Lecture Notes in Computer Science 3251

Commenced Publication in 1973
Founding and Former Series Editors:
Gerhard Goos, Juris Hartmanis, and Jan van Leeuwen

Editorial Board

David Hutchison
  Lancaster University, UK
Takeo Kanade
  Carnegie Mellon University, Pittsburgh, PA, USA
Josef Kittler
  University of Surrey, Guildford, UK
Jon M. Kleinberg
  Cornell University, Ithaca, NY, USA
Friedemann Mattern
  ETH Zurich, Switzerland
John C. Mitchell
  Stanford University, CA, USA
Moni Naor
  Weizmann Institute of Science, Rehovot, Israel
Oscar Nierstrasz
  University of Bern, Switzerland
C. Pandu Rangan
  Indian Institute of Technology, Madras, India
Bernhard Steffen
  University of Dortmund, Germany
Madhu Sudan
  Massachusetts Institute of Technology, MA, USA
Demetri Terzopoulos
  New York University, NY, USA
Doug Tygar
  University of California, Berkeley, CA, USA
Moshe Y. Vardi
  Rice University, Houston, TX, USA
Gerhard Weikum
  Max-Planck Institute of Computer Science, Saarbruecken, Germany

Lecture Notes in Computer Science 3251

Commenced Publication in 1973
Founding and Former Series Editors:
Gerhard Goos, Juris Hartmanis, and Jan van Leeuwen

Editorial Board

David Hutchison
Lancaster University, UK

Takeo Kanade
Carnegie Mellon University, Pittsburgh, PA, USA

Josef Kittler
University of Surrey, Guildford, UK

Jon M. Kleinberg
Cornell University, Ithaca, NY, USA

Friedemann Mattern
ETH Zurich, Switzerland

John C. Mitchell
Stanford University, CA, USA

Moni Naor
Weizmann Institute of Science, Rehovot, Israel

Oscar Nierstrasz
University of Bern, Switzerland

C. Pandu Rangan
Indian Institute of Technology, Madras, India

Bernhard Steffen
University of Dortmund, Germany

Madhu Sudan
Massachusetts Institute of Technology, MA, USA

Demetri Terzopoulos
New York University, NY, USA

Doug Tygar
University of California, Berkeley, CA, USA

Moshe Y. Vardi
Rice University, Houston, TX, USA

Gerhard Weikum
Max-Planck Institute of Computer Science, Saarbruecken, Germany

Hai Jin   Yi Pan   Nong Xiao
Jianhua Sun (Eds.)

Grid and Cooperative Computing – GCC 2004

Third International Conference
Wuhan, China, October 21-24, 2004
Proceedings

Springer

Volume Editors

Hai Jin
Jianhua Sun
Huazhong University of Science and Technology, Cluster and Grid Computing Lab
430074 Wuhan, China
E-mail: {hjin, jhsun}@hust.edu.cn

Yi Pan
Georgia State University, Department of Computer Science
34 Peachtree Street, Suite 1450, Atlanta, GA 30302-4110, USA
E-mail: pan@cs.gsu.edu

Nong Xiao
National University of Defense Technology, School of Computer
Changsha, 410073 China
E-mail: xiao-n@vip.sina.com

Library of Congress Control Number: 2004113939

CR Subject Classification (1998): C.2, D.2, D.4, I.2.11, H.4, H.3, H.5, K.6.5

ISSN 0302-9743
ISBN 3-540-23564-7 Springer Berlin Heidelberg New York

This work is subject to copyright. All rights are reserved, whether the whole or part of the material is concerned, specifically the rights of translation, reprinting, re-use of illustrations, recitation, broadcasting, reproduction on microfilms or in any other way, and storage in data banks. Duplication of this publication or parts thereof is permitted only under the provisions of the German Copyright Law of September 9, 1965, in its current version, and permission for use must always be obtained from Springer. Violations are liable to prosecution under the German Copyright Law.

Springer is a part of Springer Science+Business Media

springeronline.com

© Springer-Verlag Berlin Heidelberg 2004
Printed in Germany

Typesetting: Camera-ready by author, data conversion by Olgun Computergrafik
Printed on acid-free paper   SPIN: 11323662   06/3142   5 4 3 2 1 0

Preface

Welcome to the proceedings of GCC 2004 and the city of Wuhan. Grid computing has become a mainstream research area in computer science and the GCC conference has become one of the premier forums for presentation of new and exciting research in all aspects of grid and cooperative computing. The program committee is pleased to present the proceedings of the 3rd International Conference on Grid and Cooperative Computing (GCC 2004), which comprises a collection of excellent technical papers, posters, workshops, and keynote speeches. The papers accepted cover a wide range of exciting topics, including resource grid and service grid, information grid and knowledge grid, grid monitoring, management and organization tools, grid portal, grid service, Web services and their QoS, service orchestration, grid middleware and toolkits, software glue technologies, grid security, innovative grid applications, advanced resource reservation and scheduling, performance evaluation and modeling, computer-supported cooperative work, P2P computing, automatic computing, and meta-information management.

The conference continues to grow and this year a record total of 581 manuscripts (including workshop submissions) were submitted for consideration. Expecting this growth, the size of the program committee was increased from 50 members for GCC 2003 for 70 in GCC 2004. Relevant differences from previous editions of the conference: it is worth mentioning a significant increase in the number of papers submitted by authors from outside China; and the acceptance rate was much lower than for previous GCC conferences. From the 427 papers submitted to the main conference, the program committee selected only 96 regular papers for oral presentation and 62 short papers for poster presentation in the program. Five workshops, International Workshop on Agents, and Autonomic Computing, and Grid Enabled Virtual Organizations, International Workshop on Storage Grids and Technologies, International Workshop on Information Security and Survivability for Grid, International Workshop on Visualization and Visual Steering, International Workshop on Information Grid and Knowledge Grid, complemented the outstanding paper sessions.

The submission and review process worked as follows. Each submission was assigned to three program committee members for review. Each program committee member prepared a single review for each assigned paper or assigned a paper to an outside reviewer for review. Given the large number of submissions, each program committee member was assigned roughly 15–20 papers. The program committee members consulted 65 members of the grid computing community in preparing the reviews. Based on the review scores, the program chairs made the final decision. Given the large number of submissions, the selection of papers required a great deal of work on the part of the committee members.

Putting together a conference requires the time and effort of many people. First, we would like to thank all the authors for their hard work in preparing submissions to the conference. We deeply appreciate the effort and contributions of the program committee members who worked very hard to select the very best submissions and to put together an exciting program. We are also very grateful for the numerous suggestions we received from them. Also, we especially thank the effort of those program committee members who delivered their reviews in a timely manner despite having to face very difficult personal situations. The effort of the external reviewers is also deeply appreciated. We are also very grateful to Ian Foster, Jack Dongarra, Charlie Catlett, and Tony Hey for accepting our invitation to present a keynote speech, and to Depei Qian for organizing an excellent panel on a very exciting and important topic. Thanks go to the workshop chairs for organizing five excellent workshops on several important topics in grid computing. We would also like to thank Pingpeng Yuan for installing and maintaining the submission website and working tirelessly to overcome the limitations of the tool we used.

We deeply appreciate the tremendous efforts of all the members of the organizing committee. We would like to thank the general co-chairs, Prof. Andrew A. Chien and Prof. Xicheng Lu for their advice and continued support. Finally, we would like to thank the GCC steering committee for the opportunity to serve as the program chairs as well as their guidance through the process. We hope that the attendees enjoyed this conference and found the technical program to be exciting.

Hai Jin and Yi Pan

Conference Committees

Steering Committee

Guojie Li (Institute of Computing Technology, CAS, China)
Xiaodong Zhang (National Science Foundation, USA)
Zhiwei Xu (Institute of Computing Technology, CAS, China)
Xianhe Sun (Illinois Institute of Technology, USA)
Jun Ni (University of Iowa, USA)
Hai Jin (Huazhong University of Science and Technology, China)
Minglu Li (Shanghai Jiao Tong University, China)

Conference Co-chairs

Andrew A. Chien (University of California at San Diego, USA)
Xicheng Lu (National University of Defense Technology, China)

Program Co-chairs

Yi Pan (Georgia State University, USA)
Hai Jin (Huazhong University of Science and Technology, China)

Workshop Chair

Nong Xiao (National University of Defense Technology, China)

Panel Chair

Depei Qian (Xi'an Jiaotong University, China)

Publicity Chair

Minglu Li (Shanghai Jiao Tong University, China)

Tutorial Chair

Dan Meng (Institute of Computing Technology, CAS, China)

Poster Chair

Song Wu (Huazhong University of Science and Technology, China)

Program Committee Members

Mark Baker (University of Portsmouth, UK)
Rajkumar Buyya (University of Melbourne, Australia)
Wentong Cai (Nanyang Technological University, Singapore)
Jiannong Cao (Hong Kong Polytechnic University, Hong Kong)
Guihai Chen (Nanjing University, China)
Xiaowu Chen (Beihang University, China)
Xuebin Chi (Computer Network Information Center, CAS, China)
Qianni Deng (Shanghai Jiao Tong University, China)
Shoubin Dong (South China University of Technology, China)
Xiaoshe Dong (Xi'an Jiaotong University, China)
Dan Feng (Huazhong University of Science and Technology, China)
Ning Gu (Fudan University, China)
Yadong Gui (Shanghai Supercomputer Center, China)
Minyi Guo (University of Aizu, Japan)
Yanbo Han (Institute of Computing Technology, CAS, China)
Yanxiang He (Wuhan University, China)
Jinpeng Huai (Beihang University, China)
Chun-Hsi Huang (University of Connecticut, USA)
Liusheng Huang (University of Science and Technology of China, China)
Kai Hwang (University of Southern California, USA)
Weijia Jia (City University of Hong Kong, Hong Kong)
Francis Lau (The University of Hong Kong, Hong Kong)
Keqin Li (State University of New York, USA)
Minglu Li (Shanghai Jiao Tong University, China)
Qing Li (City University of Hong Kong, Hong Kong)
Qinghua Li (Huazhong University of Science and Technology, China)
Xiaoming Li (Peking University, China)
Xiaola Lin (City University of Hong Kong, Hong Kong)
Xinda Lu (Shanghai Jiao Tong University, China)
Zhengding Lu (Huazhong University of Science and Technology, China)
Junzhou Luo (Southeast University, China)
Dan Meng (Institute of Computing Technology, CAS, China)
Xiangxu Meng (Shandong University, China)
Xiaofeng Meng (Renmin University of China, China)
Geyong Min (University of Bradford, UK)
Jun Ni (University of Iowa, USA)
Lionel Ni (Hong Kong University of Science and Technology, Hong Kong)
Depei Qian (Xi'an Jiaotong University, China)
Yuzhong Qu (Southeast University, China)

Hong Shen (Japan Advanced Institute of Science and Technology, Japan)
Ke Shi (Huazhong University of Science and Technology, China)
Ninghui Sun (Institute of Computing Technology, CAS, China)
Yuzhong Sun (Institute of Computing Technology, CAS, China)
David Taniar (Monash University, Australia)
Huaglory Tianfield (Glasgow Caledonian University, UK)
Weiqin Tong (Shanghai University, China)
David W. Walker (Cardiff University, UK)
Cho-Li Wang (The University of Hong Kong, Hong Kong)
Xingwei Wang (Northeastern University, China)
Jie Wu (Florida Atlantic University, USA)
Song Wu (Huazhong University of Science and Technology, China)
Zhaohui Wu (Zhejiang University, China)
Nong Xiao (National University of Defense Technology, China)
Cheng-Zhong Xu (Wayne State University, USA)
Baoping Yan (Computer Network Information Center, CAS, China)
Guangwen Yang (Tsinghua University, China)
Laurence Tianruo Yang (St. Francis Xavier University, Canada)
Qiang Yang (Hong Kong University of Science and Technology, Hong Kong)
Shoubao Yang (University of Science and Technology of China, China)
Zhonghua Yang (Nanyang Technological University, Singapore)
Pingpeng Yuan (Huazhong University of Science and Technology, China)
Weimin Zheng (Tsinghua University, China)
Yao Zheng (Zhejiang University, China)
Luo Zhong (Wuhan Univeristy of Technology, China)
Aoying Zhou (Fudan University, China)
Wanlei Zhou (Deakin University, Australia)
Xinrong Zhou (Åbo Akademi University, Finland)
Jianping Zhu (University of Akron, USA)
Mingfa Zhu (Lenovo Research, China)
Hai Zhuge (Institute of Computing Technology, CAS, China)

Local Arrangements Chair

Pingpeng Yuan (Huazhong University of Science and Technology, China)

Exhibition Chair

Qin Zhang (Huazhong University of Science and Technology, China)

Financial Chair

Xin Li (Huazhong University of Science and Technology, China)

Industry Chair

Xia Xie (Huazhong University of Science and Technology, China)

Publication Chair

Jianhua Sun (Huazhong University of Science and Technology, China)

Conference Secretary

Cong Geng (Huazhong University of Science and Technology, China)

Reviewers

Rashid Al-Ali
Jeff Dallien
Zhiqun Deng
Jonathan Giddy
Ian Grimstead
Zhengxiong Hou
Yanli Hu
Ajay Katangur
Yunchun Li
Na Lin
Zhen Lin
Hui Liu
Tao Liu
Xinpeng Liu
Sanglu Lu
Zhongzhi Luan

Yingwei Luo
Wendy MacCaull
Praveen Madiraju
Shalil Majithia
Zhongquan Mao
Stephen Pellicer
Weizong Qiang
Ling Qiu
Shrija Rajbhandari
Omer Rana
Geoffrey Shea
Praveena Tayanthi
Ian Taylor
Baoyi Wang
Guojun Wang
Hui Wang

Xianbing Wang
Xiaofang Wang
Xiaolin Wang
Xingwei Wang
Yuelong Wang
Mark Wright
Guang Xiang
Bin Xiao
Xia Xie
Shaomin Zhang
Yang Zhang
Ran Zheng
Jingyang Zhou
Cheng Zhu
Deqing Zou

Table of Contents

The Grid: Beyond the Hype 1
   Ian Foster

High Performance Computing Trends, Supercomputers, Clusters and Grids 2
   Jack Dongarra

e-Science and Web Services Grids 3
   Tony Hey

Making Grids Real: Standards and Sociology 4
   Charlie Catlett

About Grid 5
   Greg Astfalk

Many Faces of Grid Computing 6
   Greg Rankich

From Science to Enterprise – Intel's Grid Activities 7
   Karl Solchenbach

Session 1: Grid Service and Web Service

A QoS-Enabled Services System Framework for Grid Computing 8
   Yang Zhang, Jiannong Cao, Sanglu Lu, Baoliu Ye, and Li Xie

State Management Issues and Grid Services 17
   Yong Xie and Yong-Meng Teo

Transactional Business Coordination and Failure Recovery
for Web Services Composition 26
   Yi Ren, Quanyuan Wu, Yan Jia, Jianbo Guan, and Weihong Han

Using Service Taxonomy to Facilitate Efficient Decentralized
Grid Service Discovery 34
   Cheng Zhu, Zhong Liu, Weiming Zhang, Zhenning Xu, and Dongsheng Yang

QoS Analysis on Web Service Based Spatial Integration 42
   Yingwei Luo, Xinpeng Liu, Wenjun Wang, Xiaolin Wang, and Zhuoqun Xu

Engineering Process Coordination Based on a Service Event Notification Model . 50
   Jian Cao, Jie Wang, Shensheng Zhang, Minglu Li, and Kincho Law

A Fault-Tolerant Architecture for Grid System 58
   Lingxia Liu, Quanyuan Wu, and Bin Zhou

A Workflow Language for Grid Services in OGSI-Based Grids 65
    Yan Yang, Shengqun Tang, Wentao Zhang, and Lina Fang

Membrane Calculus: A Formal Method for Grid Transactions 73
    Zhengwei Qi, Cheng Fu, Dongyu Shi, Jinyuan You, and Minglu Li

Workflow-Based Approach to Efficient Programming and Reliable Execution
of Grid Applications 81
    Yong-Won Kwon, So-Hyun Ryu, Ju-Ho Choi, and Chang-Sung Jeong

Achieving Context Sensitivity of Service-Oriented Applications
with the Business-End Programming Language VINCA 89
    Hao Liu, Yanbo Han, Gang Li, and Cheng Zhang

Mapping Business Workflows onto Network Services Environments 97
    Wenjun Wang, Xinpeng Liu, Yingwei Luo, Xiaolin Wang, and Zhuoqun Xu

Session 2: Grid Middleware and Toolkits

GridDaenFS: A Virtual Distributed File System
for Uniform Access Across Multi-domains 105
    Wei Fu, Nong Xiao, and Xicheng Lu

Ad Hoc Debugging Environment for Grid Applications 113
    Wei Wang, Binxing Fang, Hongli Zhang, and Yuanzhe Yao

Distributed Object Group Framework with Dynamic Reconfigurability
of Distributed Services 121
    Chang-Sun Shin, Young-Jee Chung, and Su-Chong Joo

Design and Implementation of Grid File Management System Hotfile 129
    Liqiang Cao, Jie Qiu, Li Zha, Haiyan Yu, Wei Li, and Yuzhong Sun

pXRepository: A Peer-to-Peer XML Repository for Web Service Discovery 137
    Yin Li, Futai Zou, Fanyuan Ma, and Minglu Li

Design Open Sharing Framework for Spatial Information in Semantic Web 145
    Yingwei Luo, Xinpeng Liu, Xiaolin Wang, Wenjun Wang, and Zhuoqun Xu

A Metadata Framework for Distributed Geo-spatial Databases
in Grid Environment 153
    Yuelong Wang, Wenjun Wang, Yingwei Luo, Xiaolin Wang, and Zhuoqun Xu

Research and Implementation of Single Sign-On Mechanism for ASP Pattern 161
    Bo Li, Sheng Ge, Tian-yu Wo, and Dian-fu Ma

Design of an OGSA-Based MetaService Architecture 167
    *Zhi-Hui Du, Francis C.M. Lau, Cho-Li Wang, Wai-kin Lam, Chuan He,
    Xiaoge Wang, Yu Chen, and Sanli Li*

Facilitating the Process of Enabling Applications Within Grid Portals 175
  Maciej Bogdanski, Michal Kosiedowski, Cezary Mazurek, and Maciej Stroinski

Towards a Distributed Modeling and Simulation Environment for Networks 183
  Yueming Lu, Hui Li, Aibo Liu, and Yuefeng Ji

Approximate Performance Analysis of Web Services Flow
Using Stochastic Petri Net 193
  Zhangxi Tan, Chuang Lin, Hao Yin, Ye Hong, and Guangxi Zhu

Complexity Analysis of Load Balance Problem
for Synchronous Iterative Applications 201
  Weizhe Zhang and Mingzeng Hu

Session 3: Advanced Resource Reservation and Scheduling

Incentive-Based P2P Scheduling in Grid Computing 209
  Yanmin Zhu, Lijuan Xiao, Lionel M. Ni, and Zhiwei Xu

Hybrid Performance-Oriented Scheduling of Moldable Jobs
with QoS Demands in Multiclusters and Grids 217
  Ligang He, Stephen A. Jarvis, Daniel P. Spooner, Xinuo Chen, and Graham R. Nudd

An Active Resource Management System for Computational Grid 225
  Xiaolin Chen, Chang Yang, Sanglu Lu, and Guihai Chen

Predicting the Reliability of Resources in Computational Grid 233
  Chunjiang Li, Nong Xiao, and Xuejun Yang

Flexible Advance Reservation for Grid Computing 241
  Jianbing Xing, Chanle Wu, Muliu Tao, Libing Wu, and Huyin Zhang

Club Theory of the Grid 249
  Yao Shi, Francis C.M. Lau, Zhi-Hui Du, Rui-Chun Tang, and Sanli Li

A Runtime Scheduling Approach with Respect to Job Parallelism
for Computational Grid 261
  Li Liu, Jian Zhan, and Lian Li

A Double Auction Mechanism for Resource Allocation
on Grid Computing Systems 269
  Chuliang Weng, Xinda Lu, Guangtao Xue, Qianni Deng, and Minglu Li

Research on the Virtual Topology Design Methods
in Grid-Computing-Supporting IP/DWDM-Based NGI 277
  Xingwei Wang, Minghua Chen, Qiang Wang, Min Huang, and Jiannong Cao

De-centralized Job Scheduling on Computational Grids
Using Distributed Backfilling 285
   Qingjiang Wang, Xiaolin Gui, Shouqi Zheng, and Bing Xie

Session 4: Grid Security

A Novel VO-Based Access Control Model for Grid 293
   Weizhong Qiang, Hai Jin, Xuanhua Shi, and Deqing Zou

G-PASS: Security Infrastructure for Grid Travelers 301
   Tianchi Ma, Lin Chen, Cho-Li Wang, and Francis C.M. Lau

Protect Grids from DDoS Attacks 309
   Yang Xiang and Wanlei Zhou

Trust Establishment in Large Scale Grid Settings 317
   *Bo Zhu, TieYan Li, HuaFei Zhu, Mohan S. Kankanhalli, Feng Bao,
   and Robert H. Deng*

XML Based X.509 Authorization in CERNET Grid 325
   Wu Liu, Jian-Ping Wu, Hai-Xin Duan, Xing Li, and Ping Ren

Alerts Correlation System to Enhance the Performance
of the Network-Based Intrusion Detection System 333
   Do-Hoon Lee, Jung-Taek Seo, and Jae-Cheol Ryou

Security Architecture for Web Services 341
   Yuan Rao, Boqin Feng, and Jincang Han

A Hybrid Machine Learning/Statistical Model of Grid Security 348
   Guang Xiang, Ge Yu, Xiangli Qu, Xiaomei Dong, and Lina Wang

A Novel Grid Node-by-Node Security Model 356
   Zhengyou Xia and Yichuan Jiang

Research on Security of the United Storage Network 364
   Dezhi Han, Changsheng Xie, and Qiang Cao

Session 5: Information Grid and Knowledge Grid

Q3: A Semantic Query Language for Dart Database Grid 372
   Huajun Chen, Zhaohui Wu, and Yuxing Mao

Knowledge Map Model 381
   Hai Zhuge and Xiangfeng Luo

Kernel-Based Semantic Text Categorization
for Large Scale Web Information Organization 389
   Xianghua Fu, Zhaofeng Ma, and Boqin Feng

Mobile Computing Architecture in Next Generation Network 397
   Yun-Yong Zhang, Zhi-Jiang Zhang, Fan Zhang, and Yun-Jie Liu

A Study of Gridifying Scientific Computing Legacy Codes 404
   Bin Wang, Zhuoqun Xu, Cheng Xu, Yanbin Yin, Wenkui Ding, and Huashan Yu

Aegis: A Simulation Grid Oriented to Large-Scale Distributed Simulation 413
   Wei Wu, Zhong Zhou, Shaofeng Wang, and Qinping Zhao

An Adaptive Data Objects Placement Algorithm for Non-uniform Capacities 423
   Zhong Liu and Xing-Ming Zhou

Resource Discovery Mechanism
for Large-Scale Distributed Simulation Oriented Data Grid 431
   Hai Huang, Shaofeng Wang, Yan Zhang, and Wei Wu

Research and Design on a Novel OGSA-Based Open Content Alliance Scheme . . 440
   ZhiHui Lv, YiPing Zhong, and ShiYong Zhang

An Ontology-Based Model for Grid Resource Publication and Discovery 448
   Lei Cao, Minglu Li, Henry Rong, and Joshua Huang

A New Overload Control Algorithm of NGN Service Gateway 456
   Yun-Yong Zhang, Zhi-Jiang Zhang, Fan Zhang, and Yun-Jie Li

Nodes' Organization Mechanisms on Campus Grid Services Environment 464
   *Zhiqun Deng, Zhicong Liu, Hong Luo, Guanzhong Dai, Xinjia Zhang,
   Dejun Mu, and Hongji Tang*

A Resource Organizing Protocol
for Grid Based on Bounded Two-Level Broadcasting Technique 472
   Zhigang Chen, Anfeng Liu, and Guoping Long

Session 6: P2P Computing and Automatic Computing

Peer-Owl: An Adaptive Data Dissemination Scheme
for Peer-to-Peer Streaming Services 479
   Xiaofei Liao and Hai Jin

WCBF: Efficient and High-Coverage Search Schema
in Unstructured Peer-to-Peer Network 487
   Qianbing Zheng, Yongwen Wang, and Xicheng Lu

EM Medical Image Reconstruction in a Peer-to-Peer Systems 495
   Jun Ni, Tao He, Xiang Li, Shaowen Wang, and Ge Wang

End Host Multicast for Peer-to-Peer Systems 502
   Wanqing Tu and Weijia Jia

Janus: Build Gnutella-Like File Sharing System over Structured Overlay 511
  Haitao Dong, Weimin Zheng, and Dongsheng Wang

Gemini: Probabilistic Routing Algorithm in Structured P2P Overlay 519
  Ming Li, Jinfeng Hu, Haitao Dong, Dongsheng Wang, and Weimin Zheng

SkyMin: A Massive Peer-to-Peer Storage System 527
  Lu Yan, Moisés Ferrer Serra, Guangcheng Niu, Xinrong Zhou, and Kaisa Sere

A Mobile Agent Enabled Approach for Distributed Deadlock Detection 535
  Jiannong Cao, Jingyang Zhou, Weiwei Zhu, Daoxu Chen, and Jian Lu

Maintaining Comprehensive Resource Availability in P2P Networks 543
  Bin Xiao, Jiannong Cao, and Edwin H.-M. Sha

Efficient Search Using Adaptive Metadata Spreading in Peer-to-Peer Networks .. 551
  Xuezheng Liu, Guangwen Yang, Ming Chen, and Yongwei Wu

Querying XML Data over DHT System Using XPeer 559
  Weixiong Rao, Hui Song, and Fanyuan Ma

Efficient Lookup Using Proximity Caching for P2P Networks 567
  Hyung Soo Jung and Heon Y. Yeom

Content-Based Document Recommendation
in Collaborative Peer-to-Peer Network 575
  Heung-Nam Kim, Hyun-Jun Kim, and Geun-Sik Jo

P2PENS: Content-Based Publish-Subscribe over Peer-to-Peer Network 583
  Tao Xue, Boqin Feng, and Zhigang Zhang

Joint Task Placement, Routing and Power Control
for Low Power Mobile Grid Computing in Ad Hoc Network 591
  Min Li, Xiaobo Wu, Menglian Zhao, Hui Wang, Ping Li, and Xiaolang Yan

Scheduling Legacy Applications with Domain Expertise
for Autonomic Computing 601
  Huashan Yu, Zhuoqun Xu, and Wenkui Ding

E-Mail Services on Hybrid P2P Networks 610
  Yue Zhao, Shuigeng Zhou, and Aoying Zhou

Implementation of P2P Computing in Self-organizing Network Routing 618
  Zupeng Li, Jianhua Huang, and Xiubin Zhao

HP-Chord: A Peer-to-Peer Overlay to Achieve Better Routing Efficiency
by Exploiting Heterogeneity and Proximity 626
  Feng Hong, Minglu Li, Xinda Lu, Jiadi Yu, Yi Wang, and Ying Li

MONSTER: A Media-on-Demand Servicing System Based on P2P Networks ... 634
  Yu Chen and Qionghai Dai

Research of Data-Partition-Based Replication Algorithm
in Peer-to-Peer Distributed Storage System 642
  Yijie Wang and Sikun Li

Local Index Tree for Mobile Peer-to-Peer Networks 650
  Wei Shi, Shanping Li, Gang Peng, Tianchi Ma, and Xin Lin

Distributed Information Retrieval
Based on Hierarchical Semantic Overlay Network 657
  Fei Liu, Fanyuan Ma, Minglu Li, and Linpeng Huang

An Efficient Protocol for Peer-to-Peer File Sharing with Mutual Anonymity 665
  Baoliu Ye, Jingyang Zhou, Yang Zhang, Jiannong Cao, and Daoxu Chen

Session 7: Innovative Grid Applications

IPGE: Image Processing Grid Environment
Using Components and Workflow Techniques 671
  Ran Zheng, Hai Jin, Qin Zhang, Ying Li, and Jian Chen

Distributed MD4 Password Hashing with Grid Computing Package BOINC 679
  Stephen Pellicer, Yi Pan, and Minyi Guo

A Novel Admission Control Strategy with Layered Threshold
for Grid-Based Multimedia Services Systems 687
  Yang Zhang, Jingyang Zhou, Jiannong Cao, Sanglu Lu, and Li Xie

Applying Grid Technologies to Distributed Data Mining 696
  Alastair C. Hume, Ashley D. Lloyd, Terence M. Sloan, and Adam C. Carter

Research and Implementation of Grid-Enabled Parallel Algorithm
of Geometric Correction in ChinaGrid 704
  Haifang Zhou, Xuejun Yang, Yu Tang, Xiangli Qu, and Nong Xiao

Mixed Usage of Wi-Fi and Bluetooth
for Formulating an Efficient Mobile Conferencing Application 712
  Mee Young Sung, Jong Hyuk Lee, Yong Il Lee, and Yoon Sik Hong

UDMGrid: A Grid Application for University Digital Museums 720
  *Xiaowu Chen, Zhi Xu, Zhangsheng Pan, Xixi Luo, Hongchang Lin,
  Yingchun Huang, and Haifeng Ou*

The Research on Role-Based Access Control Mechanism
for Workflow Management System 729
  Baoyi Wang and Shaomin Zhang

Spatial Query Preprocessing in Distributed GIS 737
  Sheng-sheng Wang and Da-you Liu

Aero-crafts Aerodynamic Simulation and Optimization by Using "CFD-Grid"
Based on Service Domain 745
  Hong Liu, Xin-hua Lin, Yang Qi, Xinda Lu, and Minglu Li

Power Management Based Grid Routing Protocol
for IEEE 802.11 Based MANET 753
  Li Xu and Bao-yu Zheng

On Grid Programming and MATLAB*G 761
  Yong-Meng Teo, Ying Chen, and Xianbing Wang

A Study on the Model of Open Decision Support System Based on Agent Grid .. 769
  Xueguang Chen, Jiayu Chi, Lin Sun, and Xiang Ao

Poster

Memory Efficient Pair-Wise Genome Alignment Algorithm –
A Small-Scale Application with Grid Potential 777
  Nova Ahmed, Yi Pan, and Art Vandenberg

Research on an MOM-Based Service Flow Management System 783
  Pingpeng Yuan, Hai Jin, Li Qi, and Shicai Li

Reliability Analysis for Grid Computing 787
  Xuanhua Shi, Hai Jin, Weizhong Qiang, and Deqing Zou

Experiences Deploying Peer-to-Peer Network for a Distributed File System 791
  Chong Wang, Yafei Dai, Hua Han, and Xiaoming Li

LFC-K Cache Replacement Algorithm
for Grid Index Information Service (GIIS) 795
  Dong Li, Linpeng Huang, and Minglu Li

The Design of a Grid Computing System for Drug Discovery and Design 799
  Shudong Chen, Wenju Zhang, Fanyuan Ma, Jianhua Shen, and Minglu Li

A Storage-Aware Scheduling Scheme for VOD 803
  Chao Peng and Hong Shen

A Novel Approach for Constructing Small World in Structured P2P Systems 807
  Futai Zou, Yin Li, Liang Zhang, Fanyuan Ma, and Minglu Li

Design of a Grid Computing Infrastructure Based on Mainframes 811
  Moon J. Kim and Dikran S. Meliksetian

SHDC: A Framework to Schedule Loosely Coupled Applications
on Service Networks 815
  Fei Wu and Kam-Wing Ng

Tree-Based Replica Location Scheme (TRLS) for Data Grids 819
  *Dong Su Nam, Eung Ki Park, Sangjin Jeong, Byungsang Kim,
  and Chan-Hyun Youn*

Evaluating Stream and Disk Join in Continuous Queries 823
  Weiping Wang, Jianzhong Li, Xu Wang, DongDong Zhang, and Longjiang Guo

Keyword Extraction Based Peer Clustering 827
  Bangyong Liang, Jie Tang, Juanzi Li, and Kehong Wang

A Peer-to-Peer Approach with Semantic Locality to Service Discovery 831
  Feng Yan and Shouyi Zhan

Implicit Knowledge Sharing with Peer Group Autonomic Behavior 835
  Jian-Xue Chen and Fei Liu

A Model of Agent-Enabling Autonomic Grid Service System 839
  Beishui Liao, Ji Gao, Jun Hu, and Jiujun Chen

MoDCast: A MoD System Based on P2P Networks 843
  Yu Chen and Qionghai Dai

Maintaining and Self-recovering Global State
in a Super-peer Overlay for Service Discovery 847
  Feng Yang, Shouyi Zhan, and Fouxiang Shen

Towards the Metadata Integration Issues in Peer-to-Peer Based Digital Libraries .. 851
  Hao Ding

Service-Oriented Architecture of Specific Domain Data Center 855
  Ping Ai, Zhi-Jing Wang, and Ying-Chi Mao

Extension of SNMP Function for Improvement of Network Performance 859
  Chunkyun Youn and Ilyong Chung

A New Grid Security Framework with Dynamic Access Control 863
  Bing Xie, Xiaolin Gui, Yinan Li, and Depei Qian

A Dynamic Grid Authorization Mechanism with Result Feedback 867
  Feng Li, Junzhou Luo, Yinying Yang, Ye Zhu, and Teng Ma

A Dynamic Management Framework for Security Policies
in Open Grid Computing Environments 871
  Chiu-Man Yu and Kam-Wing Ng

A Classified Method Based on Support Vector Machine
for Grid Computing Intrusion Detection.................................. 875
  Qinghua Zheng, Hui Li, and Yun Xiao

Real Time Approaches for Time-Series Mining-Based IDS 879
  Feng Zhao, Qing-Hua Li, and Yan-Bin Zhao

Anomaly Detection in Grid Computing Based on Vector Quantization 883
  *Hong-Wei Sun, Kwok-Yan Lam, Siu-Leung Chung, Ming Gu,
  and Jia-Guang Sun*

Integrating Trust in Grid Computing Systems 887
  Woodas W.K. Lai, Kam-Wing Ng, and Michael R. Lyu

Simple Key Agreement and Its Efficiency Analysis
for Grid Computing Environments 891
  Taeshik Shon, Jung-Taek Seo, and Jongsub Moon

IADIDF: A Framework for Intrusion Detection 895
  Ye Du, Huiqiang Wang, and Yonggang Pang

Seoul Grid Portal: A Grid Resource Management System
for Seoul Grid Testbed 899
  Yong Woo Lee

PGMS: A P2P-Based Grid Monitoring System 903
  Yuanzhe Yao, Binxing Fang, Hongli Zhang, and Wei Wang

Implementing the Resource Monitoring Services
in Grid Environment Using Sensors 907
  Xiaolin Gui, Yinan Li, and Wenqiang Gong

InstantGrid: A Framework for On-Demand Grid Point Construction 911
  *Roy S.C. Ho, K.K. Yin, David C.M. Lee, Daniel H.F. Hung,
  Cho-Li Wang, and Francis C.M. Lau*

Scalable Data Management Modeling and Framework for Grid Computing 915
  Jong Sik Lee

A Resource Selection Scheme for Grid Computing System *META* 919
  KyungWoo Kang and Gyun Woo

An Extended OGSA Based Service Data Aggregator
by Using Notification Mechanism 923
  YunHee Kang

Research on Equipment Resource Scheduling in Grids 927
  Yuexuan Wang, Lianchen Liu, Cheng Wu, and Wancheng Ni

A Resource Broker for Computing Nodes Selection
in Grid Computing Environments 931
  Chao-Tung Yang, Chuan-Lin Lai, Po-Chi Shih, and Kuan-Ching Li

A Model of Problem Solving Environment
for Integrated Bioinformatics Solution on Grid by Using Condor 935
  Choong-Hyun Sun, Byoung-Jin Kim, Gwan-Su Yi, and Hyoungwoo Park

MammoGrid: A Service Oriented Architecture Based Medical Grid Application.. 939
  Salvator Roberto Amendolia, Florida Estrella, Waseem Hassan, Tamas Hauer, David Manset, Richard McClatchey, Dmitry Rogulin, and Tony Solomonides

Interactive Visualization Pipeline Architecture
Using Work-Flow Management System on Grid for CFD Analysis 943
  Jin-Sung Park, So-Hyun Ryu, Yong-Won Kwon, and Chang-Sung Jeong

Distributed Analysis and Load Balancing System for Grid Enabled Analysis
on Hand-Held Devices Using Multi-agents Systems 947
  Naveed Ahmad, Arshad Ali, Ashiq Anjum, Tahir Azim, Julian Bunn, Ali Hassan, Ahsan Ikram, Frank Lingen, Richard McClatchey, Harvey Newman, Conrad Steenberg, Michael Thomas, and Ian Willers

WSRF-Based Computing Framework of Distributed Data Stream Queries 951
  Jian-Wei Liu and Jia-Jin Le

Mobile Q&A Agent System Based on the Semantic Web in Wireless Internet 955
  Young-Hoon Yu, Jason J. Jung, Kyoung-Sin Noh, and Geun-Sik Jo

Semantic Caching Services for Data Grids 959
  Hai Wan, Xiao-Wei Hao, Tao Zhang, and Lei Li

Managing Access in Extended Enterprise Networks Web Service-Based 963
  Shaomin Zhang and Baoyi Wang

Perceptually Tuned Packet Scheduler for Video Streaming
for DAB Based Access GRID 967
  Seong-Whan Kim and Shan Suthaharan

Collaborative Detection of Spam in Peer-to-Peer Paradigm
Based on Multi-agent Systems 971
  Supratip Ghose, Jin-Guk Jung, and Geun-Sik Jo

Adaptive and Distributed Fault Monitoring of Grid by Mobile Agents 975
  Kyungkoo Jun and Seokhoon Kang

Grid QoS Infrastructure: Advance Resource Reservation
in Optical Burst Switching Networks 979
  Yu Hua, Chanle Wu, and Jianbing Xing

A Scheduling Algorithm with Co-allocation Scheme
for Grid Computing Systems 983
  Jeong Woo Jo and Jin Suk Kim

A Meta-model of Temporal Workflow and Its Formalization 987
  Yang Yu, Yong Tang, Na Tang, Xiao-ping Ye, and Lu Liang

Commitment in Cooperative Problem Solving 993
  Hong-jun Zhang, Qing-Hua Li, and Wei Zhang

iOmS: An Agent-Based P2P Framework for Ubiquitous Workflow 997
  Min Yang, Zunping Cheng, Ning Shang, Mi Zhang, Min Li, Jing Dai,
  Wei Wang, Dilin Mao, and Chuanshan Gao

MobiMessenger: An Instant Messenger System
Based on Mobile Ad-Hoc Network 1001
  Ling Chen, Weihua Ju, and Gencai Chen

Temporal Dependency for Dynamic Verification
of Temporal Constraints in Workflow Systems 1005
  Jinjun Chen and Yun Yang

Integrating Grid Services for Enterprise Applications 1009
  Feilong Tang and Minglu Li

A Middleware Based Mobile Scientific Computing System – MobileLab 1013
  Xi Wang, Xu Liu, Xiaoge Wang, and Yu Chen

Towards Extended Machine Translation Model
for Next Generation World Wide Web 1017
  Gulisong Nasierding, Yang Xiang, and Honghua Dai

Author Index 1021

The Grid: Beyond the Hype

Ian Foster

Argonne National Laboratory, University of Chicago
foster@cs.uchicago.edu

Abstract. Grid technologies and infrastructure support the integration of services and resources within and among enterprises, and thus allow new approaches to problem solving and interaction within distributed, multi-organizational collaborations. First developed and applied within eScience, Grid tools are now increasingly being applied in commercial settings as well. In this talk I discuss the current state and planned future directions for one particular collection of Grid technologies, namely the open source Globus Toolkit. I examine, in turn, the underlying requirements that motivate its design; the components that it provides for security, resource access, resource discovery, data management, and other purposes; the protocol and interface standards that underlie its design and implementation; the ecosystem of complementary tools, deployments, and applications that build on and/or complement these components; and the nature of the most urgent challenges that must be overcome to expand its utility and breadth of application.

High Performance Computing Trends, Supercomputers, Clusters and Grids

Jack Dongarra

University of Tennessee and Oak Ridge National Laboratory
dongarra@cs.utk.edu

Abstract. In this talk we will look at how High Performance computing has changed over the last 10-year and look toward the future in terms of trends. In addition, we advocate the 'Computational Grids' to support 'large-scale' applications. These must provide transparent access to the complex mix of resources – computational, networking, and storage – that can be provided through aggregation of resources.

e-Science and Web Services Grids

Tony Hey

University of Southampton, UK
Tony.Hey@epsrc.ac.uk

Abstract. The talk will introduce the concept of e-Science and briefly describe some of the main features of the £250M 5 year UK e-Science Programme. This review will include examples of e-Science applications not only for science and engineering but also for e-Health and the e-Enterprise. The importance of data curation will be emphasized and the move towards 'Web Services Grids' – Grid middleware based on open standard Web Services.

Making Grids Real: Standards and Sociology

Charlie Catlett

Argonne National Laboratory
catlett@mcs.anl.gov

Abstract. Grid technologies attempt to answer several fundamental needs. Complex applications often require resources (instruments, computers, information...) that exist in multiple locations under the administration of multiple organizations. Grid technology attempts to address the combining of these resources to fulfill the needs of these advanced applications. However, there is a second fundamental need that Grid technology aims to meet-providing for these organizations to work together without loss of autonomy or security. Because effective Grid systems require software and information technology from multiple sources, standards are essential. Because Grid systems require resources from multiple organizations, careful process and sociology structures are necessary. These include process, policy, and organizational incentives. Catlett will talk about lessons learned as Chair of Global Grid Forum and as Executive Director of the US National Science Foundation's TeraGrid project in these areas.

About Grid

Greg Astfalk

Hewlett-Packard's High Performance Systems Division (HPSD), USA
wliu@hp.com

Abstract. This talk's title is purposely simplistic. We want to use the opportunity to discuss Grid in a very realistic, broad-reaching, and practical way. Grid's most important role is largely misunderstood, as is its ETA (Estimated Time of Arrival).

We'll consider the key enablers for Grids and the mega-trends that are acting as forcing functions on Grid. A key point is that Grid, in the future, is largely about SOAs (Service Oriented Architectures) and less so about scientific computing.

We'll discuss its fit with commercial IT, with robust management of IT resources, including services themselves, and of the need for open standards.

Above all, it is hoped this talk will be provocative, illuminating, and thought-provoking.

Many Faces of Grid Computing

Greg Rankich

Windows Server Division, Microsoft Corporation

Abstract. High Performance Computing and Grid Computing are among the fastest growing server workloads worldwide, driven by enterprises and academic adoption. HPC features some of the most demanding and exciting application scenarios that drive innovation in distributed system development, large scale management, parallel computing, networking and storage. This talk will take a hard look at the various forms of HPC and Grid Computing and attempt to distill hype from reality. The talk will compare and contrast high performance computing clusters, desktop scavenging, "data-grids", and cooperative grids and how each approach can be used. Microsoft's view and efforts in this space will also be discussed.

From Science to Enterprise – Intel's Grid Activities

Karl Solchenbach

Intel Parallel & Distributed Solutions Division
Hermuelheimer Str. 8a 50321 Bruehl
karl.solchenbach@intel.com

Abstract. The presentation addresses the current status of UNICORE and will describe Intel's concepts for a new Grid Programming Environment (GPE). UNICORE is now heavily used in academia and industry; the presentation will show some examples of successful application enabling with UNICORE. This includes results of the EUROGRID project as well as work of other Grid projects in and outside Europe. The UNICORE architecture is based on three layers: the client interfaces to the end-user, providing functions for creating jobs to be run and monitoring their progress, the gateway is the single entry point for all users at a centre and cooperates with the firewalls to only allow legitimate users in, and the server schedules UNICORE jobs for execution, translating the specification to the commands appropriate for the target platform. The server also controls transmission of files, including output and result files that can be later retrieved by the user.

Currently UNICORE is modified to support the new Grid standards OGSA and WS-RF. In addition, Intel extends the successful graphical UNICORE client to a next generation Grid Programming Environment (GPE). When the focus of the Grid applications will change from science to enterprise applications, Grid-specific environments and development tools will be needed. GPE will make it very easy to use the Grid, it will support all Grid standards and Grid middleware systems like Globus, UNICORE and others. With the GPE development Intel will accelerate the grid-readiness of scientific, technical and enterprise applications in various vertical markets.

A QoS-Enabled Services System Framework for Grid Computing

Yang Zhang[1,2], Jiannong Cao[2], Sanglu Lu[1], Baoliu Ye[1], and Li Xie[1]

[1] State Key Laboratory for Novel Software Technology,
Department of Computer Science and Technology, Nanjing University, P.R. China 210093
{sanglu,yebl,xieli}@nju.edu.cn
[2] Department of Computing, Hong Kong Polytechnic University,
Hung Hom, Kowloon, Hong Kong
{csyzhang,csjcao}@comp.polyu.edu.hk

Abstract. Grid computing is an efficient paradigm to aggregate diverse and geographically distributed computing resources and provide a transparent and services-sharing environment. The high-performance applications in grid have high demand on quality of service (QoS). In this paper, we present a comprehensive study on the QoS requirements of such high-end grid-based applications, propose a new QoS-enabled services system framework, and describe a prototype implementation of the framework that aims to provide stream media transmission services in grid. Our goal is to provide useful, transparent and convenient QoS for high-end applications in the grid environment.

1 Introduction

"Grid" is such a kind of infrastructure that provides an efficient paradigm to aggregate disparate computing resources in a flexible and controlled environment [1]. It focuses on large-scale resource sharing, innovative applications and, in some cases, high-performance orientation [2]. Because of highly strict *Quality of Service* (QoS) requirements, not only on network but on other resources including CPU, memory and so on, these new breed of grid-based applications (e.g. remote visualization, distributed data analysis, large scale stream media service, etc) are called high-end applications [3,4]. If to be transmitted over Internet, data or media flows of such high-end applications demand a high degree of QoS insurance on the whole end-to-end path. Traditional mechanisms that deal with QoS issues only on the network level are obviously inadequate to satisfy the end-to-end QoS requirements. An advisable approach is to address such QoS problems through a method that integrates different QoS mechanisms from both the network level and the end-system level.

OMEGA [5] may be the first QoS architecture that aims to provide end-to-end QoS guarantees. But it focuses on multimedia systems and the setup is only traditional Internet environment. Foster et al. propose a framework named GARA [6] in Globus Project for end-to-end QoS guarantee in Grid. It provides a unified mechanism to make QoS reservations for different types of grid resources including both network parameters and system resource parameters. However, GARA is mainly for computation grid which mainly supports computation-intensive applications but is not suitable for other dataflow-intensive applications. Moreover, GARA does not take user classification into consideration. G-QoSM [7] is a service-oriented grid framework building on the Open Grid Services Architecture (OGSA). It is able to distinguish different classes of users and treat them differently. G-QoSM, however, is not a complete system framework. It only gives the QoS metrics for describing the execution environment and depends on the third party software (e.g. Globus) to create such environment and provide management and control of services.

In this paper, we present a complete QoS-Enabled Services System (QESS) framework that aims to guarantee the end-to-end QoS in grid from both network level and end-system level. Furthermore, we introduce a preliminary prototype implementation based on the QESS framework. Our goal is to provide useful, transparent and convenient QoS for high-end applications in the grid environment.

The main contributions of this paper include:

1. A study on key issues of guaranteeing end-to-end QoS in grid-based services system;
2. A framework, "QESS", which integrates both network and end-system QoS mechanisms to achieve the QoS guarantee in grid computing; and
3. A preliminary grid services system prototype, "ND-SMGrid", which provides stream media transmission services in the grid environment.

The remainder of this paper is organized as follows. Section 2 analyses the QoS requirements of the high-end applications in the grid environment. In Section 3, a QoS-enabled service system framework called QESS is proposed. Based on QESS, a prototype system "ND-SMGrid" is implemented and described in Section 4. Finally, we conclude this paper in Section 5.

2 QoS Requirements for High-End Applications in Grid

In a traditional services system, end-to-end QoS have been perceived to be more or less governed by conventional network parameters like bandwidth, packet loss, end-to-end delay and jitter [3,8]. Previous researches mainly focus on the network level: how to enhance the network performance, how to optimize the transmission control, so as to maximize the number of served users. The existing works are able to provide QoS guarantees for common applications to some extent. However, considering the high-end applications in a grid computing environment, for the complexity of the application system, the heterogeneity of flows and the huge amount of data transmission, end-to-end QoS requirements are somewhat more diverse and harder to satisfy. It is inadequate to address such QoS issues only through the network level. End-system level QoS, which mainly aims to improve the system performance by proper services scheduling, resource reservation, service nodes status monitoring, and etc, is also a determinative factor to the end-to-end QoS guarantee in a grid service system. Therefore, a more comprehensive description on QoS consideration in grid environment should cover both the QoS of network level and end-system level.

2.1 Network Level QoS

In the grid computing environment, heterogeneous and dynamic resources are distributed over the Internet. Some important issues that influence network level QoS should be considered:

Bandwidth Management and Reservation: The whole system should use some schemes to manage and allocate network bandwidth, reserve the required resources for the user, thus achieve higher available network QoS.

Routing Stability: Because of some instabilities in the Internet (e.g. route fluttering, routing looping, etc) that induce the degradation of network QoS, some proper routing mechanisms should be provided to keep the stable routing.

Link Sharing: Different applications should be able to share the transmission links. It is the prerequisite to the generic grid applications.

Congestion Control: The network should take some congestion control techniques to avoid congestion, thus to keep the transmission unblocked, assure the transmission efficiency and improve QoS on the network.

2.2 End-System Level QoS

From the aspect of end-system level, special mechanisms still should be provided to satisfy the QoS requirements of high-end applications in grid environments. Nowadays, these issues are mainly discussed as follows [1,2]:

Service Discovery: In the grid system, resources are abstracted to some kinds of services. The system should be able to discover services in the grid for the serving users based on their QoS attributes.

Service Scheduling: It is of vital importance to the services system in grid, since it faces multi-classes of users with different significance and QoS requirements. The system should be able to select the most suitable service node for the arrival user.

QoS Mapping and Monitoring: Logical QoS parameters of diverse applications should be transformed into particular resource capacity thus help the grid system monitor the resource status.

Collaborative Allocation and Advance Reservation: For some scarce resources, special advance reservation and collaborative allocation mechanisms will help to realize the well-balanced service and resource scheduling.

Application-Level Control: High end performance requires a set of transparent APIs provided to application level, by which, the given applications will be able to control the resources level below, to monitor service status, and to modify QoS requests and application behavior dynamically.

Figure 1 summarizes an abstract model that addresses QoS requirements for a generic grid-based services system.

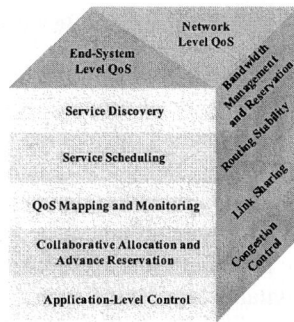

Fig. 1. QoS Requirements Model for Grid Services System.

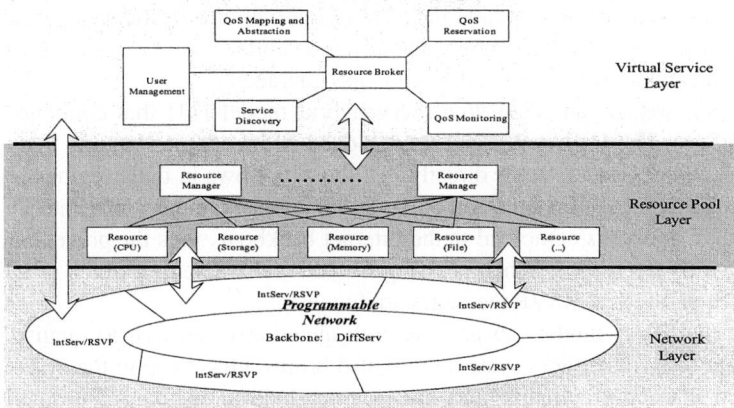

Fig. 2. The QESS Framework.

3 QESS: A QoS-Enabled Services System Framework

In this section, we propose a QoS-Enabled Services System Framework (QESS), which provides end-to-end QoS guarantee for high-end application services system in the grid environment. QESS has four main functions that satisfy the QoS requirements discussed above: 1) QoS integration from both network level and end-system level; 2) QoS abstraction and QoS mapping; 3) dynamic service selection and monitoring; and 4) service discovery and management.

3.1 QESS Framework Overview

As illustrated in figure 2, QESS is a hierarchical framework with three layers from bottom to top: network layer, resource pool layer and virtual service layer.

In the network layer, an IntServ/DiffServ hybrid network control model is used to maintain the network level QoS for the Internet transmission. Furthermore, the upper resource pool layer and virtual service layer cooperate together to provide end-system level QoS for grid service applications.

3.2 IntServ/DiffServ Model with Programmable Network Technology

IETF presents the Integrated Services (IntServ) model [9] as the extension to the traditional best-effort model. IntServ uses Resource Reservation Protocol (RSVP) as its working protocol. By explicit signaling negotiation, the transmission QoS can be guaranteed. However, because of the totally distributed control, all the intermediate routers have to keep the reservation information for each connection, which limits the scalability of this model. In the grid environment, a larger amount of services and resources are provided and the potential users are more complicated and dynamic, the signaling out-of-band will inevitably bring huge additional burden for network transmission and make it difficult to control. Differentiated Services (DiffServ), which completely eliminates storage of connection states in routers, is an alternative [10]. DiffServ uses in-band signaling that transmitted with the data in data-stream together. The data flows can be classified at the edge of Internet thus decrease the transmission burden to the networks. DiffServ has good expansibility but, for lack of explicit reservation, it cannot provide the end-to-end QoS guarantee.

In QESS, we use an IntServ/DiffServ hybrid model [11] that combines IntServ with DiffServ. The IntServ regions are deployed at the edge of Internet that connects with grid services end-systems directly. All the data flows in these regions use RSVP to exchange QoS requirements and make resource reservations. Here, IntServ regions are all small networks compared to the Internet backbone, thus the processing burden can be greatly decreased. When crossing the Internet backbone, which is a DiffServ region, the end-to-end RSVP messages are tunneled.

IntServ/DiffServ model can provide both quantitative as well as qualitative QoS guarantees for grid services system but it also brings a handicap to the system implementation. The programmable network [12] is a helpful technology to facilitate and simplify the network deployment. In programmable network, the packets can be active in the sense that they can carry executable code. This code can be executed at designated routers to achieve some particular functions. Hence, QoS parameters and QoS control information can be kept during delivery. Also, using programmable network, some other network QoS management policies, congestion control policies and routing optimization policies can be deployed and configured dynamically.

3.3 Virtual Service Layer

Virtual service layer is the upper layer of the grid services system. This layer defines APIs and provides communication interfaces for the users. It is the "access point" of all services provided in the grid system.

All users attempting to land on this grid services system will register at the *user management unit (UM)* first, which stores user information at a special *User Information Library* and carries out user identity authentication. *UM* also negotiates *Service Level Agreement (SLA)*, which sets the service level and QoS requirements of the user, with the landing users. The *SLA* will also be used to execute data packets marking at the edge routers when data flows are delivered over the network layer.

Resource Broker (RB) is the "heart" of the virtual service layer. After authenticated, the service request is relayed to *RB*, which is responsible for understanding of the request content, discomposing the request, applying and selecting resources on behalf of the user. During this procedure, other four units called: *QoS Mapping and Abstraction Unit (QMA)*, *QoS Reservation Unit (QR)*, *Service Discovery Unit (SD)* and *QoS Monitoring Unit (QM)* are working collaboratively.

Because of the diversity of resources (e.g. CPU, memory, disk storage, etc), the service capacity of different nodes in the grid system is hard to describe. *QMA* uses an abstract model to abstract and map the resource/service relation thus help to scale the capacity of different nodes dispersed in the grid.

Immediate or advance resource reservation/allocation is managed by *QR* to keep the QoS guarantee for the served users during service period. Once the *SLA* is approved, *RB* sends it to *QR* and sets up a resource reservation.

In the runtime, the QoS capacity status of the whole system changes dynamically. Here, in QESS, *QM* can be used to monitor the grid QoS status in time.

SD is mainly objective to discover services based on QoS attributes, such as CPU speed, memory capacity or storage requirement. Some other kinds of particular service types can also be the service discovery criterion.

RB collects required service information from above four units in the virtual service layer and select one or several best service node(s) to provide service for the user.

3.4 Resource Pool Layer

Resource pool layer is built below the virtual service layer. All nodes that provide service resources are deployed at this layer. Here, regardless of their service capacity, we take those resource units, which are not only close enough but have the identical service scheme, as a service node. Each node can be a single server, a multi-server cluster or other storage devices and can has its own management control mechanism.

Resource Manager (RM) here manages and monitors the status of the nodes in the grid system and submits its collected information to the *QM* at the upper layer periodically. There could be only a single *RM* or several *RMs* in the resource pool layer. Single *RM* is easily implemented and need not take synchronization and collaboration of multi-*RMs* into consideration while multi-*RMs* can also has some advantages. For example, single-point-of-failure problem can be avoided and the management burden of each *RM* can be decreased.

In addition, users do not contact to their serving nodes until the *RB* at virtual service layer successfully selects the suitable service node(s) for them.

4 "ND-SMGrid": A Prototype Implementation

Based on the QESS framework proposed above, we are developing a prototype named "ND-SMGrid", which aims to provide online stream media transmission services in a grid environment. Its architecture can be described as figure 3.

In "ND-SMGrid" architecture, the virtual service layer consists of a user management server, a central processing server and a Metacomputing Directory Service (MDS) server. The arrival users first execute identity authentication and submit ser-

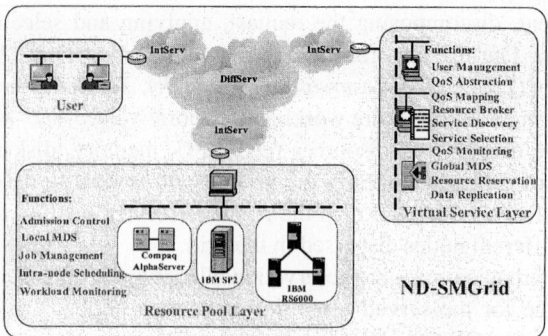

Fig. 3. The "ND-SMGrid" Architecture.

vice requirements through the user management server. This kind of requirements are relayed to the central processing server, which deals with QoS managements including QoS parameters (e.g. media file and bandwidth) mapping and abstraction, QoS reservation, information collection from MDS server, and service selection. Also at this stage, the SLA is negotiated with the user.

MDS [13] is an independent function component in Globus Toolkit. It uses LDAP mechanism to describe and discover service/resource as well as search information.

In our implementation, we use an MDS server to execute global service discovery and periodical QoS monitoring.

If this service selection phase finishes successfully, the most suitable node is assigned to the users, to which they can connect to obtain their services. In "ND-SMGrid", the resource pool layer is made up of three service nodes. The first node consists of two Compaq AlphaServer and the second is an IBM SP2 workstation. Other three IBM RS6000 servers form the third node. All the media files are stored at these three nodes. In addition, we also use a single resource manager in the resource pool layer to manager, monitor and coordinate these nodes.

When the service session is terminated, users will return their resources and the central server at virtual layer will update the record of available resources.

In "ND-SMGrid", resource objects are all media files. So, to improve the availability, we also take data replication into consideration. That is to say, as to those scare and frequently visited files, some data replication schemes are used to make data replications at different nodes, which can increase the possibility to serve more users. This function is performed at the MDS server.

Till now, we haven't completed our whole implementation. Current work is just at the end-system level. But we provide the interfaces for the future implementation on the network level and a preliminary grid stream media services system prototype is coming into being. "ND-SMGrid" can shield the differences of resource nodes and provide a transparent and convenient QoS interface to users.

5 Conclusion and Future Work

To address the QoS issues in a grid environment, we make a comprehensive study on the QoS requirements for high-end applications from both network level and end-system level in this paper. We also propose a QoS-enabled services system framework named QESS that can satisfy these requirements. Based on this QESS framework, a prototype called "ND-SMGrid" that intends to provide large-scale stream media transmission services in a grid environment is being developed and its architecture and implementation are also described in this paper.

"ND-SMGrid" is only a preliminary prototype implementation. Much work, especially at the network level, is needed to be carried out. In the immediate future, our focus will be on the implementation of the whole system and taking more other related complicated issues at network level into consideration. Furthermore, we are to broad our application range from the single media transmission service to general-purpose services thus to form a generic grid services platform that caters to more kinds of applications.

Acknowledgement

We would like to acknowledge our colleagues at Distributed and Parallel Computing Lab, of Nanjing University, for their implementation of "ND-SMGrid" to realize our idea. This work is partially supported by the National Grant Fundamental Research 973 Program of China under Grand No.2002CB312002 and the National High Technology Development 863 Program of China under Grand No.2001AA113050.

References

1. I. Foster, C. Kesselman, "The Grid, Blueprint for a New Computing Infrastructure," San Francisco, Morgan Kaufmann Publishers Inc, 1999
2. I. Foster, C. Kesselman, S. Tuecks, "The Anatomy of the Grid: Enabling Scalable Virtual Organizations," *International Journal of High Performance Computing Applications, 15(3): pp200-222,* 2001
3. D. Miras, B. Teitelbaum, A. Sadagic, et al. "A Survey on Network QoS needs of Advanced Internet Applications", *Working Document of Internet 2,* Dec 2002.
4. I. Foster, A. Roy, V. Sander, et al, "End-to-End Quality of Service for High-End Applications", *IEEE JSAC Special Issue on QoS in the Internet,* 1999.
5. K. Nahrstedt, J. Smith "Design, Implementation and Experiences of the OMEGA End-Point Architecture," *IEEE JSAC, Special Issue on Distributed Multimedia Systems and Technology, 14(7): pp1263-1279,* Sept 1996.
6. I. Foster, C. Kesselman, C. Lee, et al "A Distributed Resource Management Architecture that Supports Advanced Reservations and Co-allocation," In *Proc. IWQoS'99, pp27-36,* 1999.
7. R. Al-Ali, O. Rana, D. Walker, S. Jha, and S. Sohail, "G-QoSM: Grid Service Discovery using QoS Properties," *Computing and Informatics Journal, Special Issue on Grid Computing, vol. 21, no. 4, pp.363–382,* 2002.
8. P Nanda, A Simmonds, "Providing end-to-end guaranteed quality of service over the Internet: a survey on bandwidth broker architecture for differentiated services network," In *Proc. 4th International Conference on IT (CIT 2001), pp. 211-216,* Berhampur, India, Dec 2001.
9. B. Braden, D. Clark, S. Shanker "Integrated Services in the Internet Architecture: An Overview," *Internet RFC 1633,* June 1994.

10. S. Blake, D. Black, M. Carlson, et al. "An Architecture for Differentiated Services", *Internet RFC 2475*, Dec 1998.
11. Y. Bernet, R. Yavatkar P. Ford, et al., "A Framework for Integrated Services Operations Over DiffServ Networks", *Internet RFC 2998,* Nov 2000.
12. A. T. Campbell, H.G. Meer, M.E. Kounavis, et al., "A Survey of Programmable Networks," *Computer Communication Review, vol.29, no.2, pp.7-23,* Apr.1999
13. G. Laszewski, S. Fitzgerald, I. Foster, et al. "A Directory Service for Configuring High-Performance Distributed Computations," In *Proc. 6th IEEE Symposium on High-Performance Distributed Computing, pp.365-375, 1997*
14. K. Yang, X. Guo, A. Galis, et al. "Towards Efficient Resource on-Demand in Grid Computing," *ACM SIGOPS Operating Systems Review, v.37 n.2, p.37-43,* April 2003.

State Management Issues and Grid Services

Yong Xie[1] and Yong-Meng Teo[1,2]

[1] Singapore-Massachusetts Institute of Technology Alliance, Singapore 117576
xieyong@mit.edu
[2] Department of Computer Science, School of Computing,
National University of Singapore, Singapore 117543

Abstract. Recent development on Web/Grid services have achieved some success but also exposes several open issues such as state management. In this paper, we analyze the requirements for state management. We compare different ways of achieving state management in *Web Service*, *Grid Service* and the recent *Web Service Resource Framework*, and articulate the costs and benefits of each approach in terms of fault tolerance, communication cost, etc. We propose a new approach for grid service state management and its prototype implementation.

1 Introduction

Since the invention of the World Wide Web, one of the biggest challenges for computer scientists is to define the ways how components collaborate to carry out applications distributed over the internet. The concept of Web Service is proposed in the year 2000, together with a group of Technologies, such as Simple Object Access Protocol (SOAP) [1], Web Services Description Language (WSDL) [2], and Universal, Description, Discovery and Integration (UDDI) [3]. There has been some success in deploying Web Service in industry such as IBM's WebSphere, BEA's WebLogic, and Microsoft's .NET platform.

Recent works [4] on standardizing of Grid computing focus on the idea of Grid Services, which is defined by the Open Grid Service Architecture (OGSA) [5], and is specified by Open Grid Service Infrastructure (OGSI) [6]. The propose Grid service extends the Web service by introducing the concept of a service factory to differentiate service instances, so that *states* can be encapsulated inside the instance. However, the introduction of states open up a number of issues. For example, in the Globus Toolkit [7], deploying new Grid services to the hosting environment requires the restart of the service container, which causes all ongoing services and persistent services to be stopped.

Arising from the recent collaboration between the Web service community (mainly industry) and the Grid service community (mainly academia) in the evolution of a grid standard called Web Service Resource Framework (WSRF) [8], which separates state management from the stateless Web service.

In this paper, we first define services' state, and analyze the requirements for state management. We compare different models and state management such as in Web service, Grid service (OGSI) and WSRF, and analyze the costs and benefits in terms of fault tolerance, scalability, etc. We propose a new approach for state management and implemented a prototype demonstrate the functionality in managing states using the Globus Toolkit Version 3.

The rest of the paper is organized as follows. Section 2 overviews Web/Grid service and articulated the requirements of state management in the lifecycle of services and service instances. In Section 3, we survey existing approaches in modeling and managing states namely Web service, Grid service (OGSI) and WSRF, and compare them in terms of targeted applications, fault tolerance, communication cost, etc. We then propose a new approach in Section 4. Section 5 presents the prototype we implemented using Globus Toolkit 3. Our concluding remarks are in Section 6.

2 Fundamentals

Figure 1 shows the lifecycle of web/grid services. A service requestor or client is the software component that consumes a Web/Grid service supplied by the service provider. A service requestor creates instances (sessions) of the service and start execution of the application. The instance/session of service can be destroyed either by a destroy request from the requestor or by the expiry of its lifetime. The service provider maintains the service once it is deployed. Undeploying takes it out of service and ends the service lifecycle. In the event of failure, recovery mechanisms are needed for both the service provider and service requestor for restoring state information, etc.

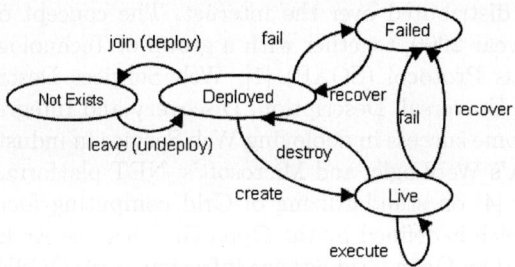

Fig. 1. Web/Grid Service Lifecycle

2.1 State and State Management

There have been many discussion on the "state" and related "stateless/stateful web services", but the definition of state is still elusive. One of the major causes of the confusion in today's Web/Grid service community regarding state management. The state of a web service is rather general, and it comprises of one or more of the following:

– *Service's internal data or attributes needed to persist across Multiple invocations.* In the example of a simple calculator, the state of the calculator is the internal integer's value which is being used for addition and subtraction in every invocation of the service [9].

- *Web service's internal data or attributes needed to maintain for multiple clients.* This is typically related to the clients' account Information, like client's bank account number in a banking web services. [10].
- *Web service's internal data or attributes needed to deal with the dependencies among multiple services.* Information such as other services' address, status, and version numbers are needed to be kept persistent.
- *Service's status at system level.* Information such as service's life cycle, system workload, is needed for monitoring and maintenance purposes.

Based on the definition of state, we found that existing discussions of stateless/stateful service [8–11] only address certain aspects of the state. In general, most of the web services today are stateful, i.e., state information is kept and used during the execution of applications.

State management deals with how state information is stored, retrieved and maintained. State management is required throughout the life cycle of services, which includes *Joining (deployment)* of services, *Creation* service instances, *Execution* of the services, *Destruction* of services, *Leaving (undeployment)* of an existing service, *Failure (instance failure and service failure)*, and *Recover*: recovery mechanisms are needed to restore states after failure.

3 Analysis

Table 1 shows the comparisons of three approaches for modeling and managing state information and our proposed method.

3.1 Web Service

Web service has been widely regarded as stateless and non-transient. However, some vendors proposed the idea of "stateless implementation, stateful interfaces" [12]. Typical implementation requires the use of client side HTTP cookie/session and/or a dispatcher-like software component at the Web service site to decide which internal service component to invoke, and where the saved state can be retrieved.

So far, Web services are targeted mainly at industrial applications, which are usually short-lived and require little state with simple data structure or even none [10]. States are typically stored at a fixed location, i.e. either at requestor side or Web service site as shown in Figure 2(a).

With the stateless implementation of Web service, the join of a new service should not affect existing services. In a typical stateless Web services, the communication overhead is just the round trip cost between the requestor and the service provider. The cost for transferring states depends on where the state information is kept, and the size of the state information. Storing states at the service side incurrs less communiation overhead, because no transfer of states is needed, but managing state information requires more scalable mechanism than simple database or filesystem. A stateless Web service can be restarted after

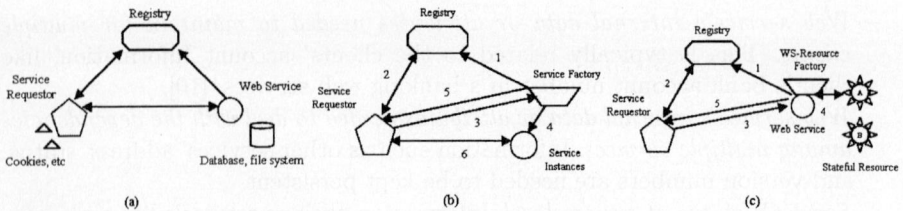

Fig. 2. (a) Web Service Model; (b) Grid Service Model; (c) WS-Resource Framework Model

failure without concerning its history of prior interactions. Even with the stateful interface, as state information is being communicated through invocation of the service or accessing of some other system components, failure of the service should not cause any loss of state information. The leaving of a service is similar to the joining process, which does not affect other services.

If state information is maintained by the requestor, the state is secure in the sense that the requestor has full control. However, if the state information is stored at the service provider side, extra work by the provider is needed to ensure the security. Both ways of maintaining state information would limit the scalability for the requestor and service provider, if the number of requestor, the number of invocation per requestor, the size of state information increase. To our knowledge, so far, there is no standard naming scheme for states to support sharing of states.

3.2 Grid Service

The Grid computing community has developed Grid standards: OGSA [5] and OGSI [6]. The Globus Toolkit 3 (GT3) [13] is a reference implementation of the OGSI specification, and has become an emerging standard for Grid middleware.

When a new service joins, it is deployed on the service provider's host, and registered with the runtime environment. In GT3, this requires a restart of the service container, which will affect all existing services in the same container. The detailed steps involved are shown in Figure 2(b).

As the state information that is associated with the instance itself, the communication cost in Grid service is smaller comparing to the Web service case. However, all state information will be lost together with the instance when the service fails. The leaving of a service is similar to the joining process, which involves update to the runtime environment. In GT3, this involves restart of the service container. Naming of the state information can be achieved using the service instance's Grid Service Handle (GSH). Although the OGSI specifies that the service instance can be remotely created on the Grid, such feature is not supported in the current version of Globus Toolkit, which would limit its scalability.

3.3 WS-Resource Framework (WSRF)

The WS-Resource framework was proposed as a refactoring and evolution of OGSI aiming at exploiting new Web services standards by a team composed of major players from both Grid service and Web service communities [8].

To create a new stateful service, besides the stateless Web service providing the function, a WS-Resource factory, which is a Web service to create the instance of a stateful resource, is also required. The steps involved in WSRF are shown in Figure 2(c).

WSRF aims to fulfill the requirements from both industry and academic applications in an interoperable way [8]. As both the Web service and the WS-Resource factory are stateless Web services, the joining and leaving of a service is the same as it is in Web services, which will not affect other services. The additional communication cost depends on how the stateful resource is implemented. As the resources can exist independently of the service, failure of Web services will not cause loss of the state information. The uniqueness of the state id can be achieved by using the Web service end point reference [8]. Moreover, as the WS-Resource factory is specific to a given web service, all states are managed by the provider. As a result, the service provider must be trusted, and it should not be compromised. The tight coupling between the factory and service may lead to some scalability issues. For example, there is a limit to the number of states that a service provider can handle. It is not clear whether the requestor has the flexibility to change the policy of state management dynamically for reasons like the requestor no longer trusts the service provider to store its states.

4 Proposed Grid Service Model

From detailed analysis, we note that although WSRF separates state management from the stateless web service and provided a mechanism for state management, it still has some limitations. Firstly, the service-specific WS-Resource factory and state management increases the complexity of developing a new service. The tight-coupling between the resource factory and the service restricts the scalability. More importantly, in the WSRF, it is the service provider who decides where to store the stateful resource and where the state management is carried out, which may introduce security problems, as the requestor and provider may have different security policies. For example, in a critical application involving a company's confidential information, the requestor of a web service does not want to expose the state information to others. In other words, although the service itself is trusted by the requestor, the security policy (like access control) of the service provider under a different administrative domain may not be trusted. Moreover, the flexibility of choosing the location of the state repository is not supported in WSRF either.

To overcome these limitations, we propose a new state management framework based on the WSRF. In our framework, we extend the idea of "separation of state management from Web service" by introducing the concept of a generic state management service. This service takes in the initial state in the form of

Table 1. Comparison of State Management Schemes

		Web Service	OGSI Grid Service	WSRF	Proposed Model
	Target Application	Industry, simple state, short-lived	Academic, complex state, long-lived	Industry and academic	Industry and academic
Schemes	State Repository	Client or Server	Service Instance	WS-Resource	Stateful Resource
	Service Join	No effect on others	Restart Container	No effect on others	No effect on others
	Service Failure	No loss of State	Loss of State	No loss of State	No loss of State
	Service Leave	No effect on others	Restart Container	No effect on others	No effect on others
Other Issues	Communication Cost	Client – service, depending on where state is stored	Client - Service	Client – Factory, Client – Service, Service – WS-Resource	Client – Service (for initialization), Client – State Management Service, Client – Service (for service interaction), Service – State Management Service – Resource
	State Security	• Non-transient for stateless service • State on client: secure • State on server: require extra work from dispatcher	Encapsulated in service instance: secure	Secure, given access control to resources and server is trusted by requestor; otherwise, not secure as server can access state resource	Client controls where to store state information: secure with access control
	Scalability of State Information	Both ways of maintaining state information have restriction on scalability.	Remote deploy service instance: scalable (GT3 does not support)	Tight coupling between factory and service, only allow one copy of factory: not scalable	Loose coupling between state management service and web service, allow multiple copies of state management service: scalable
	Naming of states	Not supported	Using GSH	WS end point reference	Many ways: e.g. combine service id, client id, and application id.

a XML document, stores it as a stateful resource, and returns the URI of the resource to the requestor. The generality makes the state management and its service loosely-coupled by setting the identifier of the stateful resource to be universally unique. Different implementation of the generic state management service could support different mechanisms for handling states such as security policy. Also interoperability can be achieved by standardizing the interface for state management service to interact with the requestor, service provider and other state management services as well.

The process of using the Web service in our framework is illustrated in Figure 3, and two steps are involved, namely the initialization step and the interaction step. The initialization step is to create the initial state resource: (1) a requestor sends the request to the desired Web service; (2) the Web service returns the initial state to the requestor; (3) the requestor combines with the initial state information provided by its own, then contacts the state management service and passes the initial state with an optional parameter to specify the type of stateful resource (e.g. XML file); (4) the State Management Service stores the initial information as a stateful resource; (5) URI of the stateful resource is returned to the service requestor. The interaction step involves: (1) the Web service takes in the URI of the stateful resource from the requestor,

(2)–(4) executes the service by accessing/updating the stateful resource though the state management service, (5) returns the service result if applicable.

Besides the reduction of work for service developers, this framework gives the control of state management to the service requestor by allowing the requestor the ability to deploy the state management service at the desired place (such as requestors' local host). Inter-dependencies of state information can be handled by the interactions between state management services, but the detailed design is out of the scope of this paper. In addition, scalability is enhanced by the flexible deployment of the state management service. Unique naming of the states can be achieved by combining the client id, service id and application id similar to WSRF.

As compared with WSRF, an additional round trip communication is needed in our framework between the requestor and the Web service at the initialization step. Moreover, the access of state information is done through the interaction between the web service and the state management service. Though additional communication overhead is incurred, this cost can be reduced by utilizing locality for deploying the state management service. (Analysis of the proposed framework is summarized in the Table 1).

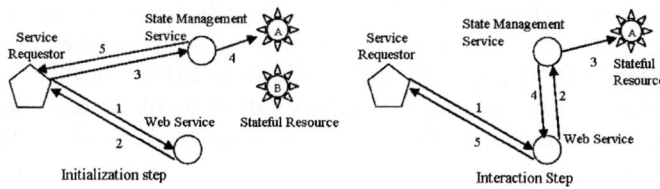

Fig. 3. Proposed Grid Service Framework

```
public interface State {                          public interface Calc {
  public String create(String doc, int cid);        public String start();
  public void set(String uri, String doc);          public int add(String uri, int b);
  public String get(String uri);                    public int subtract(String uri, int b);
}                                                   public int multiply(String uri, int b);
                                                    public int divide(String uri, int b);
                                                  }

              (a)                                                  (b)
```

Fig. 4. Interfaces for Web Service in the Prototype: (a) Interface for the state management service; (b) Interface of the sample calculator web service

5 Implementation

A prototype is implemented to illustrate our idea of separating state management from web service by using a generic state management service. We have developed two persistent services (Web Services) using the Globus Toolkit version 3. One is a calculator service (CalcService) providing the computation functionalities and the other is a general state management service (StateService). The interfaces for these two services are shown in Figure 4. In this prototype, simple state information is maintained as the current computation result.

A client program is implemented to use the `CalcService`. It first calls the 'start' method in Calc to get the initial state (e.g. 1), and then invokes the 'create' operator in `StateService`, passing the state and its client id. The `StateService` creates a XML file holding the state information and returns its *URI*. Here we model the *URI* as: host name + storage directory + 'state' + client-id + '.xml' to ensure the uniqueness. Then the client is able to call the computation functions provided by `CalcService`, such as 'add', by providing the *URI* of the state information. The information can be retrieved or updated by `CalcService` through the 'get' and 'set' operations in `StateService` respectively.

6 Conclusion and Future Work

In this paper, we have clearly defined the state of services, and articulated the requirements for state management. We analyzed different ways of modeling and managing states in Web services, Grid services and WS-Resource Framework, and compared them in terms of target applications, fault tolerance, etc. We proposed a new framework for state management, and compared it with existing state management schemes. We have also described our prototype based on the Globus Toolkit 3. More work needs to be done to reduce the additional communication cost in our framework such as to utilize locality of the state management service and the state repository. Other details such as standardizing the naming or addressing of the stateful resource and access control can be further improved in future as well.

Acknowledgment

The authors like to express their thanks to Ai Ting and Wang Caixia for discussion and comments in an earlier draft of this paper.

References

1. SOAP: Simple object access protocol. http://www.w3.org/tr/soap. (2003)
2. Christensen, E., Curbera, F., Meredith, G., Weerawarana, S.: Web services description language (wsdl). http://www.w3.org/tr/wsdl (2004)
3. UDDI: Discovery universal, description and integration. http://www.uddi.org/. (2004)
4. Foster, I., Kesselman, C., Nick, J., Tuecke, S.: The physiology of the grid: An open grid services architecture for distributed systems integration (2002)
5. OGSA: Open grid services architecture. http://www.globus.org/ogsa/. (2003)
6. OGSI: Open grid services infrastructure.
   https://forge.gridforum.org/projects/ogsi-wg. (2003)
7. Foster, I., Kesselman, C.: Globus: A metacomputing infrastructure toolkit. The International Journal of Supercomputer Applications and High Performance Computing **11** (1997) 115–128

8. WSRF: The ws-resource framework. http://www.globus.org/wsrf/. (2004)
9. Sotomayor, B.: Tutorial of globus toolkit version 3 (gt3). http://www.casa-sotomayor.net/gt3-tutorial/. (2003)
10. Shaikh, H.H., Fulkerson, S.: Implement and access stateful web services using websphere studio. http://www-106.ibm.com/developerworks/webservices/library/ws-statefulws.html (2004)
11. Vogels, W.: Web services are not distributed objects. http://weblogs.cs.cornell.edu/allthingsdistributed/archives/000343.html (2004)
12. Foster, I., Frey, J., Graham, S., Tuecke, S., Czajkowsld, K., Ferguson, D., Leymann, F., Nally, M., Sedukhin, I., Snelling, D., Storey, T., Vambenepe, W., Weerawarana, S.: Modeling stateful resource with web services. http://www-106.ibm.com/developerworks/library/ws-resource/ws-modelingresources.html (2004)
13. GT3: Globus toolkit version 3. http://www.globus.org/. (2004)

Transactional Business Coordination and Failure Recovery for Web Services Composition*

Yi Ren, Quanyuan Wu, Yan Jia, Jianbo Guan, and Weihong Han

School of Computer Science,
National University of Defence Technology, Changsha 410073, P.R. China
renyi@nudt.edu.cn

Abstract. Web services have become a new wave for Internet-based business applications. Through compositions of Web services available, value-added services can be built. More and more enterprises have ventured into this area staking a claim in their future. However, there still exists a great challenge to perform transactions in the loosely coupled environment. Traditional ACID transaction model is not suitable and failure recovery algorithm ensuring semantic atomicity is needed. In this paper, we propose a process-oriented transactional business coordination model with hierarchical structure. It allows dependencies across hierarchies and supports various transactional behaviors of different services. Based on this model, a novel failure recovery algorithm is presented. It can ensure semantic atomicity for long running Web services composition. Analysis shows it can effectively reduce compensation sphere and thus decrease costs. The correctness and implementation issues of this algorithm are also presented.

1 Introduction

Web services technology has become a new wave for Internet-based business applications. As there are more and more Web services available, Web services composition technology is proposed, which enables one business to provide value-added services to its customers by composing basic Web services, possibly offered by different autonomous enterprises [1]. The composition can be complex in structure, often with semantic atomicity requirements [2]. Moreover, its execution may take a long time to complete. Transaction is used as a common technique to ensure reliability and execution consistency since it has well known ACID properties and provides guaranteed consistent outcome despite concurrent access and failures. However, traditional ACID transaction model is not suitable for Web services environment:

- Lock mechanism is not feasible since underlying system may not support or allow locking over resources accessed by autonomous Web services, and long time locking may decrease resource utilization and lead to starvation.
- Long time interactions with other concurrent and long running services break atomicity boundaries of traditional transactions. Traditional compensation method turns to be expensive due to frequent and unpredictable failures.

---
* For detailed information, an extended version of this paper is recommended by the authors.

Thus new model and correlated mechanisms are required. As mentioned, locking mechanism is not recommended. Currently, compensation and contingency are used to handle failures caused by resource access conflicts or other reasons. The primary goal of transactional failure recovery is to meet semantic atomicity requirements for Web services composition. But existing works are mostly about the transaction model, how to ensure atomicity requirements properly is rarely addressed and remains unresolved. Since transaction model and failure recovery mechanism are closely related, in this paper we first propose a transactional business coordination model, based on which a failure recovery approach is presented.

2 Related Work

There have been efforts done for Web service transaction processing in both industrial and academic areas. The most representative ones are Business Transaction Protocol (BTP) [3], WS-C/WS-T [4], WebTransact framework [1], transactional region of eFlow [5] and Agent Based Transaction Model (ABT) [6]. According to the way to perform transactional coordination, we classify above works into strict-nested and process-oriented. The coordination mode of BTP and ABT is strict-nested. And that of WebTransact, WS-C/WS-T and eFlow is process-oriented.

Strict-nested coordination organizes participant services into a strict-nested structure. It usually adopts extended 2PC and is prone to be accepted. But it neglects the fact that compositions are often complex and with dependencies between services. Therefore, how to enact the execution of complex Web services composition and at the same time to drive transactional coordination in a 2PC fashion is difficult to resolve and remains an open problem. Moreover, autonomous and heterogeneous services may not support 2PC interfaces. Thus it needs to be driven by business logics. This is often a burden to application developers. While under process-oriented mode, the composition is a process of services. Participant services are executed and coordinated by pre-specified business rules. In case of failures, failure recovery mechanism is activated to ensure transactional requirements. It integrates execution model of Web services composition with its transactional guarantee in a natural way. Thus we believe that process-oriented mode is more practical and preferable for Web services composition.

Existing process-oriented methods have contributed a lot to transactional coordination of Web services composition. However, there remains research work to be done: WebTransact lacks failure recovery ensuring transactional properties; WS-C/WS-T describes a generalized framework and its coordination protocols, further efforts are needed to put it into practice; eFlow's ACID transactional region is too restrict for long running transactional compositions.

3 Transactional Business Coordination Model

3.1 Two Different Views of Web Services Composition

Web services composition can be represented as a directed graph, which gives the view of an execution flow. It also has semantic hierarchical view upon the execution flow. A composition is built up with elementary and composite services. It can be semantically decomposed until they are elementary services. Thus hierarchical structure is formed. Dependencies across hierarchies are possible to exist.

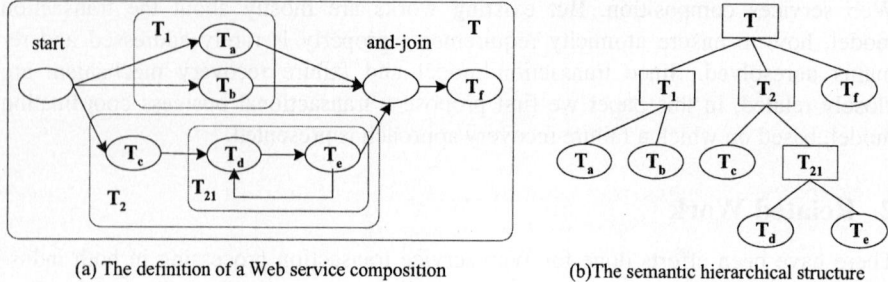

(a) The definition of a Web service composition   (b) The semantic hierarchical structure

Fig. 1. An example of Web services composition.

Fig. 1(a) illustrates flat execution view of a Web services composition. The ellipses are elementary services and blunt-squared rectangles indicate composite services. Assume T is to *plan a business travel from Changsha to Beijing*. T_1 is for *transportation and accommodation reservation*. T_2 takes charge of *payment*. T_f is to *acknowledge the customer*. T_1 and T_2 can be further decomposed, with T_a *reserving a plane ticket from Changsha to Beijing*, T_b *reserving a hotel room in Beijing*, T_c *calculating the fee* and T_{21} *charging the customer*. T_d and T_e of T_{21} is *invoicing* and *paying* respectively. Fig. 1(b) is the semantic hierarchical structure extracted from the flat view of fig. 1(a). The rectangles represent composite services, which are compositions of their lower layer services. Dependencies across hierarchies are possible to exist. In above example, T_{21} and T_1 are arranged concurrently to enhance efficiency of the composition. It is assumed that T_1 will most often success. But logically, T_{21} is control dependent on T_1. Charging can be executed only after transportation and accommodation reservation success. So the dependency must be considered when failures occur in the execution of T_1.

3.2 Semantic Atomicity

It is one of the typical characteristics of heterogeneous Web services environment that services may have diverse transactional behaviors. So we introduce the concepts of Recovery Mode and Recovery Policy.

Definition 1 Recovery Mode (RM). *RM is the property that specifies atomicity feature of a Web service. RM of a service may be atomic if it has atomicity requirement or non-atomic if it is not transactional. For instance, RM of T_1 in section 3.1 is atomic since T_a and T_b must either success or abort.*

Composite service S forms an atomic sphere if the following conditions are both satisfied: i) RM of S is *atomic*, and ii) S has no parent composite service or RM of its parent service is *non-atomic*.

Definition 2 Recovery Policy (RP). *RP is the property to describe recovery features of elementary services. The possible values of RP are non-vital, compensable, retriable, or critical. RP of a service is determined by its semantics.*

A *non-vital* service is effect free, the results of which will not influence other operations. A service is *compensable* if its effects can be undone by its compensating services. And a service is *retriable* if it can finally terminate successfully after being retried for finite times. A service is *critical* if it is not *retriable*, *non-vital* or *compensable*. RP of composite services is decided by RP of its participant services. Note that critical service is not *compensable*. Assume S is a composite service, s_c is one of the composing services of S and s_c is *critical*. If S fails at a *compensable* service after s_c has been finished and start point of alternative executing path is before s_c, then S cannot be compensated and recovered forward.

Definition 3 Semi-atomicity. *A service S meets semi-atomicity property if one of the conditions is satisfied: i) S finishes successfully (either without failures or survive by compensation and contingency[1] methods); ii) S fails, but is still compensable.*

We argue that even semi-atomicity is to be relaxed in Web service environment. Traditionally, sub transactions must to be carefully structured to be well-formed for semi-atomicity [7]. Furthermore, sometimes only one critical sub transactions is allowed. This is not realistic, since Web services environment is failure-prone, heterogeneous, autonomous and human intervention is needed when necessary. Semi-atomicity is further relaxed into semantic atomicity. Semantic atomicity is ensured so long as semantic termination guarantees of Web services are satisfied.

3.3 Web Service Transaction

According to previous analysis, we believe that Web services composition in fact can be considered as long running transaction with appropriate RM and RP defined for its participant services. Furthermore, hierarchical structure can be built upon its directed graph of execution flow, with dependencies across hierarchies specified.

Definition 4 Web Service Transaction. *Web service transaction is long running transaction corresponding to a Web services composition, which can be abstracted as a tuple $T_{WSC} = (T, \prec_t, Dep, Attr, RS, Cons)$, where:*

- **T** *is a set of sub transactions, each of its sub transaction is also a set of sub transactions if it has any. Thus T is hierarchical structured.*
- $\prec_t$ *is the set of control flows and data flows defined as business rules among leaf sub transactions of T.*
- **Dep** *is the set of dependencies on transactions across hierarchies of T.*
- **Attr** *is the set of tuple (t, rm, rp) defined on all the transactions of T, where t is a transaction in the hierarchical structure of T, rm represent its RM property and rp represent its RP property. Default RM is atomic.*
- **RS** *is the set of resources accessed by T. The resources may be memory data, database records, files, etc.*

---

[1] Semantic equivalent services are used to satisfy termination guarantees of the composition.

- **Cons** is the set of constraints on all sub transactions. For example, a transaction must be finished before Monday 12:00 pm.

Definition 4 puts forward a transactional business coordination model. T and Dep define semantic hierarchical structure with dependencies across layers. $\prec$ specifies temporal order of the elementary services and reveals execution flow. *Attr* defines transactional behaviors for participant services. The services are coordinated according to $\prec_t$. When failures occur, T, Dep and *Attr* are referred. In next section, we describe a novel failure recovery algorithm based on the proposed model.

4 Transactional Failure Recovery

Mission critical Web services compositions should have their transactional requirements, especially semantic atomicity requirements guaranteed in case of failures. Next, we will discuss transactional failure recovery algorithm TxnFailureRecover with following problem considered:

1. Web services compositions are often complex with dependencies between executing services, so identifying compensation sphere is a non-trivial work.
2. Web services may involve databases, files, applications or human intervention. So they are expensive to be compensated. It is important to minimize compensation sphere and avoid unnecessary costs.
3. Semantic equivalent Web services provided by multiple providers are available on the Web. Multi-binding and dynamic binding technologies can be used.

4.1 Algorithm Description

TxnFailureRecover is the proposed algorithm (described in fig. 2). It activates *CGGenerating* to generate compensating graph for current failure service. Compensating graph is a directed graph generated according to execution history and recovery properties. It defines a compensation sphere, or the scope of services to be compensated. When a failure occurs, services dependent on the failure service are compensated even they are not in the atomic sphere. *TxnFailureRecover* is used for transactional failure recovery. It distinguishes diverse RM and RP properties and handles the failure according to semantic hierarchical structure of the composition.

4.2 Correctness

TxnFailureRecover will terminate only under one of the following two circumstances:

1. It is automatically recoverable in the atomic sphere, i.e., the FailureService is non-vital, or can be retried, or is compensable and has contingencies, etc.
2. It is not automatically recoverable, i.e., it cannot recover forward even the top of atomic sphere is reached or critical service is found. The administrator is notified.

```
Algorithm TxnFailureRecover
Input: ID of the failure service
{  Abort executing services;
   CurrentCG:=NULL; //initial current compensating graph as NULL
   1: switch(FailureService.RP) //according to RP of failure service
   {  case Retriable: restart FailureService; break;
      case Non-Vital: break;
      case Compensable:
            if ((there are appropriate services in multi-binding ad-
                 dresses) OR (DynamicServiceAgent returns contingencies))
            {  CurrentCG:=CGGenerating(CurrentService.ID)
               enact compensation with CurrentCG; //backward recovery
               activate the optimal contingency; //forward recovery
            } //if there are alternatives
            else //is not recoverable in current layer, turn to higher
                 layer
            {  if ((FailureService has ParentService) AND (ParentSer-
vice.RM = atomic))
                     {FailureService:=ParentService; goto 1;}
               else
               {  CurrentCG:=CGGenerating(CurrentService.ID)
                  enact compensation with CurrentCG; //backward recovery
                  //is not auto-recoverable, human intervention is needed
                  notifyAdmin;
            } } break;
      case Critical: //is not auto-recoverable, human intervention
            notifyAdmin;
   } return;
}
```

Fig. 2. Transactional hierarchical failure recovery algorithm.

Therefore, when it cannot be recovered in current hierarchy, the failure will be propagated to higher hierarchy until the composition is recovered or the system administrator is notified. So what *TxnFailureRecover* ensures is relaxed atomicity.

Theorem 1. *A composition instance S meets semi-atomicity property if it is automatically recoverable in the atomic sphere or no critical service is encountered during its failure recovery.*

Proof: Omitted due to space constraints.

Thus when failed composition is not automatically recoverable in the atomic sphere and there are critical services in failure recovery, atomicity requirement is further relaxed to semantic atomicity. It is acceptable since termination guarantees can be ensured by human intervention. *TxnFailureRecover* can promote the degree of automatism since it supports both backward and forward recovery by rich transactional information and diverse recovery methods. The algorithm has the capability to determine the boundary of compensation sphere. The sphere is confined to a layer as low as possible. So the number of compensating services to be executed can be reduced. Thus the compensation costs can be potentially reduced.

4.3 Analysis of Compensation Cost

The most common used two compensation methods are complete compensation and partial compensation. The former is to compensate inversely from failure point to start point. While the latter is to rollback to a predefined save point. But it is often wasteful even with partial compensation. First, assume there no dependency across hierarchies. Then the lowest automatically service in the atomic sphere is identified as compensation sphere. Only when it is not automatically recoverable, compensation sphere will be the same as atomic sphere. Compared with complete compensation and save point based partial compensation method, the compensation sphere is reduced and unnecessary costs are avoided. Then consider the situation with dependencies between services across hierarchies. The worst case is that the compensation graph is the same as that generated with complete and partial compensation methods. Otherwise the compensation sphere is reduced.

4.4 Failure Recovery Mechanism in StarWSCop

As shown in fig. 3, failure recovery mechanism in StarWSCop prototype is mainly composed of Exception Handler, Transactional Failure Recovery Manager, Compensation Service, Log Service, Dynamic Service Agent, etc. *TxnFailureRecover* and *CGGenerating* are implemented in Transactional Failure Recovery Manager and Compensation Service respectively. *Exception Manager* receives exceptions from the Engine. Registered exceptions are handled by handlers. Unregistered exceptions are forward to Transactional Failure Recovery Manager as failures. *Transactional Failure Recovery Manager* reads information of composition definition from Service Repository in case of a failure. According to the semantic hierarchical view and transactional attributes, failure recovery is fulfilled. Compensation Service and Dynamic Service Agent will be used if necessary. *Compensation Service* can get execution history information from Log Service. It receives requests from Transactional Failure recovery Manager to generate compensating graph for services. *Log Service* records the execution events during Web services composition is in process. It provides interfaces to return relevant persistent history information. *Dynamic Service Agent* is responsible for dynamic service binding requests. It searches for matching Web services in the UDDI Registry and chooses the suitable one for forward recovery.

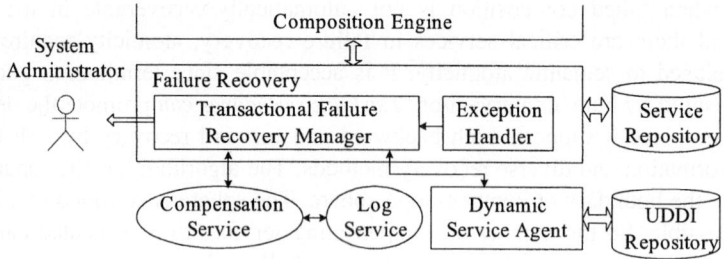

Fig. 3. Failure recovery in StarWSCop.

5 Summary

The transactional business coordination model and failure recovery mechanism are proposed to build reliable Web services compositions. However, there are other application domains where they are useful, such as Business Grid. Since Grid is moving towards Web service concepts and technologies, representing by WSRF [8], we believe the proposed ideas can be used into Grid platforms.

Acknowledgements

This work is supported by National Natural Science Foundation of China Under grant NO. 90104020 and China National Advanced Science & Technology (863) Plan under Contract No. 2002AA116040.

References

1. P.F. Pires: WebTransact: An Infrastructure for Building Reliable Web Services Compositions. Doctoral Thesis, Rio de Janeiro, Brazil. (2002)
2. H. Garcia-Molina: Using Semantic Knowledge for Transaction Processing in a Distributed Database. ACM Transactions on Database Systems, 6 (1983) 186-213
3. A. Ceponkus et al.: Business Transaction Protocol Version 1.0. OASIS Business Transactions Protocol Technical Committee. (2002)
4. F. Cabrera et al.: Web Services Coordination/Transaction specification. BEA Systems, International Business Machines Corporation, Microsoft Corporation, Inc. (2002)
5. F. Casati et al: Adaptive and dynamic service composition in eFlow. In Proceedings of International Conference on Advanced Information Systems Engineering, Stockholm (2000)
6. Tao Jin, Steve Goschnick: Utilizing Web Services in an Agent-based Transaction Model. In Proceedings of International workshop on Web Services and Agent-based Engineering, Melbourne (2003) 1-9
7. A. Zhang et al: Ensuring Relaxed Atomicity for Flexible Transactions in Multidatabase Systems. In Proceedings of the ACM SIGMOD International Conference on Management of Data, Minneapolis (1994) 67-78
8. K. Czajkowski, D. F. Ferguson, I Foster et al: The WS-Resource Framework, Version 1.0. Blobus Alliance, IBM Inc et al. (2004)

Using Service Taxonomy to Facilitate Efficient Decentralized Grid Service Discovery

Cheng Zhu, Zhong Liu, Weiming Zhang, Zhenning Xu, and Dongsheng Yang

Department of Management Science and Engineering,
National University of Defense Technology, Changsha, Hunan, P.R. China
{zhucheng,zhliu,wmzhang,xzn,dsyang}@nudt.edu.cn

Abstract. In this paper, an efficient decentralized Grid service discovery approach is proposed, which utilizes service taxonomy to address both scalability and efficiency. Grid information nodes are organized into different community overlays according to the types of the services registered, and a DHT (Distributed Hash Table) P2P based upper layer network is constructed to facilitate the organization of community overlays and provide efficient navigation between different communities. As a result, service discovery will be limited into related communities, which improves efficiency. A simple and lightweight greedy search based service location (GSBSL) method is also introduced to identify service providers with high QoS efficiently within the same community. Simulation results show that, the efficiency is improved compared with existing decentralized solutions, and the overhead is acceptable and controllable.

1 Introduction

Fundamental to Grid service virtualization and integration is the ability to discover, allocate, negotiate, monitor, and manage the use of network-accessible capabilities in order to achieve various end-to-end or global qualities of service. Grid service discovery is one of the basic components of Grid system, which discovers qualified service providers for different Grid users. The desirable features of Grid service discovery should be both scalability and efficiency. By scalability, we hope service discovery would not rely on only a few centralized servers, which could be potential performance and security bottlenecks; By efficiency, we refer to the capability of searching and locating qualified service providers with high QoS rapidly without much network and processing overhead.

Service taxonomy provides standard classification for types of services, which is potentially helpful to achieve more efficient service discovery and sharing. In fact, various service taxonomies, for example UNSPSC (United Nations Standard Products and Services Code) and NAICS (North American Industry Classification System), have been supported in centralized service discovery mechanism like UDDI. In this paper, we study how to use service taxonomy to achieve efficient decentralized Grid service discovery. The basic idea is to organize Grid information nodes into different community overlays according to the types of the services registered, and construct a DHT (Distributed Hash Table) P2P [1] based upper layer network to provide efficient navigation between different communities. In this way, service discovery request will be forwarded only to the related communities, which controls the network overhead and improves efficiency. In addition, a simple and lightweight greedy search based service location (GSBSL) method is also introduced to identify service providers with relatively higher QoS among qualified providers efficiently within the same community.

2 Related Work

Scalable and efficient service discovery has always been an important problem in Grid, Web Service and P2P. Resource (service) discovery is mainly centralized in existing Grid systems. Although interactions between information nodes are supported, the general-purpose decentralized service discovery mechanism is still absent [2].

Node organization protocols and resource discovery request processing approaches of resource discovery in P2P networks has been introduced into Grid resource discovery. For example, information nodes are organized into a flat unstructured network in [2] and various request-forwarding policies are studied. A DHT based service discovery approach is also studied in [3], where information nodes are organized into a DHT P2P network, and Hilbert curve is introduced to support fuzzy matching. However, there would be huge overhead to support attribute-based search, especially when attribute values of resources (services) change frequently over time. Another study [4] organizes information nodes into hierarchical topologies to reduce redundant messages. However, the global hierarchical topologies are hard to maintain in dynamic environment, and a well-defined hierarchy does not always exist in Grid environment. An information node grouping technique is studied in [5], in which information nodes are randomly grouped together. Though the searching space is successfully reduced, service publishing and update overhead increases vastly. A common problem in above decentralized service discovery solutions is that, they only address "finding" qualified service providers, but lacks mechanism to further selecting and locating service providers with high QoS for users.

3 System Overviews

To facilitate service discovery, service providers will publish information of services then can offer, which can be accessed via standard interfaces. The nodes implementing interfaces of service (resource) information registry and access defined in Grid specifications are called information nodes in this paper. In decentralized Grid service discovery, there would be multiple information nodes in the system, which forms overlay network so that information and discovery requests can be propagated to the whole Grid.

3.1 Assumptions

Assumption 1. There is a standard taxonomy on Grid services, and each service can be classified into one and only one type under this taxonomy.

Taxonomy is often in form of classification tree, and we consider only the atomic type that cannot be further divided. We treat a service with multiple types as multiple services under this assumption, each having only a single type.

Assumption 2. For each Grid service, there is only one information node as its registry node, and a service is treated to be the local service of its registry node.

Assumption 2 is mainly for the convenience of expression. If a service is registered at multiple information nodes at the same time, we randomly choose one of them as the registry node, and treat others as replicating nodes. Under this assumption, information nodes can be treated as service providers.

Assumption 3. Service discovery request specifies the type of the required service, as well as the constraints on its attribute values.

Under this assumption, a simple example of service discovery request would be (TYPE: sensor; ATTRIBUTE: location = XX region, network bandwidth > 1MB/s ...). In general, various types of services are integrated to form into a VO for an incoming Grid task, and the whole service discovery process can be logically divided into multiple steps where only one type of service is searched at each step. We believe that even the type information is absent in the service discovery request, it would not be difficult to infer the possible type(s) of the required service according to the description on the attributes under the help of service taxonomy.

3.2 Community Based Information Node Organization

Just like grouping nodes sharing similar files in file sharing P2P networks, we group the information nodes with the same type of service together. E.g. for each type t in the service taxonomy, all information nodes with the service of type t are organized together to form an overlay network, which is called community overlay (CO) and represented as $CO(t)$. That's to say, a community overlay $CO(t)$ includes all information nodes that can provide information on service of type t. Given a service discovery request with type t, as long as an entry node of the community overlay $CO(t)$ is obtained, the searching would be confined into a much smaller scope rather than searching all information nodes as a whole.

To facilitate self-organization of communities and efficient request forwarding among different communities, we introduce an upper layer DHT P2P based overlay network called bootstrap network. It is formed by a small number of information nodes, serving as a community entry node registry and lookup network. Each community will have a registry node in bootstrap network, which is the node taking charge of the hash value of the community ID in the DHT node topology space. For example, the successor node of the community ID hash value will become registry node for the community, if Chord protocol is adopted [6]. Entry nodes for each community overlay, usually those powerful nodes, will register and update their address information on bootstrap network periodically. On the other hand, lookup is based on the community ID. With DHT lookup protocol, the lookup initiated from any node on the bootstrap network will finally reach the registry node of the required community, which is guaranteed to succeed within $\log N$ hops with high probability, where N is the size of the bootstrap network.

New information nodes can be bootstrapped into the corresponding community overlays once entry nodes are found, and a lot of overlay network construction and maintenance protocols are available to maintain community overlays. Both theoretic analysis and practical observation show that [7,8], it's possible to construct and maintain connected overlay topology or topology with a huge connected component even though each node knows only a few other nodes in the system, and adjusts its neighbors independently.

The load of bootstrap network comes from both register/update and lookup operations. To control the overall register/update overhead in bootstrap network, the register period can be adaptively adjusted. To control the overall lookup overhead on bootstrap network, lookup request will be forwarded to bootstrap network only after the requested entry node cannot be obtained by querying local cache and neighbors. To address load balance in bootstrap network, there are many load balance techniques available [9,10]. We also put forward a load balance technique based on rumor mongering in [11] to balance lookup request processing load in bootstrap network.

In addition to efficiency and scalability, the DHT P2P based bootstrap network makes our approach adaptive to the possible changes on service taxonomy, because the DHT takes charge of the mapping between community ID and registry node.

3.3 Greedy Search Based Service Location

After an entry node of the community related to the required service is found, various request forwarding strategies can be used to propagate request inside the community, such as flooding, random walk and etc [12]. A main drawback of existing request forwarding strategies is that, there is no comparison between the QoS of qualified service providers, and user may be inundated by the retuning results when there are lots of qualified service providers. Under such case, if the number of results returned is explicitly controlled by limiting the scope of request propagation, it is not guaranteed to find service providers with high QoS.

To deal with this problem, we introduce a simple greedy forwarding strategy: greedy search based service location (GSBSL). The idea behind GSBSL is very simple: when a qualified service providers is found, it will always forward the service discovery request to the neighbor with the best QoS until no neighbor has better service quality. A simple example is shown in figure 1.

GSBSL can be combined with other request forwarding strategies to identify "good" service providers efficiently. For example, limited flooding can be used first to identify qualified service providers inside the community, and GSBSL is used to push the request further. If the discovery results returned by GSBSL are not satisfying, flooding or its variations can be used in the worst case. Implementation of GSBSL is convenient under the help of mobile agent technology. Serving for user's intention and interest, mobile agents can travel from one information node to the other, and decide the next target according to the local information of the docking site.

If the qualified service providers inside a community form a connected overlay topology, we have proved the following results in [13]: 1) the performance of GSBSL improves significantly with the increase of the average degree of the overlay network; 2) it is very likely that this method can find service providers with relatively high QoS in small number of hops in a big community.

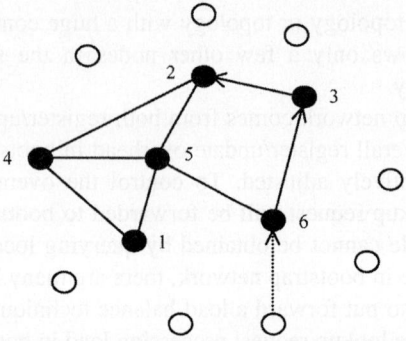

Fig. 1. A simple example on greedy search based service location. Suppose black nodes are qualified service providers in a community, and number reflects the reverse order of the node according to the QoS at the moment. Node will always forward location request to the neighbor with best QoS until no neighbor has better service quality. A request initiated in node 6 will be forwarded to node 2 eventually.

4 Simulation Studies

4.1 Simulation Settings of Decentralized Service Discovery in Grid Environment

In simulation, we ignore the dynamic change of the overlay topology of the information nodes during a service discovery process, and assume all service information to be valid. Similar to [2], we set the size of information nodes to be 10,000, size of total services to be 50,000, and size of service types defined in service taxonomy to be 1,000. In order to evaluate the influence of community size on service discovery performance, we consider 2 cases, in which the requested service falls into the largest community with 5,000 nodes and the 10th largest community with 500 nodes respectively. As to the distribution of services on each node, we have evaluated both balanced and unbalanced cases, which is similar to [2]. However, due to the space limit, we only present the result under balanced service distribution over nodes. There is no much difference between the 2 cases in our simulations.

We compare the performance of our solution with others under the same service density. By service density, we refer to the ratio of number of information nodes owning services satisfying discovery requirement to the population of the whole information nodes.

We use PLOD [14] to generate initial topologies with different size, including each community topology and other topologies needed in simulation with parameter α = 0.3 and average node degree of 4, which simulates topology of large-scale networks with power-law characters. To take the bootstrap network into account, we use evaluation results of Chord system [6] in our simulations directly. For each lookup request forwarded to bootstrap network, we set the hops needed for a successful return to be 5. This is because we expect the size of bootstrap network to be about 10^3 according to the experience of GWebCache [15] in Gnutella, and the average lookup hops in Chord system is $1/2(log_2 N)$. We also set the failure rate of a lookup in bootstrap network to be 5%, which is average value observed in Chord system [6]. If the entry node cannot be looked up, the request will be randomly forwarded between communities. Node will only forward lookup request to bootstrap network when the needed entry node cannot be found in the local host cache of all neighboring nodes within 2 hops.

In community overlay maintenance, we simulate ping/pong schemes described in [15], except that up to 5 community IDs the node belongs to also be included in the pong message. Service discovery requests are initiated randomly from nodes without satisfied service, 10 minutes are given to the randomly constructed network to stabilize before discovery requests are launched.

4.2 Performance of Service Discovery

We first compare our solution with solution in [2] where information nodes form a flat topology under the same random walk request forwarding strategy. The walker is set to be 1, and random walk terminates when all the neighbor of the request-sending node have received the request. We use hops needed to find the first satisfied service and the success rate in the service discovery to evaluate the performance, and the results are shown in figure 2. In the figures, "CO-RW(10)" means: community overlay with random walk request-forwarding policy, with the qualified service falling into the 10th largest community. The meanings of the other symbols in this paper are similar. Symbol "r" stands for the service density, and its unit are one out of one thousand (1/1000). From figure 3, we can see that, both performance on time and success rate are improved under our approach, and smaller the community is, better the performance would be.

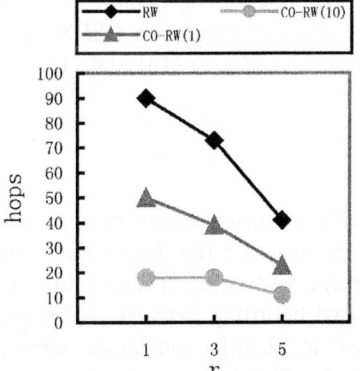 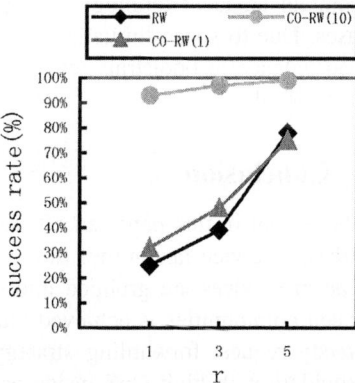

Fig. 2. Performance of community based service discovery vs. simple random walk on flat topology.

We also compare our community based service discovery with another information node grouping approach – NEVRLATE. We set the grouping scheme in NEVRLATE according to the optimal setting indicated in [5], e.g. 100 groups with 100 nodes in each group under our simulation settings. Because a service would register on at least one node in each group in NEVRLATE, we take the population of nodes covered in a discovery process and a service information update process into account as an index of network overhead. The results are shown in figure 3, in which the total number of nodes covered is represented as N. It's clear that, though the hops needed in NEVRLATE are smaller, the network overhead is much bigger.

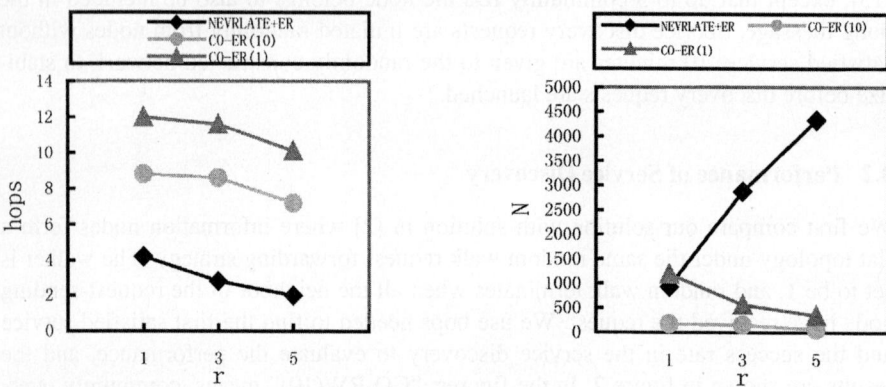

Fig. 3. Performance of community based solution vs. NEVRLATE under expanding ring request-forwarding policy.

Because node may join several community overlays at the same time, the topology maintenance overhead will increase in our service discovery approach. We observe less than 1K byte/sec average bandwidth consumption in the simulation for a node to maintain adjacent topology in a community overlay, which is acceptable in most cases. Due to space limitation, we do not include evaluation results on GSBSL and load balance in bootstrap network. Interested readers can refer to [13] and [11] for more details.

5 Conclusion

We present in this paper an efficient decentralized Grid service discovery approach utilizing service taxonomy. In our solution, information nodes registering the same type of services are grouped into community overlays, and efficient navigation between communities is achieved via a DHT P2P based bootstrap network. A Simple greedy request forwarding strategy GSBSL is used to identify and locate service providers with high QoS inside community efficiently. Simulation study shows the improvement of performance, and overhead is acceptable.

References

1. Ratnasamy S, Stoica I, Shenker S. Routing Algorithms for DHTs: Some Open Questions. In: Proceedings of 1st International Workshop on Peer-to-Peer Systems. (2002) 45-52
2. Iamnitchi A, Foster I, On Fully Decentralized Resource Discovery in Grid Environments. In: Lee, Craig A. (eds.): Proceedings of 2nd IEEE/ACM International Workshop on Grid Computing. (2001)
3. Andrzejak A, Xu Z. Scalable, Efficient Range Queries for Grid Information Services. In: Proceedings of 2nd IEEE International Conference on Peer-to-Peer Computing. (2002) 33-40
4. Huang L, Wu Z, Pan Y, Virtual and Dynamic Hierarchical Architecture for e-Science Grid, In: Proceedings of International Conference on Computational Science (2003) 316-329
5. Chander A, Dawson S, Lincoln P et al.. NEVRLATE: Scalable Resource Discovery. In: Proceedings of IEEE/ACM CCGrid (2002) 382-388.
6. Stoica I, Morris R, Karger D, et al.. Chord: A Scalable Peer-To-Peer Lookup Service for Internet Applications In: Proceedings of ACM SIGCOMM (2001) 149-160
7. Saroiu S, Gummadi P, Gribble S. A Measurement Study of Peer-to-Peer File sharing Systems. In: Proceedings of SPIE Conference on Multimedia Computing and Networking, (2002). 156-170
8. Jelasity M, Preusb M, Steen M. Maintaining Connectivity in a Scalable and Robust Distributed Environment. In: Proceedings of 2nd IEEE/ACM CCGrid (2002) 389-394
9. Byers J, Considine J, Mitzenmacher M. Simple Load Balancing in Distributed Hashing Tables. In: Proceedings of the 2nd International Workshop on Peer-to-Peer Systems (2003). 80-87
10. Rao A, Lakshminarayanan K, Suranaet S et al.. Load Balancing in Structured P2P Systems. In: Proceedings of the 2nd International Workshop on Peer-to-Peer Systems (2003). 68-79
11. Zhu C, Liu Z, Zhang W, et al, Load Balance based on Rumor Mongering in Structured P2P Networks, Journal of China Institute of Communications, 2004, 25(4): 31-41
12. Lv Q, Cao P, Cohen E. Search and Replication in Unstructured Peer-to-Peer Networks. In: Proceedings of ACM SIGMETRICS 2002.
13. Zhu C, Liu Z, Zhang WM et al. Analysis on Greedy Search based Service Location in P2P Service Grid. In: Proceedings of 3rd IEEE International Conference on Peer-to-Peer Computing. (2003) 110-117
14. Palmer C, Steffan J, Generating Network Topologies that Obey Power Laws, In: Proceedings of the IEEE Globecom. (2000)
15. http://rfc-gnutella.sourceforge.net/src/rfc-0_6-draft.html

QoS Analysis on Web Service Based Spatial Integration

Yingwei Luo, Xinpeng Liu, Wenjun Wang, Xiaolin Wang, and Zhuoqun Xu

Dept. of Computer Science and Technology, Peking University, Beijing 100871, P.R. China,
lyw@pku.edu.cn

Abstract. Web Service sets service-oriented development, deployment and integration as its basic jumping-off place, and has become the primary publishing mechanism and organization framework in the environment of distributed computing and services among Internet. Currently, especially in distributed applications of large scaled spatial information, due to user's strong dependence on data formats and platforms of certain GIS manufacturer, there are obvious difficulty and complexity in inter-platform spatial information exchange and system development. In this paper, the concrete design and implementation of an integration and interoperation platform that aims to solve heterogeneity and on-line service for geographical information is presented first. This platform concentrates on implementation to shield heterogeneity among spatial data sources and services through a common spatial data transmission and exchange format-GML. In addition, QoS control for WebGIS Services of the integration and interoperation platform is expatiated from different aspects: cache mechanism, GML data compression, map on-line generation, etc.

1 Introduction – Web Service and GIS

Web Service is a standard-based software technology, which provides programmers and integrators with possibility to have all existing and developing systems bind together in a refreshing mode. Web Service allows the existence of interoperation between softwares that are provided by different providers, running on different operating systems and written by different programming languages [1]. The key features of Web Service lie in that it is based on industrial standards, emphasizes interoperation, and implements inter-platform as well as inter-language fast deployment.

Currently, problems of QoS also become more and more popular in sorts of distributed application. As a fairly popular kind of distributed service providing mechanism, QoS in Web Service is a fatal factor in appreciating the success of service providers. Due to dynamic and unpredictable features of Web connection in Internet, implementation of QoS is confronted with great challenge. Thus, beginning with the analysis of Web Service requirement, setting up QoS-supported Web Service, comprehensively considering bottle-neck of QoS, and providing corresponded QoS methods and transaction service become a main process in the construction of Web Service based QoS model [2].

In the domain of GIS, each dominant GIS platform (e.g., ArcInfo, MapInfo) and several database manufacturers (e.g., Oracle) provide data, services and product series based on their own spatial data models, so it is very hard to make for inter-platform data transmission. But with the development of contemporary urgent need of inter-domain geographical data brought about by large scaled spatial projects as Digital Earth, Digital city, geographical information integration has already become exigent problems to be solved.

Existing tactic focuses on the writing of application-related adapter softwares for each participating platform to shield the heterogeneity of platforms. But accounting for diversity of application processes and complexity of disposition, each alteration to applications will induce redevelopment to related softwares. In certain situation, programmers are restricted by data and service security of relevant industries.

Web Service concentrates on fast conformity and deployment of legacy systems, so construction of Web Service based geographical information interoperation platform not only could make full use of legacy systems, but also can separate system implementation and function interfaces by function publishing as services. The latter permits flexible function replacement in implementation aspect, with the precondition of accordance among service interfaces participating integration satisfied. Additionally, because of the separation of interface and implementation, security of sensible data and relevant services inside some industries are well protected. This paper will detail on the application of Web Service in geographical information integration.

2 Geographical Information Integration Platform

2.1 Geographical Information Integration Model

According to the technical standards and implementation specification of Web Service, we design and implement Geographical Information Integration and Interoperation Platform (as shown in figure 1, abbreviated as Platform in the following) [3]. The design aim of the platform is to provide service integration which have layers as the basic geographical data organization units, accomplish data retrieval from multiple GIS data sources based on different platforms, proceed on-line overlapping, analysis, visualization of map data, and perform user intercommunication. Organization.

Geographical Information Integration and Interoperation Platform includes four kinds of basic services:

(1) AIS (Application Integration Service), which is the core service of business process scheduling to implement interoperation. The main functions of it includes:

◆ Set up connections with clients, parse user requests, and perform task division;
◆ Interact with available metadata services, and acquire data description and service description needed by business process;
◆ Start data requesting thread, accomplish interaction with TS (Transform Service), which adopts GML as the uniform geographical data format to retrieve spatial data of each layers;
◆ Start data processing thread, and accomplish GML based information disposition as coordinate transformation, spatial analysis, etc;
◆ Accomplish coordination and conformity of a map's multiple layers, and on account of QoS service features of clients, respond to client requests through several effective ways;

✧ Provide PAIToolkits to extended development users who based on the Platform. This toolkits provides a series of API packages and Web Services of spatial information and services and shields heterogeneity, distribution and semantic collision of information and services.

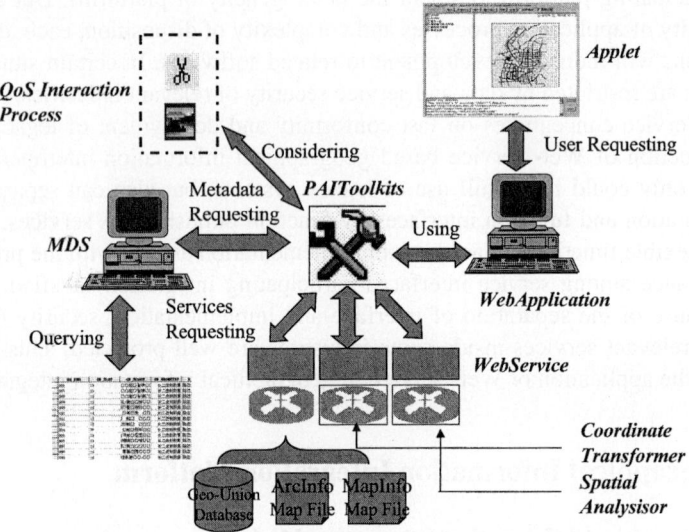

Fig. 1. Geographical Information Integration Platform.

(2) MDS (Metadata Service). Inside MDS, service information of Web Services in the Platform is maintained. It also describes exposed query interfaces using WSDL so that services can be identified and bound dynamically during the business process.

(3) TS (Transform Service), it can be set into each GIS service provider's platform to accomplish the function of data retrieval from all sorts of GIS platforms and data format transformation into uniform GML (Geography Markup Language) - the corresponding standard of XML in geographical information domain.

(4) SAS (Spatial Analysis Service), it can receive GML formatted spatial data, and provides GML based common spatial analysis abilities as spatial data clipping, projection, coordinate transformation and overlapping, etc.

As shown in figure 1, for a map request, the map requested is formed by multiple layers that are stored in heterogeneous GIS platforms. A typical service stream is that: Web Service receives a map request from client, and it calls corresponding API in PAIToolkits to start AIS, with the latter accomplishing following Web Service flow:

✧ Request MDS for layer data and metadata information of these data relevant to current client request. MDS queries for above information from the backend databases it maintains and returns it to AIS;

✧ AIS locates TS according to metadata of layer-providing services, and has above layer metadata wrapped as parameters to send parallel requests to multiple TS for data transformation, and finally it will collect all the transformation results - GML formatted layer data;

✧ AIS takes QoS into consideration according to user service features of the client end and current web performance (Detailed discussion of QoS is proposed in section 3), and performs suitable clipping or on-line lattice figure generation to available GML data, with the results sent back to the web client.

2.2 Platform Implementation – TS Implementation as an Example

In the Platform, the service structure of TS is shown in figure 2:

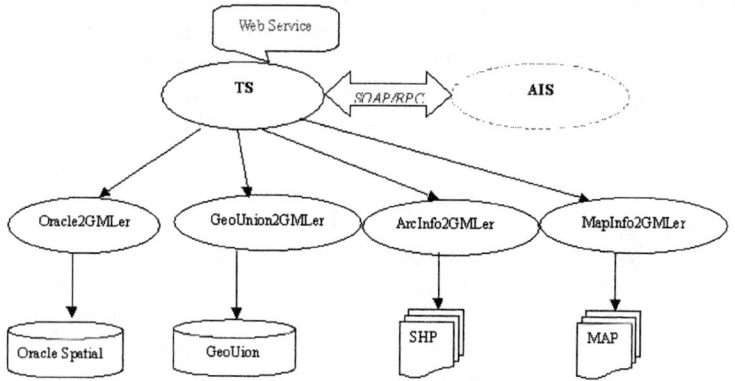

Fig. 2. TS Service Structure in Integration Platform.

TS mainly provides 4 kinds of GML data transformation abilities of GIS data: ArcInfo's SHP files, MapInfo's Tab files, spatial data stored in Oracle spatial and spatial data stored in Geo-Union – a GIS platform developed by ourselves. Through Web Service encapsulation to existing or under developing data transformation modules, TS provides common transformation interfaces. In the Platform, TS mainly interacts with AIS to accomplish service scheduling of data transformation by AIS.

The primary implementation of TS is:

(1) Implementation of transformers due to different spatial data storage types:

Map Data storage Format	Transformers
Oracle Spatial Database Storage	Oracle2GMLer
Geo-Union Database Storage	GeoUnion2GMLer
ArcInfo File Storage	ArcInfo2GMLer
MapInfo File Storage	MapInfo2GMLer

(2) Usage of Apache "Tomcat + AXIS" to implement Web Service. TS and AIS perform communication through SOAP protocol:

Message Type	Direction	Message Parameters
Request	AIS→TS	Layer name, storage format, other necessary parameters (including database connection parameter information, etc)
Reply	TS→AIS	Corresponding layer data (GML format)

3 QoS in Interoperation Platform

3.1 QoS of Web Service

QoS determines availability and efficiency of services, which greatly affect popularity of certain services. Accounting for the dynamical and unpredictable feature of web connection, discussion of QoS in Internet faces tough challenge. All these ask for the requirement of QoS standard in Internet, and they mainly include [2]: availability, accessibility, integrity, performance, reliability, regulatory and security.

(1) Bottle-neck of Web Service [2]. The performance of a Web Service mainly lies in the restriction of lowest message transmission. Web Service broadly depends on protocols like HTTP, SOAP. HTTP's main problem is that it does not ensure the arrival of packages to destination and their sending sequence. SOAP's problem mainly comes from XML format expression and parsing mechanism. Although SOAP uses legible text, XML parser is also high time-consuming. Currently asynchronous message queues or private LANs can be used as alternates to HTTP, and suitable XML parsers (such as SAX) can be adjusted to parser SOAP protocol. What's more, SOAP text may be compressed during transmission.

(2) Methods to Provide QoS. There are sorts of ways to provide QoS for a service provider, which are summarized as follows [2]: a) cache mechanism and overload balance; b) service provider can proceed Capacity Modeling to set up a top-down model for request communication volume, current performance usage and result QoS; c) service providers can classify Web Service communication volume according to the capacity of communication volume; d) Service providers can provider "Differentiated Services". That is, performance of different clients and service types is determined by performance models; and e) as to business processes that lay more emphasis on maintenance of Web Service integration, transaction QoS is used.

3.2 QoS Controls in Interoperation Platform

Because of feature of geographical information, in Web Service based Geographical Information Integration and Interoperation Platform, efficient expression of QoS is an imperative part of platform construction, and it affects platform availability.

The necessity to provide QoS in the Geographical Information Integration and Interoperation Platform lies in the following aspects:

◆ The biggest bottleneck to apply geographical data in Internet origins from the massive data transmission. What's more, although GML that is expressed in elemental XML has a common data format to normalize data process, the redundancy of its expression increases the data transmission volume sharply;
◆ Because of the diversity of web ranges where the client ends locate and habits in users' map manipulation, forms in which they acquire map data become the key to satisfy application requirements.

In Geographical Information Integration and Interoperation Platform, QoS based implementation strategies are divided into the following 3 aspects:

(1) GML Data Compression. GML data compression is the most direct but simple way to reduce geographical data transmission volume between each Web Service. It

focuses on the settlement of performance requirement of Web Service, and is the leading method to resolve bottleneck brought about by SOAP. GML data compression has a level hierarchy in the interoperation environment:

- ✧ GML data compression is accomplished immediately during the process in which TS generates GML. This is feasible when layer data themselves are stable and massive. But when layers are in a frequent alteration, the compression is not likely to be done earlier, or a new bottle-neck will emerge from the recompression of GML data of new editions;
- ✧ Compression data editions produced by TS are maintained in AIS caches. Since AIS has corresponding cache edition control mechanism (see following), this strategy is feasible. At the same time if frequent map dynamical generation by AIS is needed, single compression data editions cached are not enough, with the reason same as the above. In this case AIS should always uncompress GML data to complete data parsing;
- ✧ On-time GML data compression is used when GML data are provided to the client. This strategy has a higher efficiency when cached map data editions maintained by AIS are stable and user requests for these data are not frequent. In other case the improvement in efficiency by data compression is not ensured.

(2) Cache Mechanism. The Platform can perform on-line GML data clipping according to map usage habits of the client, map application precision and current web speed. Construction of caches is also in a hierarchy, which is reflected not only in the Platform, but also in the granularity of geographical data model. In the platform:

- ✧ In order to avoid retransformation of highly reusable layer data as to consume over computing and IO resources, caches for layer data having stable editions by TS can gain huge efficiency profit;
- ✧ Construction of caches by AIS is key to implementation of data integration. AIS's storage framework is now organized according to GML layer files. It uses directory identified by map name to organize subordinate physical layers, marking necessary editions and time stamps to ensure data validity allowed by QoS;
- ✧ Client web browsers use IE's self-carried Internet cache directories to proceed data on-line caching. If data caches of higher performance and elaboration are required, sessions of web pages can be maintained to pertinently implement geographical data process.

From the aspect of data model's granularity, caches can be arranged in three levels as map, layer and entity. Now the platform uses physically independent layers, considering the following advantages:

- ✧ Layers are independent of concrete applications, and are convenient for multiple applications to share the same layer caches. Large cache data redundancies as well as the resulting inconsistency of data editions are thus avoided;
- ✧ GML files give complete mapping to layers, so intact cache for layers avoids many time-consuming operations as construction of caches for data resulting from entity division, modification and deletion. Additionally, the choice to have application logics after GML-parsing query suitable entities is more effective than parsing large number of scattered GML files including only single entity.

(3) Other QoS-related Differentiated Services. In the Platform, provision of differentiated services based on user identity and performance requirement is a strategy to accomplish QoS control. Platform mainly supports following 2 differentiated services:

(A) Map Transformation and Data Integration Service

◇ GML formatted map clipping: the Platform can perform on-line GML data clipping according to users' map manipulation habits, map application precision and current web speed, and improve GIS service's QoS with the aid of map cache management.

◇ On-line map generation service: the Platform supports lattice figure's on-line generation and transmission to a thin client. Map cache service also supports edition controls and renovation of lattice figures.

Map clipping and on-line map generation services usually function cooperatively. For a client with lower speed, less map refreshment and lower time requirement, screen-ranged clipped lattice figures are provided. For a client with high speed, frequent map refreshment and strict time requirement, on-line painting of large volume of vector data is provided.

(B) Metadata Service. Metadata service records information of all the transformation services and data sources on their base platforms. All the metadata query interfaces are published as Web Services. The content of Metadata includes:

◇ Map metadata description: Platform provides basic metadata description information of maps to flexibly express member layers, storage locations, and access modes of maps.

◇ Map transformation service metadata description: Platform provides metadata description for services to accomplish layer transformation. It includes service ports, service interfaces, service parameter information and service privileges.

◇ Application based spatial metadata semantic modeling and querying: In order to resolve semantic collision brought about by application domains to spatial information, Platform presents a simple spatial metadata semantic model, based on the ontology specification of W3C. This model has RDF as the fundamental semantic framework, and provides convenient semantic-oriented query functions.

Metadata service also records hierarchy information of QoS, which can judge from current users and web conditions, regulate quality of metadata service, and select suitable map sources, transformation services and semantic exchange contents for specific users.

4 Conclusion

This paper takes Web Service as the basic background, discusses the its usage as information publication framework to dynamically implement application integration, sets up the Geographical Information Integration and Interoperation Platform, and further illustrates basic strategies to implement map service QoS in the Platform.

As a distributed service publication technique, Web service has lots of similarities to agent and grid service techniques. They all provide transparent operations or actions to the outside, while the implementation of them can be well encapsulated. They focus on different development direction, so differences among them are obvious:

- ♦ As an intellectual entity, Agent focuses on the conception of changes in the objective world through outer behaviors. So agent is not simply a objective entity to be called, it is also an positive and active entity to fit into its environment and affect its environment;
- ♦ Calling of services is passive. Except for the inner implementation of service interfaces are altered, Web Services usually cannot change interface functions they publish. Notwithstanding, the same interface published can have different implementations;
- ♦ Grid Service extends the functions of Web Service, with the infusion of service publication into life cycles of service instances. Grid Service publishes operations through the standard Web Service publication, but permits users to create service instances by factory methods of services and finally kill service instances. Different Grid Services can perform inner state information exchange by Service Data provided by service instances.

Now, in addition to our concentration on data renovation mechanism of the Platform, we are implementation domain ontology based platform information modeling and publication to make for semantic level sharing of application-restricted geographical information. In order to express platform service architecture overload balancing and service scheduling based on dynamical performance parameters, construction of Grid Service based service architecture using Globus Toolkit 3.0 is under consideration.

Acknowledgement

This work is supported by the National Research Foundation for the Doctoral Program of Higher Education of China (20020001015); the 973 Program (2002CB 312000); the NSFC (60203002); the 863 Program (2002AA135330, 2002AA 134030); the Beijing Science Foundation under Grant No.4012007.

References

1. Simon Watson: A Definition of Web Service, http://www.integra.net.uk/services, 2002.7.
2. IBM Developer Works: Web Services: Understanding Quality of Service for Web Services, http://www-106.ibm.com/developerworks/library/ws-quality.html, 2002.1.
3. Wang Wenjun, *et al*: Spatial Applications Integrating Infrastructure, The Seventh Annual Conference of China Association of Geography Information System, Beijing, 2003.12.
4. Mark Volkmann: AXIS - An Open Source Web Service Toolkit for Java, Object Computing, Inc., 2002.9.

Engineering Process Coordination Based on a Service Event Notification Model

Jian Cao[1], Jie Wang[2], Shensheng Zhang[1], Minglu Li[1], and Kincho Law[2]

[1] Department of Computer Science, Shanghai Jiaotong University,
200030, Shanghai, P.R. China
{cao-jian,sszhang,li-ml}@cs.sjtu.edu.cn
[2] Department of Civil and Environment Engineering, Stanford University,
Stanford, CA 94305, USA
{jiewang,law}@stanford.edu

Abstract. Due to the complexity and uncertainties, the engineering process requires dynamic collaborations among the heterogeneous systems and human interactions. In this paper, we propose a service event notification model based on grid service to coordinate different applications and human. The model takes advantage of the grid infrastructure and reduces the need for ad hoc development of middleware for supporting process coordination. In this model, an event notification server composed of a group of grid services can capture events from other grid services and generate process events. When an event occurs, event notification server decides to whom it should send the event according to an awareness model that keeps the states of the underlying coordination policies and business rules. The awareness model and the methodology for building an event notification system, together with the infrastructure of the notification server are presented in the paper. An example in applying the model to an engineering project coordination scenario is presented to illustrate the potential application of the event notification model.

1 Introduction

For complex and knowledge intensive projects such as engineering/construction and design/manufacturing, multiple participants residing in different locations often need to work together throughout the lifecycle of the project. The dynamic nature of project requirements and the inevitable collaboration among multiple organizations and participants pose many challenges from both technological and management perspectives.

The first challenge is the diverse heterogeneous software and hardware environments that are used in the engineering processes. A fully integration solution imposing on homogeneous software and hardware platforms is infeasible. We need an alternative approach that can coordinate heterogeneous applications during the project lifecycle process.

Grid-based engineering service is a potentially useful technology for process coordination. Grid concepts and technologies were first developed to enable resource sharing and to support far-flung scientific collaborations [1]. The Open Grid Services

Architecture (OGSA) [2] integrates key Grid technologies with Web services mechanisms [3] to create a distributed system framework.

Another challenge is the coordination of the participants in an engineering project. Because engineering process is often highly dynamic in nature, it is difficult to lay out an exact plan detailing all the necessary tasks, their interdependency and interactions.

To address these two challenges, a platform that can support both system coordination and human coordination is important. Current grid service technology itself does not support human coordination. However, it does provide a notification mechanism to publish the status of changes in events, which can be forwarded to interested parties to support human coordination. Thus we propose a Grid service based event notification model to support engineering process.

This paper is organized as follows. Section 2 defines an awareness model for engineering process. Section 3 discusses how to capture and transform events based on business requirements for a complex process. Section 4 introduces the event notification mechanism. In Section 5, the structure of an event notification server supporting engineering process coordination is proposed. In Section 6 we provide an example in building design and construction management to demonstrate the grid-based event notification approach described in the paper. Section 7 discusses related works and Section 8 concludes the research and points out some future works.

2 An Awareness Model for Engineering Process

Dourish and Bellotti [4] define awareness as "an understanding of the activities of others, which provides a context for your own activity." There are many types of awareness information that can be provided to a user about other users' activities [5]. We focus and categorize the awareness information of engineering process into two main types: (1) Awareness Information related to Project Artifact Sharing, and, (2) Awareness Information related to Process Logic.

We propose to build the awareness model based on the technique of integrated cooperative process modeling.

2.1 Artifact Structure Model

The artifact itself can be used as a "shared representation" through which people can communicate with one another and coordinate their activities.

An engineering artifact can be represented by a hierarchical tree, which is called an artifact tree. An artifact tree is a triple $<C, r, R>$, where C is a finite set of components, $r \in C$ is the root component, $R \subseteq C \times C$ such that $(c_1, c_2) \in R$ if $c_1 \in sub(c_2)$, where sub denotes the "sub-components of" relationship).

An artifact type can be defined as $<P, A>$, where P is a set of parameters that can characterize this artifact, A is a set of operations to manipulate this artifact. The artifacts produced during an engineering process are not independent, and are often interdependent. That is, if an artifact a depends on another artifact b, whenever b is changed, *a* must at least be checked to ensure consistency and if necessary, need to be changed accordingly. Moreover, the dependency is transitive.

To model an artifact structure, two relationship types are defined among the components: they are the *sub* relationship and, the *depend* relationship, as shown in Figure 1.

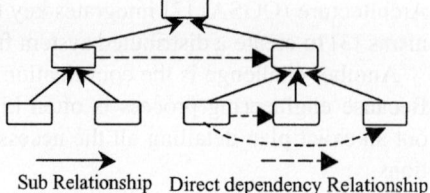

Fig. 1. Artifact Structure Model.

2.2 Process Structure Model

To model a dynamic engineering process, we partition the process model into two layers. The top layer is a project plan model and the bottom layer represents a library of workflow models with a set of tasks.

A project is denoted as $P=<T_p, R_p, SP, R_{dp}>$, where T_p is the anticipated time schedule of the project, R_p is the organizational property that will be defined in section 2.3, SP is a set of activities, R_{dp} are ordering relationships among activities which are defined according to a general project model such as CPM (Critical Path Method).

An activity can be complex or simple. The definition of a complex activity is similar to a project. A simple activity is denoted as $a_p=<T_a, R_a, TA>$, where T_a is the time scope defined for the activity, R_a is the organizational property. $TA=\{ta_{p1}, ta_{p2}, ..., ta_{ph}\}$, in which ta_{pi} is a task of activity a_p and a_p is also called the parent of ta_{pi} (denoted as $a_p=\uparrow(ta_{pi})$). A task can be expressed as $ta=<a_t, c_t, I, O, R_t>$, where a_t is an operation (defined for artifact type of c_t), c_t is an artifact, I is the input artifact set of this task, O is the output set of this task and R_t is the organization property of this task. There are no ordering relationships defined among the tasks of an activity.

In an artifact structure model, for each operation defined for an artifact type, we should specify it's content as:

(1) applications or services that are possibly conducted by a person,
(2) invoking a service by the system automatically, or
(3) workflow models that can fulfill the operation.

The process structure model is shown in Figure 2. A project model consists of activities and their relationships and it provides high level ordering constraints for the entire engineering process. While at the lower level description, tasks are fulfilled by invoking applications and services by a person or by a structured workflow.

2.3 Organizational Structure Model

A project, activity or task is usually allocated to an organizational unit for execution. Organizational structure reflects the project process structure and management structure.

Each participant can be assigned several roles. Team $te=<TR, TM, TF>$ is defined as a triple, in which TR represents a role set, TM is an actor set, and $TF \subseteq TR \times TM$ represents the enabled roles of the members TM in team te.

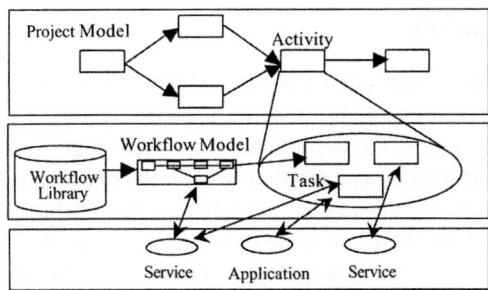

Fig. 2. Process Structure Model.

2.4 Resource Model

The resource managed by a notification server includes applications and services. Applications and services are registered in the notification server. For a public service in an engineering process, its address and the embedded methods should be registered so that other applications can find this service.

3 Event Capturing and Transforming

Notification mechanism has been defined in OGSA [2]. An important aspect of this notification model is the tight integration with service data: a subscription operation is a request for subsequent "push" delivery of service data.

In addition to capture the notification of grid service, we should capture the context of the notification, i.e., project name, the operations and the artifact affected. A service can specify the context information by adding a special service data type as follows:

```
<complexType name="EventDataType"><sequence>
        <element name="ProjectName" type="string"/>
        <element name="ArtifactName" type="string"/>
        <element name="OperationName" type="string"/>
</sequence></complexType>
```

When a method of this service is invoked, the values of the related service data elements (EventData) should be set and pushed to the notification sink.

Activities and tasks have different states. When an activity changes from one state to another, events will be triggered. These events are generated by a process engine and distributed by the notification server.

The events captured can be transformed into other events according to the business requirements. We provide a transformation rule in the following form:

On *Event Expression* **If** *Condition* **Then** *RaiseEvent* (*e*)

An *event expression* is composed of a set of event filters. A *Condition* is a conjunction of the constraints, which define the relationships among parameters of different event types. Action *RaiseEvent* will produce a new event.

4 Event Notification

After an event has been captured, the concerned individuals or applications should be notified to deal with the event. We assume that each event is related to at least an artifact or an activity, i.e., in the definitions of an event type, there is an attribute that indicates at least an artifact or an activity to which this event targets upon. If the event is related to an artifact c_0, then we can find other related artifacts through the dependency relationships among them. Suppose they compose a set C. To receive the necessary notifications for application services of a task, they should be registered at the notification server regarding the locations where these applications reside. Events related to the artifact c_0 will be broadcasted to these applications if their related artifacts are in C.

In order to notify the individuals, who are the task owner of particular tasks, when events related to artifacts happen, we should find the tasks directly related to these artifacts and notify the task owners. Because we have defined the input for each task, it is quite straightforward to find those tasks that are waiting for the events.

5 System Structure of Notification Server

The system architecture based on grid service platform for a notification server is shown in Figure 3. A user joins an engineering cooperative process through a personal workspace. In the personal workspace, a user can invoke different applications. These applications in turn can invoke services that are running in different grid service containers. A notification server consists of a set of grid services. An event receiver service is running to gather events from distributed services and it will store all the events gathered and recorded the events into an event history. Once an application starts running, it will create an application broker within the notification server. An application broker monitors the event history and determines whether the server should notify the application based on the awareness model. Similarly, a personal broker is created by each personal workspace. A personal broker will also monitor the event history and notify the personal workspace based on the awareness model. As

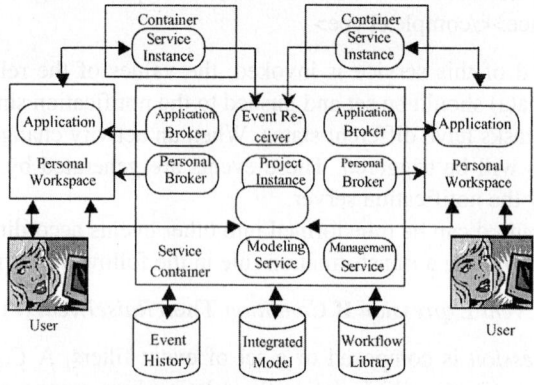

Fig. 3. Notification Server Structure based on Grid Service Platform.

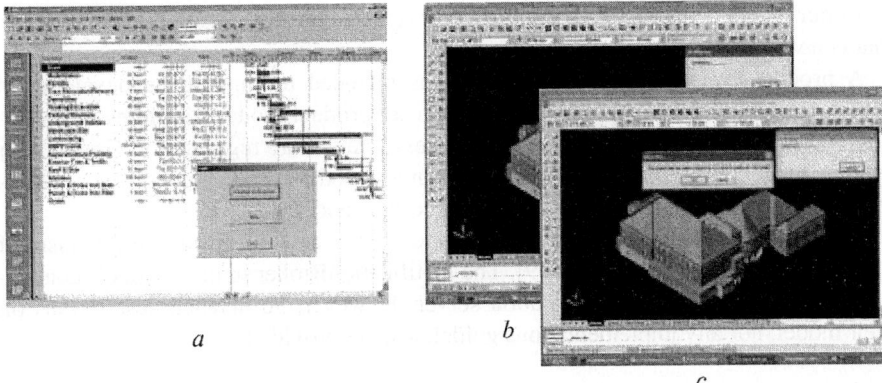

Fig. 4. A Case Study.

for each project, a project service instance will be created. The project service instance provides methods to be invoked by the personal workspaces, generates the events and coordinates the tasks according to the process logics. A set of facility services is also provided by a notification server. These services include modeling service and management service. A user can revise the project model, initiate a project and manage the project process through these facility services.

6 A Case Study

We have built a simple demonstration to illustrate the engineering service notification model. This case example demonstrates a grid based coordination system for facilitating a building design process. The project manager can define or update process plan through MS Project software. The process plan can be checked in to a grid repository service (Figure 4 a). The repository service will generate an event to the notification server. Notification server sends this event to AutoCAD software, which can be used to design and display different phases of the facility (Figure 4 b, c).

The distributed coordination framework for this example demonstration is developed based on Globus Toolkit 3.0. Specific gateways are developed to connect the software applications into a grid service. For example, we developed gateway plugins for MS Project and AutoCAD using Microsoft's .NET framework.

7 Related Works

There are many event notification servers [6]. Three representative examples are CASSIUS, Elvin and Siena [7]. CASSIUS was tailored to provide awareness information for groups. Elvin was originally developed to gather events for visualizing distributed systems, but it evolved later into a multi-purpose event notification system. Siena emphasized on event notification scalability in distributed environments. Our notification server model aims to coordinate and support engineering process in which an awareness model is designed with built-in knowledge. Another characteristic of our model is that the notification server is based on service computing paradigm that is not investigated by other approaches.

A product awareness model, Gossip, was intended to support the development process [8]. Gossip included a shared composite product model with rules of awareness information. Product object and awareness relation are registered in Gossip. Our artifact model is similar except our awareness model also includes process sub-model, organizational sub-model and resource sub-model.

Recently, a white paper describing a method based on a notification mechanism of a web service has been proposed [9]. The NotificationBroker in that model is conceptually quite similar to our notification server. However, no detailed descriptions on their model, nor any implementations guidelines, are provided.

8 Conclusions and Future Work

This paper proposes a solution based on an event notification model of grid services. Current grid service technology only emphasizes on cooperative computation. The proposed model extends the grid service to support cooperative work of individual human participants during a complex engineering process. Based on the awareness model, a notification server can not only broadcast events to the interested parties to support cooperative work, but also support process control that is important for an engineering process. Our future work includes developing a notification server to support more expressive event transformation rules in a dynamic engineering process.

Acknowledgement

This work was performed while the first author was visiting Stanford University. The authors would like to thank Mr. Jim Cheng for providing the data and the building model.

References

1. Foster, I., Kesselman, C., The Grid: Blueprint for a New Computing Infrastructure, Morgan Kaufmann (1999)
2. Foster, I., Kesselman, C., Nick, J., Tuecke, S., The Physiology of the Grid: An Open Grid Services Architecture for Distributed Systems Integration, Globus Project, http://www.globus.org/research /papers/ogsa.pdf, (2002)
3. Graham, S., Simeonov, S., Boubez, T., et. al. , Building Web Services with Java: Making Sense of XML, SOAP, WSDL, and UDDI, SAMS (2001)
4. Dourish, P., Bellotti, V., Awareness and Coordination in Shared Workspaces, Conference Proceedings of Conference on Computer Supported Cooperative Work, Toronto, Ontario, Canada, (1992) 107-114
5. Steinfield, C., Jang, C. Y., Pfaff, B., Supporting Virtual Team Collaboration: The Team-SCOPE System. Proceedings of GROUP Conference, Stockholm, Sweden, (1999) 81-90

6. Cugola, G., Nitto, E., Fuggetta, A., The JEDI Event Based Infrastructure and Its Application to the Development of the OPSS WFMS. IEEE Transactions on Software Engineering, Vol. 27(9) (2001) 827–850
7. Cleidson, R. B., Souza, D., Santhoshi, D., et. al., Supporting Global Software Development with Event Notification Servers, Proceedings of the ICSE 2002, http://citeseer.ist.psu.edu /desouza02 supporting.html, (2002)
8. Farshchia B. A., Gossip: An Awareness Engine for Increasing Product Awareness in Distributed Development Projects, http://www.idi.ntnu.no/~ice/publications/caise00.pdf, (2000)
9. IBM, Publish-Subscribe Notification for Web services, http://www-106.ibm.com/developerworks/ library/ws-pubsub/ WS-PubSub.pdf, (2004)

A Fault-Tolerant Architecture for Grid System

Lingxia Liu[1,2], Quanyuan Wu[1], and Bin Zhou[1]

[1] School of Computer Science,
National University of Defence Technology,
Changsha 410073, China
lingxia_liu@163.net
[2] The Telecommunication Engineering Institute,
Air Force Engineering University
Xi' an 710077, China

Abstract. Support for the development of fault-tolerant applications has been identified as one of the major technical challenges to address for the successful deployment of computational grids. Reflective systems provide a very attarctive solution for the problem. But in the traditional systems, all the binding relationships between the function part and the non-function part of the application are fixed. A new reflective architecture which uses events to achieve dynamic binding is proposed in the article. The two best known fault-tolerance algorithms are mapped into the architecture to set forth the dynamic binding protocol.

1 Introduction

Failure in large-scale Grid systems is and will be a fact of life. Hosts, networks, disks and applications frequently fail, restart, disappear and behave otherwise unexpectedly. Support for the development of fault-tolerant applications has been identified as one of the major technical challenges to address for the successful deployment of computational grids [1, 2].

Forcing the programmer to predict and handle all these failures significantly increases the difficulty of writing reliable applications. Fault-tolerant computing is a known, very difficult problem. Nonetheless, it must be addressed, or businesses and researchers will not entrust their data to Grid computing.

Reflective systems provide a very attractive solution for the problem. The benefits of reflective technology have been discussed in many works [3, 4]. Reflection enables fault tolerant transparency to application programmers. Application programmers care the function part of the application. Fault tolerant experts care the non-function part of the application. How to integrate the function part and the non-function part of the application is a special care in reflective system.

Fabre proposed a compile-time binding protocol [4] and a run-time binding protocol [5] based on the reflective programming language. They can't implement dynamic binding between the function part and the non-function part of the application. All the binding relationships between the function part and the non-function part of the application are fixed. We propose a new reflective architecture which uses events to achieve the dynamic binding between the function part and the non-function part of the application.

The remainder of this paper is organized as follows. In Section 2, we narrate the basic concepts including reflection and fault tolerant. In Section 3, we present the reflective architecture for grid system. Next, we set forth the architecture by mapping the fault tolerant algorithms to the architecture. Finally, in Section5, we conclude the paper.

2 Basic Concepts

2.1 Reflection

Reflection [6] can be defined as the property by which a component enables observation and control of its own structure and behavior from outside itself. This means that a reflective component provides a meta-model of itself, including structural and behavioral aspects, which can be handled by an external component. A reflective system is basically structured around a representation of itself or meta-model that is causally connected to the real system.

A reflective system is structured in two different levels: *the base-level*, in charge of the execution of the application (functional) software and *the meta-level*, responsible for the implementation of observation and control (non-functional) software. The architecture of a reflective system is shown in Figure1.

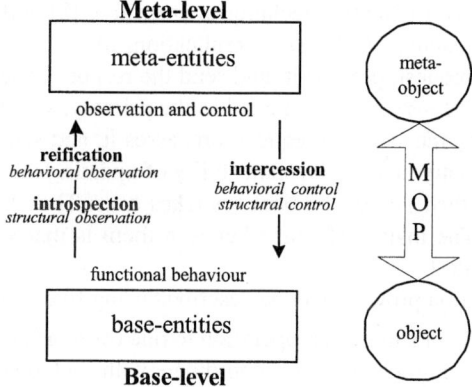

Fig. 1. The architecture of reflective system.

The reflective mechanisms of the system provide either behavioral or structural reflection. The reification process exhibits the occurrence of base-level events to meta-level. The introspection process provides means to the meta-level for retrieving base-level structural. The intercession process enables the meta-level to act on the base-level behavior (behavioral intercession) or structure (structural intercession). Behavioral reflection refers to both reification and behavioral intercession mechanisms. Structural reflection refers to both introspection and structural intercession mechanisms.

Each reflective computation can be separated into two logical aspects [7]: computational flow context switching and meta-behavior. A computation starts with the computational flow in the base level; when the base entity begins an action, such action is trapped by the meta-entity and the computational flow raises at meta-level (shift-up action). Then the meta-entity completes its meta-computation, and when it allows the base-entity to perform the action, the computational flow goes back to the base level (shift-down action).

The notion of a reflective architecture has often been used in object-oriented systems. The base-level is composed of objects, the meta-level is composed of meta-objects. The reification and intercession processes between base-level objects and meta-level objects correspond to a protocol is called MetaObject Protocol(MOP).

2.2 Fault-Tolerant

Fault tolerance is the ability of an application to continue valid operation after the application, or part of it, fails in some way. Such failure may be due to, for example, a processor crashing.

Replication is a key mechanism for developing fault-tolerant and highly available applications. Redundancy is a common practice for masking the failures of individual components of a system. With redundant copies, a replicated component can continue to provide a service in spite of the failure of some of its copies, without affecting its clients.

The two best known replication techniques are active [8] and passive [9] replication in distributed systems. With active replication, all copies play the same role. They all receive the request, process it, and send the response back to the client. With passive replication, one server is designated as the primary, while all others are backups. Only the primary receives the request, processes it, and sends the response back to the client. The primary takes the responsibility of updating the state of the backups. If the primary fails, then one of the backups takes over. They have own advantages and disadvantages. The main difference between them is that whether one or all of them process the request.

In general, replication protocol can be described using five generic phases:

1. Client Request: client submits an operation to one (or more) replicas.
2. Server Coordination: replica servers coordinate with each other to synchronize the execution of the operation.
3. Execution: operation is executed on replica servers.
4. Agreement Coordination: the replica servers agree on the result of the execution.
5. Client Response: the outcome of the operation is transmitted back to the client.

3 System Architecture

The fault tolerant architecture for the gird system named FTGA (Fault-tolerant Gird Architecture) is shown in Figure 2.

FT Services provide basic services required in fault-tolerant computing, such as failure detectors, group communication (provide atomic multicast), replication domains management facilities, event service and a stable storage support.

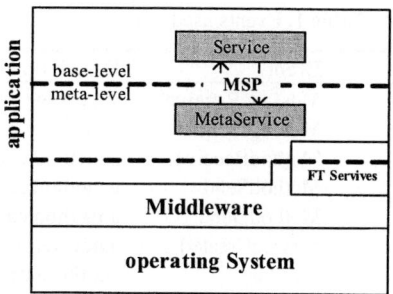

Fig. 2. Architecture of FTGA.

Application is composed of two sub-layers. The service sub-layer implements the functional part of the application and the metaService sub-layer implements the non-functional part of the application. Service and metaService are bound dynamic (at run time) using MetaService Protocol (MSP). MSP provides flexible mechanisms for structuring applications and specifying the flow of information between services and metaServices that comprise an application. The event model is used to implement MSP mechanism. Events specify interactions between services and metaServices and between metaServices and middleware. Events provide a uniform infrastructure to bind each level together.

As we all know, the middware level is responsible for the transport, coding, decoding, filtrating and transform of the message. Portable Interceptor (PI) is hook into the middware level through which the messages can be intercepted. PI is one part of the metaService.

Building a reflective system involves defining these two processes [10]:

- What is the needed information to observe the internal structure and operations of the component?
- What are the required actions that can be applied to control the component operations and structure?

Answering these questions depends very much on the final objective. From the analysis of the generic phases of the general replication protocol, we can conclude that the metaService should implement these functions:

- setting and getting state;
- the control of the client invocations;
- exchanging protocol information between participants of an algorithm;
- creating a new replica.

We also point out and classify the major event kinds that needed to implement the protocol. We classify these events into three categories: message-related events, method-related events and management-related events. The events and their descriptions are shown in table 1.

The events are raised or caught by the base-level or meta-level of the application. We should insert handlers to the appropriate events for associating applications with events.

Table 1. Events used to the protocol.

Category	Event	Description
Message-related	MessageReceive	a message is received
	MessageSend	a message is sent
Method-related	MethodReceive	a method call is received
	MethodSend	a method call is sent
	MethodDone	a method call is done
Management-related	ServiceCreated	a new replica is created
	GetState	get the state of the service
	SetState	set the state of the service

For the programmer, the most difficult aspect of program is processing the state of the service. We can use a sophisticated stub generator to generate the functions to set and get the state automatically [4]. Otherwise, programmers must define two functions for a service. One is setState(State s) which is responsible for setting the state of a service. The other is getState() which is responsible for getting the state of a service.

4 Mappings of Algorithms

We will map the two best known fault-tolerance algorithms into the architecture.

In the active replication, all the copies receive and process the request. The flow of the algorithm is shown in Figure 3:

1. The client sends a request to the service. Before it received by the service, PI intercepts the request and raises a MessageReceive event. The event will be handled by the corresponding metaService.
2. The metaService saves the request using the stable storage service (a MessageReceive event).
3. The metaService sends the request to all the copies by atomic multicast services (a MethodReceive event).
4. The service processes the request and raises a MethodDone event. The metaService handles the event, coordinates the responses from the copies.
5. The metaService saves the response using the stable storage service.
6. The metaService sends the response to the client.

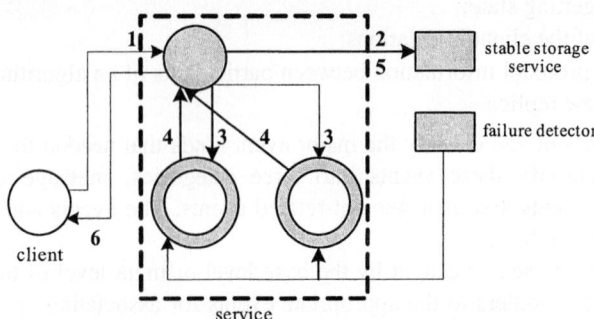

Fig. 3. The flow of the active replication algorithm.

In the passive replication, only the primary receives and processes the request. The flow of the algorithm is shown in Figure 4:

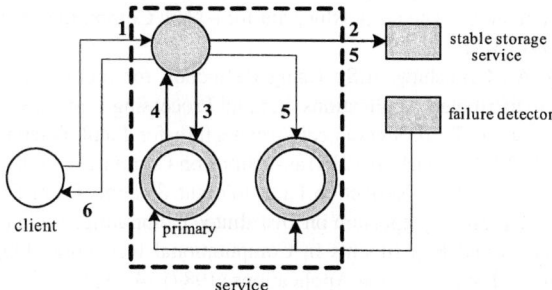

Fig. 4. The flow of the passive replication algorithm.

1. The client sends a request to the service. Before it received by the service,PI intercepts the request and raises a MessageReceive event. The event will be handled by the corresponding metaService.
2. The metaService saves the request using the stable storage service (a MessageReceive event).
3. The metaService sends the request to the primary service (a MethodReceive event).
4. The service processes the request and raises a MethodDone event. The metaService handles the event, gets the responses from the primary one.
5. The metaService saves the response using the stable storage service. The metaService raise a SetState event. The backup handles the event and set the state.
6. The metaService sends the response to the client.

The failure detector will notify the metaService when it detects a fail. The corresponding metaService will decide whether or not creating a new copy. If the answer is yes, the metaService will raise a ServiceCreated event. The corresponding metaService will create a new copy and set the state of the new one.

5 Conclusion and Future Work

In this paper, we have successfully conceptualized a fault tolerant architecture for Grid system. We have shown that reflection and event model enhances the flexibility of such system. The reflective approach described in this paper enables the dynamic binding between metaServices and services.

The FTGA has been partially implemented based on Globus toolkit3 [11]. FT Services has been implemented today. Future work consists of implementing the basic metaServices and MSP.

Acknowledgements

This work is supported by Chinese high technology plan ("863" plan) under Contract No. 2002AA116040.

References

1. Grimshaw, A.S., et al: Metasystems. Communications of the ACM. 11(1998) 46-55
2. Foster, I., Kesselman, C.: The Grid: Blueprint for a New Computing Infrastructure. Morgan Kaufmann (1999) 15-51
3. Nguyen-Tuong, A., Grimshaw, A.S.: Using Reflection for Incorporating Fault-Tolerance Techniques into Distributed Applications. Parallel Processing Letters. 9(1999) 291-301
4. Fabre, J.C., Perennou, T.: A Metaobject Architecture for Fault-Tolerant Distributed Systems: The FRIENDS Approach. IEEE Transactions on Computers. 1(1998) 78-95
5. Fabre, J.C., Killijian, M.O.: Reflective Fault-tolerant Systems:From Experience to Challenges. 14th International Symposium on Distributed Computing, Spain (2000)
6. Maes, P.: Concepts and Experiments in Computational Reflection. Object-Oriented Programming Systems, Languages and Applications (1987) 147-155
7. Yang Sizhong, Liu Jinde: RECOM:A Reflective Architecture of Middlaware. ACM Reflection'2001 Workshop on Experience with Reflective Systems, Japan (2001)
8. Guerraoui, R., Schiper, A.: Software-based replication for fault tolerance. IEEE Computer. 30(4) 68-74
9. Beedubail, G., Anish Karmarkar, Udo Pooch: Fault tolerant objects replication Algorithm. Technical Report TR95-042, Computer Science Department, Texas A&M University (1995)
10. Fabre, J.C., Killijian, M.O., Perennou, T.: Reflective Fault Tolerant Systems:Concepts and Implementation. 14th International Symposium on Distributed Computing, Spain (2000)
11. Globus toolkit3. Available online at http://www.globus.org (2003)

A Workflow Language for Grid Services in OGSI-Based Grids

Yan Yang[1], Shengqun Tang[1], Wentao Zhang[2], and Lina Fang[1]

[1] State Key Lab of Software Engineering, Wuhan University
Wuhan 430072, P.R. China
ms_yangyan@163.com
[2] Networks and Communication Lab of Computer School, Wuhan University
Wuhan 430072, P.R. China

Abstract. Workflow is a critical part of the emerging Grid Computing Environments and captures the linkage of constituent services together in a hierarchical fashion to build larger composite services. As an indispensable element in workflow systems for Grid services, Grid workflow language (and its enactor) serves as the basis for specifying and executing collaborative interactive workflows in Grids. Since Open Grid Services Architecture (OGSA) is built on Web services technology, it is natural to expect that the existing orchestration standards designed for Web services can be leveraged in OGSI based Grids. In this paper we propose a Grid workflow language called Grid Services Workflow Execution Language (GSWEL), which is based on OGSA and provides an extension to Business Process Execution Language for Web Services (BPEL4WS) in order to better support Grid characteristics.

1 Introduction

The Open Grid Services Architecture (OGSA) built on both Grid and Web services technologies is currently the most important Grid Architecture. In OGSA, everything is seen as a Grid service, which is a Web service that conforms to a set of conventions relating to its interface definitions and behaviors. The Open Grid Services Infrastructure (OGSI)[1] is one of the core elements of OGSA. It defines mechanisms for creating, managing and exchanging information among Grid services, and provides a complete specification to define interface and behavior of a Grid service.

In the Web service community, a variety of research works have been proposed in regard to how to describe and specify the collaboration and composition of Web services, such as WSFL, XLANG and BPEL4WS[2]. Similarly, Grid services can realize their full value only if there exists a mechanism to dynamically compose new services out of existing ones. In order to do so, we not only have to describe the order in which these services and their methods execute, but also present a way in which such a service collection can export itself as a service.

This paper proposes a Grid workflow language called GSWEL, which is based on OGSA and provides an extension to BPEL4WS in order to provide easy and natural support for OGSI mechanisms.

The remainder of this paper is organized as follows. Section 2 introduces some related work. In Section 3 we overview the syntax for GSWEL and discuss a few important features. Section 4 addresses some implementation details. And in Section 5 a sample Grid workflow described in GSWEL is presented. Finally, in Section 6 we conclude the paper and point out the directions for future work.

2 Related Work

GSFL[3] is probably a pioneer effort to examine technologies that address workflow for Web services and leverage this technology for Grid services. GSFL is based on WSFL and has four important elements including *Service Providers*, *Activity Model*, *Composition Model* and *Lifecycle Model*.

SCUFL is a workflow language developed as part of myGrid project[4]. The intention with it was to simplify the process of creating workflows for eScience by making the language granularity and concepts as close as feasibly possible to those that a potential user of the system would already be thinking in. The most notable feature of SCUFL is operating at a higher level of abstraction than alternatives such as WSFL.

SWFL[5] is another XML-based meta-language for the construction of scientific workflows from OGSA-compliant services. It also extends WSFL and supports a new set of conditional operators and loop constructs as well as arrays and objects. Another notable feature is its support for integrating parallel programs into workflow.

BPEL4WS results from the innovative combination of WSFL and XLANG. So it provides a sophisticated model that inherits the power of structured programming with the directness and flexibility of graph models. Because of its technical merits and industrial support from IBM and Microsoft, BPEL4WS becomes the most promising orchestration standard for Web services. It has replaced WSFL and XLANG and is now in the process of standardization in OASIS. That's exactly why we choose BPEL4WS to base our work on. Since BPEL4WS was not designed for Grids, we'll have to make some adaptions to it. GSWEL proposed in this paper, from some perspective, can be viewed as an attempt to integrate BPEL4WS concepts and elements into OGSI-based Grids.

3 GSWEL

3.1 Syntax Overview

GSWEL is an XML-based language defined by a set of XML schemas. It allows to specify Grid workflows within OGSA. The basic structure for it is shown in figure 1:

The first part is partner declarations. Unlike BPEL4WS, GSWEL allows specifying binding information for partners right in the workflow's definition document. The binding mechanism can be specified by *type* attribute (e.g. *factory*, *instance*) of the ⟨binding⟩ element. Child element *location* may be an

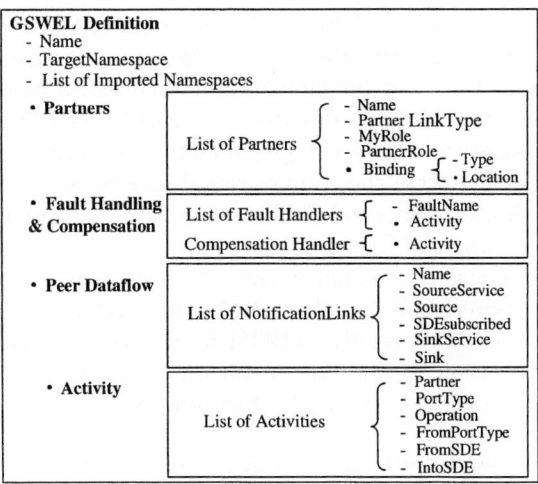

Fig. 1. GSWEL elements structure

wsa:EndpointReference from WS-Addressing (corresponding to *factory*) or a Grid Service Handle to an existing service instance. Therefore, when such a workflow is to be executed, all the partners for whom the *partnerRole* attribute is present can be resolved to actual Grid services before operations on them can be invoked.

The second part is fault and compensation handlers. The ⟨*faultHandlers*⟩ element specifies the faults that should be caught and the activities that should be executed upon a specific fault, while ⟨*compensationHandler*⟩ can specify an undo step that have already been completed.

The third part is peer dataflow model. All the peer data flows between constituent Grid services are modeled as *notification links*. Each *notification link* is used to declare a peer data flow, which is in fact a notification stream between a notification source service and a notification sink service.

Lastly comes activity declaration which addresses a Grid workflow's control flow and the implicit associated data flow. GSWEL reuses many basic BPEL4WS constructs for activities, but with slightly adapted syntax. Besides, GSWEL introduces a couple of new constructs such as ⟨*destroy*⟩ and ⟨*onNotification*⟩. ⟨*destroy*⟩ is used to explicitly destroy a Grid service instance acting as a partner. It is helpful for timely release of related resources when they have already been used and to be no more used in that workflow. ⟨*onNotification*⟩ allows the receiving end of a notification link to wait for the arrival of a specific notification before an operation can be called on exactly the same interface.

3.2 GSWEL's Features

(1) Support OGSI and be based on GWSDL. GSWEL supports OGSI. Not only the interfaces of participating Grid services are defined by GWSDL, but also the interface of the workflow itself is exposed as GWSDL document. By making its interface(s) inheriting the basic *GridService* interface defined in OGSI, a workflow acquires the most fundamental characteristics needed to become a Grid service. Thus, the workflow described in GSWEL can be viewed as a new compositional Grid service. Control logic and data flow inside are transparent to clients. So clients can access the compositional services provided by workflows the same way as they do to normal Grid services. This has also laid a foundation for recursive composition of Grid services.

Workflow monitoring and management is an important aspect of Grid workflow systems. As GSWEL is based on XML, it is relatively easy to describe workflow states in XML, especially as OGSI Service Data Elements(SDE). Thus it facilitates the creation of workflow monitoring services that can be used by workflow clients to track the execution of workflows.

(2) Explicit message routing through use of a WS-Addressing endpoint reference containing GSH. BPEL4WS uses WS-Addressing endpoint references to establish location of participating Web services for invocation of their operations. Also, endpoint references identifying the workflow are used by partners to interact with the workflow instance. However, that's not sufficient to uniquely identify the workflow instance. As BPEL4WS also introduces *correlation sets*, which need to be declared and used to correlate messages with their intended workflow instances.

GSWEL does not take in message correlation. Instead, it adopts an explicit means to route messages. As Grid workflow is also a Grid service, according to OGSI, it will be given upon creation at least one global unique *Grid Service Handle (GSH)* which globally and for all time refer to the same Grid service instance. So its GSH can uniquely identify the workflow instance. Now, with no need for correlation sets, what is only needed for message routing is a WS-Addressing endpoint reference containing GSH. With this information, (asynchronous) messages can be matched to Grid workflow instance for which they are intended.

(3) Enable Peer-to-Peer communications between services. As existing Web services workflow model do not allow data to be transmitted between services in a P2P way, all the control logic and data flow need go through the workflow engine. However, with large amounts of data often needed to be exchanged between Grid services in Grid environments, GSWEL needs to enable communications between services as depicted in figure 2. As OGSI specifies, interface NotificationSource is used for client subscription, while NotificationSink is used for asynchronous delivery of notification messages. GSWEL is required to provide a mechanism to allow connecting these two ends, thus avoiding the need for the workflow engine to mediate at every step. This requirement leads to the design of element $\langle notificationLinks \rangle$, in which every child element declaras a notification subscription link. By using $\langle notificationLinks \rangle$, GSWEL tells what direct P2P data transfers are required among constituent Grid services.

(4) Flexible lifecycle management. Unlike BPEL4WS, which needs *start activity* specified for workflow creation, GSWEL workflow instances are expected

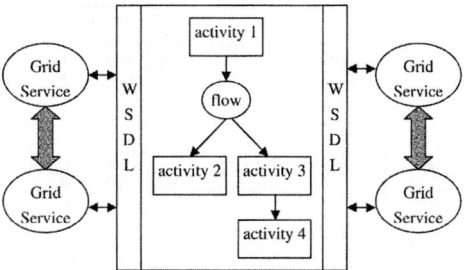

Fig. 2. Peer-to-peer communication between participating services

to be created using OGSI *factories* before they begin to execute. GSWEL reuses *terminate* activity for immediate destruction of workflow instances. And as each workflow instance implements the OGSI *GridService* interface, it has acquired the basic lifetime management capabilities that every Grid service has. Furthermore, GSWEL introduces *destroy* activity for explicit destruction of a participating Grid service instance. This helps to timely release relevant resources after the service has accomplished its task.

(5) Hold states information in SDEs. Instead of importing *containers*, one basic BPEL4WS construct, GSWEL leverages SDEs defined on the portTypes of Grid workflow to hold state information including messages that have been received from partners or are to be sent to partners, and even those that act as temporary variables for computation.

4 Implementation Framework

A GSWEL implementation may include three main components: a *GSWEL file parser*, a *GSWEL workflow container* and a *workflow interpreter*.

- GSWEL file parser

The parser is responsible for recursively parsing GSWEL document to produce a corresponding runtime workflow model which provides an in-memory process representation of a GSWEL workflow and contains complete information about the workflow definition. The workflow container will use that model as a basic template to create new workflow instances, while the interpreter will use it to control the execution of a workflow. JDOM is considered to be used for parsing GSWEL schemas, with decorations made to some generated classes.

- GSWEL workflow container

The container provides a runtime and deployment environment for GSWEL workflows. It is to be built upon GT3 with all enclosed services implemented as Grid services. This container should serve all incoming and outgoing requests. As depicted in figure 3, a *service manager* is responsible for dispatching incoming invocations to individual workflow instances of deployed workflows, or sending outgoing ones to their partners. Upon receiving an application message, the container (specifically the service manager) is required to decide whether the

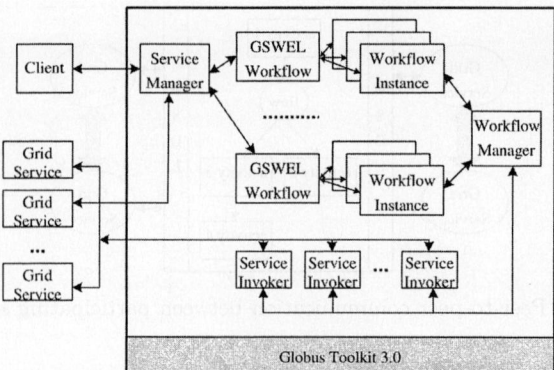

Fig. 3. GSWEL Runtime Architecture

message needs to be routed to an existing workflow instance or a new instance created. If a new instance has to be created, the in-memory runtime workflow model then will be used as template. Another service called *workflow manager* is used to control the execution of workflow instances and to request different services from the container via *service invokers*.

- Workflow interpreter

The workflow interpreter is a runtime component that executes an instance of a workflow model. Its task is to coordinate the workflow's control and data flow. Note that for all time the interpreter is not aware of the outside world. It uses the workflow container for all its external interactions.

In addition to above three main components, a *graphical workflow designer* may also be needed to create, modify and visualize GSWEL documents. It provides such features as synchronized XML source and tree views of the workflow being created, and can be built on an existing open source XML editor.

5 Case Study

This section presents a simple workflow for tour agenda planning. As figure 4 shows, it involves only two Grid services called *Planner* and *Forecaster*. *Planner* makes decisions about vehicle type according to the total number of travellers, plans travel schedule according to weather conditions, and also generates approximate fares. *Forecaster* forecasts weather conditions with the input of specific locations and dates. Executing order of this workflow is as follows:

1. *Client* invokes *sendReq* on *flow*'s *clientReqPT*, with input kept in AB.
2. Then the following two activities are executed concurrently.
   (a) Invoke *selVehicle* on *Planner*'s *planPT* with the value of AB. Once *Planner* receives notification about D from *Forecaster*, *Plan* is called. Then *Planner* returns its result to *flow* and is destroyed explicitly afterwards.
   (b) Invoke *forecast* with newly-assigned value of A. The change of D then happens and will immediately be notified to its subscriber *Planner*.
3. Finally, value of C on *plannerCallPT* is returned to *Client*.

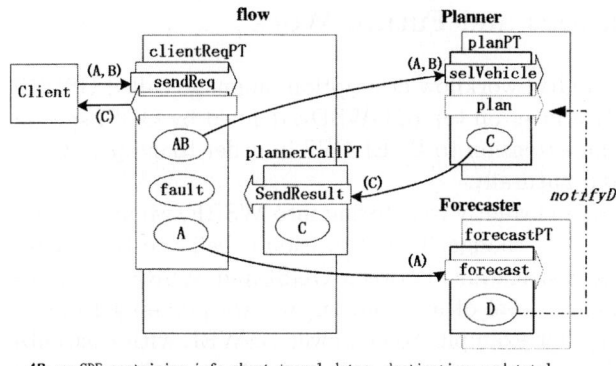

Fig. 4. A Grid Workflow example

Here is the sample Grid workflow's GSWEL document:

```
<workflow name="flow"  targetNamespace="..."  xmlns:lns="..." ...>
<partnerLinks>
    <partnerLink name="client" .../>
    <partnerLink name="Planner" ...>
      <binding type="factory">
        <location><wsa:EndpointReference><wsa:Address>http://.../Plannerfactory</wsa:Address>
        </wsa:PortType>factoryPT</wsa:PortType></wsa:EndpointReference></location></binding>
    </partnerLink>
    <partnerLink name="Forecaster" ...>
      <binding type="instance"><location>http://.../Forecasterfactory/service</location>
      </binding>
    </partnerLink>
</partnerLinks>
<faultHandlers> .... </faultHandlers>
<notificationLinks>
    <notificationLink name="notifyD" source="Forecaster" sourcePortType="forecastPT"
              sdeSubscribed="D" sink="Planner" sinkPortType="planPT"/>
</notificationLinks>
<sequence>
    <receive partnerLink="client" portType="clientReqPT" operation="sendReq" intoSDE="AB"/>
    <flow>
      <sequence>
        <assign><copy><from portType="clientReqPT" SDE="AB" element="A"/>
            <to portType="clientReqPT" SDE="A" element="A"/></copy></assign>
        <invoke partnerLink="Forecaster" portType="forecastPT" operation="forecast"
            fromPortType="clientReqPT" fromSDE="A"/>
      </sequence>
      <sequence>
        <invoke partnerLink="Planner" portType="planPT" operation="selVehicle"
            fromPortType="clientReqPT" fromSDE="AB"/>
        <onNotification name="notifyD"><call operation="plan"></onNotification>
        <receive partnerLink="Planner" portType="plannerCallPT" operation="sendResult"
            intoSDE="C"/>
        <destroy partner="Planner"/>
      </sequence>
    </flow>
    <reply partnerLink="client" portType="clientReqPT" operation="sendReq"
        fromPortType="plannerCallPT" fromSDE="C"/>
</sequence>
</workflow>
```

6 Conclusions and Future Work

In this paper, a Grid workflow description language called GSWEL is proposed. It is essentially a layer on top of GWSDL defined by OGSI specification. And it has provided an extension to BPEL4WS in order to support OGSI mechanisms more easily and naturally.

Since OGSI is currently refactoring into WSRF, we are considering to adapt our work to the new change. This transition is believed to be feasible due to the similarities and associations between OGSI and WSRF. Furthermore, in order to realize a higher level of automation, we are planning to leverage Semantic Web technology, for example, to empower GSWEL with customized automated service discovery or to constrain and guide the workflow composition. By now we are not taking performance issues into account. However, it's very important, especially for long running Grid workflows. Future GSWEL implementations are expected to have a high level of reliability and fault tolerance.

References

1. GGF OGSI-WG: Open Grid Services Infrastructure (OGSI) Version 1.0. (2003)
2. IBM, Microsoft, BEA Systems: Business Process Execution Language for Web Services Version 1.1. (2003)
3. Sriram Krishnan, Patrick Wagstrom, Gregor von Laszewski: GSFL: A Workflow Framework for Grid Services. (2002)
4. Savas Parastatidis, Paul Watson: Experiences with Migrating myGrid Web Services to Grid Services. (2003)
5. Yan Huang: JISGA: A Jini-Based Service-Oriented Grid Architecture. The International Journal of High Performance Computing Applications, 17(3): 317-327 (2003)
6. Dieter Cybok: A Grid Workflow Infrastructure. GGF10 (2004)
7. Francisco Curbera, Rania Khalaf, IBM T.J. Watson Research Center: Implementing BPEL4WS: The Architecture of a BPEL4WS Implementation. GGF10 (2004)
8. Ian Foster, Carl Kesselman: The Grid 2: Blueprint for a new computing infrastructure. Elsevier (2003)

Membrane Calculus:
A Formal Method for Grid Transactions

Zhengwei Qi, Cheng Fu, Dongyu Shi, Jinyuan You, and Minglu Li

Department of Computer Science and Engineering
Shanghai Jiao Tong University, Shanghai 200030, P.R. China
qizhwei@sjtu.edu.cn, {fucheng,shi-dy,you-jy,li-ml}@cs.sjtu.edu.cn

Abstract. The research of transaction processing in Web Services and Grid services is very active in academic and engineering areas now. However, the formal method about transactions is not fully investigated in the literature. We propose a preliminary theoretical model called Membrane Calculus based on Membrane Computing and Petri Nets to formalize Grid transactions. We introduce five kinds of transition rules in Membrane Calculus (including object rules and membrane rules) and the operational semantics of transition rules is defined. Finally, a typical long-running transaction example is presented to demonstrate the usage of Membrane Calculus.

1 Introduction

Grid is a type of parallel and distributed system that enables the sharing, selection, and aggregation of geographically distributed autonomous resources dynamically. This rapidly expanding technology proposes many challenging research problems in this area. For example, a lot of Grid applications would benefit from transactional support. A common Grid transaction processing would contribute a useful building block for professional Grid systems. The study of transaction processing in Web Services and Grid services is very active in academic and engineering areas now. Many top companies such as IBM, Microsoft, etc. and academic communities such as GGF TM-RG (https://forge.gridforum.org/projects/tm-rg) advance the research and applications in this area. There exist some transaction coordination standards or drafts to address this problem. For example, WS-Coordination/WS-Transaction [9,11] specifications delimit the two kinds of transactions in Web Services, namely, the short-time transaction called Atomic Transaction and the long-time transaction called Business Activity. OGSI-Agreement [12] defines a set of OGSI-compatible portTypes through which applications can negotiate and mediate between users and service providers within Virtual Organizations.

However, the formal method about transactions is not fully investigated in the literature. Roberto Bruni proposed one kind of process algebra called Join Calculus to prescribe Orchestrating Transactions [4]. Another semantic model, called Zero-Safe Nets, is based on the special Petri Nets and provides the unifying notion of transaction[5]. From the point of view of Grid computing, due to the static structure of these models, they are not suitable to describe the dynamic transactions. Therefore, we need a new formal method to model Grid transactions. We present a preliminary formal calculus based on Membrane Computing and Petri Nets to describe Grid transaction processing.

Membrane Computing (or P systems) were introduced by Gheorghe Păun in 1998 [1] as a distributed and parallel computable model based on biological membranes. For a general class of P systems, a high-level framework called Membrane Petri Nets [6] (MP-Nets or M-Nets in short) was introduced. MP-Nets will be suitable to model both P systems and dynamic Colored Petri Nets (CPN). MP-Nets can be a graphical as well as an algebraic modelling tool for general P systems and extend ordinary CPN by introducing the dynamic and reflective structure inspired by P systems.

The rest of this paper is organized as follows. In Section 2, we recall the definition of basic P systems and high-level MP-Nets. In Section 3, we propose Membrane Calculus to formalize Grid transactions and use one typical example to illustrate this modelling. In the last section, we conclude our work and discuss future research directions.

2 P Systems and Petri Nets

2.1 The Basic P Systems

We briefly recall the basic model of P systems from [2].

Definition 1. *A P system (of degree $m \geq 1$) is a tuple*

$$\Pi = (V, T, C, \mu, w_1, w_2, \ldots, (R_1, \rho_1), \ldots, (R_m, \rho_m)),$$

where V is an alphabet of objects; $T \subseteq V$ is the output alphabet; $C \subseteq V$, is catalyst; μ is a membrane structure consisting of m membranes, and m is called the degree of Π; $w_i, 1 \leq i \leq m$, are strings which represent multisets over V associated with the regions $1, 2, \ldots, m$ of μ; R_i is the set of evolution rules associated with membrane $1, 2, \ldots, m$ of μ; ρ_i is a partial order relation over R_i, called a priority relation.

An evolution rule is a pair (u, v), which we will usually write in the form $u \to v$, where u is a string over V and $v = v'$ or $v = v'\delta$, where v' is a string over $\{a_{here}, a_{out}, a_{in_j} | a \in V, 1 \leq j \leq m\}$, and δ is a special symbol not in V which represents that this membrane will be dissolved after the execution of this rule.

In [2], more details about this definition could be found. Generally speaking, P systems consist of several membranes, and these membranes are composed of membrane structures. In every membrane, there exist some objects and some evolving rules on these objects. There is a partial order relation providing the priority relation among these rules.

2.2 MP-Nets as the Framework of Generic P Systems and Petri Nets

In [6], a basic model of P system can be formalized by an ordinary Place/Transition Net. The membrane can be dissolved in the basic P systems, while in general P systems the membrane can be created, divided, merged and so on. We use MP-Nets to describe general P systems. MP-Nets are based on Hierarchical Colored Petri Nets(HCPN) [3]. HCPN is a kind of Colored Petri Nets(CPN) [3] containing a number of interrelated subnets called **pages**. The syntax and semantics of HCPN are similar to the definitions of non-hierarchical CPN. It is convenient to use MP-Nets to describe features and compare differences among these variants of P systems[6]. In short, MP-Nets are the generalization and integration of both P systems and HCPN.

An MP-Net is a kind of dynamic and hierarchical high level Petri Net. So the dynamic structure presents many theoretical and practical challenges in Petri Nets. MP-Nets are also the generic framework of P systems. The elementary membrane consists of the multiset of objects (every object owns its *color type*) and transitions corresponding to the rules in P systems. Each membrane (not *skin membrane*) can be *Dissolved, Created, Moved*, etc. MP-Nets can be a general framework for both P systems and Petri Nets, which is a powerful tool to model concurrent and mobile systems.

3 Membrane Calculus for Grid Transactions

3.1 Membrane Calculus

From Section 2, MP-Nets can be regarded as a powerful tool of concurrent and mobile systems based on P systems. Roughly speaking, every membrane in MP-Nets represents an entity or a transaction unit. If a membrane represents an entity, this membrane can be viewed as a page of dynamic HCPN. Otherwise, if a membrane presents a transaction unit, it denotes a transaction context. The rule sequence in every membrane represents an action sequence.

As discussed above, our model aims to express the transaction processing in Grid environments. Similar to the definition of MP-Nets, the formal Membrane Calculus (MC) is defined as follows,

Definition 2. *The Membrane Calculus (of degree $m \geq 1$) is a tuple*

$$MC = (\Sigma, \mu, P_1, P_2, \ldots, (R_1, \rho_1), \ldots, (R_m, \rho_m), C),$$

where

1. Σ *is a finite set of non-empty **types**, also called color sets;*
2. μ *is a membrane structure consisting of m membranes, and m is called the degree of MC;*
3. $P_i, 1 \leq i \leq m$, *is the set of Place which represents the Place set associated with the region $1, 2, \ldots, m$ of μ;*
4. R_i *is the set of transition rules associated with membranes $1, 2, \ldots, m$ of μ; ρ_i is a partial order relation over R_i, called a priority relation;*
5. C *is a color function. It is defined from $p \in P_i$ to Σ. This function C maps each place p in the membrane to a type $C(p) \in \Sigma$. Intuitively, this means that each Place must have a data type.*

In order to define transition rules formally, similar to Colored Petri Nets (CPN), we introduce some useful concepts as follows.

Definition 3. *(1) Let Type(v) denote the type of a variable v. (2) Var(expr) denotes the set of variables of the expression expr. (3) expr<b> denotes the expression with binding b. We often write bindings in the form $< v_1 = c_1, v_2 = c_2, \ldots, v_n = c_n >$ where $Var(r) = v_1, v_2, \ldots, v_n, r \in R_i$.*

Here we borrow the concept from CPN. The corresponding definitions in CPN can be found in Section 2 of [3]. Informally speaking, through types and bindings, we extend the P systems to a high level typed P systems, and the relationship between both systems is analogous to the relationship between CPN and Place/Transition nets.

In the definition of basic P systems, transition rules are objects evolution rules and only one operator δ is related to evolve membranes. In our MC, we adopt a general framework of P systems. That is, our membranes can not only be dissolved but be moved or created[8]. To keep our model simple, the other operations such as *Merge* and *Divide*[10] will be ignored in MC. Thus, in MC, membranes can be moved, dissolved and created. Except the *Create* operation, *Move* and *Dissolve* transition rules are similar to the *in*, *out*, and *open* operations in Mobile Ambients[13].

Definition 4. *Formally, each transition rule $r \in R_i$ has one of the following form:*

1. OR $r: P_r \to P_{here} P_{out} P_{in_j}$
2. Dissolve $r: P_r \to \delta$
3. Create $r: P_r \to [_k P_k, (R_k, \rho_k)]_k$
4. MoveIn $r: P_r \to \sigma_k$
5. MoveOut $r: P_r \to \eta_k$

The first type of rule is Object Rule (OR). All other rules are Membrane Rule, i.e., *Create, Dissolve,* and *Move,* respectively. In MC, we prescribe that the priority relation ρ of every membrane is $Dissolve > MoveOut > MoveIn > Create > OR$.

3.2 The Operational Semantics of Membrane Calculus

P systems can be referred as the high order Chemical Reaction Model(or *Cham*) [7], which is proposed to describe the operational semantics of process calculi, e.g., CCS, Join Calculus, etc. Moreover, P systems provide the mechanism of dynamic operations on membranes. Correspondingly, the semantics of MC is divided into two parts, namely, object rules and membrane rules. Informally speaking, every object rule can be presented by a transition in the term of CPN and the semantics of this rule can be represented by the changing of the marking in CPN. In a word, the operational semantics of one object rule is delimited by a transition similar to CPN.

Every membrane includes some objects, object rules, membrane rules, and submembranes. These elements (except membrane rules)compose a page in CPN. However, membrane rules can not be expressed by CPN. Meanwhile, the dynamic behaviors of MP-Nets are easy to be described by the actions on P systems. If all rules are object rules, then the structure of MC is static. So the effect of these object rules can be represented by ordinary CPN. When some membrane rules are executed, the structure of MC will be changed, similar to the reactions in *Cham*. More specifically, the static semantics of MC is similar to HCPN; the dynamic semantics is analogous with P systems.

The operational semantics of MC is listed as follows,

$$\frac{[_nP_n,\ldots[_mP_m,(R_m,\rho_m),[_jP_j,\ldots]_j,\ldots]_m]_n,\quad r\in R_m:P_r\to P_{here}P_{out}P_{in_j}, P_r\subset P_m}{[_nP_n+P_{out},\ldots[_mP_m-P_r+P_{here},(R_m,\rho_m),[_jP_j+P_{in_j},\ldots]_j\ldots]_m]_n} \quad (OR)$$

$$\frac{[_nP_n,\pi_n,(R_n,\rho_n),[_mP_m,\pi_m,(R_m,\rho_m)]_m]_n \quad r\in R_m:P_r\to\delta, P_r\subset P_m}{[_nP_n+P_m,\pi_n+\pi_m,(R_m+R_n,\rho_m+\rho_n)]_n} \quad (Dissolve)$$

$$\frac{[_nP_n,\pi_n,(R_n,\rho_n)]_n \quad r\in R_n:P_r\to[_kP_k,(R_k,\rho_k)]_k, P_r\subset P_n}{[_nP_n-P_r,\pi_n,(R_n,\rho_n),[_kP_k,(R_k,\rho_k)]_k]_n} \quad (Create)$$

$$\frac{[_nP_n,\pi_n,(R_n,\rho_n)]_n,[_mP_m,\pi_m,(R_m,\rho_m)]_m \quad r\in R_m:P_r\to\sigma_n, P_r\subset P_m}{[_nP_n,\pi_n,(R_n,\rho_n),[_mP_m,\pi_m,(R_m,\rho_m)]_m]_n} \quad (MoveIn)$$

$$\frac{[_nP_n,\pi_n,(R_n,\rho_n),[_mP_m,\pi_m,(R_m,\rho_m)]_m]_n \quad r\in R_m:P_r\to\eta_n, P_r\subset P_m}{[_nP_n,\pi_n,(R_n,\rho_n)]_n,[_mP_m,\pi_m,(R_m,\rho_m)]_m} \quad (MoveOut)$$

The first rule is object rule and this rule dose not change the membrane structure. All other rules are membrane rules and they change the membrane structure which presents the dynamic behavior of MC. In above rules, P_i represents the Place set in each membrane. π denotes the submembrane structure and (R_i,ρ_i) is transition rules and partial priority relation between transition rules, respectively. The operator +, -, denote the Multisets *Union* and *Difference*, respectively. And the operator $\subset$ denotes Multiset Inclusion[6]. Similar to CPN, every place has a type and every transition rule associates an expression binding when this rule is executed. It is easy to understand above definitions and the detail example will be given in the later part of this paper.

3.3 A Long-Running Transaction of Membrane Calculus

In this part, we discuss the model of the long-running transaction, i.e., Flight and Hotel Booking. That is an example from the section 3.3 of [14]. In this example, a transaction is composed of the sequence of two transactions: the first books a return flight ticket, the second reserves a hotel room for the nights between the arrival and the departure dates. This is a typical transaction example in Web and Grid services. In most cases, two-phase commit protocol adopted by traditional transaction processing systems is a bit too strict in general for Grid style applications. In this example, we use MC to model long running transactions in Grid environments. That is, we use the *compensation* transaction to compensate the committed transaction. In our example, if the second transaction fails, then we trigger the compensation of the first committed transaction.

According to MC, this transaction is formalized as follows,

$MC_{Flight\&Hotel} = (\{String\}, \mu, P_0, P_1, P_2, (R_0, \emptyset), (R_1, \emptyset), (R_2, \emptyset), C)$
$\mu = [_0 \quad Begin, [_1 \quad \emptyset \quad]_1, [_2 \quad \emptyset \quad]_2 \quad]_0$
$P_0 = (Begin, Success, Success1, Compensate, Fail)$,
$P_1 = (Begin1, Ticket, Compensate1)$,
$P_2 = (Begin2, Room, Fail2)$,
$R_0 =$
$\{ \quad Begin1 : Begin \rightarrow Begin1_{in_1},$
$\quad Continue : Success1 \rightarrow Begin2_{in_2},$
$\quad Compensate : Compensate \rightarrow Compensate1_{in_1} \}$
$R_1 =$
$\{ \quad Begin1 : Begin1 \rightarrow Ticket,$
$\quad Succuess : Ticket \rightarrow Success1_{out},$
$\quad Fail : Ticket \rightarrow Fail_{out},$
$\quad Compensate : Compensate1 \rightarrow Fail_{out} \}$
$R_2 =$
$\{ \quad Begin2 : Begin2 \rightarrow Room,$
$\quad Succuess : Room \rightarrow Success_{out},$
$\quad Fail : Room \rightarrow Fail2,$
$\quad Compensate2 : Fail2 \rightarrow Compensate_{out} \}$

Membrane 0 denotes the skin membrane (or the root membrane), which presents the whole transaction. From the point of view of CPN, Membrane 1 presents the behavior of booking a flight and Membrane 2 presents the second subtransaction, i.e., booking a room. In order to focus on the transaction processing, we use the rules *Success* and *Fail* in Membrane 1 and 2 to denote the booking processes in two subtransactions, and these two rules correspond to two kinds of results, *Success* or *Fail*, respectively. If we want to describe the detail booking processes in above two transactions, it is easy to use MC to denote the whole two processes, similar to Petri Nets.

In $MC_{Flight\&Hotel}$, the color set is only one type, *String*, which denotes the places such as *Begin*, *Fail*, *Success1*, etc. The function C maps each place in all membranes to the type *String*, and we omit the type mapping in all rules.

Firstly, the rule $Begin1 : Begin \rightarrow Begin1_{in_1}$ in Membrane 0 denotes that if there is a place $Begin$, this rule triggers the first subtransaction in Membrane 1. Then, in Membrane 1, the existing of $Begin1$ triggers the booking of a ticket, which is denoted by the rule $Begin1 : Begin1 \rightarrow Ticket$.

If this booking succeeds, the rule $Succuess : Ticket \rightarrow Success1_{out}$ sends the signal $Success1$ to Membrane 0, which presents the success of the first subtransactions. Otherwise, the signal $Fail$ is sent to Membrane 0 to denote the failure of the first transaction and the following failure of the whole transaction.

Secondly, if the first transaction succeeds, the rule $Continue : Success1 \rightarrow Begin2_{in_2}$ begins the second subtransaction. If this transaction succeeds, the signal $Success$ is sent to Membrane 0 to denote the Success of the second transaction and the following success of the whole transaction. Otherwise, the signal $Compensate$ is sent to Membrane 0 and the compensation of first subtransaction is triggered by the rule $Compensate : Compensate \rightarrow Compensate1_{in_1}$. This action satisfies the semantics of long-running transactions. From the above example, we can see that through the simple and concise syntax a typical long-running transaction is formalized.

4 Conclusions and Future Work

This paper presents a new calculus called *Membrane Calculus* for gird transactions based on P systems and Petri Nets. We firstly introduce P systems as a distributed and parallel computation model. Then, we use MP-Nets as the high-level framework of generic P systems. Based on MP-Nets, Membrane Calculus is proposed and the Operational Semantics of this calculus is defined. It is shown that Membrane Calculus is very suitable to describe Grid transactions through one typical example in the Grid environment.

The research about formal methods of Web/Grid Services is very hot now. In [14], an extension of the asynchronous π-calculus is used to describe long-running transactions in Web Services. The first International Workshop on Web Services and Formal Methods (WS-FM 2004, http://www.cs.unibo.it/ lucchi/ws-fm04/) have discussed many kinds of formal systems to deal with Web Services. Generally speaking, all these methods are far from mature. This paper provides a general semantic calculus of Grid transactions which has two distinct advantages: dynamic structure and location mobility. There are many properties such as data consistency, checkpoint, and weak read/write which should be explored in the future. On the other hand, we wish to implement this calculus by Maude [15] so that we can use *Model Checking* to study properties of Grid transactions. For instance, the protocols WS-Coordination/WS-Transaction [9,11] have been adopted by Grid communities, it is very meaningful to use Membrane Calculus to formalize and analyse these practical protocols.

Acknowledgments

This paper is supported by Natural Science Foundation of China (No. 60173033) and 973 project (No.2002CB312002) of China, and grand project of the Science and Technology Commission of Shanghai Municipality (No. 03dz15027 and No. 03dz15028).

References

1. Gh. Păun, Computing with membranes, *J. Comput. System Sci.* 61(1): 108-143 (2000).
2. Gh. Păun, G. Rozenberg, A guide to membrane computing, *Theoret. Comput. Sci.* 287: 73-100 (2002).
3. K. Jensen, Coloured Petri Nets. Basic Concepts, Analysis Methods and Practical Use. *Basic Concepts. Monographs in Theoretical Computer Science*, Volume 1, Springer-Verlag, 2nd corrected printing 1997.
4. R. Bruni, C. Laneve, U. Montanari, Orchestrating Transactions in Join Calculus, *Lecture Notes in Computer Science*, Vol. 2421, Springer, Berlin.
5. R. Bruni, U. Montanari, Transactions and Zero-Safe Nets, *Lecture Notes in Computer Science*, Vol. 1825, Springer, Berlin.
6. Z. Qi, J. You, H. Mao, P Systems and Petri Nets, *Lecture Notes in Computer Science*, Vol. 2933, Springer, Berlin. 2003.
7. J.-P. Banatre, P. Fradet, Y. Radenac, Higher-order chemistry, In *Proc. of Workshop on Membrane Computing 2003*, Tarragona, Spain.
8. M. Ito, C. Martin-Vide, Gh. Păun, A characterization of Parikh sets of ET0L languages in terms of P systems, In *M. Ito, Gh. Păun, S. Yu (Eds.), Words, Semigroups, and Transducers, World Scientific*, Singapore, 2001.
9. F. Cabrera, G. Copeland, T. Freund, et al., Web Services Coordination (WS-Coordination), http://www-106.ibm.com/developerworks/library/ws-coor/, August 2002.
10. M. Margenstern, C. Martin-Vide, Gh. Păun, Computing with membranes: variants with an enhanced membrane handling, In *Proc. 7th Internat. Meeting on DNA Based Computers*, Tampa, FL, 2001.
11. F. Cabrera, G. Copeland, B. Fox, et al., Web Services Transaction (WS-Transaction), http://www-106.ibm.com/developerworks/library/ws-transpec/, August 2002.
12. K. Czajkowski, A. Dan, J. Rofrano, et al., Agreement-based Grid Service Management (OGSI-Agreement), http://www.globus.org/research/papers/OGSI_Agreement_2003_06_12.pdf, June 2003.
13. L. Cardelli, A.D. Gordon, Mobile Ambients, *FoSSaCS 1998*: 140-155.
14. L. Bocchi, C. Laneve, G. Zavattaro, A Calculus for Long-Running Transactions, *FMOODS 2003*: 124-138.
15. M. Clavel, F. Durán, S. Eker, et al., The Maude 2.0 System. In *Proc. Rewriting Techniques and Applications*, 2003, *Lecture Notes in Computer Science*, Vol. 2706, 2003.

Workflow-Based Approach to Efficient Programming and Reliable Execution of Grid Applications*

Yong-Won Kwon, So-Hyun Ryu, Ju-Ho Choi, and Chang-Sung Jeong

Department of Electronics and Computer Engineering in Graduate School,
Korea University, Sungbukku anamdong 5-1, Seoul, Korea
{luco,messias,jhchoi}@snoopy.korea.ac.kr, csjeong@charlie.korea.ac.kr

Abstract. Grid computing provides a virtual organization which allows us to solve very large and complex problem with great-scale computing and data resources. But we need a powerful PSE to present easy programming and efficient execution environment for Grid application. We propose a new Grid workflow management system with abstract workflow description with integrated control and data flow and asynchronous flow control, exception for various fault handling, distributed/parallel workflow patterns, dynamic resource allocation, runtime workflow modification, and interactive graphic user interface. The architecture and implementation of our results about Grid workflow is shown in this paper for construction of efficient and powerful user environment on Grid.

1 Introduction

With Grid computing technology, a scientist will get great scale computing and data resource to solve a complex problem by constructing a VO(virtual organization)[1] which is made by selecting and gathering various geographically distributed resources. But we need a PSE(Problem Solving Environment) that provides easy Grid programming and efficient execution environment. It should enable us not only to produce new distributed applications but also to coordinate distributed services and execute them simultaneously. Also, PSE should provide a mechanism to prevent the progress of fault into serious error and system crash. Because Grid consists of heterogeneous and distributed resources under different administrators, various faults can be occurred.

Workflow is an important technology to achieve the above PSE for broad utilization of Grid because it can make multiple services to be executed at the same time as well as to be connected sequentially on various heterogeneous, distributed Grid resources. Also, it usually provides graphic user interface for us to be able to easily specify, execute, and monitor a workflow. Though several projects[3–7] have done researches about Grid workflow, it needs more advanced

---

* This work has been supported by a Korea University Grant, KIPA-Information Technology Research Center, University research program by Ministry of Information & Communication, and Brain Korea 21 projects in 2004.

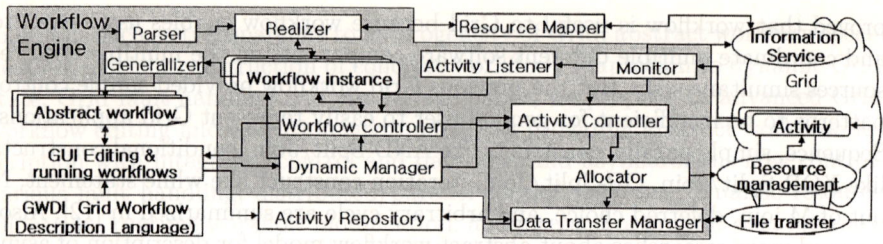

Fig. 1. Architecture

reservation without hardware fault. For example, when an activity is launched on a selected resource and if a problem occurs in the process of allocation, it is a resource utilization fault.

Efficient execution of workflow: To execute Grid workflow efficiently, we should be able to select proper resources on dynamic Grid environment. Therefore, knowledge rich information service is necessary, which provides not only the static data like speed of CPU and network, size of memory and file system, file system type, operating system and list of installed softwares, but also dynamic data like amount of CPU idle cycle, amount of free memory and free storage, available network bandwidth between point to point, and eventually, prediction of these dynamic information. But recent information service does not supply trustworthy dynamic data reliably while static data are provided. However, it should adapt dynamic Grid environment semi-automatically or manually by supporting fault handling, dynamic reallocation of activity, and runtime modification of workflow even thought not automatically. In addition, Grid workflow should execute activities as asynchronously as possible to acquire maximum profit of distributed resource.

4 New Advanced Grid Workflow System

4.1 Overall Architecture

We show our Grid workflow architecture in figure 1. GWDL(Grid Workflow Description Language)[14] is an XML-based language to specify Grid workflow. We construct an abstract workflow by using GWDL and the editing GUI. The Parser makes information about activity nodes and their interaction from a workflow. The Realizer produces a workflow instance by selecting resources on which activities are executed. The Resource Mapper helps the selection of necessary resources for activities. WC(Workflow Controller) runs workflow instances. It processes control flow, data flow, and fault handling when each event is occurred such as completion of activity, input/output data transfer, or fault generation. When an activity has to be executed, WC sends a start request with information about execution and data transfer of the activity to AC(Activity Controller). AC

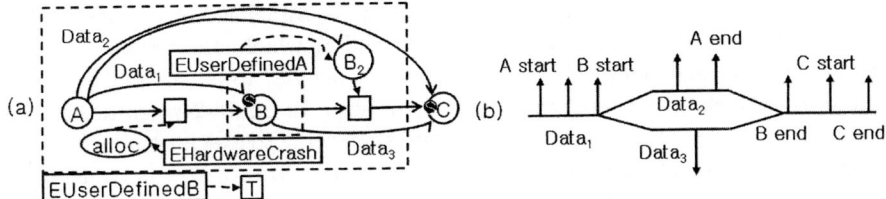

Fig. 2. Abstract workflow

transmits it to the Allocator which launches an activity on a distributed resource, and it also moves the necessary files to execute the activity by sending file transfer command to DTM(Data Transfer Manager). A running activity notifies the state into AM(Activity Monitor), and it is retransmitted again to AL(Activity Listener). AL updates the state of activity node in WC. WC proceeds control and data flow or fault handling according to the value of changed state. When an activity should be stop, WC sends a termination request to AC and it quits the running activity. When AC is aware of necessity of data transfer from activity state which is sent by AM, it commands data transfer into DTM or activity. When user want to modify a running workflow, he pause and modify it with interactive GUI, and then the changes are sent into DM(Dynamic Manager). DM reflects the modification into a workflow instance and activities through WC and AC. AR(Activity Repository) stores data of activities: name, in/out data, executable file, and etc.

4.2 Abstract Workflow Specification

Figure 2 shows our abstract workflow. A, B, B_2 and C nodes are activity nodes. All activity nodes have an activity, and each activity has input/output data, a function and exceptions. We define prenode of a node as the previous node of it and postnode as the next node. We makes an abstract workflow easily by connecting a node, the prenodes of it and the postnodes of it with control and data flow. Control flow is the execution order of activities, and data flow is movement of data between activities. To support advanced workflow pattern, we integrate control flow and arbitrary data flow. The order of execution is basically determined by control flow, however, when a node is dependant to only the first data of the prenode, executing it asynchronously is more efficient. Therefore we provide the selection of significant flow as we show in figure 2. B is dependant to the $data_1$, so B is executed when $data_1$ is generated. But C is dependant to node B, and C is executed when B is complete. Figure 2 (b) show the execution order of activities, The part of A and the part of B are calculated simultaneously because B is dependant to $data_1$ which appears faster than $data_2$. But, to adjust the execution order of A and C, C is executed after A and B is completed in order. Small circle which has letter 'S' means dependency to data or control flow. If it is connected to control flow, a node is started immediately after the prenode is complete, and If data flow, a node is started after the data is generated.

We support exception handling. Exception is attached into a dashed-line block, and the block has to contain a sub-workflow which starts from one node and finishes at one node. When an exception occurs, user can make control flow sent to back or forward, that is, retry or skip. In figure 2 when host is not available by a crash, reallocation is done and then retry to execute node B. When a user-defined exception occurs, B_2 is executed, and control flow is transmitted into the postnode of B. With exception, we detects fault, and we can use retry, reallocation, replication, alternative, and user defined activity for compensation of fault.

4.3 Workflow Patterns

We present our basic and advanced workflow patterns. In the basic patterns, there are sequence for sequential execution, XOR-Split/XOR-Join for conditional choice and merge, Loop for iteration, Multi-Split/Multi-Join for conditional multiple choice and merge, AND-Split/AND-Join for parallel execution and synchronous merge.

In the advanced patterns, there are the followings to support various parallelism: pipeline, data parallelism, and synchronization. AND-Loop is a variant of simple loop construct that AND-Split follows each iteration. Whenever an iteration of loop is complete, a new control flow is generated to execute the postnode of loop. AND-Loop can send many flows to a node N continuously. If N is bottleneck, activities that send flows to N may stop or be processed slowly. A pattern is needed to prevent this situation. In queue pattern, when control flows arrive at it several times, all control flows are stored in queue and transferred whenever the next node is idle. In node copy pattern, a node is copied up to the limited number of times and an idle node is executed in parallel. In Wait pattern, a node N blocks until some or all flows are received. we published the above advanced pattern and GWDL at [14]. There are formal definitions and detail explanation of workflow model and also GWDL which consists of Data model for data type, resource model for specifying Grid resource, activity model for defining an activity, and flow model for describing the interaction between activities. In explicit termination pattern, workflow is exited. Pause/Continue pattern allows us to stop workflow temporary and restart it. Group is a subworkflow which is specifying by Group Node in GWDL. To use predefined fault easily, we define a group with fault handling property. The value of it is retry, reallocation, replication, and alternative. Activities in Group use a specified fault handling strategy. Workflow pattern enables us to make parallel Grid application fast and easily and understand it well through explicit expression.

4.4 Exceptions for Various Fault Handling

In GWDL, we defines exceptions: EFail for unknown fault, ENotAvailable for hardware fault, EAllocation for allocation problem, EMemory for insufficient memory, EStorage for insufficient storage, ENoData for no usable input or output data, ETimeExpired for timer and user defined exceptions for software fault and so on. User can use retry, replication, reallocation, alternative, and other user-defined recovery activity. Retry is executing again, reallocation is changing resource, replication is executing copies of an activity on several distributed resources, and alternative is executing the alternative on different resources. If a user want to complete an activity to a specific time, he use the timer activity. After the activity and the timer are executed simultaneously, if timer is expired before activity is completed, ETimeExpired exception occurs and a fault handling methods can be executed according to user specification.

4.5 Concrete Workflow Execution and Dynamic Modification

We present two types of activity: loose-coupled(LC) type and tight-coupled(TC) type. LC type is an executable file which is only executed without notification of detail states, but TC type communicates with workflow engine closely. It notifies the state of itself and provides socket connection for peer-to-peer communication between TC type activities. With the LC type, we execute legacy programs. With the TC type, we make new program or it helps us to use legacy application. It also supports asynchronous execution explained in section 4.2. So before a TC type activity is launched, many parameters like communication addresses and ports, and activity identification number are specified. Of course, computing resources for activity execution of both types are determined. User can specify computing resources for activities, or workflow engine can do it statically or dynamically. If user select dynamic allocation, WC asks allocation to the Realizer. Workflow engine supports automatic parameter specification of both activity types, dynamic resource allocation, and fault handling.

User can modify a running workflow with DM and Pause/Continue pattern. Our Interactive GUI supports not only dynamic workflow modification and monitoring by presenting pause and continue buttons, but also tools for making workflow. With dynamic allocation, dynamic modification and fault handling, we can adapt workflow to dynamic Grid environment by manually, or semi-automatically.

5 Implementation

We have implemented a prototype of our Grid workflow system with object-oriented programming by using Globus toolkit[8] for Grid resource, Java CoG for Globus client library, Java for workflow engine and GUI, C++ for TC type activity, Apache Xerces for parsing and editing GWDL, Xalan for generation of TC type activity template code from activity model of GWDL and socket for peer-to-peer communication among activities and workflow engine. We also use design pattern such as Factory, Singleton, Composite and Facade. For launching an activity and notification of completion or failure of activity, we use GRAM of Globus toolkit, MDS for information service, and GridFTP for file transfer.

6 Conclusion

We have proposed a new Grid workflow system with abstract workflow model, exception, parallel workflow pattern, asynchronous flow control, dynamic allocation and modification. Abstract workflow description with integrated control/data flow enables user to utilize Grid resource without knowledge about the details of Grid such as resource management, information service, remote secondary storage access methods, and file transfer library. Asynchronous flow control maximizes the benefit of distributed resource by executing multiple activities as simultaneously as possible. Exception enables user to easily use various fault handling scheme such as retry, reallocation, replication, alternative, and user-defined fault processing for reliable execution of workflow. We provides distributed and parallel control patterns which enable us to make parallel Grid application easily and fast. To adapt dynamic Grid environment, activities are allocated at distributed resource not only statically but also dynamically. Also a running workflow can be modified by user when a exception occurs or after user pause it. Our Grid workflow system provides interactive graphic user interface to support powerful dynamic control and easy monitoring of an executed workflow. Therefore, our research provides us Grid workflow system as a powerful tool for production and execution of workflow-based Grid applications.

References

1. I. Foster, C. Kesselman, S. Tuecke, "The Anatomy of the Grid: Enabling Scalable Virtual Organizations," International J. Supercomputer Applications, 15(3), 2001.
2. G. Alonso, C. Hagen, D. Agrawal, A. E. Abbadi, C. Mohan, "Enhanced the Fault Tolerant of Workflow Management Systems" IEEE Concurrency, July-september, 2000.
3. Triana Workflow, http://www.gridlab.org
4. Sriram K., Patrick W., Gregor von L., GSFL:A Workflow Framework for Grid Services http://www-unix.globus.org/cog/projects/workflow/gsfl-paper.pdf
5. GridAnt, http://www-unix.globus.org/cog/projects/gridant/
6. myGrid workflow, 2003, http://www.mygrid.org.uk/myGrid/
7. A. Hoheisel, , U. Der, "An XML-based Framework for Loosely Coupled Applications on Grid Environments" ICCS 2003. Lecture Notes in Computer Science, Vol. 2657, Springer-Verlag
8. I. Foster, and C. Kesselman, "The Globus Project: A Status Report," Heterogeneous Computing Workshop, pp. 4-18, 1998. Globus Toolkit, http://www.globus.org
9. Unicore http://www.unicore.org
10. CoG kits, http://www-unix.globus.org/toolkit/cog.html
11. J. Kienzle, "Software Fault Tolerance: An Overview", Ada-Europe 2003, LNCS 2655, pp. 45-67, 2003.
12. W.M.P. van der Aalst, A.H.M, Hofstede, B. Kiepuszewski, A.P. Barros. (2003). Workflow Patterns, Distributed and Parallel Databases, Jule 2003, pp. 5-51
13. S. Hwang, C. Kesselman, "GridWork.ow: A Flexible Failure Handling Framework for the Grid", Proceedings of the 12th IEEE International Symposium on High Performance Distributed Computing , 2003
14. Y.W. Kwon, S.H. Ryu, C.S. Jeong, H.W. Park, "XML-Based Work.ow Description Language for Grid Applications", ICCSA, pp 319-327, 2004

Achieving Context Sensitivity of Service-Oriented Applications with the Business-End Programming Language VINCA*

Hao Liu, Yanbo Han, Gang Li, and Cheng Zhang

Institute of Computing Technology, Chinese Academy of Sciences
Graduate School of the Chinese Academy of Sciences
100080, Beijing, China
{liuhao,yhan,ligang,zhch}@software.ict.ac.cn

Abstract. This paper presents an approach to increasing the adaptability and flexibility of service-oriented applications. The focuses are on the modeling of user context information and its uses for enabling personalized service selection, composition and activation. The resulting models and mechanisms are an essential part of the novel language VINCA, which is targeted at facilitating service composition on business level in an intuitive and just-in-time manner.

1 Introduction

According to the basic principles of service-oriented architecture (SOA) [1], constructing a Web services application is simply a matter of first discovering and selecting the right services and then composing them into a solution. However, the openness and dynamism of service-oriented computing raise extra challenges, for example there can be a massive collection of constantly changing Web services, and user requirements can also be very volatile. So how to construct adaptable web service applications is becoming a key issue. This paper proposes an approach to promoting user context [2] sensitivity of service-oriented VINCA [3] applications. The main features of the proposed approach include: 1) providing adaptable personalized services to the user by user context-based service selection and composition; 2) facilitating application construction and activation by reducing interactions between the application and the user through making context as implicit inputs of the application. Essential parts of this approach include the way to model user context and the mechanism to use it, which will be discussed in detail in the paper.

The paper is organized as follows: Section 2 gives a brief overview of the VINCA language. Section 3 presents the detailed user context definition in VINCA. Section 4 describes the mechanism to achieve user context sensitivity of VINCA applications. Section 5 presents a case study to illustrate how to construct and use a VINCA appli-

---

* The research work is supported by the National Natural Science Foundation of China under Grant No. 60173018, the Key Scientific and Technological Program for the Tenth Five-Year Plan of China under Grant No. 2001BA904B07 and German Ministry of Research and Education under Grant No. 01AK055.

cation to fulfill individual spontaneous requirements in practice. Section 6 overviews related work. Finally, some concluding remarks are given in section 7.

2 An Overview of the VINCA Language

The development of VINCA is driven by a real world project FLAME2008 [4] we are undertaking. It is a Sino-German joint effort for developing a service mediation platform named CAFISE [5] to provide personalized and one-stop information services to the general public during the Beijing 2008 Olympic games. VINCA has been designed to mediate between diverse, rapidly changing user requirements and composites of individual services scattered over the Internet.

VINCA is a process-oriented language designed to capture user's needs. There are four constituent parts in VINCA: business service, process, user context and interaction. A business service is semantic abstraction of a set of concrete Web-based services providing a given business functionality from business point of view [3]. Each business service has a set of input and output parameters that are all in character form with a separate type information, a list of non-function attributes for specifying user's QoS (service quality) requirements, such as cost, responding time, location and so on. Business service semantics is based on a domain-oriented ontology [6]. They are attributed by a list of semantic descriptions, each of which refers to an ontology concept shared by all involved parties. Interaction specification mainly describes the way of service delivery. VINCA uses process to describe user's business requirements. The basic building blocks of VINCA process are abstract nodes. A node can be a service node or a control node, presenting a business service (either atomic or composite) or a control element (e.g. sequential link, concurrent link). Detailed design rationale of business service, interaction and process can refer to paper [3]. User' context is used to characterize user's situation [2]. Our recent research effort on user context definition and how to use it to promote service-oriented application adaptability will be discussed in detail in the following sections.

3 User Context

One of the most important objectives of VINCA is to make VINCA processes adaptive to web services and user requirements that are changing dynamically, as well as less demanding on user's attention. To achieve this goal, VINCA processes will need to be sensitive to user's context. Accordingly we have to define an effective context model to characterize the situation of the user.

The first step to model user context in VINCA is to identify the categories of context required. We have adopted the four categories of user context that Gross and Specht have considered in [7]: Identity, Location, Time and Activity. They can be used as indices for retrieving second level types of contextual information. For example, from a user's identity we can acquire many pieces of related information such as addresses, preference items, nationality etc. In fact these items are the basic pieces of information that characterize the situation a user in. We express these pieces of information as:

Basic structure: The atomic context. The atomic context information is the basic units that embody discrete items of user information. For example, language, country and location etc. We represent it with a four-tuple: (ContextType, OntologyURL, Operator, Value). ContextType refers to the type of context. OntologyURL is the context type's ontology file's URL. In FlAME2008, it refers to an ontology concept shared by business services to enable semantic match between user's context items and business service input and non-function property parameters. Value is the context type's value, and Operator describes the relationship between the context type and its value. It can be a comparison operator (such as =, >, or <).

Examples of atomic contexts:

(Location, http://flame2008:6888/domain/Location.daml#location, = , Stadium)

Operations on contexts: Only the atomic context is not enough to describe user's context precisely. For example, user's preference is often conditional. When the user is in the stadium during daytime, he likes fast food. While when he is in the hotel during nighttime, he may like Chuan food. So we import an abstraction layer based on first-order predicate calculus and Boolean algebra over the atomic context layer.

Boolean operations: We can construct more complex context expressions by performing Boolean operations such as conjunction, disjunction, and negation over atomic contexts.

For instance:

(Location, http://flame2008:6888/domain/Location.daml#location, = , Stadium) ∧ (Time, http://flame2008:6888/domain/Time.daml#time, = , Daytime)

refers to a situation instance that it is daytime and the user is in the stadium.

Quantification: The Value argument of the atomic context is a variable and then quantify over this variable. This lets us parameterize the context and represent a much richer set of contexts. The model allows both universal and existential quantification over this variable.

For example, to express that the user is in some location, we can write:

$\exists_{Location}$ X (Location, http://flame2008:6888/domain/Location.daml#location,=, X).

Rules: Based on Boolean operations and quantifications on atomic contexts defined above, we can easily construct various rules to express more complex context.

For instance, conditional preference can be expressed as follows:

(Location, http://flame2008:6888/domain/Location.daml#location, =, Stadium) ∧ (Time, http://flame2008:6888/domain/Time.daml#time, =, Daytime) -> (Flavour, http://flame2008:6888/domain/Restaurant.daml#taste, = ,Fast food)

refers to that when the user is in the stadium during daytime, he likes to eat fast food.

To sum up, the context model we defined above is represented to support two kinds of context information: 1) the atomic context information to state basic facts about a user. 2) Operations on the atomic context to express more complex context information such as rules, situation.

Evaluation of the context model: (1) Extensibility: Duo to our context model based on domain ontology, a user can add (delete) the atomic context items and modify the corresponding operations on it to extend his context model. While many of the existing context models for context-aware computing are restricted to narrow classes of context. In particular, several support only sensed context information and its derivatives [8][9]. (2) Strong expression power: By the operations on the atomic context, our model provides the means to express more complex context than most of existing ones [10] [11]. For example: The user can define a situation instance by Boolean operations. We extend the means to express user preference by assigning condition to user's preference.

4 The Mechanism to Achieve User Context Sensitivity of VINCA Applications

In the previous section we have defined user context model. Another important aspect in our approach addresses the question of its uses. The corresponding algorithm and supporting environments will be discussed in detail in the following subsections.

4.1 The Algorithm to Achieve User Context Sensitivity of VINCA Applications

We briefly outline the algorithm of how to use user context information for enabling personalized service selection, composition and activation.

(1) For each business service in a VINCA process, user's atomic context items will semantic match against business service's non-function property and input parameters. For each semantic matched context item, check whether it is user situation conditional. If it is, then compare condition S_i with user current situation *Situ* that includes the context types in condition S_i. If *Situ* equals S_i or S_i is a subset of *Situ*, then make it as the implicit input of the business service.

(2) For each semantic matched user's atomic context items against business service's non-function property parameters, according to the priority of them which can be defined by the user (or use platform's default configuration), use user's context item's value and default selection policy as the condition to select concrete web services. This process is a stepwise one. After the selection process, the most appropriate services will be selected.

(3) The final web service will be selected according to context parameters (e.g., cost budget) and user's default preferences (e.g. minimal execution duration) that could be applicable to the whole VINCA process. The efficient linear programming method in [12] is adopted for the optimally selecting the most appropriate web service.

4.2 The Architecture to Support User Context Sensitivity of VINCA Applications

There are three layers in supporting architecture as shown in Fig.1. End users' applications generated by VINCA application layer can be context sensitively executed by interpreter layer through invoking related resources in resource layer. We will discuss the interpreter layer in detail in the following sections.

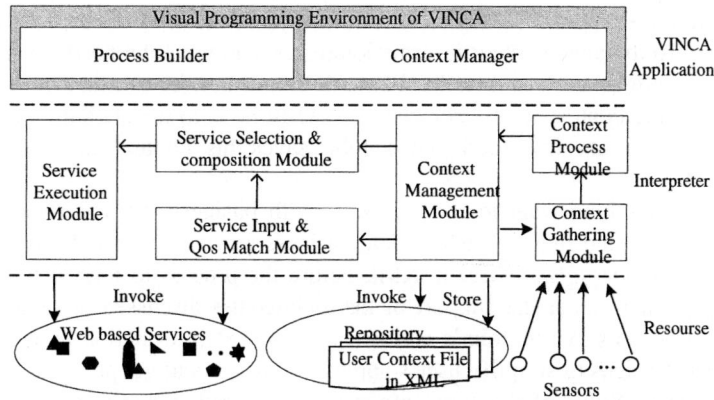

Fig. 1. The Architecture to Support User Context Sensitivity of VINCA Applications.

The interpreter layer mainly includes application execution part and context management part. Context management part is composed of three modules: context processing, context gathering and context management modules. Context management module is responsible for storing user context file into repository, sending instructions to trigger context gathering module to gather user' contextual information according to user context file. Context Gathering module is responsible for gathering some of user's context information (e.g. location) from sensors (e.g. mobile phone, PDA). Context process module processes the context data forwarded by context gathering module. Then it sends the processed data to context management module. Application execution part includes service input and Qos match module, service selection and composition module, and service execution module. The service input and Qos match module makes the semantic matched user's context parameters as the implicit inputs of the business services. It is also responsible for check whether the semantic matched context items are user situation conditional. If they are, then check whether the conditions are fulfilled. The service selection and composition module at first selects appropriate services among a possibly large set of web services according to users' context information and the constraints set by the service provider. Then further refines the set of selected services using context parameters that are relevant for the composition. After the concrete web service that satisfies all the user's context is chosen, the corresponding capsules (i.e. information about the services, addresses, ports, and invocation protocols) and semantic matched user context items' values which are made as service's input parameters are sent to service execution module to carry out the execution of the web service.

5 Case Study

To better explain how to achieve user context sensitivity of VINCA applications, the following simplified example is used: In 2008, various parties will provide a large variety of services for public uses, and VINCA helps different users to get their personalized services. Among the users is Mr. John, a tourist. With the context manager

John can build his context. To static context information such as flavor, John filled in their values. To dynamic context information (e.g. location), John needs to assign the source of this information such as his PDA. If John likes, he can also defines some rules over atomic context items. For instance, when he is in the stadium during daytime, he likes to eat fast food. Then John saves his context in a file named *john.context*.

When one day John want to run a process with business services ReserveHotel, InquiryRestaurant and InquiryBusSchedule business services, the platform will automatically add *import john.context* in XML before the process to form a VINCA application [3]. Then under the support of the architecture discussed in section 4, the process will be sensitive to John's context: (1) Context information as the implicit input of web services to simplify user's application usage and get personalized application activation. Some of the John' preference information don't need to be input each time he uses the processes. For example: When the ReserveHotel service is triggered, the service input and Qos match module will make his preference item RoomStyle value as the implicit input parameter of room style of ReserveHotel service by semantic matching between the ReserveHotel input parameters and john's atomic context items. (2) Context information as the service selection and composition criteria to enable users get personalized services. For example, when the InquiryRestaurant business service is triggered, the service selection and composition module will first select the restaurant information web services according to john's preference such as flavor and get a small list of them. If the preference is situation conditional, the conditional situation is also examined. And then a smaller set of the nearest ones will be recommended through evaluation of John's location. If John is in the stadium during day time, then several nearest fast-food restaurant information services will be selected. While when he is in the hotel during night time, several nearest Chuan food restaurant information services will be recommended. Finally John's preference item costBudget and minimal execution duration will be taken into account for further selecting the most appropriate web service from a global point of view.

6 Related Works

Our work relates to several emerging fields of research and technologies, including service selection, service composition, context-aware computing technologies and context modeling. This section reviews the related work in the former three areas. The related work about context model has been discussed in section 3.

From the user's point of view, service selection approaches can be grouped into three categories: manual, service discovery protocol and context-aware service selection. Manual selection of web services involving too much user involvement causes user inconvenience. For example, it may be tedious for a user to examine many restaurant services and compare them. Service discovery protocols may select services for a user [14]. The advantage of protocol selection is that it simplifies client programs or little user involvement is needed. On the other hand, protocol selection may not reflect the actual user's will. Predefined selection criteria may not apply to all cases. Context information is useful in selecting personalized service. But so far, only a few projects use location information as a kind of context information to help service selection [15].

There have been some works done towards dynamic service composition to suit openness environment. For example, the dynamic service composition called software hotswapping at the Carleton University, Canada [16], the eFlow [17] from HP laboratories, and the Ninja service composition at the University of California Berkeley [18]. But they all miss out the use of context when composing services. Composite web service development is still a manual activity that requires specific knowledge about the composing web services in advance and takes a lot of time and effort.

7 Concluding Remarks and Future Work

In this paper, we present an approach for increasing the adaptability and flexibility of service-oriented VINCA applications by making the application sensitive to user context information. Our contribution in this paper is the definition of an extensible user context model and the corresponding mechanism to achieve user context sensitivity of VINCA applications. We have implemented FLAME2008 prototype platform. The experiment results validate our approach's feasibility. Further enhancement of our approach includes investigations on the formal definition of context model, user situational preference generation from history data, user context information storage modes and update policy, context model usage pattern etc.

References

1. www.service-architecture.com.
2. Anind K. Dey and Gregory D. Abowd, Towards a Better Understanding of Context and Context-Awareness, 1st International Symposium on Handheld and Ubiquitous Computing (HUC '99), June 1999.
3. Y.Han, Hui Geng, Bernhaerd Holtkamp, VINCA-A Visual and Personalized Business-level Composition Language for Chaining Web-based Services. International Conference on Service-oriented Computing 2003, Trento, Italy.
4. B. Holtkamp, R. Gartmann, Y. Han, FLAME2008-Personalized Web Services for the Olympic Games 2008 in Beijing, Proceedings of Echallenges 2003, Bologna, Italy, Oct. 2003.
5. Y. Han, et al, CAFISE: An Approach Enabling On-Demand Configuration of Service Grid Applications, Journal of Computer Science and Technology, Vol.18, No.4, 2003.
6. N. Weißenberg and R. Gartmann, Semantic Web Services for Olympia 2008, Proc. IEEE/WIC Int. Conf. on Web Intelligence (WI2003), Oct 13-17, 2003, Halifax, Canada, http://wi-consortium.org /wi-iat-03/papers/Wi2009.pdf.
7. Gross, T., Specht, M. (2001), Awareness in Context-Aware Information Systems. In Mensch&Computer, Fachübergreifende Konferenz, Bad Honnef, Germany, Oberquelle, Oppermann and Krause (Eds.) Teubner, pp. 173–182.
8. Schmidt, A., et al.: Advanced interaction in context. In: 1st International Symposium on Handheld and Ubiquitous Computing (HUC' 99), Karlsruhe (1999).

9. Gray, P., Salber, Modelling and using sensed context in the design of interactive applications. In 8th IFIP Conference on Engineering for Human-Computer Interaction, Toronto (2001).
10. G. Klyne, F. Reynolds, C. Woodrow, H. Ohto, Composite Capability/Preference Profiles (CC/PP): Structure and Vocabularies, W3C Working Draft, Mar 15, 2001. (URL: http://www.w3.org/TR/2001/WD-CCPP-struct-vocab-20010315/).
11. G. Klyne, A Syntax for Describing Media Feature Sets, RFC 2533, Mar 1999. (URL: http://www.faqs.org/rfcs/rfc2533.html).
12. L. Zeng, B. Benatallah, M. Dumas, J. Kalagnanam, Q. Sheng, Quality-driven Web Service Composition, In Proc. of 14th International Conference on World Wide Web (WWW'03), Budapest, Hungary, May 2003, ACM Press.
13. CAFISE group: Service Community Specification, Technical Report, Software Division, ICT of CAS, December 2002.
14. G. G. Richard III, Service Advertisement and Discovery: Enabling Universal Device Cooperation, IEEE Internet Computing, September-October, 2000, pp. 18-26.
15. Feng Zhu, Matt Mutka, and Lionel Ni, Classificationof Service Discovery in Pervasive Computing Environments, MSU-CSE-02-24, Michigan State University, EastLansing, 2002.
16. B. Limthanmaphon and Y. Zhang.Web Service Composition with Case-Based Reasoning. Klaus-Dieter Eds: Schewe and Xiaofang Zhou Database Technologies, January 2003.
17. F. Casati, S. Lnicki, M. Krishnamoorthy, and M. Shan, Adaptive and dynamic service composition in eFlow. Technical Report, HPL-200039, Software Technology Laboratory, Palo Alto, USA, 2000.
18. UC Berkeley Computer Science Division. http://ninja.cs.berkey.edu.

Mapping Business Workflows onto Network Services Environments

Wenjun Wang, Xinpeng Liu, Yingwei Luo*, Xiaolin Wang, and Zhuoqun Xu

Dept. of Computer Science and Technology, Peking University, 100871 Beijing, P.R. China
lyw@pku.edu.cn

Abstract. In order to implement common adaptability in accessing to distributed resources, it is fatal for Web applications to deploy business workflows onto network service environments composed of Web Services and Grid Services. This paper introduces abstract resources and abstract services to express abstract workflows, and uses resources and service instances to express executable workflows that can run in service environments. Thus the deployment and execution of business workflows are transformed into two mappings: mapping of business workflow onto abstract workflows and mapping of abstract workflows onto executable workflows. The definition of relevant parameters used to implement automatic mapping is formulized as Application Template (AT). The most important component of AT is corresponding relation between business functions and abstract workflow fragments. AT also defines restrictions while each abstract services are utilized in certain application tasks, as well as associated service metadata. The latter avoids superfluous metadata by service providers, and simplifies unnecessary details in matching service instances.

1 Introduction

Workflow [1] is an important cooperation mechanism, which is used to implement automation of a whole or part business process. Workflow is being used in BPR (Business Process Reproduction), BPM (Business Process Management), and owns several specifications as XPDL, Wf-XML, BPML, etc. A workflow that is expressed by business activities is called a "business workflow".

Web Service [2] can be directly accessed through XML based message protocols, and can be seamlessly integrated across the network without consideration of programming languages and runtime environments. Based on the concepts of Web Services, OGSA brings forwards concept of "Grid Service", which is used to solve problems related to temporary services as service detection, dynamical service creation and service lifetime management. Services can form service workflows. These workflows refer to a complete operation streamline composed of Web Services and Grid Services that are used to satisfy business requirements. A series of service workflow specifications as WSFL, XLANG, BPEL4WS and GSFL [2] are being established or have already published.

In dynamical service environments, in order to implement common adaptability in accessing to distributed resources, it is fatal for Web applications to deploy business

---

* Corresponding author.

workflows onto network service environments composed of Web Services and Grid Services. Pegasus (Planning for Execution in Grids) [3], a research of ISI, is a part of the project GriPhyN and SCEC/IT, which can map complex business workflows onto grids and execute them. Workflows in Pegasus include two levels: abstract workflows and concrete workflows. Both the two level workflows use grid terms to describe themselves, without relevant mappings of terms from different domains. The project SciDAC-SDM [4] of California University makes a strict differentiation between abstract workflows described by business terminology and low-level service composed executable workflows, but the transformation between workflows is static, without the selection of suitable services in terms of real-time states of running services.

This paper presents an executable workflow formation method based on Application Template, which comprehends the advantages of the above two methods. Our method defines three-level workflows as business workflow, abstract workflow and executable workflow. With the aid of Application Template, the automatical transformation between cross-level workflows can be implemented.

2 Level of Workflow

2.1 Basic Concepts

(1) Workflow. A workflow W is a tuple <Tasks, Valuables, Successive Functions, Conditional Functions>, where:

- Tasks: Set of tasks in W, which are function modules consisting of a workflow;
- Valuables: Set of valuables in W, which includes the common valuables that must be maintained during the execution of the workflow;
- Successive functions: This kind of functions describes sequential relationships between tasks, which actually provide sides of a DAG. They are formulized as:
- $f_i(t_j) = t_k (t_j, t_k \in T)$, where T expresses the set of tasks;
- Conditional functions: This kind of functions describes the execution conditions of relevant successive functions. Only when states specified by these functions are satisfied, corresponding successive functions can be put into execution; otherwise workflows will be suspended or even canceled. They are formulized as: $f_i(a_1, a_2, ..., a_n) = true \,/\, false$, where a_i expresses possibly relevant conditions in this restriction. These restrictions can be described through terminology from above Tasks and Valuables.

A workflow is expressed as a DAG, with nodes denoting functions, sides denoting direction of control. Type of nodes in a workflow may includes: a) function nodes, nodes representing tasks; b) data nodes, data needed by function nodes which provide necessary input parameters, output results or internal data files; and c) control nodes, virtual nodes composed of successive functions and condition functions, which decide the next step in flow direction.

(2) Service Environments. Web Services and Grid Services make up of service environments, which are the execution environments of workflows.

Service environments include services and resources that are utilized by services. All the entities in service environments can be called resources, such as data files, process capabilities, storage systems, databases, etc. Services are also regarded as a special type of resources.

There are three different types of users in service environments: resources providers, service environment managers and business applicants. Resource providers provide resources, Web Services or Grid Services; service environment managers register and manage services in environments; business applicants are mainly domain experts, who specify business workflows, and implement functions of workflows through services, resources in service environments.

(3) Abstract Resources. The description of resources can be abstracted as abstract resource to express a class of resources providing the same function. It is an illustration of specifications to classified resources., or an abstraction to resources bearing the same features. There may exist multiple physical resources for a virtual resource.

(4) Abstract Services. Services are a special type of resources. They can provide entities with certain capabilities with can be used in a network environment. In this paper, the range of services is specialized as Web/Grid Services. Services of service environments are also redundant with multiple services providing the same function. The same phenomenon happens for business applicants as they make choice for certain services. The description of services is abstracted as abstract service, which provides specifications for the same class of services.

2.2 Level of Workflows

In regard of different user groups in service environment, workflows are divided as:

(1) Business Workflow (BWF), the kind of workflow that can be described by business terminology and accomplish certain business functions. Tasks in these workflows are usually set down by business applicants.

(2) Abstract Workflow (AWF), the kind of workflow that is described by abstract resources and abstract services to accomplish specified business functions. Tasks in workflows are expressed by abstract services, and usually act as transitory workflows when business workflows are deployed onto service environments.

(3) Executable Workflow (EWF), the kind of workflow that is described by resources and service instances to accomplish certain business functions. These workflows can be put into execution. In this level, the interaction between service environments and executable workflows should be considered.

The division of workflows into business workflows and service workflows makes it possible that business applicants may only need to concentrate on the design of business workflows, without caring about technical details of environments. For instance, domain experts can use "calculation of troop orders" (as shown in figure 1) directly, and the algorithms to implement this business function and the low-level technique information should not be taken into consideration.

Reasons for the subdivision of service workflows into abstract workflows and executable workflows are as follows: a) the difficulty in one-off mapping is reduced. If

BWF is directed mapped to EWF, not only the transformation of different domains' terminology should be mattered, but also the technique details relevant to the service environments should be paid attention; b) the mapping of BWF to AWF can be done in advance, while the mapping of AWF to EWF can be done in execution. Thus, not only the efficiency can be improved, but also the former mapping can be reused. Mapping from AWF to EWF need only strategies of resource selection and service selection. Problems aroused by changes of service environments are reduced, too.

Figure 1 is an example for demonstration of workflows in three levels.

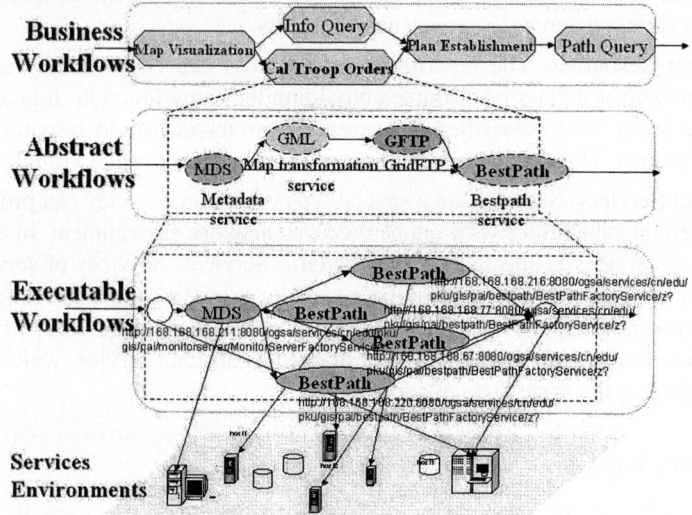

Fig. 1. Transformation Demonstration of three-level Workflows.

2.3 Description Terminology of Workflow

According to the definition of workflow, some important elements in a workflow, such as tasks, valuables, successive functions, and condition functions, should be precisely defined in all 3 levels of workflows. Ontology [5] is "the explicit formulized specification of sharing concept models", which provides common comprehension of domain knowledge and ascertains common terminology inside a domain. So when ontology is applied to the definition of workflows, terms used by workflows of different levels may come from different ontologies. Advantages for ontology introduction lie in that: a) business workflows can be intimately related to application of ontology from each levels; b) the origination of basic terminology of workflows can be traced from definite sources, which facilitates the semantic mappings; and c) the description of workflows may appear to be of more preciseness and theorization.

2.4 Transformation Between Different Levels of Workflows

Transformation between different levels of workflows can be performed through two phases: from BWF to AWF and from AWF to EWF. When BWF is transformed into

AWF, we need: a) replace each business function in BWF with corresponding abstract workflow fragments which implement the functions; b) map resources relevant to each business function onto corresponding abstract resources (see figure 2).

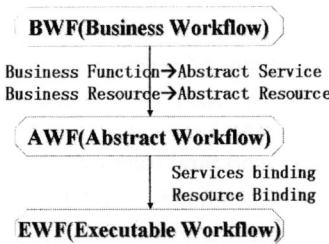

Fig. 2. Transformation Between Different Levels.

When workflows are put into execution, AWF need to be transformed into EWF, i.e.: a) replace each abstract service in AWF with a suitable service instance; and b) replace each abstract resource in AWF with a suitable resource.

3 Application Template Based Workflow Mapping Method

3.1 Application Template

Mappings from BWF to AWF and from AWF to EWF can be successfully performed only when condition of all aspects are matched. So we introduce the concept of Application Template of business functions, which is used to describe relevant assistant information (dynamic or static) needed by deployment of tasks in business workflows into service environments and associated metadata.

The soul of Application Template (abbreviated as AT in the following) is the mappings between business functions and corresponding abstract workflow fragments.

In order to obtain automatic transformation of business workflows into abstract workflows, in addition to mappings between business functions and abstract service fragments, AT also includes descriptions of function nodes, descriptions of control nodes, etc., as shown in figure 3. The formulized expression of AT is presented in the following:

$$AT =< DT, wf_A, EvaFuncs, MarkFuncs, GetVarMeta, \text{Re } sumeDesc, MetaDesc >$$

Where:

- ✧ DT (Domain Term): Domain terms for business functions and data. It also includes applicable condition, performance requirements, applicable organization and time restrictions;
- ✧ wf_A (Abstract Workflow Fragment): Abstract workflow fragments that are mapped by DT, which are the implementation of DT functions;
- ✧ EvaFuncs (Evaluation Functions of control nodes): Evaluation functions to each control node in wf_A, which are used to control the selection of tributaries in flow direction;

- *MarkFuncs* (Mark Functions of function or data nodes): Mark functions to each function node or data node in wf_A. These functions are expressed by valuables in service environments too, while they are used to control the selection of service instances from service classes and the selection of resources from resource classes;
- Re *sumeDesc* (Resume Functions of function or data nodes): The recover levels of the failure during the execution of function or data nodes include: Re-execution of corresponding services; reselection of new instances to abstract services or reselection of new workflow fragments in AT to be run; other descriptions of recover methods for each function and data node in wf_A;
- *MetaDesc* (set of Metadata Descriptions in service environments): The descriptions of service environment valuables, i.e., the metadata descriptions of available service environments to AT. They include common valuables (which are passed between tasks, not restricted to certain tasks) or metadata that should be maintained during the process of workflow mapping and execution. These descriptions can control the conditional selection and part scheduling of the whole workflows. They also include the acquisition means of the valuables and the preconditions for the execution of AT.

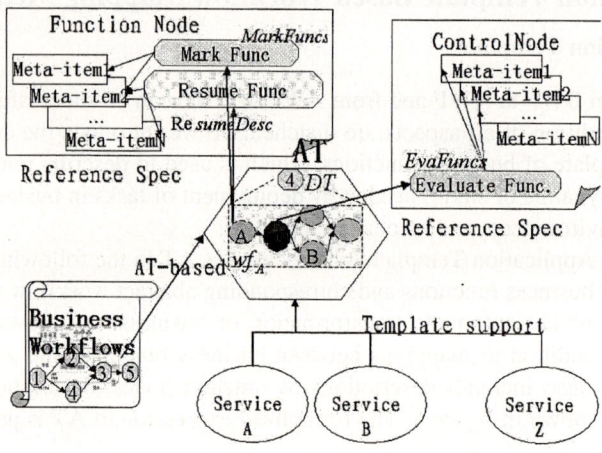

Fig. 3. Demonstration of Application Template.

3.2 Implementation of Mappings Between Different Levels of Workflow Using Application Template

There are two ways for the execution of workflows: compilation and explanation.
- The explanation way sends AWF into the workflow executor, and executes step by step according to real-time states of the environments. This kind of execution is real-time.

✧ The compilation way establishes EWF from AWF according to real-time states of the environments, and further sends the resulting EWF that is composed of Web/Grid Services into the service scheduler to be executed. This kind of execution is quasi-real-time. The execution steps of this way are list as follows (also shown in figure 4):

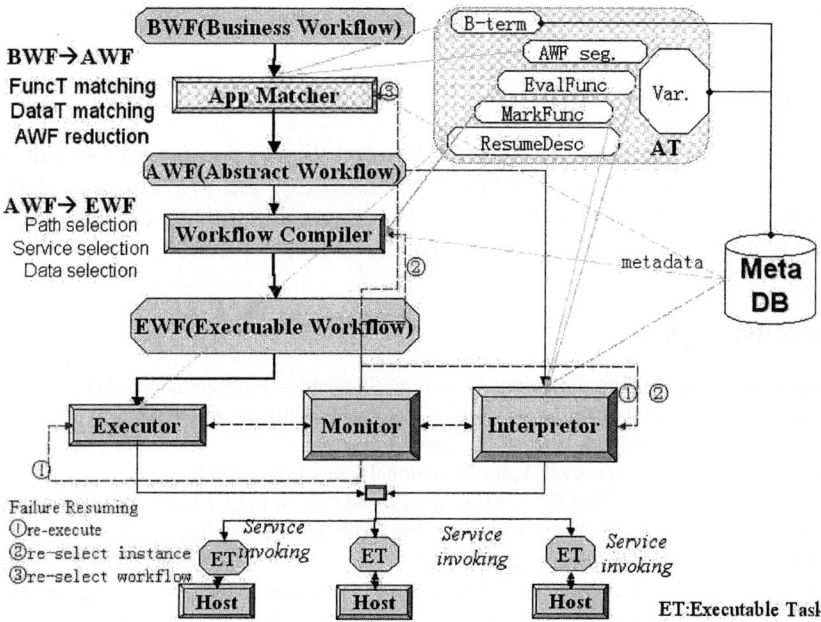

Fig. 4. Demonstration of Mapping Process.

(1) Application matching: mapping BWF in the business domains onto AWF in the service environments. This step includes detailed sub-steps as Matching function templates, Matching data templates, Reduction of AWF and Binding of instances.

(2) Judging the feasibility of execution to put AWF into service environments: the execution preconditions are examined, and the plenitude of metadata items, resources, services and data are ascertained.

(3) Workflow compiling: Mapping AWF to EWF in service environments. This step includes detailed steps as Tributary selection, Service instance selection, Data instance selection and Reduction of executable workflow.

(4) Failure recovery during execution by Ats: if a service instance fails during execution, the recover operation is performed according to the recover strategies in ATs.

4 Conclusion

This paper proposes a new method to map business workflows onto service environments - an Application Template based method to map business workflows to abstract workflows, and further to executable workflows composed of service instances.

The definition of Application Template is provided, along with an example. The example has been used as an application component in IERS of City Tianjin that has passed testing. This method provides distributed applications that consist mainly of Web/Grid Services with a dynamical mechanism in service organization, deployment and execution.

We'll further explore how to implement dynamical detection, configuration, and deployment of Web/Grid Services in the semantic level.

Acknowledgement

This work is supported by the National Research Foundation for the Doctoral Program of Higher Education of China (20020001015); the 973 Program (2002CB 312000); the NSFC (60203002); the 863 Program (2002AA135330, 2002AA 134030); the Beijing Science Foundation under Grant No.4012007.

References

1. Workflow Management Coalition: Workflow Standard – Interoperability Wf-XML Binding, Document Number WFMC-TC-1023, Version 1.0, http://www.wfmc.org (2000).
2. Sriram Krishnan, Patrick Wagstrom, and Gregor von Laszewski: GSFL: A Workflow Framework for Grid Services, Argonne National Laboratory (2002).
3. E. Deelman, J. Blythe, Y. Gil, C. Kesselman, G. Mehta, S. Patil, M. Su, and K. Vahi: Pegasus: Mapping Scientific Workflows onto the Grid, The Proceedings of the Second European Across Grids Conference (2004).
4. B. Ludäscher, I. Altintas: Compiling Abstract Scientific Workflows into Web Service Workflows, A. Gupta, In 15th Intl. Conference on Scientific and Statistical Database Management (SSDBM), Boston, Massachussets (2003).
5. Guarino, N.: Formal Ontology and Information Systems, FOIS 98, Trento, Italy (1998).

GridDaenFS: A Virtual Distributed File System for Uniform Access Across Multi-domains[*]

Wei Fu, Nong Xiao, and Xicheng Lu

School of Computer, National University of Defense Technology,
410073 Changsha, China
lukeyoyo@tom.com, xiao-n@vip.sina.com

Abstract. It is highly required to provide a uniform access and management mechanism for large-scale and distributed data across multiple administrative domains. In this paper, a virtual distributed file system called GridDaenFS is presented to provide secure, uniform and transparent file access to data resources among distributed administrative domains. Underlying heterogeneous data resources are seamlessly integrated, and data resources can be conveniently published or shared. Furthermore, suitable security and access control policies can be properly assigned to guarantee user and data safety. Experiments show that the performance is comparable to traditional network file systems and it is much suitable for data-intensive applications.

1 Introduction

Recently, data are becoming the central element in many scientific and engineering scenarios, especially in modern massive scientific researches and developments. At the same time the scale and size of data have grown explosively, and those data can be scattered at different administrative domains, being organized into different files, storing in many kinds of file systems, single ones or network ones. Therefore, how to effectively manage these files and utilize the data resources can be a big challenge, which differs from the traditional cases very much [1, 2]. Generally speaking, it requires:

- Access to large amounts of files geographically distributed in multiple domains by many applications and users;
- Transferring of massive data between storage systems, e.g. GridFTP [3];
- Methods to harmonize different administrative domains that have their own policies for connectivity, security, latency and so on.

Traditional file systems technologies cannot meet these new demands, but a newly emerging technology called Data Grid [4, 5] provides good solution. Therefore, we combine them together and present a virtual distributed file system to solve the problem described above, which we called GridDaenFS.

---

[*] This paper is supported by the National Natural Science Foundation of China (No. 60203016), the National Hi-Tech R&D 863 Program of China (No. 2002AA131010) and 973-2003CB316900.

This article is organized as follows: Section 2 introduces the framework of Grid-DaenFS. In Sections 3 the components will be respectively presented in detail. In Section 4 the performance will be evaluated. Related works are analyzed in Section 5. The paper will be ended by a brief conclusion in Section 6.

2 GridDaenFS Framework

GridDaenFS is actually a middleware system, integrating traditional file systems, such as single file systems (e.g., Linux ext3, Windows NTFS) and network file systems (e.g., NFS, HPSS). It doesn't replace the intrinsic local managing systems. On the contrary, it is built upon them, seamlessly aggregates them and provides uniform and secure access and management for them. We develop a special module, which can be embedded into native file systems, to accomplish file access and management functionalities by making use of native ones.

The framework of GridDaenFS is illustrated as Figure 1, including four parts:

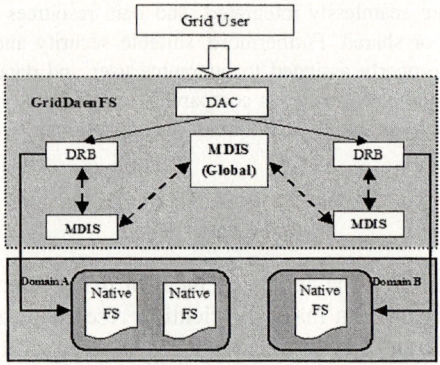

Fig. 1. GridDaenFS's four parts: DAC, MDIS, DRB and Native FS. The dashed rectangle above is the core services of GridDaenFS, and the rectangle below is local resource systems. Each DRB manages one domain, or a virtual organization.

- Meta Data Information Service (MDIS). MDIS provides catalog services for both file and directory information. The former records and maintain all kinds of file attributions (metadata). The latter includes the management of directories and provides operations that are needed to set up and query tree-like hierarchical catalog. Besides, it also keeps the mechanism of cache and replication;
- Data Request Broker (DRB). DRB is responsible for two functions: file storage and file access. File storage saves files into specific workspace at the location appointed; and file access offers POSIX file operations, such as open, close, read and write. In order to operate across different domains, DRB are wrapped into a uniform and co-operating form;
- Data Accessing Client (DAC). DAC receives file requests and deal with cooperative requirements. It provides GUI client and uniform POSIX APIs;
- Native file systems. Native file systems actually are those traditional file systems. At the present time we have realized the support to CIFS and NFS.

Since GridDaenFS is set up on the basis of native file systems, and the details of file operation are entirely accomplished by them, so this brings us two benefits:
- GridDaenFS needn't care about the underlying realization of physical storage systems. When another native file system needs to be supported, we simply plug a corresponding collaborating module into DRB;
- It endows GridDaenFS with a characteristic of autonomy. This is very important for a grid project, because local policies should be always considered.

3 GridDaenFS Components

3.1 File Naming and Global Namespace

GridDaenFS organizes a logical uniform namespace above native file systems. Users don't need to face storage details, such as physical location, access protocol and so on. Instead, GridDaenFS provides naming transparency, that is, naming a file needn't to know its location. GridDaenFS endows three names for a file, which respectively are:

1. Site File Name (SFN), an actual name which can uniquely identify the file in the native file system;
2. Internal File Name (IFN), a 128-bit hexadecimal name which is produced by MD5 algorithm. IFN is globally unique and only used in the interior of the system. It is unchanged during the entire lifetime of the file;
3. Logical File Name (LFN), the logical name that is only used in logical view.

LFN and IFN form the logical view, and SFN alone forms the physical one. Users or applications only need to face the logical view of LFN, thus GridDaenFS shields the complexity and heterogeneity of files from users. Several LFNs can be mapped to one IFN when one or more soft links are built for the same file. One IFN can only be mapped to one SFN and vice versa. All of these relationships are stored in MDIS databases.

3.2 File Addressing and Access

The naming scheme described above induces another problem: how to locate a file according to its logical name. GridDaenFS provides a fast mapping methods for file addressing between LFN, IFN and SFN. We employ two tables to keep the mapping information: one involves LFN and IFN, and the other involves IFN and SFN. Users or applications operate on LFN, and only after mapping can LFN be translated into IFN and IFN can be submitted to file access module DRB. Then another table is employed to complete the mapping between IFN and SFN at a similar way.

Hence the whole file addressing process is finished and the actual file address is obtained, the file can be really accessed. Essentially, file access envelops different kinds of access protocol to provide a uniform access interface. Once a uniform file request is submitted to any DRB, access module will check its type and then automatically transform into the correct protocol. If another protocol needs to be supported, we can plug the corresponding inlet into GridDaenFS, keeping the interface unchangeable.

GridDaenFS provides a global account for each user. In order to contact to a specific native local file system, global account should be assigned to an actual user account in this file system. Here Figure 2 illustrates the course of data access.

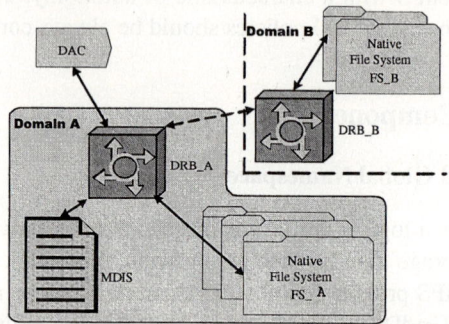

Fig. 2. GridDaenFS Data Access Process: among domain and cross domain.

1. DAC puts forward a file require to DRB_A at Domain A with a global account, then DRB_A start up an user authenticating process and decides its validity;
2. DRB_A queries its MDIS, invoking a searching process, to get necessary information from MDIS's database, including a local account at FS_A where file is held, and actual access protocol and physical file address and so on;
3. DRB_A converts and brings forward the transformed request to the native file system FS_A with the local account obtaining from MDIS;
4. FS_A gets the file and turns it back to DAC via DRB_A. This is called an access among domain;
5. If the file at FS_A is unavailable, then a cooperative access starts: DRB_A automatically finds another replica of the file, for example, a replica at Domain B. Then step 2, 3 and 4 will be repeated, except that the destination is FS_B. This process is called an access across domain.

3.3 User Management

In GridDaenFS, user management is tightly connected with file management. A use account is a global identity, which can publish data resource or get enough rights to share other user's data resources. Different users have different logical view within global uniform view if they own different data resources or have different rights on the same ones. According to some similar interesting, users are combined to form a group. Users in one group can share their resources and collaborate easily. Each group has a group administrator, who responses for activities about group management. Additionally, GridDaenFS has some system administrators, who response for activities about system initialization, maintenance, group creation and so on.

When a user is created, GridDaenFS defines a default independent workspace for him. The system manages this workspace automatically. Therefore, when he uploads or creates files, these files can be automatically copied into the workspace and saved. This can benefit him especially when user needs some storage space to hold his import files or export files for some computing applications.

3.4 Security

Large number of files across different domains may use different security mechanisms and policies. We adopt the existing grid security standard GSI [6] and develop our own security software packages. There is a Certificate Authority (CA) in the whole system, providing security functions for different entities among the domain. It concerns about the following issues:

- Signing certificates for users and system components, including DAC, DRB and MDIS;
- Producing proxy certificate for users' single-point login;
- Authentication between DAC and DRB. Users have their own certificates and can establish different authentication channels with DRB through DAC;
- Authentication between DRB and MDIS. Authentication channels between DRB and MDIS are shared by all to promote efficiency;
- Authentication between DRB and DRB. Only when a DRB is activated or a new one joins into the system is it necessary to establish a trust relationship between this DRB and other DRBs.

3.5 Access Control

GridDaenFS applies access control restrictions at the file level. It defines some basic privileges in advance, such as read, write, replicate and so on, which combine to form different roles. A role is a set of actions and privileges related to specific operations. For a specific user, it must be endowed with one or more roles to perform the operations embodied in the roles. Then GridDaenFS defines a vector:

```
<LFN, user, roles, period, local_user, local_ privs >
```

Among them, `user` stands for a grid user which may include several `roles`, `period` stands for file's valid time limit, `local_user` is a native file system user which `user` is mapped into, and `local_privs` is the local access policy for `local_user`.

The access control procedure includes three steps. Firstly, GridDaenFS examines `user` and `roles`. For each `LFN`, `roles` will be resolved and the rights should be fully validated. Secondly, `period` of `LFN` will be checked. If `LFN` doesn't expire, native file system still needs to validate `local_user` and `local_privs` by using local access control policy.

3.6 Cache and Replica

Cache is a familiar way to promote efficiency, while replica [7] is a concept emerging with data grid. Both techniques are exerted to guarantee GridDaenFS performance.

2. I. Foster, C. Kesselman, S. Tuecke: The Anatomy of the Grid: Enabling Scalable Virtual Organizations. International Journal of High Performance Computing Applications, 2001, 15(3). 200-222. Journal of Network and Computer Applications (2001)23: 187-200
3. Bill Allcock, Lee Liming, Steven Tuecke – ANL: GridFTP: A Data Transfer Protocol for the Grid. Grid Forum Data Working Group on GridFTP
4. L. Chervenak, I. Foster, C. Kesselman et al: Data Management and Transfer in High Performance Computational Grid Environments. Parallel Computing Journal, 2002, 28 (5): 749-771
5. Wolfgang Hoschek, Javier Jaen-Martinez, Asad Samar, Heinz Stockinger, and Kurt Stockinger: Data Management in an International Data Grid Project , http://www.eu-datagrid.org/, 2000
6. Nataraj Nagaratnam, Philippe Janson, John Dayka, etc: The Security Architecture for Open Grid Services. http://www.globus.org/ogsa/, 2002
7. Globus project: An Architecture for Replica Management in Grid Computing Environments. Global Grid Forum 1
8. C. Baru, R. Moore, A. Rajasekar, M. Wan: The SDSC Storage Resource Broker. In Proc. Of CASCON'98 Conference, Toronto, Canada, 1998
9. Brian S. White Michael Walker Marty Humphrey Andrew S. Grimshaw: LegionFS: A Secure and Scalable File System Supporting Across-Domain High-Performance Applications
10. R. J. Figueiredo, N. H. Kapadia, and J. A. B. Fortes: The punch virtual file system: Seamless access to decentralized storage services in a computational grid. In Proceedings of the Tenth IEEE International Symposium on High Performance Distributed Computing. IEEE Computer Society Press, August 2001
11. Osamu Tatebe, Youhei Morita, Satoshi Matsuoka, Noriyuki Soda, Satoshi Sekiguchi: Grid Datafarm Architecture for Petascale Data Intensive Computing

Ad Hoc Debugging Environment for Grid Applications

Wei Wang, Binxing Fang, Hongli Zhang, and Yuanzhe Yao

Research Center of Computer Network and Information Security Technology
Computer Science Department, Harbin Institute of Technology
Mail Box 320, Harbin Institute of Technology, China
{ww,bxfang,zhl,yyz}@pact518.hit.edu.cn

Abstract. Debugging can help programmers to locate the reasons for incorrect program behaviors. The dynamic and heterogeneous characteristics of computational grids make it much harder to debug grid applications. In this paper, we give the definition the concept of ad hoc debugging environment, which uncover the nature of debugging behavior in computational grids. Besides solving some normal problems in tradition parallel debugging, such as user interface, application instrumentation, we first address the nondeterministic dynamic behavior of computing environment in grids during the debugging session. We evaluate the similarity between computing environments, in which the grid application can be re-executed for cyclic debugging.

1 Introduction

Debugging is of paramount importance for software development, which provides the assurance of the quality of applications, including correctness, performance and reliability. Debugging is also one of the most difficult stages in software engineering. It has been estimated that the cost of providing program assurance via appropriate de-bugging, verification and testing activities can easily range from 50% to 75% of the total software development cost.

Parallel programs are much more complex than serial ones. The execution of a parallel program consists of not only the sequential computations, but also the inter-actions between processes, such as communication and synchronization. These features make it more difficult to debug parallel applications. Some errors that never happen in a serial program will be prevalent in parallel one. Deadlocks [1] and race conditions [2] are the two main unwanted behaviors. The notion of computational grids [3] is put forward for solving large-scale scientific problems. It is a much more exhausting task to debug grid applications in contrast with debugging sequential or even parallel programs.

Net-dbx [4] and P2D2 [5] are two parallel debugger for large-scale applications. Net-dbx is a web-based debugger of MPI programs. P2D2 is said to be a debugger for grid applications. The common ground of both debuggers is state-based parallel de-bugging. Net-dbx uses dbx as the base debugger and P2D2 uses gdb. The resources in the grid are diverse for its heterogeneity, not all of them support the specific serial debugger. For example, dbx and gdb are two debuggers used in UNIX-like environment, but the operating system in resources in the grid may be MS Windows. The both debuggers cannot run in Windows naturally. Another aspect is that the scale of applications executed in the grid is very large, so users will be inundate in the state data and difficult to find applications' errors. From these points of views, neither Net-dbx nor P2D2 is a true grid applications debugger. IC2D [6] is a graphical environment for remote monitoring and steering of distributed and metacomputing applications. It can be served as a building block for metacomputing and computing portals. In IC2D, only coarse-grained monitoring is provided, users cannot understand the applications' behavior in detail, especially the message passing details. Netlogger[7] monitors program behavior in distributed applications. Netlogger only cares about the performance of applications, but do not address the applications correctness analysis.

In this paper, we address the heterogeneous and dynamic characteristic of computational grids in debugging grid applications. The ad hoc concept is first introduced into grid applications debugging environment, which can describe the debugging behavior in grids exactly. With the ad hoc concept, we design and implement a grid-enable debugging environment. We also present the computing environment similarity, which support cyclic debugging in computational grids. The portal solution can interface users and the environment. We use grid services to manage resources and build the ad hoc debugging environment according users' demand. The feasibility is obvious. Since MPI has become the standard in message-passing programming, and has been implemented in computational grids as MPICH-G2[8]. So we constrain the grid application to MPICH-G2 applications. On the whole, our work enhances the debugging functionality in grid computing with high portability, usability and feasibility with the ad hoc concept.

This paper is organized as follows: Section 2 gives the definition of ad hoc computing environment and description of the architecture of the ad hoc debugging environment. In section 3 we give some implementation details of the ad hoc debugging environment. The experiment and an example will be given in section 4. Section 5 is our conclusion remarks and future work.

2 Ad Hoc Computing and Debugging Environment

2.1 Ad Hoc Environment Architecture

The nature meaning of the word "ad hoc" is "for this". The word "ad hoc" about computation is first used to describe autonomous peer-to-peer multi-hops mobile communication networks. The most particular characteristic of this kind of networks is its autonomy and provisionality.

Similar to this condition, grids include tremendous computation resources. These resources may join in and leave grids at any time. When a resource is not included in the grid infrastructure, it is only a single unit. Once a resource joins in the grid, it can be used as a computation node, a storage node, or other purpose. When no computation is submitted, these resources are idle, and have no relation with each other, except the physical network connection between each other. Some resources can be aggregated when an application is invoked. We call this aggregated computation platform an ad hoc computing environment, for it is unknown before the invocation of application. The following is its definition:

Definition 1. An ad hoc computing environment in the grid is the one that can be built by automatically aggregated the resource in the grid according to the specification of an application. It only exists temporarily during the application execution and vanishes when the application ends.

The ad hoc computing environment can describe the computation condition in grids correctly and uncover the dynamic nature of computation grids.

Definition 2. An ad hoc debugging environment is the one that provide debugging support for application executed in ad hoc computing environment.

One ad hoc debugging environment includes following components, including user interface, resource management, application instrumentation and debugging functionality. Fig. 1 is the architecture of ad hoc debugging environment.

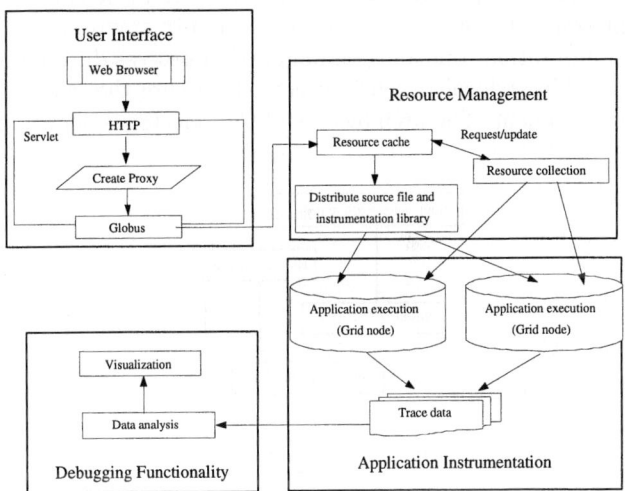

Fig. 1. Ad hoc debugging environment architecture

2.2 User Interface

The resources in ad hoc computing environment in grids are diverse. Providing a convenient interface to access this complex environment is very important. To adapt the heterogeneity of grid, the interface should be portable and can be seamless to computing environment.

Web portal technology [9] can ease users to access grid resources. A portal is an integrated and personalized Web-based interface that provides the user with a single point of access to a wide variety of data, knowledge, and services – at anytime and from anywhere using any Web-enabled client device.

2.3 Resource Management

In general, users need to execute their programs again and again to find errors. Parallel programs are always nondeterministic, which make it hard to debug them. This condition is much worse in grids. The execution environment is also nondeterministic and it can change dynamically from run to run. So it has high possibility that the re-source needed may be missed during the next debugging session of the same application. Even the same resource still exists in the next debugging process; its state may be different. So how to assure the determinism of computing environment is so important in grid applications debugging environment.

Our solution addresses the problem discussed above. How to evaluate the similarity degree between two ad hoc debugging (computing) environments is a valuable problem. The definition of this similarity in the context of debugging is following:

Definition 3. Two computing environments are said to be similar, if the same application presents almost the same behavior in both environment. The behavior includes computation time and communication process during the execution.

We design a hierarchical resource selector to assure the determinism of the computing environment. The architecture of the selector is shown in Fig.2.

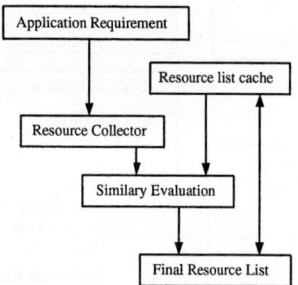

Fig. 2. The architecture of the resource selector

2.4 Debugging Functionality

The main point of this paper is of introduction of designing and building ad hoc de-bugging environment in grids. So we will give a brief introduction about the debugging functionality. In our environment, we debug MPICH-G2 grid applications by event-based post-mortem method, which has more scalability and feasibility. The trace data are visualized by space-time graph, and consistent global states detection and race condition detection are two main debugging functionalities.

3 Implementation

3.1 Building Ad Hoc Debugging Environment

When a debugging demand is submitted through the user interface, the constrained resources based on the application need should be discovered and aggregated. In our case, the constrained condition is MPICH-G2[9]. In fact not all the computing re-sources in the grid are installed MPICH-G2. So if we want to debug MPICH-G2 ap-plications, we should select the appropriate resources.

MPICH-G2 information comprises five properties, which are the hostname, MPICH-G2 installation state, and the path of MPICH-G2 library, mpicc and mpirun. We customize the core information to support MPICH-G2 information. By analyzing the structure of LDAP Schema and OID (Object IDentification), we design the OID of five properties describing MPICH-G2 information. These properties can put the MPICH-G2 object into the standard configuration of GLOBUS resources description, and can be access as other known resources in grids. It is implemented by extending the structure of OID. The MPICH-G2 resources can publish themselves by running our MPICH-G2 information provider. Once users register the MPICH-G2 information provider, their resources can be accessed by searching MDS directory.

3.2 Resource Selector

Computation Capability Similarity. The key problem to evaluate the computation capability similarity is how to quantize the CPU capability. Every resource attending the computation contributes its capability. We define the contribution as formula 1:

$$C_i = \frac{f_i \times I_i}{\sum_{j=1}^{n} f_j \times I_j} \qquad (1)$$

Here C_i represents the contribution of i_{th} resource, f_i is the CPU frequency of i_{th} resource, and I_i is its CPU idle ratio. In formula 1, we consider both static capability (CPU frequency) and dynamic information (CPU idle ratio). This strategy is correct and fairer for resource selection to evaluate the computation capability similarity.

Network Connection Similarity. The communication between processes is an important feature of parallel grid applications. So to describe the behaviors of grid applications, the communication should be considered. The network connection has a deep effect on processes communication. The network connection similarity plays an important role in evaluating the environment similarity.

Assume a resource R_m is missed in resource list cache during a new debugging session; we will select a new resource R_c from resource candidates, assuring the similar network similarity. Assume there are n resources in the resource list cache, the latencies between R_m and the other n-1 resources are also recorded, we use

$l_{m,i}$ to represent the latency between R_m and $R_i (i \neq m)$, so a latency vector, $L_{m,i}$, is constructed as follow:

$$L_{m,i} = (l_{m,1}, l_{m,2}, ..., l_{m,n})(m \neq i)$$

For every R_c, we construct a latency vector, $L_{c,i}$, we can use these vectors to select a appropriate resource. We define distance between $L_{m,i}$ and $L_{c,i}$ as $D_{c,m}$:

$$D_{c,m} = \sum_{i=1}^{n} |l_{c,i} - l_{m,i}| \qquad (2)$$

We can use formula 2 to compute $D_{c,m}$, for every R_c. We compare these values, the corresponding R_c with the minimum will be selected as a re-placement for R_m.

Resource Attributes Acquirement. In order to evaluate similarity, the data about resource attributes should be collected. Network Weather Service (NWS)[11] is a distributed system that periodically monitors and dynamically forecasts the performance of various network and computational resources. We can publish the resource information, including CPU load, memory available, network bandwidth and latency, collected by NWS, to GLOBUS MDS.

4 Experiments and Example

4.1 Test Bed Configuration

We implement an ad hoc debugging environment on the test bed in our laboratory. The test bed consists of eight heterogeneous PCs, which are partitioned three parts. Fig.3 shows our test bed configuration. Three workstations are in one LAN, whose configuration is Pentium 450M CPU and connected via 10M Ethernet. Four high performance PCs are in another LAN, whose configuration is Pentium 1G and connected via 100M Ethernet. One Dawning server, which has two Pentium 933M, is single with 100M connection. The software components comprise GLOBUS version 2.2.2, which support grid service and computation. MPICH-G2 is used as the communication library.

4.2 Test Example

Users can submit their task source code and instrumentation library through the web portal. We also provide the resource selection conditions for users. The two main aspects are the number of nodes and memory capacity users want to get.

In order to test our debugging environment, we select an actual MPICH-G2 application, whose functionality is matrices multiplication. The communication events are recorded in a trace file. We visualize the debugging data. The following results are the debugging of matrices multiplication run by 4 processes. The

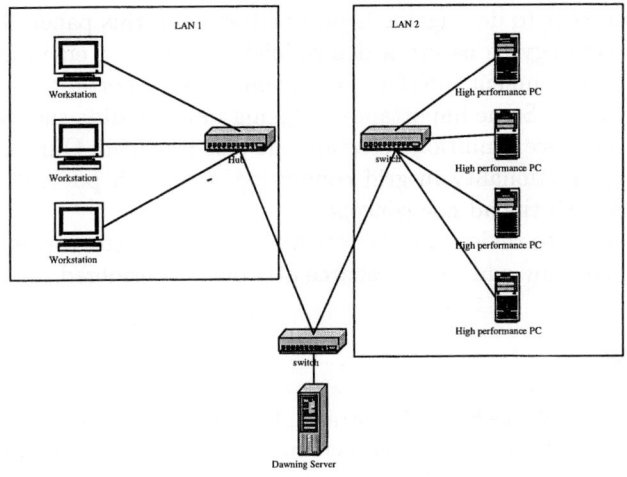

Fig. 3. Testbed configuration

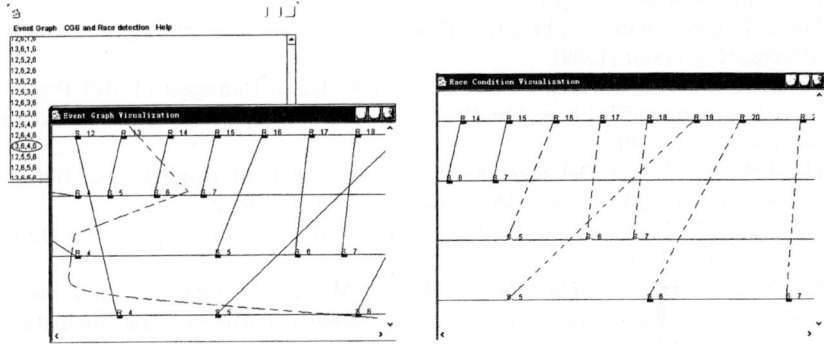

Fig. 4. CGS and race detection results

debugging functionalities are onsistent global states(CGS) detection and race condition detection. Fig.4 shows the results. The left part is the CGS detection results, CGS is represened by the cut line across the space-time graph. The right part is the results of race condition detection, the dashed lines show the messages racing with each other.

5 Conclusion

Computational grids provide computing power by sharing resources across administrative domains, but grid applications may be prone to error in this complex, dynamic environment. In this paper we have presented a debugging environment for MPICH-G2 grid applications. The ad hoc concept is given to describe the dynamic characteristic of grids. The effect of nondeterministic computing environment to de-bugging is first addressed in this paper. The resource management strategy to assure a deterministic debugging environment is put forward. The environment interface is implemented by portal technique, which ease users to access. Some important debugging functionality, such as consistent global states and race condition detection, are also provided. Our work enhances the debugging functionality in grid computing with high portability, usability and feasibility with the ad hoc concept.

Future work is to revise the evaluation of computing environment similarity to be more correct by add more resource attributes considered.

References

1. Luecke, G., Zou,Y, Coyle, J., Hoekstra, J., Kraeva, M.: Deadlock Detection in MPI Programs. Concurrency and Computation: Practice and Experience. 14(2002) 911-932
2. Netzer,R.H.B., Brennan,T., Damodaran-Kamal,S.: Debugging Race Conditions in Message-Passing Programs. Proceedings of the SIGMETRICS symposium on Parallel and distributed tools (1996) 31-40
3. Foster,I., Kesselman,C.: The Grid : Blueprint for a New Computing Infrastructure, Morgan-Kaufmann (1999)
4. Neophytou, N., Evripidou, P.: Net-dbx: a Web-based Debugger of MPI Programs Over Low-bandwidth Lines. Parallel and Distributed Systems, IEEE Transactions on. 12(2001) 986-995
5. Hood,R., Jost,G.: A Debugger for Computational Grid Applications. Heterogeneous Computing Workshop, Proceedings. 9th(2000) 262-270
6. Baude,F., Sophia,I. etc.: IC2D:Interactive Control and Debug of Distribution. Grappers. (2001) 193-200
7. Gunter,D., Tierney,B., Crowley, B., Holding,M., Lee,J.: NetLogger: a Toolkit for Distributed System Performance Analysis. Modeling, Analysis and Simulation of Computer and Telecommunication Systems, Proceedings.8th International Symposium on. (2000) 267-273
8. Karonis, N., Toonen ,B., Foster, I.: MPICH-G2: A Grid-Enabled Implementation of the Message Passing Interface. Journal of Parallel and Distributed Computing. 5(2003) 551-563
9. Chandran, A., Moutayakine, D., etc.: Architecting Portal Solutions IBM Red Book(2003)
10. Kranzlmueller, D.: DEWIZ: Event-based Debugging On the Grid, 10th Euromicro Workshop on Parallel, Distributed and Network-based Processing. (2002) 162-169
11. Wolski, R., Spring, N., Hayes, J.: The Network Weather Service: A Distributed Resource Performance Forecasting Service for Metacomputing. Future Generation Computer Systems. 5(1999) 757-768

Distributed Object Group Framework with Dynamic Reconfigurability of Distributed Services

Chang-Sun Shin, Young-Jee Chung, and Su-Chong Joo

School of Electrical, Electronic and Information Engineering, Wonkwang University, Korea
{csshin,yjchung,scjoo}@wonkwang.ac.kr

Abstract. In this paper, we constructed the Distributed Object Group Framework(DOGF) which is a reconfigurable architecture supporting dynamically adaptation of distributed services. We also developed the distributed application simulator to verify the adaptability and reconfigurability of our framework. The DOGF provides appropriate grouping of distributed objects for executing given distributed applications as the point of view of a single logical view system, and supports the adaptive distributed services for dynamic reconfiguration of distributed application. In constructing procedure of the DOGF, we designed the distributed services that consist of the objects supporting service, the load balancing service, and the real-time service. We also implemented the managing technique for dynamic reconfiguring the distributed services. Finally, for verifying whether the DOGF can support the dynamic reconfigurability of distributed services or not, we developed the Defence System against Invading Enemy Planes(DSIEP) simulator as a practical use of distributed application with real-time property on the DOGF. From this, with adaptation of given load balancing policies, we analyzed the average execution times and the deadline violation rates as the execution results for the client's request.

1 Introduction and Related Works

Applications executed on distributed environment are configured by one or more distributed objects. When considering the resource capability, we need to create the dynamically reconfigurable architecture that can reuse distributed objects used in arbitrary distributed application. Furthermore, such distributed applications must provide distributed transparency to heterogeneous distributed environments by using the state-of-the-art network and computing technologies. That is, we have to provide a standard distributed framework for distributed application. This framework supports the adaptive distributed services and dynamic reconfigurability for arbitrary distributed applications.

As some of the representative researches relating to our study, the Common Object Request Broker Architecture(CORBA)[1] and the Telecommunications Information Networking Architecture(TINA)[2] have suggested the management model of distributed object. But these models have not been supporting the concept of object group for the efficient management of distributed objects and the managing scheme of multi-duplicated objects with the same service property. In the research area of distributed real-time computing, the Real-Time CORBA[3] suggested by the Object

Management Group(OMG), the Time-triggered Message-triggered Object Support Middleware(TMOSM) developed by DREAM Laboratory at University of California at Irvine[4], and the Real-Time Object Group(RTOG)[5,6] we studied have been suggesting the models for satisfying the real-time property. However, the researches above have not been supporting the object management service for multi-duplicated objects and the load balancing service for server systems.

From a viewpoint of efficient reconfiguration of distributed resources, the Olan model by IMAG-LSR Laboratory[7] and the Configuration Management model by Hewlett-Packard Laboratory[8] have been studied. The Olan model defined the language for supporting the dynamic reconfiguration. But, they did not consider the multi-duplicated objects. The Configuration Management Model supported the load balancing by selecting appropriate server object and the efficient reconfiguration of distributed application by defining the management strategy of server objects, but they have not supported the concurrent processing of server objects for clients' requests.

In this paper, for solving the problems mentioned above, we construct a dynamically reconfigurable architecture, called Distributed Object Group Framework(DOGF), that can reconfigure the appropriate server objects as a unit of single object group for given distributed application and provide the load balancing of distributed systems. Finally, for verifying our framework, we develop the Defence System against Invading Enemy Planes(DSIEP) simulator as a practical use of distributed application on the DOGF. Then, in a viewpoint of load balancing, we analyze the execution results for the client's request and its deadline violation.

2 Backgrounds

In distributed environments, we need services that can manage the distributed objects configuring application efficiently. If we manage the distributed application as a group on this environment, we can extend distributed application through the optimal integration of distributed systems. For achieving above goal, we have been studying the distributed object group[5,6,9,10].

For executing application on distributed system environment, systems' heavy loads, complex management techniques, and communication cost are required. If we can select one of them correctly and group the appropriate objects configuring distributed application in advance, we can guarantee the optimal resource utilization and the load balancing among distributed systems. Therefore, we need architecture with dynamic reconfigurability of distributed application. For meeting above requirements, we defined the distributed object group. Our group is a logical single unit of objects for executing arbitrary distributed application, and consists of components supporting the object management services. Various distributed strategies can be adapted to each component.

Let's consider the object management scheme in the distributed object group. Our group for the distributed objects can provide the creation management, the registration/withdrawal management, the access security management, the mobility management, and the name/property management. The components of a distributed object group can be classified as the group management objects and the real-time supporting objects. For supporting group management, our model contains the Group Manager(GM) object, the Security object, the Information Repository object, and the Dynamic Binder object. For supporting the distributed real-time service, our group includes the server objects, the Real-Time Manager(RTM) objects, and the Scheduler objects. A part of server objects in an object group is multi-duplicated objects with the same service property. The Group Manager object is responsible for managing all of objects in an object group. The Security object checks access rights for the client's request by referring the access control list(ACL). The Information Repository object stores information for all server objects existing in an object group. The Dynamic Binder object implemented by an arbitrary algorithm given selects an appropriate object invoked by clients. The RTM object calculates the real-time constraints. The Scheduler object assigns the priority for service execution to the client's request. We accommodate the object registry policy of the inner group, the security policy, the load balancing policy, and the real-time service policy to the components.

3 Distributed Object Group Framework(DOGF)

As we mentioned above, the DOGF suggested in this paper provides the dynamic reconfigurable architecture supporting the logical distributed environment, the optimal utilization of system resources, and the adaptive distributed services. We construct the DOGF to manage the server objects by adapting the strategies for dynamic reconfiguration and adaptive services to object group components. In this section, we describe the distributed services provided by the DOGF and its architecture.

3.1 Distributed Services of the DOGF for Dynamic Reconfiguration

The meaning of dynamic reconfiguration of distributed application is to be satisfied with the client's request by executing the distributed services with adaptability to the change of infrastructure where the applications are configured. That is, when executing the distributed application, the DOGF provides the optimal execution environment by adding the service and relocating distributed resources. To provide the dynamic environment, the service strategies managing the change of property or function of resource executing distributed services and the dynamic load variation of network or systems are required. Hence, the DOGF provides the object group supporting service, the load balancing service, and the real-time service[6,9,10].

The Object Group Supporting Service. This service in the DOGF services supports the function which registers/withdraws server objects as a group member to an arbitrary object group, and also supports the function which inserts/deletes the access right of server object from client's request. In this step, the object registry policy and the security policy out of the reconfiguration strategies are applied to the framework.

The Load Balancing Service. The load balancing service provides client objects with binding service after dynamically selecting the server object with appropriate conditions by considering load and deadline information. The binding priority is assigned to the server object. The load balancing policy is applied to this service. Our framework balances the load among systems on which the server objects are located through the dynamic selection/binding of appropriate server object. For executing application, we adopt the binding priority algorithm to the Dynamic Binder object.

The Real-Time Service. When the clients request the service to the server object selected through the load balancing service, this service guarantees the service execution within deadline for the clients' requests. The real-time scheduling is occurred dynamically whenever the client requests the service, and task priority is assigned to the request. The real-time scheduling policy is applied for this real-time service. For executing actual application, we adopt the Earliest Deadline First(EDF) or the Rate Monotonic(RM) algorithm to the Scheduler object.

3.2 Architecture of the DOGF

We construct the DOGF to manage the server object configuring distributed application efficiently and to provide the timeliness and executability of real-time service that distributed real-time application might support. The DOGF is located on the upper layer of the Commercial-Off-The-Shelf(COTS) middleware and the platform that is composed of the heterogeneous operating system and communication network. Our framework applies the dynamic reconfiguration and adaptive services for distributed applications on the various heterogeneous platforms and COTS middlewares. The distributed applications located on the top layer are not mission critical applications, and the distributed applications supporting non real-time or real-time property of various fields can be executed on the framework based on distributed object group. Figure 1 shows the architecture of the DOGF we suggest.

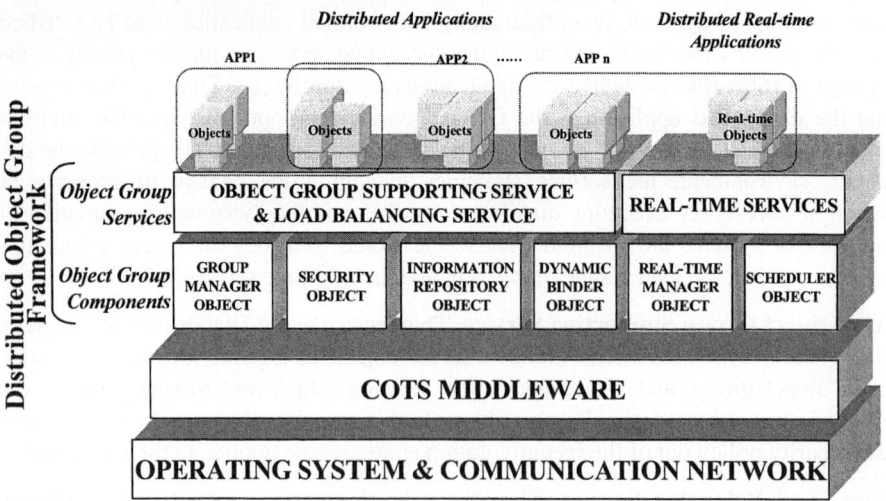

Fig. 1. Architecture of the DOGF.

4 The DOGF Based Distributed Real-Time Application

In this section, we develop the Defence System against Invading Enemy Planes(DSIEP) simulator as a distributed real-time application for verifying the DOGF. The DSIEP detects and destroys the enemy planes invading the defence domain. This simulator also has the real-time properties. The deadline is an important factor in the DSIEP. For example, when the enemy planes invade the defence domain, we must intercept them before an air attack. The server objects configuring the DSIEP are interactive objects with real-time constraints for the deadline. We developed the server objects by adopting the TMO scheme. Detailed structure and functions of the TMO scheme are described in [4]. With the component of the DSIEP, the Space TMO informs the simulation system of the enemy plane and its real-time location. The Alien TMO creates the enemy planes periodically. The Radar TMO detects the location of the enemy plane in a given defence domain, and reports the location of enemy plane to the Command Post TMO. The Command Post TMO receives the real-time detection information of enemy plane from the Radar TMOs and requests the selection of the Missile Launcher TMO that executes interception to the Group Manager object. In this step, the Group Manager object checks the access right for the Missile Launcher TMO of the Command Post TMO in the Security object through the object group supporting service. Then, the Group Manager object requests the object reference of the Missile Launcher TMO to the Information Repository object. If the enemy plane is detected in the Shared Domain where all Missile Launcher TMOs can intercept the enemy plane, the DOGF executes the load balancing service through the Dynamic Binder object for selecting the appropriate TMO out of the several duplicated Missile Launcher TMOs. After that, the Command Post TMO requests the intercepting service to the selected Missile Launcher TMO. The Missile Launcher TMO receives the intercepting request from the Command Post TMO, and intercepts the corresponding enemy plane according to the intercepting priority after executing the real-time service through the Real-Time Manager object and Scheduler object. The Figure 2 is the GUI showing the execution of the DSIEP simulator on the DOGF.

We adapt the binding priority algorithm to the Dynamic Binder object and the EDF algorithm to the Scheduler object. The real-time flying position of the enemy planes are displayed

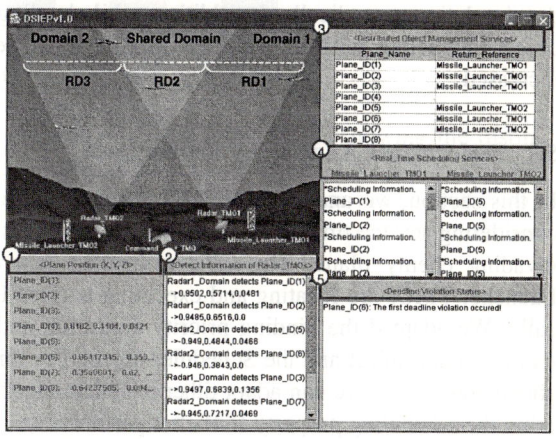

Fig. 2. Executing result of the DSIEP simulator.

in ①. The coordinates of enemy planes detecting by Radar TMOs are showing in ②. The results of the load balancing service by dynamic selection and binding of the

DVR was not showing the distinguished difference of performance. That is, we could balance the system load by applying the load balancing service. Hence, we analyzed that the DOFG supports the reconfigurability of distributed application by using the DOGF services. Therefore, our framework could adapt the various distributed service policies to the service components, and we verified that the DOGF is the object group supporting distributed framework that could improve the performance of distributed application by supporting the dynamic reconfigurability.

In future work, we are planning to suggest the adaptive distributed framework that can provide to the state-of-the-art distributed computing fields such as distributed clustering, ubiquitous, or/and Grid systems. Also, we will use this framework for developing the real-time simulator for the Public Traffic Control based on LBS.

Acknowledgements. The authors wish to acknowledge helpful discussions of the TMO Programming Scheme with Professor Kane Kim in University of California at Irvine, DREAM Lab. This research was supported by University IT Research Center Project.

References

1. Moser, L.E., Melliar-Smith, P.M., Naraimhan, P., Koch, R.R., Berket, K.: Multicast Group Communication for CORBA. In Proc. of the International Symposium on Distributed Objects and Applications (1999) 98-107
2. Axel Kupper: Locating TINA User Agents: Strategies of a Broker Federation and their Comparison. In Proc. of the 6th International Conference on Intelligence in Services and Networks(IS&N) (1999)
3. G. Cooper, L. DiPippo, R. Esibov, R. Ginis, R. Johnston, P. Kortman, P. Krupp, J. Mauer, M. Squadrito, B. Thuraisingham, S. Wohlever, V. Wolfe: Real-Time CORBA Development at MITRE, NRaD, Tri-Pacific and URI. In Proc. of IEEE Workshop on Middleware for Distributed Real-time Systems and Services (1997) 69-74
4. K.H. Kim: A TMO Based Approach to Structuring Real-Time Agents. In Proc. of the IEEE CS 14th International Conference on Tools with Artificial Intelligence(2002) 165-172
5. C.S. Shin, M.H. Kim, Y.S. Jeong, S.K. Han, S.C. Joo: Construction of CORBA Based Object Group Platform for Distributed Real-Time Services. In Proc. of the 7th IEEE International Workshop on Object-oriented Real-time Dependable Systems (2002) 229-302
6. S.C. Joo, C.S. Shin, C.W. Jeong, S.K. Oh: CORBA Based Real-Time Object-Group Platform in Distributed Computing Environments. Lecture Notes in Computer Science, Vol. 2659 (2003) 401-411
7. Rebort Cole: Application Configuration in Client-Server Distributed System. In Proc. of the International Workshop on Configurable Distributed Systems (1992) 309-317
8. Bellissard, L., Boyer, F., Riveill, M., Vion-Dury, J.-Y.: System Services for Distributed Application Configuration. In Proc. of the 4th International Conference on Configurable Distributed Systems (1998) 53-60
9. C.S. Shin, S.C. Joo, Y.S. Jeong: A TMO-based Object Group Model to Structuring Replicated Real-Time Objects for Distributed Real-Time Applications. Lecture Notes in Computer Science, Vol. 3033 (2003) 918-926
10. C.S. Shin, M.S. Kang, C.W. Jeong, S.C. Joo: TMO-Based Object Group Framework for Supporting Distributed Object Management and Real-Time Services. Lecture Notes in Computer Science, Vol. 2834 (2003) 525-535

Design and Implementation of Grid File Management System Hotfile

Liqiang Cao[1,2], Jie Qiu[3], Li Zha[2], Haiyan Yu[2], Wei Li[2], and Yuzhong Sun[2]

[1] Graduate School of the Chinese Academy of Sciences, Beijing 100080, China
`caolq@software.ict.ac.cn`
[2] Institute of Computing Technology, Chinese Academy of Sciences,
Beijing 100080, China
`char@software.ict.ac.cn`
`{Yuhaiyan,liwei,yuzhongsun}@ict.ac.cn`
[3] IBM China Research Laboratory, Beijing, China
`qiujie@software.ict.ac.cn`

Abstract. Hotfile is a user level file management system. It wraps GridFTP and GASS or any other file transfer protocol compatible with hotfile structure into a unified vegafile protocol. Based on virtual grid file layer and a set of basic grid file operations, users can access grid file without knowing the physical transport protocol of the file. Test result shows the overhead of vegafile protocol is little in file transfer and operation. Further vegafile transfer experiment shows when file size is smaller than 1M byte, Vega file copy with GASS has higher bandwidth, where as file is larger than 1M byte, Vega file copy with GridFTP has higher bandwidth.

1 Introduction

Traditionally file system runs in the kernel space of OS. It manages storage device in computers and network, manages user's data in files and supplies user with a POSIX compatible interface. However, grid environment is not like the single machine or computer network; it faces resource sharing and resource co-operations problems in dynamic environment and between multi virtual organizations [1]. To solve file transfer and file management problems in grid environments, many grid file systems and grid file management tools has been developed. The main obstacle in developing grid application is absence of standards file operation interface, Facing different grid file systems and different grid file management tools, we developed a vegafile transfer protocol and implemented it hotfile, which can utilize different file management protocol and give user a unified operation interface.

Hotfile wrapped GridFTP and GASS or any other file transfer protocol compatible with hotfile structure. Vegafile layer in hotfile abstracts the physical address and physical transfer protocol of grid file. Composed by a set of basic grid file operations, hotfile's higher level APIs are easy to use and manage grid files.

Hotfile are different from other grid file management tools in that hotfile server works as a grid services and it can dynamically deployed to the computers that has installed globus toolkit and it can work with most popular file transfer protocol such as GridFTP and GASS.

The rest of this paper is organized as follows. We first describe the structure of hotfile, and then discussed vegafile layer and basic grid file operation set in detail. We also discuss some related works. Finally we show the current status of hotfile and the future of hotfile.

2 Structure of Hotfile

Hotfile has 2 layered naming space. First layer of hotfile naming space is the name of Data Service, which is registered in router service of Vega grid or MDS of globus. The second layer of naming space is file's path in a Data Service. A data service contains one or more file or directory in it.

Hotfile is mostly like a client/server structure, but different from it in that most c/s architecture has single application server. Hotfile server can be dynamically deployed to multi servers. Files storage in multi server of one user consist his unified storage space.

Fig. 1 is the logical structure of hotfile. To single users, Hotfile is a grid file server based on his grid certificate/credential. User can access data services in any host which are trusted by CA in grid environments.

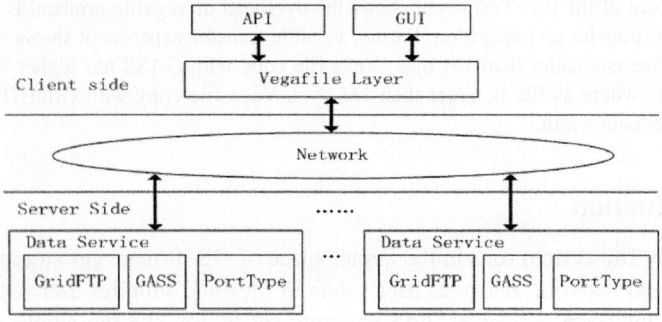

Fig. 1. Structure of hotfile.

User or application can access his data via hotfile's GUI or API. To get data service, it should has the certification which over trusted CA of hotfile server's signature. Data services that trusted by user work as part of user's storage space. All data services trusted by user compose his unified storage space.

Vegafile layer is a virtual file operation layer of hotfile. It binds Vegafile protocol to physical file address and physical file access protocol. In conjunction with discovery mechanism supported by VEGA grid or globus toolkit, vegafile layer firstly get the physical address of Data Services, and then it access physical address of data service to get physical file access protocol via PortType of Data Service. Once physical address and physical file access protocol has been obtained, physical file management channel can be set up. Because of different file transfer protocol has different file access interface. To have a unified file access protocol, we wrapped all data operation of GridFTP and GASS to a single Vega file operation interface, with which users can access file in a single storage space.

Data Services is the server side of hotfile. It can be deployed to servers installed gt3. It not only wrapper different file operation protocols, but also have PortType interface. PortType is Data service's supplement interface that management some features of Data Services, such as the physical access protocol and the base directory of Data Services, and, from PortType, client can get some services data of Data Service.

With the interface of vegafile layer, we implement an advanced grid file operation API and a graphics grid file manager.

3 Vegafile Protocol

Vegafile protocol is the core of hotfile. Files in hotfile are represented by vegafile URL, data and Meta data in file are amended by basic hotfile operations.

Hotfile has a weaker semantic compared with UNIX file systems in file lock operation. Multi-users in different client can't write to file at same time. There is two reasons why we have a weaker semantic. The first is, vegafile protocol designed to be a file transfer and management protocol, not a data I/O protocol, so there is very few concurrent write to the same files. The second is, strong semantic means more communications between client and server. Because of high latency in WAN, stronger semantic spent much time in communications, which means lower performance.

Vegafile URL is a virtual file address in hotfile system. The BNF of vegafile URL is defined as follows:

vegafile url := scheme "://" [hostport] "/" [path]
scheme := vegafile
hostport := hostport from Section 5 of RFC 1738 [5]
path := *path | (/ *([a .. z] | [A .. Z] | - | _ | .))

When hotfile work, Data Services of hotfile firstly search Resource Router in Vega grid or MDS of Globus to get all available data services of user. With the corporation of data service deployed in server, vegafile dynamically translated to physically address of physical file.

Vegafile layer firstly get the host physical address and grid services container port from MDS of globus toolkit 3 or resource router developed by our Vega Grid team, and, suppose get a Data service in 159.226.41.23:8080/OGSA/GOS/Data/DataService, then, it will utilize getURLpoint and getLocalPoint function in PortType of 159.226.41.23:8080/OGSA/GOS/Data/DataService to get the physical protocol of file transfer and the local configuration of the server, which can be used to build physical file transfer path between server and client. Once physical path was setup, Vegafile layer can operate on files using the wrapper API of physical protocol.

Though the MDS or router is a component of hotfile, it's only used when file system initialized. As part of globus toolkit and Vega Grid, it also a necessary part of other grid systems.

There are two principals in deciding which operation should and how to implement. First principal is the minimum principal, file operations can't be disassembled, and they can be used to implement other operations with few client side calculation. The second principal is the balance principal. Because of the high latency in hotfile

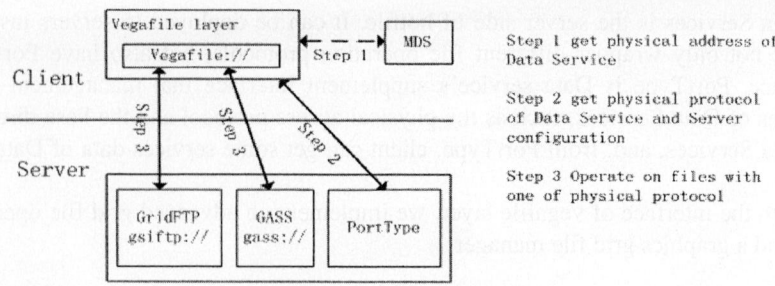

Fig. 2. Steps of vegafile layer.

environment, the main bottleneck in hotfile is communication, exist operation and list operation can be returned in almost the same time, so we choice list operation as one of basic set and implement exist operation and getFileInfo operation on the basis of list operation. With above principals, the operations in basic file operation set are: copy, initialize, open, close, delete, mkdir, and list.

GetFileInfo operation returns vegafile class, which composed mainly by physical attributes of Vegafile. Part of fields in vegafile class is follows:

private long size;
private String name;
private String date;
private byte fileType;
private String physicalpath;

With the data service and virtual vegafile layer, hotfile wrappers the current most popularly used grid file management protocol.

Fig.3 shows the Hotfile's graphic user interface. Tool bar and address lists placed in upper place of GUI, Left panel and right panel of GUI represents browser of one data service. Users can copy, move, create, delete file or directories between 2 data service.

4 Performance

We deployed Data services in Computer with 2 xeon 2.4G CPUs, 1GB memory and 1 36GB utltra3 SCSI hard disk, OS is red hat Linux 7.3, and JVM version is 1.4.2. Client is a personal computer with PIII 1G CPU, 512M memory and 60GB ATA-100 hard disk; OS is windows XP; JVM version is 1.4.1. Server and Client are connected with fast Ethernet.

4.1 The Overhead of Vegafile Layer

We test the performance of Vegafile protocol with GridFTP and Vegafile protocol with GASS. Fig 4 and Fig 5 are result of vegafile copy compared with GridFTP upload and GASS copy respectively.

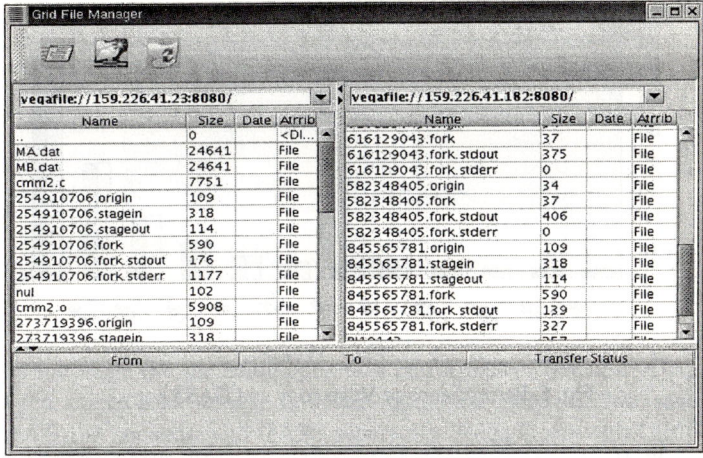

Fig. 3. GUI of hotfile.

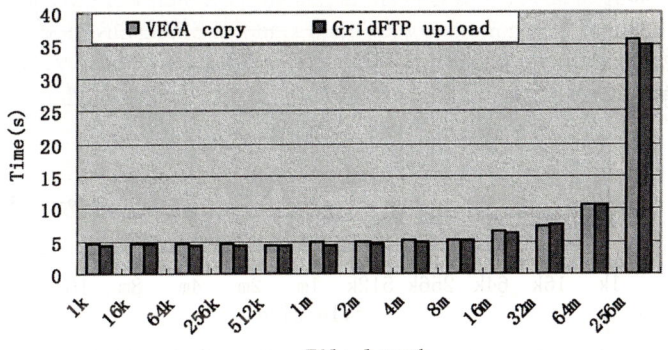

Fig. 4. Performance of Vega copy and GridFTP upload.

After protocol has been initialized, overhead in vegafile only generates in protocol transfer from vegafile protocol to physical protocol. Result shows that the overhead of vegafile protocol is minimum compared with file transfer time.

Fig 4 is the performance of coping files from 1k to 256M using Vega file copy with GridFTP; and we also upload the same set of file using GridFTP. The average overhead of Vegafile copy with GridFTP is 0.239 seconds.

In a similar performance test, file varies from 1K to 64M and we compared the performance of Vega copy with GASS and GASS copy, the average overhead of Vega copy with GASS is 0.242 seconds.

4.2 Performance of Vegafile with GridFTP and GASS

Because of certification based mechanism, Data transfer in grid environment is time cost. 1K file copy using Vega copy with GridFTP copy is 4.715 seconds, 1K file copy using Vega copy with GASS is 3.475 seconds.

environment. Hotfile server side is implemented as a service, which can be deployed dynamically to gt3 servers. In client side's vegafile layer of hotfile, we wrapped 2 widely used grid file transfer protocol to vegafile protocol with little overhead. There are still many issues to be further worked in hotfile. First is the client side cache, which can short request and answer path then accelerate hotfile operation, the second is grid file's authentication question, which is time cost now. To solve it with higher performance and same security level is quite a challenge problem.

References

1. I.Foster, C. Kesselman, J. Nick, S. Tuecke; The Physiology of the Grid: An Open Grid Services Architecture for Distributed Systems Integration, Open Grid Service Infrastructure WG, Global Grid Forum, (2002)
2. Z.Xu A Model of Grid Address Space with Applications, VGSD- 2, (2002)
3. R. Moore, A. Rajasekar, and M. Wan. The SDSC Storage Resource Broker. In Procs. Of CASCON'98, In Toronto, Canada, (1998)
4. The POSIX standards: http://www.opengroup.org/onlinepubs/007904975/toc.htm.
5. Berners-Lee, T., Masinter, L. and M. McCahill, "Uniform Resource Locators (URL)," RFC 1738 prop, (1994)
6. B. White, A. Grimshaw, and A. Nguyen-Tuong, "Grid-based File Access: the Legion I/O Model", in Proc. 9th IEEE Int.Symp. on High Performance Distributed Computing (HPDC), pp165-173, (2000)
7. Micah Beck, Terry Moore, James S. Plank. An End-to-End Approach to Globally Scalable Network Storage. ACM SIGCOMM 2002, Pittsburgh, PA, (2002)
8. Rebecca L. Collins and James S. Plank, Content-Addressable IBP -- Rationale, Design and Performance, ITCC 2004, Las Vegas, (2004)
9. J.Kubiatowicz et al., "OceanStore: An Architecture for Global-Scale Persistent Storage", Proceedings of the Ninth international Conference on Architectural Support for Programming Languages and Operating Systems (ASPLOS), (2000)
10. B. White, A. Grimshaw, and A. Nguyen-Tuong, "Grid-based File Access: the Legion I/O Model", in Proc. 9th IEEE Int.Symp. on High Performance Distributed Computing (HPDC), pp165-173, (2000)

pXRepository: A Peer-to-Peer XML Repository for Web Service Discovery*

Yin Li, Futai Zou, Fanyuan Ma, and Minglu Li

The Department of Computer Science and Engineering,
Shanghai Jiaotong University, Shanghai, China, 200030
{liyin,zoufutai,ma-fy,li-ml}@cs.sjtu.edu.cn

Abstract. The Web services are distributed across the Internet, but the existing Web service discovery is processed in a centralized approach such as UDDI, which has the limitations of single point failure and performance bottleneck. In this paper we propose a distributed XML repository, based on a Peer-to-Peer infrastructure called pXRepository for Web Service discovery. In pXRepository, the service descriptions are managed in a completely decentralized way. Moreover, since the basic Peer-to-Peer routing algorithm cannot be applied directly in the service discovery process, we extend the basic Peer-to-Peer routing algorithm with XML support, which enables pXRepository to support XPath-based queries. Experimental results show that pXRepository has good robustness and scalability.

1 Introduction

Web services are much more loosely coupled than traditional distributed applications. Current Web Service discovery employs a centralized repository such as UDDI[1], which leads to a single point of failure and performance bottleneck. The repository is critical to the ultimate utility of the Web Services and must support scalable, flexible and robust discovery mechanisms. Since Web services are widely deployed on a huge amount of machines across the Internet, it is highly demanded to manage these Web Services based on a decentralized repository.

Peer-to-peer, as a complete distributed computing model, could supply a good solution to build the decentralized repository for the Web Service discovery. Existing Peer-to-Peer systems such as CFS[5], PAST[6] and OceanStore[7] seek to take advantage of the rapid growth of resources to provide inexpensive, highly available storage without centralized servers. However, because Web Services utilize XML-based open standard, such as WSDL for service definition and SOAP for service invocation, directly importing these systems by treating XML documents as common files will make Web Service discovery inefficient. INS/Twine[8] seems to provide a good solution for building the Peer-to-Peer XML repository. However, INS/Twine does not provide a solution to provide XPath-like query.

We designed a decentralize XML repository for Web service discovery based on Peer-to-Peer network named pXRepository (Peer-to-Peer XML Repository). We have extended the Peer-to-Peer routing algorithm based on Chord[4] for supporting XPath query in pXRepository.

---

* This paper is supported by 973 project (No.2002CB312002)of China, and grand project of the Science and Technology Commission of Shanghai Municipality (No. 03dz15027 and No. 03dz15028).

2 System Overview

pXRepository is a Peer-to-Peer XML storage facility. Fig.1 illustrates the system architecture of pXRepository where there are no central servers. Each peer in pXRepository acts as a service peer (SP for simplicity), which not only provides Web service access, but also acts as a peer in the Peer-to-Peer XML storage overlay network. The architecture of the service peer in pXRepository is shown in Fig.2.

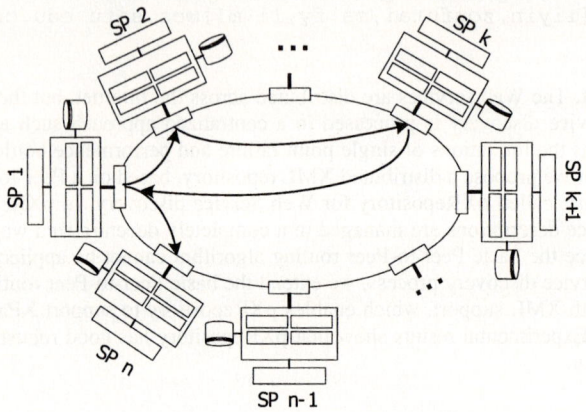

Fig. 1. System Architecture of pXRepository.

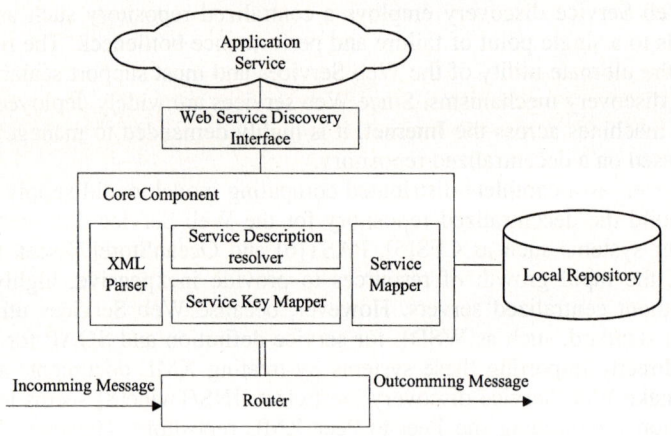

Fig. 2. The architecture of the service peer.

In pXRepository, XPath[9] is used as query language for retrieving XML documents stored over the Peer-to-Peer storage network. A SP consists of three active components called the Web Service Discovery Interface, the core component and the router, and a passive component called the local repository. Web Service Discovery Interface provides access interface to publish or locate Web services and also exposes itself as a Web service. Core component consists of the XML parser, the Service Description Resolver, a Service Key Mapper and a Service Mapper. The service description resolver is a key-splitting algorithm that extracts key nodes from a description. Each key node extracted from the description is independently passed to the service key mapper component, together with URI of the document and parent key node. The service key mapper is responsible for associating HID with each key node. More details are given in section 3. XML parser parses the XML document and XPath query, and is used by other components. Service mapper is responsible for mapping HIDs to service descriptions and will return the results to the application services through the Web service discovery interface. Local repository keeps the Web service interface, service descriptions and HIDs that SP is responsible for. The router routes query requests and return routing results.

3 Web Service Discovery Algorithm

Service locating algorithm specifies how to route the query to the service peer who satisfies the service request. In pXRepository, the service request is expressed in XPath. However, the routing algorithm, Chord, in underlying Peer-to-Peer overlay network only supports exact-match. We extend the Chord algorithm to support XPath based match. The extended Chord algorithm is called eXChord.

In pXRepository, WSDL is used to describe the Web service interface, and the service description metadata is generated based on the content of WSDL document and the description that the user inputs before publishing. An example of Web service description metadata is shown in Fig.3.

```
<?xml version="1.0" encoding="UTF-8"?>
<services><service>
  <name>ListPrice</name>
  <documentation>List the product price</documentation>
  <location>
     http://services.companya.com/product/ListProductService.wsdl
  </location></service><service>
  <name>OrderService</name>
  <documentation>Make an order to the product</documentation>
  <location>
     http://services.companya.com/product/OrderService.wsdl
  </location></service></services>
<description>
  <company>CompanyA</company>
  <industry>Manufactory</industry>
  <region>China</region>
  <keyword>Automobile Price Order</keyword>
  <comments>......</comments>
</description>
```

Fig. 3. An example of Web service description metadata in pXRepository.

To publish the Web Services, the Web Service description metadata in XML format will be first converted to a node-value tree (NVTree). Fig.4 shows an example of the NVTree converted from the Web Service description shown in Fig.3. We have to notice that only the important elements in the Web Service description will be extracted and inserted into the NVTree.

In pXRepository, eXChord extracts each node from the NVTree, and associates each node with a value. Those nodes whose values consist of several words, are further divided into single word value based nodes. Fig.5 shows the key nodes produced

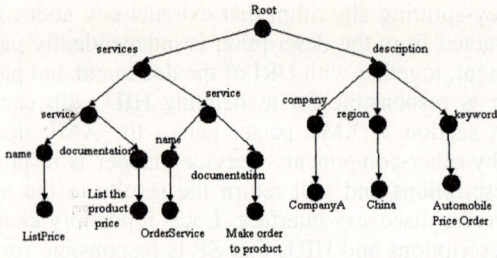

Fig. 4. A sample NVTree converted from service description shown in Fig. 3.

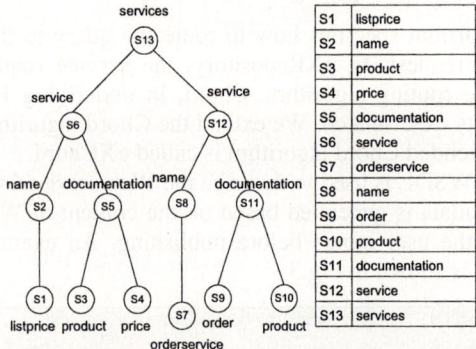

Fig. 5. Splitting a NVTree into service description key nodes.

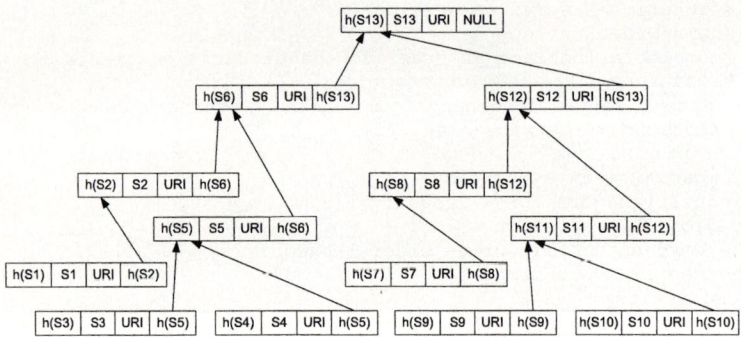

Fig. 6. Distributing node keys across pXRepository.

from the left sub-tree in Fig.4. Each node in Fig.5 is associated with a string called node key (denoted by S1,S2,...). Each node key in NVTree is then passed to the hash function to produce a hash ID, which will be used as a key to insert into the underlying Peer-to-Peer overlay network by eXChord algorithm. In pXRepository, the hierarchical relationship of the nodes in NVTree will be preserved as shown in Fig.6. Each element in Fig.6 resides in a specific service peer in pXRepository corresponding to the hash value of its node key (denoted as h(S1), h(S2),...). Each node in NVTree publishes its contents containing hash value of its node key, node key, URI (Universal Resource Identifier) of the XML document, and hash value of its parent node key, to the underlying Peer-to-Peer network. The Web Service description document is published to the pXRepository by hashing the URI and the document can be retrieved by means of Chord algorithm.

Before presenting pXRepository service publishing and locating algorithm which is named as eXChord, we first introduce some definitions:

Definition 1. Let $И$ be identifier set, and Ω be the SP set. Then the definition of node and node^{-1} are given as follows:

node: $И \rightarrow \Omega$, $И$ is HID set, and Ω is the SP set. node maps HID to corresponding node. This mapping is achieved by Chord algorithm.

node^{-1}: $\Omega \rightarrow И$, node^{-1} maps node to the HID of the node. This mapping is achieved by consistent hashing.

Definition 2. Let SD stands for a Web Service description document, then $Ü(SD)$ stands for the URI of the document, $\Gamma(SD)$ represents the NVTree of SD, and $N(SD)$ stands for the set of NVTree nodes, where $N(SD)=\{N_1, N_2,...,N_m\}$.

Definition 3. Let N stands for a NVTree node, then $P(N)$ stands for its parent node and $K(N)$ represents the node key.

The pseudocode of eXChord service description publishing algorithm is given in Fig.7. Function Publish is run on peer n, take a service description (SD) as input and publishes the SD into the Peer-to-Peer overlay network. When peer n publishes its SD, it first publishes the service description content, and computes the set of the NVTree nodes, then for each node in the set, computes its node key, which will be used as a key together with the document URI and its parent node key hash value to be distributed to the Peer-to-Peer overlay network.

```
1   n.Publish(SD){
2     key=hash(Ü(SD));  n`=n.Route(key);  n`.Insert(key, SD)
3     Compute N (SD)={N1, N2,...,Nm };
4     for each Ni in N (SD){
5       nodekey=K (Ni);   parentkey=K ( P (Ni));
6       n.Distribute(nodekey, parentkey,Ü(SD));
7     }
8   }
9   n.Distribute(nk, pk, URI){
10    id = hash(nk);  n`=n.Route(id);
11    n`.Insert(id, nk, URI ,hash(pk));
12  }
13  n.Route(id){
14    n lookups id` which is closest to id from local routing table;
15    if id<=id` and ∀ id``∈ И, id<=id`` ∧ id``<=id` → id`=id``  then
16      return node(id`)
17    else
18      node(id`).Route(id);
19  }
```

Fig. 7. The pseudocode of eXChord service description publishing algorithm.

To search for a Web Service, the client must specify the query requirements, which is expressed in XPath language. The Web Service searching can be abstracted as following problem: "Given a large collection of Web Service description documents D and an input XPath query Q, find the subset of documents in D that matches Q." pXRepository supports XPath queries, and each XPath query only contains text matching constraints. An XPath query can be converted to a tree called XPTree as shown in Fig.8. In Fig.8, the dash line connection represents ancestor/descendant relationship while solid line connection stands for parent/child relationship. An XPTree can be divided into a sub-tree set in which each sub-tree corresponds to the sub-query of the XPath query as shown in Fig.8.

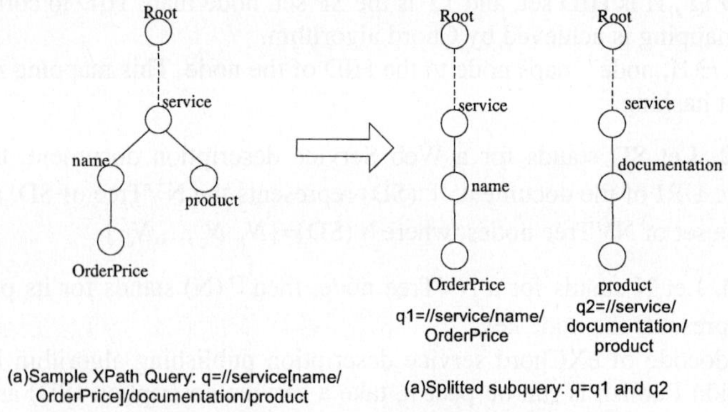

(a)Sample XPath Query: q=//service[name/OrderPrice]/documentation/product

(a)Splitted subquery: q=q1 and q2

q1=//service/name/OrderPrice

q2=//service/documentation/product

Fig. 8. A sample XPTree.

To present the Web Service locating algorithm, we should introduce some definitions:

Definition 4. Let QD stands for a XPath query, then Γ(QD) represents the XPTree of QD, Γ_s(QD) stands for the sub-XPTree set of Γ(QD), where Γ_s(QD)=$\{T_1,T_2,...,T_m\}$.

Definition 5. Let T stands for a sub-XPTree, then $X(T)$ represents the value of the leaf node of T.

The pseudocode of eXChord service locating algorithm is given in Fig.9. Function Locate is run on peer n, take query requirement QD as its input and searches the Peer-to-Peer overlay network for the services that satisfy its requirements. When node n tries to locate the services according to its QD, it first compute the set of sub-XPTrees, then for each sub-XPTree in the set, find all the URI of the service documents that match the sub-query in a recursive way. Finally, the intersection of each URI set that match sub-query is the result set. The client can get the WSDL documents according to the result URI set, and then can access the Web services by the methods described in WSDL document.

4 Evaluation and Experimental Results

In this section, we evaluate pXRepository by simulation and compare pXRepository with centralized service management approach such as UDDI. We use the Georgia Tech Internetwork Topological Models (GT-ITM)[10] to generate the network topologies used in our simulations. We use the "transit-stub" model to obtain topologies that more closely resemble the Internet hierarchy than pure random graph. An Internetwork with 600 routers and 28800 service peers (node for simplicity) are used in our experiment.

```
1    Let R= Φ // R is the result set of the query
2    n.Locate(QD){
3      Compute Γs (QD)={T1, T2,...,Tm };
4      for each Ti in Γs (QD){
5        key=hash( X (Ti));  n`=n.Route(key);
6        NV= n`.get(key); /*NV represents the set of nodes having the  node key of Κ (Ti) */
7        URIS= Φ; /* URIS is the result set that match sub-XPTree of Ti */
8        for each NVi in NV {
9          URI= n`.Match(NVi, Ti); /*Macth is a recursive process finding  matching document set*/
10         if URI!=NULL then
11           URIS = URIS ∪ URI;
12       }
13       if R= Φ then
14         R= URIS;
15       else
16         R=R∩URIS; /* R should satisfy all the sub-queries.*/
17     }
18   }
19   n.Match(N, T){
20     if T=NULL then /*exit condition that indicates a match*/
21       return N.URI; /* N.URI is URI of the service description document  that node N belongs to*/
22     if N.key= X (T) then  /*node N matches sub-XPTree at leaf level*/
23     {
24       key=hash( Κ ( Ρ (N))); // get parent node key hash value
25       n`=n.Route(key);  T= T-{leaf node of T };
26       n`.Match( Ρ (N), T); /*to see if parent node still matches.*/
27     }
28     else /*not match*/
29       return NULL;
30   }
31   n.Route(id){
32     n lookups id` which is closest to id from local routing table;
33     if id<=id` and  ∀ id`` ∈ H, id<=id`` ∧ id``<=id` → id`=id`` then
34       return node(id`);
35     else
36       node(id`).Route(id);
37   }
```

Fig. 9. The pseudocode of eXChord service locating algorithm.

4.1 Space Overhead

Experimental results in Fig.10 reveal that the space overhead of pXRepository is much better than that of UDDI.

4.2 Load

Load is an important metric to evaluate Web service management approach. This paper uses the number of messages in and out of the node as a metric to evaluate the

(2) Problem of Availability. Different GIS have definite boundaries, thus users are hard to comprehend and use heterogeneous spatial information. Users are unwilling to meet additional restrictions, and are urgent to break current phase of separation, so comes the interoperation of GIS.

One of the main reasons that lead to problem of spatial information sharing and interoperation in Internet environment is the heterogeneity. This problem can be studied from syntax and semantic levels: a) syntax heterogeneity, different spatial information resources adopt different storage formats due to different recognition models to the world (Even the same recognition model will lead to different storage formats), and there also may exist version differences in the same kind of storage format; b) semantic heterogeneity, different spatial information resources may be expressed in different concept architectures, while the same concept architecture will also take on phenomena such as same structure with different meanings or same meaning with different structures. The emergence of XML and RDF techniques [3][4] facilitates the dissolving of syntax heterogeneity. They also express structures of information that makes the elimination of semantic heterogeneity possible. Semantic Web of W3C is focusing on the disposition of such problems.

(3) Problem of Wieldy-ability. Users have sorts of requirements while using spatial information: Some want to use the common services, so GIS products provide integration environments; Some have special requirements, so GIS products provide further development tool. In Internet environment, markets cannot be monopolized by one or two powerful products. In fact, they themselves are hard to provide services due to specific applications. So an open information sharing and interoperation framework is in urgent need. Builders of framework can provide information atoms and service atoms, and users of framework can fulfill tasks through combinations among atoms. Following problems should be taken into consideration: a) what information atoms and service atoms are there in the framework; b) how can users acquaint themselves with and use information atoms and service atoms; c) what methods can be used to implement service atoms; d) how can communications be held between information atoms and service atoms; and e) how can information atoms and service atoms cooperate with each other.

2 OSISF: Open Spatial Information Sharing Framework

During the process of spatial information sharing, three concepts are referred: spatial information resources, spatial information acquirement means and spatial information applications. Thus there are three different roles: spatial information providers, spatial information disposition service providers (usually GIS product) and users. So, OSISF should perform the design according to these three different roles. OSISF has two basic targets: The first is to solve the sharing problem among spatial information, which includes three problems mentioned above; The second is to expand it into the construction of Semantic Web, which will becomes one of its components. Currently, OSISF is concentrating on the dissolving of gaps in spatial information sharing (as shown in figure 1): Information comprehension gap and information acquirement gap.

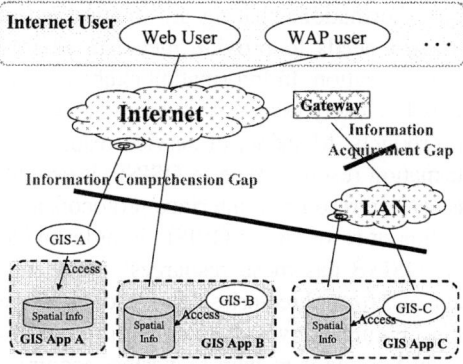

Fig. 1. Two Gaps in Spatial Information Sharing.

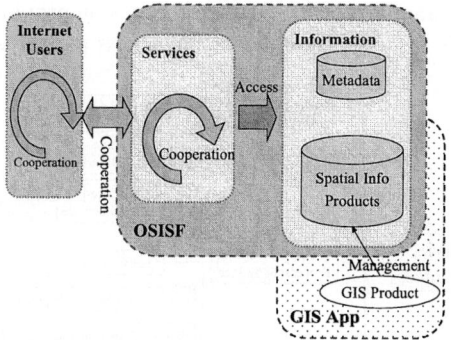

Fig. 2. Open Spatial Information Sharing Framework.

2.1 Architecture of OSISF

OSISF is the bridge between Internet users and GIS applications. In OSISF, knowledge of spatial information domain is stored into two different carriers - programs and data, thus there are two different resources there: service resource and information resource. Its organization is shown in figure 2, while Internet users themselves may have cooperation to form a concrete application system. Each Internet user can cooperate with a group of services in OSISF, and complete certain application tasks. Service resource of OSISF can acquire and comprehend all kinds of information resources in OSISF, and they can cooperate with each other according to certain mechanisms. Information resource of OSISF may be spatial information, or metadata. The framework accentuates information acquirement and comprehension, and information maintenance may be accomplished by services inside the framework, or by specific applications outside it. For example, spatial information may be managed by specific GIS products.

(1) Information Resources. Information resources include two classes of information, which are spatial information products and directories respectively. OSISF thinks that spatial information is already existed, so it mainly studies how to acquire

and comprehend spatial information. Sharing information not only needs to provide information itself, but also needs to provide metadata of information to describe meaning, syntax of the information. In information exchange, spatial information and spatial metadata are usually closely related, so we can name the combination of them as spatial information products. Metadata of service resources (service metadata) is also an important information resource type in OSISF, it describes processing ability of information such as service resources, interface protocols and accessing roads, etc. Accounting for distribution of resources in OSISF, in order to have users acquire their metadata conveniently, OSISF has these resources' metadata extracted, stored together by certain means, and form directories (or metadata databases).

(2) Service Resource. Figure 3 gives the classification of common services. Spatial Information Retrieval Service is to help requesters acquire specific spatial information and to dissolve syntax differences of heterogeneous information. Navigation Service can help users effectively and quickly index needed resources through directories. Spatial Information Processing Service focuses on providing users with spatial information computing resources, but it does not access to data resources directly.

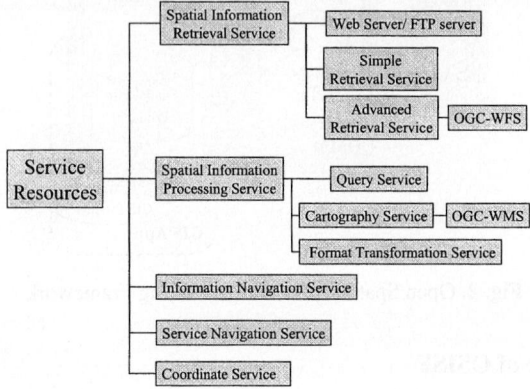

Fig. 3. Open Spatial Information Sharing Framework.

(3) Users on Internet. Generally speaking, users are representation of systems or people outside the framework, and can be divided into two types: a) browser, this kind of users are completely active while interacting with services in the framework, and the connections can only be set up from them without accepting connection requests from services. Publication of tasks and feedbacks are transmitted through user-created connections. b) gateway, this kind of users interacts with services in the framework peer to peer, with both sides originating and accepting connection requests. Publication of tasks and feedbacks can be transmitted through different connections. For example, wireless users can use median user identities to communicate with services asynchronously and without connection. They can thus rationally use resources, shorten channel occupation periods and improve channel utilization rates.

(4) Basic Relationship within Framework. Basic relations in the framework (as shown in figure 4) include two activities as acquiring of information by services and cooperation among services, and a relation of metadata's description of resources.

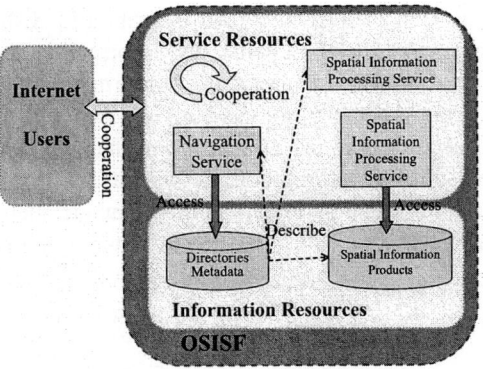

Fig. 4. Basic Relations in OSISF.

Activities of services' acquiring of information lie in two ways: between spatial information retrieval services and spatial information products, as well as between navigation services. Here, problems such as the concrete storage form, the ways to locate and access spatial information products should be considered.

Cooperation between services lies in two ways: between users and services, and among services. The aim of cooperation between a group of users and services is to complete certain tasks. Here, problems such as communication mechanisms, basic cooperation mechanisms and the way to implement cooperation should be considered.

Relations of resources described by metadata lie between metadata and all resources inside framework. Framework users can understand all resources with the help of metadata: distribution of resources, relations among resources, etc.

2.2 Maintenance and Usage of Framework

During maintenance and use of the framework, people may participate as four identities: framework designers, information providers, service providers and users (see figure 5). Framework designers are responsible to present common senses that people should conform to, such as information acquirement protocols, information comprehension protocols, service cooperation methods and task execution methods, etc. Information providers should consider providing spatial information products, both spatial information and corresponding spatial metadata registered into directories. If spatial information products cannot be accessed to from Internet, information providers should also act as service providers to provide median Spatial Information Retrieval Service. Service providers should consider providing services accessible from Internet, meanwhile they should also have service metadata registered into directories. Users should consider how to describe tasks, which resources should be used and how to cooperate services to complete tasks.

2.3 Function of Framework in Semantic Web

OSISF supports sharing of spatial information, and its structure is compatible to the Web. Information exchanged between OSISF services uses median format set by the framework, while this kind of median format is written in XML, which can be explained by XML Schema and RDF Schema, and information is organized according to their meanings. So the basic requirement of information exchanging format required by Semantic Web is met. Ontology is used to describe a group of concepts of a certain domain, and according to it, information-exchanging format is organized to have both syntax and semantic information. In this way, OSISF is not only an Ontology provider of spatial information domain in the Semantic Web, but also it provides services based on this Ontology, such as Spatial Information Retrieval Service.

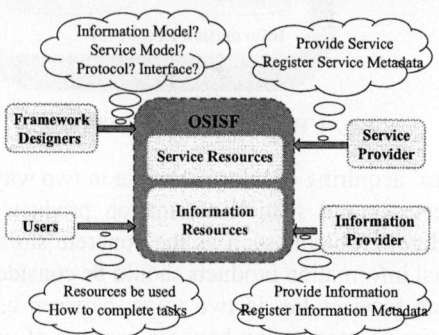

Fig. 5. Roles and Tasks of People During Maintaining and Using OSISF.

3 Implementation Techniques of OSISF

3.1 Cooperation Mechanism of Services

Information can be stored in different formats, and services can be implemented by different methods, but they should conform to basic requirements of the framework. Here how to combine and cooperate services to complete information acquirement tasks of concrete applications should be considered.

(1) Task Resolving Model. In order to fulfill a task, users firstly should divide the task into subtasks, and then send them to corresponding services, such as accessing to a certain information node, or filtrating an information segment. Information can be passed among services. Execution of subtasks can have sequent orders with one input of a service resulting from another service's output. In this way, the task is completed in steps through communication and cooperation among services, and finally the result is handed to users. Figure 6 describes task-solving model in OSISF.

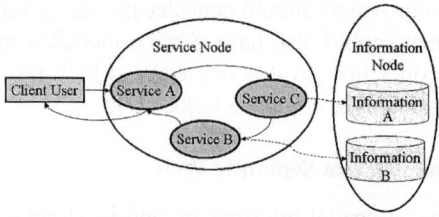

Fig. 6. Task-solving Model in OSISF.

There are three roles in the task-solving model: a) client user, brings forward tasks to the framework, and receives results of tasks. User is responsible for the division of tasks into subtasks as well as descriptions of the execution of tasks. b) service node, describes service resources used in tasks with each service responsible for a subtask. c) information node, describes information resources used in tasks.

Three relations in the model should be considered: a) relation between client user and service node; b) relation among service nodes; and c) relation between service node and information node. The former two relations are relevant to communications of application programs, while the third relation cares information acquiring way of application programs, methods of which are implemented by application programs.

(2) Information Flow Graph (IFG). The essence of task-solving procedure is the procedure to transmit information. OSISF describes information transmission through IFG. According to task-solving model, IFG adopts directed graph (Vertex, directed arc) as its representation. Roles are expressed by vertexes, which include client user (identified by red rectangles), service (identified by blue rectangles) and information (identified by green columns). Relations between roles (service - client user, service - service, information - service) are expressed by directed arcs: a directed arc links two roles, which denotes that there is information flowing from departing role to target role. Client user and service can both have multiple arcs linked to them.

Figure 7 is a graphic representation of an IFG, which has A, B as information acquirement services, C as information combination service, and D as information transformation service. It can also be expressed in literal form using Xlink [5].

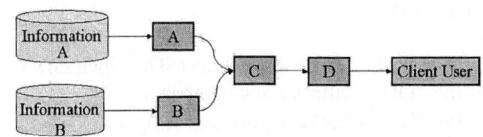

Fig. 7. Information Flow Graph.

(3) Algorithm to Accomplish Information Retrieval Task by Cooperation Services. While users have information acquirement tasks, they firstly acquire the IFG through navigation service, form task scripts by these graphs as well as their predilections, and publish these scripts to relevant Service Nodes in the framework. These Service Nodes can independently cooperate with each other to complete tasks based on the comprehension of task scripts, and finally have results sent back to users.

3.2 Information Retrieval Mechanism

In order to locate spatial information resources in Internet, metadata should be designed to describe different spatial information and services. Through metadata in directories, navigation services can assist users to find suitable spatial information and retrieval services to acquire spatial information resources. Metadata should describe resources' accessing routes clearly, such as host address and accessing control information (user names, passwords). What's more, metadata should also envision storage formats of resources and services' capacities. Navigation service should comprehend and associate metadata of service resources to those of information resources.

4 Conclusion

OSISF owns the characteristic of openness, simplicity, flexibility and automation, which provides users with great freedom to construct the framework. Service providers can use different advanced techniques to implement services, or develop domain application related proliferation services to improve the degree of automation to the framework's process, and thus improve its service ability from every aspect. Spatial information owners can conveniently add new spatial information into it to improve its information gross. Users can use abundant information and service resources in the framework to develop different applications, and they also can share development experiences to the largest extant by sharing task scripts.

Acknowledgement

This work is supported by the National Research Foundation for the Doctoral Program of Higher Education of China (20020001015); the 973 Program (2002CB3 12000); the NSFC (60203002); the 863 Program (2002AA135330, 2002AA134030); the Beijing Science Foundation under Grant No.4012007.

References

1. Buehler K, Mckee L (Editors): The Open GIS Guide (Third Edition), OpenGIS Consortium, Inc. (OGC), http://www.opengis.org
2. ISO/TC211 WG4: Geospatial services N042, Open Geographic Datastore Interface (OGDI), http://www.isotc211.org/
3. Ora Lassila, Ralph R. Swick (Editor): Resource Description Framework (RDF) Model and Syntax Specification, http://www.w3.org/TR/REC-rdf-syntax
4. Dan Brickley, R.V. Guha (Editor): Resource Description Framework (RDF) Schema Specification, http://www.w3.org/TR/rdf-schema
5. Steve DeRose, Eve Maler, David Orchard (Editor): XML Linking Language 1.0, http://www.w3.org/TR/xlink

A Metadata Framework for Distributed Geo-spatial Databases in Grid Environment

Yuelong Wang, Wenjun Wang, Yingwei Luo*, Xiaolin Wang, and Zhuoqun Xu

Dept. of Computer Science and Technology, Peking University, Beijing, 100871
lyw@pku.edu.cn

Abstract. MDS (Monitoring and Discovery Service) is a key infrastructure in grid computing, it mainly undertake registry, updating and discovery of all sorts of resource information of grid environment. This paper developed a MDS-based framework named GridMeta for coping with the problem of distributed geo-spatial metadata. First, we introduced the organization of GridMeta Servers; then we put forward a information model based on DIT, and according to the information model, geo-spatial metadata, service metadata, user metadata and all kinds of Grid resource information can be depicted and stored uniformly; and the last, we provide a set of accessing interfaces of GridMeta APIs, by the interfaces, the registering, adding, deleting, updating and (automatic) discovery of distributed metadata can be implemented.

1 Introduction

Metadata technology plays a key role in the integration and application of geo-spatial information. With the development and popularization of the Internet/Intranet technology, the metadata technology is no longer a method of data description and data index, and now it is becoming one indispensable and powerful method and tool in the domain of information sharing and interoperation all over the whole Internet, this kind of tool can help coping with a series of bottleneck problems such as data discovery, data transformation, data management and data employment [1].

It should be pointed out that metadata can be divided into common-sensed metadata and narrow-sensed metadata. The common-sensed metadata can be all sorts of descriptive information of all network or grid resources, and the narrow-sensed metadata always referred to the geo-spatial metadata. In a GIS application system, geographical information is usually a kind of leading resource, and the geo-spatial metadata always plays a key role for information sharing and interoperation. In this thesis, we put forward a framework mainly dedicated to solving the problem of distributed geo-spatial metadata, but other kinds of metadata such as service metadata, user metadata, etc, is also related to.

PAI [2] (Peking university spatial Application Integrating infrastructure) is a geographical information integration and interoperation platform, and its aim is to provide service integration, accomplish data retrieval from multiple GIS data sources based on different platforms, proceed on-line overlapping, analysis, visualization of

---

* Corresponding author.

map data, and perform user intercommunication in Internet and Grid environments. In PAI, we designed and implemented a metadata-integrating framework named GridMeta, which is based on MDS (Monitoring and Discovery Service) technology of grid computing. GridMeta focuses on the distributed, heterogeneous, and dynamic metadata in the Internet and Grid environment, effectively solves the problem of automatic discovery and integrated application of geo-spatial information, grid resources, and relative services. Figure 1 shows the role of GridMeta in the PAI infrastructure.

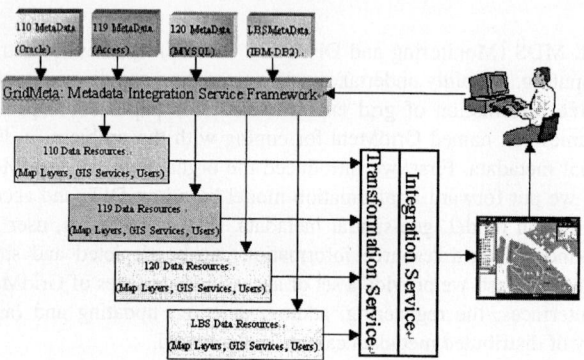

Fig. 1. The Role of GridMeta in the PAI Infrastructure.

GridMeta is designed for PAI infrastructure, but it has a universal value for the same kind of problems. This paper firstly introduces the basic principle of MDS, and addresses the organization GridMeta servers; secondly puts forward a set of universal descriptive specifications of all sorts of grid resources based on the MDS information model; at last gives a querying and searching strategy by GridMeta APIs.

2 Implement of GridMeta

2.1 Introduction of MDS

In PAI infrastructure, we adopt the MDS (Monitoring and Discovery Service) technology to implement our GridMeta framework. MDS is an information management framework serving the Grid Computing, and it is brought forward by the Globus project. MDS concentrates on solving the problem of managing, issuing and utilizing all sorts of massive, distributed and dynamic resource information in the Grid environment, and it mainly includes resource description, resource detection, resource monitoring, resource updating and so on [3].

MDS provides a configurable information provider component named GRIS (Grid Resource Information Service), and a configurable index collection component named GIIS (Grid Index Information Service). Each GRIS component takes the responsibility of registry, maintenance and searching of distributed grid resources within its domain. GIIS supports the ability to perform efficient simultaneous queries to several GRIS components. GIIS could provide whole-ranged query service by combining distributed GRIS services. And it also provides a uniform resource view to facilitate the application programs to perform searching and query operation.

The information service architecture based on GRIS/GIIS is of much scalability. MDS's resource information servers (GRIS and GIIS) can be flexibly configured in accordance with specific conditions and application requests. It should be pointed out that, GRIS and GIIS are only relative concepts, especially in the multi-leveled MDS architecture, a middle-leveled information server acts as a GIIS to the low-leveled server, on the other hand, it is regarded as GRIS by the high-leveled server.

2.2 Organization of GridMeta Servers

Based on MDS mechanism, we assemble our GridMeta servers on the basis of GRIS/GIIS architecture. First, a GRIS server is built for each GIS application system (or a Virtual Organization, namely VO). These GRIS servers are responsible for registry, maintenance and publishing of the metadata of the local systems, and the metadata here describes domain-related spatial information, service information and user information. Commonly, a GIS system (or VO) maintains a local GRIS server, nevertheless, if necessary, several VOs can share a same GRIS server. In default case, a GRIS server may be self-configurable, and performs listening on web port 2135.

Above GRIS, GridMeta designs one or more configurable GIIS servers for index collection. GIIS is responsible for maintaining and publishing each GRIS's registered information, and providing a cache service which is analogous to a Web Searching Engine. GRIS and GIIS are linked by the pointers of "referral".

Metadata of a GRIS server can be registered to GIIS through HeartBeating mechanism, or when GIIS receives a user's request, and its cached information expires, it may obtain relevant refreshing information from GRIS forwardly. The organization of GridMeta servers is illustrated as Figure 2.

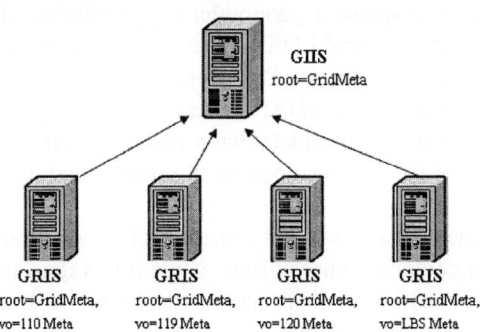

Fig. 2. Organization of GridMeta servers.

GridMeta servers are assembled based on LDAP protocol [4]. To solve the problem of load balance, the information index is maintained with redundance technology. For each GRIS server, local metadata composes the main database, and distant metadata composes the slave database. The data replication and consistency among different servers is guaranteed by the stoke mechanism of LDAP. When the main database is being modified, the mirror in the slave database is updated automatically (in-time update or timed update, while the refreshing speed may not be very quick).

3 Information Model of GridMeta

Based on LDAP protocol, GridMeta uses DIT (Directory Information Tree) to organize distributed metadata [5]. Starting from the root node, DIT maintained a hierarchical view of all resource information of a grid, and provided a searching mechanism based on the tree structure. The DIT of GridMeta is illustrated as Figure 3.

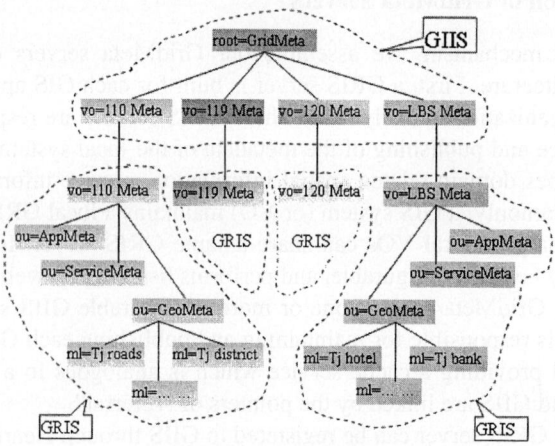

Fig. 3. DIT of GridMeta.

As shown in Figure 3, information of a DIT is stored in more than one sub-tree, and all these sub trees compose a geographically distributed directory. These sub trees are implemented with several LDAP servers (A subordinate LDAP server forms a GRIS component, while an upper LDAP server forms a GIIS), or maintained by one LDAP server if necessary. All LDAP servers are linked by referral mechanism. If a sub tree does not contain the metadata that a user needs, it can automatically redirect to other locations of the DIT by a pointer of referral, and this is transparent to the user.

DIT consists of objects that are called entries. Each object is either a container or a leaf. A container can comprise other objects, while a leaf does not contain any other objects. An entry contains one or more records depicting the real or abstract objects of a grid, such as hosts, users, services, maps and maplayers. Records are stored as <attribute/value> pairs, and are represented with the format of "name=value".

Each DIT node can be identified by one or more attributes, this attribute for node identification is called RDN (Relative Distinguished Name). RDN can be used to distinguish an entry within one container. In order to identify an entry in the whole tree, DN (Distinguished Name) is defined. A DN is a serial of concatenated RDN strings from the current entry to the root entry. A DN can identify an entry in a whole tree scope. With DN, users can access any one entry in an information tree at will.

For instance, there is an entry of maplayer, its RDN is: "lyrid=00104", and its DN is "lyrid=00104, ou=GeoMeta, vo=LBS Meta, root=GridMeta".

In GridMeta, geo-spatial metadata is organized on the basic unit of maplayer. Every one maplayer's metadata (descriptive information of a maplayer) forms a leaf node of DIT. Figure 4 shows a leaf entry that describes a maplayer.

Name	Value	Type	Size
lyrid	00104	text attribute	5
objectClass	MapLayer	text attribute	8
mapid	003	text attribute	3
mapname	lbs	text attribute	3
lyrname	bj-bank	text attribute	7
lyrkind	lounge	text attribute	6
place	beijing	text attribute	7
lyrscale	1:2000	text attribute	6
lyrdate	2003/05/05	text attribute	10
lyrsrs	longitude/latitude	text attribute	18
rng-left	117.119082	text attribute	10
rng-right	114.967047	text attribute	10
rng-top	40.470121	text attribute	9
rng-bottom	36.694378	text attribute	9
rng-zmax	0	text attribute	1
rng-zmin	0	text attribute	1
dt-fmrt	mapinfo	text attribute	7
dt-path	vectdbj	text attribute	7
dt-owner	jwzh	text attribute	4
acs-userid	vector	text attribute	6
acs-port	1521	text attribute	4
acs-passwd	vector	text attribute	6
acs-mode	002	text attribute	3
src-ip	162.105.17.111	text attribute	14
lbl-zmax	0	text attribute	1
lbl-zmin	0	text attribute	1
validfrom	20030707091038Z	text attribute	15
validto	20030708091038Z	text attribute	15
keepto	20030709091038Z	text attribute	15

Fig. 4. An entry depicting the metadata of a maplayer.

Besides geo-spatial metadata, GridMeta also integrates other kinds of metadata such as service metadata, user metadata, and software or hardware metasata or a grid. To make the upper-leveled applications more flexible and convenient, GridMeta introduced "relation" mechanism between all kinds of metadata. An instance, there is a service in LBS system, its name is "getAroundServ()", its RDN is "lbsid=12", the function of this service is to find nearby banks on a city map, it needs to integrate a maplayer (e.g. RDN is "lyrid=00104") which contains all banks' locations of the city. Here, we could add one attribute of "lyrid=00104" to this service's entry. So, when a user calls the service of "getAroundServ()", he will automatically get the service's attribute of "lyrid=00104", then by querying the maplayer's entry whose RDN is "lyrid=00104", the maplayer's metadata can be available. The service's entry is shown in Figure 5.

On the other hand, perhaps you would find a service related to a maplayer. In the same way, you could add an attribute of "serviceid=*" to the maplayer's entry.

Sometimes, a service needs to integrate more than one maplayers. To support this, you can add several values to the service's attribute, such as "lyrid=00104 & lyrid=00106", because GridMeta permits object of multi-valued attributes.

Name	Value	Type	Size
lbsid	12	text attribute	2
objectClass	LBS-Application	text attribute	15
svcname	getAroundServ	text attribute	13
lyrid	00104	text attribute	5
servtype	bank	text attribute	4
comment	03	text attribute	2
validfrom	20030707091038Z	text attribute	15
validto	20030708091038Z	text attribute	15
keepto	20030709091038Z	text attribute	15

Fig. 5. An entry depicting the metadata of a service.

4 Access to GridMeta

Because GRIS/GIIS servers run on the TCP/IP protocol, all requests on the Internet can access the information of GridMeta through the TCP/IP, regardless of platform.

GridMeta provided four kinds of principal interfaces: query, add, modify and delete. Client programs can use these APIs to implement all kinds accessing operations, and users also can extend the metadata management framework as need with the interfaces.

Because GridMeta is constructed on the base of MDS and LDAP protocol, users also can use MDS APIs or JNDI APIs to access GridMeta servers directly.

The APIs of GridMeta offer great convenience to design upper applications. Using the accessing interfaces and methods provided by GridMeta, users and upper-leveled applications could access GridMeta servers conveniently. The following java code segment utilizes GridMeta APIs to query a GridMeta server.

```
GridMeta gridmeta;
gridmeta = new GridMeta("gis.pku.edu.cn", "2135",
"ou=GeoMeta, vo=LBS Meta, root=GridMeta");
try
{
   gridmeta.connect();
   String   filter =
"(&(svcname=getArounServ)(servtype=bank))";
   String   scope  = "gridmeta.SUBTREE_SCOPE";
   String   layerMeta;
   Hashtable   maplayer;
   maplayer  = gridmeta.searchLayer(filter, scope);
   layerMeta = toXML(maplayer);
   gridmeta.disconnect();
}
catch(MDSException e)
{
   System.err.println( "Error:"+ e.getLdapMessage() );
}
```

The first step is to construct an object instance (named "gridmeta") of the class "GridMeta" with the following piece of code (namely, to employ the class of "GridMeta"'s constructing method).

gridmeta = new GridMeta("gis.pku.edu.cn", "2135",
    "ou=GeoMeta, vo=LBS Meta, root=GridMeta");

The parameters in class "GridMeta" are GridMeta server's name, port (default 2135), DN or Base Directory. Default DN is "root=GridMeta", the root node.

The next step, after running the constructing function, is to use the function of "connect()" to construct a connection. After having done this, a searching operation can be undertaken with the sub row of code:

maplayer = gridmeta.layerSearch(filter, scope);

In the preceding row of code, the parameter of "filter" defines what a searching filter should be used. When the value of "filter" is "objectclass=* ", this denotes to search all the objects of a DIT, and when the value of "filter" is "lyrname=bj-mainroad", that means to search all the maplayers whose names are "bj-mainbroad" in a DIT. One of the principal advantages of GridMeta is that it supports locating the geo-spatial metadata with relative service metadata, user metadata and others, based on the mechanism of "relation". For instance, in Listing 1, when the parameter of "filter" is given a value as following:

filter ="(&(svcname=getAroundServ)(servtype=bank))"

Above value of "filter" means to search a maplayer (or several maplayers), this (these) maplayer(s) could support the service of "getAroundServ()", and the role of this service here is to help finding the nearby banks.

Another parameter of "scope" defines the bound of searching operation, "ONELEVEL_SCOPE" denotes to search only one level that contains current entry, while "SUBTREE_SCOPE" means to search the subordinate tree of the current node.

As for the method of "searchLayer()" owned by the class of "GridMeta", its searching result is a hash table, GridMeta can transform the searching result from hash table to XML-based text with the method "toXML()", this operation can make great convenience for the upper methods to comprehend and process the searching result.

5 Conclusions and Future Work

This paper introduces an MDS-based management framework named GridMeta for distributed geo-spatial metadata, including the architecture, information model, information access, programming interfaces etc. This framework can undertake discovery, registry, query and updating of all kinds of metadata in network or grid environment, it also can describe and store the information of diversified objects (permanent and temporary) in network or grid, so it is suitable for solving the problem of distributed and dynamic geo-spatial metadata and other grid resource information.

The principal advantages of GridMeta is as following:

- Suitable for solving the problem of distributed, autonomous and heterogeneous metadata of Web/Grid GIS;
- Based on the MDS technology, regarding the metadata as a kind of grid resource that can be monitored momently by the HeartBeating mechanism, this measure solved the problem of unstable metadata in distributed environment;

- Locating geo-spatial metadata on the base of server and use information, this archived automatic discovery of geo-spatial metadata primarily;
- Distributed LDAP servers based on GRIS/GIIS, solved load balance problem;
- A GridMeta server is constructed on the base of OpenLDAP, so it has a faster respond speed than relation database; especially when more than one user access a GridMeta server simultaneously, the superiority is more outstanding;
- The entry-based security and authentication mechanism, and the encrypt transmission of LDAP can guarantee the information security in the open network and grid environment.

The future work is: First, to set up a cache mechanism for the client program, thus a searching request will not be sent to a LDAP server every time, and this can increase the respond speed further more; Second, to explore the security and authentication mechanism on the base of users' authorities; And the third, to introduce "WSDL+UDDI" technology to the framework, this can facilitate the automatic broadcasting and detecting of metadata greatly.

Acknowledgement

This work is supported by the National Research Foundation for the Doctoral Program of Higher Education of China (20020001015); the 973 Program (2002CB 312000); the NSFC (60203002); the 863 Program (2002AA135330, 2002AA 134030); the Beijing Science Foundation under Grant No.4012007.

References

1. Wang Xiaolin, Luo Yingwei, Cong Shengri, Zhang Ning, Xu Zhuoqun, Lu Zhonghui: A Study of Spatial Metadata and Its Applications (in Chinese), Journal of Computer Research and Development, Vol.38, No.3, 2001.
2. Luo Yingwei, Wang Wenjun, Wang Xiaolin and Xu Zhuoqun: Integration and interoperation of spatial information applications (in Chinese), Journal of Dalian University of Technology, Vol.43, No.SUPPL, 2003.
3. Steven Fitzgerald, Ian Foster, Carl Kesselman, Gregor von Laszewski, Warren Smith, Steven Tuecke: A Directory Service for Configuring High-Performance Distributed Computations, 6th International Symposium on High Performance Distributed Computing (HPDC '97), Portland, OR, USA, 1997.
4. Gregor von Laszewski and Ian Foster: Usage of LDAP in Globus, http://www.globus.org/mds/globus_in_ldap.html, 1998.
5. Xiao Nong, Ren Hao, Xu Zhiwei, Tang Zhimin, Xie Xianghui, Li Wei: Design and Implementation of Resource Directory in National High Performance Computing Environment (in Chinese), Journal of Computer Research and Development, Vol.39, No.8, 2002.

Research and Implementation of Single Sign-On Mechanism for ASP Pattern*

Bo Li, Sheng Ge, Tian-yu Wo, and Dian-fu Ma

Computer Institute, BeiHang University, P.O Box 9-32 Beijing 100083

Abstract. Software Services based Application Service Provider pattern is an important method in constructing enterprise applications, which integrate business systems with different authentication mechanism. So there are questions such as repeated authentication and authorization, difficulties in authorization management, difficult to describe security information interoperability. This paper proposes a method, which stores information in a uniform format, accesses it in a standard interface and exploits account federation, authentication proxy, and authorization proxy to transfer authentication and authorization results. As a result, we design and implement a single sign-on system by this method.

Keywords: ASP, Single Sign-on, LDAP, SAML

1 Introduction

Under the Internet environment, ASP platform and service is becoming the trend of E-business. Portal-based ASP platform can provide or integrate various information systems, such as OA, Email Systems, and cooperative platforms and so on. Not only does the ASP provide a standard interface for User, but provide a standard connection point for application providers.

Application always has its own authentication and authorization mechanism Heterogeneity of which result in problems [1]. ①users must remember independent accounts and login repeatedly. ②User information and security policy are unrelated, which makes the user management complicated and unsafe. ③Because of heterogeneous security mechanism, it is impossible to transmit authentication results to finish a cooperative job. Thus, to integrate the user information and security policy standardize authorization process and set up SSO system among applications becomes the key in ASP development.

So far, much SSO-related work has been done. Such as MS .Net Passport [2] and Liberty [3], but they can't meet new requirement of authorization management and sharing security information

This paper proposed a SSO solution base on LDAP [5] and SAML [6]. LDAP stores and manages user information and security. SAML descript and transmit the authentication results in a standard way.

---

* This paper is supported by China Education and Research Grid (China Grid) and the National High Technology Development Program of China under Grant No.(2001AA113030, 2002AA116050 and 2003AA115420).

an legacy application P_L, the Web application B represents a newly developed application P_N, then the single sign on in within ASP domain is as below:

1. User logged on the ASP platform and provided DS_A the account-federation information.
2. DS_A authenticated the user and stored the account- federation information in LS_A. Then this user attained a single identity and a federation – relationship.
3. The administrator of P_N customized the authorization policy, including roles and permissions; the administrator of DS_A assigns the role to the user, so the user have the corresponding permissions.
4. When User U wanted to access P_L, the authentication module of P_L requested the acount information of U, according to the established account-federation, and DSA returned the account to P_N.
5. The authentication module of P_L authenticated the user's identity.
6. When the user wanted to access P_N, according to the established security policy, DS_A constructed an authorization assertion by SAML and returned it to P_N.
7. P_N executed the SAML authorization assertion.

Through this mechanism, ASP platform cannot only support account-federation and storage for the legacy systems, but can customize access control policy for the newly developed systems. Thus, we can achieve central authentication and access control.

3.3 SSO Mechanism Among ASP Platforms

Within the enterprise alliance, to finish a transaction users need to access many applications distributed in many enterprises. The key problem is sharing and transferring the authentication and authorization results among ASPs. For example, ASP_A and ASP_B represent two independent security domains. After they release and exchange their own trust policy, a new credible relationship is established.

1. Users access ASP_A, which redirect user to authentication center (AC) to log on globally.
2. Users log on AC, AC create Http Session, SAML authentication assertion and ticket based on the authentication result (reference to the assertion) for the user.
3. AC redirect the user to ASP_A with the ticket issued by AC, which prevents repeated attack.
4. ASP_A request integrated SAML authentication assertion by ticket, AC return the assertion to the authentication module of the ASP_A.
5. Authentication module of ASP_A authenticate the identity of the user according to the SAML authentication assertion issued by AC.
6. User access ASP_B, ASP_B redirect user to AC.
7. Make use of the Http session, AC create the SAML assertion and ticket for the user, then redirect the user to the ASP_B with the ticket issued by AC.
8. ASP_B request integrated SAML authentiation assertion from AC, then AC return the assertion to the authentication module of the ASP_B.

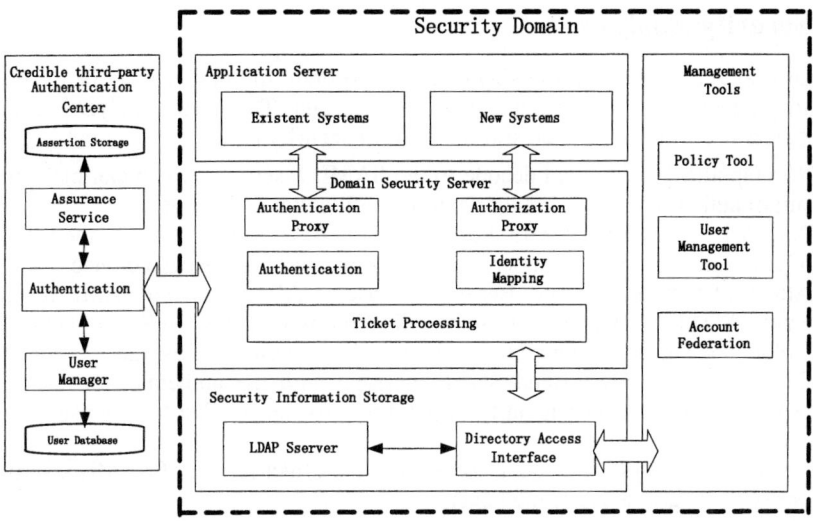

Fig. 2. Architecture of the single sign-on system for ASP pattern.

9. Authentication module of ASP_B authenticate the identity of the user according to the SAML assertion, thus the user only log on once to the AC, then he can access ASP_A and ASP_B seperately.

In above mechanism, credible third-party authentication center send authentication and authorization result by assertion and ticket to guarantee user's identity to other platforms.

4 Implementation of Single Sign-On System

In this paper, we implement a single sign-on system for ASP Pattern, (Abbreviation is ASPSSO), According to the design target, this paper design four core parts to construct the system: security information storage, domain security server credible, third-party authentication center and configuration tools, the architecture is showed as figure 2:

1. Security information storage: Introduce LDAP to centrally store user's accounts and security policy in a domain. LDAP access interface, capsulate the API to access LDAP.
2. Domain security server: Authenticate the users and map the global identity to the local identity, create the account federation for users, automatically transmit the account to the legacy application, send the authorization assertion to the new deployed application.
3. Credible third-party authentication center: Authenticate the global user, create and manage SAML authentication assertion, issue ticket relative to the assertion, respond to request of assertion.
4. Configuration tools: Customize account federation information, access control policy, and the identity mapping policy of the users.

5 Security Analysis

When user access through the single sign-on system, the system may be exposed to security attacks, the security guarantee is important. This paper analyzes security r characteristic, which can avoid several kinds of security attack:

1. Wiretap attack Code: and decode to the SAML assertion, which can make the result of authentication and authorization confidential and integrate.
2. Playback attack: Set the period of the validity of assertion, decrease the valid time of attacker, and make the assertion invalid after has been accessed once.
3. Attack of tamper message: Because we use SOAP to transmit SAML result, we extend SOAP by WS-Security standardization, in this way the SOAP message can include the signature element to guarantee the integrity of message.
4. Disguise attack: We extend SOAP message with the authentication element to prove message sender's identity, which prevents the attacker to disguise as a legal site.
5. Hostile corporation site attack: By setting the effect region of SAML assertion to prevent the hostile use of the ticket.

6 Conclusion

In order to meet the new security requirement of E-Business, this paper propose a kind of Single Sign-on means for ASP Pattern. The characteristics are as follows: ①though LDAP, the system can centrally manage the user information and security policy in a standard interface. ②though SAML, the system has standard description and transmission of authentication and authorization result. ③unified authentication and authorization, simplified complexity of authorization management: ④By Web Service, implement a reliable third party authentication center, which is interoperable and portable. ⑤the system has good security and

Problems such as reliability and performance in large-scale concurrent access should be taken into account.

References

1. Build and implement a single sign-on solution [J]. Sep.30th 2003. http://www-106.Ibm.com/developerworks/java/library/wa-singlesign/? Ca=dgr-lnxw914CASsso
2. Microsoft Corporation. Microsoft .NET Passport Technical Overview.2001 [J].
3. Jeff Hodges, Sun Microsystems, Inc. Liberty ID-FF Architecture Overview [S]. 14 April 2003.www.projectliberty.org/specs/ draft-lib-arch-overview-v1.2-04.pdf.
4. Network Working Group. Lightweight Directory Access Protocol (v3)[S]. 1997. RFC2251.http://www.ietf.org/rfc/rfc2251.txt? number=2251
5. Hallam-Baker P, Maler E.Assertions and Protocol for the OASIS Security Assertion Markup Language [S]. 2002,5.
http://www.oasis-open.org/committees/ security/docs/Cs-sstc-cor e-01.pdf

Design of an OGSA-Based MetaService Architecture*

Zhi-Hui Du[1], Francis C.M. Lau[2], Cho-Li Wang[2], Wai-kin Lam[3], Chuan He[1], Xiaoge Wang[1], Yu Chen[1], and Sanli Li[1]

[1] Department of Computer Science and Technology, Tsinghua University
100084, Beijing, China
{duzh,wangxg,yuchen,lsl-dcs}@tsinghua.edu.cn
[2] Department of Computer Science The University of Hong Kong, Hong Kong
{fcmlau,clwang}@cs.hku.hk
[3] The Hong Kong Institute of High Performance Computing, Hong Kong
lamwaikin@hkhpc.org

Abstract. This paper proposes a MetaService-based architecture for dealing with the problem of services management in grid systems. We introduce the basic idea of MetaService, and then discuss MetaService classification, new MetaService construction, and the MetaService working mechanism. A graph and a tree can be used for the optimization function of MetaService. An example from real life is used to explain the role of MetaService. We give an overview of a grid prototype system which has incorporated some of the ideas presented. A brief comparison is given between MetaService and other related work.

1 Introduction

OGSA (Open Grid Services Architecture) [1] is an open standard for grid computing [2][6]. It borrows ideas from Web Services [3], making it deployable in wide range of scientific and commercial fields. OGSA is a service-oriented architecture and any thing in OGSA should be a service. An OGSA-based grid system generally includes a large collection of distributed resources of various kinds – for example, computers, storages, networks, equipment, sensors, etc. By the requirement of OGSA, all the grid resources should be provided in the form of services in order to become a part of the grid system. So the problem of resource management in OGSA-based grid system is largely a problem of service management.

The management of the resources of an OGSA-based grid system must itself be a service. We propose a new concept called MetaService which is a service to be used to manage other grid resource services. With MetaService, we can distinguish the management service from the resource service. MetaService and resource service have the same syntax but different functions. In the same way that MetaData is data about data, so MetaService is service about service. Just as some metadata can be the data of other metadata, the difference between MetaService and resource service is also relative.

---

* This research is partly supported by Beijing Municipal NSF(No.4042018), National High-Tech Research and Development Plan of China (No. 2003AA1Z2090), Tsinghua University 985 Foundation, and IBM Basic Research Foundation.

2 MetaService Architecture

It is important that MetaService will comply with the same specification as resource service. This will make the entire OGSA system having a single uniform basis and enable all the services to interact with each other easily. The focus of resource service is to provide the resource in the form of grid service but the focus of MetaService is to manage the resource services. The different functions between MetaService and resource service explain why a MetaService architecture is necessary for an OGSA-based grid system. This paper describes the MetaService architecture and discusses some general principles for exploiting MetaService to achieve automatic management of a large grid system.

2.1 MetaService Classification

A very complex MetaService is hard to develop and in fact we cannot use one MetaService for all kinds of management tasks. In the proposed MetaServices architecture, MetaServices are divided into the three broad categories of service monitoring, optimizing and adjusting.

Certain information is necessary for an efficient management, and the MetaServices architecture must have some mechanism to obtain such information. Grid information has great variety and it is hard to use one monitoring MetaService to collect all kinds of information, so different monitoring MetaServices may be necessary for getting information from different resource providers and consumers. One MetaService should be dedicated to some special aspects of the necessary information. This can avoid building a very large monitoring MetaService which may be to clumsy to meet the dynamic requests.

An optimizing MetaService tries to provide better services and solutions to the users. Different algorithms and heuristic methods can be employed in the optimizing process. The output solution of the optimizing MetaService need not be the best, but it should be better than the one without or before optimization. The approximate best solution should be achieved after several iterations of optimization.

An adjusting MetaService can implement the effect suggested by the optimizing MetaService and it should have the ability to interact with other resource services or MetaServices. The adjusting MetaService can change the grid resource configuration or the method used, and also the behavior of different MetaServices at the same time.

The relationships among the different kinds of MetaServices are shown in Fig. 1.

2.2 MetaService Construction

Several questions need to be answered in the process of building a set of MetaServices for an OGSA-based grid system.

First, where should the MetaServices be put? We define grid services consumer as any grid entity using the other grid services, and grid services provider as any grid service providing service to other grid entities. So MetaService can be put between any grid services consumer and grid services provider if the MetaService's service is called for (see Fig. 2).

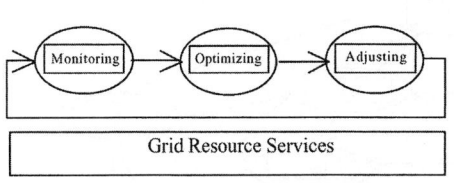
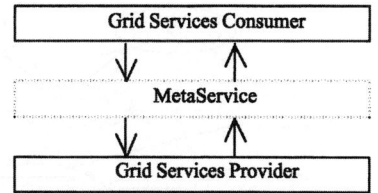

Fig. 1. MetaService classification.

Fig. 2. Location of MetaService.

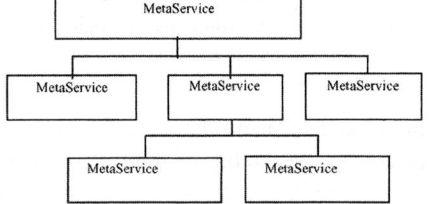
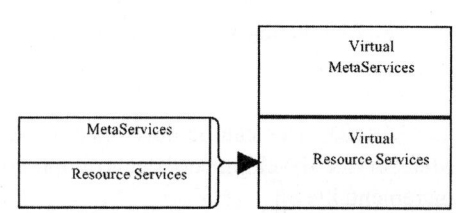

Fig. 3. MetaService layers.

Fig. 4. High-level virtual MetaServices for virual resource service being merged from low-level MetaServices and resource services.

The MetaServices on the one hand collect information from low-level grid resources providers and on the other hand receive requests from grid services consumers that are on top. MetaServices thus make decisions to match the requests from consumers with resources from providers efficiently. The dotted lines in Fig. 2 mean that the consumer in fact can use the resources provided by lower-level provider directly, bypassing the MetaService, but that may result in an inefficient use of the provider's service.

Second, how to build a powerful MetaService step by step? There are two methods. (1) MetaService layers. At the very beginning phase, some simple and low-level MetaServices are developed, and then these MetaServices are used to build higher-level and relatively more complex MetaServices (see Fig. 3). (2) Virtualization. The difference between MetaService and resource service is relative, not absolute. This means that some MetaService can be seen as resource service by other higher-level MetaServices (see Fig. 4). This abstraction method can provide the MetaService developers with different levels of well-defined and clear views. It can not only simplify the work of the overall MetaService development but also allow MetaService development to happen in different levels simultaneously.

In fact, these methods are inspired by the same in the general grid service concept. In OGSA, we can build virtual high-level services from low level services [1]. These methods can take advantage of any features of the general resource services. Any improvement in OGSA's service specification can be absorbed by MetaService smoothly because MetaService is also a service by syntax.

Although we can build MetaServices in different levels at the same time, it does not mean that they must be finished at the same time. The easiest or the most impor-

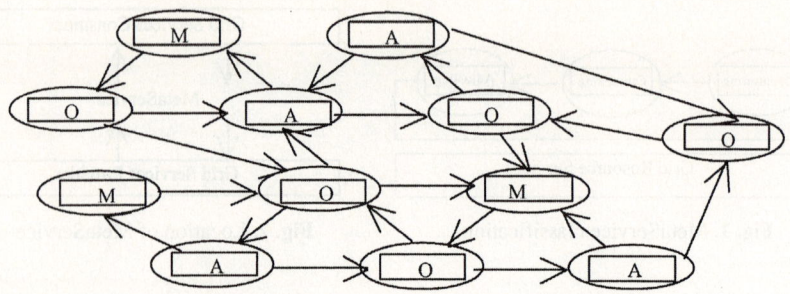

Fig. 5. Directed graph of MetaService optimization path.

M: Mornitoring MetaService; O: Optimizing MetaService;, A: Adjusting MetaService

tant MetaServices can be developed at the early stages. Then others can join in the MetaService development later to improve or add to the quality of the total system incrementally.

2.3 MetaService Working Mechanism

How do MetaServices work together to make the total grid system more efficient than before? The MetaService working mechanism can be described in two orientations, horizontal and vertical.

The relationships of MetaServices in the same layer (horizontal orientation) can be expressed as follows. Let V_i be the set of MetaServices in layer i, $i=1,2...,N$, then the MetaServices in layer i can be formulated as a directed graph

$$G_i = <V_i, E_i>, E_i = \{<v_k,v_l> | v_k,v_l \in V_i, and\ v_k \neq v_l\}$$

Any MetaService optimizing method in layer i is a subgraph of G_i. Fig. 5 is an example of an optimization subgraph in a given layer.

The relationships of MetaServices in different layers (vertical orientation) can be expressed as a directed tree T_j. Let V be the set of MetaServices in different layers, $V=V_1 \cup V_2 \cup ... \cup V_N$, V_i is the set of MetaServices in layer i, $v_i(j) \in V_i$ and it is the jth element of V_i. We assume that the MetaServices are distributed across N successive layers; then

$$T_j = <V, E>, E = \{<v_i, s_{i+1}> | v_i \in V_i, s_{i+1} \in V_{i+1}, and\ i=1,2...,N-1, if\ i=1\ then\ v_i = v_1(j)\}$$

The optimization tree T_j establishes an even more powerful method to provide high-level users with easier and more efficient services.

The framework provides us a formal approach to evaluate the efficiency of MetaServices. Different subsets of G_i and T_j can be chosen in a test to see if they can improve the performance of a given grid system (or part of a given grid system). The efficiency or performance of a grid system can have different definitions in different contexts. We will not discuss how to evaluate a grid system's efficiency in this paper and we assume that such criteria have been established.

2.4 Discussion on MetaService Architecture

The classification of MetaServices is conceptual, not physical. In other words, the functions of monitoring and optimizing can be implemented in one physical MetaService. The building method of MetaServices architecture is open and the number of MetaServices to be provided can be increased gradually. The architecture only defines the method to build MetaServices and how to make them work together, and does not impose any limit on specific MetaServices or their number. The architecture is therefore able to change its functions or behavior to adapt to different environments.

In this MetsService architecture, adaptation of MetaServices is achieved by absorbing new, better MetaServices and adjusting the behaviors of both MetaServices and resource services. In fact, only when the necessary support for MetaService has come from the resource services concerned, can the architecture then work efficiently. So the MetaService architecture also helps to improve the design of resource services.

The working mechanism of MetaServices architecture defined in the form of directed graphs and trees provides a mathematic basis on which to apply different algorithms or theories about graphs (trees). The optimization path on the tree T_j is based on a user-oriented view. It tries to provide better performance for some specific applications. The optimization path on the graph G_i is for a system-oriented view. It tries to provide better system performance by adjusting the behavior of system resources. The optimizations on G_i and T_j should be combined together when we develop MetaServices for our grid system.

3 Similar Mechanism in Real Life

To help the reader understand the basic idea of MetaService, we give a simple example here the service of which has the similar role as MetaService.

Most people have the experience to deal with personal or official business in a bank. Typically, each bank clerk would serve a queue. For a new customer, he/she must decide to join which queue by himself/herself. This mechanism cannot ensure that the first customer will always be served first (i.e., FCFS – First Come, First Served). The ranking machine is now introduced in most banks. The new working mechanism can be described as Fig. 6 which is in a services-oriented view. The bank customers can be seen abstractly as service consumers and the bank clerks as service providers. The ranking machine ranks the customers based on their order of arrival and inserts the customers into a central virtual queue. When any bank clerk becomes idle, he/she requests a new customer (task) from the head of the virtual queue. The ranking machine maintains a central virtual queue to implement a true FCFS discipline for all the customers.

It is obvious that the ranking machine has enabled a decision making which is fair to most or all of the customers without introducing significant changes to the bank clerks. With this example, we can see that a MetaService can integrate monitoring

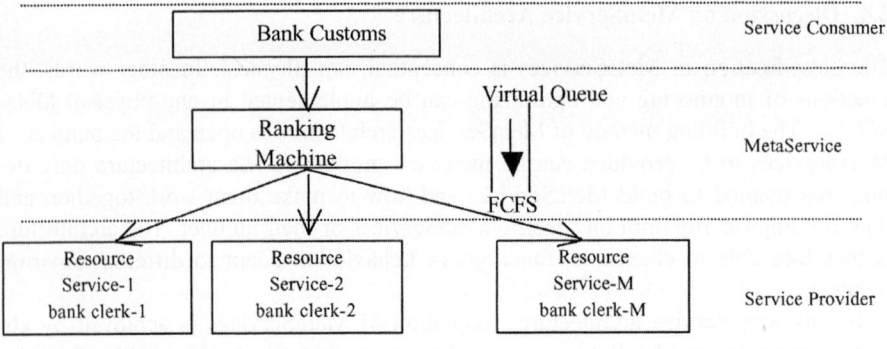

Fig. 6. Role of Ranking Machine from the view of MetaService.

and scheduling (optimizing) functions into one service, because the situation there can be as simple as and similar to this banking example.

4 Prototype Experiment

4.1 PC Grid Monitoring MetaService

The PC Grid monitoring MetaService is based on GT3[4]. This MetaService can obtain computing power information useful for subsequent optimizing and scheduling services. PC computing power contributors can test the service via the Internet. They register at a specific web address (http://166.111.68.187:8080/ csipgrid/ index.jsp), and then they can start downloading as clients. The client-side software can be installed in the Linux or Windows platform. The monitoring MetaService will collect information about the client's computer and store it into a database on the server. Fig. 7 is a snapshot of the monitoring MetaService. The scheduling and optimizing Meta-Services, which would try to provide cost-efficient computing power to grid applications based on the monitored information, are under development.

4.2 Contractor MetaService

Some applications want the grid system to provide specific QoS, for example, task ending time, computing power, bandwidth, etc. The concept of contractual computing [13] was introduced to provide such insured services in a grid environment. A Contractor MetaService is part of the design and implementation of our grid testbed. The Contractor MetaService first analyses the QoS requirement that has come from some upper layer, and then decides what kind of necessary resources and services are needed to be reserved; if the reservation requests are successful, the Contractor Meta-Service can schedule tasks onto the reserved resources and services to satisfy the request's QoS requirement. The algorithm has been tested on GridSim [14] and all the promises of the Contractor MetaService have shown to be realizable. The detailed design of the Contractor MetaService will be presented in a future paper.

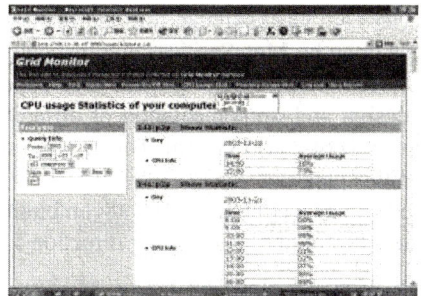 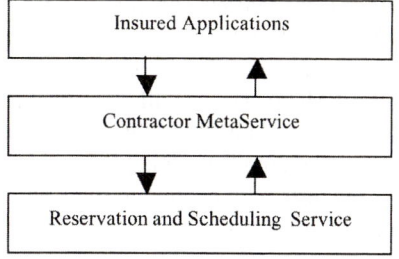

Fig. 7. A Snapshot of the monitoring Meta-Service.

Fig. 8. Contractor MetaService which can proide insured services.

5 Related Work

In IBM's autonomic computing [12], four important aspects, self-configuring, self-healing, self-optimizing and self-protecting are provided to manage a complex system. We can think that configuring, healing and protecting are some kind of adjusting, so the function of autonomic computing and that of our MetaServices architecture are closely related. The MetaServices architecture focuses on OGSA-based grid systems while autonomic computing is a more general method; so we can borrow the results of autonomic computing research and apply them to MetaServices.

The Reflection [5] mechanism has been applied to designing clever middleware components. Some functions provided by the reflection mechanism can be wrapped as monitoring MetaServices to watch the inner behavior of resource services. In fact, we can regard reflection as some kind of supporting technology for MetaServices.

There are other projects that have used the word metaservice or meta service (e.g., ICENI [9]), but their ideas are very different from that of ours. UDDI [8] Version 3 uses meta service for locating web services by enabling robust queries against rich metadata. DAPRA provides GMS (Grid Meta Service) [10] which provides a potential architecture addressing naming, white/yellow pages and directory services. It is obvious that GMS only addresses part of the problem which is being addressed by our MetaService architecture.

6 Conclusion

Although WSRF [11] is poised to replace OGSI completely in the future, the basic idea of OGSA will persist and the service-oriented architecture will become more and more important in building grid systems.

To absorb more resources into the grid system in the form of services, there will be much work of wrapping resource as services. With MetaServices, resource owners can easily and pleasantly provide their resources to the grid system, and they do not need to know how the users can use their resources efficiently and how the grid system will manage their resource services together with other services. For the grid

They require an access environment which would be much easier to use for non-advanced users. Furthermore, the portal and website operators require an easy procedure to provide the functionality of the grid services to the users of their respective portals and websites. This is why we introduce the PROGRESS Portlet Framework which allows to build reusable web components and co-operates with PROGRESS grid services to deliver a flexible environment that facilitates the process of enabling applications within grid portals and other websites. We present these solutions in this paper. First, we shortly review some of the known grid portals in section 2 and the PROGRESS grid-portal environment in section 3. The PROGRESS Portlet Framework is overviewed in section 4. Further on, in section 5, we discuss how this framework and other PROGRESS grid tools can be employed to create reusable components and facilitate the building of a specialized and user-friendly access environment to grid-enabled applications. Eventually, we draw some conclusions in section 6.

2 Grid Portals

There are numerous outstanding grid portal solutions deployed around the world. However, the architectures of many of them, unlike the architecture of the PROGRESS HPC Portal and a few others, do not facilitate the building of new portals, enabling new applications within the portals and do not make the overall portal workplace advanced and comfortable to use. While we take a closer look at the PROGRESS HPC Portal itself in sections 3, 4 and 5, let us investigate some of the other grid portals now.

To actually investigate the portals we must look for the tools utilized therewith. One of the best known grid portals is HotPage [6]. HotPage is an implementation of the Grid Portal Toolkit (GridPort) [7] infrastructure. GridPort version 3 comes as a collection of grid and web services and enables easy utilization of grid resources by multiple remote clients, for example portals, applications or other grid services. Such architecture of a grid-portal environment facilitates the building of new portals and is very close to the architecture of the PROGRESS grid-portal environment which is described in section 3.

GridPort version 3 replaces the collection of Perl modules which formed version 2 of the toolkit [8] that has been utilized by many production computing portals as it delivers a good solution to enable grid resources on the web. Another portal implemented in the Perl technology is the Legion Grid Portal [9]. The logic of LGP is a Perl CGI script, which is used to process most of the user requests. Its role is to issue Legion commands on behalf of the user. LGP may be deployed for interaction with any underlying grid infrastructure, for example Globus. It, however, cannot deliver distributed grid services.

There are a few other grid portals, like numerous deployments of the Grid Portal Development Kit [10], the portal delivered within the EnginFrame package [11], the portal from University of Lecce [12] or Sun's Grid Engine Portal [13]. These, however, are also directed towards delivering basic grid functionality, thus the portal workplaces created by these portals do not offer rich functionality and good flexibility.

We have presented a range of known computing portals herewith. Some of them provide a wide range of possibilities for the end user and make their portal workplace an excellent opportunity to process the research and scientific experiments on the grid. In the next three sections we show how the PROGRESS HPC Portal compares to these solutions and how it facilitates the work of grid users and grid portals operators.

3 PROGRESS Grid-Portal Environment

The architecture of the PROGRESS grid-portal environment, which is utilized by researchers from the bio-informatics and chemistry area and such applications as, for example, Assembling DNA Sequences [14] or Gaussian [15], is presented in Fig. 1. The PROGRESS grid computing resources are managed by a Globus Toolkit installation, and are delivered by the PROGRESS Grid Resource Broker (GRB) [16], which decides where and how to run the application associated with the job submitted by a user via the grid services grouped within the Grid Service Provider (GSP) [17, 18], based on the job requirements and the current status of the grid resources. The whole infrastructure is assisted by the Data Management System [18, 19], which is used as the source of input data and the destination for results of computing experiments performed in the PROGRESS grid. The GSP and DMS services are available for use within the PROGRESS HPC Portal.

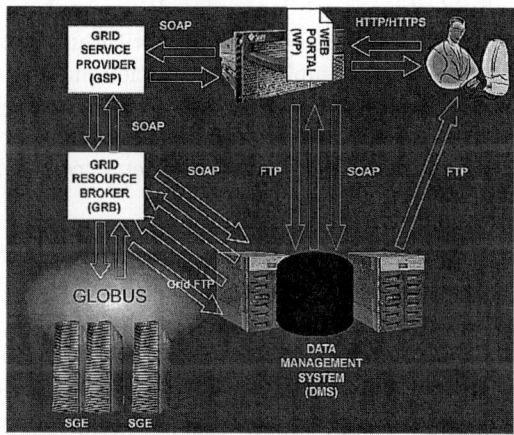

Fig. 1. PROGRESS grid-portal environment

In the PROGRESS grid-portal environment we assumed the SOAP protocol for communication between its distributed modules, thus all the modules provide Web Services interfaces. The richest Web Services interface is delivered by the GSP, which includes four independent services built using the J2EE technology. The two most important GSP services are the Job Submission and the

Application Management services. The first of these services delivers functions for building a computing job, submitting it to the grid for execution and for viewing its results; it is capable of maintaining both single-task and workflow grid jobs. The other one manages the PROGRESS application repository in a form of application descriptors, which include references to the application executables and lists of executable arguments, required environment variables and files. The main access point to the Data Management System is the Data Broker Service, which aims to deliver DMS resources and distributed modules under one external interface.

The Job Submission, Application Management and Data Broker services, which were discussed in detail in [17–19], and a few others of less importance to the grid nature of the portal, are utilized within the PROGRESS HPC Portal. The users of the portal can utilize the above-mentioned serviced and thus can manage their grid jobs, manage applications available within the application repository, and manage the data stored within the DMS.

In the testbed version of the PROGRESS HPC Portal this functionality has been provided to the portal users by a set of independent Java portlets. These were designed as separate modules, each implementing the whole stack enabling them to access the Web Services interfaces of the respective services. This has been changed for the most recent version of the PROGRESS HPC Portal and the above-mentioned portlets. The new portlets have been implemented with the use of the PROGRESS Portlet Framework and deployed for use within the PROGRESS HPC Portal; we describe the features of this framework in the next section. The use of this framework has not only enabled us to create highly reusable portlet codes, but has also provided us with an opportunity to easily reuse these parts of the already implemented portlets in new portlets that are responsible for communication with the underlying services and exposition of the obtained results as easily usable objects. This allows to easily create specialized portlets for specific applications which have been enabled on the grid; we show how this is done in section 5 of this paper.

4 PROGRESS Portlet Framework

The PROGRESS Portlet Framework features four layers in its architecture and forms a clear structure of an application accessing Web Services. The bottom layer consists of *Web Service Proxy* classes, which handle the SOAP communication with WS services. The next layer features *Request Handler* classes, which are capable of translating user's requests into a proper stream of invocations of WS methods. The third layer provides *Content Generator* classes, which prepare the content to be returned to the user. Finally, the highest layer includes *Provider* classes, which can take a form of portlets. The layered architecture of the PROGRESS Portlet Framework is presented in Fig. 2.

The Web Service Proxies extend a base proxy class, which utilizes the functionality of the Web Service Invocation Framework and uses the binding classes generated with the Axis toolkit based on the WSDL descriptions of services. The

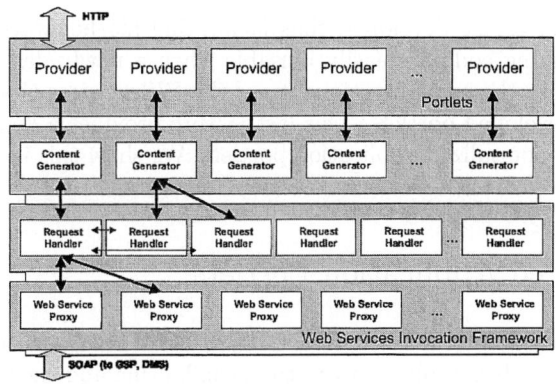

Fig. 2. Architecture of the PROGRESS Portlet Framework

Proxy classes are capable of communicating with the services using the SOAP protocol. They expose the *executeMethod()* method to enable the invocation of WS methods.

The Proxies' *executeMethod()* methods are invoked by the Request Handlers, which invoke proper WS methods based on user requests. One Request Handler method can invoke multiple WS methods. An invocation stream of Request Handler methods, which can be methods of one or of multiple Request Handlers, constitutes the action to be performed as stated in the user request. Each Http Request for portlet content contains the *action* and *page* parameters, which are analyzed by a Content Generator. Actions and pages are described in XML configuration files, in *actions.xml* and *pages.xml*, respectively. The configurations are dynamically loaded by classes referenced by Content Generators. This mechanism of dynamic load of action and page configurations allows to quickly add new pages to a portlet. When all methods necessary to deliver the content of a new page are available within the already existing Request Handlers, it is enough to add proper lines to the respective configuration files and the required actions are performed on the services, and the new page is generated on the basis of the responses returned by Request Handlers. There is no need to write any lines of Java code for such a page to be generated.

Content Generators are responsible for the preparation of the content returned by a portlet filling in a portal channel. They transform the data gathered by Request Handlers into HTML pages; in the PROGRESS HPC Portal we use the XSLT technology to build portlet contents. Each Content Generator is then utilized by exactly one Provider. Providers, which are usually portlets, can take a form of any class used to generate the content of a channel within a web portal for example the JSR 168 *Portlet* or the Sun's *ProfileProviderAdapter*, or even a generic *HttpServlet*. Providers are the top layer classes, which allow plugging the built web applications into a portal framework and the portal itself.

The layered architecture of the PROGRESS Portlet Framework enables flexible and easy development of user interfaces to programmatic Web Services. Its structure is clear to the developer, thus facilitating his/her work. The parts of the already existing code are easily reusable when a newly designed portlet is required to utilize the same functionality of the WS services that were utilized in previously developed portlets. This last feature provides a great opportunity to create highly specialized application portlets. We show this in practice in the next section.

5 Specialized Application Portlets

As it was already mentioned, the PROGRESS HPC Portal features three core portlets: "My computing jobs" for creating, configuring and executing grid jobs, "Applications" for managing the grid-enabled application, and "My data" for managing files stored within the DMS. These core portlets utilize the functionality of the GSP Job Submission and Application Management services and the DMS Data Broker.

The PROGRESS Portlet Framework has been used to rebuild these portlets to enable their reusability within other portals and websites. The other advantage is the opportunity to reuse some parts of the core portlet codes. Thus, the framework is also the environment used to create new portlets specialized to deliver an intuitive user interface for the management of grid jobs utilizing specific grid-enabled applications. Such specialized portlets carry a high value for the users of the grid portals. While the standard "My Computing Jobs" portlet allows to create and submit a grid job built on top of any available application, it is also quite complex for a non-advanced user. Most users require an easy to use and intuitive interface; otherwise, a new tool, even if it delivers a wide range of functionality, is not utilized and respected.

While developing the core portlets we have implemented all necessary classes to work within the four layers of the PROGRESS Portlet Framework. Therefore, when designing new portlets to provide user interfaces to specific grid-enabled applications, we could reuse the already existing code, because these new portlets utilize the same WS servicesas the three core portlets. The specialized application portlets are aimed to utilize the same grid infrastructure as the core portlets and to deliver a highly specialized interface to create, configure and execute grid jobs that use specific grid-enabled applications as opposed to the standard user interface available within the "My computing jobs" portlet, which allows to manage grid jobs based on any available grid application, yet it is not very intuitive for application users.

Thanks to the layered architecture of the PROGRESS Portlet Framework, a developer of a specialized application portlet, for example the developer of the "Gaussian" portlet, which enables the management of grid jobs utilizing the Gaussian chemistry application, could use the already existing Request Handlers to implement most of the actions performed by the new portlet. For example, the portlet can issue a request to list the content of a DMS directory or a request to retrieve the current status of a grid job. Both these requests can be handled by the Request Handlers, and the underlying Web Service Proxies implemented

Towards a Distributed Modeling and Simulation Environment for Networks

Yueming Lu, Hui Li, Aibo Liu, and Yuefeng Ji

The School of Telecommunications Engineering,
Beijing University of Posts and Telecommunications Beijing,
P. O. Box 128, 100876 Beijing, China
{ymlu,jyf}@bupt.edu.cn, {lihuibupt,aiboliu}@263.net

Abstract. Scalability and flexibility are key issues in the design of modeling and simulation infrastructure. In this paper, an infrastructure (called DMSE) which is dedicated to support various scalability and flexibility of modeling and simulation applications. An approach which would establish various virtual networks on the DMSE is presented. The purpose of the DMSE is to support various target networks in a common infrastructure. With help of the DBMS, the monitoring and management of the infrastructure create dynamic data in the node and view points. Thus, the infrastructure helps users to flexibly model and simulate various networks, dynamically deploy service and protocol to the node, and arbitrarily monitor and manage data from nodes.

1 Introduction

Most networks are developed with routers, optical or data switches that use routing protocols to dynamically find resource, use hardware or software to forward packets or switch data, and use signaling protocols to allocate resources [1]. Thus, the network works in a distributed environment and is difficult to be controlled.

Routing and signaling protocols are widely used in IP routers, ATM switches, and optical switches, but sometimes the protocols are not work properly in the cases of network congestion and failure, however. In practice, it's very difficult for engineers to find a failure in carries' working networks within a short time such as the failure interval per year limited by guaranteed SLA (service level agreement). Also, when carries want to deploy new services to the current working networks, they don't know how to combine new services to existing services and can't determine the quality of the new services. In addition, other issues may happen in the time of network optimization and extension.

The solution for those issues is to establish a modeling and simulation infrastructure that is equivalent to the working network (called target network). The infrastructure models and simulates failures, services and characterized features or performance which engineers or developers are interested in. The OPNET's Modeler and VINT's Ns-2 [2], VPI's Tranport Maker, and Ciena's MPS are famous modeling and simulation tools, but most of them run on a single computer. The distributed modeling and simulation systems (i.e. dedicated networks) are limited to be used within companies for special purposes and not available for open or commercial use. However, the distributed modeling environment is better to model interconnection, failure location and protocol testing, comparing with the system running in a single computer.

In this paper, we suggest a common modeling and simulation infrastructure called DMSE. The DMSE has three contributions. The first is to establish virtual networks on a fixed physical network. The second is to dynamically deploy services on the infrastructure. The third is an approach which monitors, controls and manages the infrastructure and protocol controllers.

2 Organization and Structure of the DMSE

The DMSE is designed to support modeling and simulation of various networks (i.e. IP, GMPLS [3], ASON (Automatically switched optical network)) [4] and various protocols (i.e. TCP/IP, OSPF [5], OSPF-TE [6], RSVP). However, networking, internetworking and protocols are working in terms of distributed computing environment. Previous efforts, such as NS, are based on a single computer and limited to central and single computing. In contrast, our infrastructure consists of multi computers which are identified as *programmable nodes*. The infrastructure enables the modeling and simulation of distributed services and protocols.

Organization and structure of the DMSE is depicted in the Fig. 1 (1).

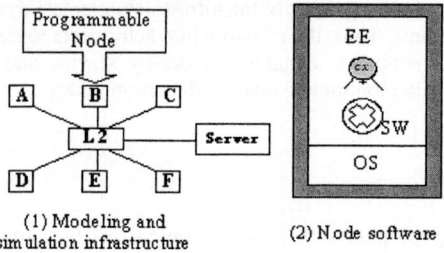

Fig. 1. Organization and structure of the DMSE.

- A,B,...,F are programmable nodes. The number of nodes is not limited to 6. In the center, a hub labeled with "L2" is an Ethernet Layer 2 switch or Ethernet bridge.
- Programmable nodes are connected in a shared Ethernet LAN (local area network). The LAN supports broadcasting, multicasting.
- The management server is also connecting to the Ethernet switch. The purpose of the server is to deploy *programmable service* and *protocol controllers* (called PC) to nodes, manage and control the PC, and monitor the *view points*.

Fig. 1 (2) presents the structure of the programmable node.

- The programmable node is formed by OS (operating system), SW (software switch), EE (execution environment) and CX (The PC NO x).
- The EE is the core of the node and loads, controls and executes PCs.
- The PC CX is a service or protocol controller and is loaded and created from the server.

3 Establishment and Maintenance of Virtual Networks

The big issue is how to dynamically model and simulate various networks, protocols and services, realize the user's various ideas on the DMSE, and meet various requirements which are

- Network node can be dynamically added and removed according to the idea of designer or user.
- Number and topology of target network can be also defined and modified.
- Protocols and services can be dynamically deployed.

Fortunately, our previous efforts focus on two parts: the programmable network and network design. The programmable network is *active network* first appeared in the DARPA and can inject mobile code to network nodes [7, 8]. Based on previous researches, the next study on the network modeling and simulation mostly shifts to distributed programmable network and *virtual network*.

- Based on the DMSE, we dynamically establish different virtual networks. Various virtual networks and applications are depicted in the Fig. 2(1), which shows the organization and structure of physical infrastructure arranged for modeling and simulation applications.
- *Target network modeling* (in Fig. 2 (2)): add and remove arbitrary virtual links between nodes and establish a virtual network which is equivalent to target network.
- *Discovery modeling* (in Fig. 2 (3)): Its purpose is to model the neighbor discovery in routing protocols like OSPF. The modeling only needs a node with neighbors. The infrastructure can establish any links between nodes.
- *Multi-network modeling* (in Fig. 2 (4)): Its purpose is to model more than one network on the DMSE at the same time.
- *Interconnection modeling* (in Fig. 2 (5)): Based on Fig. **2** (4), it also models the (traffic or signaling/routing protocols) interconnection between two networks.
- *Signaling/traffic modeling* (in Fig. 2 (6)): Model both the signaling (control plane) and traffics in a single virtual network. Multi-protocols are supported. Trace tools are available in the same time.

Virtual networks for target networks are dynamically established and some are dedicated to model one or more specified protocols or traffics, but how to establish a virtual network?

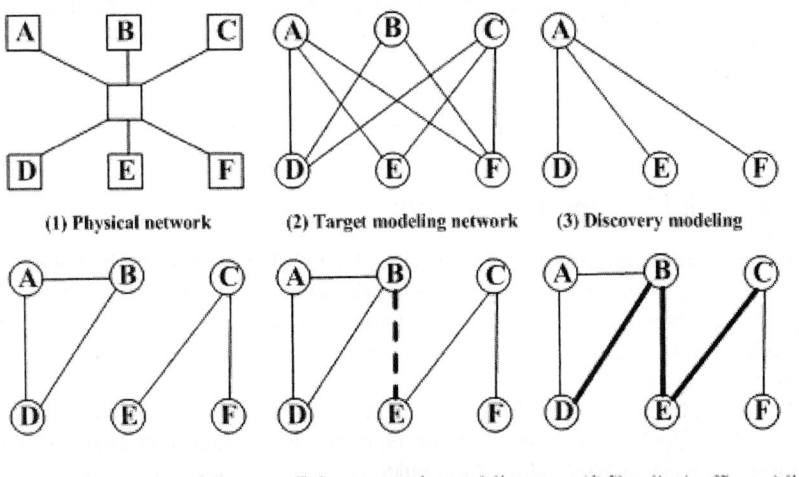

(1) Physical network (2) Target modeling network (3) Discovery modeling

(4) Multi-network modeling (5) Interconnection modeling (6) Signaling/traffic modeling

Fig. 2. Establishment and maintenance of virtual networks.

The prior experiments are conducted in a shared Ethernet network with one or more Layer 2 switches. Fig. 3 shows the establishment and maintenance of virtual links. The physical infrastructure is depicted in the Fig. 3 (1). The establishment and maintenance are processed as

- Setup Ethernet frame channel: The Ethernet frame channel (or tunnel) is a virtual link which is identified as source and destination Ethernet physical address in the Ethernet frame header.
- Maintain virtual links: The maintenance is to create records in the virtual link table at both end nodes of the virtual link. The virtual link table is depicted in the Fig. 3 (2) that records all virtual links connected to the node. It can also record the network topology if the routing protocols (i.e. OSPF) are activated.
- Maintain TE links: The virtual links may IP links which are consistent with the traffic links. However, sometimes the target network is a GMPLS network which has hundred virtual links between two nodes. The main idea is to create a TE link table in the node (depicted in the Fig. 3 (3)).

ID	name	Src address	Dst address	bandwidth
1	L01: A-B	MAC A	MAC B	10G
2	L02: B-A	MAC B	MAC A	10G
3	L03: A-B	MAC A	MAC B	2.5G
4	L04: B-A	MAC B	MAC A	2.5G
5	L05: A-D	MAC A	MAC D	1G
6	L06: D-A	MAC D	MAC A	1G
7	L07: B-D	MAC B	MAC D	10G
8	L08: D-B	MAC D	MAC B	10G

(1) Target network   (2) Link table

ID	name	Src address	Dst address	10G	2.5G	1G
1	TE01: A-B	MAC A	MAC B	1	1	0
2	TE02: B-A	MAC B	MAC A	1	1	0
3	TE03: A-D	MAC A	MAC D	0	0	1
4	TE04: D-A	MAC D	MAC A	0	0	1
5	TE05: B-D	MAC B	MAC D	1	0	0
6	TE06: D-B	MAC D	MAC B	1	0	0

(3) TE-link table

Fig. 3. Establishment and maintenance of virtual links.

4 Network Plan and Design Tool

Based on the previous establishment of virtual network, we focus on the network plan and design tool. The purpose of the network plan and design tool is to provide a graphic user interface (GUI) for users, connect to nodes' database, manage virtual link table in every node. Fig. 4 shows the network plan and design tool and the management process is.

- Network plan and design: Before modeling and simulation, user's idea must inject to a the virtual network. By network plan and design tool, user characterized the target network as nodes and virtual links.
- GUI: The GUI is a tool by which users can draw nodes and virtual links in a window. That is users can input their virtual network by mouse only.

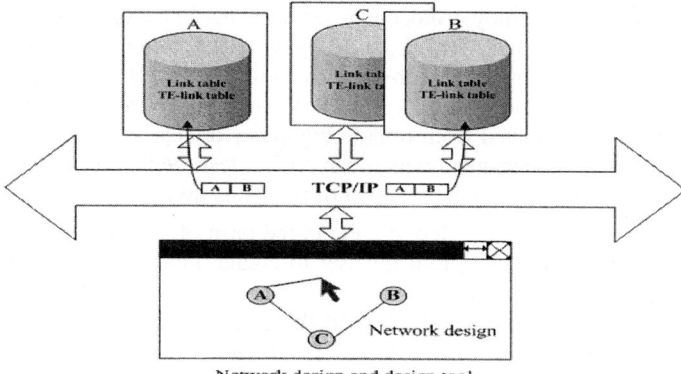

Fig. 4. Network plan and design tool and the management process.

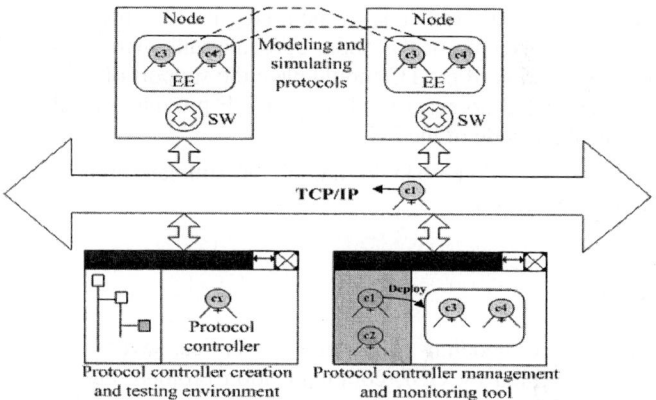

Fig. 5. Service deployment.

- Management: The management characterizes the virtual network, which users drew in the GUI window, into Ethernet frame channels and nodes. When the plan and design is finished, the tool connects to the corresponding node drawn in the GUI window and insert records to *virtual link table* or *TE link table*. That is the tool gives a message to node and tell the node how many links and how they connect to nodes.

5 Deployment of Programmable Protocol Controller

So far we concentrate on plan, design, establishment and maintenance of virtual network. Now we want to focus on the programmable nodes and service deployment of the infrastructure.

In regard to the service deployment, the DMSE consists of four parts: protocol controller (PC), protocol controller creation and testing environment (CTE), protocol controller management and monitoring tool (MMT), execution environment (EE). Fig. 5 depicts the DMSE's service deployment mechanism.

In order to dynamically deploy protocols or services, the PC, which would control the protocol or service of virtual network, is designed as mobile code (or mobile agent). That is the PC can move from one computer to another, execute in specified node and deal with dedicated protocol or service. Furthermore, the PC may not be created in the node, but can be downloaded from remote computer or servers as a file. Hence, we provide a deployment mechanism which can create, test, deploy and execute the PC. According to the sequence, the deployment of the PC is divided into 3 steps which are

- Creation and testing: In this step, the PC in the term of a file is created and tested in the CTE which provide a tool for user to create, modify, compile and test the PC.
- Deployment: The deployment is to send the dedicated PC to specified node via TCP/IP protocol. But how to transfer the PC to the specified node? We developed the MMT which is a tool enable user dynamically control the deployment of the PC in a GUI. However the TCP/IP will transfer the PC to the destination according to management and control mechanism.
- Execution: Same as the fixed code in the node, the PC downloaded from remote computer can also execute in the node, combine and share resources with fixed code. But the execution of the PC should have the support of the EE. Since the PC is dynamical code and no security check, the EE provide a mechanism to check, link, and control the code of the PC.

After the 3 steps, the PC can handle, model and simulate protocol or service, and communicate with other PCs.

6 Monitoring

In order to monitor the modeling of protocol or service, several fundament requirements have to be fulfilled by GUI, communication protocol and data organization. Normally, some sophisticated network management system should also have a performance monitoring tool. However, network management system is fixed software and nodes have to compile and maintain MIB when the PC is dynamically inserted into or removed from the node. In regard to communication protocol, the CORBA can dynamically integrated management objects from node. But it's difficult for users to develop and implement modules both in node and management system side.

In our case, we invite the DBMS (database management system) in the node side and SQL language at the server side. By SQL language, the management system dynamically creates tables or management objects in the node and establishes *view point* which shows data from a characterized performance parameter or a table in the node. The monitoring processing depicted in the Fig. 6 is

- Create a view point: First, users can define a view data or table that is concerned (i.e. the interval to find all neighbors). At last, the definition changes to SQL sentences after users submit their requirements to the *text and tree viewer* and *visual network monitor tool*.
- Create table in the node: According to the SQL sentences, the *text and tree viewer* and *visual network monitor tool* submit their requests to the DBMS in the specified node. By the DBMS, both the PC and the management system share, control and manage (read and write) data in the table.

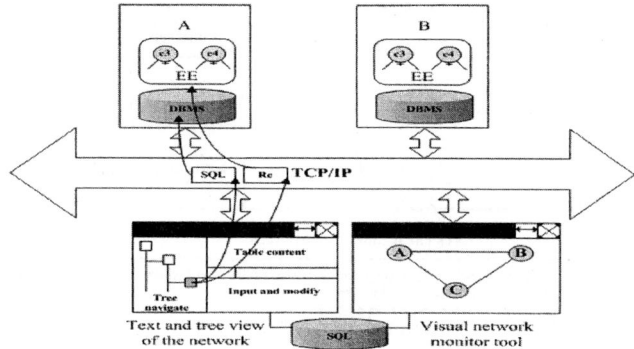

Fig. 6. Service monitoring.

- Configuration: In order to get a high performance, when the management system want to change some data in the PC, it first submits SQL sentence (labeled "SQL" in the packet) to the DBMS that changes the data in the table shared with specified PC, then sends a refresh message (labeled "Re" in the packet) to notify that PC to get data from the shared table.

7 Application Scenarios

In order to demonstrate benefits of the DMSE, we select the "hello protocol" of automatically neighbor discovery as a modeling service. The hello protocol enables node to find and remove its neighbors in fixed duration. The purpose of the scenarios is to determine how many times to send a hello packet in one minute, an how long to remove an active neighbor when the neighbor is down. Hence, we characterize the hello protocol into three view points (three parameter: hello_interval, hello_dead_interval and hello_dead_found).

Regarding the PC, the design of hello protocol is taken the flexibility and mobility into consideration. The PC for Hello Protocol and The SQL sentences for view points are illustrated in Fig. 7:

- Create the PC for Hello Protocol: As the Hello protocol works, the corresponding PC also have six or more states (i.e. Down, Init, ExState, Exchange, Loading, Full) and Hello protocol data (depicted in Fig. 7 (1)).
- Setup view point: Create table (named "hello" in this case) in the node, where the PC executes and accesses the data in the table (depicted in Fig. 7 (1)).
- Create SQL sentences: The purpose of SQL sentences (depicted in Fig. 7 (2)) is to get and set data in the "hello" table. That is data in the PC can be monitored and configured.

8 Related Work

Currently, our efforts is to support modeling and simulating routing (i.e. OSPF) and signaling (i.e. RSVP) protocols in various networks, such as MPLS-based networks and ASON (Automatically switched optical network). The networks are characterized as small networks, huge networks, hierarchical networks and various topologies. The

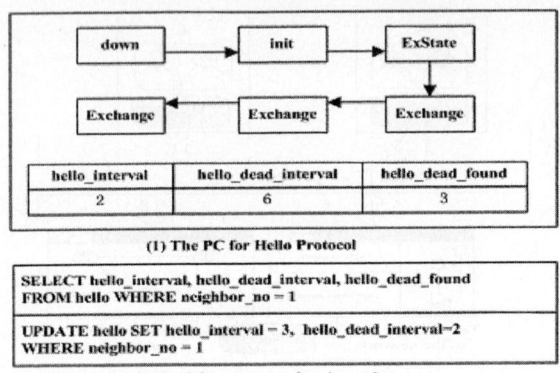

Fig. 7. The PC for Hello Protocol and the SQL sentences for view points.

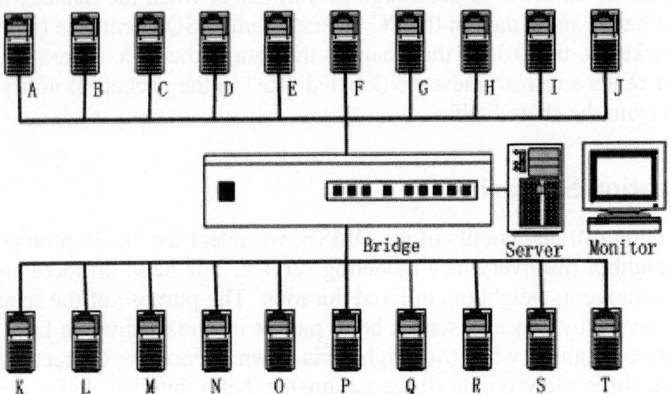

Fig. 8. The physical network of the DMSE.

protocols are extended to support various special usages which may not be standardized (i.e. the routing protocol for GMPLS and ASON). The scenarios of modeling and simulation is not fixed, thus the infrastructure must support various networks and various protocols.

The DMSE consists 6 nodes (6 high performance computers)and a server for monitoring and a Ethernet bridge, and currently is extending to more nodes(depicted in Fig. 8). Nodes of the DMSE share Ethernet frames and Ethernet frame channels (virtual links) are established dynamically.

Software in server is classified into 3 kinds: plan and design tool, PC tool, and monitoring tool.

– The purpose of the PC tool is to create, test and deploy the service and protocol controller (PC).
– The monitoring tool is designed to define view data, view point and SQL sentences.
– The plan and design tool (depicted in Fig. 9) is a GUI that gives a interface to input network topology (virtual network) by mouse.

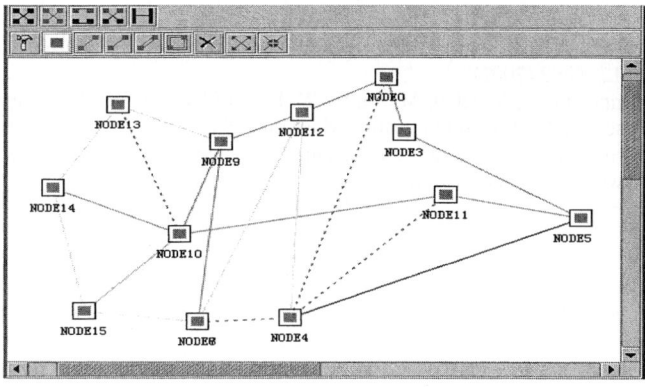

Fig. 9. Network plan and design tool.

9 Conclusion

The DMSE is designed to support dynamically planning, design and establishment of virtual networks which are equivalent to target networks, deployment of the PCs, and monitoring of the infrastructure. Hence, the features of the DMSE are flexible, scalable and dynamic. Regarding the application, the DMSE supports various networks (IP, MPLS and ASON) , various equipments (routers, switches and OXCs), various protocols ("Hello" protocol, OSPF, signaling, traffic or other standardized or unstandardized protocols) and various topologies (ring, mesh, huge or small). The benefit of the DMSE is that different modeling and simulation applications can be deployed in a common infrastructure regardless of the topologies, equipments, protocols and networks. The DMSE can help researchers to study on their new protocols or prototype of new networks, vendors to develop their new equipments, and carries to locate and fix failures in working network.

Acknowledgements

The research work is supported by the National Outstanding Youth Foundation (No. 60325104), National Natural Science Foundation (No. 90104017), National 863 Project (No. 2003AA122520), the Grand Key Science and Technology Research Program of MOE(No.0215), P. R. China. Thanks for the great help.

References

1. E. Rosen, A. Viswanathan, R. Callon: Multiprotocol Label Switching Architecture, IETF RFC 3031 (2001)
2. Flores-Lucio G., Paredes-Ferrare M., Jammeh E., Fleury M. and Reed M.: OPNET-Modeler and NS-2: Comparing the Accuracy of Network Simulators for Packet-Level Analysis using a Network Testbed. In: 3rd WEAS Int. Conf. on Simulation, Modelling and Optimization (ICOSMO 2003), Crete, Vol. 2 (2003) 700-707
3. Ashwood-Smith P. , et al: Generalized Multi-Protocol Label Switching (GMPLS) Architecture, Internet Draft, draft-ietf-ccamp-gmpls-architecture-01.txt(2001)
4. Architecture for the automatically switched optical network (ASON), ITU-T G.8080/Y.1304 Nov. (2001)

5. J. Moy: OSPF Version 2, RFC 2328, Standard (1998)
6. Greg M.Brnstein, Vishal Sharma, et al: Interdomain optical routing, Journal of Optical Networking, Vol.1, No.2 (2002)
7. David L. Tennenhouse, Jonathan M. Smith, W. David Sincoskie, et al: A Survey of Active Network Research, IEEE Communication Magazine(1997) 80-86
8. Active Networks Working Group: Architectural Framework for Active Networks Version 0.9, http://www.cc.gatech.edu/projects/canes/arch/arch-0-9.ps (1998)

Approximate Performance Analysis of Web Services Flow Using Stochastic Petri Net

Zhangxi Tan[1], Chuang Lin[1], Hao Yin[1], Ye Hong[1], and Guangxi Zhu[2]

[1] Department of Computer Science and Technology, Tsinghua University, China
{xtan,clin,hyin,yhong}@csnet1.cs.tsinguha.edu.cn
[2] Department of Electronic & Information Engineering, Huazhong University of Sci. & Tech.
gxzhu@mail.hust.edu.cn

Abstract. Web services provide a language-neutral, loosely-coupled, and platform-independent way for linking applications within organizations or enterprises across the Internet. In such a scenario, quantitative characteristics such as service execution throughput should be evaluated to measure the system performance. Usually, the first step is to define an abstract workflow model, for example Stochastic Petri Net Models. However, large system always raises the problem of state explosion. In this paper, we discuss a set of simplification rules for five basic structures of web service flow: sequential, parallel, conditional, loop and mutex. Our approach can effectively reduce the state space and is applied to the performance analysis of a web service flow management system.

1 Introduction

Services such as online purchasing, information querying or management etc, can be provided on the Internet using web services technology, which offer a conceptual foundation and a technology infrastructure for service-oriented computing [1]. It allows programs written in different languages on different platforms communicate with each other in standard-based way [2]. Besides, web services are considered as reusable software components [3] and can be integrated into every software system that is web service-aware [4].

With the rapid evolving and development of web services, web services flow includes more execution activities and interacts with activities of other web services on remote sites. To compute performance metrics for a given web services flow, many existing tools like Petri Nets [5] and newly developed Web Service Flow Language (WSFL) by IBM [6] can be employed for abstract modeling. For the sake of fast computation time, many researchers [2, 7] prefer Petri Nets, since they are well suited for capturing flows in web services, for modeling the distributed nature of web services, for representing methods in a web service and for reasoning about the correctness of the flows. In this paper, we model web service flows using Stochastic Petri Net (SPN), an extended version of Petri net. However, the state space explosion, which entails the solution of the underlying Markov Chain (MC) with a large number of states, limits the applicability of Petri Nets based methods for analyzing large-scale systems. Unfortunately, relative small efforts are put on this problem while model abstraction and hierarchical model merging have been deliberated discussed.

To solve this issue, we have developed a set of reduction techniques based on numeric compression and decomposition techniques, which focus on a set of five basic structures in web service flow: sequential, parallel, conditional, loop and mutex. The rest of this paper is organized as follows. Section 2 gives some preliminary definitions on Stochastic Petri Net modeling of web services. Section 3 gives four performance equivalent formulae for the first four structures. Section 4 introduces a new decomposition strategy for the fifth structure and a solution methodology for the entire model. Section 5 presents an illustrative example. Section 6 concludes the paper.

2 Stochastic Petri Net Modeling of Web Services

2.1 Web Services

Web services are general-purpose, standards-based, federated and modular. Web services provide a synchronous or asynchronous messaging environment that is built on the multi-layer XML foundations. The goal of web services is to offer a platform for building distributed applications, such as reliable messaging and business transactions. Web services enable XML messages to move data between heterogeneous systems. XML Schema (XSD) defines the message format which describes the type and structure of XML documents. Web Services Description Language (WSDL) explains how operations can be invoked using particular transport protocol bindings. Universal Description, Discovery and Integration (UDDI) provides a mechanism for clients to dynamically find other web services. XML message are encapsulated using lightweight XML based protocols like Simple Object Access Protocol (SOAP) for exchange of information in a decentralized distributed environment. In addition, web services cooperate with a collection of standards and protocols that allow us to make processing requests to remote systems using existing transport protocols (e.g. HTTP, SMTP).

2.2 Stochastic Petri Net Modeling

Definition 2.1 Petri Nets and Stochastic Petri Nets: A *Petri Net* (PN) is a quadruple (P, T, F, M_0). $P = \{p_1, p_2, \ldots, p_k\}$ is a finite set of places. $T = \{t_1, t_2, \ldots, t_M\}$ is a finite set of transitions ($P \cap T = \varnothing$). $F \subseteq (P \times T) \cup (T \times P)$ is a set of arcs. $M_0 = \{m_{01}, m_{02}, \ldots, m_{0k}\}$ is an initial marking. The sets of input and output places for transition t are denoted by *t and t*, respectively. Input and output transitions for place p are denoted by *p and p*. A *Stochastic Petri Net* (SPN) is obtained from a PN by associating a set of transition firing rate $\lambda = \{\lambda_1, \lambda_2, \ldots, \lambda_m\}$ (with exponentially distributed transition times) to transitions. In a *Generalized Stochastic Petri Net* (GSPN), we differentiate timed and immediate transitions. The immediate transitions are assigned firing probabilities and timed transitions are assigned firing rates. The GSPN definition has been extended in some literatures [9, 10] by adding priorities for all transitions and inhibitor arcs. The models used in this paper are based on GSPN with these extended definitions, but simply we use term SPN instead.

2.3 Stochastic Petri Net Modeling

The first step to set up a model is to map the WSDL file to SPN as follows:

a) portType, data storage -> places
b) Service-Port (Name, Binding name, Location) -> immediate transitions

c) Computational primitives -> timed transitions
d) Message data -> Tokens
e) Binding (portType, Protocol) -> arcs

3 Performance Analysis of the First Four Structures

To give general modeling and simplification techniques for the first four structures mentioned in Section 1, we exploit the definition of *subnets* as follows:

Definition 3.1 Subnets [8]: Let $N' = (P', T', F')$ and $N = (P, T, F)$ be two nets. N' is a subnet of N, $N' \subseteq N$, iff $P' \subseteq P$, $T' \subseteq T$, $F' \subseteq F \cap ((P' \times T') \cap (T' \times P'))$.

We model these structures as *subnets* and present time performance equivalent formulae that transform a *subnet* into a transition with an expected delay time equivalent to the delay time of the original *subnet*.

Sequential Structure

SPN Model and Simplification: All transitions in a *subnet* fire sequentially, one after another. We call it a *sequential subnet*, shown in Fig. 1.

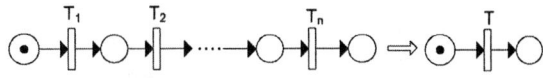

Fig. 1. A *sequential subnet* and its simplification.

Parallel Structure

SPN Model and Simplification: When every transition in a *subnet* can fire at the same time or in any order we have a *parallel subnet*. The *parallel subnet* and its performance equivalent transition are shown in Fig. 2.

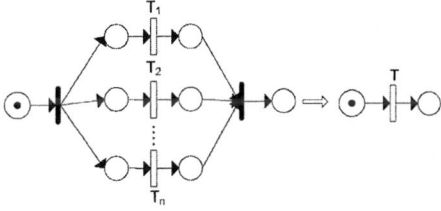

Fig. 2. A *parallel subnet* and its simplification.

Conditional Structure

SPN Model and Simplification: We identify a *conditional subnet* when every transition in a *subnet* is enabled at some time, but only one transition can fire at that time. The *conditional subnet* and its performance equivalent transition are shown in Fig. 3.

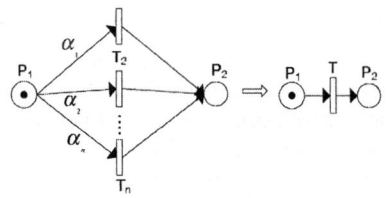

Fig. 3. A *conditional subnet* and its simplification.

Loop Structure

SPN Model and Simplification: We identify a *loop subnet* when a transition in a *subnet* fires repeatedly. The *loop subnet* and its performance equivalent transition are shown in Fig. 4.

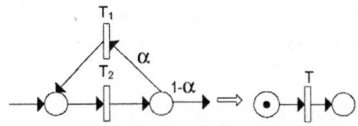

Fig. 4. A *loop subnet* and its simplification.

4 Approximate Analysis of the Mutex Structure (The Fifth Structure)

4.1 Mutex Structure and Its SPN Model

Description: A set of activities share common resources and execute in their own order, which is decided at run-time. No two activities can access the identical shared resource at the same moment.

SPN Model
A SPN model with mutex structures is composed of M *subnets*, which model parallel web service flows. These *subnets* are structurally connected through places with one initial token in each that model shared mutex resources, shown in Fig. 5. Also, the token can 'travel' through these *subnets*. We call these places 'mutex places' and their tokens 'mutex token'. Obviously, at any time only one *subnet* can consume the mutex token (access the exclusive shared resource).

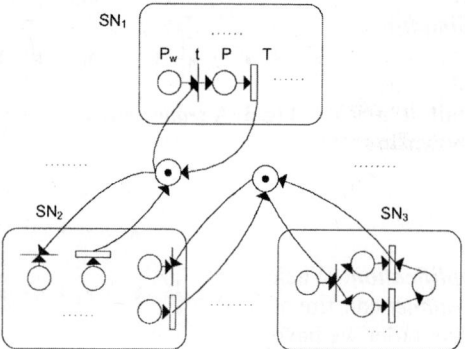

Fig. 5. The SPN model of mutex structure.

4.2 Equivalent Complementary Nets and the Decomposition Strategy

Analyzing mutex structures is the crucial one in state space reduction. In many cases, we need to employ the 'divide and conquer' principle and an auxiliary structure, namely *Equivalent Complementary Net*. Thus, the solution of mutex structures is an iterative approximate process. First, we describe our decomposition strategy and give the definition of *Equivalent Complementary Nets*. Like other 'divide and conquer' techniques, our scheme also needs to identify *subnets*, while this is straightforward for mutex and other resource sharing structures where *subnets* are clearly separated by shared mutex places. After identifying *subnets*, *complements of* these *subnets* are acquired according to the following definition:

Definition 4.1 Complement of a Subnet (CN): Let N' and N be defined as Definition 3.1. Then N^c is obtained by deleting all places and transitions which are simultaneously in N' and N. N^c is called *complement of* N'.

Definition 4.2 Equivalent Complementary Net of a Subnet (ECN): Let SN_i be a *subnet*, CN_i be the *complement of* SN_i and $Token_m$ be the mutex token. ECN_i, an *Equivalent Complementary Net* of SN_i is an aggregated version of CN_i, which guarantees the following equivalent relations: 1) The probability of mutex token $Token_m$ in ECN_i is equal to that of $Token_m$ in CN_i. 2) The throughput of mutex token $Token_m$ traveling through ECN_i is equal to that of $Token_m$ traveling through CN_i.

Definition 4.3 Frame Network (FN): Let SN_i and ECN_i be defined as above. Together they compose the Frame Network (FN_i).

4.3 Parameter Tuning of Equivalent Complementary Nets and the Iterative Solution

During the iterative solution, we should adjust parameters of each ECN to satisfy Definition 4.2. In SN_i of Fig. 6, place $P_{s_{ik}}$ is the kth serving place related with the mutex place P_{mutex} and λ_{ik} is the firing rate of *serving transition* T_{ik}. In ECN_i, $P_{s_i}^c$ and $P_{w_i}^c$ are respectively the *aggregated serving* and *waiting place*. We define $T_{a_i}^c$ of ECN_i the *aggregated arriving transition* which is enabled when there is no token in $P_{w_i}^c$. The probability of conflict immediate transition t_i^c is defined as the maximum probability that CN_i will consume the mutex token when conflicts with SN_j:

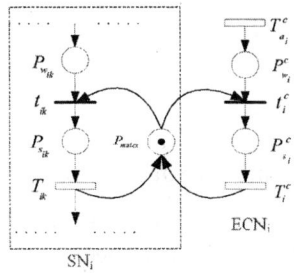

Fig. 6. Sample of frame network.

$$\text{Prob}(t_i^c) = \max\{\sum_{j \neq i} \sum_{t_{jk} \text{ is enabled}} \text{Prob}(t_{jk})\} \quad (1)$$

According to Definition 4.2, in ECN_i it requires the nonempty probability of *resource serving place* $P_{s_i}^c$ and throughput of the mutex token as follows:

$$\text{Prob}(\#(P_{s_i}^c) = 1) = \sum_{j \neq i} \sum_{k} \text{Prob}(\#(P_{s_{jk}}) = 1) \quad (2)$$

$$\text{Throughput}(T_i^c) = \sum_{j \neq i} \sum_{k} \text{Throughput}(T_{jk})$$
$$= \sum_{j \neq i} \sum_{k} \lambda_{jk} \times \text{Prob}(\#(P_{s_{jk}}) = 1) \quad (3)$$

In equation (2)(3), λ_{jk} is the firing rate of *serving transition* T_{jk} of SN_j in CN_i. Similarly, $P_{s_{jk}}$ is the *serving place* of SN_j. During the iterative solution, the probability $\text{Prob}(\#(P_{s_{jk}}) = 1)$ is acquired by solving *Frame Network* FN_j.

Thus, we get the firing rate of T_i^c:

$$\lambda_i^c = \frac{\text{Throughput}(T_i^c)}{\text{Prob}(\#(P_{s_i}^c) = 1)} = \frac{\sum_{j \neq i} \sum_{k} \text{Throughput}(T_{jk})}{\sum_{j \neq i} \sum_{k} \text{Prob}(\#(P_{s_{jk}}) = 1)} \quad (4)$$

The overall iterative algorithm solving the mutex structure is described below (the superscript of the variable indicates in which round of the iteration the result has been got):

Algorithm 4.1 Solving the mutex structure with Equivalent Complementary Nets: *Precompress the first four structures using and decompose the model according to Section 4.2. Then, order all subnets {SN_i}, equivalent complementary networks {ECN_i} and frame networks {FN_i} from i=1 to L.*

Choose proper initial probabilities $\text{Prob}(\#(P^c_{s_i}) = 1)^{(0)}$ *and calculate the firing probability of* t^c_i *for each* ECN_i *as equation (1).*

    m=1;
    do
    for i= 1 **to** L **do**
        Solve FN_i using **Algorithm 4.1** and acquire Throughput$(T_{ik})^{(m)}$ and $\text{Prob}(\#(P_{s_{ik}}) = 1)^{(m)}$ for SN_i.
    end for
    $m = m+1$
    until *(The accuracy meet the stopping criterion.)*

Algorithm 4.1 can take the advantage of the fact that the structures of SN_i, ECN_i, and FN_i are all unchanged. As a result, the build of reachability graph and elimination of vanish states do not need to be repeated.

5 An Illustrative Example

To demonstrate our approximate analysis methods, we exploit a sample web services system, which is composed of the five basic structures described above. The web services concern a company's human resource database management system, shown in Fig. 7. Two types of clients perform remote database update by calling web services. They can select different port numbers (80 for http, 443 for https) to invoke different access methods and perform different database operations.

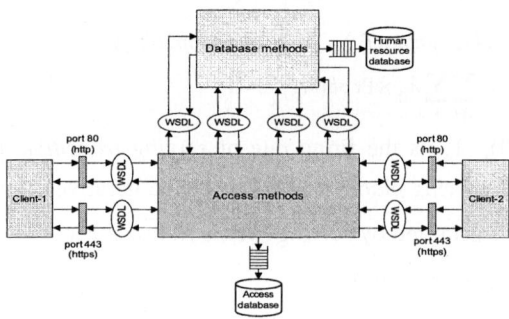

Fig. 7. Flow model of the sample web service management system.

In our experiment, we assign conflict immediate transitions of one mutex place the same priority. The firing rates of transitions are all exponentially distributed random variables. Both complete and decomposed SPN models used in this article are solved with direct Markovian analysis by SPNP 6.0 [10]. Observed in Fig. 8, our simplifica-

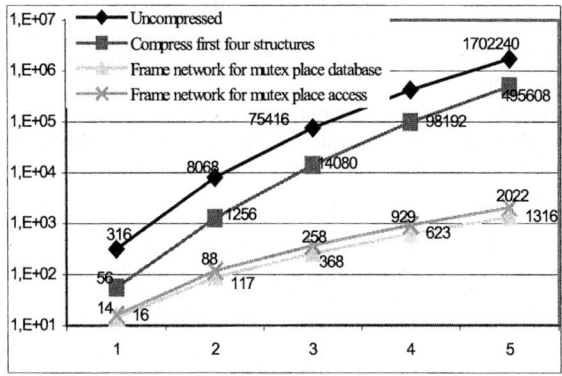

Fig. 8. Comparison of state spaces between original model and simplified model.

tion techniques effectively reduce the state space, especially the decomposition of mutex structures. In all the models we have tested, the relative error remain quite low (<3%). Moreover, the approximate results outputted from two separated *frame networks* are well converged and nearly identical.

6 Conclusions

In this paper, we identify five basic structures when modeling web services flows. We also investigate their performance analysis techniques. Our method is an iterative approximate process and its advantage is twofold:

1. It decomposes the whole model by mutex places 'cleanly' and greatly reduces the state space. The state complexity of each *frame net* is a linear value in the number of its related *subnet*.
2. In examples we have experimented, approximate results are quite near the exact ones.

Acknowledgment

This research was supported by Intel IXA University Research Plan (No.9077), the Natural Science Foundation of China (No.90104002, 60173012, 60273009 and 60372019), NSFC and RGC (No.60218003), the Projects of Development Plan of the State Key Fundamental Research (No.G1999032707 and 2003CB314804) and China Postdoctoral Science Foundation (No.2003034152).

References

1. Web Services Architecture Tam. "Web services architecture overview. The next stage of evolution for e-business", IBM Technical Document, Web Architecture Library, 2000
2. J. P. Thomas, M. Thomas and G. Ghinea, "Modeling of Web Services Flow", Proceedings of IEEE International Conference on E-Commerce 2003, pp. 391-398: June, 2003
3. Szyperski, C., Component Software, Addison Wesley, 1988.

4. Rettberg, A., Thronicke, W., "Embedded system design based on web services", Proceedings of Design, Automation and Test in Europe Conference and Exhibition, 2002.
5. J.L. Peterson, "Petri Net Theory and the Modeling of Systems", Englewood Cliffs, NJ: Prentice-Hall, Inc, 1981
6. F. Leymann, "Web Services Flow Language 1.0", IBM Software Group, 2001.
7. Hamadi, R. and Benatallah, B. "A Petri Net-based Model for Web Service Composition" Pro-ceedings of the Fourteenth Australasian Database Conference (ADC2003), Adelaide, Australia, pp. 191-200: 2003.
8. J. M. Blake, K. S. Trivedi, "Reliability analysis of interconnection networks using hierarchical composition," IEEE Trans. Reliability, vol. 38, pp. 111-120, Jan. 1989.
9. G. Chiola, M.A. Marsan, G. Balbo, G. Conte, "Generalized Stochastic Petri Nets: A Definition at the Net Level and its Implications", IEEE Transactions on Software Engineering, 19(2):89-107, February 1993.
10. G. Ciardo, J. K. Muppala, K. S. Trivedi, "SPNP: Stochastic Petri Net Package", Proceedings of 3rd International Workshop on Petri Nets and Performance Models, pp142-150, Kyoto, Japan, Dec.1989

Complexity Analysis of Load Balance Problem for Synchronous Iterative Applications

Weizhe Zhang and Mingzeng Hu

School of Computer Science and Technology, Harbin Institute of Technology, P. R. China
zwz@pact518.hit.edu.cn,mzhu@hit.edu.cn
http://pact518.hit.edu.cn/index.html

Abstract. Load balance technologies for tightly coupled applications on the large scale heterogeneous cluster systems are paid more and more attentions nowadays with the emergence of the Meta computing and Grid computing environment. Focusing on a class of representative tightly coupled applications, synchronous iterative applications, we formulate their load balance problem into a combination optimization problem. Meanwhile, we establish a complexity result that accesses the difficulty of the problem. The theory analysis result will do great help for the design of approximate algorithms.

1 Introduction

More and more large-scale grand applications today make use of heterogeneous computing systems to obtain the cycles they need. Currently, heterogeneous computing systems are classified by power-level as computing grid, Meta-computing system and heterogeneous cluster system (HCS) [1]. Computing grid [2] links the most powerful supercomputers of the largest supercomputer computing centers through dedicated high-speed networks. Meta-computing system [3] connects medium-size servers through fast but non-dedicated links. However, heterogeneous cluster system, consisting of network of workstations or PCs with the highest performance-to-price ratio, is most widely adopted by university departments and companies.

The major limitation to parallelize grand applications on the HCS arises from the difficulty of balancing the load using different speed processors and heterogeneous networks. It is convenient to classify classic parallel problems into embarrassingly parallel, synchronous, loosely synchronous, asynchronous, and meta problems [4]. The class of embarrassingly parallel problems requires no communication or synchronization and loosely synchronous applications are data parallel but data points evolve differently. Almost all kinds of the traditional load balancing strategies (such as domain decomposition, recursive bisection and round robin algorithm) can be brought into play for such applications on HCS because there is no or little interprocess communication. However, fully synchronous applications with higher communication and computation ratio require all the operations are synchronized at regular points and computation/communication capabilities of each resource must be taken into account at one time. Therefore, it is an actual challenge to balancing the load for this class of tightly coupled applications on HCS. Moreover, asynchronous and meta problems have two or more subtasks; each of the subtasks belong to one of the above categories, which demands more sophisticated meta scheduling strategies [5, 6] that beyond this article.

We focus on the important subclass of fully synchronous applications – synchronous iterative (or multiphase) applications, which encompass elliptic PDE solvers [7], finite element equations solvers [8], atmosphere and ocean circulation simulation, heat distribution problem and cellar automata [9]. In such applications, parallel iterative algorithms repeatedly execute a computation on a large collection of application data, with an explicit synchronization of the tasks and exchange of data performed at the end of each iteration step. Each processor reaches a barrier synchronization after every iteration and the next iteration cannot begin until all processes have finished the previous iteration. Several authors have derived performance models and load balance strategies for synchronous iterative algorithms running on homogeneous resources [10-12]. However, the study of load balance strategies for the use of heterogeneous resources has been sparse. In this paper, we develop a performance modeling for synchronous iterative applications executing on shared, heterogeneous resources. Also, we prove the NP-complexity of full processor load balance problem of synchronous iterative applications and reduce the FPSILBP problem into Traveling Salesman Problem.

The rest of this paper is organized as follows. In Section 2, the system, task and performance model are presented. We formally state the combination optimization problem in Section 3. We show in Section 4 the decision problem is NP-complete and TSP based heuristic algorithm may perform well. The paper ends with a brief conclusion.

2 Problem Description

2.1 Heterogeneous Computing System Model

The HCS is represented by a completely connected, undirected graph $G = (V, E)$, where set V denotes heterogeneous machines and set E denotes communication link. Let $v_i \in V$ denotes a machine and $e_{i,j} \in E$ denotes the link between machines v_i and v_j, where $1 \leq i,j \leq |V|$. The heterogeneity of v_i and $e_{i,j}$ is depicted by a six-tuple (α_i, β_i, δ_i, ξ_i, $\omega_{i,j}$, $\varphi_{i,j}$), which are defined as follows:

1. α_i represents the processor number of machine v_i.
2. β_i represents the weighted factor of processor, which are determined by the processor architecture and type of applications.
3. δ_i represents the load status (e.g. CPU utilization rate) of processors.
4. ξ_i represents the computation startup time of processors.
5. $\omega_{i,j}$ represents the transfer rate between v_i and v_j. The symmetric cost ($\omega_{i,j} = \omega_{j,i}$) is assumed in most situation.
6. $\varphi_{i,j}$ represents the communication latency time between v_i and v_j.

2.2 Task Model

The core computation of a synchronous iterative application consists of a sequence of iterations with a set of variables being evaluated at each iteration step as a function of some or all of the values from previous iterations. The variables are distributed over heterogeneous processors, with each processor being responsible for evaluating the variables allocated to it. All processors must complete an iteration step and exchange variable values before the next iteration can begin. A generic abstract model is shown as follow:

```
partition( )                      /*data decomposition*/
for(k=0 ; k<n ; j++)              /*for each synchronous
                                    iteration*/
   forall(i=0 ; i<v ; i++)        /*|V| processes execute in
                                    parallel*/
   { compute(u_i)                 /*compute local data u_i*/
     update(u_i)                  /*communication and update
                                    data u_i*/

     barrier(mygroup)             /*explicit or implicit
                                    synchronization*/
}
```

Data decomposition phrase, which plays the key role for load balance in the HCS, determines how to divide the problem data into different parts. Let U be the total size of the problem data to be performed at each step of the algorithm. Processor V_i will accomplish a share $u_i = \lambda_i U$ of this total work, where $\lambda_i \geq 0$ for $1 \leq i \leq |V|$ and $\sum_{i=1}^{|V|} \lambda_i = 1$. The input data can be partitioned into strips, blocks, recursive or arbitrary rectangles. Strip partition is straightforward for most of synchronous iterative applications on homogeneous resources. However, as will be shown in this paper; it is a tough problem to split the data into strips on the HCS.

Traditionally, computing phrase of iterative methods adopts two parallel schemes: Jacobi parallel iterative scheme and Gauss-Seidel parallel iterative scheme. In the Jacobi method, we used the old values $u_i^{(k)}$ for computing $u_i^{(k+1)}$. If instead we use $u_i^{(k+1)}$ in the algorithm as soon as it is available, we have the Gauss-Seidel method. Though the convergence of the Gauss-Seidel scheme is faster for large problems than Jacobi scheme, data dependences of Gauss-Seidel scheme incur more communication unsuitable for the distributed computing environment [13] such as the HCS. Thus, we assume computing phrase adopts the Jacobi parallel iterative scheme and the problem data update only once at each iterative step.

A circle of processors is not too complicated but can be very effective for many synchronous iterative applications [14-16]. Also, the circle topology can be embedded perfect on other structures such as mesh and hypercube. Therefore, we assume that update phrase introduces a virtual circle for the HCS and the processor only communicates with its predecessor and successor.

2.3 Performance Model

The accuracy of predicting the execution time of synchronous iterative applications is determined by its performance model. Let T_{total} denotes the overall predicted execution time of synchronous iterative applications; $T_{iter,i}$ denotes execution time of one iteration of the i-th processor v_i; N denotes the iterative numbers; $T_{comp,i}$ and $T_{comm,i}$ represent the calculation and communication time on one iteration of the i-th processor, respectively. U and L are computation and communication data scale of each iteration, which are const on different iteration. Now the execution time model of synchronous iterative applications is presented as follow:

$$T_{total} = T_{iter} \times N \tag{1}$$

$$T_{iter} = \max_{1 \le i \le |V|} T_{iter,i} \tag{2}$$

$$T_{iter,i} = T_{comp,i} + T_{comp,i} \tag{3}$$

$$T_{comp,i} = (\lambda_i \times U) \times \alpha_i \times \beta_i \times (1-\delta_i) + \xi_i \tag{4}$$

$$T_{comm,i} = L \times (\omega_{i,i-1} + \omega_{i,i+1}) + \varphi_{i,i-1} + \varphi_{i,i+1} \tag{5}$$

Based on the equation (1), the execution time of the application will be proportional to the application iteration time. Due to the synchronization of the processors each iteration step, the iteration time will be equal to the time of the slowest participating processor, which is shown on equation (2). The synchronous iterative algorithm is both calculation-intensive and communication-intensive so that the startup time and communication latency can be neglected. Let $\pi_i = \alpha_i \times \beta_i \times (1-\delta_i)$, the equation(2), (4), (5) can be rewritten as (6), (7), (8).

$$T_{comp,i} \approx (\lambda_i \times U) \times \pi_i \tag{6}$$

$$T_{comm,i} \approx L \times (\omega_{i,i-1} + \omega_{i,i+1}) \tag{7}$$

$$T_{iter} \approx \max_{1 \le i \le |V|}((\lambda_i \times U) \times \pi_i + L \times (\omega_{i,i-1} + \omega_{i,i+1})) \tag{8}$$

Our purpose is to minimize T_{iter} by partitioning and distributing program data among a virtual circle of k processors out of $|V|$. The reason that we choose k processors but not all of the $|V|$ is because the additional computation power does not pay off the communication cost. Now, we are ready to formulate the general definition of load balance of synchronous iterative applications:

3 Formulation

Definition 1 (SILBP) Load Balance Problem of Synchronous Iterative applications:

Given a set V of n processors $v_1,....v_n$ with computation power π_i and a communication link (v_i,v_j) between each processor with bandwidth $\omega_{i,j}$, given the total workload U and the communication volume L at each step of the synchronous iterative application, determine q processors, a circle permutation of the processors $(v_1,....v_q)$ and λ_i, where $1 \le q \le n$, $1 \le i \le q$, $0 \le \lambda_i \le 1$, $\sum_{i=1}^{|q|} \lambda_i = 1$, so that

$$T_{iter} = \min_{1 \leq q \leq |V|} \{ \max_{1 \leq i \leq |q|}((\lambda_i \times U) \times \pi_i + L \times (\omega_{i,(i-1) \bmod q} + \omega_{i,(i+1) \bmod q})) \} \quad (9)$$

However, [17] has proved that if the ratio U/L is large enough, the optimal solution of similar problems will involve all the processors. In that case, the impact of communications will become small in front of the cost of computation, and these computations should be distributed to all resources. The special case of SILBP is formulated as follow:

Definition 2 (FPSILBP) Full Processor Load Balance Problem of Synchronous iterative applications

Given a set V of n processors $v_1,....v_n$ with computation power π_i and a communication link (v_i,v_j) between each processor with bandwidth $\omega_{i,j}$, given the total workload U and the communication volume L, $U/L \geq$ CONST, can we find a circle permutation of the processors $(v_1,....v_n)$ and λ_i, where $1 \leq i \leq n$, $1 > \lambda_i > 0$, $\sum_{i=1}^{n} \lambda_i = 1$, so that the resulting time T_{iter} of each iteration step

$$\max_{1 \leq i \leq n}((\lambda_i \times U) \times \pi_i + L \times (\omega_{i,(i-1) \bmod n} + \omega_{i,(i+1) \bmod n})) \leq D ?$$

4 Complexity

The decision problem associated to the general SILBP optimization problem has been proved to be NP-complete by using a reduction from the Hamiltonian Cycle Problem [18]. In this section, we will prove that full processor load balance problem of synchronous iterative applications also has the intrinsic difficulty.

Theorem 1. FPSILBP is NP-complete.

Proof. We first show that FSILBP belongs to NP. Given an instance of the problem, we use as a certificate the sequence of n processors in the circle. The verification algorithm checks that this sequence contains each processor exactly once, compute and compare ($\lambda_i \times U) \times \pi_i + L \times (\omega_{i,i-1} + \omega_{i,i+1}$) for each processor, selects the max value, and checks whether the value is at most D. This process can certainly be done in polynomial time.

To prove that FPSILBP is NP-hard, we show that TSP $\leq_p$ FPSILBP. Consider an arbitrary instance M_1 of TSP: TSP = {<G, $c_{i,j}$, K>: $G = (V, E)$ is a complete graph, $c_{i,j}$ is a function from $V \times V \to Z$, $K \in Z$, and G has a traveling-salesman tour $(v_1,v_2,...v_n,v_1)$ with cost $(c_{1,2}+c_{2,3}+,...+c_{n-1,n}+c_{n,1}) \leq K$}. We construct the following instance M_2 of FPSILBP: FPSILBP= {<G', π_i, $\omega_{i,j}$, U, L, λ_i, C, D>: $G' = (V', E')$ is a complete graph, $n=|V'|$, $\pi_i =1$, $\omega_{i,j} = \dfrac{c_{i,j}}{2}$, $\dfrac{U}{L}=C$, $L=1$, $\sum_{i=1}^{n}\lambda_i =1$, $D=\dfrac{C}{n}+K$ and G has a circle permutation of n processors with $T_{iter} \leq D$}.which is easily formed in polynomial time.

We now show that graph G has a tour $(v_1, v_2,...,v_n, v_1)$ so that $(c_{1,2}+c_{2,3}+,...,+c_{n-1,n}+c_{n,1}) \le K$ if and only if graph G' has the same processor permutation such that $T_{iter} \le \frac{C}{n}+K$. In order to perfectly balance load among all the n processors, iteration time of different processors should be equal, e.g. $T_{iter} = T_{iter,i} = T_{iter,0} = (\lambda_i \times U) \times \pi_i + L \times (\omega_{i,i-1}+\omega_{i,i+1})$. Since $\sum_{i=1}^{n}\lambda_i = 1$, we derive that $\sum_{i=1}^{n}\frac{T_{iter}-L\times(\omega_{i,i-1}+\omega_{i,i+1})}{U \times \pi_i} = 1$. Thus, $T_{iter} = \frac{U}{\sum_{i=1}^{n}\frac{1}{\pi_i}} + L \times \sum_{i=1}^{n}(\frac{\omega_{i,i-1}+\omega_{i,i+1}}{\pi_i})$. According to the definition of instance M_2 of FPSILBP, $T_{iter} = \frac{C}{n}+\sum_{i=1}^{n}(\omega_{i,i-1}+\omega_{i,i+1}) = \frac{C}{n}+\sum_{i=1}^{n}(\frac{c_{i,i-1}+c_{i,i+1}}{2}) = \frac{C}{n}+\sum_{i=1}^{n}c_{i,i+1} \le \frac{C}{n}+K$

therefore the sufficient condition is proved.

Conversely, suppose that graph G' has a permutation $(v_1', v_2'... v_n', v_1')$ and $T_{iter} \le D = \frac{C}{n}+K$. If $(v_1', v_2', ..., v_n', v_1')$ is not the solution of instance M_1, then $(c_{1,2}+c_{2,3}+,...,+c_{n-1,n}+c_{n,1})>K$. Then, the iteration completion time of instance M_2 $T_{iter} = \frac{U}{\sum_{i=1}^{n}\frac{1}{\pi_i}} + L \times \sum_{i=1}^{n}(\frac{\omega_{i,i-1}+\omega_{i,i+1}}{\pi_i}) > \frac{C}{n}+K$, a contradiction.

Proposition 1 Finding the optimal circle permutation for FPSILBP is equivalent to solving the TSP problem in the weighted graph G' = (V, E'), where $e_{i,j} = \frac{\omega_{i,j}}{\pi_i}+\frac{\omega_{j,i}}{\pi_j}$.

Proof. Since $\sum_{i=1}^{n}\lambda_i = 1$ and Equation (8), we derived that $\sum_{i=1}^{n}\frac{T_{iter}-L\times(\omega_{i,i-1}+\omega_{i,i+1})}{U \times \pi_i} = 1$. Then $T_{iter} = \frac{U}{\sum_{i=1}^{n}\frac{1}{\pi_i}} + L \times \sum_{i=1}^{n}(\frac{\omega_{i,i-1}+\omega_{i,i+1}}{\pi_i})$, where U, $\sum_{i=1}^{n}\frac{1}{\pi_i}$ and L are constant. Therefore, T_{iter} will be minimal when $\sum_{i=1}^{n}(\frac{\omega_{i,i-1}+\omega_{i,i+1}}{\pi_i})$ is minimal. This can be achieved by computing the shortest Hamiltonian Cycle in the graph G' = (V, E'), where $e_{i,j} = \frac{\omega_{i,j}}{\pi_i}+\frac{\omega_{j,i}}{\pi_j}$. Once we have derived the path, we can determine the T_{iter} and derive the load λ_i from Equation (8).

Frederic Vivien tried to prove this proposition in [18]. The correct method was adopted while an error occurs that he mistaken $e_{i,j} = \dfrac{\omega_{i,j}}{\pi_i} + \dfrac{\omega_{j,i}}{\pi_j}$ for $\dfrac{\omega_{i,j} + \omega_{j,i}}{\pi_i}$.

Since SILBP and FPSILBP are both NP-complete, we don't expect to find a polynomial-time algorithm for finding a minimum iteration time of a circle processor permutation. The polynomial-time "approximation algorithm", which produces "approximate" solutions for these problems, should be designed later.

5 Conclusion

In this paper, the Load Balance Problem of Synchronous Iterative applications (SILBP) and its special case, Full Processor Load Balance Problem of Synchronous (FPSILBP), are formulated into the combination optimization problems. Their performance model and NP-Complexity is discussed in detail. According to the theorem and proposition, the next step in the research is to design TSP based and Hamiltonian Cycle Problem based heuristic algorithm to facilitate solving the SILBP and FPSILBP.

References

1. Vincent Boudet, Fabrice Rastello, and Yves Robert. Algorithmic issues for (distributed) heterogeneous computing platforms. In Rajkumar Buyya and Toni Cortes, editors, Cluster Computing Technologies, Environments, and Applications (CC-TEA'99), pages 09–712. CSREA Press, 1999.
2. I.Foster and C. Kesselman, editors. The Grid: Blueprint for a New Computing Infrastructure. Morgan Kaufmann, San Francisco, CA, 1999
3. Matyska, L.and Ruda. M. Metacomputing. New direction in high performance computing. Information Technology Applications in Biomedicine, 1997. ITAB '97. Proceedings of the IEEE Engineering in Medicine and Biology Society Region 8 International Conference, 7-9 Sept. 1997 Pages:106 – 108
4. Fox, G.C., Williams, R.D., and Messina, P.C. Parallel Computing Works! Morgan Kaufmann, San Francisco, 1994.
5. Vadhiyar, S.S and Dongarra, J.J. A metascheduler for the Grid. High Performance Distributed Computing, 2002. HPDC-11 2002. Proceedings. 11th IEEE International Symposium on 23-26 July 2002
6. Weissman, J.B. Metascheduling: A scheduling model for metacomputing systems. High Performance Distributed Computing. 1998. Proceedings. The Seventh International Symposium on 28-31 July 1998 Pages: 348 – 349
7. Daoqi Yang. A Parallel Iterative Domain Decomposition Algorithm for Elliptic Problems, Journal of Computational Mathematics, 16(1998) 141-151.
8. K.-A. Mardal and H. P. Langtangen. An efficient parallel iterative approach to a fully implicit mixed finite element formulation for the Navier-Stokes equations, ECCOMAS CFD 2001, Computational Fluid Dynamics Conference Proceedings
9. Barry Wilkinson and Michael, Allen Parallel Programming: Techniques and Applications using Networked Workstations and Parallel Computers, Prentice Hall, 1999

10. Brochard L, Prost J-P and Faurie F. Synchronization and load unbalance effects of parallel iterative algorithms Proc. Int. Conf. on Parallel Processing (St Charles, IL) vol 1 (University Park, PA: The Pennsylvania State University Press) pp 153–60
11. Dubois M and Briggs F A. Performance of synchronized iterative processes in multiprocessor systems IEEE Trans. Software Engng SE-8 419–31, 1982
12. Kruskal C P and Weiss A. Allocating independent subtasks on parallel processors. IEEE Trans. Soft. Engng SE-11 1001–16, 1985
13. Mark F. Adams. A distributed memory unstructured gauss-seidel algorithm for multigrid smoothers, Conference on High Performance Networking and Computing archive, Proceedings of the 2001 ACM/IEEE conference on Supercomputing, Pages: 4-4, 2001
14. Chung-Ming Chen and Soo-Young Lee. On parallelizing the EM algorithm for PET image reconstruction,Parallel and Distributed Systems, IEEE Transactions on ,Volume: 5, Issue: 8, Aug. 1994 Pages:860 - 873
15. Yang, L.T. Data distribution and communication schemes for IQMR method on massively distributed memory computers, Parallel Processing, 2000. Proceedings. 2000 International Workshops on, 21-24, Aug. 2000, Pages:299–306
16. Walker, E.and Morgan, G.. Pipeline ring data-flow architecture for solving large iterative structures, Computers and Digital Techniques, IEE Proceedings,Volume: 141, Issue: 4, July 1994, Pages: 212–220
17. Arnaud Legrand, Hélène Renard, Yves Robert and Frédéric Vivien. Load-balancing iterative computations on heterogeneous clusters with shared communication links. In PPAM-2003: Fifth International Conference on Parallel Processing and Applied Mathematics, LNCS, 2003. Springer Verlag.
18. H.Renard, Y.Robert and F.Vivien. Static load-balancing techniques for iterative computations on heterogeneous clusters. Euro-Par 2003. August 26 - 29, 2003 in Klagenfurt/Austria

Incentive-Based P2P Scheduling in Grid Computing

Yanmin Zhu[1], Lijuan Xiao[2], Lionel M. Ni[1], and Zhiwei Xu[2]

[1] Department of Computer Science
Hong Kong University of Science and Technology
Clearwater Bay, Kowloon, Hong Kong
`{zhuym,ni}@cs.ust.hk`
[2] Institute of Computing Technology
Chinese Academy of Sciences, Beijing, China
`xiaolijuan@software.ict.ac.cn,`
`zxu@ict.ac.cn`

Abstract. Grid computing has emerged as an attractive computing paradigm recently. In typical grid environments, there are two distinct parties, resource consumers and resource providers, which have different optimization objectives. Enabling an effective interaction between the two parties (i.e., scheduling jobs of consumers across resources of providers) is particularly challenging due to the distributed ownership of grid resources. In this paper, we propose an incentive-based P2P scheduling for grid computing, with the goal of building a practical and robust computational economy. The goal is realized by building a computational market supporting fair and healthy competition among consumers and providers. To build the healthy computational market, we propose the P2P scheduling infrastructure to efficiently support the scheduling, and the incentive-based algorithms for consumers and providers, respectively.

1 Introduction

With the rapid development of high-speed wide-area networks and powerful yet low-cost computational resources, grid computing [1] has emerged as an attractive computing paradigm. Computational grids strive to aggregate the computational power of heterogeneous, geographically distributed and dynamic computational resources. These resources are usually administrated by different domains and owned by various instances. Therefore, they are highly autonomous and differ from each other in many aspects, such as scheduling policy, security requirement, performance strategy, and desired objective. Effective scheduling is of fundamental importance. However, due to unique characteristics described above in grid computing, scheduling in grid environments is particularly challenging.

In typical grid environments, on one hand, some users (*resource consumers*) have computational jobs to execute, but they may lack computational resources for their jobs. On the other hand, some resource owners (*resource providers*) have relatively underutilized resources. It is highly desirable for consumers to schedule jobs across those resources, but the scheduling is significantly complicated by the distributed ownership of grid resources. Consumers and providers are independent from each other, each having its own access policy, scheduling strategy, and optimization objec-

The goal is to build a computational economy to enable efficient interaction between consumers and providers. It is to be realized by building a healthy computational market. The health of the market means every player in the market can have sufficient incentive for joining in the market such that the market is stable and lasts.

4 Incentive-Based P2P Scheduling

The large scale of the virtual market implies that simplicity, self-organizing, robustness and scalability are important features that the system should possess. To this end, we propose the P2P scheduling infrastructure for organizing the resource consumers and resource providers into a P2P alike network. Taking the advantages of P2P networks, the scheduling infrastructure greatly facilitate the scheduling operation over the distributed grid system.

The basic idea is that we try to form a complete competition among all participants. On one hand, given a job request, enough providers will actively compete for the job request. On the other hand, given a provider, enough consumers will compete for the provider's resource. We expect that the incentives of both consumers and providers can be automatically achieved through the complete competition mechanism.

In Figure 1, the main steps for executing a job in the computational market are listed. Every job will experience the same steps until its completion.

4.1 P2P Scheduling Infrastructure

P2P networks, such as Gnutella [7], and Kazza [8], have been widely used in file sharing for their simplicity, scalability and robustness. In general, there is no centralized controller in P2P networks and each peer is autonomous. To take the advantages of P2P networks, we organize resource providers into a P2P network. The P2P network forms the scheduling infrastructure on which job announcements are forwarded. A consumer initiates a job announcement and sends it into the P2P network. On receiving the job announcement, every provider is required to forward the job announcement to its neighboring providers except the one which forwards this job announcement to it.

4.2 Incentive-Based Consumer Algorithm

The behavior of a consumer is characterized by two operations: *budget assigning* and *job offering*. Given a job, the budget assigning algorithm is responsible for assigning a proper budget to the job. And the job offering scheme determines which provider the job is offered to from the candidates that replied.

Many factors are involved in deciding the budget of a job. These factors can be generally classified into two categories: internal ones and external ones. Internal factors include the job length and the urgent level of the deadline, and external factors include the overall system load and the budget assigning schemes of other consumers in the system.

```
step 1:  the consumer creates a job and sends out a job an-
         nouncement over the P2P scheduling infrastructure;
step 2:  the provider gets the job announcement and replies
         to the consumer with estimated completion time;
step 3:  the consumer selects the provider providing the
         earliest completion time for the job, and offers
         the job to that provider;
step 4:  the provider gets the job offer and inserts it into
         the pending offered job queue;
step 5:  the provider executes the job;
step 6:  the result is sent back to the consumer;
```

Fig. 1. The General Scheduling Procedure.

The urgent level of a job is defined as follows. According to the definition, a higher *urgLev* value means more urgently the job is required to be completed.

$$urgLev = \frac{jobLength}{deadline - creationTime}$$

It is intuitive that a longer job length requires more computing cycles and therefore a job with longer job length needs more budget. And a higher urgent level implies that the job should precede many other jobs to be executed, and therefore it requires higher priority. To obtain a higher priority, the job has to be given more budget. Thus, the budget is proportional to the job length and job urgent level. The following is the proposed budget assigning scheme for consumers.

$$jobBudget = \lambda \times (a \times jobLen + b \times urgLev)$$

- jobBudget : job budget
- λ : job importance factor
- a, b : constant coefficient

In addition to the internal factors, the budget assigning algorithm should be aware of the external factors. One basic observation is that when the deadline missing rate increases, a consumer should conclude that jobs were assigned relatively less budget such that the jobs were inferior while competing with other jobs. Therefore, the budget assigning algorithm has to be adaptive to the deadline missing rate. When the deadline missing rate increases, the algorithm should increase importance factor accordingly; otherwise, it should decrease λ.

After sending out a job announcement, the consumer waits for a short while and expects to receive a number of replies from those providers meeting the job's deadline. A reply includes the completion time when the provider claims to complete the job. The job offering scheme is responsible for determining to which provider the job is going to be offered. Different consumers may have different optimization objectives, and therefore have different job offering schemes. We have implemented the job offering scheme that a consumer will offer the job to the provider who claims the earliest completion time for the job.

4.3 Incentive-Based Provider Algorithm

The behavior of a resource provider is characterized by two operations: *competing for jobs* and *local scheduling*. The former operation is responsible for deciding how to compete for jobs, and the later operation is to schedule the local offered jobs, with the aim of maximizing the profit.

Once a resource provider is offered with a job, it will perform scheduling for its own benefit only, without taking into account the job's performance. In order to prevent the provider from not keeping the promise it made to the consumer, we propose a penalty model. The basic idea is that a provider will be penalized with an amount of money if the provider could not complete the job before the completion time it promised to the consumer. The amount of penalty is related to the budget of the job and the length of exceeding time. To some extent, the penalty model helps force the provider to keep its promise, but still allows the freedom of local scheduling such that it is able to maximize its profit.

The following expression summarizes the penalty model mathematically. p_1B and p_2B are the slopes of the penalty lines, where p_1 and p_2 are constant coefficients. In general, p_1 is less than p_2 because for consumers exceeding deadline is more serious than exceeding the claimed completion time. *MaxPen* represents the maximal penalty that could be posed.

$$penalty = \begin{cases} 0 & T \leq CT \\ p_1B(T-CT) & CT < T \leq DL \\ p_2B(T-DT) + p_1B(DL-CT) & DL < T \leq T_0 \\ MaxPen & T > T_0 \end{cases}$$

In the computational market, each provider competes actively for jobs, and tries to maximize its profit. One basic operation of providers is how to respond to job announcements. Ideally, the provider could try to compete for those jobs that will result in the maximal profit. But it is hardly possible because the provider is not able to predict the possible future job announcement arrivals, and also cannot determine whether it will be offered with a specific job. We propose an aggressive competing algorithm for providers. By this algorithm, each provider tries to complete for every job whenever it can satisfy the job's deadline.

Whenever a job is completed, the provider has to make a decision which job to execute next. Suppose that there are n offered pending jobs. It is a well-known NP complete problem to schedule the jobs such that the resulting profit is maximized [9]. One simple optimal solution is to investigate each of the $n!$ permutations, compute the profit, and select the permutation which produces the maximal profit. But it is certainly computationally infeasible when n is large. We propose a heuristic local scheduling algorithm, which is computationally efficient and produces near-optimal profit. The basic idea is that for each provider a sorted list of offered pending jobs is maintained with respect to the profit. The ordering of these jobs results in a near maximal profit. The heuristic is that after a new offered job is inserted, the relative order of the original jobs is probably not changed. According to the heuristic, only $n+1$ *possible* positions for the new job should be investigated. The new offered job will be inserted to the position which produces the maximal profit out of the n+1 positions. With the availability of the sorted list of jobs, it is simple to decide which job to execute next. The provider will select the job on the head of the list to execute whenever a previous job is completed.

5 Simulation Results

We design the first experiment to study the incentives obtained by consumers with different job importance factor. In this experiment, simulations are performed using the following parameters. There are in total 20 consumers and 80 providers. 10% of the consumers are assigned with job importance factor $\lambda=1.5$, 10% of consumers with $\lambda=0.5$ and the rest with $\lambda=1.0$. The system load is varying from 0.3 to 0.7.

Figure 2 shows the resulting incentives in terms of the deadline missing rate with respect to the system load. As seen in this figure, the consumers with higher importance factor really experience less deadline missing rate. Those jobs with relatively higher importance factor will gain relatively higher priority and consequently be executed earlier. This trend becomes sharper when the system load is increasing. Thus, it makes sense that consumers with importance factor $\lambda=1.5$ experience minimum deadline missing rate. This experiment demonstrates that our approach achieves to guarantee the incentive of consumers.

The second experiment is designed to study the incentive of providers. Before talking about the experiment results, we define the terminology *fairness scale* of an individual provider. The fairness scale is defined by the following expression.

$$fairness\ scale_i = \frac{normalized\ profit_i}{normalized\ capability_i}$$

A fairness scale reflects the individual incentive for a provider. The ideal case is that the fairness scale of every provider is one. A fairness scale less than one means that the provider does not make the profit conforming to its computational capability, which leads to the situation that the provider does not get enough incentive. If the fairness scale of a provider remains much less than one, it possibly quits the market. The standard deviation (SD) of fairness scales of providers is employed to study the overall incentive situation of providers.

As shown in Figure 3, the SD of fairness scales is fairly good, which is less than 25% of the ideal fairness scale. When the system load is increasing, the SD will be further reduced. This experiment demonstrates that our approach really guarantees the incentive for every provider.

6 Conclusion

Distributed ownership of resources greatly complicates scheduling in grid computing. Enabling the interaction between consumers and providers is highly challenging. In this paper, we have proposed the incentive-based P2P scheduling, aiming at building a decentralized, scalable and robust computational market. In this market, each participant behaves for its own benefit only. However, the computational market is proved to be healthy since each participant is guaranteed to obtain sufficient incentive

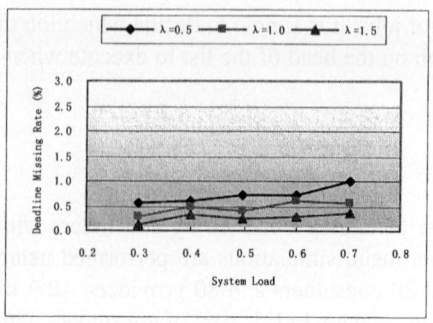
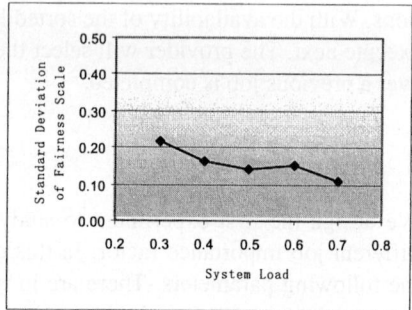

Fig. 2. Resource Consumer Incentive.   **Fig. 3.** Resource Provider Incentive.

for joining the market. Detailed simulation results demonstrate that our approach is successful in building a healthy and scalable computational economy.

References

1. Foster, I. and C. Kesselman, The Grid 2: Blueprint for a New Computing Infrastructure. 2003: Morgan Kaufmann Publishers.
2. Shetty, S., P. Padala, and M. Frank, A Survey of Market Based Approaches in Distributed Computing. 2003.
3. Buyya, R., et al., Economic Models for Resource Management and Scheduling in Grid Computing. Special Issue on Grid Computing Environments, Journal of Concurrency and Computation: Practice and Experience, 2002. 14(13-15): p. 1507-1542.
4. Buyya, R. and S. Vazhkudai. Compute power market: Towards a market-oriented grid. in the First International Symposium on Cluster Computing and the Grid. 2001.
5. Malone, T.W., et al., Enterprise: A market-like task scheduler for distributed computing environments, in The Ecology of Computation, B.A. Huberman, Editor. 1988, Amsterdam: north-Holland. p. 177-205.
6. The Standard Performance Evaluation Corporation (SPEC) Home Page, http://www.specbench.org/.
7. Gnutella Homepage, http://gnutella.wego.com, April, 2004.
8. Kazaa Homepage, http://www.kazaa.com, April, 2004.
9. Gonzalez, M.J., Deterministic Processor Scheduling. ACM Computing Surveys, 1997. 9(3): p. 173-204.

Hybrid Performance-Oriented Scheduling of Moldable Jobs with QoS Demands in Multiclusters and Grids*

Ligang He, Stephen A. Jarvis, Daniel P. Spooner, Xinuo Chen, and Graham R. Nudd

Department of Computer Science, University of Warwick
Coventry, UK CV4 7AL
liganghe@dcs.warwick.ac.uk

Abstract. This paper addresses the dynamic scheduling of moldable jobs with QoS demands (soft-deadlines) in multiclusters. A moldable job can be run on a variable number of resources. Three metrics (over-deadline, makespan and idle-time) are combined with weights to evaluate the scheduling performance. Two levels of performance optimisation are applied in the multicluster. At the multicluster level, a scheduler (which we call MUSCLE) allocates parallel jobs with high packing potential to the same cluster; MUSCLE also takes the jobs' QoS requirements into account and employs a heuristic to achieve performance balancing across the multicluster. At the single cluster level, an existing workload manager, called TITAN, utilizes a genetic algorithm to further improve the scheduling performance of the jobs allocated by MUSCLE. Extensive experimental studies are conducted to verify the effectiveness of the scheduling mechanism in MUSCLE. The results show that the comprehensive scheduling performance of parallel jobs is significantly improved across the multicluster.

1 Introduction

Separate clusters are increasingly being interconnected to create multicluster or grid computing architectures [1][2][6][7] and as a result, workload management for these architectures is becoming a key research issue. Parallel jobs that run in these domains can be classified into two categories: rigid and moldable [11]. In this paper, a mechanism is developed to schedule moldable jobs in multiclusters/grids. A moldable job is defined as a parallel job that can be run on a variable number of computers.

A job's execution time may not be inversely proportional to its size due to the presence of communication among execution components [10][12]. Consequently, the smallest product of a job's size and the corresponding execution time results in the least consumption of resources. This size is called the *preferable size* of the job.

In the multicluster architecture assumed in this paper, the constituent clusters may be located in different administrative organizations and as a result be managed with different performance criteria. In this scheduling work, we combine three metrics (*over-deadline*, *makespan* and *idle-time*) with additional variable weights; this allows the resources in different locations to represent different performance scenarios.

---

* This work is sponsored in part by grants from the NASA AMES Research Center (administrated by USARDSG, contract no. N68171-01-C-9012), the EPSRC (contract no. GR/R47424/01) and the EPSRC e-Science Core Programme (contract no. GR/S03058/01).

Over-deadline is defined as the sum of excess time of each job's finish time over its deadline; makespan is defined as the duration between the start time of the first job and the finish time of the last executed job [8][9].

In this work, the multicluster architecture is equipped with two levels of performance optimisation. A multicluster-level scheduler (called MUSCLE) is developed to allocate moldable jobs with QoS demands (deadlines) to constituent clusters. When a job is submitted to the multicluster, MUSCLE identifies the job's preferable size and corresponding execution time and then allocates jobs with high packing potential in terms of their preferable sizes to the same cluster. It also takes the QoS demands of jobs into account and exploits a heuristic to allocate suitable workload to each cluster. When MUSCLE makes scheduling decisions to distribute jobs to individual clusters, it also determines a *seed schedule* for the jobs allocated to each cluster assuming the jobs are run with their preferable sizes. These seed schedules are sent to the corresponding clusters where an existing workload manager (TITAN [12]) uses a genetic algorithm to transform the schedule into one that further improves the (local) comprehensive performance.

The rest of the paper is organized as follows. The system and workload model is introduced in Section 2. Section 3 briefly presents the genetic algorithm used in TITAN. The design of MUSLCE is proposed in Section 4. Section 5 presents the experimental results. Finally, Section 6 concludes the paper.

2 System and Workload Model

The multicluster architecture assumed in this work (shown in Fig. 1) consists of n clusters, $C_1, C_2, \ldots, C_n$; where a cluster C_i ($1 \leq i \leq n$) consists of m_i homogeneous computers (i.e., the size of cluster C_i is m_i), each with a service rate of u_i. There are two scheduling levels in the architecture; MUSCLE acts as the global scheduler for the multicluster while TITAN schedules the jobs sent by MUSCLE within each local cluster. MUSCLE and TITAN are interconnected through an agent system [3][4][5]. Users can submit parallel jobs to the multicluster through MUSCLE or through

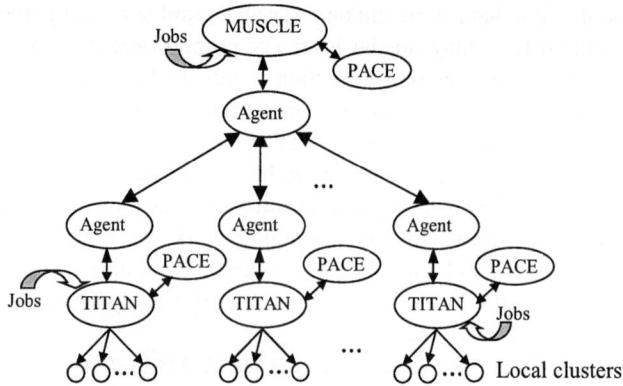

Fig. 1. The multicluster job management architecture.

TITAN. PACE (Performance Analysis and Characterisation Environment) [10][15] is incorporated into the architecture to provide the jobs' preferable size and corresponding execution time. Parallel jobs are moldable, and a parallel job, denoted by J_i, is identified by a 4-tuple (a_i, s_i, e_{ij}, d_i), where a_i is J_i's arrival time, s_i is its preferable size, et_{ij} is its execution time in cluster C_j (1≤j≤n) and d_i is its soft-deadline (QoS).

3 Local-Level Scheduling via TITAN

This section briefly describes the genetic algorithm used in the TITAN workload manager [12]. Three metrics (over-deadline, makespan and idle-time) are combined with the additional weights to form a comprehensive performance metric (denoted by *CP*), which is used to evaluate a schedule. The CP is defined in Eq.1, where Γ, ω and θ are makespan, idle-time and over-deadline, respectively; W^i, W^m and W^o are their weights. For a given weight combination, the lower the value of CP, the better the comprehensive performance.

$$CP = \frac{W^i\Gamma + W^m\omega + W^o\theta}{W^i + W^m + W^o} \quad (1)$$

A genetic algorithm is used to find a schedule with a low CP. The algorithm first generates a set of initial schedules (one of these is the seed schedule sent by MUSCLE). *Crossover* and *mutation* operations are performed to transform the schedules in the current set and generate the next generation. This procedure continues until the performance in each generation of schedule stabilizes.

4 Global-Level Scheduling via MUSCLE

The main operations performed by MUSCLE are as follows. First, MUSCLE determines which parallel jobs with preferable sizes can be packed into a *computer space* with a given size. These possible compositions of parallel jobs are organized into a *composition table*. After the composition table is constructed, MUSCLE searches the table for suitable parallel jobs to allocate to the available computer space in a cluster. When the computer space is available in multiple clusters, MUSCLE orders the processing of the space using a heuristic.

4.1 Organizing Parallel Jobs

Suppose the maximum cluster size in the multicluster is m_{MAX} and that at some time point, p parallel jobs, J_1, J_2, ..., J_p are collected (into a queue) in the multicluster. Algorithm 1 outlines the steps for constructing the composition table. The p jobs are filled into suitable rows in the table.

```
Algorithm 1. Constructing the composition table
1. for each parallel job J_i (1≤i≤p) to be scheduled do
2.     for each j satisfying 1≤j≤ m_MAX do
3.         if s_i = j
```

```
4.        Append J_i to the tail of the j-th row of the
          table;
5.      if s_i < j
6.        r←j-s_i;
7.        if the r-th row in the table is not NULL
8.          if there is such a composition of parallel
            jobs in the r-th row in which no job is
            in the j-th row of the table;
9.            Append J_i as well as the parallel jobs in
              the composition from the r-th row to
              the tail of the j-th row;
```

There are two for-loops in Algorithm 1 and Step 8 searches a row for the qualified composition. In the worst case, the time taken by Step 8 is $O(p)$. Hence, the worst-case time complexity of Algorithm 1 is $O(p^2 m_{MAX})$.

4.2 Searching the Composition Table

The algorithm for allocating jobs to a computer space with size r in a cluster proceeds as follows. First, it searches the composition table from the r-th row up to the first row to obtain the first row that is not null. Then, in this row, the algorithm selects the composition in which the number of jobs having been allocated is the least. If the number is zero, these jobs are allocated to the computer space. If a job J_i, whose size is s_i, in the composition has been allocated, a function is called to search the s_i-th row for alternative jobs for J_i. The function is called recursively if a composition cannot be found in the s_i-th row in which no job in it is allocated. The recursive call terminates when there is only one composition in a searched row (i.e. there are no alternative jobs) or when the composition consisting of unallocated jobs is found. If the algorithm fails to allocate jobs to the computer space with size r, it continues by trying to identify jobs to allocate to the computer space with size $r=r-1$. The procedure continues until r reaches 0. After the allocated jobs have been determined, the schedule for these jobs can also be computed. The time complexity of the procedure is based on the number of jobs that are allocated. The best-case time complexity is $O(1)$ while the worst-case time complexity is $O(p^2 n_{MAX})$.

As can be seen from the above description, the job allocation algorithm always attempts to identify the jobs that maximally occupy the given computer space. In doing so, the number of computers left idle is minimized.

4.3 Employing a Heuristic to Balance Performance

A performance metric ε is formed by integrating the workload attributes in clusters. ε is defined in Eq.2, where p_i is the number of jobs, $etSum_i$ is the sum of the execution times of all jobs, $sizeSum_i$ represents the total job sizes and $slkSum_i$ is the total slack (the slack of a job is its deadline minus its execution time and its arrival time). When multiple clusters offer available computer space, the cluster with the smallest ε is given the highest priority and will be allocated the jobs.

$$\varepsilon = \frac{p_i \times etSum_i \times sizeSum_i}{slkSum_i \times m_i} \qquad (2)$$

The complete scheduling procedure for MUSCLE is outlined in Algorithm 2.

Algorithm 2. MUSCLE scheduling
```
1.    while the expected makespan of the jobs yet to be
      scheduled in clusters is greater than a prede-
      fined valve
2.      Collect the arriving jobs in the multicluster;
3.      Call Algorithm 1 to construct the composition ta-
        ble for the collected jobs;
4.      do
5.        for each cluster do
6.          Calculate ε using Eq.2
7.        Get the earliest available computer space in
          cluster C_j which has the minimal ε;
8.        Allocate jobs to this space;
9.        Update the earliest available computer space and
          the values of the workload attributes in C_j;
10.     while all collected jobs have not been allocated;
11.     Go to Step 1;
```

The time of Algorithm 2 is dominated by Step 3 and Step 8 in the do-while loop. Their time complexities have been analysed in subsection 4.1 and 4.2.

5 Experimental Studies

A simulator is developed to evaluate the performance of the scheduling mechanism in MUSCLE. The experimental results focus on demonstrating the performance advantages of the scheduling mechanism in MUSCLE over the scheduling policies frequently used in distributed systems. Weighted Random (WRAND) and Dynamic Least Load (DLL) policies are two selected representatives. The DLL policy schedules jobs to the resource with the least workload. In the WRAND policy, the probability that a job is scheduled to a resource is proportional to its processing capability.

40,000 parallel jobs are generated; the submissions of parallel jobs follow a Poisson process. A job's preferable size follows a uniform distribution in [MIN_S, MAX_S]. Given a job's preferable size s_i, suppose that the corresponding execution time is $e_{ij,si}$; then when the job's size is s_i+k, the corresponding execution time $e_{ij,si+k}$ is determined by Eq.3, where k can be positive or negative integer and φ is the factor that determines the scalability of the job regarding its size.

$$e_{ij,si+k} = e_{ij,si} \times \frac{s_i}{s_i+k} \times (1+|k|\varphi) \qquad (3)$$

When jobs take preferable sizes and the service rate is the average service rate of the constituent clusters in the multicluster, their execution time follow a bounded

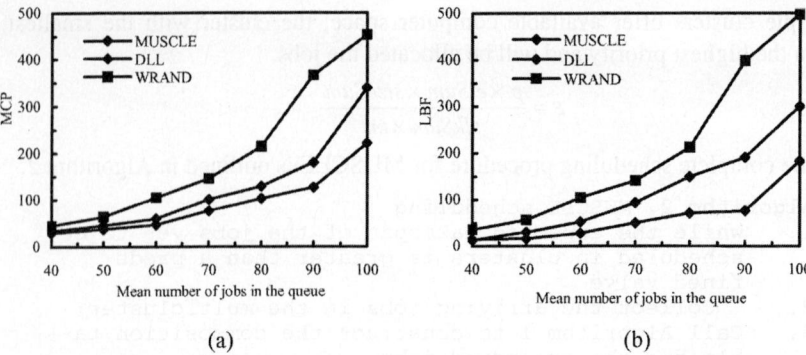

Fig. 2. The comparison of MUSCLE, DLL and WRAND under different workload levels in terms of (a) mean comprehensive performance (MCP), and (b) Load balance factor (LBF); (W^o, W^m, W^i)=(4, 3, 1); (e_l, e_u)=(5, 100); (MIN_S, MAX_S)=(1, 10); (MIN_DR, MAX_DR)=(1, 5); φ=0.05; The cluster size and service rate are shown in Table 1.

Table 1. The multicluster setting in Fig.4.

Clusters	C_1	C_2	C_3	C_4
Size	20	16	12	10
Service rate ratio	1.0	1.2	1.4	1.6

Pareto distribution, shown in Eq.4, where e_l and e_u are the lower and upper limit of the execution time x.

$$f(x) = \frac{\alpha e_l^{\alpha}}{1-(e_l/e_u)^{\alpha}} x^{-\alpha-1} \quad (4)$$

A job's deadline, d_i is determined by Eq.5, where dr is the deadline ratio. dr follows a uniform distribution in [MIN_DR, MAX_DR], which is used to measure the deadline range.

$$d_i = \max\{et_{ij}\} \times (1+dr) \quad (5)$$

The performance metrics evaluated in the experiments are the *Mean Comprehensive Performance* (MCP) and *Load Balance Factor* (LBF). MCP is the average of CP in each cluster and LBF is the standard deviation of these CP's.

5.1 Workload Levels

Fig.2.a and Fig.2.b demonstrate the performance difference among MUSCLE, DLL and WRAND scheduling policies under different workload levels. The workload level is measured by the mean number of jobs in the queue (for accommodating the collected jobs) in the multicluster. It can be observed from Fig.2.a that MUSCLE outperforms the DLL and WRAND policies under all workload levels. This is because the jobs are packed tightly in the seed schedules sent by MUSCLE to the individual clusters. Therefore, the further improvement by the genetic algorithm in each cluster is based on an excellent "seed". As can be observed from Fig.2.b, MUSCLE outperforms DLL and WRAND in terms of LBF except when the mean number of jobs is 40 (in that case, the performance achieved by MUSCLE is slightly worse than that by DLL). The reasoning behind this is as follows. When the workload is low, a small number of jobs miss their deadlines. DLL is beneficial to obtaining the balanced resource throughput and resource utilization. Therefore, DLL shows a more balanced MCP performance. However, as the workload increases further, more jobs miss their deadlines. MUSCLE takes the QoS demands into account so that the MCP performance remains balanced among the clusters.

5.2 Cluster Size

Parallel jobs have to be packed into the clusters. Cluster size is therefore an important parameter for parallel job scheduling. Fig.3.a and Fig.3.b compare the performance of MUSCLE, DLL and WRAND under different heterogeneity levels of cluster size. The heterogeneity levels of cluster size are measured by the scale of the range from which cluster sizes are selected. The multicluster consists of five clusters. Five sets of cluster sizes, all with the same average, are uniformly chosen from five ranges, [10, 42], [14, 38], [18, 34], [22, 30] and [26, 26]. The range [26, 26] means the multicluster is homogeneous in terms of cluster size. The service rates of computers in all clusters are set to be the same.

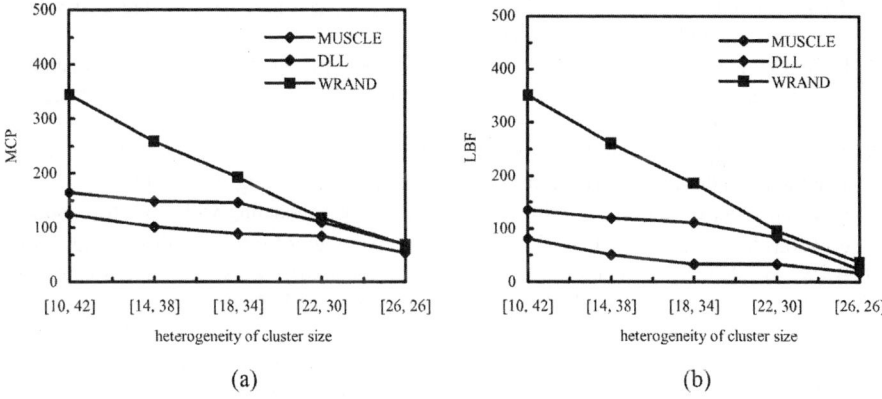

Fig. 3. Performance comparison of MUSCLE, DLL and WRAND under different heterogeneity levels of cluster size, measured by a range $[a, b]$; (e, e_u)=(5, 100); (MIN_S, MAX_S)=(1, a); (MIN_DR, MAX_DR)=(1, 5); $\varphi=(a+1)/a$; the mean number of jobs in the queue is 100.

As can be observed from Fig.3.a and Fig.3.b, MUSCLE outperforms DLL and WRAND in terms of MCP and LBF in all cases. Further, MCP and LBF performance achieved by all three policies improves as the heterogeneity level decreases. This is because the decrease in the heterogeneity of the cluster size is beneficial to achieving a balanced load for all policies. Thus, the MCP performance is also improved.

6 Conclusions

A multicluster-level scheduler, called MUSCLE, is described in this paper for the scheduling of moldable jobs with QoS demands in multiclusters. MUSCLE is able to allocate jobs with high packing potential (in terms of their preferable sizes) to the same cluster and further utilizes a heuristic to control the workload distribution among the clusters. Extensive experimental studies have been carried out to verify the performance advantages of MUSCLE.

References

1. M. Barreto, R. Avila, P. Navaux.: The MultiCluster model to the integrated use of multiple workstation clusters. Proc. of the 3rd Workshop on Personal Computerbased Networks of Workstations, 2000, pp. 71–80
2. R. Buyya, M. Baker.: Emerging Technologies for Multicluster/Grid Computing. Proceedings of the 2001 IEEE International Conference on Cluster Computing, 2001
3. J. Cao, D. J. Kerbyson, G. R. Nudd.: Performance Evaluation of an Agent-Based Resource Management Infrastructure for Grid Computing. Proc. of 1st IEEE/ACM International Symposium on Cluster Computing and the Grid, 2001
4. J. Cao, D. J. Kerbyson, E. Papaefstathiou, G. R. Nudd.: Performance Modeling of Parallel and Distributed Computing Using PACE. Proceedings of 19th IEEE Intl Performance, Computing, and Communications Conference, 2000
5. J. Cao, D. P Spooner, S. A Jarvis, G. R Nudd.: Grid load balancing using intelligent agents. To appear in Future Generation Computer Systems special issue on Intelligent Grid Environments: Principles and Applications, 2004
6. L. He, S. A. Jarvis, D. P. Spooner, G. R. Nudd.: Optimising static workload allocation in multiclusters. Proceedings of 18th IEEE International Parallel and Distributed Processing Symposium (IPDPS'04), April 26-30, 2004
7. X. He, X. Sun, G. Laszewski.: QoS Guided Min-Min Heuristic for Grid Task Scheduling. Journal of Computer Sci&Tech, Special Issue on Grid Computing, 18(4), 2003
8. B. G. Lawson, E. Smirni.: Multiple-queue Backfilling Scheduling with Priorities and Reservations for Parallel Systems. 8th Job Scheduling Strategies for Parallel Processing, 2002
9. A. W. Mu'alem, D. G. Feitelson.: Utilization, predictability, workloads, and user runtime estimates in scheduling the IBM SP2 with backfilling. IEEE Trans. Parallel & Distributed Syst. 12(6), pp. 529-543, 2001
10. G. R. Nudd, D. J. Kerbyson, E. Papaefstathiou, J. S. Harper, S. C. Perry, D. V. Wilcox.: PACE: A Toolset for the Performance Prediction of Parallel and Distributed Systems. In The Intl Journal of High Performance Computing, 1999
11. E. Shmueli, D. G. Feitelson.: Backfilling with lookahead to optimize the performance of parallel job scheduling. 9th Job Scheduling Strategies for Parallel Processing, 2003
12. D. P Spooner, S. A Jarvis, J Cao, S Saini, G. R Nudd.: Local Grid Scheduling Techniques using Performance Prediction. IEE Proc. Comp. Digit. Tech., 150(2):87-96, 2003

An Active Resource Management System for Computational Grid*

Xiaolin Chen[1], Chang Yang[1], Sanglu Lu[2], and Guihai Chen[2]

[1] Department of Computer Science, Chuxiong Normal University, Chuxiong 675000, China
{chenxl,yc}@cxtc.edu.cn
[2] State Key Laboratory of Novel Software Tech., Nanjing University, Nanjing 210093, China
{sanglu,gchen}@nju.edu.cn

Abstract. In this paper, we propose an active grid resource management system supported by active networks for computational grid. First, we construct a scalable two-level resource management architecture. In this architecture, resources in the system are divided into multiple autonomous domains. In each domain, an active resource tree (ART) is organized with resources as its leaf nodes and active routers as non-leaf nodes. Resource information is disseminated into each active router in ART, and nodes in ART work cooperatively to discover and schedule resources. Communications between domains are done via the root of ART. Second, the resource trade between consumers and providers is carried out by an improved barter marketing model. The proposed model reduces the complexity and cost of the trade, and provides desired soft quality of service as well. We illustrate that the proposed system exhibits good scalability, autonomy, flexibility, and fault-tolerance.

1 Introduction

A grid [1] is a very large-scale network computing system that can scale up to Internet size environment, in which kinds of computing, storage and data resources, as well as scientific devices or implements, are distributed across multiple organizations and administrative domains. The computing resources can be supercomputers, SMPs, clusters, desktop PCs, or even mobile computing devices such as PDA.

In the grid's architecture, grid resource management system (GRMS) is the central component, which is responsible for disseminating resources information across the grid, accepting requests for resources, discovering and scheduling the suitable resources that match the requests from the global grid resources, and executing the requests on scheduled resources [2]. As a grid is geographically distributed, heterogeneous and autonomous in nature, the design and implementation of RMS is challenging. In this paper, we focus on GRMS for computational grid.

The scalability of GRMS determines the scalability of the grid, and the scalability of GRMS is determined by the resource organization structure, resource scheduling structure and economic model for resource trade. In this paper, we propose an active resource management system (AGRMS) supported by active networks for computa-

---

* This research was supported by the National Natural Science Foundation of China under Grant No.60363001, the National High Technology Development 863 Program of China under Grant No.2001AA113050 and the National Grand Fundamental Research 973 Program of China under Grant No.2002CB312002.

tional grid. AGRMS employs a two-level resource management architecture and an improved resource barter trade model, which demonstrate good scalability, autonomy, flexibility, extensibility and fault-tolerance, and provide soft quality of service.

The remainder of this paper is organized as follows. In section 2 background and related works are reviewed. Section 3 presents the assumed application characteristics. The design of ARRMS is described in Section 4. Section 5 introduces the current status of the implementation. Conclusion and the future work are finally presented in Section 6.

2 Related Work

Most GRMSs for computational grid, such as AppLeS [3], Condor [4], Globus [5], and Legion [6], lack an economic model for resource trade. Because the motivation for contributing resources to grid has been driven by public good, prize, fun or collaborative advantage, it makes these GRMSs not scale well. [7] presents a number of arguments for the need of market or economy-based mechanisms in GRMS.

Economic Model for Resource trade can be classified into barter trade and trade with currency as mediate. The disadvantages of the former are that it is less flexible and it has limitation on single goods at a time, while the advantages of the latter is that it is flexible and it is suitable for multiple goods exchange, but its process is complicated and the cost of trade is too high, which are caused by the introduction of electronic cash, electronic bank or trade market due to pricing, accounting and payment mechanisms in the process of trade. Distributed computing systems that adopt the latter trade mode can't scale well, because there are some centralized electronic banks or trade markets existing to reduce the complication of trade. In this paper, we proposed an improved resource barter trade model for computational grid, which overcomes the disadvantages of traditional barter trade and reduces the complexity and cost of the trade.

Active network [8, 9] is a novel approach to network architecture in which customized computation can be performed on the fly at routers as active packets pass through them. The user-defined computation carried by an active packet may extend and modify the network infrastructure. Active networks are composed of active nodes, which are either active routers or active terminal nodes (hosts). Active nodes provide execution environment (EE) for user-defined computation carried by the active packets. When an active packet that contains active code or references to code arrives at an active node, the code is extracted or retrieved and executed. The code can modify the content of the packet or the state of the active node, or transmit one or more packets. The active node also provides soft-state cache for the active code to store active data and other state information.

3 Basic Assumptions

To focus our research on grid resource management, we made a simple assumption that applications running on AGRMS are long running SPMD (Single Program Multiple Data stream). In the SPMD model, each processor runs same program and operates on part of the overall data. There are many applications that are characteristic of SPMD model such as operation of two large-scale matrix, crypto attack by brute force, and so on. Furthermore we do not care for the issue of communication and

collaboration between processes that run in parallel on multiple machines to implement the tasks of application. The assumption implies the following issues:
- Only computing resources are considered, so as to simplify resource description. We use a binary (C, M) (representing estimates of compute and memory) to describe resource provided by a machine at a certain moment.
- According to the computational capacity and time the application needs to run in a single machine, the user divides application data into chunks, and requests GRMS to schedule application code and data to multiple machines with resource chunk as a unit. A resource chunk is represented by a triple (Cr, Mr, Tr), which means it takes Tr scheduling intervals to finish the processing of the application data of the chunk with the computing speed of Cr and the memory of Mr.
- Because the type of resource needed is monotonous, only the CPU time and memory are used, it is suitable to adopt the barter trade model. For the convenience of calculating the amount of resource and guaranteeing the fairness of trade, we make a unified price standard, the price of 1MIPS per 1 scheduling interval is 1, and that of 1K memory per 1 scheduling interval is 1.

4 Active Grid Resource Management System

4.1 Architecture

According to the behavior of autonomous system in the IP network, resources are divided into separate autonomous domains. In the domain, a hierarchical structure is constructed where the root node stands for the root of domain, the mediate node stands for sub-domain, and the leaf node stands for computing resources such as supercomputers, clusters, PCs, and etc. In the tree-structured domains, all the nodes are active routers (AR) except for the leaf nodes, root node can be the border router of IP autonomous system or not, and the connection between any two nodes can be either physical or logical. Thus, an active resource tree (ART) is built. Among domains, the root node is used to communicate with other domains, and circuits can exist in topology of domains.

Resource information is disseminated to soft-state cache of each AR in ART, and the cooperation of nodes in ART realizes recourses discovery and schedule, thus building a hierarchical resource management structure within the domain. As the resource broker of each user in a domain, the root of ART (root node) trades on load and repayment with other domains. In the process of trading, the trade entity is domain but not the users of domain. Thus, a scalable two-level hierarchical architecture of resource management system is formed (see Fig.1).

4.2 Resource Dissemination

In the domain, resource manager reports resource state within a week to resource active router (RAR) which neighbors to the resource by the way of resource state matrix RS[7][48]. Each element RS[i][j] of the matrix RS refers to the total number of available CPU and memory resource in a scheduling interval j in day i after today. A scheduling interval lasts for 30 minutes. In the last scheduling interval of each day, resource manager updates matrix RS which is stored in AR's soft-state cache. Each element of matrix RS is 8 bytes and the size of RS is less than 3k. Suppose that RAR

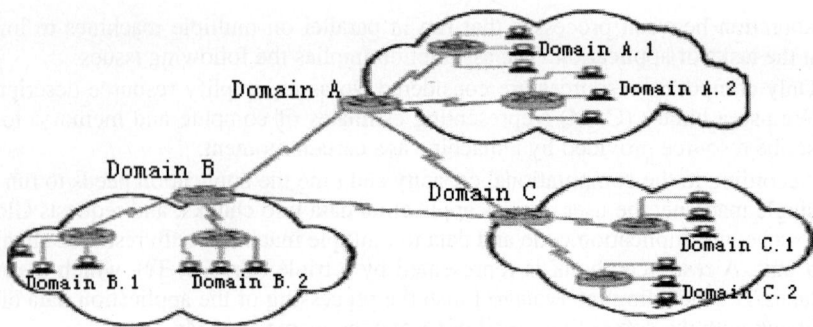

Fig. 1. The architecture of AGRMS.

has n ports, and RS_k stands for resource state matrix reported by port K. RAR computes the sub-domain's resource state matrix RS,

$$RS = (\sum_{K=1}^{n} RS_k) . \qquad (1)$$

then reports the result to parent node. Similar to the RAR, each node reports to parent node about the sub-domain's RS except for root node. According to the number of child nodes and the soft-state cache's free space, AR can store each matrix RS_k reported by each child node, or store its sub-domain's RS to save space.

4.3 Resource Trade Model

Users are taken as trade entities of both sides in traditional barter trade model, which results in the following problems:
- In order to get the debt back, debtee may communicate with debtors that distributed on multiple machines of different domains.
- Demanding thousands of debtors to schedule the job at the same time is impossible, thus the deadline of completing the job cannot be guaranteed.
- Due to the great gap of the time to schedule the job of debtee among different debtors, the communication between processes running in parallel on multiple machines is complicated. This brings difficulty in designing grid application.

In order to solve these problems, we use domain as trade entity to loan and repay resource among domains, and use user and domain as trade entities to loan and repay resource in the domain. Among different domains, root node is the resource agent of every user in the domain. It loans and repays resource from the outside for its clients, and maintains a debt database that records the transactions with other domains. When root node receives resource request from other domains, it will schedule the whole domain's idle resource to provide service; when root domain node receives repayment request from other domain, it will check the debt relation and schedule the whole domain's idle resources to repay. In the domain, each user maintains a balance which is initialized set to zero. When the user obtains resources from either inside or outside of the domain, it will subtract the price of resource from the balance, accordingly, when user provides resources for request which comes from either inside or outside of the domain, it will add the price of resource to the balance. For example, Alice and Mike are users of domain A. They borrow resources valued 100, 300 from domain B, then domain B schedules user John to provide resources valued 40, 100 to Alice and Mike respectively, and schedules user Martin to provide Alice and Mike resource valued 60, 200 respectively. Later, John from domain B requests resource valued 400 outside domain B, then domain B's root node asks domain A to pay back 400. Domain A schedules its user Alice, Mike and Merry to provide resource valued 100, 200, and 100 respectively to repay. If the balance of each above user is 0 at the beginning, when the trade ends, the balance of user Alice, Mike, Merry, John and Martin will be 0, -100, 100, -260 and 260 respectively.

The process of trade is greatly simplified by making use of the methods aforementioned. Meanwhile the cost of communication and scheduling for realizing the trade is reduced, and a large amount of idle resource inside the domain can be scheduled by root node at a scheduling interval. This improves the system's efficiency and scalability; Moreover, it can provide soft quality of service.

4.4 Resource Discovery and Scheduling

4.4.1 Resource Request

The resource request sent from the user includes the following information:

- Rr (Cr, Rr, Tr): definition of a resource chunk.
- number: the number of resource chunks.
- balance: the current balance of requester.
- income: the average daily income that can be earned by requester within a week.
- deadline: the job requested should be completed before one scheduling internal ends. If there's no requirement for deadline, deadline is 0.
- bandwidth: minimum network bandwidth of resource provider.

When receiving resource request, RAR forwards it to parent node. Nodes in ART forward the request to parent nodes until it reach to root node. As agent of user inside the domain, root node requests resource from inside and outside of the domain for its clients.

If root node requests resource in the domain with a deadline, it computes the amount of resource each child node should sell according to resource state matrix RS_k, and notifies each child node to sell a certain amount of resource between some scheduling interval and deadline. After the mediate node of ART receives notification of selling resource, the node does the same as root node does.

When the user receives the above notification forwarded by RAR, it decides how much resource to be sold according to its own selling strategy. Meanwhile, it preserves resource and replies to RAR with the amount of resource to be sold and the updated resource state matrix RS_k. Every user may independently designs its own selling strategy which defines how much resource to be sold according to such factors as risk coefficient, demand of network bandwidth and so on. Risk coefficient is defined as:

$$\text{Risk coefficient} = (\text{number} * \text{price}(Rr) - \text{balance}) / \text{income}. \tag{2}$$

When receiving the responses for all the notifications, the mediate node of ART records the amount of resource to be sold by each child node in soft-state cache, and reports to parent node about the total amount of resource to be sold and the updated resource state matrix of its child nodes. Finally, root node will know how much resource is actually sold to resource requester. If the requester is inside the domain, root node will inform the requester about the amount of resource to be sold. Otherwise, root node will calculate and record liquidated debt between the domain and the requester, and send the requester a notification which carries the information of the amount resource to be sold and the liquidated debt. After receiving the notification signed by resource provider, the requester signs an agreement about the liquidated debt and sends it to resource provider.

If the resource requester send request without deadline, root node will put priority on scheduling users whose balance is negative to sell resource; if the total amount of resource provided by all such users cannot satisfy resource request, the remains should be processed according to the way of that of deadline, and the deadline is the last scheduling interval of that week. In order to implement schedule priority of debtors, each user reports to RAR about its balance. Each AR in ART maintains a debt bitmap in soft-state cache to indicate which child nodes have debtors.

4.4.2 Resource Repayment
When receiving resource repayment notification from other domains, root node checks the validity of the notification base on the debt database. If the notification is not valid, root node demands the requester to send the last liquidated debt signed by root node to verify the sum root node owes.

If the repayment requires a deadline, root node process in the way described in 4.4.1. If there's no deadline, root node process in the similar way described in 4.4.1. The difference is: in the latter case, the user can decide the amount of resource to be sold according to its own selling strategy after receiving selling resource notification; however, in the former case, the amount of resource to be sold is the maximum between the resource that is needed to repay off the debt and the resource that can be supplied within one week.

When receiving resource request, root node first notify every debtors within the domain to repay to avoid that some debtors can't earn money for a long time and can't ask for loans.

4.4.3 Resource Scheduling
After receiving response for resource request from root node, resource requester will send active codes and data which realize computing task to root node. Root node will distribute the active codes and data to child nodes according to the amount of resource preserved by child nodes, and actions of the rest mediate node in ART is the same as root node's. After receiving active code and application data, resource provider will executes active code on assigned scheduling intervals.

4.4.4 Resource Authorization
In the domain, any sub-domain can authorize a user of another sub-domain to consume its resource free or on discount. And it is the same case between domains. If some domain or sub-domain authorizes a user of another sub-domains to consume its resource free or on discount, its administrator will send authorization statements to active router which charges that domain. The active router will store these statements in soft-state cache.

4.4.5 Fault-Tolerance

When root node fails, a new root node will be selected from its child nodes. After logically connecting all child node of the old root, the new root node asks them to report resource information matrix RS, debtor bitmap and so on. It also requests resource provider to report records of trade with other domains. After receiving the response, the new root node computes out debt relation between its domain and other domain. Thus, the state of the new root is the same as that of the old one before it fails. When the mediate node fails, process in the similar way as the previous one does. In addition, root node can store redundant resource proportion in soft-state cache. When some resource provider or mediate node fail, the nodes which detect these failures will report to root node about the amount of resource planned to be provided by these failed nodes. Root node allocates corresponding resource from redundant resource.

5 Implementation Status

We are currently implementing AGRMS on the ANTS [10] toolkit. ANTS is a java-based active network execution environment developed by MIT. It has three components: Capsule, Active Node and Code distribution System. A capsule encapsulates the application data and user customized program. When the capsules pass through active nodes that can be routers, switches or end system, user-defined codes carried in the capsule are executed and routed to destination. Code distribution System provides on-demand code distribution using mobile code.

Most components of AGRMS are implemented by ten types of capsule. Some components have been finished or nearing completion. Other parts remain being developed.

6 Conclusions and Future Work

In this paper, we propose an active grid resource management system (AGRMS) supported by active networks for computational grid. AGRMS employs a two-level resource management architecture and an improved resource barter trade model, which demonstrate good scalability, autonomy, flexibility, extensibility and fault-tolerance, and provide soft quality of service.

To simplify the model and concentrate on resource management strategy, we have not discussed the communication between processes distributed on multiple machines within a grid in this paper. As a supplement, we are doing some research work from the following two aspects. First, we are designing a layered process communication service based on MPI (Message Passing Interface) to accommodate layered scheduling. Second, network bandwidth will be further considered in the resource model, and AGRMS will guarantee soft qualify of service also on bandwidth.

References

1. I. Foster and C. Kesselman (editors), The Grid: Blueprint for a New Computing Infrastructure, Morgan Kaufmann Publishers, USA, 1999.
2. Klaus Krauter, et al, A Taxonomy and Survey of Grid Resource Management Systems, International Journal of Software, Practice and Experience, Volume 32, Issue 2, Pages: 135-164, Wiley Press, USA, 2002.

3. F. Berman and R. Wolski , The AppLeS Project: A Status Report, Proceedings of the Eight NEC Research Symposium, Germany, May 1997.
4. M. Litzkow, et al, Condor - A Hunter of Idle Workstations, Proceedings of the 8th International Conference of Distributed Computing Systems, June 1988.
5. I. Foster and C. Kesselman., Globus: A Metacomputing Infrastructure Toolkit, International Journal of Supercomputer Applications, 11(2): 115-128, 1997.
6. S. Chapin, et al, The Legion Resource Management System, Proceedings of the 5th Workshop on Job Scheduling Strategies for Parallel Processing, April 1999.
7. R. Buyya, et al D. Abramson, J. Giddy, An Economy Driven Resource Management Architecture for Global Computational Power Grids, The 2000 International Conference on Parallel and Distributed Processing Techniques and Applications (PDPTA 2000), Las Vegas, USA, June 26-29, 2000.
8. D.L.Tenenhouse and D.Wetherall, Towards an Active Network Architecture, Proc. Multimedia Comp and Networking 96, MMCN' 96, San Jose, CA, Jan 1996.
9. D.L.Tenenouse et al, A Survey of Active Network Research, IEEE Communications Manazine, 1997, 35(1): 80 - 86.
10. D.Wetherall et al, ANTS: A toolkit for building and dynamically depoloying network protocols, Proc. IEEE OpenArch' 98[C] ,San Francisco, CA, Apr,1998.

Predicting the Reliability of Resources in Computational Grid*

Chunjiang Li, Nong Xiao, and Xuejun Yang

School of Computer,
National University of Defense Technology,
Changsha, 410073, P.R. China
+86 731 4575984
lcj@hnxinmao.com

Abstract. The computational grid built on wide-area distributed computing systems is a more variable and unreliable computing environment, hence it is undoubtedly necessary to predict the reliability of the resources before allocating them to a grid application in order to ensure certain availability for the application. In this paper, we present an algorithm for predicting the reliability of resources based on time series of performance data and status data generated by the Grid Monitoring Service. The predicted reliability information can augment Grid Information Services. Simulation results show that with the reliability information, the grid resource allocation service can guarantee the high availability of the applications in the computational grid.

1 Introduction

The computational grid [1, 2] enables the coupling and coordinated use of geographically distributed and substantially heterogeneous resources for such purposes as large-scale computation, distributed data analysis, and remote visualization. Achieving large-scale distributed computing in a seamless manner introduces a number of difficult problems. One of the most critical problems is resource unreliability, which seriously degrades the availability of grid applications. Researches [3, 4] show that a large wide-area system that contains hundreds to thousands of machines and multiple networks has a small mean time to failure. The most common failure modes include machine faults in which hosts go down, and network faults where links go down. When some resources fail, the applications using such resources have to stall. Additionally, the performance of the resource in grid environment is more variable, can not satisfy the requirement of the application all the time during its lifetime. Moreover, it is very difficult to distinguish a failed resource with a resource with very low performance in distributed systems. So, when a resource fails or its performance can not meet the requirements of the application, the resource seems unavailable for the application.

---

* This work is supported by the National Science Foundation of China under Grant No.60203016; the National High Technology Development 863 Program of China under Grant No.2002AA131010 and 973-2003CB316900

The goal of reliability prediction is to estimate in what percent a resource could serve an application. In our former work [5], a measurement model for the availability of a grid application was presented. The calculation of application availability requires knowledge about the reliability of the resources used by the application. In order to guarantee certain availability for grid applications, it is absolutely necessary to query the Grid Monitoring Service for the reliability of resources when scheduling. But, by now, in typical Grid Monitoring Systems such as Monitoring and Discovery Service (MDS) [6] in the Globus Toolkit which distributed by the Globus Alliance [7], Network Weather Service system [8] and R-GMA [9] et al., such kind of information about the resources is not presented. In this paper we propose the algorithm for predicting the reliability of resources using the data from the Grid Monitoring System. The general Grid Monitoring Service covers two complementary areas: application monitoring and infrastructure monitoring. Both can be subdivided into performance monitoring and status monitoring [10]. This paper focuses on the performance monitoring and status monitoring of Grid infrastructure(resources). Our algorithms use the monitoring data series generated by performance monitoring and status monitoring of Grid resources to predict the reliability of resources.

In the next section, related work about performance and reliability prediction is discussed. In section 3, the predicting algorithm is presented. Section 4 introduces the resource allocation framework which takes the reliability prediction into consideration. The experiments for evaluation is presented in Section 5. In section 6,we summarize our work.

2 Related Work

Much work has been done in the area of performance prediction including host load [11–13] and network performance [14, 15]. But reliability prediction is substantially different with performance prediction. Generally, system-level reliability predictions are developed based on a system model and component-level reliability prediction. Reliability estimation may be based on extrapolated accelerated life testing, an analytical physics-of-failure model, or some other method. Different sources of reliability predictions include company-specific field data, accelerated life test results, physics-of-failure models, public available data (e.g., Reliability Analysis Center (RAC)) and empirical models [16].

In the grid environment, however, it is very difficult to fetch the reliability data sources of so many resources. A feasible way is to estimate the reliability of a resource by the historical status data and performance data generated by the Grid Monitoring Service during its life time.

3 Resource Reliability Prediction in Grid

In this section, firstly, we present the reliability assessment model used in the algorithm for predicting reliability of resources. Then we give detailed implementation of this algorithm in the predictor of resource reliability.

3.1 Reliability Assessment Model

The status of a resource during its lifetime is given in 1.

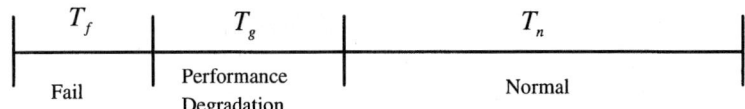

Fig. 1. Status and Time Duration of a Resource in a Time Interval when Applications Running on It

As shown in 1, a time interval in a resource lifetime can be divided into three parts: resource failure, total time duration is T_f; performance degradation that can not support the running the application, total time duration is T_g; normal state that can normally support the running of the application, total time duration is T_n. Then the reliability in this time interval of the resource for the application can be assessed by the following equation:

$$R(Res, app) = \frac{T_n}{T_f + T_g + T_n}$$

3.2 Interval Reliability Assessment

Predicting the reliability of a resource during its whole lifetime is not only difficult and time consuming but also means little for the application running on this resource for the application uses the resource only for a time interval. What an application cares most is the reliability of the resource the application will encounter during its execution. So, we aggregate the historical monitoring data of a resource into a series of time interval, and assess the reliability of the resource in each interval then we get an interval reliability series based on which we can predict the reliability of the resource in the future interval. And we use the expected execution time of the application as the time interval.

We suppose T_e is the expected execution time of an application, and T_i is the time interval of monitoring data, for NWS, is 10 seconds. Then the number of monitoring data points (K) aggregating a time interval can be calculated as $K = \lceil \frac{T_e}{T_i} \rceil$.

Suppose the historical monitor data series of a resource is:

$$H = h_1, h_2, \ldots, h_N$$

we aggregate them into a series containing M interval, and $M = \lceil \frac{N}{K} \rceil$.

For a time interval l which contains K monitoring data points ($h_l, h_{l+1}, \ldots, h_{l+k-1}, (1 \leq l \leq M)$), we use the following method to estimate the reliability of a resource in this time interval.

Firstly, we define the following functions:

$Num_{fail}(h_l, h_{l+1}, \ldots, h_{l+k-1})$: number of data points that the resource is in failure state;

$Num_{deg}(h_l, h_{l+1}, \ldots, h_{l+k-1})$: number of data points that the resource is in performance degradation state;

$Num_{normal}(h_l, h_{l+1}, \ldots, h_{l+k-1})$: number of data points that the resource is in normal state.

Then the estimated reliability of the resource of the interval can be calculated as:

$$R_l = \frac{Num_{normal}}{Num_{fail} + Num_{deg} + Num_{normal}}$$

3.3 Interval Reliability Prediction

Using the interval reliability assessment method in the upper subsection, we can get a series of assessed reliability $R = R_1, R_2, \ldots, R_M$ from the historical status and performance data series. Based on this reliability series, we use a tendency-based time series predictor similar with which presented in [17,18] to predict the reliability of a resource in the future interval. The predicting process is illustrated in 2.

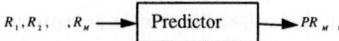

Fig. 2. Function of the Predictor

We use the tendency-based predicting algorithm in 3 in this predictor:

```
       if    ((R_{t-1}-R_t)<0)
                 Tendency="Increase";
       else if   ((R_t-R_{t-1})<0)
                 Tendency="Decrease";
       if   (Tendency="Increase")   then
                 PR_{t+1} = R_t +IncrementConst;
                 IncrementConst adaptation process
       if   (Tendency="Decrease")   then
                 PR_{t+1} = R_t -DecrementConst;
                 DecementConst adaptation process
```

Fig. 3. Tendency-based Predicting Algorithm

In this algorithm, R_t is the reliability value at the t_{th} assessment, PR_{t+1} is the predicted value for the reliability of the next interval R_{t+1}. At each time interval, we measure the real data R_{t+1} and calculate the difference between the current measured value and the last measured value R_t to determine the real increment (decrement) we should have used in the last prediction in order to get the actual value. We adapt the value of the increment (decrement) value accordingly and use the adapted *IncrementConst* (or *DecrementConst*) to predict the next data point. And initially, the *IncrementConst* and the *DecrementConst* are both set to 0.05.

4 Resource Allocation with Reliability Prediction

In this section, we present the framework of resource allocation which takes resource reliability into consideration. There is much work about performance prediction when allocating resources to an application. But in computational grid, not only performance but also reliability must be taken into consideration in order to ensure certain availability for the applications.

In computational grid, before being allocated some resources, the user of the application usually submit a job script which describe the resource requirement of the application such as number of computing node, processor architecture, CPU load etc.

When allocating resources to an application as well as taking resource reliability into consideration, the resource allocation procedure should include following steps:

(1) Choose the resources which meet the architecture requirements of the application;
(2) Measure the current performance of these resources, such CPU load and net-work bandwidth, select the ones that meet the requirement of the application;
(3) Predict the performance of selected resources, choose the ones which could meet the performance requirements with high possibilities;
(4) If takes availability into consideration, predict the reliability of the resources chosen in the upper step, then select the resources with high reliability.

In the following section, by simulation, we will show that resource allocation with reliability consideration is absolutely necessary in computational grid in order to provide high availability service to the applications.

5 Experiments

We evaluate the reliability prediction method by simulation with monitored data from a NWS system running on a cluster of LINUX machines. The NWS records a trace of CPU including CPU load and status (on or fail) for each machine. We use a bag-of-tasks style application for simulation. There are totally 12 machines in the cluster, and the application needs 4 machines among them for execution.

In each trace, a data point is recorded as a $< t, s >$ pair, t denotes the time of monitoring and s denotes the result. If the machine is on, s is the available CPU capacity; else, s is equal to zero. In order to reduce the complexity, if the available CPU capacity is very low (less than 5%), we also set to zero. For a machine with very low available capacity can never meet the application requirement, moreover in distributed computing environment, it is very difficult to distinguish a failed machine from a very slow one.

In the prediction, we suppose the performance requirement of the application is 20% CPU time, and the expected execution time of the application is 5 minutes. We used 80 minutes trace data for simulation, and repeat running the application 10 times. We compared the execution time and stall time of the application under three resource allocation policies: Random, High Performance First, and Performance with Reliability. The definition of these policies is given below:

Random (RD): Randomly select the hosts which current spare CPU time is more than 20%.

High Performance First (HPF): Select the hosts with highest spare CPU time.

Performance with Reliability (PR): Select the hosts with high performance and high reliability.

The execution time of these three policies in our experiment is shown in 4. It is obvious that **PR** policy can substantially improve the performance of the applications.

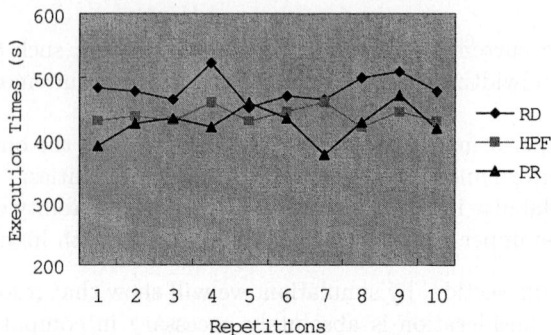

Fig. 4. Comparison of the three resource allocation policies: **RD**, **HPF**, and **PR**

The stall time the application encountered in each repetition is shown in 5. Under **PR** policy, the average stall time of the applications is very low.

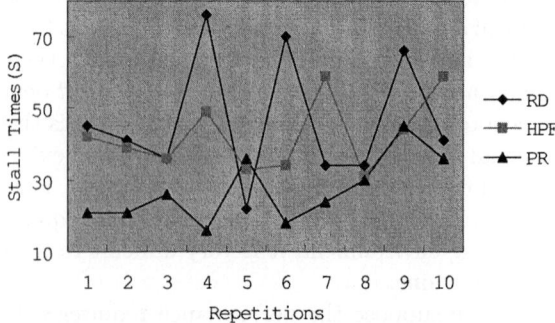

Fig. 5. Comparison of the stall time in each execution of the three resource allocation policies: **RD**, **HPF**, and **PR**

6 Conclusion

We have argued that reliability prediction of resources in computational grid is needed absolutely, and proposed a reliability prediction mechanism in this paper. Taking reliability into consideration, resource allocation service (grid scheduler) can provide high availability guarantee to the applications in grid. Elementary experiments show that the method is encouraging. And the reliability prediction mechanism proposed in this paper can also used to predict the reliability of services which are abstract resources in the service grid environment.

References

1. Foster, I., Kesselman, C.: The Grid: Blueprint for a New Computing Infrastructure. Morgan Kaufmann Publishers (1999)
2. Foster, I.: The grid: A new infrastructure for 21st century science. Physics Today **54(2)** (2002)
3. Bolosky, W.J., Douceur, J.R., Ely, D., Theimer, M.: Feasibility of a serverless distributed file system deployed on an existing set of desktop pcs. Proceedings of the ACM international conference on Measurement and modeling of computer systems, SIGMETRICS (2000) 34–43
4. Bosilca, G., Bouteiller, A., et al: Mpich-v: Toward a scalable fault tolerant mpi for volatile nodes. Proceedings of the International Conference on SuperComputing 2002 (SC2002) (2002)
5. Li, C., Xiao, N., Yang, X.: Application availability measurement in computational grid. Proceedings of the 2nd workshop on Grid and Cooperative Computing (GCC2003), Springer LNCS **3032** (2003) 151–154
6. Czajkowski, K., Fitzgerald, S., Foster, I., Kesselman, C.: Grid information services for distrib-uted resource sharing. Proceedings of the Tenth IEEE International Symposium on High-Performance Distributed Computing (HPDC-10) (2001)
7. WWW: (http://www.globus.org)
8. Wolski, R., Spring, N.T., Hayes, J.: The network weather service: A distributed re-source performance forecasting service for metacomputing. Future Generation Computer Systems **15(5-6)** (1999) 757–768
9. Byrom, B., et al: Datagrid information and monitoring service architecture: Design, re-quirements and evaluation criteria. Technical Report DataGrid-03-D3.2-334453-4-0 (2002)
10. Holub, P., Kuda, M., Matyska, L., Ruda, M.: Grid infrastructure monitoring as reliable information service. In The 2nd European Across Grids Conference, Nicosia, Cyprus, January (2004)
11. Liu, C., Yang, L., Foster, I., Angulo, D.: Design and evaluation of a resource selection framework for grid applications. Proceedings of the 11th IEEE International Symposium on High-Performance Distributed Computing (HPDC 11) (2002)
12. Dinda, P.A., O'Hallaron, D.R.: Host load prediction using linear models. Cluster Computing **3** (2000)
13. Dail, H.J.: A modular framework for adaptive scheduling in grid application development environments. Technical Report CS2002-0698, Computer Science, University of California San Diego (2002)

14. Frogner, B., Cannara, A.B.: Monitoring and prediction of network performance. International Workshop on Advance Issues of E-Commerce and Web-Based Information Systems (1999)
15. Psounis, K., Pan, R., Prabhakar, B., Wischik, D.: The scaling hypothesis: Simplifying the prediction of network performance using scaleddown simulations. ACM SIGCOMM Computer Communications Review **33(1)** (2003)
16. Coit, D.W.: System reliability prediction prioritization strategy. Proceedings of Annual Reliability and Maintainability Symposium (2000)
17. Yang, L., Foster, I., Schopf, J.M.: Homeostatic and tendency-based cpu load predictions. International Parallel and Distributed Processing Symposium (IPDPS2003) (2003)
18. Yang, L., Schopf, J.M., Foster, I.: Conservative scheduling: Using predicted variance to improve scheduling decision in dynamic environments. Proceedings of SuperComputing 2003 (SC2003) (2003)

Flexible Advance Reservation for Grid Computing[*]

Jianbing Xing[1,3], Chanle Wu[2], Muliu Tao[3], Libing Wu[3], and Huyin Zhang[3]

[1] The State Key Laboratory of Software Engineering, Wuhan, 430072, China
[2] National Engineering Research Center of Multimedia Software, Wuhan, 430072, China
[3] School of Computer, Wuhan University, Wuhan, 430079, China
netlab@whu.edu.cn

Abstract. Requests of advance reservation with fixed parameters, i.e. start time, end time and resource capability, may be rejected due to instantaneous peaks of resource utilization. Gaps between these peaks are too narrow for additional requests to fit in. As a result, the call acceptance rate of reservation would decrease dramatically, and the performance of resource may be reduced. In fact, many resource reservations for grid applications don't need fixed parameters. In this paper, a flexible advance reservation is introduced. Its parameters can be modified according to resource status in order to fill the gaps of resource. Particular admission control algorithm for this new type of reservation is provided too. Simulation shows that it can improve performance of resource reservation in terms of both call acceptance rate and resource utilization.

1 Introduction

Grid applications need guarantees of Quality of Service (QoS). Resource reservations, which include advance and immediate reservations, are the effective technologies to provide grid QoS. An advance reservation [1] is a possibly limited or restricted delegation of a particular resource capability over a defined time interval, obtained by the requester from the resource owner through a negotiation process. Immediate reservation can be viewed as advance reservation that starts at "now". Examples of reserved resource capabilities are number of processors, amount of memory, disk space, software licenses, network bandwidth, etc. In this paper we will use network bandwidth as a representation of these resource capabilities.

As more requests of grid resource reservation are admitted, at certain times the resource utilization shows "peaks" increasing over time. Although gaps between these peaks have great resource capability, some reservations with fixed parameters, such as start time, end time and resource capability, cannot utilize these idle resource fragments in gaps. The rejection of reservation request reduces user's satisfaction. And the unused resource fragments decrease the utilization rate of resources.

Demands on resource capability of grid applications are always variable. For example, large-scale data transfer must be finished before a deadline. It is concerned about the total bytes that the network transferred other than the instantaneous transfer rate or the duration of it. As a result, resource reservations for this application could

---

[*] Supported by the fund of State Key Lab of Software Engineering and National High Technology Development 863 Program of China under Grant (No.2003AA001032).

request alterable network bandwidth. Fewer resources will be reserved at daytime when network is busy, while more will be reserved from vacant network at night. The overall transfer rate is kept quite high in order to complete the transfer task in time.

Flexible advance reservation is introduced for these grid applications in this paper. The flexible advance reservation doesn't have all parameters fixed. Some of these parameters are appointed by reservation request, while the resource owner determines the others by taking into account the resource state. Flexible advance reservation makes full use of grid resources, and increases call acceptance rate of requester. A new admission control algorithm for this type of advance reservation is provided in this paper too.

The paper is structured as follows. Section 2 provides an overview of related works and section 3 introduces the conception of flexible advance reservation. Admission control algorithm is presented in section 4. Performance of flexible advance reservation is verified by simulation in section 5. We conclude the paper with a summary of conclusion and future works.

2 Related Works

Early researches on advance reservation were concentrated on reservation protocols, such as RSVP [2] and ST-II [3], admission control mechanisms [4][5], and routing algorithm for network with advance reservations [6][7].

Advance reservations in grid involve not only network resources but also computing and storage resources. GARA [8] supports reservation and allocation of multi-type grid resources. But performance of reservation was seldom considered.

As described in the previous section, fixed parameters in advance reservation may deteriorate the performance of reservations. Burchard [9] proposed a malleable reservation to meliorate this problem to a certain extent. Start time, duration time and required network bandwidth of malleable reservation can be accommodated to avoid the peaks of resource utilization while remaining the total transfer bytes unchanged. These measurements improve the admission probability of reservation requests. But there is only a single bandwidth reserved for one request. So it cannot make full use of resource fragments.

3 Flexible Advance Reservation

In this section, the conception of flexible advance reservation is introduced, which can improve the performance of resource reservations greatly.

3.1 Advance Resource Reservation

In the environment of advance reservation, the period for which requests can be submitted is called *book-ahead period*. Usually, this period is divided into slots of fixed size, e.g. minutes (see Fig.1). There are two phases of an advance reservation: the intermediate phase between the request and the start of the transmission and the actual usage phase where the transmission takes place.

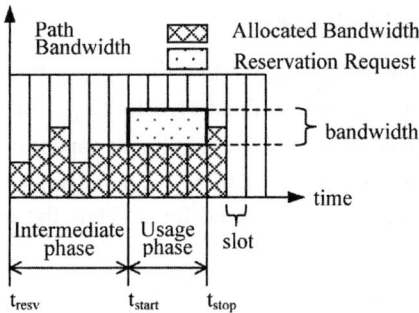

Fig. 1. Advance reservations: slotted time and status storage. A bandwidth advance reservation request is denoted by the dotted rectangle. The position of rectangle means the start time of reservation. The length of dotted rectangle is the duration of reservation request, and the height is the requested bandwidth. The area of rectangle is the total transfer bytes of the reservation.

At least three basic parameters should be defined for a reservation request, i.e. start time t_{start}, stop time t_{stop} and resource-specific parameter, such as network bandwidth or number of CPUs.

3.2 Attribute of Flexible Reservation

We define *reserved resource block* as a constant resource capability, which starts from and stops at a determinate time. It can be denoted as $b=(start, stop, bandwidth)$. Its total transfer bytes, i.e. capability, are $bandwidth*(stop-start)$.

A flexible reservation is defined as $r=(t_{start}, t_{stop}, capability, bw_{min}, bw_{max}, n)$. The parameter t_{start} is the earliest start time of the flexible reservation, t_{stop} is the latest finish time, *capability* is the total bytes need to be transferred, bw_{min} and bw_{max} are constrains to the bandwidth that should be reserved, and n is the maximum number of reserved resource blocks that flexible reservation will have. A complete network reservation request is composed of the flexible reservation and other information, such as IP address of end nodes. It may be denoted as $r_{net}=(node\text{-}a, node\text{-}b, t_{start}, t_{stop}, capability, bw_{min}, bw_{max}, n)$.

Suppose m reserved resource blocks are booked for a flexible reservation. $b(i)=(start, stop, bandwidth)$, $i=1$ to m. Value of the three parameters in each $b(i)$ may be different. It is obvious that:

$$m \leq n. \qquad (1)$$

$$t_{start} \leq b(i).start < b(i).stop \leq t_{stop}, \quad i=1 \text{ to } m. \qquad (2)$$

$$\sum_{i=1}^{m} b(i).bandwidth \cdot (b(i).stop - b(i).start) = capability. \qquad (3)$$

Malleable reservation in [9] is in fact a flexible reservation that has only one reserved resource block. The default value of n is 1. The bandwidth of successfully reserved request can vary for more times as n becomes greater, and the reservation can use more bandwidth fragments.

3.3 Realization of Flexible Reservation

For a flexible reservation $r=(t_{start}, t_{stop}, capability, bw_{min}, bw_{max}, n)$, the transmission needs to be started after t_{start} and be finished before t_{stop}. According to the resource state, at most n resource blocks that have maximum capability between t_{start} and t_{stop} are selected. If the total capabilities of these resource blocks can meet the required capability of reservation, they are reserved for it. Thus the request can be admitted successfully.

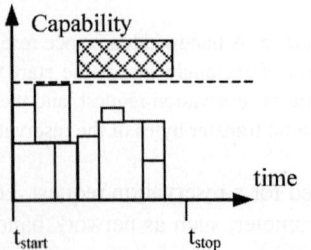 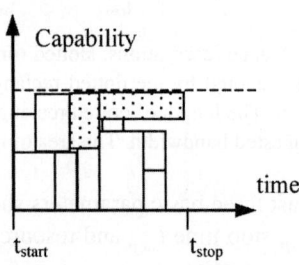

(a) Rejected request of normal reservation   (b) Admitted request of flexible reservation

Fig. 2. Contrast between flexible and normal advance reservations. A flexible reservation (b) is admitted while the normal advance reservation with same capability (a) is denied.

It is shown in the figure that the flexible reservation can reserve more bandwidth when the network is vacant, and reserve less bandwidth when the network is in heavy payload. In this way the request will not be rejected for instantaneously busy network, and can make full use of network bandwidth.

4 Admission Control Algorithm for Flexible Reservation

In the environment of advance reservation, state of each path in book-ahead period is recorded. We use $Bw(t)$ to denote a path state. It is a single-value function of time, which means the vacant bandwidth at time t of the path.

The inputs of admission control algorithm are the flexible reservation request $r_f=(u, v, t_{start}, t_{stop}, capability, bw_{min}, bw_{max}, n)$ and the topology of network $G(V, E)$. u and v are two end nodes of transmission. $G(V, E)$ depicts the topology of network. Dijkstra's shortest path algorithm can be used to find path between nodes u and v. It can be implemented with complexity $O(E*log(V))$. As a common algorithm, its detail is ignored in this paper.

Once the path is found and the state of its bandwidth is available, the following admission control algorithm answers for whether to reject or admit the flexible reservation request. If admitted, the output result is in the form of $r_{out}=(u, v, b(1) \sim b(m))$, as described in section 3.2. The total transfer bytes of all blocks equal the requested one.

AdmControl_Flex ()
Input:
    Network topology $G(V, E)$
    Flexible reservation request $r_f=(u, v, t_{start}, t_{stop}, capability, bw_{min}, bw_{max}, n)$
Output:
    Admitted flexible reservation with m reserved resource blocks $b(m)$.

```
(1).   Bw(t)=Find_path(G(V, E), u, v)
(2).   If ∫_{t_start}^{t_stop} Bw(t)<capability then return NULL
(3).   For each m∈ [1,n] do
(4).       Initiate b(m)
(5).       For each rd∈ [bw_max, bw_min] do
(6).           (t1, t2)=find_max_duration(Bw(t),rd,t_start,t_stop)
(7).           If capability_of(b(m)) ≤ rd*(t2-t1) then
                   b(m)=(t1, t2, rd)
(8).           Done
(9).           Subtract resource block b(m) from Bw(t)
(10).          capability = capability-
                   b(m).bandwidth*(b(m).stop-b(m).start)
(11).          If capability ≤ 0 then break
(12).      Done
(13).  If capability ≤ 0 then
(14).      Revise redundant capability of b(m)
(15).      Reserve b(1) ~ b(m) for the flexible
               reservation request
(16).      return (u,v,b(1) ~ b(m))
(17).  Else return NULL
```

Admission control for flexible reservation can be performed in polynomial time. The algorithm requires at most n cycles of the outer loop (line 3), and $bw_{max}-bw_{min}$ cycles of the middle loop in line 5. The function *find_max_duration* in line 6 needs $t_{stop}-t_{start}$ cycles to find maximum duration of the path in bandwidth rd. Hence, besides the function *find_path* in line 1, which complexity is $O(E*log(V))$, the complexity of the admission control algorithm is $O(n*(bw_{max}-bw_{min})*(t_{start}-t_{stop}))$, which is linear with the number of reserved resource blocks n. In the environment of advance reservation, the spent time of admission control algorithm can be tolerant for that the reservation is made a very long time before it is actually used.

5 Simulation and Result Analysis

A simulation and its results are presented in this section. The simulation shows the advantage of flexible reservation.

5.1 Simulation Environment

A simulation is carried out to contrast the performance of flexible reservation with that of normal advance reservation. The network in the simulation has a single path between node u and v. In other words, path needs not to be found for reservation request in the admission control algorithm. The path provides a bandwidth capacity (denoted by *bw*) of 100 Mbit/s for advance reservations. Array is used as the data structure of bandwidth in this simulation.

We designed a reservation request generator to produce a set *R* of advance reservation requests on the path between node *u* and *v*. Each request includes start time, duration and bandwidth requirement. And the requested capability is the bandwidth multiplying duration.

The simulation period has a length (denoted by *length*) of 20000 slots. One slot is one second. Within this period, requests are generated with Poisson distribution with a mean of 0.9 requests per slot. The duration time is exponentially distributed with a mean value of 150 slots and the bandwidth requirement is uniformly distributed between 2 Mbit/s and 6 Mbit/s. The start time is uniformly distributed between 200 slots and 700 slots after the request arrival time.

In this high payload situation the call acceptance rate (see below) is very low in order to obtain meaningful results, which show differences between normal and flexible advance reservations. Otherwise, both reservation types behave identically, for all of them would be accepted when network is under light burthen.

5.2 Performance Metrics

The performance metrics used for examinations are call acceptance rate and bandwidth usage rate. The call acceptance rate is defined as

$$Rate_{accept} = Number(A) / Number(R) . \qquad (4)$$

where *A* denotes the set of accepted requests and *R* denotes the whole set of requests as mentioned before. The bandwidth usage rate is defined as

$$Rate_{bw} = \sum_{r \in A} capability\ (r) / (bw \cdot length) . \qquad (5)$$

5.3 Simulation Scenario

Flexible reservations are randomly chosen form the set *R*. The minimum and maximum bandwidth are allowed to differ at most 50% from the originally requested bandwidth. This means a normal advance reservation with 2 Mbit/s bandwidth would allow bandwidths of reserved resource blocks of a flexible reservation to change between 1 and 3 Mbit/s. The earliest start time and the latest stop time are at most 25% of the originally defined duration of normal reservation earlier and later, respectively. This means that a request with a given duration of 100 slots is allowed to commence 25 slots earlier than originally specified. The percentage of flexible reservations steps up from 0% to 50%. We repeat the simulation for 3 times with reserved resource block number *n* from 1 to 3 respectively.

5.4 Results and Analysis

The results are shown as below.

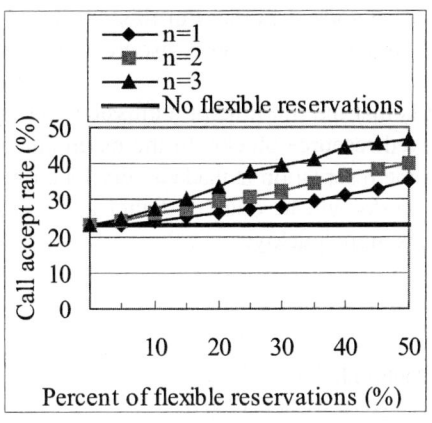 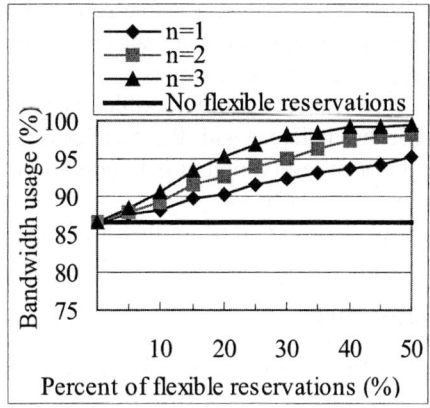

Fig. 3. Call acceptance rate and bandwidth usage rate using varying percentage of flexible reservations.

In Fig.3, the call acceptance rate and bandwidth usage rate are outlined. The heavy black line shows the case where all advance reservations are normal ones. The call acceptance rate in this case is 22.95% and the bandwidth usage rate is 86.54%. With the rising percentage of flexible reservations, both of the two performance metrics increase. When half of normal advance reservations are flexible reservations, the call acceptance rate in the case of $n=3$ is 46.52%, and the bandwidth usage rate is 99.5%.

In the case of $n=1$, the flexible reservation is equivalent to malleable reservation in [9]. When the parameter n have changed from 1 to 3, which means that more reserved resource blocks are permitted in flexible reservations, performance metrics increase accordingly.

It can be concluded from the figure that using flexible reservations is an opportunity to make use of resource fragments and improve the performance of advance reservations. On one hand it can admit more reservations to satisfy the requester. On the other hand grid resource owners can benefit from this mechanism by increasing the resource usage rate.

6 Conclusion and Future Works

In the advance reservation environment of grid computing, a lot of resource fragments are produced while the resource is reserved. The duration of these fragments are often too short to be used by reservation request, although these fragments have great resource capability. A flexible advance reservation mechanism is introduced in this paper. Such type of advance reservation can have variant resource capability reserved, in the form of reserved resource blocks, during its lifetime according to the resource states. Grid resource owners can benefit from making full use of resource fragments. Also grid resource users can be satisfied for more reservation requests are accepted by applying this mechanism. It is verified by simulation that the flexible advance reservation mechanism can improve the performance of advance reservation greatly.

We have discussed flexible reservation on network resource. But this mechanism can really be applied to advance reservations on other types of grid resources, such as computing, storage and others.

A trouble brought by flexible reservation is that the run time of admission control algorithm is linear with the number of reserved resource blocks. In the environment of advance reservation, it is tolerant because the reservation is made a very long time before it actually uses the resource. Future works will be done on better admission control algorithm to admit flexible reservations more quickly.

References

1. The Definition of Advance Reservations. Available online at http://www.fz-juelich.de/zam/RD/coop/ggf/graap/graap-wg.html (2003)
2. Schill, A., Breiter, F.,Kuhn, S.: Design and Evaluation of an Advance Reservation Protocol on Top of RSVP. In: IEIP 4th International Conference on Broadband Communications (BC '98), Stuttgart, Germany. IFIP Conference Proceedings 121, Chapman & Hall, (1998) 23–40
3. Reinhardt, W.: Advance Resource Reservation and its impact on Reservation Protocols. In: Proc. of Broadband Islands '95, Dublin, Ireland, 9 (1995)
4. Ferrari, D., Gupta, A., Ventre, G.: Distributed Advance Reservation of Real-Time Connections. Multimedia Systems, Vol.5. Springer-Verlag, Berlin Heidelberg New York (1997) 187-198
5. Wischik, D., Greenberg, A.: Admission Control for Booking Ahead Shared Resources. In: Proc. Of IEEE INFOCOM '98. 2 (1998) 873–882
6. Guerin, R., Orda, A.: Networks with Advance Reservations: The Routing Perspective. In: Proc. of IEEE INFOCOM 2000. 1 (2000) 118–127
7. Burchard, L.-O.: Source Routing Algorithms for Networks with Advance Reservation. Technical Report No. 2003-2, TU Berlin.
   Available online at http://kbs.cs.tu-berlin.de /publications/res_mgnt/tr-2003-03.pdf
8. Foster, I., Kesselman, C. Lee, C., et al.: A Distributed Resource Management Architecture that Supports Advance Reservations and Co-Allocation. In: 7th International Workshop on Quality of Service (IWQoS '99). (1999) 27–36
9. Burchard, L.-O.: On the Performance of Computer Networks with Advance Reservation Mechanisms. In: 11th IEEE Intl. Conference on Networks (ICON 2003). (2003) 449–454

Club Theory of the Grid

Yao Shi[1], Francis C.M. Lau[2], Zhi-Hui Du[1], Rui-Chun Tang[1], and Sanli Li[1]

[1] Department of Computer Science and Technology,
Tsinghua University, Beijing 100084, China
shiyao00@mails.tsinghua.edu.cn
{duzh,tangrc}@tsinghua.edu.cn
lsl-dcs@mail.tsinghua.edu.cn
[2] Department of Computer Science and Information Systems,
The University of Hong Kong, Hong Kong, SAR, China
fcmlau@csis.hku.hk

Abstract. Grid is a new type of resource sharing and distributed infrastructure. Because of limitations of software and hardware, the service that a certain grid can offer is finite, and so is the number of users. If the number of users is too small, much of the planned resources could turn out to be wasteful; on the contrary, if there are too many users, the excessive loading could substantially reduce the benefit of each user and also the efficiency of the grid service. Therefore, the following are two central problems for grid design: 1. How many users should the grid serve so that each user can receive the maximum benefit? 2. To a certain group of users, how much resources should be invested so that the construction and maintenance the grid become viable? Based on the economic theory of clubs, this paper makes a quantitative analysis of the quasi-optimal size of the grid by regarding grid services and resources as club goods. The results are two important conclusions. We use an experiment involving GridFTP to verify the theory.

1 Introduction

Grid is a new type of resource sharing and distributed infrastructure [1], integrating resources and services in different places and using standard open implements and protocols [2]. It is widely used in the fields of science portal, distributed computing, large-scale date analysis and collaborative work [3].

Since grid is usually applied to distributed computing of a large group of users, its reliability of performance and quality of service are quite important. However, in application, it is difficult to measure dynamic and heterogeneous grid with reliable performance. So in a shared environment, description of service quality and measure of performance become an urgent problem to be solved [4](Chapter 21).

Obviously, one of the crucial factors of the service quality of a certain grid is the number of users it serves. With the increase of users, the quality of service (benefit) that each user gains decreases. Meanwhile, the cost of grid construction and other costs that each user bears also decrease.

In this paper, the number of users who use the services and resources of the grid is defined as the size of the grid. By referring to the club theory in economics [5], we regard the group of users as a typical club, and then study the quasi-optimal size of the grid.

The remainder of this paper is organized as follows.

Section 2 describes the economic theory of clubs and how the grid can be modeled as a club and its resources and services as club goods.

Section 3 constructs two models of the grid and provides some solutions for their optimal sizing. The solutions are summarized as two sets of conclusion and deduction, which show the relationship between various key factors in grid computing including cost, benefit, and number of users.

Section 4 verifies the above conclusions and deductions with an experiment using GridFTP [6].

Section 5 gives a summary.

2 Club Theory

2.1 Economic Theory of Clubs

In economics, goods are divided into private consumption goods and collective consumption goods [7][8][9]. The difference lies in that the former has excludability of benefits and rivalry of consumption while the latter has non-excludability of benefits and non-rivalry of consumption. Goods whose benefits can be withheld costlessly by owner or provider display excludable benefits. Rivalry of consumption means that one person's using some goods or service can prevent others from getting benefits from the goods or service [10].

In 1965, Buchanan put forward his famous economic theory of clubs [5], which claims that there is one kind of goods with excludability of benefits and partially non-rivalry of consumption, which is called club goods, existing between private consumption goods and public consumption goods. Take swimming pool for instance. When it is used by only several people, the entry of somebody else hardly disturbs the use of others. However, when the number of users rises to a certain point, people in the swimming pool begin to feel uncomfortable, and then the increase of users will cause low efficiency of others.

2.2 Characters of Grid Computing

Grid computing has the following two characters.

Excludability of Benefits. The service of grid and sharing of resources are organized. In the definition of virtual organization, Ian Foster claims that "This sharing is, necessarily, highly controlled, with resource providers and consumers defining clearly and carefully just what is shared, who is allowed to share, and the conditions under which sharing occurs. A set of individuals and/or institutions defined by such sharing rules form what we call a virtual organization" [11]. Therefore, only after joining the organization of the grid can a user enjoy the relevant service. Those outside the virtual organization cannot enjoy the services and resources of the grid. In our view, it has two reasons: firstly, the present realization of Grid Computing Environment (GCE) [4](Chapter 20) is based on middleware (Globus Toolkits [12, 13, 14], Legion [15, 16], Java/CORBA CoG Kit [17, 18], etc.), but far from the extent that every terminal serves as a member of the grid; secondly, due to the limitation of the resources of a practical grid, if the grid has no excludability, the infinite increase of users will necessarily lead to a radical drop in the quality of service.

Partially Non-rivalry of Consumption. For a grid with limited resources, when there are only a small number of users, the increase of users will exert little influence on others, i.e. non-rivalry of consumption. However, when the number of users reaches a certain extent, the increase of users will cause obvious decline in service quality, and thus disturb the use of others with congestion. Non-rivalry of consumption does not work at this moment.

Conclusion of the Characters. The two characters mentioned above determine service and resources of the grid as typical service and goods of clubs.

3 Solution and Conclusion of the Model of Club

3.1 Cost and Benefit of Grid Computing

As is mentioned above, grid has excludability of benefits. So each member should pay some cost when joining the grid, and he can obtain benefits at the same time.

To study this problem, we refer to the method used before. Suppose there is a currency of grid called $G [4] (Chapter 3), then cost and benefit have a united standard of measure.

There is no need to give further explanation of considering the cost of gird-construction in the form of currency, since it is very clear. As for transforming benefits of CPU resource, storage resource and media resource into the form of currency (i.e. establishing a mapping from information amount to the grid currency), it is a more reasonable description way than using only information amount. And thus, benefits vary in terms of importance. For instance, there are two grids with the same amount of information. One provides military information while the other provides entertainment information. Obviously, their importance differs. The former is more valuable than the latter, because it has a stricter requirement for the quality of service. If we measure benefits only by information amount, the two grids will generate the same benefit, which goes against the practical situation.

We have only explained the rationality of mapping information amount to grid currency. In application, the mapping from information amount to grid currency $G depends on the specific circumstance, which is beyond the scope of this paper.

3.2 Basic Assumptions of the Model

The model is based on the following two assumptions:

Assumption 1. (Rational user): Every user pursues the maximum benefit.

Assumption 2. (Comparability of cost and benefit): Since both cost and benefit can be measured by the grid currency $G, they are comparable and addible.

3.3 Simplified Model of Remote Service

To make our model simple, we make the following assumptions:

Assumption 3. (No difference among users). All users enjoy the same amount of service. (If the amount of service is not the same, they can be regarded as different users.)

Assumption 4. (Only one service exists). In the model, there is only one type of grid service, and no other service is taken into consideration.

Suppose the number of users is N; total cost is C; the benefit a typical user can get from the service is TB; when N changes, the change of cost of a user is MC and the change of benefit is MB.

$$MB = \frac{dTB}{dN}. \tag{1}$$

If MB>0, benefit increases. If MB<0, benefit decreases; in other words, user has suffered loss, or there generates the cost, called congestion cost.

From Assumption 3, the cost that every user bears is C/N, therefore,

$$MC = \frac{d(C/N)}{dN} = -\frac{C}{N^2}. \tag{2}$$

According to Assumption 3, since every user is the same, if one more user enters, total congestion cost for all the users increases. The net benefit of user U(N) is benefit minus cost, that is,

$$U(N) = TB - C/N. \tag{3}$$

Maximum of U(N) equates maximum of the net benefit of a single user,

$$U_{max}(N) = MAX\{TB - C/N\}. \tag{4}$$

From (3), we get the solution of (4),

$$U'(N) = \frac{dU(N)}{dN} = \frac{dTB}{dN} - \frac{d(C/N)}{dN} = MB - MC = 0$$
$$\Rightarrow MB = MC = -C/N^2 \Rightarrow -N \cdot MB = C/N. \tag{5}$$

Conclusion 1. When total congestion cost equates the average cost every user bears, each user achieves the maximum net benefit, and the size of the sharing service is optimal.

In this conclusion, we should pay attention to the following two points:
1. Rule out the non-maximum points.
2. The utility function U(N) may have several extrema. In this case, we can regard the several larger values as the quasi-optimal ones.

Generally, the trend is that MB decreases with the rise of N, so,

Deduction 1. The optimal size of users increases with the rise of the total cost and decreases with the fall of the total cost. Especially when we take no notice of cost (C=0, i.e. regard the grid as an ideal public infrastructure that requires no payment), each user gains maximum benefit when the change of his benefit is 0 (i.e. the turning point between the rise and fall of benefit).

3.4 General Model

If we do not make the two consumptions in 3.3, i.e. take the difference of users and variety of grid services into account, and suppose each user receives local service (local service can be considered as the gird service that can be used by only one person).

The utility function of the ith user is,

$$U^i[(X_1^i,N_1^i),(X_2^i,N_2^i)\cdots(X_n^i,N_n^i)] . \qquad (6)$$

In this function, X_j^i is the amount of jth service or resource, and N_j^i indicates the number of users sharing jth service or resource. Obviously, $\partial U^i/\partial X_j^i > 0$, which shows that if other conditions stay the same, the increase of service leads to the increase of individual benefit; $\partial U^i/\partial N_j^i > 0$, which indicates that if other conditions stay the same, the increase of users sharing the service makes the individual benefit decrease.

As the total cost of the system is fixed, the distribution of service is restricted. Suppose the restriction of ith user's service is,

$$F^i[(X_1^i,N_1^i),(X_2^i,N_2^i)\cdots(X_n^i,N_n^i)]=0 . \qquad (7)$$

In this function, X_j^i is the amount of jth service or resource, and N_j^i indicates the number of users sharing jth service or resource. $\partial F^i/\partial X_j^i > 0$, i.e. when other conditions remain unchanged, the more service there is, the more restriction the ith user has in using resource. $\partial F^i/\partial N_j^i > 0$, i.e. when other conditions remain unchanged, the more people share the service, the less is the cost of service assigned to each user, and thus F decreases.

According to (6) and (7), this model can be thought as a problem about the maximum of a non-linear programming.

$$\max\{U^i[(X_1^i,N_1^i),(X_2^i,N_2^i)\cdots(X_n^i,N_n^i)]\} . \qquad (8)$$

$$s.t. \ F^i[(X_1^i,N_1^i),(X_2^i,N_2^i)\cdots(X_n^i,N_n^i)]=0 . \qquad (9)$$

By using Lagrange multiplier method [19] on (8) and (9), we get the Lagrange function,

$$L^i[(X_1^i,N_1^i),(X_2^i,N_2^i)\cdots(X_n^i,N_n^i)]=$$

$$U^i[(X_1^i,N_1^i),(X_2^i,N_2^i)\cdots(X_n^i,N_n^i)]+\lambda F^i[(X_1^i,N_1^i),(X_2^i,N_2^i)\cdots(X_n^i,N_n^i)] . \qquad (10)$$

In (10), suppose the partial derivatives of X_j^i and N_j^i equate 0, we have the following system of equations,

$$\left.\begin{array}{l}\dfrac{\partial L^i}{\partial X_j^i}=\dfrac{\partial U^i}{\partial X_j^i}+\lambda\dfrac{\partial F^i}{\partial X_j^i}=0\\[6pt]\dfrac{\partial L^i}{\partial N_j^i}=\dfrac{\partial U^i}{\partial N_j^i}+\lambda\dfrac{\partial F^i}{\partial N_j^i}=0\end{array}\right\}\Rightarrow\left\{\begin{array}{l}\dfrac{\partial U^i/\partial X_j^i}{\partial U^i/\partial X_k^i}=\dfrac{\partial F^i/\partial X_j^i}{\partial F^i/\partial X_k^i}\cdot\quad(11)\\[6pt]\dfrac{\partial U^i/\partial X_j^i}{\partial U^i/\partial N_j^i}=\dfrac{\partial F^i/\partial X_j^i}{\partial F^i/\partial N_j^i}\cdot\quad(12)\end{array}\right.$$

(11) indicates that for any two services, their marginal rate of substitution (MRS) equates their marginal rate of transformation (MRT) [20]. Here, marginal rate of substitution shows how many units of service B a user is willing to give up so as to achieve one unit of service A, i.e. marginal benefit of using service A. Marginal rate of transformation indicates under the vested benefit and cost, how many units of service B a user should give up so as to increase one unit of service A, i.e. marginal cost of using service A; this cost can also be measured by service B. If all the services are the same, in other words, there is no difference in quality or speed, (11) stays true undoubtedly.

(12) indicates that for any service and the number of people sharing that service, their marginal rate of substitution equates marginal rate of transformation. Here, marginal rate of substitution shows the number of users that can be increased when there is one more unit of service offered (the benefit of each user remains unchanged). Marginal rate of transformation shows the number of users that should be increased when there is one more unit of service offered.

From the derivation above, we get the following conclusion.

Conclusion 2. In the general model of the grid, individual user achieves the maximum net benefit when the following two conditions are satisfied:

1. For any two services, their marginal rate of substitution equates marginal rate of transformation.
2. For any service and the number of users sharing that service, their marginal rate of substitution equates marginal rate of transformation.

The size of the grid that meets the requirement of both (11) and (12) is called the club size of the grid.

Deduction 2. In a general model of the grid, if all the services are the same, meeting condition 2 in Conclusion 2 is enough for a user to gain the maximum net benefit.

4 A Case for the Model of Club

4.1 Environment of Testing

The computer used in this test is DeepSuper-21C supercomputer [21], which ranks the 163rd in the Nov. 2003 list of the Top 500 supercomputers [22], developed by Institute of High Performance Computing, Tsinghua University. The test uses 31 nodes in the cluster. The configuration of each machine is as follows,

Table 1. Hardware configuration.

CPU	Daul 3.06G Xeon processors
Memory	1 giga bytes DDR memory
Hard disk	80 giga bytes IDE hard disk
Network adapter	Myrinet network adpter(2Gbps bandwidth)
Operating system	Red Hat Linux 8.0
File system	Ext 3

Globus Toolkit 3.2 with all services [11] installed on each machine. One machine is the server and the other 30 are the clients. Each client can run several processes, each of which simulates one user who visits the server.

The test uses the GridFTP [6] tool originally brought by Globus Toolkit 3.2. In the test, each client runs 1 to 20 processes. Since there are 30 clients in all, there are 30, 60 ... 600 connections. All the processes download different files from the server during the same period (download different files so that the cache will not be used too often). The size of each file is 16 mega bytes (128 mega bits).

4.2 Result of Test and Analysis

The result of test is described in Fig. 1 to Fig. 3, the symbols "*" in these figures show the measured data.

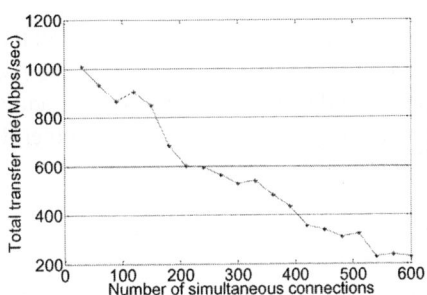

Fig. 1. Total transfer rate versus number of simultaneous connections.

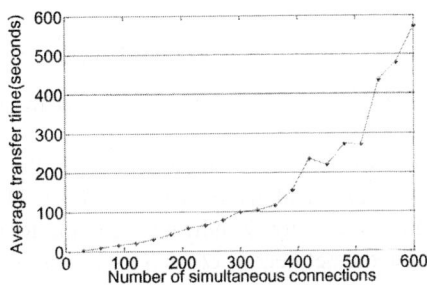

Fig. 2. Average transfer time versus number of simultaneous connections.

We analyze the result of the test according to the simplified model of 3.3.

Fig. 1 shows that when the number of simultaneous connections rises, the total transfer rate at which all the users download the files is not always the same, but declines. There are two causes: 1. The increase of users enhance the load of the server, so IO rate decreases; 2. The number of processes increases and each process occupies some resources of the server.

Fig. 2 shows that when the number of simultaneous connections increases, the average transfer time at which each user download the file is on the rise. Fig. 3 shows that with the increase of simultaneous connections, the average transfer rate decreases. The result of Fig. 2 and Fig. 3 is obvious.

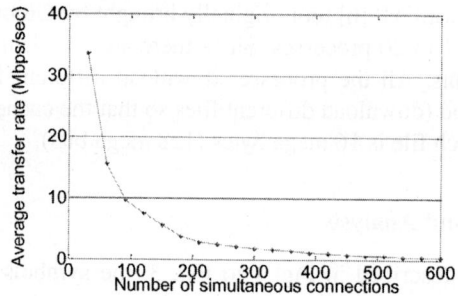

Fig. 3. Average transfer rate versus number of simultaneous connections.

In order to get the approximate average transfer rate (ATR) corresponding to each user, we use interpolation on the 20 points in Fig. 3, in the range of 30 to 600 users. To keep the data monotonic and as exact as possible, we adopt piecewise cubic Hermite interpolation [23] and thus get Fig. 4, which is smoother than Fig. 3.

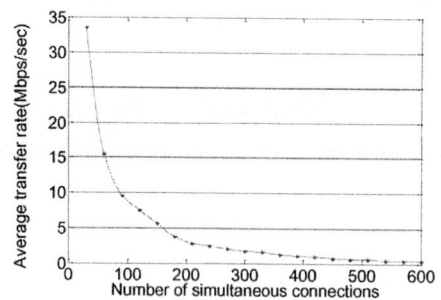

Fig. 4. Adopt piecewise cubic Hermite interpolation on the points in Fig. 3.

Now, suppose the value of each mega bits data is K$G, the benefit of each user in each unit of time is TB= K·ATR, and his net benefit is TB-C/N=K·ATR-C/N.

When K=2, Fig. 5, Fig. 6 and Fig. 7 indicate when the total cost equates 500$G, 2200$G and 4500$G, how the net benefit of each user changes with the changing the number of users.

When drawing the curves of Fig. 5-7 in one diagram, we get Fig. 8.

Seen from Fig. 5, when the total cost is very low, the fewer users there are, the more net benefit each user enjoys. It is because when the cost that each individual bears is very low, the benefit shared by extra users is far more than the cost shared by them.

Seen from Fig. 6, when the total cost increases, the function, which shows the relation between individual user's net benefit and the number of users, vibrates in a small range. Net benefit of each user when the number of users is 50, 150, 350 and 500 is much at one.

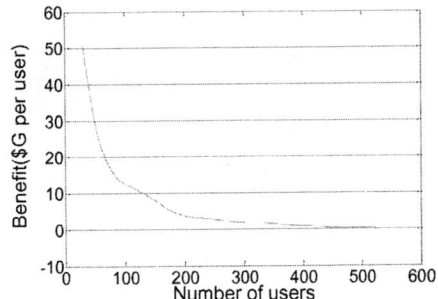

Fig. 5. Benefit curve (2$G/Mbits, total cost = 500$G).

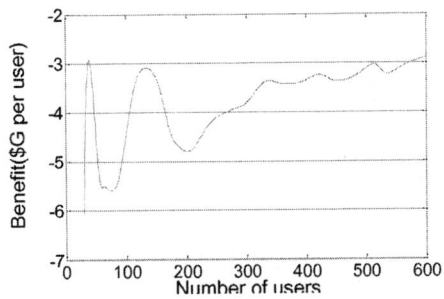

Fig. 6. Benefit curve (2$G/Mbits, total cost = 2200$G).

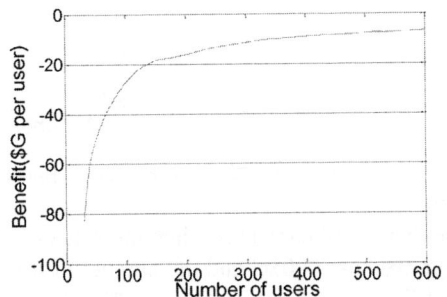

Fig. 7. Benefit curve (2$G/Mbits, total cost = 4500$G).

According to Conclusion 1, from Fig. 6, we can get several extrema, and the larger ones, where the total congestion cost of all users equates the cost shared by each user, are the quasi-optimal size of users.

Fig. 7 shows the case of high total cost. In this case, the more users there are, the more net benefit each user has. Since the cost is too high, more users are needed to share the cost.

Take Fig. 5 to Fig. 8 into account, we arrive at the conclusion that the optimal size of users increases with the increase of the total cost, and decreases with the fall of the total cost, as is predicted in Deduction 1.

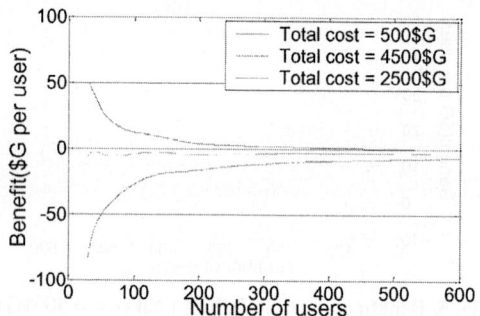

Fig. 8. Curves of benefits (2$G/Mbits).

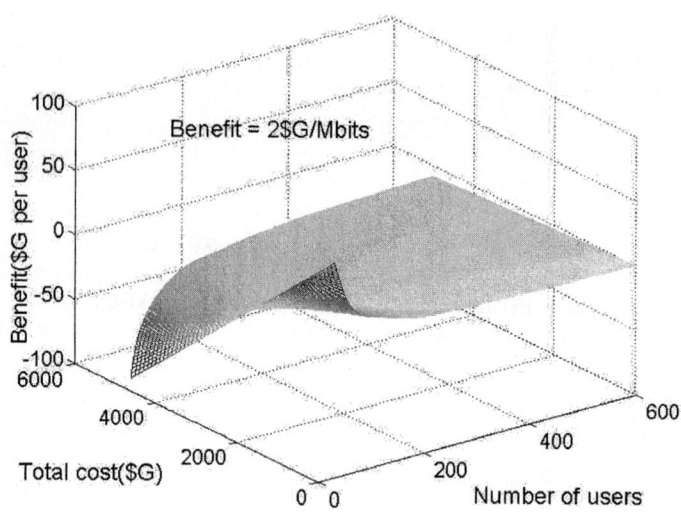

Fig. 9. Statistic of benefit, total cost and number of users (Benefit = 2$G/Mbits).

Fig. 9 shows the net benefit of each user when the value of each mega bits data is 2$G, cost changes from 0$G to 5000$G and the number of users changes from 30 to 600. Fig. 5, Fig. 6 and Fig. 7 are three sections of Fig. 9, which shows the relation between the number of users and net benefit.

The camber in Fig. 9 is representative. To demonstrate this point, we generate Fig. 10, which shows the net benefit of each user when the value of each mega bits data is 5$G, the cost changes from 0$G to 10000$G, and the number of users changes from 30 to 600. Fig. 9 and Fig. 10 are similar in shape, only different in the range of values.

If we use the value of benefits from other test data, the shape of camber may vary that in Fig. 9. However, the fundamental trend of changing is the same, corresponding with the description in Conclusion 1 and Deduction 1.

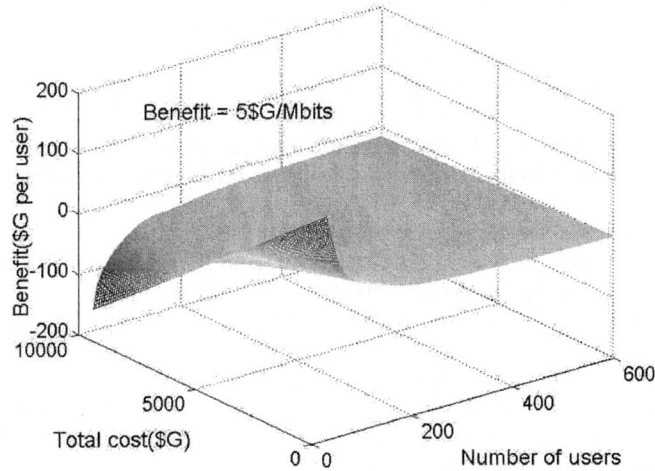

Fig. 10. Statistic of benefit, total cost and number of users (Benefit = 5$G/Mbits).

5 Summary

This paper presents the club theory in grid computing. In this theory, grid resources and services are regarded as club goods, and users as club members. The theory is meaningful because it facilitates the derivation of the most suitable number of uses when the cost for constructing and maintaining a grid is fixed, and it can be used to work out the optimal size of the grid and the optimal cost incurred.

The club theory of the grid can be applied to measure whether an existing grid is being used adequately or excessively. In planning for a new grid, this can prevent inappropriate or over-investment.

The club theory can be applied not only to grid computing, but also to other forms of distributed computing.

References

1. Foster, I.: Grid Computing. AIP Conference Proceedings. 583 (2001) 51-56
2. Foster, I.: What Is The Grid? Three Point Checklist.
   http://www.globus.org/research/papers.html
3. Foster, I.: THE GRID: A New Infrastructure for 21st Century Science. Physics Today. 55 (2002) 42-47

4. Berman F., Hey A., Fox. G.(eds.): Grid Computing – Making the Global Infrastructure a Reality. John Wiley & Sons, Ltd., Chichester (2003)
5. Buchanan, J. M.: An Economic Theory of Clubs. Economica. 32 (1965) 1-14
6. Allcock, W., Bester, J., Bresnahan, J., Chervenak, A., Liming, L., Meder, S., Tuecke, S.: GridFTP Protocol Specification. GGF GridFTP Working Group Document. September 2002.
7. Bowen, H. R.: The International of Voting in the Allocation of Economic Resources. Quarterly Journal of Economics. 58 (1943) 27-48
8. Samuelson, P. A.: The Pure Theory of Public Expenditures. Review of Economics and Statistics. 36 (1954) 387-389
9. Samuelson, P. A.: Diagrammatic Exposition of a Theory of Public Expenditure. Review of Economics and Statistics. 37 (1955) 350-356
10. Cornes R., Sandler T.: The Theory of Externalities, Public Goods, and Club Goods. Cambridge University Press, Cambridge (1986) 6-7
11. Foster, I.: The Anatomy of The Grid: Enabling Scalable Virtual Organizations. Cluster Computing and the Grid. 2001. Proceedings. First IEEE/ACM International Symposium on. 6-7
12. The Globus Project: http://www.globus.org
13. Foster I., Kesselman, C.: The Globus Project: A Status Report. Proc. IPPS/SPDP '98 Heterogeneous Computing Workshop. (1998) 4-18.
14. Foster I., Kesselman C.: Globus: A Metacomputing Infrastructure Toolkit. Intl J. Supercomputer Applications. 11 (1997) 115-128
15. The Legion Project: http://legion.virginia.edu
16. Grimshaw, A., Ferrari, A., Knabe, F., Humphrey, M.: Legion: An Operating System for Wide Area Computing. Dept. of Comp. Sci., U. of Virginia, Charlottesville, Va. Available at ftp://ftp.cs.virginia.edu/pub/techreports/CS-99-12.ps.Z.
17. von Laszewski, Foster, I., Gawor J., Lane, P.: A Java Commodity Grid Kit. Concurrency and Computation: Practice and Experience. 13 (2001) 643-662
18. von Laszewski, G., Parashar M., Verma, S., Gawor, J., Keathey, K., Rehn N.: A CORBA Commodity Grid Kit. 2nd International Workshop on Grid Computing in conjunction with Supercomputing 2001.
19. Hadley, G.: Nonlinear and Dynamic Programming. Addison-Wesley Publishing Company, Inc., Reading, Mass. (1964)
20. Hicks, J. R.: Value and Capital: An Inquiry Into Some Fundamental Principles of Economic Theory. The Clarendon Press, Oxford (1941)
21. http://hpclab.cs.tsinghua.edu.cn/~top500/
22. http://www.top500.org/lists/2003/11/
23. Davis, P. J.: Interpolation and Approximation. Blais Dell Publishing Company, New York (1963)

A Runtime Scheduling Approach with Respect to Job Parallelism for Computational Grid

Li Liu, Jian Zhan, and Lian Li

Computer Science Department, Lanzhou University, Lanzhou, 730000, China
liliu03@st.lzu.edu.cn

Abstract. Researches have demonstrated that runtime scheduling with respect to parallelism can provide a low-overhead load balancing with global load information. In this paper, we present a new approach that is with respect to job parallelism for computational grid. The approach has the capability of redressing the amount of parallel jobs automatically through grain size and other significant factors we specified. As all the jobs parallelized have been divided by an aptotic grain size before being scheduled, the approach can analyze and manage the affiliation of parallel jobs automatically. The approach is so polished that scheduling can be adjusted in real time with little delay. It provides high-quality load balancing for parallel jobs and is universal for computational grid making every grid resource work efficiently, not only as a good scheduling scheme but also as a scheme for scheduling and managing job parallelism.

1 Introduction

Computational grids[1][2] are becoming more attractive and promising platform for solving large-scale computing intensive problems. Various geographically distributed resources are logically coupled together and presented as a single integrated resource[3]. Resource scheduling becomes very important for a computational grid. Besides, in our environment, we introduced a job parallelism scheme on fine-grain level for the computational grid. And how to manage the jobs dispatched parallel to match resources efficiently is a pivotal issue. Parallel scheduling activity is performed at run time. Therefore, it can deal with dynamic job submission. The possible load imbalance caused by the grain size of current turn would be corrected by the next turn of scheduling[4]. So we defined a set of estimation formulas to appraise the parallelism scheduling and redress the grain size and other parameters automatically.

This paper presents an integrated resource management and parallelism job runtime scheduling scheme for computational grid. In section 2, we take a formulation to educe a parallelism algorithm based on instruction level that we developed and other parallelism algorithms. In section 3, computational grid architecture and job flow for the approach are introduced in detail. Section 4 some significant formulas and estimation standards of the approach are take into account. Section 5 focuses on the implementation of the approach and the performance evaluation of this scheme. Finally, we draw the conclusion and give out future work in section 6.

- Sends them in parallel from the holding container to a donor where they can be executed
- Combines parallel jobs in a correct order and form a new job into the job queue
- Judges the time when result shall be send back to users
- Manages them during job parallelism and job execution
- Logs a record of their execution when they are finished, including success and failure.

When a set of jobs are submitted to the job queue, firstly they are sent to be parallelized. We use Common Parallelism Approach (CPA for short) to parallelize the dependent jobs into independent ones. The parallelism approach satisfies "once a job is executable, run it promptly". Then the jobs can be sent to the scheduler which takes into account the order in which the job was submitted, what donors are available, and the priority of the job. The scheduler places all the independent jobs in a single pending list and continuously re-evaluates the availability of resources and priority of all parallelized jobs in the grid. If a donor has the resources available, this approach will automatically move the highest priority job to the queue on that donor to begin execution immediately.

4 The Runtime Scheduling Approach with Parallelism

In this section we combine the fine-grain size parallelism and scheduling theories together to model Runtime scheduling computational grids.

$$E = \frac{\int_{t_1}^{t_2} \frac{Ncpa(t)}{Np(t)} dt}{t_1 - t_2} \quad (1)$$

E is the average parallelism efficiency between execution time t_1 and t_2. $N_{cpa}(t)$ presents the number of independent executable jobs parallelized by CPA at point t. $N_p(t)$ presents the number of independent executable jobs parallelized at most in theory at point t. So we can find easily that $E \leq 1$. If at time t $N_{cpa}(t) = N_p(t)$, we say CPA has reached its peak at t. E can estimate the degree of parallelism at a period of time. However, considering the capacity and the throughput, it is not the best one that E equals 1 at any given period of time because of load balance and throughput. Grid Manager can adjust the capacity and throughput of parallelism by E.

As we parallelize jobs on the fine-grain level, we must specify the grain size we parallelize those jobs by, e.g. the job

$$Greater(Add(A, B), Multi(A, C))$$

has three operations, and it can be decomposed three sub jobs as Greater(E1,E2), Add(A,B) and Multi(A,C). Each of them has one operation. We do parallelizing on the level of sub jobs which have the same number of operations.

$$\left\lceil \frac{OP_{job}}{gsize} \right\rceil \leq N_{sub}(gsize) \leq OP_{job} \quad (2)$$

$N_{sub}(gsize)$ means the number of sub jobs that have been divided by the given grain size $gsize$. It is actually decided by the location of operations. OP_{job} presents the operation number of jobs.

$$N_{cpa}(t) = \sum_{dj(t)} N_{sub}(gsize) \qquad (3)$$

dj(t) presents the divided job at t. If $gsize$ is too small, $N_{cpa}(t)$ will be so large that it would take more time to process the dependence among sub jobs. On the contrary, the parallelism degree would be too low if the $gsize$ is too large. So grid manager should adjust $gsize$ at any time to satisfy the load balance.

Let's consider the relationship between the number of parallelized jobs and the number of completed jobs at a period of time.

$$perI(t_1, t_2) = \frac{\sum_{t1}^{t2} N_{COMPLETE}(t)}{\sum_{t1}^{t2} N_{cpa}(t)} \qquad (4)$$

$$perII(t_1, t_2) = \frac{\sum_{t1}^{t2} N_{COMPLETE}(t)}{\sum_{t1}^{t2} N(t)} \qquad (5)$$

$$\Delta perI(t_1, t_2) = perI(t_1 + \Delta t_1, t_2 + \Delta t_1) - perI(t_1, t_2) \qquad (6)$$

$$\Delta perII(t_1, t_2) = perII(t_1 + \Delta t_1, t_2 + \Delta t_1) - perII(t_1, t_2) \qquad (7)$$

$perI(t_1, t_2)$ means the percentage of the amount of finished jobs compared to the amount of executable jobs between t_1 and t_2. $per(t_1, t_2)$ means the percentage of the amount of finished jobs compared to the amount of the jobs submitted. $N_{COMPLETE}(t)$ denotes the amount of the finished jobs at t. $N(t)$ denotes the amount of the jobs at t. Grid Manager would adjust $gsize$, f and C automatically to reach a better $perI$ and $perII$.

For each job, resource matching should be done after parallelizing. Find out the satisfied resource to the job. Estimate the total processing time T_{total} on each job.

$$T_{total} = FRT + T_{transfer} + Wait + T_{run} \qquad (8)$$

FRT means the delay time finding out the matching resource. $T_{transfer}$ denotes the transfer time on the net. It is dependent on the transmitted data size and the network bandwidth the job passes by. $Wait$ is the waiting time for getting computation on the donor's buffer queue. Every job may have its $Wait$ own because of different priority. And each donor has its particular buffer queue size, so $Wait$ is different on different donor for different job. T_{run} is the execution time on the job. Besides the difficult of the job, this parameter is dependent on the capability of the donor the job run on. The donor can be a single processor, multiple processor group or a cluster even another grid. So T_{run} may be discrepant because of choosing different donor.

Because all the submitting jobs to a matching donor are formed of a fixed size $gsize$, we can presume all the jobs have the same computation difficulty. So we can specify

$$T_{deadline} = C \times f(gsize) \tag{9}$$

f is an incremental function that can be defined by grid manager. C is a constant that relies on concrete system. If a job occupies the computation resource on a donor all along, it shall be detected by the manager. So we specify a parameter $T_{deadline}$ to present the deadline which constraint a job's execution time. Compute as following

$$T_{out} = T_{total} - T_{deadline} \tag{10}$$

Once $T_{out} > 0$ we deem the job is dead and we would resubmit the job or abandon it.

For a given set of jobs, because of the parallelization and resource matching, every time to execute the same jobs may confront that E and *per* are different. So we need average data to find the statistical optimal $gsize$, f and C.

5 Implementation of the Approach Based on MICE

5.1 MICE

MICE is the abbreviation of Mathematics Internet Computing Environment. It is a new Grid Computing Model based on Specification Oriented Programming method and also is an application of Grid Computing technologies. This architecture enables the mathematics software resources to be deployed as services on the internet and advances the integration and utilization of mathematics software resources.

As a runtime scheduling approach with respect to job parallelism the approach is implemented in the MICE[6] prototype system. MICE provides a simple programming model and an execution vehicle for applications with unstructured and(or) dynamic parallelism on network computing system. The MICE runtime system allows dynamically changing set of computing resources and scheduling resources runtime controlled by MICE manager. All jobs waiting to be sent to a matching resource are executable and any pair of them is independent to each other. In particular, these jobs get the same amount of operations. Based on the adjustment parameters, MICE can adjust the parallelism grain size and control the throughput of jobs which are waiting for relative resources.

5.2 Experimental of Results

In this section, we evaluate the parallelism grid scheduling approach from the viewpoints of grid jobs. The process of parallelizing, matching, scheduling, and executing is performed so that some metric delivered to the requestor is maximized[7].

We have tested a set of groups under the same jobs and the same environment. We choose three of them to produce utilization percentage and as Fig.2 shows.

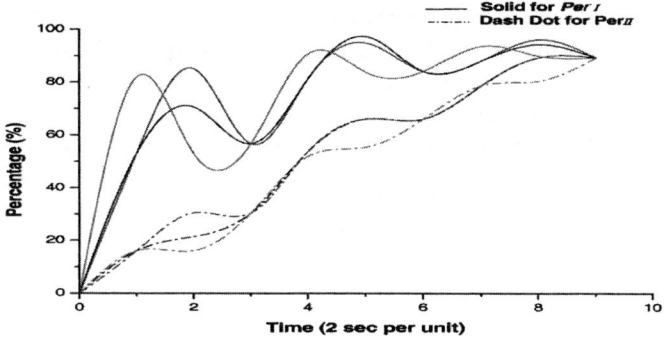

Fig. 2. perI(0,t) and perII(0,t) for a set of parallelism jobs

Because of the indetermination of the executing order of parallel jobs, these groups got some differences. At the beginning, it is generally high as there are a lot of parallel jobs. Because the relationships between two jobs are very exiguous in the beginning, the acceleration amplitude is great. And the dithering is so sharp on account of that there are many executable jobs at a period of time because of the completion of a job and at another period of time amount of jobs are waiting for the completion of a dependent job, so it is mild when it reaches its partial nadir and ascends rapidly to arrive at its partial vertex.

Analyzing the relationship between grain size and completion percentage as Fig.3 and Fig.4 shows.

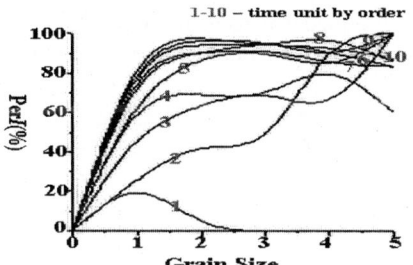
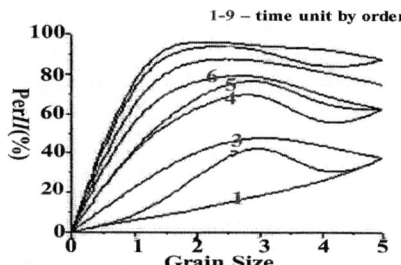

Fig. 3. Relationship between perI(0,t) and grain size using B-Spline

Fig. 4. Relationship between perII(0,t) and grain size using B-Spline

In order to gain a better resource utilization, Grid Manager would choose an appropriate grain size with which each pair of neighbor time units get the minimal differences. In our test, grain size 2 or 3 obtains a minimal difference relative to other ones.

6 Conclusions

In this paper, we discussed an effective technique for computational grid with respect to job parallelism on a given grain size which is exactly not greater than statement or instruction level. Although the speedup is modest, it should be noted that this runtime scheduling succeeds where other methods fail to detach different jobs (or programs) parallel executing. The variety of architectures benefiting from this approach underscores its importance. Although an NP-Complete scheduling problem, this approach has numerous effective heuristics that have been developed. Resources can be automatically assigned to a job under the estimated standard we specified and Grid Manager can adjust those regulable parameters to let scheduler reach a better resource utilization.

References

1. I. Foster and C. Kesselman (ed.):The Grid Blueprint for a New Computing Infrastructure. Morgan Kaufmann Publishers. (1998)
2. M. Baker, R. Buyya, and D. Laforenza: The Grid: International Efforts in Global Computing. Proc. of International Conference on Advances in Infrastructure for Electronic Business, Science, and Education on the Internet, Rome, Italy. (2000)
3. Ran Zheng, Hai Jin: An Integrated Management and Scheduling Scheme for Computational Grid, The second international workshop on grid and cooperative computing, Shanghai, China. (2003)
4. Min-You Wu: On Runtime Parallel Scheduling for Processor Load Balancing. (1997)
5. Hui C C, Chanson S T: Improved strategies for dynamic load balancing. IEEE Concurrency. (1999) 58-67
6. Zhan Jian, Li Lian: Mathematical Network Computing based on Specification Oriented Programming. Master thesis, Computer Science Department, Lanzhou University. (2003)
7. M. Maheswaran: Quality of Service Driven Resource Management Algorithms for Network Computing. Proc of International Conference on Parallel and Distributed Processing Technologies and Applications. (1999)

A Double Auction Mechanism for Resource Allocation on Grid Computing Systems

Chuliang Weng, Xinda Lu, Guangtao Xue, Qianni Deng, and Minglu Li

Department of Computer Science and Engineering, Shanghai Jiao Tong University,
Shanghai 200030, P.R. China
{weng-cl,lu-xd,xue-gt,deng-qn,li-ml}@cs.sjtu.edu.cn

Abstract. Considering dynamic, heterogeneous and autonomous characteristics of computing resources in grid computing systems and the flexibility and effectivity of economics methods applied to solve the problem of resource management, a double auction mechanism for resource allocation on grid computing systems is presented. Firstly, a market model of double auction is described, in which agents are utilized to represent the computational resource traders in the grid environment and are equipped with the reinforcement learning algorithm. Secondly, a double auction mechanism is presented, where the uniform-price auction is adopted aiming at CPU resources, and the transaction fee can be adjusted flexibly. Finally, the efficiency of the presented double auction mechanism is analyzed through experiments, and experimental results show that the presented mechanism is efficient, and the transaction price varies mildly.

1 Introduction

One characteristic of resources on grid systems is autonomy [1], that is, resources belong to different organizations that have different usage strategies for their own resources in order to purse their own maximal interest. It is a nature way that resource allocation on emerging grid systems could be studied based on the economic mechanism.

One kind of effective economic mechanisms is double auction [2]. In a double auction, sellers and buyers submit asking price and biding price respectively to the auctioneer, and a trade is made if a buyer's bid exceeds a seller's ask. Typically, a buyer wants to purchase multiple units and a seller has more than one unit for sale, which is a multiple-unit double auction (MDA). Therefore, a buyer's bid may satisfy several sellers' asks and a seller's ask may match several buyers' bids. It is the auctioneer that deals with this sort of matching between multiple sellers and multiple buyers involving multiple units. In addition to MDA, a buyer or a seller only transacts a single-unit resource in a single-unit double auction (SDA).

In this paper, we focus on a multiple-unit double auction mechanism for resource allocation on grid computing systems, which will be tested with the reinforcement learning algorithm for studying the mechanism efficiency and the price variation of the computational resource market.

The paper is organized as follows. In Section 2, a brief overview is given for current research on resource allocation based on the economic mechanism for computer systems. The market model is discussed in Section 3. The double auction mechanism is presented in Section 4. In Section 5, the efficiency of the mechanism is discussed. In Section 6 we test the presented mechanism with experiments, and show the experimental result. Finally, we conclude the paper in Section 7.

2 Related Work

The Grid Architecture for Computational Economy (GRACE) [1] is presented, which mainly focuses on incorporating economic models into a grid system with existing middlewares, such as Globus and Legion. It demonstrates that the economic viewpoint is better suited than other mechanisms to run a large-scale distributed computing system. However, the resource prices are given to them instead of determining in accordance with economic mechanisms in its experiments. The price of resource usually is determined with two kinds of economic mechanisms: auction model and commodity market model. Research efforts on commodity market model include [3, 4, 5], although the commodity market model is effective mechanism, the lack of flexibility stands out in the large-scale distributed systems such as grid computing systems. Study on applying auction model to price resource in distributed systems include [2, 6, 7, 8, 11], the double auction model and the one-side auction model are applied to determine the resource price in many situations, which include E-commerce and the computational electricity market, etc. In economics research domain, research efforts on double auction include [9, 10].

3 The Market Model

In this section, we will describe the market model from the two aspects of the trader behavior and the double auction rule, which is the basis of the following discussions and experiments.

3.1 Trader Behavior

In this paper, Traders are equipped with the modified version of the Roth-Erev reinforcement learning (hereafter, referred as to MRE) algorithm, which was described in [11]. The MRE algorithm is calibrated with four parameters, a scaling parameter $s(1)$, a recency parameter r, an experimentation parameter e, and a parameter K denoting the number of possible actions that will be taken by the learner. For simplicity, each buyer and seller is assumed to learn in accordance with the MRE algorithm characterized by the same four values for these parameters.

The main idea of the MRE algorithm is as follows. At the beginning of the first auction round 1, each trader i assigns an equal propensity $q_{ik}(1) = s(1)X/K$, where X is the average profit that traders can achieve in any given auction round. In addition, an equal choice probability $p_{ik}(1)=1/K$ is assigned to each of its feasible auctions k which total is K. Each trader i probabilistically selects a feasible action k' to submit to the auctioneer in accordance with its current choice probabilities. After transaction, each trader i updates the corresponding parameters for next round auction according to the profit achieved in this auction round. After the nth auction round, traders i updates its existing action propensities $q_{ik}(n)$ based on its newly earned profit $R(i,k',n)$ as follows:

$$q_{ik}(n+1) = (1-r)q_{ik}(n) + ME(i,k,k',n,K,e) \qquad (1)$$

Where, ME is an updating function reflecting the experience gained from past trading activity, and it takes the form:

$$ME(i,k,k',n,K,e) = \begin{cases} R(i,k',n)(1-e), & k = k' \\ q_{ik}(n)\dfrac{e}{K-1}, & k \neq k' \end{cases} \qquad (2)$$

Given the updated propensities $q_{ik}(n+1)$ for auction round $n+1$, updated choice probabilities $p_{ik}(n+1)$ of trader i which is for selecting k' among its feasible action k in auction round $n+1$ is as follows:

$$p_{ik}(n+1) = \frac{q_{ik}(n+1)}{\sum_{m=1}^{K} q_{im}(n+1)} \qquad (3)$$

3.2 Auction Rule

One kind of double auction is the continuous double auction (CDA), which matches buyers and sellers immediately on detection of compatible bids. A periodic version of the double auction instead collects bids over a specified interval of time, and then clears the market at the expiration of the bidding interval. In addition, double auction can be classified to uniform-price auction and discriminatory-price auction. In uniform-price auctions, all exchanges mandated by the auction clearing policy are to occur at the same price; in contrast the prices are set individually for each matched buyer-seller pair in discriminatory-price auctions.

For simplicity we adopt the periodic continuous version of the double auction in this paper. That is, in each *round*, all traders are solicited for their bids and asks. While submitting a bid or ask, a trader does not know the bids and asks submitted by other traders in this round. And then the auctioneer clears the market and determines the buyer-seller matches according to the bids and asks, and determines the trading volume and the transaction value for the buyer-seller matches. One argument in favor of the uniform-price mechanism is its perceived fairness, and the other benefit is that it simplifies the auctioneer's task of calculating transaction value [12]. Therefore we adopt the uniform-price mechanism in this paper.

Further, we assume that the interval of auction rounds is fixed and equal, and restrict computational resources to one kind of resource, i.e., CPU resource, and the unit of the computational resource is Mflops (Mega floating point operations per second).

4 Auction Mechanism

The number of sellers is denoted by NS, and the number of buyers is denoted by NB. In each auction round, each trader should submit the two values to the auctioneer, which include the bidding or asking price for unit computational resource and the demand or supply of computational resource. Let a_i denote the ith asking price (in the non-decreasing order) of sellers and b_j denote the jth bidding price (in the non-increasing order) of buyers, i.e., $0 \leq a_1 \leq a_2 \leq \ldots \leq a_{NS}$ (*supply curve*) and $b_1 \geq b_2 \geq \ldots \geq b_{NB} \geq 0$ (*demand curve*). In this paper, we mainly focus on how to determine the price of computational resources through the double auction, so we assume that the demand of each buyer and the supply of each seller for the computational resource are fixed individually at a certain value for a period. S_i denotes the supply of computational resource seller i and D_j denotes the demand of computational resource buyer j.

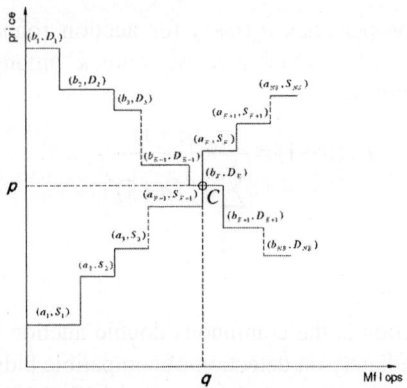

Fig. 1. Illustration for determining the E and F.

We can determine the point of intersection $C(q, p)$ illustrated as Fig.1, and when the intersection of the demand curve and the supply curve is a line segment, we take $C(q, p)$ to be the midpoint of the line segment.

There are two kinds of situations for $C(q, p)$, which take the form individually as follows:

$$a_F \geq b_E \geq a_{F-1}, \text{ simultaneously, } \sum_{j=1}^{E} D_j \geq \sum_{i=1}^{F-1} S_i \geq \sum_{j=1}^{E-1} D_j \qquad (4)$$

$$b_{E-1} \geq a_F \geq b_E, \text{ simultaneously, } \sum_{i=1}^{F} S_i \geq \sum_{j=1}^{E-1} D_j \geq \sum_{i=1}^{F-1} S_i \qquad (5)$$

The situation of $C(q, p)$ can determine the value of E and F illustrated as Fig.1, and seller i ($i<F$) and buyer j ($j<E$) will have a transaction for computational resources. The transaction price for buyers is defined as $p^b = \max\{a_{F-1}, b_E\}$, and the transaction price for sellers is defined as $p^a = \min\{a_F, b_{E-1}\}$.

The transaction fee for seller i ($i<F$) is defined as ϕ^a, and it satisfies:

$$p^a - p \leq \phi^a \leq p^a - a_{F-1} \qquad (6)$$

The transaction fee for buyer j ($j<E$) is defined as ϕ^b, and it satisfies:

$$p - p^b \leq \phi^b \leq b_{E-1} - p^b \qquad (7)$$

In each auction round, the transaction volume of sellers and buyers take the forms as situation 1 and situation 2, corresponding to equation (4) and (5) respectively.

1. The transaction volume V_j^b of buyer j ($j<E$) equals to the volume of computational resource submitted by the buyer, i.e., $V_j^b = D_j$; the transaction volume of seller i ($i<F$) is $V_i^a = S_i \times (\sum_{k=1}^{E-1} D_k / \sum_{k=1}^{F-1} S_k)$.

2. The transaction volume V_i^a of seller i ($i<F$) equals to the volume of computational resource submitted by the seller, i.e., $V_i^a = S_i$; the transaction volume of buyer j ($j<E$) is $V_j^b = D_j \times (\sum_{k=1}^{F-1} S_k / \sum_{k=1}^{E-1} D_k)$.

In addition, r_j represents the marginal revenue received by buyer j for unit computational resource; the marginal cost c_i denotes how much it cost seller i to provide unit computational resource.

5 Mechanism Efficiency

For seller i ($i \geq F$) who participates in one auction round and is unsuccessful, its utility is zero, and for seller i ($i<F$) who participates in the auction round and is successful, the utility received by it is:

$$U_i^a = (p^a - c_i - \phi^a) \cdot V_i^a \tag{8}$$

For buyer j ($j \geq E$) who participates in the auction round and is unsuccessful, its utility is zero, and for buyer j ($j<E$) who participates in the auction round and is successful, the utility received by it is:

$$U_j^b = (r_j - p^b - \phi^b) \cdot V_j^b \tag{9}$$

Efficiency is one of the most important goals people usually pursue when designing a market mechanism. An efficient market can achieve the goal of maximizing the total profit obtained by all traders including sellers and buyers. By stronger efficiency definition, the efficiency is defined as a measure of how much all traders actually obtain compared to the maximum total market value [2].

The maximal potential total profit for one auction round is:

$$T = \sum_{i=1}^{F-1}(p^a - c_i) \cdot S_i + \sum_{j=1}^{E-1}(r_j - p^b) \cdot D_j \tag{10}$$

The practical achieved profit of all sellers and buyers is:

$$U = \sum_{i=1}^{F-1} U_i^a + \sum_{j=1}^{E-1} U_j^b = \sum_{i=1}^{F-1}(p^a - c_i - \phi^a) \cdot V_i^a + \sum_{j=1}^{E-1}(r_j - p^b - \phi^b) \cdot V_j^b \tag{11}$$

The *efficiency* of the auction mechanism is defined as:

$$EA = \frac{U}{T} \times 100 \qquad (12)$$

In the following section, we will analyze the efficiency of the presented auction mechanism with experiments based on the reinforcement learning algorithm.

6 Experiments and Results

In this paper, the traders including sellers and buyers are equipped with the MRE algorithm for adjusting their biding or asking price of computational resource. And we investigate efficiency outcomes and price variations for a computational resource market with double auction pricing and with traders who continually update their price offers on the basis of past profit experiences.

In the experiment, the number of buyers, NB, is 100, the number of sellers, NS, also is 100, and there is one auctioneer. As illustrated in the section 3, the interval of auction rounds is fixed and equal, the auction resource is CPU resource, and the unit of the computational resource is Mflops. During this short-run auction, the demand and the supply is fixed for each auction *run*, and uniformly distributed in the range [1000, 10000] (Mflops), and marginal costs of sellers and marginal revenues of buyers are both distributed in the range [40, 80] (Grid$/1000Mflops). According to equation (6) and (7), The transaction fees are set individually as follows:

$$\phi^a = \max\{\frac{2}{3}(p^a - a_{F-1}),\ (p^a - p)\},\ \phi^b = \max\{\frac{2}{3}(b_{E-1} - p^b),\ (p - p^b)\} \qquad (13)$$

The parameter values for the MRE algorithm are as follows. The number of possible price offers $K = 20$, the recency parameter $r = 0.1$, the experimentation parameter $e = 0.20$, a scaling parameter $s(1) = 20.0$. Each auction *run* consists of 10000 auction *round*s. The experiment results are the average value of the 100 auction runs.

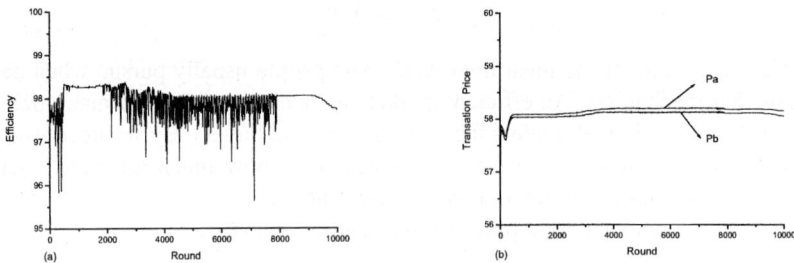

Fig. 2. The distribution of the buyer's marginal revenue and the distribution of the seller's marginal cost are both [40,80].

According to the above experiment parameters, we develop a simulator for testing the performance of the presented double auction mechanism. The experimental results are showed as Fig.2. According to the efficiency outcomes of the experiment, we can find that the efficiency of the presented double auction mechanism is larger than 90% during the 10000 rounds, although the value varies in different rounds, which indicates that the mechanism is efficient.

According to Fig.2 (b), it can be noticed that the transaction price varies mildly in different rounds. In the practical economic market, it is important that the price of resources changes gradually instead of changing discontinuously. The experimental result indicates that the mechanism can satisfy this requirement. In addition, although the transaction price of buyers p^b is less than the transaction price of sellers p^a, the budget balance can be achieved through the transaction fee in the presented auction mechanism.

7 Conclusion

In this paper, we present a double auction mechanism for resource allocation on grid computing systems. Agents are utilized to represent the computational resource traders in the grid computing system, and they are equipped with the reinforcement learning algorithm for analyzing the efficiency outcomes of the mechanism in the short run. Experimental results indicate that the presented mechanism is efficient, which is studied with uniform-price, multi-unit, continuous periodic auction, which is different from the existing relative research. Also the transaction price varies mildly, which is helpful to keep the computational resource market stable.

Acknowledgements

This research was supported by the National Natural Science Foundation of China, No. 60173031.

References

1. Buyya, R.: Economic-based Distributed Resource Management and Scheduling for Grid Computing. PhD Dissertation, Monash University (2002)
2. Huang, P., Scheller-Wolf, A., Sycara, K.: Design of a Multi-unit Double Auction E-market. Computational Intelligence, Vol.18, No.4 (2002) 596-617
3. Cheng, J., Wellman, M.: The WALRAS Algorithm: A Convergent Distributed Implementation of General Equilibrium Outcomes. Computational Economics, Vol.12, No.1 (1998) 1-24
4. Subramoniam, K., Maheswaran, M., Toulouse, M.: Towards a Micro-economic Model for Resource Allocation in Grid Computing System. In: Proceedings of the 2002 IEEE Canadian Conference on Electrical & Computer Engineering (2002) 782-785
5. Wolski, R., Plank, J., Brevik, J., Bryan, T.: Analyzing Market-based Resource Allocation Strategies for the Computational Grid. The International Journal of High Performance Computing Applications, Vol.15, No.3 (2001) 258-281
6. Waldspurger, C., Hogg, T., Huberman, B., Kephart, J., Stornetta, W.: Spawn: a Distributed Computational Economy. IEEE Transactions on Software Engineering, Vol.18, No.2 (1992) 103-117
7. Regev, O., Nisan, N.: The Popcorn Market – Online Markets for Computational Resources. In: Proceedings of the 1st International Conference on Information and Computation Economies (1998) 148-157

8. Lalis, S., Karipidis, A.: JaWS: an Open Market-based Framework for Distributed Computing over the Internet. In: Proceeding of the 1st IEEE/ACM International Workshop on Grid Computing (2000) 36-46
9. Yoon, K.: The Modified Vickrey Double Auction. Journal of Economic Theory, Vol.101, No.2 (2001) 572-584
10. Code, D., Sunder, S.: Allocative Efficiency of Markets with Zero-intelligence Traders: Market as a Partial Substitute for Individual Rationality. Journal of Political Economy, Vol.101, No.1 (1993) 119-137
11. Nicolaisen, J., Petrov, V., Tesfatsion, L.: Market Power and Efficiency in a Computational Electricity Market with Discriminatory Double-auction Pricing. IEEE Transactions on Evolutionary Computation, Vol.5, No.5 (2001) 504-523
12. Wurman, P., Walsh, W., Wellman, M.: Flexible Double Auctions for Electronic Commerce: Theory and Implementation. Decision Support Systems, Vol.24, No.1 (1998) 17-27

Research on the Virtual Topology Design Methods in Grid-Computing-Supporting IP/DWDM-Based NGI*

Xingwei Wang[1], Minghua Chen[1], Qiang Wang[1], Min Huang[2], and Jiannong Cao[3]

[1] Computing Center, Northeastern University, Shenyang, 110004, China
wangxw@mail.neu.edu.cn
[2] College of Information Science and Engineering, Northeastern University,
Shenyang, 110004, China
[3] Department of Computing, Hong Kong Polytechnic University, Hong Kong, China

Abstract. Grid computing is one of the key applications in NGI. Virtual topology design is considered to be one of the key problems in the IP/DWDM-based NGI. Since the virtual topology design problem aiming at minimizing the sum of traffic-weighted hop count is NP-hard, three methods are presented, adopting heuristic algorithms and intelligent algorithms respectively. A simulation environment is developed and the analysis has been done on the effects of the primary parameters on the results of the proposed methods. Performance comparisons among the proposed methods have also been done. Simulation results have shown that the methods presented here are feasible and effective. Thus, it can help to provide grid computing with high performance network support.

1 Introduction

Grid computing is one of the key applications in NGI (Next Generation Internet) and IP/DWDM is one of the critical networking techniques of NGI backbone [1]. Therefore, research on virtual topology design in IP/DWDM-based NGI backbone – IP/DWDM optical Internet is very important. It can help to provide grid computing with high performance network support.

Virtual topology [2] is also called logical topology, which is consisted of a series of end-to-end lightpaths. Performance of IP/DWDM optical Internet can be improved substantially by constructing an optimal virtual topology for it.

A lot of investigations have been done on the problem of virtual topology design, most of which oriented to regular topologies [3-4]. In this paper, based on mesh networks, methods to design the virtual topology are presented, aiming at minimizing the sum of the traffic-weighted hop count. Due to its NP-hard nature [5], heuristic algorithms, GA (Genetic Algorithms) and SAA (Simulated Annealing algorithm) are adopted respectively. To verify the feasibility and the effectiveness of the proposed algorithms, a simulation environment is developed, and the analysis has been done on the effects of the primary parameters on the results of the proposed methods.

---

* This work is supported by the National Natural Science Foundation of China under Grant No. 60003006 (jointly supported by Bell Lab Research China) and No. 70101006; the National High-Tech Research and Development Plan of China under Grant No. 2001AA121064; the Natural Science Foundation of Liaoning Province in China under Grant No. 20032018 and No. 20032019; Modern Distance Education Engineering Project by China MoE.

2 Basic Assumptions

Network nodes are connected together by optical fibers. The maximum amount of available wavelengths (called wavelength number below) in each optical fiber is W. One or more concatenated optical fibers form an optical channel, on which light signals are transmitted without optical-to-electric or electric-to-optical conversions. Each optical channel occupies a certain wavelength. One or more concatenated optical channels between two nodes form a lightpath. When optical signals pass through optical channels with different wavelengths, wavelength conversion occurs.

In this paper, (s, d) represents a pair of traffic nodes, s is the source node and d is the destination one; (i, j) represents an optical channel from node i to node j; (m, n) represents an optical fiber from node m to node n.

Physical degree of a node is the number of optical fibers, through which the node connects with others directly, and it can be classified into physical in-degree and physical out-degree. Virtual degree of a node is the number of lightpaths, through which the node connects with others directly, and it can be classified into virtual in-degree and virtual out-degree, if they are equal, call them virtual degree. Virtual degree is limited by a lot of factors, such as the wavelength number in each optical fiber, the number of optical receivers and transmitters on each node and the network connectivity. The physical hop count is the number of the optical fibers that form an optical channel. The virtual hop count is the number of optical channels a lightpath covers. $[\lambda_{sd}]_{N \times N}$ is a matrix, describing the amount of traffic flowing from one node to another. Here, $\lambda_{ss}=0$, and N is equal to the amount of nodes in the network. λ_{ij} represents the total amount of traffic passing through the optical channel (i, j). λ_{ij}^{sd} indicates the amount of traffic passing through the optical channel (i, j) from s to d.

3 Problem Formulations

Given the physical topology G and the specific traffic matrix $[\lambda_{sd}]_{N \times N}$, generate a corresponding virtual topology G_v, and route the traffic over it.

A physical topology can be modeled as a graph $G(V, E)$, where V is the set of nodes representing optical nodes and E is the set of edges representing optical fibers that connect the nodes. Graph $G(V, E)$ contains $|V|$ nodes and $|E|$ edges. Each node $v_i \in V$ is associated with the following parameters: available optical transmitters, $t_i=t(v_i)$, $t:V \rightarrow R^+$; available optical receivers, $r_i=r(v_i)$, $r:V \rightarrow R^+$. Each edge $e_{ij}=(v_i,v_j) \in E$ is associated with the following parameters: available wavelengths, $w_{ij}=w(e_{ij})$, $w:E \rightarrow R^+$; traffic, $\xi_{ij}=\xi(e_{ij})$, $\xi:E \rightarrow R^+$, $\xi_{ij}=\xi_{ji}$, $1 \le i,j \le n$.

To reduce the processing time of optical signals in the electronics, and thus to alleviate the speed bottleneck problem in the electronic domain, three methods to design virtual topology are presented, attempting to minimize the sum of traffic-weighted hop count. The objective function is defined as follows: $F = \min \left(\dfrac{\sum_{sd} \lambda_{sd} H_{sd}}{\sum_{sd} \lambda_{sd}} \right)$.

Here, H_{sd} is the count of hops, which lightpath (s, d) has traversed. The following constraints have to be satisfied: virtual degree constraints, the virtual in-degree and out-degree of each node in the virtual topology are no more than their physical counterparts respectively; traffic constraints, the total amount of traffic passing through any optical fiber is at most equal to its possible maximum traffic load, traffic conservation needs to be satisfied at each node, that is, the amount of traffic introduced by any node should be equal to the amount leaving it minus the amount entering it; wavelength constraints, the number of wavelengths available on a lightpath is at most equal to the number of wavelength converters on it, and the wavelength continuity constraint needs to be met with between any two converters; optical channel constraints, the length of an optical channel is at most equal to D, and the traffic routed on an optical channel is no more than C, due to virtual topology restriction, not all traffic between s and d can be routed, here, D is the maximum length of the optical channel, which is determined by the attenuation of the optical signal, and C is the maximum cap of the optical channel.

4 Heuristic Algorithms for Virtual Topology Design

4.1 Heuristic Virtual Topology Construction Algorithm

This algorithm is to determine a virtual topology upon the given physical topology and to determine the source and destination node pairs and their corresponding lightpaths. The basic idea is to set up lightpaths at first for those traffic node pairs with the largest amount of traffic and to attempt to route the largest amount of traffic within one hop. Optical channel and virtual degree constraints are both to be met with here. The proposed algorithm is described as follows:

(1) Sort all traffic node pairs by their amount of traffic in descending order.
(2) Choose a node pair (s, d) that has not been marked routed. If the current lightpath can route the traffic between s and d, compute its hop count h_1.
(3) If node s and node d both meet with the virtual degree constraints, compute the shortest path, and denote the sum of distance it spans as l; otherwise, go to (6).
(4) If $l<D$, establish the optical channel (s, d); otherwise, introduce a node m_1 with the largest amount of traffic on the shortest path from node s within the distance of D, and then introduce a node m_2 with the largest amount of traffic on the shortest path from node m_2 within the distance of D. Just do as the above, till node d is reached.
(5) Compute the lightpath from s to d, denote its hop count as h_2. If $h_1 < h_2$ release all optical channels established on the shortest path between s and d.
(6) Mark node pair (s, d) routed, go to (2), till all traffic node pairs have been routed, the algorithm ends.

4.2 Heuristic Wavelength Assignment Algorithm

The proposed algorithm is to determine the optical fibers that make up an optical channel and to determine which wavelength each optical channel will be assigned. To ensure that the wavelength continuity constraint be satisfied between any two wavelength converters, the following measures are taken: number all wavelengths available on the optical fibers; every time, try to find out the wavelength with the smallest number, which has not yet been assigned; whenever such a wavelength is not available, release the whole lightpath. The proposed algorithm is described as follows:

(1) Number all the wavelengths available on the optical fibers. If no wavelength is available, the algorithm ends; otherwise, sort all wavelengths by their number in ascending order.
(2) Select a lightpath to which the wavelength has not yet been assigned as the current one, if such a lightpath does not exist, the algorithm ends; otherwise, $s = i$.
(3) Search for the next node with wavelength converter after node s along the selected lightpath. If found, name it m, let $w = 0$, go to (4); otherwise, $m = j$, $w = 0$, go to (7).
(4) Assume that L is the first optical fiber on the optical channel (s, m).
(5) If L is the last optical fiber on the optical channel (s, m), go to (7); otherwise, if $w=W$, go to (8).
(6) If wavelength w is not used on L, select the next optical fiber as L, go to (5); otherwise, increase w by 1, go to (4).
(7) Assign wavelength w to the optical channel (s, m).
(8) If $m = j$, mark the wavelength assignment to the current lightpath succeeded, go to (2); otherwise, mark the wavelength assignment to the current lightpath failed. Release the lightpath, go to (2).

4.3 Heuristic Traffic Routing Algorithm

The proposed algorithm tries to find the shortest paths for the traffic node pairs by greedy algorithm upon the constructed virtual topology and to assign traffic as much as possible to the path with hops as less as possible without exceeding the optical channel capacity:

(1) Sort all the traffic node pairs by their amount of traffic in descending order. Initialize the allowed amount of traffic of each lightpath to be the maximum capacity of the optical channel C.
(2) Have all traffic node pairs been routed, the algorithm ends; otherwise, select the node pair with the largest traffic amount and compute the shortest path P_{sd} from s to d.
(3) Compute the MAT (Maximum Allotable Traffic) of P_{sd}, and then add traffic $\lambda=\min(C-MAT, \lambda_{sd})$ to all optical channels that P_{sd} covers. ($MAT =\min(\lambda_{ij})$, λ_{ij} is the allotable capacity of C_{ij}, and C_{ij} is the optical channel that P_{sd} covers.)
(4) If $\lambda_{sd}-\lambda>0$, compute the next shortest path P_{sd} from s to d, go to (3); otherwise, mark (s, d) routed, select the node pair with the largest amount of traffic from the not-routed ones as (s, d), go to (2).

5 Intelligent Algorithms for Virtual Topology Design

5.1 Intelligent Virtual Topology Design Method Based on GA

Numerical coding is adopted to denote the solution, which is also called chromosome here. Each gene of a chromosome is a cluster of integers, and each integer of the cluster corresponds to one node. The integer cluster corresponding to the ith gene represents the nodes which node i connects with directly through optical channels.

The length of chromosome is equal to the amount of nodes in the given network. The length of gene *i* is equal to the smaller one of the amount of nodes in the network minus 1 and the amount of optical transmitters at the node. Validity is one of the problems to be considered. In the *i*th gene cluster, there must be no *i*, that is, there is no optical channel going from node *i* to itself. The initial population is generated randomly according to the above coding rules, and the population size is equal to *P*.

Generate the son chromosome by having the crossover operation upon two father chromosomes. Three kinds of crossover schemes are used: parallel, exchanging and odd-even crossover. In the proposed algorithm, one of the above three schemes will be chosen randomly to do the crossover operation. The genes of the chromosomes generated through parallel crossover are all valid. However, the genes of the chromosomes generated through exchanging or odd-even crossover may be invalid and their validity has to be checked. Generate new valid genes randomly to replace those invalid ones. Random single-gene mutation is adopted. Select son chromosomes randomly according to the mutation probability. Then, choose a gene in each chosen chromosome randomly to have the mutation operation on it, generating a new gene, checking its validity, and replacing the old gene with the new and valid one.

Fitness function is defined as follows: $F_t = MAX - \left(\dfrac{\sum_{sd} \lambda_{sd} H_{sd}}{\sum_{sd} \lambda_{sd}} \right)$. Here, *MAX* is a large enough positive number to ensure that all possible values of F_t are positive. The problem of computing the fitness value is decomposed into two subproblems. Each chromosome corresponds to a virtual topology. Once the virtual topology is constructed, assign the wavelength to it and route the traffic on it, and then, get the value of λ_{sd} and H_{sd} to compute the value of F_t. Here, wavelength assignment and traffic routing are performed, using the algorithms in section 4.2 and 4.3 respectively.

To speed up the population evolution, select the $P \times (1-p_c)$ best chromosomes from the *P* father chromosomes and reserve them directly as son chromosomes. And then, have crossover operation on the remaining $P \times p_c$ chromosomes in the father population to generate the new son population. Here, p_c is the crossover probability. Father chromosomes in the crossover operation are selected from the father population according to the roulette wheel.

To ensure the solution quality, a safety valve is established to reserve the elitist (the chromosome with the largest fitness value), which need not take part in the crossover and selection operation. Only when the best chromosome of the current population is better than the reserved elitist, does the former replace the latter. Set the maximum evolution times, if the elitist remains unchanged after the maximum evolution times of generations, the algorithm ends. The proposed algorithm is as follows:

(1) Clear the evolution times and the fitness value of the elitist, and then generate the initial population randomly as father one.
(2) If the evolution times is not less than the predetermined maximum one, the algorithm ends; otherwise, increase the evolution times by 1.
(3) Compute the fitness value of each chromosome in the father population, and reserve the largest fitness value and its corresponding chromosome as the best one.
(4) If the current largest fitness value is not more than the fitness value of the elitist, that is, no evolution occurs in current population, go to (6).

(5) Replace the elitist in safety valve and clear the evolution times.
(6) According to the crossover probability, save a certain proportion best father chromosomes directly as son chromosomes.
(7) Select parent chromosomes from the remaining father ones according to the roulette wheel, and then do the crossover operations on them to generate son chromosomes.
(8) Do the mutation operations on all son chromosomes according to the mutation probability.
(9) Use all son chromosomes as the new father population ones, go to (2).

5.2 Intelligent Virtual Topology Design Method Based on SAA

This method is different from that GA-based one mainly on the fitness function definition and the annealing mechanism. Its fitness function is defined as follows:

$F_t = \left(\dfrac{\sum_{sd} \lambda_{sd} H_{sd}}{\sum_{sd} \lambda_{sd}} \right)$. A suitable initial temperature is the one that can bring the higher

average initial state accepted probability [5]. Generate two solutions randomly, and then denote their fitnesses by *fitness*1 and *fitness*2 respectively. If these two values are equal, generate a new solution, till they are unequal. Set the initial temperature according to the following formula: *temperature*=-|*fitness*2-*fitness*1|×log(e)/log(β), 0<β<1. Use exponential cooling scheme to control the temperature with the following formula: $t_k = \alpha t_{k-1}$, 0<α<1.

6 Performance Evaluations

Simulations have been done over some actual network topologies, including NSFNET [5].

6.1 Heuristic Virtual Topology Design Method

Set virtual degree to 6, and the effect of wavelength number on average hops is shown in Fig.1. Set the wavelength number to 6, and the effect of virtual degree on average hops is shown in Fig.2.

Set the virtual degree to 6, and the effect of the wavelength number on the established optical channels is shown in Fig.3. Set the wavelength number to 6, and the effect of virtual degree on the established optical channels is shown in Fig.4.

6.2 Intelligent Virtual Topology Design Method

Set virtual degree to 6, the wavelength number to 6 and the population size to 20. The effect of the crossover probability is shown in Fig.5 and Fig.6.

Set virtual degree to 6, the wavelength number to 6 and the population size to 20 and 30 respectively. The effect of the mutation probability is shown in Fig.7 and Fig.8.

Set virtual degree to 6 and the wavelength number to 6. The effect of the population size is shown in Fig.9 and Fig.10.

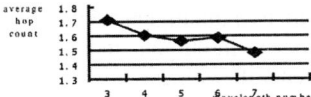

Fig. 1. Effect of wavelength number on average hop count.

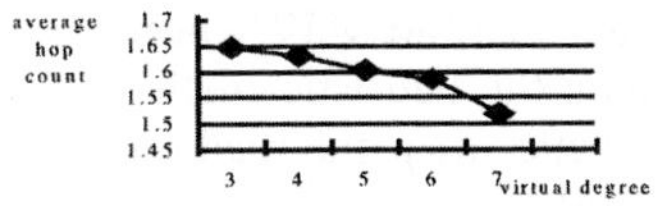

Fig. 2. Effect of virtual degree on average hop count.

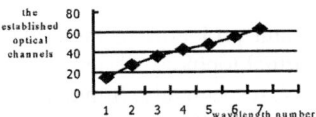
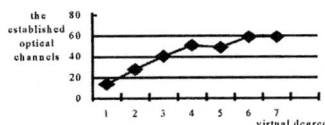

Fig. 3. Effect of wavelength number on the established optical channels.

Fig. 4. Effect of virtual degree on the established optical channels.

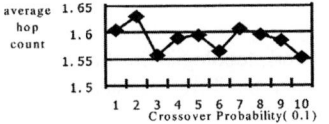
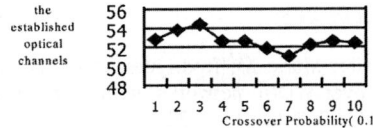

Fig. 5. Effect of crossover probability on average hop count.

Fig. 6. Effect of crossover probability on the established optical channels.

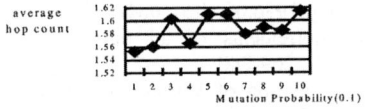
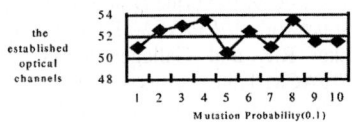

Fig. 7. Effect of mutation probability on average hop count.

Fig. 8. Effect of mutation probability on the established optical channels.

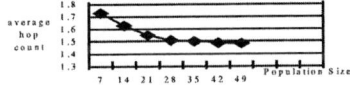
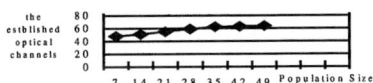

Fig. 9. Effect of mutation probability on average hop count.

Fig. 10. Effect of population size on the established optical channels.

With the MTTD (Maximum Temperature Declined Times) increased, the produced solutions are getting better and better. However, MTTD cannot be increased without limitation, causing the running time unbearable. After MTTD has become large enough, its effect on algorithm performance changes little. Thus, a proper value of MTTD has to be selected, in order to improve the algorithm performance further and to keep the running time bearable.

6.4 Comparisons Between the Three Proposed Methods

According to the analysis and comparisons of the performance evaluation results, the following conclusions can be drawn: by properly setting the parameter values, all the above three proposed methods can get better even optimal solutions; and when the population size and MTTD become large enough, the two intelligent methods are both better than the heuristic method, at the same time, the intelligent method based on SAA is better than the method based on GA in general, with the cost of the longer runtime [5].

7 Conclusions

Virtual topology design is one of the most important problems in the IP/DWDM-based NGI. Oriented to the mesh networks, three virtual topology design methods are proposed, adopting heuristic algorithm, GA and SAA respectively. Simulation results have shown that methods presented here are feasible and effective. Thus, it can help to provide grid computing with high performance network support.

References

1. Freire, Mario M, Rodrigues, *et al*. The role of network topologies in the optical core of IP-over-WDM networks with static wavelength routing. Telecommunication Systems, 2003.02, 24(2): 111-122.
2. Bregni S, Janigro U and Pattavins A. Optimal allocation of limited optical-layer resources in WDM networks under static traffic demand. Photonic Network Communications, 2003.01, 5(1): 33-40.
3. Mukherjee B, Banerjee D and Mukherjee A. Some principles for designing a wide area optical network. IEEE/ACM Transactions on Networking, 1996.05, 4(5): 684-695.
4. Pankaj R.K and Gallager R. G. Wavelength requirements of all optical networks. IEEE/ACM Transactions on Networking, 1995.03, 3(3): 269-280.
5. CHEN Ming-Hua. Research and Simulated Implementation of Virtual Topology Design Algorithms in IP/DWDM Optical Internet [D]. Shenyang: Northeastern University, 2003.06.

De-centralized Job Scheduling on Computational Grids Using Distributed Backfilling[*]

Qingjiang Wang, Xiaolin Gui, Shouqi Zheng, and Bing Xie

Department of Computer Science and Technology, Xi'an Jiaotong University,
Xi'an 710049, China
qjwang@mailst.xjtu.edu.cn
{xlgui, sqzheng}@mail.xjtu.edu.cn
xiexiebing@sohu.com

Abstract. To dynamically improve node selections of the waiting jobs under de-centralized scheduling, each node together with its neighbors is assumed to compose a subgrid, using distributed backfilling to optimize grid scheduling on each subgrid. Whenever a job terminates, distributed backfilling is triggered to rebackfill all waiting jobs on the corresponding subgrid in order of their submittal. Each subgrid is overlapped with some other, so the waiting jobs may be migrated around the grid. A simulated grid is established, while grid workload is modeled by extending workload models of parallel systems. Job speedup is used to evaluate scheduling strategies. Results show the dynamic optimization of node selections brought by distributed backfilling is grid-wide, and can improve scheduling performance remarkably as long as grid load is not too light and the job migration costs are not too high.

1 Introduction

A computational grid aggregates an abundance of computation resources, also called grid nodes. User jobs may be single-node jobs or multi-node jobs, and only the grid scheduling of single-node jobs are studied here.

The implementations of grid scheduling may be centralized or de-centralized. In centralized scheduling, grid schedulers select the appropriate node for each job without collaborations with other grid schedulers [1, 2, 3, 4, 5]. In de-centralized scheduling, the appropriate nodes are found by collaborations between grid schedulers [6, 7]. De-centralized scheduling facilitates the expansion of grid resources, so it is the reasonable implementation of grid scheduling. For de-centralized grid scheduling, most papers published in this area are dealing with static scheduling, and these static scheduling usually optimizes job assignments not in the whole grid. So, the performance of static scheduling is not always satisfying.

This paper assumes each node together with its neighboring nodes composes a subgrid, using a novel strategy called distributed backfilling to improve node selections of the waiting jobs on each subgrid. Whenever a node finishes a job, distributed backfilling on the subgrid with the node as the center is triggered, which makes the waiting jobs be migrated around the grid, so grid-wide dynamic grid scheduling may be implemented. Simulated experiments evaluate the impacts of various components

---

[*] This work was sponsored by the National Science Foundation of China (NSFC) under the grant No. 60273085, the state High-Tech Research and Development project (863) under the grant No. 2001AA111081, and the ChinaGrid project of China.

of grid workload on the efficiency of distributed backfilling, including job arrival interval, job submittal location, job migration cost, runtime overestimation, and migration cost overestimation.

In Section 2, the topology and performance metrics of grid scheduling are discussed. In Section 3, distributed backfilling is described. In Section 4, distributed backfilling is evaluated by simulated experiments. Finally, conclusions and the future works are given.

2 Topology and Metrics of Grid Scheduling

2.1 Topology of Grid Scheduling

Definition 1 Neighboring grid schedulers

If two grid schedulers can exchange their scheduling-related information with each other, then one scheduler is called a neighbor of the other.

Definition 2 Topology of grid scheduling

Suppose grid schedulers are represented by vertexes, and the neighbor relations are represented by edges, then the graph is called the topology of grid scheduling.

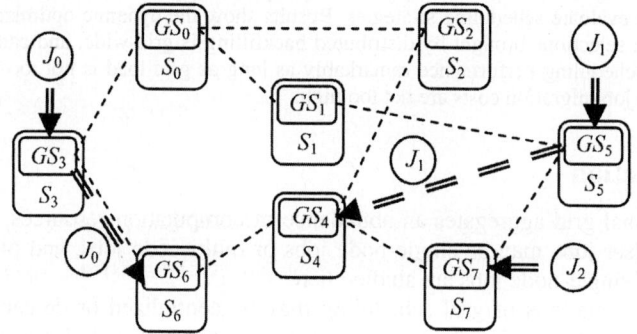

J_i: a job GS_i: a grid scheduler S_i: a node Dashed line: neighbor relation
Real double line: job submittal Dashed double line: job migration

Fig. 1. The topology of de-centralized grid scheduling.

The topology in [6] is a non-complete graph, while the topology in [7] is a tree. In a tree, the failure of any non-leaf node will result in a sub-tree being separated from the tree, but the problem does not occur to a non-complete graph. So, non-complete graphs are the reasonable topology of de-centralized scheduling, illustrated as Fig. 1, where one grid scheduler resides at each node.

2.2 Performance Metrics of Grid Scheduling

Suppose the time of submitting J_i, executing J_i, and finishing J_i is denoted by t_i^a, t_i^s, and t_i^f respectively. Then, the waiting time (t_i^w) is equal to $t_i^s - t_i^a$, the runtime (t_i^e) is equal to $t_i^f - t_i^s$, and the response time (t_i^r) is equal to $t_i^f - t_i^a$.

t_i^e may vary when J_i is assigned to different nodes, so the slowdown [8] of J_i is not always consistent with t_i^r, and the mean slowdown of jobs does not exactly indicate the performance of grid scheduling. By scheduling an identical segment of grid workload, two kinds of scheduling methods can be compared with each other.

Definition 3 Job speedup

Suppose the jobs submitted to a grid during a period are $\{J_i \mid i \in [j,k]\}$. These jobs are scheduled by method A and B respectively, and the total of response time are $T_j^k(A)$ and $T_j^k(B)$ respectively. Then, $\dfrac{T_j^k(A) - T_j^k(B)}{T_j^k(A)} \times 100\%$ is called job speedup, and denoted by η_A^B.

Where $\{J_i\}$ is a segment of grid workload, and $k - j + 1$ should be big enough to eliminate the impact of individual job on η_A^B. $\eta_A^B > 0$ means that B is generally more efficient in shortening the response time of jobs than A.

3 Distributed Backfilling

Consecutive backfilling is competent to accomplish parallel job scheduling on parallel systems [8, 9, 10]. Without loss of generality, suppose consecutive backfilling implements the local scheduling on each grid node.

Suppose l_i^k is the runtime of J_i on S_k, $e(J_i, S_k)$ represents the estimating accurateness of l_i^k, and $e(J_i, S_k) \times l_i^k$ is the runtime estimate of J_i on S_k, where $e(J_i, S_k) \geq 1$. After J_i is backfilled on S_k, the planed starting time (t_i^{an}) is called the anchor of J_i [8]. Thus, the predicted response time of J_i on S_k is $t_i^{an} + e(J_i, S_k) \times l_i^k$. Suppose $m(J_i, S_r, S_k)$ is the cost of migrating J_i from S_r to S_k, $e_m(J_i, S_r, S_k)$ reflects the estimating accurateness of $m(J_i, S_r, S_k)$, and $e_m(J_i, S_r, S_k) \times m(J_i, S_r, S_k)$ is the migration cost estimate, where $m(J_i, S_r, S_r) = 0$. If $e_m(J_i, S_r, S_k) < 1$, the migration cost estimate is shorter than the reality, some reserved CPU cycles would be wasted, and the remaining cycles might not be enough for completing J_i. So, the migration cost estimate should satisfy $e_m(J_i, S_r, S_k) \geq 1$. Migrating J_i from S_r to S_k, the anchor of J_i on S_k should satisfy $t_i^{an} \geq e_m(J_i, S_r, S_k) \times m(J_i, S_r, S_k)$.

Suppose N_k represents the neighboring nodes of S_k, and $\{S_k\} \cup N_k$ compose the subgrid P_k. Distributed backfilling is proposed to improve node selections of the waiting jobs on each subgrid, which will be triggered on P_k whenever a job is submitted to GS_k or terminated on S_k.

Algorithm Distributed Backfill

1. If J_i is submitted to GS_k, assign J_i to its appropriate node:
   a. For $\forall S_x \in N_k \cup \{S_k\}$, try to get the predicted response time:
      If $t_i^{an} \geq e_m(J_i, S_k, S_x) \times m(J_i, S_k, S_x)$, put J_i into the job queue on S_x to obtain $t_i^{an} + e(J_i, S_x) \times l_i^x$. Then, remove J_i from S_x.
   b. Suppose S_m has the predicted response time shortest. If S_m is not S_k, migrate J_i from S_k to S_m, and put J_i into the job queue on S_m.
2. If a job terminates on S_k, rebackfill the waiting jobs on the subgrid:
   a. For $\forall S_x \in N_k \cup \{S_k\}$, remove the waiting jobs from the job queue on S_x, and insert them into the queue *tempQueue*.
   b. Sort the jobs in *tempQueue* according to the ascending submittal time.
   c. Loop over *tempQueue* until it is empty:
      i. Pop the first job (J_i) from *tempQueue*. Suppose J_i currently resides at S_r.
      ii. For $\forall S_x \in N_r \cup \{S_r\}$, try to get the response time predict:
         If $t_i^{an} \geq e_m(J_i, S_r, S_x) \times m(J_i, S_r, S_x)$, put J_i into the job queue on S_x to obtain $t_i^{an} + e(J_i, S_x) \times l_i^x$. Then, remove J_i from S_x.
      iii. Suppose S_m has the predicted response time shortest. If S_m is not S_r, migrate J_i from S_r to S_m, and put J_i into the job queue on S_m.

Compressing queue [8] on each subgrid will become much complicated owing to the node differences, and it does not conform to FCFS. So, rebackfilling is adopted here.

Apparently, distributed backfilling on two overlapped subgrids should be mutually exclusively triggered, by which a waiting job may be migrated many times.

4 Experiments

4.1 Simulated Grid

The simulated grid includes 12 nodes. Its network topology is randomly established as Fig. 2. Suppose the grid scheduling topology is the same as the network topology.

Suppose there are 128 processors in each node, $P_{comp}(S_j)$ denotes the computation performance of S_j, and $P_{comm}(S_j, S_k)$ denotes the communication performance between S_j and S_k. To roughly represent the differences between computation performances, let $P_{comp}(S_j) = j+1$, $j \in [0,11]$. The backbone network is more powerful in communication performance than the region networks. If S_j is connected with S_k by backbone network, let $P_{comm}(S_j, S_k) = 1$. Otherwise, $P_{comm}(S_j, S_k) = 10$.

4.2 Grid Workload Model

Single-node jobs are usually traditional parallel applications, so grid workload can be modeled by extending workload models [11, 12, 13] of parallel systems.

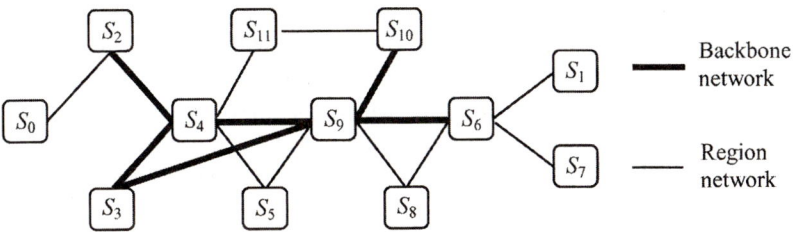

Fig. 2. The network topology of simulated grid.

According to [13], the job parallel degrees are divided into three categories, namely 1, power of 2, and not power of 2. Suppose the probability of serial job is p_1, the probability of parallel job (power of 2) is $p_2 \times (1-p_1)$, and the probability of parallel job (not power of 2) is $(1-p_2) \times (1-p_1)$, where $p_1=0.21$ and $p_2=0.89$. When parallel degree is power of 2, parallel degree conforms to two-phase-uniform distribution (0.8, 4.5, 7, 0.86). The job arrival interval conforms to Gamma distribution (0.21, 0.89).

Let $l_i^k = \dfrac{P_{comp}(S_k) \times l_i^j}{P_{comp}(S_j)}$. Thus, only l_i^0 needs to be modeled. According to [13], l_i^0 after a logarithmic transformation conforms to hyper-Gamma distribution (4, 45, 6, 0.3, 0.3). To shorten the experimental cycles, l_i^0 is shortened as $1.5^{\ln l_i^0}$.

The waiting jobs may be migrated, so the migration cost should be modeled. Let $m(J_i, S_j, S_k) = \dfrac{P_{comm}(J_i, S_j, S_k) \times m(J_i, S_u, S_v)}{P_{comm}(J_i, S_u, S_v)}$. Thus, only the migration cost between two certain nodes needs to be modeled. The migration cost of a job is mainly spent on transferring the executable and the data for processing, so $m(J_i, S_j, S_k)$ may be supposed as $d(J_i) \times P_{comm}(S_j, S_k)$, where $d(J_i)$ is the amount of data transferred. Here, suppose $d(J_i)$ conforms to the uniform distribution.

Besides, the job submittal location where jobs are submitted to should be also modeled. To simulate the distribution uniformity, the probabilities of submitting a job to various grid schedulers are supposed to follow the geometric proportion (α), where schedulers are sorted randomly, such as (GS_0, GS_1, GS_8, GS_{11}, GS_7, GS_5, GS_6, GS_3, GS_2, GS_9, GS_4, GS_{10}), and $\alpha \geq 1$.

Combining the models of parallel degree, runtime, migration cost, arrival interval, and submittal location, the model of grid workload is obtained.

4.3 Result Analysis

DB denotes distributed backfilling, and N denotes the static scheduling implemented in DB. The effect of dynamic scheduling can be examined by job speedup.

time, et al, is moved. When the job is no longer migrated, its executable and data for processing are transferred from the submittal location to the final node. The smart distributed backfilling will be studied in the future.

References

1. Di Martino, V. Sub Optimal Scheduling in a Grid Using Genetic Algorithms. International Parallel and Distributed Processing Symposium (IPDPS'03), 2003. 148-154.
2. Ernemann, C., Hamscher, V., Schwiegelshohn, U., et al. On advantages of grid computing for parallel job scheduling. 2nd IEEE/ACM International Symposium on Cluster Computing and the Grid (CCGRID2002), 2002. 31-38.
3. Ranganathan, K., Foster, I. Decoupling Computation and Data Scheduling in Distributed Data-Intensive Applications. 11th IEEE International Symposium on High Performance Distributed Computing, 2002. 352-258.
4. Subramani, V., Kettimuthu, R., Srinivasan, S., et al. Distributed Job Scheduling On Computatinal Grids Using Multiple Simultaneous Requests. 11th IEEE International Symposium on High Performance Distributed Computing, 2002. 359-366.
5. Hongtu Chen, Muthucumaru Maheswaran. Distributed Dynamic Scheduling of Composite Tasks on Grid Computing Systems. 16th International Parallel and Distributed Processing Symposium (IPDPS'02), 2002. 88-97.
6. Arora, M., Das, S.K., Biswas, R. A De-Centralized Scheduling and Load Balancing Algorithm for Heterogeneous Grid Environments. 2002 International Conference on Parallel Processing Workshops, 2002. 499-505.
7. Junwei Cao, Spooner, D.P., Jarvis, S.A., et al. Agent-Based Grid Load Balancing Using Performance-Driven Task Scheduling. 2003 International Parallel and Distributed Processing Symposium (IPDPS'03), 2003. 49-58.
8. Mu'alem, A.W., Feitelson, D.G. Utilization, predictability, workloads, and user runtime estimates in scheduling the IBM SP2 with backfilling. Parallel and Distributed Systems, IEEE Transactions on, 2001, 12(6): 529 -543.
9. Srinivasan, S., Kettimuthu, R., Subramani, V., et al. Characterization of backfilling strategies for parallel job scheduling. 2002 International Conference on Parallel Processing Workshops, 2002. 514 -519.
10. Zhang, Y., Franke, H., Moreira, J.E., et al. Improving parallel job scheduling by combining gang scheduling and backfilling techniques. 14th International Parallel and Distributed Processing Symposium (IPDPS 2000), 2000. 133 -142.
11. Windisch, K., Lo, V., Feitelson, D., et al. A comparison of workload traces from two production parallel machines. 6th Symposium on the Frontiers of Massively Parallel Computation, Annapolis, MD, USA, 1996. 319-326.
12. Feitelson, D. Workload modeling for performance evaluation. http://citeseer.nj.nec.com/531320.html, 2003-12-3.
13. Lublin, U., Feitelson, D. The Workload on Parallel Supercomputers: Modeling the Characteristics of Rigid Jobs. http://citeseer.nj.nec.com/lublin01workload.html, 2003-12-3.

A Novel VO-Based Access Control Model for Grid*

Weizhong Qiang, Hai Jin, Xuanhua Shi, and Deqing Zou

Cluster and Grid Computing Lab
Huazhong University of Science and Technology, Wuhan, 430074, China
{wzqiang,hjin}@hust.edu.cn

Abstract. As an important aspect of grid security, access control model gets more and more attention. Entities in *virtual organizations* (VOs) must establish a dynamic, secure and cooperative trust mechanism. This paper analyses the cross-organization, dynamic, cooperative and multilevel characteristics of access control problem in grid, and proposes a novel VO-based access control framework. The multilevel access control model is introduced for multilevel requirements and delegation concept is also introduced for permission delegation across organizations.

1 Introduction

Heterogeneity, dynamic and organization self-governed challenge grid technology. Grid security, as the focus of the paper, includes many aspects, such as encryption transferring, authentication, authorization and audit. *Grid Security Infrastructure* (GSI) [3] is a security architecture for grid computing. It provides single-sign-on approach, cross-domain protocol and some convenient security APIs for grid applications. It also presents a simple authorization mechanism, which is very inflexible and coarse-grain by using a mapping table in each grid node that only maps the global name (a ticket or certificate) [2] into a local name.

To solve cross-organization cooperative problems, the conception of *"virtual organizations"* (VOs) is coming into being, which comprises grid entities that unite logically together for a special purpose and spread around many administration organizations or domains. Security requirements within the VO-supporting grid environment have changed to accord with the need to support scalable, dynamic, distributed VOs [1].

In this paper, we present a VO-based access control model and provide a multilevel formalization framework for access control in grid computing. The framework stratifies the access control mechanism into three levels and describes the access control models respectively. The principle and innovation of the paper is based on the two aspects. First, the motivation of each organized VO is presented in the upper level model with the components, projects and functions. Second, we introduce delegation concept for the permission delegated from real organizations to virtual organization.

The paper is organized as follows. First, we present some related works about some access control models. Then we describe access requirements for grid computing. Our multilevel access control components and model are also presented. Finally, we conclude our paper and present some future considerations.

---

* This paper is supported by ChinaGrid project from Ministry of Education.

2 Related Works

Many access control models have been developed in order to enforce the system security, such as *access control list* (ACL), *discretionary access control* (DAC), *mandatory access control* (MAC), *Role Based Access Control* (RBAC) [3][4].

Team based access control model (TMAC) [5] is a primitive attempt to apply role based access control for collaborative environments. It integrates team concept and RBAC [6] for cooperative hypermedia environments. *Task based authorization control* (TBAC) [7] model is specialized to support secure workflow management. TBAC allows the permissions for an executable module changing with the overall progress of a task. *Workflow authorization model* (WAM) [8] is also for workflow environment as TBAC, and permissions is linked to a time interval in authorizations. *Coalition-based access control* (CBAC) [9] incorporates the aspects of RBAC, TMAC and TBAC, and gives some formalization abstractions. Our paper utilizes some of the methods in CBAC for reference.

Some access control attempts in grid computing are listed below. Akenti [10] is an access control architecture that can address issues that all the resources are controlled by multiple authorities. Permis [11] is a policy driven RBAC *Privilege Management Infrastructure* (PMI), in which the policies are written in XML and stored in X.509 *attribute certificates* (AC) which may be widely distributed. In *Community Authorization Service* (CAS) [12], the owners of resources grant access to a community account as a whole. The CAS server is responsible for managing the policies that govern access to a community's resources. It maintains fine-grained access control information and grants restricted GSI proxy certificates to the users of the community. A fine grain authorization system [13] is proposed in grid by modifying GRAM of Globus. An authorization infrastructure by providing authorization at the component interface is also presented [14].

3 Access Control Challenges in Grid Services Supporting VOs

A virtual organization is formed with some participants including users, services and resources, each of which is from an independent classical organization which has its own security requirements, as Fig.1 illustrates.

As to authentication, some organizations may adopt *public key infrastructure* (PKI), other organizations may adopt Kerberos. To solve the interoperability problem, Kerberos-to-PKI and PKI-to-Kerberos translation mechanism are needed.

As Fig.1 illustrates, the participant users or services are governed independently by classical organizations which have different local access control policies. In order to establish authorization and access control mechanism, each VO must coordinate access control policies delegated from classical organizations and then establish VO's access control policy repository to restrict the security in the VO.

The establishment and security management of a VO is a challenging mission. The traditional methods that manually edit VO policy repository and issue assert authorization credential can not meet the dynamic characteristic. A user-driven security model is proposed in [15] that allows user to create entities and policy domains in order to coordinate and utilize services within VOs.

In fact, user-driven security model is not practical and operateable, because the security restriction to each user can not be determined by the user itself. It is the responsibility of each domain's administrator.

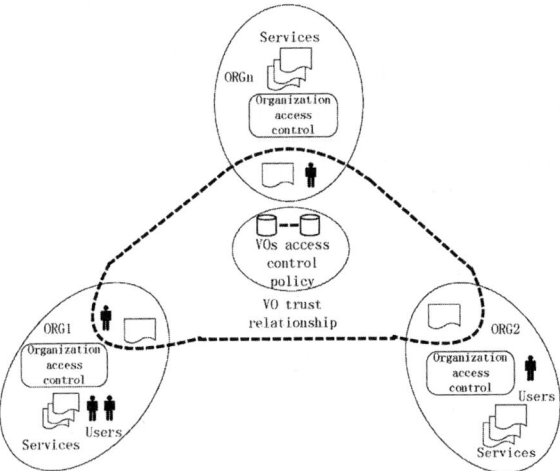

Fig. 1. The architecture of a virtual organization.

The motivation of establishing a VO is to assemble the services from independent security domains and complete a special mission. For example, for the image grid, the processes include 3-D reconstruction service, parallel romancing service and remote displaying service, and these services are dominated by separate organizations that include classified services and users. An engineering laboratory has the function of some types of 3-D reconstruction services, and users can utilize and manage the reconstruction services, also services or users in the other laboratories can call the reconstruction services through some authorizations. In order to easily cooperate, a virtual organization can be established for the image processing grid. From this point of view, the VO is not as dynamic as mentioned above. The lifecycle of a VO is relatively static, which is the repetition of the aggregation of services, which includes all the functions that must be carried out in order to complete the target.

4 Components of Virtual Organizations

Before building an effective and accurate access control model for services sharing in a virtual organization, we must distinguish and classify the security components and the relationship between them. In the subsections we will describe the components in three levels: virtual organization level, organization level, and permission level. The multilevel component models express all the security entities in VO-based grid.

4.1 Virtual Organization Level Access Control Components

At the virtual organization level, four types of components exist: VOS, ORGS, PROJECTS, and FUNCTIONS.

For the application in the image processing grid, data fetching service, 3-D reconstruction service, organ identifying service, parallel romancing service and remote displaying service participate in a virtual organization (VO). Each service comes from an independent organization (ORG) which is an individual security domain. Data fetching service is from body slice database, which belongs to the hospital and has its security mechanism. 3-D reconstruction service is from a special research department which specializes image reconstruction. Organ identifying service is from an artificial intelligent laboratory in a university. Parallel romancing and remote displaying service are from a high performance computing center. These four organizations participate and contribute in the virtual organization.

Each organization (ORG) has its special functions and offers part of them to the project. In this virtual organization, the project is virtual body application.

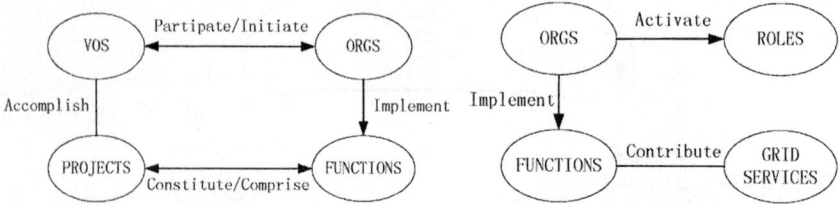

Fig. 2. Virtual Organization Level.   **Fig. 3.** Organization Level.

4.2 Organization Level Access Control Components

At the organization level, also four types of components exist, ORGS, ROLES, GRID SERVICES, and FUNCTIONS, as described in Fig.3.

An organization undertakes some functions, such as organ identifying, face identifying and remote sensing image identifying, which are from the artificial intelligent laboratory (ORG). All the functions are contributed by the grid services that belong to the ORG. The ORG may activate some roles for the convenience of permission management and assignment.

4.3 Permission Level Access Control Components

At the permission level, also four types of components exist, ORGS, USERS, ROLES, and PEMISSIONS, as described in Fig.4.

The permission level access control components describe the role-based access control mechanism in an organization or a virtual organization.

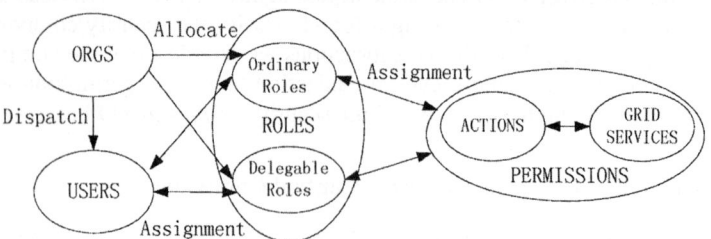

Fig. 4. Permission Level.

In our paper, we introduce the delegation concept in ROLES component. The ROLES set is separated into two types: ordinary roles and delegable roles. The ordinary roles are special for access control inside an organization (ORG). The delegable roles are special for being delegated by other organizations (ORG) or virtual organizations (VO).

The USERS component is dispatched by the ORGS or VOS component and assigned some roles from ROLES component.

5 VO-Based Access Control Model

5.1 Virtual Organization Level Access Control Model

Definition 1. The following is a list of Virtual Organization Level Access Control Model:

1. VOS, ORGS, PROJECTS, and FUNCTIONS, are sets for virtual organization, classical organization, projects, and functions, respectively.
2. $PARTICIPATE \subseteq ORGS \times VOS$, is a many to many *org* to *vo* participation relation.
3. $Orgs: VOS \rightarrow 2^{ORGS}$, is a function derived from *PARTICIPATE* mapping each *vo* to a set of *orgs* where $Orgs(vo) = \{orgs \mid (org, vo) \in PARTICIPATE\}$. This function expresses the *orgs* that participate in a *vo*.
4. $Vos: PROJECTS \rightarrow VOS$, is a mapping of each *project* to an *vo* where $accomplish(project) = vo$. This function expresses the *project* that accomplished by a *vo*.
5. $CONSTITUTE \subseteq FUNCTIONS \times PROJECTS$, is a many to many *function* to *project* constitute relation.
6. $Funtions: PROJECTS \rightarrow 2^{FUNCTIONS}$, is a function derived from *CONSTITUTE* mapping each *project* to a set of *functions* where
   $Functions(project) = \{functions \mid (function, project) \in CONSTITUTE\}$.
   This function expresses the *functions* that constitute into a *project*.
7. $FUNCTIONS = FUNCTIONS_{class1} \cup FUNCTIONS_{class2} \cup \cdots \cup FUNCTIONS_{classn}$, this expression means *functions* are classified into some classes, just like the example in section 4.2.
8. $Functions: ORGS \rightarrow FUNCTIONS_{classi}$, is the mapping of each *org* to a set of classified *functions*. This mapping expresses the *functions* implemented by an *org*.

5.2 Organization Level Access Control Model

Definition 2. The following is a list of Organization Level Access Control Model:

1. ORGS, ROLES, FUNCTIONS, SERVICES, are sets for classical organization, roles, functions, grid services, respectively.
2. $Roles: ORGS \rightarrow 2^{ROLES}$, is the mapping of each *org* to a set of *roles*. This mapping expresses the *roles* that activated by an *org*.
3. $Services: FUNCTIONS \rightarrow SERVICES$, is a mapping of each *function* to an *service* where $contribute(function) = service$. This function expresses the *function* that contributed by a *grid service*.

5.3 Permission Level Access Control Model

Definition 3. The following is a list of Permission Level Access Control Model:
1. ORGS, USERS, ROLES, PERMISSIONS, are sets for classical organization, users, roles, permissions, respectively.
2. The $PERMISSIONS_{org} \subseteq ACTIONS_{org} \times GRIDSERVICES_{org}$ set corresponds to the set of *Actions* which might be applied to *grid service* in an *org*.
3. $Users: ORGS \rightarrow 2^{USERS}$, is the mapping of each *org* to a set of *users*. This mapping expresses the *users* that dispatched by an *org*.
4. $ROLES_{org} = OR_{org} \cup DR_{org}$, *ROLES* in an *org* are separated into two classes, ordinary roles (OR_{org}) and delegable roles (DR_{org}). The subscript *org* means the component set inside an *org*, just like the second item in Definition 2.
5. $OR_{org} \cap DR_{org} = \phi$, this means OR_{org} and DR_{org} are mutually exclusive, which avoids the confusion of responsibility.
6. $UAO_{org} \subseteq USER_{org} \times OR_{org}$, is a many to many mapping of *users* to *ordinary roles* assignment relation inside an *org*.
7. $UAD_{org} \subseteq USER_{org} \times DR_{org}$, is a many to many mapping of *users* to *delegable roles* assignment relation inside an *org*.
8. $UA_{org} = UAO_{org} \cup UAD_{org}$.
9. $PAO_{org} \subseteq PERMISSION_{org} \times OR_{org}$, is a many to many *permission* to *ordinary roles* assignment relation inside an *org*.
10. $Permission\_o: OR_{org} \rightarrow 2^{PERMISSIONS_{org}}$, is a function mapping a ordinary role to a set of permissions, where $Permission\_o(role) = \{p:P | \exists r' \leq r \cdot (r',p) \in PAO_{org}\}$.
11. $PAD_{org} \subseteq PERMISSION_{org} \times DR_{org}$, is a many to many *permission* to *delegable roles* assignment relation.
12. $Permission\_d: DR_{org} \rightarrow 2^{PERMISSIONS_{org}}$, is a function mapping a delegable role to a set of permissions, where $Permission\_d(role) = \{p:P | \exists r' \leq r \cdot (r',p) \in PAD_{org}\}$.
13. $PA_{org} = PAO_{org} \cup PAD_{org}$.

Rule1. This rule illustrates the constraint related to the delegation of authorizations inside an *org*:

$can\_delegate \subseteq DR_{org} \times Pre\_con$, where DR_{org} means the set of *delegable roles* in an *org*, Pre_con means set of prerequisite conditions.

$(r, cr) \in can\_delegate$ means a user in an *org* with the membership of delegable role *r* or a role senior to *r* in an *org* can delegate the role *r* or a role junior to *r* to the other entities (for example a *vo*) which satisfies the prerequisite condition *cr*. In fact, the purpose of delegation of a *role* is to delegate the *permission* of it.

$(r, \phi) \in can\_delegate$ means a user with the membership of delegable role *r* or a role senior to *r* in an *org* can delegate the role *r* or a role junior to *r* to the other entities without any constraint.

For example, $(r_1, cr_1) \in can\_delegate$ (r_1 expresses one of the *roles* in the *data storage organization* with the *permission* of *read action* on *data fetching service*, *cr1* expresses the *vo* which can accomplish the *virtual body application project* of image processing grid), means a user (a database manager for *virtual body department* in the *data storage organization*) with r_1 role assignment can delegate the *permission* belong to the role r_1 to the *vo*, *virtual body application project*.

6 Conclusion and Future Works

In this paper, we propose a VO-based framework for access control in service grid environment. We introduce hierarchy concept to clarify access control components and their relationship, and also introduce access control model for each level. Major contributions of the paper include the following two aspects: The objective and functional composites of each *vo* are introduced with *PROJECTS* and *FUNCTIONS* components in the VO level access control model. Delegation concept is introduced for the expression of permission delegated from real organizations to virtual organization, from which we conclude that the delegation model integrates the Permission level and the VO level by the delegation of permission in each delegable role.

There remain many challenges to be explored. The first is a specification language to specify and enforce policies for the access model proposed in the paper. We also need an administrative model for the access control model in order to manage the creation and management of the policies, just as the definition in ARBAC [16].

References

1. Foster, C. Kesselman, and S. Tuecke, "The Anatomy of the Grid: Enabling Scalable Virtual Organizations", *International Journal of High Performance Computing Applications*, 15 (3). 200-222. 2001.
2. Foster, C. Kesselman, G. Tsudik, and S. Tuecke, "A Security Architecture for Computational Grids", *Proceedings of the 5th ACM Conference on Computer and Communications Security*, pp.83-92, San Francisco, CA, USA, 1998.
3. R. Sandhu, E. Coyne, H. Feinstein, and C. Youman, "Role-based access control models", *IEEE Computer*, Vol.29, No.2, February 1996.
4. D. F. Ferraiolo, R. Sandhu, et al., "Proposed NIST Standard for Role-Based Access Control", *ACM Transactions on Information and System Security*, 4(3): pp.224-274, 2001.
5. R. K. Thomas, "Team-based access control (TMAC): a primitive for applying role-based access controls in collaborative environments", *Proceedings of the 2nd ACM workshop on Role-based access control*, pp.13-19, Fairfax, VA, USA, October 1997.
6. W. Wang, "Team-and-Role-Based Organizational Context and Access Control for Cooperative Hypermedia Environments", *Proceeding of ACM Hypertext'99*, pp.37-46, Darmstadt, Germany, February 1999.
7. R. K. Thomas and R. S. Sandhu, "Task-based Authorization Controls (TBAC): A Family of Models for Active and Enterprise-oriented Authorization Management", *Proceedings of the IFIP WG11.3 Workshop on Database Security*, Lake Tahoe, California, August 1997.
8. V. Atluri, W. K. Huang, "An authorization model for workflow", *Proceeding of the Fourth European Symposium on Research in Computer Security*, pp44-64, Sept. 1996.
9. E. Cohen, R. K. Thomas, W. Winsborough, and D. Shands, "Models for coalition-based access control (CBAC)", *Proceedings of the seventh ACM symposium on Access control models and technologies*, Monterey, CA, USA, June 2002.

10. M. Thompson, W. Johnston, S. Mudumbai, G. Hoo, K. Jackson, and A. Essiari, "Certificate-based Access Control for Widely Distributed Resources", *Proceedings of the Eighth Usenix Security Symposium*, Aug. 1999.
11. D. Chadwick, A. Otenko, "The Permis X.509 Role Based Privilege Management Infrastructure", *Proceedings of SACMAT 2002 Conference*, ACM Press, pp.135-140.
12. L. Pearlman, V. Welch, I. Foster, C. Kesselman, and S. Tuecke, "A Community Authorization Service for Group Collaboration", *Proceedings of the IEEE 3rd International Workshop on Policies for Distributed Systems and Networks*, 2002.
13. K. Keahey, V. Welch, S. Lang, B. Liu, and S. Meder, "Fine-Grain Authorization Policies in the GRID: Design and Implementation", *Proceedings of the 1st International Workshop on Middleware for Grid Computing*, 2003.
14. L. Ramakrishnan, et al., "An Authorization Framework for a Grid Based Component Architecture", *Proc. of the 3rd International Workshop on Grid Computing*, 2002.
15. V. Welch, F. Siebenlist, I. Foster, J. Bresnahan, K. Czajkowski, J. Gawor, C. Kesselman, S. Meder, L. Pearlman, and S. Tuecke, "Security for grid services", *Proceedings of 12th International Symposium on High Performance Distributed Computing (HPDC-12)*, IEEE Computer Society Press, 2003.
16. R. Sandhu, V. Bhamidipati, E. Coyne, S. Ganta, and C. Youman, "The ARBAC97 model for role-based administration of roles: preliminary description and outline", *Proceedings of the 2nd ACM workshop on Role-based access control*, pp.41-50, October 1997.

G-PASS: Security Infrastructure for Grid Travelers*

Tianchi Ma, Lin Chen, Cho-Li Wang, and Francis C.M. Lau

Department of Computer Science
The University of Hong Kong, Hong Kong
{tcma,lchen2,clwang,fcmlau}@cs.hku.hk

Abstract. Grid travelers are special mobile processes responsible for coordinating resources that are distributed across multiple virtual organizations (VOs). We propose a security infrastructure called G-PASS to provide security support for grid travelers during their trip and credential mapping when crossing VO boundaries. We demonstrate the power and feasibility of G-PASS with a bio-informatics application running on multiple VOs. We report and analyze the overheads incurred in migration decisions and the actual process migrations. G-PASS can be installed with GSI as the base, thus making it compatible with existing grid middleware.

1 Introduction

In grid computing, a virtual organization (VO) is a group of individuals or institutions who share some computing resources for a common goal. As grid technologies become more mature and widely adopted, more and more VOs are formed and various types of resources and modes of information sharing are supported. It then becomes inevitable that the future development of Grid applications will move towards large-scale deployment, and they need to access services and resources that are scattered among multiple VOs. To facilitate such deployment, several new features should be developed within the grid security infrastructure, which include: (1) support for VO crossing; (2) features to guarantee the security of the coordinating agents and their hosts; and (3) ways to maintain the trust relationship between multiple autonomous VOs.

In this paper, we propose a new type of mobile process, called *grid traveler*, which has the special ability to move across VO boundaries to coordinate the use of resources and access control under different protection domains. An accompanying security infrastructure named G-PASS is proposed for the credential management of grid travelers. G-PASS provides two useful functions: (1) G-PASS implements a new trust model for supporting simple credential verification and transfer, as well as the creation and atomicity of security transactions. The core

---

* This research is supported in part by HKU Foundation Seed Grant 28506002, the China 863 National Grid project, and a HKU grant for the HKU Grid Point.

of this trust model is the concept of "security instance". A security instance includes a security transaction and its constraint specification. One can accomplish the delegation by binding his/her identity onto the security instance instead of binding onto some special host. (2) To support VO crossing, G-PASS provides an RBAC2 [2] qualified role-based privilege mapping with the granularity of security instance. Different from traditional role-based access control (RBAC), a gateway service called G-custom is imported to map the original credentials (recorded in a credential carrier called G-passport) to a locally recognized approval table. The local resource publisher can then work with the normal access control directly without having to install the role-mapping mechanism.

The rest of the paper is organized as follows. Section 2 gives the background of this research. In section 3, an overview of G-PASS and its features are given. The instance-oriented trust model and role-based privilege mapping are discussed in Section 4. Section 5 presents the performance test of G-PASS. Sections 6 and 7 discuss related work and conclude the paper.

2 Background

GSI is the generally used security system in current grids. It maintains basic trust relationships for resource sharing and job submission. However, GSI cannot satisfy the complex security requirements of grid travelers during VO crossings.

Firstly, GSI is too simple to deal with grid travelers. In general, each VO has its own security policy space. There is a set of identity bindings on access rights. A delegation issued from outside of the policy space will be regarded as an invalid request because the identity binding on it cannot be recognized by the local access control policy. Although GSI has implemented a simple role-based mapping mechanism, it is just a simple one-to-one mapping and can hardly deal with complex situations when crossing VOs. For example, for a traveler with multifarious identities, each of the identities provides only a partial approval that will be valid only under some specific constraints. The one-to-one mapping cannot support such complex role mapping relationships.

Secondly, GSI uses the X.509 delegation model and the delegation is bound to the target host. As the grid traveler travels over multiple hops, the delegation will be transferred by continuously issuing new delegation documents. The verifier at grid entry point needs to check all the signatures created in the delegation chain, thus introducing large overheads. The whole security system becomes less scalable.

Thirdly, the atomicity of security transactions cannot be ensured in GSI. For example, a modification to a bank account will include a read and a write operation. There is no safe state between the two operations. If the two operations are approved by two identities, it is difficult to decide which operation should take the responsibility upon the occurrence of an exception during the modification - known as "separation of duty" problem. GSI allows only simple approvals from single identity. It can not deal with the separation of duty problem well.

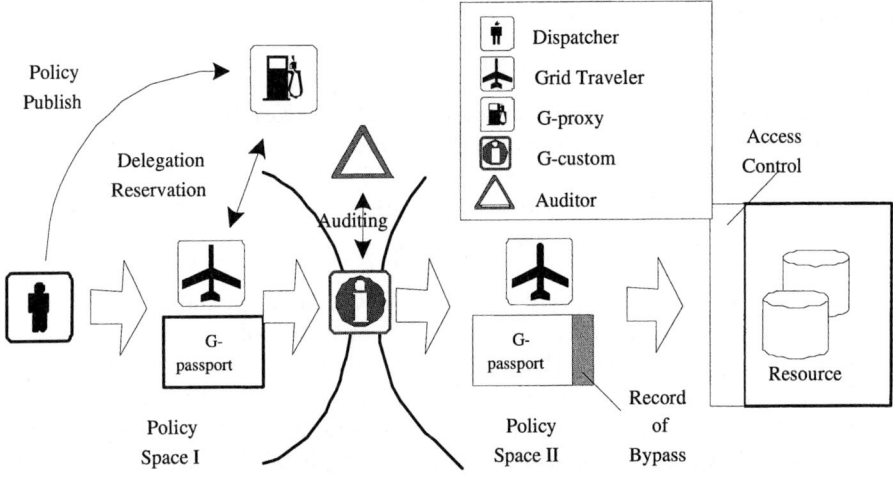

Fig. 1. G-PASS Architecture

3 System Architecture

Figure 1 shows the basic working mechanism in G-PASS. G-PASS consists of various security-related components to support grid travelers.

The *G-passport* resembles a passport in real life with continuous passport pages. G-passport keeps several types of page content: (1) *G-dispatch* declares the delegation from the traveler's dispatcher. It records a privilege set asserted legal by original dispatcher. (2) *G-warrant* claims the intent of warranting a subset of the privileges declared in G-dispatch by a certain warrantor. At the same time, the warrantor promises to be responsible for the traveler's corresponding behavior. (3) *G-exception* records the security exceptions thrown by the hosts or resource access controllers. It is used for security monitoring. Systems can adjust their policy according to such records. (4) *G-event* records the user-defined security events.

With G-passport we design an instance-oriented trust model [3], which ensures the atomicity of security transactions. A chained signature technology [3] is adopted to ensure the integrity of the G-passport. Thus, the attacker cannot find a way to substitute, modify, or delete a page in the G-passport, nor to insert a page. Each page is defined as a contract, in which at least two identities are required to provide a digital signature, and to claim their responsibilities. This complies with Clark-Wilson's principle of separation of duty [4]. An auditor, if it exists, may be required to reserve a copy of the contract and to accuse any concerned identity who attempts to deny the approval or event.

G-custom is a border checker at the entry of a policy space. The role-based mapping is performed in G-custom. *G-proxy* grants new delegation in the name of the user. It enables single sign-on by the user and keeps his/her policy active even the user is currently offline. This procedure is called *delegation reservation*.

4 Trust Management

4.1 Instance-Oriented Trust Model

The traditional trust model sits on top of the Public Key Infrastructure (PKI). Supposed user U holds the keypair $KP = (Pk, Sk)$, where Pk is the public key and Sk is the private key. A digital signature $S_{KP}(v)$ proves the correctness of statement v in U's name. We define a delegation

$$Del(U', p, U) = ((U' \parallel p), S_{KP}(U' \parallel p), Pk), \ p = \{r_i, c_i\} | i = 1, \ldots, n\}, \ n \geq 1$$

to claim that U' can have privileges in the set p in the name of U, where r_i represents the detailed privilege and c_i is the constraint of r_i.

From the above, we can see that the traditional delegation is identity-oriented. In the migration of a grid traveler, the target identity is the identity of the target host. So the delegation is also a host-oriented one. It has the following disadvantages for supporting grid travelers. Firstly, it fails to achieve separation of duty. The Clark-Wilson's principle regulated that the states before and after the transaction must be safe and verifiable, so that the responsibility is clear once any exception is raised. It is the issuer of the delegation who can define transactions; however, the privileges are defined by the resource provider. So new delegation mechanism should be developed to enforce the recording of privileges in the form of transactions. Secondly, it incurs a large overhead. Supposed a grid traveler gets a delegation of p_1 from identity U_1; after $k-1$ times of migration, it arrives at the host with identity U_k. A delegation chain is generated during the trip. The host will need to check at least $k-1$ signatures to assert the validity of this delegation document.

To overcome the above drawbacks, we adopt an instance-oriented approach [3]. A security instance includes a security transaction and its validity specifications. Identities will be simply delegated onto the transaction instead of the privilege operations. This provides the atomicity of security transactions by which the separation of duty can be achieved.

Let $T(r_1, \ldots, r_k)$ be a transaction including a sequence of k operations $(o_1, \ldots, o_k)$, where the operation o_i is performed according to the defined privilege r_i, for $1 \leq i \leq k$. During the delegation granting, the issuer can specify the constraint set C for the transaction. Let $Ins(T, C) = (\{r_1 \ldots r_k\}, C)$ be a security instance of $T(r_1, \ldots, r_k)$ under C. Let $req(r, S)$ represent a request for operating p under the system state S. When it satisfies $r \in \{r_1 \ldots r_k\}$ and $S \in C$, the request is said to be covered by the instance $Ins(T, C)$.

When U wants to grant delegation to $Ins(T, C)$ with keypair $KP = (Pk, Sk)$, it can simply issue a capability, which is a signed document with permitted privileges to serve as the delegation of the instance; that is,

$$Del(Ins(T, C), U) = (Ins(T, C), S_{KP}(Ins(T, C)), Pk)$$

Note the target identity in the traditional host-oriented delegation has been removed from the new delegation document. Thus there needs not be a delegation chain to implement the one-by-one security guarantee, and the overhead in verifying the delegation can be greatly reduced from $k-1$ to 1.

4.2 Role-Based Privilege Mapping

To support VO crossing, the security credentials should be made effectively in the target VO's local policy space. G-PASS imports a role-based privilege mapping, which can proceed in two phrases: (1) role-based privilege mapping, and (2) normal access control. Figure 2 shows the main operations performed in phase (1). The credentials recorded in the G-passport are transformed into a privilege table that can be fully recognized in the target VO's local policy space. The privilege table is formed with an array of instances. Each instance is approved by several locally recognizable roles. As an assistance of the transformation, a role table is imported in which roles and their corresponding global identities are recorded. The role table can be published onto a G-custom service. In phase (2), because the role-based mapping has been done when the grid traveler entered the VO, the local resource provider need not perform RBAC again on the G-passport. It will firstly select an instance according to the given requests. Then it will check if there exists a role that is granted to use all the privileges recorded in the instance by the local access control policy.

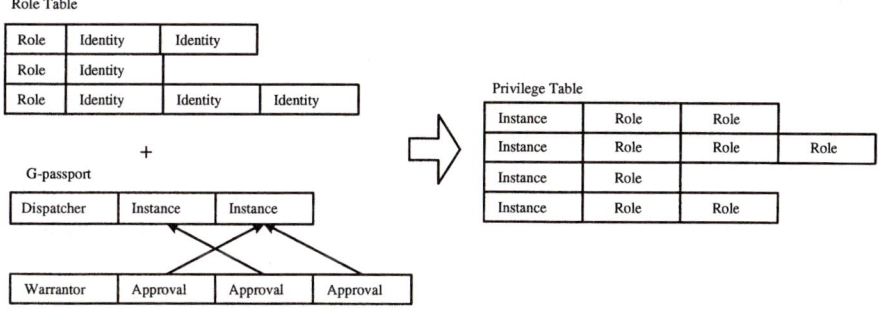

Fig. 2. Role-based privilege mapping

The advantage of this two-phase procedure is that the local resource provider needs not provide a role table themselves. This makes a policy adjustment easier because no synchronization and consistency problem need to be considered.

5 Sample Application and Performance Evaluation

We evaluate the efficiency of G-PASS using a distributed bio-informatics application, BLAST (Basic Local Alignment Search Tool [5]), written in G-JavaMPI [8]. G-JavaMPI is a new MPI middleware which supports parallel and distributed Java computing in a grid environment. A special feature of G-JavaMPI is its support of transparent Java process migration which can facilitate dynamic load sharing and resource-driven task migrations. In G-PASS, the migratable JavaMPI process is regarded as a kind of grid traveler which needs security protection when it migrates across VOs.

7 Conclusion

In this paper, a security infrastructure called G-PASS is proposed. The goal is to support VO crossing and information gathering for grid travelers. G-PASS works as an infrastructure, providing protocols and documents (G-passport) as well as primary establishments (G-custom). It is compatible with the GSI in terms of key preparation and GRAM plug-in. Therefore, G-PASS can be used together with existing grid middleware, especially the Globus Toolkit. We envision that with more large-scale applications taking advantage of the grid environment, mobile travelers will be more common and demand more capabilities, and thus mobility support and role-based privilege mapping will be under the limelight in the grid security field.

References

1. I. Foster, C. Kesselman, G. Tsudik, and S. Tuecke. "A Security Architecture for Computational Grids," Proc. 5th ACM Conference on Computer and Communications Security Conference, pp. 83-92, 1998.
2. R. S. Sandhu, E.J. Coyne, H.L. Feinstein, C.E. Youman, "Role-Based Access Control Models," IEEE Computer 29(2): 38-47, IEEE Press, 1996.
3. T. Ma, S. Li, "An Instance-Oriented Security Mechanism in Grid-based Mobile Agent System," IEEE International Conference on Cluster Computing, pp. 492-495, 2003.
4. D.D. Clark. and D.R. Wilson, "Non Discretionary Controls Commercial Applications," IEEE Symposium on Security and Privacy, pp. 184-194, April 1997.
5. S. F. Altschul, W. Gish, W. Miller, E.W. Myers, and D.J. Lipman. "Basic local alignment search tool," J. Mol. Biol., 215:403-410, 1990.
6. L. Pearlman, V. Welch, I. Foster, C. Kesselman, S. Tuecke. "A Community Authorization Service for Group Collaboration," IEEE 3rd International Workshop on Policies for Distributed Systems and Networks, 2002.
7. S. Tuecke, et.al. "Internet X.509 Public Key Infrastructure Proxy Certificate Profile," IETF, 2003.
8. L. Chen, C.L. Wang, and F.C.M. Lau. "A Grid Middleware for Distributed Java Computing with MPI Binding and Process Migration Supports," Journal of Computer Science and Technology, Vol. 18, No. 4, July 2003, pp. 505-514.

Protect Grids from DDoS Attacks

Yang Xiang and Wanlei Zhou

School of Information Technology, Deakin University
Melbourne Campus, Burwood 3125, Australia
{yxi,wanlei}@deakin.edu.au

Abstract. Recently a number of highly publicised incidents of Distributed Denial of Service (DDoS) attacks have made people aware of the importance of providing available securely the grids' data and services to users. This paper introduces the vulnerability of grids to DDoS attacks, and proposes a distributed defense system that has a mixture deployment of sub-systems to protect grids from DDoS attacks. According to the simulation experiments, this system is effective to defend grids against attacks. It can avoid overall network congestion and provide more resources to legitimate grid users.

1 Introduction

Grid computing has been identified as one of the emerging technologies that will change the world [13]. In [10], the Grid can exist everywhere in our daily lives, such as PDAs, cell phones, appliances, Game Boys, cash registers and so on. Several emerging technologies are combining to create a Great Global Grid (GGG) infrastructure that will enable universal connectivity to large computing resources [9]. Many technologies are converging to create these future Internet architectures including utility grids, web services, wireless devices, peer-to-peer collaboration, virtual clusters, business process automation, and enterprise portals, etc. To develop these grid models, security and privacy are always the major challenges.

Many security systems are researched to secure the grids in authentication and encryption areas. However, availability and access control in grid computing still need more research. Recently the notorious Distributed Denial of Service (DDoS) attacks make people aware of the importance of providing available grids' data and services securely to users. A DDoS attack is characterized by an explicit attempt from an attacker to prevent legitimate users of a service from using the desired resource [1]. Although today the Great Global Grid (GGG) infrastructure has not been maturely built like the World Wide Web, it also faces the challenges of DDoS attacks.

In this paper we show a system how to protect grids from DDoS attacks. Our contribution is that we analyze the vulnerabilities in grid computing related to DDoS attacks, propose a novel distributed system to protect grids from DDoS attacks by using statistical analysis, and conduct experiments to test the validity of our idea. Our system is one of the first systems that have a mixture deployment, which is to deploy sub-systems in both victim end and attack source end, thus achieve better control results over other mechanisms.

2 Related Work

A number of grid platforms with security features exist today. In [3], a security system for computational grids is introduced. A distributed Grid Security (GridSec) architecture is developed in University of Southern California [4]. This architecture is built with distributed firewalls, packet filters, security managers, attack databases, and automated intrusion detection and response subsystems running over multiple platforms. To detect intrusion, distributed Micro-Firewalls are developed by Hwang [6] to protect the grids. However, most of the current research in grid computing does not concern the DDoS problem, or does not propose effective DDoS defense mechanisms.

A typical DDoS attack involves sending a large number of packets to a destination, thus causing excessive amounts of endpoint, and possibly transit, network bandwidth to be consumed [5]. The real attacker is always trying to hide himself from detection, for example, by providing spoofed IP addresses.

Traditional defending mechanisms include detecting mechanism, and reacting mechanism, detailed analysis of current research can be found in [14]. After detecting the malicious actions of DDoS attacks, the defense system turns into the reacting stage, which includes filtering, congestion control and traceback. Most of the current systems are not automated and often with high detection false positive. Moreover, these systems are deployed only at victim end, which cause the lagging response behind the attack; therefore it would not be a potent defense method inherently. It cannot effectively avoid network congestion, since the flood has already arrived. Our defense system proposed in this paper is deployed in both victim end and attacker source end, so it greatly enhances the defense capability.

3 DDoS Defense System for Grids

3.1 Vulnerabilities of Grids

The vulnerabilities of DDoS in grid computing arise from both the network environment that it relies on and the algorithms that are applied. For the first case, flood attacks are the most possible DDoS attack in grid computing. For the second case, the algorithmic complexity attack is another class of low-bandwidth Denial of Service attacks [2]. In this paper, we mainly talk about flood attacks and the countermeasures.

The characteristic of the Internet comprising of limited and consumable resources is an inherent reason to attract attacks. Bandwidth, processing power, and storage capacities are all targets of attacks. Each grid has limited resources that can be exhausted by a sufficient number of users. Thus when the attacks are successful, the grid can be out of service. The DDoS attacker keeps sending malicious packets in a certain long period of time to launch an attack, and this kind of attacks is called flood attacks. Some examples of flood attacks include TCP-based attacks, ICMP-based attacks, and request attacks [1].

3.2 Distributed Defense System

The following figure shows the defense system to protect grids from DDoS attacks. This system is deployed in both victim end and source end, which means all the networks that are near the attacked grid side and near the attacker side. This system is distributed, and each sub-system is coordinated autonomously with others.

This system includes a protecting communication network, which is used to communicate with other coordinated defense systems located at different grids. Because the DDoS attacks usually come from different networks, this part is essential to make a successful defense system. Other components of this defense model are intrusion surveillance system, attack control system and traceback system.

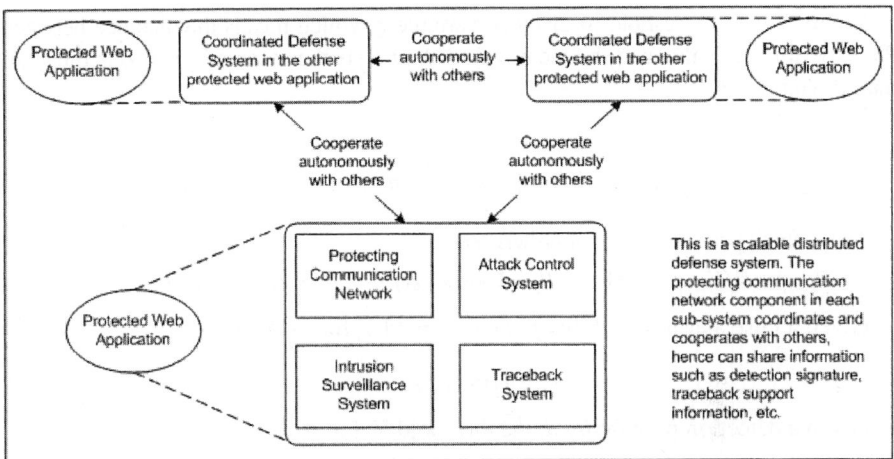

Fig. 1. Distributed defense system.

The intrusion surveillance system automatically monitors the potential intrusion actions. It is very important for an effective defense system. In our work, we apply the statistical method to analyze the network characteristics. When its sensors find the malicious scanning, propagation, communication actions or symptoms by flood of possible attackers, it alerts the attack control system and traceback system, and sends the alarm message to coordinated defense systems. The traceback system also plays a key role in the defense system. We find that an effective traceback system relies on an ingenious intrusion surveillance system. Simply passively dropping the malicious packets is not enough to be a strong protection system against DDoS attacks. So the attack control system not only blocks the source attack traffic, but also record the crime actions for later forensic purposes.

3.3 Statistic-Based Detection

To detect the abnormal network characteristics caused by DDoS attacks, two main mathematical methods are the statistical method [8] and the Cumulative Sum (CUSUM) algorithm [11]. Here we use statistic-based detection in order to obtain a simple and fast process, and a high-accurate hit rate.

One assumption of this method is that the network traffic variables obey a Normal Distribution. The other assumption is that the attacker uses spoofed source IP addresses in the attacking packets to avoid traceback. So we apply this statistic method [8] to detect the abnormal changes of new IP addresses when networks are attacked. Although network traffic volume monitoring may be an easy way to detect DDoS attacks, it cannot differentiate flash crowds from the real DDoS attacks [7]. Hence by observing the normal network IP addresses pattern and comparing it with the current traffic, the possible attacks can be detected. Therefore the detection probability P_d, false alarm probability P_f, and missing probability P_m can be written as follows.

$$P_d = \int_{\frac{V-\mu_\xi}{\sigma}}^{\infty} \frac{1}{\sqrt{2\pi}} e^{-\frac{t^2}{2}} dt \quad P_f = \int_{\frac{V}{\sigma}}^{\infty} \frac{1}{\sqrt{2\pi}} e^{-\frac{t^2}{2}} dt \quad P_m = \int_{-\infty}^{\frac{V-\mu_\xi}{\sigma}} \frac{1}{\sqrt{2\pi}} e^{-\frac{t^2}{2}} dt \quad (1)$$

Where V is the threshold of distance variable ξ, which means the distance between the experimental value and the mean value. And μ_ξ is the mean of ξ and σ is the standard deviation of ξ.

Let

$$\phi(t) = \int_{-\infty}^{t} \frac{1}{\sqrt{2\pi}} e^{-\frac{t^2}{2}} dt \quad (2)$$

These three probabilities can be written by

$$P_d = 1 - \phi[(V-\mu_\xi)/\sigma] \quad P_f = 1 - \phi(V/\sigma) \quad P_m = \phi[(V-\mu_\xi)/\sigma] \quad (3)$$

Given a false alarm probability f, the threshold V_f for $P_f(V_f) \le f$ is

$$V_f \ge -\sigma\phi^{-1}(f) \quad (4)$$

Given a detection probability d, the threshold V_d for $P_d(V_d) \ge d$ is

$$V_d \le \mu_\xi - \sigma\phi^{-1}(d), \quad if \quad \mu_\xi - \sigma\phi^{-1}(d) > 0 \quad (5)$$

For $d=1$ and $f=0$, we get

$$P_d = 1, P_f = 0, \quad if \quad V \in [4\sigma, \mu_\xi - 4\sigma], \mu_\xi - 4\sigma > 0 \quad (6)$$

So by adjusting the threshold variable the best detection probability P_d and least false alarm probability P_f can be achieved.

3.4 Simulation

To build a real testing defense environment for grids is expensive. So we did simulation work by a network simulator, SSFNet (Scalable Simulation Framework) [12], and gathered the experimental data for analysis. Our test results show this system can efficiently protect grids from DDoS attacks.

With this capability to build models of large scale network environments by using the SSFNet, we did many experiments to simulate our system. Figure 2 is the grid topology for our experimental analysis of DDoS attacks and defense. We setup four grids, which are linked by 5 routers. Attacking hosts are deployed in Grid 0 and Grid 3, while the victim server is host 1:4 in Grid 1. Two packages in the SSFNet simulator are exploited in our experiments, one is DDoS package and the other is tcpdump package. We observe the number of new IP addresses within a time period and apply statistic method to detect the attack pattern.

3.5 Results and Analysis

In the experiments, we use the TCP-based flood attacks to simulate the real situation. The attacking hosts keep sending SYN packets with spoofed source IP addresses to exhaust the target's resources. The intrusion surveillance systems and attack control systems of our distributed defense system are deployed at each router.

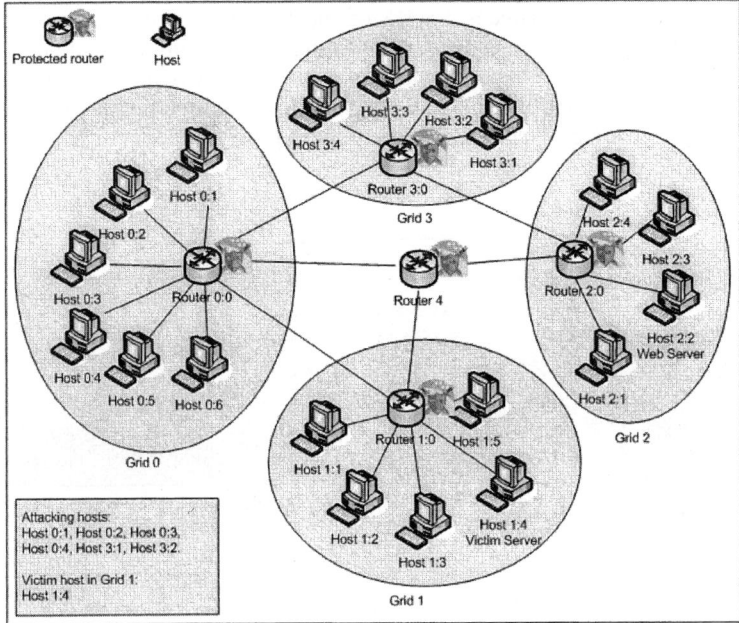

Fig. 2. Grid topology in simulation.

The record of new IP address numbers in the routers are shown in figure 3. In Router 1:0, the number rises dramatically during the attack. But in the other routers, the rises are not so evident. Because every upper router contributes a small part of change, then the overall change can be obviously shown at the victim end router. The original traffic volumes in Router 1:0 and Router 0:0 are shown in figure 4. The abrupt change traffic volume can be easily detected at the victim end. We can see both the victim end router and upper routers suffer network congestion.

By using the statistical method described above, the normal traffic pattern can be modeled. Then in each router, attack control sub-system can filter the malicious packets by comparing the ongoing traffic with the normal one. If we only deploy the defense system at the victim end as the traditional research work discussed above, that is, activate only the attack control sub-system at the victim end in our experiments, then lots of network traffic still will be consumed as it is shown in figure 5. From figure 4(b) and figure 5(b) we find there is not much difference in traffic volume of the upper routers between the protected case and the unprotected one. And the malicious packets can not be filtered well as it is shown in figure 5(a).

When the defense system is deployed as we proposed, then the network traffic is normal as shown in figure 6. From figure 5(a) and figure 6(a) we can see in the later deployment less malicious packets arrive at the victim end. These two figures obviously show that the amount of filtered malicious packets in our proposed deployment is greater than current traditional systems' deployment. In figure 6(a), the victim end traffic resembles to normal condition, because most of the attacking traffics are controlled in the source end. Figure 6(b) shows the traffic volume in Router 0:0 is shaped by the attack control sub-system activated in all routers. It also demonstrates the global deployment of the defense system is better than other deployments.

In this simulation experiment, 98.3% of the malicious packets are filtered. We achieve the active defense objectives by detecting and filtering the malicious packets as early as possible. The experiments prove our approach is a fast and accurate method to detect and control the DDoS attacks. Moreover, the mixture of both victim end and source end deployments is a more powerful way to protect grids from DDoS attacks.

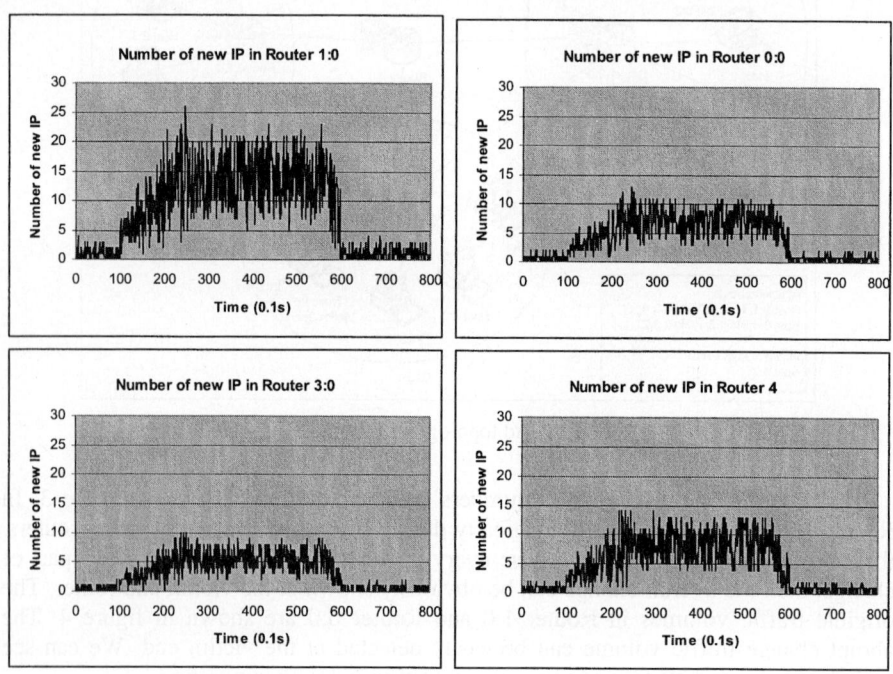

Fig. 3. Number of new IP addresses in different routers.

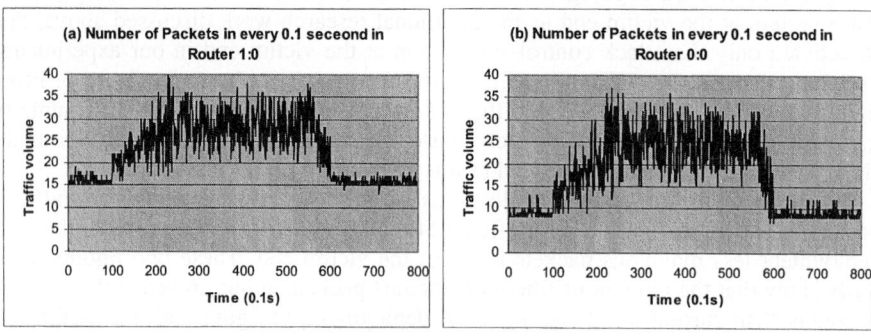

Fig. 4. Original traffic volume in Router 1:0 and Router 0:0.

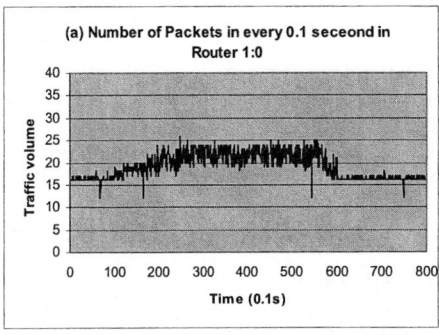
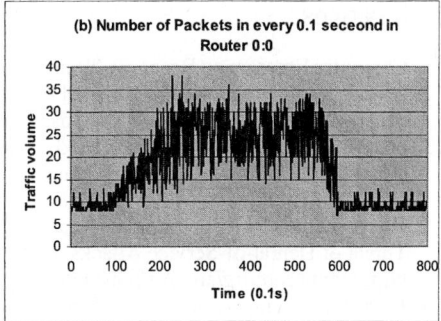

Fig. 5. Traffic volume in Router 1:0 and Router 0:0 if defense system deployed at Router 1:0.

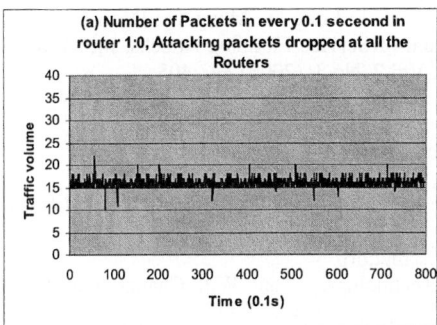
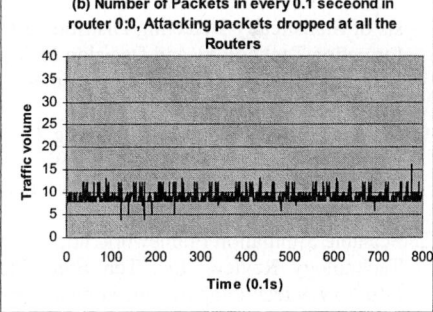

Fig. 6. Traffic volume in Router 1:0 and 0:0 when attacking packets are dropped at all the routers.

Compared with the current DDoS defense systems, the benefit of this deployment is not only a good filtering, but also avoiding the network congestion from the attacker side to target side. If the malicious packets are dropped as early as possible in the upstream routers, then more network resources can be available to legitimate users. Moreover, since the sensors are deployed near the attack sources, it offers more opportunities to traceback the real attackers.

4 Conclusions

In this paper, a distributed system to protect grids from DDoS attacks is presented. The simulation experiments show this system can detect DDoS attacks quickly and safeguard protected grids effectively. However, more work toward realizing the secure gird computing still need to be done. In the next step, experiments in real network environment will be designed and deployed. Coordination between the distributed sub-systems for secure grid computing and an accurate traceback system will be the main points. Robust grid security system will benefit all societies and countries, and more research should be done in this area.

References

1. Computer Emergency Response Team, CERT. http://www.cert.org
2. Crosby S. A., Wallach, D. S.: Denial of Service via Algorithmic Complexity Attacks. Proc. of the 12th USENIX Security Symposium (2003) 29-44
3. Foster, I., Kesselman, C., Tsudik, G., Tuecke, S.: A Security Architecture for Computational Grids. 5th ACM Conf. on Computer and Communication Security (1997)
4. Grid Security Project. University of Southern California, http://gridsec.usc.edu
5. Householder, A., Manion, A., Pesante, L., Weaver G. M., Thomas, R.: Managing the Threat of Denial-of-Service Attacks. CERT, http://www.cert.org/archive/pdf/Managing_DoS.pdf (2001)
6. Hwang K., Gangadharan, M.: Micro-Firewalls for Dynamic Distributed Intrusion Detection. IEEE Int'l Symp. of Network Computing and Applications (2001)
7. Jung, J., Krishnamurthy, B., Rabinovich, M.: Flash Crowds and Denial of Service Attacks: Characterization and Implications for CDNs and Web Sites. Proc. of the International World Wide Web Conference 2002 (2002) 252-262
8. Li, M., Chi, C., Jia, W., Zhao, W., Zhou, W., Cao, J., Long D., Meng, Q.: Decision Analysis of Statistically Detecting Distributed Denial-of-Service Flooding Attacks. Int'l J. of Information Technology and Decision Making, Vol.2. No.3. (2003) 397-405
9. Marcus, R.: Great Global Grid: Emerging Technology Strategies, Trafford Publishing, ISBN 1-55369-884-3 (2002)
10. Malone, M. S.: Internet II: Rebooting America. Forbes ASAP, http://www.forbes.com/asap/2001/0910/044_5.html (2001)
11. Pollak, M.: Optimal Detection of A Change in Distribution, Ann. Statist., Vol. 13 (1986) 206-227
12. Scalable Simulation Framework. http://www.ssfnet.org.
13. Technology Review, Inc. Ten Emerging Technologies That Will Change the World, http://www.technologyreview.com/articles/ emerging2003.asp (2003)
14. Xiang, Y., Zhou, W., Chowdhury, M.: A Survey of Active and Passive Defence Mechanisms against DDoS Attacks, Technical Report, TR C04/02, School of Information Technology, Deakin University, Australia (2004)

Trust Establishment in Large Scale Grid Settings

Bo Zhu[1,2], TieYan Li[2], HuaFei Zhu[2],
Mohan S. Kankanhalli[1], Feng Bao[2], and Robert H. Deng[2]

[1] National University of Singapore
{zhubo,mohan}@comp.nus.edu.sg
[2] Institute for Infocomm Research
{zhubo,litieyan,huafei,baofeng,deng}@i2r.a-star.edu.sg

Abstract. Trust establishment is hard in grid architecture by the *ad hoc* nature. To set up trust in large scale of network is more difficult. In this paper, we propose an automatic key management (AKM) model and corresponding key construction schemes. The hierarchical structure is formed automatically and scale seamlessly in arbitrary network sized. Regions are configured differently according to various levels of risks faced. The novel model provides an integrated solution for self-organized trust establishment, upon which rich appliances are securely supported. It is automatic, flexible, and scalable. Furthermore, simulation results show that computation costs due to the variations are very small under common threshold and region size settings.

1 Introduction

Computational grids are a collection of heterogeneous computers and resources spread across multiple administrative domains with the intent of providing users easy access to these resources. Typical characteristics of grids include simultaneous use of large numbers of resources with dynamic requirements, use of resources from multiple administrative domains, complex communication structures, and stringent performance requirements. Under easier intentional attacks, security in this settings becomes more crucial and is eagerly needed.

To protect the grid infrastructure, Globus Toolkit [1] has been developed (current version GT3.0) and Grid Security Infrastructure (GSI) [4] is the *de facto* security standard in grid community. However, as indicated in [9], GSI suffers from many potential security drawbacks such as uncontrolled delegation, leaky infrastructure and insecure services. Thus, more security mechanisms must be developed to complement GSI and finally ensure inter-grid trust establishment. In the literature, several well known trust models have already been proposed [5, 3, 12]. All of these trust models are designed for certain scenarios, however, none of them is suitable to be used in large scale of grid environment efficiently.

In this paper, we proposed an automatic key management (AKM) model, which is hierarchical and formed dynamically. We carefully study the joining and leaving behaviors to design a scalable scheme. In addition, to achieve flexibility and adaptability, a new concept-*Regional Trust Coefficient* (RTC) is introduced

3 Node-Based and Region-Based Operations

Common node-based and region-based operations include "Join", "Leave", "Merge", "Partition", "Expansion", and "Contraction". The first two are executed in the same way as [6], and here we concentrate on region-based operations only.

3.1 "Merge" and "Partition"

"Merge" operation happens when the number of nodes within a region drops under its threshold. Then, the region is divided into a few parts and each part is combined into one nearby region. Since the thresholds of the target regions are invariable, "Merge" operation can be viewed as a series of "Join" operations.

"Partition" operation happens when RTC of a region drops under GTC or is lower than the security level expected. A straight-forward solution is to partition the region into two regions with almost the same size. For example, region B_i with size $2n$ and the threshold k is partitioned into two regions B_i and B_{m+1}, each of which has n nodes and keeps its threshold as k. For the n nodes remaining in region B_i, they just need to renew their secret shares after the "Partition" operation. In order to assign new secret shares to the n nodes in region B_{m+1}, firstly, region B_{m+1} chooses k regions at level 1, and then selects k nodes from each of the k regions. Without lost of generality, we denote the group of the k regions, the group of the k nodes from region B_j $(j = 1, \cdots, k)$, and the group of all these k^2 nodes as $G_B = \{B_1, \cdots, B_k\}$, $G_j = \{C_{j1}, \cdots, C_{jk}\}$, and $G = \{C_{11}, \cdots, C_{1k}, \cdots, C_{k1}, \cdots, C_{kk}\}$, respectively. Then, a "Partition" request signed by the secret key of region B_i is generated and multicasted to all the nodes in group G. The IDs of region B_i, B_{m+1}, and the k regions are sent together with the request.

By Lagrange interpolation, we have $SK_{B_i} = \sum_{j=1}^{k} SK_{B_j} l_{B_j}(ID_{B_i}) \pmod{q}$ and $SK_{B_j} = \sum_{h=1}^{k} SK_{C_{jh}} l_{C_{jh}}(0) \pmod{q}$, where $l_{B_j}(ID_{B_i}) = \prod_{r=1, r \neq j}^{k} \frac{ID_{B_i} - ID_{B_r}}{ID_{B_j} - ID_{B_r}} \pmod{q}$ and $l_{C_{jh}}(0) = \prod_{r=1, r \neq h}^{k} \frac{ID_{C_{jr}}}{ID_{C_{jr}} - ID_{C_{jh}}} \pmod{q}$. Combining them, we have $SK_{B_i} = \sum_{j=1}^{k} \sum_{h=1}^{k} SK_{C_{jh}} l_{C_{jh}}(0) l_{B_j}(ID_{B_i}) \pmod{q}$. Similarly, we get $SK_{B_{m+1}}$. Therefore, we know that $SK_{B_{m+1}} - SK_{B_i} = \sum_{j=1}^{k} \sum_{h=1}^{k} SK_{C_{jh}} l_{C_{jh}}(0) R_j \pmod{q}$, where $R_j = l_{B_j}(ID_{B_{m+1}}) - l_{B_j}(ID_{B_i})$.

To help a node with identity $C_{(m+1)i}$ in region B_{m+1} generate the new shares of $SK_{B_{m+1}}$, each node C_{jh} in group G computes the partial share $SK'_{C_{jh}} = SK_{C_{jh}} l_{C_{jh}}(0) R_j + \sum_{r=1}^{k-1} a_{jhr} ID^r_{C_{(m+1)i}} \pmod{q}$, where $a_{jhr} \in Z_q$, for $r = 1, \ldots, k-1$. Then distributes the partial share to node $C_{(m+1)i}$. After receiving k^2 partial secret shares from the nodes in group G, node $C_{(m+1)i}$ adds them together to get a share of $SK_{B_{m+1}} - SK_{B_i}$, denoted as $SK'_{C_{(m+1)i}}$. According to the homomorphic property, each node in region B_{m+1} can compute its new share of $SK_{B_{m+1}}$ by adding $SK'_{C_{(m+1)i}}$ to its original share of SK_{B_i}.

3.2 "Expansion" and "Contraction"

"Expansion" Operation happens when RTC of a region drops under GTC or when AKM reaches its capacity upper limit (namely, RTCs of all the regions in AKM

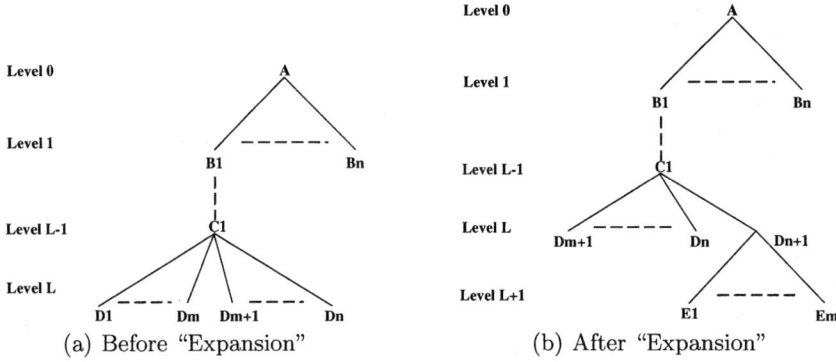

Fig. 2. "Expansion" Operation

are equal to GTC). For example, as shown in Figure 2(a), before "Expansion" the level of AKM is L. We assume that, original secret sharing of region C_1 is executed by $SK_{D_i} = a_0 + a_1 \cdot ID_{D_i} + \cdots + a_{k-1} \cdot ID_{D_i}^{k-1} \pmod{q}$.

Now there is a new node that wants to join, but the RTC of region C_1 is already equal to GTC. Therefore, "Expansion" operation is executed.

Firstly, chooses a group of m nodes from region C_1 which will be degraded to Level $L+1$, where $k \leq m \leq n-k+1$. Without loss of generality, let the group be $G = \{ID_{D_1}, \cdots, ID_{D_m}\}$. Then, selects k nodes from group G. Without loss of generality, let the group be $R = \{ID_{D_1}, \cdots, ID_{D_k}\}$. Following that, chooses a new identity denoted as $ID_{D_{n+1}}$ for the master node of all the nodes degraded. According to Shamir's threshold scheme [8], the secret share for node D_{n+1} can be calculated by the k nodes in group R. However, during "Expansion" operation, this secret share would be never calculated out or recovered explicitly, since we do not assume the existence of a TA at this stage.

For simplicity, the same (n, k) threshold scheme is employed in the newly created region D_{n+1}. Without loss of generality, we assume that a new identity ID_{E_i} is assigned to the node whose old identity is ID_{D_i}, or the identity is chosen by the node itself. Then each node in group R calculates the partial secret share denoted as SK'_{E_i} by $SK'_{E_i} = SK_{D_j} \cdot l_{D_{n+1}} + \sum_{r=1}^{k-1} b_{jr} \cdot ID_{E_i}^r \pmod{q}$, where $l_{D_{n+1}} = \prod_{h=1, h \neq j}^{k} \frac{ID_{D_{n+1}} - ID_{D_h}}{ID_{D_j} - ID_{D_h}}$, and distributes it to the node whose new identity is ID_{E_i}. Finally, each node recovers its new secret share in the new region by combining all the secret shares: $SK_{E_i} = b_0 + b_1 \cdot ID_{E_i} + \cdots + b_{k-1} \cdot ID_{E_i}^{k-1} \pmod{q}$, where $b_0 = SK_{D_{n+1}}$, $b_r = \sum_{j=1}^{k} b_{jr}$ $(r = 1, 2, \ldots, k-1)$. Therefore, all the coefficients of the secret sharing polynomial of the new region are cooperatively determined by the k nodes.

"Contraction" Operation happens when AKM reaches its capacity lower limit. To ensure that the number of nodes in any region in AKM is not less than the threshold set before, we have to decrease the level of the structure. Similar to "Merge" operation, "Contraction" Operation can be viewed as a series of "Join" operations as well.

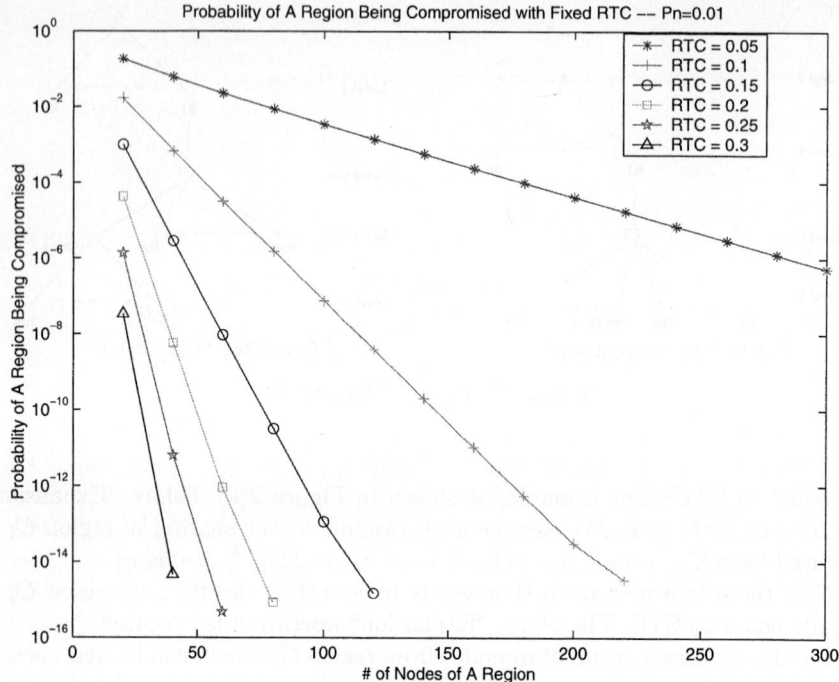

Fig. 3. P_r with fixed RTC

4 Simulation Analysis

Intuitively observed, as in highly dynamic environment, lots of regions with small size result in rapid transform of the structure of the key tree. On the other hand, if their sizes are too big, we may have problems with intra-region routing. Current on-demand routing protocols, such as AODV [7] and OLSR [2], handle well when the size is around 100 to 250 nodes. Thus, it is suitable to set the region size within this range.

4.1 Regions with the Same RTC

As shown in Figure 3, say a single curve, we can see that the higher the RTC, the smaller the value of p_r. Therefore, it is suitable to use RTC as an approximate index of the security condition of a region. In addition, observing the bunch of curve, the higher the RTC, the faster the value of p_r decreases. Due to the two properties, we say that a higher RTC is good for the sake of security. However, it leads to a higher threshold which consumes more computation power.

At the same time, we find that, when both RTC and the region size are small, this region is easy to be compromised. For example, in Figure 3, given that the p_n is 0.01 and the RTC of a region is 0.05, p_r may be higher than 0.01 when the size of the region is less than 75. Fortunately, we need to be aware of this only during the initialization of a region. Thereafter, if nodes leave the region, compared to the decreasing of the region size, the increase of the RTC can bring more positive effects on the security condition of the region. For instance, suppose a region has 40 nodes at the beginning, and its RTC is 0.05. According to Figure 3, p_r is 6.07%. After 20 nodes' leaving the region, the region size drops to 20. However, because its RTC increases to 0.1 at the same time, we find that p_r decreases to 1.69%. It means that this region is more secure than before.

4.2 Computational Costs of Region-Based Operations

Since both "Merge" and "Contraction" operations can be viewed as a series of "Join" operations, our simulation focuses on "Partition" and "Expansion" operations. We run the simulation on a Pentium III 800 laptop.

Table 2 shows the computation cost of "Partition" and "Expansion" operations under different ORSs and thresholds. In the tables, GPSS stands for time for generating partial secret shares for all the nodes in the newly generated region, while GNSS stands for time for generating the new secret share. Simulation results show that the computation cost of both "Partition" and "Expansion" operations is quite small under common threshold and region size settings. For example, when the ORS of a region is 100 and its threshold is 15, it only takes 31 miliseconds to complete the "Partition" operation. As to the "Expansion" operation, the total cost is less than 8 miliseconds.

Table 2. Computation Costs of "Partition" and "Expansion" Operations

ORS	Threshold	"Partition" Operation (msec)			"Expansion" Operation (msec)		
		GPSS	GNSS	Total	GPSS	GNSS	Total
100	5	8.87	0.03	8.90	1.59	0.01	1.60
100	10	18.59	0.11	18.70	4.52	0.02	4.54
100	15	30.96	0.28	31.24	7.91	0.02	7.93
100	20	46.24	0.60	46.84	11.69	0.02	11.71
100	25	62.00	0.80	62.80	16.42	0.03	16.45
250	10	47.63	0.11	47.74	11.00	0.03	11.03
250	20	124.20	0.68	124.88	27.67	0.03	27.70
250	30	205.27	1.29	206.56	54.71	0.07	54.78
250	40	302.27	2.24	304.51	90.16	0.07	90.23
250	50	422.73	3.77	426.50	135.78	0.07	135.85

5 Related Works

X.509 [5], SPKI [3], and PGP [12] are three of the most well-known trust models. The X.509 trust model [5] is centralized that each participant has a certificate signed by a central CA. Since GSI employs X.509 certificates, this trust model can be used within a grid domain. SPKI trust model [3] is more flexible by supporting delegation certificate. But how to control the proxy/delegation certificates is still an unsolved problem. PGP [12] is a distributed trust model that builds one's trust from its neighbors.

Technically similar with our approach, in [11], Zhou and Haas focused on establishing a secure key management service in an ad hoc networking environment. They proposed to use threshold cryptography to distribute trust among a set of servers. Their solutions are only suitable for a small group of servers and are inefficient given large scale networks. In [6], Kong et al. extended the scheme in [11] to normal nodes. It minimizes the effort and complexity for mobile clients to locate and contact the service providers. One major weakness of this scheme is that it becomes either inefficient or insecure with increasing participants. In [10], a more extreme case, where there is no CA at all, was considered. Each user is its own authority domain and issues public-key certificates to other users.

6 Conclusion

In this paper, we proposed a hierarchical trust establish model, which is efficient, flexible and scalable. By building a dynamic key management scheme, we achieve strong security without much tradeoff on efficiency. Furthermore, simulation results show that our scheme is very efficient under common threshold and region size settings.

References

1. Globus toolkits v3.0 of the globus project. http://www.globus.org.
2. Thomas Clausen Et. Al. Optimized link state routing protocol, September 2001. Internet draft, draft-ietf-manet-olsr-06.txt.
3. C. Ellison, B. Frantz, B. Lampson, R. Rivest, B. Thomas, and T. Ylonen. SPKI certificate theory, September 1999. RFC 2693.
4. I. Foster, C. Kesselman, G. Tsudik, and S. Tuecke. A security architecture for computational grids. In *Proceeding of 5th ACM Conference on Computer and Communications Security Conference*, 1998.
5. R. Housley, W. Polk, W. Ford, and D. Solo. Internet x.509 public key infrastructure certificate and certificate revocation list (CRL) profile, 2002. RFC 3280.
6. Jiejun Kong, Petros Zerfos, Haiyun Luo, Songwu Lu, and Lixia Zhang. Providing robust and ubiquitous security support for mobile ad-hoc networks. In *IEEE 9th International Conference on Network Protocols (ICNP'01)*, 2001.
7. Charles E. Perkins and Elizabeth M. Royer. Ad-hoc on-demand distance vector routing. In *WMCSA'99*, 1999.
8. A. Shamir. How to share a secret. *Communications of the ACM*, Vol. 22(11):612–613, November 1979.
9. Mike Surridge. A rough guide to grid security, 2002.
10. Srdjan Čapkuny, Levente Buttyán, and Jean-Pierre Hubaux. Self-organized public-key management for mobile ad hoc networks. Technical Report 2002/34, EPFL/IC, May 2002.
11. Lidong Zhou and Zygmunt J. Haas. Securing ad hoc networks. *IEEE Network Magazine, Special Issue on Network Security*, Vol. 13, No.6, 1999.
12. Phil Zimmermann. Pretty good privacy (PGP), October 1994.

XML Based X.509 Authorization in CERNET Grid

Wu Liu[1,2,3], Jian-Ping Wu[1], Hai-Xin Duan[1], Xing Li[2], and Ping Ren[1,4]

[1] Sichuan Agricultural University, Dujiangyan, 611830, Sichuan, China
{liuwu@cernet.edu.cn}
[2] Network Research Center of Tsinghua University, 100084 Beijing, China
[3] Computer Software Lab. of Sichuan Normal University, 611168 Chengdu, China
[4] Business Management School of Sichuan University, 610064 Chengdu, China

Abstract. This paper presents an authorization solution for resource management and control developed as part of the China Education and Research Network (CERNET) to perform fine-grained authorization of job and resource management requests in a Grid environment which meets the Fusion-Grid's security needs in large scale networks such as CERNET. It integrates the GT2 job manager and X.509 authorization and this model can be extended to other authorization decision functions. It allows the system to evaluate a user's resource specification language request against authorization policies on resource usage. Furthermore, based on XML integrated authorization policies, it allows other virtual organization members to manage the user's resources.

1 Introduction

Users from different organizations who are geographically dispersed but are working together to solve a common problem or related problems typically organize themselves into Virtual Organizations (VO)[5]. The VO not only defines who its members are and possibly assigns roles or attributes to the members, but also arranges with the owners of various resources for VO member access. The resources may consist of computing platforms, storage elements, scientific instruments, data or services.

The China Education and Research Network (CERNET) [3] is an example of such a VO. The CERNET is building a Fusion-Grid to provide computational and data services to its members. Being so widely used as Grid middleware, the Globus Toolkit (GT2) [4] has been chosen by CERNET for remote job submission and secure access to its common data servers. While object-oriented distributed programming frameworks such as CORBA provide very fine-grained access-control at the level of object methods, GT2 provides a coarse-grained "admission control" facility and leaves fine-grained access control up to the resource provider. This simple approach is entirely acceptable for the initial stages of a Grid, when there is a limited set of potential users who negotiate access directly with the resource providers, but it does not scale to large numbers of resource hosts and users. Hence, the GT2 access control mechanisms must be extended to meet the Fusion-Grid's security needs. The solution we present here is to integrate the X.509 authorization service [6] with GT2.

Section 2 is a brief overview of how authorization is currently handled in GT2. Section 3 introduces the X.509 authorization service. Section 4 describes our integration of the GT2 job manager and X.509 authorization and how this model can be extended to other authorization decision functions. Section 5 presents our conclusions.

2 GT2 Authorization

We assume the following model for job submission and control. An interaction is initiated by a user submitting a request to start a job, including the job description, accompanied by the user's Grid credentials, in the form of an X.509 certificate [8]. In the current case, this is just an identity certificate and asserts no other attributes about the user. This request is then evaluated by an ACDF which may be called from several different Access Control Enforcement Functions (ACEF) located in the resource management modules. If the request is authorized, it is started under a local credential (namely User-ID).

During the job execution, a VO user may submit management requests composed of a management action (e.g., request information, suspend or resume a job, cancel a job). The resource manager may decide to perform the action or to pass it on to the locally executing job.

In order to perform these transactions, the Globus Resource Acquisition and Management (GRAM) [2] system is used, which has two major software components: the **gatekeeper** and the **job manager**.

The gatekeeper is responsible for translating Grid credentials to local credentials (e.g. mapping the user to a local account based on their Grid credentials) and creating a job manager instance to handle the specific job invocation request. The job manager is a Grid service which instantiates and then provides for the ability to manage a job. Figure 1 shows the interaction of these elements.

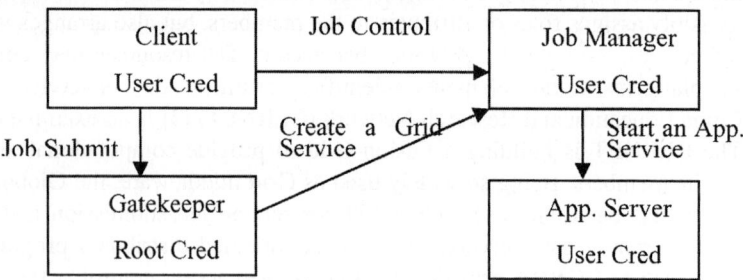

Fig. 1. Interaction of the main components of GRAM.

The GRAM gatekeeper is responsible for authenticating the requesting Grid user, authorizing a job invocation request, and determining the account in which the job should be run. Authentication, done using GT2's Grid Security Infrastructure (GSI) [1], verifies the validity of the presented Grid credentials, the user's possession of those credentials, and the user's Grid identity as indicated by those credentials. Authorization is based on the user's Grid identity, the site's trust policy, and the site grid-mapfile, which maps each allowed Grid identity to a local User-ID.

The gatekeeper then starts a job manager instance, executing with the user's local credential. This mode of operation requires the user to have an account on the resource and implements fine-grained access enforcement by privileges of the account.

The GRAM job manager parses the user's request, including the job description, and calls the resource's job control system (e.g., exec, LSF, PBS etc.) to initiate the user's job. During the job execution, the job manager monitors its progress and handles job control requests (e.g., suspend, stop, query etc.) from the user. Since the job manager instance is run under the user's local credential, as defined by the user's account, the operating system, and local job control system are able to enforce local policy on the job manager and user job by the policy tied to that account.

The job manager does no authorization on job startup because the gatekeeper has already done so. Once the job has been started, however, the job manager accepts, authenticates, and authorizes management requests on the job.

3   X.509 Authorization Model

As noted in Section 1, the authorization provided by GT2 is coarse-grained. Because of the large user community, the CERNET needed to add fine-grained authorization for job execution and management. Rather than writing an authorization function from scratch, the CERNET decided to use the X.509 authorization service [7].

The X.509 model consists of resources that are being accessed via a resource gateway (the ACEF) by clients. These clients connect to the resource gateway using the TLS [9] handshake protocol, or something equivalent, to present authenticated X.509 certificates. The stakeholders for the resources express access constraints on the resources as a set of signed certificates, a few of which are self-signed and must be stored on a known secure host (probably the resource gateway machine), but most of which can be stored remotely. These certificates express the attributes a user must have in order to get specific rights to a resource, identify the stakeholders who are trusted to create use-condition statements and determines the attribute authorities who can attest to a user's attributes. At the time of the resource access, the resource gatekeeper (ACEF) asks a trusted X.509 server (ACDF) what access the user has to the resource. The X.509 server finds all the relevant certificates, verifies that each one is signed by an acceptable issuer, evaluates them, and returns the allowed access.

In the application shown in Figure 2, the pull model is used to allow applications to transparently use X.509 authorization over standard GSI/TLS connections that transport and verify X.509 certificates. X.509 can also be used in a push model because it returns its authorization decision as a signed capability certificate containing the subject's distinguished name (DN), public key, the certification authority (CA) that signed for this name, the name of the resource, and the subject's rights. These capability certificates are short-lived in order to avoid revocation problems.

In GT2, the gatekeeper acts as the resource gateway: it allows access only to Grid users who appear in the grid-mapfile. In our current work we make the job manager an ACEF as well, by enabling it to enforce policy about fine-grained job access.

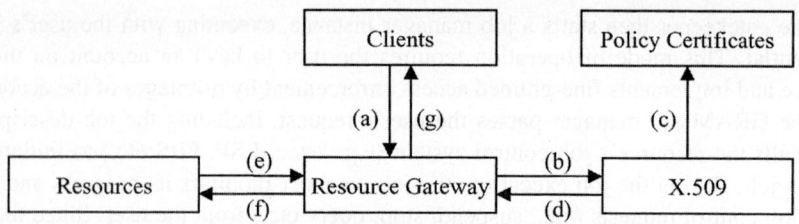

Fig. 2. The X.509 authorization model in pull mode.

4 Integration of X.509 and Authorization via XML

In this section we describe how we integrate the GT2 job manager with X.509.

4.1 Code Integration

While the Globus gatekeeper currently acts as the ACEF and ACDF for job submission, we decided to add our callout for fine-grained access control to the GT2 job manager [6] for two reasons. First, the job manager is the component that parses the Resource Specification Language (RSL) [2] of the job request. RSL consists of attribute value pairs specifying job parameters such as executable description (name, location, etc.), and resource requirements (number of CPUs to be used, maximum allowable memory, etc.). These are the attributes that we want to control. Second, the job manager decides and enforces access policy for job control. Requests to terminate, signal or query a job go directly to the job manager via the job handle URL that is returned on job creation. In GT2 the job manager allows these actions only if the requestor has the same Grid-ID as the job initiator. These were the other actions we wanted to control.

Specifically, our additions consisted of the following:

(1) **Authorization callout API.** We designed a callout API to integrate an ACDF with the job manager. The callout passes to the ACDF module all the information relevant to access control, such as the credential of the user requesting a remote job, the credential of the user who originally started the job, the action to be performed (such as start or cancel a job), a unique job identifier, and the job description expressed in RSL. The ACDF responds through the callout API with either success or an appropriate authorization error. This call is made whenever an action needs to be authorized, that is, before instantiating a job and before canceling, querying, or signaling a running job.

(2) **Policy-based authorization for job management.** As discussed in Section 2, each job management request other than job start is currently authorized by the job manager so that only the user that started a job is allowed to manage it. We modified the authorization in GRAM to enable Grid users other than the job initiator to manage the job based on policy with decisions rendered through the authorization callout API. In addition to changes to the authorization model, this modification also required extensions to the GRAM client to allow one user to signal a job manager instance owned by another user.

(3) **RSL parameters.** We extended RSL to add the "job-tag" parameter allowing the user to submit a job to a specific job management group. If the user does not provide a job-tag on start, a default one will be assigned to the job.
(4) **Error reporting.** We further extended the GRAM protocol to return authorization errors describing reasons for authorization denial as well as authorization system failures.
(5) **Callouts to be configurable at run time.** In order to provide for easy integration of third-party authorization solutions, the job manager allows callouts to be configurable at run time. Callouts can be configured through either a configuration file or an API call. Configuration consists of specifying an abstract callout name, the path to the dynamic library that implements the callout, and the symbol for the callout in the library. Callouts are invoked through runtime loading of dynamic libraries. Arguments to the callout are passed using the C variable argument list facility. The insertion of callout points into job manager required defining a GRAM authorization callout type, that is, an abstract callout type, the exact arguments passed to the callout and a set of errors the callout may return. These callout points are configured by parsing a global configuration file.

4.2 Authorization Policy

When the job manager calls X.509, the access decision is based on the X.509 authorization policy. X.509 organizes policy according to the resources that are being controlled. Hence, the first step in writing policy is to determine the set of resources. In the case of fine-grained control of GT2 job submission, the things that can be controlled are the right to execute a job on a machine in which binaries may be executed, RSL parameters such as requested CPU time, requested scheduling queue, and the rights to stop, resume, cancel, or query currently executing jobs.

From the viewpoint of the Fusion-Grid resource provider, some of these are more important than others and some are hard to enforce:

(1) **Right to submit any job to a machine** – already enforced by gatekeeper
(2) **Right to start a specific binary** – important and can be enforced by the job manager
(3) **Right to limit CPU cycles for a specific job** – currently not important, would need to be enforced by the run queue manager (PBS)
(4) **Right to restrict a user or group to a total CPU limit per month** – may be important, requires an accounting system
(5) **Right to choose an execution queue** – may be important for service guarantees
(6) **Need for at least one class of administrative users who can kill any job** – important
(7) **Need for multiple administrative classes that can kill a restricted set of jobs** – possibly useful but requires users to understand job classes.

4.3 XML Based Policy for the Fusion-Grid

The policy we designed running at CERNET has two levels, with several branches at the lower level. There is a site-wide level where policy specifies the CAs that will be trusted to issue X.509 certificates, the stakeholders for other resources, and the location of the use-condition and attribute certificates. There is also a subordinate level that contains separate policies for each class of executables, for example, the production code, test utilities, a development version of the code, and policies for each job category (at the moment we have only one job category).

```
<?xml version="1.0" encoding=
"GB-2312"?>
 <X.509Certificate xmlns:xsi="http://
www.cernet.edu.cn/XMLSchema-instance"
 xsi:noNamespaceSchemaLocation=
'http://www.cernet.edu.cn
/X.509/docs/X.509Certificate.xsd'>
  <SignablePart>
   <Header Type="Policy" SignatureDi-
gestAlg="RSA-MD5" Canon-
Alg="Ak1CanAlg" Version="2">
    <UID>"rocky.lbl.gov#104b8965#Thu May 03
17:15:30 PDT 2004"</UID>
    <Issuer>
     <UserDN>
/O=CERNET.edu.cn/OU=People/CN=Zhang
Shan
     </UserDN>
     <CADN>
/DC=net/DC=es/OU=Certificate Authori-
ties/OU=CERNET Grid/CN=PKI1
     </CADN>
    </Issuer>
    <ValidityPeriod Begin="010504001529Z"
End="101005001529Z"/>
   </Header>
   <PolicyCert>
    <ResourceName>
CERNET
    </ResourceName>
    <CAInfo>
     <CADN>
     </CADN>
    </Principal>
<URL>file:/p/Fusion-Grid/certs</URL>
   </UseCondIssuerGroup>
   <AttrDirs>
<URL>file:/p/Fusion-Grid/certs</URL>
   </AttrDirs>
```

```
/DC=net/DC=es/OU=Certificate Authori-
ties/OU=CERNET Grid/CN=PKI1
     </CADN>
     <X509Certificate>
MIICvzCCAiigAwI-
BAgIBETANBgkqhkiG9w0BAQUFADBbMRkwFw
YDVQQKExBET0Ug...
     </X509Certificate>
     <IdDirs> <URL>
file:/p/Fusion-Grid/idCerts
     </URL></IdDirs>
     <CRLDirs> <URL>
ldap://ldap.doegrids.org
     </URL></CRLDirs>
    </CAInfo>
    <UseCondIssuerGroup>
     <Principal>
      <UserDN>
/O=cernet.edu.cn/OU=People/CN=Zhang Shan
      </UserDN>
      <CADN>
/DC=net/DC=es/OU=Certificate Autho-
rities/OU=CERNET Grid/CN=PKI1
      </CADN>
     </Principal>
     <Principal>
      <UserDN>
/O=CERNET.edu.cn/OU=People/CN=Li Shi
      </UserDN>
      <CADN>
/DC=net/DC=es/OU=Certificate Authori-
ties/OU=CERNET Grid/CN=pki>
      <CacheTime>3600</CacheTime>
    </PolicyCert>
  </SignablePart>
  <Signature>
This is a fake signature
  </Signature>
 </X.509Certificate>
```

Fig. 3. Top-level policy certificates for CERNET.

The complete policy certificate at the top level is shown in Figure 3. It specifies the trusted CAs and where they publish certificates and CRLs, <CAInfo>; the stakeholders and where they publish their use-conditions, <UseCondIssuerGroup>; directories to be searched for attribute certificates, <AttrDirs>; and the maximum caching time for any certificates used in an authorization decision, <CacheTime>. The header of this certificate, and all X.509 certificates, has the type of the certificate, a unique ID for the certificate, the issuer who signed the certificate, and a validity period.

Four user groups are granted specific rights:

(1) **General** – used for middleware testers
(2) **Clients** – users who are allowed to run the production code
(3) **Developers** – who can run experimental versions of the code
(4) **Administrators** – who can control other users' jobs

Use-conditions are written for each class of executables and job category. A portion of a use-condition that grants users in the client group to start the production code is shown in Figure 4. Note that the Attribute Info element includes the authority that is allowed to assert that a user is in the client group.

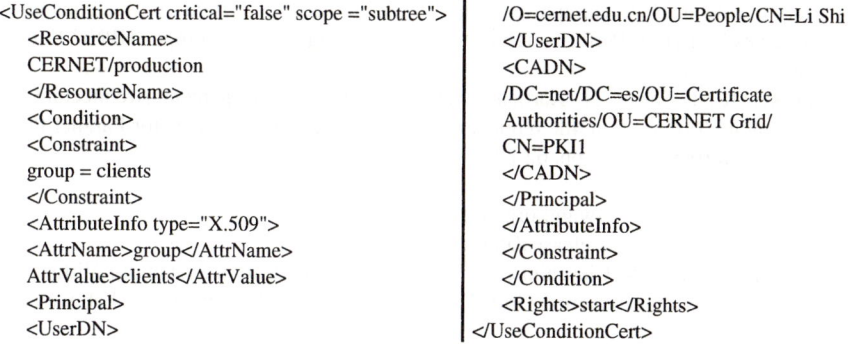

Fig. 4. Use-condition fragments for production code.

Figure 5 shows the portion of an attribute certificate that asserts a user's membership in the client group. This certificate had to have been issued and signed by Li Shi for it to be accepted by the X.509 policy engine. Note that more than one attribute authority can be specified in a use-condition.

5 Conclusion

This paper presents an authorization solution for resource management and control developed as part of the China Education and Research Network (CERNET) to perform fine-grained authorization of job and resource management requests in a Grid environment which meets the Fusion-Grid's se-curity needs in large scale networks such as CERNET. It integrates the GT2 job manager and X.509 authorization and this model can be extended to other au-thorization decision functions. It allows the system to evaluate a user's resource specification language request against authorization policies on resource usage. Furthermore, based on XML integrated authorization policies, it allows other virtual organization members to manage the user's resources.

In the future, CERNET members may want to control access to data located at several repositories. In this case there will be two stakeholders for the data, the owner

association rules, outlier detection, and classification. Frequent episodes and association rules are most frequently used. The latter approach finds temporal relationships among huge amount of data, and the former is used to analyze relationships of elements in intrusion detection alerts. It is useful to separate the critical threats from other intrusion alerts. However, real time analysis using data mining is not possible until now.

Stanford CIDF Correlator developed to reduce false alerts and to find large-scale attack such as DDoS through relationship analysis among the intrusion detection information [2]. Relationship analysis is processed by Event Processing Agents (EPA) according to the pre-defined scenario with time-based causal relationship. If the scenarios are perfect, the rule based approaches and planning process model guarantee the more efficient than other approaches. However, because it is difficult to define all attack scenarios, it analyzes the relationships only against limited attack.

Herve and Andreas suggest ACC (Aggregation and Correlation Components) based on object oriented concept [3]. This approach uses hierarchical structure to represent intrusion detection information. It classifies 'intrusion detection alerts set' into 7 situations by using attributes such as source/destination IP address, attack class, and security level. However, the system generates different event at the same situation, if it is ambiguous what classification rules are used to classify level.

Probabilistic approach aggregates intrusion detection alerts by similarity [4]. It is calculated by using some attributes such as source IP, target IP, attack class, and timing information. This approach practically has low performance.

M-Correlator uses conditional priority table (CPT) when analyze intrusion detection alerts [5]. CPT has prioritized parameter such as importance value of resource/service and seriousness value of attack. According to these values, it makes relevance score, 0 to 255. At last, it calculates priority of each incident by using Bayes Calculation with CPT, relevance score, and output value of sensor.

3 The Proposed System Architecture

3.1 System Architecture

Fig. 1 shows a schematic view of proposed system. Management Center performs event correlation. Sensor is a network-based intrusion detection system to detect possible intrusion in a system. Being applied to individual sub-network, the sensor captures intrusion packets originated from appropriate network. It generates and delivers an event to management center. Management center decides whether received events should be correlated or not. Depending on the decision, it generates a new event for the attack. To accomplish this, the proposed system employs 4 modules such as filter, aggregator, correlator, and situator.

3.2 Event Structure

The proposed system has four kinds of events. The hierarchical relationship between two events is the right side of Fig.1.
- thread event: it is generated by filtering in sensor
- aggregation event: it is generated by aggregation in center
- correlation event: it is generated by correlation in center
- situation event: it is generated by situation in center

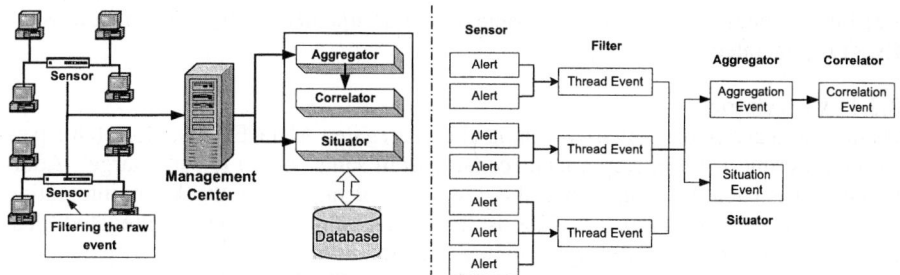

Fig. 1. A schematic view of proposed system and Events Hierarchy.

3.3 Architecture of Each Module

3.3.1 Management Center

The Management Center (MC) aggregates and correlates received events from sensors. On initialization process, it sets environment variables related to analyze events and connect to database. The MC module contains viewer and data schema. Viewer shows each hierarchical event. Data structure has the information of each event.

3.3.2 Filter

Filter consists of following three modules. Alert Receiver is a process. ThreadEvent Maker and ThreadEvent Sender module operate as multi-thread in a process.

Alert Receiver is the module that receives intrusion detection information from a sensor. In this proposed system, it receives information formatted Alertpkt structure. The structure is log information structure of snort known as network intrusion detection system.

ThreadEvent Maker receives alert information from Alert Queue. It compares the alert information with the information in current ThreadEvent Table. If the information is overlapped in ThreadEvent table, ThreadEvent Maker combines the received alert information with the information in ThreadEvent table. Otherwise, it creates new ThreadEvent.

ThreadEvent Sender transmits collected ThreadEvent in periodical time to Management Center in order to reduce burdens of processing events. Using timer module, ThreadEvent Sender stop the task of updating ThreadEvent table. After stopping, it transmits ThreadEvent information to Management Center.

3.3.3 Aggregator

Aggregator performs the integration of roles to upper level Meta Event, Aggregation Event, by analyzing the similarity among the ThreadEvents. The ThreadEvents means the transferred events from the sensor deployed in a network. Since the lifetime of events is longer than filter's one, some ThreadEvents can not be integrated. Therefore, the events are integrated to Meta Events by the aggregator.

When ThreadEvent is sent to management center, communication management modules call Aggregator function. The communication management module manages the communication with network and sensor in the management center. The Aggregator function selects the existing Meta Event in database table and compares with the ThreadEvent. Next, it integrates to the Meta Event. If there is the Meta Event that satisfies a condition, the existing Meta Event is updated. Otherwise, a new Meta Event is generated.

3.3.4 Correlator
Using the analysis of similarity among the aggregated Meta Events, Correlator performs grasping of multi-step attack scenario from same source to same target. Additionally, it is capable of analyzing the new types of multi-step attacks by integrating of events which have temporal relationship. Correlator only manages aggregated Meta Events. In analysis of similarity, it focuses on finding multi-step attacks by applying high minimum expectation at source and destination IP address. These attacks are that an intruder attacks a target step by step.

When transferred Meta Event is updated, correlated Meta Event is just updated. In the other case, when new aggregated Meta Event is transferred, temporal similarity was evaluated using the events occurred in a specific time interval. According to the result of evaluating, it combines into existing correlated Meta Event with time information or generates a new Meta Event.

3.3.5 Situator
Situator analyzes the relationship between 'source ip' and 'destination ip' field among ThreadEvents, and it identify the types of attack at network level. According to the analysis, each attack is classified by 4 types of attack such as 1:1, 1:N, N:1, and M:N.

The type '1:1' is a situation that a host is attacked by one source host, such as DoS(Denial of Service) attack and CGI(Common Gateway Interface) scan attack. '1:N' attack is a situation that several hosts are attacked by a host, such as network/service scan attack. This kind of attack aims to gather information related to target host. 'N:1' attack is a situation that a host is attacked by several different hosts, such as DDoS(Distributed Denial of Service) attack. This phenomenon can be detected by analyzing change of event's number. The type 'M:N' is a situation that the whole networks or some hosts are attacked by many hosts, such as Internet worm.

The situator of the proposed system treats three types of attack except for '1:1' attack, since '1:1' types of attack are controlled by the correlator. The system employs three detectors such as worm detector, DDoS detector, and scan detector. When worm detector receives the ThreadEvent from the sensor, it relay the event to two other detectors. DDoS detector and scan detector stores it candidates buffer. If some of candidates in the buffer are decided as an attack, they send the event to the worm detector. The worm detector should analyze the similarity between events from other two detectors. If there is similar events from different detector, the worm detector generates a event for 'M:N' attack.

4 Evaluations

We implements the sensor based on snort NIDS. It runs on the Linux-based system. The management center is implemented on the UNIX system using C language. The database system is the Oracle for the UNIX. To evaluate proposed system, we use 7 scenarios such as Stealth scan (scenario 1), Attack against FTP server (scenario 2), CGI attack against Web server (scenario 3), Attack against RPC service vulnerability (scenario 4), Network scan attack (scenario 5), DDoS (scenario 6), and Worm/Virus (scenario 7).

In this paper, we will show scenario 4 and scenario 6. These two scenarios evaluate correlator and situator. Additionally, to evaluate filter and aggregator, we measure three features such as event reduction rate of filter/aggregator and execution time of each module.

◆ **Thread Event Table** ◆

Num	EventID	EventName	CreatedTime	LastModifiedTime	Count	SensorID	AttackClass	Protocol	SIP	DIP	SPort	DPort	Signature
1	88532	Misc activity	2003-10-22 20:36:30	2003-10-22 20:36:34	13	1	29	UDP	143.248.204.35	More	More	More	More
2	88533	Misc activity	2003-10-22 20:36:34	2003-10-22 20:36:34	3	1	29	UDP	143.248.82.95	More	More	More	More
3	88531	Misc activity	2003-10-22 20:36:24	2003-10-22 20:36:31	13	1	29	UDP	143.248.222.30	More	More	More	More
4	88526	Detection of a non-standard protocol or event	2003-10-22 20:35:50	2003-10-22 20:36:23	3	1	25	ICMP	143.248.137.1	More	More	More	More
5	88530	Misc activity	2003-10-22 20:36:06	2003-10-22 20:36:11	13	1	29	UDP	143.248.201.167	More	More	More	More
10	88516	Misc Attack	2003-10-22 20:35:40	2003-10-22 20:35:46	31	1	30	TCP	143.248.206.62	More	More	More	More
11	88525	Attempted Administrator Privilege Gain	2003-10-22 20:35:46	2003-10-22 20:35:46	1	1	12	Unknown	143.248.137.40	More	More	More	More
12	88524	Decode of an RPC Query	2003-10-22 20:35:46	2003-10-22 20:35:46	1	1	14	TCP	143.248.137.40	More	More	More	More
13	88520	Attempted Information Leak	2003-10-22 20:35:43	2003-10-22 20:35:43	1	1	4	Unknown	143.248.137.40	More	More	More	More
14	88523	Attempted Information Leak	2003-10-22 20:35:43	2003-10-22 20:35:43	1	1	4	Unknown	143.248.137.40	More	More	More	More
15	88522	Attempted Information Leak	2003-10-22 20:35:43	2003-10-22 20:35:43	1	1	4	Unknown	143.248.137.40	More	More	More	More
16	88521	Attempted Information Leak	2003-10-22 20:35:43	2003-10-22 20:35:43	1	1	4	Unknown	143.248.137.40	More	More	More	More
17	88517	Attempted Information Leak	2003-10-22 20:35:42	2003-10-22 20:35:42	1	1	4	UDP	143.248.137.40	More	More	More	More
18	88519	Attempted Information Leak	2003-10-22 20:35:42	2003-10-22 20:35:42	1	1	4	Unknown	143.248.137.40	More	More	More	More
19	88518	Attempted Information Leak	2003-10-22 20:35:42	2003-10-22 20:35:42	1	1	4	Unknown	143.248.137.40	More	More	More	More
20	88514	Misc activity	2003-10-22 20:35:35	2003-10-22 20:35:40	13	1	29	UDP	143.248.201.213	More	More	More	More
21	[illegible]	Misc Attack	2003-10-22 20:31:59	2003-10-22 20:31:59	5	5	30	TCP	[illegible]	More	More	More	More

Fig. 2. ThreadEvents for RPC service vulnerability.

Correlator Table

Num	EventID	EventName	CreatedTime	LastModifiedTime	AttackCount	SIP	DIP	SPort	DPort	LowerEvent	Attack	Signature
1	2	Attack steps of 143.248.137.40-> 143.248.137.64	2003-10-22 20:35:42	2003-10-22 20:35:46	3	More	More	More	More	More	More	More

Correlator Lower Events with ID = 2

EventID	EventName	CreatedTime	LastModifiedTime	Count	SIP	DIP	SPort	DPort	LowerEvent	Signature
50004	Attempted Information Leak	2003-10-22 20:35:42	2003-10-22 20:35:43	7	More	More	More	More	More	More
50005	Decode of an RPC Query	2003-10-22 20:35:46	2003-10-22 20:35:46	1	More	More	More	More	More	More
50006	Attempted Administrator Privilege Gain	2003-10-22 20:35:46	2003-10-22 20:35:46	1	More	More	More	More	More	More

Correlator Signature Lists with ID = 2

EventID	AttackClass	Signature	Count
2	4	ICMP PING NMAP	1
2	12	RPC STATD TCP stat mon_name format string exploit attempt	1
2	14	RPC portmap status request UDP	1
2	4	SCAN Proxy (8080) attempt	1
2	4	SCAN SOCKS Proxy attempt	1
2	4	SCAN Squid Proxy attempt	1
2	4	SNMP AgentX/tcp request	1

Fig. 3. The result of correlation on the scenario 4: Correlation Event and the details of it.

4.1 Scenario 4 – Attack Against RPC Service Vulnerability

For the simulation, we perform the attack by using following three steps.

- Host scan using NMAP
- Perform the Buffer overflow attack using vulnerabilities of RPC service
- Take important system file in the Root Shell.

Fig. 2 shows the ThreadEvent and Fig. 3 shows the result of correlation. As shown in second table of Fig. 3, the proposed system identifies correctly that a set of events are occurred with causal and temporal relationship. The system also shows the details of the attack events.

◆ Thread Event Table ◆

Num	EventID	EventName	CreatedTime	LastModifiedTime	Count	SensorID	AttackClass	Protocol	SIP	DIP	SPort	DPort	Signature
1	[illegible]	Potentially Bad Traffic	2003-10-22 21:10:25	2003-10-22 21:11:25	311	7	3	UDP	[illegible]	More	More	More	More
2	[illegible]	Potentially Bad Traffic	2003-10-22 21:10:25	2003-10-22 21:11:25	61	7	3	UDP	143.248.131.73	More	More	More	More
3	[illegible]	Potentially Bad Traffic	2003-10-22 21:10:25	2003-10-22 21:11:25	60	7	3	UDP	143.248.131.34	More	More	More	More
4	[illegible]	Potentially Bad Traffic	2003-10-22 21:10:25	2003-10-22 21:11:25	60	7	3	UDP	143.248.131.32	More	More	More	More
5	[illegible]	Misc activity	[illegible]	[illegible]	5	7	29	UDP	[illegible]	More	More	More	More
6	[illegible]	Detection of a non-standard protocol or event	[illegible]	[illegible]	3	7	25	ICMP	143.248.137.1	More	More	More	More
7	[illegible]	Misc activity	[illegible]	[illegible]	15	7	29	UDP	[illegible]	More	More	More	More
8	[illegible]	Potentially Bad Traffic	[illegible]	[illegible]	[illegible]	7	3	UDP	[illegible]	More	More	More	More
9	[illegible]	Misc activity	[illegible]	[illegible]	41	7	29	UDP	192.249.31.100	More	More	More	More
10	[illegible]	Misc activity	[illegible]	[illegible]	15	7	29	UDP	[illegible]	More	More	More	More
11	[illegible]	Misc activity	[illegible]	[illegible]	55	7	29	UDP	[illegible]	More	More	More	More
12	[illegible]	Misc activity	[illegible]	[illegible]	[illegible]	7	29	UDP	[illegible]	More	More	More	More
13	[illegible]	Potentially Bad Traffic	[illegible]	[illegible]	[illegible]	7	3	UDP	[illegible]	More	More	More	More
14	[illegible]	Potentially Bad Traffic	[illegible]	[illegible]	406	7	3	UDP	[illegible]	More	More	More	More
15	[illegible]	Misc activity	[illegible]	[illegible]	[illegible]	7	29	UDP	192.249.31.100	More	More	More	More
16	[illegible]	Potentially Bad Traffic	[illegible]	[illegible]	[illegible]	7	3	UDP	[illegible]	More	More	More	More
17	[illegible]	Potentially Bad Traffic	[illegible]	[illegible]	60	7	3	UDP	143.248.131.32	More	More	More	More
18	[illegible]	Potentially Bad Traffic	[illegible]	[illegible]	60	7	3	UDP	143.248.131.34	More	More	More	More
19	[illegible]	Detection of a Network Scan	[illegible]	[illegible]	[illegible]	7	[illegible]	[illegible]	[illegible]	More	More	More	More

Fig. 4. ThreadEvents for DDoS attack.

Situation N to 1 Table

Num	EventID	EventName	CreatedTime	LastModifiedTime	AlertCount	DIP	SIP	SCount	AttackClass	Signature
1	216	Potentially Bad Traffic	2003-10-21 22:16:45	2003-10-22 21:09:15	3427	143.248.137.40	More	10	3	More
2	188	Detection of a Network Scan	2003-10-21 23:26:42	2003-10-22 21:09:14	11388	239.255.255.250	More	1292	23	More
3	215	Misc Attack	2003-10-22 15:26:59	2003-10-22 21:06:27	4550	239.255.255.250	More	13	30	More
4	138	Misc Attack	2003-10-21 16:17:16	2003-10-22 15:25:39	20186	239.255.255.250	More	26	30	More
5	107	Detection of a Network Scan	2003-10-20 23:16:49	2003-10-21 23:25:56	10569	239.255.255.250	More	1250	23	More
6	185	Misc Attack	2003-10-20 15:08:04	2003-10-21 15:03:39	16177	239.255.255.250	More	27	30	More

Situation N1 Source IPs with ID = 216

EventID	SourceIP
216	143.248.109.234
216	143.248.137.179
216	143.248.137.32
216	143.248.137.34
216	143.248.137.37
216	143.248.137.73
216	143.248.137.84
216	143.248.137.88
216	192.249.31.1
216	192.249.31.100

Situation N1 Signature Lists with ID = 216

EventID	Signature	Count
216	ATTACK-RESPONSES id check returned root	7
216	ICMP Large ICMP Packet	3783
216	TELNET login incorrect	3

Fig. 5. The result of situator's processing for the scenario 6.

4.2 Scenario 6 – DDoS Attack

DDoS is not detected by most of IDS, while the proposed system can detect it in almost real-time. To carry out DDoS attack, we send large packet (3000 bytes) to the target host repeatedly by using ICMP Flooder. Fig. 4 and Fig. 5 show the result of the situator's process. Since many ThreadEvents has the same signature and target system, the situator decides these events are a DDoS attack.

4.3 Event Reduction Rates of Filter/Aggregator

Event reduction rates are calculated with the number of total events and the number of reduced events by filter and aggregator. Table 1 shows the results. If we set the period as 1 minute, average reduction rates of filter is about 11%. In the case of 'Nachi' worm by ICMP Ping CyberKit, approximately a hundred events are aggregated in a ThreadEvent. So, the reduction rate is down to 1%.

Table 1. Event reduction rates of Filter and Aggregator.

Modules	Event type	Numbers of events	Reduction Rates
Sensor	Raw Event	599403	–
Filter	ThreadEvent	66775	**11.1%** (# of ThreadEvent/# of Alert)
Aggregator	AggregationEvent	33173	**5.5%** (# of AggEvent/# of Alert)

4.4 Execution Time of Modules

We measured execution time in which one ThreadEvents are performed. Table 2 shows the results. Execution time for a ThreadEvent is 0.8 seconds approximately. The event processing process of IBM Tivoli needs a second to process one high-level event and general network management environments are often capable of handling hundreds of alerts per second [7]. Therefore 0.8 seconds is acceptable time and applicable to real-time system.

Table 2. Execution time of modules.

Modules	Execution time (sec.)
Storing ThreadEvent	0.0097
Aggregator	0.5398
Correlator	0.0874
UpdateCorrelator	0.0013
Situator	0.1691
Total	**0.8103**

5 Conclusions and Future Works

This paper proposed a flexible and effective system to analyze correlation of intrusion detection system. The system analyzes the relationship among intrusion detection information, and generates high-level events. It helps the administrator manage whole system and network easily and effectively. Additionally, the system detects large-scale attacks such as DDoS and Internet worm timely. Since it performs all process in acceptable times, the system also applies to the real-time system.

Since there is no data set for evaluating this kind of system, we use some scenario to evaluate it. So, it is need to construct more detailed scenarios for evaluating. And the system should treat the host based IDS to achieve more secure system.

References

1. Lee, W.: A Framework for Constructing Features and Models for Intrusion Detection System. PhD thesis, Columbia University (1999)
2. Perrochon, L., Jang, E., Luckham, D.C.: Enlisting Event Patterns for Cyber Battlefield Awareness. DARPA Information Survivability Conference & Exposition (DISCEX'00), Hilton Head, South Carolina, USA (2000)
3. Debar, H., Wespi, A.: Aggregation and Correlation of Intrusion-Detection Alerts. Proceedings of 2001 International Workshop on Recent Advances in Intrusion Detection, Davis, CA, USA (2001)
4. Valdes, A., Skinne, K.: Probabilistic Alert Correlation. Fourth International Workshop on the Recent Advances in Intrusion Detection, Davis, CA, USA (2001)

5. Phillip, A., Porras, et al.: A Mission impact-Based Approach to INFOSEC Alarm Correlation. Fifth International Workshop on the Recent Advances in Intrusion Detection, Zurich, Switzerland (2002)
6. Cuppens, F.: Managing alerts in a multi intrusion detection environment. 17th Annual Computer Security Applications Conference (ACSAC), New Orleans, USA (2001)
7. Porras, P., Neumann, P.: Emerald: Event Monitoring Enabling Responses to Anomalous Live Disturbances. National Security Conference (1997)

Security Architecture for Web Services

Yuan Rao, Boqin Feng, and Jincang Han

Department of Computer Science and Technology, Xi'an Jiaotong University,
710049 Xi'an, China
yuanrao@163.com

Abstract. A security architecture model for web services, Security extended-Resources, Services, Roles, Protocols and Methods (SX-RSRPM) model, was proposed based on the conceptions about security of web services and Degree of Safety for Web Services (DoS4WS), which is a qualitative analysis index of Web Services security, in the paper. This model could effectively increase the value of DoS4WS by the extended emerging XML-based security of protocol stacks and a new affiliated role as security CA center. In addition, an integrated security solution has been developed and the results in practice show that this solution can build a confidence security environment for all roles and provide a flexible and extensible security manage mechanism.

1 Introduction

The emerging Web Services technology, which is based on a series of XML-based protocols (e.g., SOAP, WSDL, WSFL, UDDI, WS-Security) and traditional Internet standard protocols, has the capabilities to support the development of Internet-based loosely coupled applications that can integrate heterogeneous and distributed systems between different organizations [1]. Although this application-to-application method can greatly improve the flexibility and integration of business processes, it also increases the exposure of critical business resources. Hondo[2] considered that a Web Service security model must support protocol-independent policies that service providers could enforce, and descriptive security policies attached to the service definitions that clients could use in order to securely access the service. Therefore, how to build a securing service-oriented architecture and assess the securing service performance is a vital problem.

This paper introduces some conceptions about Security and Security architecture of Web Services. A qualitative analysis index of Web Services security, e.g., Degree of Safety for Web Services (DoS4WS), which was influenced by the degree of credible environments around the WS-based applications and the long-duration of the transaction times, was analyzed. From the DoS4WS perspective, a securing Web Services architecture should be designed to improve the credible environment by affiliated a security role and enforced the securing communication process based on some emerging security technologies and existing polices.

The rest of this paper is organized as follows: Section 2 presents an overview of Web Services security and related works. Section 3 presents the conception of DoS4WS and the security architecture model of WS, e.g., SX_RSRPM model. A securing WS integrated solution is introduced in Section 4. The conclusion is presented in last Section.

ating to composing can be divided into publish, discover and bind phases, SCAC plays a critical role to provide CA to other roles and guarantee security enforcement and credibility of the executing environment around WS-based applications.

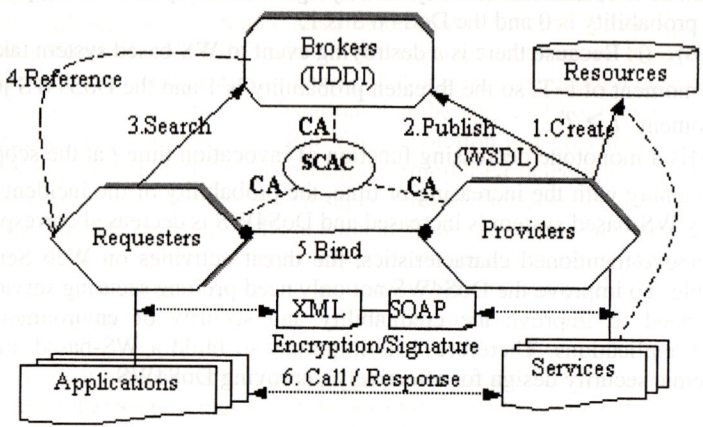

Fig. 1. SX_RSRPM Web Services Security Architecture.

Figure 2 illustrate the process of security enforcement can be applied at publish, discover and bind phases of the SX-RSRPM model. Each role acquires the CA from SCAC before interoperating with others.

In publish phase, after encrypting their data resource, creating the web service and inserting the security assertion into WSDL, the service provider can apply the registration request to UDDI. As soon as received the provider's registration request, the UDDI decipher the encrypted SOAP header and acquire the provider's digital signature and CA information. Then, UDDI check the provider's identification and CA by exchanging the SOAP messages with SCAC. After received the confirmation information from SCAC, UDDI will send the service registration acknowledgement to provider for registering and publishing their service. Finally, UDDI classify and manage the registration service for other authorization user to invoke.

In discover phase, when UDDI received the searching service request from the service requestor, it also need decipher the encrypted SOAP header to acquire the requestor's digital signature and CA. Then, UDDI checks the requestor's identify and CA information by exchanging the SOAP messages with SCAC. Once it received the confirmation information from SCAC, UDDI will send the searching acknowledgement to service requestor. While the requestor can search the service from the UDDI center by quickly searching engine and get the services list. After selected the corresponding service from the list, the requestor will receive the whole information of the service included white page (e.g., service owner information), yellow page (e.g., service information) and green page (e.g., detail technology information included WSDL with security assertion). The provider can decipher the security SOAP header and acquire the service interface parameters information from WSDL. In bind phase, after acquired the service interface information from WSDL, the requestor can reference and instance the web service and corresponding methods as well as send the message of binding request to service provider. The provider deciphers the encrypted SOAP header and acquires the requestor's digital signature and CA information, as well as

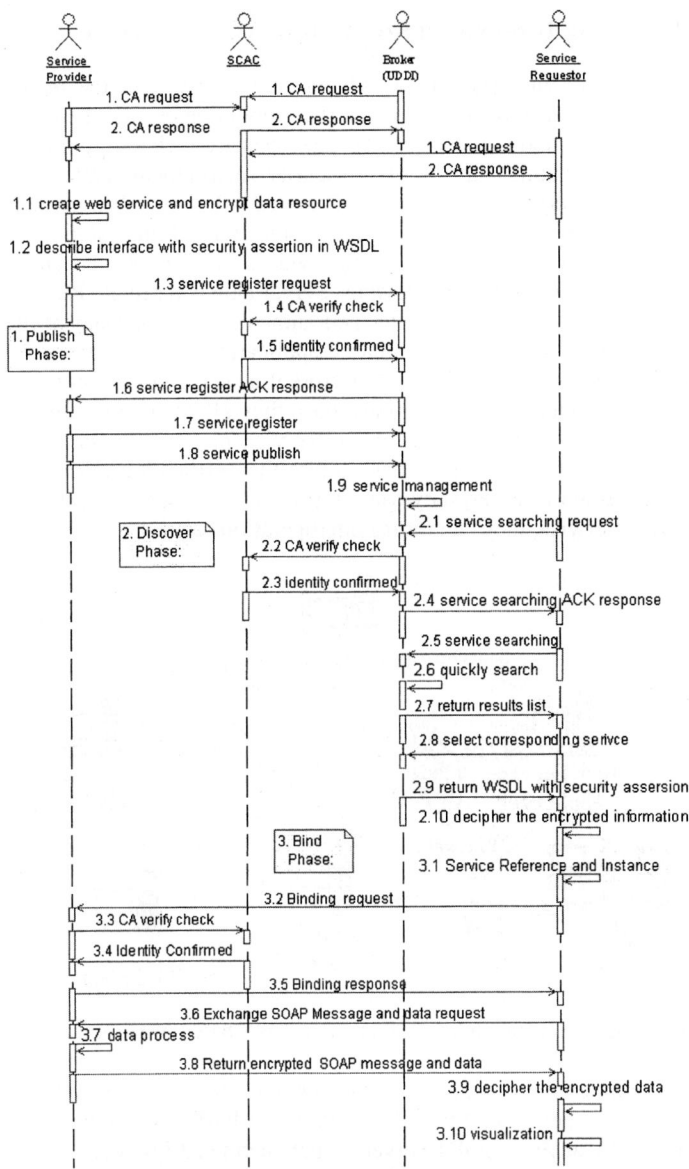

Fig. 2. Security Publishing, Discovering and Binding Phases in the SX-RSRPM Model.

checks the requestor's identify and Certification and Authentication by exchanging the SOAP message with SCAC. The provider will send the binding service acknowledgement to requester after received the confirmed information from SCAC and built a communication channel between each other. When the provider received the data access request, they will parse the SOAP message, encrypt the data contents and return it to the requestor. The requestor receives and deciphers the encrypted data and visualizes these data in program.

4 The Securing Integrated Solution of Web Services

The requirements of enterprise always change constantly with the modification of business process, the developers of applications should determine the proper security rank in programming and could insert the new security technology into the original framework without rebuilt it again when security requirements have been changed. Therefore, this paper proposed a securing integrated solution of WS application based on the OPEN-CLASS prototype system. The architecture of this solution illustrates in figure 3, where the private UDDI server provide service published and discovered mechanism for different users. The Passport Server also is a private Register server to provide a Single Sign-on logon, register and authentication mechanism for all users. Each user, who wants to enter the WS-based application environment, needs to register on this server firstly and obtain a CA. Before the users in different roles interoperate each other over Internet by SOAP messages, they all need verify the mutual certification (CA) in both service requestor and service provider by Passport Server at first. When both requestor and provider acquire the verification information of each other from the Passport Server, they can interoperate and communicate over Internet to each other. Otherwise, they can refuse interoperation request.

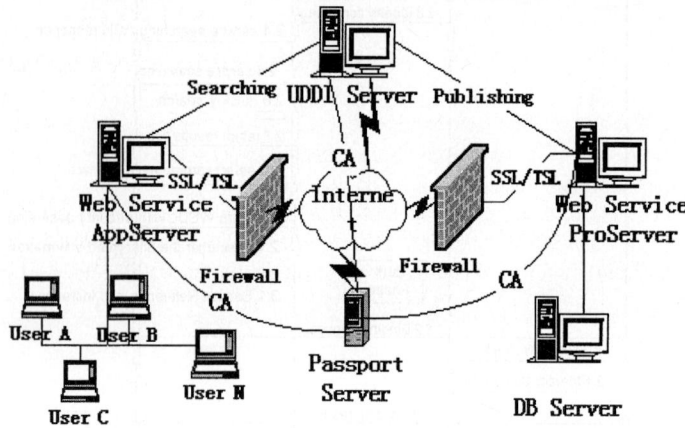

Fig. 3. Securing Integrated Solution of WS-based Application.

In service publish phase, when the provider publishes his own web service from PorServer(e.g., provider server) to UDDI server, the PorServer should logon to Passport Server for CA at first. UDDI Server, which also need login and acquire the CA from the Passport Server before providing service, receives the SOAP message request from the PorServer and deciphers the encrypted CA information by RSA algorithm and digital signature by SHA-1 algorithm in SOAP header and send these critical information to the Passport Server for validation. Moreover, UDDI can accept the service interface description (WSDL) registration provided by ProServer until it received the identity authentication of ProServer from Passport Server.

In service Discovery phase, the service requestor also need identify authentication from Passport Server before it send the searching request from PorServer to UDDI. After acquired the certification from the Passport Server, the UDDI Server provides corresponding access control privilege to AppServer for discovering web services.

In service Binding phase, the AppServer acquired the service interface parameters information in WSDL document from UDDI Server, while it also need authenticate the ProServer's identification by Passport Server before referencing and binding with ProServer. On the one hand, the SOAP messages transmit by SSL/TSL and through firework with access control and security policy for providing a point-to-point transport security between ProServer and AppServer. On the other hand, this solution introduces two kinds of data encryption mechanisms for exchanging message in creating web service process, one is directly encrypting the critical data and storing them into database, another is extracting the recordset from database and mapping them into XML document as well as encrypted the critical data information by RSA algorithm, the public key can be deployed to other users by Passport Server. Otherwise, the timestamp and digital signature with SHA-1 algorithm can be use to provide the Integrity of the SOAP messages, proof of message origin and avoid the message replay attack. But the application program in AppServer need parse and decipher SOAP message and validate the certification and authentication of ProServer for implementing the dynamic binding with web services to provide the individual services selection for location users.

5 Conclusion

This paper introduced the conception of DoS4WS and built an SX-RSRPM security architecture model, which investigated some vital problems should be resolved in security paradigm of WS-based application. In order to improve the credibility environments around of WS-based applications and to prolong the duration of web service invoking times, an integrated security solution integrated some emerging XML-based security techniques and based on the SX-RSRPM model have been developed. This solution builds a concentrated security management method of web services (in UDDI Server), users identification and CA information (in Passport Server) and provides a flexible and extensible security integrated mechanism.

References

1. U. Ogbuji, The Past, Present and Future of Web Services (Part I): What is Web Services?, http://www.webservices.org/index.php/article/articleview/663/1/61/2001.9
2. M. Hondo, N. Nagaratnam, A. Nadalin, Securing Web Services, IBM Systems Journal, 2002, 41(2), Pp.228
3. Seokwon Yang, Herman Lam, Stanley Y.W. Su, Trust-Based Security Model and Enforcement Mechanism for Web Service Technology, Lecture Notes in Computer Science, Vol. 2444:151-160 2002
4. S.Yang, H.Lam, and S.Y.W.Su, Trust-Based Security Model and Enforcement Mechanism, Lecture Notes in Computer Science, Vol. 2444:150, 2002
5. G. Denker, L. Kagal, T. Finin et al., Security for DAML Web Services: Annotation and Matchmaking, Lecture Notes in Computer Science, Vol. 2870:335, 2003
6. Lee SM, Kwon OS, Lee JH, et al., Ty*SecureWS: An Integrated Web Service Security Solution Based on Java, Lecture Notes in Computer Science, Vol.2738: 186, 2003

A Hybrid Machine Learning/Statistical Model of Grid Security

Guang Xiang[1], Ge Yu[1], Xiangli Qu[2], Xiaomei Dong[1], and Lina Wang[3]

[1] School of Information Science and Engineering,
Northeastern University, Shenyang, 110004, China
xguang80@hotmail.com
[2] School of Computer Science, National University of Defence Technology,
Changsha, 410073, China
[3] State Key Lab of Software Engineering,
Wuhan University, Wuhan, 430072, China

Abstract. Most current Grid security techniques concentrate on traditional security aspects such as authentication, authorization, etc. While they have shown their usefulness, the significance of the information hidden in the historical data corpus denoting the user-Grid interactions has been largely neglected. In fact, such information provide great insight into Grid security and if properly harnessed, will help better protect the Grid against potential attacks.

To utilize these hidden information in a service-oriented Grid environment, we propose a hybrid machine learning and statistical model. The machine learning component predicts the security of a service by considering the probability distribution of the past services, while the statistical component evaluates a service's security statistically based on its own past behaviors and users' opinions. We construct an overall architecture based on this hybrid model and demonstrate through examples its effectiveness and potential to offer stronger security to the Grid.

1 Introduction

The Grid first emerged in the mid 1990s to facilitate coordinated resource sharing and problem solving in dynamic and multi-institutional virtual organizations. In this era of information explosion, while more people are enjoying the convenience brought by the Grid, an increasing number of security problems came as well.

Much work has been done with respect to Grid security. In the groundbreaking work [1], Grid computing and virtual organization (VO) are defined and Grid security is discussed in the framework of a layered Grid architecture. [2] analyzes many unique security requirements of a large-scale distributed Grid environment and develops security policies solving such issues as single sign-on, protection of credentials, etc. A large-scale authentication infrastructure based on the Grid Security Infrastructure (GSI) is described in detail in [3]. In [4], an online credentials repository called MyProxy is proposed to enable Grid Portals to access GSI protected resources in a secure and scalable way. To address the problems related to scalability, flexibility and expressibility in distributed VOs, [5] introduces a new approach named the Community Authorization Service, a trusted third party that performs fine-grained control of community policy while leaving the ultimate control of resource access to the resource owners. [6] defines and standardizes X.509 Proxy Certificates to offer restricted proxying and delegation within a PKI framework. In another work, [7] applies the approach of trust negotiation to Grid security which establishes trust iteratively and bilaterally by the gradual disclosure of certificates thus minimizing the sensitive information exposed to others through some unnecessary exchange of credentials. Toolkit.

Most current Grid security techniques concentrate on traditional security aspects. While they are useful, the significance of the information hidden in the historical data corpus has been largely neglected. In fact, such information provide great insight into Grid security and can help better protect the Grid.

To fully utilize these information, we propose in this paper a novel hybrid machine learning/statistical model of Grid security. Although there are many candidate machine learning algorithms, bearing Occam's Razor[1] in mind, we favor the simple and efficient naive Bayes classifier and use it to classify an incoming service into corresponding security level. A statistical method, which fully captures the inherent behavioral continuity of a service, is also applied to evaluate Grid security. By combining these two approaches, we obtain a stronger hybrid Grid security model upon which we build our overall architecture.

The rest of the paper is organized as follows. In Sect.2, we investigate our machine learning model. Section 3 describes the statistical model. Section 4 presents our hybrid security model and overall Grid security architecture. Finally, we draw our conclusions in Sect.5.

2 Machine Learning Model

2.1 Naive Bayes Classifier

The naive Bayes classifier is a highly practical Bayesian learning method in machine learning. In many domains its performance is comparable to that of the decision tree and neural network classifiers.

The naive Bayes classifier applies to learning tasks where each instance X is represented by a conjunction of attributes $\langle x_1, x_2, \ldots, x_n \rangle$ and a class label which can take on any value from some finite set $V = \{C_1, C_2, \ldots, C_m\}$. Given a new instance X, the classifier will assign the most probable target value, V_{MAP}.

$$V_{MAP} = \arg\max_{C_i \in V} P(C_i | x_1, x_2, \ldots, x_n) \ . \tag{1}$$

Based on the Bayes theorem and the naive Bayes assumption that the values of the attributes are conditionally independent of one another, given the target class value, the approach used by the naive Bayes classifier is

$$V_{NBC} = \arg\max_{C_i \in V} P(C_i) \prod_{j=1}^{n} P(x_j | C_i) \ . \tag{2}$$

---

[1] Occam's Razor: Prefer the simplest hypothesis that fits the data ([8]).

In the service-oriented Grid framework, each web service can be represented by a set of features and a class label. The naive Bayes classifier can then be applied to classify an incoming service by comparing it with the training samples.

2.2 Feature Selection

Irrelevant features and weakly relevant features will reduce the accuracy of supervised learning algorithms and may lead to overfitting the training data. In addition, two phenomena, the so called curse of dimensionality and peaking phenomenon, may negatively impact the classification performance. These lend justification to feature selection in our model.

Two kinds of features are used in our model: the original features and the additional statistical features calculated over the historical data.

Original Feature. We construct a set of original features after considering the Grid specialty and the tradeoff between accuracy and computational overhead. Feature values are in the second parenthesis after the feature name.

1. TYPE_OF_SERVICE (TOS) (PRIVATE, COM, HARDWARE) denotes the function of the services.
2. DOMAIN (DOM) (LOCAL, MEDIUM, FAR) denotes the distance between the remote service and the service users. Note this feature is not fixed if resource and code mobility is enabled.
3. SYS_LOAD (SLD) (IDLE, MODERATE, BUSY) is a measure indicative of the system running.
4. SYS_SECURITY (SSC) (WEAK, NORMAL, STRONG) denotes how secure the underlying infrastructure of the services is, such as OS, etc.
5. NET_RANK (NER) (WEAK, NORMAL, STRONG) is a comprehensive measuring of the quality of the network infrastructure.

Additional Feature. For the complex, large-scale and dynamic Grid environment, static features alone are not enough. Hence we extend our set of features by adding some continuous and intensity measures.

1. NET_ERRORS (NEE) (SMALL, MEDIUM, LARGE) denotes the average network errors occurring on the host where the requested service exists.
2. AVE_CONS (AVC) (SMALL, MEDIUM, LARGE) is the average number of connections to the same host.

These temporal-statistical features offer additional information from a continuous perspective and improve the accuracy of our machine learning model.

2.3 Case Study

Suppose the class label set is $V = \{$WEAK, NORMAL, STRONG$\}$. For brevity, we use the first letter to represent the feature values. Table 1 lists a set of 15 training samples for the naive Bayes classifier.

Table 1. Training samples

No.	TOS	DOM	SLD	SSC	NER	NEE	AVC	LABEL
1	P	L	I	S	S	S	L	S
2	P	L	M	N	N	M	M	N
3	C	F	B	N	W	L	S	W
4	H	M	M	W	W	M	M	W
5	H	M	I	S	N	S	M	S
6	H	F	I	N	N	S	L	N
7	C	F	B	W	N	M	S	W
8	P	M	M	S	S	S	L	S
9	P	F	B	W	N	L	S	W
10	H	L	I	N	S	S	M	S
11	P	L	M	S	N	S	M	S
12	C	L	B	N	S	M	M	N
13	C	M	M	S	S	M	L	N
14	P	F	M	W	W	L	S	W
15	C	L	M	S	S	L	L	S

Scenario 1:

Suppose the incoming service is $X = \langle P, F, M, N, W, M, S \rangle$. We adopt a Bayesian approach to estimate the conditional probabilities in (2) to avoid the zero probability problem.

m-estimate of probability:

$$P(x_j|C_i) = \frac{n_c + mp}{n + m}. \tag{3}$$

where n is number of training samples of class C_i. n_c is the number of training samples of class C_i that has attribute value x_j. p is the prior estimate of the probability and is chosen to be $1/3$ here. m is called the equivalent sample size and is assigned 15 here.

Based on (2) and (3), we have the probabilities after normalization $P(W|X) \approx 66.67\%$, $P(N|X) \approx 23.1\%$, $P(S|X) \approx 10.23\%$. Hence the security level predicted by the naive Bayes classifier is WEAK and this fits our expectation.

Scenario 2:

Service Y is $\langle H, L, I, N, N, M, M \rangle$. The posterior probabilities after normalization are $P(W|Y) \approx 13.99\%$, $P(N|Y) \approx 47.85\%$, $P(S|Y) \approx 38.16\%$. Y is then classified as NORMAL and this is reasonable considering its feature values.

Our machine learning model is accurate and robust since it turns to the past service corpus for information and thus does not introduce bias by concentrating on any single service overly.

3 Statistical Model

The performance of the machine learning approach in Sect.2 depends to a large extent on the size and quality of the training data. In this section we propose a

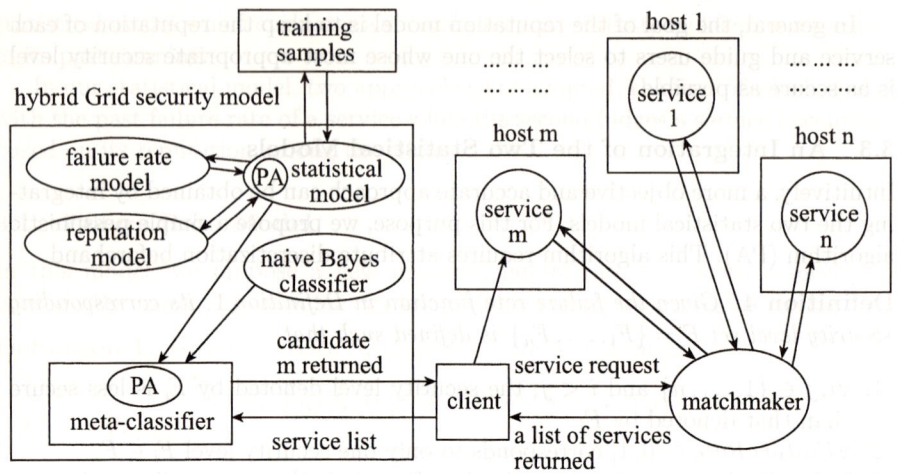

Fig. 1. Overall Grid security architecture

There are times when the machine learning model and statistical model do disagree. This may be due to the fact that a service's reputation does not match its serving records. Anyway, PA never gives an overoptimistic prediction because it always outputs the lowest security level of its input. In general, by emphasizing both the service corpus and service-users' opinions, our hybrid security model is more flexible and accurate and can thus provide finer-grained security guarantee to the Grid.

5 Conclusions

This paper presents a hybrid machine learning and statistical model of Grid security. In a web service based Grid environment, our approach is based on the observation that the behaviors of a service tend to be continuous. For a target service, a final prediction regarding its security is given based on the output of the machine learning model and the statistical model. The comprehensive evaluation of security in our hybrid model provides much useful information otherwise unknown to the clients and thus further decreases their risks.

The contribution of this paper is that it proposes a novel Grid security model based on machine learning and statistics. Exploiting the valuable information hidden in the historical data corpus, our Grid security model has demonstrated the effectiveness and potential of offering stronger security to the Grid.

Our model is still preliminary and further extension is needed. Due to the scarcity of publicly available data, large-scale experiment was currently not conducted. Several issues deserve future research, in particular, a more appropriate feature selection algorithm to capture the Grid-specific characteristics and a more efficient agent mechanism to decrease the realtime computational overhead and network traffic to the minimum.

Acknowledgements

This research was partially funded by the National Research Foundation for the Doctoral Program of Higher Education of China under Grant No.20030145029, the National High Technology Research and Development Program under Grant No.2003AA414210 and the Teaching and Research Award Program for Outstanding Young Teachers in Higher Education Institution of the Ministry of Education.

References

1. Foster, I., Kesselman, C., Tuecke, S.: The Anatomy of the Grid: Enabling Scalable Virtual Organizations. International Journal of Supercomputer Applications. **15**(3) (2001) 200–222
2. Foster, I., Kesselman, C., Tsudik, G., Tuecke, S.: A Security Architecture for Computational Grids. Proceedings of 5th ACM conference on Computer and communications security. ACM Press New York (1998) 83–92
3. Butler, R., Engert, D., Foster, I., Kesselman, C., Tuecke, S., Volmer, J., Welch, V.: A National-Scale Authentication Infrastructure. IEEE Computer **33**(12) (2000) 60–66
4. Novotny, J., Tuecke, S., Welch, V.: An Online Credential Repository for the Grid: MyProxy. Proceedings of 10th International Symposium on High Performance Distributed Computing. IEEE Press New York (2001) 104–115
5. Pearlman, L., Welch, V., Foster, I., Kesselman, C., Tuecke, S.: A Community Authorization Service for Group Collaboration. Proceedings of the IEEE 3rd International Workshop on Policies for Distributed Systems and Networks (2002) 50–60
6. Welch, V., Foster, I., Kesselman, C., Mulmo, O., Pearlman, L., Tuecke, S., Gawor, J., Meder, S., Siebenlist, F.: X.509 Proxy Certificates for Dynamic Delegation. 3rd Annual PKI R&D Workshop (2004)
7. Basney, J., Nejdl, W., Olmedilla, D., Welch, V., Winslett, M.: Negotiating Trust on the Grid. 2nd Workshop on Semantics in P2P and Grid Computing at the Thirteenth International World Wide Web Conference. New York (2004)
8. Mitchell, T.M.: Machine learning. McGraw-Hill (1997)
9. Han, J.W., Kamber, M.: Data mining: Concepts and Techniques. Morgan Kaufmann (2000)

A Novel Grid Node-by-Node Security Model

Zhengyou Xia[1] and Yichuan Jiang[2]

[1]Department of computer, Nanjing University of Aeronautics and Astronautics, Nanjing, China
zhengyou_xia@yahoo.com
[2]Department of Computer Information & Technology, Fudan University, Shanghai, China
jiangyichuan@yahoo.com.cn

Abstract. In this paper, we present a novel grid Node by Node security model. We partition the numerous grid nodes into different autonomy areas and each area is designated a server. The key, integrity and encrypt algorithm between the grid nodes are negotiated through the autonomy area server. The algorithms and keys exchange among the different autonomy area servers is described, At last, we use SPI calculus to verify security negotiation process, which proves that our solution is correct.

1 Introduction

Grid computing has been widely accepted as a promising paradigm for large-scale resources sharing in recent years [1, 2]. While the security architectures of specific Grids such as Globus [3, 4, 5], are well understood, the general authorization framework for shared resources in grids is not. Research and development efforts within the Grid community have produced protocols, services, and tools that address the challenges arising when we seek to build scalable VOs. Security is one of the characteristics of an OGSA-compliant component. The basic requirements of an OGSA security model are that security mechanisms be pluggable and discoverable by a service requestor from a service description. This functionality then allows a service provider to choose from multiple distributed security architectures supported by multiple different vendors and to plug its preferred one(s) into the infrastructure supporting its Grid services [6][7]. We are interested in ensuring authenticating and preserving the integrity of a message that travels among grid nodes, each of which can modify the message during the packet forwarding process. It is likely that a solution of this type would involve wide scale sharing of secrets thus rendering the solution weak from a security perspective. We may adopt adequate solution: authenticate between each pair of nearest active neighbors, and iterate the Node-by-Node trust relationship over the entire path. The Node-by-Node authentication and integrity problem can be subdivided into two parts. Authenticating the origin, protecting the message during the transmission and reception phase to assure that no message modification has occurred, and preventing replay attacks using old messages. Protecting the message from being processed by un-trusted forwarding message.

In this paper, we present a new kind of method for grid Node-by-Node security. We partition the grid into different autonomy area and designate an autonomy area server for per different autonomy area. In one autonomy area, when the grid node joins the autonomy area, the grid node must first registers into the area autonomy area server. After registering into the autonomy area server, the grid node is legal one. The autonomy area server will check and maintain the grid nodes of its autonomy area.

The grid node negotiates algorithm and key of grid Node by Node through autonomy area server.

2 Basic Definition and Autonomy Area

We are interested in ensuring authenticating and preserving the integrity of a message. We propose to partition grid into different autonomy areas. The three autonomy areas are composed of large-scale grid in figure 1. Some definitions are described as the following:

Definition 1: Grid Autonomy Area (GAA): the algorithms and keys of different grid nodes are negotiated through management server. We called it autonomy area that is composed of the some grid nodes and server. We generally assign number to GAA. For example GAA 10, which means that the number of grid autonomy area is 10.

Definition 2: Bordergrid Node: we call grid node as the border grid node that is belong to at least two autonomy areas. The border node can negotiate algorithms and keys with different autonomy area server.

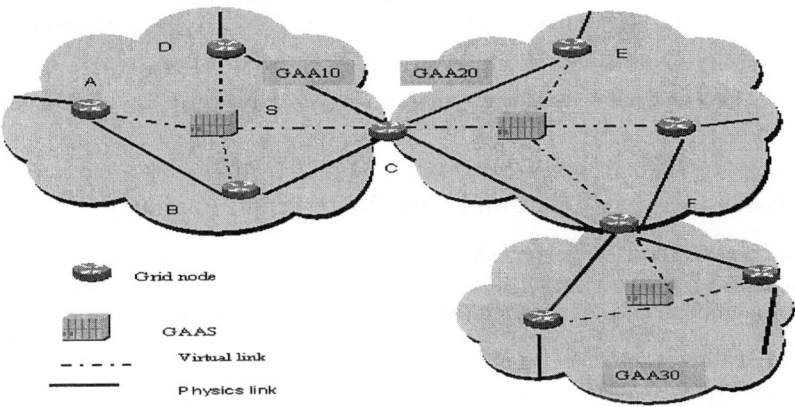

Fig. 1. Grid autonomy area.

Definition 3: Neighbor Grid Node: if the two grid nodes hold the interface of same network, we called the two grid nodes as neighbor grid nodes.

Register Process of Grid Node: when an grid node A joins GAA 10, the grid node A first sends register request to the autonomy area server (GAAS) of GAA 10. The Key_ID_A, authentication information and public key of the grid node A are included into the register request information. The Key_ID_A is an unsigned 64-bit number that MUST be unique for each given sender. Locally unique Key Identifiers can be generated using some combination of the address (IP or MAC) of the sending interface and the key number. The request information is encrypted by public key of GAAS. When the GAAS receives the request that comes fro the grid node A, the

GAAS first uses the private Key to decrypt the request, draws the authentication information from the quest and check the authentication information. If the authentication information cannot be passed, the GAAS will return register fail information. If the authentication is passed by GAAS, the GAAS will send response of register request to grid node A. the register sequence number (64 bit unsigned number) and Key_ID_A are included in the response of register request. The response of register request is encrypted by private key of GAAS and public key of grid node A. The GAAS will save the Key_ID_A, register sequence number and IP address of grid node A. When the grid node A receives the response of register request, the grid node A decrypts the response by private key and public key of GAAS and saves the register sequence number.

Logout Process of Grid Node: the GAAS will periodically check the status of the grid nodes that belong to the same GAA with it. For example, in GAA 10, the GAAS sends hello information to the registered grid node A per 30 second. If the GAAS sends the three times hello information and cannot receive the response of hello from A, the GAAS will consider the grid node A that the grid node A has left off the autonomy area. The GAAS will delete register information of the grid node A from the GAA 10. That is the logout process of grid node A from GAA 10. If the grid node A receives the hello information from the GAAS of GAA 10, the grid node A will send response of hello to GAAS. The response of hello includes Key_ID_A and resister sequence number and the response is encrypted by private key of grid node A and public key of GAAS.

3 Process of Algorithms and Keys Negotiation Protocol

When the grid node A joins the GAA 10 and registers to autonomy area server S, the grid node A begins to negotiate algorithms and keys with grid node B. the negotiation process is through autonomy area server S. we describe the process is as the following figure 2(supposing K_n^+ and K_n^- respectively express using public key and private key of n to encrypted and decrypt).

$1: A \rightarrow S : \{key\_ID_A, \{[\text{Re}gNum_A]\}_{K_A^-}\}_{K_s^+}$

The first step: the grid node A sends challenge request. The request information includes Key_ID_A and $\text{Re}gNum_A$ (register sequence number) assigned by autonomy area server S.

$2: S \rightarrow A : \{key\_ID_A, \{[N_s]\}_{K_s^-}\}_{K_A^+}$

When the server S receives the challenge request from grid node A. the server S parses the Key_ID_A and $\text{Re}gNum_A$ from the request information. If the Key_ID_A and $\text{Re}gNum_A$ of the request information is the same with them which is saved in server S, the server S will send response of challenge request to grid

5. Foster, I., Kesselman, C., Intl J.: Globus: A Metacomputing Infrastructure Toolkit. Supercomputer Applications. 11(2) (1997) 115–128.
6. Foster, I. and Kesselman, C. Globus: A Toolkit-Based GridArchitecture. Kaufmann, 1999, 259-278.
7. Foster, I., Kesselman, C., Nick, J. and Tuecke, S. The Physiology of the Grid: An Open Grid Services Architecture for Distributed Systems Integration, Globus Project, 2002. http://www.globus.org/research/papers/ogsa.pdf.
8. M. Abadi and A. D. Gordon. A calculus for cryptographic protocols: The spi calculus. Information and Computation, 148:1-70,1999. An extended version with full proofs appears as Digital Equipment Corporation Systems.

Research on Security of the United Storage Network

Dezhi Han[1], Changsheng Xie[2], and Qiang Cao[2]

[1] Department of Computer Science, JiNan University, Guangzhou,
510632 Guangdong, P.R. China
dezhihan88@sina.com
[2] Huazhong University of Science and Technology, Wuhan,
430074 Hubei, P.R. China

Abstract. The USN made by integrating a NAS and a SAN has the advantages of NAS and SAN, but its security has become more complicated. To solve this problem, this paper presents the MUVFS, which offers a storage volume view for each authorized user, who can only access the data in his own storage volume. And it introduces a security scheme, with which all users can encrypt and decrypt the data of the themselves storage volume at clients, and the USN server only need to check the users' identities and the data's integrity. As shown in the experiments, the security of the USN has been improved greatly with little influence on the system performance.

1 Introduction

An ideal storage architecture would provide strong security, data sharing across platforms (i.e., operating systems), high performance, and scalability in terms of the number of devices and clients. The three storage architectures in common use today are Direct-attached Storage (DAS), Storage Area Networks (SANs), and Network-attached Storage (NAS). The fourth architecture is often called the Object Storage [1].

The DAS connects block-based storage devices directly to the I/O bus of a host machine (e.g., via the SCSI or the ATA/IDE), but its connectivity is limited. The SAN is a switched architecture that provides a rapid, scalable interconnection for large numbers of hosts and storage devices. But its cross-platform data sharing is limited. The NAS is just another name for file serving, which is introduced to share data across platforms, but clients will be limited by the performance of the file server. The object storage needs a modified OS and a special object interface instead of the block object, which holds back its popularization [1].

The United Storage Network (USN) provides a cross-platform data sharing (files) as well as a high performance (blocks). In a USN, the file server and clients are all connected to it, including a NAS device, an object storage device and a SAN device. Given this connectivity, the file server can share the file metadata with the clients, which enables the clients to directly access the storage devices. Without any mechanism for the I/O authorization, the USN security is decreased. And we have designed and implemented a reliable and practical security system with low costs to improve it.

2 The USN Model and Security Problem

The USN is made up of the NAS devices, the iSCSI devices, the object devices and the SAN devices. As shown in Figure 1, the USN includes two channels: a server channel and a high-speed network-attached channel. The metadata and small files are transferred by the server channel, and the big files are transferred by the high - speed network-attached channel [2].

The file server can share the file metadata with clients, allowing the clients to directly access the storage devices. Figure 1 illustrates a USN model used to share files among a number of clients. In order to improve the security for the USN, firstly, we have designed the Multi User View File System (MUVFS) which provides a storage volume view for each authenticated user; secondly, we have developed a security algorithm to the USN system to prevent it from many types of attacks, providing security choices for users at clients.

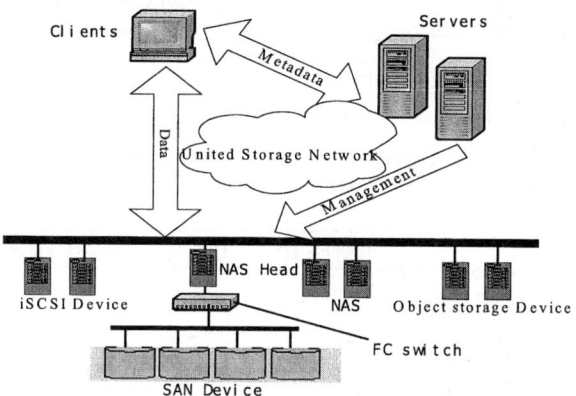

Fig. 1. A USN used to share files among clients.

3 Design of the MUVFS

The goal of the MUVFS is to improve the security of the previous standard distributed file systems while preserving the flexibility and performance of them. The MUVFS is a stackable file system layer based on the inode structure. It is placed on the top of the native file systems. A stackable file system structure is discussed in [3].

3.1 The Design Principle

The MUVFS is illustrated in Fig.2. First, the MUVFS introduces the concept of storage volume views. Each storage volume contains a special storage mode which contains a metadata architecture for a certain file system. In the storage volume, there are a logic storage space allocation, an operating interface, a function collection, a storage space allocation and reclamation and a data structure. In the storage volume, the storage mode of the file volume contains the operating function and the directory structure for implementing the POSIX semantic; the storage mode of the zoning volume must face special zoning types, such as the NTFS and the ext3. The zoning volume can derive a corresponding file volume when it is formatted by a special file system. All storage modes have registered in the MUVFS.

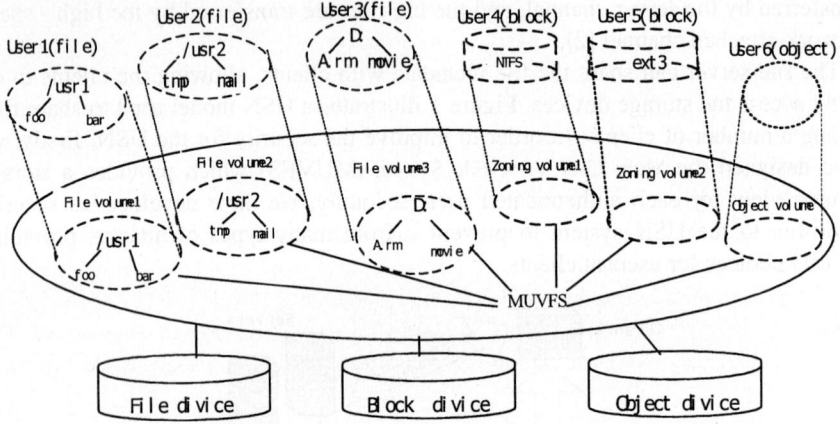

Fig. 2. Concept of storage volume in MUVFS.

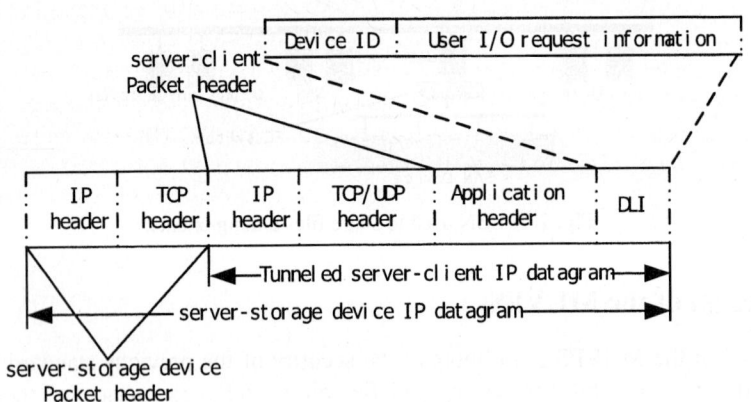

Fig. 3. Redirected packet structure.

Second, the MUVFS introduces one new file system interface of the new_read. The server's application that supports the third-party transfers the call of the new_read instead of the normal read for those data directly sent back to clients by the storage device. The main difference between the new_read and the normal read is that: the new_read only provides the location of the data and writes data location information (DLI) in the front of the supplied user's buffer. The DLI normally contains the device ID and the user's I/O request information. The size of a DLI is typically a few dozens of bytes and is usually much smaller than the size of the requested data. Third, the MUVFS uses a packet redirector. The packet redirector is placed on the USN servers, monitoring all outgoing packets and intercepts packets that contain the I/O request information. These requesting packets are redirected to the storage device. Fig. 3 shows the redirected packet structure.

3.2 Cache Consistency

A multi-user file view leads to the following question: if different views of the file may contain different data, how should these views be cached? The MUVFS solves this problem by a locking mechanism which is similar to the locking policy of the Windows/CIFS. The MUVFS implements the integrated locking function as follows: (1) The MUVFS implements the file locking and the record locking function simultaneously and supports the allowing and denying mode at the same; (2) In the MUVFS, the file locking and record locking forms a hierarchy architecture. When the client accesses the record in the MUVFS, he must obtain corresponding file lock; (3) The MUVFS implements the compelled locking semantic, and all users who access the same data are compelled to comply with each locking regulation; (4) The MUVFS supports the compatibility checking and implements the locking collision checking.

3.3 Packet Filtering

To minimize the impact on the other network traffic, the MUVFS uses a two-level filtering scheme in its packet redirector. At the first level, packets are filtered based on their UDP/TCP port numbers in a way similar to that of a firewall. The outgoing packets of applications which don't support the third-party transfer are filtered out at this level. In the second level, packets are filtered based on the information provided by the applications. The outgoing packets that belong to applications supporting third-party transfer instead of requiring data transfer are filtered out.

4 Design of the Security Scheme

4.1 Basic Features

The MUVFS can provide different storage volume view for different users or groups based on their accessing authorization, which can provide security for users' information. If someone filches a user's password and imitates the user, he can obtain the user's storage volume view and can access the user's information. To solve the problem, we have designed a security scheme associated with the MUVFS, with several important features. The first one is the end-to-end encryption at a client. The user uses a standard algorithm such as the RC5 and the Blowfish to encrypt the data at client, ensuring that the data is unreadable by anyone until it is decrypted at the client.

Monitoring the data's integrity is a second goal. Both the client and the USN server share the HMAC key. When the user at a client sends the data to the USN, at first, the user needs to calculate the HMAC of the data with the HMAC key, then he sends the data and its HMAC to the USN server. The USN server calculates and compares the HMAC of the above-mentioned data and authenticates the user and verifies the data. The valid data and its HMAC are stored in the user's storage volume of the MUVFS. When the user reads any data from his storage volume of MUVFS, this data and its HMAC are sent to the requiring client, then the user verifies the integrality of the data at a client.

Thirdly, providing the Access Control List[4], the USN shall have a sharing function at least as powerful as that of the standard Unix group, and the multi - users or the UNIX group users shall share the same private key.

Preventing invalid users from accessing the storage device is a fourth feature for this security system. A user is authenticated by his password. In case that the password or the HMAC key may fail to work in a long run, a time stamp is fixed to control the lifetime of the user's password and the HMAC key.

4.2 The USN Certificate File

The USN contains a single certificate file, as shown in Fig. 4, which contains both the administrative and cryptographic information about each USN user. The USN server uses the information in the certificate file to authenticate users and to arrange basic storage management.

User ID	HMAC key	User password	User public key	Times tamp
User ID	HMAC key	User password	User public key	Times tamp
....				
User ID	HMAC key	User password	User public key	Times tamp

Fig. 4. Certificate file.

The certificate file contains a list of tuples, each of which includes a user's ID, a HMAC key, a public key, a user's password and a time stamp. The user's ID identifies the user or group to which the remainders of the tuple pertain. The public key is stored on the disk as a convenience so that the user needs not to consult a centralized key server. The password is used for the USN server to authenticate the user's identification easily. The HMAC key is used to verify the integrity of the data written by a user and the user's identity. The HMAC key and the user's password are encrypted and stored in the non-volatile memory on the USN metadata server. When the USN metadata server is rebooted, the certificate file is loaded into the memory, and the user HMAC key and the user's password are decrypted and cached in the volatile memory. The time stamp field is updated each time when a user changes his password and the HMAC key with the help of the administrator.

4.3 USN Security Schemes

The encryption and decryption in the USN is done at clients. Though the time spent on it cannot be easily reduced, the symmetric algorithms are relatively fast. There are two different schemes for users' authentication, varying from the security and the speed.

The first scheme is more secure with some effects on the speed of the system. When writing data to the USN storage device, the user encrypts the file by using the

RC5 key. He generates the HMAC of the encrypted file by using a HMAC key, then sends the encrypted file and its HMAC to the USN server, and the USN server verifies the integrity of the encrypted file by the user's HMAC key. When the verification is done, the encrypted file together with its HMAC are put into the user's storage volume in the MUVFS. Otherwise, the encrypted block will be discarded. When these stored data is read, the user's authorization is done before the encrypted file and its corresponding HMAC are provided, allowing the user to verify their integrity and decrypt it at the client.

The Scheme 2 reduces the load on the client's CPU by replacing the ciphertext with a cleartext. The client only generates the HMAC for a cleartext data file by using the user's HMAC key and sends the cleartext and its HMAC to the USN server. The USN server verifies the integrity of the cleartext file by using the user's HMAC key. When the verification is passed, the cleartext file and its HMAC are written into the user storage volume in the MUVFS. Otherwise, the cleartext block will be discarded. This scheme is considerably high in speed without any encryption and decryption done at the client, but it is at the expense of a loss in security.

5 Performance Testing and Analyzing

5.1 Testing Environment

The experiment environment is illustrated in Figure5. Host 1 is a general PC, CPU(Intel Pentium4 1.6G), memory(256MB), OS(Linux 7.1), Hard disk(Maxtor 91020D6 (IDE)), NIC/HBA(AGE-1000SX (NIC)). Host 2 is a USN server, CPU (Intel Pentium4 1.7G), OS(Linux 7.1), memory(256MB), Hard disk (ST318437LW (SCSI)), NIC/HBA(AGE-1000SX (NIC) and SANbladTM 300). The NAS is a ELAKE NAS1000 developed by ourselves, CPU(Intel Pentium3 730MB), memory (256MB), OS(Linux 7.1) Hard disk(ST318437LW(SCSI)), NIC/HBA(AGE-1000SX (NIC)) There are two ELAKE NAS1000 used in the experiment. The FC60-RAID is a RAID based on a fiber channel.

To evaluate the influence of the security system on the file system performance for the USN server, the test is divided into two groups. In the first group, the tested file system is the NFS and the NFS+MUVFS. In the second group, the tested file system is the NFS and the NFS+MUVFS+secure_module, the security module is a software module that implements the security scheme made by ourselves. In the experiment, we use the Bonnie++, which may test the transferring rate of the file system and the occupation rate of the CPU by three modes, which are sequential read, sequential write, and random read/write. In the test, the parameters are set as follows: the file size is set for 200M, the memory is set for 50M.

5.2 Testing Data

The testing results of the NFS (A group) and the NFS+MUVFS (B group) are shown in Figure.6 while the testing results of the NFS (AA group) and the NFS+MUVFS+ secure_module (BB group) are shown in Figure 6.

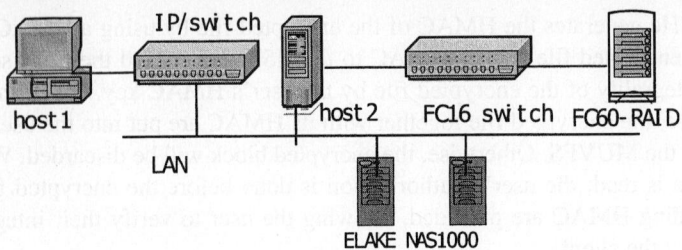

Fig. 5. Experiment configuration.

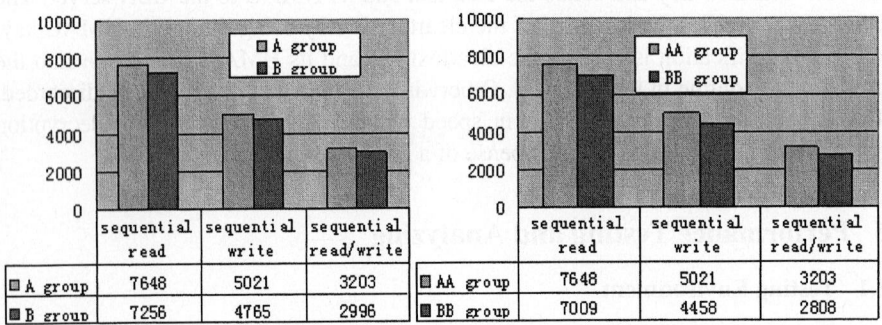

Fig. 6. The transferring rate of the file system.

5.3 Data Analyzing

The first experiment is illustrated in Fig.6. The sequential read rate of the NFS is 7648 KBps while the sequential read rate of NFS+MUVFS is 7256 KBps, with the transferring rate dropped 4.1%; the sequential write rate of the NFS is 5021 KBps while the sequential write rate of the NFS+MUVFS is 4765 KBps, with the transferring rate dropped 5.1%; the sequential read/write rate of the NFS is 3203 KBps while the sequential read/write rate of the NFS+MUVFS is 2996 KBps, with the transferring rate dropped 6.4%. And the transferring rate of NFS+MUVFS has dropped 4.1%~6.4%.

The second experiment is illustrated in Fig.6. The sequential read rate of NFS is 7642 KBps while the sequential read rate of the NFS+MUVFS+secure_module is 7009 KBps, with the transferring rate has dropped 8.4%; the sequential write rate of the NFS is 5021 KBps while the sequential write rate of the NFS+ MUVFS+ secure_module is 4458 KBps, with the transferring rate has dropped 11.2%; the sequential read/write rate of NFS is 3203 KBps and the sequential read/write rate of the NFS+MUVFS+secure_module is 2808 KBps, with the transferring rate has dropped 12.3%. And the transferring rate of NFS+MUVFS has dropped 8.4%~12.3%.

When the MUVFS and the secure _module are integrated to the USN server, a better security for the USN is achieved. However, the performance of the USN server drops 8.4%~12.3%, which is accepted in the practical application.

6 Conclusions

The presented security USN system has shown that the cryptographic security is possible for the USN integrated by the NAS and the SAN. And it is adaptable to today's computing technology, and will become more effective as clients' processors have a higher speed. This security mechanism for the USN has solved many problems concerning the system performance, providing the users with a good data confidentiality and integrity the moment it leaves the client's computer. And its decentralized security function can improve the performance and scalability of the system. Furthermore, its distributed security can remove the singe point of failure that plagues many proposed centralized security schemes. Compared with the SSDFS[5], this security scheme can provide strong security for clients of the block-oriented, the file-oriented and the object-oriented requests.

References

1. Mike Mesnier, Gregory R. Ganger, Erik Riedel. Object-Based Storage. IEEE Communications Magazine, August 2003, pages 84-90.
2. FU Xianglin, XIE Changsheng, HAN Dezhi.. The study and implementation of a iSCSI-based SAN. Research and development of computer, 2003(5).
3. J.S. Heidemann and G.J. Popek, "File System Development with Stackable Layers," ACM Trans. Computing Systems, vol. 12, no. 1, pp. 58-89, Feb. 1994.
4. J.Howard, et.al.,"Scale and Performance in a Distributed File System", ACM Trancations on Computer System 6(1), February 1988, page 51-81.
5. M.Ethan, L.Darrell, F.William and R.Benjamin,"Strong Security for Distributed File Systems", IEEE Micro, 20(1), May 2001, page 34-40.

Q3: A Semantic Query Language for Dart Database Grid*

Huajun Chen, Zhaohui Wu, and Yuxing Mao

College of Computer Science, Zhejiang University, Hangzhou, 310027, China
{huajunsir,wzh,zzzgz,maoyx}@zju.edu.cn
http://grid.zju.edu.cn/dart/q3

Abstract. In presence of Grid where a huge amount of databases can be involved in sharing cycle, database tools and middlewares should be well suited for schema mediation and query processing in a semantically meaningful way. *Dart* is an implemented prototype system whose goal is to provide a semantic solution capable of deployment in grid settings. This paper particularly concerns the problems of semantic query answering in DartGrid. Our first contribution is a semantically enriched query language, called Q3, for specifying complex queries using RDF/OWL semantic. We describe its implementation in Dart Database Grid, and also introduce a semantic interface for user to graphically browse RDF semantic, and visually construct Q3 queries.

1 Introduction

The Grids is emerging as a building infrastructure that support coordinated management and sharing of inter-connected hardware and software resources. That raises the question as to how database resource can be deployed and integrated in such a new paradigm where a huge amount of decentralized, independently administrated databases can be involved in the grid-scale sharing cycle. In such a new setting, database tools and middlewares should be well suited for schema mediation and query processing in a *semantically meaningful* way.

Dart[1] database grid [1] is an implemented prototype system whose goal is to provide a semantic solution capable of deployment in grid-settings. Dart is built upon several semantic web standards and grid technologies. This paper considers, more particularly, the problems of query answering in Dart database grid. We introduce a semantically enriched query language, called Q3, for specifying complex queries using RDF/OWL semantic and present its implementation in Dart Grid. We also present a semantic interface for facilitating the user to graphically browse RDF semantic, and visually construct Q3 queries.

This paper is structured as follows. We first articulate on the design of Q3 language in Section 2. Section 3 introduces the implementation. Section 4 mentions some key problems remained unresolved.

---

* This work is supported in part by China 973 fundamental research and development project: The research on application of semantic grid on the knowledge sharing and service of Traditional Chinese Medicine; Intel / University Sponsored Research Program: DartGrid : Building an Information Grid for Traditional Chinese Medicine; and China 211 core project: Network-based Intelligence and Graphics.
[1] Dart Home Page: http://grid.zju.edu.cn/dart, demo is available at our web site.

2 Q3 Language

2.1 Overview

Q3 is a sub language of N3 [2], a compact and readable alternative to RDF's XML syntax. Q3 carries the following aims:
- To improve readability. Q3 inherits this characteristic from N3.
- To accurately capture the semantic of queries. In Q3, every query can be viewed as an OWL class definition; and query processing is reduced as computing all instances satisfying the query concept definition.
- Designed specially for formulating query on database resources using RDF/OWL semantic.

2.2 Design Details of Q3

This section gives a informal introduction on Q3 by some examples.

1. Specify the QName
Being similar to N3, the @prefix directive binds a prefix to a namespace URI, as the following example displays.
@prefix q3:<http://grid.zju.edu.cn/q3#>.

2. Q3 Operators and Specifying a query pattern
A Q3 query begins with a Q3 operator. Currently, two operators has been implemented in Dart, they are: q3:Query & q3:Update. We plan to support all SQL operations including select, update, insert & delete, all at a semantic level.

Q3 query pattern can be view as a concept definition with which each result returned must satisfy. Each query has only one pattern definition, and this definition is enclosed by a pair of square brackets "[...]". As a matter of fact, this part can be seen as an anonymous object (remember: in N3, we use "[...]" to denote anonymous nodes). The following example indicates that the result returned must be an instance of the class of nw:Product: q3:Query   q3:pattern   [a nw:Product].

Note: in N3, the property *rdf:type* can be abbreviated in N3 to just "a". So we can introduce an instance of the class of nw:Product as:
nw:cellphone001   a   nw:Product.

To introduce an anonymous instance of the class of nw:Product , we can use : [a nw:Product].

3. Specify comparison constraints
Q3 allow user to specify comparison predicates in a query like SQL, for an example as below, "nw:ProductName "Cell Phone"" indicates that the instance returned must be a product with the name of "Cell Phone"; "nw:UnitPrice [>= 2000]" indicates that this product's unit price should be larger than or equal to 2000. This can be viewed as the "where" part of a SQL query and can also be viewed as owl:hasValue and owl:hasSomeValue restriction in OWL representation.

   q3:Query q3:pattern [a   nw:Product;
            nw:ProductName "Cell Phone";
            nw:UnitPrice [>= 2000]
         ].

4. Specify what attribute values that should be returned

This can be seen as the "select" part of a SQL query, i.e., it specifies what specific attribute values that the result returned should include. We use a "[]" following an attribute name to indicate that the user wants to query the data value corresponding to that attribute. The following example indicates that the user want to query the product's name and unit price.

```
q3:Query q3:pattern [ a         nw:Product;
            nw:ProductName [];
            nw:UnitPrice   []
        ].
```

Note: in N3, "[...]" denotes an anonymous node or blank node. Since in RDF semantic, bland nodes are treated as simply indicating the existence of a thing, it is reasonable to view "[...]" as an existential variable. Therefore, we can represent the example above by introducing some existential variables as below.

```
q3:Query q3:pattern {
  this log:forSome ?x1,?x2,?x3.
  ?x1  a         nw:Product;
     nw:ProductName ?x2;
     nw:UnitPrice   ?x3.
}.
```

The above example would mean in mathematic logic:

$$\exists x, y, z. Product(x) \cap productName(x, y) \cap unitPrice(x, z)$$

5. Specify join constraints

In some case, we need to specify a relationship between two classes in a query, for example, we maybe want to find out all products supplied by a specific supplier. In that query, two classes are involved, they are *Product* and *Supplier*. The property of *suppliedBy* poses a new constraint on the product instances returned. In relational model, this can be viewed as a join constraint between two tables.

The following example means: find out those products supplied by a Supplier that locates in China. As you can see, we introduce a new anonymous node, denoted by "[...]", for the Supplier class.

```
q3:Query q3:pattern [ a       nw:Product;
        nw:ProductName [];
        nw:suppliedBy [ a nw:Supplier;
                nw:Country   "China"
                ]
    ].
```

6. Aggregation Predicate: q3:count

q3:count is a reserved predicate of Q3. While it is very similar to the count(*) aggregation function in SQL, it provide us with some more advanced functionalities that we can use it to specify cardinality restriction in a semantic query.

The following example indicates: find out those products that supplied by at least 3 suppliers and list their product name.

```
q3:Query
    q3:pattern [
            a           nw:Product;
            nw:ProductName [];
            nw:Supply_by  [ a     nw:Supplier;
                        q3:count ">3"
                    ]
        ].
```

If we represent the upper example using description logic syntax, it would be:

$Product \cap \exists_{=1} productName \cap \exists_{>3} suppliedBy.(Supplier)$

7. Universal Quantifier

Sometime we need to express a query using universal quantifier. For example, we want to query those product whose supplier are all from China. In this case, we need to declare universal variable in a Q3 query. In N3, we can use "this log:forSome ?x" to declare existential variable, and "this log:forAll ?x" to declare a universal variable, as the example displayed below.

```
q3:Query q3:hasPattern {
        this log:forAll ?y.
        ?x  a    nw:Product;
            nw:productName [];
            nw:suppliedBy ?y.
        ?y  a    nw:supplier;
            nw:locatedIn "China".
    }.
```

With mathematical logic, the universal quantifier set a specific constraint that the result should satisfy:

$$\forall y.(x,y) \in R \Rightarrow y \in C^I$$

The C^I denotes the semantic interpretation of concept C. That is to say, if y has a relationship of R with x, then y must be an instance of C. If we represent the example above using description logic's syntax, it can be clearer:

$Product \cap \forall suppliedBy.(Supplier \cap locatedIn:"China")$

2.3 Abstract Syntax and Formal Semantic of Q3

We use description logic syntax to define the abstract syntax of Q3. The following definition specifies the expressivity of current version of Q3.

Definition 2. Let **CN** be set of atomic concept names, and **RN** be set of role names, and **AN** be set of attribute names, and **ON** be set of individual names or literal values. The set Q for Q3 is the set of formula defined by the following abstract syntax:

$C | \exists A.L | \exists A.D | \exists R.D | \leq_n A.L | \geq_n A.L | \forall A.L | \leq_n R.D | \geq_n R.D | \forall R.D | Oneof(o_1, o_2,...) | A:o | C \cap D$

Where C, D is concept expression, A is attribute name in **AN**, L is basic literal datatype, R is role name in **RN**, and $o, o_1, o_2...$ are individual names in **ON** or literal values.

We do not present the formal semantic of Q3, since it is very similar to DL descriptions.

3 Implementation in Dart Database Grid

3.1 DartGrid Prototype

The principal technical notions of DartGrid are highlighted as below:

- The base development and deployment platform are Globus 3.0 . And, all service definitions are OGSI[2]-compliant;
- Source data semantic and Ontologies are represented by RDF/OWL[3], the W3C semantic web standards.
- We use Protégé to build the ontologies and HP Jena2 toolkit to store, read, process RDF/OWL data.

We would like enumerate all base elements of DartGrid. Figure 1 illustrates how these elements can be organized in two different deployment modes: hierarchical mode (H-Mode), and P2P mode (P-Mode). DartGrid is intended to provide a generic framework that is flexible to accommodate different application requirements.

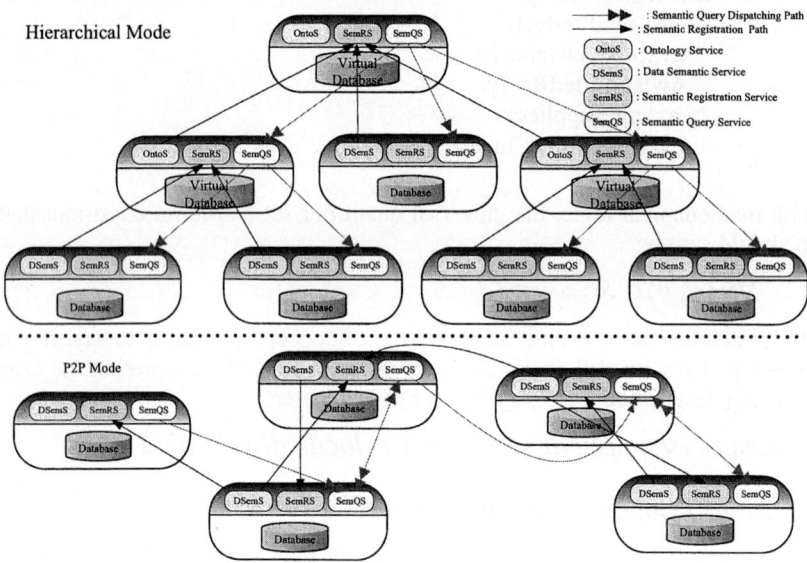

Fig. 1. Abstract Architecture.

1. Source Data Semantic Service (DSemS):
DSems is used to export source database's relational schema as RDF/OWL semantic description. Typically, local DBA publishes source data semantic of a specific database by DSemS. Others can inquire of this service to fulfill some tasks such as brows-

---

[2] GGF's OGSI Working Group: https://forge.gridforum.org/projects/ogsi-wg
[3] W3C Web-Ontology (WebOnt) Working Group: http://www.w3.org/2001/sw/WebOnt/

ing the source semantic, using the source semantic to define semantic queries or do semantic mapping. We've developed a tool to automatically convert relational schema to its corresponding RDF/OWL description. This tool is downloadable at our website.

2. Ontology Service (OntoS):
Ontologies can be viewed as the mediated schema. Ontologies are published by OntoS for user browsing or posing queries. DSemS can be viewed as a special case of OntoS.

3. Semantic Registration Service (SemRS):
Semantic registration establishes the mapping relationship from source data semantic to mediated ontologies, or from ontologies to other ontologies (Hierarchical mode), or from local source data semantic to other remote target data semantic (P2P mode). Database service provider can register his/her database to a SemRS, which possibly locates in other database node (P2P mode) or upper-level mediated node (Hierarchical mode). To understand the detailed registration strategy he police, please refer to our web site.

4. Semantic Query Service (SemQS):
SemQS accepts Q3 semantic queries, inquires of SemRS to determine which lower-level SemQS are capable of providing the answer, then rewrites the query in terms of lower-level ontologies and generates a series of sub semantic queries, which are then dispatched to those SemQS involved. Finally, the queries will be delivered to the SemQS of a specific database where the queries will be ultimately converted into SQL queries. The results of SQL queries will be wrapped by RDF/OWL semantic again and what SemQS returns is always in RDF/XML format.

3.2 Semantic Browser and Visual Semantic Query

DartGrid offers a semantic browser [8] enabling user to interactively specify a semantic query. Typically, user first visits a DSemS or OntoS and browses the source data semantic or ontologies graphically; then user selects concepts of interest, specifies the constraints, and a Q3 query string will be generated simultaneously; thirdly, user submits this Q3 query string to the SemQS relative to the DSemS or OntoS he/she visit; when the result is returned, the user can browse the result graphically again. Figure 2 and 3 illustrate an example from our TCM application[7]. It showcases how user can step-by-step specify a semantic query to find out those Chinese *Compound-Formulas* constituted by some Chinese *UnitMedicine* that can help *influenza*. In this query, three concepts are involved, they are: *tcm:CompoundMedicine, tcm:Unit Medicine* and *tcm:Disease*.

3.3 Rewriting Q3 Using Relational Tables and Views

When a Q3 query is dispatched to the SemQS of a specific database, the SemQS needs to rewrite the Q3 query in terms of the source relational schema, i.e., convert the Q3 query into a SQL query. We specify a set of rules to fulfill this task. A Q3 query can be presented as a query tree ,for example, the following query can be translated into a query tree as Figure 4 depicts. In Table 1, L(x) denotes the description label on the node.

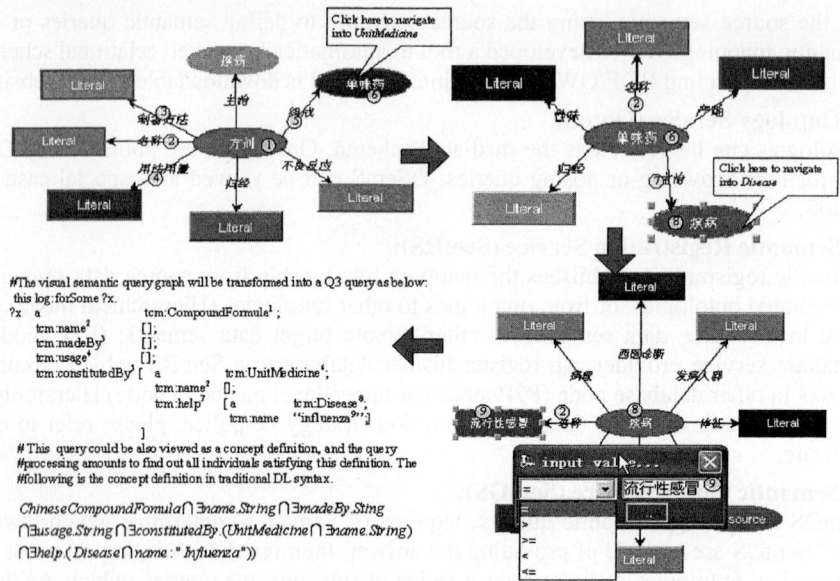

Fig. 2. An example of visual semantic query.

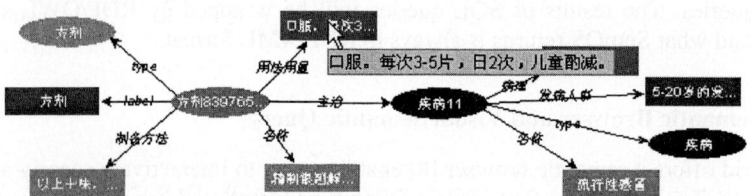

Fig. 3. The query result.

$Student \cap \exists name.String \cap \exists enterYear.(>: 2002) \cap \exists hasDept.(Department \cap name:"CS") \cap \forall select.(Course \cap hasCreditHour : 3 \cap \geq 20 selectedBy.Student)$

4 Summary and Future Work

Database management and integration has recently received significant attention for both Grid development [5, 6] and Semantic Web research [3,4,9]. Dart is an implemented prototype system aimed to provide such a kind of data sharing toolkit. Q3 is a semantically enriched query language implemented in Dart database grid. The salient characteristics of Q3 include: a) it conforms to RDF/OWL semantic; b)the query semantic can be accurately captured; c) it follows

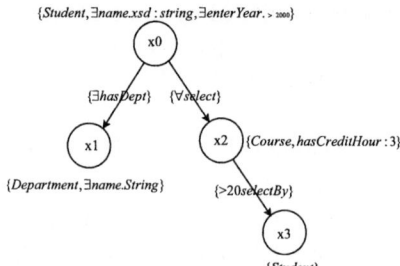

Fig. 4. The Q3 query tree.

Table 1. Rules for converting Q3 to SQL.

	Description
1	If $C \in L(x_i)$ then replace C with its corresponding table name, we denote it by $tn(C)$; If $\exists\ A.L \in L(x_i)$ where A is a owl:DatatypeProperty, then rewrite it as a *projection*, i.e., $\prod_{cn(A)}(tn(C))$ where $cn(A)$ is the column name corresponding to attribute A, C is the domain class of A; If $A{:}o \in L(x_i)$ or $\exists\ A.D \in L(x_i)$ where A is a owl:DatatypeProperty, o is a value, D is a *value comparison expression* such as ">: 18", then rewrite it as a *selection*, i.e., $\sigma_{cn(A)=D}(tn(C))$;
	For example: a query $Student \cap \exists name.String \cap \exists age.>18 \cap sex{:}"female" \xrightarrow{convertedTo}$ Select student_id,student_name,student_age from student where student_age>18 and sex='female'
2	If $\exists\ R.D \in L(x)$ where R is a owl:ObjectProperty, then: Rewrite it as an *equi-join*, i.e., $tn(C_1)$ join $tn(C_2)$ where C_1 is the domain class of R, C_2 is the range class of R
	$\exists hasDept.Department \xrightarrow{convertedTo}$ Select student.student_id, department.dept.id from student, department where student.dept_id=Department.dept_id
3	If $\leq n\ R.D \in L(x)$ or $\geq n\ R.D \in L(x)$ where R is a *owl:ObjectProperty* then Rewrite it using *"group by ... having"* as the example below shows.
	Find out those courses that selected by more than 20 students: $\geq 20 selectedBy.Student \xrightarrow{convertedTo}$ select cid, count(*) from select_course group by cid having count(*) >=20 (cid is the course id)
4	If $\forall\ R.D \in L(x)$ where R is a *owl:ObjectProperty*, then Rewrite it using *"exists"* as the example below shows. Note: $\neg \exists R . \neg D \Rightarrow \forall R . D$
	For example, find out those students for which all courses selected by him/her have a credit hour of 3: $\forall select.(Course \cap hasCreditHour : 3) \xrightarrow{convertedTo}$ Select sid from Student where not exists (select course.cid from course, select_course where Course.credit_hour<>3 and select_course.sid=Student.sid and select_course.cid=course.cid))
5	If $\exists\ A.ONE\text{-}OF(\ldots)| \leq n\ A.ONE\text{-}OF(\ldots)| \geq n\ A.ONE\text{-}OF(\ldots)|\forall\ A.ONE\text{-}OF(\ldots) \in L(x)$ Rewrite it using *"in"* as the example below shows.
	Find out those students who select course 1 or 2, $\exists select.ONE\text{-}OF(1,2) \xrightarrow{convertedTo}$ Select student_id from select_course where course_id **in** ("1","2").

the syntax feature and inherits good readability from N3. One key problem remains unresolved is to develop efficient algorithm for distributed semantic query where a huge number of semantic resources are involved. That is our next work focused in the future.

References

1. Zhaohui Wu, Huajun Chen, et al. DartGrid: Semantic-based Database Grid. Lecture Notes in Computer Science 3036/2004, pp. 59 - 66
2. Tim Berners-Lee, Dan Connolly. Semantic Web Tutorial Using N3: http://www.w3.org/2000/10/swap/doc/.
3. Alon Y.Halevy, Zachary G. Ives, Peter Mork, Igor Tatarinov, Piazza: Data Management Infrastructure for Semantic Web Applications, in: Proc. of The Twelfth International World Wide Web Conference, 2003, 556-567.
4. Frank Manola, A Database Perspective on the Semantic Web: A Brief Commentary, Bulletin of the IEEE Computer Society Technical Committee on Data Engineering 26(4) (2003) 5-11.

5. Ian Foster, Jens Vöckler, Michael Wilde, Yong Zhao, The Virtual Data Grid: A New Model and Architecture for Data-Intensive Collaboration, in: Proc. of Conference on Innovative Data System Research, 2003.
6. I. Foster, S. Tuecke, J. Unger, OGSA Data Service, Global Grid Forum Documents. http://www.cs.man.ac.uk/grid-db/papers/draft-ggf-dais-dataservices-ggf9.pdf, 2003.
7. Huajun Chen, Zhaohui Wu, Chang Huang: TCM-Grid: Weaving a Medical Grid for Traditional Chinese Medicine. Lecture Notes in Computer Science, 2659 1143- 1152.
8. Mao Yuxin, Wu Zhaohui, Chen Huajun, SkyEyes: A Semantic Browser for the KB-Grid, Lecture Notes in Computer Science , 3032 (2003) 752-759.
9. Tim Berners-Lee, Relational Databases on the Semantic Web, http://www.w3.org/ DesignIssues/RDB-RDF.html.

Knowledge Map Model[*]

Hai Zhuge[1] and Xiangfeng Luo[2]

[1] Knowledge Grid Research Group, Key Lab of Intelligent Information Processing,
Institute of Computing Technology, Chinese Academy of Sciences, 100080, Beijing, China
zhuge@ict.ac.cn
[2] Hunan Knowledge Grid Lab, Hunan University of Science &Technology, Hunan, China
luoxf@kg.ict.ac.cn

Abstract. Knowledge representation and inference is a key issue of Knowledge Grid. This paper proposes Knowledge Map (KM) model for representing and reasoning causal knowledge as an overlay in Knowledge Grid. It extends Fuzzy Cognitive Map (FCM) to represent and infer not only cause-effect causal relations, but also time-delay causal relations, conditional probabilistic causal relations and sequential relations. Mathematical model and inference rules of KM are presented. Simulations show that KM is more powerful than FCM in emulating real world.

1 Introduction

Fuzzy cognitive maps aim to mimic human reasoning for causal knowledge [1, 2]. They have the following characteristics [1-5]:

1. Domain knowledge is in the form of concepts and directional connections;
2. Relation between concepts is in the form of production rule;
3. The inference of FCMs can be computed by numeric matrix operations instead of explicit production rules;
4. FCMs can express tacit knowledge of experts;
5. Augmented FCM's matrix sum makes the addable characteristic.

Though FCMs have many desirable properties, they have the following limitations:

1. FCMs do not support the inference of sequential relations or time-delay causal relations. The characteristic of inference process of FCMs is that all the interaction of FCMs' concepts is synchronous. Therefore, it does not reflect sequential relations or time-delay causal relations which widely exist in real world.
2. FCMs do not support the inference of conditional probabilistic causal relations.

Fig.1 is an FCM representing an expert's knowledge about terror events. The *casualty* and *explosion* are the subsequences of *terrorists*, and the *terrorists* is the subsequence of *the friendly foreign policy* and *the power of strike*. The friendly foreign policy has not an immediate effect and it needs days or months to make a full impact on *terrorists*. *The power of strike* also needs hours or days to make a full im-

---

[*] The research work was supported by the National Science Foundation of China and National Grand Fundamental research 973 programs (2003CB316900).

pact on *terrorists*. But, FCMs do not support this type of inference. Their inference results in some intelligent systems are usually distorted.

Hagiwara extends FCMs to solve time-delay causal relations and conditional causal relations [1], but he does not consider probabilistic causal relations. Miao et al. put forward Dynamical Cognitive Network (DCN) to define dynamic causal relation between concepts [2], but DCN still does not distinguish different types of causal relations. Neural Cognitive Map (NCM) is presented to solve complex causal relations by Obata and Hagiwara [6], but NCM needs much training data that are difficult to be obtained in some intelligent systems, and time-delay causal relations as well as conditional probabilistic causal relations are difficult to be found by neural networks. Based on the object-oriented paradigm for decision support, contextual Fuzzy Cognitive Map (cFCM) is proposed by Satur and Liu [7]. The cFCM considers contextual causal relations and cause-effect causal relations without probabilistic causal relations and time-delay causal relations.

This paper classifies causal relations into cause-effect causal relations, time-delay causal relations, and conditional probabilistic causal relations, and considers the sequential relation as a special time-delay causal relation.

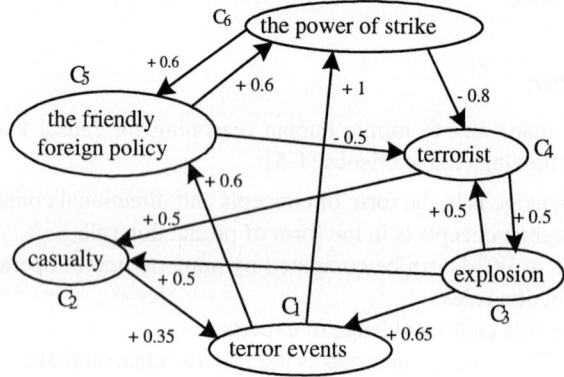

Fig. 1. An expert's knowledge about terror events represented by fuzzy cognitive map.

2 The Mathematical Model of Knowledge Map

Knowledge Map (KM) is the map that can represent cause-effect causal relations, time-delay causal relations, conditional probabilistic causal relations and sequential relations. It can be formally defined as follows:

Let V_{ci} and V_{cj} be the state values of C_i and C_j; R_{ij} and w_{ij} be the relation type and the weight from C_i to C_j respectively; $\phi(x)$ be the reasoning function of C_j. R_{ij} and w_{ij} are the elements of the KM's relation matrix **R** and the adjacency matrix **E** respectively. The mathematical model of KM is as follows:

$$V_{cj}(t+1) = \phi(\sum_{\substack{i=1 \\ i \neq j}}^{N} g(V_{ci}(t), R_{ij}, w_{ij}), \sum_{\substack{i=1 \\ i \neq j}}^{N} V_{ci}(t) w_{ij}).$$

The operator function $g(V_{ci}(t), R_{ij}, w_{ij})$ is determined by the following operations of Rule 1- 4:

Rule 1: If there exist conditional probabilistic causal relations from different concepts C_i to C_j, $V_{cj}(t+1)$ should be computed first, and then set $w_{ij}=0$.

$$V_{cj}(t+1) = \begin{cases} V_{cj}(t)+u(\sum_{i=1}^{N}V_{ci}(t)w_{ij}|C_p,C_q,...)) & -1 \leq V_{cj}(t)+u(\sum_{i=1}^{N}V_{ci}(t)w_{ij})) \leq 1 \\ 1 & V_{cj}(t)+u(\sum_{i=1}^{N}V_{ci}(t)w_{ij})) > 1 \\ -1 & V_{cj}(t)+u(\sum_{i=1}^{N}V_{ci}(t)w_{ij})) < -1 \end{cases}$$

where $C_p, C_q,...$ are conditions of weight w_{ij}, $u(x)$ is a computing function of all the concepts C_i occurrences leading to the increase / decrease probability of concept C_j. Its function can be defined as follows:

$$u(|x|) = \begin{cases} +u(|x|) & \text{if } x \geq 0 \\ -u(|x|) & \text{if } x < 0 \end{cases}$$

here $u(|x|) = \tanh(|x|) \in [0,1]$.

Rule 2: If there exists time-delay causal relation from C_i to C_j, then reserve the primary value of C_i during the time delay.

Rule 3: If all values of the ith row are zero in the KM's adjacency matrix **E**, then set $V_{ci}(t)$ equal to the original value of C_i.

Rule 4: The $(m-1)^{th}$ sequential relation should be reasoned before the reasoning of the mth sequential relation.

After the operations of Rule 1-4, infer each concept of KM one times by the following inference rule:

Rule 5: The effect concept's state value of KM at time $(t+1)$ should partly depend on its own state value at time t [4].

The computing of effect concept's state value is as follows:

$$V_{cj}(t+1) = \mu f(\sum_i w_{ij} V_{ci}(t)) + \lambda V_{cj}(t) \in [-1, 1],$$

where $f(x) = \tanh(x)$, μ and λ are allotted coefficient, and $\mu + \lambda = 1$.

In KM, causal knowledge is in the form of concepts, relations, directional connections and weights. Concept and its state value represent knowledge and its existent degree respectively. The relation matrix **R** and the adjacency matrix **E** describe the relation types and the weights between concepts of directional connections respectively. And all of their interactions among concepts, relations, directional connections and weights compose a dynamic network and form a track that corresponds to a flow in the causal relation space of p dimensions (p is the number of relations). The terror events knowledge represented by Fig.1 can be represented as the knowledge map (denoted as KM1) shown by Fig.2. The relation matrix and the adjacency matrix of KM1 correspond to **R1** and **E1** respectively. In the relation matrix, the cause-effect causal relation and the conditional probabilistic causal relation are denoted as R_{sc} and $R_{p/condition}$ respectively. If there does not exist causal relation between concepts, it is denoted as N in **R**. The mth subsequence is denoted as mS and the time-delay causal relation is denoted as kD (k is the delay times). For example, **R1** (5, 4) = ($1S, nD$) represents that there exists first sequential relation and the time-delay causal relation (delay time is n) from C_5 to C_4. The denotation C_4 ($1S56, nD56, P34/normal$) in Fig.2 represents that there exists first sequential relation and the time-delay causal relation (delay time is n) from C_5 and C_6 to C_4, and there also exists conditional probabilistic causal relation from C_3 to C_4.

$$R1 = \begin{bmatrix} N & N & N & N & R_{sc} & R_{sc} \\ R_{sc} & N & N & N & N & N \\ R_{sc} & R_{p|normal} & N & R_p & N & N \\ N & (2S,kD) & (2S,kD) & N & N & N \\ N & N & N & (1S,nD) & N & R_{sc} \\ N & N & N & (1S,nD) & R_{sc} & N \end{bmatrix} \quad E1 = \begin{bmatrix} 0 & 0 & 0 & 0 & +0.6 & +1 \\ +0.35 & 0 & 0 & 0 & 0 & 0 \\ +0.65 & +0.5 & +0.5 & +0.5 & 0 & 0 \\ 0 & +0.5 & +0.5 & 0 & 0 & 0 \\ 0 & 0 & 0 & -0.5 & 0 & +0.6 \\ 0 & 0 & 0 & -0.8 & +0.6 & 0 \end{bmatrix}$$

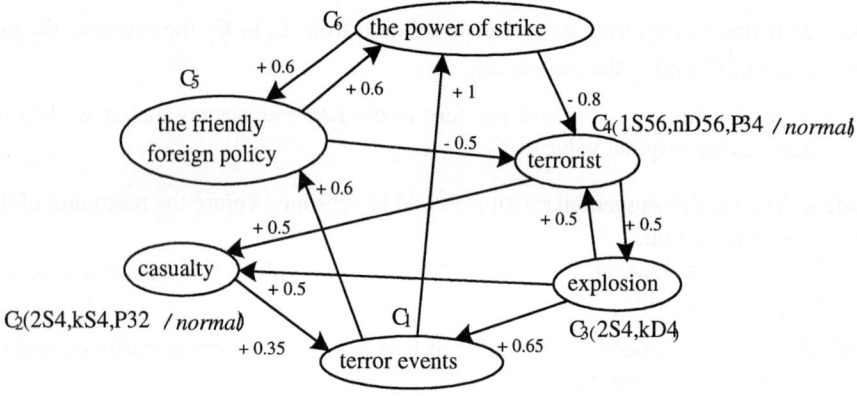

Fig. 2. An expert's knowledge about terror events represented by knowledge map.

3 Simulations

3.1 Terror Events Inference by Fuzzy Cognitive Map

The FCM1 shown by Fig.1 about terror events is constructed by expert. The inference results of FCM1 are shown by Fig.3. Fig.4 is the zoom of Fig.3's anterior iterative times. The reasoning results show that the terror events exhibit limit-cycle equilibrium behaviors (T ≈ 20). We know that if the casualty and the explosion occurrence degree are very little, then the terror events occurrence degree is also very little. The inference results of the 3rd iteration (in Fig.4) show that if the government executes

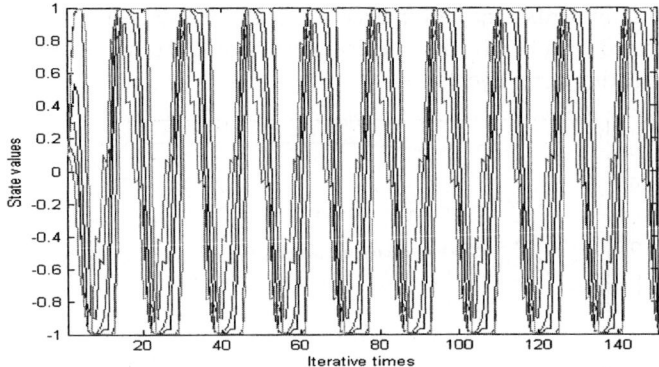

Fig. 3. Terror events reasoning results by fuzzy cognitive map.

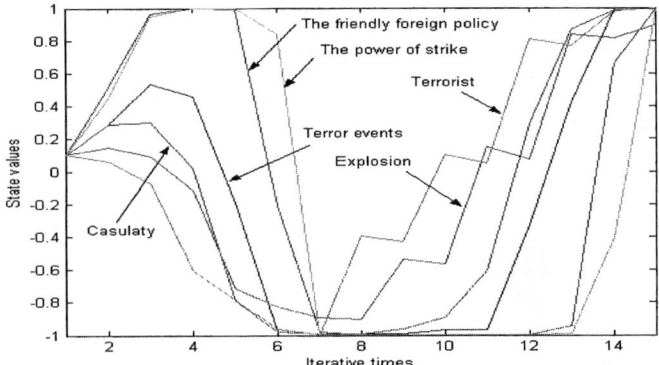

Fig. 4. The zoom of Fig.3's anterior iterative times.

very friendly foreign policy (0.9973) and very strong powers of strike (0.9974), then the occurrence degree of terror events becomes 0.4569, and the casualty and explosion occurrence degree become 0.0228 and -0.1113 respectively. So the reasoning results of the 3rd iteration are not reasonable. The reasoning results of the eleventh iteration show that if the government does not strike the terrorists (-0.9998), only executes wild foreign policy (-1.0000), then the terror events would hardly occur (-0.9669). This situation is impossible to occur in real world.

3.2 Terror Events Inference by Knowledge Map

Knowledge map of KM1 is shown by Fig.2. The relation matrix and the adjacency matrix are **R**1 and **E**1 respectively. The inference results of KM1 are shown by Fig.5. The inference results of KM1 show that if the government executes very friendly foreign policy (0.9260) and very strong power of strike (0.9253), then the terror events occurrence degree is -0.0165, and the casualty and the explosion occurrence degree are -0.0058 and -0.0056 respectively. Reviewing all of the results of KM1 we can observe the tendency that with the strengthening of the friendly foreign policy and the power of strike, the terror events' occurrence degree becomes lower and lower. Compared with FCM1, the inference results of KM1 are more reasonable than FCM1 in emulating terror events.

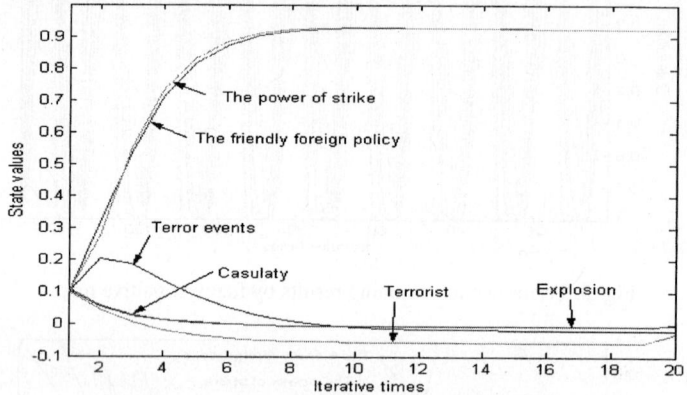

Fig. 5. Terror events reasoning results by knowledge map.

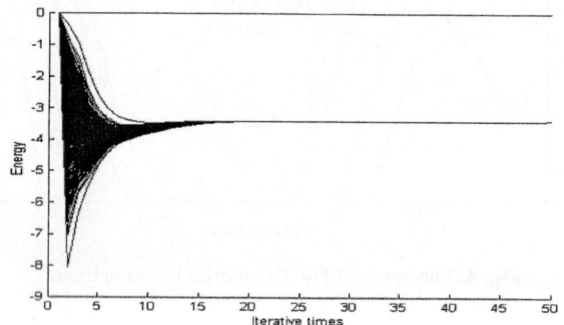

Fig. 6. 1000-times repeated stability experiments on the inference process of KM1.

3.3 The Stability of KM1

1000-times repeated stability experiments on the inference process of KM1 shown in Fig.6 show that the output of concept is controlled by KM1 and it can achieve steady states.

4 Analyses

The following characteristics make KMs have the following capabilities over FCMs in emulating real world.
1. KMs consider not only cause-effect causal relations, but also sequential relations, time-delay causal relations and conditional probabilistic causal relations at the same time when emulating real world.

2. The effect concept's state value of KM at time $(t+1)$ depends on not only causal concepts' state values at time t but also its own actual state value. FCMs only consider the actual causal relations; they do not include the actual state of the effect concept. So, the reasoning values of FCMs often change sharply. This does not accord with real world. According to Rule 5, the value of the next effect concept of KM includes not only the actual state values of the causal concepts but also its own actual value.
3. If a concept does not have any causal concept, it can maintain its own original value in the inference process of KMs. In FCMs, if a concept does not have any causal concept, the next inference value of the concept would be zero according to the inference rules of FCMs. In the inference process of KMs, concepts can keep their original values in the inference process of KMs according to Rule 3, so KMs can avoid this type of distortion in the inference process.
4. Before the reasoning of the cause-effect causal relations, time-delay causal relations and sequential relations, conditional probabilistic causal relations can be reasoned. In FCMs, if one concept's occurrence increases / decreases another concept's existent degree, then the increased / decreased value would be used in the inference process. For example, we know that there may exist terrorists in a town (assuming that its existent degree is 0.6) if one day we hear from news that the town was attacked by explosion (shown in Fig.7). At that moment, the terrorists' existent degree may jump to a high value. To take Fig.7 (a) and (b) as an example, assume that $V_{C4}(0) = V_{C3}(0) = 0.6$. *FCM inference out:* $V_{C4}(1) = f(w_{ij} * V_{C3}(0)) = tanh(0.5*0.6) = 0.2913$; *KM inference out:* $V_{C4}(1) = V_{C4}(0) + u(w_{34} * V_{C3}(0)) = 0.8913$. Common sense tells us: when an explosion occurs, the terrorists' existent degree may increase. But the reasoning result of FCM shows that the existent degree of terrorists decreases. It is not true clearly. So the inference of KM is more reasonable than FCM.

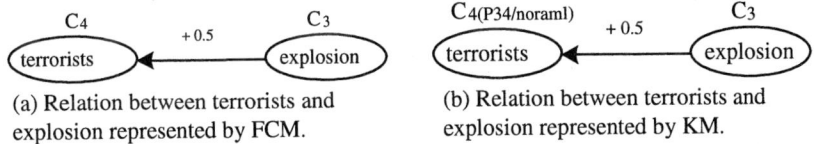

(a) Relation between terrorists and explosion represented by FCM.

(b) Relation between terrorists and explosion represented by KM.

Fig. 7. Different representation of relations between terrorists and explosion.

5 Conclusions

Fuzzy cognitive map only describes cause-effect causal relations of causal knowledge. This paper presents a knowledge map model that can describe not only cause-effect causal relations, but also time-delay causal relations, conditional probabilistic causal relations and sequential relations. The results of simulations indicate that knowledge map can emulate real world more natural than fuzzy cognitive map. The ongoing work is to use knowledge maps to share and understand causal knowledge in Knowledge Grid environment [8].

References

1. M. Hagiwara. Extended Fuzzy Cognitive Maps. In: the proc of IEEE International Conference on Fuzzy System FUZZ-IEEE, San Diego, 1992, 795-801.
2. Y. Miao, Z.Q. Liu, et al. Dynamic Cognitive Network, IEEE Transactions on Fuzzy System. 9(5), 2001, 760-770.
3. Y.Miao, Z.Q.Liu. On Causal Inference in Fuzzy Cognitive Maps. IEEE Transactions on Fuzzy System.8 (1), 2000, 107-119.
4. C. D. Stylios, P. P. Groumpos. Fuzzy Cognitive Maps: A Soft Computing Technique for Intelligent Control. In: the proc of IEEE International Symposium on Intelligent Control, Patras, 2000, 97-102.
5. B.Kosko. Fuzzy Engineering. Prentice Hall, 1997.
6. T. Obata, M. Hagiwara. Neural Cognitive Maps. http://citeseer.ist.psu.edu/ .
7. R. Satur, Z.Q. Liu. Contextual fuzzy cognitive map for decision support in geographic information systems. IEEE Trans. on Fuzzy Systems, 7(5), 1999, 495-507.
8. H. Zhuge. China's E-Science Knowledge Grid Environment. IEEE Intelligent Systems, 19 (1) 2004, 13-17.

Kernel-Based Semantic Text Categorization for Large Scale Web Information Organization

Xianghua Fu, Zhaofeng Ma, and Boqin Feng

Department of Computer Science and Technology, Xi'an Jiaotong University,
Xi'an, 710049, China
csdnfxh@yahoo.com.cn

Abstract. The volume of digital documents increases rapidly in recent years, so fast and accurate methods to classify text documents are needed imminently. In this study a new kernel-based semantic method is proposed to reduce the dimensionality of the documents and extract the semantic concept of the words simultaneously using kernel principal component analysis (KPCA). With a new semantic concept kernel function, text documents are mapped into semantic concept space that consists of independent semantic concepts of words directly, and then classified by support vector machine. Moreover, a kernelized Hebbian algorithm is applied to improving the computing process of KPCA. In comparison to BASE-SVM, the experiment results on Reuters-21578 data set show that the new method is effective on enhancing the performance of the text categorization. Furthermore in the case the training data set is not efficient, the improvement is much clearer.

1 Introduction

In recent years, there is a tremendous growth in the volume of text documents available on the Internet, digital libraries, and company-wide intranets. This had led to an increased interest in developing methods that allow users to quickly and accurately classify these types of information. Traditional text categorization methods will lead to poor categorization performance because most terms may have multiple meanings and the same concept can be described by multiple terms. For the concept-centric nature of the documents, an effective method to classify documents should relate to the semantic concepts. Many existing methods attempts to extract the latent semantic concept of the documents based on the statistical analysis, such as latent semantic indexing (LSI)[1,2], concept indexing (CI)[3], distributional clustering (DC)[4] et al.

Kernel methods (KMs) are a state-of-the-art class of learning algorithms [5], which have been applied to the fields of the supervised learning, unsupervised learning and regression problem. The best known example of KMs is support vector machine (SVM). The combination of SVM and text categorization has been pioneered by Joachins [6], and a variety of document representing models were tested to compare their effect of the performance of SVM [7]. But the approaches in [6,7] only based on the words and did not considered the semantic correlation in the words of documents. They were extended by Siolas in [8] and Cristianini in [9].

In this study, integrating kernel principal component analysis (KPCA) with SVM, it introduced a new kernel-based semantic method for text categorization. KPCA is used to reduce the dimensionality of the input space of documents, transform many co-related terms of documents to independent semantic concepts, throw off some unimportant terms and get the approximate map from the input space to a semantic concept space (SCS). And then the text categorization task was performed in SCS with SVM classifier. A similarity measurement function between the vectors in SCS was defined as the kernel function of SVM classifier, with which it need not to implement the map from the original input space to SCS explicitly.

2 Kernel-Based Semantic Text Categorization

In this section we will discuss the principle of Kernel-based Semantic Text Categorization in details.

2.1 Semantic Concept Space

There are some dependent relations between the terms and the semantic concepts in documents. The semantic concept can be inferred through the context of the words. Given a documents set $D=\{d_1,d_2,\ldots,d_n\}$, a category set $Y=\{y_1,y_2,\ldots,y_l\}$, and vocabulary $T=\{t_1,t_2,\ldots,t_m\}$, with vector space model (VSM), all the documents constitute a term by document matrix $\mathbf{X}=(\mathbf{d}_1,\mathbf{d}_2,\ldots,\mathbf{d}_n) =(\mathbf{t}_1^T,\mathbf{t}_2^T,\ldots,\mathbf{t}_m^T)^T$, where a column $\mathbf{d}$ is a document, and a row $\mathbf{t}$ is a term, every element x_{ij} is the word frequency of the term t_i in document d_j. The similarity between two documents is defined as $s(\mathbf{d}_i,\mathbf{d}_j)= \mathbf{d}_i^T\mathbf{d}_j$, and the similarity between two words is defined as $s(\mathbf{t}_i,\mathbf{t}_j)=\mathbf{t}_i\mathbf{t}_j^T$, so the matrix $\mathbf{X}^T\mathbf{X}$ contains similarities between all pairs of documents, and the matrix $\mathbf{X}\mathbf{X}^T$ contains similarities between all pairs of words. If decorrelate the correlation of the $\mathbf{X}\mathbf{X}^T$, we can get independent semantic concepts of the words.

As we known, the covariance matrix of the document set D is $\mathbf{C}=E(\mathbf{X}-E(\mathbf{X}))(\mathbf{X}-E(\mathbf{X}))^T$, where $E(\cdot)$ represents to compute the expectation. To simplify the problem, we assume $E(\mathbf{X})=\dfrac{1}{n}\sum_{k=1}^{n}\mathbf{d}_k = 0$, then $\mathbf{C}= E(\mathbf{X}\mathbf{X}^T)= \dfrac{1}{n}\sum_{j=1}^{n}\mathbf{d}_j\mathbf{d}_j^T$.

Definition 1: Given a documents set D, its vector space is $\mathbf{X}$ and its covariance matrix is $\mathbf{C}$, $\lambda_1 \geq \lambda_2 \geq \ldots \geq \lambda_q \geq 0$ are the eigenvalues of $\mathbf{X}$, $\mathbf{a}_1, \mathbf{a}_2, \ldots, \mathbf{a}_q$ are corresponding normalized eigenvector, then the ith principal component of D is $\mathbf{v}_i=\mathbf{a}_i^T\mathbf{X}=a_{1i}\mathbf{d}_1+a_{2i}\mathbf{d}_2+\ldots+ a_{ni}\mathbf{d}_n$. The vector space composed of $\mathbf{V} = (\mathbf{v}_1,\mathbf{v}_2,\ldots,\mathbf{v}_q)$ is called the semantic concept space of the document set D.

In real problem, only the first kth ($k \leq q$) eigenvectors are selected to constitute the SCS S, namely $S = \mathbf{V}^k = (\mathbf{v}_1,\mathbf{v}_2,\ldots,\mathbf{v}_k)$.

2.2 Kernel-Based Text Categorization

The basic idea of text categorization using SVM is mapping documents to a linear feature space F through a map function φ and then classifying documents by constructing separating hyperplane.

Definition 2: R is space of real numbers, and C is space of complex numbers. Given a document set D and a kernel function $k(\mathbf{d}_i,\mathbf{d}_j):D\times D \to R$, the $n\times n$ matrix $\mathbf{K}:=(k(\mathbf{d}_i,\mathbf{d}_j))_{ij}$ is the kernel matrix of $k(\mathbf{d}_i,\mathbf{d}_j)$ with respect to D. If $\sum_{ij} c_i c_j K_{ij} \geq 0$ for all $c_i \in C$, then the kernel matrix $\mathbf{K}$ is called positive, and the function $k(\mathbf{d}_i,\mathbf{d}_j)$ is called a positive definite kernel.

Proposition 1: Given a document set D and a positive defined kernel function $k(\mathbf{d}_i,\mathbf{d}_j)$, there always exists a map function $\varphi:D\to F$, which maps the data set D to a high dimensional linear feature space F and satisfies $k(\mathbf{d}_i,\mathbf{d}_j)=(\varphi(\mathbf{d}_i)\cdot\varphi(\mathbf{d}_j))$, where $(\varphi(\mathbf{d}_i)\cdot\varphi(\mathbf{d}_j))$ is the dot product in the feature space F.

Apparently the kernel function computes the similarity of all the vector pairs in the feature space F.

Suppose the separating hyperplane $f(\mathbf{d})=\mathbf{w}^T\varphi(\mathbf{d})+b$ in F split the document data set correctly, where $\mathbf{w}=\sum_{i=1}^n w_i\varphi(\mathbf{d})$ is a weight vector of the map $\varphi(\mathbf{d})$, and b is a constant. The process of training SVM classifier is to solve the optimum problem:

$$\max_{\mathbf{w}} \sum_{i=1}^n w_i - \frac{1}{2}\sum_{i,j=1}^n w_i w_j y_i y_j k(\mathbf{d}_i,\mathbf{d}_j) \qquad (1)$$

$$\text{s.t. } 0\leq w_i \leq C, i=1,\dots,n, \; \sum_{i=1}^n w_i y_i = 0.$$

where $C>0$ is a constant, y_i is the class label of the document d_i. Then we get the decision function to classify text documents

$$f(\mathbf{d})=\text{sgn}\left(\sum_{i=1}^n y_i w_i k(\mathbf{d}_i,\mathbf{d})+b\right). \qquad (2)$$

2.3 Extract Semantic Concept with Kernel Principal Component Analysis

We extract the semantic concepts in the linear feature space F to form SCS, and then classify documents in the SCS using SVM.

In the feature space F, the matrix $\Phi=(\varphi(\mathbf{d}_1),\dots,\varphi(\mathbf{d}_n))$ is corresponding to the term by document matrix $\mathbf{X}$, and the covariance matrix of the documents is:

$$\overline{\mathbf{C}}=1/n\sum_{i=1}^n (\varphi(\mathbf{d}_i)\cdot\varphi(\mathbf{d}_i)).$$

To get independent semantic concepts, we need to find Eigenvalues $\lambda\geq 0$ and $\mathbf{v}\in F\backslash\{0\}$ satisfying $\lambda\mathbf{v}=\overline{C}\mathbf{v}$.

By the same argument as above, the solution $\mathbf{v}$ lies in the span of $\varphi(\mathbf{d}_1),...,\varphi(\mathbf{d}_n)$. So we consider the equivalent equation $\lambda(\varphi(\mathbf{d}_k)\cdot \mathbf{v}) = (\varphi(\mathbf{d}_k)\cdot \overline{\mathbf{C}}\mathbf{v})$ for all $k=1,...,n$, to which there exists coefficients $\mathbf{a}^T = (a_1,...,a_n)$ such that $\mathbf{v}^T = \mathbf{a}\cdot \Phi^T = \sum_{i=1}^{n} a_i \varphi(\mathbf{d}_i)$. Let $\mathbf{K} = \Phi^T \Phi$ by $K_{ij} := (\varphi(\mathbf{d}_i)\cdot \varphi(\mathbf{d}_j))$, then $n\lambda \mathbf{K}\mathbf{a} = \mathbf{K}^2 \mathbf{a}$, namely $n\lambda \mathbf{a} = \mathbf{K}\mathbf{a}$ gives us all solutions $\mathbf{a}$ and λ.

The Eigenvectors $\mathbf{V}^k = (\mathbf{v}_1,...,\mathbf{v}_k)$ corresponding to the first kth $(k \leq q)$ maximal Eigenvalues in F constitute SCS S. Every dimension in S is a semantic concept. So the semantic concepts $\mathbf{s}$ of the document d can be extracted using following equation:

$$s_t = (\mathbf{v}_t \cdot \varphi(\mathbf{d})) = \sum_{t=1}^{n} a_t(\varphi(\mathbf{d}_t)\cdot \varphi(\mathbf{d})) \tag{3}$$

Eq. (3) projects the image $\varphi(\mathbf{d})$ on the Eigenvectors $\mathbf{v}_t$.

Now documents can be classified in S with SVM. Because the nonlinear features of documents have been transformed to the F, we only need to select a linear kernel function to compute the similarity of the vectors in S. Suppose a function $\theta: D \rightarrow S$ maps D to S, i.e. $\mathbf{s} = \theta(\mathbf{d}) = \mathbf{V}^k \cdot \varphi(\mathbf{d})$, then the kernel function used by SVM is:

$$\begin{aligned} k_S(\mathbf{s}_i, \mathbf{s}_j) &= \theta(\mathbf{d}_i)\cdot \theta(\mathbf{d}_j) \\ &= (\mathbf{V}^k \cdot \varphi(\mathbf{d}_i))(\mathbf{V}^k \cdot \varphi(\mathbf{d}_j)) \\ &= \sum_{t=1}^{k}\sum_{l=1}^{n} a_{tl}^2 k(\mathbf{d}_l,\mathbf{d}_i)k(\mathbf{d}_l,\mathbf{d}_j) \end{aligned} \tag{4}$$

It is obviously that $k_S(\mathbf{s}_i,\mathbf{s}_j)$ is a positive defined kernel, and satisfies the proposition 1. We called $k_S(\mathbf{s}_i,\mathbf{s}_j)$ as semantic concept kernel (SCK). Using the new SCK, SCS can be accessed directly in the original document vector space, and documents are classified based on the semantic concepts.

2.4 Kernelize GHA to Compute Semantic Concept Kernel

To compute SCK, the key problem is to compute the Eigenvalue λ and Eigenvector $\mathbf{a}$ of the kernel matrix $\mathbf{K}$. The large-scale Web documents which are updated continuously make it inappropriate to perform KPCA using static methods such as Eigenvalue decomposition or singular value decomposition. Through kernelizing the Generalized Hebbian Algorithm (GHA) [10], we can get the Kernelized Hebbian Algorithm (KHA) [11], which need not to restore the kernel matrix $\mathbf{K}$ and has good time and space performance to solve KPCA problem.

Let $\mathbf{U}(t) = (\mathbf{u}_1(t)^T,...,\mathbf{u}_k(t)^T)^T$, where $\mathbf{u}_i(t) \in R$. Given a random initialization of $\mathbf{U}(0)$, the GHA update rule is represented in the feature space F as

$$\mathbf{U}(t+1) = \mathbf{U}(t) + \eta(t)(\mathbf{y}(t)\varphi(\mathbf{d}(t))^T - LT[\mathbf{y}(t)\mathbf{y}(t)^T]\mathbf{U}(t)) \tag{5}$$

where the rows of $\mathbf{U}(t)$ are vetors in $\mathbf{F}$ and $\mathbf{y}(t) = \mathbf{U}(t)\varphi(\mathbf{d}(t))$, $\varphi(\mathbf{d}(t))$ is a pattern presented at time t which is randomly selected from the mapped data points $\{\varphi(\mathbf{d}_1),...,\varphi(\mathbf{d}_n)\}$, $LT[\cdot]$ sets all elements above the diagonal of its matrix argument to zero, thereby making it lower triangular. Then $\mathbf{U} \to \mathbf{V}^T$ as $t \to \infty$.

Based on KPCA, $\mathbf{u}(t)$ can be expressed as an expansion of the mapped data points $\varphi(\mathbf{d}_i)$, i.e. $\mathbf{u}_i(t) = \Phi \mathbf{a}_i(t)$. Thus we get $\mathbf{U}(t) = \mathbf{A}^T(t)\Phi^T$, where $\mathbf{A}(t) = (\mathbf{a}_1(t),...,\mathbf{a}_k(t))$ is the expansion coefficients. The ith row $\mathbf{a}_i = (a_{i1},...,a_{il})$ of $\mathbf{A}(t)$ corresponds to the expansion coefficients of the ith eigenvector of $\mathbf{K}$ in the $\varphi(\mathbf{d}_i)$. Using this representation, the update rule becomes

$$\mathbf{A}^T(t+1)\Phi^T = \mathbf{A}^T(t)\Phi^T + \eta(t)(\mathbf{y}(t)\varphi(\mathbf{d}(t))^T - LT(\mathbf{y}(t)\mathbf{y}(t)^T)\mathbf{A}^T(t)\Phi^T). \tag{6}$$

Representing mapped data points $\varphi(\mathbf{d}(t))$ as $\varphi(\mathbf{d}(t)) = \Phi \mathbf{b}(t)$ with a canonical unit vector $\mathbf{b}(t) = (0,...,1,...,0)^T$ ($b_j = 1$ represented that the mapped data point $\varphi(\mathbf{d}_j)$ is selected at time t), the update rule is

$$\mathbf{A}^T(t+1) = \mathbf{A}^T(t) + \eta(t)(\mathbf{y}(t)\mathbf{b}(t)^T - LT(\mathbf{y}(t)\mathbf{y}(t)^T)\mathbf{A}^T(t)). \tag{7}$$

Representing it in component-wise form gives:

$$a_{ij}(t+1) = \begin{cases} a_{ij}(t) + \eta(t)(y_i(t)) - \eta y_i(y)\sum_{k=1}^{i} a_{kj}(t)y_k(t) & \text{if } b_j = 1 \\ a_{ij}(t) - \eta y_i(y)\sum_{k=1}^{i} a_{kj}(t)y_k(t) & \text{if } b_j \neq 1 \end{cases} \tag{8}$$

where $y_i(t) = \sum_{k=1}^{l} a_{ik}(t)\varphi(\mathbf{d}_k) \cdot \varphi(\mathbf{d}(t)) = \sum_{k=1}^{l} a_{ik}(t)k(\mathbf{d}_k,\mathbf{d}(t))$.

3 Experiment Results

In this paper, experiments have been conducted to evaluate the classifying performance of the kernel based semantic methods in term of the measurement F1.

3.1 Experiment Setup

Two kinds of text categorization classifiers were implemented using the LIBSVM package [12]. BASE-SVM served as the baseline classifier, in which the linear kernel function was selected and all the words in the documents were used to classify documents. KPCA-SVM was implemented with the kernel-based semantic method discussed in this paper. The most frequently used Reuters-21578 collection was used in this experiment. To divide the collection into a training set and a test set, the modified Apte ("ModApte") split was applied. After preprocessing, the document number of every class used in the experiments was listed in Table 1.

Table 1. The document number of the training set and test set in six categories.

	Earn	acq	money-fx	crude	grain	trade
Training set	2709	1488	460	349	394	337
Test set	1014	630	133	160	130	106

The cosine similarity of documents was used, i.e. $s(\mathbf{d}_i, \mathbf{d}_j) = \dfrac{\sum_{k=1}^{n} x_{ki} x_{kj}}{\sqrt{\sum_{k=1}^{n} x_{ki}^2 \sum_{k=1}^{n} x_{kj}^2}}$, where $x_{ij} = tf_{ij} \cdot \log(n/df_i)$, tf_{ij} is the number of the word t_i that occurs in the document d_j, and df_i is the number of documents where the word t_i occurs. If a is the number of documents correctly assigned to this category, b is the number of documents incorrectly assigned to this category, and c is the number of documents incorrectly rejected from this category, then precision $R_p = \dfrac{a}{a+b}$ and recall $R_r = \dfrac{a}{a+c}$. The measurement $F1 = \dfrac{2R_r R_p}{R_r + R_p}$ was used to evaluate the performance of the classifiers.

3.2 Result and Analysis

In experiments, the word frequency threshold h is set to be 0.2, and the argument C in SVM is set to be 10. The polynomial kernel function $k_p((\mathbf{d}_i,\mathbf{d}_j)=(\mathbf{d}_i^T \mathbf{d}_j + b)^p$, $(p>1, p \in N)$ is applied to KPCA-SVM. One SVM classifier is induced according to each class and then is combined to form a multi-class SVM classifier. The experiment results are averaged over the results of 10 runs and shown in Table 2 and 3. The best F1 value of the KPCA-SVM and the BASE-SVM were shown in Fig. 1.

Table 2. The F1 value of six categories in the experiment with $p=3$ in polynomial kernel and different k.

	Earn	acq	money-fx	crude	grain	trade
100	0.912	0.889	0.62	0.738	0.631	0.668
200	0.928	0.895	0.633	0.741	0.658	0.683
300	0.939	0.917	0.667	0.769	0.674	0.718
400	0.941	0.916	0.673	0.769	0.671	0.723
500	0.940	0.917	0.674	0.767	0.671	0.721
600	0.937	0.914	0.671	0.769	0.672	0.721
BASE-SVM	0.935	0.889	0.656	0.747	0.66	0.704

Table 3. The F1 value of six categories in the experiment with k=300 and different p in polynomial kernel.

	Earn	acq	money-fx	crude	Grain	trade
p=1	0.934	0.884	0.657	0.751	0.661	0.7
p=2	0.935	0.901	0.659	0.757	0.665	0.709
p=3	0.939	0.917	0.667	0.769	0.674	0.718
p=4	0.934	0.914	0.675	0.763	0.691	0.733
p=5	0.936	0.915	0.673	0.766	0.687	0.73
p=6	0.936	0.913	0.675	0.766	0.684	0,731
BASE-SVM	0.935	0.889	0.656	0.747	0.66	0.704

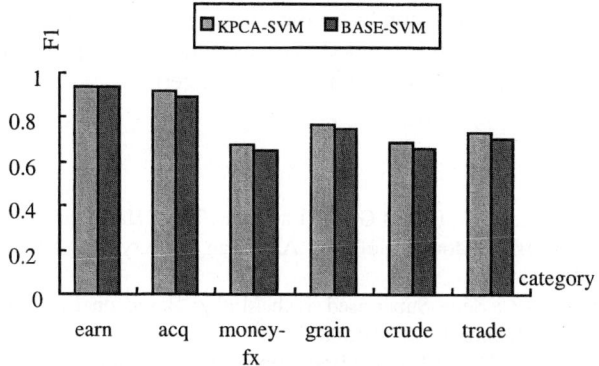

Fig. 1. The best F1 value of KPCA-SVM and that of BASE-SVM.

Table 2 shows the experiment results with different k, which indicates that a few dimensions can represent all the information of documents, and through increasing the dimensions of SCS, the F1 value rises to reach a maximum and then falls to a value equivalent to that of the BASE-SVM. In most of categories, the F1 value of the KPCA-SVM outperforms that of the BASE-SVM. But when the dimensionality is very small, the performance of the KPCA-SVM is poorer than that of the BASE-SVM because of some lost information of the documents.

Table 3 shows the results of different k for the polynomial kernel function, which indicates that using the nonlinear kernel function can get better performance than the linear kernel function. When the order of the polynomial kernel function p is 3 or 4, the performance of the SVM classifier is good, and then the increase of the order cannot improve the performance clearly. Moreover, when the training data set is small, nonlinear kernel function can enhance the performance of classification much greatly.

The best results of the KPCA-SVM and the BASE-SVM are compared in Fig.1, which shows that the results of the KPCA-SVM are much better than that of the BASE-SVM in all the categories. So the kernel-based semantic method presented in this study can improve the traditional method based on the word space. Moreover, in the case that the training data set is not efficient, the new method enhances the classification performance much greatly.

4 Conclusion and Future Work

The volume of digital documents increases rapidly in recent years, so fast and accurate methods to classify text documents are needed imminently. But the high dimensionality and the synonymous and multi-vocal words of the documents make the traditional method ineffective. In this study we proposed a new kernel-based semantic text categorization method and a new semantic concept kernel function to solve this problem. Experiment results show that this method is available and can improve the performance of the text categorization.

In future work we intend to conduct experiments at a larger scale on web data to evaluate the effectiveness of this method. We also intend to explore faster and more efficient algorithms to compute the semantic concept kernel.

References

1. Derwester S., Dumais S. T., Furnas G. W., Landauer, T. K., Harshman, R.: Indexing by latent semantic indexing. Journal of the American Society for Information Science. 41(1990)391–407
2. Gong X-J., Shi Zh-Z.: Semi-Supervised Web Mining Based on Bayes Latent Semantic Model. Journal of software. 13(2002)1508–1514
3. Karypis G. and Han E-H.: Fast Supervised Dimensionality Reduction Algorithm with Applications to Document Categorization & Retrieval. In proceeding of 9th International Conference Information and Knowledge Management. ACM press, New York, USA (2000)12–19.
4. Baker L. and McCallum A.: Distributional clustering of words for text classification. In proceedings of the 21st annual international ACM SIGIR conference on Research and development in information retrieval. ACM press, New York, USA (1998)96–103
5. Müller K.-R., Mika S., Rätsch G., Tsuda K.and Schölkopf B.: An Inroduction to kernel-Based Learning Algorithms. IEEE Transaction on Neural Networks. 12(2001):181–202
6. 6. Joachims, T.:Text Categorization with support vector Machines. In proceedings of the 10th European conference on Machines learning. Springer-Verlag, Berlin Heidelberg New York (1998)137–142.
7. Leopold E. and Kindermann J.: Text Categorization with Support Vector Machines. How to Represent Text in Input Space?. Machine Learning. 46(2002)423–444
8. Siolas G. and d'Alché-Buc F.: Support Vector Machines based on a Semantic Kernel for Text Categorization. In proceeding of the International Joint conference on Neural Networks, IEEE Computer Society Washington, DC, USA, 5(2000): 205–209.
9. Cristianini N., Shawe-Taylor J., and Lodhi H.: Latent Semantic Kernels. Journal of Intelligent Information Systems. 18(2002.):127–152
10. Haykin S.: Neural Networks : A Comprehensive Foundation (second edition). Tsinghua University Press, (2002.)392–442
11. Kim K I, Franz M O, Schölkopf B. Kernel Hebbian Algorithm for Iterative Kernel Principal Component Analysis.Technical Report No. 109. Max Planck Institute for Biological Cybernetics, (2003)
12. http://www.csie.ntu.edu.tw/~cjlin/libsvm

Mobile Computing Architecture in Next Generation Network

Yun-Yong Zhang, Zhi-Jiang Zhang, Fan Zhang, and Yun-Jie Liu

Postdoctoral Programme, Technology Department, China Unicom,
100032 Beijing, China
{zhangyy,zhangzhj,zhangf,liuyj}@chinaunicom.com.cn
http://www.chinaunicom.com.cn/

Abstract. Next generation network faces many challenges. In this paper, the new mechanism based on mobile agent is focused and the core architecture called common agent request broker architecture (CARBA) is established in order to embody intelligent collaboration and high performance in the next generation network. Then mobile agent based computing architecture (MABCA), extension of MABCA in nomadic environment and validation of MABCA openness is discussed. In the end, experiment results are given out.

Keywords: Next Generation Network (NGN), Mobile Agent, Intelligent Agent, Agent Request Broker, CARBA, MABCA

1 Introduction

With the rapid development of network and the quick increase of information, next generation network (NGN) faces new challenges: how to organize and share information efficiently, how to let the network support dynamic and changeable service, how to let network support real-time continuous media and how to support mobile computing in network. New technology is needed to meet these challenges and mobile agent technology is suitable.

Mobile agent is a program that can move from a host to another autonomously in heterogeneous network environment, it is the mixture of agent technology and distributing computing technology. While the interaction between traditional RPC client and server needs continuous communication, mobile agent can move to server and communicate with server locally and the local communication needs not occupy network resources.

If we use mobile agent to design NGN mobile computing middleware and mobile agent system communicates each other using MASIF specification [1], interoperability, mobility and intelligent collaboration will be embodied in next generation network.

We take deep researches of distributing processing systems and collaborative mobile computing, and then build highly efficient collaborative middleware – CARBA (Common Agent Request Broker Architecture) based on interoperability middleware and develop many intelligent, efficient and highly available network applications with CARBA. In the end, we discuss the CARBA in nomadic environment.

2 Mobile Computing Architecture and Its Core CARBA in NGN

As shown in fig1, four main parts are included in MABCA: agent request broker (ARB) that is soft bus, agent common facilities, agent domain interface, and agent services. MABCA also includes knowledge library that saves the knowledge got from migration process and task solving module. ARB is the kernel of MABCA, through it, mobile agent can send request and receive reply transparently. Agent facilities provide high-level services. Agent domain interface builds specific interface according to the application interface and agent services provide some necessary low-level agent service.

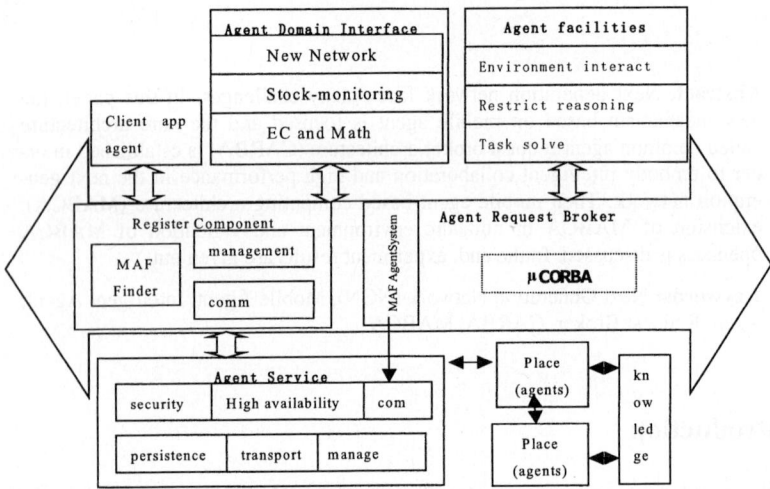

Fig. 1. MABCA Architecture.

2.1 Agent Request Broker (ARB)

ARB is the kernel of CARBA; it transfers the agent's request and reply. In order to achieve interoperability in heterogeneous network environment, MAFAgentSystem and MAFFinder interfaces are included.

MAFAgentSystem interface is the main interface of ARB; it is responsible for receiving agent, listing agents/places, getting MAFFinder interface, getting agent system type and getting agent status.

MAFFinder interface is responsible for managing all registered agent information. It includes the following services:

- Management: Provides the location of agent in region.
- Communication: Provides agent's communication between regions.
- MAFFinder: Provides services such as registering, unregistering and querying agent.

2.2 Agent Facilities

Agent Facilities provide some high-level shareable agents such as task solving agent, environment interacting agent and restrict reasoning agent to build application.

Task solving agent and restrict reasoning agent include task's reasoning method and rules. Environment interacting agent implements the semantic of ACL (agent communication language) and ensures right communication between agent and agent facilities.

2.3 Agent Domain Interface

Agent domain interface is the specific application oriented interface such as the new generation network, electronic commerce, remote teaching and stock-monitoring field.

2.4 Agent Services

Agent services provide some basic low-level services such as:
- security: Uses access control to support internal security.
- high availability: Provides services such as fault tolerance, Quality of Service (QoS) and load balance.
- persistence: Enables the storage of agents and places on a persistent medium. In this way, it is possible to recover agents or places when needed, e.g. when an agency is restarted after a system crash.
- management: Provides the services such as creating agent, deleting agent, suspending agent, activating agent and copying agent.
- transportation: Provides the services such as the agent's serialization, migrating agent to destination region and restarting agent in destination region.
- communication: Provides some communication modes such as synchronous, asynchronous, dynamic and multicast.

2.5 Agent Request/Service

There are some steps between client agent's request and server's reply:
- initialization: MAFFinder interface uses register_agent () function to register agent's information.
- MAFAgentSystem interface uses receive_agent() function to receive client agent's request, uses get_agent_info() function to get agent's type, and uses get_authinfo() function to authenticate. If the authentication fails, error will be returned, otherwise MAFAgentSystem interface will use get_MAFFinder () function to get the reference of MAFFinder interface and goto the following steps.
- MAFFinder interface uses the information in the register library and uses lookup_place(), lookup_agent_system(), lookup_agent() functions to get the destination information in the client agent's request.
- MAFAgentSystem interface moves client agent to the destination, interacts with the server and returns reply.

All the functions listed above can be defined with IDL (Interface Definition Language) language.

3 MABCA's Intelligent Collaboration

To implement the intelligent collaboration, ACL should be imported in CARBA. Two solutions are put forward: implementation of ACL module in CARBA to embody ACL directly and construction of IDL/ACL gateway to embody ACL in CARBA indirectly. The former is complete and the later is convenient while has low efficiency.

3.1 ACL Based Approach

The ACL module in CARBA consists three sub modules: common basic performative module, network performative module and facilitation performative module [2].

3.2 IDL/ACL Gateway Based Approach

In the agent research field, FIPA [3] is a very important organization besides OMG, specification of FIPA embody intelligence well. The function of IDL/ACL gateway is mapping the CARBA services to FIPA specification's format. Following mappings between CARBA and FIPA must be identified:

- CARBA region corresponds to FIPA domain
- CARBA agent system corresponds to FIPA agent platform
- CARBA MAFFinder corresponds to FIPA directory facilitator
- CARBA MAFAgentSystem corresponds to FIPA agent management system
- CARBA resource corresponds to FIPA agent wrapper

4 MABCA High Availability Services

4.1 Fault Tolerance Service

Fault tolerance is very important in MABCA system. Possible agent failure because of network failure and long-time computer shutdown must be considered in MABCA. Three fault tolerance strategies are included in MABCA: (1) creating many backup of one task in network and compare them after the end of task. (2) Central fault tolerance: one server keeps the original backup of mobile agent and resends original backup to disabled mobile agent. (3) Distributing fault tolerance: dividing the responsibility of fault tolerance into whole network.

4.2 QoS Service

MABCA provides many services needed in distributed computing, but with the application of MABCA in multimedia environment and real-time NGN environment, the end–to-end QoS guarantee is the key. We use QoS and its stream specification; plug a QoS agent in CARBA ARB to be responsible for parameter translation between application level, OS level and network level. Also, the QoS agent interacts with RSVP protocol.

4.3 Load Balance Service

To increase the throughput of MABCA, it is necessary to import load balance. Similar to the QoS service, we plug load balance agent in the CARBA ARB. When receive client agent request, the load balance agent forwards the quest to other service agents dynamically using ant colony system algorithm.

5 CARBA in Nomadic Environment

With the rapid development of wireless technology, mobile computer network emerges these years. Many companies use these wireless networks to carry through distributing transaction. But in nomadic environment, congestion control and recovery of the loss package mechanism of TCP/IP are not efficient because of the burst error and delay [4]. To solve the above problem, OMG and some companies publish the wireless CORBA specification [5]. Hard mobile middleware μCORBA is developed based on this specification. If we import mobile agent technology into the hard mobile middleware, it will decrease the transfer delay and increase the network performance further. MABCA and the idea of μCORBA is integrated to build the mobile MABCA architecture.

6 Validation of MABCA Openness

How about the openness of MABCA? ODP specification is used to validate the openness of MABCA.

6.1 Static Relation Between CARBA System and ODP

The static relation between CARBA system and ODP is the relation between CARBA system architecture and ODP engineering viewpoint. The relation is shown in table1.

Table 1. Relation between CARBA architecture and ODP engineering viewpoint.

CARBA System architecture	ODP engineering viewpoint
Region	IRMD
MAFFinder	
CoreAgency	Nucleus
MAFAgentSystem	
Agent System	Node
Place	Capsule
Agent	Objects and modules

6.2 Dynamic Relation Between CARBA System and ODP

The dynamic relation between CARBA system and ODP is the relation between CARBA system action and ODP function. The relation is shown in table2 and table3.

7 Experiment and Summary

CARBA prototype is developed and tested with chain stores model. Supposing agent1 is the depot director of ChengDu branch and agent2 is the depot director of BeiJing

Table 2. Relation between CARBA Finder System method and RM-ODP function.

CARBA MAFFinder method	RM-ODPfunction
Register_agent	Engineering
Register_agent_system	Interface Reference
Register_place	Tracking Function
Lookup_agent	Trading Function
Lookup_agent_system	
Lookup_place	
Unregister_agent	Engineering
Unregister_agent_system	Interface Reference
Unregister_place	Tracking Function

Table 3. Relation between CARBA Agent system method and RM-ODP function.

CARBA MAFAgentSystem method	RM-ODPfunction
Create_agent	Cluster Management Function
Fetch_class	RM-ODP technology viewpoing
Find_nearby_agent_system_of_profile	Trading Function
Get_agent_status	Cluster Management Function
Get_agent_system_info	Node_Management
Get_authinfo	Authentication Security Function
GetMAFFinder	Node Management
List_all_agents	Trading Function
List_all_agents_of_authority	Trading Function
List_all_places	Trading Function
Receive_agent	Migration Function
Resume_agent	Cluster Management Function
Suspend_agent	Cluster Management Function
Terminate_agent	Cluster Management Function
Terminate_agent_system	Node Management Function

branch. Now, the ChengDu store lacks of some goods and wants to query that whether BeiJing store has these goods. Firstly, agent1 sends ask-If performative to agent2 to query whether agent2 has these goods. If agent2 has these goods, agent2 will send reply performative to agent1 and agent1 will create a new agent to look for goods (agent2 will authenticate agent1 for the security reason and if authentication fails, ageent2 will send deny performative to agent1). Finally the new agent sends the goods information to agent1. If agent2 does not have these goods, it will send sorry performative to agent1 and agent1 can contact the city using the "reply-with id2" field of the sorry performative. We focus on the difference of the tow systems and validate the efficiency, high availability and intelligence.

In order to test the scalability of mobile CARBA, increment test is carried through. The network environment is wireless network (2M bandwidth and 11M bandwidth). We add the subsystems from 2 to 6 dynamically; table4 shows the information querying time.

Table 4. Increment test results.

n	Time of returning first data (s)						Time of returning all data (s)					
n	mBCBA		wCORBA		CORBA		mBCBA		wCORBA		CORBA	
m	11M	2M	11M	2M	11M	2M	11M	2M	11M	2M	11M	2M
2	1.25	1.39	1.71	2.89	1.89	3.23	1.56	1.70	1.84	7.93	4.67	10.44
3	1.27	1.48	1.88	3.01	2.10	4.87	1.74	1.78	4.65	12.37	6.99	15.67
4	1.31	1.53	1.93	5.67	2.38	7.44	2.02	2.85	6.97	17.44	8.14	21.88
5	1.34	1.59	2.44	8.86	2.79	10.89	2.37	3.88	9.10	25.55	10.21	30.52
6	1.38	1.71	3.11	13.95	3.20	15.87	2.63	4.01	11.34	38.22	12.68	45.14

Results show that mobile MABCA has better performance than μCORBA and CORBA, with the network bandwidth decreasing, the performance of CARBA does not change obviously but the performance of μCORBA and CORBA changes evidently. With the increment of subsystems, mobile MABCA has the ability of scalability, but the time of querying information in μCORBA and CORBA increases rapidly. Also we can find that the difference between time of returning all data and time of returning first data in mobile CARBA is not big, but in μCORBA and CORBA, the difference is big.

Mobile agent is different from RPC; its unique property brings tremendous renovation to mobile computing in NGN. Now, most of researchers research mobile agent from AI so it is hard to make it applicable. We research mobile agent from distributing computing and hope that through the discussing of MABCA, researchers can pay more attention to the utilization of mobile agent.

References

1. Zhang Yunyong, Mobile agent and its application, Tsinghua university press, Beijing, January 2002
2. Zhang Yunyong, Liu Jinde, Research of Mobile Agent Interoperability, Computer Science, Vol28, No.9, 2001 74-77
3. FIPA: FIPA98 Draft Specification: Part11: Agent Management Support for Mobility, FIPA8415, Version 0.3, Foundation for Intelligent Physical Agents, April 1998
4. H.Balakrishnan, Improving Reliable Transport and Handoff Performance in Wireless Networks. Wireless Networks, 1(4), 1998 4-10
5. OMG, Wireless Access and Terminal Mobility in CORBA, Revised Submission, OMG Document telecom/2001-01-1

A Study of Gridifying Scientific Computing Legacy Codes*

Bin Wang[1], Zhuoqun Xu[1], Cheng Xu[1], Yanbin Yin[2],
Wenkui Ding[1], and Huashan Yu[1]

[1] Department of Computer Science and Technology,
School of Electronics Engineering and Computer Science,
Peking University, 100871, Beijing, P.R.China
wbin@pku.edu.cn

[2] Center of Bioinformatics (CBI),
College of Life Sciences,
Peking University, 100871, Beijing, P.R.China

Abstract. In pre-Gridservice times, conventional way of pooling domain expertise is by sharing legacy codes. With the introduction and wide adoption of OGSA framework and OGSI specification, Grid computing has entered a new era of service-orientation, in which implementation of sharing domain expertise resorts to portable, interoperable Grid services. So comes the challenge to gridify existing scientific computing legacy codes (to convert to OGSI--compliant Grid services). This paper analyzes features of scientific legacy codes and proposes a design scheme of gridificaiton. The design scheme consists of three parts: gridified computing services; scientific computing Operation Providers(OPs); "on-demand" scientific computing factories. The design solution has been successfully applied to Bioinformatics field. Application scope tests show that the design solution applies to most types of scientific computing legacy codes. Performance experiments find that the proposed gridification mechanisms bring about only slight efficiency loss.

1 Introduction

Grid computing has its origin from the scientific community. It is one of the major motivators and missions for Grid computing to resolve e-science problems [1]. In pre-Gridservice times(before the introduction of OGSI-compliant Grid services), major way of realizing sharing in scientific computing community is to share legacy code applications and data. With the popularizing of OGSA framework [2] and OGSI specification, Grid computing has entered a new stage, which is featured by service-orientation. In the form of Grid services, Grid developers are able to implement Grid functionality on the framework of Web services, characterized by good interoperability, portability, and language neutrality. Web services provide a standard means of interoperating between different software applications, running on a variety of plat-

---

* The work is supported by National Natural Science Foundation of China under contract 60173004 and 60303001.

forms and/or frameworks. After the transfer to service-oriented architecture framework, means of achieving sharing to build a Grid moves to creation and invoking of Grid services.

There are a variety of legacy codes in fields of scientific computing. These codes have seen wide application and distribution, playing a vital role in scientific researches. At present, a big obstacle facing scientific domain grid developers is how to gridify existing scientific computing legacy codes (to convert to OGSI-compliant Grid services).

Regarding the subject of "gridifying existing scientific computing legacy codes", this paper proposes a design scheme to gridify existing scientific computing legacy codes. According to the design solution, gridified scientific computing services implement three interfaces(or "portTypes" in WSDL): OGSI GridService interface, LegacyCodePortType interface and ManipulationPortType interface. GridService interface ensures OGSI-compliance. LegacyCodePortType interface is a wrapper which encapsulates implementations of legacy codes. It invokes requested legacy code and detects running state of computing processes. ManipulationPortType interface is for manipulating computing processes, such as terminate or restart a computing process.

The gridification design has been implemented and successfully applied to the field of Bioinformatics. With this design solution, experiments show that it applies to most types of scientific computing legacy codes. Performance tests indicate efficiency loss is puny.

This paper makes the following contributions:

- fulfils a key step for Grid computing moving towards service-oriented era, which implements an advancement in sharing means from sharing of legacy code applications to creation and invoking of legacy code Grid services;
- proposes a common interface for all the scientific computing legacy codes which abstracts main function interfaces of all the scientific computing legacy codes, so different legacy code can be used in a uniform way, hiding varied implementations;
- proposes and implements "on-demand" factory, which is able to produce different Grid service instances according to service requestors' demands;
- proposes and implements a solution to service information advertisement in which a factory is responsible for publishing information about gridified computing service instances the factory can produce;
- provides a uniform monitor and manipulation interface for Grid services in a Grid container, so as to coordinate and control the running of services under a common framework

It is noted that the proposed gridification solution in this paper applies not merely to scientific computing legacy codes and can be easily extended to applications in other fields.

The rest of the paper is structured as follows: Sect. 2 reviews background related to conventional way of scientific computing and Grid computing in the service-oriented era. Sect. 3 presents a design solution to gridify existing scientific computing legacy codes. Sect. 4 describes application of our design scheme to Bioinformatics field. Sect. 5 deals with experiments on applicability scope and performance evaluation. Sect. 6 concludes the whole paper and expects future work.

2 Background

2.1 Conventional Way of Conducting Scientific Computing

Conventional way of conducting scientific computing is sharing legacy codes in the form of library-style toolkit packages.

One common practice is that widely used computing legacy codes of a science domain are put in a Web/FTP server known to the domain community for download. The community members can download, install legacy codes and then use them to do computing locally.

But there are many situations where local computing power is insufficient. So another common practice is that a group of domain legacy codes are installed and configured in a highly capable computational resource (e.g. a super computer server, a Beowulf cluster) and then the computational resource are granted access to users in a certain range. For example, users within a university are authorized to log on via Telnet or SSH protocol to the computing server of the university. The users upload source data files, manually start computing and download the result data files on the computing server.

2.2 Grid Computing in the Service-Oriented Era

Since its birth, Grid computing has undergone several changes and evolutions. Not least of these is the move towards a service-oriented architecture, which was based on the OGSA(Open Grid Service Architecture) [3].

The concept of Grid Service is at the core the service-oriented OGSA framework and OGSI specification. A Grid Service is a Web Service that conforms to a set of conventions that provide for the controlled, fault resilient and secure management of stateful services. OGSA (Open Grid Service Architecture) addresses architectural issues related to the requirements and interrelationships of Grid Services. In this way, OGSA further refines the Web Services model to focus on the features required by Grid infrastructures and applications. OGSI (Open Grid Service Infrastructure) addresses detailed specifications of the interfaces that a service must implement in order to fit into the OGSA framework [4].

Grid services are characterized (typed) by the capabilities that they offer. A Grid service implements one or more interfaces, where each interface defines a set of operations that are invoked by exchanging a defined sequence of messages. Grid service interfaces correspond to portTypes in WSDL. Grid services can maintain internal state for the lifetime of the service. The existence of state distinguishes one instance of a service from another that provides the same interface. The term Grid service instance is referred to as a particular instantiation of a Grid service.

3 Design

3.1 Characteristic Analysis of Scientific Computing Legacy Codes

Prior to gridifying scientific computing legacy codes, an analysis is necessary to study characteristics of these legacy codes in various scientific domains. The legacy codes have some characteristics in common:

- single interface: most scientific computing legacy codes are written in C, Fortran, or parallelized with MPI, OpenMP. They have but one main function as an entry into specific computing implementation. Users invoke the main function with input parameters and get exitValue of the function as output. Input parameters usually specify input and output data files as well as other configuration parameters.
- "dark box" operation: from the perspective of users, internals of legacy codes are invisible, just like a "dark box". Users know the functionality of legacy codes, but, in many cases, have no idea about the interior implementation. Legacy codes are just like processes which produce a data transformation from a set of inputs to a set of outputs.
- computationally intensive: involves much floating calculations, and occupies considerable CPU time [5].
- time-consuming: takes a long time to run their courses, ranging from hours to days, even as long as weeks, months.
- operate on large, complex, and heterogeneous data [6]:
  - input and output data files may consume considerable amount of disk space
  - often need data transformation beforehand
  - may produce large amount of scratch result data needing to archive when computation in progress

3.2 Goals of Gridifying Legacy Codes

The course of gridifying legacy codes is to convert scientific computing legacy codes to Grid services. The following goals are set for the gridification design:

- sacrifice no original functionality:
  - retains full scientific computing capability without any modification to legacy codes internals
  - minimizes efficiency losses derived from service gridification mechanism
- OGSI-compliant:
- user-friendly and fault-tolerant:
  - state monitoring: support state monitoring. Scientific computing grid services keep watching status of computing processes. Notification of status would sent to users if any change occurs or at a certain interval
  - runtime manipulation: allows for runtime control over computing processes, such as termination, restart etc
- wide applicability: can be applied to most scientific computing legacy codes

3.3 A Design Scheme

Based on the above analysis and consideration, a design scheme of gridifying scientific computing legacy code is proposed, which consists of three parts:
1) Gridified Computing Services
2) Scientific Computing Operation Providers
3) "On-demand" Scientific Computing Factory

3.3.1 Gridified Computing Services

We call Grid services we get after gridifying existing scientific computing legacy codes "gridified computing services". A gridified computing service implements three interfaces(or "portTypes" in WSDL): OGSI GridService interface, LegacyCodePortType interface and ManipulationPortType interface(optional).

- GridService interface ensures OGSI-compliance.
- LegacyCodePortType interface

This interface serves as a wrapper which encapsulates different implementations of scientific computing legacy codes. It invokes requested legacy code, detects running state of computing processes and sends status notification to parties interested. The interface consists of an operation, execlegacycode() and a ServiceData, ComputingInstanceStatus:

  - execlegacycode(): abstracts main program interfaces of all the scientific computing legacy codes and provides consistent operation for invocation.
  - ComputingInstanceStatus: a customized ServiceData, which tracks runtime information about the associated scientific computing process.

- ManipulationPortType interface

This interface encapsulates manipulation methods of scientific computing processes, consisting of two operations, terminatejob() and rerunjob():

  - terminatejob(): terminate the scientific computing process
  - rerunjob(): restart the scientific computing process

3.3.2 Scientific Computing Operation Providers

Delegation approach is employed to implement scientific computing legacy code Grid services. Scientific computing legacy codes are wrapped as different scientific computing OPs. In addition to OperationProvider interface, scientific computing OPs implement LegacyCodePortType interface and ManipulationPortType interface(optional).

Then scientific computing OPs can be loaded in a modular way to different computing services. For example, Laplace legacy code is wrapped as Laplace OP, BLAST as BLAST OP. As a result, scientific OPs, derived from different legacy codes, constitue a LegacyCode OPs' repository.

3.3.3 "On-Demand" Scientific Computing Factory

Considering scalability, convenience and common features of various scientific computing legacy codes, a customized "on-demand" scientific computing factory is proposed. "On-demand" denotes that the factory can create different gridified computing Grid service instances requested by users, in contrast to factories in OGSA framework, which can create only one type of service instance.

The "on-demand" factory performs the following two jobs:

- create different service instances on demand:

According to the scientific computing type specified by a user, the factory service loads corresponding scientific computing OP from OPs' repository, together with GridServiceImpl, creates a computing service instance desired by the user and return the GSH of the instance to the user.

- information advertisement

The factory service publishes to a service Registry information on all the gridified scientific computing services hosted by the computational resource where the factory resides in. The registered information, materialized in XML file style, consists of the following two aspects of information:

- information on factory service and hosting computational resource: name, handle, host name, host information
- information on gridified computing services: name, description, portTypes, arguments, invocation sample, etc.

3.4 Running Scenario

As shown in Fig.1, after the implementation of gridification, a typical running scenario can be described as follows:

(1) A factory service publishes service information to a Service Registry
(2) A service requestor discovers services wanted in a Service Registry
(3) The service requestors sends createservice request to the factory
(4) The factory service creates a scientific computing service instance as requested
(5) The factory returns the GSH of the created service instance to the service requestor
(6) The service requestor invokes the computing service instance to perform computation

4 Application in Bioinformatics

On the platform of Globus Toolkit 3.02, the design scheme(in section 3) is implemented with Java and C language as main development tools.

After the prototype implementation, we applied our gridification solution to Bioinformatics and obtained satisfactory results.

Bioinformatics is a newly emerging interdisciplinary research area involving biology, computer science, and applied mathematics [9]. One of the major concerns of Bioinformatics is the comparison of nucleotide or protein sequences from the same or different organisms. By finding similarities between sequences, scientists can infer the function of newly sequenced genes, predict new members of gene families, and explore evolutionary relationships. Now that whole genomes are being sequenced,

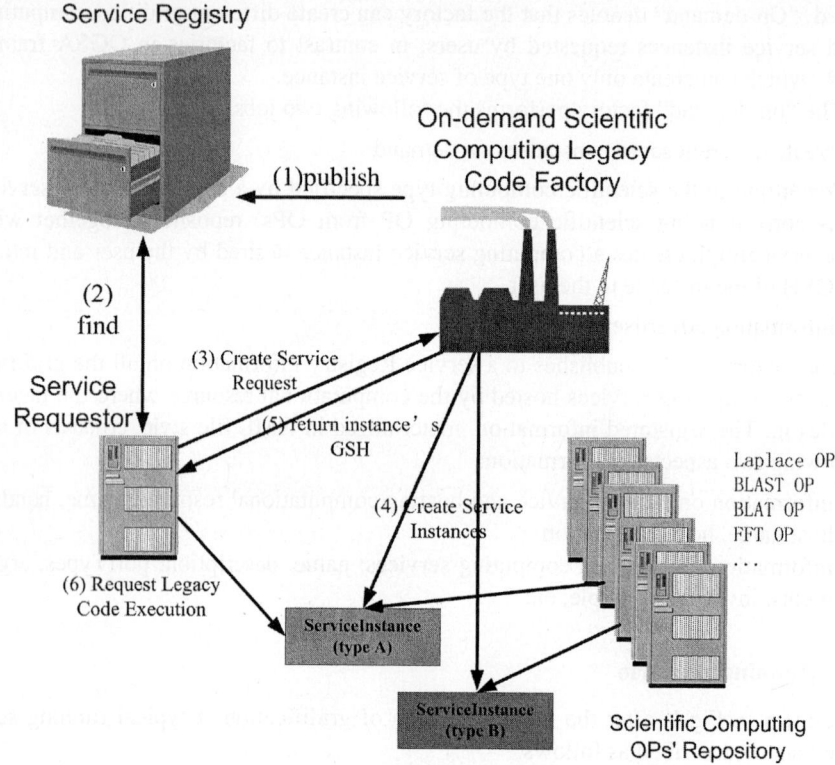

Fig. 1. Running Scenario.

sequence similarity searching can be used to predict the location and function of protein-coding and transcription-regulation regions in genomic DNA.

BLAST(Basic Local Alignment Search Tool) is the tool most frequently and popularly used for calculating sequence similarity. BLAST provides a method for rapid searching of nucleotide and protein databases. Since the BLAST algorithm detects local as well as global alignments, regions of similarity embedded in otherwise unrelated proteins can be detected. Both types of similarity may provide important clues to the function of uncharacterized proteins. BLAST comes in variations for use with different query sequences against different databases.

BLAT(The BLAST-Like Alignment Tool) is another bioinformatics software tool for sequence comparison. BLAT performs rapid mRNA/DNA and cross-species protein alignments. BLAT is more accurate and faster than popular existing tools for mRNA/DNA alignments and for protein alignments at sensitivity settings typically used when comparing vertebrate sequences.

Our application work in Bioinformatics consists of gridification of BLAST legacy code and BLAT legacy code. BLAST is gridified as BLAST Grid service, and BLAT as BLAT Grid service, which can be deployed into OGSI-compliant containers for invocation from service requestors.

5 Experiment Result and Evaluation

5.1 Application Scope

In order to examine the applicability scope of our design solution, tests have been conducted with various kinds of scientific computing legacy code applications. Test result is shown in Table 1.

Table 1. Application Scope Test Result.

Legacy Code Type	Applicable
C/C++ Application	Yes
Fortran Application	Yes
MPI+C Parallel Application	Yes
MPI+Fortran77 Parallel Application	Yes
Bash Shell Script Application	Yes
Perl Script Application	Yes

Based on the test result of Table 1, it can be concluded that our gridification design solution applies to most kinds of scientific computing legacy codes.

5.2 Performance Evaluation

Experiments are conducted to evaluate the performance of our proposed gridification mechanisms. Legacy codes are run in two ways: directly and as gridified Grid services. Then time used in either way is recorded and compared.

Fig. 2 depicts change of portion variation of service overhead with total time. Here service overhead corresponds to difference between time used by a legacy code run

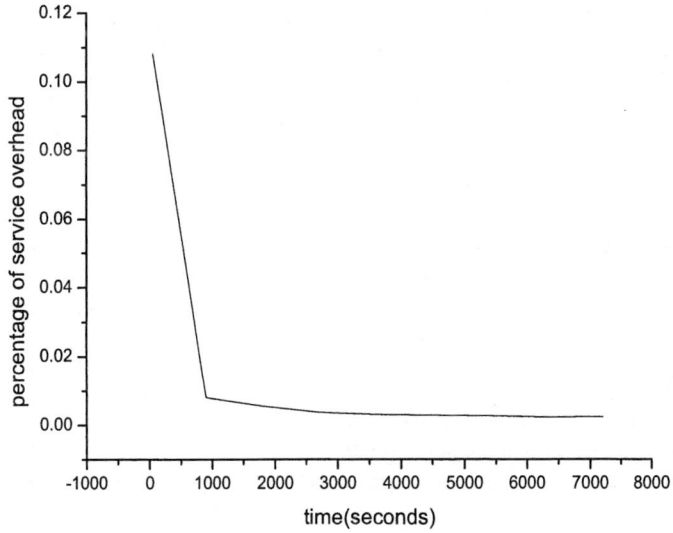

Fig. 2. Percentage Variation of Service Overhead with Time.

as a Grid service and time used by the same legacy code run directly. As shown in Fig. 2, as total running time of a legacy codes increases, portion of service overhead decreases sharply. It is also noticed that the portion of service overhead is lower than 0.3 percent when the whole running time of a legacy code exceeds an hour.

Based on performance tests, it can be concluded that the proposed gridification mechanisms brings about trifling time loss, and the portion of service overhead decreases as whole running time increases. The efficiency loss can be ignored when the whole running time is longer than a certain amount.

6 Conclusion and Future Work

Our work solves "the last mile" challenge during the course of Grid deployment and spreading, bridging a gap between Grid service platforms and Grid service providers. Scientific domain Grid builders, who may be unfamiliar with Grid technologies, can take advantage of our work to gridify domain-specific legacy codes and establish their domain Grids.

Legacy code grid services are building blocks which can be used alone or with other services to form composite services or scheduled in a workflow.

In the future, more interfaces(or portTypes) can be added to incorporate more functionality of fault-tolerance and workflow.

References

1. I. Foster.: The Grid: A New Infrastructure for 21st Century Science," Physics Today, vol. 55, no. 2, 2002, pp. 42-47.
2. I. Foster, C. Kesselman, J. Nick, S. Tuecke.: The Physiology of the Grid: An Open Grid Services Architecture for Distributed Systems Integration. June 22, 2002.
3. Mark Baker.: Ian Foster on Recent Changes in the Grid Community. IEEE Distributed Systems. Feb. 2004 http://dsonline.computer.org/0402/d/o2004a.htm
4. S. Tuecke, K. Czajkowski, I. Foster, J. Frey, S. Graham, C. Kesselman, T. Maguire, T. Sandholm, P. Vanderbilt, D. Snelling.: Open Grid Services Infrastructure (OGSI) Version 1.0. Global Grid Forum Draft Recommendation, 6/27/2003.
5. Wang Bin et al.: A Grid-computing-based Solution to Science&Engineering Computing and its Implemenation. ISTM/2003: 5th International Symposium on Test and Measurement, June 2003, Vol.1, pp 581-586
6. Bin Wang, Ping Chen, Zhuo-qun Xu.: An Improved Solution to I/O Support Problems in Wide Area Grid Computing Environments. GCC2003: 2nd International Workshop on Grid and Cooperative Computing, LNCS 3032,.Springer-Verlag Berlin Heidelberg, December 2003, pp. 677–684
7. Carl Kesselman.: Open Grid Services Architecture: A Tutorial. GCC2003: Shanghai, Dec.7, 2003 (POWERPOINT)
8. Borja Sotomayor.: The Globus Toolkit 3 Programmer's Tutorial. http://www.casa-sotomayor.net/gt3-tutorial/
9. National Center for Biotechnology Information(NCBI). http://www.ncbi.nlm.nih.gov/

Aegis: A Simulation Grid Oriented to Large-Scale Distributed Simulation*

Wei Wu, Zhong Zhou, Shaofeng Wang, and Qinping Zhao

The Key Laboratory of Virtual Reality Technologies, Ministry of Education
BeiHang University, Beijing 100083, P.R. China
{wuwei,zz,wangsf}@vrlab.buaa.edu.cn

Abstract. This paper focuses on the resource management technology of large-scale distributed simulation based on grid service. As a data grid, Aegis is established to provide computing service environment and uniform view of various data resources for simulation applications. This paper starts from demands of distributed simulation, introduces the architecture of Aegis simulation grid, discusses the realization of main modules, and establishes a platform of resource management oriented to distributed simulation. Finally, it describes the design and implementation of simulation computing service with the example of tactical notification based on strategy rules, and makes experiments of distributed simulation supported by Aegis simulation grid.

1 Introduction

Distributed simulation has been widely applied in military field due to its characteristic of supporting mass users in shared environment. Since the 1990s of last century, HLA[1] technology has been followed universally as a de-facto standard for modeling and simulation. As a new developing technology, grid has become one of the hottest topics in computer science after 2000. Some research institutions attempt to adopt grid technology in HLA simulation to solve some problems of current HLA simulation.

Early in 1998, California Institute performed SF Express[2] warfare simulation integrating Globus with subsidy of DARPA (Defense Advanced Research Projects Agency). Based on DIS (Distributed Interactive Simulation) protocol, it enhanced flexibility and robustness of the system by means of dynamic management function of Globus in aspects of scene distribution, resource deployment, information service, log service, supervision and fault tolerance. However, it was based on old DIS protocol and supported only a small quantity of super computers. Comparing with it, modern distributed simulation has to bear high extensibility and caters for the requirements of large-scale simulation.

As part of European CrossGrid[3], Katarzyna Zajac etc. enhanced HLA RTI architecture on the basis of Grid at three levels: RTI, federation and federates after analyzing the drawbacks of RTI 1.3-NG supported by US DMSO (Defense Modeling and Simulation Office) [4]. Grid applications at these three levels are mainly based on modifications to the drawbacks of RTI 1.3-NG proposed by the author and such research fields as data transfer, creating grid service for HLA-based simulation, etc. still need further validation.

---

* This paper is supported by National 973 Program of China under the grant No.2002CB312105 and China Education and Research Grid (ChinaGrid) under the grant NO.CG2003-GA004.

Wentong CAI etc. proposed to provide HLA-based distributed interactive simulation with load balance service by establishing LMS (Load Management System) over grid. This method employs grid for load management among simulation applications, in which Globus performs connection authentication, resource discovery and task assignment while RTI still serves data transfer between federates without affecting transfer efficiency [5]. Because little computation occurs in processing single message and the main bottleneck of systematic resource consumption lies in large quantities of messages in usual HLA distributed simulation, the advantages of load balance for distributed simulation may be restricted to special applications and this method need further studies in practice.

The emergence of large-scale distributed simulation in China has brought out many key technology researches. Regular means of resource replication in each host and the means of central resource database cannot meet the requirements for accessing, storing, transferring and managing such massive data resources as terrain or models. Simulation computation, which demands data from a lot of simulation nodes, brings additional computation and communication overhead to simulation nodes, and thus affects the simulation advance and run-time efficiency of the whole system.

This paper proposes a data grid oriented to distributed simulation, which provides HLA-based distributed simulation with resource management service and computing service to meet the requirements of large-scale simulation and improve the scalability. In this paper a simulation grid called Aegis is established, whose main modules and the realization of simulation computing service are introduced in detail. Finally experiment results are given to validate the system.

2 Aegis Simulation Grid Architecture

The terrain and model data in distributed simulation increase with the rise in scale and entity types. However, terrains in schedule and entity models loaded by a single federate during a period of time are only part of the whole. There are two modes at present: host replication and central database mode. Ordinarily distributed simulation adopts host replication mode. The simulation scale is unavoidably limited by such restrictions as the data maintenance and consistency checking. The central terrain and model database limits the number of clients and the server usually becomes the bottleneck. Standard access interface which can accommodate different simulation applications is an important factor for system scalability.

Large-scale simulation involves numerous hosts and applications locating at different nodes far apart. The collaboration in simulation run-time period is complicated and the initialization process even more. The management of consistency of resource distribution, logs, configuration files, etc. requires careful confirmation and consumes a lot of manpower and time. The simulation computation of large-scale simulation, which is based on the global information or requires data information of a large amount of simulation nodes as input, is a troublesome issue.

The simulation grid supported and HLA-based distributed simulation concept model is shown as Fig. 1. Federates join the federation execution, perform communication of simulation data and get simulation advance via. RTI. The simulation architecture still accords with HLA standard. Aegis simulation grid integrates various resources required in simulation and provides uniform data access methods to support simulation computing service of specified application with dynamic data.

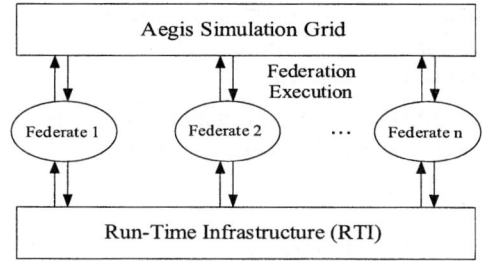

Fig. 1. Conceptual model of grid-supported distributed simulation.

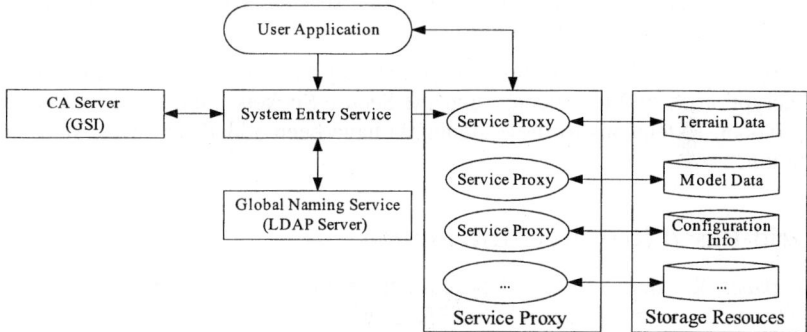

Fig. 2. Aegis simulation grid architecture.

Aegis simulation grid integrates static resources, such as terrain data, entity model, multiple media information and configuration information, etc. in the distributed simulation, and provides uniform data access methods via. grid service. At the same time, it can collect and integrate the diversiform dynamic data required in simulation computation through the registration of metadata computing directory service.

The Aegis simulation grid architecture is shown as Fig. 2. Aegis simulation grid comprises system entry service, global naming service, replication management service, security certification center (CA, Certificate Authority), service proxy and storage resources. The system entry service is the entry of the whole system via which user applications request for service. The global naming service is responsible for the management of metadata information of the system. The service proxy screens the diversity of storage resources and provides uniform storage access interface for the system. In addition, it supports file replication management and serves efficient transfer. Offering security authority, CA is mainly responsible for security and control of access to data of the system, and at the same time records user's registration information. The storage resources mainly comprise blocked terrain model and texture, entity model library, configuration information of simulation application, logs, global or common files, dynamic information database, etc.

System entry service is the uniform access interface for users, providing access to CA authority center, MDS service and replication management service.

Global naming service defines metadata information, including a series of files such as stored resources, terrain, rules, etc., as well as user information. Global naming service also provides such functions as locating metadata information and updating metadata information, and adopts LDAP metadata computing directory service (MDS) to manage metadata information.

Replication management service of Aegis mainly comprises replica creation, replica catalog, replica selection and replica management. Replication management service helps to improve the overall performance of the system and balance the load of the system.

Service proxy provides users with various data service. System entry service searches metadata information server according to user's request, gets the metadata information about data storage information and corresponding access method, and then generates service proxy. The data operation among service proxy, storage resources and user application is supported by GridFTP.

GSI supported security is an important aspect in grid service. GSI is based on public key encryption, X.509 certificates, and the Secure Sockets Layer (SSL) communication protocol. Extensions to these standards have been added for single sign-on and delegation.

3 Grid-Based Resource Distribution

In this paper, the shared terrain data required by the simulation entities are managed by means of resource service and GridFTP service in Grid technology. It adopts a new grid-based transfer and exhibition method for models. The method needs two scene segmentations in practice. The first segmentation occurs in the stage of modeling. The second scene segmentation is performed on grid servers. The scene should be segmented into nodes ease to transfer in the network, and the priority among nodes should be considered [6].

The basic scheduling unit for 3-D information is 3-D scene. The model for large-scale grid-based shared terrain distribution is shown as Fig. 3.

The 3-D scene is constructed as this: the VirtualUniverse on top level with Locales below and Locale is the root of a whole scene; each Locale may contains several BranchGroups, i.e. sub-scenes; and each sub-scene contains multiple nodes including View, Canvas3D, Screen3D, etc. The outputs of first scene segmentation are: sub-scene in 3-D model with coordinates and the visible-distance parameters of the model for the purpose of scene prediction.

The scheduling tasks of grid server are to load 3-D sub-scene from storage, such as database or hard disk, into memory and to prepare for data transfer. The basic scheduling algorithm is shown as Fig. 4.

The client adds node data continuously received to the current scene. In the grid-supported distributed simulation system, the scheduling of 3-D scene is accomplished by both user and grid server. However, the scheduling is launched by user since only user can finally determine which information is desired.

Suppose the user is in the scene as Fig. 5.The four geometry areas that may be loaded are A, B, C and D respectively and suppose the distance between the user and the center is R, while $R1$ and $R2$ are two systematic parameters. The algorithm to predict and schedule scenes on client is as following:

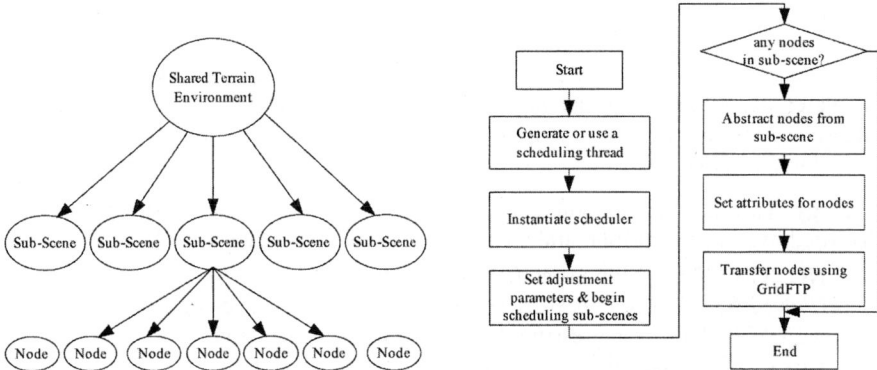

Fig. 3. Terrain distribution model.   **Fig. 4.** Scheduling algorithm for 3-D sub-scene.

```
While navigation {
   (1)Calculate the distance from the center: R;
   (2)Selection scheme:
   If R<R1: continue;
   Else If R1<R<R2: select LOD1;
   Else select LOD2 (finer resolution)
   (3) load geometry according to the current position of the
   navigator(A,B,C,or D)
   (4) cull the invisible tiles
}
```

Such transfer and exhibition method for 3-D model has high operability, high transfer efficiency and ease to use with combination of geometric information and texture information.

4 Implementation of Grid-Based Computing Service

Along with the development in domestic distributed simulation, tactical simulation has developed from concept demonstration of small scale to campaign scale gradually. The strategy and campaign simulation need to be combined with tactical simulation more closely and offer guidance for tactical simulation more directly.

Aegis simulation grid provides such a simulation computing service that calculates the optimal tactics for a federate at that time according to its current status, and notifies a tactical notification to the federate.

The entire terrain is plotted into two-dimensional grids. Each grid is given a specific number and the distribution information of terrain numbers is recorded in a two-dimensional array, *raw_grid[i][j]*, where i and j are the division numbers of the terrain in x and y directions respectively.

Correspondingly, the forces of red and blue sides are maintained in a two-dimensional array, *force[i][j]*, and *force* is defined as following:

```
struct force{
int red_count,    // forces of red side
int blue_count    // forces of blue side
};
```

In tactical notification system, strategy declarer determines the strategy, selects or imports new constraint rules from the constraint rule library according to strategy types, and then issues notification to corresponding federates on the basis of data calculated according to the parameters of the federates. The constraint rules are the core of tactical notification, involving strategy establishment, grammar of rule and tactical notification.

Strategy types are the embodiment of strategy and it's what the constraint rules are classified by. At present four strategy types are defined in Aegis as following:

> NO_STRATATEM: no specified strategy type.
> CONSERV_STRA: conservative strategy aiming at least casualties.
> ATTACK_STRA: aggressive strategy aiming at aggression.
> DEFENSE_STRA: defensive strategy aiming at protection.

Constraint rules are the base of performing tactical policy computation. It can be imported from script or be added to the library of strategy constraint rules manually.

> <Strategy constraint rules> ::=< Strategy type>::<Constraint rules>

Vocabularies defined at present include THISSIDE, OTHERSIDE, COUNT and AROUND. Logical and arithmetic operation symbols includes le, ge, eq, +, -, * and /.

At present Aegis defines four types of **tactical notification** as following:

> NO_NOTE: no enemy, go on according to scenario.
> ATTACK_NOTE: besiege the enemy.
> WARNING_NOTE: warning, be careful of the enemy and do not act rashly.
> RETREAT_NOTE: enemy is in the dominance, retreat to avoid the enemy's main forces.

5 Validation of Grid Service-Based Distributed Simulation

"World 2003", which originated from "World" system - an HLA-based large-scale distributed simulation program, is a prototype system developed to validate the functions of Aegis simulation grid. Supported by simulation grid, each federate starts a local tank instance in "World 2003" after the initialization of simulation resources, and the red and blue sides perform tactical counterwork in the virtual environment. Aegis provides tactical notifications to corresponding fighting federate on the basis of strategy guidance of red and blue sides. In the simulation, "World 2003" reached the anticipated goal. The concrete implementation is introduced below.

5.1 Overall Architecture and Simulation Process

The overall architecture of "World 2003" prototype system is shown as Fig. 6. Two kinds of grid clients are involved: tAgent requests for services to Aegis simulation grid and schedules required data resources, while confAdmin is responsible for the adjustment of simulation parameters.

The basic thought of "World 2003" is that "World" regularly outputs its situation information; the grid client tAgent regularly reads such information and requests the server for tactical notification service; the server performs rules computation based on such information, compares military forces of federates in adjacent areas, and returns

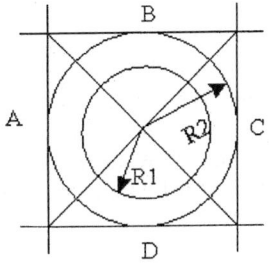
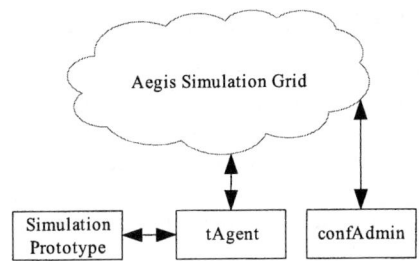

Fig. 5. Prediction and scheduling algorithm.　　**Fig. 6.** "World 2003" architecture.

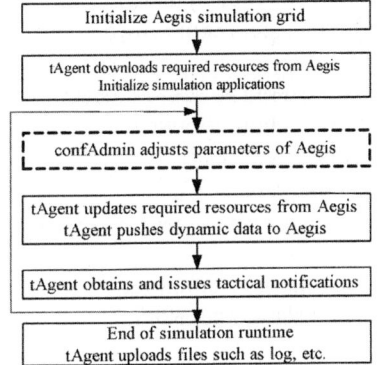
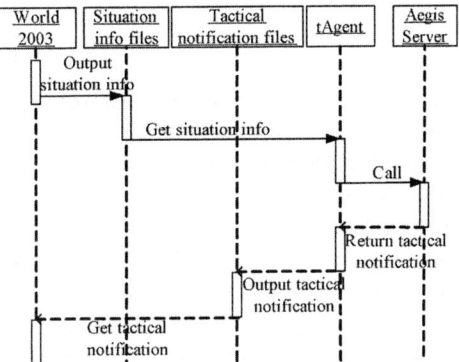

Fig. 7. Execution of grid-based simulation.　　**Fig. 8.** Interaction between tAgent and "world".

strategy rules-based tactical notification to tAgent; "World" periodically obtained tactical notification as the guidance to the next operation. The execution process of distributed simulation supported by simulation grid can be shown as Fig. 7.

The grid client confAdmin adjusts simulation parameters by means of access to service proxy. ConfAdmin is mainly responsible for strategy definition and adjustment at present. Each side can adjust strategy rules dynamically and obtain real-time tactical notifications with the support of computing service.

5.2 Interactive Process Between Grid Client tAgent and "World"

The communication between "World" and grid client tAgent can adopt several approaches, such as shared memory, shared files, TCP/IP communication, process communication, remote procedure call, etc., among which the approach of shared files is used in "World 2003".

The interactive process is shown in Fig. 8. "World 2003" outputs the situation information to specified location. Then tAgent traverses the situation files in the specified location, calls simulation computing service to deal with them orderly, gets the tactical notification and returns it to tAgent. Then tAgent outputs the tactical notification to the notification file at specified location from which "World 2003" obtains the tactical notification. In this course, both of the fighting sides can perform strategy adjustment to grid server and obtain tactical feedbacks through the grid client confAdmin.

5.3 Establishment of Constraint Rules

In the strategic counterwork of "World 2003", the constraint rules for conservative strategy and aggressive strategy are defined in Table 1 and Table 2 respectively. The primary goal of conservative strategy is to conserve the forces and keep casualty to the minimum. The aggressive strategy is characterized by initiative attack. If enemy's forces are less than or close to ours, start initiative attack; retreat would be suggested only if our forces are in an absolutely inferior status.

Table 1. Conservative strategy rules.

//conserve. There is no enemy around
if COUNT(OTHERSIDE) AROUND eq 0
return NO_NOTE;
//Enemy's forces around are less than half of ours. In absolute dominance, attack!
if COUNT(OTHERSIDE) AROUND le (COUNT(THISSIDE) AROUND/2)
return ATTACK_NOTE;
//Enemy's forces around are greater than double of ours. Retreat!
if COUNT(OTHERSIDE) AROUND ge (2*COUNT(THISSIDE) AROUND)
return RETREAT_NOTE;
//Enemy's forces around are close to ours. Warning!
return WARNING_NOTE;

Table 2. Aggressive strategy rules.

//aggressive. There is no enemy around
if COUNT(OTHERSIDE) AROUND eq 0
return NO_NOTE;
//Enemy's forces around less than ours. Attack!
if COUNT(OTHERSIDE) AROUND le COUNT(THISSIDE) AROUND
return ATTACK_NOTE;
//Enemy's forces around are greater than double of ours. Retreat!
if COUNT(OTHERSIDE) AROUND ge (2*COUNT(THISSIDE) AROUND)
return RETREAT_NOTE;
//Enemy's forces around are close to ours. Warning!
return WARNING_NOTE;

5.4 Experimental Environment and Test Analysis

The experiments are carried out from two aspects: functional and scalability. The functional experiment is mainly used to validate the functions of resource service and computing service that Aegis provides, including resource download, update, tactical notification, etc. The simulation process is shown as Fig. 9.

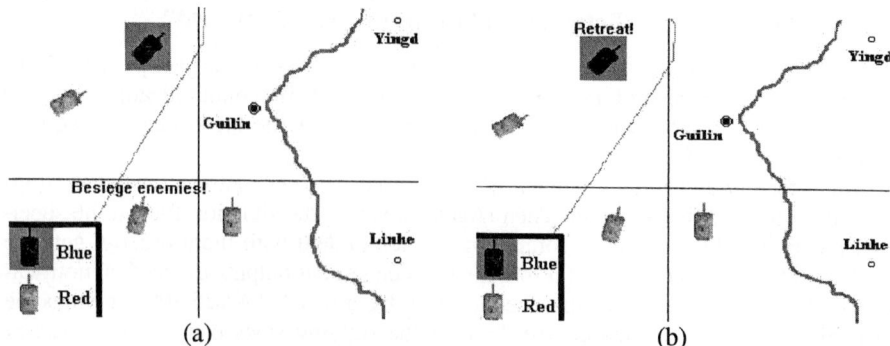

Fig. 9. Partial scene of "World 2003" simulation.

Aegis: A Simulation Grid Oriented to Large-Scale Distributed Simulation 421

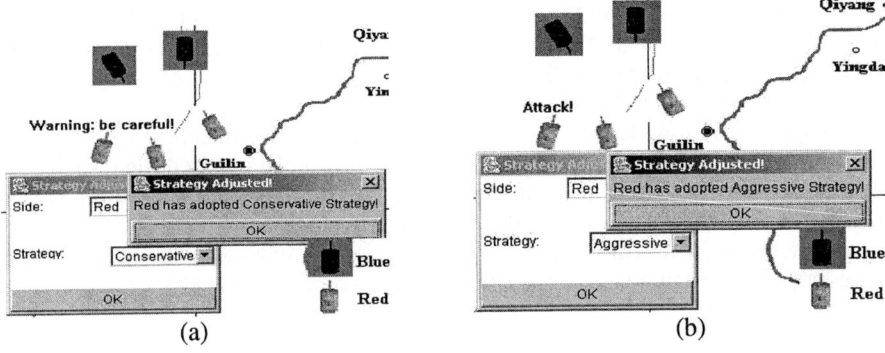

Fig. 10. Scene before and after dynamically strategy adjustment.

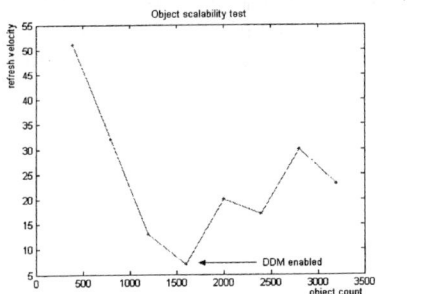

Fig. 11. Result of object scalability test. **Fig. 12.** Result of federate scalability test.

The computing service of tactical notification can dynamically respond to the strategy adjustment of the simulation grid in different notifications. The simulation scene after dynamic strategy adjustment is shown as Fig. 10. The client confAdmin adjusts the strategy parameters of simulation grid to set strategy for the red side in the figures. Fig. 10 (a) stands for the scene of conservative strategy and Fig. 10 (b) is of aggressive strategy.

Because grid service is independent to RTI data transfer and neither of them affects the other, the simulation efficiency does not decline. On the other hand, because each federate is supported with tactical notifications on basis of strategy of its side, the actual combat is simulated better.

In order to find out the usability of grid service-based distributed simulation, object and federate scalability test is performed, that is, the scale of objects and federates that the grid service-based distributed simulation can support is tested.

The test environment is as following: 100M Ethernet, DVE_RTI 2.0, 8 hosts with frequencies from P2 450 to P4 2.4G and memory from 128M to 512M. All the hosts are installed with windows 2000/XP and 4 hosts have the frequency above 1G. The result of object scalability experiment is shown as Fig. 11.

In Fig.11, refresh velocity is the frequency graphics of "world 2003" refresh and it represents the processing load of the host and the runtime efficiency in some aspects. "World" in each host instantiates 400 objects. Objects in the first four hosts congregate in close areas and they begin to disperse to different areas since the next four hosts begin to join. Because Data Distributed Management of DVE_RTI 2.0 is used and irrelevant messages are filtered, the amount of processed messages decreases and the refresh velocity increases. The result of testing initialization of mass federates using 8 hosts is shown as Fig. 12.

The application used for test is helloworld - a basic HLA application. Each application instantiates one object and notify the client to download required file resources to local from Aegis server at the initialization stage. StartApp, an application to start federate at a single host, is developed to start helloworld every other second. Since each host starts 64 federates in the test, there are 512 federates started totally. All these federates are initialized in ten minutes,

6 Conclusion

As a newly emerged technology, grid service is not mature yet. However, the grid-enhanced distributed simulation may lead to some advancements. Relevant research is booming throughout the world so that it is worth further investigation and exploration. This paper proposed a data grid oriented to distributed simulation and the grid offers resource management service and computing service for HLA-based distributed simulation so as to both meet the demands of large-scale simulation and improve the scalability. Some experiments are made to validate the simulation grid based distributed simulation and the results shows the usability of Aegis.

References

1. Simulation Interoperability Standards Committee (SISC) of the IEEE Computer Society. IEEE Standard for Modeling and Simulation (M&S) High Level Architecture (HLA) - IEEE std 1516-2000, 1516.1-2000, 1516.2-2000. New York: The Institute of Electrical and Electronics Engineers Inc., 2000.
2. S. Brunett, D. Davis, T. Gottschalk, P. Messina, and C. Kesselman. Implementing distributed synthetic forces simulations in metacomputing environments. Proceedings of the Heterogeneous Computing Workshop, pp. 29-42. IEEE Computer Society Press, 1998.
3. http://www.eu-crossgrid.org/
4. K. Zajac, A. Tirado-Ramos, Z. Zhao, P. Sloot, M. Bubak. Grid Services for HLA-based Distributed Simulation Frameworks. European Across Grids Conference, 2003.
5. W.T.Cai, S.J.Turner, H.F.Zhao. A Load Management System for Running HLA-based Distributed Simulations over the Grid. Proceedings of the sixth IEEE International Workshop on Distributed Simulation and Real-Time Applications (DS-RT'02) , pp.7-14, Fort Worth, Texas, USA, October 11-13, 2002.
6. Qian Xueping, Hao Aiming, Zhao Qingping. Research And Implementation Of Distributed Virtual Environment Server. The Sixth International conference for Younger Computer Scientists (icycs'2001), 2001.

An Adaptive Data Objects Placement Algorithm for Non-uniform Capacities

Zhong Liu and Xing-Ming Zhou

National Laboratory for Parallel and Distributed Processing, Changsha, 410073

Abstract. The capacities of storage nodes usually are non-uniform and storage nodes are dynamically changed in large-scale distributed storage systems. In this paper, a novel dynamic data objects placement algorithm is proposed; data objects are always distributed among the storage nodes according to their capabilities. When storage nodes are changed, it affords to immediately rebalance data objects distribution according to weight of storage nodes. Simulation results indicates that data objects are always distributed among the storage nodes according to their capabilities and data objects are migrated throughout all storage nodes in parallel, resulting in minimum amount of replacement of objects.

Keywords: distributed storage, non-uniform capacity, dynamic interval mapping, balanced distribution and adaptive.

1 Introduction

As the use of large distributed systems and large-scale clusters of commodity computers has increased, significant research has been devoted toward designing scalable distributed storage systems. Its applications now span numerous disciplines, such as: higher large-scale mail system, online numeric periodical, digital libraries, large online electric commerce system, energy research and simulation, high energy physics research, seismic data analysis, large scale signal and image processing applications, data grid application and peer-to-peer storage application, etc. Usually, it will no longer be possible to do overall upgrades of high performance storage systems. Instead, systems must grow gracefully over time, adding new capacity and replacing failed units seamlessly-an individual storage device may only last five years, but the system and the data on it must survive for decades. Since the capacities of storage nodes usually are non-uniform and storage nodes are dynamically changed in large-scale distributed storage systems, systems must distribute data objects among the storage nodes according to their capabilities and afford to immediately rebalance data objects distribution according to weight of storage nodes when storage nodes are changed. So we study the problem of designing flexible, adaptive strategies for the distribution of objects among a heterogeneous set of servers. Ideally, such a strategy should be able to adapt with a minimum amount of replacements of objects to changes in the capabilities of the servers so that objects are always distributed among the servers according to their capabilities.

Previous techniques are able to handle these requirements only in part. For example, standard hashing techniques use hash function $h(id) \equiv id \pmod{n}$ where id is object ID and n is the number of storage nodes in system. If storage nodes have the uniform capabilities, it can be used to distribute data objects evenly among n servers. However, but they usually do not adapt well to a change in the capabilities. Moreover, If a new server is added, approximately the fraction $n/(n+1)$ of the data objects must be replaced before the data objects can be accessed using the new mapping. In contrast, the minimum fraction that must be relocated to obtain a balanced mapping is approximately $1/(n+1)$. Litwin, et al. [1] has developed many variations on Linear Hashing (LH*), the LH* variants are limited in two ways: they must split buckets, and they have no provision for buckets with different weights. LH* splits buckets in half, so that on average, half of the objects on a split bucket will be moved to a new empty bucket, resulting in suboptimal bucket utilization and a "hot spot" of bucket and network activity between the splitting node and the recipient and the distribution is unbalanced after replacement. Other data structures such as DDH[2] suffer from similar splitting issues. Choy, et al. [3]describes algorithms for perfect distribution of data to disks that move an optimally low number of objects when disks are added. However, these algorithms do not support weighting of disks and removal of disks. Brinkmann, et al.[4][5]proposes a method for pseudo-random distribution of data to multiple disks using partitioning of the unit range. This method accommodates growth of the collection of disks by repartitioning the range and relocating data to rebalance the load. However, this method does not move an optimally low number of objects of replacement.

In our algorithm, data objects are always distributed among the storage nodes according to their capabilities. When storage nodes are changed, it affords to immediately rebalance data objects distribution according to weight of storage nodes. Moreover, our algorithm almost always moves a statistically optimal number of objects from every storage node in the system to each new storage node, rather than from one storage node to one storage node. The rest of the paper is organized as follows. Section 2 describes the balls into bins system model. Section 3 presents an adaptive data objects placement algorithm for non-uniform capacities based distributed dynamic interval mapping. Section 4 gives simulation results and performance analysis. Section 5 summarizes the paper.

2 The System Model

We adopt and extend the standard balls into bins model similar to [4], as follows:

1. A bin represents a storage node and a ball represents a data object.
2. Let $\{0, 1, \ldots, N\}$ be the set of all possible bins and $\{0, 1, \ldots, M\}$ be the set of all possible balls in the system. Suppose $M \gg N$.
3. Suppose that we have a random function $H : \{0, 1, \ldots, M\} \to [0, 1)$, H maps the balls uniformly at random to real numbers in the interval [0, 1).
4. The capability weight of bin i (which represents storage node i) is w_i

3 The DDIM Algorithm

The basic idea of DDIM (Distributed Dynamic Interval Mapping) is to create one interval chain for each bin and establish one to one mapping relation between interval chain and bin. The interval chains may have many intervals; each data objects, which are mapped into the interval chain, are distributed into the same bin. All the interval chains form an interval chain table. Initially, the interval [0, 1) is divided into different length intervals according to weight of storage nodes; these intervals are assigned into corresponding interval chain. The way to compute the mapping given a ball id x is first to find the interval d to which x belongs, and find the interval chain to which d belongs, and then to compute the mapping using one to one mapping relation between interval chain and bin. When bins are changed, current intervals are divided into more small intervals rather than the interval [0, 1) is redefined and different intervals are reassigned into interval chains, resulting in balls replacement. But the one to one mapping relation between interval chain and bin remain unchanged. The rest of construction of the dynamic interval chains table consists of the following.

3.1 The Initial Interval Chains Table L_0

In general, we suppose that there are k($\geq$1)bins in the system initially, which are numbered $0, 1, 2, \ldots, k--1$. For arbitrary i, let $a_i = \dfrac{\sum_{j=0}^{i} w_j}{\sum_{j=0}^{k-1} w_j}$, for $0 \leq i \leq k-1$, and define $a_{-1} = 0$, then we get real numbers a_i where $0 = a_{-1} < a_0 < a_1 < a_2 < \ldots < a_{k-1} = 1$ The ith interval is $d_i = [a_{i-1}, a_i)$. We construct the initial interval chains table $L_0 = \{l_0, \ldots, l_{k-1}, \ldots l_N\}$, where $l_j = \begin{cases} \{d_j\} & 0 \leq j \leq k-1 \\ \phi & k \leq j \leq N \end{cases}$ (Φ represents empty). We define interval mapping function $F_0 : L_0 \rightarrow \{0, 1, \ldots, N\}$, let $F_0(l_j) = j$ for $0 \leq j \leq N$. For arbitrary ball x, the way to compute the mapping bin is find the interval d, to which x belongs, and then to find the interval chain l_j, to which d belongs. The function computing the mapping bin follows:

int GetBin(ball x, interval chains table L_i) {
    if(query interval chains table L_i and find interval chain l_j, make $H(x) \in d$ and $d \in l_j$ then return bin j
    else return -1 }

3.2 The Dynamic Interval Chains Table L_i

Suppose the ith interval chains table is $L_i = \{l_0, \ldots, l_{k-1}, \ldots l_N\}$, where $l_j = \begin{cases} \{d_{j1}, d_{j2}\ldots\} & 0 \leq j \leq k-1 \\ \phi & k \leq j \leq N \end{cases}$. We construct the (i+1)th interval chains table according to L_i. We have to consider the following two cases:

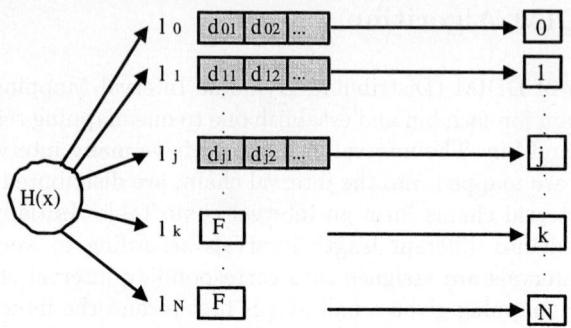

Fig. 1. Dynamic interval mapping.

(1) Adding bin. We know all the interval chains to the right of the kth interval chain have empty interval by the define of L_i. So the next bin to add is the k+1th bin and numbered k, let its capability weight w_k. For arbitrary interval chain l_j, a scan is sent to all available interval $d_{j1}, d_{j2} \ldots$, until getting an interval aggregate D_j (if need, dividing one interval) and making the sum of all interval length $\sum_{d_j \in D_j} |d_j| = \frac{w_j}{\sum_{i=0}^{k} w_i}$. Remain these interval in the jth interval chain l_j, but migrate the other intervals into jth interval chain l_k, consequently we get $L_{i+1} = \{l_0, \ldots, l_k, \ldots l_N\}$, where $l_j = \begin{cases} \{d_{j1}, d_{j2}\ldots\} & 0 \leq j \leq k \\ \phi & k+1 \leq j \leq N \end{cases}$.

(2) Removing bin. Suppose the system remove the bin numbered $z (0 \leq z \leq k-1)$. For interval chain l_z, let the sum of interval length $|D_z| = \sum_{d_j \in l_z} |d_j|$, a scan is sent to all available interval $d_{z1}, d_{z2} \ldots$, until getting k-1 interval aggregate D_j (if need, dividing one interval). For arbitrary j, if $0 \leq j < z$, the sum of interval length of aggregate $D_j = \sum_{d_j \in D_j} |d_j| = \frac{w_j}{\sum_{i=0, j \neq z}^{k-1} w_i} * |D_z|$, then migrate the intervals of aggregate D_j into jth interval chain l_j; if $z \leq i \leq k-2$, the sum of interval length of aggregate $D_j = \sum_{d_j \in D_j} |d_j| = \frac{w_{j+1}}{\sum_{i=0, j \neq z}^{k-1} w_i} * |D_z|$, then migrate the intervals of aggregate D_j into (j+1)th interval chain l_{j+1}. After removing the bin z, in order to continuity of bins number, let the number of bin to the right of the zth bin reduces one, the corresponding interval chain is forward one row. Consequently we get $L_{i+1} = \{l_0, \ldots, l_k, \ldots l_N\}$, where $l_j = \begin{cases} \{d_{j1}, d_{j2}\ldots\} & 0 \leq j \leq k-2 \\ \phi & k-1 \leq j \leq N \end{cases}$.

Lemma 1. *the number of balls placed in a bin is proportional to the total length of intervals contained in the corresponding interval chain.*

Proof. Since H is a uniform random function, it maps the balls uniformly at random to real numbers in the interval $[0, 1)$, so the number of balls mapped to

An Adaptive Data Objects Placement Algorithm for Non-uniform Capacities

different intervals is probabilistically proportional to the length of target interval. On the other hand, all the balls mapped to intervals contained in the same interval chain are placed in the same bin, so the number of balls placed in a bin is proportional to the total length of intervals contained in the corresponding interval chain.

Theorem 1. *the number of balls placed in any bin is proportional to its capacity.*

Proof. By Lemma 1, we just have to prove that: for arbitrary two bins, the total length of intervals contained in the corresponding interval chain is proportional to their capacities. For arbitrary two bin p and q, assuming that the total length of intervals contained in the corresponding interval chain is $|D_p|$ and $|D_q|$, respectively. The proof is by induction on i.

(1) Initially $i = 0$, by the construction algorithm of initial interval chains table L_0, then

$$\frac{|D_p|}{|D_q|} = \frac{|d_p|}{|d_q|} = \frac{|a_p - a_{p-1}|}{|a_q - a_{q-1}|} = \frac{\sum_{j=0}^{p} w_j}{\sum_{j=0}^{k-1} w_j} - \frac{\sum_{j=0}^{p-1} w_j}{\sum_{j=0}^{k-1} w_j} = \frac{w_p}{w_q},$$

So the total length of intervals contained in the corresponding interval chain is proportional to their capacities.

(2) Assuming that the conclusion holds for the ith expansions, we prove it for the (i+1)th expansions. Consider the following two cases in the (i+1)th expansions:(a) Adding bin. By the construction algorithm of dynamic interval chains table, if $0 \le p, q < k$ then $|D_p| = \frac{w_p}{\sum_{i=0}^{k} w_i}$, $|D_q| = \frac{w_q}{\sum_{i=0}^{k} w_i}$, $\frac{|D_p|}{|D_q|} = \frac{w_p}{w_q}$. If p or q is the new k+1th bin numbered k, assume $q = k$. by the assumption, in the ith interval chains table L_i, the total length of intervals contained in the any interval chain is proportional to its capacities, then for any interval chains l_j, $|D_j| = \frac{w_j}{\sum_{i=0}^{k-1} w_i}$, after the (i+1)th expansions, the total length of intervals migrated from l_j to l_k is equal to $(\frac{w_j}{\sum_{i=0}^{k-1} w_i} - \frac{w_j}{\sum_{i=0}^{k} w_i})$, then $|D_q| = |D_k| = \sum_{j=0}^{k-1}(\frac{w_j}{\sum_{i=0}^{k-1} w_i} - \frac{w_j}{\sum_{i=0}^{k} w_i}) = \frac{w_k}{\sum_{i=0}^{k} w_i}$, $\frac{|D_p|}{|D_q|} = \frac{|D_p|}{|D_k|} = \frac{w_p}{w_k}$, So the total length of intervals contained in the corresponding interval chain is proportional to their capacities.(b) Removing bin. By the construction algorithm of dynamic interval chains table, for any interval chains l_p, $|D_p| = |D_{old}| + |D_{new}|$, where $|D_{old}|$ is the total length of intervals contained in the old interval chain, and $|D_{new}|$ is the total length of intervals migrated from

l_z to l_p, by the assumption, $|D_{old}| = \frac{w_p}{\sum_{i=0}^{k-1} w_i}$, $|D_{new}| = \frac{w_p}{\sum_{i=0, j\neq z}^{k-1} w_i} * |D_z|$, then for any

$$p, q, \frac{|D_p|}{|D_q|} = \frac{\frac{w_p}{k-1} + \frac{w_p}{k-1}}{\sum_{j=0}^{k-1} w_j + \sum_{j=0, j\neq z}^{k-1} w_j} * |D_z| = \frac{w_p}{w_q},$$ So the total length of intervals contained

in the corresponding interval chain is proportional to their capacities. By induction we show that the total length of intervals contained in the corresponding interval chain is always proportional to their capacities.

Theorem 2. *When bins are changed, the number of balls migrated is the minimum.*

Proof. By Lemma 1, we just have to prove that the total length of intervals migrated is the minimum. Consider the following two cases. (a) Adding bin. By Theorem 1, the number of balls placed in any bin is proportional to its capacity, then the minimum total length of intervals migrated to l_k is $\frac{w_k}{\sum_{j=0} w_j}$, By the construction algorithm of dynamic interval chains table, the total length of intervals migrated to l_k is: $\sum_{j=0}^{k-1} (\frac{w_j}{\sum_{i=0} w_i} - \frac{w_j}{\sum_{i=0} w_i}) = \frac{w_k}{\sum_{i=0} w_i}$. (b) Removing bin. The intervals migrated is contained in l_z, that is just the minimum. So the total length of intervals migrated is always the minimum.

4 The Simulation Results Analysis

In our algorithm, we need a random function H, which maps the balls uniformly at random to real numbers in the interval [0, 1). We select the Mersenne Twister[6] as the random function H in the implementation of our algorithm. The simulation system includes four storage nodes, which capability weight is assigned to 1, 2, 3, 4 respectively. The 10000, 40000, 80000, 120000 data objects from four clients are sent respectively to storage nodes.

Table 1 lists the distribution of data objects sent from four clients and the total data objects in all different storage nodes. Table 1 show that data objects

Table 1. The distribution of data objects for non-uniform capacities.

Client\Nodes	Node 1	Node 2	Node 3	Node 4
Weight	1(10%)	2(20%)	3(30%)	4(40%)
Client1	1008(10.08%)	1956(19.56%)	3022(30.22%)	4014(40.14%)
Client2	4075(10.19%)	7929(19.82%)	11931(29.83%)	16065(40.16%)
Client3	8020(10.03%)	15704(19.63%)	24166(30.21%)	32110(40.14%)
Client4	12108(10.09%)	23674(19.73%)	36040(30.03%)	48178(40.15%)
Total	25211(10.08%)	49263(19.71%)	75159(30.06%)	100367(40.15%)

sent from four clients and the total sum are always distributed among the storage nodes according to their capabilities weight.

Add three storage nodes with capabilities weight 5, 6 and 7. Table 2 lists the redistribution of data objects sent from four clients and the total data objects in all different storage nodes. Remove a storage node with capability weight 4. Table 3 lists the redistribution of data objects sent from four clients and the total data objects in all different storage nodes.

Table 2 and Table 3 show that data objects sent from four clients and the total sum are always redistributed among the storage nodes according to their capabilities weight and data objects are migrated throughout all storage nodes in parallel after add or remove storage nodes, resulting in near-minimum amount of replacement of data objects.

5 Conclusions

In this paper, we propose a novel dynamic data objects placement algorithm based on pseudo-random hash function and dynamic interval mapping. In our algorithm, data objects are always distributed among the storage nodes accord-

Table 2. The redistribution of data objects according to nodes capabilities weight after adding three storage nodes.

Client\Nodes	Node 1	Node 2	Node 3	Node4	Node 5	Node 6	Node 7
Weight	1 (3.57%)	2 (7.14%)	3 (10.71%)	4 (14.29%)	5 (17.86%)	6 (21.43%)	7 (25%)
Client1	369 (3.69%)	710 (7.10%)	1107 (11.07%)	1406 (14.06%)	1777 (17.77%)	2132 (21.32%)	2499 (24.99%)
Client2	1462 (3.66%)	2830 (7.08%)	4201 (10.5%)	5694 (14.24%)	7241 (18.1%)	8599 (21.5%)	9973 (24.93%)
Client3	2911 (3.64%)	5700 (7.13%)	8602 (10.75%)	11336 (14.17%)	14290 (17.86%)	16975 (21.22%)	20186 (25.23%)
Client4	4371 (3.64%)	8434 (7.03%)	12844 (10.7%)	16984 (14.15%)	21674 (18.06%)	25637 (21.36%)	30056 (25.05%)
Total	9113 (3.65%)	17674 (7.07%)	26754 (10.7%)	35420 (14.17%)	44982 (17.99%)	53343 (21.34%)	62714 (25.09%)

Table 3. The redistribution of data objects according to nodes capabilities weight after removing one storage node.

Client\Nodes	Node 1	Node 2	Node 3
Weight	1(16.67%)	2(33.33%)	3(50%)
Client1	1716(17.16%)	3237(32.37%)	5047(50.47%)
Client2	6702(16.76%)	13275(33.19%)	20023(50.06%)
Client3	13453(16.82%)	26311(32.89%)	40236(50.29%)
Client4	20109(16.76%)	39557(32.96%)	60334(50.28%)
Total	41980(16.79%)	82380(32.95%)	125640(50.26%)

ing to their capabilities weight. When storage nodes are changed, it affords to immediately rebalance data objects distribution according to weight of storage nodes. Simulation results indicates that data objects are always distributed among the storage nodes according to their capabilities weight for non-uniform capability and data objects are migrated throughout all storage nodes in parallel, resulting in near-minimum amount of replacement of objects.

References

1. W.Litwin, M.A.Neimat,and D.A.Schneider. LH*-a scalable, distributed data structure. ACM Transactions on Database Systems, 1996, 21(4): 480-525.
2. Devine, R.Design and Implementation of DDH: Distributed Dynamic Hashing. Int. Conf. on Foundations of Data Organizations, FODO-93.Lecture Notes in Comp. Sc., Springer-Verlag (publ.), Oct. 1993.
3. D.M.Choy, R.Fagin, and L.Stockmeyer. Efficiently extendible mappings for balanced data distribution. Algorithmica, 1996, 16:215-232.
4. A.Brinkmann, K.Salzwedel, and C.Scheideler. Efficient, distributed data placement strategies for storage area networks. In Proceedings of the 12th ACM Symposium on Parallel Algorithms and Architectures (SPAA), ACM Press. Extended Abstract. 2000, 119-128.
5. A.Brinkmann, K.Salzwedel, and C.Scheideler. Compact, adaptive placement schemes for non-uniform capacities. In Proceedings of the 14th ACM Symposium on Parallel Algorithms and Architectures (SPAA), Winnipeg, Manitoba, Canada, Aug. 2002. 53-62.
6. M.Matsumoto and T.Nishimura, "Mersenne Twister: A 623-dimensionally equidistributed uniform pseudorandom number generator", ACM Trans. on Modeling and Computer Simulation Vol. 8, No. 1, January pp.3-30 1998.

Resource Discovery Mechanism for Large-Scale Distributed Simulation Oriented Data Grid[*]

Hai Huang, Shaofeng Wang, Yan Zhang, and Wei Wu

School of Computer Science and Engineering
BeiHang University, Beijing 100083, P.R. China
{huanghai,wangsf,zy}@vrlab.buaa.edu.cn

Abstract. A new type of resource discovery mechanism is proposed for simulation oriented data grid. By introducing distributed resource brokers at edge routers of each domain, a majority of resource discovery queries can be processed solely by resource brokers at edge routers, instead of accessing remote global resource broker. The response time and network overhead of resource discovery is reduced and system scalability is enhanced. Experiment results show that the proposed resource discovery mechanism is feasible.

1 Introduction

Computer military operation simulation attracts more and more attention all over the world. With the rapid growth of simulation scale, the traditional simulation method has exposed more and more drawbacks. In recent years, the Grid technology, aiming at resolving resource sharing and collaboration, has become a hot topic and some research institutions attempt to adopt Grid technology into simulation to solve the problems in large-scale distributed interactive simulation [1, 2].

In large-scale distributed simulation supported by Grid environment, simulation applications usually frequently access data resources distributed in disparate areas. Furthermore, they not only expect high usability, but also have strict requirement on the response time of resource access request (RAR). The entire process of accessing resource, from sending out a RAR to obtaining the desired result, comprises resource discovery and resource access. Correspondingly, the response time of a RAR comprises the time of resource discovery and the time of resource access, that is, the time spent on obtaining the resource metadata about storage information and corresponding access method and the time spent on accessing the desired resource according to the obtained metadata. The fact that distributed simulation needs high real-time performance indicates it is significant for research on resource discovery and resource access mechanisms to curtail their response time.

2 The Contemporary Mechanisms

A lot of work in resource discovery has been carried out. Within Globus, the GIIS, which is called Index Service in GT 3.0, may accepts registration messages from "child" GRIS or GIIS instances and merges these information sources into a unified

---

[*] This paper is supported by the National Grand Fundamental Research 973 Program of China under Grant No. 2002CB312105.

information space, forming a tree-like hierarchy model of metadata [3, 4]. There exists a global center, the tree root, affecting the system scalability. Furthermore, in a large-scale grid environment, the GIIS hierarchy is extraordinarily complex. In the resource discovery process, the leaf node is first queried, and if the metadata of required resource is not found the direct upper node is queried. The rest may be deduced by analogy until the node is tree root. Therefore, there is no a priori formula for predicting the performance of resource discovery. The more complex the GIIS hierarchy, the more unpredictable the performance, and in general, the longer a query might take to answer. The contradiction between system scalability and real-time performance of resource discovery mechanism should be considered carefully in large-scale distributed simulation.

The Web Services technologies have achieved great progress in resource discovery. Among which, UDDI [5, 6] and WS-Inspection [7, 8] are representative methods. UDDI is a centralized discovery method, comprising a UDDI center and a set of APIs used to access the center by users. However, users can only find services listed in the UDDI center. Furthermore, because of its characteristic of centralization, excessive access requests may lead to failure of the server. Similarly, there exists the contradiction between scalability and real-time performance. The WS-Inspection provides a distributed discovery method, overcoming the shortcomings lying in the centralized ones. However, a WS-Inspection document is generally made available at the point-of-offering for the services that are referenced within the document. So, WS-Inspection cannot work independently if the point-of-offering is not known.

In China, [9] proposed a consumer-oriented resource discovery mechanism, which expedites resource discovery and reduces the frequency of accessing remote server by caching. However, it is merely complementary of existing service discovery methods, such as UDDI, WS-Inspection, and needs further study in application. [10, 11] proposed resource management structures similar to Fig. 1. The whole system comprises a global resource broker (RB) server and several local RBi (i = 1,2,3), forming a hierarchical tree structure. The global RB server is responsible for metadata management of the whole system while each RBi, which is a partial copy of RB, is only responsible for the management of local metadata. The resources managed by RBi form a domain and neighboring domains are connected via edge routers (ER). When a RAR arrives at ER, it queries local RBi via the ER. If the metadata of desired resource is not found in RBi, it is redirected to RB for further query.

Such system structure is simple and easy to implement. Part of the burden of global RB is transferred onto RBi of each domain, bearing distributed characteristic. However, for each RAR, it is a must to access RBi at least once, even to access global RB

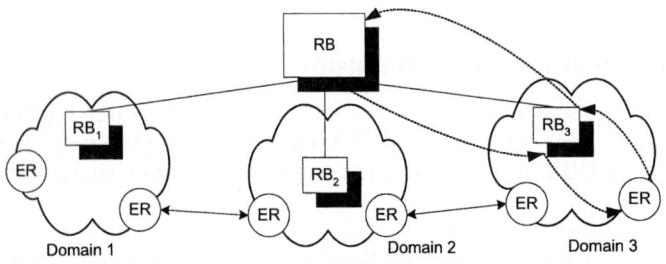

Fig. 1. Hierarchical Tree Structure.

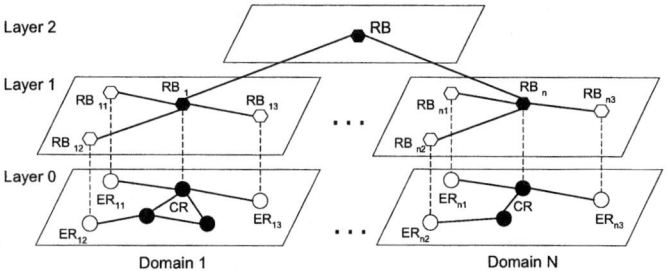

Fig. 2. System Architecture of Aegis Resource Discovery.

on some occasions. When the scale of domain gets larger, querying RBi for resource metadata would take a relatively long period of time, which not only prolongs the response time but also increases network overhead. On the other hand, excessive RARs might result in the fact that RB becomes the bottleneck of system performance.

3 Aegis Resource Discovery Architecture

Aegis is a data Grid oriented to large-scale distributed simulation that is being established by BHU. It is expected to apply the proposed mechanism to Aegis in the near future. Therefore, it is called Aegis resource discovery mechanism (Fig. 2.).

Aegis resource discovery mechanism is based on the following considerations.

Firstly, querying local RB for metadata costs less time and less network overhead than querying remote RB.

Secondly, a simulation application usually need to access resources distributed in multiple domains. The reason is obvious. On the one hand, resources in a grid environment are extensively distributed, involving complex resource types, spanning in an extensive area. It is common for several domains to collaboratively serve a simulation application since one domain cannot meet its demands solely. On the other hand, resources in a grid environment are dynamic, which means that resources in a domain may be available at some moment and become unavailable later, or new resources may join in as time goes on. In such cases as the Grid resources decrease dynamically or a failure occurs, simulation applications usually access resources in other domains.

Finally, although the mode of accessing resources is not very clear, yet simulation applications have their own favor of different resources. For example, a Tank simulator may fire continuously which leads to frequent access to resource regarding exploding effect of bomb and almost no access to that regarding effect of missile flying.

In Fig. 2, Layer 0 is physical layer, which indicates that the Grid comprises N domains, and each Domain i (i=1, 2, ..., N) comprises Mi ERij (j=1, 2, ..., Mi) and several core routers (CR). Layer 1 and Layer 2 are the views of resource layers. For each Domain i, there is a RBi, managing resources in local domain, and for each ERij, there is a corresponding RBij. RBi and RBij have different roles. First, RBij stores the metadata of resources that might be accessed in all probability by RARs arriving at ERij in the near future, including resource providers' information, resource location, access method, access probability, etc. Second, the metadata stored in RBij is not just related to resources in Domain i, it is related to resources distributed in different domains that might be frequently accessed. As a result, RBij is not just a partial copy of

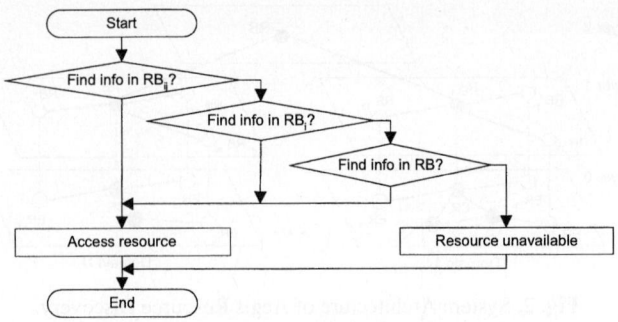

Fig. 3. Flowchart of Resource Discovery.

RBi. To manage resources in different domains collaboratively, a global RB is established on Layer 2, which stores the resource metadata of the entire Grid. The process of querying metadata is described as following and its flowchart is shown as Fig. 3.

(1) Suppose a RAR arrives at ERij, then query RBij for metadata of desired resource. If get the right metadata, then goto (4), else goto (2);
(2) Query RBi for metadata. If get the right metadata, then goto (4), else goto (3);
(3) Query RB for metadata. If get the right metadata, then goto (4), else return the exception of resource unavailable and goto (5);
(4) Access resource according the information described in the metadata, goto (5);
(5) Carry out post-exception handling, such as take some action according to policy when resource is unavailable, and end the process.

The mechanism weakens the role of remote global RB. It is possible that a majority of resource discovery queries can be processed only at edge routers, instead of accessing the global RB. The response time and network overhead is reduced, and the burden of global RB is alleviated. Accordingly, performance bottleneck is resolved to some degree and system scalability is enhanced, which can better meet the requirements of simulation.

4 Key Technologies of Aegis Resource Discovery Mechanism

According to Aegis resource discovery architecture and RAR process flow, the key to guarantee this mechanism effectively lies in forecasting the frequency of resource access as well as choosing the appropriate metadata of resources from numerous metadata to store in RBij. These two strategies will be discussed below.

4.1 Resource Access Frequency Forecast

The metadata of resources, which will most possibly be accessed by resource access requests via ERij in a future period, is stored in RBij. Time series analysis [12] is employed at every edge router (such as ERij) to make statistics and to forecast the access frequency for each resource accessed by RARs arriving at ERij in the next period in order to determine which resources will most possibly be accessed.

The fundamental idea of time series analysis is to regard the data series formed over time by the forecasted object as a stochastic series. That is to say, time series is a

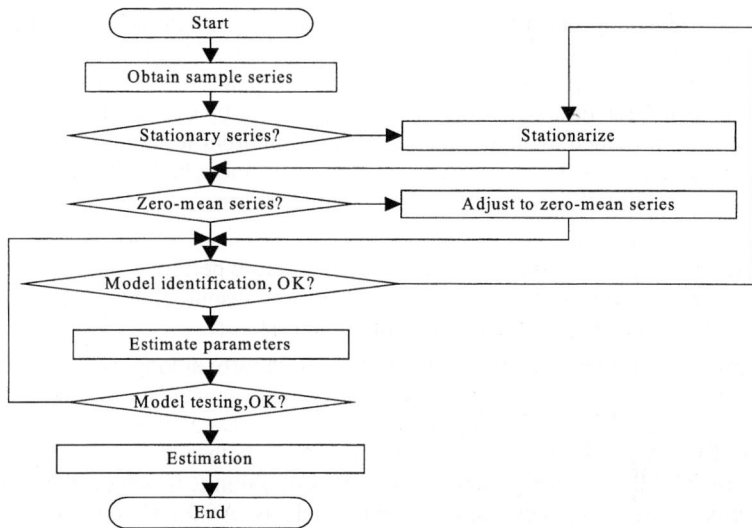

Fig. 4. Flowchart of Resource Access Frequency Forecast.

series of stochastic variables dependent on time t except for a few observed values caused by occasional factors. Although individual value forming the time series bears uncertainty, changes of the whole series bear certain regularity, which can be described in corresponding mathematical model approximately. The interdependent relationship or self-reference of this group of stochastic variables indicates developmental continuity of the forecasted object. Furthermore, once the autocorrelation is described in relevant mathematical model, their future values can be forecasted according to their past values and present values of the time series.

There are three fundamental models for real-number stationary time series with mean value of zero: Auto Regression Model - AR (p), Moving Average Model - MA (q) and Auto Regressive Moving Average model - ARMA (p, q). However, a lot of phenomena generally show a certain ascending or descending trend as time goes on, composing non-stationary time series. Solution to this is to carry on difference stationarizing process and zero-mean process before building the model, i.e. the so-called Integrated Auto Regressive Moving Average Model - ARIMA (p, d, q). ARIMA is a short-term forecast method with comparatively high accuracy. The process of forecasting resource access frequency with ARIMA model is shown as Fig. 4.

4.2 Strategy for Metadata Storage Selection in RBij

Grid resources are dynamically changing. What are stored in RBij should be metadata of resources that is most possibly accessed and available in a future period. In Ageis data grid, both availability and efficiency of data resources are improved by creating multiple replicas of the same data. As for the replicas of resource provided by the same provider, their availability may be left out of account in RBij, while selection of replicas from multiple replicas is achieved by Aegis replica selection mechanism. On the other hand, different resource providers may provide the same resource, so not only access frequency but also availability of resource should be considered when determining whether to store the metadata of resource provided by a certain provider in RBij. In addition, metadata stored in RBij should guarantee that RAR arriving at ERij could spend less time accessing resources provided by one service provider than accessing those provided by other service providers in order to minimize the response time of resource access. Consequently, we can determine whether the metadata of a resource is stored in RBij according to the principles below:

1. Resource is available;
2. Resource may be most frequently accessed in a future period;
3. If there are multiple service providers of the same resource, only metadata of resource that is accessed with the minimum time is stored;

The priorities of three conditions above decrease in turn, among which Condition 1 is necessary but not sufficient condition. That is to say, metadata of unavailable resources is not stored in RBij. RBij queries for Service Data Elements (SDEs) every period of TTR (Time To Request) in order to judge whether resources are available, among which TTR is a configurable parameter. Condition 2 is achieved by forecasting resource access frequency via time series analysis. As for Condition 3, RBij records response time of accessing each resource in order to determine the resource provider that has minimal access response time. Suppose the recorded response time of accessing a resource is Tcur, then after an access with new response time of Tnew, the recorded response time would be updated to Tcur=(1-α)Tcur+αTnew, among which α is weight. Its value scope is 0<α<1, representing the influence degree of the latest response time of accessing resource, aiming at eliminating impacts of response time jitter. The system compares Tcur of every service provider, and selects the service provider with the minimal Tcur value.

5 Experimental Environment and Test Analysis

Suppose the response time of a resource discovery is T, querying RBij spends T1, querying RBi spends T2, and querying RB spends T3. Then their relationship is T=T1+T2+T3.

If the metadata of desired resources can be found in RBij, then T1≠0, T2=0, T3=0; if the metadata is found in RBi, then T1≠0, T2≠0 ,T3=0; and if RB is required, then T1≠0, T2≠0, T3≠0.

Because the operation of querying RBij is approximately equivalent to a local query, querying RBi is performed in a domain, and querying RB needs remote operation across multiple domains. So the relationship among T1, T2 and T3 is T1<<T2<T3.

Therefore, there exists much difference among response time. The method of reducing such difference as well as shortening response time is to ensure that the metadata of desired resources are stored in RB_{ij} in all probability, which in turn is exactly the basic idea of the proposed resource discovery mechanism.

The test of Aegis resource discovery mechanism should include two parts. The first part is to test its capability of reducing response time and network overhead of resource discovery. Aegis minimizes the response time to T1 (T1<<T) by storing the metadata of resources that may be accessed most frequently in the near future. The higher the hit rate of querying RBij is, the lower the average response time and network overhead of resource discovery is.

The second part is to test its capability of alleviating the burden of global resource brokers and enhancing the system scalability. As for traditional resource mechanism, such as GRIS&GIIS in GT 2.0, it is necessary to query the centralized GIIS every time accessing resources distributed across multiple domains. However, in Aegis, a majority of RARs can be processed by RBij at local ERij no matter the desired resources are in a single domain or in multiple domains. Only when the metadata is not stored in RBij is it necessary for the remote RBi or the global centralized RB to be queried. Therefore, compared to GRIS&GIIS, Aegis resource discovery mechanism is more "distributed", and resolves the performance bottleneck existing in the centralized GIIS in a sense. Furthermore, the system scalability is improved.

Aegis data grid is at the stage of establishment. The proposed Aegis resource discovery mechanism has not been integrated into it. Therefore, large-scale test validating its scalability has not been carried out. A simple experiment was conducted to validate its capability of reducing the response time and network overhead of resource discovery and thus to prove the feasibility.

Due to the restriction of experiment environment, the experiment was carried out in local area network to simulate the wide area network. The purpose is to test the hit rate of querying RBij, not to calculate the value of response time. It is based on the basic fact that the response time of querying RBij is much smaller than that of querying remote server. If the hit rate of querying RBij is relatively high, the average response time of resource discovery would be reduced greatly.

In Aegis, the physical characteristic and logical characteristic of a file are inter-independent, which means changes of any side do not affect the other. Therefore, the hit rate of querying RBij is independent of the physical location of resource and merely related to how the metadata stored in RBij reflects resource access frequency. Based on considerations above, the experiment neither strictly follows Aegis architecture, nor involves multiple domains. Instead, a network topology as Fig. 5 is used.

In the experiment, four PCs form Domain i, among which three PCs not only ran Tank, Plane and Cannon simulator respectively, but also acted as their own RBij (j=1, 2, 3). The fourth PC stored the terra data, entity model, multimedia, and configure information used by the simulators. Although these four PCs had no routing function, they corresponded to routers described above from the view of RBs. Fig. 6 shows the result that Tank simulator queried the metadata of configure file tankFile and exploding effect model bombModel during the simulation, among which the x-axis is time with 30 seconds per unit and y-axis is the ratio that the metadata of desired resource was found in RB11 and RB1 respectively. The curves A and B are the ratios that bombModel metadata were found in RB11 and RB1 (Fig. 6 (a)); C and D are the ratios that tankFile metadata were found in RB11 and RB1 (Fig. 6 (b)) respectively.

As shown in Fig. 6, the metadata of tankFile were always stored in RB1 because the file was accessed only once by Tank simulator in the simulation process. As for bombModel, in the first 120 seconds, its metadata were stored in RB1, which was the initial place its metadata was stored in. At 120th second, enough sample data were produced and time series analysis' forecast showed that bombModel would be accessed frequently, so the metadata of bombModel were copied to RB11 and could query RB11 for its metadata. At about 210th second, the Tank simulator stopped firing and began to patrol, that is, bombModel was no longer accessed. Since the 300th second, the metadata of bombModel could only be obtained from RB1 because they were overwritten by metadata of other resources. This is expected.

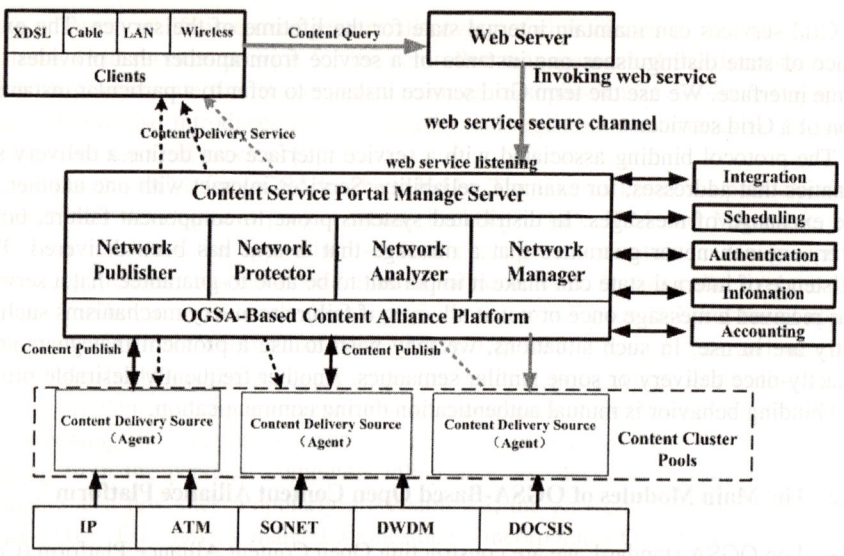

Fig. 1. Broadband Multimedia Content Alliance Platform based on OGSA.

the content registration is out of order and they are processed in the isolated environment, content can not be discovered easily and widely, including all kinds of content resources and meta-material including picture, image, audio, and video and more, provided by different content communities.

The caPlatform MServer performs a key function of integration and registration. The integration and registration service includes not only content directory, but also content services. Different hierarchical content directories can be collected into a total directory in Content Alliance Platform. At the same time, different content delivery sources can encapsulate their own content entities to become standard grid service using GWSDL, and register as a member of Content Alliance Platform to provide multimedia service. After registration, a content delivery source can be uniformly managed and checked in and out of the pool managed by caPlatform MServer according to the system's publishing standards. Checking in implies that the content delivery source can be used by caPlatform to provide multimedia service for other clients. By this platform, all kinds of content services can be discovered more easily.

(2) Content Discovery Service

Following that, the user might want to search for content services matching certain criteria or browse the resources available in the local caPlatform installation. By implementing a hierarchical content discovery platform using OGSA, users can easily discover the proper content from rich resources and be served by the proper servers, including on-demand server or real-time server. In a word, the goal is to provide multimedia content service as close to the user as possible to minimize content latency and jitter and to maximize available bandwidth speed.

The discovery platform interfaces to the directory service to search for content resources matching the requested criteria. The discovery target is not only to search static content directory, but also to search the matching dynamic service. In order to track the state of services, the discovery platform might also have to communicate with the database service to obtain more dynamic information such as system load over the past few hours or days on the selected computers. The latter information is needed when a user requests streaming service from a streaming server in resource pool for a certain period. The main information platform (including directory service and database service) is Globus Toolkit 3 Monitoring and Discovery Service (MDS), consisting of Grid Resource Information Service (GRIS) and Grid Index Information Service (GIIS).

(3) Content Cooperating Service
The most-promising aspect of OGSA is their ability to resolve the differences among shared, networked applications. Various web applications can be stitched together: Applications from different vendors, of various vintages; written in different languages; running on disparate platforms can easily communicate and cooperate. OGSA is a set of standards and technologies to integrate applications within the enterprise and also enable standards based integration with partner applications. We have argued that within internal enterprise IT infrastructures, SP-enhanced IT infrastructures, and multi-organizational Grids, computing is increasingly concerned with the creation, management, and application of dynamic ensembles of resources and services (and people) – of Virtual Organizations (VO) [3]. Depending on context, these ensembles can be small or large, short-lived or long-lived, single institutional or multi-institutional, and homogeneous or heterogeneous. Individual ensembles may be structured hierarchically from smaller systems and may overlap in membership. Open Grid Services Architecture supports the creation, maintenance, and application of ensembles of services maintained by VO. Virtualization allows for consistent resource access across multiple heterogeneous platforms with local or remote location transparency, and enables mapping of multiple logical resource instances onto the same physical resource and management of resources within a VO based on composition from lower-level resources. Virtualization allows the composition of services to form more sophisticated services – without regard for how the services being composed are implemented. Virtualization of Grid services also underpins the ability to map common service semantic behavior seamlessly onto native platform facilities [2]. One target of using VO is to realizing smooth service cooperation of all kinds of heterogeneous distributed content servers. The different content entities at the same VO or different VO can easily communicate and cooperate to provide uniform content service, while they are running on disparate platforms, provided from different vendors, containing different content resources. For example, they can realize striped parallel media file download.

(4) Authentication Service
Before any service can be obtained on behalf of a user, authentication must be done. The authentication platform is based on Grid Security Infrastructure (GSI) of OGSA. The PKI-based GSI protocol provides single sign-on authentication, communication protection, and some initial support for restricted delegation. In brief, single sign-on

allows a user to authenticate once and thus create a proxy credential that a program can use to authenticate with any remote service on the user's behalf. GSI uses X.509 certificates, a widely employed standard for PKI certificates, as the basis for user authentication.

(5) Scheduling Service
The caPlatform MServer will have to identify one that best suits the user's needs. For multi-thread task requests submitted by the user such as streaming service, the caPlatform MServer will forward the request to the scheduler, which will schedule the task on the local caPlatform cluster if possible. If needed, it will interact with a higher-level scheduler to request resources from other clusters that participate in the grid.

(6) Accounting and Payment Service
Accounting and payment service gathers resource and content usage information for the purpose of accounting, payment, and/or limiting of resource usage by community members.

3.2.4 Resource Pool (Members: Content Alliance Platform Server, caPlatform Server) and Agents

The main members of the resource pool are content alliance platform server (caPlatform Server), including all kinds of media center, from different vendors, of various vintages; running on disparate platforms. After authentication, caPlatform Servers can register themselves as a member of resource pool to provide multimedia service. After registration, a caPlatform Server can be checked in and out of the pool managed by caPlatform. Checking in implies that the caPlatform Server can be used by caPlatform to provide multimedia service for clients. At the same time, users can register their own workstations as a microserver to provide multimedia service such as file sharing like delivery servers. Each member will have a caPlatform agent running on it. The agent receives requests from the caPlatform Mserver, such as a request to launch the VOD server (one kind of caPlatform Servers) to provide content service. The agent also monitors use of the VOD server it is running on, and periodically sends status information to the caPlatform MServer. The agent can also ensure that the quality of service criteria used in selecting the system initially is maintained over time. In case the criterion gets violated the agent can aid in taking corrective action. When the user disconnects from the VOD server, the agent informs the caPlatform MServer, which returns the VOD server to the pool of available resources.

3.2.5 The Data Management Tools
The data management tools in the resource pool include Grid Access to Secondary Storage (GASS), GridFTP and Globus Replica Management provided by OGSA. GASS is usually used to transfer media files within the same media cluster. The GridFTP facility provides secure and reliable data transfer between various media clusters, such as large-scale media files. At the same time, with Globus Replica Management, overall system can store copies of the most hot media files on previous local storage for faster access, while keeping track of these replicated media files.

4 Conclusions and Future Work

OGSA-based multimedia content alliance service is a new distributed media content service scheme. The new scheme further being realized in global Internet will necessarily make this technology closer to business field and obtain more industry space. Before realizing delivery scheme on a worldwide scale, researchers must overcome many key technologies and carry on many experiments in test bed with carefully controlled trials, continuously validating its scalability and evaluating its actual performance.

References

1. Foster, I., Kesselman, C., Nick, J. and Tuecke, S.: The Physiology of the Grid: An Open Grid Services Architecture for Distributed Systems Integration. Globus Project, 2002. www.globus.org/research/papers/
2. Tuecke, S., Czajkowski, K., Foster, I., Frey, J. et al.: Grid Service Specification. www.globus.org/research/papers/, 2002
3. Foster, I., Kesselman, C., and Tuecke, S.: the Anatomy of the Grid: Enable Scalable Virtual Organizations. Int J. Supercomputer Applications, 2001
4. ZhiHui L., ShiYong Z., YiPing Z.: Research on Service Model of Content Delivery Grid. In the Proc. of APWeb'04, Lecture Notes in Computer Science, Vol. 3007. Springer-Verlag, Hangzhou (2004), 321-330
5. Muthucumaru M.: Main Topic- Part B: Next-Generation Middleware Platforms. Canadian Conference on Electrical & Computer Engineering2002, CCECE'02, May 2002
6. Zhihui L., Yiping Z., Shiyong Z., Jie W.: Study of Main Technology in Rich Media Grid Delivery. In the Proc. of 2003 International Conference on Computer Networks and Mobile Computing, ICCNMC'03, IEEE Computer Society Press, 2003
7. Gartner Group Research Note.: Delivery grid Surfaces in Enterprise: A Tutorial. January 2003
8. Sven, G., Winfried K., Carsten R.: Modeling and Simulation of Media-On-Demand Services – Evaluating a Digital Media Grid Architecture. HP Laboratories Technical Report HPL-2002-192, July 2002

An Ontology-Based Model for Grid Resource Publication and Discovery*

Lei Cao[1], Minglu Li[1], Henry Rong[2], and Joshua Huang[2]

[1] Department of Computer Science and Engineering, Shanghai Jiao Tong University,
Shanghai 20030, China
{clcao,mlli}@sjtu.edu.cn

[2] E-Business Technology Institute, The University of Hong Kong, Hong Kong, China
{hrong,jhuang}@eti.hku.hk

Abstract. Resource management system is the core component of a Grid system. It has two important functions: resource publication and discovery. This paper presents an ontology-based model for Grid Resource Publication and Discovery(GRPD). We adopt multiple domain-specific registries to manage corresponding resources of a Virtual Organization(VO) in order to obtain high GRPD efficiency. Resource descriptions and resource requests are all based on domain-specific ontology. The ontology-based matchmaker of the domain-specific registry plays the important role in resources selection. The "Index" node of a VO hosts the general registry. Other domain-specific registries are distributed in the VO. This is a two-level registry mechanism. A large-scale Grid system may contain many VOs. "Index" nodes from various VOs connect to each other in the peer-to-peer mode instead of the hierarchical mode.

1 Introduction

Grid technology is one of the most important technologies coming forth in recent years. It has emerged to enable large-scale flexible resources sharing among dynamic Virtual Organizations(VOs) in a networked environment. A basic service in Grid is resource discovery: given a description of resources desired, the resource discovery mechanism will return a set of (contact addresses of) resources that match the description[1]. We also call this process resource matching. Resource discovery in a Grid is a challenging task because of the following Grid features[2]: heterogeneous, dynamic, autonomic and numerous. These characteristics create significant difficulties for traditional centralized and hierarchical resource discovery services. Furthermore, existing resource description and resource selection in the Grid are highly strained because traditional resource matching is done based on symmetric, attribute-based matching. The exact matching and coordination between providers and consumers make such system inflexible and difficult to extend to new characteristics or concepts.

---

* This paper has been supported by the 973 project (No.2002CB312002) of China, grand project of the Science and Technology Commission of Shanghai Municipality (No.03dz15027)

In this paper, we present an ontology-based model for Grid Resource Publication and Discovery(GRPD). Resource descriptions and requests are all fundamental in GRPD. We use separate ontologies to describe resources and requests respectively. Instead of exact syntax matching, our ontology-based matchmaker performs semantic matching using terms defined in those ontologies. The loose coupling between resources and requests removes the tight coordination requirement between resource providers and resource consumers.

A large-scale Grid system may contain many VOs. Each VO has its own "Index" node that hosts the general registry. We build up multiple domain-specific registries in a VO, in accordance that ontology is domain specific. All heterogeneous resources in a VO will register themselves with those domain-specific registries in soft-state way via the general registry. This is a two-level registry mechanism. "Index" nodes from various VOs connect to each other in the peer-to-peer mode instead of the hierarchical mode.

The rest of this paper is organized as follows. We discuss related work in Section 2. Section 3 presents our ontology-based model in detail. Section 4 presents an ontology-based matchmaking example. We conclude in Section 5 with lessons learned and future research plans.

2 Related Work

Condor-G[3] combines the inter-domain resource management protocols of the Globus Toolkit and the intra-domain resource management methods of Condor to allow users to harness multi-domain resources. Condor-G agent formulates resources information and user requests in the Classified Ads resource specification language, and then uses the matchmaker to make brokering decisions based on symmetric, attribute-based matching. Obviously, the matchmaker becomes the system bottleneck especially when the system's scale becomes larger.

In Globus[4], resource information is managed in the Information Services (MDS3) that consist of resource-layer services and some of higher-level services (a collective-layer Index Service). There is typically one Index Service per VO but, in large organizations, several Index Services can be hierarchically included in a higher-level Index Server. The Grid community agrees that it is not easy to devise scalable Grid resource discovery based on centralized or hierarchical mechanism when a large number of Grid hosts, resources, and users have to be managed[5].

Legion[6,7] is a reflective, object-based operating system for the Grid. Proper scheduler objects use information from the collection and resource owners in making scheduling decision. Objects are used as the main system abstraction throughout. Co-allocation of resources is not supported.

EU Data Grid[8] was designed to provide distributed scientific communities access to large sets of distributed computational and data resources. A job request is expressed in the Classified Ads of Condor. Resource discovery is done by queries and employs periodic push for dissemination. It does not support advanced reservation or co-allocation of resources.

Nimrod-G[9] is a Grid-enabled resource management and scheduling system based on the concept of computational economy. It uses Globus middleware services for dynamic resource discovery and dispatching jobs over computational Grids. Nimrod-G has the same shortcomings as Globus by using Globus MDS.

3 Ontology-Based Model

3.1 Architecture

Figure 1 shows our model architecture in a VO. Domain-specific registries register their metadata with the general registry. Each one is responsible for resources registration of this domain and implements resource discovery by ontology-based matchmaking. We apply various domain ontologies in different specific registries. The mechanism strengthens both the efficiency of resources discovery and the scalability of the Grid system. We use redundant technology to enable the reliability of the Grid system. Each registry has its own online substitute. In our model we also use the P2P philosophy and techniques among VOs of the Grid system.

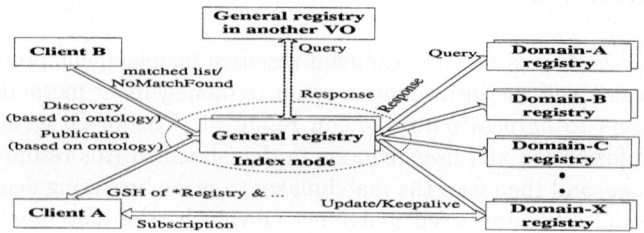

Fig. 1. Model architecture

3.2 Grid-Related Ontologies

The term ontology has been in use for many years. Today's use of ontology on the Web has a different slant from the previous philosophical notions. One widely cited definition of an ontology is "A specification of a conceptualization"[10]. It is domain specific so that there are many kinds of ontologies in the world. Using RDF-Schema[11], each ontology defines objects, properties of objects, and relationships among objects, which belong to one domain. We have designed and prototyped our matchmaker using existing semantic web technologies to exploit ontologies and rules for Grid resource matching. Resources in Grid environment are heterogeneous. We can simply group Grid resources into five domains. They are: (1)Computational resources (Cluster, PC, Supercomputer, Operating System, etc.), (2)Database and storage resources (Magnetic disk array, Optical disc library, Magnetic tape library, Oracle, Sybase, etc.), (3)Application resources

(Online game, Specific computing software, etc.), (4)Instrument resources (Telescope, Spectral analyzer, etc.), (5)Network resources (Rooter, Switcher, etc.).

Besides this, we have built five domain ontologies for Grid resources using Protege-2000[12]. Each ontology includes three parts: resource ontology, resource request ontology and resource policy ontology[13, 14]: The resource ontology provides an abstract model for describing resources, their capabilities and their relationships. The majority of our resource vocabularies are taken from the Common Resource Model(CRM)[15]. We extend it to fit our abstract description requirements. The resource request ontology focuses on a request, properties of the request, characteristics of the request and the resource requirements. The ontology supports requests of multiple independent resources. The resource policy ontology describes the resource authorization and usage policies.

3.3 Two-Level Registry Mechanism Used in a VO

A large-scale Grid system is usually comprised of a large number of heterogenous resources located in different organizations. To have a registry that holds the handles for every resource in the Grid would be impractical. On one hand it will be too large. On the other hand it will be updated frequently after resources are created or removed from the system. Figure 2 shows two kinds of registry structures in our two-level registry mechanism. The general registry contains two parts: the Super-index neighbour set and the Domain-specific registry set. The former will be introduced in the next section. The later contains the metadata (e.g., GSHs) of all domain-specific registries in a VO. We use domain-specific ontology vocabularies to describe Grid resources. Those metadata of various Grid resources from the same domain are stored in a resource registration database of a domain-specific registry. The database is based on soft-state updates from resource providers. There are three parts in a domain-specific registry (See Figure 2). The resource registration database has been mentioned above. The ontology-based matchmaker is responsible for resource selection. It also consists of three components[13]: The Domain ontologies contains the domain model and vocabularies for expressing resource publications and job requests. The Domain background knowledge contains additional knowledge about the domain that is not captured by the ontology. TRIPLE[16], a rule system based on deductive database techniques is used to implement the background knowledge.

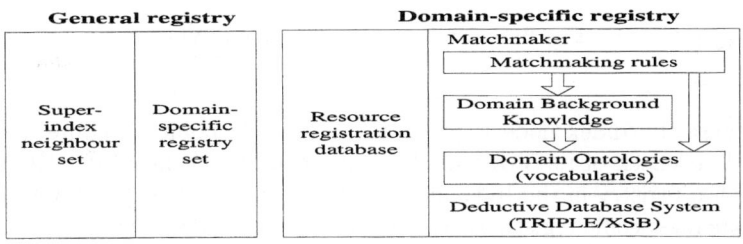

Fig. 2. Registry structures

The matchmaking rules define the matching constraints between requests and resources. We also use the TRIPLE rule language to implement these rules.

Domain background knowledge uses the ontology vocabularies to capture background information. Matchmaking uses both domain ontology and domain background knowledge to match a request to resources. Our ontology-based matchmaker is built on top of TRIPLE/XSB deductive database system.

Benefits of this two-level registry mechanism are: (1)Decreasing the centralized registry's workload and increasing resource discovery efficiency by distributing various domain-specific resource queries to corresponding registries; (2)Strengthening the system's adaptability because resources adopt soft-state method to register themselves and may only submit their keep-alive messages to domain-specific registry periodically. They can join or leave at any moment. (3)Improving the system's scalability. The system doesn't need to do much work but to remove (or add) the resource metadata from (or to) the domain-specific registry on the resource's departure (or entrance).

3.4 Super-index Network

In large-scale Grid environment, we may use the P2P techniques to implement non-hierarchical decentralized Grid systems. Use of P2P protocols is expected to improve the efficiency and scalability of large-scale Grid systems[17]. The Grid system contains many VOs. Every VO has its own Index node to host the general registry we mentioned above. We connect those Index nodes from different VOs into a P2P network like "Super-peer Network"[18]. Certainly, the Index node will contain the neighbor set of other VOs' Index nodes. Therefore we have two different scenarios: a typical Grid one in a VO, a traditional P2P one among VOs.

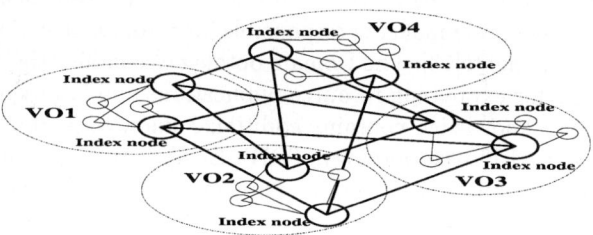

Fig. 3. Super-Index network among VOs

Because the Index node is prone to be a single point of failure and a potential bottleneck of its VO, we take some measures to avoid it: (1)Only host the general registry in the Index node. After Grid resources register themselves with domain-specific registries, they can exchange messages with them directly. (2)Use those nodes having the largest capabilities within each VO as the Index nodes. (3)Introduce redundancy into the design of the Index node to provide more reliability to the VO and less load on the Index node. However, doing so will add additional costs, so it's important to balance reliability and cost. In our model we have used two nodes as the "virtual" Index node to ensure good reliability.

The same reliability problem is faced by those nodes that host domain-specific registries. We have also used the same measures to solve them.

3.5 Grid Resource Publication and Discovery

Grid Resource Publication(GRP) is depicted as follows:

1. The resource provider wants to join the VO. It submits the publication request to the VO's general registry.
2. The general registry will verify the domain that the forwarded request corresponds with and deliver it to a domain-specific registry that is responsible for registration in the verified domain.
3. The domain-specific registry then returns success message to the general registry if the registration is successful.
4. The general registry returns the GSH of the domain-specific registry to the resource provider.
5. The resource provider will periodically send out the living messages to the domain-specific registry. Thus it is called soft-state registration.
6. The domain-specific registry may subscribe with the resource provider for some metadata. When subscribed metadata changes, the resource provider will notify the domain-specific registry in time.

Grid Resource Discovery(GRD) is described as follows:

1. The requester submits the resource query to the general registry.
2. As step 2 in GRP.
3. The ontology-based matchmaker in the domain-specific registry makes matching between resources and request.
4. The domain-specific registry will return the matched resource list (or NoMatch-Found) to the general registry.
5. If no requested resource can be found, the general registry will use some policies to select another general registry from its neighbor set of VOs. The search will continue until the requested resources can be found or the Time-to-Live(TTL) expires.
6. If the requested resources can be found, the matched resource list will be sent back to the requester via the general registry.

Due to the paper space, the sequence diagrams of GRP and GRD are omitted.

4 Ontology-Based Matchmaking Example

Here we present a matching example that cannot be done easily by attribute-based matchmakers. We utilize our ontology-based matchmaker in the example.

Property Name	Property Values
MDiskArray.Name	"MDA1.cs.sjtu.edu.cn"
MDiskArray.AuthorizedGroup	"cs@sjtu.edu.cn"
MDiskArray.NumberofAvailableDisks	20
MDiskArray.RAIDLevel	3
MDiskArray.Cost-perGB	$30
MDiskArray.IORate	200MB/S
MDiskArray.MaxCapacity	10TB
MDiskArray.UsingControler.type	"Hard"
MDiskArray.UsingControler.name	"Disk-controling card"

(a) A Magnetic Disk Array with 10TB MaxCapacity

Property Name	Property Values
MTapeLibrary.Name	"MTL1.cs.sjtu.edu.cn"
MTapeLibrary.AuthorizedGroup	"cs@sjtu.edu.cn"
MTapeLibrary.NumberofAvailableTapes	10
MTapeLibrary.LinkType	"SAN"
MTapeLibrary.Cost-perGB	$10
MTapeLibrary.IORate	35MB/S
MTapeLibrary.MaxCapacity	50TB

(b) A Magnetic Tape Library with 50TB MaxCapacity

Property Name	Property Values
JobRequest.Name	"Request1"
JobRequest.Owner	"User1"
JobRequest.JobType	"Save data online or nearline"
JobRequest.NumberofResources	1
JobRequest.RequestResource.ResourceType	"large-capacity storage system"
JobRequest.RequestResource.RankBy	"Cost-perGB"
JobRequest.RequestResource.MinStorageSpace	9TB
JobRequest.RequestResource.MinIORate	25MB/S

(c) Job Request

Fig. 4. Job request and available resources

Figure 4(a, b) shows examples of two instances of resources: a 10TB-MaxCapacity Magnetic Disk Array and a 50TB-MaxCapacity Magnetic Tape Library. We only list some relevant properties to the example. In the example, both storage resources belong to the Department of Computer Science and Engineering, Shanghai Jiao Tong University. They only allow users in the "cs@sjtu.edu.cn" group to access the resources. Figure 4(c) shows an example of a job request specifying that it wants one "Large-capacity storage system" resource for a "Save data online or nearline" job. The resource requirements are also specified in the list. As our background knowledge indicates that a "Save data online or nearline" job can be done by MDiskArray or MTapeLibrary systems, so two storage systems both are candidate resources. Assuming that User1 has an account that belongs to the "cs@sjtu.edu.cn" group, User1 is authorized to access both storage systems. The matchmaker then checks the capabilities of both resources against the resource requirements. Again, since our background knowledge specifies that these two storage systems are "large-capacity storage system", both resources pass the "RequestResource.ResourceType" requirement criteria. Because both resources are compatible with the resource requirements, the "RankBy" is used to select the match. Since the "Cost-perGB" of "MTL1.cs.sjtu.edu.cn" is lower than that of "MDA1.cs.sjtu.edu.cn", the matchmaker returns "MTL1.cs.sjtu.edu.cn" as a match.

5 Conclusion and Future Work

This paper has described an ontology-based model for GRPD. We focus on the grouping of Grid resources to build some domain-specific registries in a VO and give a two-level registry mechanism. The ontology-based matchmaker that locates in every domain-specific registry is responsible for the semantic matching between resources and requests. Doing so increases both the efficiency of resource discovery and the scalability of the Grid system. Using P2P techniques we construct a Super-Index network among VOs of the Grid system. In the near future, we plan to do a practical performance evaluation to show that our ontology-based model can be efficiently used in the Grid environment.

References

1. A. Iamnitchi, I.F.: On fully decentralized resource discovery in grid environments. In: Proceedings of International Workshop on Grid Computing. (2001)
2. Y. GONG, e.a.: Vega infrastructure for resource discovery in grids. Computer Science and Technology **18** (2003) 10
3. J. Frey, e.a.: Condor-g:a computation management agent for multi-institutional grids. In: Proceedings of the 10th IEEE International Symposium on High Performance Distributed Computing. (2001)
4. S. Tuecke, e.a.: Open grid services infrastructure (ogsi) version 1.0. In: Specification of GGF by OGSI-WG. (2003)
5. C. Mastroianni, D. Talia, O.V.: P2p protocols for membership management and resource discovery in grids. (2004)
6. S.J. CChapin, e.a.: The legion resource management system. In: Proceedings of Job Scheduling Strategies for Parallel Processing. (1999)
7. A. Natrajan, M.A. Humphrey, A.G.: Grid Resource Management In Legion. Kluwer Academic Publishers, Virginia (2004)
8. W. Hoschek, e.a.: Data management in an international data grid project. In: Proceedings of the 1st IEEE/ACM International Workshop on Grid Computing. (2000)
9. R. Buyya, D. Abramson, J.G.: Nimrod/g: An architecture for a resource management and scheduling system in a global computational grid. In: Proceedings of the 4th International Conference on High-Performance Computing in the Asia-Pacific Region. (2000)
10. Gruber, T.: A translation approach to portable ontology specification. In: Proceedings of the Knowledge Acquisition Workshop. (1992)
11. D. Brickley, R.G.: Rdf vocabulary description language 1.0. In: RDF. (2004)
12. Protege: Protege2000. In: http://protege.stanford.edu/. (2004)
13. H. Tangmunarunkit, S. Decker, C.K.: Ontology-based resource matching in the grid-the grid meets the semantic web. In: Proceedings of the 1st Workshop on Semantics in Peer-to-Peer and Grid Computing. (2003)
14. P. Pothipruk, P.L.: An ontology-based multi-agent system for matchmaking. In: Proceedings of the 1st International Conference on Information Technology and Applications. (2002)
15. E. Stokes, N.B.: Common resource model(crm). In: Specification of GGF by CMM-WG. (2003)
16. M. Sintek, S.D.: Triple-a query,inference,and transformation language for the semantic web. In: Proceedings of the 1st International Semantic Web Conference on The Semantic Web. (2002)
17. D. Talia, P.T.: Toward a synergy between p2p and grids. IEEE Internet Computing **7** (2003) 3
18. B. Yang, H.G.M.: Designing a super-peer network. In: Proceedings of the 19th International Conference on Data Engineering. (2003)

A New Overload Control Algorithm of NGN Service Gateway

Yun-Yong Zhang, Zhi-Jiang Zhang, Fan Zhang, and Yun-Jie Li

Postdoctoral Programme, Technology Department, China Unicom,
100032 Beijing, China
{zhangyy,zhangzhj,zhangf,liuyj}@chinaunicom.com.cn
http://www.chinaunicom.com.cn/

Abstract. New requirements to the overload control of Parlay gateway are analyzed. And the maximum revenue based overload control model of Parlay gateway is proposed. Furthermore, the network-based overload control architecture based on the agent linear programming theory is proposed. And network overload control architecture based on linear programming is given out. Simulation results show that using agent linear method, it is quick to find out maximum revenue based overload control algorithm; also it is suitable for large-scale softswitch network.

Keywords: Next Generation Network (NGN), Softswitch, Overload Control, Mobile Agent, Linear Programming, Maximum Revues

1 Introduction

With the development of network, current fixed network, mobile network and Internet will be converged based on IP protocol [1]. Next Generation network is service driven and service development and deployment is very important so as to utilize the network capacity. The services can be opened up to third party using OSA/Parlay [2].

Overload control mechanism must be included in Parlay gateway so as to let Parlay gateway serve as NGN telecommunication level device. Overload control methods have been researched in SCP. In SCP, overload control can be divided into two types, which are access control and overload detection. Access control methods include windows based method, callgap method, ACG method, percent method and token bucket method [3~5]. Overload detection methods include queue length based method, number of calls based method, average response time based method and timeout of queued message based method.

Comparing with SCP, Parlay service has more types such as basic two-party call, multi-party call, message service and mobile service. Overload control mechanism is more difficult than SCP due to the factor that different service has different time limit and Parlay gateway must listen to the request of both softswitch and application server.

Current overload control methods in Parlay gateway are based on nodes and lack of flexibility. In this paper, network based overload control methods are put out so as to satisfy maximum service revenue requirement, improve network efficiency, divide subnets according to regions and enhance the sensitivity.

2 Overload Control Model

As shown in Fig.1, service requests are classified into different priorities by service classifier, requests in different priority queue are scheduled according to schedule algorithm. Rate of CPU utilization, resource and service arrival rate of softswitch are measured by system. When system is detected to be overloaded, overload control method works.

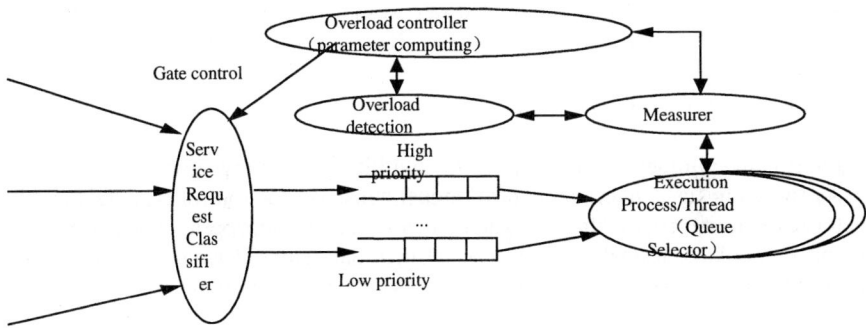

Fig. 1. Parlay overload control model.

Definition 1: Overload Condition
The load condition is described by three levels. Load level 0 corresponds to normal load, load level 1 corresponds to overload and load level 2 corresponds to severe overload. Different Parlay gateway can have different threshold values for the load levels.

Definition 2: Priority
Each priority corresponds to a guaranteed rate of application calls per second, where $p_j(j=1,2,...M)$ guarantees a higher rate than p_j+1. Priority is independent with time constraint and each application can has different priority.

Definition 3: Time Constraint
A time constraint T_k corresponds to the maximum delay a message should experience each time it passes the Gateway. Time constraint is independent with priority and each application can has different time constraint. The time constraints are set such that $T_k < T_k+1(k=1,2...N)$. The set of applications with time constraint k is denoted $A(T_k)$, The total guaranteed rate of applications with time constraint k, is given by

$$\lambda_k = \sum_{l \in A(T_k)} d_l .$$

We denote with N(t) the number of messages in the gateway with a remaining time less than t before their deadline expires. x_{GW} is an estimation of the execution time of a random message in the gateway.

Definition 4: Utility Function

Utility function describes the relation of income revenue and consumed resource. Utility function $U(l) = \dfrac{r(l)}{x_{tot(l)}}$, where r(l) is the revenue. The above expression gives a good *estimation of the utility seen from the aspect of revenue in the short run*. But in the long run it should be an advantage for the operator to have an extra goodwill parameter, G, that could be set individually and be added to the above expression.

An application, A_l, has a guaranteed rate of d_l calls per second, and a total execution time in the Gateway of $x_{tol(l)}$ seconds. Of course the system must be stable when all applications face their guaranteed rate, which implies that $\sum_{i=1}^{R} d_l . x_{tot}(l) < 1$, where R is the number of applications. Each utility corresponds to a priority.

As shown in fig.2, overload control model consists of a controller, a gate and a selector (execution process/thread). The *controller* makes appropriate measurements on the gateway. Also, it analyses the measurement data and determines what action that has to be taken by the *gate*, which regulates the acceptance of new application calls.

Algorithm of controller:
  while(controller examines the overload state of Parlay gateway)
  {
  Call agent linear programming algorithm;
  if(selector can pick the messages in some order that none of the time constraints is expired)
   {controller tell the gate to increase the acceptance of new application calls}
   else
   { controller orders the gate to decrease the acceptance of new application calls
   }
  }

Algorithm of selector:
As different applications can have different time constraints the selector has to decide in which order the messages in the gateway should be served. The selector uses an Earliest Deadline First (EDF) scheduling algorithm.

Algorithm of gate:
  if(Parlay gateway is overloaded)
  {
  let the gate use a call gapping method to reject application calls. The time is divided into small intervals of a certain length, and then the first application call in the interval is accepted. The interval lengths depend on the guaranteed rate the application has.
   If(in level 1 overload status)
   {
   gate introduces call gapping on the lowest priority applications;
   if(the overload condition remains after X seconds)
   {call gapping is introduced on application calls of the next priority level;}
   }

else if (in level2 overload status)
{ all the priority levels are blocked at once, only letting the guaranteed amount of application calls through
}
}

3 Linear Programming Based Network Overload Control

3.1 Model

For each arriving or departing message the controller checks if the time constraints for the messages waiting in the queue may fail if the message is admitted. The following condition should of course always be fulfilled:

$N(T_k) \cdot x_{GW} <= T_k (1<=k<=N)$, If not fulfilled, application calls with time constraint T_k will most probably fail even if the gate starts to reject arriving application calls at this stage.

Through further discussion, above condition should be improved to also include the guaranteed rate of new application calls. This condition can be expressed as

$$[N(T_k) + \sum_{j=1}^{k-1} \lambda_j (T_k - T_j)] \, x_{GW} < Tk (1<=k<=N) \quad (1)$$

Also, controller should also check that the condition (1) is not violated in the future by admitting too many calls from applications with less tough time constraints. This new condition can be described as:

$$[N(T_i) - N(T_i - T_k) + \sum_{j=1}^{k-1} \lambda_j (T_k - T_j) + \sum_{j=k}^{i-1} \lambda_j T + \sum_{j=1}^{k-1} \lambda_j T_j] \, x_{GW} < T_k \quad (2)$$

$(2<=I<=N, i>k)$

To get a more calmly behavior, the controller uses a marginal when signaling for overload. This marginal is created by multiplying the right hand of the conditions with a marginal factor, $f<1$. If any of the conditions are violated when the right hand side is multiplied with f, the controller signals overload (load level 1). If any of the conditions are violated without the marginal factor, the controller signals severe overload (load level 2) to the gate.

Definition 5: Total Utility Function

The total utility can then be defined as $U_{tot} = \sum_{l=1}^{R} s(l) \cdot x_{tot(l)} \cdot U(l)$, where s(l) is the number of served application calls from application l.

From the point of view of maximum running revenue, network based overload control is a optimization question with constrained conditions. The total running revenue of all services can be used as object function with constrained conditions (1) and (2) and linear programming theory can be introduced to solve this optimization problem.

3.2 Agent Based Linear Programming Solution

The concept of mobile agent [6] is brought forward by General Magic corporation in the 1990's when it distributed Telescript. Mobile agent is a program that can move from a host to another autonomously in heterogeneous network environment, it is the mixture of agent technology and distributing computing technology. While the interact between the traditional RPC client and server needs continuous communication, mobile agent can move to server and communicates with server locally and the local communication need not occupy network resources.

Mobile agent has many advantages: mobile agent migrates service request to server dynamically to avoid transferring of the large data in network, so it reduces the dependence of network bandwidth, it dos not need central schedule and agent can run in different network nodes simultaneously and asynchronously, it is autonomous and it can routes intelligently.

Characters of parallel computing, goal driving and less time complicacy in mobile agent are very suited for dealing with the problem of integer linear programming.

Definition 6: Agent
Agent is a quaternion set with object, clone, action and collaboration and can be expressed with G. The behavior of an agent can be described as the following:

3.2.1 Clone
Each agent G can be cloned into two sub agent G_1, G_2, which has the same object but different context environment with G. G is called father agent, while G_1, G_2 is called son agent. One agent can be both father agent and son agent the same time.

3.2.2 Action
Agent G will always computing its object according some manner until the object is achieved or some termination condition is met. In the above tow conditions, agent will be disposed.

3.2.3 Collaboration
High collaboration of mobile agent is achieved using agent communication language that is based on speech act theory. The best known ACL is KQML (Knowledge Query and Manipulation Language), developed by the ARPA knowledge sharing initiative and FIPA(Foundation of Physical Intelligent Agent) ACL.

G is cloned into G_1, G_2, where goal of son agent is the same as G, while environment is divided into two sub environments(say E_1 and E_2), which is different from the environment of G(say E). G_1 and G_2 computes its goal in individual environment, then send the goal to father agent using agent communication language KQML and the father agent G determines whether the goal is achieved and whether the action of G_1 and G_2 should be terminated according the sub goal of G_1 and G_2.

3.3 Solving Algorithm

Above agent model can be used to solve the linear programming problem using relaxion B as the goal and environment of G. G is solved according to the simplex method. G is cloned into G_1, G_2(corresponding to the sub question B_1, B_2) and agent computing the solution according the Branch and Bound steps until the integer solution are met. The whole algorithm is shown as the following:

(1) Let the object and environment of agent G be the object function and constrained conditions of linear programming problem. Define a variant U(Utility function) that is assigned value NULL.
(2) Agent G solves relaxation problem
if(all optimized solutions are integer (U has achieved optimized value)||relaxation problem has no solution)
    {agent will goto (6)}
    else
    {agent goto (3)}.
(3) If the value of variant λ_j is a fraction (say b_j){
Computing the maximum integer $[b_j]$ which is less than b_j and the minimum integer $[b_j]+1$ which is lager than b_j
    Starting the clone of agent G(the son agent are G_1, G_2), where object(goal) of son agent is the same as G, while environment is the environment of G plus the constrained conditions $x_j <= [b_j]$ and $x_j >= [b_j]+1$.}
(4) Agent G_1, G_2 computes its individual object;
    if optimized solution(integer) has been solved{
        If(U!=NULL)
        Then
        {U is the optimized value}
        Else
        {U=max(U, optimized value)}}
(5) For the non-integer optimized value Gi in G_1, G_2
    if(U is NULL || U< G_i){then G=G_i;
    goto(3)}
(6) If(U!=NULL){
    printf(U);}
    else{Display that linear programming conditions is not right}

In the above algorithm, the whole network is divided into different subnets according the factors such as distance, time delay. Each agent communicates others using KQML language and detect the overload using Advertise KQML performative. Each agent is only responsible for the overload controlling in its local subnet, so it can reduce the numbers of variants in the algorithm and the execution time of algorithm.

3.4 Implement

Aglet and jKQML are used to implement above model. Context is used by aglet to manage the behaviors of mobile agent, which include agent creation'agent clone, agent dispatch, agent retraction, agent deactivation, agent activation and agent disposal.

    Firstly, Aglet creates agent G'then aglet clone is executed when agent need to clone agent G_1 and G_2, aglet dispatch is executed to dispatch agents to specific environment and aglet disposal is executed to dispose agent.

Time complexity of pure linear programming algorithm is $O(n)$ while time complexity of agent based algorithm is $O(\log^n)$. Also agent-based algorithm has the following advantages:
- Problem solving process is collateral due to simultaneous clone in agent;
- Problem solving process is object driven due to the BDI (Belief, desire and Intention) architecture in agent;
- and algorithm has flexibility and scalability due to the openness of agent.

4 Experiment

Three applications, two different time constrains and three different priorities are used in the experiment. Each application has different execution time and different numbers of Parlay messages in Parlay gateway and different delay in application server and softswitch network. Only when continuous five messages are overloaded, the controller changes overload status. Parlay call sequences are createCall, routeReq, routeRes, deassignCall, CAPeventReportBCSM, CAPeventReport and configuration in experiment is shown in Table 1.

Table 1. Experiment configuration.

APP	Parlay execution time (s)	AS execution time(s)	SS execution time(s)	Priority	Time limit (s)	X	f factor
1	0.001, 0.002, 0.002,0.002, 0.002, 0.001	0.001, 0.001, 0.001	0.008, exp(2,0)	1	0.1	50	0.9
2	0.002, 0.003, 0.003,0.003, 0.003, 0.002	0.002, 0.002, 0.002	0.01, exp(0.01)	2	1.0		
3	0.0001,0.0002, 0.0002,0.0002, 0.0002, 0.0001	0.001, 0.001, 0.001	0.01, exp(2.0)	3	0.1		

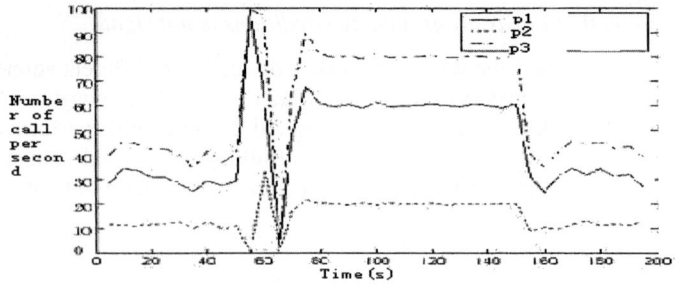

Fig. 2. Number of calls according to different priority.

As shown in Fig.2, after overload control algorithm started up and initialized, number of calls for high priority service is 80 while for low priority service is 20. The overload control algorithm is both impartial and efficient. Supposing the individual utility function of priority 1, priority 2 and priority 3 is 3, 2 and 1, then the total utility function is 2280, which approach the theory value 2350.

5 Further Work

Overload control mechanism can be affected due to the self- similarity of network signal messages. The distribution of above signal message arrival is expressed by $R(\tau) \sim \tau^{-\alpha} (0 < \alpha < 1)$, which is different with traditional Possion distribution. On the other hand, in the next generation network, softswitch network, application server, data server, media gateway, media resource server are used to run service, so these devices must be included in overload control algorithm. Due to the more complex operation, ant colony based agent can be used in its overload control computing.

References

1. Hubaux. The Impact of the Internet on telecommunication architecture. Computer Networks, 1999,(31):257~273.
2. Parlay Group, "Parlay 4.1 Specification," http://www.parlay.org/, 2002.
3. Donald E. Smith. Ensuring Robust Call Throughput and Fairness for SCP Overload Controls. IEEE/ACM Transactions on Networking, 1995, (5):538-548.
4. M.Kihl, C.Nyberg. Investigation of overload control algorithms for SCPs in the intelligent network. IEE Proc.-Commun., 1997, (6):419-424.
5. A.Karmouch, Special Section on Mobile Agents, IEEE Communications, 1998,(7): 1-10.

Nodes' Organization Mechanisms on Campus Grid Services Environment

Zhiqun Deng[1], Zhicong Liu[1], Hong Luo[1], Guanzhong Dai[1],
Xinjia Zhang[1], Dejun Mu[1], and Hongji Tang[2]

[1] Control & Networks Institute, College of Automation,
Northwestern Polytechnical University, Xi'an, 710072, China
zhiqundeng@tom.com
[2] Department of Computer Science,
East China Normal University, Shanghai, 200062, China
yb02241001@student.ecnu.edu.cn

Abstract. The Campus Grid Services Environment provides many services such as the computing service, storage service, communication service, and research and education services. For the storage service, the multilevel linked lists are built to organize the storage resources contributed by the nodes of the campus network for effective utilization of the storage resources. Additionally, the mechanisms of nodes' joining and leaving, nodes' storage resource scheduling and allocation, and nodes' unallocated storage spaces management algorithms are presented in detail. The total nodes number aggregated is 16,843,261. The storage space aggregated is attained to 17 PB, much larger than those of other current campus grids. This architecture is scalable and more nodes can join this storage pool to contribute a more huge storage space.

1 Introduction

Unlike WAN, the campus network is not too complex and most users are students and staffs of the university. Based on the campus network, campus grids are constructed in many universities and institutes. Although there are still discussions whether campus grids belong to Grid Computing, they indeed provide a means to share computing, storage, and information. Campus grids coordinate all of PCs, supercomputers and clusters in the campus network to deliver nontrivial qualities of services.

Hiroshi Arikawa [1] presented the node selection mechanism based on the node usage pattern on the campus computational grid. Campus grids for computing are being constructed in China. This kind of campus grids offers the computing service with a high computing efficiency.

Campus grids for computing are different from the high performance centers. Most high performance centers consist of supercomputers, which are very expensive. But the cost of the campus grids is relatively low. In addition, campus grids deliver other services as well, such as storage service, communication service, and etc.

In this paper, we give a new construction of the campus grid by aggregating all the resources of the voluntary nodes within the university. The present work is to organize the storage of the voluntary computers, by which users can store the data in the storage pool aggregating by these computers. The objective is to link all the nodes in the network layer. In a sense, nodes' organization mechanisms being built are below the application level. While, the EzGrid [2] campus storage grid is in the application level.

In the following of the paper, the Campus Grid Services Environment is presented in Section 2. Section 3 proposes nodes' organization mechanisms and the multi-level linked lists. Section 4 presents nodes' joining and leaving mechanisms. Section 5 presents the nodes' storage resources allocation. Section 6 discusses nodes' unallocated storage space management. Section 7 gives the conclusion and future work.

2 Campus Grid Services Environment

The Campus Grid Services Environment (CGSE) is to facilitate collaboration and sharing of resources and services. Campus grid services consist of computing service, storage service, communication service, education and research service, and etc.

The computing service is mainly the high performance computing applying to Aeronautics, Astronautics, and Marine science, which are the main subjects in the university. The Center for High Performance Computing in the university generally provides such a service. In a sense, this service belongs to the education and research service, however, due to the much attention paid on the high performance computing, this computing service is distinct from other education and research services.

The storage service, which we are implementing, is to aggregate the storage resources in tens of thousand voluntary computers in the university.

The communication service is another killer application in the CGSE. The CGSE is a platform for information exchanging. Any user anywhere anytime in the campus can communicate with each other at a low cost. Typically, a medium size university has more than thirty thousand people. Thus it is not necessary to buy extortionate communication devices to communicate with a sky-high price in the campus. The communication service aims to build an internal communication environment at a very cheap communication price. The communication nodes may be the special cheap devices similar to the cell phone, or the computers (see Fig.1.)

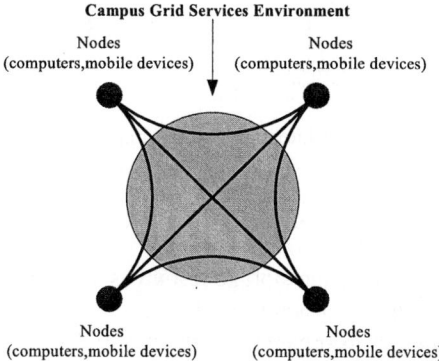

Fig. 1. Communication service based on the Campus Grid Services Environment.

The education and research services are the main services in the CGSE. And they are also the aims to construct the campus grid. In the CGSE, there are all kinds of services about the education and research, such as software, lecture videos, real time video courses (including the courses of remote education), etc.

Of course, there are still other services need to be further explored in the future.

The CGSE is based on the campus high-speed network. Although computers are geographically distributed, they are in the same region. The distances between multi-institutes in the same university are approximately ignored while constructing the campus grid. The transmission delay will not be considered (but this must be considered in WAN).

3 Nodes' Organization Mechanisms

3.1 Campus Network Environment

Firstly, we should understand the campus network environment. In the campus network, assuming there are at least twenty colleges or institutes. Nowadays, each college or institute usually has an enterprise server with a huge storage capacity. For example, there are one SUN FIRE v880 server (with a capacity of 73GB*6), SUN Ultra Enterprise 450(36GB*4), SUN StorEdge 3500(876GB) in Control & Networks Institute of the College of Automation. Thus the total storage capacity is at least 1TB in Control & Networks Institute. Consequently the storage capacity that all of the colleges' or institutes' servers can contribute will be huge. It will be the main storage capacity in the CGSE. Additionally, in the Campus Grid Center (CGC), there is a HP rx2600 Cluster (416G Flops with a storage capacity: 1TB). While, there still are at least ten thousand PCs or workstations in the campus network. If each host provides 1GB storage capacity, the overall capacity aggregated will be more than 10TB (see Fig.2.).

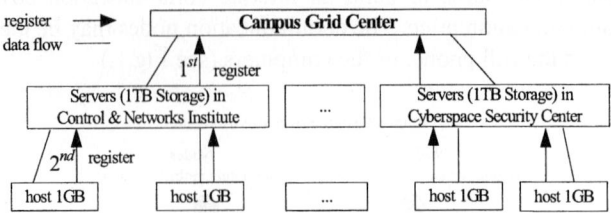

Fig. 2. Nodes' organization architecture in the Campus Grid Services Environment.

3.2 Nodes' Organization Mechanisms

How to effectively organize the storage resources contributed by all the nodes? In this paper, a multilevel index structure (see Fig.3.) similar to the UNIX files index mechanism [3] is proposed to organize all the hosts. There is a global index in the CGC and a local index in the server of each college or institute. Such organization mechanisms have some merits: It is easy to append and delete the node in the index.

Also, the nodes can be accessed directly and randomly. In addition, the speed to locate the host with specified contents is fast.

In Fig.3 the Host Node Index (global index) is stored in the CGC, a high available system. The CGC administrates the entire grid services environment. The total nodes number is $253+256+256^2+256^3=16,843,261$.

The Direct Nodes 0~252 mean the servers in each college or institute. There are 253 servers in Direct Nodes. These servers possess huge capacities and provide reliable consistent services (running 24 hours). So their storage capacities can be accessed directly all the time. Each server provides at least 1TB storage capacity.

While, Single Indirect (253rd pointer) means that nodes are linked to an index node. There will be 256 nodes by this means. These nodes are always on the campus grid network, and can provide services within a less time than the servers. They are the extended servers of the Direct Nodes. Similarly, each node provides at least 1TB storage capacity.

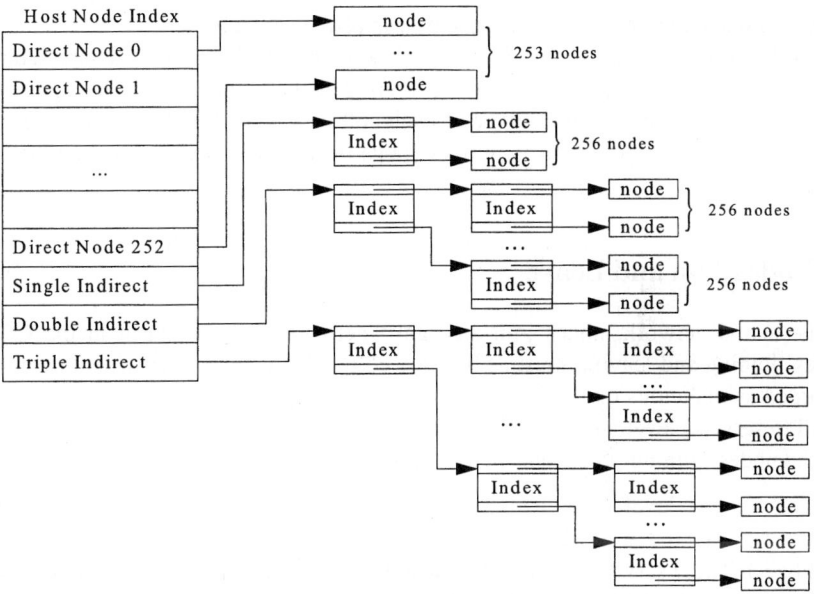

Fig. 3. Nodes' organization mechanisms.

More nodes can use the Double Indirect (254th pointer) to be linked in the CGC. These nodes are often in the campus grid. For example, the nodes are alive on the working hours (from 9:00AM to 5:30 PM). And the storage resources provided are only available in the daytime. These computers mainly refer to those in offices. Every node contributes at least 1GB storage capacity.

More computers can use the Triple Indirect (255th pointer) to join the pool of the CGC. These nodes are temporarily in the CGSE. They join and leave randomly, but still are alive at some time.

The total storage capacity is as follows:
The storage capacity of Direct Nodes 0~252 is 1 TB*253=253TB (Terabytes).
The storage capacity of Single Indirect is 256 TB.
The storage capacity of Double Indirect is 256*256 GB=65 TB.
The storage capacity of Triple Indirect is 256*256*256 GB $\approx$ 16777 TB $\approx$ 16.7 PB (PetaBytes).
And the total storage capacity is approximately 17351TB (i.e.17.351PB).

In fact, the voluntary computers can contribute more storage capacities, not only 1GB, and may be 2GB, or more. So the actual maximum storage summed is more than 17.351PB. The storage pool aggregated will be huge enough, as can compare with the campus grid in University of Houston, which owns more than 4000GB storage [4].

If the nodes' number is large enough, the linked list can be extended to the larger one. Nodes' number in this work is 256. Actually, the direct nodes number can be 512, and more according to the CGC computers' throughput. This means the whole system is scalable and the storage capacity can also be extended.

All the computers in the campus network will be linked by this structure. This index structure is stored in the CGC and is referred as the global index. The same index structure can be stored in the server of each college or institute in a similar manner. This index is to mange the hosts in the college or institute and is termed as the local index.

4 Nodes Join and Leave

In the actual network, nodes join and leave network randomly. But in the CGSE, nodes' joining and leaving mustn't be allowed too frequently; otherwise they will be regarded as malicious ones. The processes of nodes' joining and leaving can be classified to servers' level and hosts' level.

The nodes' joining processes are as follows:

- The servers' level refers to the storage servers in the colleges or institutes (see Fig.2.), which are the Direct Nodes 0~252 and the Single Indirect Nodes (see Fig.3.). These servers are directly linked to the CGC. They directly send the register requests to the Certificate Center in the CGC. This is the first level register. They provide consistent storage services, and these servers are also the proxies for the nodes in the hosts' level.
- The hosts' level refers to the hosts in the colleges or institutes, which are the Double Indirect Nodes and the Triple Indirect Nodes (see Fig.3.). These hosts submit register requests to the proxy servers, the servers in the colleges or institutes. In our scheme, we adopt the proxy mechanism like MyProxy [5]. Due to computers in the network of each college or institute, their joining and leaving messages are only to be informed to the servers in the college or institute. These servers as proxies will send these hosts' register information to the CGC periodically. In Fig.2, the register process can be seen.

In general, the nodes' joining must pass the authentication, and claim that how many storage capacities they can provide and how long time they can serve. If the storage capacity contributed is not enough or the service time is too short, the nodes will not be allowed to join the CGSE.

How to encourage the nodes to join the campus grid? We build a prize mechanism for the voluntary computers. If the node joins the CGSE, and contributes at least 1 GB storage space (the user cannot read or write it), it can get two size of the same storage capacity, thus the node can get at least 2GB storage capacity. (The storage capacity contributed varies at users' desire; he may contribute more than 1GB storage capacity.)

Another prize mechanism is based on the storage economy according to the service time and the storage capacity. This mechanism needs to be further explored.

The nodes' leaving processes are follows:

- If the servers want to leave the CGSE, due to the servers directly linked to the CGC, they should send the logoff requests to inform the CGC, who will know the servers' dynamic storage capacities in time. At the same time, the CGC will periodically check whether the servers are alive. Thus even if the servers leave the system abruptly, they can also be detected.
- If the hosts in each college or institute leave the system, they will send the logoff requests to the servers in the colleges or institutes that they belong to. The servers will periodically report the hosts' storage capacities to the CGC. Similarly, the servers also detect whether the hosts are alive.

5 Nodes' Storage Resource Scheduling and Allocation

Since there are huge storage spaces aggregated, how to effectively schedule them and allocate the data to the right capacity in the node decentralized in the CGSE?

The CGC is responsible for the servers' storage nodes scheduling; while the servers manage the hosts' storage nodes scheduling. We adopt the local scheduling first policy.

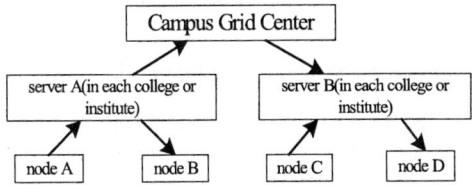

Fig. 4. Storage nodes scheduling and data allocation.

In Fig.4, Node A submits the request to the server A. The server A will firstly search its own storage capacity: whether there is a block to store the data. Otherwise, it will search in the local index to lookup whether there is a proper storage capacity for the data. If there is a node with an enough storage capacity, the server A will inform node A about the node address. Then node A will store the data in that node. If there is no node to store the data, the server will submit this request to the CGC. The CGC will further continue this scheduling.

6 Nodes' Unallocated Storage Space Management

In the CGSE, the storage capacities and the data are dynamically changed with time. When the data are being written, new storage block in the host must be allocated to them. Additionally, the old data will be deleted and the allocated space will be freed. There must be monitors to administrate the unallocated storage space. Similarly, the unallocated storage spaces management is also based on the multilevel: the servers' level and the hosts' level (see Fig.5.).

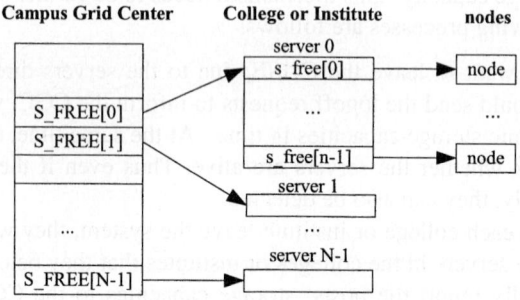

Fig. 5. Unallocated storage space linked list.

6.1 Servers' Level Unallocated Storage Management

There are 509(253+256) servers in the CGSE. (The servers' number can be added or reduced according to the requirement of the campus network). In fact, the current total number of servers is much less than 509. Each server's storage capacity is at least 1TB. The minimum block size allocated to each requester is 1GB.

The unallocated storage space is managed by linked lists. Each server's storage contexts such as the unallocated storage size and server's address, and etc, are stored in the CGC. And the linked lists are also stored in the CGC (see Fig.5).

6.2 Hosts' Level Unallocated Storage Management

The data storage space requested is allocated according to the block. The block size that each requester is allocated is 1GB, that is, the storage capacity that each host contributes. A user may request 0.5GB storage space, but the system still allocates him 1GB. Similarly, the storage context of each host is stored in the server, and the server manages the linked list of hosts.

7 Conclusion and Future Work

Our work focuses on aggregating the storage resources of all the voluntary computers in the Campus Grid Services Environment and providing the storage services. The multilevel linked lists are proposed to organize the storage nodes of the campus network for effective utilization of the storage resources. The total nodes number aggregated is 16,843,261.The storage space aggregated is scalable and attained to 17 PB, much larger than those of other current campus grids.

However, there are still some problems to need to be solved in our future work. Firstly, nodes' joining and leaving is independently a stochastic process. The number of nodes' storage space is a birth and death process. How is formally this process described mathematically? And how can nodes' joining and leaving affect the performance of the whole system? There is much work to do in this research. Secondly, in our system, we just adopt the authentication to ensure the user identity. Security problems such as virus protection, malicious attack behaviors are also needed to consider. Thirdly, to effectively encourage users to contribute their storage space, the prize mechanism based on the economy need to be further studied.

We will further extend the work to construct a practical Campus Grid Services Environment.

Acknowledgements

Our work has been supported by the Graduate Innovation Seed Foundation of Northwestern Polytechnical University under Grant No. Z20030051. We very much appreciate Mr. Daowu Zhou in Cambridge University for his numerous useful suggestions. We are grateful to Mr. Jun Yao for many helpful discussions.

References

1. Hiroshi Arikawa, Kazutoshi Fujikawa and Hideki Sunahara: A Node Selection Mechanism Based on the Node Usage Pattern on Campus Grid. Proceedings of 2003 IEEE Pacific Rim Conference on Communications, Computers and Signal Processing (PACRIM '03) Vol. 1. (2003) 217-220
2. The EzGrid project, the campus storage grid. Description available at http://hpclab.cs.tsinghua.edu.cn/~duzh/project/ogsa/ogsa.htm
3. Gray Nutt: Operating System: A Modern Perspective. 2nd edition, Lab Update. Pearson Education North Asia Limited and Posts & Telecommunications Press, Beijing (2002)
4. The campus grid in University of Houston. Description available at http://www.grid.uh.edu/resources.html (2003)
5. J. Novotny, S. Tuecke, V. Welch. An Online Credential Repository for the Grid: MyProxy. Proceedings of the Tenth International Symposium on High Performance Distributed Computing (HPDC-10), IEEE Press (2001)

A Resource Organizing Protocol for Grid Based on Bounded Two-Level Broadcasting Technique

Zhigang Chen, Anfeng Liu, and Guoping Long

College of Information Science and Engineering, Central South University,
ChangSha 410083 China

Abstract. Aiming at the peculiarities of the resource organization in distributed, heterogeneous grid environment, a new protocol based on resource tree is proposed in this paper. The resource tree is constructed through bounded two-level broadcasting technique. We investigate the data structures used by this protocol and the resource tree construction algorithm in detail in this paper. Our simulation results show that the protocol proposed in this paper has a number of advantages, including: low node storage overhead, supporting dynamic resource addition & deletion, effectively hiding the heterogeneity of the resources in wide-area networks, high scalability of resource organization, fast construction of resource trees and high resource searching performance, etc. Although there is some broadcasting overhead involved, we can still get relative high performance due to the limited number of control packets required in this protocol.

1 Related Work

The grid resources organization[1-4] we discussed in this paper aims at the globe scale wide-area networks. In grid environment, each kind of resources has various properties and it is not easy for a client to get exactly the grid resources he/she needs. Since different clients requests different resources, we think we can categorize the resources according to the resources' properties. Based on this, we can organize all resources with the same category into a spanning-tree, which is called resource tree. Each spanning-tree represents one kind of resources, which is similar to the widely used resource model: the resource pool model[5-6]. Here each spanning-tree corresponds to a resource pool, the combination of many resource trees forms a back-bone system in grid environment, as Figure 1 illustrates. In Figure 1, different bold lines represent different kinds of resource trees. When there is some kind of new resources added into the grid, it simply adds itself into a proper resource tree; when this resource is deleted by some node in the grid, it is also removed from the resource tree. Every node in the spanning-tree has a copy of the information about the entire spanning-tree. Thus if there is a client who wants to submit some kind of resources to the resource tree, it needs only submit the resources to any node of the resource tree. Obviously this is very convenient for distributed resource searching.

2 ROBBTLB: Introduction and Overview

The Resource Organization Based on Bounded Two Level Broadcast (ROBBTLB) framework proposed in this paper is based on the improvement of Spanning-join algorithm and the routing algorithm proposed by D.Ghosh[7], etc. The main features

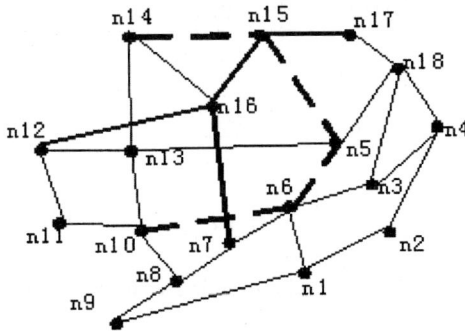

Fig. 1. The back-bone grid system formed by resource trees.

of the protocol proposed in this paper, which is based on the basic concepts of the two-level broadcasting, are as follows: Firstly, since the grid resource management is in fact the construction of resource trees, the design of the ROBBTLB protocol should not only consider the registration of new resource nodes, but also the de-registration and miscellaneous management work of resource trees. Secondly, since ROBBTLB establishes resource tree through probing technique, we don't need a central node in the grid. If a source node want to find some resources in the grid, it needs not know the address of the destination resource tree, because the only thing it does have to know is the resource category number of the resource tree. Thirdly, after the probing packets are sent out by the source node, they may reach multiple on-tree-nodes. So if there exists multiple on-tree-nodes who all satisfy the condition raised by the source node, multiple ACKs will be sent back to the source node. Then the source node will choose the best path to the destination resource tree. Thus ROBBTLB is a multi-path routing protocol.

3 Protocol Description

3.1 Network Model

Definition 1. The grid resource model could be defined as a weighted no-directed graph, $G = (V, E)$. $V = n$, the number of nodes and $E = m$, the number of edges in the graph. V is the set of the nodes in grid and E is the set of network links in grid.

3.2 The Two-Level Forwarding Table for Path Searching

Let's consider the node v. Suppose N1(v) is the set of neighbor nodes, E1(v) is the set of links which connect v and the nodes in N1(v), N2(v) is the set of all the neighbors of N1(v), and E2(v) the set of links which connect N1(v) and N2(v).

The two-level forwarding table of the node v contains the state information of all the links included in E1(v) and E2(v); we say all the entries corresponding to E1(v) are called the-first-level-entries, denoted as R_v^1; while all the entries corresponding to E2(v) are called the-second-level-entries, which is denoted as R_v^2.

Figure 2 illustrates the two-level forwarding table of the node v:

Links	Nodes	Path Delay	Properties
L_1	V_1	D_1	*
⋮	⋮	⋮	⋮
L_n	v_n	D_n	CAddr
(L_1,l_1)	u_1	D_1+d_1	*
⋮	⋮	⋮	⋮
(L_n,l_m)	u_m	D_n+d_m	CAddr

Fig. 2. The two-level forwarding table of the node v.

3.3 Data Structures

The following data structures are required by ROBBTLB protocol:
 1, *Join_Probe*(Jioning Probe packet)

Definition 2: *Join_Probe* is a set contain five elements which can be defined as (C_Addr, Req_DelayC, Call_ID, CurNToDestCacD, Eligible_paths). In this definition, C_Addr is the category address, Req_DelayC is the maximum delay that can be accepted by the source node after it sends out the probing packets to the destination resource tree, Call_ID is call identifier(the IP address of the source node which sends the probing packets could be used as this identifier), and CurNToDestCacD is the delay between the moment the source node sending probing packets and the moment the destination node receiving the probing packets. Eligible_paths is the path set of the resource trees that are reachable.
 2, *Ack_For_Probe*(ACK packet)

Definition 3: Ack_For_Probe is defined as (C_Addr, Call_ID, t_Delay).
 3, *Nack_For_Probe*(negative ACK packet)

Definition 4: Nack_For_Probe is defined as *(C_Addr, Call_ID)*. In this definition, C_Addr is the category address, and *Call_ID* is call identifier
 4, *Prune*(prune tree packet)

Definition 5: *Prune* is defined as *(Call_ID, C_Addr)*. In this definition, *Call_ID* is call identifier, C_Addr is the category address.

3.4 The Resource Tree Construction Algorithm

When a node v receives a *Join_Probe* packet, it first checks the *Eligible_paths* set of this packet. If it is not NULL, this means the node v is simply an intermidiate node during the fowarding process. Thus the node v simply forwards Join_Probe packets according to the path specified in Eligible_paths. Of couse, before it forwards Join_Probe packets it must first clear the Eligible_paths set, while at the same time it must store the *Call_ID*, the input port *(In_Interface)*, the output port *(Out_Inter-face.O1,OutInter_Interface.O2,...)* and the category address *(C_Addr)* of the packets. Otherwise if the Eligible_paths set is NULL, this means this node is not an intermidiate node, but a second-level node(neighbor's neighbor). If v is an on-tree-node, it first checks if the bealoon expression *t_Delay + Join _Probe. CurNToDestCacD <= Req_DelayC* is true. If so, an *ACK* packet is sent back to the source node. If v is not an on-tree-node, it checks its two-level forwarding table. If there exists some on-tree-nodes in this table, v then finds out the best available path, and changes the *CurNTo-DestCacD* value of the Join_Probe packet to the sum of the current value and the best available path delay. After this, the node v forwards the packet to the node of the resource tree, while at the same time it stores the *Call_ID*, the input port (*In_Interface*), the output port (*Out_Interface.O1, OutInter_Interface. O2 ,…*) and the category address (*C_Addr*) of the packets. If there isn't any on-tree-node on the two-level forwarding table, v will find out a group of available paths to its second level neighbor(neighbor's neighor) based on its forwarding table, and then after v modifies the *CurNToDestCacD* valuse of the *Join_Probe* packet it sends it to the second level neighbors(neighbor's neighbors) following the paths selected before. After v's second level neighbors receive the Join_Probe packet, they will repeat the process discussed before until the *Join_Probe* packet is received by the resource tree or the timer is expired.

Figure 3 illustrates the *Join_Probe* processing algorithm:

```
Algorithm : Join probe message processing
Join _probe _process (join _probe )
begin
Switch ( Join _probe.path )
  Case Join_probe.path ≠ NULL
        R.in = In_Interface;  // In_Interface is the input port of //Join_probe
        R.out =
        Join _probe.age = Join _probe – 1;
        Forward Join _probe to R.out ;

  Case Join_probe.path = NULL
        If ( this node is on-tree-node )
          if  (t_Delay  +  Join  _probe.  CurNToDestCacD  <= R_DelayConstraint)
              M.out=M.out + In_Interface ;
              Send  ACK to In_Interface ;
              Return ;
          else
              Send NACK to In_Interface ;
              Return ;
          endif
        Else
          if ( this node's two-level forwarding table contains on-tree-nodes)
             Join _probe.path = available paths ;
             if ( Join _probe.path ≠ NULL)
                Forward Join _probe to on-tree-nodes ;
                R.in =
                R.out =
                R.status =
             else
                Send NACK to In_Interface ;
             end if
          else
             Join  _probe.path  =  all  available  paths  to  second-level nodes(neighbor's neighbors) ;
             if ( Join _probe.path ≠ NULL )
```

```
            Forward Join _probe sccording to Join _probe.path ;
            R.in = In_Interface ;
            R.out = R.out + Join _probe.path ;
            R.status = unfixed ;
            Return ;
         Else
            Send NACK to In_Interface ;
            Return ;
         endif
      Endif
      EndSwitch of age
EndSwitch
End the processing
```

Fig. 3. *Join _probe* packets processing algorithm.

```
      Algorithm : ACK message processing
      ACK_procee ( ACK )
      if (theExistingPath.t_delay better than ACK.t_delay )
         send a prune to ACK's coming interface ;
         Return ;
      else
         R.status = fixed ;
         R. t_delay = ACK. t_delay ;
      endif
      if (v is the starting probe node )
         Mark its forwarding table ; // the resource node is
      added into the resource tree successfully
      else
         ACK.t_delay = ACK.t_delay + this node's delay ;
         Send ACK to join_probe coming interface ;
      endif
      Return ;
```

Fig. 4. *ACK* packets processing algorithm.

When the node v receives an *ACK* packet, it checks the value of *ACK.t_Delay*. If there is already an accepted path which is better than the current path(the accepted path's *ACK.t_Delay* is lower than the current path), v sends back a prune packet to the port which sent the ACK. Otherwise if no path is accepted before the current path arrives or if the current path's *ACK.t_Delay* is lower than the previously accepted paths, v thinks the port which sent the *ACK* is the best interface and marks the corresponding entry which was recorded when the *join_probe* packet arrived at v in forwarding table R as fixed, meaning this path is the best so far. If v is the node which sent the original *join_probe* packet, then v is successfully added into the resource tree; otherwise if v is not the node which sent the original *join_probe* packet, then v will send back *ACK*, with the node's delay value been added on the *ACK* packet's *t_Delay* field, through the input port(In_Interface) of the previous *join_probe* packet. Figure 4 illustrates the *ACK* packets processing algorithm.

When v receives an *NACK* packet, it deletes the input port(In_Interface) *NACK* of the packet from its forwarding table; If its forwarding table corresponding to *Call_ID* is NULL(doesn't exist an available path), all the entries in *Call_ID* are deleted. If v is not the node which sent the original *Join_Probe* packet, it must send the *NACK* packet to the node which sent the original *Join_Probe* packet. Figure 5 illustrates the *NACK* processing algorithm.

```
Algorithm : NACK message processing
NACK_process (NACK)
      R.out = R.out – NACK_coming_interface ;
   If  R.out = NULL
      Delete the entry for the Call_ID
   Endif
   If  v isn't the starting probe node
      Sned the ACK to In_Interface of the Probe coming
   Endif
   Return ;
```

Fig. 5. The *NACK* packets processing algorithm.

When v receives a *prune* packet, it first deletes the input port of the *prune* packet from the forwarding table *M*. If its forwarding table corresponding to *Call_ID* is NULL, all the entries in *Call_ID* are deleted, while at the same time v sends a *prune* packet to the input port through which resource tree packets arrived. Figure 6 illustrates the *prune* packets processing algorithm.

```
Algorithm : prune message processing
prune_process (NACK)
      M.out = M.out – prune_coming_interface ;
   If  M.out = NULL
         Delete the entry in M  for the Call_ID ;
         Sned the prune to In_Interface of the packet coming;
   Endif
   Return ;
```

Fig. 6. The *prune* packets processing algorithm.

4   Simulation Results

The ROBBTLB protocol involves extra broadcasting overhead, since this grid resources management scheme is based on two-level broadcasting technique. But as [7] points out, the extra overhead is not disastrous because the network traffic spurred by control packets required by this protocol is low.

Figure 7 illustrates the results of our simulation on different scale networks. Under different grid environment(with different number of resource nodes), the time required to build the resource tree seldom changes when the resource node ratio(number of resource nodes/number of nodes in grid) is 1/50 or 1/100. Simulation results show that our protocol scales well with respect to network scales and resource ratio.

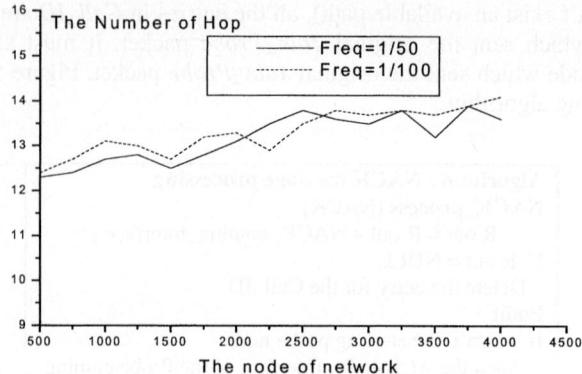

Fig. 7. The resource tree building time under different network scales and different resource ratio.

5 Conclusion and Future Work

In this paper, we designed an entire set of data structures and a resource tree building and management algorithm, while at the same time a new grid resource organization protocol, ROBBTLB, which support wide-area grid networks is implemented.

References

1. LI Wei,XU Zhi-Wei, A Model of grid adress space with applications, journal of computer research and development vol40 2003(12) p1756-1762
2. CAO Hong-Qiang XIAO Nong etc a market-based approach to allocate resources for computational grids journal of computer research and development vol39 2002(8) p913-916
3. Wei Li, Zhiwei Xu, Fangpeng Dong, Jun Zhang, Grid Resource Discovery Based on a Routing-Transferring Model, 3rd International Workshop on Grid Computing (Grid 2002)
4. I Foster,C Kesselman,S Tuecke The Anatomy of the Grid: Enabling Scalable Virtual Organizations, International Joural of Supercomputer Application,2001,15(3):200~222
5. Parabon Computation Inc'2000.http://www.parabon.com
6. Entropia Inc'2002.http://www.entropia.com
7. Donna Ghosh , Venkatesh , and Raj Acharya , Quality-of-Service Routing in IP Networks , IEEEE TRASACTION ON MULTIMEDIA. VOL. 3 , NO. 2, pp. 200-208, JUNE 2001

Peer-Owl: An Adaptive Data Dissemination Scheme for Peer-to-Peer Streaming Services*

Xiaofei Liao and Hai Jin

Cluster and Grid Computing Lab
School of Computer Science and Technology
Huazhong University of Science and Technology, Wuhan, 430074, China
{xfliao,hjin}@hust.edu.cn

Abstract. More and more researchers have put their emphases on peer-to-peer streaming services, which can provide massive and cheap video-on-demand services. Data dissemination is one of the most important open problems when providing peer-to-peer streaming services, including live streaming and video-on-demand. Considering the performance and fault-tolerance of streaming systems, content should be distributed and replicated onto peer nodes according to some kind of strategy. In this paper, a novel data dissemination scheme is proposed. According to the proposal, media files are recoded into injected objects with several segments according to the characteristics of VCR operations' frequency, without changing the media compressing formats. Then, a peer who wants to get one media file can acquire the corresponding segmented object files from many source peers, not from only one peer. The new scheme does not need any extra network bandwidth. Results from simulations have proved that our Peer-Owl scheme has good performances.

1 Introduction

In the last few years, due to the increasing demand for streaming services on the Internet, many studies are being undertaken to find an efficient scheme. The new network model, peer-to-peer, can aggregate abundant resources from thousands of computers and address the above problems. Now many researchers provide new overlays on p2p networks to support file-sharing services, game services, especially the media streaming services [6].

But the peer-to-peer streaming networks have many open problems [1], such as fault-tolerant schemes and data replication schemes. The key to a p2p system, and one of the most challenging design aspects, is efficient techniques for data dissemination. There are several advantages using a good data dissemination scheme. First, services can be more stable than ever because there are many supplying peers to support the same media objects in p2p networks. Second, QoS can be improved. Third, load balance performance also can be improved. Hot movies have enough replicas, which are distributed onto almost all peers. Requests for the same hot movie can spread around the p2p networks.

Some researchers put their emphasis on the data dissemination schemes, but not for streaming services and only to support data sharing overlays. Other researchers

---

* This paper is supported by National 863 Hi-Tech R&D project under grant 2002AA1Z2102.

focus on data dissemination of the streaming services. But they only support the broadcast services and living streaming services, not on-demand services. It is obvious that data dissemination schemes in different streaming patterns, including living streaming and on-demand streaming, are not the same.

In [2], a novel data-splitting scheme *Owl* has been proposed to improve the performance of cluster video servers. In this paper, we extend the idea of *Owl* scheme to p2p environments. We provide a data dissemination scheme, called *Peer-Owl*, for p2p streaming services. The new scheme can encode movie objects into segments and transmit them onto ultra-peers when a new stream is created. Without any extra network bandwidth, the scheme can obtain a high success ratio of data dissemination because it takes full consideration of the characteristics of the frequency of VCR operation and the roles of ultra-peers.

The paper is organized as follows. We present some statistics about p2p streaming network and describe some properties of streaming services in section 2. Section 3 focuses on the method of how to segment media objects. In section 4 *Peer-Owl*, a novel data dissemination scheme for p2p streaming networks, will be described in detailed. Section 5 focuses on the methodology and results of simulations. In section 6 we survey some related works. Finally, section 7 closes with conclusions and future works.

2 Modeling Peer-to-Peer Streaming Networks

There are several models to describe the p2p networks, such as Erdos-Renyi (ER) model, BA model [1], and EBA model [7]. But when we consider a data dissemination scheme for a p2p streaming network, it is difficult to find a compatible model to describe our needs.

2.1 Topological Properties of Large-Scale P2P Networks

We look at two aspects of a p2p network: network topology, distribution of degrees of all peers. All these topological properties of p2p networks are based on Gnutella networks. All experimental data are calculated from our crawler on Gnutella network and the source codes are rewritten from *Limewire* open source client. Our crawler can discover over ten thousands peers and their connections. There are three data sets, calculating with three different time lengths, 20 minutes, 30 minutes and 40 minutes, called Data<200404031026>, Data<200404031426> and Data<200404022126>, respectively. All our conclusions are from these data sets. To describe a stable topological graph of p2p network, we only run our crawler for no longer than 40 minutes.

From Tab.1, we can conclude that though the number of ultra-peers is little, most edges are connected with them.

We have made the following conclusions. First, ultra-peers take more resources, such as routing paths, in hand than that of leaf peers. Second, ultra-peers are the most important agents for almost all shortest paths of pairs of peers. If we want to provide an effective data dissemination scheme, it is important to make more replicas onto ultra-peers.

Table 1. Statistical results about degrees of Gnutella networks. Properties measured are: the number of peers (or called nodes) *v*; total ultra-peers number *uv*; total edges number *e*; the number of edges connected between two ultra-peers *u2u*; the number of edges connected between one ultra-peers and one leaf peer *u2l*; the number of edges connected between two leaf peers *l2l*.

Items	Data<200404031026>	Data<200404031426>	Data<200404022126>
v	17395	28119	38115
uv	3302	5287	6257
e	20426	35181	52531
u2u	2803	4649	5788
u2l	17425	30163	46128
l2l	198	369	615

2.2 Topological Properties of Large-Scale P2P Networks

Streaming services based on p2p networks have some characteristics that other applications overlays do not have, such as file-sharing overlays and data storage overlays. VCR operations will make great impact on data dissemination. When one requesting peer accepts the video-on-demand service, it should store the media data onto its own storage systems and its neighbors'. But if the requesting peer does seek operation, media data accepted before will be dropped. To make the VCR operations frequency clear, we give two definitions about *MTTVCR* (called Mean Time To VCR) and *ttvcr* (called time to VCR) as in Equation 1. In Eq.1, t_{i+1} means the time position of the current VCR operation and t_i means the time position of last VCR operation.

$$MTTVCR = \frac{\sum ttvcr_i}{n}; ttvcr_i = t_{i+1} - t_i; \tag{1}$$

In the subsection, we use the log files from the video server located in the CCRNC (*Center China Regional Network Center*) of CERNET (*China Education and Research Network*). Log files of the video server record historical data. Figure 1(a) shows the *ttvcr* (time to VCR) of different users who choose the movie "*Thelma & Louise*" in March 17, 2001. The value of *MTTVCR* means that there are about three or four VCR operations in viewing one movie with 120 minutes length. Figure 1(b) shows the number of VCR operations in different time range. It is obvious that most VCR operations have been done in the first 20 minutes.

We can make the following conclusions about dada dissemination for p2p streaming services: first, media data should be cut into segments according to the above statistical results of VCR operations; second, when one requesting peer is accepting media data, it should store media onto local storage systems to build new replicas.

3 Media Encoding and Decoding

We have proposed a novel data splitting scheme, called *Owl*, to improve the performance of clustered video servers [3]. In clustered video servers, media data should be

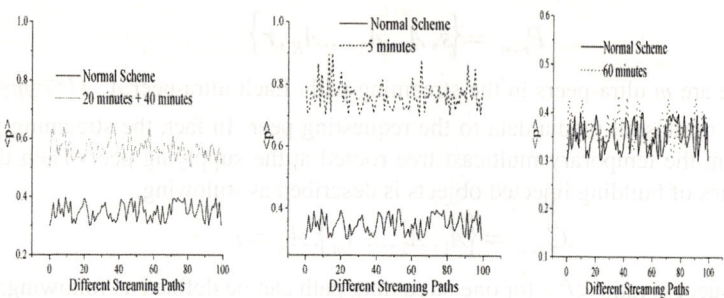

Fig. 2. (a) Success Ratio: Normal Scheme *vs Peer-Owl* with the first Clip's length 20 minutes and other clips' length 40 minutes; (b) Success Ratio: Normal Scheme *vs Peer-Owl* with the clip's length 5 minutes; (c) Success Ratio: Normal Scheme *vs Peer-Owl* with the clip's length 60 minutes.

0.865. When selecting two peers s, r randomly, there is a streaming path $P_{s \to r}$ between them. Then the number of ultra-peers on the path can be described as $0.865 * | C_{s \to r} |$. Second, the characteristics of the frequency of VCR operations can be described in mathematical model. For one requesting peer, the probability of VCR operations on one time position can be described as following:

$$f = \frac{1}{t} * c; \sum_t \frac{1}{t} * c = 1; t = 10 * \tau, 1 <= \tau <= 11 \qquad (5)$$

There is an assumption that one movie is only 120 minutes long. After computing, c equals to 3.3. When the frequency of VCR operations and the selection of streaming paths have been decided, we can build a static environment to simulate the data dissemination. In our simulation, we select one pair peers from Data<200404031026> for 100 times and generate 100 different streaming paths randomly.

In the following simulations, there are several analyses in Fig.2. In these figures, there are four types of schemes: normal scheme without movies segmentations, *Peer-Owl* scheme with the first clip's length 20 minutes and other clips' length 40 minutes, *Peer-Owl* scheme with the clip's length 5 minutes, *Peer-Owl* with the clip's length 60 minutes.

From these figures, the success ratio of the third scheme is better than that of other three types of schemes. The success ratio of second scheme is middle in the four schemes. The normal scheme and the fourth scheme have almost the same performances of success ratio. But in the third scheme, clip size is too small and the length of tasks list is too longer. Systems need too resources to process the fragmentized injected objects.

6 Related Works

Most researches about data disseminations focus on mobile ad-hoc networks and structured p2p networks. Its main functions are to support the popular data sharing and improve the efficiency of p2p services.

Popular file sharing and living streaming are the primary topics [5] when researchers mention the data dissemination and prefacing techniques. It is obvious that these two services have much comparability. Because of no VCR operations to interrupt the normal data streams, data transmission of these two types of services are stable and consecutive. When systems begin data streams, data channels will be created and keep stable. A useful way to make data dissemination with the help of Tornado coding [3][4] is provided. The paper first proposed a new model Street-and-Building to describe the mobility of peers in mobile ad-hoc networks. Then, it provided a data dissemination protocol to disseminate Tornado encoded file segments (packets). But the beautiful scheme only supports file-sharing services. The middleware only supports file downloading services.

Researches on the statistics of VCR operations for on-demand streaming services are not hot topics. Intra-movie skewness from logs of traditional video servers is studied [9]. In fact, due to users' interest, there are some VCR operations when the movie is being viewed. From these logs, we can make some interesting conclusions, described in subsection 2.2. These conclusions will be helpful in designing a good data dissemination scheme.

Media objects segmentation techniques are mentioned at [4][6]. Tornado coding use the coding methods to support file sharing in normal and p2p environments [4]. A stream is segmented into *blocks* only for the efficient use of storage space [6] and the retrieval of necessary parts of a media stream. To apply segmentations of media objects in data disseminations for on-demand streaming services is one contribution.

7 Conclusions and Future Work

In this paper, we first give some statistics from Gnutella networks and traditional video servers; then a novel data dissemination scheme for P2P streaming networks is presented. From simulations, the success ratio of our *Peer-Owl* scheme is good according to the frequency of VCR operations. But the scheme *Peer-Owl* has great rooms to improve. In the future work, our focus will be concentrated on the media data splitting scheme for P2P networks, the bulk data transmitting technique and other open issues.

References

1. L. Barabasi and R. Albert, "Emergence of scaling in random networks", *Science*, Vol.286, pp.509, 1999.
2. E. Korpela, D. Werthimer, D. Anderson, J. Cobb, and M. Lebofsky, "SETI@home: Massively Distributed Computing for SETI", *Scientific Programming*.
3. H. Jin and X. Liao, "Owl: a new multimedia data splitting scheme for cluster video server", *Proceedings of 28th Euromicro Conference*, 2002, pp.144-151.
4. J. Byers, J. Considine, M. Mitzenmacher, and S. Rost, "Informed content delivery across adaptive overlay networks", *Proceedings of ACM SIGCOMM 2002*, August 2002.
5. J. Byers, M. Luby, and M. Mitzenmacher, "Accessing multiple mirror sites in parallel: Using tornado codes to speed up downloads", *Proceedings of IEEE INFOCOM'99*, pp.275-83, March 1999.

6. N. Leibowitz, M. Ripeanu, and A. Wierzbicki, "Deconstructing the Kazaa Network", *Proceedings of 3rd IEEE Workshop on Internet Applications (WIAPP'03)*, 2003, Santa Clara, CA.
7. Q. Wang, Y. Yuan, J. Zhou, A. Zhou, "Peer-Serv: A Framework of Web Services in peer-to-Peer Environment", *Proceedings of the Fourth International Conference on Web-Age Information Management (WAIM 2003)*, Springer-Verlag, 2003.
8. S. Wu and H. Jin, "Symmetrical Pair Scheme: a Load Balancing Strategy to Solve Intra-movie Skewness for Parallel Video Servers", *Proceedings of 16th International Parallel and Distributed Processing Symposium (IPDPS'02)*, Florida, USA.

WCBF:
Efficient and High-Coverage Search Schema in Unstructured Peer-to-Peer Network

Qianbing Zheng, Yongwen Wang, and Xicheng Lu

School of Computer Science, National University of Defense Technology, Changsha
410073, P.R. China
nudt_zhengqianbing@hotmail.com

Abstract. Random walks is a typical search schema in unstructured peer-to-peer network. It generates partial coverage problem and makes some resources indiscoverable. An efficient search schema *WCBF* is presented. WCBF utilizes compressed Bloom filter for representing the path information to enlarge the coverage of random walks and alleviates the problem with a low overhead. Experimental results show the coverage of WCBF is almost 6 times that of random walks in a Gnutella-like overlay network.

1 Introduction

Gnutella-like unstructured P2P file sharing systems cause significant impact on Internet traffic[1–3]. These systems use flooding-based search policy and TTL-based termination condition. Such search schema has a high coverage, but produces vast redundant messages and wastes enormous bandwidth[4, 5].

Efforts have been made to reduce redundant messages generated by flooding. Related works can be classified into two categories:

- Breadth First Search(BFS)-like methods, including modified-BFS[6], Directed BFS[7], Expanding Ring[8]
- Depth First search(DFS)-like methods, including DFS in Freenet[9], random walks[8] as well as its variations APS[10] and GIA[11].

The main idea of prior work is to reduce the number of peers that receive query messages. That is to say, a peer selects a fixed number of its neighbors to forward query message instead of all its neighbors. However, this leads to *partial coverage problem*[12] which leaves a large percentage of the peers unreachable no matter how large the TTL value is set. As a result, some resources will never be discovered.

The direct cause of partial coverage problem is the query path loops generated by these solutions. Zhuang selects variable number of neighbors based on TTL to avoid generating query path loop[12]. This approach can resolve this problem to some extent, but it can only be used in BFS-like search methods and is not suitable to DFS-like search methods for a peer only selects one neighbor to forward query message. In fact, DFS-like methods can get better performance than BFS-like methods in reducing bandwidth. It's important to solve the partial coverage problem in DFS-like methods.

In this paper, we propose an efficient search schema – *WCBF(Walker with Compressed Bloom Filter)* for unstructured P2P network. The search schema is based on random walks[8]. Using compressed Bloom filter representing the path information, WCBF enlarges the coverage of random walks and alleviates partial coverage problem with a low overhead.

The rest of this paper is organized as follows. In section 2, we introduce a typical DFS-like method: random walks. In section 3, we describe our schema in details. In section 4 and 5, we make some experiments and discuss the results. Finally, we end with conclusions in section 6.

2 Random Walks

Random walks[8] is a typical representative of DFS-like search methods. It forwards a query message to a randomly chosen neighbor at each step. And this message is called a "walker". To get user-perceived delay, it increases the number of "walkers" (typically 16 to 64). The requesting peer sends k query messages and each query message takes its own random walk. The termination condition is checking method that walkers periodically contact the query source asking whether the user demand has been satisfied.

Random walks is an excellent improvement of flooding-based search schema. It can achieve significant message reduction compared with the standard flooding scheme[8]. Furthermore, random walks is identified that it achieves better results than flooding for searching when the overlay topology is clustered and it is an excellent candidate to simulate sampling for P2P networks[13].

However, Random walks still suffers the query path loops. A query path loop is defined as follow: the path information which a walker traverses is denote as a peer set $L = \{v_{i1}, v_{i2}, ..., v_{in}\}$, a query path loop forms if a peer v_j which is selected to forward the walker belongs to the L. We will utilize compressed Bloom filter[14] to avoid it.

3 Walker with Compressed Bloom Filter

To avoid query path loops, we consider to add path information to a walker. If a peer selects a neighbor not belonging to L to forward the walker, it will not generate a loop. According to Gnutella protocol, it needs 16 bytes to identify one peer. Then, the overhead of representing L is too high. So it is better to use a space efficient data structure to represent the path information.

3.1 Bloom Filter

Bloom filter(BF)[15] is space efficient data structure representing a set $S = \{s_1, s_2, ..., s_n\}$ of n elements to support probabilistic membership queries. S is described by a vector V of m bits, initially all set to 0. Bloom filter uses k independent hash functions $h_1,...,h_k$ with range $\{0,...,m-1\}$. For each element $s_i \in S$, the bits at positions $h_1(s_i), h_2(s_i),..., h_k(s_i)$ in V are set to 1(A particular bit might be set to 1 multiple times). To check if an item x is in S, we check all $h_i(x)$. If any of them is 0, then certainly x is not in the set S. Otherwise x is conjectured to be in the set although there is a certain probability that it is wrong. This is called a *false positive*. If a bloom filter is used to represent the path information, the probability of *false positive* is the probability of detecting loop in error. Suppose that hash functions are perfectly random, the probability of a *false positive* is

$$\left(1-\left(1-\frac{1}{m}\right)^{kn}\right)^k \approx \left(1-e^{\frac{-kn}{m}}\right)^k$$

We can get a conclusion that the more the value of m is, the less the probability of *false positive* is. That is to say, the smaller the size of bloom filter is, the less the probability of detect looping behavior rightly is.

3.2 Compressed Bloom Filter and WCBF

Since the transmission size is more important than the bloom filter size, the size of bloom filter must be small. Thus, we decide to use Compressed Bloom Filter(CBF) to represent path information to strike a good balance between overhead and the loop detection accuracy. Bloom filter was used to represent path information to avoid short forwarding loops[16]. We apply the method in [16] to random walks and call it *walker with BF*. Our search schema is called *Walker with Compressed Bloom Filter*, or WCBF for short.

The main idea of WCBF is each walker carries a CBF which logs path information. When a walker arrives at one peer, the peer will select one neighbor which will not cause a loop according to its CBF to forward this walker. Furthermore, each peer v_i keeps two state sets. v_i will not select its neighbor belonging to the two state sets to forward a walker. One state set S_1 comprises the peers which v_i has selected to forward the walkers. This state set is proposed in[8] to accelerate random walks. The other state set S_2 we propose contains the peers which forward the walkers to v_i. Because one walker is less likely to visit the peers which other walkers visit, it will cover more peers. Meanwhile, when a peer can't find a neighbor to forward, it will randomly select a neighbor which is not in S_1 and S_2. This method is called *false allowed* which is equal to infinite *reprieves*[16]. *False allowed* reduces the probability of detecting loop in error. Simulation results later will confirm these improvement.

Choosing a good hash function and a compression program is the critical part of implementing compressed Bloom filter. Mitzenmacher shows that $k < ln2 \cdot (m/n)$ can achieve improved performance for compressed Bloom filter and proposes the arithmetic coding is a near-optimal compression program[14]. Let the max value of m be 512. And n varies in the range of the size of P2P network. So, the value of k should be 1 at most. Thus, we only use one hash function.

Each peer in P2P network is uniquely identified by a 16-byte string r, and the binary representation of r is d. The hash function $H(d)$ is as follow:

$$H(d) = d \bmod m$$

where m is the length of BF. From the above equation, we can know $H(d) \in [0, m)$. Let m be the power of 2, the hash function can be finished very easily by the operation of right shift. We choose PPM[17] which is one of the arithmetic coding model as the compression program.

The pseudo-code of WCBF is shown in Algorithm1. s is the request peer. The number of walkers is 32 and the neighbor set of a peer u is $N(u)$.

Algorithm 1: The pseudo-code of *WCBF*.
WCBF(s)
foreach walker w_i
    s sets all bits in BF of W_i to 0
    $d \leftarrow$ the binary representation of s'id
    $d' \leftarrow H(d)$
    s set the $d'th$ bit in BF to 1 and compress BF to CBF.
    s add CBF to w_i and select u_i randomly to forward w_i.
if one peer v_i receives one walker w_i
    if (v_i has received w_i before) **or** (the demand of s has been satisfied)
        stop
    else
        decompress CBF of w_i to BF
        foreach $u_i \in N(v_i)$ **and** $u_i \notin S_1$ **and** $u_i \notin S_2$
            $d \leftarrow$ the binary representation of u'_iid
            $d' \leftarrow H(d)$
            if the $d'th$ bit in $BF \neq 1$
                set the $d'th$ bit in BF to 1 and compress BF to CBF.
                add CBF to w_i and select u_i to forward w_i.
                stop
        if v_i can't find a neighbor to forward w_i
            select u_i not in S_1 and S_2 randomly to forward w_i
            stop

4 Simulation Methodology

Coverage and *query finish rate* are defined to compare the performance of three algorithms: random walks, walker with BF and WCBF. *Coverage* of one algorithm is the number of peers which the algorithm visits. The more the coverage is, the more the number of resource which algorithm can find is. If the number of one user's request resources is x, *query finish rate* is y/x after a search algorithm discovers y resources.

Let the size of path information be l bits. There are several questions we are interested in:

- How does the increasing of l affect the coverage of these algorithms? And with the same l, how does the coverage of WCBF as compared with that of other algorithms?
- How many peers does WCBF visit under $false\ positive = 0$? And what value of l can get the approximate coverage to the coverage under $false\ positive = 0$?
- How does the increasing of l affect the query finish rate of these algorithms with different query set?

Recent measurements and researches[4, 18, 19] show that Gnutella-like unstructured networks have power-law degree distribution. Furthermore, they have the similar characteristics to the scale-free model[20]. So we use the scale-free model to simulate P2P overlay network. We generate a P2P overlay graph OL with 10,004 nodes whose properties approximate to the real Gnutella overlay network by using Barabais Graph Generator[21]. The graph generator is based on the scale-free model. The average node degree of OL is 3.99 which approximates to the actual value 3.4[4]. Furthermore, two topologies can be assessed to have similar properties through power-law index k. The power-law index k of OL is -2.343 which approximates to the actual value -2.3[19].

Three distributions are built to simulate the P2P system environment.

- replication distribution: it controls the number of each resource in P2P network. A Zipf-like distribution with parameter $\alpha = 1$ is used in the simulated environment. The form of Zipf-like distribution is as follow:

$$f(x) \propto \frac{1}{x^\alpha}$$

- resource popularity distribution: it represents how many queries are made for each resource. According to [22], the distribution follows a Zipf-like distribution.
- user demand distribution: it represents how many numbers is the user demand for each requested resource.

Two query sets are generated: one query set QS_1 comprises queries to the top 50 unpopular resources and user demand is the half of the number of the resource in P2P network for each requested resource; the other query set QS_2 consists 50 queries whose resource popularity distribution and user demand distribution both follow a Zipf-like distribution with parameter $\alpha = 1$.

5 Experimental Results

We design a set of experiments to compare the performance of three algorithms in OL. $l = 8, 16, 64, 128, 192, 256, 320, 384, 448, 512$ respectively.

Firstly, three algorithms with 32 walkers run with the same request peer. After that, the coverage of each is computed. Figure1 shows the coverage of the algorithms with the increasing of l.

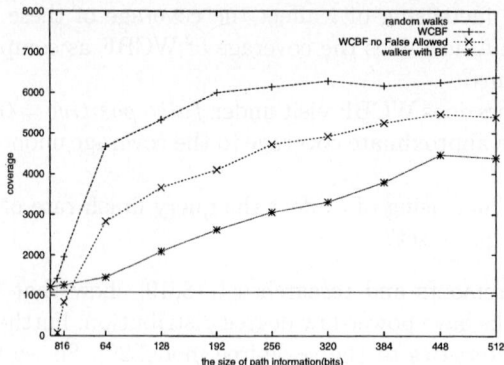

Fig. 1. The coverage vs. the size of path information

If we let l equal to the size of OL, *false positive* equals 0. Thus, the max coverage of WCBF under *false positive*= 0 is 6540. With the increasing of l, the coverage of WCBF rises sharply within $l = 192$ at first, then grows close to the max coverage slowly. At the best condition, the coverage of WCBF is almost 6 times that of random walker and over 3 times that of walker with BF. Moreover, the coverage of WCBF under $l = 64$ approximates that of walker with BF under $l = 512$. The benefits of S_2 and *false allowed* also can be seen from Figure1. WCBF under $l = 0$ equals random walks with S_1 and S_2, its coverage increases by 11%, as compared with random walks with S_1. When WCBF doesn't use *false allowed* method, its coverage shows different degrees of drop. Especially under *small l*, its coverage is even smaller than that of random walks.

Secondly, for each query in the query sets a peer which doesn't have the requested resource is randomly chosen to start the query. Query finish rate of each algorithm is computed. Figure2 compares the average query finish rate of these algorithms with the increasing of l for two query sets QS_1 and QS_2.

The results show that the partial coverage problem weakens the ability of random walks discovering resources, especially for the unpopular resources. WCBF improves the ability of random walks for it enlarges the coverage of random walks.

6 Conclusions

We have proposed an efficient and high-coverage search schema – WCBF. This search schema utilizes compressed Bloom filter for representing the path information to alleviate the partial coverage problem of DFS-like methods and increases the coverage of random walks greatly. Experimental results show the coverage of WCBF is almost 6 times that of random walks. As a result, WCBF can find more resources than random walks in a Gnutella-like overlay network. From the experimental results, WCBF with the size of path information=64 can get a good trade-off between overhead and query path loops detection accuracy. Fur-

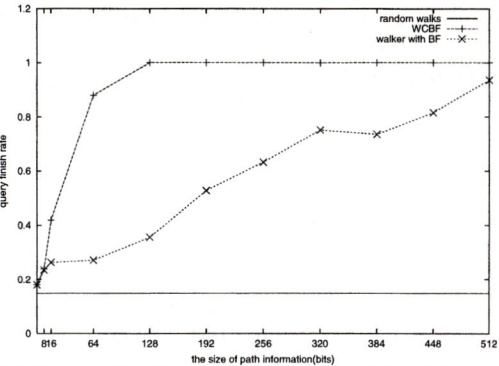

(a) the results of query set QS_1 consists of the top 50 unpopular resources

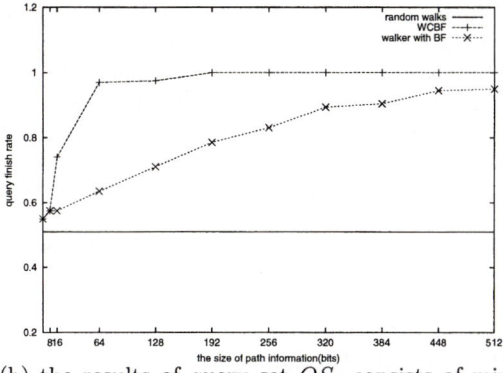

(b) the results of query set QS_2 consists of mix resources

Fig. 2. Query finish rate vs. the size of path information

thermore, WCBF also can be used to improve the performance of other DFS-like search methods such as APS and GIA.

Acknowledgement

This work is supported by NSFC project 90204005 and 863 project 2003AA121510.

References

1. http://www.caida.org/analysis/workload/oc48.
2. http://eddie.csd.sc.edu.
3. WISC: (traffic statistics) http://wwwstats.net.wisc.edu.

[17], such as VLSI specific systems, pipeline/vector machines, shared memory supercomputers, and multi-processor platforms and clusters [18-24]. Due to the rapid development of networking, people have begun to deploy image reconstruction algorithms on distributed computer systems, which are throughput on a fast local network (LAN) [25-26] or on large-scale distributed computational Grids [27-30].

The authors propose a simple economic and efficient Internet-based distributed parallel system for iterative X-ray CT EM medical image reconstruction using P2P technology. The proposed system is feasible and applicable to research and clinical use. Section 2 briefly reviews and discusses distributed computing models for medical image reconstruction. Section 3 presents the parallelism of EM algorithm utilized in the system. Section 4 addresses system architecture and application implementations. Section 5 presents imaging results computed by the proposed system, followed by a conclusion and plans for future exploration in this area.

2 TCP/IP-Based Distributed P2P Model

The proposed alternative approach integrates fully network-enabled distributed systems, referred to as "throughput" or "P2P-enhanced" architecture. The P2P distributed system has a decentralization paradigm that allows image reconstruction among computing peers. The formation of such a system can be built at "run-time." Each computer has its own IP address and serves as both client and server, depending on demand and system configurations. The system is quite scalable and portable, low cost, and easy to set up and maintain. It can aggregate many distributed computer resources. The "virtual" parallel system can be built within a local network or be linked globally through a fast network. In such a system, the client directly connects with all other computing nodes. In our system, computing peers run a peer establishment program before a client searches available nodes through the Internet. The client can therefore submit part of the image reconstruction jobs to selected nodes for intensive computation. A major disadvantage is that performance of the system depends heavily on network characteristics such as bandwidth, job loading and balancing, fault tolerance, and concrete parallelism of data decomposition. Regarding computational mechanisms, such a model would serve as an innovative and core component in application layers in computational Grids for medical imaging.

3 The Parallel Algorithms for Iterative EM

The cross-sectional image is divided into n square pixels. Each pixel has constant X-ray-attenuation factors, x_j ($j=1,...,N$). The x represents the corresponding N-dimensional column vector. Projection data b_i ($i=1,...M$) is the line integral along the ray-path. b is the system matrix of dimension $M\times N$. Let $A=(a_{ij})$ be the matrix mapping x to b, where each element a_{ij} measures the contribution of x_j to b_i. This results in the following linear system, $Ax=b$. The CT image reconstruction problem is to find the vector x. The parallel iterative reconstruction procedure are described as follows. Each computational node has an individual IP address. There is no distinction between client and server. A client initiates node selection, domain-decomposition, task-distribution, and the sending of computational sub-tasks to servers to perform computation. Once the servers accomplish forward/backward projections, the client receives the image data and performs global updating before the next iteration. The algorithm are listed as follows

1. The client decomposes total projection data into smaller groups and distributes the sub-grouped projection data to computational nodes.
2. The client makes an initial guess about the initial value of the image to be reconstructed and sets the iteration index;
3. The servers perform re-projection by estimating projection data.
4. The servers transfer the modified back-projection data and weighting-factors to the client node.
5. The client compares real projection data with re-projected ones.
6. The client modifies the current image by incorporating weighted back-projection in a pixel-specific fashion.
7. The client broadcasts the image data to all servers for next-step reconstruction.

Repeat Steps 2 to 6 until a pre-defined error between two consecutive iterations is sufficiently small.

4 System Implementations

The system aggregates many legacy heterogeneous computers through a 3Com switch (Superstack II 3300-24 ports). The software is developed in Java to preserve platform independency. In the current study, each client or server node has its own TCP/IP address. In the initialization of client service, a scope of TCP/IP address (for example, "128.122.*.*" or "128.155.120.*") as an IP string is needed to search available IPs of potential servers. The IP address such as "197.*.*.*." can be input to simulate a localized PC cluster.

Once the IP scope is defined, the client creates an instance of Collector class (responsible for communication) for collecting available servers that recognize the image reconstruction portal. The successful IP addresses are stored in a backup data file for the next system configuration. The system allows users interactively edit the backup file to adjust the configuration. Once the system is dynamically established, image projection data are transferred to all the participating servers. After passing projection data to all the servers, the system perform back/forward iterative process for medical image reconstruction.

Each computational node has a server socket with a specified port number. Through the socket, multiple connection sockets for data I/O can be established. The server socket is always available so that it can continue receiving jobs from, or returning data to, the client. Once a server has been called, it dynamically generates threads. Each thread handles specific data I/O and performs a sub-tasked image forward and backward projection. After a server begins its computation, it accepts all the communications from potential clients that have the identical communication protocol. Once the server starts to function, its listener detects the communication from client.

5 Performance and Results

We conducted several experiments to study system performance. The sample image size was 252x252. The voxel size was 0.4. The projection data was obtained from an XCT scanner. The parameters of the XCT scanner are listed. The total number of intervals per translation was 310. In the scanner, there were translations of 15 in 12-

degree increments. The total number of detectors was 12. The between two neighboring detectors was 1 degree. The offset angle for the first detector was 5.5 degrees. We selected 10 heterogeneous distributed computers as computational nodes.

Figure 1 plots the performance in terms of speedup T^1_{iter}/T^n_{iter}. The T^1_{iter} is the total computational time to perform complete iteration using one server; and T^n_{iter} is the total computational time to perform complete the iteration using multiple servers. The total computational time is defined as the time for projection data transfer and total iterative computation time, i.e., $T^n_t = T^n_{comp}(T^n_{iter}) + T^n_p$. The speedup linearly increases as the number of deployed nodes increases, up to five nodes. After that, the speedup increases moderately with added nodes. With about eight nodes, the speedup begins to remain constant and no improvement in performance is observed. As the size of the dataset increases, the increase in performance becomes more pronounced. The major detractors to performance are network latency and heavy data communications.

Figure 2 plots the total job time vs. the time required to perform server-side iteration $T^n_s = T^n_{iter} - T^n_f$, where T^n_f is the time used to transfer reconstructed image data during iterations. It can be simply calculated as $T^n_f = aNT^n_p$. The calculation assumes the image transfer time equals to the projection data transfer time within the data communication. The proportional constant a=3 accounts for the times of data transfer within each iteration, basically the transfer of accumulated backward projected data, weighted factors from each server to client, and image data from client to server after image assembly on client system. N is the total number of servers (computational nodes). It is a fact due to the present parallel EM algorithm. Figure 2 clearly shows that with the increase of number of computational nodes, the desired iterative computation on the servers reduces to zero, which gives no benefit to the parallelism. The reason of such a reduction is due to the networking loss during data transfer, since T^n_f is proportional to the number of nodes used and the individual data transfer rate on Internet. It indicates the loss is strongly upon the number of nodes in a non-linear relationship, since $T^n_p = bN + c$ is can be interpolated as a linear function of N, as the first approximation (see Fig. 3), where b is dependent on the concurrent network bandwidths and other effects. Therefore, the data transfer cost is proposal to the bandwidth and the second power of numbers of servers. $T^n_f = abN^2 + ac = aN^2 + B$ where the coefficient b (mainly accounts for bandwidth, in an inverse relation) and c are time-dependent variables and the coefficient a is prefixed based on specific parallel algorithm, both coefficients A and B are time-dependent functions.

Fig. 4 plots the comparison between server-site contribution to reconstruction vs. the data transfer cost during the iteration process. The crossing point indicates the server-site gained computing balances with the data transfer cost. It is critical that reflects where the parallel performance (speedup) is not significant. This study tells that the corresponding optimal number of nodes is about eight nodes.

Fig. 5 plots the experimental results by selecting bandwidths of 100 MBPS using local centralized network system, of 70 MBPSB using 10/100MBPS local switch system, and of 11MB Wireless service, respectively. The bandwidth data are measured before submitting jobs from client. We assume the bandwidths maintain the same during whole computation. Fig. 5 indicates clearly that with a low bandwidth the performance is worse. That is due to the data transfer. The projection data transfer vs. different bandwidth can be seen in Fig. 6. The coefficient A is the product of 3 times b value. A determines the total cost during iterations. Therefore the data transfer cost strongly depends on the bandwidth of networking between client and servers. Increase the averaged bandwidth, the system gains high benefit in performance.

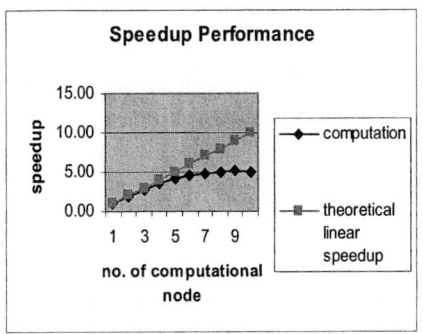

Fig. 1. Performance (speedup) vs. number of computational nodes (servers).

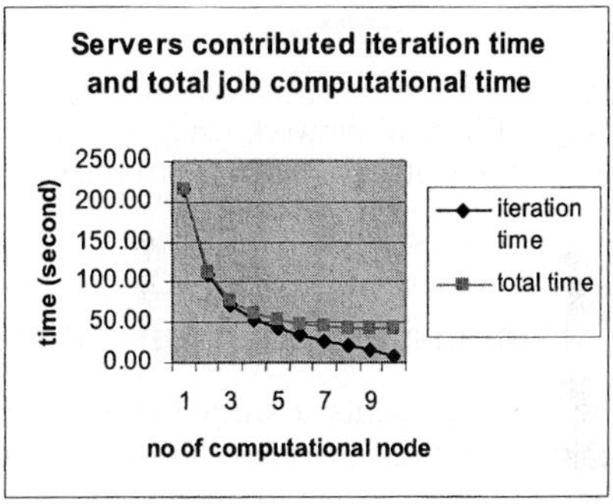

Fig. 2. Total iteration time and total computation time vs. number of computational node (servers).

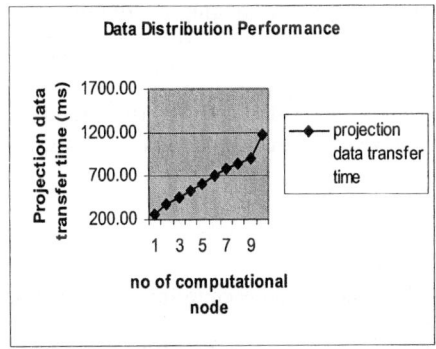

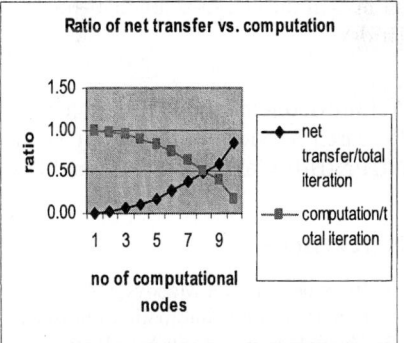

Fig. 3. Time to transfer the projection data to all the computational nodes.

Fig. 4. The total ratio of time required to data transfer data during iteration vs. computation time.

The bandwidth increase, the coefficient b decreases, and thus the data transfer rate decreases. B can be modeled as $b = \dfrac{1}{B_{max}} e^{-B_{avg}}$, where $B_{max} = \max\{B_{1,max}, B_{2,max}, \ldots, B_{n,max}\}$ and $B_{i,max}$ is possible maximum bandwidth for server i, and B_{avg} is the average bandwidth between client and selected servers.

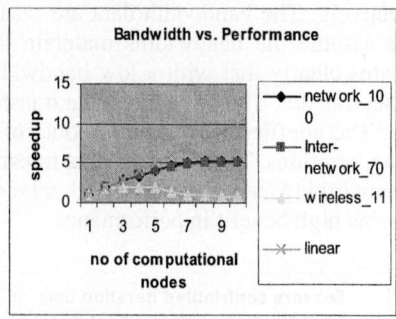
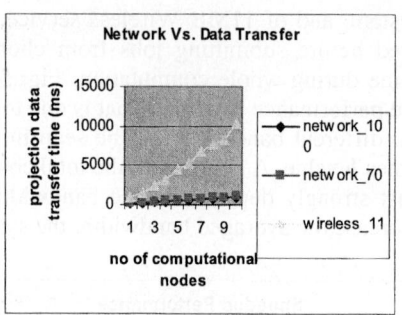

Fig. 5. Effect of network using various network bandwidths on parallel performance.

Fig. 6. Bandwidth speed vs. data transferring within the computation.

6 Conclusion and Future Work

The study presents a simple P2P-enhaced distributed system for iterative EM algorithm in medical imaging. The experimental results are discussed with simple evaluation models in order to analyze performance. The future work will focus on (1) the deployment of various parallelism in image reconstruction to the system; (2) study the effect of image size on the performance; (3) develop an optimal available IPs for node selection; (4) study load balance and fault tolerance; and (6) conduct comparisons of same parallel algorithm using GlobusToolKit (GT2/3) enhanced computation on TeraGrid. The proposed Internet-based P2P system and its implementation provides an insight into large-scale distributed computing of medical imaging in computational Grids.

Acknowledgements. The project is financially supported by internal funds at Research Services, Information Technology Services of the University of Iowa, and NIH/NIBIB grants EB001685 and EB002667.

References

1. Rockmore J. and Macovski A.: A Maximum Likelihood Approach to Image Reconstruction, Proc Joint Automatic Control Conf, 782-786 (1977)
2. Shepp L.A. and Valdi Y.: Maximum Likelihood Reconstruction for Emission Tomography, IEEE, Trans. Med. Imag., **MI-1** (1982) 113-122
3. Lange K. and Carson R.: EM Reconstruction Algorithms for Emission and Transmission Tomography, Journal Computer Assist. Tomography, 8(2), (1984) 302-316
4. Andersen A. H.: Algebraic Reconstruction in CT from Limited Views, IEEE Trans. Med. Imag., **8** (1989) 50-55
5. Andersen A. H. and Kak A. C.: Simultaneous Algebraic Reconstruction Technique (SART): A Superior Implementation of the ART Algorithm, Ultrasonic Imaging, **6** (1984) 81-94
6. Jiang M. and Wang G.: Convergence of the Simultaneous Algebraic Reconstruction Technique (SART), IEEE Trans. Image Processing, **12** (2003) 957-961
7. Synder D.L. et al.:Deblurring Subject to Nonnegativity Constraints, IEEE Trans. Signal Processing, **40** (1992) 1143-1150

8. Hutton B. F. et al: A Clinical Perspective of Accelerated Statistical Reconstruction, European Journal of Nuclear Medicine, 24 (1997)
9. Blocket D. et al: Maximum-Likelihood Reconstruction with Ordered Subsets in Bone SPECT, J. Nucl. Med, **40** (12), (1999) 1978-1984
10. Bai C.Y.et al: Postinjection Single Photon Transmission Tomography with Ordered-subset Algorithms for Whole-body PET Imaging, IEEE Trans. Nucl. Sci., **49** (1) (2002) 74-81
11. Kamphuis C. et al: Dual Matrix Ordered Subsets Reconstruction for Accelerated 3D Scatter Compensation in Single-photon Emission Tomography, Eur. J. Nucl. Med, **25** (1) (1998) 8-18
12. Manglos S.H. et al: Transmission Maximum-likelihood Reconstruction with Ordered Subsets for Cone-beam CT, Phys. Med. Biol. **40** (7) (1995) 1225-1241
13. Narayanan M.V. et al: Optimization of Regularization of Attenuation and Scatter- Corrected Tc-99m cardiac SPECT Studies for Defect Detection Using Hybrid Images, IEEE Trans. Nucl. Med., **48**, (3) (2001) 785-789
14. Wang G., et al: Iterative Deblurring for CT Metal Artifact Reduction, IEEE Trans. Med. Imag., **15** (1996) 657-664
15. Leahy R. and Byrne C.: Recent Developments in Iterative Image Reconstruction for PET and SPECT, Editorial, IEEE Trans. Med. Imag., **19** (4) (2000).
16. Hudson H.M. and Larkin R.S.: Accelerated Image Reconstruction Using Ordered Subsets of Projection data, IEEE Trans. Med. Imag., **13**, (1994) 601-609
17. Kamphuis C. and Beekman F.J.: Accelerated Iterative Transmission CT Reconstruction Using an Orderd Subsets Convex Algorithm, IEEE Trans. Med. Imaging, **17**(6) (1998)
18. Jones W. F. et al: Design of a Super Fast Three-dimensional Projection System for Positron Emission Tomography, IEEE Trans. Nucl. Sci., **37** (1990) 800-804
19. Guerrini C. and Spaletta G.: An image Reconstruction Algorithm in Tomography: A Version for the CRAY X-MP Vector Computer, Computers and Graphics, **13** (1989) 367-372
20. Miller M. and Butler C.: 3-D Maximum a Posteriori Estimation for Single Photon Emission Computed Tomography on Massively-parallel Computers, IEEE Trans. Med. Imag., **12** (1993) 560-565
21. McCarty A. and Miller M.: Maximum Likelihood SPECT in Clinical Computation Times Using Mesh-connected Parallel Computers, IEEE Trans, Med. Imag., **10** (1991) 426-436
22. Atkins M. et al: Use of Transputers in a 3-D Positron emission Tomography, IEEE Trans. Med. Imag., **10** (3 (1991) 276-283
23. Chen C. M. et al: A Parallel Implementation of 3-D CT Image Reconstruction on Hypercube Multiprocessor, IEEE Trans. Nucl. Sci., vol. **37** (3) (1990) 1333-1346
24. Johnson C. and Sofer A.: A Data-parallel Algorithm for Iterative Tomographic Image Reconstruction, www.nih.gov/publications
25. Shattuck D. et al: Internet2-based 3D PET Image Reconstruction Using a PC Cluster, Phys. Med. Biol., **47** (2002) 2785-2795
26. Chen C-M.: An Efficient Four-connected Parallel System for PET Image Reconstruction, Parallel Computing, **24** (1998) 1499-1522
27. Kontaxakis G. et al: Grid Processing Techniques for the Iterative Reconstruction of Large Clinical Positron Emission Tomography Sonogram Datasets, Health Grid 2003, www.die.upm.es/im.
28. Vollmar S. et al, "HeinzelCluster: Accelerated Reconstruction for FORE and OSEM3D, Phys. Med. Biol. **47** (2002) 2651-2658
29. Bevilacqua A.: Evaluation of a Fully 3-D BPF Method for Small Animal PET Images on MIMD Architectures: International Journal of Modern Physics C, **10** (4) (1991) 1-17
30. Backfrieder W. et al: Web-based Parallel ML_EM Reconstruction for SPECT on SMP Clusters, Proceeding of the International Conference on Mathematics and Engineering Techniques in Medicine and Biological Science, Las Vegas, Nevada, USA (2001)
31. Bevilacqua A.: A Dynamic Load Balancing Method on a Heterogeneous Cluster of Workstations, Special Issue: Parallel Computing on Networks of Computers, **23** (1) (1999) 49-56

End Host Multicast for Peer-to-Peer Systems*

Wanqing Tu and Weijia Jia

Department of Computer Engineering and Information Technology,
City University of Hongkong, Kowloon, Hongkong, P.R. China
tu.wanqing@student.cityu.edu.hk, itjia@cityu.edu.hk

Abstract. Multicast is an effective means for conducting the cooperative P2P communications. This paper studies an algorithm to construct a scalable and efficient end host multicast tree – *Shared Cluster Tree* (SCT). Compared with the previous work, by utilizing the underlying network properties, the novelty and contributions of our multicast algorithms are: (1) Scalability: by fully utilizing the underlying network properties to layer and cluster the group members, SCT decreases the transfer of the data traffic in the *backbone domain* links and therefore reserves more *backbone domain* resources to more *local domains* which include many more end hosts; (2) Efficiency: by searching the cluster cores based on the *cluster core selection* method, packets are multicasted to the cluster members by the cluster cores in the minimum delay. Our simulation results indicate that SCT is scalable and efficient for the end host multicast in P2P systems...

1 Introduction

Peer-to-peer (P2P) computing is the system of mutually exchanging information and services directly between the sender and the receiver. Grids are systems that integrate of instruments, displays, computational and information resources, managed by diverse organizations in potentially geographically widespread locations. In P2P systems, any two peers are of comparable capabilities and permitted to communicate with each other in such a way either ought to be able to initiate the communications. Thus, P2P computing is a powerful tool to organize grid and cooperative computing. A number of grid and cooperative applications (e.g. audio-video conferences, network games and distance learning) require multicast transmission to enable efficient many-to-many communications. Because multicast is an efficient means for the distribution of information that should be transmitted to a group of members. As end hosts in most P2P systems (e.g. Content-Addressable Network (CAN)) construct an overlay network, such overlay structure is in the sense that each of its edges corresponds to a unicast path between two end hosts in the underlying Internet. In this paper, our motivation is to develop a scalable and efficient multicast in the overlay network for P2P networking. Multicast in the overlay network is often called as the end

---

* The work is supported by Research grant council (RGC) Hong Kong, SAR China under CityU strategic grant nos. 7001587.

host multicast by many researchers. Compared with unicast, although the end host multicast communication has more scalability and efficiency, the following analysis shows that the end host multicast is less scalable and efficient than the network layer multicast.

One of the dominant differences between the end host multicast and the network layer multicast is that the multicast packets are replicated and forwarded by the end hosts instead of the intermediate routers. End host replicating and forwarding incurs the identical packets traversing the same underlying links more than one times. Excessive identical packets occupy the network resources. It influences the scalability of the end host multicast.

As for efficiency, our studies show that the end hosts instead of intermediate routers forwarding can cause not only worse scalability but also longer transmission delay. This is mainly because of the excessive delay that packets transmit from the application layer to the network layer as shown in Figure 1. We briefly analyze the delay difference as follows: Let the delay from the application layer to the network layer be $d_{a->n}$, the delay from the network layer to the physical layer be $d_{n->p}$, and the delay that packets transmit in the physical links be $d_{p->p}$. The total end-to-end delay in (a) is $d_{nt} = 2d_{a->n} + 4d_{n->p} + 2d_{p->p}$, however, the total end-to-end delay in (b) is $d_{et} = 4d_{a->n} + 4d_{n->p} + 2d_{p->p}$. The more the number of end hosts that forward the packets, the longer end-to-end transmission delay is required for the multicast. Moreover, the end hosts have lower capacities to deal with and forward packets than the intermediate routers, which incurs longer processing delay. Hence, the end host multicast is costly in terms of delay. Although the end host multicast cannot overtake the network layer multicast in terms of the performances, the end host multicast's independence of the underlying network architecture makes it feasible and popular. It is thus worthwhile to improve the scalability and efficiency of end host multicast. Since the underlying network is the fundamental of the overlay network, we fully utilize the underlying network characteristics to develop the end host multicast, which is called as *Shared Cluster Tree* (SCT). *"Shared tree"* refers to the multicast data path, and a *"cluster"* refers to a "mini-group". Unlike the

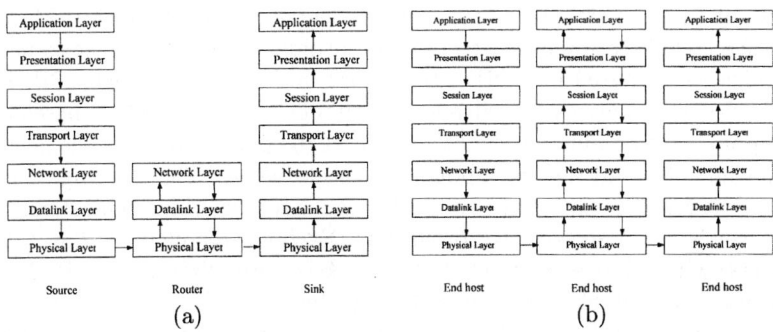

Fig. 1. Delay difference between the router forwarding architecture (a) and the end host forwarding architecture (b)

previous work, the novelty and contributions of the paper are: *1) Scalability*: by fully utilizing the underlying network properties to layer and cluster the group members, our *cluster construction, layer construction* and *shared tree creation* regulations decrease the transfer of the data traffic in the *backbone domain* links and therefore reserve more *backbone domain* resources to more *local domains* which contain many more end hosts. (The *local domain* and the *backbone domain* will be defined in Section III); *2) Efficiency*: by applying the *cluster core selection*, each cluster selects a core. The core guarantees to multicast the data packets to the cluster members with the minimum delay; by utilizing the hierarchy and cluster, SCT limits the flow transmission in the links among routers, which also improves the efficiency.

The rest of the paper is organized as follows: Previous related work is discussed in Section 2. Section 3 provides the detailed SCT design. Section 4 evaluates the well-known end host multicast protocols: CAN-multicast, NICE, and our DSCT, and gives the simulation observations and performance comparisons. Section 5 concludes the paper.

2   Related Work

In terms of the sequence of constructing the data and control topologies, the current end host multicast protocols are classified into three categories: *1) Mesh-first multicast*. NARADA [8] is a *mesh-first* end host multicast protocol. It demonstrates the feasibility of implementing the multicast functionality at the application layer. It distributedly organizes group members into an overlay mesh topology, and runs a protocol similar to the conventional DVRMP to accomplish multicast routing among these members. To guarantee robust, each group member in NARADA periodically sends refresh packets to each of other group member, which compromises the scalability of NARADA; *2) Tree-first multicast*. YOID [9] is one of the earliest end host multicast protocols. It constructs a shared data delivery tree among members first. The use of tree routing is to get logarithmic scaling behavior with respect to the number of receivers. But YOID would not be efficient and simple, because it has a direct control over various aspects of the tree structure (e.g. the out-degree at the members). Moreover, expensive techniques of loop detection and avoidance, and resilience to network partitions are required in YOID; *3) Implicit multicast*. Well-known NICE [10] and CAN-based multicast [11] all adopt the implicit approach. CAN-based multicast employs the flooding method to multicast the packets to all of the neighbors who haven't received the corresponding packets before. The flooding scheme is simple, but resource-consumed and inefficient. NICE is a hierarchical multicast, each layer is composed of a set of clusters. Each cluster includes $k \sim (3k-1)$ members, where k is a constant. In NICE, k is normally set as 3. Packets are multicasted to each cluster member by the cluster leader, except in the source cluster where packets are sent to each cluster member by the source. Compared with other protocols, NICE is more scalable for the hierarchy and cluster. But the hierarchy and cluster is independent of the underlying network. Actually, by utilizing the underlying network properties, as described in this paper, we can improve the scalability further.

3 SCT Design

We first make the following definitions. *1) Local domain*: It is composed of group members that attach to the same router directly or through several local network components (e.g. the hubs) and other local network resources (e.g. physical links, the hubs) used to connect these members. In this definition, the router mainly refers to the local router, not the router in the backbone network. *2) Backbone domain*: It is composed of all the routers used to connect the multicast group members and other network resources (e.g. physical links) among these routers. The routers in this definition include the local routers and the backbone routers. *3) Unassigned end hosts*: Define the end hosts in a multicast group that haven't been assigned into any cluster as the *unassigned end hosts*.

Apart from employing the hierarchy scheme, our basic idea to design SCT is to limit the transfer of packets in the *backbone domain* links, which reserves the *backbone domain* resources to carry more end hosts and decreases the chance that packets transmit in the long delay *backbone domain* links.

3.1 Cluster Construction

In SCT tree, each *local domain* has a *local core*. The selection of *local core* is introduced in the *layer construction* regulation. Members in the same *local domain* form the "intra-clusters". *Local cores* in different *local domains* form the "inter-clusters". Different cluster sizes exist for these two kinds of clusters.

As for the "intra-cluster", the size is expressed as:

$$s_{ina} = \begin{cases} (k, 3k-1) &, r_n > 3k-1 \\ r_n &, r_n \leq 3k-1 \end{cases} \quad (1)$$

where r_n is the number of *unassigned end hosts* in the same *local domain*, and the expression $(k, 3k-1)$ represents a random constant between k and $3k-1$. Like NICE, k is a constant, and in our experiment, we also use $k = 3$. The cluster size bound, $(k, 3k-1)$, is the same as the one in NICE, which avoids the frequent cluster splitting and merging (see [10]). The way to cluster *local domain* members is to partition s_{ina} *unassigned end hosts* who are the closest ones to one another among all *local domain* members into the same cluster. Expression 1 guarantees that each "intra-cluster" only contains members of the same *local domain*. Hence, the packets' traversing over the *backbone domain* links is limited.

Similarly, as the "inter-cluster" contains end hosts of different *local domains*, the size is denoted as:

$$s_{ine} = (k, 3k-1) \quad (2)$$

3.2 Cluster Core Selection

In SCT, for the efficient communications, the core of each cluster is selected *in terms of the minimum sum of unicast distance from the core to all the members*.

Let s be the cluster size, $d(i,j)$ be the unicast distance from member i to member j, $i,j \in [1,s]$, d_i be the sum of $d(i,j)$, d_i can be expressed as:

$$d_i = \sum_{j=1, j \neq i}^{s} d(i,j) \qquad (3)$$

Denote the cluster core as c. Then, the cluster core satisfies:

$$d_c = min\{d_i, i \in [1,s]\} \qquad (4)$$

3.3 Layer Construction

Local domain members are mapped to different layers using the following scheme: All members in the same *local domain* d, $d \in [0, n_{ld} - 1]$, belong to the lowest layer L_0. The *Cluster construction* regulation assigns them into different clusters in L_0. By using the *cluster core selection*, a cluster core is selected for each cluster in layer L_i. Cluster cores in L_i join in layer L_{i+1}. Also, members in L_{i+1} are partitioned into different clusters. Then, the cluster cores of clusters in L_{i+1} become the members of layer L_{i+2}. The hierarchy in the *local domain* d terminates until the number of members is not greater than $3k - 1$, and the core of these members is the *local core*, which constructs the uppest layer $L_{LHL(d)}$ of *local domain* d. $LHL(d)$ is the uppest layer number of *local domain* d. For the layer is labelled from 0, the number of layers in *local domain* d is $LHL(d)+1$. In this layer method, each member in layer L_i is actually the cluster core of cluster in layer L_j, $j \in [0, i-1]$ that the member belongs to. Formula 5 illuminates the relationship among the number of members n_d, the uppest layer number $LHL(d)$, and the size of the cluster s_{ina} in *local domain* d.

$$\frac{n_d}{(s_{ina})^{(LHL(d)-1)}} \leq 3k - 1 \qquad (5)$$

Local cores are mapped to different layers using the similar scheme: As different *local domains* contain different numbers of end hosts, the local uppest layers $L_{LHL(d)}$ may be different from one another. Denote the maximum value of the local uppest layer numbers LHL as:

$$LHL = max\{LHL(d), d \in [0, n_{ld} - 1]\} \qquad (6)$$

All *local cores* belong to layer L_{LHL}. *Local cores* go on to be layered into several layers that are not lower than L_{LHL}, then partitioned into several clusters with the size of $(k, 3k - 1)$. The hierarchy terminates until the number of *local cores* is not greater than $3k - 1$, and the core of these members is the *SCT core*, which constructs the highest layer L_{HL} of SCT. HL is the highest layer number of SCT tree. The total layer number of SCT is $(HL+1)$. Formula 7 illuminates the relationship among the number of *local cores* n_{lc}, the number of layers ($HL - LHL + 1$) containing the "inter-clusters", and the size of the clusters s_{ine}.

$$\frac{n_{lc}}{(s_{ine})^{(HL-LHL)}} \leq 3k - 1 \qquad (7)$$

Formula (5) and (7) show that the cluster size is the inverse proportion to the total layer number $HL + 1$ with respect to the fixed number of group members $n = \sum_{d=0}^{n_{ld}-1} n_d$.

3.4 Shared Tree Creation

The main idea of the tree generation algorithm can be sketched as follows: SCT tree is a hierarchical shared tree. The *layer construction* assigns the group members into different layers. The *cluster construction* partitions the members of the same layer into different clusters. Each cluster is a minor shared tree on the SCT tree. The cluster core of either the "intra-cluster" or the "inter-cluster" coordinate the communications between the members inside and outside the cluster. Namely, packets from the outside of the cluster are forwarded to each of the cluster member by the cluster core. And, packets sent by a cluster member forward to the cluster core first, then, the cluster core multicasts the packets to: a) other members in the source cluster; b) group members in other clusters in which it is also the cluster core; c) its cluster core of the cluster in the uppest layer that it belongs to. Similarly, these receivers go on to forward the packets along the shared tree until all group members receive the packets. The setting of the *local core* (i.e. the cluster core of "inter-cluster") decreases the packet transmission in the *backbone domain* links greatly. One packet transmitted from *local domain* a to *local domain* b only utilizes the same *backbone domain* links between these two *local domains* one time. The following is the *SCT hierarchical shared tree creation algorithm*.

Algorithm 1 SCT Hierarchical Shared Tree Creation Algorithm
Input: Group $G = \{g_1, g_2, ..., g_n\}$;
Output: SCT hierarchical shared tree T_{SCT};

---

1. $T_{SCT} = \{\}$;
2. For $j = 0$ to $n_{ld} - 1$ do
3.   Apply the *layer construction* and the *cluster construction* regulations to layer and cluster group members of the *local domain* j, select the cluster cores according to the *cluster core selection* method, and add the operation results to T_{SCT};
4.   Select the member in $L_{LHL(j)}$ as the *local core* of the *local domain* j, add it to T_{SCT};
5. Apply the *layer construction* and the *cluster construction* regulations to layer and cluster *local cores* from layer L_{LHL}, select the "inter-cluster" cores by using the *cluster core selection* method, and add the operation results to T_{SCT};
6. Select the member in the highest layer L_{HL} as the *SCT core* of T_{SCT}, and add it to T_{SCT}.

---

4 Experimental Evaluation

We use *ns*-2 [12] to run our simulations on a group of SUN SOLARIS workstations. The simulation backbone network adopts the MCI ISP backbone as shown in Fig. 2. End peers directly or indirectly attach to the routers in the backbone. The link bandwidth in the backbone network and the *local domain* are 1000Mbps and 100Mbps respectively. Random integers between 20 and 40 and between 4 and 8 are set as the link costs of *backbone network* and *local domain* respectively. Three criteria are used to evaluate the performances of different end host multicast protocols. 1) Average end-to-end delay (AD): AD of source s is define as the ratio of the sum of end-to-end delay from s to each group member to the number of group members; 2) Average cost stress (ACS): Define ACS as the ratio of the sum of the consuming costs of identical packet copies in each link to the sum of each link's inherent cost; 3)Average link stress (ALS): ALS is defined as the ratio of the sum of numbers that identical packet copies traverse over the same underlying links to the number of links in the group.

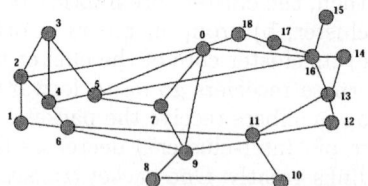

Fig. 2. The Experimental Backbone Network

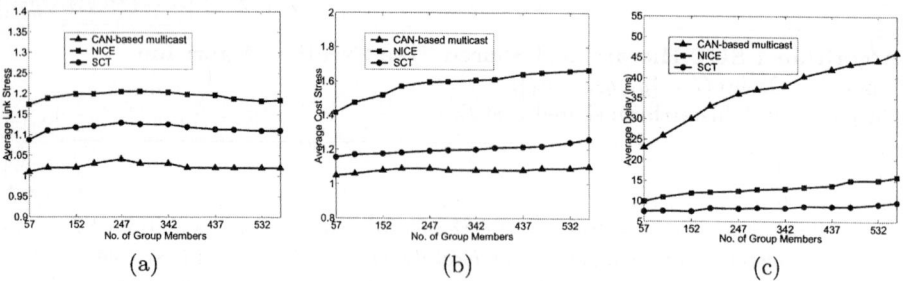

Fig. 3. The performance results. (a) is for the average link stress, (b) is for the average cost stress, (c) is for the average end-to-end delay

In this simulation, there are 5 sources, we observe the performance of proposed multicast protocols as the group members increase from 57 to 570. Fig. 3 (a) illuminates that SCT achieves lower link stress than NICE. In SCT and NICE, the worst link stress appears in the links around some cluster core (or some cluster leader in NICE). The *cluster construction* in SCT disperses the worst link stress by using different cluster sizes. For CAN-based multicast, the flooding distributes the link stress much more evenly over all links, hence, the worst link stress of CAN-based multicast is much lower than SCT and Nice. It introduces lower ALS of CAN-based multicast. However, the flooding scheme occupies all of links in the multicast group, which makes it resource-consumed. Fig. 3 (b) illuminates that SCT achieves much lower cost stress than NICE. As we have analyzed, SCT avoids the packets' unnecessary visit to the costly links. The curves prove that the construction of SCT tree meets the demand of limiting the data traffic in the *backbone domain* links. Hence, SCT is more scalable and efficient than NICE. For the same reason of ALS, the ACS of CAN-based multicast is lower than SCT and NICE. The *AD* curves shown in Fig. 3 (c) illuminate that SCT is the most efficient one in these three multicast mechanisms in terms of *average end-to-end delay*. NICE achieves relatively much lower *AD* than CAN-based multicast. The flooding scheme in CAN-based multicast incurs much longer delay. The simulation results show that the core selection and the hierarchy and cluster in SCT are effective for improving efficiency and scalability of end host multicast in P2P systems.

5 Conclusion

Unlike previous related studies, in this paper, we utilize the underlying network properties to cluster and layer the group members, to select the cores and to construct the multicast tree. Our main contribution is to improve the scalability and efficiency of end host multicast by our novel data structure – SCT (Shared Cluster Tree). Theoretic analysis and performance data have demonstrated that our SCT is indeed scalable and efficient for multicasting in the overlay networks of P2P systems. It is our target to implement SCT multicast in practical P2P cooperative systems.

References

1. C.Lv, P.Cao, ECohen, K.Li, and S.Shenker, "Search and replication in unstructured peer-to-peer networks", Preprint, Optinal 2001.
2. E.Cohen and S.Shenker,"Optimal replication in random search networks", Preprint, Optional 2001.
3. Q.Lv, S.Ratnasamy, and S.Shenker, "Can heterogeneity make gnutella scalable?", *Proc. of the 1st Internetional Workshop on Peer-to-Peer Systems (IPTPS '02)*, MIT Faculty Club, Cambridge, MA, USA, March 2002.
4. I.Stoica, R. Morris, D.Karger, M.Kaashoek and H.Balakrishnan, "Chord: A scalable content-addressable network," In *Proc. Of ACM SIGCOMM'01,* Aug. 2001.
5. Sylvia Ratnasamy, Mark Handley, Richard Karp and Scott Shenker, "Application-level multicast using content-addressable networks", *Proc. Of ACM SIGCOMM'01,*,Volume 8 Issue 4. Aug. 2001.
6. Rowstron, A. and Druschel, P., "Pastry: Scalable, distributed object location and routing for large-scale peer-to-peer systems", Available at http://research.mecrosoft.com/antr/PAST/, 2001.

7. Zhao, B.Y., Kubiatowicz,J. and Joseph, A, "Tapestry: An infrastructure for fault-tolerant wide-area location and routing", Available at http://www.cs.berkeley.edu/ravenben/tapestry/, 2001.
8. Y.H.Chu, S.G.Rao and H.Zhang, "A case for end system multicast", In *Proc. of ACM SIGMETRICS,* June 2000.
9. P.Francis, "Yoid: Extending the multicast internet architecture", White paper http://www.aciri.org/yoid/,1999.
10. S.Banerjee, B.Bhattacharjee and C.Kommareddy, "Scalable application layer multicast," *Proc. of ACM SIGCOMM,* August 2002.
11. Sylvia Ratnasamy, Mark Handley, Richard Karp and Scott Shenker, "Application-level multicast using content-addressable networks", *Proc. of NGC,* 2001.
12. UC Berkeley,LBL,USC/ISI,and Xerox PARC, *"ns* Notes and Documentation," October 20, 1999.

Janus: Build Gnutella-Like File Sharing System over Structured Overlay

Haitao Dong, Weimin Zheng, and Dongsheng Wang

Computer Science and Technology Department, Tsinghua University, Beijing, China
dht02@mails.tsinghua.edu.cn, {zwm-dcs,wds}@tsinghua.edu.cn

Abstract. How to build an efficient and scalable p2p file sharing system is still an open question. Structured systems obtain $O(\log(N))$ lookup upper bound by associating content with node. But they can not supporting complex queries. On the other hand, Gnutella-like unstructured systems support complex queries, but because of its random-graph topology and its flooding content discovery mechanism, it can not scale to large network systems. In this paper, we present Janus, which build unstructured file sharing system over structured overlay. Different from previous approaches, Janus keeps bidirectional links in its routing table. And with one-hop replication and biased random walk Janus make it possible to implement complex queries in the scalable manner. The experimental results indicate that, when the system running over a network of 10,000 peers, it only needs 100 hops to search half of the total system.

1 Introduction

In recent years, P2P overlay has become one of the most important systems over the Internet, which have consume more bandwidth than http access. The rapid development of P2P system owe much to its function as substrate for large data sharing and content distribution application. In such serverless system, a node looking for files must find the peers that have the desired content at first. Napster was one of the first peer-to-peer systems, which recognizes that requests for popular content need not be sent to a central server but instead could be handled by the many hosts, or peers, that already possess the content. It adapts a hybrid design, which consists of a directory server as centralized search facility based on file lists provided by each peer. But because RIAA's lawsuit, these centralized systems have been replaced by new decentralized systems that have no such centralized search capabilities. There are two kinds of decentralized P2P overlay networks: unstructured overlay and structured overlay.

Unstructured overlay, such as Gnutella and KaZaA, distribute both the download and search because of the lack of centralized directory server. In these systems, peers establish a randomly connected overlay network and the placement of files within them has no constrains. Consequently, each query should be flooded(or randomly dispersed) throughout this overlay with a limited scope(by setting the TTL of query). If a peer receives such a query message, it will check locally if it has the file matching the query. If having such content, it sends the list of all the appropriate content back to the original node. And if the TTL is not zero, the node will decrease the TTL by 1, and resends it to all its neighbors. So we can make out that the load on each node grows linearly with the total number of queries, which grows with system size. Obviously, this approach supports arbitrarily complex queries but not scale to large size systems.

To solve the scalability problem of unstructured overlay, several approaches have been simultaneously but independently proposed, all of which support a distributed hash table (DHT) functionality; among them are Tapestry[10], Pastry[6], Chord[8], and Content-Addressable Networks (CAN)[5]. In these DHT systems, files are associated with a key, which is produced, for instance, by hashing the file name, and each node in the system is responsible for storing a certain range of keys. There is one basic operation in these DHT systems, lookup(key), which returns the identity (e.g., the IP address) of the node storing the object responsible for that key. By allowing nodes to put and get files based on their key with such operation, DHTs support the hash-table-like interface. The core of these DHT systems is the routing algorithm. The DHT nodes form an overlay network with each node having several other nodes as neighbors. When a lookup(key) operation is issued, the lookup is routed through the overlay network to the node responsible for that key. Then, the scalability of these DHT algorithms is dependent on the efficiency of their routing algorithms. Each of the proposed DHT systems listed above – Tapestry, Pastry, Chord, and CAN – employ a different routing algorithm. Though there are many details different between their routing algorithms, they share the same property that every overlay node maintains O(log n) neighbors and routes within O(log n) hops (n is the system scale). To achieve this routing performance, the graph formed by peers is structured to enable efficient discovery of data items given their keys at the cost of not supporting complex queries.

[2] points out that complex searches based on keyword are more prevalent, and more important, than exact-match queries. Can we design a system scalably supports complex query?[11] proposes a newly thought that it is plausible to build Gnutella like unstructured overlay on a structured overlay. They replace the random graph in Gnutella by a structured overlay while retaining the content placement and discovery mechanisms of unstructured overlays to support complex queries. Structella also uses either a form of flooding or random walks to discover content. Though, in flooding model, it takes advantage of structure to ensure that nodes are visited only once during a query and to control the number of nodes that are visited accurately, it does not radically solve the scalable problem. It is because when a query is initiated, many nodes irrelevant to the query have to re-route the query message. And in random model, because Structulla simply walk along the ring formed by neighbouring nodes in the id space. In worst case, its query path length is O(n).

In this paper, we propose Janus, a new file sharing system, which builds unstructured P2P system (Gnutella) on DHT with bidirectional fingers. The motivation of modifying conventional structured overlay's unilateral finger into bidirectional is that we observe the fact that though conventional DHT overlay has balanced fan-out fingers(typically O(log$n$)), the number of their fan-in fingers is not homogeneous at all. By maintaining bidirectional routing table, peers have different size of neighbors, which make the biased random walk strategy possible. And the degree of the imbalance between peers' routing table size largely influence the efficiency of the performance biased random walk. We also introduce one-hop replication (neighboring peers replicate file information with each other), which improve query performance greatly based on our simulation results.

The rest of this paper is organized as follows. Section 2 introduces related works in the field of P2P file sharing overlay networks research. Section 3 presents the design of Janus protocol. Experimental results are presented in section 4. Final conclusion is given in section 5.

2 Background

Janus protocol is based on two ongoing research interests: unstructured file sharing and structured overlay design. In this section, first, we generally introduce representative DHT systems, then give a brief description of Gnutella, the most prevalent unstructured file sharing system.

DHTs: Chord takes, as input, a key and, in response, route a message to the node responsible for that key. The keys are strings of digits of some length. Nodes have identifiers, taken from the same space as the keys (i.e., same number of digits). Each node maintains a routing table consisting of a small subset of nodes in the system. When a node receives a query for a key for which it is not responsible, the node routes the query to the neighbor node that makes the most "progress" towards resolving the query. The notion of progress differs from algorithm to algorithm, but in general is defined in terms of some distance between the identifier of the current node and the identifier of the queried key.

Gnutella: Here we mainly introduce the Gnutella 0.4 protocol[1]. Gnutella is based on a random graph. Each node in the overlay maintains a neighbour table with the network addresses of its neighbours in the graph. The neighbour tables are symmetric; if node has node in its neighbour table then node has node in its neighbour table. The neighbour tables are designed to be symmetric in order to reduce maintenance load. There is an upper and lower bound on the number of entries in each node's neighbour table, typically the lower bound is 4 and upper bound is 8.

Structella: Structella replaces the random graph in Gnutella by a structured overlay but it does not use the structure to organize the content. Structella supports complex queries using variants of flooding and random walks like Gnutella but it takes advantage of structure to ensure that nodes are visited only once during a query and to control the number of nodes that are visited accurately. Structella also leverages the structured overlay to reduce the maintenance overheads. But it is only a primary approach. Janus adopts some idea of Stuctella, such as loosing the constraint in DHT that couple content to nodes according their key. But Janus introduces several mechanisms to improve system performance of Structella.

3 The Janus System

This section describes the Janus protocol. The Janus protocol specifies how to how to construct the routing table, how to maintain the structure of the system, how deal with query operation, and how to use the k-hop replication strategy to improve system performance.

While Janus can be implemented based on most of existing O(log n) structured overlays, in the following part of this paper, we only discuss Janus based on Chord (Janus on other DHTs has the similar result). Like Chord, Janus assigns each node an m bit identifier using a base hash function such as SHA-1. But, different from Chord, the placement of the content in Janus has no constraint with where it is located, which is similar to Gnutella.

In the following part, we introduce how nodes in Janus will be organized. A node's identifier is chosen by hashing the node's IP address. The term "node" will refer to both the node and its identifier under the hash function. Consistent hashing assigns

keys to nodes as follows. Nodes are ordered in an identifier circle modulo 2m by its identifier. A node (with identifier) K is called node L's successor node if and if only
1. K > L;
2. there is no such node M satisfying L<M<K;

The first node whose identifier is equal to or follows (the identifier of) key in the identifier space is called the *successor node* of key, denoted by *succ(key)*.

3.1 Routing Table

A Janus node has a routing table consists of two parts. The first one is identical with Chord's routing table, which we denote as *finger list*: let m be the number of bits in the key/node identifiers. Each node n, maintains a routing table with (at most) m entries, called the *finger table*. The m entry in the table at node n contains the identity of the *first* node, s, that succeeds n by at least 2^{i-1} on the identifier circle, i.e., s=succ(n+2^{i-1}) where 1≤i≤m(and all arithmetic is modulo 2^m). We call node s the ith *finger* of node n. A finger table entry includes both the Chord identifier and the IP address (and port number) of the relevant node. Note that the first finger of n is its immediate successor on the circle; for convenience we often refer to it as the *successor* rather than the first finger.

The second part of node n's routing table contains all the nodes having n in its finger list, we call it as n's *reverse finger list*. All the neighbors in Janus' routing table can be used to discover content. We define size of node n's routing as n's *degree*, denoted as degree(n). Figure 2 show degree distribution of Janus. To implement biased random walk (detail in section 3.3), Janus must also record all the neighbors' degree.

Figure 1 shows an example, the identifier circle has an m=3, and there are 3 nodes in the circle with node identifier 0, 1, 3.

By keeping reverse fingers, Janus establish bidirectional link between peers in the overlay, because the imbalance of fan-in of Chord protocol, the degree distribution of Janus is not uniformed distributed. From the chart, we can see that the degree distribution is similar to power-law distribution.

3.2 Janus Protocol

Janus protocol preserves two invariants:
1. Each node's finger table is identical to Chord's finger table.
2. For every node n in a node m's reverse routing table, m is in n's finger table.
   This section shows how to maintain these invariants when node joins.

To preserve the first invariant, like Chord, Janus perform following tasks when a node joins into the ovelay:
1. Initialize the predecessor and fingers of node n.
2. Update the fingers and predecessors of existing nodes to reflect the addition of n.

Both of the two tasks are identical with Chord, so we do not discuss them due to the limit of space.

To preserve the second invariant, when a node n initializes its fingers, its finger will insert n into its reverse finger list.

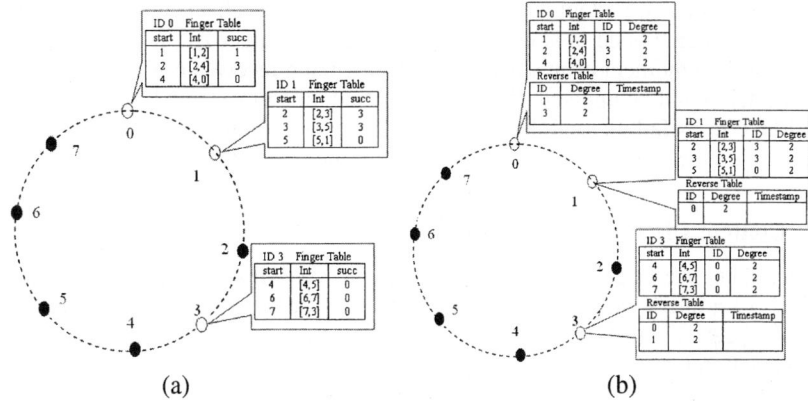

Fig. 1. (a) State of Chord's Routing Table. (b) State of Janus's Routing Table.

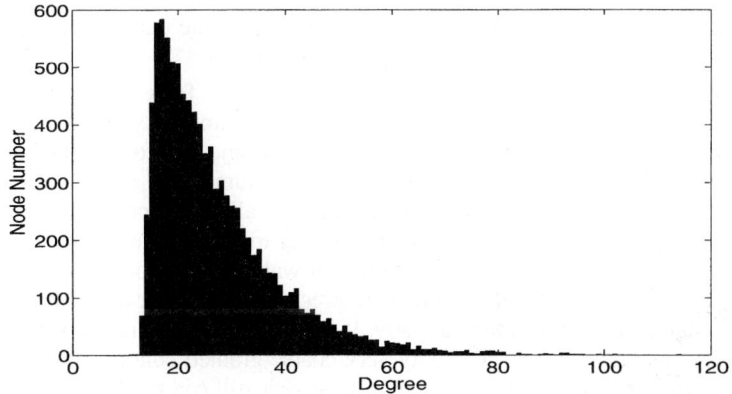

Fig. 2. The degree distribution of Janus with the scale of 10000 nodes.

Join: When a new node n joins Janus system, firstly it should contact an existing Janus node b, which is named as n's bootstrap node. On receiving n's join request, b initiates a lookup of n's to find succ(n), say s. After find s, n insert s as its successor, and inform s's form predecessor, say p, to update its successor as n. Then n begin to initiate its finger list. When successfully find a finger, the finger node will insert n into its reverse finger list.

Maintain: In a dynamic network, nodes can join, fail or leave voluntarily at any time. To handle these situation, Janus must introduce maintain mechanism.

To avoid introduce more maintenance cost than Chord, Janus choose different mechanism to update finger list and reverse finger list. Because Janus node's finger list is identical to Chord node's finger table, Janus adopt the same update protocol as Chord. To see the detail, please refer[8].

For reverse finger list, Janus introduce timestamp for every item in the list. In finger list's maintaining protocol, node n probe the node in its finger list periodically. So, for node m in n's finger list, if node n is alive, m will periodically receive n's probe message. That is to say, when n fails, m will never receive n's probing message any more. In that, we adopt a timestamp-based update mechanism for node's reverse finger list. When a node m is inserted into n's reverse finger list, it will be attached a timestamp (the moment of m's insertion). And when n receives m's probing message, the timestamp will be updated. If in several period, m's timestamp has not be updated, n will take m as failure, and delete m from its reverse finger list. We can see that to maintain reverse finger list, Janus need not introduce extra maintenance cost.

3.3 Content Discovery

Structella[11] use finger table to flood query message. For a general instance, given the TTL of a query flood operation is t, and every peer has m neighbor, and there is only one peer contains the content which the query is looking for, the efficiency of this query flood is:

$$Factor = \frac{1}{m(m-1)^{l-1}} \quad (1)$$

That to say, in the $m(m-1)^{l-1}$ query messages, only one query message is effective. Obviously the flooding query model is not scalable.

Structella also modify the random walk search process. To search for an object using a random walk, a node chooses a neighbor at random and sends the query to it. Each neighbor in turn repeats this process until the object is found. Random walks avoid the problem of the exponentially increasing number of query messages that arise in flooding, but the response time of random walk is not predictable. Structella implement random walk by walking along the ring formed by neighbouring nodes in the id space. One of the advantages of random walks over flooding in unstructured overlays is that random walks provide more precise control over the number of overlay nodes that is visited to satisfy a query. It is possible to constrain floods using the TTL mechanism but this provides only very coarse-grained control. Though this approach ensures that each node is visited only once, it still can resolve the inefficiency of query process. If the popularity of the content in the system is p%, that is to say p% nodes in the system have the content, in Structulla, it need

$$hops = N \times p\% \quad (2)$$

on average to find the content. If N is 10^5, and p% = 1%, it need 1000 hops to get the content!

To improve the efficiency of the content discovery process, each Janus node actively maintains an index of the content of each of its neighbors (nodes in finger list and reverse finger list as well). These indices are exchanged when neighbors establish connections to each other, and periodically updated with any incremental changes. Thus, when a node receives a query, it can respond not only with matches from its own content, but also provide matches from the content offered by all of its neighbors. When a neighbor is lost, either because it leaves the system, or due to topology adaptation, the index information for that neighbor gets flushed. This ensures that all index information remains mostly up-to-date and consistent throughout the lifetime of the node. And the timestamp mechanism mentioned in last section makes this process easy and effective.

The fact that degree unevenly distributed in Janus and one-hop replication (whereby nodes keep an index of their neighbors' shared files) ensures that high de-

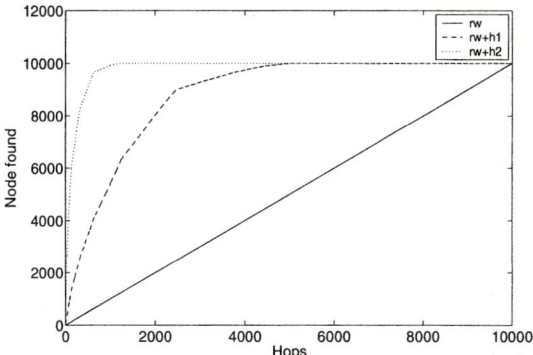

Fig. 3. Biased random walk with different replication strategy(rw is biased random walk without replication, rw+h1 is biased random walk with 1-hop replication, rw+h2 is biased random walk with 2-hop replication).

gree nodes can greatly accelerate the query process. Hence, the Janus search protocol uses a biased random walk: rather than forwarding incoming queries to randomly chosen neighbors, a Janus node selects the highest degree neighbor and sends the query to that neighbor.

4 Simulations

In this section, we present experimental results obtained with a simulator for Janus protocol. The simulator was implemented on ONSP, an overlay networks simulation platform, which can parallel simulating the function of most off-the-shelf peer-to-peer protocols. By implementing the event logic according to the protocol's definition, the user can easily simulate various protocols. The detailed description of ONSP exceeds the scope of this paper, we will present it in another paper.

Our experiments were performed on a 32 processors cluster (Pentium IV CPU and 2G memory), running Linux Redhat 7.0.

To simulate the real network topology and latency, we implemented a network topology generator based on GT-ITM[9]. The generator produces a transit-stub network topology model. The average living time of a node over the trace was about 2.3 hours. Experiment result is shown as figure 3.

5 Conclusion

We started this paper with the thought that if we can build unstructured file sharing system on structured overlay and get scalable query performance. Building on the work in[1][2][4], we proposed a design that appears to achieve significant scalability by harnessing peer's heterogeneity.

References

1. The Gnutella protocol specification, 2000. http://dss.clip2.com/GnutellaProtocol04.pdf.
2. Y. Chawathe, S. Ratnasamy, L. Breslau, N. Lanham, and S. Shenker. Making gnutella-like p2p systems scalable. In Proc. ACM SIGCOMM, Aug. 2003.

Although its expectation can be calculated, no fixed upper limit lies. This makes room for scalability design: we can raise the expectation of hops to keep a low overhead by adjusting system arguments.

Kelips [2] is another p2p protocol with $O(1)$-hop lookup efficiency. We will compare Gemini and Kelips after a 5,000,000-node case study in section 2. Then in section 3 we give the formal description of Gemini and explain how to determine system arguments. Section 4 proposes some novel methods to gain even high scalability. Final conclusion is in section 5.

Case Study: 5,000,000-nodeGemini

To make our presentation more tangible, we first describe how Gemini works in a network of 5,000,000 nodes. Formal description and analysis are left to the next section.

Like Pastry and Chord, Gemini assigns every node a *nodeId*. Commonly a nodeId is 64-bit long, being hash result of the node's IP address.

To a node, say node M[1], we call the first 10 bits of M the node's *hat*, and the last 10 bits the node's *boot*.

Routing table. A Gemini node has a routing table consists of two parts, the first one containing all the nodes having the same cap with M, called *hatcase*, while the second one containing all the nodes having the same boot with M, called *bootcase*.

Obviously if two nodes have a same hat, their hatcases must also be the same. In this way, all the nodes are split into $2^{10}=1024$ groups according to their different haps. We call these groups *hat clubs*. Nodes within the same hat club are fully interconnected via their hatcases. Because nodeIds scatter evenly, every hat club contains about 5,000,000/1024≈4883 nodes. The *boot clubs* are defined symmetrically.

Routing. Every message has a destination address that is also 64-bit long. The first 10 bits of the address is also called its *hap*, and the unique hat club with this cap is called its *hat club*.

Slightly different with Pastry, root node of a message satisfies the following two conditions:

1) It is in the hat club of the destination address.
2) Its nodeId is numerically closest to the destination address among the whole club.

Considering that nodes in a hat club are fully interconnected, the basic task of message routing turns to forward the message to the correct hat club. After that the message will directly reach its root node in one hop.

In a probability of 99%, Gemini accomplishes this task using only one hop. This is because that a node's bootcase contains 4,883 items that is almost evenly distributed in the nodeId space whilst only 1,024 hat clubs exists.

When routing a message with destination address D, node M first checks whether M and D have the same cap. If so, M directly forwards it to the root node, which must be in M's cap chest; otherwise M inspects its bootcase, tries to seek out a node who has D's cap and forwards the message to it. The probability that M successfully finds the proper node from its bootcase is called *hit ratio*, which is determined by the size of M's bootcase (noted Bc) and the number of hat clubs (noted H). Figure 1 shows

---

[1] Node M means this node has a nodeId M.

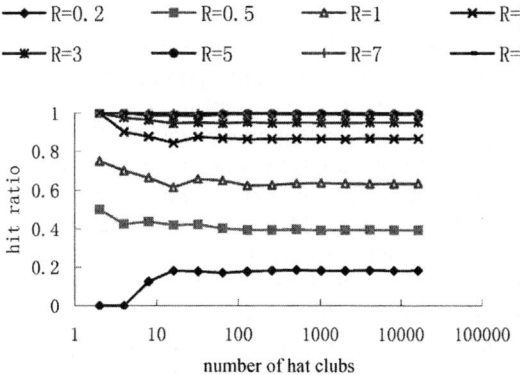

Fig. 1. Hit ratio vs. number of hat clubs. R is the ratio of bootcase size (Bc) to number of hat clubs (H). It shows that when H exceeds 50, hit ratio is almost a constant.

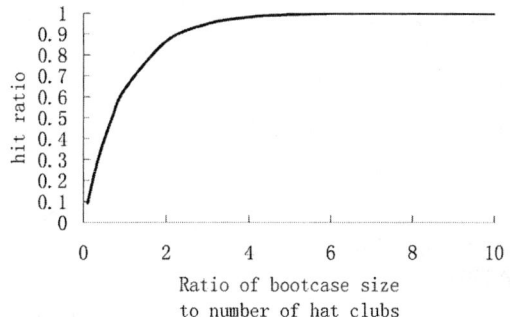

Fig. 2. Hit ratio vs. R, ratio of bootcase size to number of hat clubs. Here number of hat clubs fixes at 1024.

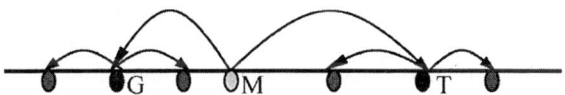

Fig. 3. An instance of tree-based multicast.

how hit ratio varies with respect to H under different ratios of $R=Bc/H$. We can see that when H exceeds 50, hit ratio is almost a constant value. For clearly illustrating, Figure 2 fixes H at 1024. In our case, $R=4883/1024\approx4.77$, so hit ratio is beyond 0.99. The equation supporting Figure 1 and 2 will be introduced in section 3.

If there is no node with D's cap existing in M's hatcase, M forwards the message randomly to a node in its hatcase whose boot is different with M.

Maintenance. Like one-hop overlay, Gemini uses broadcast (or say multicast) to maintain nodes' routing table. Note that hat clubs are independent to one another, and so do boot clubs, which allows us only to consider how to maintain nodes' hatcase/bootcase within a hat/boot club.

Nodes in a club are thought of as a sorted list in nodeId order. Every node sends heartbeats to the next node in the list every 10 seconds, while the last node sends heartbeats to the first one. If a node does not respond to heartbeats continuously for 3 times, the sender then multicasts its death to all the nodes in this club. The message used to inform other peers a member-changing event is called an event message.

Noting that nodes within a club are fully interconnected, we have many ways to realize such multicast of events. Here we only employ a simple tree-based multicast, illustrated in Figure 3. When a node M initiates a multicast, it sends the event to a node before it (say G) and another behind it (say T). Then alike, G and T each sends the event to two nodes, one ahead and the other behind. This procedure continues. At each step every node should ensure that once it sends the event to another node, there is no other node between them that has already received this event.

In this manner, except for the changing member's nodeId, ip address and port, an event message should additionally include nodeId of the first node before the receiver who has already received the message, as well as nodeId of the first such node behind it. Plus UDP header (64 bits) and ip header (160 bits), an event message will not exceed 500 bits.

If more efficient multicast is desired, the multicast tree can be modified to be 2^b-based. And if more reliable multicast is desired, response-redirect mechanism can be deployed, which will double the maintenance cost.

Maintenance cost. Assuming that the average lifetime of nodes is 1 hour, which is the statistic result from [6], all the items in the routing table have to be refreshed in a period of 1 hour. It means that on average every node receives (4883+4883)*2=19532 even messages per hour, namely 5.43 messages per second. Plus heartbeats and their responses, the total message count per second will not exceed 6, i.e., the bandwidth cost is lower than 3kbps.

Joining of new node. When a new node X joins Gemini, it first contacts an existing Gemini node E. E picks a node H with X's hat from its bootcase, as well as a node B with X's boot from its hatcase. X collects its hatcase from H and bootcase from B. If E does not find appropriate H or B, it asks another node in its hat or bootcase for help.

Remarks. From this case we can see that 1) the reason why Gemini can route messages efficiently is that nodes in bootcase are sufficient to cover all the hat clubs in a very high probability; 2) the low maintenance cost comes from the maintaining manner of event multicast.

Comparison with Kelips. Another p2p protocol Kelips [2] also ensures that data lookups be resolved in $O(1)$ hops. There are four major differences between Gemini and Kelips:

First, Kelips divides nodes into groups only in one manner, while Gemini divides nodes in two symmetric manners.

Second, Kelips nodes explicitly keep pointers to other groups, while Gemini nodes employ items in the other dimension of groups as such pointers.

Third, in Gemini whether a message can successfully reach its hat club in one hop is in some probability. Reducing this probability can trade hop count for bandwidth overhead, making Gemini more scalable. (Detail discussion is in section 4) Kelips has no such character.

Fourth, Kelips is designed for file lookups, so it replicates file pointers all over a group. Gemini is just a substrate, leaving data operations to its application designers.

Formalization and Analysis

In this section, we describe how to determine the length of hat and boot in a given system environment, and then estimate routing table size and maintenance cost.

Assuming that in a Gemini overlay network of N nodes, every node has a hat with h bits and a boot with b bits. Then there are $H=2^h$ hat clubs and $B=2^b$ boot clubs totally. On average, each hat club contains $Hc=N/2^h$ nodes and each boot club contains $Bc=N/2^b$ nodes.

We hope that message routing can be accomplished in 2 hops, which is directly determined by hit ratio. To a certain node M, it has a bootcase with about Bc items. M wishes that these items could cover all the hat clubs. But unfortunately it is not the reality: there must be some hat clubs that are not pointed to by any items in M's bootcase. We note number of such hat clubs as S. Expectation of S can be calculated as follows:[2]

$$S = \sum_{k=0}^{H-1} k \times \frac{\binom{H}{k}\binom{Bc-1}{H-k-1}}{\binom{H+Bc-1}{H-1}}$$

They occupy a fraction of $s=S/H$ in the bootcase, so hit ratio is $hr = 1-s = 1-\frac{S}{H}$.

Figure 1 and 2 are based on this inequation. We can see from Figure 2 that when H is larger than 50, choosing R as 4.7 will ensure a hit ratio more than 0.99. So in common cases we demand that $R>R_0=4.7$. That is to say, $N/(2^b \cdot 2^h)>R_0$, namely

$$b+h \leq \log_2(N/R) \quad (1)$$

Next we estimate maintenance cost. Heartbeat cost is a fixed small value, so we simply ignore it. The substantial cost is for maintaining hatcase and bootcase, with a size of $Hc+Bc$. Assuming that nodes' average lifetime is L seconds, then each node triggers two events in a period of L seconds on average. So every node receives $2\cdot(Hc+Bc)$ events every L seconds. If redundancy of employed multicast algorithm is f, then the number of messages a node receives per second is

$$m = (Hc+Bc) \times 2 \times f/L = \left(\frac{N}{2^h}+\frac{N}{2^b}\right) \times 2f \Big/ L$$

When $h=b$, m reaches its minimum[3]

$$m_0 = \frac{2N}{2^{\frac{1}{2}(h+b)}} \times 2f = \frac{4N \cdot f}{2^b \cdot L} \quad (2)$$

---

[2] This problem is equivalent to the scenario randomly throwing r identical balls into n different boxes. The total number of possible results is $\binom{r+n-1}{n-1}$.

[3] Choosing h and b that holds $|h-b|=1$ is also feasible, but analysis will be more complicated. Considering that its result is of only slight difference with the scenario where $h=b$, we omitted it in this paper.

Simultaneously considering (1) and (2), we should set $b = h = \left\lfloor \dfrac{\log_2(N/R)}{2} \right\rfloor$ to ensure a 2-hop-routing probability larger than 0.99 using a minimal maintenance cost. Then we can get the following results:

a) Routing table size R_{size} satisfies

$$2\sqrt{N \cdot R} \leq Rsize < 4\sqrt{N \cdot R} \qquad (3)$$

b) Maintenance cost m_0 satisfies

$$\dfrac{\sqrt{N \cdot R} \times 4f}{L} \leq m_0 < \dfrac{\sqrt{N \cdot R} \times 8f}{L} \qquad (4)$$

Figure 4 show the variation of b, R_{size} and m_0 as functions of N using $f=1$ and $L=3600$.

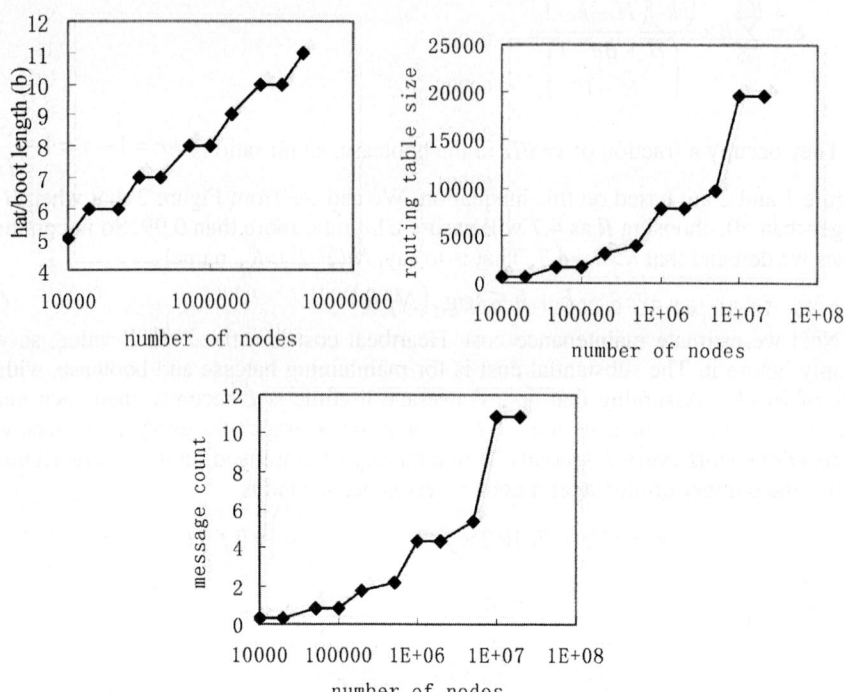

Fig. 4. Hat length vs. N, Routing table size vs. N and Routing table size vs. N.

Scalability

Inequation (3) and (4) show that a Gemini node has a routing table with $O(\sqrt{N})$ items and consumes $O(\sqrt{N})$ bandwidth to maintain it. This allows for a good scalability of Gemini.

In addition, when the maintenance cost is not acceptable by peers, Gemini also can trade hops for bandwidth consumption like other overlay protocols. To illustrate it, we put Gemini into a stricter environment where $N=10,000,000$, $f=2$ and $L=2,400$. Using inequation (4) we know that if keeping 2-hop routing, a node should receive at least 22 messages per second.

Gemini can reduce this cost by decreasing R, which will raise b and h, increase H, shrink Bc, and reduce hr. Expectation of hop count can be calculated as:

$$hops = 2 \cdot hr + 3 \cdot (1 - hr) \cdot hr + \cdots = 1 + \frac{1}{hr}$$

Even when hr drops to 0.25, average hop count only rises to 5. How message cost and hop count vary in relation to R is shown in Figure 5.

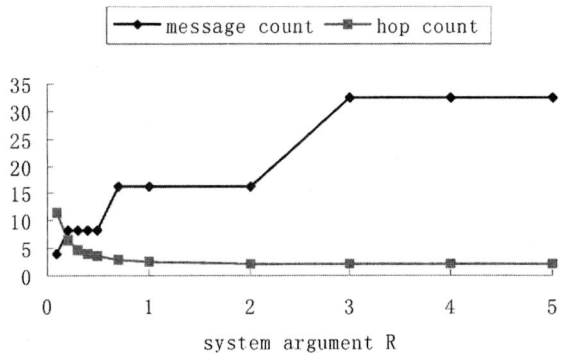

Fig. 5. Message cost and hop count in relation to the argument R, where $N=10,000,000$, $f=2$ and $L=2,400$.

Conclusion

Apparently, in structured overlay design, larger nodes' routing table is, fewer routing hops are needed. Traditional Pastry-like protocols can only afford a small number of entries for the reason that they use heartbeats to maintain routing tables. Recent one-hop overlay employs broadcast for maintaining, significantly decreasing bandwidth cost per entry. But letting every node keep a panoramic sight does burden most weak nodes.

References

1. Anjali Gupta, Barbara Liskov, Rodrigo Rodrigues. One Hop Lookups for Peer-to-Peer Overlays. HOTOS IX. May 2003.
2. Indranil Gupta, Ken Birman, Prakash Linga, Al Demers, Robbert van Renesse. Kelips: building an efficient and stable P2P DHT through increased memory and background overhead. IPTPS '03. February 2003.
3. Kazaa. http://www.kazaa.com. November 2003.
4. Petar Maymounkov and David Mazieres. Kademlia: A Peer-to-peer Information System Based on the XOR Metric. IPTPS '02. March 2002.

5. S. Ratnasamy, P. Francis, M. Handley, R. Karp, and S. Shenker. A Scalable Content-Addressable Network. SIGCOMM 2001. August 2001.
6. A. Rowstron and P. Druschel. Pastry: Scalable, distributed object location and routing for large-scale peer-to-peer systems. Middleware 2001. November 2001.
7. Saroiu, S., Gummadi, P. K., and Gribble, S. D. A Measurement Study of Peer-to-Peer File Sharing Systems. MMCN '02. January 2002.
8. Stoica, R. Morris, D. Karger, M. F. Kaashoek, and H. Balakrishnan. Chord: A scalable peer-to-peer lookup service for internet applications. SIGCOMM 2001. August 2001.
9. Ben Zhao, John Kubiatowicz, and Anthony Joseph. Tapestry: An infrastructure for fault-tolerant wide-area location and routing. Technical Report UCB/CSD-01-1141, U. C. Berkeley. April 2001.

SkyMin: A Massive Peer-to-Peer Storage System

Lu Yan[1], Moisés Ferrer Serra[2], Guangcheng Niu[2], Xinrong Zhou[1], and Kaisa Sere[1]

[1] Turku Centre for Computer Science (TUCS) and Department of Computer Science,
Åbo Akademi University, FIN-20520 Turku, Finland
{lyan,xzhou,kaisa}@abo.fi

[2] Department of Computer Science, Åbo Akademi University, FIN-20520 Turku, Finland
{mferrer,gniu}@abo.fi

Abstract. The aim of the project *SkyMin* is to design a large-scale, Internet-based storage system providing scalability, high availability, persistence and security. Every node serves as an access point for clients. Nodes are not trusted; they may join the system at any time and may silently leave the system without warning. Yet, the system is able to provide strong assurance, efficient storage access, load balancing and scalability. Our approach to construct a massive storage file system on Internet is to implement a layer on top of existing heterogeneous file systems. The architecture is as follows: it consists of lots of FS (File Server) and one or some NS (Name Server). NS is the control center. Users can access the file system from every node.

1 Introduction

The aim of the project *SkyMin* is to build a global storage system, based on peer-to-peer (p2p) technology, and running on Internet [1] [2] [3] [4]. In that system we encounter a central node (Name Server) and a lot of peers (File Servers). The relationship between the peers and the name server is modeled as client-server paradigm, while between the peers themselves is pure p2p (they work as client and server at the same time). Every node can join the system at any time, and leave it silently (without warning).

The system is specially tailored for groups with limited budgets that need to store and share information in a secure way, making it easily available for everyone (the peers will store the information, and all the shared space will be available for the whole group). No special skills are required for the peer user, since the application values a user-friendly design; only some administrative skills are required for running the name server since the correct behavior of the system depends on the appropriate configuration and administration of the name server.

The remainder of the paper is organized as follows. We start with the requirement specification in Section 2. We present the architecture of *SkyMin* in Section 3 and more design details in Section 4. We conclude the paper in Section 5.

2 Requirement Specification

As mentioned earlier, the system is based on the implementation of two different applications: the File Server (from now on just "peer") and the Name Server (from now on just "server"). While the server program will not be used "directly" by most end users (except the administrators), but the peer program will be.

2.1 Functionality

The functionality requirement of the system is intended to hide all the complexity of the system. Therefore, although more sophisticated functionality is provided (such as login/logout etc.), the main functionality of the system is the one described in Table 1.

Table 1. Main Functionality.

Feature	Functionality that provide
Add User	Add users to a group (p2p network group)
Remove User	Remove an existing user from a group (p2p network group)

Server application functionality

Feature	Functionality that provide
Search	Allow the user to lookup files
Download	Get (read) files from the shared space
Upload	Put (write) files to the shared space
Delete	Delete a file from the shared space

Peer application functionality

2.2 More Specification Points

Although we have introduced the main functionality of the system from the user's point of view, we have to regard that the server side holds a very important role determining the communication between peers (each peer will be connected to the server, and it decides how the connections will be done). Therefore, we need to refine our functionality requirement to better specify the behavior of the system. We present a list with these important (and/or clarifying) points:

- Nodes can join and leave silently. There is no change in the data availability. Neither is there any change in the behavior of other nodes.
- Storing data means copying a file from a node (located in a user computer, but not in the shared space), to the storage space, if there is enough space in the storage system. If there is no space available, the system prompts an error message.
- The system must not store duplicated files in the same node.
- A name server may store information of several groups. A user should not be able to access information that is out of the scope of his/her own group.
- The system does not need to support a user running multiple instances of the program at the same time (for example, over different groups).

2.3 Non-functional Requirements

Non-functional requirements are required for the system. Some of them are contradictory, so we have to make some tradeoff among these requirements:

- Scalability: there is no restriction on the number of nodes that can simultaneously join a group (and the performance must not be affected).
- High availability: redundancy in the storage system (allow duplicates and provide mechanisms/algorithms to guarantee the availability of resources).
- Persistence: the system should be capable of nonstop running.
- Security: a very important feature, since the information traveling over Internet might be private and should not be accessible to people out of the p2p group.
- Reliability: strongly related to persistence; the server side must be error tolerant. (Some kind of recovering mechanism must be provided for that.) The peer side should also be reliable (for example, the server crashes but the peer continues the communication with the other peers).
- Load balancing: store information in an efficient way, not randomly.

3 System Architecture

The *SkyMin* framework is a global storage system, as shown in Fig. 1, based on p2p technology. This system consists of one (or more) central server and many peers which are connected to Internet. The server works as a control point to grant the peer's access to the storage system, to manage the indexes of the peers' resources (files can be accessed by other peers who are members of the whole storage system), etc. On the other hand, every joined peer contributes a slice of their local storage space to the system, and the peer can save and retrieve files to/from the system.

3.1 Peer-Side Architecture

There are several tasks that the peer should be able to perform. These tasks can be summarized as:
- Receive and analyze the user's commands.
- Send its resource information to the server and maintain it periodically.
- Send user's requests to the server and process the server's response.
- Work as a server itself to respond to other peer's download or upload request.

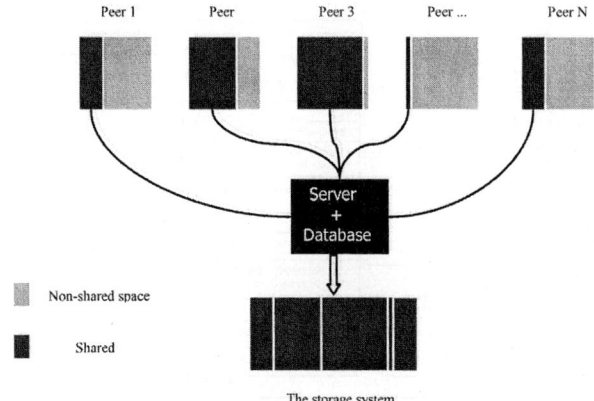

Fig. 1. System Architecture.

Note that some of these tasks are more complicated than the other ones. Therefore several classes are used for them. According to the requirement specification of our peer system, we built the following classes to accomplish these tasks in Fig. 2:

The "Peer" class holds the main logic of the peer program in the peer application. It starts the system, (taking care of program variables, server in the background, login, etc) and accepts commands from the user and process them (doing corresponding actions). The "Shell" class handles the user input; it parses the input. If the user's command makes no sense, this class will prompt an error message. The "ProtocolMsg" class holds the information of building different messages for different purposes (e.g. search, upload, etc.). This class keeps the protocol information of within messages (each message has a unique identifier, and this class keeps that information). The "PeerResource" class is used for constructing a message for sending the peer resources. The class allows sending different kinds of resources, including peer space (maximum available space) and files.

It is the "MessageTransport" class that takes the responsibility of the message transportation over Internet. Possible exceptions that may occur during this transportation are built on the "MessageTransportException" class which extends the "WrappedThrowable" class. So far, all messages are sent in a homogeneous way, thanks to the protocol. Messages are sent identically; with the identifier provided with the "ProtocolMsg" class, they can be properly interpreted.

The peer also runs a server thread in the background, which is an object of the "PeerServer" class. This class implements a listener to a specific port; when a message arrives (a peer wants to connect), the process will be controlled by a "ConnectionHandler" object.

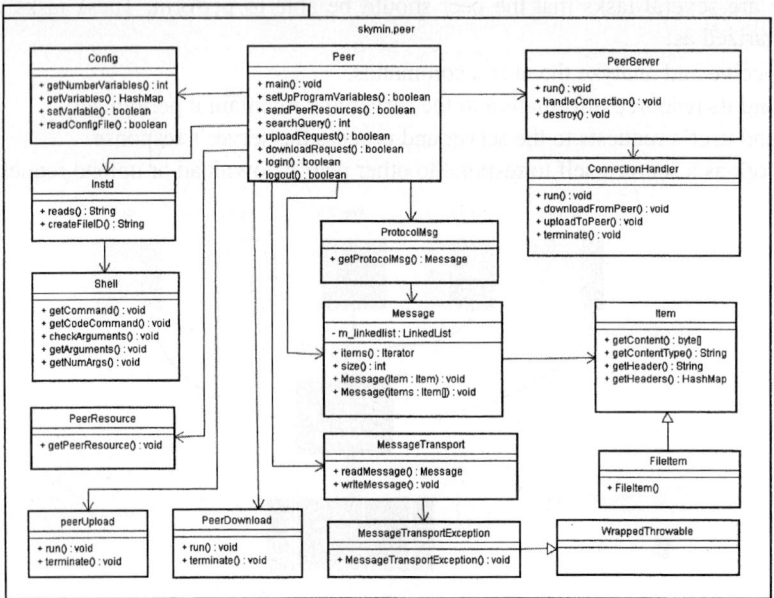

Fig. 2. Peer Class Diagram.

The "Item" class is used to build the content of messages. Items are text ones, but can be extended to support files. This is done by the class "FileItem". Items are finally encapsulated in a message (the "Message" class) and therefore can be sent. The "Instd" class provides several functions with different purposes. It is like a utility class which builds the identifier of files and also takes care of reading from the standard input.

The "PeerDownload" class is responsible to manage the process when the peer wants to download a file. It connects the peer and updates the pertinent information in the server. The "PeerUpload" class has the same purpose as the previous class except that the peer wants to perform an upload. The "Config" class is responsible for reading the configuration file and load program variables.

3.2 Server-Side Architecture

There are also several tasks that the server should be able to perform. These can be summarized as:

- Receive and analyze the user's commands.
- Listen to a port for incoming connections.
- Handle incoming connections.
- Interpret peer messages and send proper answers.
- Add or remove users to/from a group.

As before, some functions are more complex than the other ones. Therefore they will be implemented in several classes. The server class diagram is shown in Fig. 3.

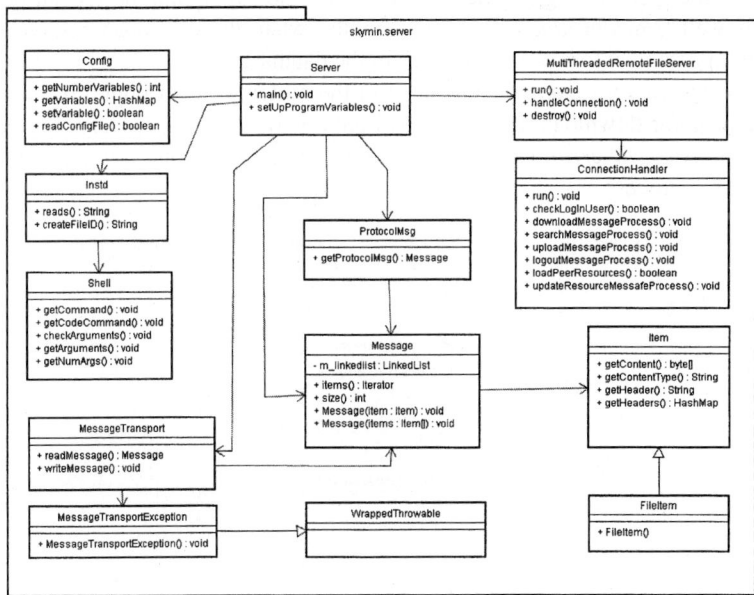

Fig. 3. Server Class Diagram.

Some classes have the same behavior here as in the peer side, even when working in a little different way (one clear example is the shell, which carries out the same work but process different commands). Therefore "Shell", "Config", "Instd", "Item", "FileItem", "Message", "MessageTransport", "MessageTransportException", "WrappedThrowable" and "ProtocolMsg" do not need further explanation.

The "Server" class is like the "Peer" class in the peer side. It performs all the initializations and prepares the system for working. The "MultithreadedRemoteFileServer" is analogous to the "PeerServer" in the other package. It implements the listener to a specific port for incoming connections. The "ConnectionHandler" is used to handle the incoming connections.

4 System Design

The behavior of the system can be explained by sequence diagrams. In the sequence diagrams there are some descriptions like *Msg(keyword)*. This defines the protocol of our system; it means that the system expects some messages and reacts to the received one and/or the expected one. That is the reason between peer and server and between peer and peer all the descriptions are of the type *Msg(x)*.

Download. In the download scenario in Fig. 4, first of all a search is needed (before the user downloads a specific file, he/she needs to know where it is). The following steps 1 to 4 are the same as a normal search. After that, the user selects one of the files provided by the search operation to download, and sends the information to the server in a *downloadrequest* message. The server searches the database for the best available peer, and answers with an *okdownloadrequest* message. Then the peer connects the other peer asking for the file (*download* message), and the second peer answers as well (*startdownload* message). Finally, when the download completes, the peer will update its information resources in the database.

Different variations can be appreciated in the system. For example, if there is no suitable peer for download, the server will answer with an *emptyquery*. So the peer realizes it. Also an *error* message can come from the second peer, aborting the downloading process.

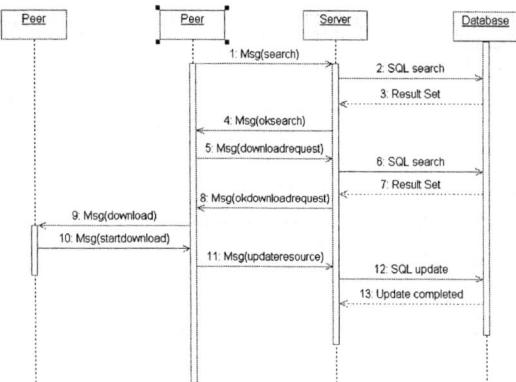

Fig. 4. Download sequence diagram.

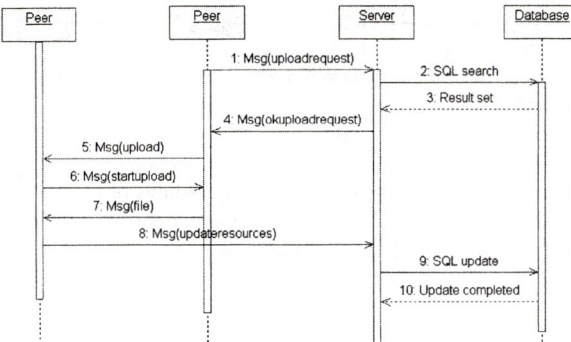

Fig. 5. Upload sequence diagram.

Upload. Here the situation is shown in Fig. 5. The peer sends an *uploadrequest* message, telling the ID and the size of the file to upload. Then the server checks the database which peer is the best to upload, and answers an *okuploadrequest* (note that the server can also answer an *uploadnotrequired* message), specifying the best peer to upload (note that the own peer that makes the request can also be the peer chosen). Then the peer contacts the chosen peer by an *upload* message, and waits for the answer (*startupload*). Finally, after the file is sent, the peer informs the server about its new resources, and the server updates the database.

5 Concluding Remarks

The increasing demand for massive storage systems has spawned an urge for a large-scale storage solution with scalability, high availability, persistence and security. Nowadays, Internet has become cheap and widespread, which makes it possible to build an economical massive storage solution over Internet.

In this paper, we proposed *SkyMin*, a global storage system, based on peer-to-peer (p2p) technology, and running on Internet, specially tailored for groups with limited budgets to store and share information in a secure way. This paper presented a prototype implementation of the architecture and design elements of *SkyMin*; several design elements still need fine-tuning (e.g. in the current release, there is no support for transport-level security; more sophisticated protocols like SSH will be incorporated in the next release). The rise of pervasive computing has brought new innovative design ideas for our architecture. In the future work, more design elements will be further refined in favor of ubiquity and mobility to make *SkyMin* a massive storage solution for the pervasive computing era.

References

1. L. Yan and K. Sere: Developing Peer-to-Peer Systems with Formal and Informal Methods. In Proceedings of the 2nd International Workshop on Refinement of Critical Systems: Methods, Tools and Developments (RCS'03). Turku, Finland, June 2003.

to a request whenever the resource is available, hoping that this will not lead to a deadlock. Since it attempts to find and resolve actual deadlocks concurrently with normal computing activities of the systems, more flexibility is gained with little negative influence on the system performance. In this paper, we are only concerned with deadlock detection.

Many papers on distributed deadlock detection can be found in the literature [Bra83, Cha83, Lee01, Mit84, Rub03, Sin89]. Traditional distributed deadlock detection algorithms are mostly implemented with message passing, which is more suitable in a closely coupled distributed system. In a wide-area network environment like the Internet, there is a larger delay in message passing and greater variance of network topology. A pure message-passing based algorithm may not work well. In this paper, we propose an alternative approach using mobile agents [Lan99, Pha98]. It has been found that the mobile agent is especially suitable for structuring and coordinating wide-area network and distributed functions that require intensive remote real-time interactions [Cao01]. The proposed Mobile Agent Enabled Deadlock Detection (MAEDD) framework uses cooperating mobile agents as an aid for distributed deadlock detection. Because mobile agents can take the advantages of being in the same site as the peer site, interacting with the peer locally and autonomously, it allows us to design algorithms that make use of up to date system information for deadlock detection and reduce remote communications. Using the mobile agent technology also allows us to provide clear and useful abstractions in designing distributed systems through the separation of concerns. In MAEDD, the function of deadlock detection and network communication are filtered out from the server's logic and combined into one single layer, where mobile agents are used as a means to achieve the dual goals.

The remainder of this paper is organized as follows: section 2 presents a brief overview of traditional solutions for distributed deadlock detection. In section 3, the generic MAEDD framework for mobile agent enabled deadlock detection is proposed. A specified algorithm for MAEDD is introduced at section 4. The experimental results for performance evaluation are reported in section 5. Section 6 concludes the paper.

2 Related Works

Distributed deadlock detection algorithms can be classified into four categories [Kna87]: *Path-Pushing (WFG-based)* [Ash02], *Edge-Chasing (Probe-based)* [Cha82, Mit84], *Diffusing Computation* [Cha83] and *Global State Detection* [Bra83]. The former two types of algorithm are widely adopted in database and distributed systems.

Path-pushing algorithms maintain an explicit Wait-for-Graph (WFG). A WFG consist of a set of processes $\{P_1, P_2, ..., P_n\}$ as the node set. An edge (P_i, P_j) exists in the graph if and only if P_i is waiting for a resource held by P_j [Wu99]. Each site periodically builds a local WFG by collecting its local wait dependencies, then searches for a cycle in the WFG and tries to resolve these cycles. After that every site sends its local WFG to its neighboring sites. Each site updates its local WFG by inserting wait dependency received, and detects cycles in the updated WFG. The updated WFG is passed along to neighboring sites again. This procedure will repeat until some sites finally detect the deadlock or announce the absence of deadlock. The most famous algorithm in this category is Obermarck's algorithm [Obe82] implemented in System R*. In a newly proposed algorithm targeting at handling deadlocks in mobile agent systems [Ash02], a "Detection Agent" is dispatched to all resources held by its target "Consumer Agent" for collecting deadlock information. The gathered information is finally returned to the "Shadow Agent" which is responsible for monitoring that Consumer Agent and detecting deadlock cycles.

Instead of explicitly building the WFG, Edge-chasing algorithms send a special probing message to detect deadlocks. A process (initiator) sends probes to processes holding the locks it is waiting for. A process receiving a probe message forwards it to all the processes it is waiting for. The probe message contains information to identify the initiator. If the initiator receives a probe sent by itself, it can announces a deadlock because the probe must have traveled a cycle. This idea was originally proposed in [Cha82] with the correctness proof presented in [Ksh97]. Similarly in the algorithm of [Mit84], a probe consists of a single number that uniquely identifies the initiator. The probe travels along the edges in the opposite direction of global WFG. When it returns to its initiator, a deadlock is detected.

Our algorithm adopts the ideas of both *Path-pushing* type and *Edge-chasing* type algorithms. It is based on WFG and mobile agents are working like the probes.

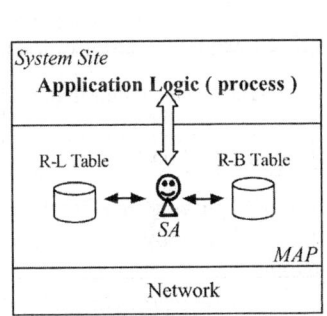
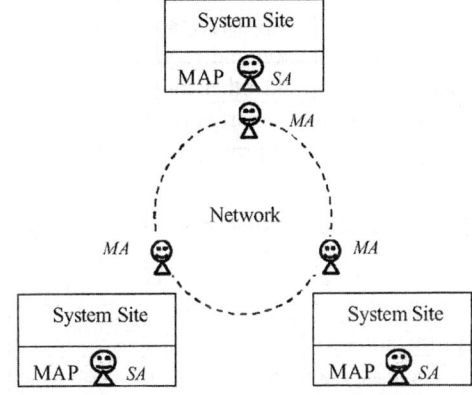

Fig. 1. The Framework of MAEDD (I).   **Fig. 2.** The Framework of MAEDD (II).
SA: Static Agent
MA: Mobile Agent
MAP: Mobile Agent Platform
R-B Table: Request-Block Information Table
R-L Table: Resource Location Information Table

3 The MAEDD Framework

We consider a group of mobile agent enabled server sites in a distributed system. In traditional distributed systems using message passing, the process coordination functions have to be integrated into the server code. The use of mobile agents allows us to develop a flexible architecture by having mobile agents carry out information collection and process coordination functions, which are functionally separated from the server application logic. Fig. 1 and Fig. 2 illustrate the MAEDD framework.

Mobile agents embedded with specified algorithm are capable of navigating through the network and performing tasks at the sites they visit. Mobile agents are initialized and dispatched by Static Agents (SA) reside on the Mobile Agent Platform (MAP) of each system site. The SA is responsible for managing a resource location table which records where the available resource is. It also provides standard primitives for interacting with applications running on that server, such as informing the application process that the request for some resource is involved in a deadlock cycle so that the process must abort to release the resource it occupies. At the same time, applications can request the SA to apply for a certain resource. When a deadlock is suspected, the SA will dispatch a mobile agent which encapsulates the deadlock detection strategy/algorithm, together with necessary data, and then dispatch it to the network. The deadlock detection algorithm to be embedded in the mobile agent maybe adaptively selected based on the current server status and network condition.

Mobile agents interact with the SA on each server they visited for exchanging information. If necessary, several cooperating mobile agents can be dispatched to work together. Once a deadlock cycle is detected by a mobile agent, it will select a victim process according the embedded deadlock resolution algorithm by informing the corresponding SA to terminate the application process.

Issues need to be addressed within the framework include which distributed deadlock detection and resolve algorithm should be employed, who and how to manage the resource location and request-block information table (the static agent or the application itself), and other mobile agent interaction details. In the following section, we will describe a particular mobile agent enabled distributed deadlock detection algorithm under the proposed framework.

4 A Mobile Agent Enabled Deadlock Detection Algorithm

In this section, we first describe the system model and data structures. Then we present the mobile agent enabled deadlock detection algorithm. Finally, we introduce an optimization to the algorithm by using two mobile agents.

There is a collection of n sites, each is identified as S_i (i=1, ..., n). On each site S_i, one or more application processes reside, each has a global unique identifier, such as P_{i1}, P_{i2}, etc. The processes request resources autonmoulsy. If the requested resource is held by other processes, the requesting process will be blocked until the requested resource is released. At any time, a process can be in only one of the following two states: *active* when it performs tasks normally or requests an idle resource; *blocked* if it requests a resource currently held by others. On granting the requested resource, the state of the process will change back to *active*.

Different deadlock models can be used, including *single-resource model, and model, or model,* and *p-out-of-q model* [Kna87]. The most widely adopted model is the single-resource model, in which a process has only one outgoing request at a time. Hence, it can be involved in just one deadlock cycle. Algorithms designed for deadlock detection under this model can be easily extended to other models [Kri99]. In this paper, we only consider the single-resource model.

The set of processes that hold resource for which a process P is waiting for is denoted by *Block (P)*. The set of processes that are waiting for resources hold by process P is denoted by *Request (P)*. The processes in the same set can reside in one site or over several sites. In fact, there is only one process in the set under the single-source model. The dependency relationship is often denoted by a directed graph, known as *Wait for Graph (WFG)*, where a node represents a process and an edge depicts the Request-Block relation between a pair of processes. For each process P_{Ki} in the set of Request (P_K), there is a corresponding edge initialized from P_{Ki} to P_K.

Every site accommodates a static agent responsible for maintaining a local WFG for that site. The WFG maintained is actually a list of entries, each describes a dependency edge. The data structure of the WFG is shown as follows:

Request Process ID	Blocking Process ID	Remote Host	Time Elapsed

The "Remote Host" filed records the site where the blocking process resides. If the value is the same as the local site's address, it means that both the requesting process and the blocking process are local to this site. The "Time Elapsed" variable is used to track how long the edge has been formed by increasing its value at intervals.

In our algorithm, a Mobile Deadlock Detection Agent (MDDA) is used mainly for detecting and resolving the deadlock. When roaming in the network, the MDDA carries two types of data, both implemented as a linked list. One is *S-List* containing sites that the agent plans to travel. The other is *P-List* recording IDs of processes that are suspected to be involved in a deadlock. If we consider the P-List as a sequence of processes $<P_1, P_2,..., P_k>$, $k \geq 0$, then for every P_i in the list (i<k), P_{i+1} belongs to the set of Block (P_i).

From time to time, the SA on a site checks the states of all local processes and resources to keep the local WFG up to date. When the value of the field "Time Elapsed" in some entry exceeds the predefined value, i.e. upon timeout, the SA will first analyze the local WFG to see whether the requesting process is involved in a local wait-for cycle. If so, the agent will resolve this deadlock in a way similar to handling deadlocks in a centralized system. If the suspicious deadlock cannot be resolved, the SA initializes a Mobile Deadlock Detection Agent (MDDA) and dispatches it. The algorithm for MDDA to execute comprises of two parts: one is for detecting the deadlocks and the other is for resolving them.

Upon being initialized, the MDDA will first process the WFG of its origin site, by simply adding the ID of the involved local processes to the P-List and the local site's address to S-List. That is, for a requesting process P_i, the process P_{i+1} that blocks P_i (corresponding to the timeout entry in the WFG) will be added to the P-List. The MDDA then searches the WFG for entries recording a process P_{i+2} that blocks P_{i+1}, and append P_{i+2} to the P-List. This procedure is repeated until an entry is found in the local WFG where the value of "Remote Site" field is not equal to the local site's address. At this time, the remote address will be appended to the S-List and set as the next migration target. Once a site is appended to the S-List, its ID will not be deleted no matter the agent has visited it or not. By doing so, the MDDA is sure that it makes no redundant visits.

On arriving at a new site, the MDDA will analyze the local WFG and update the P-List and S-List if necessary. If the ID of some process to be appended to the P-List is already in that list, the MDDA knows that a deadlock cycle is detected and tries to resolve it. Otherwise the MDDA will continue searching for the deadlock cycle until there is no unvisited remote site in the S-List. In this case, the MDDA announces that no deadlock exists for the requesting process and all the processes on the sites along the agent's traveling path. The MDDA finally disposes itself. Note that if two-phase locking (2PL) is not adopted, for avoiding detecting false deadlocks, a SA should inform the mobile agent that has visited its site that some process on that site has been granted the requested resource. On receiving the notification, the agent will delete related processes' and sites' ID from the P-List and S-List.

In our algorithm, when a MDDA detects a cycle, it resolve the deadlock by choosing the last process appended to the P-List as the victim. The MDDA passes that process's ID to its corresponding SA, which will first kill the selected process directly or inform it to abort, and delete all entries in the WFG that contain the victim process's ID. The resource held by the victim is then released.

When processes distributed over many different sites are suspected to be involved in a deadlock cycle, multiple mobile agents can be used to improve the performance. Here, we consider the use of only two mobile agents. To support this optimization, the SA should maintain an extra data set for each local process P_i. This set contains all processes that belong to the Request (P_i), but don't reside on the same site as P_i. Also, the addresses of the sites on which these processes reside are recorded. The SA will dispatch a pair of mobile agents, FMDDA (Forward MDDA) and BMDDA (Backward MDDA), which will travel the network in reverse directions. The BMDDA travels backward along the WFG, that is, appends the ID of a process that in the set of Request (P) to its P-List for a requesting process iteratively. Similarly, relevant remote sites' addresses are appended to the S-List reversely. The two mobile agents know each other's agent ID and will set a variable "flag" on every site they visit. The value of the flag can be the ID of the agent who sets the flag and will be cleared by the SA after a period of time. Once one of the mobile agents finds that the flag in the site it is currently visiting has already been set by another, it will move back to the first site in its S-List. On the site where they are originally decided to meet, the two agents will combine their P-List and determine whether a cycle exists (a single process appears twice in the combined list). If the flags of every site one agent visits are all unset, the agent will keep moving along the S-List until it detects a deadlock by itself or no more site is appended to the S-List indicating that no deadlock cycle exists.

5 Performance Evaluation and Comparison

Experiments have been carried out to evaluate the performance of the proposed algorithm against exisitng message passing algorithms such as Obermarck's Algorithm [Obe82]. Twenty-five inter-connected workstations are organized as a network system. The IBM Aglet platform [Agl] is deployed for each workstation. Network nodes host different number of processes but the number of processes on each site is a constant throughout the experiments. Processes randomly request resources or block other processes by informing the SA to update the corresponding entries in the local WFG. The time for a process to hold a resource follows a passion distribution with the expected value?

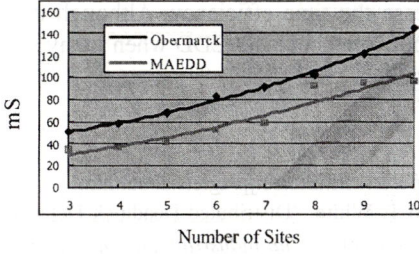

Fig. 3. Comparison of Response Time (I).   Fig. 4. Comparison of Response Time (II).

We choose *Response Time* as the main performance metric. The response time is defined as the time interval between MDDA initiation and termination, which reflects how quick a deadlock can be handled. Fig. 3 illustrates the average response time, as a function of sites involved in a distributed deadlock. We can see that our MAEDD algorithm using single MDDA detects an existing distributed deadlock 40% faster than then Obermarck. As shown in Fig. 4, when more sites are involved, MAEDD gains higher efficiency by employing two MDDAs.

Fig. 5 shows the network traffic required for detecting a given deadlock by the two algorithms. Network traffic is measured by a network inspecting software called "Ethereal" [Eth03]. We record the size of all related TCP packages. To minimize the byte code transmitted with the aglet, we implement our algorithm by using several small classes, such as class "Work", "Graph", "FIFO" and "Node". Each time an aglet arrives at

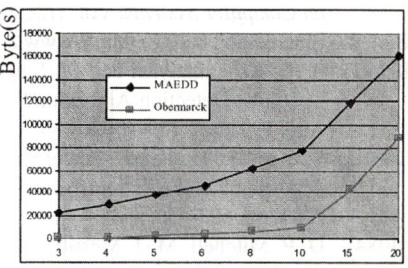

Fig. 5. Comparison of Network Traffic.

a new site, objects of these classes are created. When leaving the site, these objects are destroyed. Hence, the byte code of these classes does not need to be transferred with the Aglet. We can further combine them into one JAR file. But the MAEDD still requires more network traffic. However, the traffic increases greatly for Obermarck when more than 10 sites are involved in a deadlock because many message packages are required to be sent between every pair of sites. We woule like to point out that, in a network environment where bandwidth is always sufficient, the network traffic is not a serious issue. Also, the result reported here is based on the Aglet platform. Using other MAP platforms with smaller overhead in size, such as the Naplet [Xu02], we believe that the performance of our algorithm should be better.

one or several files. During a file sharing session for a task, a peer holding part of content would receive data from and send data to others in a P2P network. However, a peer containing complete content only takes part in uploading data to others, which is referred to a *seed* in the paper. The success for a peer in a file sharing session is identified by the file totally incorporated in its local storage. The shortage of a small piece of content can make its previous work become useless since the integrity of the file is not reached. Much work has been done as how to efficiently discover the required resources (e.g., particular files) [1,9,11]. However, further investigation on the content sharing reveals the equal importance of complete resource available in P2P networks.

Upon receiving various requests in a P2P network, a peer faces the challenge as how to assign its resources. For instance, which peer would get service and which fragment of the shared content would be delivered first. An effective resource assignment method can yield more seeds within a constant time for a P2P infrastructure. Large number of peers with the total content lead to the resource availability increased, and much more fault tolerance when some seed peers depart. Furthermore, the distribution of generated seeds can be more balancing inside the whole system. The network traffic in terms of content transferring is evenly distributed since the data can be delivered from many places. Thus, the resource availability issue for content sharing is addressed in this paper in order to maintain more comprehensive resources in a P2P network.

The contribution of this paper is two fold. First, we analyze the probability of a peer to become a seed consisting of all content. The probability is formalized by the duration time of a peer in a P2P network modeled by uniform, exponential and normal random variables respectively. Second, a simple, distributed comprehensive algorithm for resource management is proposed. This algorithm can be deployed in each peer only with local information known inside a P2P system. In order to achieve more available resources, the peer as well as the pieces of content are deliberately allocated during the content sharing process in the described algorithm. We believe that the presented method can outperform randomized resource assignment techniques in most cases, such as longer session time, large number of peers that successfully contain all content.

In the next section, an example will be given to illustrate the request-based content sharing in a P2P system. The state of a peer is denoted by a 6-tuple. In Section 3, after the presented model for content request, the probability for a peer to become a seed is analyzed. The duration time of a peer is formulated by continuous uniform, exponential and normal random variables respectively. The comprehensive resource management algorithm is described in detail in Section 4. Concluding remarks are given in Section 5.

2 Request-Based Content Sharing

A P2P network can be represented by an undirected graph $G = (V, E)$ where each peer is denoted by a node $v \in V$. An edge $(u, v) \in E$ represents a bidirectional communication link between peer u and v. Let the size of a shared

system. The IBM Aglet platform [Agl] is deployed for each workstation. Network nodes host different number of processes but the number of processes on each site is a constant throughout the experiments. Processes randomly request resources or block other processes by informing the SA to update the corresponding entries in the local WFG. The time for a process to hold a resource follows a passion distribution with the expected value?

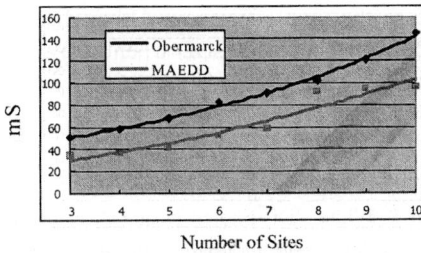

Fig. 3. Comparison of Response Time (I). **Fig. 4.** Comparison of Response Time (II).

We choose *Response Time* as the main performance metric. The response time is defined as the time interval between MDDA initiation and termination, which reflects how quick a deadlock can be handled. Fig. 3 illustrates the average response time, as a function of sites involved in a distributed deadlock. We can see that our MAEDD algorithm using single MDDA detects an existing distributed deadlock 40% faster than then Obermarck. As shown in Fig. 4, when more sites are involved, MAEDD gains higher efficiency by employing two MDDAs.

Fig. 5 shows the network traffic required for detecting a given deadlock by the two algorithms. Network traffic is measured by a network inspecting software called "Ethereal" [Eth03]. We record the size of all related TCP packages. To minimize the byte code transmitted with the aglet, we implement our algorithm by using several small classes, such as class "Work", "Graph", "FIFO" and "Node". Each time an aglet arrives at

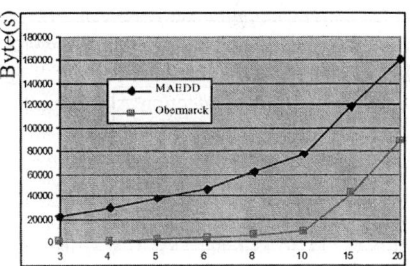

Fig. 5. Comparison of Network Traffic.

a new site, objects of these classes are created. When leaving the site, these objects are destroyed. Hence, the byte code of these classes does not need to be transferred with the Aglet. We can further combine them into one JAR file. But the MAEDD still requires more network traffic. However, the traffic increases greatly for Obermarck when more than 10 sites are involved in a deadlock because many message packages are required to be sent between every pair of sites. We woule like to point out that, in a network environment where bandwidth is always sufficient, the network traffic is not a serious issue. Also, the result reported here is based on the Aglet platform. Using other MAP platforms with smaller overhead in size, such as the Naplet [Xu02], we believe that the performance of our algorithm should be better.

6 Conclusions

We have proposed MAEDD, a distributed deadlock detection framework using mobile agents. A specific deadlock detection algorithm is also presented under this framework. A system prototype is built on IBM Aglet, based on which experiments are conducted to evaluate the proposed framework and algorithm. The proposed approach has several advantages over the message-passing algorithm. For performance, it reduces the response time and so to increase the time efficiency. Although it requires larger network traffic, it is still economic to use our MAEDD when many sites in the distributed system are involved in a deadlock.

References

[Agl]   Aglets Software Development Kit, http://www.trl.ibm.co.jp/aglets.about.html
[Ash02] B. Ashfield, D. Deugo, F. Oppacher and T. White, "Distributed Deadlock Detection in Mobile Agent Systems", In *Proc. 15th Inte'l Conf. on Industrial and Eng. Application of AI and Expert Sys.*, Caims, Australia, Jun. 17-20, 2002, pp. 146-156.
[Bra83] G. Bracha, S. Toueg, "A Distributed Algorithm for Generalized Deadlock Detection", *Tech. Rep. TR 83-558*, Cornell University, Ithaca, N.Y., 1983.
[Cao01] J. Cao, G. H. Chan, W. Jia, T. Dillon, "Checkpointing and Rollback of Wide-Area Distributed Applications Using Mobile Agents", In *Proc. 15th International Parallel and Distributed Processing Symposium*, Apr. 23-27, 2001, pp. 1-6.
[Cha82] K. M. Chandy and J. Misra, "A Distributed Algorithm for Detecting Resource Deadlocks in Distributed Systems", In *Proc.1st ACM Annual Symp. on Principles of Distributed Computing*, Ottawa, Canada, Aug. 18-20 1982, pp. 157-164.
[Cha83] K. M. Chandy, J. Misra, L. M. Haas, "Distributed Deadlock Detection", *ACM Trans. on Computer Systems*, Vol. 1(2), May. 1983, pp. 144-156.
[Eth]   Ethereal (V. 0.10.0), http://www.ethereal.com/
[Ksh91] A. D. Kshemkalyani and M. Singhal, "Invariant-based Verification of a Distributed Deadlock Detection Algorithm", *IEEE Trans. on Software Engineering*, Vol. 17(8), Aug. 1991, pp. 789–799.
[Lee01] S. Lee, J. L. Kim, "Performance Analysis of Distributed Deadlock Detection Algorithms", *IEEE Trans. on Knowledge and Data Engineering*, Vol. 13(4), Apr. 2001, pp. 623-636.
[Mit84] D. P. Mitchell, M. J. Merritt, "A Distributed Algorithm for Deadlock Detection and Resolution", In *Proc. ACM Symposium on Principles of Distributed Computing*, New York, USA, 1984, pp. 282-284.
[Pha98] V. A. Pham, A. Karmouch, "Mobile Software Agents: An Overview", *IEEE Communications*, Vol. 36(7), Jul. 1998, pp. 26-37.
[Rub03] J. M. M. Rubio, P. Lopez, J. Duato, "FC3D: Flow Control-based Distributed Deadlock Detection Mechanism for True Fully Adaptive Routing in Wormhole Networks", *IEEE Trans. on Parallel and Distributed Systems*, Vol. 14(8), Aug. 2003, pp. 765-779.
[Sin89] M. Singhal, "Deadlock Detection in Distributed Systems", *IEEE Computer*, Vol. 22(11), Nov. 1989, pp.37-48.
[Wu99]   J. Wu, "Distributed System Design", CRC Press, USA, 1999.
[Xu02]   C. Z. Xu, "Naplet: A Flexible Mobile Agent Framework for Network-Centric Applications", In *Proceedings of the 16th International Parallel and Distributed Processing Symposium*, Florida, USA, Apr. 15-19, 2002, pp. 219-226.

Maintaining Comprehensive Resource Availability in P2P Networks[*]

Bin Xiao[1], Jiannong Cao[1], and Edwin H.-M. Sha[2]

[1] Department of Computing
Hong Kong Polytechnic University
Hung Hom, Kowloon, Hong Kong
{csbxiao,csjcao}@comp.polyu.edu.hk
[2] Department of Computer Science
University of Texas at Dallas
Richardson, Texas 75083, USA
edsha@utdallas.edu

Abstract. In this paper, the resource availability issue has been addressed with respect to the intermittent connectivity and dynamic presence of peers in a P2P system. We aim to maintain more comprehensive resource available. The contribution of this paper is two fold. First, we analyze the probability of a peer to become a seed consisting of all content. Second, a simple, distributed comprehensive algorithm for resource management is proposed. In order to increase the availability of resources in a P2P network, the peer as well as the fragment of shared content are deliberately allocated during the content sharing process.

1 Introduction

P2P (peer-to-peer) systems have drawn much attention in recent years. The infrastructure of a P2P network makes a peer to act as both a server and a client. A peer (e.g., a computer in a network) can build a direct connection with another one without the involvement of a server. Such a network architecture avoids the drawbacks of conventional client-server model, such as the computation bottleneck on servers, and thus greatly improved the efficiency of system computing performance [6]. The P2P computing encompasses applications in a P2P network that can be categorized as distributed computing, content sharing and collaboration [2,7]. The popularity of content sharing (or file sharing) systems (e.g., Gnutella [8], Napster and Freenet [3]) in turn encourages the continuous research for advanced technologies in P2P networks.

Sharing content [5] in the P2P context addresses the techniques and technologies for effective ways of discovering the existence and locations of content, and furthermore getting it transferred. The shared content can be files or a storage. For the more popular file sharing applied in Internet, the content can be

---

[*] This work is partially supported by HK Polyu ICRG A-PF86 and COMP 4-Z077, and also by TI University Program, NSF ETA 0103709, Texas ARP 009741-0028-2001 and NSF CCR-0309461

one or several files. During a file sharing session for a task, a peer holding part of content would receive data from and send data to others in a P2P network. However, a peer containing complete content only takes part in uploading data to others, which is referred to as a *seed* in the paper. The success for a peer in a file sharing session is identified by the file totally incorporated in its local storage. The shortage of a small piece of content can make its previous work become useless since the integrity of the file is not reached. Much work has been done as how to efficiently discover the required resources (e.g., particular files) [1,9,11]. However, further investigation on the content sharing reveals the equal importance of complete resource available in P2P networks.

Upon receiving various requests in a P2P network, a peer faces the challenge as how to assign its resources. For instance, which peer would get service and which fragment of the shared content would be delivered first. An effective resource assignment method can yield more seeds within a constant time for a P2P infrastructure. Large number of peers with the total content lead to the resource availability increased, and much more fault tolerance when some seed peers depart. Furthermore, the distribution of generated seeds can be more balancing inside the whole system. The network traffic in terms of content transferring is evenly distributed since the data can be delivered from many places. Thus, the resource availability issue for content sharing is addressed in this paper in order to maintain more comprehensive resources in a P2P network.

The contribution of this paper is two fold. First, we analyze the probability of a peer to become a seed consisting of all content. The probability is formalized by the duration time of a peer in a P2P network modeled by uniform, exponential and normal random variables respectively. Second, a simple, distributed comprehensive algorithm for resource management is proposed. This algorithm can be deployed in each peer only with local information known inside a P2P system. In order to achieve more available resources, the peer as well as the pieces of content are deliberately allocated during the content sharing process in the described algorithm. We believe that the presented method can outperform randomized resource assignment techniques in most cases, such as longer session time, large number of peers that successfully contain all content.

In the next section, an example will be given to illustrate the request-based content sharing in a P2P system. The state of a peer is denoted by a 6-tuple. In Section 3, after the presented model for content request, the probability for a peer to become a seed is analyzed. The duration time of a peer is formulated by continuous uniform, exponential and normal random variables respectively. The comprehensive resource management algorithm is described in detail in Section 4. Concluding remarks are given in Section 5.

2 Request-Based Content Sharing

A P2P network can be represented by an undirected graph $G = (V, E)$ where each peer is denoted by a node $v \in V$. An edge $(u,v) \in E$ represents a bidirectional communication link between peer u and v. Let the size of a shared content be f_s. In this paper, we concern the sharing of a file unit F among a given P2P system in a session. For each node $v \in V$, its bandwidth resource for data transfer is limited. A 6-tuple $(B_{uu}, B_{ul}, B_{du}, B_{dl}, f_l, t_p)$ is associated to denote its current state, where B_{uu} describes the bandwidth used for node v to upload data in F while B_{ul} represents the left bandwidth, B_{du} shows the occupied bandwidth to download F while B_{dl} means the left bandwidth for content downloading, f_l is the size of the left content, and t_p provides the time passed for peer v connected in the system in a session of content sharing of F. The size of f_l is calculated by the number of missed blocks of file F, which is denoted by the percentage of f_s. It is possible that node v is downloading file F from other peers while uploading pieces of F concurrently in a distributed P2P system. For a node v with $f_l = 0$, which means peer v contains the whole file F, v is defined as a *seed* in the system. The seed node only sends data in a particular file F to non-seed nodes. For those nodes that are with $f_l \neq 0$, they are attempting to derive data fragment to minimize f_l and as a source to provide data to others.

A P2P system is shown in Figure 1 for the sharing of a file F among 5 peers. The network topology represents the request connections in a file sharing session. Along with each engaged node, a table provides its current state. The state information is free to others whenever a connection is built. For node A, its $f_l = 0$ reveals that A can be treated as a seed. Nodes B, C and D are in the data transmission process and part of the content is attained. Node E is at the beginning of file downloading since f_l is the same as f_s. Suppose that the uploading bandwidth of A is free of occupation ($B_{uu} = 0$). Upon the request from node B and C of sharing of file F, node A can clearly view their tables as in Figure 1. Thus, node A needs to make a decision as how to assign its limited bandwidth for package delivery to those 2 peers (B and C). Moreover, peers, such as B, C and D, face the same problem to handle requests from others.

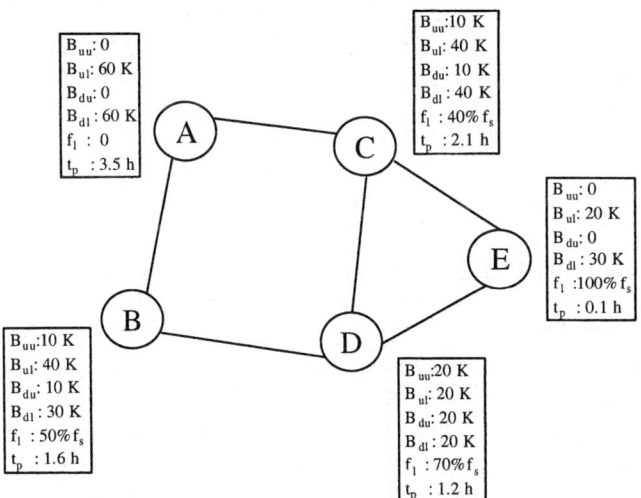

Fig. 1. The state of a file sharing in a P2P system.

3 Probability Analysis

3.1 A Model for Content Request

In Figure 2, a graph depicts the case as n peers (from b_1 to b_n) send their requests as content sharing to peer a in a P2P distributed system. Peer a is assumed to be a seed that consists of the entire content. Although a P2P setting is loosely organized and fully decentralized, a participant can be aware of the configurations of those peers connecting to it. Hence, graph in Figure 2 to peer a is transparent after message exchange.

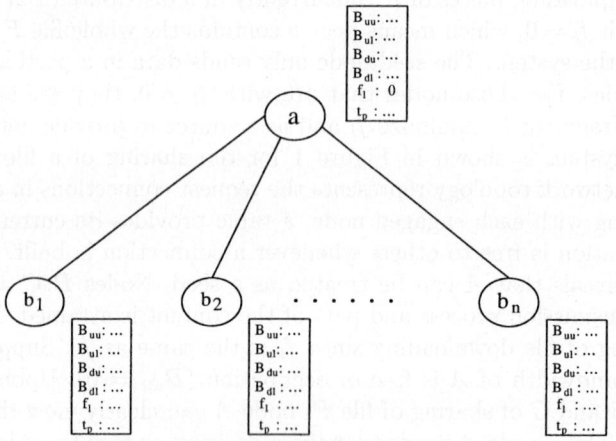

Fig. 2. n peers send their requests to peer a.

More seeds available in a P2P system, it is more fault tolerance. Given that peer b_i is supported by peer a, let B_i ($i = 1, 2, \ldots, n$) be the event that peer b_i becomes a seed. The value of B_i is either 1 or 0 where 1 represents the defined event to be true. Let t_i be the time interval from its request to peer a to the time that B_i is true. Thus, we have

$$t_i = \frac{f_{l_i}}{min(B_{ul_a}, B_{dl_i}) + B_{du_i}}$$

where the subscript i denotes the parameter of peer b_i while a for peer a in this paper. The content obtained by peer b_i can either from other peers, which is represented as through a channel by B_{du_i} bandwidth (assumed to be consistently existing), or from the seed a. The second path with the source node a has the maximum bandwidth supported by $min(B_{ul_a}, B_{dl_i})$ for packet delivery. f_{l_i} means the content contained by a but not in peer b_i (even if peer a is not a seed, the above equation is valid following the same definition of f_{l_i}). Let T_a, T_{b_i} be the event that peer a, b_i will last t_i time longer respectively. Then

$$B_i = T_a \cdot T_{b_i} \tag{1}$$

In a P2P distributed system, peer a and peer b_i can be assumed to be two independent hosts. It follows that T_a and T_{b_i} are two independent events. From Equation 1, the probability of event B_i to be true can be depicted as

$$P(B_i) = P(T_a) \cdot P(T_{b_i}) \qquad (2)$$

Let X_a be the random variable that denotes the time for peer a online. Because peer a has already been in the session for a t_{p_a} time period, the probability of event T_a to be true is under the conditional probability case. Hence, we have:

$$P(T_a) = \frac{P\{X_a > t_{p_a} + t_i\}}{P\{X_a > t_{p_a}\}} \qquad (3)$$

Let X_{b_i} be the random variable that denotes the time for peer b_i online. From Equation 2 and 3, we can rewrite $P(B_i)$ as

$$P(B_i) = \frac{P\{X_a > t_{p_a} + t_i\}}{P\{X_a > t_{p_a}\}} \cdot \frac{P\{X_{b_i} > t_{p_{b_i}} + t_i\}}{P\{X_{b_i} > t_{p_{b_i}}\}} \qquad (4)$$

3.2 Uniform Random Variable

Although the topology of a P2P network is constantly changed, after an initial setting up process for content sharing of a particular file unit, the network can be treated as in a steady state. In other words, the number of peers arrived equalizes those departed and the network keeps in a balanced situation. Let X be a uniform random variable to denote the connecting time for a peer on the interval (α, β). Its probability density function is given by

$$f(x) = \begin{cases} \frac{1}{\beta - \alpha}, & \text{if } \alpha < x < \beta \\ 0, & \text{otherwise} \end{cases}$$

For a given $t \in (\alpha, \beta)$, $P\{X > t\} = 1 - P\{X \leq t\} = 1 - \frac{t-\alpha}{\beta-\alpha} = \frac{\beta-t}{\beta-\alpha}$. Thus

$$P(B_i) = \frac{\beta - t_{p_a} - t_i}{\beta - t_{p_a}} \cdot \frac{\beta - t_{p_{b_i}} - t_i}{\beta - t_{p_{b_i}}} \qquad (5)$$

3.3 Exponential Random Variable

The duration time for a peer stays connecting to a network can be deployed by the exponential random variable with parameter λ. It is said the behavior of a P2P system can be better modeled by a stochastic and memoryless setting [10]. The arrival and departure of peers comply with a Poisson process. Each peer is independently and exponentially distributed online for content sharing. The probability density function for an exponential random variable X with parameter λ is given by:

$$f(x) = \begin{cases} \lambda e^{-\lambda x} & x \geq 0 \\ 0 & x < 0 \end{cases}$$

The memoryless property of the exponential random variable makes that the probability of a peer surviving for another t time longer does not involve any previous performance. Thus

$$P(B_i) = P\{X_a > t_i\} \cdot P\{X_{b_i} > t_i\} = e^{-\lambda t_i} \cdot e^{-\lambda t_i} = e^{-2\lambda t_i} \qquad (6)$$

3.4 Normal Random Variable

Suppose that the connection time for a peer in a P2P system approximately satisfies the normal distribution with parameters μ and σ^2. Let X be the modified normal random variable to represent the duration time of a peer and the probability density function can be represented as following

$$f(x) = \begin{cases} \frac{1}{\sqrt{2\pi}\sigma} e^{-(x-\mu)^2/2\sigma^2} & x \geq 0 \\ 0 & x < 0 \end{cases}$$

Thus, we have

$$P(X > t) = 1 - P(X \leq t) = 1 - \frac{1}{\sqrt{2\pi}\sigma} \int_0^t e^{-(x-\mu)^2/2\sigma^2} dx$$

$$= 1 - \frac{1}{\sqrt{2\pi}\sigma} \int_0^\mu e^{-(x-\mu)^2/2\sigma^2} dx - \frac{1}{\sqrt{2\pi}\sigma} \int_\mu^t e^{-(x-\mu)^2/2\sigma^2} dx$$

$$= 1 - \frac{1}{\sqrt{2\pi}\sigma} \int_0^\mu e^{-y^2/2\sigma^2} dy - \frac{1}{\sqrt{2\pi}\sigma} \int_0^{t-\mu} e^{-y^2/2\sigma^2} dy \quad (\text{let } y = x - \mu)$$

$$= 1 - \frac{1}{\sqrt{\pi}} \int_0^{\mu/\sqrt{2}\sigma} e^{-t^2} dt - \frac{1}{\sqrt{\pi}} \int_0^{(t-\mu)/\sqrt{2}\sigma} e^{-t^2} dt \quad (\text{let } t = \frac{y}{\sqrt{2}\sigma})$$

$$= 1 - \frac{erf(\mu/\sqrt{2}\sigma)}{2} - \frac{erf((t-\mu)/\sqrt{2}\sigma)}{2}$$

where $erf(z)$ is an "error function" from the normal distribution, which is defined as $erf(z) = \frac{2}{\sqrt{\pi}} \int_0^z e^{-t^2} dt$ [4]. Hence, from Equation 4 it follows

$$P(B_i) = \frac{1 - \frac{erf(\mu/\sqrt{2}\sigma)}{2} - \frac{erf((t_{p_a}+t_i-\mu)/\sqrt{2}\sigma)}{2}}{1 - \frac{erf(\mu/\sqrt{2}\sigma)}{2} - \frac{erf((t_{p_a}-\mu)/\sqrt{2}\sigma)}{2}} \cdot \frac{1 - \frac{erf(\mu/\sqrt{2}\sigma)}{2} - \frac{erf((t_{p_{b_i}}+t_i-\mu)/\sqrt{2}\sigma)}{2}}{1 - \frac{erf(\mu/\sqrt{2}\sigma)}{2} - \frac{erf((t_{p_{b_i}}-\mu)/\sqrt{2}\sigma)}{2}} \qquad (7)$$

4 Comprehensive Resource Management Algorithm

Suppose that there are m pieces together for a shared file. They comprise a group $P = \{p_1, p_2, \ldots, p_m\}$. In a session whenever a peer receives any content sharing requests from others, it doesn't need to know the global network topology information. Its decision for the shared pieces assignment among all applicants depends on its local knowledge as peer a in Figure 2. Let the peer that receives n content sharing requests be peer a. Those n peers form a group $B = \{b_1, b_2, \ldots, b_n\}$. For a peer b_i, part of the resource file is stored in its local storage area. Thus, each peer b_i may have different content request upon the source peer a. Let the pieces of content requested by peer b_i constitute the group $P_{b_i} = \{p_{b_{i_1}}, p_{b_{i_2}}, \ldots, p_{b_{i_k}}\}$ with $b_{i_1}, b_{i_2}, \ldots, b_{i_k} \in \{1, 2, \ldots, n\}$. A comprehensive resource management algorithm for file block assignment is shown in Figure 3.

---

while $B_{ul_a} \neq 0$
 $b_i = Extract_{peer}(B)$
 $p_j = Extract_{block}(P_{b_i})$
 $B = min(B_{ul_a}, B_{dl_i})$
 Block P_j is sent to peer b_i from the source peer a through a channel with bandwidth B
 $B_{ul_a} = B_{ul_a} - B$
end while

Fig. 3. The comprehensive resource management algorithm.

Whenever a peer has part of content available and some requests from others, the proposed algorithm can be performed inside the peer. The resource assignment algorithm ends when all its outgoing channel bandwidth consumed. Two functions are conducted to select the right peer as the receiver and the block to be sent:

- $Extract_{peer}(B)$ is an instruction that will select a peer b_i from the group B and remove it. The selected peer has the maximum value of $P(B_i)$ among the group, which is defined in Equation 4. More specifically, the value of $P(B_i)$ can be calculated according to the employed model for the duration time of a peer as discussed in Section 3. Equation 5, 6 and 7 define the probability of a peer b_i to retrieve all needed data from the requested peer. If there is a tie, a peer has smaller f_l is selected.
- $Extract_{block}(P_{b_i})$ indicates which piece of content is chosen. After the decision of peer b_i to be the one served, the block among P_{b_i} that has been mostly requested in peer a is removed and assigned to p_j. In other words, compared with other blocks in the group P_{b_i}, p_j has the maximum number of peers that would like to copy it. If there is a tie, the block is arbitrarily selected.

In the proposed complete resource management algorithm, the peer that has the high probability to become a potential seed is the one to get service firstly, which is implemented by the message $Extract_{peer}(B)$. The maximum bandwidth for data transfer is $min(B_{ul_a}, B_{dl_i})$. The block of content to be copied first is the one most popularly requested. If there is still some abundant B_{ul_a} left, another loop of peer and block selection is executed.

5 Conclusion

In this paper, the resource availability issue has been addressed for a P2P network in terms of content sharing. When more peers containing the whole file, the P2P system is more stable in the sense that the data delivery traffic is evenly distributed and fault tolerance property is well achieved. The comprehensive resource management algorithm is proposed in order to maintain more seeds available. In the presented algorithm, the peer and the piece of content selection problem is solved. The peer with a higher probability to become a potential seed has the priority to be firstly served. The piece of content that has been mostly requested will be delivered first. We believe that the presented resource allocation algorithm can outperform randomized resource assignment techniques in most cases.

References

1. K. Aberer, M. Punceva, M. Hauswirth, and R. Schmidt. Improving data access in P2P systems. *IEEE Internet Computing*, 6(1):58–67, Jan.-Feb. 2002.
2. D. Barkai. Technologies for sharing and collaborating on the net. In *Proceedings of First International Conference on Peer-to-Peer Computing*, pages 13–28, Aug. 2001.
3. I. Clarke, S.G. Miller, T.W. Hong, O. Sandberg, and B. Wiley. Protecting free expression online with freenet. *IEEE Internet Computing*, 6(1):40–49, Jan.-Feb. 2002.
4. Eric W. Weisstein. "Erf." From MathWorld-A Wolfram Web Resource. http://mathworld.wolfram.com/Erf.html.
5. A. Grimshaw, A. Ferrari, F. Knabe, and M. Humphrey. Wide area computing: resource sharing on a large scale. *IEEE Computer*, 32(5):29–37, May 1999.
6. Xiaohui Gu and K. Nahrstedt. A scalable QoS-aware service aggregation model for peer-to-peer computing grids. In *Proceedings of 11th IEEE International Symposium on High Performance Distributed Computing*, pages 73–82, July 2002.
7. K. Kant, R. Iyer, and V. Tewari. A framework for classifying peer-to-peer technologies. In *Proceedings of 2nd IEEE/ACM International Symposium on Cluster Computing and the Grid*, pages 338–345, May 2002.
8. R. Matei, A. Iamnitchi, and P. Foster. Mapping the gnutella network. *IEEE Internet Computing*, 6(1):50–57, Jan.-Feb. 2002.
9. D.A. Menasce. Scalable P2P search. *IEEE Internet Computing*, 7(2):83–87, March-April 2003.
10. G. Pandurangan, P. Raghavan, and E. Upfal. Building low-diameter peer-to-peer networks. *IEEE Journal on Selected Areas in Communications*, 21(6):995–1002, Aug. 2003.
11. S. Waterhouse, D.M. Doolin, G. Kan, and A. Faybishenko. Distributed search in P2P networks. *IEEE Internet Computing*, 6(1):68–72, Jan.-Feb. 2002.

Efficient Search Using Adaptive Metadata Spreading in Peer-to-Peer Networks*

Xuezheng Liu, Guangwen Yang, Ming Chen, and Yongwei Wu

Department of Computer Science and Technology
Tsinghua University, Beijing P.R. China
{liuxuezheng00,cm01}@mails.tsinghua.edu.cn
{ygw,wuyw}@tsinghua.edu.cn

Abstract. Search is until now a difficult problem in peer-to-peer (P2P) file-sharing systems. In this paper, we propose to use adaptive metadata spreading to make search in P2P networks efficient and scalable to large-scale systems. We model the search process in unstructured P2P networks obtain the optimized metadata populations for performance optimization. Based on the model, we propose adaptive metadata spreading approach which can adapt metadata populations to variational environment and achieve the optimized search performance. To implement our approach in fully decentralized P2P system, we employ self-organized fault-tolerant overlay trees, through which peers can easily cooperate with each other to perform adaptive metadata spreading with minor overhead.

1 Introduction

Peer-to-peer (P2P) networks have become one of the most popular Internet applications, and are widely used for sharing files among Internet users (Napster [1], Gnutella [2], KaZaa [3]). Currently, the effectiveness of P2P file-sharing applications largely depends on the efficiency and scalability of its search mechanism, which is a vital factor in exploiting the shared resources from all peers. Although centralized P2P systems (e.g., Napster [1]) are efficient in small scales, they have been abandoned due to non-scalability, vulnerability and lawsuit problems. Unstructured decentralized P2P systems (also called "Gnutella-like" systems) use blind search by forwarding query messages among peers, which severely suffer from both poor efficiency and heavy bandwidth burden, since a single query must traverse numerous peers and generate large amount of querying messages to find a less-popular file. So, they can hardly scale to very large P2P systems. Structured P2P designs like Chord [4] and Pastry [5] use distributed hash tables (DHTs) to guarantee routing convergence and is competent for precise search, but they are insufficient for practical cases where users do not have the precise filename and can only search with partial-match queries, e.g., keywords.

In this paper, we propose the adaptive metadata spreading approach to implement efficient decentralized search in P2P networks. We find that the metadata populations (i.e., number of valid metadata in the system) act as a crucial role in controlling search efficiency and bottleneck bandwidth, and the optimal metadata populations can greatly improve the efficiency and scalability of blind search. So, different from exist-

---

* Supported by NSFC under Grant No. 60373004, No. 60373005, and 973 project numbered 2003CB3169007.

ing methods which randomly spread metadata [7, 8] or only cache metadata along reverse paths [9, 10], we deliberately spread metadata and adapt them to variational environment, so as to keep the metadata populations at the optimal point that leads to best search performance. To achieve this goal in fully decentralized manner, we employ the Pastry infrastructure [5] as underlying peer organization, on top of which we link the peers that share a common file into self-organized fault-tolerant overlay trees. By means of overlay trees peers can effectively estimate needed environment parameters for calculating optimized populations, and we can apportion the tasks of controlling and spreading metadata to all peers, so that each peer has only a light burden. The simulations show that our approach is very efficient, and can greatly improve search performance and guarantee scalability for very large systems, with only minor overhead in maintaining Pastry and overlay trees.

2 Model for Metadata Populations and Search Performance

First of all, we consider a simple case where there is only one file f in the P2P network. Some peers store and share a replica of f, and other peers may want to find and download f. When a peer originates a search for f (e.g., sends a query containing some keywords in f's filename), the querying messages are forwarded from a peer to another until it gets to some of the f's owner (search succeeds) or reach the upper bound of steps (search fails). When f has few replicas, the search process will need many steps to hit one owner. Therefore, we must use metadata to accelerate search process and reduce querying messages. A metadata of f is a pointer pointing to some of f's owner, consisting of f's descriptor and the IP address of a peer that currently shares f. Since the size of metadata is much smaller than f, we can spread many metadata in the network to accelerate search. However, we cannot simply draw the conclusion that the more metadata the better performance is achieved. Metadata have inherent overhead: due to the dynamics of P2P networks, metadata continually expire and become invalid, and thus the system has to keep spreading new valid metadata in place of expired ones to maintain a certain metadata population. More metadata means more bandwidth consumption for maintaining them. So, there is a trade-off of bandwidth between search and maintenance, which is determined by metadata population. Since the peers' narrow bandwidth is the main constraint of scalability in P2P networks, we must deliberately choose metadata population so as to achieve the highest search performance, i.e., find files as quickly as possible while not overuse the available bandwidth.

Now we give an analytic model for metadata population and search performance. Consider a system consisting of N peers (N is around 10^6) and sharing U unique files (we don't count file replicas in U), denoted by $f_1, f_2, \ldots f_U$. Each unique file may have some replicas shared by peers who download the file. Suppose that there are totally C_i metadata of file f_i in the system (for $i=1\ldots U$), i.e., C_i is f_i's metadata population (we also count f_i's owners in C_i). Since on average the success likelihood of a search probe can not be better than that of probing random peers [7], the *search size* (number of probed peers) for resolving a query of f_i is a random variable with the expectation equal to N/C_i [11]. We use Q_i to denote the query load on f_i, i.e., number of search queries for f_i issued in the system per second. Thus, the aggregate bandwidth consumption for search is:

$$BW_{search} = \sum_{i=1}^{U} Q_i \cdot \frac{N}{C_i} \cdot m_{search} \qquad (1)$$

where m_{search} is the size of querying message (in bits).

Then we estimate the maintenance costs. Generally, peers in P2P networks have short session time, and peer variations are usually modeled with Poisson process with parameter $\lambda = 1/T_{session}$ [12], meaning that the rate of peer departures and joins in a duration is rather steady and can be measured. After a peer P go offline, both metadata stored in P and metadata pointing to P become invalid. We emphasize that the rate of metadata invalidation (i.e., the fraction of invalid metadata out of all metadata in a certain time interval) is also steady, and independent of which file the metadata is pointing to. In other words, metadata of different file has the same invalidation rate. This is because for a certain metadata M stored in peer P_A and pointing to peer P_B, the probability that M becomes invalid during an interval is equal to the probability that P_A or P_B goes offline in that interval, which is determined by the average session time of peers, no matter which file M is correlated. Therefore, we can use a global parameter R_{inv} to denote the invalidation rate of metadata, meaning that on average there are R_{inv} of the metadata become invalid per second. Thus, for file f_i there are $C_i \cdot R_{inv}$ metadata of f_i being invalidated per second which should be eliminated from the system. Due to this persistent metadata loss, the system must provide equivalent number of valid metadata as complement. Thus, we have the aggregate bandwidth consumption for metadata maintenance is:

$$BW_{maintain} = \sum_{i=1}^{U} C_i \cdot R_{inv} \cdot m_{maintain} \qquad (2)$$

where $m_{maintain}$ is the size of updating message. The total bandwidth should be no less than:

$$BW_{total} = BW_{search} + BW_{maintain} = \sum_{i=1}^{U} \left(Q_i \cdot \frac{N}{C_i} \cdot m_{search} + C_i \cdot R_{inv} \cdot m_{maintain} \right) \qquad (3)$$

And the bandwidth consumption per peer (which is the main bottleneck constraint) is:

$$bw_{peer} = \frac{BW_{total}}{N} = \sum_{i=1}^{U} \left(Q_i \cdot \frac{1}{C_i} \cdot m_{search} + \frac{C_i}{N} \cdot R_{inv} \cdot m_{maintain} \right) \qquad (4)$$

For large P2P systems, U and Q_i become enormously large, so bw_{peer} is inclined to grow unboundedly. To guarantee scalability, i.e., making peers capable to afford search and maintenance demand from the large system, we need to constrain bw_{peer} within peer's capability. This goal can be achieved by deliberately choose C_i. Besides, the search efficiency is characterized by average response time for search, which is proportional to the average number of probed peers (or "hops") to resolve a query, as:

$$Hops = \frac{total\ hops}{number\ of\ queries} = \frac{\sum_{i=1}^{U} Q_i \cdot \frac{N}{C_i}}{\sum_{i=1}^{U} Q_i} \propto \sum_{i=1}^{U} Q_i \cdot \frac{N}{C_i} \qquad (5)$$

Based on the model, we can optimize performance by using best metadata parameters, in order to achieve minimized bandwidth consumptions or minimized search hops under certain bandwidth constraints. Smaller bandwidth consumptions indicate fewer burdens in each peer, while smaller search hops imply less waiting time for results. Both metrics are critical to system performance.

2.1 Minimizing Bandwidth Consumption

First, we can minimize bw_{peer} in (4) to reduce peer's bandwidth burden and achieve scalability. The optimized choice of C_i for minimum bw_{peer} should be:

$$\frac{\partial bw_{peer}}{\partial C_i} = 0 \Rightarrow C_i = \sqrt{\frac{Q_i \cdot N \cdot m_{search}}{R_{inv} \cdot m_{maintain}}} \qquad (6)$$

From (6) the optimal C_i equalizes the bandwidth for search f_i and maintenance f_i's metadata. Therefore, if all C_i (i=1…U) are appropriately chosen with (6), the P2P network will utilize the same amount of bandwidth for searching for files and maintaining metadata. Note that N, R_{inv}, m_{search} and $m_{maintain}$ are independent of i, and so the optimal C_i is proportional to $Q_i^{1/2}$, the square root of f_i's query load.

2.2 Minimizing Search Hops Under Bandwidth Constraint

If peer's available bandwidth (denoted by BA) is sufficient and our purpose is to improve search efficiency, we need to minimize (5) under constraint of (4), i.e., minimizing $Hops$ subjecting to $bw_{peer} \leq BA$, as the following:

$$\{C_1,...C_U\} = \arg\min_{C_1,...C_U} Hops = \arg\min_{C_1,...C_U} \sum_{i=1}^{U} Q_i \cdot \frac{N}{C_i}$$

$$\text{subject to } \sum_{i=1}^{U} \left(Q_i \cdot \frac{1}{C_i} \cdot m_{search} + \frac{C_i}{N} \cdot R_{inv} \cdot m_{maintain} \right) \leq BA \qquad (7)$$

Here we omit the constant ΣQ_i in the denominator of the first equation.

The equation (7) is easily solved using Lagrange multiplier, and the optimal C_i is

$$C_i = \xi \cdot \sqrt{\frac{Q_i \cdot N \cdot m_{search}}{R_{inv} \cdot m_{maintain}}}, \text{ where } \xi \text{ is a constant equal to the bigger real root the quadratic:}$$

$$(\xi + \frac{1}{\xi}) = \frac{BA}{\sqrt{N \cdot R_{inv} \cdot m_{maintain} \cdot m_{search}}} \Big/ \sum_{i=1}^{U} \sqrt{Q_i} \qquad (8)$$

From (8) we have the optimal C_i for minimized hops should also be proportional to $Q_i^{1/2}$, which is the same with (6). This is not surprising, because (6) is indeed a special case of (8). If we use the minimized bandwidth derived from (6) as bandwidth constraint BA, the right part of the second equation in (8) will be equal to 2, which is the minimum value that guarantees ξ has real root solution. In this case, ξ is equal to 1 and the first equation in (8) becomes equivalent to equation (6). So, (6) is the extreme case of the general correlation of (8), and we obtain that in order to optimize search performance the metadata population should be in proportion with the square root of the file's query load.

3 Adaptive Metadata Spreading Using Self-organized Overlay Trees

Now we explain how to implement above optimized metadata populations in fully decentralized P2P networks. To calculate the optimal populations we need to sample the parameters (Q_i, C_i, N and R_{inv}) from environment. After that we control metadata populations to be optimal by actively and adaptively spreading new metadata to the system. Since there's no global knowledge, peers need cooperation to achieve the goals. We propose self-organized overlay trees built on top of Pastry [5], the prefix-based distributed hashing tables (DHTs), so as to organize owners of a unique file into an efficient and fault-tolerant tree structure, in which owners can easily communicate and cooperate with each other, sharing both information of environment parameters and the burden for spreading metadata. Due to space limitation, we omit the introduction of Pastry system. For more knowledge, please refer to [5].

3.1 Self-organized Overlay Trees

On top of Pastry, we build for each unique file f_i an overlay tree, namely f_i-tree, which consists of one root peer and all f_i's owners. We assign each unique file a uniformly distributed "fileId" generated by hash function, just as the peerIds of Pastry. For each f_i, the peer whose peerId is currently closest to f_i's fileId is employed as f_i-tree's root, taking the charge of a landmark for f_i's owners to join the f_i-tree. Each owner of f_i is a member of f_i-tree. When a peer P goes online in P2P network, it firstly joins the corresponding trees for each of its shared files. For each file f shared in P, P routes towards f's fileId to the corresponding root of f-tree, and locates a parent to link itself in the tree by going down along the f-tree and looking for an appropriate position. The join operation (see below for details) guarantees that each member of a tree has bounded number of children (no more than the radix k in Pastry's ID space), and that the tree has a small tree-height no more than the length of peerId.

The organization of an overlay tree is directed by prefix-matching of members' peerIds. The *level* of a member is defined as the number of hops between the member and the root, and we demand that, for a j-level member P, its child must share at least j-digit-long prefix with P in peerId. Thus, children in deeper levels have longer common prefixes with their parents. For join operation, consider peer P which is a newly-coming owner of file f. As mentioned above, P firstly reaches f-tree's root by routing to the peer with closest peerId to f's fileId, though $\log_k N$ hops. Then, P goes down along the tree from the root, until it finds an appropriate member to which P links itself and becomes a child. In every step, P examines the current tree member Z to see whether Z has less than k children. It is held that at j^{th} step ($j=0,1,...l$ where l is the length of peerId) P always shares at least j-digit-long prefix with the current member Z, e.g., at first step ($j=0$) P has a "0-digit-long" common prefix with the root, etc. Thus, if Z has less than k children, P stops going down the tree and sends a "join" message to Z to become Z's child. If Z has exact k children (recall that no member has more than k children), P investigates the peerIds of these k children and looks for a peer whose peerId has a longer common prefix with P (i.e., at least (j+1)-digit-long). If such peer exists (e.g., peer Z_1) then P goes down to Z_1 in the next step (i.e., (j+1)th step) and continues the above procedure with Z_1. In the case that none of Z's child has

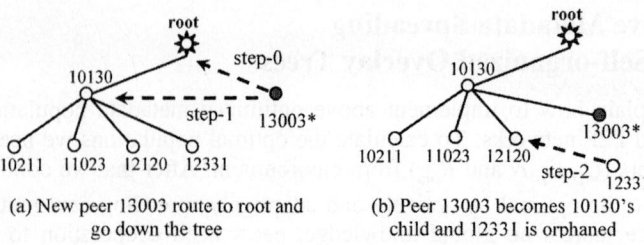

Fig. 1. Join operation to an overlay tree.

(j+1)-digit-long prefix with P, since these k children has common j-digit-long prefix with Z and P in peerId, we know at once: 1) none of Z's child has the same $(j+1)^{th}$ digit with P's peerId, and 2) there must be two children of Z (denoted by Z_A and Z_B) has the same $(j+1)^{th}$ digit in their peerIds, due to the pigeon hole principle. In this case, we let Z discard Z_B and adopt P as its child. Thus, Z still has k children and P finds its parent. The orphaned child Z_B then turn to Z_A to see whether Z_A can adopt it, and this is consistent with our law in constructing tree, since Z_A is in $(j+1)$-level and Z_B has at least (j+1)-digit-long prefix with Z_A. Thus, Z_B goes down along the tree from Z_A, and the procedure continues recursively until all peers are well settled. Note that peerIds have only l digits, so it takes no more than l steps to finish the join procedure.

As exampled in Fig.1, a peer with peerId 13003 is to join an overlay tree. In step-0 of Fig.1.(a), 13003 checks the root and finds that the root has already k (k=4) children (we omit the other children of the root in Fig.1). So, 13003 goes down the tree to a deeper level and asks the child of root which has a 1-digit-long prefix with itself. In step-1 the 13003 asks 10130 to be taken as a child (10130 and 13003 has a common prefix "1"). However, 10130 also has k children, and 13003 checks 10130's children for a longer prefix. But there's no child of 10130 has the prefix "13" in peerId, so there must be two pees having the same 2-digit-long prefix (12120 and 12331, having prefix "12"). Thus, in Fig.2.(b) 13003 takes the place of 12331 and becomes 10130's child, while 12331 is orphaned and in step-2 it asks 12120 to take it (recall that 12120 is in the 2nd level of the tree, and 12331 has a 2-digit-long prefix with 12120).

For fault-tolerance, each peer periodically probes its parent in order to maintain the tree link, as in common structured overlays. When some peer fails or leave the system, its children will soon find the failure and reconnect to the tree by routing to the root peer and performing a join operation. Since a failure can at most influence k other peers (its children) and has no effect on the child's descendants (for they are well connected), the recovery of failure is very efficient with minor cost.

3.2 Report and Notification Along Overlay Trees

A tree is commonly used to perform statistical tasks on its members by adopting the report and notification messages. Suppose that each peer in a tree has some local information (e.g., states or locally measured values for environment parameters), and all peers want to know the aggregation of such information (e.g., the sum or average of the states or values). Thus, each peer employs *report messages* to tell its parent the aggregate states collected from the subtree rooted from it, and also receive from its parent and distribute to its children the *notification messages*, which contains global aggregation that each peer wants to know. After that, all tree members get to know the global aggregation knowledge.

3.3 Parameter Estimation and Adaptive Metadata Spreading

Based on the overlay trees and report-and-notification messages, peers are easy to estimate the necessary parameters for calculating optimized metadata populations. To estimate querying load Q_i, each peer in f_i's overlay tree (i.e., f_i's owners) has a locally measured querying load, obtained by counting number of locally-received queries for f_i in a certain interval. The local querying loads are reported and aggregated along the tree, and an accurate Q_i is obtained in the root. Thus, after the notification phase every f_i's owner knows the estimated Q_i. To estimate peer number N, we take advantage of Pastry's uniformly distributed peerId and Leaf-set. Peer is aware of the peerIds in its Leaf-set, i.e., a number of adjacent peerIds, so it can thus estimate the density of peers in ID space, in other words, the total peer number N. To alleviate the imprecision due to fluctuation in local density, the estimated values of N are also reported and averaged in each tree, like Q_i. For metadata population C_i, we turn to measure N/C_i which is equal to the expectation of search hops for f_i. Each owner of f_i can locally measure the average search hops of f_i from the arriving query messages, which is further averaged along with f_i-tree and regarded as estimation of N/C_i. From the N/C_i and N, peers obtain estimation of metadata population C_i. The metadata invalidation rate R_{inv} is also locally measured in each peer and globally averaged along trees.

After estimation, the owners of file f_i are aware of Q_i, C_i, N and R_{inv}. Then, it calculates the optimized population of f_i's metadata (i.e., C_i) and further control the population. If we want to reduce bandwidth consumptions, equation (6) is used, and the peer investigates whether current C_i is in this optimized level. If C_i is less than demand, metadata spreading is employed that f_i's owners actively spread f_i's metadata to random chosen peers. For example, if current population is δ metadata less than the optimized population, and there are r owners of the file, then each owner have the task of spreading δ/r metadata to the system. Each owner randomly chooses δ/r peerIds, routes to these peerIds, and put a new metadata of the file to the closest peer of the peerId. So, the environment variations are quickly answered by adjusting metadata population levels, and we always guarantee minimized bandwidth consumption. If we want to minimize search hops under bandwidth consumption, the metadata populations should be calculated with equation (8). However, we meet a parameter ζ which needs global knowledge. Here we simply use adaptive adjusting method, i.e., peers measure current bandwidth cost bw_{peer} and communicate it as well as the current ζ with others. If bandwidth cost is less than the constraint, peers will enlarge ζ to improve search efficiency, and vice versa. Thus, the best search performance under bandwidth constraint is also achieved.

4 Performance Evaluations

We first study the overhead of overlay trees. A peer needs to join an overlay tree for each of its shared file. In each overlay tree, an intermediate member keeps at most

(k+1) tree links (i.e., one to its parent and k to its children). From previous researches on Pastry, it is sufficient to use one heart-beating message per minute to maintain a P2P link [13], and so a peer sharing less than 100 files (peers in practical P2P file-sharing system typically share only 10~20 files) need to send less than $100 \cdot (k+1)/60$ messages per second for tree maintenance. In our system the radix k is equal to 4, and thus in the extreme case a peer sends 9 messages every second for maintaining overlay trees. This overhead is very low for current Internet applications, and is even less than Pastry's overhead. So, overlay trees are very lightweight approach and consume minor system costs.

We then evaluate the search performance of our approach using simulations. Our simulation results prove the correctness of the model and system efficiency, and we outperform Gnutella by 2~3 magnitude in search efficiency and bandwidth consumption. Due to paper limitation, we omit the detailed reports of our results.

5 Conclusions

The contribution of this paper is in the following aspects. First, we propose a model of correlations between search performance and metadata populations, and show that metadata populations have significant effects on search performance in unstructured P2P systems. Second, we obtain the analytical solution of optimized metadata populations, which is a universal result and can be used to improve any unstructured P2P system. Third, we propose the overlay tree structure so that peers can easily cooperate with each other to implement adaptive metadata spreading. Our design is fully decentralized and very efficient, and greatly improves search performance in unstructured P2P systems with negligible additional overhead.

References

1. Napster, the napster homepage. In http://www.napster.com/
2. Gnutella, In http://www.gnutell.com
3. KaZaA, file sharing network. In http://www.kazaa.com
4. I. Stoica, R. Morris, D. Karger, F. Kaashoek, and H. Balakrishnan. Chord: A scalable peer-to-peer lookup service for internet applications. In Proceedings of SIGCOMM'2001, 2001.
5. A. Rowstron and P. Druschel. Pastry: Scalable, decentralized object location, and routing for large-scale peer-to-peer systems. In IFIP/ACM Middleware, Nov. 2001.
6. Beverly Yang, Hector Garcia-Molina. Improving search in peer-to-peer networks. In ICDCS'02, 2002
7. Q, Lv, P Cao, E. Cohen, K. Li, S. Shenker. Search and Replication in Unstructured Peer-to-Peer Networks . In Proceedings of 16th ACM International Conference on Supercomputing (ICS'02), 2002.
8. Yatin Chawathe, Sylvia Ratnasamy, Lee Breslau, Nick Lanham, Scott Shenker. Making Gnutella-like P2P Systems Scalable, In Proceeding of ACM Sigcomm'03
9. Bobby Bhattacharjee, Sudarshan Chawathe, Vijay Gopalakrishnan, Pete Keleher, Bujor Silaghi. Efficient Peer-To-Peer Searches Using Result-Caching, In Proceedings of IPTPS'03
10. Freenet, Open Source Community. The free network project – In http://freenet.sourceforge.net/
11. E. Cohen, A. Fiat, and H. Kaplan. Associative Search in Peer to Peer Networks: Harnessing Latent Semantics. In Proceedings of the IEEE INFOCOM'03 Conference. 2003
12. Z. Ge, D. R. Figueiredo, S. Jaiswal, J. Kurose, D. Towsley. Modeling Peer-Peer File Sharing Systems. In Proc of Infocom, 2003

Querying XML Data over DHT System Using XPeer

Weixiong Rao, Hui Song, and Fanyuan Ma

Computer Science and Engineer Department, Shanghai Jiaotong University,
HuaShan Road 1954, Shanghai, China, 200030
{rweixiong,songhui,fyma}@sjtu.edu.cn

Abstract. DHT systems like Chord, Pastry, CAN and Tapstry can only handle semantics-free, large-granularity requests for objects by identifier (typically a name). How to implement the content-based query in DHT system is a challenge. Query the XML data using XPath language can provide expressiveness for the DHT system. In this paper, we propose an XML-based content query system, termed as XPeer, built on DHT systems like Chord, Pastry or Tapstry. Besides the inherent properties provided by DHT, XPeer has several unique features. First, XPeer can utilize XML to implement content-based query using XPath as the query language; Second, XML data in XPeer can be totally heterogeneous, without conforming to the same XML schema or DTD. Third, XPeer can support range query over DHT. Finally, the XPeer can be easily extended to real P2P application like Napster, Gnutella and Freenet.

1 Introduction

Recently Peer-to-peer computing has become popular with the free file sharing service provided by Napster, Gnutella and Freenet. Many researchers in distributed system and network communities have been encouraged by the success of P2P systems, begun to deeper study of P2P, and many research systems like Tapstry[1], called structured P2P, have been implemented using distributed hash table (DHT). However, most of DHT systems focus strictly on handling semantics-free, large-granularity requests for objects by identifier (typically a name). They are limited to caching, prefetching, or pushing of content at the object level, and know nothing of overlap between objects. Moreover, they are highly ineffective at content-based query. Even the real P2P application like Naspter, Gunetella and FreeNet can only provide the keyword search. How to implement the content-based query in both real P2P systems and structured P2P systems is a high challenge. XML can provide a solution to such a challenge. The recent emergence of XML (eXtensible Markup Language) has become the standard for information exchange on the Internet. The XPath language, which is a W3C proposed standard for addressing parts of an XML document, has been adopted as a filter-specification.

In this paper, we propose an XML-based query system, termed XPeer, built on DHT systems. Besides the inherent properties provided by DHT, XPeer has several distinguishing features. First, XPeer can utilize XML to implement XML-based content-based query using XPath as the query language; Second, XPeer can support range query over DHT. To support range query rather than an exact value query over DHT is an open question. Finally, the XPeer can be easily extended to any real P2P application. By exporting any data that can be described in XML format as meta-data into XPeer, the content-based query service, not just the simple keyword searching, can be implemented by XPeer. To built XPeer, we build it based FreePastry, a DHT system developed at Rice university. According to XPeer's evaluation data in environment, XPeer can provide a preliminary solution to a semantic P2P network, and show high scalability, expressive and flexible content-based query over DHT.

Remainder of the paper is organized as follows. Section 2 presents the related work. Section 3 introduces XPeer's background work XISS and our XISS extension. Section 4 gives an overview of XPeer. In section 5 we give XPeer query process over DHT. Our primary prototype built on FreePastry and experimental result is shown in section 6. Section 7 gives the discussion. Finally the conclusion is given in section 8.

2 Related Work

Recent work on the scalable design of structured P2P overlay networks has introduced a new class of structured networks called Distributed Hash Tables (DHT). Well known representatives include Tapstry [1] etc. All of these systems were built to allow efficient key lookup. Nodes in P2P network send messages to each other based on a unique name, generated from a secure one-way hash of some unique string. Assigned as unique ID for node, messages are delivered to the destination host in a fault-resilient fashion. The P2P routing and locating infrastructure can provide scalability, fault-tolerance, self-maintenance and adaptation for upper applications.

Paper [2] presents the design and implementation of PIER, a structured query system intended to run at Internet scale. PIER system is targeted at querying the data that pre-exists in the wide area. The authors aptly relaxed the standard database system design requirements to achieve extreme scalability in a wide area network like the Internet, and also presented a novel architecture combining traditional database query processing with recent P2P network technologies. The simulation of the scalability of PIER system of over 10,000 nodes shows that PIER is a scalable distributed database querying system. The limitations of PIER include: (1) Except the equal join, the PIER provides litter solution about complex query like range query or fuzzy query; (2) For non-structured data, forcing the non-structured data to structured data in PIER is unpractical.

3 XISS and Our Extension

For an XML document, element name and text are the basic data components (for simple, we do not discuss attribute). Besides the data components, the structural nesting relationship between data components is also needed to model the XML document. In this paper, we use an extended XISS's numbering scheme[3] to address the structural nesting relationship between data components of an XML data. We extend XISS's numbering scheme by coupling the depth of each element in an XML with <order, size>. In the extended scheme, the depth of root element in an XML is 0; the depth of root element's son element is 1…In this way, an element, termed by U, can be formalized by U: =<DId, N, T, O, S, D >, where DId is the XML Document Id, N is element name, T is element text, O is element order, S is element size and D is depth. For those elements, which are not the leaf node of XML tree, their texts are null, and their T can be termed as **NIL**. Using the element-tuple defined by U: =<DId, N, T, O, S, D >, we address an element's both the content and structural component, and an XML data M can be simply modeled as multiple element-tuples, without building an element-tuple tree.

We use XPath as our query language, which provides an expressive query over XML data. The path structure in XPath can be absolute path like E1/E2 using parent-child operator '/' or relative path like E2//E3 using ancestor-descendant operator '//'. Also XPath allows the use of a wildcard operator '*' to match any element name. Besides the path structure query, XPath can allow one or more element text selection operation. In XPeer, we decompose an XPath expression into multiple subXPath and these subXPath are connected by operators like '/', '//' or '/*/'. Operator '/' shows the parent-child elements relationship, '/*/' shows the grandfather-grandson elements relationship; and '//' shows the ancestor-descendant elements relationship. In this way, the XPath expression can be shown by **XPath := subXPath$_1$ ∧…∧ subXPath$_k$** where ∧ can be '/', '//' or '/*/'. The subXPath is consisted of the element name, predicate and operator like '>', '<', '='… etc. Using the element predicate, subXPath can be used as value filtering of XML document, and the '∧' is used to address the structure filtering of XML document.

Table 1. Extended XPath model.

SubXPath	Level	Depth	P-Connector	S-Connector
book	0	0	NIL	/
chapter	1	1	/	//
Section	2	>=2	//	/*/
table	3	>=4	/*/	/
figure[@textlike '*.jpg']	4	>=5	/	NIL

In our model, we further model a subXPath as subXPath : = <N, P, O, D, L, P-Con, S-Con> by introducing subXPath depth and subXPath level. In <N, P, O, D, L, P-Con, S-Con>, N is subXPath's element name; P is predicate, O is operator, L is subXPath level, D is subXPath depth, P-Con is connector the between current subXPath with its parent subXPath, and S-Con is the connector between current subXPath and its child subXPath. SubXPath depth shows the depth of a subXPath in XPath expression; and subXPath level is the element depth in XML document where the XML document is the query result of XPath. When the '∧' is '/' or '/*/', the calculation of XPath level is easy; while the '∧' is '//', the XPath level can only be based on the relative value. For XPath expression **XPath := subXPath$_1$∧…∧ subXPath$_k$** where each '∧' is '/', both the **subXPath$_i$** level and depth are (i-1). For **XPath := subXPath$_1$ ∧…∧ subXPath$_k$** where each '∧' equals to '/*/', **subXPath$_i$** depth is (i-1), and **subXPath$_i$** level is (i-1)*2. For the XPath expression **XPath := subXPath$_1$∧…∧ subXPath$_k$** where each '∧' equals to '//', the **subXPath$_i$** level is only denoted as a relative value: **subXPath$_i$** level >= **subXPath$_{i-1}$** +1. Besides the subXPath level and depth, the structure connectors between subXPath with its parent subXPath and its child subXPath can be shown by **P-Connector** and **S-Connector**. In Table 1 where the XPath: /book//chapter/*/table/ figure[@text like '*.jpg'], where book's level is 0, chapter's level is 1, and section's level > = chapter's level +1 = 2, table's level=section's level + 2>= 4.

4 Overview of the XPeer System

In XPeer, a large number of nodes are organized into the XPeer overlay to offer the XML-based query service using XPath as query language. All nodes inside XPeer overlay cooperatively form the XPeer query engine. Clients intending to use XPeer connect to an XPeer access node to publish XML documents or submit XPath queries. XPeer offers querying services on top of DHT. Different some other systems, XPeer can be built over any DHT systems offering functions, not devoting a particular underlying DHT system. Figure 1 shows a picture of XPeer system architecture.

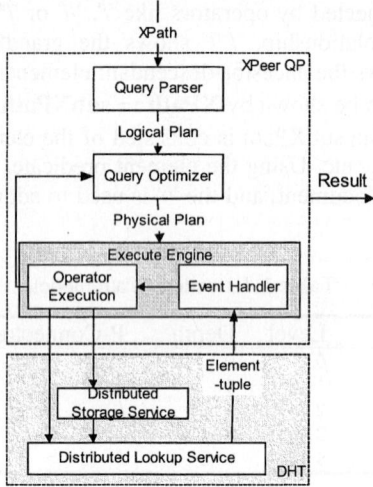

Fig. 1. XPeer Architecture.

In XPeer architecture, distributed hash tables (DHTs) can be used to distribute, store and retrieve data among many nodes in a network. The core of DHT is the dynamic content routing, mapping a key into the node responsible for that key. For XPeer query processor, we currently use XPath as the query language. Different from SQL or XQuery query language, XPath does not involve the join operation between XML documents, and most of XPath query processing is the XML document filtering: the structure join between elements in the same XML and the element text value filtering. Upper applications interact with XPeer query process, which make utilize of the underlying DHT. If XPeer QP lookups and find the data the data in local cache or index, XPeer QP will directly retrieve the result from local storage service; otherwise, XPeer QP depends on DHT's lookup service to retrieve the data from another proper node given a key. In fact, the data in XPeer may be any type of object - documents, audio, video, or any format of data that can be identified and described by XML, called as meta-data. Data type in XPeer does not affect XPeer behaviour as XPeer works with XML. This XML data is simply composed by XML element, which can be modeled as the element-tuple <DId, N, T, O, S, D > in section 3.

5 XPeer Query Process over DHT

Before the discussion of how XPeer implements the basic query process, we make the definitions:

- **ElementID:** a fix-bit Hash ID generated by hashing an element name using DHT;
- **DocID:** a fix-bit Hash ID generated by hash a whole XML document using DHT;

- **XPathID:** a fix -bit Hash ID generated by hashing an XPath expression using DHT.
- **HostID:** any node in XPeer overlay is assigned a unique fix -bit ID using DHT;
- **ElementHost:** a node in XPeer overlay whose HostID is the closet to an ElementID
- **DocHost:** a node in XPeer overlay whose HostID is the closet to a given DocID;
- **XPathHost:** a node in XPeer overlay whose HostID is the closet to an XPathID.

Based on our XML extension model in section 3, the XML document is modeled as multiple element-tuples. To publish the XML document to XPeer, The whole XML document is first hashed into **DocID**, and then the XML document is stored in **DocHost** whose **HostID** is the closet to **DocID**. The element name in an element-tuple is hashed into **ElementID**, and then the element-tuple is stored at **ElementHost** whose **HostID** is the closet to **ElementID**. When an element-tuple is stored in **ElementHost**, the element-tuple is inserted into a B+ tree index based on element-tuple's value. The use of B+ tree index can provide the efficient rang query function without scanning the whole element-tuples in **ElementHost**. When upper applications submit an XPath query into XPeer overlay, the XPath expression is hashed into **XPathID** and routed to **XPathHost** whose HostID is the closet to **XPathID**. In XPeer QP, **XPathHost** acts as query coordinator, who is responsible for XPeer query process for the XPath expression. First, the whole XPath expression is parsed and decomposed by XPeer analyzer into multiple **subXPaths**, based on our XPath extension model in section 3. After the query plan is optimized, the physical query plan is executed.

5.1 XPeer Query Operators

In XPeer QP, we define four basic operators:

- **PathAccess**(String subXPath) operator: Given subXPath, PathAccess operator returns the element-tuples Iterator whose element name is equals with the given element name in subXPath and whose element depth satisfies subXPath's depth condition as described in setion 3.
- **ValueAccess**(String subXPath) operator: Given an subXPath, **ValueAccess** operator returns the element-tuples Iterator whose element name is equals with the given element name in subXPath, element depth satisfies subXPath's depth condition as described in section 3, and element text also meets subXPath's value filtering constraints. By using the B+ tree index, the ValueAccess operation can be accelerated. The returned element-tuples operator of **ValueAccess** is actually a subset of the returned element-tuples of **PathAccess**.
- **StructureJoin**(Iterator i1, Iterator i2, String connector): executes the structure joins between the element-tuples from Iterator1 and Iterator2 where tuple1 in Iterator1 and tuple2 in Iterator2 satisfy the given connector. Using the extended XISS numbering scheme in section 3, we can immediately determine the structure relationship between element-tuples.
- **ActiveJoin**(Iterator i1, String subXPath, String connector): ActiveJoin first sends the element-tuples of Iterator1 to subXPath's ElementHost, and execute the StructureJoin between element-tuples of Iterator1 and the returned Iterator of **PathAccess**(String subXPath). Different from StructureJoin, ActiveJoin actively requests ElementHost for structure join between element-tuples; while **StructureJoin** itself

lazily waits for the incoming element-tuples for structure join. For the case that the size of Iterator is small, the size of **PathAccess**(subXPath), and the size of StructureJoin is samll, ActiveJoin operation can decease the communication cost.

5.2 XPeer Cost Metric and Query Execution

About the cost-based query optimizer, we need to establish a metric by which we can estimate the execution cost for a given physical plan or sub-plan. In DHT, any node can know only part of global knowledge of the whole overlay, and it is hard in advance to determine the cost of an operation exactly. As a result, the execute plan based on the metrics chosen in XPeer is the best of all possible plans, but not the most optimal. XPeer, an application built over DHT, chose the communication cost as the metric because the disk I/O or CUP cost of query operations locally in **ElementHost** is trivial compared with the communication cost to lookup the element-tuples from a given ElementHost to another **ElementHost** given a key. To estimate the communication cost in XPeer QP, we use the overlay hops between nodes in XPeer overlay (Although overlay hops can not determine the IP hops of a node to another, we thought the overlay hop can be considered as the cost metric of lookup in DHT because the routing table entries in underlying DHT are chosen to provide locality properties), and the size of result element-tuples in the destination **ElementHost**. The overlay hops between nodes in XPeer overlay can be determined by the numerical close of the node **HostIDs**. Without looking up the **ElementHost**, the size of element-tuples in the **ElementHost** is hard to determine exactly. However, since we know an XML document is a hierarchy shape, the size of the root element-tuple in the XML document is the least, and the size of the root element-tuple is the most. For the leftmost subXPath **subXPath_L** and rightmost subXPath **subXPath_R** of an XPath expression, the size of **PathAccess(subXPath_L)**<**PathAccess(subXPath_R)**) because the element-tuples of **PathAccess(subXPath_L)** is located at the top of XML hierarchy while the element-tuples of **PathAccess(subXPath_R)** is located at bottom.

The difficulties of XPeer QP includes: (1) **XPathHost** knows little about the statistic of the element-tuples in **ElementHost**; (2) Element-tuples in **ElementHost** are only accessible via DHT's lookup, which results in the fact that the communication cost in DHT is high, unpredictable and variable. Fortunately, XPeer QP is based on XPath, which make the QP is relatively simple compared with XQuery because XPath does not execute the join between XML documents. As a result, we need an adaptive query execution engine to interleave the planning and execution of XPeer in the query execution. The operator tree is executed using the down-top "iterator" model. Control flows from the root node and makes its way down to the tree leaves. The leaves are the **PathAccess** or **ValueAccess**, which make use of DHT to lookup the element-tuples at ElementHost. The basic pipelined unit in XPeer query engine is subXPath. At the end of each subXPath query execution, results are materialized. The execution engine executes the query plan produced by the query optimizer based on XPeer cost model, on the other hand, it gather the statistics about the element-tuples Iterator returned by the operator execution, and handle the event for operator execution.

6 Experimental Results

Currently we have made an initial implementation of the XPeer simulator, which is built on FreePastry. We experimented with publicly available XML data including

SIGMOD Record, DBLP Computer Science Bibliography and The Plays of Shakespeare in XML. All Experiments were conducted on a 1.5GHz Intel Pentium 4 machine with 1024MB of main memory running Windows 2000 Server platform. For the performance limit, the simulation is up to 8192 nodes. For time urgency, we currently just make an initial performance simulation of XPeer including the scalability. The XPath element text filtering including >, <, = or between operation can be effective because the operations are accelerated by using B+ tree index locally in ElementHost. In XPeer, the query efficiency of /, // and /*/ are the same by using the extended XISS numbering scheme because three structure operations are all done locally. So we just define the / operation as above.

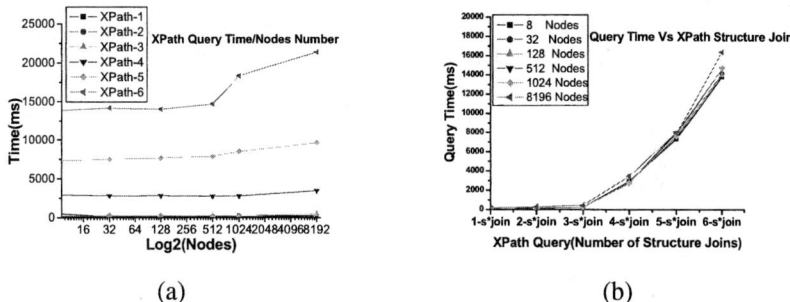

(a)   (b)

Fig. 2. (a) Query Time/Node Number (b) Query Time/XPath structure Join.

To address the scalability of XPeer, we calculate the XPath query time under different node number and XPath queries. For the XPeer optimisation, we still continue the simulation experiment. The XPath query time can defines as the time between the XPeer client issuing the XPath query and the query result arriving at the XPeer client. For XPath query, we make six different XPath query expressions including: XPath-1(*/SigmodRecord*), XPath-2(*/SigmodRecord/issue*), XPath3(*/SigmodRecord/issue/articles*), XPath-4 (*/SigmodRecord/issue/articles/article*), XPath-5(*/SigmodRecord/issue/articles/article/authors*), XPath-6(*/SigmodRecord/issue/articles/article/authors/author*).

Figure 2 (a) shows the XPath query time vs. node number. The query time of 6 XPath queries under different node number is relative smooth. From the figure, we know the query time can work well even the node number grows. Figure 2 (b) shows the query time vs. different XPath query expressions. With the growth of the number of structure join in XPath queries, the query time grows, especially when the sub-XPath number is over 3, the XPath query time grows in linear. Based on the structure join algorithms in section 5.1, the more of sub-XPaths in XPath query expression, the more structure-join between sub-XPaths in XPathHost.

7 Conclusion

This paper outlines our underlying ideas of our under-developing system, termed XPeer, which provides both scalability and expressiveness for content-based publish/subscribe system. The idea mentioned in above sections will be more and more concrete in the future development of the system. We have begun to implement our

XPeer prototype based on Pastry, and expect to deploy it in both a simulation environment and real wide are network like PlanetLab platform very soon. The initial experiment data shows the scalability and efficiency of XPeer.

References

1. B. Y. Zhao, J. D. Kubiatowicz, and A. D. Joseph. Tapestry: An infrastructure for fault-tolerant wide-area location and routing. Technical Report UCB/CSD-01-1141, UC Berkeley, Apr. 2001.
2. Ryan Huebsch Joseph M. Hellerstein Nick Lanham Boon Thau Loo Scott Shenker_ Ion Stoica, Querying the Internet with PIER, Proceedings of the 29th VLDB Conference, Berlin, Germany, 2003
3. Quanzhong Li Bongki Moon, Indexing and Querying XML Data for Regular Path Expressions, Proceedings of the 27th VLDB conference, Roma, Italy, 2001.

Efficient Lookup Using Proximity Caching for P2P Networks

Hyung Soo Jung and Heon Y. Yeom

School of Computer Science and Engineering,
Seoul National University,
Seoul, 151-744, South Korea
{jhs,yeom}@dcslab.snu.ac.kr

Abstract. In peer-to-peer(P2P) network, the overall efficiency heavily depends on the performance of the lookup procedure. So, we propose a simple caching protocol, which intuitively obtains an information about physical network structure. Our caching protocol utilizes the internet address(IP) i.e. first 16 bits of IP address (can be adaptive), The metadata used in our caching protocol is exchanged using piggy-back mechanism, and we extract useful IP prefix set by using round trip time threshold value. We have deployed our caching protocol into Chord, a well-known distributed hash table-based lookup protocol. And our result showed genuine relationship between physical and logical network structure.

1 Introduction

Lots of current peer-to-peer(P2P) lookup protocols (such as *Chord*[1], *Pastry*[2], and *CAN*[3]) are contructed on the logical network structure where peer ID and data ID should be mapped through a hash-based technique, and there is little similarity between real distance and logical distance. Benefits of such P2P protocol architecture over the centralized lookup protocol are load balancing, fault tolerance, and the possibility to be used in cooperative system service. On the other hand, weak knowledge of data location information in P2P network is the inherent cause of slowness in accessing data compared to the centralized system. Therefore, finding the data or key (hashed value of IP or contents) in P2P network must be fast and have a delay bound to overcome its lookup latency.

Our main concern is about the efficiency of lookup procedure in P2P systems. There are number of related works which focused on the fast lookup, or added additional features to the number of well known P2P protocols to reduce the structural inhomogeneity between structure of physical network and that of logical network. Most of research works mentioned above are concentrated on finding physically near-located peers so their proposed protocols are based on the caching either peer ID or data itself for the redundancy.

In this paper, we propose a simple caching protocol for the P2P networks to reduce the overhead of lookup latency. To extract appropriate metadata from various types of messages exchanged between peers, our protocol takes advantage of the inherent information in the IP address. In our caching protocol, each peer should send or receive its own or other peer's network topology metadata over the existing P2P protocol's control packet by piggy-backing. After receiving metadata, a peer should compare its IP prefix set with other peer's IP prefix set. The peer should cache other peer's IP only if any of other peer's prefixes match one of its own prefix set. The length of IP prefix is not static. As the network diameter grows, the length of IP prefix should also increase. Another benefit of our caching protocol is that it is possible to predict the number of lookups probabilistically. This prediction is possible if we look at the difference between bit sequence of node id and that of key id, since both node id and key id are generated using MD5 or SHA-1 hash function. However, if the spread of logical network is not done evenly, the effect of prediction value is reduced. We will discuss more about this topic in section 4.

2 System Architecture

Although our protocol can be easily applied to any current working peer to peer routing protocol to improve both lookup efficiency and total delay, we have chosen Chord as the testbed. The main reason is that the routing protocol of Chord is still primitive without any topology-aware optimization. Let us first look at the routing protocol of Chord. *Chord* uses a 160-bit circular id space. Each Chord node has a unique m-bit node ID, obtained by hashing a node's IP address and *virtual node index*. Chord views the IDs as occupying a circular identifier space. Keys are also mapped into this ID space, by hashing them to m-bit key IDs. Chord defines a node responsible for a key to be the *successor* of that key's ID. The *successor* of an ID j is the node with the smallest ID that is greater than or equal to j, much as in consistent hashing.

Unlike Pastry, Chord forwards messages only in clockwise direction into the circular ID space. Instead of the prefix-based routing table in Pastry, Chord nodes maintain a routing table consisting of up to 160 pointers to other live nodes(called "finger table"). The ith entry in the finger talbe of node n refers to the live node with the smallest node ID clockwise from $n+2^{i-1}$. The first entry points to n's successor, and subsquent entries refer to nodes at repeatedly doubling distances from n. Each node in Chord also maintains pointers to its predecessor and to its n successors on the node ID space. Lookups performed only with successor lists would require an average $N/2$ message exchanges, where N is the number of the servers.

2.1 Proximity Metric

Our caching protocol uses network proximity metric to obtain useful vicinal node IDs. In other word, network proximity metric is used to gather nearest peer nodes so it first measures the existing gap between physical network distance and overlay network distance among peer nodes then selects vicinal nodes and cache them in local caching storage. One simple method to obtain network distance information is using ping or traceroute-like system. There have been quite a bit of research on distance measurements and several tools are available such as IDMaps[9], Global Network Positioning(GNP)[10], and Internet Iso-bar[11]. These systems are concentrating on the finding the distance in terms of metrics such as latency or bandwidth between Interent hosts. For simplicity, we are using Internet topology i.e., Internet address assignement structure. Although measuring network distance using traceroute like system looks quite brute-force, this direct measuring scheme dosen't need any external helper like landmark node in GNP or Tracer in IDMaps.

Our caching protocol requires cirtain length of IP prefixes of each peer node. We have decided to use 16 bit prefixes since there are still lots of research groups and universities using class B addresses. For the accuracy we extract variable length of IP address according to the round-trip time. Every node should maintain its own vicinal IP prefix set which represent close AS(Autonomous System) router's address. We use traceroute-like method to construct IP prefixes; nodes preodically send ICMP packet to randomly selected IP addresses, continues to querry that address until current querry's round-trip time exceeds pre-defined RTT value(threshold). This scheme guarantees that the rate change of number of IP prefix set per each round stabilizes after initial fluctuation. The threshold value can be adjusted so that nodes do not spend too much time in IP gathering.

Once those IP prefix set is obtained, each node stores the IP prefixes in stable storage. Except the mobile hosts, the Internet connectivity remains the same over time and this information can be used afterwards. Next time the node joins overlay again it does not have to go through the prefix-gathering procedure. When a peer node respond to a query message, it should send its prefix set including its own IP. This mutual exchange can accelerate the prefix gathering procedure, and this exchanged information can be used for fast lookup procedure. Detailed explanation of using this cached node information is shown in next section. To make lookup procedure efficient, every lookup should be performed in local node itself as much as possible, i.e., peer node first searches requested ID in its cache storage then forward that query to next hop node in iterative way. Reducing the number of forwarding messages and choosing appropriate next peer node are one of the better ways of improving lookup procedure.

3 Lookup Algorithm Using Proximity Caching

Our lookup procedure is based on the existing P2P protocol. *Chord*, a well-known distributed lookup protocol, guarantees that the lookup can be resolved in $O(log\ N)$ messages to other peers. In *Chord* network every peer should maintain a finger table whose interval is expanded with 2^i ($0 \leq i \leq 159$).

3.1 Lookup Algorithmn

When a peer chooses the next hop, it should select two candidate nodes. One is from the cache, and the other is from the finger table. Then our lookup-time prediction procedure is applied for both candidate nodes. Between candidate nodes, the node with better prediction value would be selected as the next lookup node. The lookup-time calculated using candidated nodes' 160-bit hashed ID and implicitly- gained logical network size. Although node ID is determined by hashing the physical IP address of the node in *Chord*, the hashed ID value is not analyzed thoroughly during the lookup procedure. Each intermediate node just compares its own finger node's ID with the requested data key. We have found that there is certain relationship between 160-bit ID and the number of lookups. The number of bits which are set among 160-bit array is approximately close to the number of lookups really performed.

When node n constructs its finger table in *Chord*, the ith finger node is selected as the first node, s, that succeeds node n by at least 2^{i-1} on the identifier circle, i.e., $s = \text{successor}(n+2^{i-1})$, where $1 \leq i \leq m$ (all arithmetic is modulo 2^m). A node who has a key to find its successor should find key's successor node using *find_successor* function according to original Chord algorithm. In original *Chord* protocol, *find_successor()* just returns the closest preceding node in its finger table. Usually the number of lookups which is taken to find the key's successor is $logN/2$. The main difference of our caching protocol lies in the *closest_preceding_finger()* routine. Since every local lookup is accomplished through this routine regardless of the lookup type, it has a profound effect on the overall lookup efficiency. What follows is our modified version of *closest_preceding_finger()*.

```
//return closest finger preceding id
n.closest_preceding_finger(id)
  for i = m downto 1
   if(finger[i].node ∈ (n, id))
   {
        cache = find_in_cache(id);
     n' = proximity_metric(finger[i].node, cache, id);
        return n';
   }
   return n;

//select among cached nodes
n.find_in_cache(id)
  node = n' = cached_node_list;
  while (node)
    if(node ∈ (n', id))
      n' = proximity_metric(n', node, id);
    node = node.next;
  return n';

//return node who has better rtt prediction
n.proximity_metric(node_finger, cache, id)
```

$hop_{finger} = \alpha * METRIC(node_{finger}, id);$
$hop_{cached} = \beta * METRIC(cache, id);$
$rtt'_{finger} = RTT_{finger} + RTT_{avg} * hop_{finger};$
$rtt'_{cache} = RTT_{cache} + RTT_{avg} * hop_{cache};$
if$(rtt'_{finger} > rtt'_{cache})$
   return *cache*;
return $node_{finger}$;
(where $0 \leq \alpha \leq 1,\ 0 \leq \beta \leq 1$)

//caculate hop value prediction
$n.\textbf{METRIC}(node, id)$
  $bit\_array = id - node;$
  $bit\_array = bit\_array - AVERAGE/2;$
  for $i = 160$ **downto** $(160 - SIZE\_logical)$
    if(i^{th} *position of bit_array is set*)
      $bit\_array = bit\_array - AVERAGE/2;$
      $result\_hop++;$
  return $result\_hop;$

//average node distance
$\textbf{AVERAGE} = 2^{160 - SIZE_{logical}};$

Closest finger replica method was proposed in [4] where each node maintains additional *r-1* immediate successors of its initial finger. To route messages, a node selects the closest node in terms of network distance amongst the finger with the largest identifier preceding the message's identifier and the *r-1* immediate successors of that finger. We compared this heuristics with our caching protocol through the simulation.

3.2 Algorithm Validation

In our proximity caching protocol, we can predictict the value of number of the lookups. Like other peer-to-peer lookup protocol *Chord* protocol also assigns unique 160-bit hashed value to each node and the assigned 160-bit value can be a good binary representation of total *Chord* ID space. To predict the number of lookups, the size of the logical network should be known in advance. The logical network size can be inferred from each node's finger table index. When a node joins the P2P network, the node should send the largest finger table index N up to which all *i-th* finger node points to the same node ID. That means in logical network each node is mapped onto the ID space with about 2^N interval to each other. It implies that the number of nodes joined in current P2P network is about 2^{160-N}.

The main routine of our proximity caching is *proximity_metric()*. In *proximity_metric()*, it calculates lookup value predictions for both the finger node and the cached node and multiply weight factors(α, β) to these values. The weight factors should be changed dynamically because the calculation of *METRIC()* is

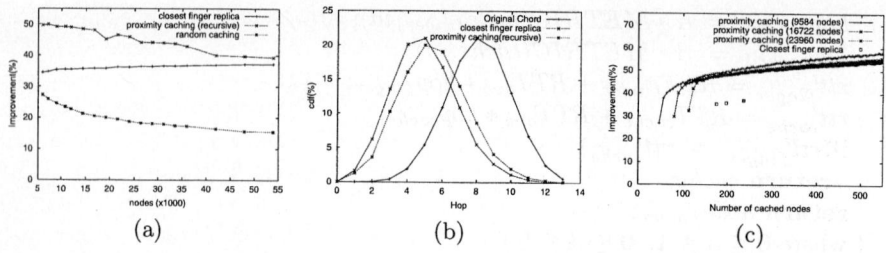

Fig. 1. The percentile delay ratio of our protocol to default Chord protocol(a). Cumulative density function to complete a routing procedure(b). Improvement over different size of proximity caches(c).

heavily affected by implicitly gained logical network size and **AVERAGE**. If the logical network size is small, the average predicted hop has low value, then difference percentile between real hop count and predicted hop count evidently becomes large. For example, in small size network consisting couple of thousand nodes, even *METRIC()* predicted 4 hop, if real hop has 6 hop, the difference percentile is almost $\frac{6-4}{6} = 33.3\%$. In that case, the weight factor should be changed so that the dependability on the result of *METRIC()* can be minimized, and local node should prefer to choose physically nearest cached node. Otherwise if the network size is quite large which is over several thousand nodes, the difference percentile can be lower than previous environment.

In *METRIC()*, we get the hop value prediction from the **bit_array** which is the difference between the nodeID and the key. During the successive bit position checking procedure, the **bit_array** is changed by being subtracted by average node distance. The reason we should subtract average node distance is that the **bit_array** can not be the exact representative of the finger node location because when a node finds the finger node, the finger node is selected as a node whose node id is the first one which is not exceeding the designated finger start id. That makes our metric calculation need some compensation value. Here the compensation value can be interpreted as the average node distance, to be exact, half of the average node distance. Therfore each bit position can tell the probabilistic existence of the finger node. Whenever we find i^{th} position of **bit_array** is set, we should subtract average node distance from the **bit_array** for compensation.

4 Simulation Results

In this section, we evaluate the routing improvement of our caching protocol using simulation. The performance metric we applied is *delay ratio* which is the ratio of the inter-node latency on our protocol to the inter-node latency on the underlying IP network. The simulator is based on the *Chord* protocol which uses recursive style routing. Our underlying network topology is composed of 5400 AS routers and the number of nodes varies from 6000 to 48000. For the simulation topology, we used Tiers[8] topology generator. Tiers topology model can provides maximum 3-level hierarchy of routing domains i.e, inter-network domain, intra-network domain, local-network domain. To these *Tiers* topology, The round trip time threshold value we used in our simulation is decided by the simple heuristics to gain better performance in that 10 percentile value of average direct inter-node round trip time value. And another factor which is set by the simple rule of thumb is α, β. In this simulation, we configured α/β is greater than 0.75 so we increased the probability of cached node to be selected as a next lookup node.

In our caching protocol, two types of cache can be defined. One is the real vicinal cache node whose RTT is within the threshold value. And the other is normal random cache node which is randomly selected to supplement node's cache. Whether a node has enogh number of vicinal cache or not, some meaningful number of random cache is necessary for synergy effect which we will explain later. Figure 1(a) plots the percentile *delay ratio* of our protocol to default *Chord* protocol. The closest finger replica method is proposed in [4] and is similar to our protocol. In that scheme, node should maintain r-1 immediate successor of each finger node. To route message, node selects the closest node in terms of network distance amongst finger set. While increasing the number of nodes, the percentile ratio of our protocol is decreasing due to the static number of cached nodes(we used 128 cached nodes because of comparison with random cache model which dose not have dynamic control of cache size). But that of closest finger replica is slowly increasing because its finger set is growing proportionally to the size of logical network and it can benefit from large size of potential node. Important feature is the decreasing rate of the percentile ratio of both rou protocol and random cache model whose cache size is 128 also. Our protocol shows slowly decreasing feature while random cache model showed decreasing rate reflect that the expected number of hop using random cache model is *(1+0.5log(M/N))*(where M is logical network size, N is the number of cahed nodes).

In Figure 1(b), we ploted cumulative density function of th number of hops to complete routing. The distribution of our protocol and closest finger replica is not largely different but the probability to exploit network proximity is higher in our protocol than that of closest finger replica so the real effect showed like Figure 1(a). Figure 1(c) displayed performance improvement with increasing cache entry in our caching protocol. The reason of different start point of each graph is that in each simulation environment the number of cached node is not same since the density of simulation network is not uniform. 3 points showes under line graph indicate the performance improvement of closest finger replica. It tells our caching protocol show better performance which is simulated based on the same size of finger set in closest finger replica and same size of cached nodes in proximity caching. And the range of number of cache size occupied by initial steep curve shown in each graph is almost same regardless of entire node size. This implies that the minimum number of additional cache nodes to reach the so called stable point is not dependent on the P2P network size so in any network condition, each node can easily determine cache size dynamically according to its own current really vicinal cache size. This is the synergy effect we mentioned earlier. Random cache can also be an increasing factor of performance with vicinal cache.

5 Conclusion

In this paper, we proposed proximity caching protocol to improve *Chord* routing protocol. Our simulation result showed our caching protocol exploits network proximity effectively. As future work we should find more accurate model of prediction of hop values to achieve optimal improvement. Since the spread of node id can not be pefectly uniform in *Chord* id space like other P2P protocols as well and the degree of uniform spread is very important factor to predict hop value as close as real hop value.

References

1. Ion Stoica, Robert Morris, David Karger, M. Frans Kaashoek, Hari Balakrishnan, "Chord: A Scalable Peer-to-peer Lookup Service for Internet Applications", *ACM SIGCOMM*, Aug. 2001.
2. Rowstron, A., Druschel, P., "Pastry: Scalable,distributed object location and routing for large-scale peer-to-peer systems", *ACM ICDSP*, Nov. 2001.
3. Ratnasamy, S., Francis, P.,Handley, M.,Karp, R.,Shenker, S., "A scalable content-addressable network", *ACM SIGCOMM*, Aug. 2002.
4. Frank Dabek, M. Frans Kaashoek, David Karger, Robert Morris, Ion Stoica, "Wide-area cooperative storage with CFS", *ACM SOSP*, Oct. 2001.
5. Sylvia Ratnasamy, Mark Handley, Richard Karp, Scott Shenker, "Topologically-Aware Overlay Construction and Server Selection", *IEEE INFOCOM* 2002
6. Ben Y. Zhao, John Kubiatowicz and Anthony Joseph, "Tapestry: An Infrastructure for Fault-tolerant Wide-area Location and Routing", *UCB Tech. Report UCB/CSD-01-1141*
7. The Gnutella Protocol Specification v0.4
8. Matthew B. Doar, "A Better Model for Generating Test Networks", *Proceedings of Global Internet* November 1996
9. Paul Francis, Sugih Jamin, Vern Paxson, Lixia Zhang, Daniel F. Gryniewicz, Yixin Jin "IDMaps: An Architecture for a Global Internet Host Distance Estimation Service", *IEEE/ACM Trans. on Networking*, Oct. 2001.
10. E. Ng and H. Zhang, "Predictiong Internet Network Distance with Coordinate-based Approach", *IEEE INFOCOM* June 2002.
11. Yan Chen, Randy Katz, Chris Overton, "Internet Iso-bar: A Scalable Overlay Distance Monitoring", 2002.

Content-Based Document Recommendation in Collaborative Peer-to-Peer Network

Heung-Nam Kim[1], Hyun-Jun Kim[1], and Geun-Sik Jo[2]

[1] Intelligent E-Commerce Systems Lab, School of Computer Science & Engineering,
Inha University, 253 Younghyun-dong, Incheon, Korea
{nami,dannis}@eslab.inha.ac.kr
[2] School of Computer Science & Engineering, Inha University, Incheon, Korea
gsjo@inha.ac.kr

Abstract. As the Internet infrastructure has been developed, many diverse and effective applications attempt to gain the potential of that infrastructure. Peer-to-Peer, one of the most representative systems for sharing information on a distributed environment, is a system can helps peer users to share their files with other peer users easily. But Peer-to-Peer network includes not only uncountable files but also plenty of duplicates, which is, bring about increase of network traffic. To solve this problem, we suggest an effective information sharing system supporting collaboration among distributed users with similar interests, or who are part of the same workgroup. In this paper, we exploit the techniques of association rules in deriving peer user profiles represented as a prefix tree structure called PTP-*tree* (**P**ersonalized **T**erm **P**attern tree). In addition, we employ content-based filtering approach to search documents that are similar to personalized term patterns. For the performance evaluation, we formed a simple Peer-to-Peer network to make experiments on real data and 10 users. Experimental results show that the proposed system helps users to reduce time for gathering documents relevant to users' needs. In addition, PTP-*tree* structure of a user profile saves the memory usage.

Keywords: Peer-to-Peer, Content-based filtering, Personalization

1 Introduction

With the explosive growth of the Internet, a wealth of contents available for direct and easy access on the users' computers. However, as the Internet has experienced continuous growth, users have to face a variety and a huge amount of documents, and often waste a lot of time on gathering documents that are relevant to their interests. For overcoming this problem, the recommendation systems and Peer-to-Peer have been issued as a solution for information retrieval [1].

Peer-to-Peer, one of the most representative systems for sharing information on a distributed environment, is a system can helps peer users to share their files with other peer users easily [2]. However, Peer-to-Peer network includes not only uncountable files but also plenty of duplicates, which is bring about increase of network traffic. To reduce duplicate documents, we suggest an effective information sharing system supporting collaboration among distributed users with similar interests, or who are part of the same workgroup.

H. Jin, Y. Pan, N. Xiao, and J. Sun (Eds.): GCC 2004, LNCS 3251, pp. 575–582, 2004.
© Springer-Verlag Berlin Heidelberg 2004

In this paper, we exploit the techniques of association rules in deriving peer user profiles represented a prefix tree structure called PTP-*tree* (**P**ersonalized **T**erm **P**attern tree). PTP-*tree* presents term patterns of each user by means of discovering associations between the terms existing in collected documents. In addition, we employ content-based filtering approach to search documents that are similar to personalized term patterns. For the performance evaluation, we formed a simple Peer-to-Peer network to make experiments on real data and users. The rest of this paper is organized as follows: The next section contains a brief overview of some related work. In section 3, we describe our approach for learning of user profiles using association rules mining. In addition, we show how to search similar documents based on the content-base filtering approach. Then, the performance evaluation is presented in section 4. Finally, we remark the conclusions and future works.

2 Related Work

Recommendation systems, regarded as a part of personalization technique, are mainly used in personal information systems [1]. Two approaches for recommendation systems have been discussed in the literature, *i.e.*, the content-based filtering approach and the collaborative filtering approach. Collaborative filtering utilizes user ratings to recommend items liked by similar user. PHOAKS [4] recommends web links mentioned in newsgroup, and BISAgent [5] shares bookmark information by collaborative filtering. Instead of computing the similarities between the users, the content-based filtering recommends only the data items that are highly relevant to the user profiles by computing the similarities between the data items and the user profiles [7], [8]. Some systems, such as OTS [10], use both content-based and collaborative filtering approaches.

Personalization is one of very active ongoing research. Most personalization systems are based on a user profile. SiteIF [9] and ifWeb [11], which aim at personalized search and navigation support, stored user profiles in the form of weighted semantic networks. OTS [10] employs the techniques of association rule mining for user profiles similar to our system.

3 Learning User Profile for Collaborative P2P

The capability to learn peer users profile is at the heart in a personalized information filtering system. In this section, we describe our approach to designing a profile representation, which is mined semi-automatically from the peer user's collected documents. In this paper, each user profiles are generated by analyzing the collected documents of users and are represented a prefix tree structure called **PTP-tree** (**P**ersonalized **T**erm **P**attern tree).

3.1 Extraction Terms with Weights of Section

In general, a collected document in a peer user's DB consists of sections, such as 'Title', 'Authors', 'Abstract' and 'References'. In this paper, the term support value

$sup_{t,D}$ is measured based on a term frequency (TF) and a weight of the each section instead of using the TF×IDF weighting approach. The terms are extracted from each section of collected documents that has been preprocessed by: removing stop words and stemming words. After a term t is extracted with the term frequency (TF), the term support value $sup_{t,D}$ is measured based on a TF and a weight of the each section, as defined in (1).

$$Sup_{ti,D} = \frac{sup'_{ti,D}}{MAX\{sup'_{ti,D}\}} \qquad Sup'_{ti,D} = \sum_{Sj} tf_{ij} \cdot w_{Sj} \qquad (1)$$

where tf_{ij} is the term frequency of t_i in the section S_j and W_{Sj} is the weighted value of section S_j. The term support value $sup_{t,D}$ is normalized in the range of [0, 1] and indicates the importance of a term in representing the document.

After extracting weighted terms, each collected document CD_j is represented as a vector of attribute-value pairs like the following.

$$CD_j = \{ (t_1, Sup_{t1,Dj}), (t_2, Sup_{t2,Dj}), ..., (t_n, Sup_{tm,Dj}) \} \qquad (2)$$

where t_n is extracted terms in CD_j and $sup_{t,D}$ is the weighted term support of t_n. Based on the above representation, the personalized term patterns are mined by association rules.

3.2 Mining Term Patterns for Personalization

Once collected documents in peer user's DB are represented as weighted term vectors, we extract association rules that associate the terms of a collected document. In this paper, each transaction corresponds to a collected document and items in transaction are terms selected from the weighted term vector. We adopt a FP-growth method to mine frequent term patterns [12]. A lot of the term patterns can be discovered through FP-growth method. To make the personalization effective and efficient, we need to sort and prune the term patterns to delete redundant and noisy information [13]. For the purpose of that, we define a *Term Pattern Cohesion* measure (*TPC*) as the following Eq. (3).

$$TPC(t_1,...,t_n) = \frac{C(t_1,...,t_n)}{\sqrt[n]{C(t_1) \times ... \times C(t_n)}} \times \mu(Sup_{t1}) \times ... \times \mu(Sup_{tn}) \qquad (3)$$

where $C(t_1,...,t_n)$ is a number of documents where the term patterns occur, $C(t_i)$, $i=1,...,n$, is a number of documents containing t_i, $\mu(Sup_{ti})$, $i=1,...,n$, is the mean of the *term support value* for t_i. $\mu(Sup_{tn})$ is measured as defined in the Eq. (4).

$$\mu(Sup_{tn}) = \frac{1}{C(t_n)} \times \sum_{1}^{j} Sup_{tn,Dj} \qquad (4)$$

where $C(t_n)$ is a number of documents containing t_n, $Sup_{tn,Dj}$ is the term support of t_n to document D_j. For example, given two patterns P_1, P_2, if the *TPC* of P_1 is greater than that of P_2, P_1 has higher rank than P_2.

3.3 Personalized Term Pattern Tree (PTP-Tree)

The generated term patterns are stored in a PTP-*tree*, which is a prefix tree structure. PTP-*tree*, which is the modification of CR-*tree* (Compressed Rule tree), can save memory space and explore relationship of terms [13]. All the mined terms are stored in header table and sorted according to descending their frequency. If four personalized term patterns are found as shown in Table 1 after mining collected documents, PTP-*tree* is then constructed as shown in Fig.1.

Table 1. After mining collected documents of a user, four term patterns are found

Patten-id	Term Patterns	TPC	length
1	{T1, T2, T3}	0.56	3
2	{T1, T2, T3, T4}	0.51	4
3	{T1, T2, T5}	0.47	3
4	{T2, T3, T4}	0.32	3

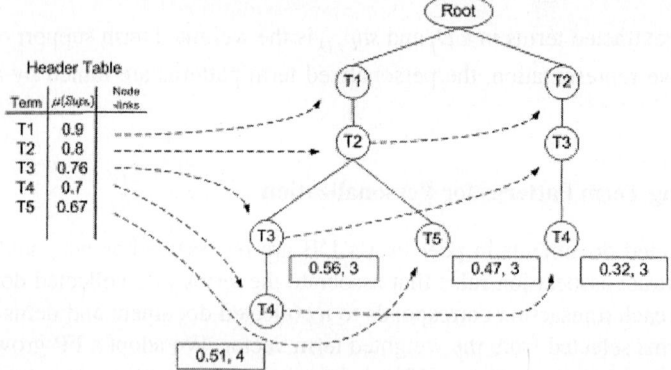

Fig. 1. A PTP-*tree* for Personalized term patterns in Table 1. Only the lowest node of ever term pattern has a *pattern property* such as the *Term Pattern Cohesion* or the *length*

In this paper, user profiles are represented by PTP-*tree* and each user has this PTP-*tree*. Based on PTP-*tree*, the server services recommendations of documents that are collected by other users using content-based filtering approach. The proposed system adopts hybrid decentralized P2P technique such as Napster [2], [3]. Our central server keeps PTP-*tree* of each user and helps to find users who are relevant to document. In addition, the central server monitors users' activity continuously. Whenever a user collects a new document, the system extracts terms with weights and calculates Term Support, $sup_{t,D}$, as mentioned in Section 3.1. Once the document, *CD*, is represented as a vector, the system searches users who are interested in this document based on PTP-*tree*. The similarity between the personalized term patterns PTP_i and the collected documents of the other users CD_j is computed by *Cosine similarity* [1]. If the similarity is greater than some threshold, the system recommends the document to the user.

4 Experimental Evaluation

An application of a peer was implemented using Visual C++ 6.0. And our central server was implemented using MS-SQL 2000 and ASP (Active Server Page) on IIS 5.0 environments. In addition, we ran our experiments on Window 2000 Server with Intel Pentium IV processor having a speed of 2.4GHz and 1GB of RAM.

4.1 Data Set and Evaluation Methods

Our experiments are based on P2P environment and real data/users. For the experiments, we collected documents from ACM Digital Library [14]. ACM has defined a hierarchical categorization of scientific documents called CCS (Computing Classification System). All of collected documents are classified into I.2.11 Distributed Artificial Intelligence of CCS that is divided into four field: *Coherence and coordination, Intelligent agents, Languages and structures, and Multiagent systems*.

For gathering documents, we invited 10 users to do the experiments. They collected documents for about 3 weeks (2 hours per day) to learn their profile. Whenever they found the document related to their own interests, they collected that document. After 3 weeks, total 533 documents were collected which may contain double counts. To evaluate the performance of our systems, we used *F1-measure* [6] widely used in information retrieval community to find the optimal threshold of *Term Support* and *Term Pattern Cohesion*. And also, to evaluate the quality of recommendation, we use the hit ratio, middle ratio, and the miss ratio as the following Eq. (5) [10].

$$hit = \frac{D_g}{D_r}, \quad middle = \frac{D_m}{D_r}, \quad miss = \frac{D_b}{D_r} \tag{5}$$

where D_r is the number of recommended documents to the user, D_g, D_m, and D_b is the number of good, moderate, and bad answers of users.

4.2 Experimental Results

In this section, we present our experimental results of the proposed system. Our experimental results are mainly divided into two parts - effectiveness of user profile and quality of performance. Before running the main experiment, we first determined the sensitivity of the two parameters: *Term Support* threshold and *TPC* threshold.

4.2.1 Parameters Tuning Experiments

Term Support and *Term Pattern Cohesion* control the size of PTP-*tree*. In general, if the size of PTP-*tree* is too small, some information may be lost. On the other hand, if too large, some noise patterns may be included.

Fig. 2 (a) shows the variation of average precision (AVG_P), recall (AVG_R), and F1 measure (AVG_F1) as changing *Term Support* threshold. For mining the personalized term patterns, we set minimum support to 5% and minimum confidence to 50%. And we did not take *TCP* threshold into consideration. As result, when *Term Support* threshold is from 0.5 to 0.6, the performance is stable as shown in Fig. 2.

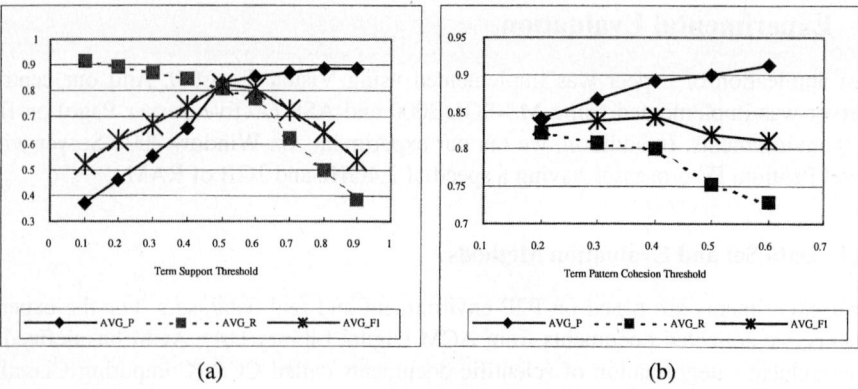

Fig. 2. Average of Precision, Recall, and F1 measure as changing *Term Support* (a) and *Term Pattern Cohesion* (b)

We do experiment again to find the effective value of *TCP* threshold. From the previous experiment, we choose 0.5 as the value of *Term Support* threshold. In general, with the growth of *TCP* threshold, the average precision grows but the average recall decreases. When we set 0.4 as the value of *TCP* threshold, this leads to the best performance as shown in Fig. 2 (b).

4.2.2 Performance Evaluation

Once we obtain the optimal values of the parameters, we evaluate the effectiveness of user profile, which is represented by PTP-*tree*. To test the effectiveness of PTP-*tree*, we check the memory usage of five users when their profiles were constructed.

As can be seen from Table 2, the memory usage is saved 63.93% when the personalized term patterns are stored in PTP-*tree*. This result can be explained as followings. In PTP-*tree*, many terms in the personalized term patterns can be shared. In addition, redundant and noisy patterns can be deleted by *Term Pattern Cohesion* while it is constructing. Thereby, PTP-*tree* contributes to the saving of memory usage.

If a user profile is learned, we assume that each user does not change his interests during our experiments. For evaluating the quality of performance, 10 users gathered documents from ACM DL based on their own profile during 28 hours (2 hour for 2 weeks). As a result, total 447 documents, which may contain double counts, were collected during this period. After resetting the collected documents of all users, they began to collect documents again with recommendation of the system based on peer-to-peer environment. Of course, the central server keeps PTP-*tree* to search the user who is relevant to the document. And according to the documents recommended, the user has to respond to the system with his rating. As can be seen from Table 3, when users collect documents with recommendation of the system, each user is able to collect more documents about 39.4 % as compared without recommendations. In addition, total document were collected for 17.8 hours with recommendations and 36.4% of the time could be save as compared without recommendations. The average acceptance rate of our recommendation is 91.3% as shown in Table 4.

Table 2. The comparison of the memory usage

User	# CD (number)	Without *PTP-tree* (MB)	With *PTP-tree* (MB)	Saving (%)
A	57	253	87	65.61
B	48	231	89	61.47
C	52	249	89	65.25
D	63	304	103	66.11
F	59	254	96	62.20
Average	55.8	258.2 MB	92.8 MB	63.93 %

Table 3. The number of average collections for each user and time for collecting

	Average Collections for each user (numbers)	Time for collecting of total documents (hours)
Testing data	44.7	28
With recommendation	73.8	17.8

Table 4. The average of hit, middle, and miss ratio for recommendations

Hit ratio (%)	Middle ratio (%)	Miss ratio (%)	Acceptance ratio (%)
54.8 %	36.5 %	8.7 %	91.3 %

5 Conclusions and Future Works

Recommendation systems are a powerful technology for users to find information relevant to users' needs. We have presented, in this paper, a recommendation system to provide a personalized service of documents recommendation in collaborative peer-to-peer network. And we propose a new effective method for learning and constructing user profile. The major advantage of our system is that it supports collaboration among distributed users with similar interests, or who are part of the same workgroup. In addition, the proposed user profile, PTP-*tree*, contributes to the saving of memory usage. To evaluate the effectiveness of our approach, we make experiment on real data and users. Our experimental results show that the proposed method saves the memory usage and reduces time consuming of users for gathering documents relevant to their interests.

However, our proposed system dose not consider about the changes in user interests despite changing interests are an undeniable fact in real life. Therefore, we will research into an adaptive learning for quickly changing of user interests in future work.

Acknowledgements

This research was supported by University IT Research Center Project (31152-01).

References

1. Hyun-Jun Kim, Jason J. Jung, and Geun-Sik Jo, "Conceptual Framework for Recommendation System Based on Distributed User Ratings," Lecture Notes in Computer Science, Vol. 3032, Springer-Verlag, 2003.
2. D. S. Milojicic, et. Al., "Peer-to-Peer Computing," HP Technical Report, HP Laboratories, Mar. 2002.
3. K. Aberer and M. Hauswirth, "An Overview on Peer-to-Peer Information Systems," Workshop on Distributed Data and Structures(WDAS-2002), Paris, France, 2002.
4. L. Terveen, W. Hill, B. Amento, D. McDonald, and J. Creter, "PHOAKS: a system for sharing recommendations," Communications of the ACM, 40(3):59-62, March, 1997.
5. Jason J. Jung, Jeong-Seob Yun, and Geun-Sik Jo, "Collaborative Information Filtering by Using Categorized Bookmarks on the Web," Lecture Note in Artificial Intelligence, Vol.2543, Springer-Verlag, 2003.
6. Y. Yang, and X. Liu, "A Re-examination of Text Categorization Methods," In Proceedings of ACM SIGIR'99 conference, 1999.
7. Raymond J. and M. L. Roy, "Content-based Book Recommending Using Learning for Text Categorization," In Proceedings of the 5th ACM conference on Digital libraries, p195-204, 2000.
8. C. Yang, "Peer-to-Peer Architecture for Content-Based Music Retrieval On Acoustic Data," In Proceedings of the twelfth international conference on World Wide Web, p376-378, 2003.
9. A. Stefani and C. Strappavara, "Personalizing Access to Web Sites: The SiteIF Project," In Proceedings of the 2nd Workshop on Adaptive Hypertext and Hypermedia HYPERTEXT '98, June 1998.
10. Y. Wu , Y. Chen , A. L. P. Chen, "Enabling Personalized Recommendation on the Web Based on User Interests and Behaviors," In Proceedings of the 11th International Workshop on research Issues in Data Engineering, April 2001.
11. F. Asnicar and C. Tasso, "ifWeb: A Prototype of User Model-Based Intelligent Agent for Documentation Filtering and Navigation in the World Wide Web," In Proceedings of the 6th International Conference on User Modeling, June 1997.
12. J. Han, J. Pei and Y. Yin. "Mining Frequent Patterns without Candidate Generation," In Proceeding of 2000 ACM-SIGMOD, May 2000.
13. We. Li, J. Han and J. Pei. "CMAR: Accurate and Efficient Classification Based on Multiple Class-Association Rules," In Proceeding of ICDM 2001, December 2001.
14. The ACM Digital Library, http://www.acm.org/dl.

P2PENS: Content-Based Publish-Subscribe over Peer-to-Peer Network

Tao Xue, Boqin Feng, and Zhigang Zhang

School of Electronics and Information Engineering,
Xi'an Jiao Tong University,
710049 Xi'an, China
Xt73@163.com

Abstract. Content-based publish/subscribe systems have recently received an increasing attention. Efficient routing algorithms and self-organization are two challenges in this research area. Peer-to-peer network topologies can offer inherently bounded delivery depth, load sharing and self organization. In this paper, we present a content-based publish/subscribe system built on top of a dynamic peer-to-peer overlay network. A scalable routing algorithm using Pastry network is presented that avoids global broadcasts by creating rendezvous nodes. Self-organization and fault-tolerance mechanisms that can cope with nodes and links failures are integrated with the routing algorithm resulting in a scalable and robust system. The experiment results reveal that our system has better routing efficiency, lower cost and less routing table size at the event brokers.

1 Introduction

Content-based publish/subscribe (pub/sub) middleware differs significantly from traditional subject-based pub/sub system, in that messages are routed on the basis of their content rather than pre-defined subjects or channels. The flexibility of content-based systems comes at the expense of added challenges in design and implementation. The open issue is how to efficiently route events to subscribers interested in them through the overlay network of event brokers. The next challenge is dynamic reconfiguration of the topology of the distributed dispatching infrastructure. That is, the overlay network should dynamically adapt to the changing of lower network topology raised by node or link failure, requiring no manual tuning or system administration. Some approaches for content-based pub/sub [1] [2] [3] [4] [5] [6] have been presented, but to the best of our knowledge, none of them provides any special mechanism to support the self-organization and fault-tolerance addressed by this paper.

On the other hand, peer-to-peer (P2P) systems have recently received significant attention in both academia and industry. P2P systems are distributed systems without any centralized control or hierarchical organization, where the software running at each node is equivalent in functionality. Recent work on peer-to-peer overlay networks offers a scalable, self-organizing, fault-tolerant substrate for decentralized distributed applications such as CAN [7], Chord [8], Pastry [9] and Tapestry [10]. Essentially, these protocol all adopt DHT(Distributed Hash Table)strategy, which provides a mapping of keys in some keyspace to machines in the network and a lookup protocol to allow any searcher to find the particular machine responsible for any key.

This paper presents a content-based publish/subscribe middleware P2PENS that combines the P2P system and content-based publish/subscribe system. In the bottom layer, Pastry is used to realize an overlay network of arbitrary topological structure; in the top layer, a content-based routing protocol with fault-tolerance and self-organizing mechanism is used to construct event dispatcher trees. Furthermore, dis-patcher trees will be reconstructed automatically, which vary with the interests of subscribers.

2 System Architecture

A novel pub/sub architecture we present is illustrated in Fig. 1. Pastry is selected as lower layer communication mechanism in the event broker network. Every event broker will run two protocols. One is Pastry's P2P protocol that handles join or leaving of nodes and routes messages to a rendezvous point. The other protocol is our content-based routing protocol that constructs event dispatcher trees and forwards events along the trees. Event brokers are interconnected with arbitrary topological structure, and communicate in point-to-point mode.

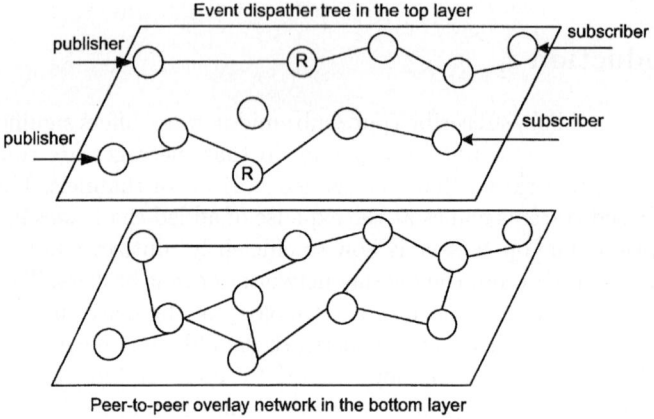

Fig. 1. System Architecture

Unlike the state of the art content-based publish/subscribe systems, which have only one event dispatcher tree for all of events, our system borrows rendezvous point mechanism from Herald [11] and uses it to constructs multi-dispatcher trees. In our system, events are classified by event type. Every event type corresponds to a rendezvous point in the network. Rendezvous point is a special event broker; it is the root of an event dispatcher tree. Events and subscriptions firstly meet in their corresponding rendezvous point. Then events are filtered and propagated along the dispatcher tree stemmed from the rendezvous point. Fig.1 illustrates two dispatcher trees that represent two different event types respectively. The rendezvous point identified with R. Clearly the mapping between an event type and its rendezvous point is a key problem we must resolve, which will be discussed in following section.

3 Routing Algorithm

The design of routing algorithm plays a decisive role for the scalability of pub/sub middleware. The routing algorithm presented here is suited to overlay networks with arbitrary topology, and need not globally broadcast events or subscriptions. Furthermore, it featured in self-organization and fault-tolerance, which distinguished it from existing routing algorithms.

3.1 Rendezvous Point

The purpose of introduction of rendezvous points is to have related events and subscriptions met and construct a dispatcher tree. There can be a lot of rendezvous points in the network; each rendezvous point takes charge of one event type. Similar to Core-Based Multicast Trees [12], multi-rendezvous points share network flow and the burden of matching events against subscriptions, which increases the scalability largely. We use Rt to indicate the rendezvous point of event type t.

Obviously, we should map event types to rendezvous points. Our system uses Pastry P2P network at the bottom layer to set up or search rendezvous point of an event type. All Events published or subscriptions subscribed contain an event type t. The result of Pastry Hash computing on t is the key of the event or subscription. Pastry will route the event or subscription to the node with the nodeId that is numerically closest to the key. This node is the rendezvous point of event type t, namely Rt.

3.2 Propagation of Events and Subscriptions

A subscriber sends its subscriptions to a connected event broker. Every event broker got subscriptions works according to the following steps:

1: Extract event type t from the subscription and perform Hash computing on t to get the key.
2: Save the subscription and the nodeId of the node who sent this subscription to routing table.
3: Determine if this event broker is Rt, if not, call Pastry's routing operation to route the subscription to the next hop according to the key.

A publisher sends events to a connected event broker. Every event broker got events works according to the following steps:

1: Determine whether the event has past Rt by checking the event's Rt field, if yes, then go to step 4.
2: Extract the event type t from the event and perform Hash computing on t to get the key.
3: Determine if this event broker is Rt, if not, then go to step 5; otherwise set the event's Rt field.
4: Match the event against subscriptions in routing table, and forward the event along reverse propagation path of every subscription matched successfully, then the algorithm stop.
5: Call Pastry's routing operation to route the event to the next hop according to the key, then the algorithm stop.

Events and subscriptions of the same type meet at their rendezvous point, and then events are forwarded along reverse propagation path of each subscription matched successfully. Fig. 2 shows the algorithmic process. Publisher P1 publishes event e1 and subscriber S1, S2 subscribe subscription s1 and s2 respectively. Without loss of generality, suppose the event type of e1 and that of s1, s2 are the same and they are matched. Firstly, in term of subscriptions propagation algorithm, s1 and s2 are routed to the rendezvous point Rt(node 4 in Fig. 2) and their propagation path are also recorded by every node along the path. Then e1 is routed to the rendezvous point Rt according to event propagation algorithm. Rt matches e1 against s1and s2, if successful, e1 is forwarded along the reverse propagation path of s1 and s2 until e1 gets to s1 and s2 respectively.

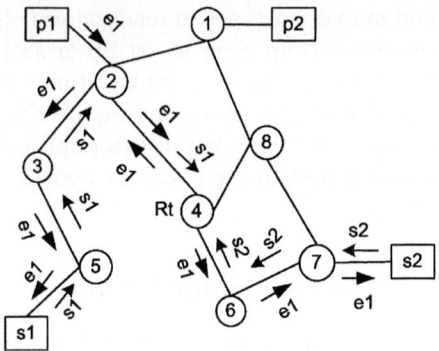

Fig. 2. Pastry-based Routing Algorithm

3.3 Optimization of Routing Algorithm

The main problem of the routing algorithm described above is that all of the events and subscriptions with the same event type must firstly reach their rendezvous point, which may causes the rendezvous point overload in some cases.

With the introduction of advertisement and coverage relation between subscriptions [4] [5], our algorithm can be further optimized.

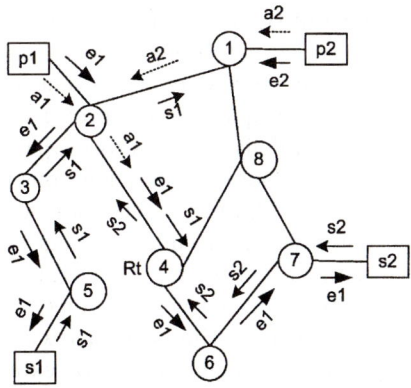

Fig. 3. Pastry-based Routing Algorithm with Advertisement

Fig. 3 illustrates an example of routing optimization. Suppose that event e1, e2, subscription s1, s2 and advertisement a1, a2 have the same event type. E1 matches s1 and s2, but e2 and s1, s2 are mismatched. Moreover a1 covers a2 and s1 covers s2. Firstly, a1 and a2 are sent by P1 and P2 respectively. They will be routed to their rendezvous point, but a2 will not be forwarded at node 2 because a1 covers a2, so only a1 will reach the rendezvous point at last. S1 is routed according to the subscription propagation algorithm. When it reaches node 2, due to advertisement a1 and a2 are the same event type with s1, so s1 is forwarded along reverse propagation path of advertisement (s1 is forwarded to node 1 in Fig. 3) on the one hand. On the other hand, s1 is routed to the rendezvous point. S2 is also routed to the rendezvous point and continue to be forwarded along reverse propagation path of a1. At node2 because s1 covers s2 and s1 has been forwarded to node 1, so the propagation of s2 will be stopped. Then e1 and e2 are published; they are forwarded along reverse propagation path of subscriptions. When e2 reaches node 1, it's discarded because it's not matched with s1. When e1 reaches node 2, since e1 matches both s1 and s2, so e1 is forwarded along the reverse propagation path of s1 and s2 until e1 gets to s1 and s2 respectively.

From the illustration above it's clear that, due to the introduction of advertisement, the propagation of mismatched events are restricted (e.g. e2 is discarded at node 1 in Fig. 3), and events need not be definitely routed to rendezvous points at first (e.g. e1 is forwarded from node 2 to the direction of S1 source in Fig. 3). Furthermore, coverage relation reduces unnecessary propagation of subscriptions and advertisements (e.g. node 2 only forwards a1 and s1 in Fig. 3).

References

1. B. Segall, D. Arnold, J. Boot, M. Henderson, and T. Phelps. Content based routing with elvin4. In Proceedings AUUG2K, Canberra, Australia, June 2000.
2. IBM. Gryphon: Publish/subscribe over public networks. Technical report, IBM T. J. Watson Research Center, 2001.
3. Banavar. G, Chandra. T, Mukherjee. B, Nagarajarao. J, Strom. R, Sturman. D. An Efficient Multicast Protocol for Content-based Publish-Subscribe Systems. Proceedings of IEEE International Conference on Distributed Computing Systems 99, Austin, TX, 1999.
4. A. Carzaniga, D. S. Rosenblum, and A. L. Wolf. Design and evaluation of a wide-area event notification service. ACM Transactions on Computer Systems, 19(3):332-383, 2001.
5. A. Carzaniga and A. L. Wolf. Content-based networking: A new communication infrastructure. In NSF Workshop on an Infrastructure for Mobile and Wireless Systems, Scottsdale, AZ, Oct. 2001.
6. G. Cugola, E. DiNitto, and A. Fuggetta. The JEDI event-based infrastructure and its application to the development of the OPSS WFMS. IEEE Transactions on Software Engineering, 27(9), 2001.
7. S. Ratnasamy, P. Francis, M. Handley, R. Karp, and S. Shenker. A scalable content-addressable network. In Proc. SIGCOMM (2001)
8. I. Stoica, R. Morris, D. Karger, Kaashoek, H. Balakrishnan. Chord: A scalable peer-to-peer lookup service for Internet applications. In Proc. SIGCOMM (2001)
9. A. Rowstron and P. Druschel. Pastry: Scalable, distributed object location and routing for large-scale peer-to-peer systems. In Int. Conf. on Distributed Systems Platforms (Middleware), Nov. 2001, LNCS 2218, pp. 329-350.
10. B. Zhao, J. Kubiatowicz, and A. Joseph. Tapestry: An infrastructure for fault-tolerant wide-area location and routing. Tech. Rep. UCB/CSD-01-1141, Computer Science Division, U. C. Berkeley, Apr. 2001.
11. L. F. Cabrera, M. B. Jones, and M. Theimer. Herald: Achieving a Global Event Notification Service. In Proc. of the 8th Workshop on Hot Topics in Operating Systems (HotOS-VIII), 2001.
12. T. Ballardie, P. Francis, and J. Crowcroft. Core Based Trees (CBT). In ACM SIGCOMM 93, Ithaca, N.Y., USA, 1993.
13. Rice University, Houston, USA. Pastry project. http://freepastry.rice.edu/.

Joint Task Placement, Routing and Power Control for Low Power Mobile Grid Computing in Ad Hoc Network

Min Li[1,2], Xiaobo Wu[1], Menglian Zhao[1], Hui Wang[2], Ping Li[2], and Xiaolang Yan[1,2]

[1] Institute of VLSI Design, Zhejiang University
[2] School of Information Science and Engineering, Zhejiang University

Abstract. Nowadays, almost all mobile devices in the market have Wireless Network Interface Cards (WNICs) attached that enable them to construct Ad Hoc network and communicate flexibly with each other. Most emerging mobile applications, which are usually computation and data access intensive, have different power consumption on different devices. Hence, with aim at lower power consumption, we can apply the on-demand grid computing to mobile devices, migrating task among mobile devices or, if available, stationary computers. Unavoidably, the task migration introduces additional communication activities, which result in significant power consumption on WNIC. Most existing papers use oversimplified communication energy model, and none of them fully exploits the important features of wireless Ad Hoc network to aggressively reduce communication energy consumption. Taking a cross-layer approach, we present a general modeling framework for joint task placement, routing and power control in the context of low power mobile grid computing. The modeling framework can serve as the basis of various network aware optimizations targeting on energy consumption, energy dissipation rate, network life time, etc.. Moreover, we propose an algorithm for minimizing global energy consumption, and study how various factors, such as WNIC, device density, communication penalty, etc., affect the potential of energy saving and the communication pattern in the mobile grid computing.

1 Introduction

For most emerging mobile applications, such as image processing, multimedia streaming, automatic target recognition, 3D game, video phone, face recognition, etc., computation and data access related energy is the dominating part of total energy consumption. Hence, reducing the computation related energy consumption is very important. The idea of low power mobile grid computing is motivated by a number of facts. The first fact is that, for a computation and data access intensive application such as MPEG4 encoding/decoding, 3D rendering, etc., the power consumption varies on different mobile devices. The second fact is that the mobile devices are usually connected to a number of stationary computers that are powered by AC and have far more abundant resources than the mobile counterpart.

Although the task migration can save computation and data access related power consumption, unavoidably it will introduce additional communication energy cost on Wireless Network Interface Card (WNIC). The cost is by no means neglectable.. Most existing papers do not explicitly consider the communication energy cost [1].

Besides the absence of convincing communication energy cost model, they do not put the problem in the context of the multi hops Ad Hoc network, none of them present a general modeling framework addressing the delicate tradeoff between computation energy saving and additional communication energy cost. Finally, none of existing paper takes a cross-system view and fully delivers the greatest potential of energy saving.

In order to minimize computation energy and communication energy as much as possible, we advocate the cross layer design that fully exploits features of all involved layers in the protocol stack by utilizing their interaction. The motivation of cross layer design is that wireless network differs from the wired one in a number of important aspects, which results in much more complexities and constraints. The cross layer design philosophy has been proved to be very effective in wireless network research, especially in wireless multimedia, Ad Hoc network, etc. [2]

The contributions of our work are as following:
- Adopt a computation-efficient model for the power consumption of a realistic WNIC.
- Propose a general modeling framework for joint task placement, routing and power control. To our knowledge, we are the first to consider the cross layer design and give the general formulation.
- Analyze the energy saving rate, number of hops, Return of Investment (ROI) of energy with varying device density, communication overhead and different WNICs.

The rest part of the paper is organized in four sections. Section 2 gives the basic model of WNIC, RF channel, link and routing. Section 3 introduces the model of joint task placement, routing and power control in Ad Hoc network. Section 4 gives the simulation result. Section presents the conclusion and briefs the future work.

2 Model of Joint Power Control and Routing

2.1 WNIC Power Model

As discussed above, the totally energy consumption of WNIC should be considered, because it is indeed the energy cost of communication. There are some existing works on power model of WNIC [3], but none of them takes into account the power control capability of WNIC. The energy model adopted is $p^{WNIC\_T} = p^{TB} + p^{AB} + p^{RFT}/\alpha$, $p^{WNIC\_R} = p^{RB}$, $p^{WNIC\_I} = p^{ID}$, where $T/R/I/S$ means transmitting, receiving and idle state, p^{TB}, p^{AB}, p^{RB} and p^{ID} are constant, α is the direct energy efficiency [4]. Hence, any WNIC can be described by a tuple $I_i = (p_i^{TB}, p_i^{AB}, p_i^{RB}, p_i^{ID}, \alpha_i)$. More details of the model, including derivation and verification, can be found in [4].

2.2 RF Channel, Link and Routing

The purpose of the RF channel model is to generate Bit Error Rate(BER) according to the distance, transmission power, bandwidth, data rate and modulation scheme.

In most environments, the radio signal strength falls as some power of the distance, called the power distance gradient or path loss gradient $p^T / p^R = (4\pi d / \lambda)^n = (4\pi df / c)^n$, where p^T is the transmission power of transmitter, p^R is the corresponding power received by receiver, d is the distance between two nodes, λ is the free space wave length, c is the speed of light, f is the frequency of carrier, n is a constant related to the environment and carrier. For narrow band system and indoor environment, $n \in (3,4)$. The BER is

$$BER \propto erfc(\sqrt{\beta p^R /(NR)}), \qquad (1)$$

where β is a constant for given modulation scheme, p^R is the received power, N is the noise spectrum density, R is the channel data rate.

The expectation of retransmission times when MAC layer retransmission limit is t.

$$E(n) = \sum_{i=0}^{t} FER^i (1-FER)i = \sum_{i=1}^{t} FER^i (1-FER)i. \qquad (2)$$

The possibility of retransmission fails is $p_f = FER^{t+1}$. Denote the protocol overhead by δ, the total number of bits transmitted is $(1+E(n))(1+\delta)s$. Assuming the time for ACK is relatively trivial compared with frame transmission, the expected value of total time for delivering the frame is $E(t_f) = (1+E(n))(1+\delta)s/R$. The expected value of the throughput observed at the Service Access Point (SAP) of MAC is $E(\phi) = s/E(t_f)$. The expected value of the energy consumption of the wireless interface on source node is $E(\varepsilon_{i,j}^{TN}) = (1+E(n_{i,j}))(1+\delta)(p_i^{TB} + p_i^{AB} + p_i^{RFT}/\alpha_i)s/R$, where $n_{i,j}$ is the number of retransmission over link $i \to j$. Ignoring the trivial ACK related transmission activities on destination node, the expected value of the energy consumption of the wireless interface on destination node is $E(\varepsilon_{i,j}^{RN}) = (1+E(n_{i,j}))(1+\delta)p_j^{RB}s/R$. The expected value of the total energy consumed on link (d_i, d_j) is $E(\varepsilon_{i,j}^{Link}) = E(\varepsilon_{i,j}^{TN}) + E(\varepsilon_{i,j}^{RN})$

The QoS requirement is defined as

$$\Xi \equiv \begin{cases} E(\phi) \geq \phi_{min} \\ p_f \leq p_{f,max} \end{cases}. \qquad (3)$$

Formally speaking, the optimal transmission power level $p_{i,j,opt}^{RFT}$ for link $i \to j$ is defined as

$$p_{i,j,opt}^{RFT} \equiv \underset{p_{i,j}^{RFT} \in [p_{i,min}^{RFT}, p_{i,max}^{RFT}]}{\arg\min} E(\varepsilon_{i,j}^{Link}), \text{ s.t., } \begin{cases} E(\phi) \geq \phi_{min} \\ p_f \leq p_{f,max} \end{cases} \qquad (4)$$

In the following, we use $\varepsilon_{i,j}^{LINK}$ to denote the expected value of per frame energy consumption on link $i \to j$ when using the optimal transmission power level $p_{i,j,opt}^{RFT}$ as defined above. The power level optimization can be performed offline. During runtime, it can be lookup from table.

With the defined optimal transmission power level $p_{i,j,opt}^{RFT}$, we can annotate link $i \to j$ with the energy cost of delivering a frame when using the optimal transmission power level. Some time the signal transmitted from i may not be able to reach j because of the upper limit of transmission power of i, or the significant path loss between i and j, in this case, the routing metric is ∞. The routing metric can serve for a number of existing routing algorithms, including shortest path, PARO, MPRC, etc.

3 Model of Mobile Grid Computing in Ad Hoc Network

3.1 Model of Device, Tasks and Environment

There are n devices in the system, and each device is described by a tuple:

$$d_j = \{c_j, r_j, \hat{p}_j^{TK}, \hat{p}_j^{MD}, p_j^{TB}, p_j^{AB}, p_j^{TM}, p_j^{RB}, p_j^{IB}, p_j^{DB}, p_j^{RTH}\}$$

Where c_j is battery capacity ($V \cdot mA \cdot Hour$); r_j is data rate of wireless interface ($Mbps$). Define $r(d_j) \equiv r_j$, $\hat{p}_j^{TK}$ is power consumption budget (maximum) for tasks (mw), $\hat{p}_j^{MD}$ is power consumption budget (maximum) for the whole device (mw), p_j^{TB} is background WNIC power of transmission state (mw). Define $p^{TB}(d_j) \equiv p_j^{TB}$, p_j^{AB} is background WNIC power RF power amplifier (mw). Define $p^{AB}(d_j) \equiv p_j^{AB}$, $\hat{p}_j^{RF}$ is maximum WNIC transmission power (mw). Define $\hat{p}^{RF}(d_j) \equiv \hat{p}_j^{RF}$, p_j^{RB} is WNIC power consumption for receiving state (mw), define $p^{RB}(d_j) \equiv p_j^{RB}$; p_j^{IR} is NIC Power consumption for idle state (mw), define $p^{IR}(d_j) \equiv p_j^{IR}$; p_j^{MB} is background power of the rest part of the mobile device (mw), define $p^{MB}(d_j) \equiv p_j^{MB}$.

The set of all devices is $D = \{d_1, d_2, ..d_j.., d_m\}$, D has two subsets, namely D^{AP} and D^M, $D^{AP} \bigcup D^M = D$, $D^{AP} \bigcap D^M = \phi$. D^{AP} is the set of Access Points (APs) which joint Ad Hoc network with the fixed infrastructure.

From the viewpoint of mobile devices, AP is the gateway to stationary computers, when a task is migrated to certain stationary computer in network, we can deem that the task is mapped onto AP. Since the collection of stationary computers have infinite power supply and infinite processing capability. Hence, $D^{AP} = \{d_j \mid c_j = +\infty \wedge p_j^{TK} = +\infty\}$. D^M is the set of mobile devices: $D^M = D / D^{AP}$.

A task is partitioned into two blocks, one is named as interaction block, and the other is named as computation block. Major functionality of the interaction block is to perform various I/O operations with user via I/O devices such as keyboard, microphone, camera, LCD, speaker, etc... The computation block is to performance the intensive computation according to user input and feed back the result. The interaction block can only run on the original host device, while computation block can migrate among different platforms based on virtual machine or middleware technology.

Each task can be described by a tuple $t_k = \{h_k, r_k^{IC}, r_k^{CI}, r_k^{CP}, r_k^{PC}, t_k^{PR}, p_k^{IR}, \vec{p}_k^{C}\}$. Where h_k is the original host, define $h(t_k) \equiv h_k$ as the original host of t_k; r_k^{IC} is data rate of the flow from interaction block to computation block (bps); r_k^{CI} is data rate of the flow from computation block to interaction block (bps), define $r_k^{CI} + r_k^{IC}$ as the communication overhead/penalty of task migration; p_k^{IR} is power consumption of interaction block (mw). $\vec{p}_k^{C} = \{p_{k,1}^{C}, p_{k,2}^{C}, ..., p_{k,n}^{C}\}$ is power consumption of computation block (vector), define $p_i^{C}(t_k)$ as the power consumption of t_k on i^{th} device: $p_i^{C}(t_k) \equiv p_{k,i}^{C}$.

The environment in which the Ad Hoc network resides can be described by a matrix L with dimension being $n \times n$. Define $L(d_i, d_j) \equiv l_{i,j}$, $l_{i,j} = p_i^{RFT} / p_j^{RFR}$, where p_i^{RFT} is the RF transmission power of device, and p_j^{RFR} is the received RF power. Hence, $p_{i,j}^{RFT} = l_{i,j} p_j^{RTH}$.

3.2 Model of Joint Task Placement, Routing and Power Control

The joint task placement, routing and power control is to give: $\Omega = \{\psi, \Theta, \chi\}$, where $\psi : T \rightarrow D$ is to map tasks onto mobile devices or Aps, $\Theta : D \times D \rightarrow R$ is to set the RF transmission power level among devices, and χ is to construct paths for all data flows existing in the system. In detail, $\psi(t_i) = d_j$ means the computation block of t_i is arranged to run on d_j; $\Theta(d_i, d_j) \equiv p^{RFT}(d_i, d_j) = p_{i,j}^{RFT}$ means the RF transmission level of d_i is set on $p_{i,j}^{RFT}$ when communicating with d_j. After tasks mapping, there are l data flow in the system: $F = \{f_1, f_2, ..., f_l\}$, each flow can be represented by a three tuple $\{d_k^{FS}, d_k^{FD}, r_k^{F}\}$, where d_k^{FS} is the source device, d_k^{FD} is the destination device and r_k^{F} is the data rate.. The formal definition of χ is:

$$\chi : F \rightarrow \bigcup_{i=2}^{n} D^i, D^2 = D \times D, D^i = D^{i-1} \times D, \chi(f_k) = \vec{u}_k, \chi(F) = U = \{\vec{u}_1, \vec{u}_2, ... \vec{u}_l\}.$$

For the kth flow f_k, the $\vec{u}_k$ is an ordered set: $\vec{u}_k = (u_{k,1}, u_{k,2}, ... u_{k,p}, ..., u_{k,d}), u_{k,p} \in D$, $|\vec{u}_k| = 2$ means direct transmission (one hop). Define $f(\vec{u}_k) \equiv f_k$ as the flow corresponding to $\vec{u}_k$ and define $r^F(f_k) \equiv r_k^F$ as the data rate of flow f_k.

First we can study the energy consumption from the viewpoint of individual mobile device. Power consumption of running tasks on mobile device d_k:

$$p_k^{TK} = \sum_{\{t_i | \psi(t_i) = d_k\}} p_k^{C}(t_i) + \sum_{\{t_i | h(t_i) = d_k\}} p^{IR}(t_i) \qquad (5)$$

As source point of data flows, the time during which the WNIC works in transmission state is:

$$\tau_k^{FS} = \sum_{\{u_i|u_{i,1}=d_k\}} \frac{r^F(f(\vec{u}_i))(1+\delta)}{\min(r(u_{i,1}),r(u_{i,2}))}. \tag{6}$$

Since WNIC of mobile device works in idle state when there is no transmission or receiving activities, the additional energy consumption introduced by aforementioned flows:

$$\varepsilon_k^{FS} = \sum_{\{u_i|u_{i,1}=d_k\}} \frac{r^F(f(\vec{u}_i))(1+\delta)(p_k^{TB}+p_k^{AB}+p^{RFT}(r(u_{i,1}),r(u_{i,2}))/\alpha_k - p_k^{TD})}{\min(r(u_{i,1}),r(u_{i,2}))}. \tag{7}$$

As destination point of data flows, the time during which the WNIC works in receiving state is:

$$\tau_k^{FD} = \sum_{\{u_i|u_{i,|\vec{u}_i|}=d_k\}} \frac{r^F(f(\vec{u}_i))(1+\delta)}{\min(r(u_{i,|\vec{u}_i|}),r(u_{i,|\vec{u}_i|-1}))}. \tag{8}$$

The additional energy consumption introduced by the flows is:

$$\varepsilon_k^{FD} = \sum_{\{u_i|u_{i,|\vec{u}_i|}=d_k\}} \frac{r^F(f(\vec{u}_i))(1+\delta)(p_k^{RB}-p_k^{ID})}{\min(r(u_{i,|\vec{u}_i|}),r(u_{i,|\vec{u}_i|-1}))}. \tag{9}$$

As a relay point of data flows, the time during which the WNIC works for relaying is:

$$\tau_k^{FR} = \sum_{\{u_{i,j}|\exists u_{i,j}, u_{i,j}=d_k \wedge j\neq 1 \wedge j\neq|\vec{u}_i|\}} \frac{r^F(f(\vec{u}_i))(1+\delta)}{\min(r(u_{i,j-1}),r(u_{i,j}))} + \sum_{\{u_{i,j}|\exists u_{i,j}, u_{i,j}=d_k \wedge j\neq 1 \wedge j\neq|\vec{u}_i|\}} \frac{r^F(f(\vec{u}_i))(1+\delta)}{\min(r(u_{i,j}),r(u_{i,j+1}))}. \tag{10}$$

The additional energy consumption introduced by the flows:

$$\varepsilon_k^{FR} = \sum_{\{u_{i,j}|\exists u_{i,j}, u_{i,j}=d_k \wedge j\neq 1 \wedge j\neq|\vec{u}_i|\}} \frac{r^F(f(\vec{u}_i))(1+\delta)(p_k^{RB}-p_k^{ID})}{\min(r(u_{i,j-1}),r(u_{i,j}))}$$
$$+ \sum_{\{u_{i,j}|\exists u_{i,j}, u_{i,j}=d_k \wedge j\neq 1 \wedge j\neq|\vec{u}_i|\}} \frac{r^F(f(\vec{u}_i))(1+\delta)(p_k^{TB}+p_k^{AB}+p^{RFT}(r(u_{i,j}),r(u_{i,j+1}))/\alpha_k - p_k^{TD})}{\min(r(u_{i,j}),r(u_{i,j+1}))}. \tag{11}$$

If $\tau_k^{FS}+\tau_k^{FD}+\tau_k^{FR} \leq 1$, the period of time during which the device is in idle state is $(1-\tau_k^{FS}-\tau_k^{FD}-\tau_k^{FR})$. The energy consumption of the WNIC if it is always in idle state is $\varepsilon_k^{DB}=p_k^{DB}\times 1s$. The total energy consumption of the WNIC in one unit of time (1s) is $\varepsilon_k^{MD} = \varepsilon_k^{TK}+\varepsilon_k^{ID}+\varepsilon_k^{FS}+\varepsilon_k^{FD}+\varepsilon_k^{FR}+\varepsilon_k^{DB}$. Average power consumption is $p_k^{MD}=\varepsilon_k^{MD}/1s$, Energy dissipation rate is $\gamma_k^{MD}=p_k^{MD}/c_k$. Moreover, we can study the energy consumption from the viewpoint of individual flow and task. For the kth flow f_k, the $\vec{u}_k$ is an ordered set: $\vec{u}_k = (u_{k,1},u_{k,2},..u_{k,p}.,u_{k,d}), u_{k,p} \in D$

Additional energy consumption introduced by the flow:

$$\varepsilon^F(f_k) = \sum_{i=1}^{|\overline{u}_k|-1} \frac{r^F(f_k)(1+\delta)(p^{RB}(u_{k,j+1}) - p^{ID}(u_{k,j+1}))}{\min(r(u_{k,j}), r(u_{k,j+1}))}$$
$$+ \sum_{i=1}^{|\overline{u}_k|-1} \frac{r^F(1+\delta)(p^{TB}(u_{k,j}) + p^{AB}(u_{k,j}) + p^{RFT}(r(u_{k,j}), r(u_{k,j+1}))/\alpha(u_{k,j}) - p^{ID}(u_{k,j}))}{\min(r(u_{k,j}), r(u_{k,j+1}))} \quad (12)$$

For the kth task t_k, denote the data flow from computation block to interaction block as $f^{CI}(t_k)$, denote the data flow from interaction block to computation block as $f^{IC}(t_k)$. If $\psi(t_k) = h(t_k)$, then $\{f^{CI}(t_k), f^{IC}(t_k)\} = \phi$. If $\psi(t_k) \neq h(t_k)$, then $\{f^{CI}(t_k), f^{IC}(t_k)\} \neq \phi$.

The energy consumed by k^{th} task t_k is:

$$\varepsilon^{TK}(t_k) = \begin{cases} p_k^{IR} \times 1s + p_{k,k}^C \times 1s, \psi(t_k) = h(t_k) \\ p_k^{IR} \times 1s + p_{k,j}^C \times 1s + \varepsilon^F(f^{CI}(t_k)) + \varepsilon^F(f^{IC}(t_k)), \psi(t_k) = d_j \neq h(t_k) \end{cases} \quad (13)$$

Finally, if we aim to lower down the energy consumption or energy dissipation rate of the whole system, the optimization problem can be formulated as:

$$\min_{\{\psi,\Theta,\chi\}} \sum_{k=1}^n \varepsilon_k^{MD}, \text{ or } \min_{\{\psi,\Theta,\chi\}} \sum_{k=1}^n \gamma_k^{MD}, \text{ or } \min_{\{\psi,\Theta,\chi\}} \left(\frac{1}{n-1} \sum_{i=1}^n (\gamma_i^{MD} - \frac{1}{n} \sum_{j=1}^n \gamma_j^{MD})^2 \right)^{\frac{1}{2}}$$

s.t., (1) $(1 - \tau_k^{FS} - \tau_k^{FD} - \tau_k^{FR}) \geq 0$; (2) $p_k^{TK} \leq \hat{p}_k^{TK}$; (3) $p_k^{MD} \leq \hat{p}_k^{MD}$; (4) $\Theta(d_i, d_j) < \hat{p}_j^{RF}$.

3.3 An Suboptimal Algorithm

Apparently the formulated problem is fairly complex. It is easy to prove that it is a NP-hard problem. When ignoring the complexity brought by joint power control and routing, the task placement problem itself is already NP-hard, because it has higher complexity than Traveling Salesman Problem (TSP). Also, when ignoring the complexity of task placement, the routing problem is a specialized case of Bin-Packing.

In this paper we propose a suboptimal algorithm to solve the joint task placement, routing and power control problem. The algorithm is described as following:

A suboptimal algorithm for joint task placement, routing and power control

```
Adjust RF transmission power for node pairs
For each task
    For each task placement option
            Choose energy optimal path based
            Evaluate total energy consumption
    End For
    Rank all task placement options
End For
Rank all tasks according to the maximum energy saving
Do from the highest rank task to the lowest rank task
    Place the task and arrange path for data flow
```

```
            If any constraints are violated
                     Choose another task placement option
                     Rank all tasks again
            End If
    End Do
```

4 Simulation Result

The simulation is to study how various factors, such as WNIC, device density, communication penalty, etc., affect the potential of energy saving and communication pattern in the Ad Hoc network.

In the simulation mobile devices are randomly arranged on a square with size being $100m \times 100m$. The power consumption of tasks on original hosts follows the uniform distribution between 400mw to 800mw and the power consumption on other devices follow the uniform distribution between 20mw to 1200mw. We are to study the behavior of system when communication penalty varies from 1Mbps to 25Mbps, and the number of devices ranges from 10 to 50. The first metric we want to study is the computation energy saving rate, the rate is defined upon task as: $\eta(t_k) = (p_{k,j}^C \times 1s - p_{k,j}^C \times 1s - \varepsilon^F(f^{IC}(t_k)) - \varepsilon^F(f^{IC}(t_k)))/(p_{k,j}^C \times 1s)$. The second metric is the number of hops, which is defined upon flow as $\lambda(\vec{u}_k) = |\vec{u}_k| - 1$.

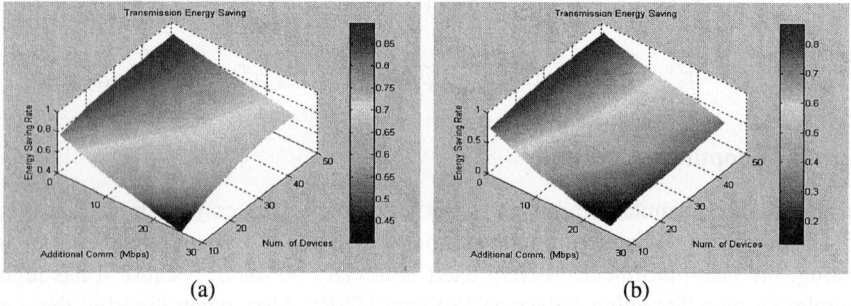

Fig. 1. Average Energy Saving Rate for UWB and 802.11b.

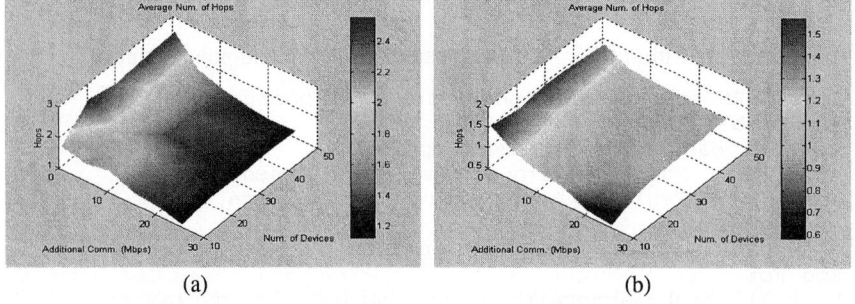

Fig. 2. Average Number of Hops for UWB and 802.11b.

In order to gain insightful view on the global behavior, we calculate average result over 1000 samples. Two typical WNICs are considered in the simulation, the one is IEEE 802.11a/b WNIC, and the other is Ultra Wide Band (UWB) WNIC. We parameterize the power model according to the standard specs and datasheet.

Fig. 1 is the colored surface of average computation energy saving rate versus number of devices and communication penalty. Fig. 1 (a) is for UWB WNIC, while Fig. 1 (b) is for 802.11 WNIC. It is apparent that the $\overline{\eta}$ increases with the increment of devices density and the decrement of communication penalty. The maximum value of $\overline{\eta}_{UWB}$ and $\overline{\eta}_{802.11}$ is almost the same, while the minimum value of $\overline{\eta}_{UWB}$ is significantly larger than that of $\overline{\eta}_{802.11}$. Moreover, the average amplitude of the surface of $\overline{\eta}_{UWB}$ is higher than that of $\overline{\eta}_{802.11}$. The result shows that the deployment of UWB is very promising for mobile grid computing.

Fig. 2 is the colored surface of average number of hops versus number of devices and communication penalty. Fig. 2 (a) is the result for UWB WNIC, while Fig. 2 (b) is the result for 802.11 WNIC. The average amplitude of the surface of $\overline{\lambda}_{UWB}$ is higher than that of $\overline{\lambda}_{802.11}$ because that, with 802.11a/b WNIC, additional hops will introduce more much energy than UWB WNIC. When the communication penalty is relatively small, both $\overline{\lambda}_{UWB}$ and $\overline{\lambda}_{802.11}$ show large value, because computation energy saving may compensate the energy consumption introduced by multi hops. When communication penalty is very large, $\overline{\lambda}_{802.11}$ falls rapidly below 1, because the computation energy saving is not worth the significant communication energy cost, so that computation block tends to run locally. While using UWB WNIC, multi hops are always promising because of the low power consumption on UWB WNIC. The results show that mobile grid computing can indeed benefit from multi hop Ad Hoc network.

5 Conclusion and Future Work

In this paper, we propose a general modeling framework for joint task placement, routing and power control. The modeling framework can serve as the basis for various optimizations, such as global energy consumption optimization, global energy dissipation rate optimization, per task optimization and so on. We propose a suboptimal algorithm for the global energy consumption optimization, which is actually NP-Hard. Finally, we study how various factors, such as WNIC, device density, communication penalty, etc., affect the potential of global energy saving and the communication pattern in the Ad Hoc network. We find that the mobile grid computing can indeed benefit from the multi hops Ad Hoc network.

Acknowledgement

This work is supported by the National Natural Science Foundation of China under grant No. 90207001.

References

1. Flinn, J., Narayanan, D., and Satyanarayanan, M. Self-tuned remote execution for pervasive computing. Proceedings of the 8th Workshop on Hot Topics in Operating Systems (HotOS-VIII), Schloss Elmau, Germany, May 2001.
2. S. Banerjee and A. Misra, Energy Efficient Reliable Communication for Multi-hop Wireless Networks, to appear in the Journal of Wireless Networks (WINET).
3. Feeney, L.M.; Nilsson, M., Investigating the energy consumption of a wireless network interface in an ad hoc networking environment, INFOCOM 2001. Twentieth Annual Joint Conference of the IEEE Computer and Communications Societies. Proceedings. IEEE, Volume: 3, 22-26 April 2001 Pages:1548 - 1557 vol.3
4. Min Li, Xiaobo Wu, Menglian Zhao, Hui Wang, Ping Li, Xiaolang Yan, Power Consumption of Wireless NIC and Its Impact on Joint Routing and Power Control in Ad Hoc Network, to appear in Journal of Embedded Computing

Scheduling Legacy Applications with Domain Expertise for Autonomic Computing*

Huashan Yu, Zhuoqun Xu, and Wenkui Ding

School of Electronic Engineering and Computer Science, Peking University
Beijing, 100871, China
yuhs@ailab.pku.edu.cn, {zqxu,ding}@pku.edu.cn

Abstract. In the domain of scientific computation, there exist a large number of legacy applications that are too valuable to give up and too complex to rewrite. Enabling them on the computational grid plays an important role for realizing the Grid potential. This paper presents a component-based approach for managing and accessing legacy applications on the computational grid. It encapsulates a collection of complementary and competitive legacies to be an independent entity, which automatically schedules these distributed legacies with domain expertise to perform computation. Computing jobs are specified in domain terms and dynamically mapped to appropriate resources. Transparently, competitive legacies are utilized to improve load balance and reliability, and complementary legacies are scheduled to meet the requirements of jobs with different characteristics. To evaluate this approach, a prototype has been implemented, and preliminary experiment results are also presented.

1 Introduction

In almost every scientific and engineering domain, there exist a large number of legacy applications distributed among different institutions. Due to the complexity of scientific problems, most of them are very time-consuming and restricted in applicability. Let us consider the multi-elastic body problem [1] in geological dynamics for example. The total of constitute bodies varies greatly from one simulated geological system to another, and different elastic bodies may differ hundredfold or more in size. To simulate the dynamic process of geological systems with different tectonic characteristics, different algorithms are required. Every simulation may involve several gigabytes of data, and the problem size and complexity are multiplied as its fidelity increases. Generally, legacy applications were developed for specific target platforms. They are unchangeable and very strict in resource-requests for execution environments. At the same time, these applications are too valuable to give up or too complex to rewrite. Some of them are complementary in resolvable problem characteristics, and others are compatible or even substitutable. The idea of scheduling legacies with Grid technologies [2,3] for cross-institutional sharing is straightforward. And the potential advantages are obvious, including the improvements in efficiency, robustness and availability, etc.

Legacy application management has been recognized as one foremost system property requirement for a Grid infrastructure [4]. However, dealing with legacy is-

---

* This work was supported by National Natural Science Foundation of China (No. 60303001, No.60173004).

sues remains one of the major technical hurdles standing in the way of realizing the Grid potential, despite of many technical advances achieved in recent years. To model and access legacies on the computational grid, we have devised a service-oriented component architecture *AOD* on top of OGSA [4]. It is an attempt to aggregate legacy applications and their target platforms with Grid technologies for autonomic computing. The ultimate goal is to enable legacies on the computational grid, and utilize their complementariness and substitutability to provide an on-demand computing environment for daily scientific and engineering work. In such an environment, all legacies are ubiquitously accessible and automatically scheduled as a whole. Computing requests are specified in domain terms and submitted uniformly. Every request is mapped to appropriate resources dynamically, according to resource availability and problem characteristics. The desired computation is automatically completed. The environment is also responsible for monitoring its process and recovering it from resource failures by re-mapping to other competitive resources. Concurrent requests are mapped to different resources, so as to balance the workloads.

In the next section, we first present a service-oriented component framework for modeling legacy applications and their target platforms. The AOD is detailed in section 3. Section 4 introduces a prototype implemented for AOD, which is evaluated with preliminary experiment results. Related works are overviewed in section 5, followed by a conclusion.

2 Scheduling and Monitoring Distributed Resources

To address the challenge of legacy application management, we have proposed the concept of *hyper-resource* for resource virtualization. Based on the software component technology, it combines Grid technologies with domain expertise to automatically schedule and monitor distributed resources on the computational grid. We first introduce this concept, then a component framework for developing hyper-resources and its underlying mechanisms are discussed. This framework provides an approach for incorporating domain expertise into the computational grid, so as to deal with the legacy issues. The most important advantage of this framework is that it allows distributed complementary and competitive legacies to be scheduled as a whole for grid computing, so as to be fault-tolerant and self-optimizing.

2.1 Concept of Hyper-resource

A *hyper-resource* is an autonomic and extensible entity, aggregating a collection of legacy applications augmented with domain expertise, and providing a set of consistent functions for scientific computing on the computational grid. Every function implies some computability of the hyper-resource, and its interface is domain-termed. Combining the augmented expertise with the Grid technologies, the hyper-resource automatically schedules the aggregated legacies to implements its functions. These functions are adaptive to problem characteristics. When a function is requested to perform some computation, the computation characteristics are automatically analyzed, and a most appropriate host is dynamically selected to run an instance of the function. The hyper-resource maps concurrent instances of the same function to different hosts, so as to improve load-balance and every job's efficiency. It is also responsible for monitoring every function instance's execution. When the process is stopped by some failure occurred on the selected host, another substitutable host is selected to recover it automatically. At the same time, a hyper-resource can be enhanced later with additional legacies, so as to provide new functions or improve service quality of existent functions. And the extension poses no effect on the interface of any existent function. Since functions provided by hyper-resources are consistent, adaptive, self-optimizing and self-healing, we call every function as an *on-demand computing service* (OD service).

Generally, all legacies encapsulated in a hyper-resource adhere to a common concept of some scientific domain. Image processing and linear equation solving are examples of such domain concepts. The legacies were independently developed and optimized, and each runs on a local platform to perform some specific computation. Utilizing the augmented domain expertise, the hyper-resource automatically schedules them to implement its OD services. Some legacies are complementary or competitive in function and resolvable problem characteristics. And most legacies neither are substitutable to nor can interoperate with others directly, due to the difference in performance, syntax and semantic of arguments, programming languages and target local platforms. In order to make the legacies compatible with each other, a hyper-resource usually encapsulates a set of specially developed executables to transform arguments of the legacies into domain-standardized data objects. Therefore, one OD service often has multiple candidate implementations. Different candidates are provided by different local platforms independently. They are substitutable or applicable to problems with different characteristics. And they are utilized to improve the OD service's dependability, efficiency and adaptability to problem characteristics.

To specify arguments of the OD services, a hyper-resource declares a set of IO ports on its interface. Every OD service employs a subset of IO ports to input and output its arguments, and every IO port corresponds to one different argument. An IO port is either an IN port for inputting or an OUT port for outputting some argument, and is associated with a *file-composing descriptor* to describe the transferred argument's semantics and syntax in domain terms. A *file-composing descriptor* (FC descriptor) is a high-level description of some complex data object type in a scientific or engineering domain. It implies the decomposing of a complex data object into several relatively small data components, and defines a set of data files for these data components. Every data component is independently specified by one data file, and its semantic interpretation is also presented in the description. Generally, a FC descriptor is independent of any hyper-resource and defined by domain experts.

With the concepts of FC descriptor and IO port, every job submitted to an OD service is formally specified with a list of triplets; and each triplet is in the form of < *port.arg, file, loc*>, where *port* is an IO port employed by the OD service for transferring some argument, *arg* is a file defined by *port*'s FC descriptor, *file* is a data file at the URL location *loc*. Triplet < *port.arg, file, loc*> specifies that *file* is provided as the argument component denoted with *port.arg*.

2.2 A Component Framework for Resource Virtualization

As discussed above, every hyper-resource automatically schedules a collection of legacy applications to implement a set of OD services for performing some complex computation on data files located anywhere on the computational grid. The legacies are augmented with necessary domain expertise for the hyper-resource to manage and enable them automatically, and the OD services are consistent, self-optimizing, self-healing and adaptive to problem characteristics. All of the legacies and the domain expertise are encapsulated in a configuration descriptor that's interpreted by AOD. A hyper-resource is registered by submitting its configuration descriptor to AOD. AOD then configures its OD services on the computational grid, and equips it with the mechanism for conducting its behaviors according to the augmented expertise. When an OD service is invoked, the desired computation is efficiently and dependably completed with the most appropriate resources. With the automatically equipped mechanisms and the expertise augmented to the underlying legacies, the hyper-resource is responsible for all implementing details, including execution environment provisioning, remote data transporting, legacy scheduling and executing, load balancing among competitive resources and troubleshooting, etc.

```
hyper-resource name
IO-port-list
service-name₁
   IN-port-list
   OUT-port-list
   service-body
service-name₂
   IN-port-list
   OUT-port-list
   service-body
......
```

Fig. 1. Hyper-resource's configuration

The configuration framework of a hyper-resource is illustrated in fig.1. It begins with an identical and domain-termed name. Next is it's interface, consisting of a list of IO ports and OD services. Every OD service is identified by a service name, employing a subset of IO ports to transfer its arguments. The employed IN ports and OUT ports are listed in *IN-port-list* and *OUT-port-list* respectively. Its service-body specifies all the underlying legacy applications and the augmented domain expertise with a formal and generic structure, which is interpreted by AOD to configure it and conduct its behaviors on the computational. The expertise augmented to every encapsulated legacy includes its syntactic interface, functionality, applicability restriction, usage, semantic relationships with others, and methods for monitoring its target platform and its running process, etc.

To specify formerly the augmented domain expertise, an OD service's underlying legacy applications is divided into several independent subsets in the service-body, and the service-body also declares a set of *inspectors*, *analyzers* and *evaluators*. Each subset contains one or more legacies installed on the same local platform, corresponding to one candidate implementation. An *inspector* is a module for querying some specific information from problem data or computational resources. It is either a set of rules or a pre-developed executable. There are two kinds of inspectors. One is to detect problem domain characteristics like problem size and time complexity, involving some simple but professional testing on a problem domain's data objects. Another is to inquire some host for dynamic status of its resources, like inquiring the available storage and idle CPU cycles. The inquiring often requires support from the hosts or their operating systems. An *analyzer* is a module for acquiring a problem's resource requirements by synthesizing its characteristics with domain expertise, resulting in a criterion value for judging whether a candidate resource is applicable. And an *evaluator* synthesizes a problem's characteristics and a host's resource status with domain expertise, resulting in a numeric value to evaluate whether resources provided by the host are optimal for the problem. Examples of evaluating criteria include efficiency, price and reliability, etc. Both analyzers and evaluators can be implemented by either rule-sets or executables. Combination of the inspectors, analyzers and evaluators specifies all domain expertise required for automatic candidate implementation selecting. When the OD service is invoked, it dynamically selects one optimal candidate to perform the desired computation.

In a service-body, every candidate implementation is specified independently, detailing the involved executable modules and their dependency. All involved modules are installed on the candidate's underlying host. They are either legacies or specially developed executables for transforming arguments of legacies into standard data objects described by FC descriptors. The service-body also annotates every candidate with a list of criterion-ranges to specify its restriction on resolvable problem characteristics, and associates each with a resource-booker. The resource-booker is an executable responsible for provisioning an execution environment on the underlying host when the candidate is selected, including reserving essential resources such as storage and processors, etc.

3 A Grid Environment for Autonomic Computing

Built on top of OGSA, AOD equips every OD service with mechanisms for autonomously conducting its behaviors on the computational grid, according to the expertise augmented to its underlying legacy applications. With the concept of hyper-resource, it utilizes distributed and heterogeneous legacy applications to provide an autonomic computing grid environment for scientific problems. Its architecture is illustrated in fig. 2, consisting of *repository, scheduler* and *broker*. The repository is responsible for incorporating domain expertise into the computational grid. The scheduler automatically maps computations specified in domain terms to requests for physical resources on the computational grid. The broker deals with resource requests of every submitted computation. Both the scheduler and the broker require the support of grid middleware like Globus Toolkit [5]. AOD requires that both the repository and the scheduler should be specified a host respectively. On every host that provides some resources for grid computing, there exists a local broker instance.

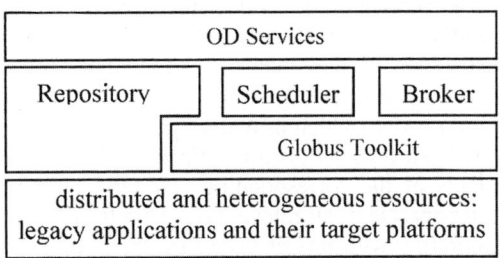

Fig. 2. Illustration of AOD

The repository provides an environment to develop hyper-resources and configure OD services on the computational grid. It interprets every hyper-resource's configuration descriptor, and equips it with the mechanisms to implement its OD services on the computational grid. With the concept of hyper-resource, it supports augmenting legacy applications with domain expertise, aggregating complementary and competitive legacy applications, and encapsulating heterogeneous and distributed legacy applications to be consistent OD services. Every registered hyper-resource's configuration can be retrieved from the repository, in order to query its interface for specifying computation requests.

The scheduler provides an identical and universal entry for users to perform computations on the computational grid. Every job is described in domain terms with an OD service to specify the desired computation. Different jobs are submitted concurrently. For every submitted job, the scheduler dynamically creates a light and temporary environment on local host, according to the specified OD service and its hyper-resource. This environment provides all mechanism required by the OD service for completing the job with the underlying resources and the augmented expertise, including analyzing the job's characteristics, deducing its resource requirements, selecting optimal resources, preparing candidate implementation's executing scheme, interacting with the submitter and remote broker instances, and recovering from resource faults, etc.

For every OD service, its behaviors on the underlying hosts are conducted by the local broker instance. According to domain expertise specified in the OD service's service-body, the instance queries the dynamic status of local resources as required by the scheduler, and implements the OD service when the local host is selected for any submitted job. It is responsible for provisioning local executing environment for the OD service, transfers its arguments on the computational grid, carrying out the executing scheme and monitoring the executing process.

4 Implementation and Experiment

We have implemented a prototype for AOD. In this prototype, every hyper-resource is configured to be a set of XML documents and Grid services conforming to OGSI [6]. Every Grid service implements a candidate implementation of some OD service, associating with an XML document to describe its restriction on resolvable problem characteristics and the methods for monitoring its underlying resources. According to the hyper-resource's configuration descriptor, AOD creates an additional XML document for every OD service, so as to describe expertise specified in the service-body for dynamic candidate implementation selecting, including the declared analyzers, evaluator and inspectors. All documents are stored in the repository.

The prototype is illustrated in fig. 3. Every broker instance is implemented as an OGSI-conforming Grid service, which provides operators for the scheduler to execute any inspector that has been installed locally. When a job is submitted, the scheduler gets relevant documents of the corresponding OD service from the repository and creates a light environment. Next, all relevant documents are passed to the light environment. Then collaborating with broker instances, the light environment selects one candidate and invokes its Grid service to perform the desired computation. The job's arguments are transferred on the computational grid with GridFTP.

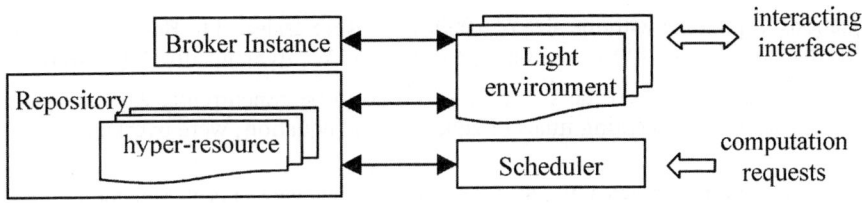

Fig. 3. Implementation of AOD

This prototype provides three XML schemas for its users. The first schema is for domain experts to define FC descriptors. The second one is for developing the configurations of hyper-resources, and the last one is for specifying computing jobs.

4.1 Experiment and Evaluation

In the rest of this section, we present a preliminary experiment result for evaluating both AOD and the prototype. The experiment was performed on a test-bed that consists of five workstations and a 4-node cluster. All of the six hosts are connected to the college network of Peking University, running Linux and Globus Toolkit. One workstation is specified for the scheduler, another is specified for the repository. Every job performed on the test-bed is completed on either one of other three workstations or the cluster. For convenience, we denote these three hosts with the notations W_1, W_2, W_3 respectively, and the cluster is notated as symbol Beo.

Our experiment is to solve Laplace equation on the test-bed and no restriction on the problem size is assumed. There three legacy applications for the problem, installed on W_2, W_3 and Beo respectively. Every legacy reports its execution process by outputting the matrix value every 5-iterations. The legacy on W_2 was developed with C and that on W_3 was developed with F77. Both require that matrix size of an applicable Laplace equation cannot be larger than 500×500. The legacy on Beo was developed with HPF. We have the experience that the speedup achieved by the HPF program will less than 2 when matrix size of the solved Laplace equation is less than 500×500. We also have the experience that the loss of efficiency is very serious when one legacy is employed for more than one Laplace equations at the same time.

To evaluate AOD and the prototype, we developed a hyper resource HR-Lap that provides one OD service. Its arguments are two text files, one for input and one for output. For every argument, the OD service assumes that the first two words are the Laplace equation matrix's storage order in the file and the total of its rows respectively. The three legacies and the above expertise are encapsulated in HR-Lap. For every submitted job, the OD service allows its submitter to set an up-bound for the number of iterations that the job requires. If the execution has not completed after the pre-set iterations have been executed, the OD service will stop the execution and return the submitter a Failure message. We have prepared five Laplace equations for the experiment. Three of them have the size of 300×300 and others have the size of 1000×1000. Every submitted job's arguments are stored on W_1.

Table 1 is the experimental result. It shows that HR-Lap is able to detect every job's size automatically. And it has scheduled all of the three legacies and their platforms as expected. W_2, W_3 and Beo were scheduled to run concurrently for different equations. Equations with the size of 1000×1000 were assigned to Beo, and equations with the size of 300×300 were assigned to W_2 and W_3 dynamically. HR-Lap did not allow two equations with the size of 1000×1000 to be concurrently executed on the test-bed, while the executing times of equation$_3$ and equation$_4$ were overlapped.

Table 1. Experimental Results for Resource Scheduling and Fault Monitoring

	size	I_{pre}	s-time	s-msg	e-time	e-msg	I_{exe}
equation$_1$	1000	500	9:18:17 PM	Beo	9:18:49 PM	Succ	120
equation$_2$	1000	500	9:25:10 PM	Null		Fail	
equation$_1$	1000	100	10:08:01 PM	Beo	10:08:23 PM	Fail	100
equation$_3$	300	400	10:29:40 PM	W_2	10:29:45 PM	Succ	100
equation$_4$	300	400	10:29:42 PM	W_3	10:29:47 PM	Succ	100
equation$_5$	300	400	10:29:44 PM	Null		Fail	

size: the job's size detected by the HR-Lap, retuned to the submitter by the scheduler;
I_{pre}:   the maximal number of iterations that the job's submitter sets;
s-time: the time when the job is submitted;
s-msg: the message returned by the scheduler to notify which host is allocated for the submitted job. A Null is returned if there are no resources available;
e-time: the time when the job's execution is stopped;
e-msg: the message returned by the scheduler when the job's execution is stopped, notifying whether the job is completed. If the job has been completed within less than I_{pre} iterations, the message is *Succ*; otherwise, it is *Fail*;
I_{exe}:   number of iterations that have been executed for the job, returned retuned to the submitter by the scheduler after the job's execution has stopped.

5 Related Works and Conclusion

OGSA is the first effort to standardize Grid functionality and produce a Grid programming model consistent with trends in the commercial sector. It integrates Grid and Web services concepts and technologies. In this architecture, resources are encapsulated to be Grid services with standard interfaces and behaviors. It does not take account of complementary or competitive resources in the computational grid, and no uniform approach is provided to schedule competing Grid services universally for load balance and reliability of grid computing.

H. Shan and L. Oliker have proposed a resource scheduling architecture in [7] for grid computing. The architecture assumes that any host can satisfy a submitted job's resource requirements. It also assumes that the grid middleware is always able to evaluate an approximate waiting time for every submitted job, according to the job's resource requirements and the local system's workload. However, these two assumptions are too strict for scientific computation, especially that most legacies are heterogeneous. ICENI [8] seeks to annotate the programmatic interfaces of Grid services using WEB Ontology Language, and allows syntactically different but semantically equivalent services to be autonomously adapted and substituted.

AOD provides the mechanism for scheduling complementary and competitive resources universally, according to problem characteristics and dynamic resource statuses. It abstracts distributed and heterogeneous legacy applications to be OD services that are adaptive, consistent, self-optimizing and self-healing, so as to improve the dependability and efficiency of grid computing. Another important advantage of AOD is that it is domain-customizable and extensible. For different scientific domain, a customized Grid environment can be constructed by registering domain-specific hyper-resources in the repository. And the customized Grid environment is improved in both computing power and performance, when additional legacies are aggregate, new hyper-resources are registered or new OD services are added.

References

1. Yongen Cai, Tao He and Ren Wang. Numerical Simulation of Dynamic Process of the Tangshan Earthquake by a New Method–LDDA. Pure and Applied Geophysics, Vol. 157(2000) 2083~2104.
2. I. Foster, C. Kesselman, S. Tuecke. The Anatomy of the Grid: Enabling Scalable Virtual Organizations. International J. Supercomputer Applications, 15(3), 2001.
3. I. Foster, C. Kesselman, J. Nick, S. Tuecke. The Physiology of the Grid: An Open Grid Services Architecture for Distributed Systems Integration. Open Grid Service Infrastructure WG, Global Grid Forum, June 22, 2002.
4. The Open Grid Services Architecture. http://forge.gridforum.org/project/ogsa-wg. Mar. 10, 2004.
5. Globus Toolkit. http://www.globus.org/toolkit/default.asp
6. S. Tuecke, K. Czajkowski, I. Foster, J. Frey, S. Graham, C. Kesselman, T. Maguire, T. Sandholm, P. Vanderbilt, D. Snelling. Open Grid Services Infrastructure (OGSI) Version 1.0. Global Grid Forum Draft Recommendation, 6/27/2003.
7. H. Shan, L. Oliker, Job Superscheduler Architecture and Performance in Computational Grid Environments. http://www.sc-conference.org/sc2003/paperpdfs/pap267.pdf
8. hJ. Hau, W. Lee, and Steven Newhouse, Autonomic Service Adaptation using Ontological Annotation. In 4th International Workshop on Grid Computing, Grid 2003, Phoenix, USA, Nov. 2003.

E-Mail Services on Hybrid P2P Networks*

Yue Zhao, Shuigeng Zhou, and Aoying Zhou

Dept. of Computer Sci. and Eng., Fudan University, Shanghai, 200433, China
{zhaoyue,sgzhou,ayzhou}@fudan.edu.cn

Abstract. As one of the most important services of Internet, e-mail is widely used in both commercial business and personal communication. In this paper, we explore ways to develop e-mail systems on P2P networks to overcome the weakness of traditional server-centric e-mail systems. We propose a framework of e-mail system based on hybrid P2P network, and present three implementations of the proposed framework, each of which has different capability, reliability. Our system is more flexible and reliable than the traditional e-mail systems.

1 Introduction

As one of the most important services of Internet, e-mail is widely used in both commercial business and personal communication. The traditional e-mail system employs a typical Client/Server (or simply C/S) design which has some inherent drawbacks. Once the server crashes all data will be lost and is hard to be recovered. Moreover, the rapidly growth of users and services will degrade the performance of e-mail service because of overloading.

Recently, *Peer-to-Peer* (P2P) computing technology is getting more and more popular. In a P2P system, the resources of all peers can be brought together to yield large pools of information, computing power, storage and bandwidth. Thus, the P2P paradigm offers more flexibility, robustness and aggregated resources and computing capability than the traditional centralized systems. By applying P2P technology to e-mail system, we can extend e-mail service to a larger scale with lower cost and better robustness and reliability.

However, to provide e-mail services based on P2P computing, the underlying P2P system need to support a special P2P styled e-mail mechanism with at least the following features: (1) sending and receiving mails in P2P style, instead of through a single centralized mail server; (2) providing a proper protocol to support mail service between users of inside and outside the P2P network, i.e., being compatible to the existing server-centric e-mail protocol.

So far, research on P2P e-mail systems is mostly based on structured P2P systems[1]. In this paper, we develop P2P e-mail system based on *hybrid* P2P network where some functionalities are still centralized. We choose hybrid P2P network [2] because (1) hybrid file-sharing P2P systems are now very popular. Some mature systems or platforms have been established, such as JXTA[3],

---

* The work was supported by the Natural Science Foundation of China under grant No. 60373019 and partially supported by IBM Software Research Award.

OpenNap[4], and (2) hybrid P2P system is a kind of trade-off between structured and unstructured P2P systems (also a kind of trade-off between centralized and pure distributed systems), it is natural and reasonable to take hybrid P2P network as the startup of establishing P2P e-mail systems.

The rest of this paper is organized as follows. Section 2 gives a basic framework of e-mail systems based on hybrid P2P network. Section 3 introduces three implementations of the proposed framework in detail. Section 4 describes the approach to transferring and replicating mails with the network by using trust relationship between nodes. Section 5 presents the system recovery mechanism. Section 6 gives the performance evaluation and Section 7 concludes the paper.

2 A Framework for E-Mail Services on Hybrid P2P Networks

Suppose the campus network of a certain university is a hybrid P2P network. Computers of each department form a community. Nodes within a community are hidden behind a fire-wall and access to the Internet through the local server (usually the super node takes the role of local server) of their own community (for security consideration). In such a network environment, there are three different e-mail services: inter-community service, intra-community service and out-of-P2P service. Inter-community and intra-community services are mail-services between nodes within the P2P environment. The former serves nodes in different communities while the latter serves nodes within the same community. Out-of-P2P service is the mail service between two nodes that one is inside the P2P network and the other is outside the P2P network. For this purpose, there must be at least one node (termed as *broker node* or simply *broker*) within the P2P network that takes the role of traditional mail server to exchange e-mails with mail servers outside the P2P network. All regular nodes within the P2P network can access to the *broker* through their super nodes. Figure 1 shows the framework of e-mail system based on hybrid P2P network.

When sending a mail, the sender first sends a SENDING_REQUEST (including the mail addresses of the sender and the receiver, and a timestamp) to its super node (or called requesting super node). The super node then assigns a mail-ID to uniquely identify the mail and parses the destination address of the sent mail.

- If the destination is outside the P2P network, the super node retrieves the mail body from the sender, wraps the mail according to RFC 2822 and transfers it to the broker that will replay the mail to outside mail server.
- If the destination is an inside P2P node but not in the local community, the super node broadcasts the request to other super nodes. Receiving the request, the super node whose community includes the destination node send a reply to the requesting super node. Then the mail can be sent based on the techniques in section 3.
- If the destination is within the local community, the mail will be sent based on the techniques in section 3.

For the out-of-P2P service, when the broker receives a mail from outside mail server, it transfers the mail to the super node whose community includes the destination node and then deletes the replication of the mail on the broker. The super node then works as sending an intra-community mail.

3 Implementing Schemes

According to the situations of network environments of different capabilities, we present three kind of implementing schemes. Here, our focus is on e-mails transferring inside P2P environment.

3.1 The Pseudo-cooperative Scheme

Suppose each super node is powerful and stable enough to play the role of an application server which can support various services at the same time and has enough storage to keep replications of user data. In such a situation, we use the super nodes as mail servers each of which is responsible for the mail service of its community.

In this scheme, all the information of users and replications of mails are stored on the super nodes. We called this scheme *pseudo-cooperative scheme* because each super node is a centralized mail server and regular nodes do not share their resource to support the mail service. Essentially, this scheme is still a typical C/S model except that multiple mail servers (super nodes) replace the single mail server in traditional e-mail system. Obviously, the new P2P mail system is more reliable and robust because the breakdown of some super nodes will not lead to services' stop of the whole system.

3.2 The Simple Cooperative Scheme

In this scheme, super nodes neither save replications of mails nor transfer mails to destination nodes directly. They only provide look-up service and assign both replicating and sending tasks to regular nodes in their communities.

When sending a mail, the sender sends a SENDING_REQUEST to its super node first. The super node then saves the meta-data which containing the from-to addresses and the identifier of the mail. If the receiver is online at this time, the super node informs the sender of the receiver's IP address with which the sender can build a connection to the receiver and transfers the mail (encrypted version) to it directly.

If the receiver node is not online, the super node uses a hash function $H(s,r)$ to assign k_{min} online nodes to keep the replications of the mail. In the function $H(s,r)$, the parameter s represents the sender's ID and r represents the receiver's ID. k_{min} is the minimal number of replications that will be kept in the community to guarantee the rate of successful sending. When the receiver logs on the community, the super node picks up a proper online node from the replication table and informs it to transfer the mail to the receiver. The picked node is called a *transferrer*. If the transferrer fails to transfer the mail to the receiver successfully, the super node will choose a new transferrer. When the transferrer completes the transferring task and responses the super node with an SEND_OK message, the super node then informs the nodes who have kept the replications of the mail to delete it. If the receiver is in another community, the notification of the receiver's online should be done between the corresponding super nodes.

As described above, for the purpose of transferring mails to the offline nodes, the information that must be kept on the super nodes includes: (1)replication tables, one for each unsent mail, to record the nodes that keep the replications; (2)a sending notification table, to record the nodes which have new mails to receive and the mail identifiers which should be sent; (3)a deleting notification table, to record the currently offline nodes which should delete replications of sent mails on them.

3.3 The Advanced Cooperative Scheme

The simple cooperative scheme does not store mail replications after delivering successfully. In this section, we will present a more powerful mail service scheme with storage capability so that it can be used with no difference from the traditional centralized mail systems.

For each user in a community, the super node will keep an inbox list and an outbox list which store meta data of incoming and outgoing mails respectively. The meta data contains essentially the information of mail's identifier, from-to address, subject and its physical locations where the replications are kept. When a mail is to be sent, there are two cases to handle: (1) For intra-community service, the sending super node uses the hash function $H(s,r)$ to assign k_{min} online regular nodes in its community to store the replications first. (2) For inter-community service, both the sending and the receiving super nodes use the hash function $H(s,r)$ to assign $\lceil k_{min}/2 \rceil$ online nodes in their corresponding communities. Then the information of replications' physical locations will be appended to both the sender's outbox and receiver's inbox. When a user wants to read his mails, either incoming or sent, he accesses his inbox or outbox on the super node. If the mail he wanted is available (at least one of the nodes which keep the mail's replications is online), the super node will direct the user to get a replication to read. However, even the user has connected to the node where the target mail is replicated, he still possibly fails to get the mail because of *misrouting*. Then the super node will direct the user to another node which keeps the replication of the target mail. Particularly, the user will access to the node in other communities to get mails when the mail is not available in its own community. In this case, the super nodes of the two communities will exchange messages to get the location information of the mail replications.

The mail-delete operation is simple. When a user sends a DELETE_REQUEST to the super node, the super node removes the meta data of the mail in the user's inbox or outbox. However, this is only a logical deletion. The replications of the mail are still kept because they are shared by both the receiver and the sender

5 Recovery

In hybrid P2P networks, super nodes are relatively stable. However, the super nodes still possibly get down. Once a super node does not work, all services attached to it such as lookup service, mails routing become unavailable in its community. A feasible way to deal with this situation is to vote another super node. However, the new super node has no information about the e-mail services of that community. So we need a recovery mechanism for our P2P e-mail system.

Recovery from local nodes. When replicating a mail, the node also stores the meta data of the mail, including mail identifier, sender and receiver information, timestamp and unread flag (only for advanced cooperative scheme). When recovering, the new super node gets all these meta data on the regular nodes in its community, rebuilding all necessary information. For simple cooperative scheme, the super node will rebuild the replication tables and the sending notification table mentioned in section 3.2. For advanced cooperative scheme, the super nodes have to rebuild the inbox and outbox information for each node. In addition, the contribution to the e-mail service and the usage of public space of each node will also be recalculated according to its mailbox information.

Recovery from other super nodes. For inter-community mail service, there are information exchanges between super nodes. In simple cooperative scheme, the new super nodes will request other super nodes for the inter-community receivers in their local community so that they can remind these old super nodes to send the mails when the receivers join the P2P network.

Notice that the deleting notification table in simple cooperative scheme cannot be recovered because no one except the downed super node knows which replications should be deleted. This will cause some mails to be transferred twice or even more times. However, this problem does not matter very much. It can be solved on the receiver side by checking the mail identifier before receiving to guarantee that no mail will be received more than one time.

6 Evaluation

We establish a simulator of the advanced cooperative scheme in Java. The hybrid P2P network in our simulation consists of 5 communities each of which contains 20 regular nodes.

6.1 Space Usage

To test the space usage, we randomly choose 5 nodes to send mails periodically (1 minute interval in our experiments). We set $k_{min} = 5$, and the space used to store mail replications is 10 MB on each node at the very beginning. The default size of mailbox for each peer is 2MB. However, the space each user can use is not limited within 2MB. Figure 3 shows the relationship between public space a user used for his/her mailbox and local space he contributes to store replications of other peers' mails.

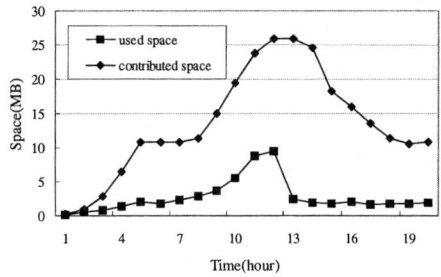
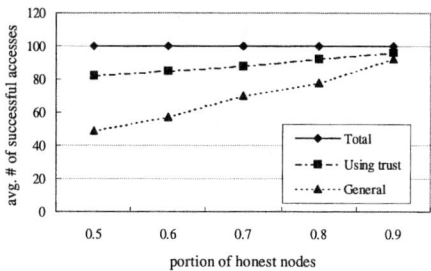

Fig. 3. The Space Used to Replicate Mails ($k_{min} = 5$)

Fig. 4. Performance of Schemes using Trust Relationship

6.2 Effectiveness

To demonstrate the effectiveness of the replicating and transferring mechanism using trust relationship, we compare it with the case using no trust information. We assume there are some malicious nodes that will misroute files (including mails) in P2P environment. We denote r the ratio of honest nodes in the network. Let $p_{down} = 0.2$ and $k_{min} = 5$. Each of the total 100 nodes accesses 100 mails in its mailbox. Figure 4 presents the average number of successful accesses from the first selected transferrers by the super node in networks of different reliability (corresponding to different r).

7 Conclusion

In this paper, we propose a framework of e-mail system based on hybrid P2P network, and present three implementation schemes. We discuss implementing techniques in detail, including mail replicating and transferring. We also give a recovery mechanism for the e-mail system so that the mails will not be lost when a super node is down. Preliminary simulated experiments are carried out to evaluate the performance of the proposed schemes.

References

1. J. Kangasharju, K.W. Ross, and D.A. Turner. Secure and Resilient Peer-to-Peer E-Mail: Design and Implementation. In *Proc. of P2P'03*, 2003.
2. B. Yang, and H. Garcia-Molina. Comparing Hybrid Peer-to-Peer Systems. In *Proc. of VLDB'01*, 2001.
3. JXTA homepage http://www.jxta.org/.
4. OpenNap homepage. http://opennap.sourceforge.net/.
5. S. Marti, P. Ganesan and H. Garcia-Molina SPROUT: P2P Routing with Social Network In *Proc. of IPTPS'04*, 2004.
6. The Gnutella protocol specification, 2000 http://dss.clip2.com/GnutellaProtocol04.pdf.
7. I. Stoica, R. Morris, D. Karger, M.F. Kaashoek, and H. Balakrishnan Chord: A scalable peer-to-peer lookup service for Internet applications In*Proc. ACM SIGCOMM 01*, 2001.

Implementation of P2P Computing in Self-organizing Network Routing

Zupeng Li[1,2], Jianhua Huang[2], and Xiubin Zhao[1]

[1] Telecommunication Engineering Institute, Airforce Engineering University,
NO. 1 Fenggao Road, 710077, Xi'an, Shanxi, P.R. China
Lizp@mail.ndsc.com.cn, Zupengli@hotmail.com
[2] National Digital Switching System Engineering & Technological R&D Center, No. 783
P.O.Box 1001, 450002, Zhengzhou, Henan, P.R.China

Abstract. We find lots of similarities between P2P and self-organizing network, and how to implement P2P technology into self-organizing network is the main object of our research. By building a virtual network topology named P2P overlay network on top of internet's physical topology layer based on P2P computing mode, we can effectively build a full-decentralized internet based self-organizing network routing model- Hierarchical Aggregation Self-organizing Network (HASN). The target and architecture of HASN are described in this paper, as well as a detailed description of a P2P decentralized naming, route discovering and updating algorithm- HASN_Scale. The network simulation results testify that the new routing model can effectively increase the network scalability and routing efficiency by reducing the consumption of network bandwidth.

1 Introduction

How to implement P2P technology into self-organizing network is the main object of our research. During our research, we find that self-organizing network has lots of similarities with P2P network. In a self-organizing network, each node is acting as a router and a common host at the same time, and there doesn't exist such a place like a network administration center. Each node is peer to each other, and the routing protocol is usually full decentralized to keep the network robust and flexible. All of these characters are just the core characters of P2P technology [1].

This paper provides a Hierarchical Aggregation Self-organizing Network (HASN) that efficiently combines structured overlay network based P2P routing technology and self-organizing network routing technology [2]. HASN model is a self-organizing network routing model based on Internet's infrastructure. It aims to construct an efficient, load balanced and dynamic self-organizing network model for WAN. This model lies in application layer. By building a P2P overlay network on top of internet's physical topology layer and implementing P2P routing algorithm on it, we can effectively build a full-decentralized internet based self-organizing network routing model.

2 HASN Model Architecture

In fact, HASN model is made up of two levels of different layers. One is network topology layer (NTL) (see Fig.1); the other is network discovery layer (NDL) (see Fig.2).

2.1 Network Topology Layer (NTL)

There are two basic components in this layer:

- Global Register Server (GRS): Function of GRS is similar to that of root servers in Domain Name System (DNS). And it is a global database, which manages to store and maintain different ranks of IP address of Hierarchical Agent.
- Hierarchical Agent (HA): HA acts as core of each Cluster. Its function is to provide guide service for initializing new node's routing table. There are many HA in each level in HASN model, which may be located at different places in the world.

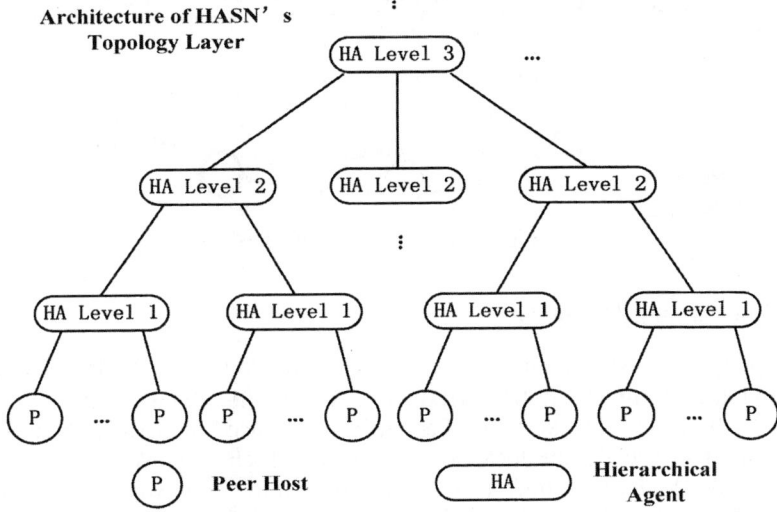

Fig. 1. The architecture of NTL

2.2 Network Discovery Layer (NDL)

By constructing NDL, HASN realizes the routing of discovering messages. According to the different levels of Cluster on NTL depicted in Fig.1, we introduce the concept of cluster level in NDL:

- Cluster Level k (CLk): CL corresponds to different level of clusters in HASN model. For example, HA1 and all nodes it manages are in Cluster Level 0 (CL0). And all HA1 merge into Cluster Level 1 (CL1), etc.
- Full Cluster (FC): FC corresponds to the set of all node and HA in the network.

3 HASN_Scale Routing Algorithm

In a P2P computing mode based self-organizing network, any node may join and leave the network at any time. HASN_Scale is the P2P based routing algorithm implemented in NDL. By adopting DHT based routing algorithm and improving Chord[3], HASN_Scale provides a highly scalable node naming, route discovering and updating mechanism for HASN.

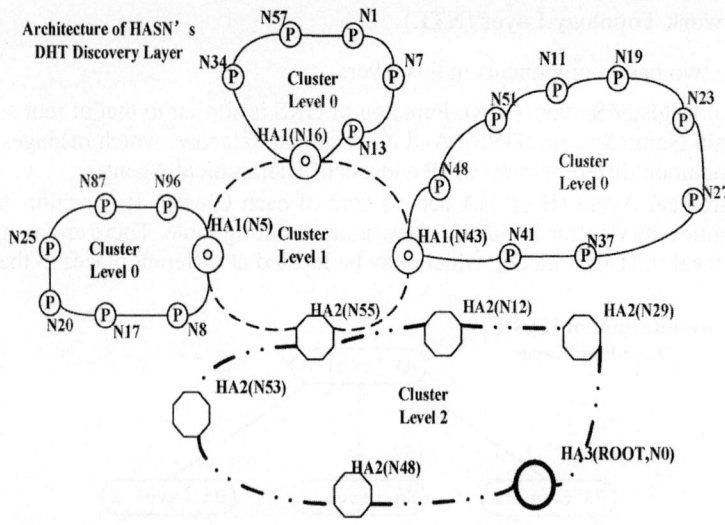

Fig. 2. The architecture of NDL

3.1 Node Naming Mechanism

By adopting a consistent hashing algorithm [4] similar to Chord, HASN assigns each node and key an m-bit identifier (NID) (N ≤ 2^m). While NID is chosen by hashing the node's IP address, the key identifier (KID) is produced by hashing the key. NID can be used to specify the location of node in a hash ring. When a node joins HASN_Scale for the first time, a NID ranges from 0 to (2^m-1) will be automatically assigned to him based on hash algorithm. The value of each m-bit long NID can be computer as follows:

$$Valueof(NID) = \sum_{\Gamma=1}^{M} (NID_\Gamma \times 2^\Gamma) \qquad (1)$$

We introduce the concepts of global ring zone (GRZ) and individual ring zone (IRZ) into HASN_Scale. GRZ corresponds to the whole data scope in $[0, 2^M - 1)$. When a new node joins the system, it will divide the GRS according to its NID. And a node's IRZ corresponds to the private hash ring zone assigned to him. IRZ can be depicted as $(NID\_Predecessor, NID]$, where NID_Predecessor represent the predecessor of current node. As for a network with N nodes, IRZ of each node satisfy the following formula:

$$IRZ_1 \cup IRZ_2 \cup IRZ_3 ... \cup IRZ_N = GRZ \qquad IRZ_i \cap IRZ_j = \varphi \quad i \neq j, 1 \leq i, j \leq N \qquad (2)$$

3.2 Node State

Every node in HASN_Scale keeps H Hierarchical Routing Tables (HRT), where H is the height of NTL. If there are n nodes in cluster i, each node in cluster i will hold a level i HRT with O(logN) entries. HRT is an improvement on Chord's finger table. But since the node naming mechanism and network architecture are different between these two models, the structure of HRT is greatly different from finger table, too. The following concepts are related to a node's HRT in an m-bit namespace.

- Current Cluster Level (CCL): the cluster level of NTL which a node belongs to in current HRT;
- Predecessor: the first node in the counterclockwise direction of hash ring from current node;
- Successor: the first node in the clockwise direction of hash ring from current node;
- HRTEntry[K].start: the starting location of the ring range that the Kth HRT entry covers.

$$HRTEntry[K].start = (n + 2^{k-1}) \bmod 2^m \quad 1 \leq K \leq m \quad (3)$$

- Next Hop Ring Interval (NRI): the ring range that the Kth HRT entry covers.

$$NRI[k] = [HRTEntry[K].start, HRTEntry[K+1].start) \quad 1 \leq K \leq M \quad (4)$$

- Next Hop IRZ (NIRZ): the IRZ of the first active node G in the clockwise direction of NRI.

$$HRTEntry[K].start \in IRZ(G) \quad 1 \leq K \leq M \quad (5)$$

The content of HRT can be depicted as follows:

Current Cluster Level (CCL)	HRTEntry[K].start	Next Hop IRZ (NIRZ)
<CCL i>	<HRTEntry X>	<NIRZ Y>

Fig. 3. Structure of Hierarchical Routing Tables

3.3 HASN_Scale Route Discovering Algorithm

The route discovering algorithm of HASN_Scale can be described as follows:

1. When node A initiates a process of route discovery (or node A receives a transmitted route discovering message), it will first pick up the querying KID from the message and compared KID with its IRZ. If KID belongs to IRZ, a Queryhit message is constructed and returned to the query initiating node, and the route discovering procedure ends here. Otherwise go into step 2.
2. Node A searches for KID in HRT according to its cluster level. In each HRT, A compares KID with HRT entries and searches for NIRZ which is closest to KID and then transmit message to the next hop node in the corresponding HRT entry. By this means, the message is finally transferred to a node whose NID is closest to KID. If the target node is found, a Queryhit message is constructed and returned to the query initiating node, and the route discovering procedure ends here. Else if the query procedure failed in the current level of HRT, the discovering process will go to a higher level of HRT. And this process continues until the target node is found or the highest level of HRT is reached.
3. If the discovering process has already reached the highest level of HRT and the target node is still not found, the route discovering procedure ends here and a failure message is returned.

3.4 Node Join and Departure

As for a routing algorithm for self-organizing network, HASN_Scale must sustain dynamic join and departure of nodes. When a new node joins the network, HASN_Scale will assign an IRZ to him by dividing a member node's IRZ according to his NID. And when a node leaves the network gracefully, HASN_Scale will reclaim his IRZ by combining it with a member node's IRZ.

3.5 Route Updating Mechanism

The route updating process caused by new node's joining the network can be described as follows:

1. The route updating process is triggered by event and started by the new node M. The route updating process works in a manner of retrieval. The retrieval works in a counterclockwise direction and each step of retrieval is 2^{i-1}, i=1,2,...,m. In each step, the route updating process will compare and update the i_{th} entry in HRT.
2. In the i_{th} step of the retrieval, the route updating process will compare the NID of the new node with that of the next hop node in the i_{th} HRT entry of the retrieved node. If the former one is smaller, the next hop node in the i_{th} HRT entry will be replaced by the new node. And then the route updating process continues to walk in the counterclockwise direction on the hash ring until it encounters a node whose i_{th} entry in HRT precedes n.
3. After this, the route updating process will continue to retrieve the (i+1) step node until value of i reaches m.

The route updating process caused by node departure is similar to the process caused by new node's joining the network. But there are two key points of difference, which can be described as follows:

- The route updating process is triggered by the departing node or successor of the departing node.
- The route updating process also works in a manner of retrieval. But in each step of the retrieval, the route updating process will just compare whether the NID of the departing node is equal to that of the next hop node in the ith HRT entry of the retrieved node. If they are equal, the next hop node in the ith HRT entry will be replaced by successor of the departing node.

4 Experiment Evaluation

Under the environment of Intel P4 1.6G CPU and Linux Redhat 8.0, we make our simulating experiment by using NS2 as network simulator [5]. And by OTCL and C++ programming, we realized the HASN and Chord routing algorithm respectively.

1. Average Message Transfer Delay (AMTD)

Average Message Transfer Delay (AMTD) is an important guideline in evaluating the performance of a routing algorithm. The relationship of AMTD in Chord, HASN and optimized HASN with N (number of nodes) is depicted in Fig. 4. As we can see, although AMTD in each routing algorithm demonstrates an O(logN) relationship with N, HASN and optimized HASN's performance is much better than Chord. This is because HASN has taken the network proximity Character into account.

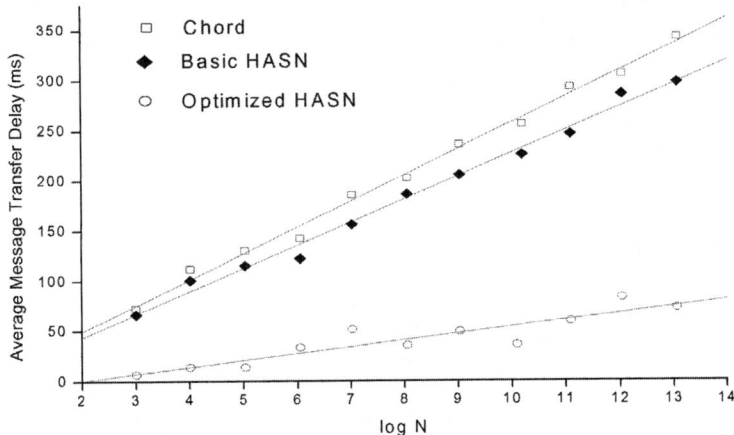

Fig. 4. Relationship of AMTD with N

2. Number of sent messages for node join
Number of sent messages for node join is an important guideline in evaluating the performance of a self-organizing network. The relationship of number of sent messages for node join in Chord and HASN with N is depicted in Fig. 5. As we can see, both routing algorithm demonstrate an $O(\log^2 N)$ relationship with N, and the performance of Chord and HASN is similar to each other.

3. Number of sent messages for node departure
Number of sent messages for node departure is also an important guideline in evaluating the performance of a self-organizing network. The relationship of number of sent messages for node departure in Chord and HASN with N is depicted in Fig. 6. As we can see, both routing algorithm demonstrate an $O(\log^2 N)$ relationship with N, and the performance of Chord and HASN is similar to each other.

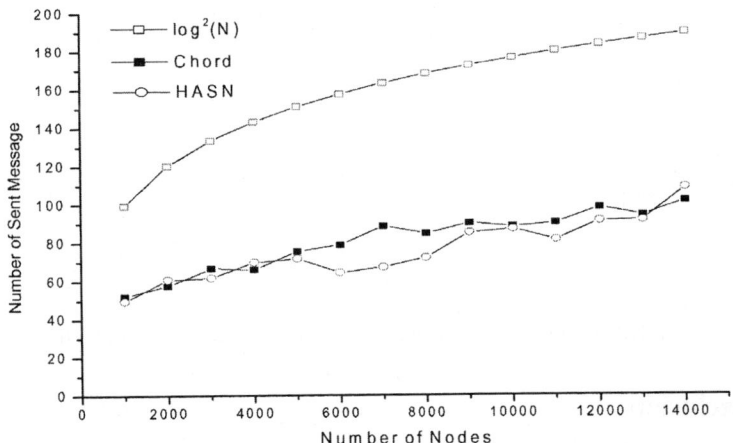

Fig. 5. Relationship of number of sent

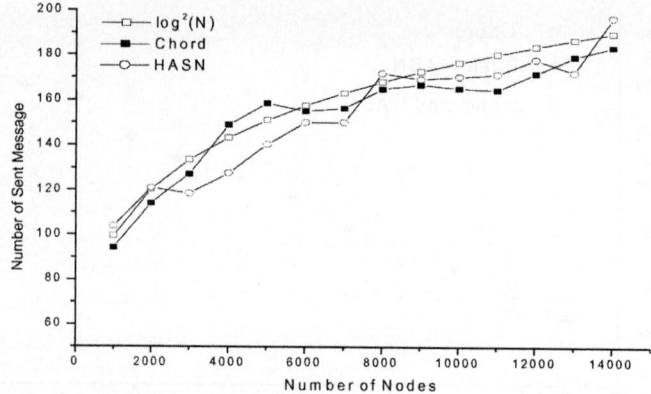

Fig. 6. Relationship of number of sent message for node departure with N

4. Path locality

Another advantage of HASN is that it has utilized the network locality character by introducing data item replication mechanism. By this means, HASN realizes localized placement of data items. Besides improving the routing efficiency, HASN has also guaranteed the network security. Since query message of local data item will not be sent outside of local network, HASN turns localized routing into reality.

The relationship of average number of physical network hops for lookup requests in Chord and HASN with fraction of local lookups is depicted in Fig. 7. Since Chord has not taken network locality character into account, average number of physical network hops for lookup requests will not change while the fraction of local lookups increases. In contrast, average number of physical network hops in HASN will reduce quickly while the fraction of local lookups increases.

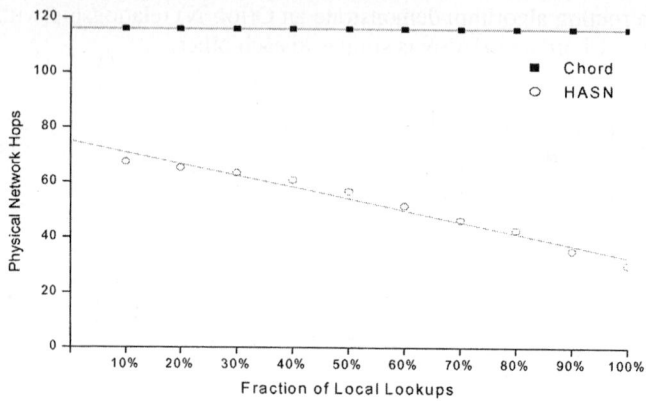

Fig. 7. Relationship of physical network hops with fraction of local lookups

5 Conclusion

By adopting DHT based routing algorithm, it has inherited all advantages of structured P2P overlay network and guaranteed network's scalability. And by introducing cluster grouping mechanism and hierarchical structure, HASN has improved the routing performance by full utilizing Internet's network proximity character and network locality character.

References

1. Manoj Parameswaran, Anjana Susarla, P2P Networking: An Information-Sharing Alternative, IEEE Computing Practices, July, 2001.
2. Sushant Jain, Ratul Mahajan, David Wetherall. A Study of the Performance Potential of DHT-based Overlays. 4th USENIX Symposium on Internet Technologies and Systems (USITS 2003). March 2003.
3. I. Stoica, R. Morris, D. Karger, M. F. Kaashoek, and H. Balakrishnan. Chord: A scalable peer-to-peer lookup service for Internet applications. Annual conference of the Special Interest Group on Data Communication (SIGCOMM 2001). August 2001.
4. Glewin, D. Consistent hashing and random trees: Algorithms for caching in distributed networks. Master's thesis, Department of EECS, MIT, 1998. Available at the MIT Library, http://thesis.mit.edu/.
5. The NS-2 Homepage. http://www.isi.edu/nsnam/ns/.

HP-Chord: A Peer-to-Peer Overlay to Achieve Better Routing Efficiency by Exploiting Heterogeneity and Proximity*

Feng Hong, Minglu Li, Xinda Lu, Jiadi Yu, Yi Wang, and Ying Li

Department of Computer Science & Engineering
Shanghai JiaoTong University, China
{hongfeng,li-ml,lu-xd,yujd,wangsuper,liy}@cs.sjtu.edu.cn

Abstract. Routing efficiency is the critical issue when constructing peer-to-peer overlay. A peer-to-peer overlay, called HP-Chord, is illustrated in this article which aims to achieve better routing efficiency. HP-Chord is built on the basis of Chord overlay, which makes use of the heterogeneity of the bandwidth of the nodes and exploits the proximity of the underlying network. The simulation shows that HP-Chord has achieved better routing efficiency than standard Chord overlay on two aspects, which are lower number of hops and lower RDP per message routing.

1 Introduction

Structured peer-to-peer overlay like Tapestry, Pastry, Content-Addressable Networks(CAN) and Chord provide a substrate for building large-scale distributed applications. These overlays allow applications to locate objects stored in the system in a limited number of overlay hops.

Routing is the core operation of structured peer-to-peer overlay, which can be summarized as two problems: the number of overlay hops and the RDP per message routing. The number of overlay hops calculates the number of intermediate nodes it passes when a message goes through several nodes till reaching its destination. This aspect weighs the logical aspect of routing of a peer-to-peer overlay. RDP is the relative delay penalty, which is defined as ratio of the actual delay before a node receives the message and the unicast delay if the the source sends the message directly to the recipient on the physical network. RDP evaluates the relative speed of message routing on the physical network.

Moreover, observations in [1] have shown that there is great heterogeneity among the bandwidth of the nodes. An open question to peer-to-peer overlay has been brought forward as Question 14 in [2] which is "Can one redesign these routing algorithms to exploit heterogeneity?". At the same time, it is another important issue how to exploit network proximity in the underlying Internet [3].

---

* This research is supported by National Basic Research Program of China(grant No.2002CB312002), ChinaGrid Program of MOE of China and Grand Project of the Science and Technology Commission of Shanghai Municipality(grant No.03dz15027)

There is little consideration of proximity in standard Chord overlay. Another question is brought forward how to exploit proximity in Chord to achieve better routing efficiency.

HP-Chord is the answer to both question above, which is built on Chord, but achieves better routing efficiency by exploiting heterogeneity and proximity. The core design can be concluded as follows:

1) HP-Chord makes use of the heterogeneity of the bandwidth of the nodes to decrease the number of overlay hops. The idea of virtual nodes [4] has been expanded here. The number of virtual nodes of a real node is built proportion to the bandwidth of the real node.

2) HP-Chord exploits the proximity by holding a list of proximate real nodes, which describes the proximity relationship between nodes of the underlying network. RDP is decreased with the help of the proximity list.

3) Special routing algorithm has been introduced into HP-Chord, which can make full use of the virtual nodes and the proximity list to achieve both low overlay hops and low RDP.

The rest of the paper is organized as follows. Section 2 illustrates the core design of HP-Chord comparing to Chord overlay. A simulation is given out in section 3, which shows the result of routing efficiency of HP-Chord. Conclusion is discussed in Section 4.

2   Core Design of HP-Chord

2.1   Preliminary

Chord [5] is one of the typical DHT peer-to-peer overlay. Chord uses a one-dimensional circular key space. The node responsible for the key is the node whose identifier most closely follows the key (numerically); that node is called the key's *successor*. Routing correctness is achieved with the pointer *successor*. Routing efficiency is achieved with the *finger list* of $O(logN)$ nodes spaced exponentially around the key space, where N is the number of nodes in Chord. Routing consists of message forwarding to the node which is closest to the key and not past the key; routing path lengths are $O(logN)$ hops. The routing algorithm of Chord is described in Fig.1.

HP-Chord's modifications of Chord are essentially modifications of the routing algorithm, i.e. HP-Chord inherits unchanged Chord's *successor* and *finger list* to use in HP-Chord's routing algorithm. And Chord's stabilization operation is inherited directly in HP-Chord to handle the churn of nodes. For the remainder of the paper, we ignore basic issues of maintenance of such lists and the stabilization in the following description of HP-Chord, but focus on the different routing algorithm of HP-Chord.

2.2   Virtual Nodes in HP-Chord

The idea of virtual nodes is used in HP-Chord to exploit full capability of the bandwidth of the nodes taking part into the system. When a real node joins

```
//ask node n to find the successor of id          //search the finger table to find
//successor is the node responsible for id        //the finger node closet preceding to id
n.find_successor(id)                              //m=2log(n)
    if(id ∈ (n,successor))                        n.closest_preceding_node(id)
        return successor;                             for i=m downto 1
    else                                                  if(finger(i) ∈ (n, id))
        n'= closet_proceding_node(id);                        return finger[i];
        return n'.find_successor(id);                 return n.successor;
```

Fig. 1. The pseudocode of routing algorithm in Chord

HP-Chord, a set of virtual nodes is created according to the bandwidth that the node has. The node of lowest bandwidth in the system is defined as holding only one virtual node. The number of the virtual nodes that other real node holds can be defined as the ratio of the bandwidth of the node and the lowest bandwidth. As a sequence, a node with higher bandwidth will hold more virtual nodes, take part in more message routing process and have more routing information for message delivering. Therefore the heterogeneity of the bandwidth of the nodes has been evaluated quantitatively and can be exploited in the routing process.

2.3 Proximity List in HP-Chord

A proximity list of real nodes is added to the real node to measure the underlying network proximity. An entry of the list contains the IP of the proximate real node and a list of identifiers of the virtual nodes which the proximate real node holds. Proximity is weighed by the ping value which can be easily got when communications happened between two real nodes. And the proximity list is created and modified when a node finds another node near itself by ping value. And stabilization of the proximity list is just to make sure whether the node in the list is still in the system by probing directly. Therefore it doesn't need extra operation to create and maintain this list.

Proximity list is used to evaluate the topology of the underlying network. When the number of proximity list is the same as the number of the nodes in the network cluster, the overlay will achieve its best performance. So if the number of nodes in the cluster of network can be measured beforehand, the number of entries in the proximity list should be adopted to the number of nodes in the network cluster. And the proximity list can choose $log(N)$ entries if the number of the nodes in network cluster cannot be got beforehand. N is the number of real nodes in HP-Chord which can be forecasted.

2.4 Routing Algorithm

As heterogeneity of the bandwidth has been described by the number of virtual nodes, and the underly network topology information has been evaluated by the proximity list, routing efficiency can be improved by exploiting such information.

Special routing algorithm is brought forward in HP-Chord, which can make use of such information.

Content of Routing Algorithm. The pseudocode of routing algorithm is illustrated in Fig.2.

```
//ask virtual node v to find
// the virtual node responsible for id
v.find_successor(id)
    r=v.get_realnode();
    successor= r.find();
    if(successor!=null)
        return successor;
    else
        next=r.cloest_preceding_node(id);
        return next.find_successor(id);
```

Fig. 2. The pseudocode of routing algorithm in HP-Chord

Virtual node's $v.find\_successor(id)$ is the basic function of routing algorithm, which is the beginning of the routing process and is called on every hop of the routing path. Comparing with $n.find\_successor(id)$ in Fig.1, it can be found that $v.find\_successor(id)$ has been expanded to ask help from real node r which the virtual node v belongs for routing information.

Real node's $r.find(id)$ function is to check whether the other virtual node on r holds the item id or whether id falls in the space the real node's virtual node and its successor responsible for. If true, return that virtual node or the virtual node's successor.

Real node's $n.cloest\_preceding\_node(id)$ is to find the virtual node closest preceding to id in all virtual nodes' *finger* lists and all virtual nodes' identifier in the proximity list which the real node holds. Therefore, next hop of routing can not only be a larger jump towards the target than the result $n.cloest\_preceding\_node(id)$ in Chord, but also be decided according to the network topology which eliminates the possibility of routing message coming back to the same cluster when the number of entries in the proximity list is the same as the number of nodes in the network cluster. So it can be concluded that the next hop is chosen as local optimum which will definitely increase routing efficiency.

Routing Procedure. A whole routing procedure can be explained as follows: a virtual node v try to find the virtual node responsible for item id. v first ask the real node r which it belongs to check whether id is held by another virtual node in r. And r will also check all its virtual nodes whether id falls in the space between the virtual node and its successor. If still failed, r will choose the virtual node closest preceding to *id* from the *finger* lists of all its virtual nodes and from all virtual nodes in its proximity list. The virtual node chosen is denoted as *next* which is the next hop of routing. Therefore virtual node *next* receives the call from virtual node v to find the virtual node responsible for item *id*. Virtual node *next* will deal with the same routing operation as v does. Then this process can be carried on iteratively till the virtual node responsible for *id* is found to check whether item *id* is held by it or not.

Routing Example. The key operation of routing algorithm in HP-Chord is the choose of the next hop which exploits the information of all virtual nodes and the proximity list to achieve better routing efficiency. An example of HP-Chord routing is illustrated in Fig.3.

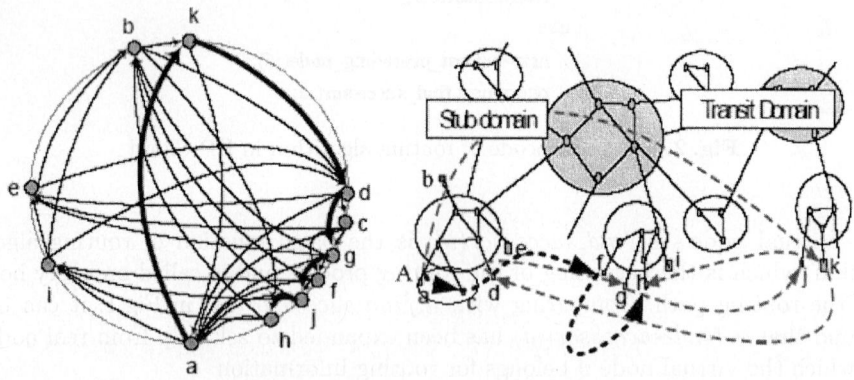

Fig. 3. An example of HP-Chord routing

Fig.3 describes a HP-Chord overlay with 10 real nodes(A, c, d,......, i) which holds 11 virtual nodes(a, b, c, ..., i), for real node A holds two virtual nodes a,c. The left part of Fig.3 shows the relationship of the virtual nodes, a line here denotes an entry in the virtual node's *finger* list. The right part of Fig.3 shows the topology of the nodes on the network. The topology is generated from Transit-Stub Internet Model [6] with two Transit domains and six Stub domains.

A routing example is illustrated in Fig.3, which is virtual node a trying to locate the virtual node h or the item with identifier h. For standard Chord overlay, the hops of routing process is the emphasized arrowhead line in the left part and the thinner arrowhead line in the right part of Fig.3, which are a,k,d,g,j,h of 5 hops. For HP-Chord, as virtual node a is held by real node A, the routing information can be got from A. A holds two virtual nodes a,c, then the closest preceding virtual node can be got as f, which is the *finger* of virtual node c. For f is more proximate to h than all other *fingers* of both a and c, and all the virtual nodes in the proximity list of a which are b,d,e. Virtual node f is chosen as the next hop of routing, which will ask the real node f to help in message routing. As h is in the proximity list of f, f will send message to h directly which ends the routing. This routing process is only 2 hops: a,f,h.

If the latency of message delivering on Transit-Stub Internet model is defined, the RDP of two routing process can be calculated. The latency of link in Stub domain is defined as 1, the one between Stub domain and Transit domain is defined as 10, the one in Transit domain is defined as 50, and the one between Transit domains is defined as 100. Then the RDP of Chord routing above is 8.2358, while the RDP of HP-Chord routing is 1.0244.

The example shows the routing path passes the same Stub domain only once. Here the unit of network cluster is Stub domain. And the routing path stretches on directly to the target in identifier space which eliminates a certain number of hops of jumping between different Stub domains. These two aspects are the key issue of improving routing efficiency.

3 Simulation

3.1 Preliminary

The simulation of HP-Chord is built with the help of GT-ITM [7] and p2psim [8]. In the simulation there are 160 real nodes with different bandwidth which is organized in Zipf distribution [9]. In Zipf distribution the bandwidth B_i of the nodes is proportion to $i^{-\beta}$, where i is the number of nodes with bandwidth B_i, i.e. $B_i = Z * i^\beta$. Z is chosen as 112 and β is chosen as 1.2 in the simulation. Fig.4 shows the distribution of the real nodes. So it can be calculated that there is total 512 virtual nodes in the simulation system. Then the number of the nodes in standard Chord overlay is chosen as 512 accordingly in the following comparison.

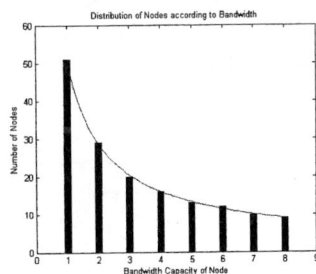

Fig. 4. Distribution of real nodes of different bandwidth capacity

The topology of the simulation is generated from GT-ITM, which can be described by the following parameters:
  T: number of Transit domain
  N_t: average number of transit nodes in each Transit domain
  S: average number of Stub domain per transit node

N_s: average number of nodes in each Stub domain
P_T: probability of edge between Transit nodes in different Transit domains
P_t: probability of edge between Transit nodes in same Transit domain
P_s: probability of edge between nodes in same Stub domain
L_T: latency of link between Transit domains
L_t: latency of link in Transit domain
L_{ts}: latency of link between Transit domain and Stub domain
L_s: latency of link in Stub domain

Table 1. Parameter/value of the simulation network topology

Parameter	T	N_t	S	N_s	P_T	P_t	P_s	L_T	L_t	L_{ts}	L_s
Value	2	4	2	10	1.0	0.6	0.42	100	50	10	1

The value of each parameter is defined in Table 1. The 160 real nodes are randomly distributed into the network topology to form the simulation environment. The number of entries in the proximity list is chosen as 9 which is the average number of nodes per Stub domain except the node itself.

3.2 Result

Fig. 5 shows the probability density function(PDF) of the number of hops per message routing. It can be found that HP-Chord decreases the number of hops greatly comparing to standard Chord overlay.

Fig.6 shows the PDF of $log_2 RPD$ per message routing. The definition of x axis illustrates that the lower of the curve when the x value get larger will increase routing efficiency greatly. It can be found that HP-Chord decreases RDR greatly comparing to standard Chord overlay.

Table 2 shows the comparison of routing efficiency between HP-Chord and Chord quantitatively to give a whole image of performance increasing in HP-

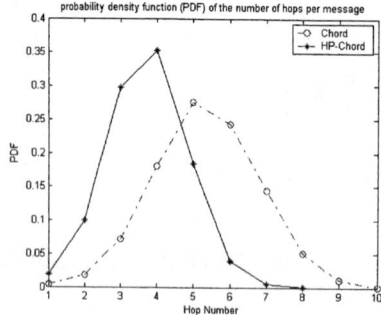

Fig. 5. PDF of the number of hops per message

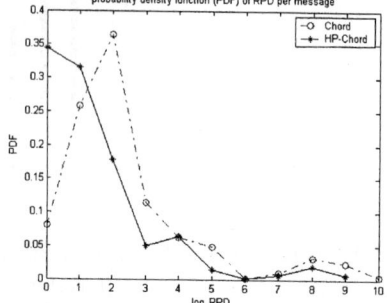

Fig. 6. PDF of $log_2 RPD$ per message

Table 2. Quantitative comparison between Chord and HP-Chord

Overlay Type	average hop number	average RDP
Chord	5.3354	33.748
HP-Chord	3.7269	18.998

Chord. It can be concluded from Table 2 that HP-Chord's routing efficiency increases 43% in the number of overlay hops and 68% in RDP per message routing in the simulation.

4 Conclusion

HP-Chord is a peer-to-peer overlay constructed on Chord aiming to achieve better routing efficiency. The main idea of HP-Chord is to make full use of the heterogeneity of node's bandwidth and the proximity relationship of the nodes of the underly network topology. The simulation result has shown that HP-Chord achieves better routing efficiency than standard Chord overlay.

References

1. S. Saroiu, P. K. Gummadi, and S. D. Gribble: A measurement study of peer-to-peer file sharing systems. In Proceedings of Multimedia Conferencing and Networking(2002)
2. S. Ratnasamy, S. Shenker, and I. Stoica: Routing algorithms for dhts: Some open questions. In First International Workshop on Peer-to-Peer Systems (IPTPS'02)(2002)
3. M. Castro, P. Druschel, Y. C. Hu and A. Rowstron: Exploiting network proximity in peer-to-peer overlay networks, Technical report MSR-TR-2002-82(2002)
4. D. Karger, E. Lehman, T. Leighton, M. Levine, D. Lewin, and R. Panigrahy: Consistent hashing and random trees: Distributed caching protocols for relieving hot spots on the world wide web. In Proceedings of the 29th Annual ACM Symposium on Theory of Computing(1997)
5. I. Stoica, R. Morris, D. Karger, M. F. Kaashoek, and H. Balakrishnan: Chord: A scalable peer-to-peer lookup service for internet, IEEE/ACM Transactions on Networking, Vol. 11, No. 1,(2003)17-32
6. E. w., Zegura, K. Calvert. , and S. Bhattacharjee: How to model an Internetwork. In Proceedings of IEEE INFOCOM (1996).
7. GT-ITM,http://www.cc.gatech.edu/projects/gtitm/
8. p2psim,http://www.pdos.lcs.mit.edu/p2psim/
9. Z. Zhang, S.M. Shi and J. Zhu: Self-Balanced Expressway: When Marxism Meets Confucian. Technical report MSRTR-2002-72(2002)

MONSTER: A Media-on-Demand Servicing System Based on P2P Networks

Yu Chen and Qionghai Dai

Broadband Network & Multimedia Research Center, Graduate school in Shenzhen, Tsinghua University, Shenzhen, Guangdong, China, 518055
{cheny,daiqh}@mail.sz.tsinghua.edu.cn

Abstract. This paper presents MONSTER, a MoD servicing system based on P2P networks. In MONSTER, the media server divides each clip into several segments; the peer hosts caching same segment consist of a cluster, and several clusters work together to transmit a whole clip to other peers. Sharing the peer hosts' resource makes the media server not the bottleneck of MoD servicing system, and system can accommodate more users concurrently. This paper introduces the model of MONSTER, and presents the clip caching strategy and cluster scheduling algorithm. The simulation results show that, with enough peer hosts and the caching time, the capacity of MONSTER is much better than the conventional MoD systems.

1 Introduction

Streaming Media on-Demand (MoD) over Internet is becoming one of the main Internet services, but the limitations of network and server resource make MoD servicing systems low capacity, low QoS and high cost. Next Generation Network will provide sufficient network resource and abundant QoS support for MoD service, and the capacity of media server becomes the bottleneck of MoD service.

This paper presents MONSTER, a MoD servicing system based on P2P networks. In MONSTER, the media server divides each clip into several segments; the peer hosts caching same segment consist of a cluster, and several clusters work together to transmit a whole clip to users. Sharing the peer hosts' resource makes the media server not the bottleneck of MoD servicing system, and system can accommodate more concurrent users. Section 2 introduces the model of MONSTER. Section 3 and 4 give out host caching strategy and host cluster scheduling algorithm, respectively. Section 5 simulates and evaluates the performance of MONSTER. Section 6 introduces some related works. In the last section, we conclude our works.

2 System Model

A basic MONSTER system is composed of a WEB server, a streaming media server and hosts that accessing clips storing on media server. This section gives out the concept model of MONSTER, and uses Finite State Machine (FSM) to introduce a host' behaviors.

2.1 Concept Model

Figure 1 gives out the concept model of MONSTER. Media server divides clip f into n segments. All hosts that access clip f are in the dotted rectangle. The host that can receive, cache, and forward data at same time is called peer host. Peer host cache one segment at most. The hosts caching the first segment of *f* consist of *cluster_1*, and the hosts caching the second segment of *f* consist of *cluster_2*, etc. The host that only receives data is called selfish host. Host sends requests to WEB server, and WED server redirects the request to media server. Media server maintains the host clusters' member information, and sends those information to host. Host tries to get media data from a member in host cluster firstly. If can't get service from a member, host gets media data from media server.

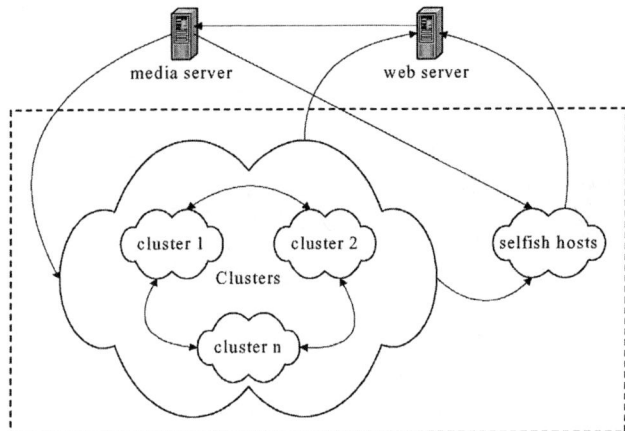

Fig. 1. Concept Model of MONSTER

Peer host will join in a host cluster after caching a segment. Web server is the rendezvous of host tree, which redirect host's request to media server. Wed server maintain and issue the basic information of the clips. The model in figure 1 is two-level (server level and host level), and is designed for a large autonomous network. This model can be scaled a three-level structure, the highest level is a central media library that stores all clips, the second level is composed of streaming media servers and cache servers that are deployed in autonomous networks and the third level includes the hosts that access clips.

2.2 Host's Finite State Machine

The potential states of a host is given below,
    init - initial state, hasn't sent request
    req - send request, hasn't connected with parent node
    recv - receive, playback and cache (for peer host) at same time
    serv - finish playback, work as a parent node, only for peer host
    recv_serv - receive, playback and forward at same time, only for peer host
    serv_end - only connect with current descendant, doesn't accept new request
    temp_cons - uplink failure temporarily, playback with playback cache
    temp_serv - uplink failure temporarily, cache a segment and provide service
    host_err - host failure permanently
    link_err - uplink failure permanently

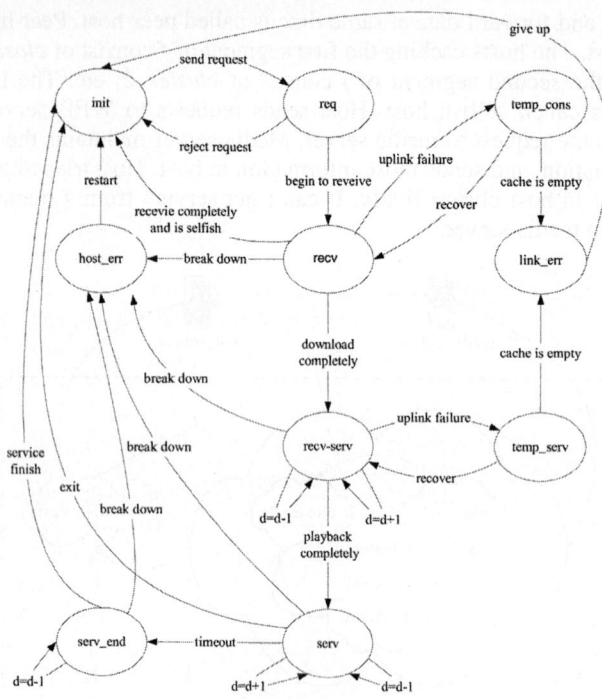

Fig. 2. The FSM of a host

Figure 2 gives out a host's finite state machine. In fact, host in *init* isn't a member of MONSTER, and host becomes a member after entering *recv*. Like conventional system, each host has playback cache, which store the data that will playback soon. Playback cache is used to keep playback smoothly when host enter *temp_cons* or *temp_serv*. Besides playback cache, peer host has cache one of some clip's segments. Compared with playback cache, the duration of a segment is much longer. So, peer host caches the segment on hard disk as a temporary file. Selfish host will change from *recv* to *init* after finishing playback. Peer host download a segment from parent node in *recv*. After download, peer host joins in the segment's host cluster and enter *recv_serv*. In *recv_serv*, peer host can playback and forwarding at same time. At the end of playback, peer host enter *serv*. When peer host is in *recv_serv* or *serv*, d is number of current descendants.

The interval that peer host stay in *recv_serv* is called peer host's life span. At the end of life span, peer host enter *serv_end*, and doesn't accept any requests, but continue providing service to current descendants. When $d=0$, peer host has no descendant, and return to *init*. Besides this, the owner of peer host can execute *exit* to make peer host return to *init* before the end of life span.

Host failure and link failure can make host enter error state. Peer host's failure changes itself from *recv*, *recv_serv*, *serv*, or *serv_end* to *host_err*. When in *serv* or *serv_end*, peer host doesn't need the support from parent nodes any more, so doesn't care about the failure of parent nodes and uplink. When in *recv* or *recv_serv*, peer host can't distinguish the difference between failure of parent node and failure of uplink. So, if can't communicate with parent node, peer host enters *temp_cons* or *temp_serv*, and use playback cache to keep playback smoothly in a short time. Peer host must connect with original parent node or create a link with new parent node before depleting playback cache. Otherwise, peer host will enter *link_err* after depleting playback cache.

3 Host Caching Strategy

Caching a segment is the foundation of providing service to other hosts. Media server has stored many clips. Each clip is divided into several segments, and each segment maps a host cluster. Dividing a clip with longer segments can reduce the number of parent nodes host connect to get a whole clip, but the connect time with each parent node is extended. Due to the unpredictableness of parent host, the longer connect time will increase the unpredictableness of the QoS, and increase the overhead of parent node. So, dividing a clip should pursue the balance between host's overhead, QoS, and parent node's overload. Besides this, a smaller segment can shorten the download time, and make peer host join host cluster earlier.

Media server specifies which segment peer host should cache. The clips' access probabilities are Zipf distribution. In order to improve the utilization of peer host's shared resource, peer host only caches the segment that belongs to the first K clips with highest accessing probability. Media server divides [0,1] into K intervals, and the width of interval i is mapped to the accessing probability of clip i. When peer host enter *recv*, media server creates a random number ε, $0 \leq \varepsilon \leq 1$, and specified the clip ID of segment that peer host will cache according to ε. Normally, the clip peer host playback isn't the clip peer host cache, so peer host creates two uplinks with two parent nodes to playback and download at same time.

Media server specify segment ID for peer host, too. The accessing probability of clip's head is higher than the accessing probability of clip's tail, so the number of peer hosts that cache clip's beginning segment should be more than other segments.

Media server uses a three-dimension list to maintain host cluster. X dimension is clip ID, and y dimension is segment ID, and z dimension is host ID. Let $i \in x$, $j \in y$, $k \in z$. The peer host caching the jth segment of the ith clip is a member of C_{ij}, and h_{ijk} is register information of the kth member of C_{ij}. The register information of peer host includes clip ID, segment ID, host address, join time, and its life span. If peer host fail silently, media server will delete its registration information at the end of its life span. Due to the space limit, this paper wouldn't give out the algorithms host caching strategy in detail.

4 Host Cluster Scheduling Algorithm

In a server cluster, a dedicate scheduler chooses the server with lowest workload to provide service to new arrival request. The merits that scheduler using to choose server include throughput, number of links, and response time, etc. Scheduler must keep track of the servers' state real time to make correct choice. In MONSTER, keeping track of the state of peer hosts real time and choosing parent node for each host increase media server's overload heavily, and will counterbalance the benefit of using peer host to reduce server's overload. So, in MONSTER, host itself gets cluster information from media server, and makes its decision by getting state information from each member in the cluster.

Host sends a *send_request* message to media server, and the return value (address of members in host cluster) is store in array *peer_host*. Then, host uses *get_status* message to get peer hosts' state. Host calculates each peer host's factor, and choose peer host with largest factor as parent node. Host connect parent node with *connect()*, if return is true, host get service from this node, otherwise connect the node with largest factor in the left peer hosts. If no peer host can provide service, host gets service from media server directly.

In order to control the overload caused by forwarding segment, peer host specify *num*, the number of descendants it can support concurrently. Peer host uses *cur_num* to record the number of current descendants, and *free_num* to record free links. It's clearly that, *num* is the sum of *cur_num* and *free_num*. When receiving get_status message, peer host responses with its *free_num*.

The other parameter related with peer host's factor is the round trip time (rtt) between sender and peer host. A smaller rtt means the distance between sender and peer host closer. Let host send request message to n peer host, and get $free\_num_i$ and rtt_i of peer host i, $i \in \{1,\cdots,n\}$. Then the factor of peer host i is calculated as below,

$$factor_i = a_1 (free\_num_i / \Sigma^n_{i=1} free\_num_i) - a_2 (rtt_i / \Sigma^n_{i=1} rtt_i) \qquad (1)$$

In equation 1, a_1 and a_2 are parameters, and $a_1+a_2=1$, $0 \leq a_1 \leq 1$, $0 \leq a_2 \leq 1$. The first item in equation 1 calculates peer host's *free_num* ratio, and the larger the ratio is, the higher the probability of peer host accepting sender is. The second item in equation 1 calculates peer host's *rtt* ratio, and the larger the ratio is, the farther the distance between peer host and sender is. The second item has side effect on peer host's factor, so its sign is negative. Function 1 uses parameters to control the effect of *free_num* and *rtt* on peer host's factor.

Host has to connect with several parent nodes to playback clip completely. To keep playback smoothly, host get the members' register information of next cluster form media server and choose new parent node before the end of playbacking current segment. If host itself cache next segment, host needn't require media server provide cluster information.

When parent host or uplink is failed, host should choose new parent node. Because the state of members in the host cluster changes during host playback, host should get the update information from media server before choosing new parent node.

5 Simulation and Evaluation

In this section, we simulate a MONSTER servicing system on NS2, and evaluate its performance. There are 200 CBR video clips with access probabilities drawn from a Zipf distribution with parameter $\theta = 0.271$ on media server. The bit rate and duration of each clip is 450Kbps and 360 time units, respectively. There are 1850 hosts in this autonomous system. The arrival time of requests is Poisson distribution, and the mean time interval between two consecutive requests is 1 time unit. Media server begins to work at 0, and the first request arrives at 0.5.

Figure 3(a) and 3(b) give out the number of rejected requests and the workload of media server under different peer host ratio. When there is no peer host, MONSTER works as a normal MoD servicing system. When media server hasn't free channel, all requests are rejected. The number of rejected requests reduces along with the increase of the number of peer hosts. At the same time, the number of free channel increases.

When the ratio of peer hosts is up to 80%, the performance of MONSTER improves remarkably. Only 22 requests rejected in [290, 420], and 1828 hosts playback correctly, and there are more that 60 free channels.

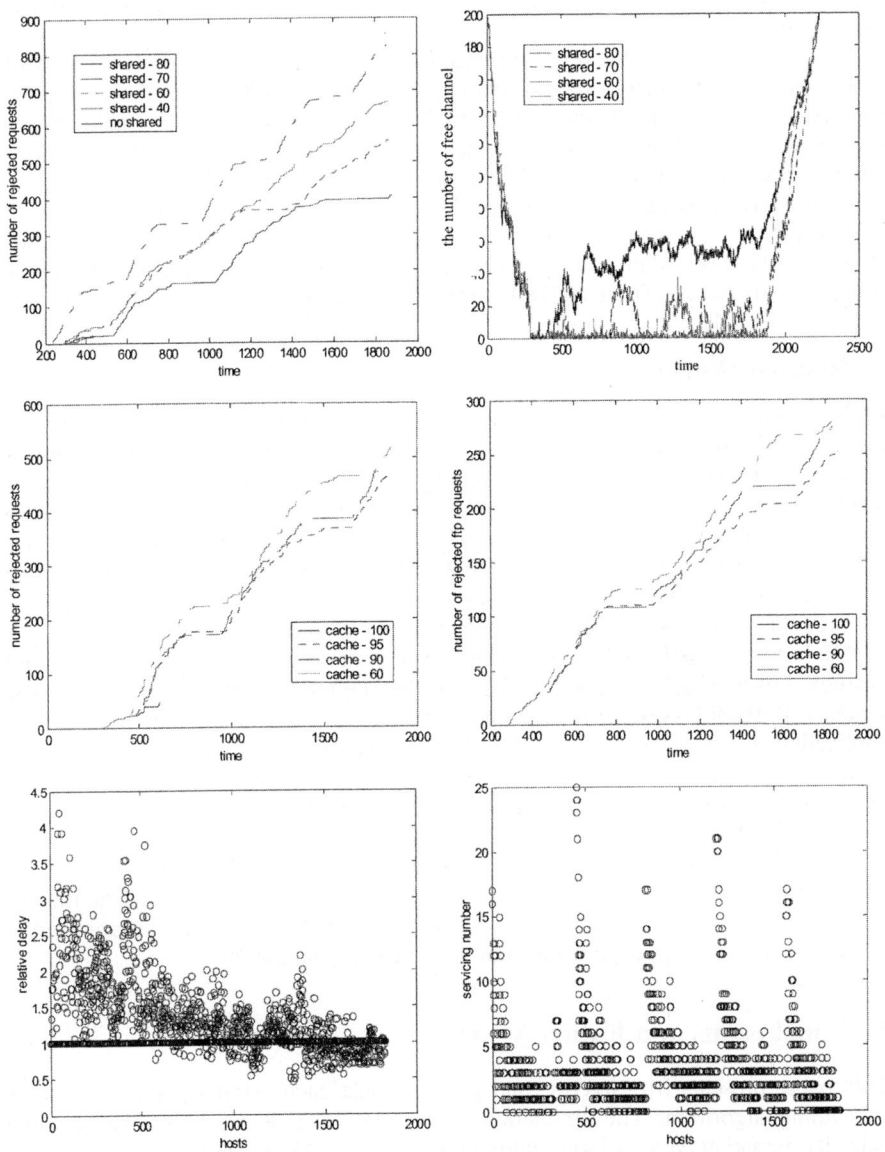

Fig. 3. The result of simulation

Figure 3(c) and 3(d) give out the number of rejected playback requests and the number of rejected download requests under different life span. The number of rejected requests increases along with the shorter life span. At the same time, the cumulative number of rejected download requests increase significantly. The shorter life span of peer host reduces the number of descendants peer host can support, and more and more hosts have to get service from media server directly. Besides this, download requests competing resource with playback requests makes the lack of system resource much seriously.

Figure 3(e) gives out the relative servicing delay. 800 hosts get service from MoD server directly, and their relative servicing delay is 1. In 1000 hosts that get service from one peer host at least, about 300 hosts' relative servicing delay is smaller than 1. Less than 150 hosts' relative servicing delay is larger that 2, in which, about 10 hosts' relative servicing delay is beyond 3, and only one is beyond 4.

Figure 3(f) gives out the number of descendants peer host connecting with in its life span. In this simulation, about 1450 hosts register as peer hosts, in which, less than 20 hosts' descendants number is beyond 20, about 80 hosts' descendants number is between 10 and 20. There are 180 peer hosts that never provide any service.

6 Related Works

IP multicast improves server and networks' resource utilization remarkably, but hasn't been deploy wildly because of the problem of management, dependability, security, and servicing time, etc [1]. Application layer multicast makes hosts to forward data they receive to other host, and needn't dedicated support provided by lower layer. So, its flexibility, adaptation, and dependability are better, and control overhead is lower [2]. Typical application layer multicast protocols include Narada [3], Gossamer [4], Yoid [5] and NICE [6], etc. Based on NICE, Tran, et al, develop Zigzag [7] for streaming media broadcast with lower control overhead.

P2P has been used wildly in file transferring systems and instant message communication, but its usage in MoD system is at the beginning stage. Hefeeda, et al, presented a P2P MoD system [8]. In this system, host cache some data, and become application server at the end of playback. Root server chooses application server for host, which increase root server's overhead significantly. AllCast is a streaming media broadcasting system produce by Pipe Dream [9]. Hosts receive and forward data at same time, but don't cache data for forwarding later. P2Cast is VoD servicing system based on patching [10]. The hosts access same clip construct a multicast tree, and hosts receive and forward data at same time. Because hosts don't cache data, the host join the tree later have to get the data that have transmit on the tree from root server, so root server has to manage many additional patching streaming.

7 Conclusions and Future Works

This paper presents MONSTER, and give out host cache strategy and host cluster scheduling algorithm. The simulation shows, MONSTER improves the servicing capacity remarkably in a large autonomous system. MONSTER needs more peer hosts with longer life span to share their resource with other hosts. So, MONSTER should develop a rewarding strategy to encourage host work as peer. The other works in the future is optimizing cache strategy and cluster scheduling algorithm.

References

1. C. Diot, B. N. Levine and B. Lyles, et al, Deployment Issues for the IP Multicast Service and Architecture. IEEE Network, Jan 2000.
2. J. Jannotti, D. K. Gifford and K. L. Johnson, et al, Overcast: Reliable multicasting with an overlay network. In Proceedings of the 4th Symposium on Operating System Design and Implementation, October 2000, pp. 197--212.
3. Y. Chu, S. Rao, and H. Zhang. A Case For End System Multicast. In Proc. ACM Sigmetrics, June 2000.
4. Y. Chawathe, S. McCanne and E. Brewer, An architecture for internet content distribution as an infrastructure service. http://www.cs.berkeley.edu/ yatin/papers, 2000.
5. P. Francis, Yoid: Extending the internet multicast architecture. Unpublished paper, http://www.aciri.org/yoid/docs/index.html, Apr. 2000.
6. S. Banerjee, B. Banerjee and C. Kommareddy, Scalable Application Layer Multicast, ACM SIGCOMM 2002.
7. D. A. Tran, K. A. Hua and T. T. Do, Zigzag: An Efficient Peer-to-Peer Scheme for Media Streaming, IEEE INFOCOM, 2003.
8. M. M. Hefeeda and B. K. Bhargava, On-Demand Media Streaming Over the Internet.
9. Allcast, http://www.allcast.com.
10. Y. Guo, K. Suh and J. Kurose, et al. P2Cast: Peer-to-Peer Patching Scheme for VoD Service, WWW 2003.

Research of Data-Partition-Based Replication Algorithm in Peer-to-Peer Distributed Storage System[*]

Yijie Wang and Sikun Li

Institute of Computer, National University of Defense Technology, Changsha, China, 410073
wwyyjj1971@vip.sina.com

Abstract. Replication is the key technology of distributed storage system. In this paper, according to the intrinsic characteristic of distributed storage system, based on the peer-to-peer model, the data-partition-based replication algorithm is proposed. In the data-partition-based replication algorithm, the data object is partitioned into several data blocks, and then these data blocks are encoded in order that there is data redundancy between data blocks. Compared with the traditional replication algorithm, the data-partition-based replication algorithm has fine granularity of replication, less bandwidth cost and storage cost, and can provide higher availability, durability and security. The results of performance evaluation show that the encode time and decode time is proportional to data size, and that the irregular cascade bipartite graphs are of great advantage to improve the success ratio of data recovery, and that if the number of data blocks used to recover data object is greater than a certain value, the success ratio of data recovery approaches to 100%.

1 Introduction

At present, along with the explosive growth of network bandwidth, computing power and storage capacity, how to aggregate the geographically distributed heterogeneous storage resources to form the virtual storage space and provide secure efficient data storage service is becoming a challenging research topic in the worldwide.

The hierarchical model and peer-to-peer model can be used to build distributed storage systems. Hierarchical model, including the new trend toward storage virtualization, use layers of control and abstraction to stitch together distributed and disparate storage providers into a single virtual whole. Peer-to-peer computing has emerged as a significant social and technical phenomenon. In the peer-to-peer model, many peers work together in a symmetrical way to provide storage services, clients need not install any new software to consume basic storage services. The distributivity, autonomy, dynamic, scalability and flexibility of peer-to-peer model can be utilized to improve the efficiency of distributed heterogeneous storage resources. Therefore, many research projects([1],[2],[3],[4],[5],[6],[7]) on distributed storage system are based on peer-to-peer model, for example, the goal of Freenet project ([1]) is to develop a distributed information storage and retrieval system; the goal of Oceanstore project ([2]) in University of California is to implement persistent data storage in the worldwide; the goal of Chord project ([3]) in MIT is to build a scalable distributed

---

[*] This work is supported by the National Grand Fundamental Research 973 Program of China (No.2002CB312105), A Foundation for the Author of National Excellent Doctoral Dissertation of PR China (No.200141), and the National Natural Science Foundation of China (No.69903011, No.69933030).

storage system, most of its research work is about data location; the goal of StarFish project ([4]) in Bell Labs is to build a transparent geographically replicated storage system.

In the most research projects on distributed storage system, the traditional replication technology is utilized to achieve the high availability and durability of distributed storage system ([4],[5],[8],[9],[10]). In the traditional replication technology, the replica size is equal to data size; in order to guarantee the availability and durability, the number of replicas need to be increased while the system size increased; however, to some extent, more replicas need more network bandwidth and more storage capacity, especially for large data object; on the other hand, the increasing replicas also influence data security. Therefore, the traditional replication technology is not adequate for data management in distributed storage system very well.

In this paper, according to the intrinsic characteristic of distributed storage system, based on the peer-to-peer model, the data-partition-based replication algorithm is proposed. In the data-partition-based replication algorithm, firstly, the data object is partitioned into several data blocks; secondly, these data blocks are encoded in order that there is data redundancy between data blocks, so that the data replication is realized indirectly, the high availability and durability of distributed storage system are achieved, and the data security is guaranteed to a certain extent. Section 2 describes the data-partition-based replication algorithm. Section 3 presents the results of performance evaluation. Section 4 provides a summary of our research work.

2 Data-Partition-Based Replication Algorithm

In the data-partition-based replication algorithm, data object D is partitioned into m data blocks which are of the same size b, then the m data blocks are encoded into n encoded data blocks of size b by Tornado code ([11],[12]), the n encoded data blocks are distributed among several storage nodes; if some read requests about data object D are received, any r (r>m) data blocks of n encoded data blocks can recover the data object D. e_r=m/n named as code rate. Tornado code is systematic code, so the first m data blocks of n encoded data blocks are the same as the m data blocks of data object D, which are named as information blocks, the other n-m data blocks of n encoded data blocks are named as check blocks.

The key components of data-partition-based algorithm are encoding and decoding. The bipartite graph B is utilized to encode the data blocks $d_1, d_2, \ldots, d_m$ into $(1+\beta)$m encoded data blocks which include $d_1, d_2, \ldots, d_m$, so m information blocks and βm check blocks are produced. In the bipartite graph B, the mapping between information blocks and check blocks is defined, each check block can be derived from the exclusive or computation of its neighbor information blocks (Figure 1(a)). Therefore, the cost of encoding is proportional to number of edges in the bipartite graph B. For a certain check block, if one of its neighbor information blocks missed, then the missing information block can be found by exclusive or computation of the check block and the existing neighbor information blocks (Figure 1(b)).

It is clear that the missing information blocks can be found by all check blocks and some other information blocks, thus the data object D can be recovered. However, if some check blocks are missed, perhaps the data object D cannot be recovered. So the cascade bipartite graphs are used to encode the data blocks. The cascade bipartite

3.2 Relation Between the Structure of Cascade Bipartite Graphs and the Performance of Data-Partition-Based Replication Algorithm

In the data-partition-based replication algorithm, the data object D is encoded into n encoded data blocks, part of the encoded data blocks are used to recover D. The success ratio of data-partition-based algorithm is 100% ideally. But in the practical situation, the success ratio is related to the structure of cascade bipartite graphs such as the number of levels, the number of check blocks in each level, and the data block degree (number of edges attached to data block). If the number of levels is greater, and the number of data blocks in each level is greater, the success ratio may be greater, but the computation cost and storage cost are increased. Commonly, the 3-level cascade bipartite graphs are appropriate, the number of information blocks is decided by the data size and data block size, and the number of the check blocks in each level is decided according to the number of information blocks. If the data block degree is greater, the number of neighbor data blocks is greater, if the data block exists, there are more data blocks can be recovered by it; if the data block does not exist, there are more other data blocks can be used to recover it.

In this section, the relation between the data block degree and the performance of data-partition-based replication algorithm is analyzed in detail. In the data-partition-based replication algorithm, the data object D is partitioned into m data blocks, the m data blocks are encoded into n encoded data blocks, any $(1+\alpha)m$ encoded data blocks can recover D, in the following experiment, α is set to 0.05. The size of data object is set to 8MB, the size of data block is set to 1KB. The 3-level cascade bipartite graphs (Figure 3) are used to replicate the data object, the number of information blocks in B_0 is 8000, the number of check blocks in B_0 is 4000, the number of check blocks in B_1 is 2000, the number of check blocks in B_2 is 1000, the number of check blocks in B_3 is 1000. For each level of cascade bipartite graphs, the sum of information block degrees is equal to the sum of check block degrees.

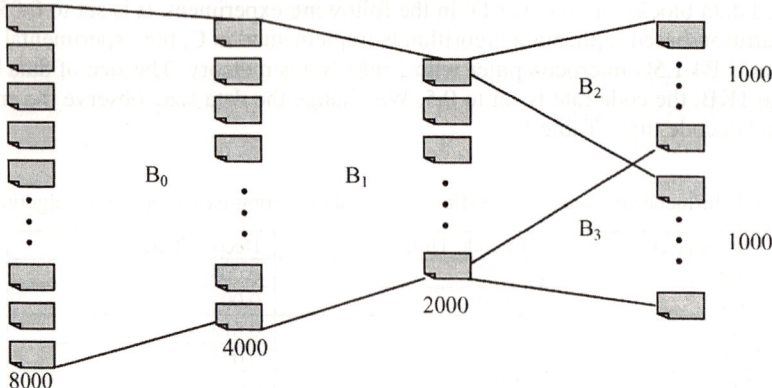

Fig. 3. 3-level cascade bipartite graphs

Regular Cascade Bipartite Graphs

In the regular cascade bipartite graphs, the data block degrees are uniform. In the following experiment, the information block degrees in 3-level cascade bipartite graphs are set to the same value. For a certain cascade bipartite graphs, the success ratio of data recovery lies on the data blocks selected. Therefore, for each value of information block degree, the different data blocks are selected to recover data object for 1000 times, and the success ratio is computed.

In Figure 4, while the information block degree is set to minor value, the success ratio of data recovery is increased as the information block degree increases; while the information block degree is set to 4, the success ratio of data recovery is the greatest; while the information block degree is set to greater value, the success ratio of data recovery is decreased as the information block degree increases. Therefore, the value of information block degree is crucial for the regular cascade bipartite graphs.

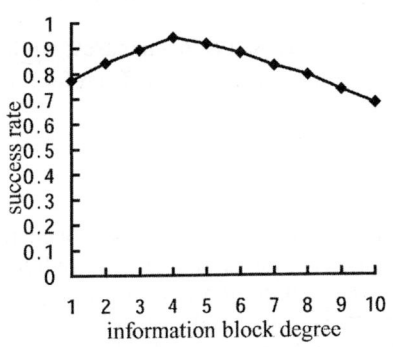 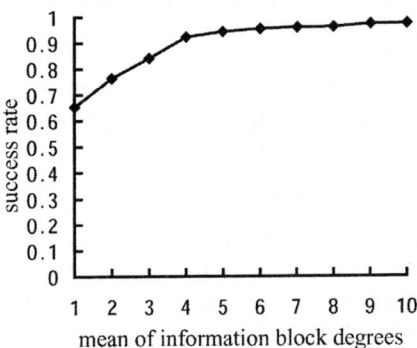

Fig. 4. Test result of regular cascade bipartite graphs

Fig. 5. Test result of irregular cascade bipartite graphs

Irregular Cascade Bipartite Graphs

In the irregular cascade bipartite graphs, the data block degrees are different. In the following experiment, the data block degrees of 3-level cascade graphs are of Poisson distribution, and the means of information block degrees are set to the same value. For each value of mean of information block degrees, the different data blocks are selected to recover data object for 1000 times, and the success ratio is computed.

In Figure 5, the success ratio of data recovery is increased as the mean of information block degrees increases. Therefore, for the irregular cascade bipartite graphs, if the information block degrees are greater, the success ratio of data recovery may be greater.

3.3 Relation Between the Number of Data Blocks Used to Recover Data Object and the Performance of Data-Partition-Based Replication Algorithm

In the data-partition-based replication algorithm, the data object D is partitioned into m data blocks, the m data blocks are encoded into n encoded data blocks, any $(1+\alpha)m$ encoded data blocks can recover D. In this section, the relation between α and the performance of data-partition-based replication algorithm is analyzed. In the following experiment, the data block degrees of the 3-level cascade bipartite graphs (Figure 3) are of Poisson distribution, the mean of information block degrees is set to 6. For each value of α, the different data blocks are selected to recover data object for 1000 times, and the success ratio is computed.

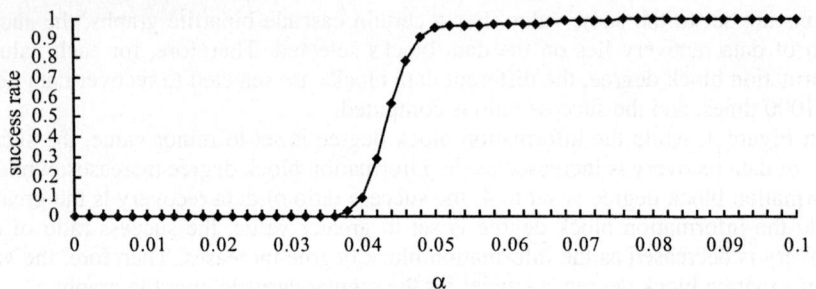

Fig. 6. The relation between α and performance of data-partition-based replication algorithm

In Figure 6, the success ratio of data recovery is increased as α increases; while α is set to 0.048, the success ratio of data recovery is greater than 90%. Therefore, if α is greater, the success ratio of data recovery may be greater.

4 Conclusion

In the data-partition-based replication algorithm, the data object is partitioned into data blocks, and then the data blocks are encoded. Compared with the traditional replication algorithms, the data-partition-based replication algorithm has fine granularity of replication, less bandwidth cost and storage cost, and can provide higher availability, durability and security. The results of performance evaluation show that the encode time and decode time is proportional to data size, and that the irregular cascade bipartite graphs are of great advantage to improve the success ratio of data recovery, and that if the number of data blocks used to recover data object is greater than a certain value, the success ratio of data recovery approaches to 100%.

References

1. Ian Clarke, Oskar Sandberg, Brandon Wiley, Theodore W. Hong: Freenet: A Distributed Anonymous Information Storage and Retrieval System. Lecture Notes in Computer Science, vol 2009 (2001)46-59
2. John Kubiatowicz, David Bindel, Yan Chen, Steven Czerwinski: OceanStore: An Architecture for Global-Scale Persistent Storage. Proc. Conf. Architectural Support for Programming Languages and Operating Systems (ASPLOS-IX), ACM Press, New York (2000) 190-201
3. Ion Stoica, Robert Morris, David Karger, M. Frans Kaashoek, and Hari Balakrishnan: Chord: A Scalable Peer-to-peer Lookup Service for Internet Applications. ACM SIGCOMM 2001, San Deigo, CA (2001)160-177
4. Eran Gabber, Jeff Fellin, Michael Flaster, Fengrui Gu: StarFish: highly-available block storage. 2003 USENIX Annual Technical Conference, San Antonio, TX, USA (2003)151-163
5. P. Druschel and A. Rowstron: PAST: A large-scale, persistent peer-to-peer storage utility. Proc of HotOS VIII, Schloss Elmau, Germany (2001)75-80
6. Lu Xicheng, Li Dongsheng, Wang Yijie: Research of Peer-to-Peer Distributed Storage System. Journal of Computer Research and Development, 40(Suppl.) (2003)1-6
7. LI Dong-sheng, WANG Yi-jie, LU Xi-cheng: A Scalable Peer-to-Peer Network with Constant Degree. Proceedings of 5th International Workshop, APPT 2003, Xiamen, China (2003)414 – 424

8. Edith Cohen, Scott Shenker: Replication strategies in unstructured peer-to-peer networks. The ACM SIGCOMM'02 Conference, Pittsburgh, USA (2002)308-321
9. Wang Yijie, Jiang Xueyang: Research of Replication Techniques in Internet Distributed Storage System. Journal of Computer Research and Development, 40(Suppl.) (2003)30-35
10. Jussi Kangasharju, James Roberts, Keith W. Ross: Object Replication Strategies in Content Distribution Networks. Journal Computer Communications, vol. 25, n. 4 (2002) 367-383
11. Michael Luby: Tornado Codes: Practical Erasure Codes Based on Random Irregular Graphs. 2nd. International Workshop on Randomization and Approximation Techniques in Computer Science (RANDOM '98), Barcelona, Spain (1998)171-175
12. D.Spielman: Linear-Time Encodable and Decodable Error-Correcting Codes. IEEE Transactions on Information Theory, 42(6) (1996)1723-1731

Local Index Tree for Mobile Peer-to-Peer Networks[*]

Wei Shi, Shanping Li, Gang Peng, Tianchi Ma, and Xin Lin

College of Computer Science, Zhejiang University, Hangzhou, P.R. China 310027
shiwei@zj165.com, shan@cs.zju.edu.cn, e_pglmary@hotmail.com,
tcma@csis.hku.hk, alexlinxin@163.com

Abstract. An effective algorithm for searching and retrieving information is proved to be critically relevant to the performance of mobile Peer-to-Peer (MP2P) systems. However, existing searching algorithms in MP2P environment are inefficient in respect of user response time and network traffic, due to the nature of limited bandwidth and mobility of the electronic devices. A searching algorithm Local Index Tree (LIT) is proposed in this paper. By building a index which keeps track of the relative locations of the peers near by, LIT reduces the number of query requests and accelerates the lookup process especially when the frequency of queries is relatively high. Simulations on NS2 show that LIT greatly reduces the user response time compared with existing searching mechanisms in MP2P systems while saving bandwidth of the network.

1 Introduction

Recently more research work has been focused on mobile peer-to-peer computing (MP2P) [2][5][8]. In MP2P environment, each mobile device plays the same role of functionality. They roam at different speed and each of them has different processing capability and network bandwidth. As a result, MP2P has to take into account the typical restrictions of mobile devices, such as frequent disconnections and limited network bandwidth. Like in P2P computing, one of the most important aspects of MP2P computing is locating and routing. In mobile P2P environment, each peer knows very little about the location of the searching target and the topology of the network. Mechanisms are seriously required to quickly locate the desired information to improve the users' experience as well as alleviate the load of network traffic.

Conventional searching mechanisms in P2P system can be classified into three categories – centralized directory model, flooded requests model and document routing model, each with their own advantages and disadvantages. However, these technologies are not suitable or efficient enough for mobile environments in which peers change their shares or locations much frequently because of their mobile and decentralized nature. Centralized directory model used in Napster [10] can't be deployed on MP2P since there exists a single point whose failure will lead to a sudden stop of the search operation, hence will deduct the stability of applications. The document routing model, used by FreeNet [7], is the most recent approach. By matching the file ID and the peer's ID, searching process moves intelligently towards the required resources. But the bandwidth used to deploy the document routing model is beyond what MP2P can offer. However the flooded requests model is more suitable for addressing the issues arising in MP2P environment such as dynamic location, real-time

---

[*] This paper is sponsored by Zhejiang Provincial Natural Science Foundation of China (No.602032).

searching and so on since it provides a pure decentralized and up-to-date searching mechanism. There are no critical nodes whose failure will result in a panic of the whole system. Each request from a peer is flooded (broadcast) to directly connected peers, which continue the process until the results are found. But this model, deployed in Gnutella, requires a lot of network bandwidth when queries are launched frequently and hence is not very scalable.

In [3] the author put forward some improved searching methods and tries to address the issues mentioned above. The local indices mechanism presented in that paper caches the metadata of all the peers within static R hops around. However this method does not take into account the nature of the mobility of MP2P network and restrictions on diverse wireless devices. In mobile settings the cached metadata about neighbors may be invalid in a short time. Thus, it is not appropriate to be deployed in MP2P environment.

In this paper, a searching mechanism, which is called "Local Index Tree (LIT)", for MP2P environment is proposed based on the flooded requests model. For each peer, a tree structure is introduced to keep track of its neighbors near by. Peers cache as many metadata about their stable neighbors as possible. To avoid too many peers being involved as in flooded requests model, the structure of LIT changes dynamically according to the topology of the network. If the metadata of the desired resource can not be found in the index tree, the searching peer just sends the query requests to the peers corresponding to the leaf nodes of its index tree. In that way, one can quickly locate the needed resources by only consulting limited number of peers, and thereby, improving the users' experience and smoothing the network traffic. By sending metadata requests to its neighbors periodically, each node maintains the index tree. And according to the mobility of the network, the tree would be modified to keep consistent with the topology of current network.

The paper is organized as follows. In the next section we put forward our dynamic searching mechanism using index tree. Section 3 describes the settings used in our experiments. Simulation results and analyses are presented in section 4 to demonstrate the performance of our mechanism. In section 5 the related work is discussed. Finally we conclude the study in section 6.

2 The Local Index Tree Mechanism

Our main goal is to reduce the user response time as much as possible, which is to say that a peer needs to cache more metadata of the peers around it. After submitting a query to the MP2P network, users may quickly get the results by consulting the cached metadata without bothering too many other peers. Every peer builds a local index tree which provides reference information about resources shared by the neighbors in the range of N hops (index radius), where N is changeable according to the mobility of the nearby network. Different from the behavior in the local indices algorithm, when the peer finds that there are nodes that change their location rapidly and the whole topology of current network changes slightly, it simply deletes these nodes in its local index tree rather than just reducing the index radius.

2.1 Initializing the Local Index Tree

When a peer joins a MP2P network, it has no idea about its neighbors. We call this new peer "participant" and the peers already in the network "veterans". So the participant broadcasts a *"Get"* message with the value of time to live (TTL) which equals to the default index radius. The message also contains the metadata over its shared resources. The veteran receiving that *"Get"* message will send a *"Reply"* message containing metadata over its shared resources as well as the predecessor address (the address from which it get the *"Get"* message) back directly to the participant. Then the veterans forward the *"Get"* message to its neighbors if current TTL of the *"Get"* message is greater than zero. The TTL is decremented each time the message is forwarded. When the TTL becomes 0, the message is no longer forwarded. In this way, the participant knows the metadata about its neighbors and learns exactly about the nearby network topology. At the same time, veterans directly connecting to the participant will insert the metadata of the participant into their local index trees. Other veterans can also be aware of the join of the participant by broadcasting update requests (described in the next section) later and add it to their own index trees. For example, in Fig.1, when peer A wants to take part in a specific MP2P network, it sends *"Get"* messages with default index radius R=2. When peer F gets the message, it simply sends back its metadata and its predecessor B's address directly to A. Using the information contained in the *"Reply"* messages, A will know the network topology and be able to build its local index tree (see the nodes connected with the thick lines in Fig.1) to represent it. Note that, in Fig.1, peer E has two predecessors. In this case, A will just choose the first *"Reply"* message coming from E. We also employ the method introduced in [14] to build the index tree without cycles.

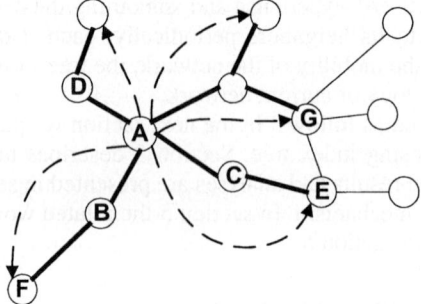

Fig. 1. The topology of network. The network connected with thick lines is the tree built by peer A with the initial index radius of 2. The requests that A sends are showed by the dashed lines

2.2 Maintaining the Local Index Tree

To get the fresh metadata about neighbors, the peer broadcasts *"Get"* messages frequently and builds a new index tree. For each peer, by comparing the nodes in the newly constructed tree and the old one, it decides to modify the index radius or just delete some nodes in its own new tree while leaving the index radius unchanged. The decision of deleting the node depends on the "moving speed" in the index tree. We adopt the ratio between the number of deleted and all peers in the newly built index, marked as P, to evaluate the mobility of current network. The index radius should be decreased when P is beyond a predefined value Q, which means the peers around move quickly and the metadata about them should not be indexed any more. On the contrary, if P remains less than Q for a certain period of time (say after T times *"Get"* requests launched), that is to say the network is somewhat stable and that the peer can safely increase the index radius R to cache more peers' metadata and provide more powerful searching capability.

In the trees building after two sequential "*Get*" requests, the level of each node will be used to calculate the node's "moving speed". Assuming after peer A gets the new index tree, the formula listed as following is introduced to calculate peer B's speed:

$$V = \frac{clevel - flevel}{time\_unit} .\qquad(1)$$

In formula 1, *clevel* is the level of node B in the newly built tree of A and *flevel* is the one in the old tree. Then A can decide to delete or keep node B in its tree according to the value of V on B, since V_B reflects the movement speed and direction of peer B from A's viewpoint. The peer with great value of V means that it is leaving away quickly. And its metadata will be quickly out of date, thus it should be deleted from the tree and its metadata will not be cached in certain period of time. Otherwise, it should be kept. We don't take into account when V is negative, which means the peer becomes closer to the requesting peer. We consider it is the positive behavior and should not be thought of change in order to guarantee the accuracy of the metadata on the peer's sharing resources.

If the node appears only alternatively in one of the newly/oldly built trees, the value of the node's level will be set to a large number for the tree in which the node is absent. So when peer B quits at the interval of A's two issues of "*Get*" request, the value of V will be great, resulting in the deletion of B in the newly built tree. And if the peer finds after broadcasting the "*Get*" message that there are peers which were not in its tree previously answering its request, that peer's value of V will be a great negative one and the participant will be indexed in the tree.

With respect to the performance of the peer, the index radius should not grow infinitely when the whole network is steady. Since the peers might be poor performance devices such as PDA and sensors that can't offer the computing power and bandwidth the mechanism needs. To address this problem, each peer can define a upper limit U beyond which index radius should not grow and the quantity of memory the algorithm can use.

2.3 Searching with LIT Technique

The main advantage of LIT is that it dynamically caches as much metadata as possible about the stable peers near by. Hence it needn't consult all its neighbors to find the file requested. When a request comes, it just forwards the request directly to the leaf node of its index tree. In this way, user response time will be greatly reduced and the bandwidth will be saved.

When a peer wants to find certain number of copies of a file, at first, it looks up its local index tree to see whether there are enough results. If there exist, it simply return the results to the user. And if there do not exist, the peer will send the queries directly to the peers corresponding to the leaf node of its local index tree and tell them the number of results still needed (see the dashed line in Fig.1). The leaf peers will check their index tree and forward the queries to the leaf node of their own index tree if necessary. To avoid the request message propagating infinitely, the peer also sets TTL value for its request message. If at last the user still can't get the enough results he needs, it seems that there are not so many copies of that file currently in the network. In this case, the peer should take down its expectation.

3 Experimental Settings

We have developed a simulated MP2P environment by modifying NS2 [1] to evaluate the newly proposed searching algorithm. The broadcast range of each peer is randomly selected from 20m to 100m. Peers all broadcast the *"Get"* requests to their neighbors every one time unit of 5 seconds. If the tree remains unchanged for 4 time units, the peer will expand its index scope; otherwise the unstable nodes of the local index tree will be deleted. We emphasize here that in our searching algorithm, the caching range, updating request time and index tree rebuilding time are decided completely by each peer according to its broadcast range, processing capability and network bandwidth.

In the experiments, we initially put 50 peers in an area of 1000m*1000m and gradually add peers until the total number of peers reaches 250. At beginning, 100 files are created from which each peer randomly selects 10 to share when it joins the MP2P network. During the experiments, each peer continuously issues queries and the query interval is randomly selected from 40 seconds to 60 seconds. The time spent to locate each file is recorded respectively. The setdest utility developed by CMU in NS2 is used to generate scene (topology) in the area every second. The network bandwidth is also monitored to evaluate the performance of our mechanism. We also record the user response time when the peers' highest speed becomes higher and higher while the number of peer remains 200.

For comparison, we also implement the pure flooded searching mechanism and local indices searching mechanism. The implement of pure flooded searching mechanism is easy, and in fact when the caching radius remains unchanged, LIT will be degraded to local indices.

4 Results and Discussion

In this section, we present the results of our experiments and analysis. As shown in Fig.2, the average time needed to search a file (user response time) is calculated using LIT, local indices and pure flooded request mechanism respectively. According to Fig.2 we find that as the peers join to the area, the user response time becomes longer. It occurs because when the number of peers increases, the desired file would scatter among these peers and thus one will spend more time to locate the file. However, as we can see, the LIT mechanism provides relatively shorter period to query the file. Fig.3 shows the network traffic recorded after a file query is launched. As the figure shows, the more peers join the area, the heavier network traffic is. Nevertheless the network traffic of LIT mechanism is lighter than the other two since each peer caches the metadata of its neighbors when it is not busy and tries to preserve each stable peer in its local index tree. The traffic line of local indices mechanism shown in the figure is closer to the one of LIT at beginning, but higher when the number of peers became large. It is so because as the peers join the area, the metadata cached by peers becomes relatively small. Fig.4 shows user response time of three mechanisms with different highest movement speed. We can figure out that the line of pure flooded query mechanism only fluctuates in a limited range because the peers using pure flooded query mechanism don't cache any metadata of their neighbors and thus the

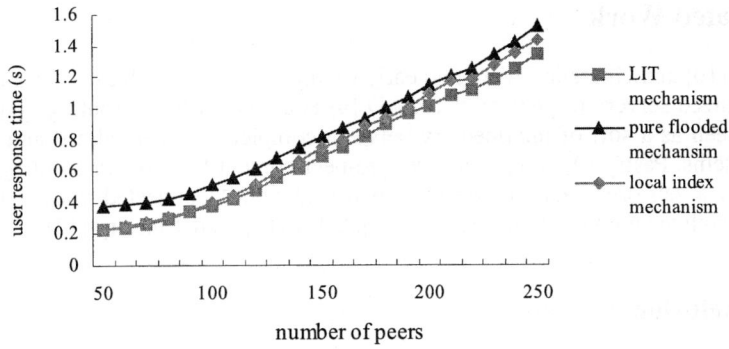

Fig. 2. User response time of the three mechanisms with different number of peers

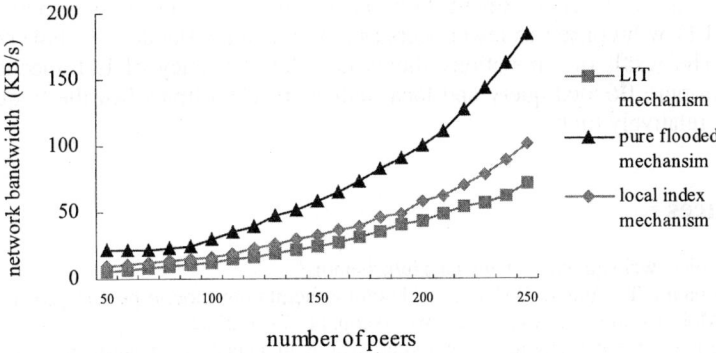

Fig. 3. Network traffic of three mechanisms with different number of peers

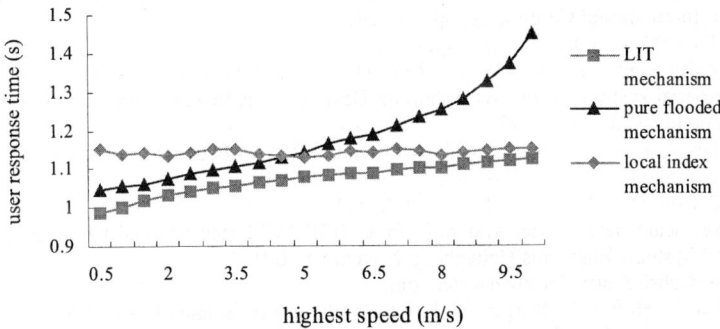

Fig. 4. User response time of three mechanisms with different highest speed

change of the topology impacts little to its user response time. When the locations of some peers change rapidly, a big part of the cached metadata by peers using local indices mechanism goes invalid soon, then the searching peer has to re-launch the query when it finds the peer which has the required files no longer reachable, thus making its user response time longer and network traffic heavier. The LIT algorithm, however, by caching the stable peer's metadata, gains the greatly improved results.

5 Related Work

Napster [10] and Gnutella [6] are two early routing systems which use centralized and decentralized servers respectively. TRIAD [4] is a content-based routing system that can be seen as a sort of intermediary between completely decentralized and centralized systems. Pastry [9] is a generic peer-to-peer content location and routing system based on a self-organization overlay network of nodes connected via the Internet. Other search techniques include Chord [11], CAN [12], and Tapestry [13].

6 Conclusions

Peer-to-peer networks offer several advantages such as simplicity of use, robustness and scalability. In this paper, we propose and simulate an information retrieval mechanism in MP2P environment. In order to provide as short user respond time as possible, LIT will consult as fewer peers as possible to accelerate the whole searching process. The result of simulations shows that the efficiency of LIT mechanism is better than pure flooded query and local indices mechanisms when the frequency of queries is relatively high.

References

1. NS2 project website. http://www.isi.edu/nsnam/ns/
2. N. Maibaum, T. Mundt. "JXTA: a technology facilitating mobile peer-to-peer networks", Proc. Mobility and Wireless Access Workshop, pp. 7-13, 2002.
3. B. Yang, H. Garcia-Molina. "Improving Search in Peer-to-Peer Networks", Distributed Computing Systems, Proc. 22nd International Conference, pp. 5-14, 2002.
4. D. Cheriton et. al. http://www.dsg.stanford.edu/triad, July 2000.
5. M. Wiberg. Folkmusic. "A Mobile Peer-to-Peer Entertainment System", Proc. 37th Annual Hawaii International Conference, pp. 290-297, 2004.
6. Gnutella website. http://www.gnutella.com.
7. I.Clarke, O.Sandberg, and B.Wiley. "Freenet: A distributed anonymous information storage and retrieval system", Proc. Workshop on Design Issues in Anonymity and Unobservability, Berkeley, California, 2000.
8. H. Hsiao, C. King. "Bristle: A Mobile Structured Peer-to-Peer Architecture", Proc. Parallel and Distributed Processing Symposium, pp. 22-26, 2003.
9. A. I. T. Rowstron, P. Druschel. "Pastry: Scalable, decentralized object location and routing for large-scale peer-to-peer systems", Proc. IFIP/ACM International Conference on Distributed Systems Platforms Heidelberg, November 2001.
10. Napster website. http://www.napster.com.
11. I. Stoica, R. Morris, D. Karger, M. F. Kaashoek, and H. Balakrishnan. "Chord: A Scalable Peer-to-Peer Lookup Service for Internet Applications", ACM SIGCOMM, San Diego, CA, August 2001.
12. S. Ratnasamy, P. Francis, M. Handley, R. Karp, S. Shenker. "A Scalable Content-Addressable Network", ACM SIGCOMM, San Diego, CA, August 2001.
13. J. Kubiatowicz, D. Bindel, et al. "OceanStore: An Architecture for Global-Scale Persistent Storage", Proc. 9th International Conference on Architectural Support for Programming Languages and Operating Systems, November 2000.
14. J. R. Sack and J. Urrutia, eds., Elsevier. "Spanning trees and spanners", Handbook of Computational Geometry, pp. 425-461, 1999.

Distributed Information Retrieval Based on Hierarchical Semantic Overlay Network

Fei Liu, Fanyuan Ma, Minglu Li, and Linpeng Huang

Department of Computer Science and Engineering, Shanghai Jiaotong University, Shanghai,
P.R. China, 200030
{fliu,ma-fy,li-ml,huang-lp}@cs.sjtu.edu.cn

Abstract. One fundamental problem that confronts information retrieval is to efficiently support query with higher accuracy and less logic hops. This paper presents HSPIR (Hierarchical Semantic P2P-based Information Retrieval) that distributes document indices through the P2P network hierarchically based on documents semantics generated by Latent Semantic Indexing (LSI) [1]. HSPIR uses CAN [2] and Range Addressable network organize [3] nodes into a hierarchical overlay network. Comparing with other P2P search techniques [4, 5] those are based on simple keyword matching, HSPIR has better accuracy for it considers the advanced relevance among documents. We use Agglomerative Information Bottleneck (AIB) [6] to cluster documents and train Directed Acyclic Graph Support Vector Machines (DAGSVM) based on these clustered documents. Owning to the hierarchical overlay network, the average number of logical hops per query is smaller than other flat architectures.

1 Introduction

Content-based full-text search and Per-to-peer (P2P) networks have become, in a short period of time, fast growing and popular Internet applications. In recent year the digital data added each year exceeds 10^{18} bytes and is estimated to grow exponentially. This trend calls for equally scalable infrastructures capable of indexing and searching rich content such as HTML, plain text, music, and image files. So P2P is adapt to content-based full-text search system. A number of P2P systems provide keyword search, including Gnutella [7] and KaZaA [8]. These systems use the simple and robust technique of flooding queries over some or all peers. Another class of P2P systems achieves scalability by structuring the data so that it can be found with far less expense than flooding. These are commonly called distributed hash tables (DHTs). Most of DHTs are based on simple keyword matching, ignoring advanced relevance ranking algorithms devised by the IR community through decades of refinement and evaluation. Without effective ranking, queries consisting of popular words may return a superfluous number of documents that are beyond the user's capability to handle. We use the most popular IR algorithms in our system such as Agglomerative Information Bottleneck (AIB). Directed Acyclic Graph Support Vector Machines (DAGSVM) and latent Semantic Indexing (LSI). AIB cluster the unlabelled documents, DAGSVM classify the new added documents base on the clustered documents, VSM and LSI represent documents and queries as vectors in a Cartesian space. The cosine of the angle between documents vectors and queries vectors are the similarity between a query and a document.

2 The Agglomerative Information Bottleneck Method

The Information Bottleneck Method (IB) [9] has been presented to find a short code for one vector that preserves the maximum information about other vector. In this paper, Let T_m be the current m-partition of X and T_n denote the new n-partition of X after the merge of several components of T_m. Obviously, n<m. Let $\{t_1, t_2,..., t_3\} \in T_m$ denote the set of components to be merged, and $t_{nk} \in T_n$ the new component that is generated by the merge, so. n=m-k+1 For each $t \in T_n$, $t \neq t_n$ its probability distributions $(p(t), p(y|t), p(t|x))$ (remains equal to its distributions in T_m. For the new component, $t_{nk} \in T_n$.

We assume t_* is the cluster which comes from merging cluster t_i and cluster t_j. The decrease in the mutual information $I(T,Y)$ due to the above merger is defined by

$$\delta I(t_i, t_j) = I(T_{before}; Y) - I(T_{after}; Y) \quad (1)$$

From [10] we can get equation as follows

$$\delta I(t_i, t_j) = r_i^j = (p(t_i) + p(t_j)).D_{JS}[p(y|t_i), p(y|t_j)] \quad (2)$$

where the functional D_{JS} is the Jensen-Shannon (JS) divergence [11],[12] defined as

$$D_{JS}[t_i, t_j] = \pi_i D_{KL}[p_i | \tilde{p}] + \pi_j D_{KL}[p_j | \tilde{p}], \quad (3)$$

The JS divergence is non-negative and is equal to zero if and only if both its arguments are identical. It is upper bounded and symmetric. N. Slonim presented Agglomerative Information Bottleneck (AIB) which clusters documents based on a greedy bottom-up merging and r_i^j.

3 The Directed Acyclic Graph Support Vector Machines

SVMs method has been introduced in automated text categorization (ATC) by Joachims [13] and subsequently extensively used by many other researchers in information retrieval community. It has shown to yield good generalization performance in both text classification problems and hypertext categorization tasks. Let's consider the original primal optimization problem describing the principle of SVMs:

$$\text{minimize}: V(\vec{\omega}, b, \vec{\xi}) = \frac{1}{2}\vec{\omega} \cdot \vec{\omega} + C\sum_{i=1}^{n}\xi_i$$

$$\text{s.t.} \quad \forall_{i=1}^{n}: y_i[\vec{\omega} \cdot \vec{x}_i + b] \geq 1 - \xi_i \quad (4)$$

$$\forall_{i=1}^{n}: \xi_i > 0$$

Instead of solving the above optimization problem directly, one can also consider the following dual program:

$$\text{maximize}: W(\vec{\alpha}) = \sum_{i=1}^{n}\alpha_i - \frac{1}{2}\sum_{i=1}^{n}\sum_{j=1}^{n}y_i y_j \alpha_i \alpha_j (\vec{x}_i \cdot \vec{x}_j)$$

$$\text{s.t.} \quad \sum_{i=1}^{n} y_i \alpha_i = 0 \quad (5)$$

$$\forall i \in [1...n]: 0 \leq \alpha_i \leq C$$

In the training phase of DAGSVM, $k(k-1)/2$ binary SVMs are solved respectively (k is the category numbers). In the testing phase, it uses rooted binary directed acyclic graph which has $k(k-1)/2$ internal nodes and k leaves. Each node is a binary SVM of ith and jth classes.

4 Latent Semantic Indexing

LSI uses singular value decomposition (SVD) [14] to transform a high-dimensional term vector into a lower-dimensional semantic vector. Each element of a semantic vector corresponds to the importance of an abstract concept in the document or query. Let N be the number of documents in the collection and d be the number of documents containing the given word. The inverse document frequency (IDF) is defined as $IDF=log[N/d]$. The vector for document Do is constructed as $Do=(T_1*IDF_1, T_2*IDF_2,..., T_n*IDF_n)$. Where T_i takes a value of 1 or 0 depending on whether or not the word i exists in the document Do. Suppose the number of returned document is m, the documents matrix S is constructed as $S=[S_1, S_2,..., S_m]$. Based on this document matrix S, the algorithm is described as follow. Since S is a real matrix, there exists SVD of S: $S=U\sum V$ where U and V are orthogonal matrices. Suppose $rank(S)=r$ and singular values of matrix S are. $q_1 \geq q_2 \geq ... \geq q_r \geq q_{r+1}=...=q_n=0$. For a given threshold ε ($0< \varepsilon \leq 1$), we choose a parameter k such that $(q_k-q_{k-1})/q_k \geq \varepsilon$. Then we denote $U_k=[u_1, u_2,...,u_k]_{m*k}$, $V_k=[v_1, v_2,..., v_k]_{n*k}$, $\sum_k=diag(q_1, q_2,...q_k)$, and $S_k=U_k\sum_k V_k^T$. S_k is the best approximation matrix to S and contains main information among the documents. In this algorithm, the documents matching queries are measured by the similarity between them. For measuring the documents similarity based on S_k, we choose the ith row R_i of the matrix $U_k\sum_k$ as the coordinate vector of documents i in a k-dimensional subspace $R_i=(u_{i,1}q_1, u_{i,2}q_2,...,u_{i,k}q_k)$ $I=1,2,...,m$. The similarity between document i and query j is defined as $sim(R_i, R_j)=|R_i \cdot R_j|/||R_i||_2||R_j||_2$

5 Storing Indices of Documents in HSPIR

We construct semantic overlay networks based on CAN overlay network and Range Addressable network. We define P to be an m-dimensional CAN overlay network and let $P=[p_1, p_2,..., p_z]$, p_i is one of the nodes in P. Assume documents collection D contains y topics, we use AIB to cluster D. Define $D=[D_1, D_2,..., D_y]$ and $D_i=[d_{1i}, d_{2i},..., d_{wi}]$ (w is the number of documents contained by D_i). According to the result come from clustering, we train the DAGSVM. We calculate the semantic vector of each document in D using LSI and let $v(d_{ij})$ be the semantic vector of d_{ij}. From the introduction in section 2, we can set the dimension of $v(d_{ij})$ to be $10*m$ by adjusting in order to resolve match between CAN and semantic vector. For each cluster D_i we compute its directory semantic vector as follows:

$$dir(D_i) = (\frac{\sum_{j=1}^{n_i}\sum_{k=1}^{10} v(d_{ij}).x_k}{10*n_i}, \frac{\sum_{j=1}^{n_i}\sum_{k=11}^{20} v(d_{ij}).x_k}{10*n_i}, ..., \frac{\sum_{j=1}^{n_i}\sum_{k=10m-10}^{10m} v(d_{ij}).x_k}{10*n_i}) \qquad (6)$$

where m is the number of dimension of CAN. Each directory node records mapping between topics and $dir(D_i)(i=1,2,...,y)$ to supply information for further querying and storing index of document.

We assume node p_i is in charge of the space containing point $dir(D_i)$. Then nodes $p_1, p_2, ..., p_y$ manage the index of documents $D_1, D_2, ..., D_y$ respectively. An index of document includes the semantic vector of a document and a reference (URL) to the document. $index(D_i)$ denote the indices of documents among D_i. Nodes $p_1, p_2, ..., p_y$ being regarded as directory nodes in HSPIR are used as root nodes of Range Addressable network (Fig. 1):

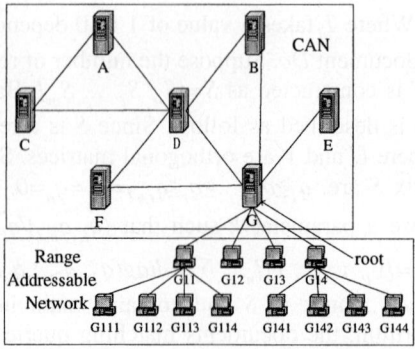

Fig. 1. The topology of HSPIR

We define $R(p_i)$ to be the Range Addressable network which root is p_i and each node in $R(p_i)$ has at most 4 nodes. We define average semantic vector as

$$avg(D_i) = \sum_{j=1}^{n_i} v(d_{ij})/n_i \qquad (7)$$

If the number of nodes in $R(p_i)$ is f then the depth of tree topology of $R(p_i)$ is $log4^f$. Indices of documents are stored in leaves of $R(p_i)$. After p_i receive $index(D_i)$, it divided $index(D_i)$ into s (s is the number of son nodes of p_i and smaller than 5) parts – $index(D_{i1}), ..., index(D_{is})$ equally. We arrange $index(D_{i1}), ..., index(D_{is})$ in increasing order of D_i's semantic vectors. p_i transmits the above s portions of indices to its son nodes. Then p_i records the mapping between $avg(D_{i1}), ..., avg(D_{is})$ and its son nodes, which can provide route information for future query. If the son node of p_i receives the indices of documents is not leaf, it does the same as p_i. Otherwise it stores the indices. According to the algorithm above, indices distribute in leaves evenly. Now we consider how to store a new document's index in our system. For example consider the process of storing document index of d in Fig. 2. Assume the client which wants to publish document d connects directory node C, then it transmits d and URL of d to C. After C receives d and URL, it uses DAGSVM to classify d into a special topic which is charged by node G in this example. C generates the index of d by combining URL and the semantic vector of d using LSI. C transmit index of d to node D and D transmit it to node G according to route algorithm of CAN.

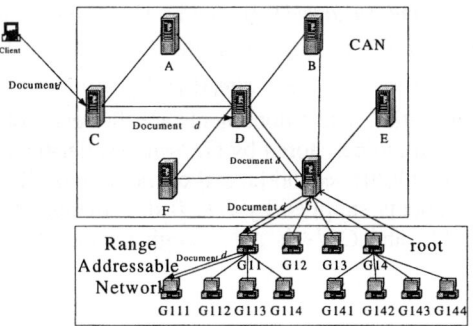

Fig. 2. The process of storing index of document

After G receive the index, it computes $sim(v(d),arg(D_{i1}))$, $sim(v(d),arg(D_{i2}))$, $sim(v(d),arg(D_{i3}))$, $sim(v(d),arg(D_{i4}))$. The largest of the four similarities is chosen to decide which son of G is in charge of index of d. In our example $G11$ is selected to receive index of d. then $G11$ calculate $sim(v(d),arg(D_{i1}))$, $sim(v(d),arg(D_{i2}))$, $sim(v(d),arg(D_{i3}))$, $sim(v(d),arg(D_{i4}))$ and find the largest of them: $sim(v(d),arg(D_{i1}))$. So it transmits index to node $G111$ which is a leaf and stores the index of d.

6 Query Documents in HSPIR

Querying documents in HSPIR is based on semantic of queries and documents. Fig. 3 shows an example of querying document. We want to explain the process of querying by this example. The client that connects directory node B sends a query q. After B receives q, it categorizes q through DAGSVM to confirm which topic q is interesting in. In this example, the topic q want to query is charged by G. Then B compute the semantic vector of q through LSI and send it to G according to CAN route algorithm.

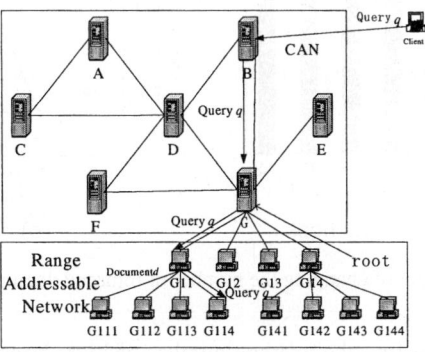

Fig. 3. The process of querying

After G receive the semantic vector of q, it computes $sim(v(q),arg(D_{i1}))$, $sim(v(q), arg(D_{i2}))$, $sim(v(q),arg(D_{i3}))$, $sim(v(q),arg(D_{i4}))$. The largest of the four similarities is chosen to decide which son node receives the query. In our example $G11$ is selected to receive query q. then $G11$ calculate $sim(v(q),arg(D_{i1}))$, $sim(v(q),arg(D_{i2}))$, $sim(v(q),arg(D_{i3}))$, $sim(v(q),arg(D_{i4}))$ and find the largest of them is $sim(v(q),arg(D_{i4}))$. So it transmits query to node $G114$ that is a leaf. Then $G114$ computes $sim(v(q),d_{i141})$, $sim(v(q),d_{i142})$,…, $sim(v(q),d_{i14m})$ and select k documents with the first k highest similarities. The k documents are the result of this query to be sent to client. If the number of indices stored by $G114$ is smaller than k, $G114$ reports $G11$. Then $G11$ select $G113$ with the second largest of the four similarities to ask for k-h (h is the number of documents supplied by $G114$) documents. If the number of documents supplied by $G113$ and $G114$ is not enough for client's request, $G113$ do the same as $G114$.

7 Experimental Results

We performed experiments using the 20 Newsgroup dataset. 20 Newsgroups consists of 20000 messages taken from 20 newsgroups. One thousand Usenet articles were taken from each of the following 20 newsgroups. We implement the DAGSVM and a hierarchical overlay simulator based on CAN and Range Addressable network. AIB is implemented and linked with each directory node in hierarchical overlay simulator. DAGSVM is trained with 20 Newsgroups datasets and we use AIB to cluster 20 Newsgroups into 20 topics to test the accuracy of our system. We use SMART [15] to index the Newsgroups corpus. In our experiment our system has 10100 nodes those contain 100 directory nodes. The dimensionality of LSI is 250 and the dimensionality of CAN is 25. Fig. 4 shows the effect of varying the number of returned documents for each query. It seems that the number of logic hops grows moderately and the accuracy increases slightly with increasing the returned documents.

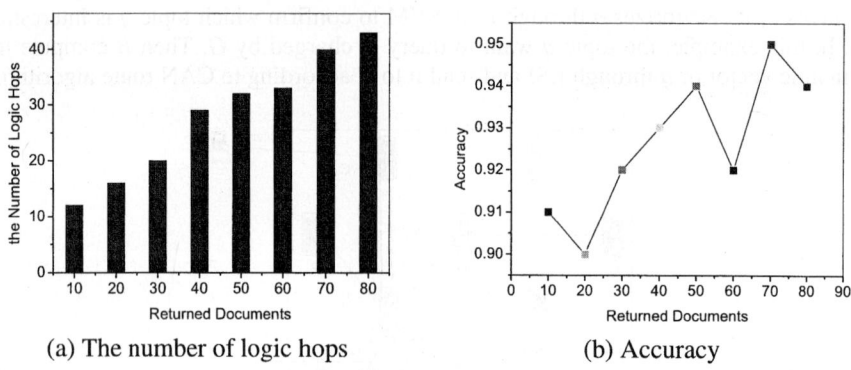

(a) The number of logic hops (b) Accuracy

Fig. 4. The effect of changing the number of returned documents in HSPIR

From Fig. 5, we can find the average number of logic hops to return one relevant document decreases drastically as we increase the number of returned documents.

When the user requests 10 documents, on average 1.2 logic hops to cost to find one relevant document. When the number of returned documents increases to 80, on average only 0.54 logic hops to cost to find one relevant document.

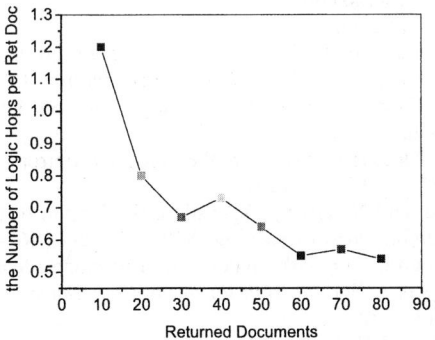

Fig. 5. The effect of changing the number of returned documents on the number of logic hops per returned documents

8 Conclusions

This paper presents HSPIR that distributes document indices through the P2P network hierarchically based on document semantics generated by Latent Semantic Indexing. HSPIR uses CAN and Range Addressable network organize nodes into a hierarchical overlay network to support query with queries and documents semantics. Comparing with other P2P search techniques those are based on simple keyword matching, HSPIR has better accuracy for it considers the advanced relevance among documents. Owning to the hierarchical overlay network, the average number of logical hops per query is smaller than other flat architectures. Both theoretical analysis and experimental results show that HSPIR has higher accuracy and less logic hops.

Acknowledgements

This paper is supported by 973 project (No.2002CB312002) of China, and grand project of the Science and Technology Commission of Shanghai Municipality (No. 03dz15027 and No. 03dz15028)..

References

1. S. C. Deerwester, S. T. Dumais, T. K. Landauer, G. W. Furnas, and R. A. Harshman. Indexing by Latent Semantic Analysis. *Journal of the American Society of Information Science*, 41(6):391–407, 1990.
2. S. Ratnasamy, P. Francis, M. Handley, R. Karp, and S. Shenker. A scalable -addressable network. In *ACM SIGCOMM'01*, August 2001.
3. A.Kothari, D.Agrawal, A.Gupta, S.Suri. Range Addressable Network: A P2P Cache Architecture for Data Ranges.
4. Q. Lv, P. Cao, E. Cohen, K. Li, and S. Shenker. Search and Replication in Unstructured Peer-to-Peer Networks. In *ICS'02*, June 2002.

5. S. Rhea and J. Kubiatowicz. Probabilistic Location and Routing. In *IEEE INFOCOM'02*, June 2002.
6. N. Slonim and N. Tishby. Agglomerative Information Bottleneck. In Proc. of Neural Information Processing Systems (NIPS-99), pages 617–623, 1999.
7. Gnutella. http://gnutella.wego.com.
8. Kazaa. http://www.kazza.com.
9. N. Tishby, F.C. Pereira, and W. Bialek. The information bottleneck method. In *Proc. of 37th Allerton Conference on communication and computation*, 1999.
10. N. Slonim and N. Tishby. The Hard Clustering Limit of the Information Bottleneck Method. In preperation.
11. J. Lin. Divergence Measures Based on the Shannon Entropy. *IEEE Transactions on Information theory*,37(1):145–151, 1991.
12. R. El-Yaniv, S. Fine, and N. Tishby. Agnostic classification of Markovian sequences. In *Advances in Neural Information Processing (NIPS-97)*, pages 465–471, 1997.
13. Jochims, T. Text categorization with support vector machines: learning with many relevant features. *Proceedings of ECML-98, 10th European Conference on Machine Learning*. Berlin: Springer. 1998. 137-142.
14. M. Berry, Z. Drmac, and E. Jessup. Matrices, Vector Spaces, and Information Retrieval. *SIAM Review*, 41(2):335–362, 1999.
15. C. Buckley. Implementation of the smart information retrieval system. Technical Report TR85-686, Department of Computer Science, Cornell University, Ithaca, NY 14853, May 1985.

An Efficient Protocol for Peer-to-Peer File Sharing with Mutual Anonymity*

Baoliu Ye[1], Jingyang Zhou[1,2], Yang Zhang[1,2], Jiannong Cao[2], and Daoxu Chen[1]

[1] Department of Computer Science and Technology, Nanjing University, Nanjing, China
yebl@vip.sina.com, {jingyang,cdx}@nju.edu.cn
[2] Department of Computing, The Hong Kong Polytechnic University, Hong Kong, China
{csyzhang,csjcao}@comp.polyu.edu.hk, cdx@nju.edu.cn

Abstract. A fundamental problem in a pure Peer-to-Peer (P2P) file sharing system is how to protect the anonymity of nodes when providing efficient data access services. By seamlessly combining the technologies of multi-proxy and IP multicast together, we propose a multicast-based protocol for efficient file sharing with mutual anonymity in this paper. Furthermore, the proposed protocol can adaptively adjust file distribution and reduce the multicast cost simultaneously. The simulations show that Mapper possesses the merits of scalability, reliability, and high adaptability with high performance.

1 Introduction

P2P provides a low-cost and highly available means of resource sharing and information dissemination over the Internet [1]. However, the open and decentralized feature of P2P also increases previously inconceivable opportunities for gathering private information and monitoring activities about individuals [2]. Consequently, anonymity is highly desired in P2P systems. The authors of literature [3] defined six types of anonymity. From the standpoint of a pair of communicating peers, three kinds of privacy guarantee can be achieved [1], namely *initiator anonymity*, *responder anonymity* and *mutual anonymity*, respectively. Due to the address information included in each packet, it's not an easy task to achieve anonymity.

In this paper, we propose a novel protocol which seamlessly combines the technologies of multi-proxy routing and IP multicast to guarantee mutual anonymity with low cost and high performance. The remainder of this paper is organized as follows: Section 2 introduces related works; Section 3 describes the proposed Mapper; Section 4 presents our initial simulation results; Finally we summarize this paper in Section 5.

2 Related Work

Most existing techniques for anonymity are mainly focus on initiator anonymity. Anonymizer [4] and LPWA [5] introduce a proxy between clients and servers to generate consistent untraceable aliases for clients. This approach relies too much on the intermediate proxy and cannot fully protect requestor privacy against the logs kept by the proxy. Onion Routing [6] and Crowds [7] use covert path to achieve initiator ano-

---

* This work is partially supported by the National 973 Program of China under Grant No. 2002CB312002 and the National High-Tech Research and Development 863 Program of China under Grant No. 2001AA113050.

nymity. However, organizing the covert path and periodical reformation introduce much resource consumption and access delay. Web Mixes [8] employs mixing technique to thwart timing analysis, but it still provide no responder anonymity. Napster [11] is regarded as the first P2P application, but it provides no anonymity. Athough the mutual anonymity is achieved in search phase by flooding-based method in Gnutella [12], this merit is unfortunately lost later by the direct HTTP connection of file transfer. Freenet [9] uses a depth-first algorithm to organize a forwarding path to providing mutual anonymity. However, the multi-proxy forwarding schema consumes many resources and inevitably leads to low efficiency. Similar to our work, APFS [10] is a mutual anonymity protocol with IP multicast. But IP multicast technology is adopted there for session management and anonymity is achieved via Onion Routing.

3 Mapper Design

3.1 Overview

We assume each node in the system offers a certain amount of local disk quota for file sharing. The file is identified with a binary file key generated by hash function [13]. Every node maintains three different tables. Among them, the File Information List (FIL) is used to record basic information of all the available files; The Dynamic Routing Table (DRT) is employed to determine the next forwarding node, and it consists of two fields: the file key and the forwarding node; The Request Forwarding Frequency Table (RFFT) indicates the forwarding frequency of a certain file through node. File retrieval is conducted through the following steps.

Step 1: Initialization.

On requesting a file specified by the user, the requestor generates a request message in the form of <grp_id, file_id, ttl, IP_m>, where grp_id is a randomly generated group identifier; file_id is the key value of the target file; ttl is a predefined hops-to-live variable; IP_m is a multicast address.

Step 2: Message Routing.

2.1. Search the file locally in the FIL table. If found, update the latest access information of the corresponding entry in the FIL table and go to Step 4, otherwise:

2.2. Determine the next forwarding node according to the local routing information. If the node itself is the initialing node or the ttl exceeds the predefined value, let current node join IP_m and go to step 5, otherwise:

2.3. Execute the Mostly Forwarding Frequency Caching (MFFC) algorithm to estimate whether let current node join the multicast group IP_m, forward the request message to the next node.

Step 3:

Repeat step 2 (the request message will be forwarded among the nodes).

Step 4: File Transfer.

4.1. The node that stores the requested file joins the multicast group IP_m, and sends the target file to all the group members through the multicast group IP_m.

4.2. File receivers update the corresponding records in both the RFFT and the routing table. A new entry in routing table is created by associating the requested file key with a node randomly selected from the group. The update of RFFT will be introduced in Section 3.3.

Step 5: Session Close.
Send an end message to all the group members notifying the end of this group session.

Actually, Mapper implements a six-state machine at every node for a particular file retrieval session (See Figure 1). For a specific session, all nodes are initially set into state I except the initiator. A node in State I changes its state to A and performs a local search process on the receipt of a REQUEST; If a copy of the desired file is found from this active node, a SUCCESS message will be returned and the state will switch to S; otherwise, the state alters to F. Active node with State F may further change to State G according to the MFFC algorithm. If the hops-to-live exceeded (ttl<0), the current active node changes its state to T and ends the message routing procedure. Upon receiving either a REQUEST or NACK message, the node previous in state F or state G could change their states temporarily to state A with its original state reserved.

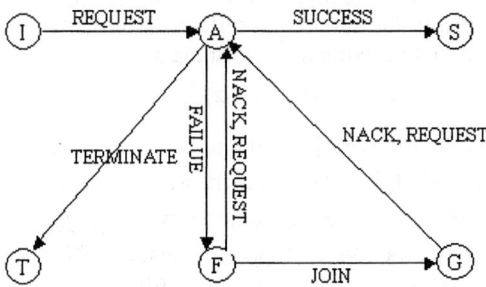

Fig. 1. State transition diagram

3.2 Request Message Routing

The procedure of request message routing is actually a dynamic directed graph traversal. Here, we propose a depth-first backtracking routing algorithm based on the key value similarity. The pseudo-code of the algorithm is as follows.

Algorithm 1: The RMR algorithm

```
RMRouting(Message msg) {
  if(msg.type==REQUEST) {
    if(curNode.state==F || curNode.state==M) {
      msg.type=NACK; forwardMsg(msg, prevNode);
      return();}
    msg.ttl= msg.ttl-1;
    if(msg.ttl<0) {
      joinGroup(curNode, msg, IP_m);  curNode.state=T;
      return();};
    if(localSearch()= =SUCCESS) {
      curNode.state=S; joinGroup(curNode, msg, IP_m);
      return();};
```

```
    else{
      curNode.state=F;
      if(memSel(msg) == true) joinGroup(curNode, msg, IP_m);
        N=fisrtRT(msg.file_id);
        if(N<>null) forwardMsg(msg, prevNode);
        else{
          msg.type=NACK; forwardMsg(msg, prevNode);};}
   else if(msg.type==NACK){
     N=nextRT(P);
     if(N<>null) forwardMsg(msg, prevNode);
     else{
       msg.type=NACK; forwardMsg(msg, prevNode); };};}
```

3.3 Group Member Selection

Economically select replication locations and multicast group members are an efficient approach for the aim of reducing file access delay and constraining bandwidth consumption. Here we propose the MFFC algorithm for this purpose. The basic idea is that if the requested file has the highest Forwarding Frequency (FF) at current node, then this node should join the group. To reflect file request patterns accurately, we establish a low-filter based FF computing formula as follows:

$$\lambda(file\_id) = \alpha \lambda_{avg}(file\_id) + (1-\alpha)\lambda_{new}(file\_id), 0 \leq \alpha \leq 1 \quad (1)$$

Where $\lambda(file\_id)$ is the filtered FF of file_id at current node, $\lambda_{avg}(file\_id)$ is the average FF over the last period, $\lambda_{new}(file\_id)$ is the average value of FF in present period. α is a constant used for coordinating the ratio between $\lambda_{new}(file\_id)$ and $\lambda_{avg}(file\_id)$. As said in Section 3.1, each node in Mapper maintains a RFFT. The table consists of four fields: $file\_id$, $\lambda_{avg}(file\_id)$, $\eta(file\_id)$, $\lambda(file\_id)$. $\eta(file\_id)$ is the total value of FF in this period. Every node updates all the records in the table periodically according to formula (2) and (3).

$$\lambda_{avg}(file\_id) = \eta(file\_id)/T \quad (2)$$

$$\eta(file\_id) = 0 \quad (3)$$

Algorithm 2 gives the pseudo-code of the MFFC algorithm.

Algorithm 2: The MFFC algorithm

```
Boolean memSel(REQUEST msg) {
  if(chkFFTable(msg.file_id)= =null)
    {addFFRec(file_id); return (false);}
  else{
    λ_new(file_id)=( η(file_id)/(Δt)
    λ(file_id)=computeFF(file_id,λ_avg(file_id),
                         λ_new(file_id));
    updateFFTable(file_id);
    if(max(λ(file_id))= = true) return (true)
    else return (false);};}
```

4 Performance Analyses

4.1 Network Convergence

We used the GT-ITM generator to create a 1000-node transit-stub graph for the overlay network topology. The simulation settings and measurements are consistent with those described in [9]. Figure 2 illustrates the evolution of the average request path length of Mapper and Freenet. This figure shows that Mapper has the similar median request path length as Freenet, but the convergence speed of Mapper is a little slower than Freenet. This is because files in Mapper are replicated selectively on the forwarding path instead of Freenet's full-path caching technique. And the later case requires more storage space, network bandwidth and transfer delay.

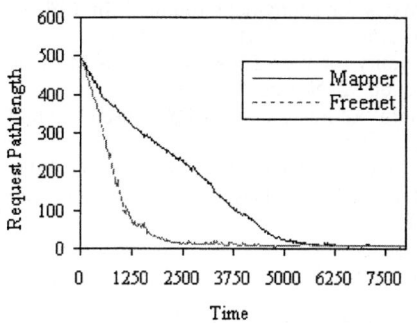 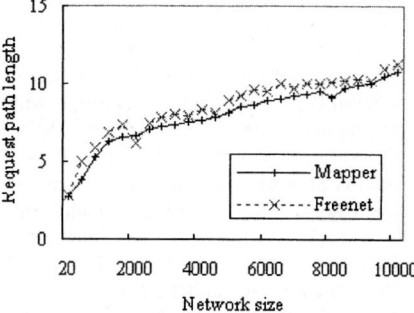

Fig. 2. Request Pathlength *vs.* Time **Fig. 3.** Request Pathlength *vs.* Network Size

4.2 Scalability

The scalability of Mapper can be demonstrated by the relationship between the request path length and the network size. Figure 3 depicts the evolution of the average request pathlength versus network size by periodically adding new nodes into the system. From this figure, we can see that both Mapper and Freenet have similar averaged requestpath length over network size. This indicates that the network size has a little impact on the overall request pathlength. Thus, Mapper scales well despite how the number of nodes increases. The average requestpath length also indicates the size of a multicast session. It is guaranteed in Mapper that the size of each multicast session won't grow too large due to the features of scalability and convergence discussed above. Meanwhile, replicating requested files inside a multicast group would further reduce the multicast scale.

5 Conclusions

In this paper, we propose an anonymous file sharing protocol for pure P2P file sharing systems by taking advantages of the multi-proxy forwarding technique and the IP multicast communication technique to locate and transfer files. Furthermore, we also present a request frequency based multicast group member selection algorithm and Mostly Forwarding Frequency Caching algorithm to optimize protocol performance. In the future, we are planning to conduct more simulations and experiments to evaluate our proposal.

References

1. Milojicic,D., Kalogeraki,S., Lukose,V. R., et. al., "Peer to Peer Computing", *Technical Report: HPL-2002-57*, Mar. 2002.
2. Shields,C., and Levine,B. N., "A Protocol for Anonymous Communication Over the Internet," In Proc. 7th ACM Conference on Computer and Communication Security (ACM CCS 2000), November 2000.
3. Dingledine,R., Freedman,M.J., Molnar,D., "The free heven project: Distributed anonymous storage service," In Proceedings of the Design Issues in Anonymity and Unobservability(LNCS 2009), July 2001.
4. Anonymizer, http://www.anonymizer.com/
5. Gabber,E., Gibbons,P., Kristol,D., et. al., "Consistent, yet anonymous, Web access with LPWA," Communications of the ACM. 42, 2. February 1999.
6. Reed,M., Syverson,P., and Goldschlag,D., "Proxies for anonymous routing," In 12th Annual Computer Security Applications Conference, Pages95-104. IEEE, December 1995.
7. Reiter,M. K., and Rubin,A. D., "Crowds: Anonymity for Web Transactions," ACM Transactions on Information and System Security, 1(1):66-92, November 1998.
8. Chaum,D., "Untraceable Electronic Mail, Return Addresses, and Digital pseudonyms," *Communications of the ACM*, Vol. 24(2), 1981.
9. Clarke, I., Sandberg,O., B. Wiley, T. W. Hong, "Freenet: A Distributed Anonymous Information Storage and Retrieval System," In ICSI Workshop on Design Issues in Anonymity and Unobservability, Berkeley, California, 2000.
10. Scarlata,V., Levine,B. N., and Shields,C., "Responder Anonymity and Anonymous Peer-to-Peer File Sharing," In Proceedings of the IEEE Intl. Conference on Network Protocols (ICNP) 2001. November 2001.
11. Shirky,C., "Listening to Napster," In Oram, A., (ed) Peer-to-Peer:Harnessing the Benefits of a Disruptive Technology, O'Reilly and Associates, Inc., Sebastopol, California. 2001.
12. Clips, "Gnutella Protocol Specifications", V. 0.4, http://www.clips2.com
13. American National Standards Institute, "American National Standard X9.30.2-1997: Public Key Cryptography for the Financial Services Industry – Part 2: The Secure Hash Algorithm (SHA-1)", 1997.

IPGE: Image Processing Grid Environment Using Components and Workflow Techniques*

Ran Zheng, Hai Jin, Qin Zhang, Ying Li, and Jian Chen

Cluster and Grid Computing Lab
Huazhong University of Science and Technology, Wuhan, 430074, China
{zhraner,hjin}@hust.edu.cn

Abstract. Computational grids have become a vital emerging platform for high-performance computing. However, some obstacles for the prevalence of grid and the grid application development are still far from mature, which is largely due to the immature grid-enabled computing environment. *Image-Processing Grid Environment* (IPGE) is a project that aims at providing high performance image-processing platform in a grid computing environment. IPGE is a combination of components and workflow techniques on which complex applications can be modelled as grid workflows with local or grid-enabled remote service components. In this paper, we discuss the infrastructure to provide the integration platform with both techniques, which provides flexible and useful mechanism and scheduling strategy to achieve series image processing operations. Components and workflow techniques provide benefits of flexibility, reusability and scalability and to construct cooperative image-processing applications easily.

1 Introduction

Image processing applications have been used widely in many scientific research areas such as biomedicine, and pattern recognition. In commercial and open community, more and more packages and applications are developed in image modeling, rendering, recognition and etc, and image data sets are growing exponentially. High performance computers and bulky spaces are needed for the increasing if image computing and storage. Grid technology [1][2] are used effectively to advance image processing and analysis. IPGE (*Image Processing Grid Environment*) is a project aiming to create a grid platform for image practitioners and researchers with access to advanced image processing services. It provides grid workflow to construct series applications, in which a grid platform is developed with components and workflow techniques.

The remainder of the paper is organized as follows: the next section introduces related works. Section 3 presents the whole image processing grid framework and the layered components. The techniques and models of components and grid workflow used in image grid environment are discussed in detail in section 4. Section 5 focuses on the scheduling strategy and evaluation for high efficiency and high performance. Finally, we draw the conclusion and give out future work in section 6.

---

* This paper is supported by ChinaGrid project from Ministry of Education

2 Related Works

There are some other projects that are using grid computing or workflow techniques for image processing or other related applications.

A Java-based tool for rapid prototyping of image processing operations is described in [3]. This tool uses a component-based framework and implements a master-worker mechanism. The design of a toolkit that allows rapid and efficient development of biomedical image analysis applications in a distributed environment is presented [4], which employs the Insight Segmentation and Registration Toolkit and Visualization Toolkit layered on a component-based framework. Other projects include European MammoGrid project for breast cancer screening [5], UK e-Science Programme MyGrid [6], Japan BioGrid project, and so on.

Workflow is the automation of a business process, during which documents, information or tasks are passed from one participant to another, according to a set of procedural rules. However, the concept extends beyond its use in business process management, and is also used more broadly in many grid projects. Early grid projects Condor [7] and UNICORE provide workflow capability to manage task dependencies. In the context of e-Science, a grid workflow is a process composed of interacting services running in a grid. WebFlow project [8] uses workflow as high level mechanism for high performance distributed applications.

IPGE targets image analysis and visualization applications in grid. It is a layered framework, uses service-oriented components and workflow techniques to specify workflow model of grid applications. A friendly programming environment is provided for deploying existing software or toolkits as service components, and building composite services by "drag-and-drop" interface.

3 Image Processing Grid Framework

Recent advances of development evolve towards resources sharing with grid computing and service-oriented computing by web services [9]. Building on concepts and technologies from these two communities, *Open Grid Services Architecture* [10] has been designed to significantly broaden the established approach of grid computing and tap into the emerging capabilities of web services. The *Image Processing Grid Platform* (IPGP) middleware is built on existing web services technologies and follows the OGSA standards.

The architecture of image processing grid, shown in Fig.1, consists of several layered parts. The IPGP middleware locates between grid basic middleware and grid-enable image processing applications, which is the most important and pivotal with grid application middleware.

Grid resources and grid basic middleware lie in the lowest layer. Grid resources are the foundation of IPGP, whose responsibilities are providing computing power. The share, integration and interoperation of all resources are accomplished by grid basic middleware. The services of grid application middleware provide generic interfaces such as indexing service, meta-data management, service management and security service, uploading or downloading, or accessing actual HPC grid resources, which hides the details of the grid resources as far as possible, and gives user transparent accessing. Based on it grid-enable image processing environment and related service toolkits are designed, which aim to provide flexible supports to various image-processing application scenarios by some APIs and SDK so that users can establish their custom-built individual applications or services.

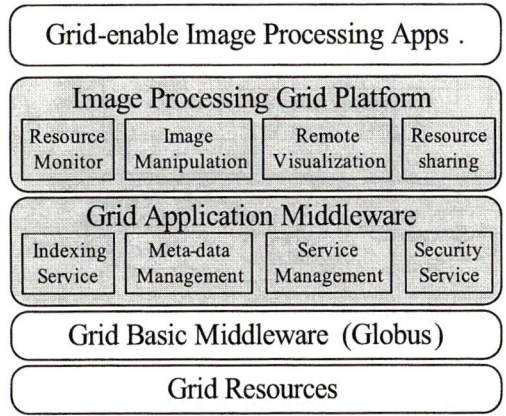

Fig. 1. Image Processing Grid Architecture

4 Service-Based Problem-Solving Environment of IPGE

The target of IPGE is to afford an image Problem-solving environment for simple or complex image applications. Every grid resource has different limited capabilities so that all image requirements can not be listed sufficiently. It is just like any other programming environment–although multifarious libraries are provided, programmers must still program their own programs to suit for his given problem.

4.1 System View

Service-based problem-solving environment is for both grid resource providers and grid clients, shown in Fig.2. For resource provider, it provides a portable environment for creating and publishing services as components. For grid client, it presents a convenient environment for browsing and querying service components, creating and running applications by composing them into application workflow.

This service-based Problem-solving environment of IPGE is designed with OGSA-compliant grid, which not only completes some image requirements independently, but also is convenient to model workflow to realize collaborative processing for complex applications. IPGE workflow engine classifies these services and assembles them as complicated components. Grid users on-line consult with the engine, customize the context of services to construct their own compatible workflow. Finally the engine invokes workflow execution module to start-up the processing, at the same time, additional services such as state browsing and querying are provided.

4.2 Component Technique of IPGE

In order to get high performance, grid-enable image applications are evolving into sets of cooperative components, which are presented as an image workflow by assembling these components in a plug-and-play fashion.

Definition 1. A is the set of all activities in IPGE, whose sum is m.

$$A=\{A_1, A_2, \ldots, A_m\} \quad (1)$$

Definition 2. R is the set of all resources in IPGE, whose sum is n.

$$R=\{R_1, R_2, \ldots, R_n\} \quad (2)$$

Definition 3. AR is a planar matrix presenting the relation between activities and matching resources.

$$AR = \begin{bmatrix} AR_{11} & AR_{12} & \cdots & AR_{1n} \\ AR_{21} & AR_{22} & \cdots & AR_{2n} \\ \cdots & \cdots & \cdots & \cdots \\ AR_{m1} & AR_{m2} & \cdots & AR_{mn} \end{bmatrix} \quad (3)$$

$$AR_{ij} = \begin{cases} 1 & R_j \text{ is one of matching resources with activity } A_i \\ & 1 \le i \le m, 1 \le j \le n \\ 0 & R_j \text{ is not matching with activity } A_i \end{cases} \quad (4)$$

Definition 4: P_{ij} is the execution price of scheduling the next activity A_j on resource R_j when activity A_i is finished on resource R_i.

$$P_{ij} = \begin{cases} 1 + \dfrac{RTT_{ij} + \dfrac{TransferSize}{BW_{ij}}}{\dfrac{Power}{V_j}} & i \ne j \\ & 1 \le i, j \le n \\ 0 & i = j \end{cases} \quad (5)$$

where RTT_{ij} denotes the round-trip time of network, *TransferSize* is the quantity of transfer data between resources R_i and R_j, BW_{ij} is the bandwidth, *Power* is the logical computational "cost" (in some arbitrary units) for the task, and V_j is performance (in units per second) of resource R_j.

The scheduling strategy is described as follows. Suppose an image-processing workflow is composed with a set of activities (service components) with sum *num*.

$$wfA=\{wfA_1, wfA_2, \ldots, wfA_{num}\} \quad (6)$$

A set of resources will be selected for the activities with sum *num* too, although same resources maybe exist in it.

$$wfR=\{wfR_1, wfR_2, \ldots, wfR_{num}\} \quad (7)$$

For each activity wfA_i ($1 \le i \le num$), there is a class of resources matching the activity, which can be got from the row vector AR_i of AR.

Compute every execution price P_{ij} for those elements whose values are not zero using Eq.5, and choose the lowest one adding into the set *wfR* orderly.

The final set *wfR* is the set of scheduling resources for this image-processing workflow, which can cooperatively satisfy the requirement of applications.

5.2 Performance Evaluation of Scheduling Algorithm

Analyzing the characteristics of grid environment and image applications, we find the inconsistent facts that the grid resources are widely distributed with low bandwidth, high latency and non-stable performance, while image-processing applications are data-intensive computing, and it leads to abundant data (almost small files) transfer between conjoint components of application workflow. In order to optimize the performance, the conjoint activities should be scheduled to one or low-price grid nodes if possible.

The scheduling algorithm described above is just conforming to the rule. Resources with the lowest price are scheduled for activities. From Eq.5, we get the conclusion that 1 is the boundary of the lowest prices. The situation of 1 can be happened when "$i=j$", namely next activity is processing on the same resource of previous activity. It avoids transferring so much small files that the transferring time is saved and optimal performance is got.

When there is no next activity on the previous resource matching activity, the next activity will be scheduled to other resources. It is inevitable that the transferring and latency of network is the main part of execution price except for the processing time. The higher the ratio of them, the larger the executing price is. Therefore, resources with high bandwidth, low latency and high processing capability will be scheduled for image-processing applications.

6 Conclusions and Future Work

High performance grid applications are evolving into sets of cooperating components, which should be constructed as grid workflow by assembling these components in grid. The research is underway. There are many factors and aspects that should thought carefully, such as QoS of applications, friendly workflow modelling tools and user interface. The support of more complex and flexible processing of grid workflow and components also should be strengthened in the future.

References

1. I. Foster and C. Kesselman (ed.), *The Grid: Blueprint for a New Computing Infrastructure*, Morgan Kaufmann Publishers, 1998.
2. I. Foster, C. Kesselman, and S. Tuecke, "The Anatomy of the Grid: Enabling Scalable Virtual Organizations", *International Journal of High Performance Computing Applications*, 15(3), 200-222. 2001.
3. E. S. Manolakos and A. Funk, "Rapid Prototyping of Component-based Distributed Image Processing Applications Using JavaPorts", *Proceedings of Workshop on Computer-Aided Medical Image Analysis, CenSSIS Research and Industrial Collaboration Conference*, 2002.
4. S. Hastings, T. Kurc, S. Langella, U. Catalyurek, T. Pan, and J. Saltz, "Image Processing for the Grid: A Toolkit for Building Grid-enabled Image Processing Applications", *Proceedings of 3rd International Symposium on Cluster Computing and the Grid*, pp.36-43, Tokyo, Japan, May 2003.
5. S. R. Amendolia, M. Brady, R. McClatchey, M. Mulet-Parada, M. Odeh, and T. Solomonides, "MammoGrid: Large-Scale Distributed Mammogram Analysis", *Proceedings of the XVIIIth Medical Informatics Europe conference (MIE'2003)*, pp.194-199, St Malo, France, May 2003.

6. R. Stevens, A. Robinson, and C. A. Goble, "myGrid: Personalised Bioinformatics on the Information Grid", *Proceedings of 11th International Conference on Intelligent Systems in Molecular Biology*, pp.302-304, Australia, July 2003.
7. J. Frey, T. Tannenbuaum, M. Livny, I. Foster, and S. Tuecke, "Condor-G: A computation management agent for multi-institutional grids", *Proceedings of the Tenth IEEE Symposium on High Performance Distributed Computing (HPDC10)*, Aug 2001.
8. D. Bhatia, V. Burzevski, M. Camuseva, G. Fox, W. Furmanski, and G. Premchandran, "WebFlow-a visual Programming Paradigm for Web/Java Based Coarse Grain Distributed Computing", *Concurrency: Practice and Experience*, Vol.9, No.6, pp.555-566, 1997.
9. S. Graham, S. Simeonov, T. Boubez, D. Davis, G. Daniels, Y. Nakamura, and R. Neyama, *Building Web Services with Java: Making Sense of XML, SOAP, WSDL, and UDDI*, SAMS publishing, 2001.
10. I. Forster, C. Kesselman, J. Nick, and S. Tuecke, "The Physiology of the Grid: An Open Grid Services Architecture for Distributed Systems Integration", *Open Grid Service Infrastructure WG, Global Grid Forum*, June 22, 2002.
11. H. Zhuge, "Timed Workflow: Concept, Model, and Method", *Proceedings of 1st International Conference on Web Information Systems Engineering (WISE2000)*, Hong Kong, June 2000.
12. W. M. P. van der Aalst and T. Basten, "Inheritance of Workflows: An approach to tackling problems related to change", *Theoretical Computer Science*, January 2002.

Distributed MD4 Password Hashing with Grid Computing Package BOINC

Stephen Pellicer[1], Yi Pan[1], and Minyi Guo[2]

[1] Department of Computer Science, Georgia State University, 34 Peachtree Street, Suite 1450
Atlanta, GA 30302-4110, USA
spellicer1@student.gsu.edu, pan@cs.gsu.edu
[2] School of Computer Science and Engineering, The University of Aizu, Tsuruga, Ikki-machi
Aizu-Wakamatsu City, Fukushima 965-8580
minyi@u-aizu.ac.jp

Abstract. Distributed computing on heterogeneous nodes, or grid computing, provides a substantial increase in computational power available for many applications. This paper reports our experience of calculating cryptographic hashes on a small grid test bed using a software package called BOINC. The computation power on the grid allows for searching the input space of a cryptographic hash to find a matching hash value. In particular, we show an implementation of searching possible 5 character passwords hashed with the MD4 algorithm on the grid. The resulting performance shows individual searches of sections of the password space returning a near linear decrease in calculation time based on individual participant node performance. Due to the overhead involved of scheduling these sections of the password space and processing of the results, the overall performance gain is slightly less than linear, but still reasonably good. We plan to design new scheduling algorithms and perform more testing to enhance BOINC's capability in our future research.

1 Introduction

Cryptographic password hashing is a standard security mechanism in operating systems and applications that is impacted by improvements to parallel computing. When viewed alongside heavily distributed computation architectures such as grid computing, the strength of many cryptographic hashing techniques comes into question. The raw processing power available to grid computers makes circumventing certain cryptographic calculations both inexpensive and viable in regard to computing time. As grid computing techniques become more popular, the cryptographic techniques and security measures surrounding these techniques must adapt to computationally powerful and widely available grid resources.

This paper examines a small-scale implementation of a publicly available grid computing system for computing a subset of the MD4 password hashing problem. The project uses the Berkeley Open Infrastructure for Network Computing (BOINC) framework to implement a distributed algorithm for searching the password keyspace used in systems such as the Microsoft Windows NT password system. While the final results for the implementation show a less than linear improvement in computation time, detailed analysis shows that this can be attributed to the small problem space exacerbating inefficiencies with the scheduling algorithm.

The paper begins with an introduction to the MD4 hashing problem and a sequential approach to calculating the password keyspace. The next section introduces the

BOINC architecture server complex implementation. This is followed by the client portion of the BOINC system. Finally, the results of the distributed implementation are analyzed and conclusions are presented.

2 MD4 Password Hashing Problem

MD4 is the cryptographic hashing algorithm used to store and validate passwords on Windows NT and interoperable systems such as the open source SMB system SAMBA. The system takes the Unicode encoded user password, pads the password to fill a single MD4 input block, and processes the block with the MD4 algorithm [1]. The result is a 128-bit value. This value is stored in order to validate the user password for authentication. Given the 128-bit hash value, there is no reasonable computation to recover the user password. As such, in order to compute a matching 128-bit hash, the original password must be used as input to the MD4 algorithm. If the password is not known, all possible input passwords must be computed to find the matching hash value.

The possible inputs for the MD4 hashing problem are all valid passwords for the system. A reasonable set of passwords are all 4 to 14 character passwords that can be input on a typical keyboard. This paper uses all upper and lower case letters, 10 digits, and all punctuation available on a standard keyboard for a total of 94 possible characters in each position. With 4 to 14 character passwords, the possible passwords are the sum of $94^4+94^5+94^6+\ldots+94^{13}+94^{14}$. This paper explores only 5 character passwords for a total password space of 94^5.

MD4 is a cryptographic hashing function and as such it is designed to foil pre-computation and sharing of computation given similar inputs. Given this design goal, parallelization of the algorithm is restricted to parallel computation of different sections of the possible inputs. There have been many studies attacking the uniqueness of hashes given different inputs to the hash function, however, as a one-way hash function, the integrity of MD4 and similar cryptographic hashes remains intact [2].

3 Sequential Algorithm

The sequential algorithm used in this paper implements a straightforward search and calculation of MD4 in portable C++. Bosselaers et al provide detailed optimizations to the sequential algorithm that show a substantial increase in performance for the algorithm on the Pentium architecture [3]. Since a nearly identical algorithm is used in the distributed implementation, optimizations to the calculation of MD4 were not considered since these improvements can be implemented in both the sequential and distributed approaches with similar results.

An iteration of the main loop of the algorithm processes a single password guess from the possible passwords. The loop begins by incrementing the password guess to be processed. Next, the password is given as input to a single block iteration of the MD4 hashing algorithm. The resulting hash value is compared to a target value for a match. If there is a match, this password guess is recorded and the loop continues.

The sequential algorithm was run with two different sets of password inputs for performance measurement. The first run searched the entire 5 character password space with a running time of 46 minutes and 13 seconds. The second run searched only a slice of the 5 character password space containing only $94^4 \times 2$ possible combinations with a running time of 1 minute and 1 second. The second run was performed to compare the sequential algorithm performance to the single node performance in the distributed implementation. The distributed algorithm was also run in a standalone mode outside of the BOINC infrastructure with a running time of 1 minute and 24 seconds. This result shows there is a nontrivial amount of overhead associated with the supporting code included with the distributed algorithm over the sequential algorithm.

4 BOINC Server Complex and Scheduling

The Berkeley Open Infrastructure for Network Computing provides a generic framework for implementing distributed computation applications within a heterogeneous environment [4]. This system is currently deployed in a beta test for the SETI@home and Astropulse distributed applications [5]. The system is designed as a software platform utilizing computing resources from volunteer computers.

Computing resources are allocated to the computational problem by the assignment of workunits. Workunits are a subset of the entire computational problem. Workunits for the MD4 hashing problem were chosen to create workunits small enough to evenly distribute work among the participating nodes and large enough to minimize startup and reporting times. For the four participating nodes in the project, the entire password space of all five character passwords were divided into 47 even workunits. This allowed for an even division of workunits containing $94^2 \times 2$ possible passwords. Workunit input files contained three fields. The first field contained the first password to check. The second field contained the last password to check. The final field contained the matching hash target.

The BOINC architecture consists of a server complex handling scheduling, central result processing, and participant account management, a core software client running on participant nodes, and project specific client software running on participant nodes. The server complex consists of a database, web server, and five BOINC specific processes. Communication between participant nodes and the server complex is handled via standard HTTP.

Database functions in BOINC are implemented for MySQL; however few, if any, MySQL specific functions are used so it can be easily replaced by a comparable database. A straightforward schema is used to store information on application versions, participant hosts, user accounts, workunits, and results. Additional tables are included for tracking user participation level and managing teams.

BOINC uses HTTP for both interaction with the user for account management and RPC communication between client nodes and the server complex. User account management is accessed using a standard web browser to interact with a web site implemented mainly in PHP. This web site allows users to manage preferences for job execution of BOINC projects. RPC communication between the BOINC core client and the server complex is handled by two cgi programs. The first cgi program, named cgi, handles work scheduling based on client requests for work. The second program, named file_upload_handler, receives completed work results from clients for processing by the server complex. The MD4 password hashing project used the standard web interface, cgi, and file_upload_handler from a standard BOINC installation.

The first BOINC process is named feeder. Feeder is responsible for accessing the database and selecting candidate workunits to be processed by clients. The next

BOINC process is the assimilator. Assimilator processes workunit results and integrates them into the project. As valid results become available, assimilator does any project specific processing including storage of results in the project database. Transitioner is the third BOINC process. This process handles state transistions of workunits. The fourth BOINC process is the validator. Validator handles testing the validity of workunit results. Result validation comes into play when redundant computing is used to compare the results of the same workunit returned by two or more clients. The final BOINC process is the file_deleter. File_deleter examines the state of workunits to determine which input and output files are no longer needed.

Scheduling workunits is handled by the cgi interfaces and the BOINC server processes. Participating nodes contact scheduling server through the web server interface named cgi. The participant node identifies itself and amount of work effort available based on high and low water marks for processing time, available memory, and disk space. Cgi then selects viable workunits from the shared memory segment filled by the feeder process. Cgi bases workunit selection based upon client capabilities and also favoring workunits that have been evaluated as infeasible by other participating nodes. The available workunits returned by feeder will be affected by workunit configurations requiring redundant computing and result validation. Workunits that have already been processed may need additional results from other participating nodes in order to arrive at a quorum of identical results in order to accept a result unit. The MD4 password hashing project did not implement redundant computing.

Configuration of the MD4 password hashing project manipulated the estimated running time of the individual workunits in order to accommodate the scheduling algorithm. Since the entire MD4 password search required a running time shorter than the default low water mark for the default BOINC project client, the first node requesting work would consume all available workunits. To circumvent this scheduling problem, the estimated time of the workunits reported running times far greater than the actual running time. This caused the scheduling algorithm to distribute the workunits more evenly across the participating nodes.

5 Participant Nodes

MD4 password search runs were performed with 4 participating nodes of varying processor speed and host operating system (Table 1). These nodes were set to run client computations even when not idle with a typical set of unrelated processes running alongside the BOINC client. The benchmarks reported are based on the benchmark run by the standard BOINC core client. These benchmarks seem to be somewhat inaccurate since the highest score actually belongs to the slowest node. This is either due to the difference in CPU architecture or host operating system. Final results show the slowest performance exhibited by the host with the highest benchmark. The other three systems with similar CPU and OS reported results correlating to their benchmark.

All clients were connected on the same local area network. The topology was a switched Fast Ethernet (100 Mbs) network. BOINC traffic shared the network with light day to day unrelated network traffic.

The BOINC core client software handles communication with the server complex from the participant node. Participating nodes are configured by pointing the BOINC client software to the BOINC project URL and providing an authentication key as-

Table 1. Participant CPU Nodes

Host	OS	CPU	FLOPS
Psyduck	Windows XP	AMD Athlon	234228795
Registeel	Windows XP	AMD Athlon	207524271
Jynx	Windows 2003	AMD Athlon	107413010
Hoothoot	Linux	Intel Pentium III	282608695
		Total Flops	831774771

signed to the user when volunteering for the project. The BOINC client software then contacts the project URL and receives the URL for the scheduler. This scheduler URL points to the cgi program of the server complex. The scheduler is contacted and work is requested based upon user configurable preferences for the amount of work to perform. The scheduler assigns workunits and instructs the client to download client software for executing the workunit and any input files necessary for processing the workunit.

The client application used to perform the MD4 hash search is nearly identical to the sequential algorithm. The most significant change to the sequential algorithm is the inclusion of BOINC client API calls to handle work check pointing and file mapping in order to accommodate workunit processing. Since scheduling in BOINC uses workunits to divide work among participating nodes, the client application requires a mechanism for specifying the slice of the MD4 password search space to process. The workunit input file specifies this slice of passwords to calculate. The project client application uses the BOINC client API to open a generic 'input' file handle. The core BOINC client receives this API call and maps the file operation to the specific input file for this workunit. The algorithm can then be called using the workunit parameters for this workunit. Similarly, all output from the client calculation is sent to an output file opened using the BOINC client API to map the output to a file specific to the result unit.

Another task of the BOINC client API is handling work checkpointing. Periodically, the client application can signal readiness for checkpointing. According to user preferences, the BOINC core client will report if checkpointing is ready. If it is ready, the client application can save the current work state. This saved state should contain all necessary information for continuing work from the given checkpoint. Checkpointing aides work interruption on the participating nodes. Because BOINC targets volunteer resources, this mechanism can save processing time if computing resources suddenly become unavailable. This can be due to a number of external events such as power loss and termination of the BOINC client. Additionally, the client can be configured to terminate workunit processing if the system state changes based on idle user activity or running on battery power in the case of portables. Experimental results in the MD4 password hashing application showed a significant amount of overhead due to the checkpointing mechanism. Decrease of the checkpointing period showed a significant increase in runtimes. While analysis of the checkpointing function showed a small amount of code necessary for the checkpoint, the impact to cache use, register use, and pipeline level due to branch prediction are likely the primary culprit in the significant increase in runtime on a small computational problem such as

MD4 password search. For larger computational problems that greatly exceed these architectural factors, the impact of the checkpointing function is likely to be insignificant.

The design of the BOINC scheduling algorithm pays close attention to the communication overhead imposed by a restricted interconnection network. In a large deployment, slow and unreliable network links can be greatly impacted by the polling nature of the BOINC client software communication with the scheduling server. Communication can bottleneck due to three major factors:

- Large amounts of workunit data
- Large amounts of result data
- Large number of participant nodes

While the MD4 password hashing project did not suffer from these three factors, the scheduler design introduced significant delay in client polling of the scheduling server and reporting results. This delay resulted in significantly longer per workunit times compared to the sequential algorithm. While this delay could have been altered for the project, it would require a significant change to the BOINC system. It is not known how significant such changes would impact larger projects that exhibit the above three factors.

6 Distributed Algorithm Performance

Based upon the standalone algorithm performance for a workunit-sized search, the theoretical speed increase should approach 3 times the sequential algorithm. The total running time of the project was 18 minutes and 44 seconds resulting in a speed increase of only 2.4 times the sequential algorithm (Figure 1). Total computation time was measured from the first workunit assigned to the final result received by the server complex. A plot of the completion time for each workunit shows a slow start for the distributed algorithm followed by a roughly linear time to completion (Figure 2). Speedup of the algorithm shows performance approaching the theoretical speedup as the number of workunits completed increases (Figure 3). Dips in the speedup curve show bursts of performance decrease attributed to inefficiencies in work scheduling. These decreases in performance are likely to occur as batches of workunits are completed and the distributed algorithm incurs scheduling and communication overhead.

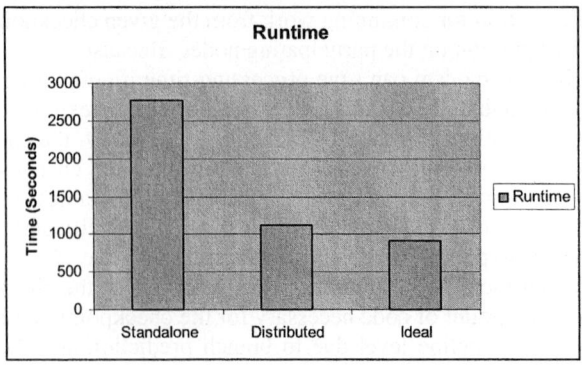

Fig. 1. Running Time Comparison

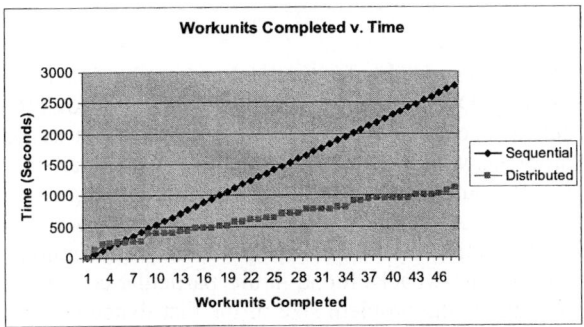

Fig. 2. Sequential and Distributed Computing Times with Different Number of Workunits

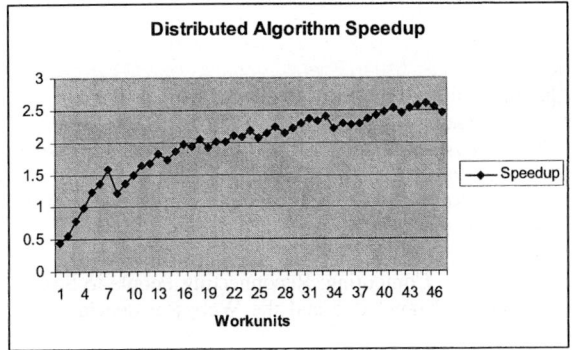

Fig. 3. Speedup with Different Number of Workunits

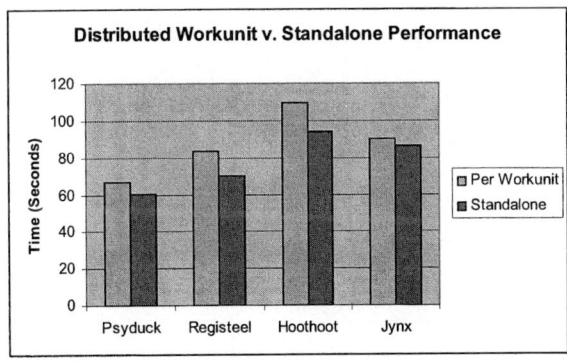

Fig. 4. Distributed Workunit vs. Standalone Performance

A closer look at distributed node per workunit average performance versus node performance for a standalone workunit shows runtimes closer to the expected speedup (Figure 4). The average wallclock time of sending the first workunit to a node and receiving the final result from a node was 16 minutes and 47 seconds. The overall average for a workunit was 1 minute and 27 seconds. These results show a performance approaching a per workunit performance on par with the sequential algorithm.

7 Conclusions

The overall less than linear speedup for the total processing time is likely due to inefficiencies in scheduling. Due to the design of the BOINC scheduling system, there is no relationship between scheduling and overall size of problem space. The system is geared towards problems with tremendous amounts of work available. Large amounts of work and excess computing capacity can compensate for long polling and reporting delays in the scheduler. As a whole, the BOINC framework provides a powerful foundation for implementing distributed computations with minimal changes to sequential algorithms. The drawbacks found in this particular computation would likely become less significant as the problem size of the distributed computation grows. In the future, we plan to design several new scheduling algorithms to enhance BOINC's scheduling service.

References

1. Auer, Karl and Jeremy Allison. Smbpasswd man page.
   http://us1.samba.org/samba/docs/man/smbpasswd.5.html
2. Dobbertin, H. *"Cryptanalysis of MD4."* Fast Software Encryption. 1996. pp. 53--69.
3. Bosselaers, Antoon et al. Fast Hashing on the Pentium. 25 May 1996.
4. BOINC contact page. 28 November 2003. http://boinc.berkeley.edu/contact.php.
5. Berkeley Open Infrastructure for Network Computing (BOINC). 26 November 2003. http://boinc.berkeley.edu/.
6. David P. Anderson. "Public Computing: Reconnecting People to Science," Proceedings of the Conference on Shared Knowledge and the Web, Residencia de Estudiantes, Madrid, Spain, Nov. 17-19, 2003.

A Novel Admission Control Strategy with Layered Threshold for Grid-Based Multimedia Services Systems*

Yang Zhang[1,2], Jingyang Zhou[1], Jiannong Cao[2], Sanglu Lu[1], and Li Xie[1]

[1] State Key Laboratory for Novel Software Technology,
Department of Computer Science and Technology, Nanjing University, P.R. China, 210093
{jingyang,sanglu,xieli}@nju.edu.cn
[2] Department of Computing, Hong Kong Polytechnic University,
Hung Hom, Kowloon, Hong Kong
{csyzhang,csjcao}@comp.polyu.edu.hk

Abstract. There is always a trade-off between system profits and the Quality of Service (QoS) in service-oriented systems. Especially in the case of grid-based large-scale multimedia services systems which serve more classes of requests with different significance, the larger number of the requests admitted doesn't necessarily lead to a higher profit rate. This paper presents a novel layered threshold-based admission control strategy for such system. This strategy provides more guarantees to higher-priority requests by assigning them more resources reservation. Meanwhile, by the layered setting of the resources, the service system can also keep the agility of the admission control procedure and achieve higher profits. Theoretical evaluation and simulations are carried out to demonstrate the effectiveness and flexibility of the proposed strategy.

1 Introduction

Given the rapid development of high-speed networks and communication technologies, people can enjoy nearly all desired services conveniently and economically. The contents run from simple text data to complex pictures, audio files or even video media. Of all the services, online real-time multimedia service, including Video-on-Demand (VoD), Video Conference and other services with highly strict timing constraints, is more and more important and becomes the focus of researches [1,2]. Furthermore, Grid technology, which aims to provide a robust paradigm to aggregate disparate resources in a controlled environment, prompts such kind of services to a higher level. Resources at different places can be organized to form a huge service grid thus scale up the capacity of the grid service system to a great extent [3].

In a generic online multimedia services system, multimedia streams consist of video frames and audio samples, which are arranged in a specific sequence. Because of the highly strict timing constraints on the online service, the *Quality of Service (QoS)* is satisfied and kept only when all the video frames and audio samples are played steadily and sequentially in a determinative period [4,5]. In a grid services

---

* This work is partially supported by the National High Technology Development 863 Program of China under Grand No.2001AA113050 and the National Grant Fundamental Research 973 Program of China under Grand No.2002CB312002.

system, the system capacity is greatly improved compared with traditional services system. But facing such a huge number of potential requests, it is always not enough. The system has to reserve the resources for the admitted requests in advance to guarantee their future QoS. This procedure is known as *Resource Reservation*. In grid, resource reservation is highly needed. Meanwhile, a commercial grid system allows management of diverse kind of requests with different priority. If more requests with higher priority are satisfied, the system can achieve more profits. Therefore, a well selective control mechanism is sure to have great impact on the efficiency and the profit of the whole service system.

In this paper, we discuss three essential admission control strategies with threshold on *resource reservation* and indicate their weakness in the grid-based large-scale multimedia services system. Meanwhile, we propose a novel admission control strategy with layered threshold. It's more flexible and efficient than those existing classic strategies after some special changes are taken on the threshold setting. Through theoretical analyses and simulations, we prove that the novel admission control strategy with layered threshold is especially potent when the requests with higher priority arrive more frequently and can bring more profits to the grid services system.

The remainder of this paper is organized as follows. Section 2 gives an overview over admission control strategies based on resource reservation. In Section 3, a novel admission control strategy with layered threshold is introduced and in Section 4, comprehensive performance evaluations including both theoretical analyses and simulations based on the reward/penalty model are carried out. Finally, we conclude the paper in Section 5.

2 Admission Control Strategy on Resource Reservation

Three types of admission control strategy on reservation admission are defined [6,7]:

None-threshold strategy: it allows all requests share the whole resources without concerning service priority. For unawareness of the service priority-type, the number of the higher-priority requests, which can bring more profits to the online service provider than lower-priority ones, cannot be assured. Hence, this kind of strategy is not appropriate for the profit-emphasizing online multimedia service systems.

Fixed-threshold strategy: by an explicit reservation, resources that originally assigned to higher-priority requests will never be consumed by lower-priority requests. Thus, the number of the higher-priority requests can be assured to some extent. But this kind of strategy is poor in flexibility because the threshold partition is fixed and the requests with a higher priority can not use the resources exclusively assigned to lower priority requests.

Dynamic-threshold strategy: there exists an extra part of resource acting as a sharable part for all types of requests. When the resources assigned to the requests of one class are used up, the sharable resources can be taken into account. For the sharable part of resources, the flexibility of admission control is greatly improved. But, since there are still some resources only kept for each fixed class of requests, with the increment of the number of service class, this kind of strategy is obviously not so efficient and the resource utilization factor is still low.

To overcome the limitation of the three classic strategies, we propose a novel strategy with layered threshold, which is also based on resource reservation.

3 Admission Control with Layered Threshold: A Novel Strategy

The none-threshold strategy neglects the service priority of requests; the fixed-threshold strategy is not flexible; and in a large-scale multimedia system with several classes of requests, the dynamic-threshold strategy is still not so efficient. Here, we propose a novel strategy with layered threshold, which takes the impact of the differences in service priority on the system profits into consideration, thus helps to improve the system performance and keeps the admission flexibility as well.

3.1 System Model

The layered-threshold-based admission control strategy proposed in the paper is based on resource reservation. When a new request comes, the system will reserve resources for it if the request can be admitted. Otherwise, it will be rejected. Also, we assume that an arriving request that is not admitted immediately will be blocked or dropped, i.e., a request is never buffered.

The system tags the request with its *Service Level*(SL, $SL \in [0,1,2,......SL_{max}]$). A request with a larger SL value indicates a higher service priority. The request marked with SL_{max} has the highest priority. The requests arrive as *Poisson Flow* with mean arrival rate λ_{SL}. Once a request is accepted, it is assured of the service capacity until its task is completed. The *Service Time(ST)* of each request is independent of the SL. Hence, all the requests of any SL follow the same ST distribution. Here, we assume it accords to negative exponential distribution with an average time of μ.

System resources are divided into basic *Resource Units(RU)* with the same size. RU is the minimum resource unit and several RUs can be combined to form a *Resource Block(RB)*.

3.2 The Layered Threshold Strategy

In the proposed admission control strategy with layered threshold, we also use *resource layers (RL)* to denote the system resource according to the corresponding serving SL. The RL of the request with SL i is denoted as R_i. Meanwhile, each RL consists of two kinds of RBs: the RBs shared with the requests in a higher SL (RB_i^u) and the RBs shared with the requests in a lower SL (RB_i^d). So, as for SL i, we get $RB_i^d = R_{i-1}$ (in the extreme condition, we set RB_0^d as 0). Then we can deduce the equation (1) below:

$$R_i = RB_i^u + RB_i^d = RB_i^u + R_{i-1} = \sum_{j=0}^{i} RB_j^u \qquad (1)$$

The resource allocating relationship can be described as Figure 1.

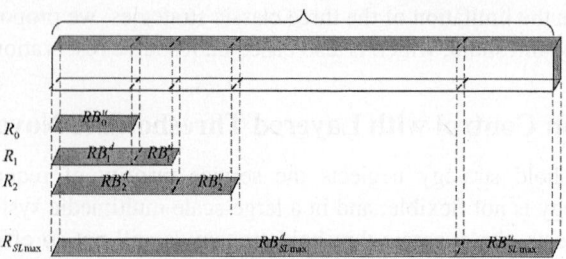

Fig. 1. Resource Allocation in Layered Threshold Strategy

In this strategy, the system will first determine the *SL* for an arrival request. Here we still suppose it as *i* and its corresponding *RL*, say, R_i. Then, the system will check the status of RB_i^u. If it is available, the request will be admitted and system will reserve resources for it; otherwise, it will turn to RB_{i-1}^u. The decision-making procedure will be repeated in the same way until the request is admitted or it reaches the last level, say, level 0. If RB_0^u is unavailable yet, the request will be refused.

We observe that, in this strategy, with the layered threshold setting, the class with a higher *SL* can use more resources and its request may have a higher possibility to be admitted. More over, through the layered resource sharing, the admission flexibility can also be achieved and the resource utilization factor and system performance can be greatly improved.

4 Performance Evaluation

In this section, we present an elaborate performance evaluation with (a) theoretical evaluation with pay-off rate function expressions and (b) simulations. The advantage of the new strategy over other three classic strategies will be demonstrated.

4.1 The Reward/Penalty-Based Evaluation Method

We consider that every request can be assigned with a reward indicating its value to the system when the request is successfully served and conversely a penalty indicating the negative value imparted to the system when the request is rejected [8]. With this method, the performance metric takes both rewards and penalties of the requests into consideration. It is called the system's total pay-off value. In other words, under a particular admission control strategy if the system on average serves *N* requests while rejecting *M* requests, the system total pay-off value *P* is

$$P = Nv - Mq \qquad (2)$$

where *v* and *q* denote the reward value and penalty value, respectively. Now, the problem we are interested in solving is to identify the best admission control strategy under which this performance metric is maximum.

4.2 Theoretical Evaluation

Here, we consider the case, in which the system serves a mix of $SL_{max}+1$ classes requests with different kinds of service priority, the highest priority is tagged with SL_{max} and the lowest with 0. Each request characterizes its own arrival/departure rate as well as its reward/penalty value. We also predefine the total resource units of the system as n and each request may ask for one unit of resource. Table 1 summarizes the notations used in this paper.

Table 1. Notation Parameters

λ_i	Arrival rate of the requests with SL i, $i \in [0,1,2,....SL_{max}]$
μ	Departure rate of the requests admitted
v_i / q_i	Reward/Penalty value of the request with SL i when served/rejected, $i \in [0,1,2,....SL_{max}]$
RB_i^u	Resource units reserved for requests with SL i, (in the layered threshold scheme, it denotes the upside-shared resource units), $i \in [0,1,2,....SL_{max}]$
RS	Totally sharable resource (only exist in the dynamic threshold scheme)
n	Maximum number of resource units that the system provides
P	The system total pay-off rate (initiated as 0)

In the case of none-threshold strategy, the system can be modeled as a classic $M/M/n/n$ hybrid queue system with the arrival rate $\Lambda = \sum_{i=0}^{SL\,max} \lambda_i$. Then we get the traffic intensity of the system as $\rho = \sum_{i=0}^{SL\,max} \lambda_i / \mu$. According to the Finite-State Limit Theorem of Birth-Death Process [9], the system's total loss probability can be given in terms of p_0,

$$p = p_n = \frac{1}{n!} \rho^n p_0 \qquad (3)$$

where $p_0 = [\sum_{i=0}^{n} \frac{1}{i!} \rho^i]^{-1}$,

When there are i requests in the system, its reward rate equals to

$$\hat{v} = \sum_{k=0}^{SL\,max} \left(i\mu \times \frac{v_k \lambda_k}{\Lambda} \right) \qquad (4)$$

Therefore, in the case of none-threshold and fixed-threshold strategy, the system total pay-off rate can be expressed as a multiple-parameter function expression as:

$$P_{none}(n, \lambda_0,...,\lambda_{SL\,max}, \mu, v_0,...,v_{SL\,max}, q_0,...,q_{SL\,max})$$
$$= \sum_{i=0}^{n} \left(i\mu \times \left(\sum_{j=0}^{SL\,max} \frac{v_j \lambda_j}{\Lambda} \right) \times \frac{1}{i!}\left(\frac{\Lambda}{\mu}\right)^i \left[\sum_{k=0}^{n} \frac{1}{k!}\left(\frac{\Lambda}{\mu}\right)^k \right]^{-1} \right) - \left(\sum_{j=0}^{SL\,max} q_j \lambda_j \right) \times \frac{1}{n!}\left(\frac{\Lambda}{\mu}\right)^n \times \left[\sum_{k=0}^{n} \frac{1}{k!}\left(\frac{\Lambda}{\mu}\right)^k \right]^{-1} \qquad (5)$$

$$P_{fixed}(RB_0^u,...,RB_{SL\max}^u,\lambda_0,...,\lambda_{SL\max},\mu,\nu_0,...,\nu_{SL\max},q_0,...,q_{SL\max})$$

$$= \sum_{j=0}^{SL\max}\left[\sum_{i=1}^{RB_j^u}i\mu\times v_j\times\frac{1}{i!}\left(\frac{\lambda_j}{\mu}\right)^i\times\left(\sum_{k=0}^{RB_j^u}\frac{1}{k!}\left(\frac{\lambda_j}{\mu}\right)^k\right)^{-1}\right] - \sum_{j=0}^{SL\max}\left[\lambda_j q_j\times\frac{1}{RB_j^u!}\left(\frac{\lambda_j}{\mu}\right)^{RB_j^u}\times\left(\sum_{k=0}^{RB_j^u}\frac{1}{k!}\left(\frac{\lambda_j}{\mu}\right)^k\right)^{-1}\right] \quad (6)$$

Similarly, in the case of dynamic-threshold strategy, we get

$$P_{dynamic}(RB_0^u,...,RB_{SL\max}^u,RS,\lambda_0,...,\lambda_{SL\max},\mu,\nu_0,...,\nu_{SL\max},q_0,...,q_{SL\max})$$

$$= \sum_{j=0}^{SL\max}\left[\sum_{i=1}^{RB_j^u}i\mu\times v_j\times\frac{1}{i!}\left(\frac{\lambda_j}{\mu}\right)^i\times\left(\sum_{k=0}^{RB_j^u}\frac{1}{k!}\left(\frac{\lambda_j}{\mu}\right)^k\right)^{-1}\right] + P_{none}(RS,\lambda_0,...,\lambda_{SL\max},\mu,\nu_0,...,\nu_{SL\max},q_0,...,q_{SL\max}) \quad (7)$$

where

$$\lambda_i' = \lambda_i\times\frac{1}{RB_i^u!}\left(\frac{\lambda_i}{\mu}\right)^{RB_i^u}\left(\sum_{j=0}^{RB_i^u}\frac{1}{j!}\left(\frac{\lambda_i}{\mu}\right)^j\right)^{-1} \quad (8)$$

In the case of the layered threshold admission strategy, the pay-off rate expression is more complex. In the $RB_{SL\max}^u$ field, it is an $M/M/RB_{SL\max}^u/RB_{SL\max}^u$ queue system and in the next $RB_{SL\max-1}^u$ field, the system equals to a compound queue system serving two kinds of requests with the SL of $SL_{\max}$ and $SL_{\max-1}$. Thus, the rest can be deduced in the same way and in the last resource field RB_0^u. We can get the expression of $P_{layered}$ as

$$P_{layered}(RB_0^u,...,RB_{SL\max}^u,\lambda_0,...,\lambda_{SL\max},\mu,\nu_0,...,\nu_{SL\max},q_0,...,q_{SL\max})$$

$$= \sum_{i=1}^{RB_{SL\max}^u}i\mu\times v\times\frac{1}{i!}\left(\frac{\lambda_{SL\max}}{\mu}\right)^i\times\left(\sum_{j=0}^{RB_{SL\max}^u}\frac{1}{j!}\left(\frac{\lambda_{SL\max}}{\mu}\right)^j\right)^{-1}$$

$$+ P_{none}(RB_{SL\max-1}^u,\lambda_{SL\max-1},\lambda_{SL\max}',\mu,\nu_{SL\max-1},\nu_{SL\max},q_{SL\max-1},q_{SL\max})$$

$$+ P_{none}(RB_{SL\max-2}^u,\lambda_{SL\max-2},\lambda_{SL\max-1}',\lambda_{SL\max}'',\mu,\nu_{SL\max-2},\nu_{SL\max-1},\nu_{SL\max},q_{SL\max-2},q_{SL\max-1},q_{SL\max}) \quad (9)$$

$$\cdots\cdots\cdots\cdots$$

$$+ P_{none}(RB_0^u,\lambda_0,\lambda_1',...,\lambda_{SL\max}^{(SL\max)},\mu,\nu_0,...,\nu_{SL\max},q_0,...,q_{SL\max})$$

After the detailed deduces above, we have the pay-off rate function expressions of four different strategies as $P_{none}, P_{fixed}, P_{dynami}$ and $P_{layered}$, with which, we can make a keen numerical performance comparison of these strategies on admission efficiency.

We assume the number of total resource units is 20 and the system servers 5 kinds of requests. All requests arrive in the same rate as 2 per minute. If a request is admitted, its service time is 5 minutes. Since a well served request with a higher priority can always bring more profits to the system, but if it is rejected, the profit loss of the system will also be larger, we assume that a request with a higher SL has a higher reward/penalty value than the request with a lower SL. Here, we assume the penalty value of each class requests is 1,2,3,4, and 5, respectively. The computed numerical results according to the expressions (5), (6), (7) and (9) with variations of the profits and arrival rate of each request are demonstrated in Figure 2 and Figure 3.

Figure 2 shows that there is only a small difference among the system pay-off values of these four strategies when the reward-penalty rate is low. But with the increment of the ratio, the performance advantage of our layered threshold strategy will be more obvious. In Figure 3, we observe that the system total pay-off under our layered threshold strategy is much higher than those under other three strategies. More over, with the increment of the arrival rate of higher-priority requests, the layered strategy shows more superiority in gaining more system total pay-off, all of which indicate that our new admission strategy is more efficient and competent.

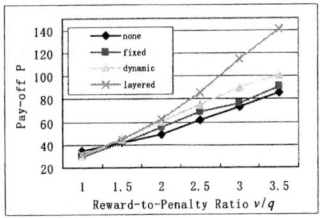

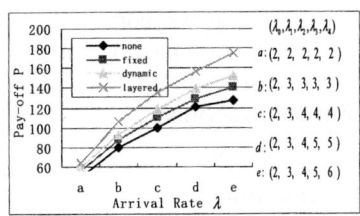

Fig. 2. Impact of Profits Differences   **Fig. 3.** Impact of Arrival Rate Differences

4.3 Simulation

Besides the elaborate theoretical evaluation, we also make some simulations to indicate the validity of our strategy. The simulations are carried out in the environment of GPSS (General Purpose Simulation System) [10].

Here, we can assume the total number of system resource unit much larger, say, 100, and they are still requested by 5 kinds of requests, which will arrive for services. If a request is admitted by the system, the average service time will be 30 minutes. We still fix the penalty values as 1,2,3,4 and 5, respectively and in each simulation we will initiate 10,000 requests.

Table 2. Impact of Profit Differences

	P_{none}	P_{fixed}	$P_{dynamic}$	$P_{layered}$
$v = q$	7873	11206	11696	12634
$v = 1.5q$	17374.5	21492	22125	23316
$v = 2q$	26876	31778	32554	33998
$v = 2.5q$	36377.5	42064	42983	44680
$v = 3q$	45879	52350	53412	55362
$v = 3.5q$	55380.5	62636	63841	66044

Table 3. Impact of Arrival Rate Differences

$(\lambda_0,\lambda_1,\lambda_2,\lambda_3,\lambda_4)$	P_{none}	P_{fixed}	$P_{dynamic}$	$P_{layered}$
(1,1,1,1,1)	26876	31778	32554	33998
(1,1.5,1.5,1.5,1.5)	11983	16554	16985	19812
(1,1.5,2,2,2)	3730	7183	7758	9457
(1,1.5,2,2.5,2.5)	-930	2090	2388	3053
(1,1.5,2,2.5,3)	-4308	-1864	-1266	1071

By increasing the ratio of reward value to penalty value and arrival rate differences step by step we get the simulations results as Table2 and Table3. In Table3, because of the limitation of the system resource capacity, when higher-priority requests arrive more frequently, more requests are refused by the system and more penalty values are produced. But even in this situation, our layered threshold based admission control strategy still has a better performance metric than other strategies.

Based on the simulation results above, we can see the fact that when higher-priority requests arrive more frequently and higher-priority requests can provide more profits, the system deploying this new strategy can get larger pay-off value, and thus, will achieve more profits. These results are consistent with the previous theoretical performance evaluation.

5 Conclusion

We propose a novel admission control strategy with layered threshold in this paper. Both our theoretical analyses and simulations reveal that the grid-based services system under the new strategy yields an improvement in performance and profit over other three existing strategies. We also get the result that with the increase of the arrival rate and profit of higher-priority clients, the new strategy can not only keep the flexibility but give the higher-priority clients more resource priority.

This new strategy is being implemented at "ND-SMGrid", a grid-based prototype system providing stream media transmission services at Distributed and Parallel Computing Laboratory, of Nanjing University. And in the future, we will pay more attention to the admission control in advance resource reservation for Grid-based multimedia service management [11] and measurement-based adaptive admission control in grid service system [12].

References

1. B. Leiner, V. Cerf, D. Clark, et al "The Past and Future History of the Internet", *Commun. ACM 40, No. 2, pp.102–108*, Feb. 1997.
2. D. Wu, T. Hou, W. Zhu, Y.-Q. Zhang, J. Peha, "Streaming Video over the Internet: Approaches and Directions", *IEEE Transactions on Circuits and Systems for Video Technology, Special Issue on Streaming Video, vol. 11, no. 3, pp. 282-300*, March 2001.
3. I. Foster, C. Kesselman, J.M. Nick, S. Tuecke, "The Physiology of the Grid, An Open Grid Services Architecture for Distributed Systems Integration", *www.globus.org* 2002
4. K. Cleary, "Video on Demand - Competing Technologies and Services", *In Proceedings of International Broadcasting Convention, pp. 432-437*. Sept. 1995.
5. S. Lim, C. Lee, C. Ahn, "An Adaptive Admission Control Mechanism for a Cluster-Based Web Server System", *IPDPS 2002 Workshops*, April 15 - 19, 2002
6. R. Simon, P. Mundur, A. Sood, "Access Policies for Distributed Video-on-Demand Systems" *Journal of the WAS, Special Issue on Communications, 85(2)*, Dec. 1998.
7. Ing-Ray Chen, Sheng-Tun Li, "A Cost-Based Admission Control Algorithm for Handling Mixed Workloads in Multimedia Server Systems", *the Eighth International Conference on Parallel and Distributed Systems*, June 26 - 29, 2001
8. I.R. Chen, T.H. Hsi, "Performance Analysis of Admission Control Algorithms Based on Reward Optimization for Real-Time Multimedia Servers", *In Performance Evaluation 2(33), pp.89-112*, Jul. 1998.

9. K.S. Trivedi, "Probability and Statistics with Reliability, Queuing and Computer Science Applications. Second Edition", Wiley, New York, 2002
10. http://www.minutemansoftware.com/
11. D. Bruneo, M. Guarnera, A. Zaia, et al, "A grid-based architecture for Multimedia services management", *2003 Annual Crossgrid Project Workshop & 1st European Across* Grids Conference, Feb. 2003
12. I.H. Kim, J.W. Kim, S.W. Lee, et al "Measurement-Based Adaptive Statistical Admission Control Scheme for Video-on-Demand Servers", *In proceedings of The 15th International Conference on Information Networking (ICOIN'01), pp. 472-478,* Feb. 2001.

Applying Grid Technologies to Distributed Data Mining

Alastair C. Hume[1], Ashley D. Lloyd[2,3],
Terence M. Sloan[1], and Adam C. Carter[1]

[1] EPCC, The University of Edinburgh, Edinburgh, UK
{A.Hume,T.Sloan,A.Carter}@epcc.ed.ac.uk
http://www.epcc.ed.ac.uk
[2] Curtin Business School, Curtin University of Technology, Perth, Australia
ashley@curtin.edu.au
http://www.cbs.curtin.edu.au
[3] University of Edinburgh Management School, The University of Edinburgh, Edinburgh, UK
http://www.ems.ed.ac.uk

Abstract. The Grid promises improvements in the effectiveness with which global businesses are managed if it enables distributed expertise to be efficiently applied to the analysis of distributed data. We report an ESRC-funded collaboration between EPCC in Edinburgh and Curtin University of Technology in Perth, Australia, that is applying public-domain Grid technologies to secure data mining within a commercial environment. We describe this Grid infrastructure and discuss its strengths and weaknesses.

1 Introduction

Data mining projects often require distributed analysts to submit jobs to distributed compute resources that process data from distributed data resources. These requirements, along with others such as secure communications and access control, make data mining an ideal application of Grid technologies.

The INWA project [1] has investigated the suitability of existing grid technologies for secure commercial data mining. This project has been funded under the Pilot Projects in E-Social Science programme [2] of the UK's Economic and Social Research Council (ESRC). The full title of the project is 'Informing Business and Regional Policy: Grid-enabled fusion of global data and local knowledge' but for ease of communication this has been abbreviated to INWA.

The project is a collaboration between academic, commercial and government organisations drawn from the UK and Australia. EPCC, the University of Edinburgh Management School (UEMS) and Lancaster University Management School (LUMS) [3] are the academic partners in the UK with Curtin Business School from the Curtin University of Technology the academic partner in Perth, Australia. The commercial and government partners are from the UK and Australia

Financial, telecommunications and property data have been provided by the commercial and government partners. These partners have also helped formulate requirements for mining of this data. The major requirement being to ensure that any data supplied can only be accessed by trusted parties. The data from UK partners is sited at EPCC with the Australian data sited at Curtin. Sun Microsystems in Australia provided the project with the compute servers for the data located at Curtin.

Such a collaboration between multiple data services in multiple jurisdictions tests acceptance of the grid - a pre-requisite for anyone to adopt this technology.

This paper elaborates on a initial project report presented in [4].

2 The INWA Grid Infrastructure

The project has designed and implemented a Grid infrastructure using existing publicly available Grid technology. This allows analysts at Edinburgh or Perth to submit batch jobs securely that are run on a compute resource local to the data being processed. The results from the batch jobs are automatically transferred back to the user. The infrastructure also allows analysts to interact with the relational data sources via SQL queries.

To submit and transfer batch jobs and their results between local and remote sites the infrastructure uses Grid Engine V5.3 [5] as the compute resource manager, and Transfer-queue Over Globus (TOG) [6] with Globus Toolkit V2 [7] for the Grid middleware.

Grid Engine is an open source distributed resource management system that allows the efficient use of compute resources within an organisation. The Globus Toolkit is essentially a Grid API for connecting distributed compute and instrument resources. TOG integrates Grid Engine V5.3 and Globus Toolkit V2 to allow access to remote resources. This allows an enterprise to access remote compute resources at any collaborating enterprises [8].

The infrastructure uses OGSA-DAI Release 3.1 [9] and the FirstDIG browser [10] to provide interactive access to the relational data resources via SQL queries. This allows Grid technologies to be applied at the crucial data cleaning and preparation stages that precede any data mining.

OGSA-DAI (Open Grid Services Architecture-Data Access and Integration) is grid middleware software designed to assist with the access and integration of data from separate sources via Grid Services. OGSA-DAI Release 3.1 uses the Globus Toolkit V3.0.2. The FirstDIG browser allows a user to interact with OGSA-DAI grid service enabled data sources using SQL queries via a GUI.

This paper describes the INWA Grid infrastructure and discusses its strengths and weaknesses.

3 Batch Job Submission and Execution

The infrastructure is used to run data cleaning and mining batch jobs on the data located at EPCC and Curtin. The data cleaning batch jobs generally use GritBot [11] and Perl scripts whilst the data mining batch jobs use C5.0 [12].

Batch job submission and execution using the infrastructure are illustrated in Figure 1. The infrastructure has two compute resources, one at EPCC and the other at Curtin. Both these compute resources are running Grid Engine. Grid Engine provides job queues into which all jobs to be executed on the compute resource must be placed. Grid Engine will cause the jobs to be executed when the required compute resource is available. The analysts do not interact with the compute resources directly. Analysts submit jobs to personal versions of Grid Engine running on their local system.

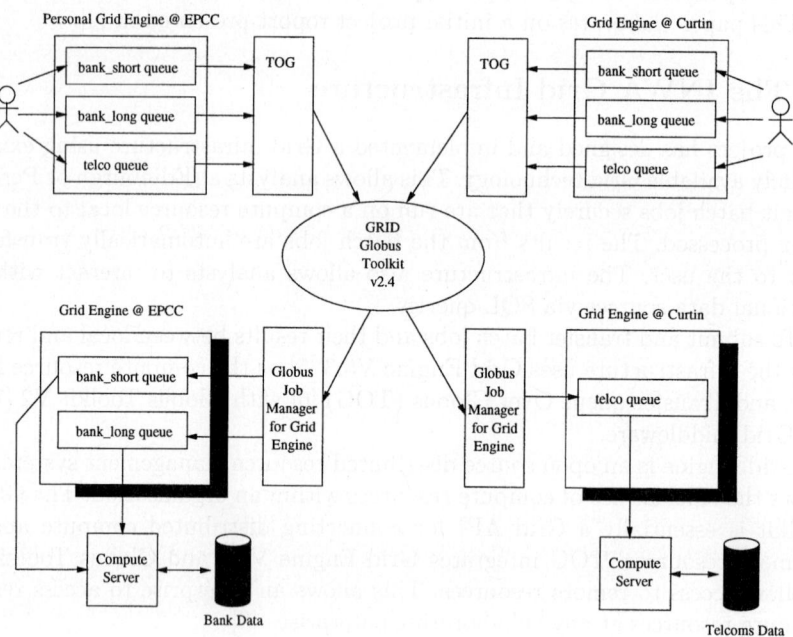

Fig. 1. Job submission and execution using the INWA Grid infrastructure

The queues in these personal Grid Engine installations are configured to use TOG to forward the job to a Grid Engine queue on a remote compute resource. TOG is configured to ensure that the Globus Toolkit securely encrypts a job and its results during transfer to and from the remote compute resource.

This infrastructure has several advantages for the analyst over the non-Grid approach of simply having a secure shell connection to each of the compute resources:

- The Globus toolkit provides a certificate-based authentication mechanism that allows the jobs to be sent to the remote resources without the need to remember a username and password for each resource.
- Analysis can use multiple compute resources without any need to know where they are located. This information is held within the personal Grid Engine configuration that maps queue names (which are based on data set names) to queues on remote resources.

– All input and output files associated with jobs are stored on the analyst's account on his local Unix system. There is no need for the analyst to remember where files are and explicitly transfer them between systems.

There are also a few potential disadvantages of the infrastructure:

  – The analyst has to submit all jobs via Grid Engine. If the analyst is not familiar with Grid Engine this will be an additional learning curve. In our case the only way to run jobs on the EPCC compute resource is through Grid Engine so it is something analysts are already familiar with and understand.
  – There is no simple way for an analyst to determine the state of a job running on a remote compute server. Querying the state of a job submitted to a personal Grid Engine will return the state of the job on that Grid Engine. There will be a corresponding job, and hence job state, on the compute resource's Grid Engine. It is this job state that is difficult to determine without logging on to the remote compute resource.

Several problems were encountered while setting up this infrastructure. The first problem discovered was that TOG did not provide any mechanism to configure the Globus security settings. All file transfers carried out by TOG were sending the data over the Internet in clear text. This was not suitable for a commercial data mining environment where the output files will contain commercially sensitive information. The project was able to get access to the TOG source code and write an enhancement to configure the data channel protection level used by Grid FTP when transferring files. This enhancement has since been given to the TOG project.

The second problem was a bug in the Java CoG [13] software that prevented the use of Grid FTP in a secure manner. This problem took weeks to track down because it initially looked like a server bug. Eventually it was traced down to the Java CoG software and a unit test written that displayed the problem. When this information and the unit test were passed to the Java CoG team the problem was resolved very quickly.

All the analysts working on the project were able to obtain Grid certificates for the UK e-Science certification authority (CA). This was convenient because this CA was already trusted by the EPCC compute resource. Had the analyst in Australia been unable to obtain a certificate in the UK he would have had to obtain a certificate from an Australian CA and this CA would have had to be added to the EPCC compute resource. At the time there did not appear to be a high profile, trustworthy provided of Grid certificates in Australia.

4 Interactive Data Access

A secure OGSA-DAI configuration and the FirstDIG browser were installed at EPCC to allow interactive access to the relational data resources. The FirstDIG browser allows the user to interact with the suite of OGSA-DAI Grid data services and send SQL queries to the data resource. The query results are displayed in a result window and can be saved to a CSV file if desired.

This secure OGSA-DAI configuration ensures that only authorised and authenticated users can access the grid data services and that SQL queries and their results are encrypted during transfer. OGSA- DAI Release 3.1 uses Globus Toolkit V3.0.2 to provide the authorisation, authentication and encryption facilities.

Unfortunately, no general-purpose software packages, such as interactive statistics packages, currently support the OGSA-DAI interface and it is likely to require great leaps forward in terms of Grid technology and standardization before they do so. If the Grid community wishes to use existing data processing and visualization tools with Grid-enabled data resources it ought to be possible to develop ODBC and JDBC bridges for OGSA-DAI that would allow applications to access these data resources over the Grid using standard interfaces.

Given that there are so many existing applications that access data resources using ODBC and JDBC drivers it is worth addressing what advantages would be gained by using OGSA-DAI for simple access to remote data resources. OGSA-DAI offers the following advantages:

- Most JDBC and ODBC drivers pass login details in clear text and hence highly compromise security. OGSA-DAI allows the use of the Globus Toolkit's more secure certificate-based authentication mechanism instead.
- There is no need for the user to obtain and remember login details to every data resource they wish to access. All authentication and authorization will be carried out using the using the Globus Toolkit's certificate based mechanism.
- OGSA-DAI provides an extra level of access control mapping that eases the task of managing user access at the server.
- OGSA-DAI can encapsulate many diverse data resources (e.g. Oracle, MySQL and DB2 databases) behind a single interface. Thus the client application does not require a driver for each data resource type. These drivers are required only at the server.

5 Integrating External Data Using OGSA-DAI

Grid technologies offer exciting opportunities for collaborative research or business ventures. OGSA-DAI provides a means by which data can be shared between several parties in a controlled and secure way. This creates many exciting possibilities for data mining in particular. Grid technologies may allow an analyst to find an external data source, confirm its suitability and integrating it with his existing data all within a very short time.

The commercial and governmental organisations whose data we wished to integrate do not provide access to their data using OGSA-DAI. In order to investigate data integration issues we obtained these external data sets as database exports and created our own database containing this data. These database were then exposed through OGSA-DAI Grid Services.

The task of integrating external data with existing data is a test of OGSA-DAI's data integration functionality. The OGSA-DAI data browser has a window that presents a standard data integration pattern for performing an SQL join operation across distributed data resources. The pattern involves three Grid Data Services (GDS). Two of these GDSs, termed the source GDSs, correspond to the two distributed data resources. The third GDS, termed the scratch GDS, provides a place to store temporary results and execute the SQL join operation. A unique SQL query is sent to each of the source GDSs. The results of these queries are obtained by the scratch GDS using OGSA-DAI's data transfer mechanisms and loaded into temporary tables created at the scratch GDS. The two temporary tables now contain data and are at the same data resource so it is possible to execute an SQL join query over these tables. The result of this SQL join query is returned to the client and the temporary tables are destroyed.

This data integration pattern was used to integrate external data with our internal data using the postcode as a common field for the SQL join. Unfortunately applying this pattern using the OGSA-DAI data browser is not a trivial exercise. The user has to enter seven SQL queries. Two of these queries select data from each of the source GDSs and a third query is the join operation to the performed at the scratch GDS. The other four queries create and drop the temporary tables at the scratch GDS. In theory, it ought to be possible to automatically construct the four queries that create and drop the temporary tables, but OGSA-DAI's current set of activities do not provide the required functionality. In particular the OGSA-DAI bulk load activity, which loads data from an input stream into a specified table, cannot automatically create a table that matches the format of the input data. It is important the user of the pattern appreciates the importance of reducing as much as possible the amount of data transferred between the GDSs. It is hence important is be as restrictive as possible within the SQL select statements that are sent to the source GDSs.

The data integration task was made slightly more complicated because the format used for the postcode was slightly different at the two data resources. This was easily overcome using simple SQL string processing functions supported by the DMBS. It some cases the differences may be less trivial and require more than simple string processing. In these cases it may be possible to use OGSA-DAI's XSL transform activities to transform the data from one format to the other. The XSL transform activities may also be useful for converting date formats, which are a typical cause of data integration difficulties.

To give an example of the performance of OGSA-DAI for data integration this distributed join took 239 seconds to complete. Approximately 98% of this time was spent transferring and inserting data to the temporary tables on the scratch GDS. One temporary table consisted of 2,881 records and the other 12,749 records. The OGSA-DAI activity code used to insert the data into the temporary tables was developed as a proof of concept and was not optimised for speed. It ought to be possible to re-write this activity such that the overall time for this distributed join is considerably reduced. Deelman et al. [14] have analysed the performance of OGSA-DAI is greater detail.

Several problems were encountered while using the OGSA-DAI data integration pattern. The first problem was the Grid Data Transform activities, which

Research and Implementation of Grid-Enabled Parallel Algorithm of Geometric Correction in ChinaGrid*

Haifang Zhou[1], Xuejun Yang[1], Yu Tang[2], Xiangli Qu[1], and Nong Xiao[1]

[1] Institute of Computer, National University of Defense Technology, Changsha, China
[2] Institute of Electronic Technology, National University of Defense Technology
{haifang_zhou,yutang18}@sina.com

Abstract. ChinaGrid is an important project sponsored by China ministry of education. In this work, one of the ChinaGrid applications, parallel remote sensing image processing, is described. Geometric correction is a basic step during the processing of remote sensing image, which is traditionally a computation-intensive and communication-intensive application if in parallel mode. In order to moving this application into grid, a new grid-enabled parallel algorithm of geometric correction GPGC is proposed. Experimental results show that our algorithm is more suitable for grid platform, excelling old method in performance and salability. In the end, the overall architecture of service system of ChinaGrid and the portal of our application are given.

1 Introduction

With the rapid innovations of remote sensing technology, remote sensing image processing is facing three big challenges: a great demand of huge storage resource, an increasing complexity of computation and an intensive requirement of fast even real-time processing [1][2]. The size of the image, for example, is up to gigabytes; and only a process of geometric correction for one 10000×10000 image needs over ten billion float operations [3]. Obviously, single processor system would not meet all these demands. Therefore, massively parallel processing (MPP) or high performance computing (HPC) becomes the best way to speed up the processing and analyzing of remote sensing image. However, considering allowed cost of production, not all application or research units own costly supercomputers. On this background, how to utilize low cost computing resources and idle computing capacity in collaborative research and development environment receives much attention.

Grid [4], which builds on the accessibility of the Internet to allow effective use of geographically distributed resources, offers a solution of cooperative computing. Computational grid [5], for example, could support traditional parallel applications, aiming to offer consistent and inexpensive access to resources irrespective of their physical location or access point. Standard MPI programs can be directly executed in computational grid by compilation with MPICH-G2 [6]. However, in grid, many existing parallel algorithms could not achieve same efficiency as them executing on

---

* This work is partially supported by the National 863 High Technology Plan of China under the grant No. 2002AA1Z201, 2002AA104510 and 2002AA714021, and the Grid Project sponsored by China ministry of education under the grant No. CG2003-GA00103.

H. Jin, Y. Pan, N. Xiao, and J. Sun (Eds.): GCC 2004, LNCS 3251, pp. 704–711, 2004.
© Springer-Verlag Berlin Heidelberg 2004

traditional supercomputers; especially for those communication-intensive and fine-grain parallel algorithms that are suitable to closely-coupled multiprocessor system. The reason is the nodes in grid are coupled loosely and network bandwidth and speed is limited. Therefore, to get benefit from cooperative computing, grid-enabled parallel algorithms must be proposed and discussed in depth.

In this paper, we focus on *geometric correction*, which is a basic step in remote sensing image processing requiring intensive computing power and normally taking long time to complete. Traditional parallel geometric correction algorithm is communication-intensive only suitable for closely-coupled parallel system. Thus, facing grid environment, we will propose a new parallel algorithm for geometric correction.

Our research is supported by ChinaGrid project that is an important project granted by China ministry of education. *Grid-enabled remote sensing image processing* is one of applications of ChinaGrid. Our work in this paper has been integrated into related service system of image processing in ChinaGrid.

2 Geometric Correction and Its Traditional Parallel Method

When gathering the information of earth objects to generate images on satellite, the sensors, remote sensing platform, the earth itself and other factors can cause geometrical distortion in images. Geometric correction [7] (also called geometry rectification) is to correct this distortion when the image is formed. Geometric correction is an indispensable process during the preprocessing of remote sensing image [1]. Figure 1 shows the main flow of geometric correction: 1) the forward mapping functions $(u,v) = f(x,y)$ from input image space to output image space must be given (see figure 2). Different correction models make different functions [2], such as polynomial transform, collinear equation, random field interpolation etc. 2) Because of using inverse mapping method [8] for resampling, the second step is to find the scope of the output image according to forward mapping functions (see the area surrounded by the real line in figure 2(b)), which is the area of resampling. 3) The third step is to get the inverse mapping functions $(x,y) = g(u,v)$ which map the output space to the input space. 4) The fourth step is geometric transformation, that is to transform the pixel's coordinates (u,v) in the scope of the output image to the conjugated position (x,y) in the input image by the inverse mapping functions. 5) The last step is to compute the intensity (gray value) of pixel (u,v) according to the intensities of pixels around conjugated pixel (x,y), i.e. resampling. The interpolation method often used is cubic convolution, bilinear interpolation or nearest neighbor method.

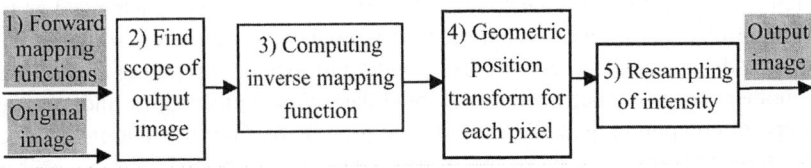

Fig. 1. The main flow of serial algorithm of geometric correction

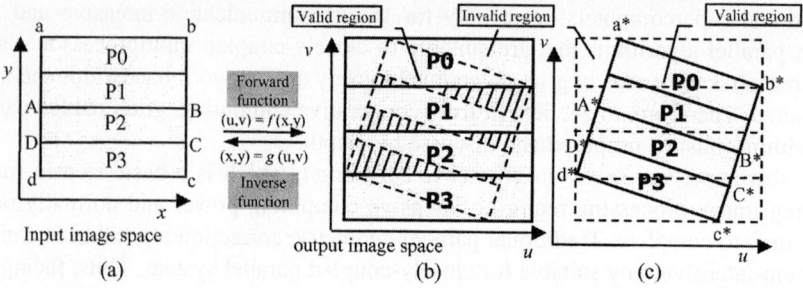

Fig. 2. The principle of the parallel algorithms

Because of the complexity of the inverse function and the demand of high-precision resampling, most computation is concentrated on the last two steps. In the traditional parallel algorithms [2][3][9], the input data is divided into regular strips or blocks, and distributed to each processor. For example in figure 2(a), the original data is divided into 4 strips evenly and each strip data is stored in the local storage of four processors. And then correction starting from the output image space, the scope of resampling for each processor is also regular partitioned, seeing the area surrounded by the black real line in figure 2(b). This method is straightforward and simple, but it needs data communication and synchronization between processors during resampling: when computing intensities of pixels (u,v) in output image space, a part of needed data in the input image space are not stored locally and must be got from other processors. This leads to frequent small data transfer among processors. See figure 2(b), the part indicated with the bias is the part needing communication to complete resampling. Complex communication mode makes this parallel method only suitable closely-coupled parallel system like SMP. It is poorly programmable and applicable in computational grid. In addition, because of geometric distortion, the corrected image is larger than source image. Thus there is blank area (set to zero directly) in output image besides the valid content (see figure 2(b)). But the traditional parallel algorithm did same computation for blank area not distinguishing it with the valid area, so it not only introduces much redundant computation but also causes the load imbalance between the processors. For these reasons, in order to migrate this application into grid, the parallel strategy must be reconsidered.

3   A Grid-Enabled Parallel Algorithm of Geometric Correction

The critical point is how to keep data locality during the resampling to avoid communication between grid nodes. One way is to keep computing area unchanged as traditional algorithm (partition of output space is regular) but to dispatch irregular dataset to every node (the needed dataset corresponding to regular computing area is irregular); another way is to keep partition of input dataset is regular but to find out irregular scope of computing area. To simplify the process of data distribution in grid, we choose the latter way. Therefore, we aim to make the work load of every grid node that participates in computing limited in its regular dataset that distributed from beginning, but not need remote data in other grid nodes during the resampling.

Based on above analysis, we proposed a new *Grid-enabled Parallel algorithm for Geometric Correction* (GPGC). Its basic idea is: correction starting from the input image space (not output space), the edges of each output image strip (computing area) that corresponds to the input image strip in each node are computed using the forward transform (these edges of output image strip is often irregular). For example, in figure 2(c) the black real lines show the edges of each output image strip. Each node implements the position transform and resampling in its own area surrounded by these edges, so all the computation only uses the local data and needs no communication.

There is another problem that the final output data strips only includes the valid area (see figure 2(c)) and can not be directly jointed into an output image. Therefore, in the last step, we add a process that regularizes the output data into a regular image (that is redistributing the final data). Although this step still needs communication, but the delayed and concentrated communication not only eliminates the synchronization during the whole computation and also makes the most of network bandwidth. Moreover, all the computation is valid, so it eliminates effect of the load imbalance.

Some technical details about GCPC will be described in the following sections. Before that, delicate data partition should be designed firstly. The regular 1-D or 2-D data partition of the input image is straightforward. But there are another two problems. Firstly, to seamless joint the different parts of output image, some redundant data need to be stored along the edges of each part in each node. Secondly, resampling along the local edges needs some extra data allocated to other nodes, and the amount of needed data is related to the width of the resampling template. The extra data can be obtained by two ways. The first way needs some communications with the neighboring nodes, and the second utilizes redundant-stored data along the edges and no communications happen. When the redundant-stored data is used for these two problems, no communications happen during the computation and it is convenient for parallel computing.

3.1 Compute the Edges of Local Output Data Space

The key point of our algorithm is to find out the irregular area edges of local computing area corresponding to the regular area of input image for each node. See figure 2(a) and (c): the edges of regular strip ABCD in the input image are mapped to the edges of irregular area A*B*C*D* in the output image through forward functions, and then intensity resampling of all pixels contained in the area A*B*C*D* is completed using local data in strip ABCD through the inverse functions.

The mapping function probably makes the shape of the irregular area very complex, so suppose that it has the limited connectivity and no extremum exists along the edges besides four corner points. The concrete computing method is showed as figure 3, which can be described as following four steps: 1) The four irregular edges in the output image correspond to four regular edges of local area in the input image, and they are computed through the forward mapping functions and stored. 2) The relative positions of four irregular edges A*B*, B*C*, C*D* and D*A* can be obtained according to the four corner points A*, B*, C* and D*. 3) The four irregular edges obtained at step 1 are dealt with as follows: the coordinates v of the edges intersected

be tradeoff, including image size, image distortion extent, number of grid nodes and network status.

Geometric correction is one of basic functions in remote sensing image processing. Based on study of the algorithms, we have developed a Parallel Remote-sensing Image Processing Service System for Grid (PRIPSS-G), which is a sub-project of Image Processing Grid project that is one of five important applications supported by ChinaGrid. Details about PRIPSS-G are presented in another paper. GPGC proposed in this paper has been encapsulated as a service integrated in PRIPSS-G. The portal of this service is showed as figure 6.

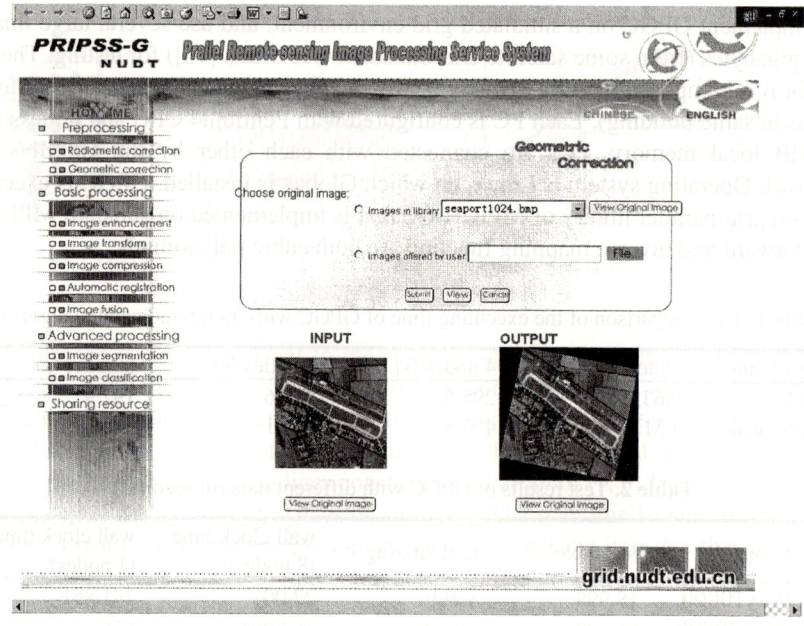

Fig. 6. The portal of our application

5 Conclusion

Grid offers a solution to utilize idle computing resources in network. Through grid, we can realize parallel computing to fasten the execution of our application programs even if there is no supercomputer available at hand. But not all existing parallel algorithms could get same efficiency in grid as them in traditional parallel systems. Therefore, research on grid-enabled parallel algorithms for large-scale application is significant.

In this paper, a new grid-enabled parallel algorithm of geometric correction GPGC is proposed for remote sensing application in ChinaGrid. GPGC changes the frequent and fine-grain communication mode of existing parallel method into a delayed but concentrated exchanging mode by computing irregular local output area. This changing makes no communication and synchronization happened during the resampling step that occupies most execution time. Experimental results show that our algorithm has better performance and scalability than traditional algorithm in grid environment. Moreover, our research is being integrated into related service system of ChinaGrid.

For the future, more parallel image processing algorithms for remote sensing application will be integrated into PRIPSS-G. And with the development of ChinaGrid, real grid service for the image processing applications will be open to all users who need it.

References

1. Zhu, S. L., Zhang, Z. M.: The Acquisition and Analysis of Image for Remote Sensing. Publishing House of Science, Peking, China. (2000) 109-132
2. Su, G. D.: Technology of Parallel for Image Processing. Publishing House of Tsinghua University, Peking, China. (2002) 6-56
3. Zhou, H. F.: Study and Implementation of Parallel Algorithms for Remote Sensing Image Processing. PhD thesis. National University of Defense Technology, Changsha, HuNan, China (2003) 1-20, 21-39
4. Foster, I., Kesselman, C., Tuecke, S.: The Anatomy of the Grid: Enabling Scalable Virtual Organizations. Int'l J. High-Performance Computing Applications, Vol.15.3 (2001) 200-222
5. Foster, I., Kesselman, C.: The Grid: Blueprint for a New Computing Infrastructure. Morgan Kaufmann Publishers, Inc., San Francisco, USA. (1999)
6. Karonis, N. T., Toonen, B., Foster, I.: MPICH_G2: A Grid-Enabled Implementation of the Massage Passing Interface. In Journal of Parallel and Distributed Computing (JPDC), Vol. 63.5 (2003) 551-563
7. Zhang, Y. S., Wang, R. L.: Dynamic Inspect in the Remote Sense. Publishing House of Liberation, Peking, China. (1999) 65-80
8. Wolberg, G., Sueyllam, H. M., Ismail, M. A., Ahmed, K. M.: One-dimensional resampling with inverse and forward mapping functions. Jounal of Graphics Tools. Vol. 5.3 (2001) 11-33
9. Chalermwat, P.: High performance automatic image registration for remote sensing. PhD. thesis. George Mason University, Fairfax, Virginia, USA (1999) 10-17, 86-96
10. http://geogratis.cgdi.gc.ca/download/landsat_7/ortho/geotiff/lcc/063006/063006_0100_990 813 _l7_08_lcc00.tif. Download at 2002.12.

Mixed Usage of Wi-Fi and Bluetooth for Formulating an Efficient Mobile Conferencing Application

Mee Young Sung, Jong Hyuk Lee, Yong Il Lee, and Yoon Sik Hong

Department of Computer Science & Engineering, University of Incheon
{mysung,hyuki,yongil,yshong}@incheon.ac.kr

Abstract. This paper proposes a mobile conferencing system which allows the mixed usage of Wi-Fi and Bluetooth and the performance analysis of it. The performance analysis includes the analysis of interference due to the mixed usage of Wi-Fi and Bluetooth. The objective of this study is to propose a method for formulating the best networking structure of a mobile conferencing system in a given situation. We developed a mobile conferencing system which works for both types of mobile networking methods; they are Wi-Fi for configuring infrastructure networking and Bluetooth for configuring ad hoc networking. We performed some experiments for comparing three mobile networking methods; Wi-Fi only, Bluetooth only, and the mixed usage of Wi-Fi and Bluetooth. These experiments lead us to learn that the mixed usage of Wi-Fi and Bluetooth for a mobile conferencing system can be a good solution for formulating an efficient mobile conferencing application. Our mobile conferencing system can benefit greatly the advantages of both types of networking methods (Wi-Fi for infrastructure networking and Bluetooth for ad hoc networking). Through our experiments, we experienced that the interference of Wi-Fi and Bluetooth degrades the system performance to some degree, however, it occurs randomly and it is tolerable if the application is carefully implemented to avoid the effect of interference.

1 Introduction

Currently, two types of mobile networking methods are distinguished. One is ad hoc networking [1], [2], [3] and the other is infrastructure networking [1]. Bluetooth [4], [5], [14] is a technology typically used for ad hoc networking, even though Bluetooth can be used for infrastructure networking. Bluetooth is secure and it allows devices to communicate even in areas with a great deal of electromagnetic interference, but it is slow and has a low communication range. In comparison, Wi-Fi (Wireless-Fidelity) certifies the IEEE 802.11 standard [1], [6] that is a technology typically used for infrastructure networking, even though Wi-Fi (IEEE 802.11) also allows to be used for ad hoc networking. Wi-Fi is much faster and has a wider communication range, but it is much less secure and more susceptible to interference. There are many studies on the comparison of Wi-Fi and Bluetooth [7], [8], [9], [10], [11], [12], [13].

This paper provides a method to operate the most appropriate and efficient networking structure for a mobile conferencing system in mixed usage of Wi-Fi and Bluetooth depending on given situations. In addition, this paper also addresses the problem of interference in mixed usage of Wi-Fi and Bluetooth. There are two types of mobile networking structures; infrastructure and ad hoc networking, and they have advantages and disadvantages. We intend to develop a mobile conferencing system which profits the advantages of both types of networking structures. For this, we developed a mobile conferencing system which works for both mobile networking structures. Our mobile conferencing system is realized in two ways; infrastructure networking is implemented using Wi-Fi technology and the ad hoc networking is implemented using Bluetooth technology. We performed some experiments for verifying which mobile networking method is desirable for formulating an efficient mobile conferencing application in three cases; Wi-Fi (infrastructure networking) only, Bluetooth (ad hoc networking) only, and the mixed usage of Wi-Fi and Bluetooth. We also analyzed the interference in mixed usage of Wi-Fi and Bluetooth in addition to the performance of the system in given situations.

We will briefly present the comparison of Wi-Fi and Bluetooth in the following section. In section 3, our mobile conferencing system will be examined. Then we will discuss the experiment settings and the results of the experiment in section 4. Finally, the last section will provide conclusions and the future work.

2 Comparison of Wi-Fi and Bluetooth

It is remarked that the maximum speed of the IEEE 802.11b WLAN (Wireless Local Area Network) is up to 11Mbps, which is very advantageous. In addition, Wi-Fi allows for the easy construction of a wired and wireless network. However, it is required that the specification of users is changed for constructing an ad hoc network in a WLAN, and this is a tedious job. The interference of wireless radio waves can be resolved by using WEP (Wired Equivalent Privacy), however, intentional hacking cannot be resolved.

In the Bluetooth wireless environment, a Piconet, an ad hoc network can be composed of a master and up to 7 active slaves within 10m. A slave can become a master of another Piconet. Therefore, multiple Piconets can form a Scatternet that is an extended network of Piconets. It is reported that coupling up to 10 Piconets would not degrade the network performance. A mobile application can be designed for enabling a conference of users with mobile devices within 10m and allows up to 70 users using a Scatternet. A convocator of a conference becomes the master of a Piconet, and eventually the master of the Scatternet. Bluetooth allows for recognizing devices using a 48-bit addressing scheme. It also provides a secret key implemented by a frequency-hopping schema, which allows only for decoding the data sent to the master. Moreover, it is secure because authentication is required for communicating with another device. Bluetooth is safer than Wi-Fi from a security point of view. However, it is disadvantageous that the communication range of the Bluetooth is up to 10m and the transmission speed is up to 1Mbps.

Both of Wi-Fi and Bluetooth allow for configuring an ad hoc networking, however, Wi-Fi is commonly used for WLAN (infrastructure networking) and Bluetooth is appropriate for WPAN (Wireless Personal Area Network; ad hoc networking). Note that Wi-Fi is more advantageous than Bluetooth as far as concerned the initial delay for connection.

The Wi-Fi is a DSSS (Direct Sequence Spread Spectrum) system operating in 2.4GHz band [1]. The Wi-Fi radios operate in the 79MHz frequency band, which is divided into 3 channels spanning 22MHz each. Bluetooth is a FHSS (frequency hopping spread spectrum) operating at 1Mbps in the same frequency band as Wi-Fi (2.4GHz) [2]. The hop frequency is 1600 hops per second; the frequency spectrum is divided into 79 hops of 1MHz bandwidth each. Figure 1 presents the frequency bands where interference may occur in mixed usage of Wi-Fi and Bluetooth.[7]

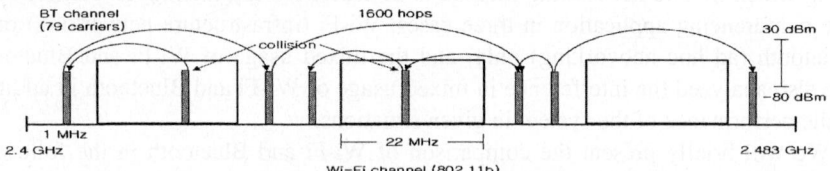

Fig. 1. Interference between Wi-Fi (white) and Bluetooth (gray) System

Theoretically, there exists the problem of interference between Wi-Fi and Bluetooth. Wi-Fi was developed for fast transmission and it is advantageous than Bluetooth in transmission speed. However, Wi-Fi is not immune to collisions. On the contrary, Bluetooth is collision-free because of its FHSS mechanism. However, the transmission capability of Bluetooth is poor [7], [8], [12]. Our system uses both mobile networking technologies, therefore we need to analyze and resolve the problem of interference caused by the coexistence of Wi-Fi and Bluetooth.

3 Mobile Conferencing System

A conference will start when a client connects to the server. A server (a presenter) can run on any device, such as a desktop PC, notebook PC, or a PDA. However, we prefer to run the server on a desktop PC or a notebook PC for the ease of measuring performance. Even a client can also run on any device the same as the server, we focused to run the client on mobile devices, because it is more probable in many cases. A conference server can be defined as either the infrastructure (Wi-Fi) or the ad hoc networking structure (Bluetooth), as well as the mixed usage of Wi-Fi and Bluetooth. The client (the user) can connect to the server through the appropriate networking method for a given situation. Figure 2 presents the network configuration of our system in mixed usage of Wi-Fi and Bluetooth. Our system is designed to avoid the effect of interference by implementing that the data transmission of critical data is realized using TCP/IP which is reliable, while the data transmission of large data such as video or audio is transmitted using UDP/IP which is not reliable, however, some packet losses of those data are not critical.

Our mobile conferencing system provides the built-in viewer (white board) and the audio-visual and textual communication, as well as the file sharing. We implemented the built-in viewer which allows the server to display shared documents. Our system allows users to mark a specific region of a shared document or to annotate on the built-in viewer using a mouse or a stylus pen. In addition, users can transfer images and files with each other through the system. The conference starts with sending a shared image to every client. Participants of the conference can see the shared image on their screens or on the beam projectors controlled by the server. Figure 3 illustrates the sharing between a desktop PC (server) and a PDA (client). A presenter can use a PDA as a remote controller which controls diverse functions for the conference. For example, the slide pages of the presentation can be moved up and down using two hardware buttons of PDA.

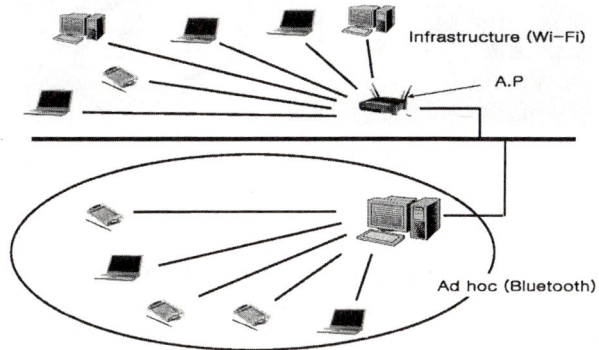

Fig. 2. Network Configuration in Mixed Usage Wi-Fi and Bluetooth

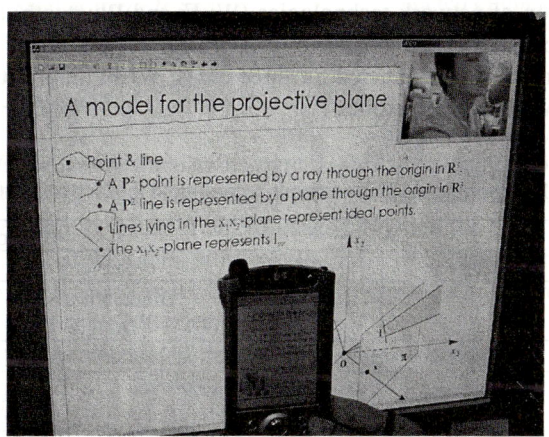

Fig. 3. A Screen Capture of our Mobile Conferencing System

For communicating in a remote conference, a real time communication facility is required. Our system provides a real-time audio-visual and textual communication facility as shown in Figure 4.

4 Experiments

We performed some experiments for verifying which mobile networking method is desirable for formulating an efficient mobile conferencing application in given situa-

Table 2. Throughput of each station calculated with the packet of 2,000 bytes

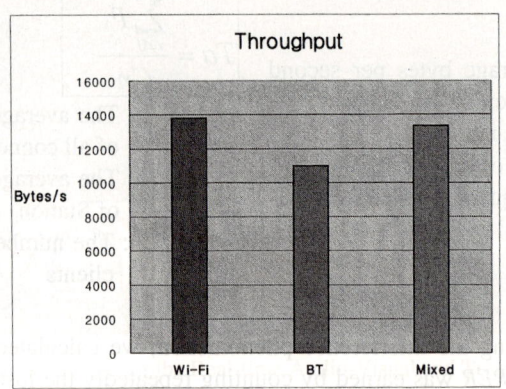

mobile conferencing applications. The results of our experiments illustrate the following:

- We calculated the Throughputs of three cases (Wi-Fi, Bluetooth, and mixed usage of both) in respecting the *PER* using our system. Through the experiment, we recognized that *PER* in the mixed usage case was the greatest and the *PER* in Bluetooth only was the smallest. However, we concluded that the *PER* due to the interference is negligible and the throughput depends mostly on the transmission rates. In the prior study, we leaned that threshold of usefulness of Bluetooth is around 8192 bytes. There is no great difference between the performance of Wi-Fi and Bluetooth if the average packet size of the application is less than 8192 bytes, and Bluetooth can be a good alternative for the Wi-Fi, in the situation where communication speed or bandwidth are not critical and the ad hoc networking is inevitable (if there is no accessible APs). In addition, Bluetooth is useful in the case of users are gathered within 10 meters, in this case, users practically do not need audio-visual communications. Therefore, we propose to configure a network structure as illustrated in Figure 2 for an efficient mobile conferencing application.
- Bluetooth is more preferable than Wi-Fi in most criteria except the transmission rates. Bluetooth is more advantageous in formulating an ad hoc networking. Wi-Fi needs to use private IPs for configuring an ad hoc network and this can inconvenience the user. However, Bluetooth only requires an adjustment of the channel for the connection to an ad hoc network. In addition, Bluetooth is safer than Wi-Fi from a security point of view and collision free.
- In conclusion, the mixed usage of Wi-Fi and Bluetooth for a mobile conferencing system can be a good solution for formulating an efficient mobile conferencing application. We propose to configure the mixed usage of Wi-Fi and Bluetooth for mobile conferencing applications and to profit from the advantages of both types of networking methods (Wi-Fi and Bluetooth). In the future, we plan to explore an agent system that automatically configures the best networking environment for on-going mobile application in real-time.

Acknowledgement

This work was supported by the Korea Science and Engineering Foundation (KOSEF) through the Multimedia Research Center at the University of Incheon.

References

1. Mattbew S. Gast, 802.11 Wireless Networks The Definitive Guide, O'Reilly, 2002. 4.
2. Kumar A., Karnik A. Performance Analysis of Wireless Ad hoc Networks, in Ad Hoc Wireless Network, CRC Press, 5-1~5-17, 2003.
3. Perkings C.E., Ad hoc Networking, Addison-Wesley, 2000. 12.
4. Bray J. and Sturman C.F., Bluetooth, Prentice Hall PTR, 2001.
5. Dreamtech Software Team, WAP, Bluetooth, and 3G Programming, Hungry Minds, 2002.
6. Wi-Fi overview. http://wi-fi.org/OpenSection/why_Wi-Fi.asp
7. Salazar, A.E.S.: Positioning Bluetooth® and Wi-Fi™ Systems. IEEE Transactions on Consumer Electronics. Vol. 50. No. 1. Feb. (2004) 151–157
8. Bluetooth performance in the presence of 802.11b WLAN Howitt, I., IEEE Transactions on Vehicular Technology, Volume: 51, Issue: 6, pp.1640–1651, Nov. 2002.
9. Performance evaluation of multiple IEEE 802.11b WLAN stations in the presence of Bluetooth radio interference Jung-Hyuck Jo; Jayant, H.; Proceedings on the IEEE International Conference on Communications ICC '03. May 11-15 2003, Volume: 2, pp1163–1168, 2003.
10. Bluetooth and IEEE 802.11 coexistence: analytical performance evaluation in fading channel Andrisano, O.; Conti, A.; Dardari, D.; Masini, B.M.; Pasolini, G.; Proceedings on the 13th IEEE International Symposium on Personal, Indoor and Mobile Radio Communications, Sept. 15-18 2002, Volume: 4, pp1752–1756, 2002.
11. Performance of a Bluetooth piconet in the presence of IEEE 802.11 WLANs Wang Feng; Arumugam, N.; Krishna, G.H.; Proceedings on the 13th IEEE International Symposium on Personal, Indoor and Mobile Radio Communications 2002, Sept. 15-18 2002, Volume: 4, pp1742–1746, 2002.
12. Iyer, A., Desai, U.B.: A Comparative Study of Video Transfer over Bluetooth and 802.11 Wireless MAC. Wireless Communications and Networking Conference (WCNC) 2003. IEEE. Volume: 3. 16-20 Mar. (2003) 2053–2057
13. Analysis of the interference between IEEE 802.11b and Bluetooth systems Fainberg, M.; Goodman, D.; Proceedings on the 54th IEEE Vehicular Technology Conference (VTC) 2001 Fall, Oct. 7-11 2001, Volume: 2 , pp967-971, 2001.
14. http://www.korwin.co.kr
15. http://www.widcom.com/Products/bluetooth_comm_software_bte.asp

UDMGrid:
A Grid Application for University Digital Museums*

Xiaowu Chen, Zhi Xu, Zhangsheng Pan, Xixi Luo, Hongchang Lin,
Yingchun Huang, and Haifeng Ou

The Key Laboratory of Virtual Reality Technology, Ministry of Education
School of Computer Science and Eng., Beihang University
Beijing 100083, P.R. China
{chen,xuzhi,panzs}@vrlab.buaa.edu.cn

Abstract. Because the eighteen online university digital museums of China confront a problem that the multi-discipline resources at these digital museums are isolated and dispersed without sufficient interconnection, ChinaGrid (China Education & Research Grid) supports a project named UDMGrid (University Digital Museum Grid), which studies a grid application for University Digital Museums using grid technology, especially information grid technology. According to the analysis on the problem, UDMGrid focuses on the resource sharing, information management and information service about those university digital museums. This paper presents research work on UDMGrid's framework, metadata, replica management, application server, etc.

1 Introduction

There are eighteen featured university museums having already been digitized, which mainly relate to Geology & Geography, Archaeology, Humanities & Civilization, and Aeronautics & Astronautics [1][2]. These digital museums play an important role in the fields of education, scientific research, as well as specimen collection, preservation, exhibition, and intercommunication.

But, these digital museums dispersed on different nodes in CERNET (China Education and Research Network) have several shortcomings: first, the present digital museums lack sufficient interconnection, and only connected by web links. This brings problems on management such as information redundancy and storage resources waste. And, the stored information is isolated in the so-called information island, which makes it difficult to intercommunicate and share resources among museums; second, these digital museums are unable to filter, classify, and process the stored raw information to fit for different kinds of requirements of visitors, and all the visitors having different knowledge background now access information through the same channel [3].

On the other hand, the information grid technology gives a better solution to these digital museums, since information grid is good at resource sharing, information management, and information service. In an information grid, the raw data from various sources is standardized, classified, and organized to be information; then, the grid digs

---

* This paper is supported by a project of China Education and Research Grid (ChinaGrid) under the grant No.CG2003-GA004, and a project of National 973 Program of China under the grant No. 2002CB312105.

out the information's internal relations; finally, uniform and transparent information service is presented to users [4][5]. The recent hot spots of information grid research focus on the grid architecture, information representation and organization, information interconnection and uniformity, security, robust, etc [6].

The San Diego Supercomputer Centre (SDSC) at UCSD is one of the original research institutes, which pays attention to the information grid and relationship among data, information and knowledge [6]. SDSC has devoted to the abstraction, organization and management of data, and advised to manage them by layers.UK e-Science concerns three levels including computational and data grid, information grid, knowledge grid [8]. It discusses the case of automating the process from raw data to information, and further to knowledge, and it plans to create new types of digital libraries for scientific data [7]. The e-Science proposes to integrate distributed data, computing and storage resources, and implement further data mining and knowledge discovery in grid environment. IBM has been actively involved in providing access to heterogeneous files, databases, and storage systems, and it concentrates on sharing of data for processing and large-scale collaboration [9]. IBM supports virtualizing data across diverse formats to solve the challenge of accessing data stored in different format, supports using of Storage Access Network (SAN) technology to solve the poor storage resource utilization, supports developing a replication solution to solve the problem of having to move a large volume of data across network to facilitate remote processing. Institute of Computing Technology (ICT), Chinese Academy of Sciences, emphasizes the information management and processing in the information grid. ICT has developed some applications based on the Vega Grid, including the Vega Information Grid (Vega-IG) [10][4] and Railway Information System [11]. The UK-wide digital museums have been built since 1999. Recently, UK planned to integrate the collections of most famous museums by Grid, so as to offer an interconnected and interactive environment for learning [12]. For example, visitors can receive the information service filtered according to their requirements, such as age and interests.

So, it is necessary and possible for ChinaGrid (China Education & Research Grid) to research on the UDMGrid [13], whose purpose is to integrate the enormous dispersed resources of various digital museums, to share the resources effectively and eliminate the information island, to filter and classify the collection information, and to provide appropriate information service to users according to their knowledge levels and motivation through unified grid portal. The UDMGrid will promote the education and scientific research in universities as well as in public, and also improve the public infrastructure facilities development in the Information Age.

The rest of this paper is organized as follows. Section 2 describes UDMGrid's system framework, metadata, replica management, and A. Server. Section 3 presents the description and analysis on a prototype of UDMGrid. Finally, section 4 draws its conclusion and outlines future work.

2 Design of UDMGrid

Using information grid technology, the UDMGrid is designed to overcome the existent shortcomings of present digital museums.

The UDMGrid is composed of three parts, as illustrated in Fig.1. The Grid Supporting Middleware offers basic grid services, such as authentication, authorization, file transfer, index service, etc. Bolstered by Grid Supporting Middleware, the Grid

Application Platform is developed and provides high level services to Grid Application. Using services through their APIs, Grid Applications can easily be developed and deployed oriented to user's requirements.

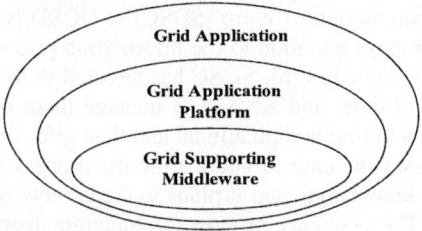

Fig. 1. The relationship of Grid Supporting Middleware, Grid Application Platform and Grid Application

Some developed grid software can be applied in UDMGrid as the Grid Supporting Middleware, such as Globus Toolkit 3.0 [14], Unicore [15], VegaGos [16] or Web-SASE4G [3].

2.1 System Framework

As shown in Fig. 2, the function modules can be divided into four parts: User Interface, Job Management, Job Monitor, and Resource Management.

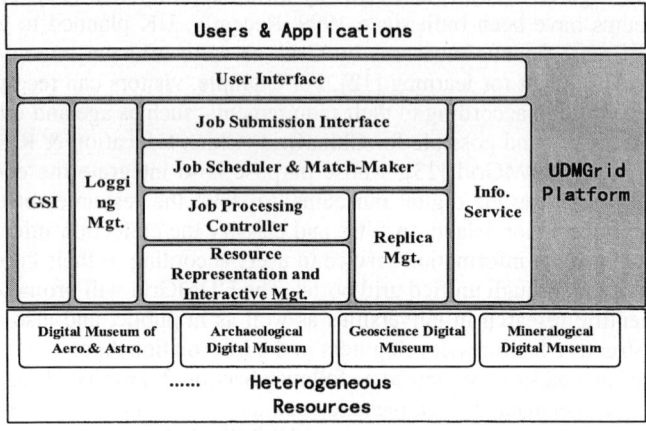

Fig. 2. The framework and function modules of UDMGrid

User Interface is the entrance for users to grid, which provides service environment for job submission and job information acquisition. Before submitting the job, user interface will pre-standardize user's request to job descriptions following specified criteria.

Job Management takes charge of job processing control and management in the workflow [17], including three modules: Job Submission Interface, Job Scheduler & Match-maker, and Job Processing Controller.

Job Monitor, composed of Grid Security Infrastructure (GSI) [18] and Logging Management [19], monitors the status of UDMGrid system and the procedure in job processing. It provides information feedback when requested by other modules.

Resource Management comprises Information Service module and Replica Management module. The functions of Resource Management part include providing resource information for match-making, metadata description and organization, replica management, etc. It is a research emphasis in UDMGrid.

2.2 Metadata

It is required to describe and standardize the mass and various data present in grid using metadata. The metadata in UDMGrid is designed according to the characters of the users, the grid it self, and the different service and various collections in digital museums, such as mass digital specimen information.

The metadata in the UDMGrid mainly includes: System Status & Configuration metadata, Replica metadata, Job metadata, Resource metadata, User metadata, which mainly service for system monitoring, replica management, job monitoring, resource accessing, and grid security. The detailed classification of the metadata in UDMGrid is depicted as following:

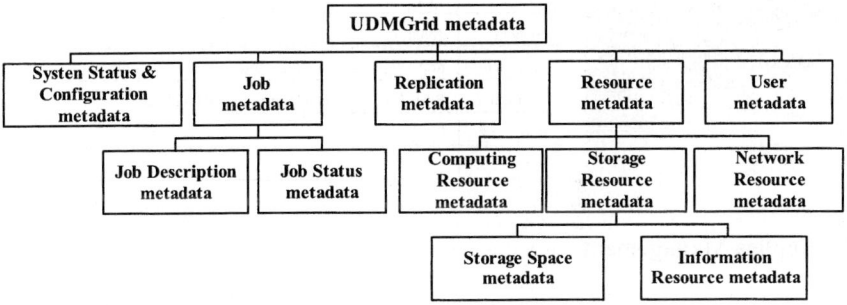

Fig. 3. the metadata in UDMGrid

System Status & Configuration metadata, stored in the Logging Management module, describes the status and configuration information of the grid system. It services for the system monitoring and provides information for cooperation of modules.

Replica metadata expresses the attributes and status of the replica. It provides the mapping information between logical files and replica files.

Job metadata describes the jobs in grid, including Job Describing metadata and Job Status metadata. The Job Describing metadata describes the content of submitted jobs, and provides information for the match-making between resources and job. Its design is suitable for data operation and accords with the characters of job in UDMGrid, such as time-sensitive and interactive.

Resource metadata describes the related information about computing resource, storage resource, and network resource. Because the computing resource in UDMGrid is generally represented as A.Server, the Computing Resource metadata is mainly designed according to the attributes of A.Server (the details are presented in part 2.4). The Storage metadata includes Storage Space metadata and Information Resource metadata. The Storage Space metadata describes the information of storage resource which can provide storage service, and the Information Resource metadata describes the attributes and some content information of the information resource. The Network resource metadata describes the status and parameters of networking.

User metadata describes information concerning users and resource providers. This information of users and providers is required in authorization, authentication, accounting, etc.

Among the metadata listed above, the Information Resource Metadata is one of the most important types in UDMGrid, since how to describe the numerous and various collections in these university digital museums is very challenging. The typical collections in university digital museums are images, text, video, etc. Take image for example, a picture about P-61 airplane in Digital Museum of Aero. & Astro. can be described as following:

Table 1. A example of Information Resource Metadata in UDMGrid

title	P-61
type	Picture
format	Jpeg
coverage	Aviation
identifier	http://digitalmuseum.buaa.edu.cn/picture/craft/black.jpg
description	P-61 is the first battle plant equipped with radar
rank	0
...	...

2.3 Replica Management

Replica Management provides transparent access to distributed data and improves the efficiency of data accessing and the quality of information service. In the UDMGrid, the Replica is managed in several domains, each domain comprise five modules: Replica Catalog, Replica Manager, Replica Catalog Query, Replica Selector and Consistency Manager (see Fig. 4).

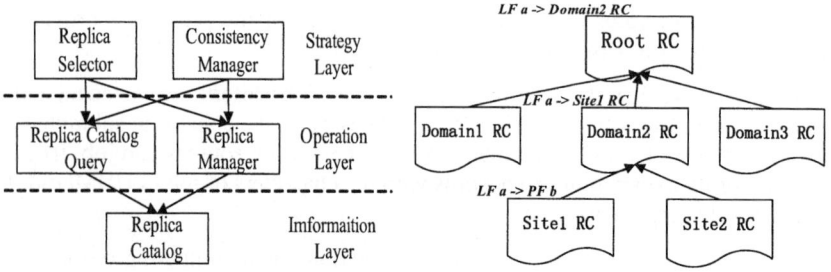

Fig. 4. Modules of replica management   **Fig. 5.** Hierarchical Replica catalog

The five modules are explained as following:

Replica Catalog (RC) stores the mapping information from logical file name (LFN) to physical file names (PFN). As illustrated in Fig. 5, the Replica metadata is stored in Site RC; the upper Domain RC contains the mappings from LFNs to the Site RCs; the Root RC contains mapping from LFNs to Domain RCs. For example, if a mapping from LF a to PF b is stored at Site1 RC, the Domain2 RC will keep a mapping record form LF a to Site1 RC and the Root RC will keep a mapping form LF a to Domain2 RC [20].

Replica Manager executes actual creation and deletion of replica within its domain and updates the RC.

Replica Catalog Query provides query service and a unified view of replicas.

Replica Selector receives request for replicas from its domain and searches the RC for replicas according to the request. Considering the result of searching, the Replica Selector will decide to select a proper replica or to create a new replica. In different storage conditions, different strategies will be used to make the decision. If create, it will inform the Replica Manager to apply the actual actions. In addition, Replica Selector maintains a table that records the accessing frequencies to logical files in this domain.

Consistency Manager detects the change of master, and informs Replica Managers to maintain the corresponding replicas. In the UDMGrid, only the master can be modified or deleted. If modified, the master will be republished and all the replicas related to it will be deleted.

2.4 Application Server

Application Server (A.Server) is a kind of deployed computing resource which provides specialized and persistent service in the UDMGrid. This specialized service is pre-designed and deployed according to the requirements of users and applications.

In UDMGrid, A.Server gets raw data about the collections from dispersed university digital museums, explores data's internal and external relationships, and processes it to information according to user's requirements. Take information classifying for example: a middle school teacher wants to query some information about "china's aeronautics history". The serving A.Server will firstly find all kinds of correlative data, such as images, video, files, etc. in the university digital museums, and filter useful information according to the job description and user's information, such as "teacher". In the end, the result will be sent back by time order.

Fig.6 illustrates the procedure of information classifying service. Firstly, A.Server providing such service must exist and have registered its service as Grid Service in the Information Service module. When a job arrived, the Job Scheduler & MatchMaker module will seek proper A.Server in the Information Service module, and deliver the job to the chosen A.Server, here is information classifying A.Server. While processing the job, A.Server may need to get a lot raw data about different digital museums' collections. The results will return to Job Processing Control module.

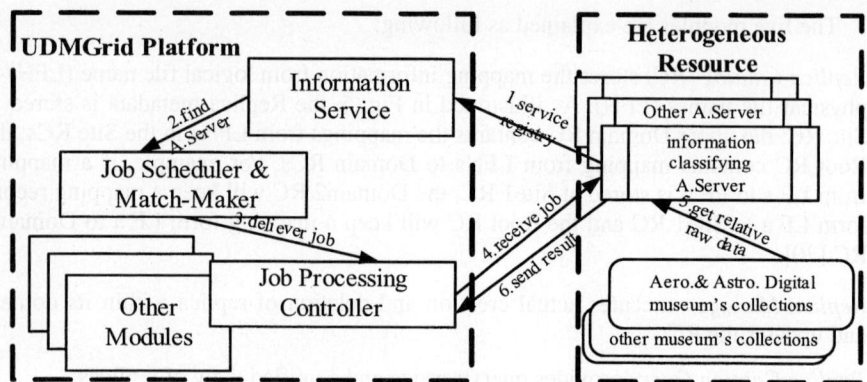

Fig. 6. A.Server in the UDMGrid

3 Prototype of UDMGrid

In this prototype, we propose to integrate the information resource of several university digital museums, and provide integrative and intelligent information searching, filtering and classifying service through grid Portal. The university digital museums involved include the Digital Museum of Aeronautics and Astronautics (BUAA) [21], the Archaeological Digital Museum (SDU) [22], the Geoscience Digital Museum (NJU) [23], and the Mineralogical Digital Museum (KUST) [24], etc.

The services and resources in each university digital museum compose one site, and the Fig.7 shows the architecture of site BUAA. The Collection provider service gives access to digital museum's collections as raw data.

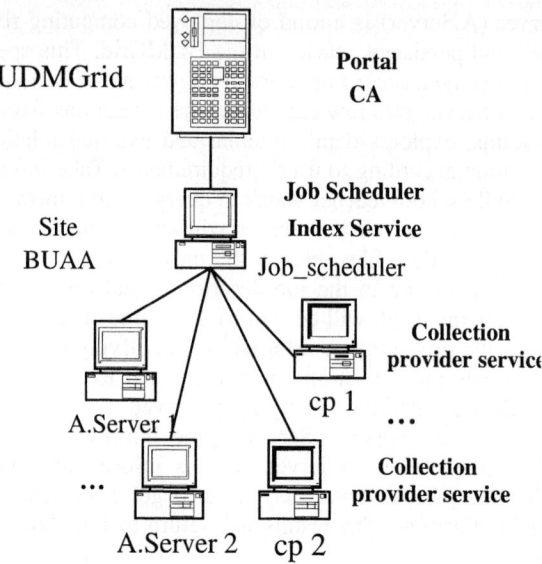

Fig. 7. The prototype in site BUAA

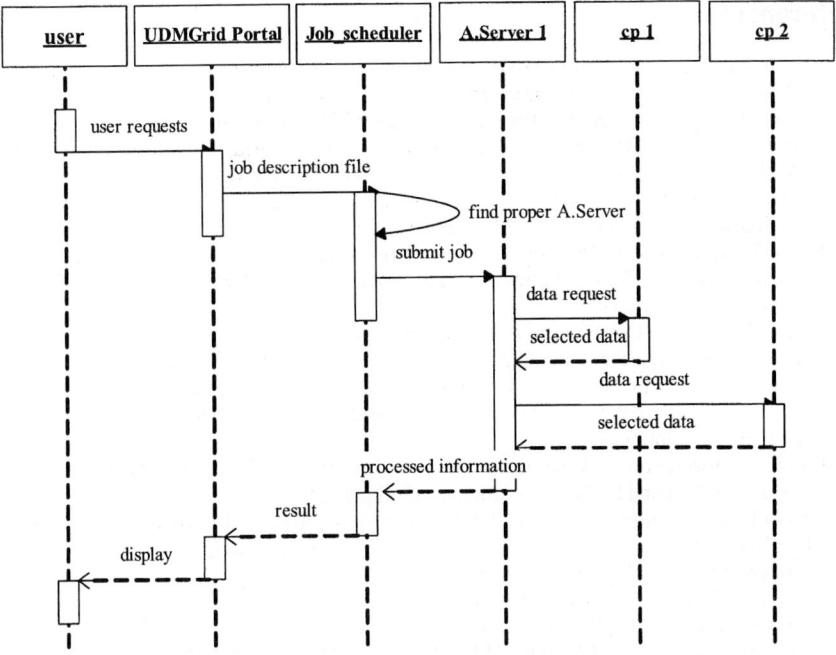

Fig. 8. Interaction in the site BUAA

The interactive process is presented in Fig.8. A user submits his request to UDMGrid from Portal. Then the portal formats user's requests to job description file and delivers it to Job_scheduler. Then the Job_scheduler searches the Index Service for proper A.Server and Collection provider services which provide materials for processing in A.Server. And the Job_scheduler creates a job according to the job description file. Then this job is submitted to the chosen A.Server. While processing, the A.Server needs to get raw data from Collection provider services as materials, and these raw data may be acquired from different sites, for example, the A.Server located in site BUAA may need resource from site NJU. After processing, the result will be send back to Job_scheduler and finally displayed to user.

4 Conclusion and Future Work

We present an information grid infrastructure for UDMGrid, and our research work concentrates on system framework, metadata, replica catalog, A.Server, etc. Its purpose is to solve the information island problem of established university digital museums, provide easy resource sharing, and offer intelligent information filtering and classifying service. As an important part of ChinaGrid Project, the UDMGrid will be the next generation information service infrastructure for museums and play an important role in the fields of learning, teaching, and public education.

The research is just at the beginning and needs continued efforts on the key technologies, such as architecture, metadata description, replica management, resource information service, A.Server, and the implementation.

References

1. The digital museum research group at China Univ. of Geosciences, Hankou Campus, The methods and steps of building university digital museum;
2. University Digital Museums, http://www.edu.cn/20020118/3018035.shtml;
3. ChinaGrid project–Beihang University Research Group, Grid Supporting Platform based on Web Services and its Typical Application, ChinaGrid Newsletter No.1;
4. Xu Zhi-Wei, Li Xiao-Lin, and You Gan-Mei, Architectural Study of the Vega Information Grid, Journal of Computer Research and Development;
5. Keith G Jeffer, EU GRIDs Activity: Information Issues;
6. Reagan W. Moore, Digital Libraries, Data Grids, and Persistent Archives, San Diego Supercomputer Center(SDSC);
7. National e-Science Centre, http://www.nesc.ac.uk/;
8. Tony Hey and Anne Trefethen, the Data Deluge: An e-Science Perspective;
9. The information grid, Rob Vrablik, Melissa Hyatt, IBM developerWorks;
10. Xu Zhi-Wei and Li Wei, Research on Vega Grid Architecture, Journal of Computer Research and Development;
11. Railway information grid Research group at Institute of Computing Technology, CAS, Information Grid and Railway Information Grid platform;
12. UK-wide digital museum-National Grid for Learning (NGfL), http://www.ngfl.gov.uk;
13. ChinaGrid Project, Ministry of Education, http://www.cergrid.cn/;
14. Globus project, http://www.globus.org/;
15. UNICORE forum, http://www.unicore.org/;
16. Vega Grid project, http://vega.ict.ac.cn/;
17. Massimo Sgaravatto, INFN-EDG, WP1: Workload Management (DataGrid-01-D1.4-0127-1_0);
18. Grid Security Infrastructure, http://www-unix.globus.org/security/;
19. Francesco Giacomini, INFN-EDG, definition of architecture, technical plan and evaluation criteria for scheduling, resource management, security and job description (DataGrid-01-D1.2-0112-0-3);
20. Ben Segal, CERN-EDG, Data Management (WP2) Architecture Report;
21. The Digital Museum of Aeronautics and Astronautics (Beihang University, BUAA) http://digitalmuseum.buaa.edu.cn/;
22. The Archaeological Digital Museum (Shandong University) http://museum.sdu.edu.cn/index/index.asp;
23. The Geoscience Digital Museum (Nanjing University)http://202.119.49.29/museum/default.htm;
24. The Mineralogical Digital Museum (Kunming Univ. of Sci. & Technol.) http://www.kmust.edu.cn/dm/index.htm.

The Research on Role-Based Access Control Mechanism for Workflow Management System

Baoyi Wang[1] and Shaomin Zhang[1,2]

[1] School of Computer, North China Electric Power University, Baoding 071003, China
wwangbaoyi@sohu.com
[2] School of Computer, Xidian University, Xi'an 710071, China

Abstract. Access control is an important protection mechanism for information system. Access control enforces subjects access restrictions to objects. Legitimate users should be allowed to access objects and illegitimate users should be detained from accessing objects. Access control between multi-user and multi-object is the key technique in security management of workflow management systems (WfMS), and the mechanism of the role-based authorization and access control is an effective way to solve the problem. This paper designs a set of practical constraint mechanism of role-based access control for WfMS. After introducing authorization constraint mechanism, a new model of role-based access control, RBACWF, is proposed. The role assignment algorithm, role-assign(tasknumber, pre_ass), is also given as a key technique in this model. After this, an application example is given to explain the execution of the algorithm. Finally, a practical WfMS based on above designs is supplied. The practice in the system's design and execution has proved that the authorization mechanism is flexible, the performance of the access security is improved greatly, and it also simplifies the task complexity of security administrator.

1 Introduction

Workflow Management Systems (WfMS) are used to run day-to-day applications in many application domains, such as finance and banking, healthcare, telecommunication, manufacturing, production and office automation. A workflow separates the various activities of a given organization process into a set of well- defined tasks. The various tasks in a workflow are usually carried out by several users in accordance with the organizational rules relevant to the process represented by the workflow. The access control between multi-user and multi-object is always the focus of security management in a workflow system. The mechanism of role-based access control can solve this problem preferably. A concept of role is introduced between users and authorizations, and access authorization of object is granted to a certain role. Because of the role, user is apart from authorization in logic. Users can get various authorizations when they are given corresponding roles. Since it can greatly simplify the operation of authorization and security management, it has been concerned and studied widely. However, there are many limitations in current role-based access control model.

RBAC96 model cluster[1] has four different models; they are RBAC0, RBAC1, RBAC2, and RBAC3. Role hierarchies and authorization constraints are added to RBAC3 on the base of primary model. But RBAC3 does not give formalized definition and systemic description of its constraint model (RBAC2) in detail, and the operations of the security administrators are very complex. NRBAC Model [2] is a new model, but it only integrates parts of RBAC96 Model, and can't solve the problem of the management of authorization and role themselves better. These models do not take characteristics of WfMS into account, either. So this paper gives a new role-based access control model that supports workflow and the model has been applied in a real office automation workflow system.

2 A Role-Based Access Control Model for WfMS

In WfMS, task is a basic unit, resources are needed to use when a task is executed. When a user obtains execution authorization of a task, it automatically gains the authorization of using necessary resources by which the task would be accomplished. Because the life cycle of a practical workflow is long, the static and dynamic constraints among tasks must be considered in this life cycle. The RBAC96 model can't satisfy the authorization constraint demands of WfMS. In order to apply role-based access control to a WfMS, authorization constraints should be introduced into it, thus a new role-based access control model RBACWF is proposed. Permission authorization constraints contain role constraints, task constraints, and authorization constraints. Combining all above constraints with role-based access control, the system can be accessed more securely. The RBACWF model is shown as Fig. 1.

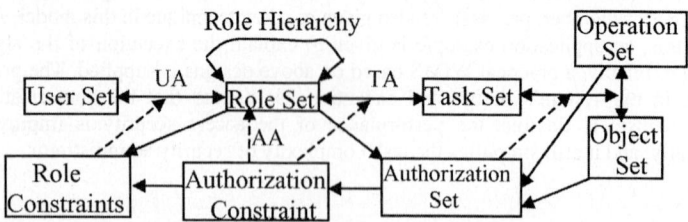

Fig. 1. Role-based access control model RBACWF for workflow system

Definition 1: Role-based access control model for WfMS = {US, RS, UA, RSH, TA, AA, TS, AS, OPS, OS, RC, AC}. By details:

US, User Set. US = { $U_i | i \in N$ }, here and in the following, N is the set of natural numbers. US is the set of users, each U_i is an actual user in US, i.e., user account number, represents a subject which can access data or other sources represented by data in computer system. A subject is one entity which executes tasks in WfMS.

RS, Role Set. RS = { $R_i | i \in N$ }, role's attributes include authorizations and constraints between them, R_i is role which represents one organization or position or work in tasks.

OS, Object Set. OS={ $O_i | i \in N$ }, each O_i is a object that can be accessed in a workflow system, it can be recognizable resource.

OPS, Operation Set. OPS={ $OP_i | i \in N$ } is the set of operations that can access the objects, for example, read, write, modify, execute , etc.

TS, Task Set. TS={ $T_i | i \in N$ } is the finite set of tasks, T_i is single task, it is a basic unit of a work.

AS, Authorization Set. AS = OS × OPS(Cartesian product). A role has the authorization $AS_i=(OS_j, OPS_k)$ means the role can execute act (OPS_k) to object (OS_j).

TA, Role-Tasks Assignment, a map function from RS to TS in workflow.

AA, Authorization Assignment, represents granting role with authorization.

UA, User-Roles Assignment, a map function from US to RS in workflow. It represents that one user possesses RS, denotes as: $US \rightarrow 2^{RS}$.

RSH, Role Set Hierarchy, is a duality relationship between RS and RS. It is semi-order relationship, represents the hierarchical relationship among the roles in an organization. Role Set Hierarchy can be marked as RSH, RSH = (RS, $\succ$). For example, the role of purchase department manager in an enterprise is higher than purchase employee in hierarchy, so the manager inherits the authorization of the employee, i.e., the manager role possesses more authorizations than employee role.

RC(Role Constraints) and AC(Authorization Constraints) are defined in section 3 of this paper.

3 Constraint Mechanism of Role-Based Access Control for Workflow Management System

Definition 2: Workflow process $W=\{TRS_1, TRS_2, ..., TRS_n\}$ is a finite set of TRS_i, here $TRS_i=(T_i, RSH_i, act_i, C_i)$, i=1,...n。 Among it, T_i is a task, RSH_i which possesses hierarchy relationship $RSH =(RS_i, \succ_i)$ is the set of roles authorized to execute T_i, $act_i \in N$ is the number of possible activations of task T_i, and C_i is constraint conditions authorizing to execute task T_i. The workflow tasks are sequentially executed in order of appearance in the workflow role specification. For example, W={(T_1, ({refund clerk}, { }), 1, C_1), (T_2, ({refund manager, general manager}, { }), 2, C_2), (T_3, ({refund manager, general manager}, { }), 1, C_3), (T_4, ({refund clerk}, { }), 1, C_4)}.

In a workflow system, there are many constraints among tasks, roles and authorizations. The Separation Rule of Duties is a kind of constraints, demands that a role be not able to undertake two tasks belonging to the Separation Rule of Duties. So the concept of collision is used to described the Separation Rule of Duties.

(1) Authorization Constraints

RCTS[3] and RCTM[3] are two different types of authorization constraints. RCTS = {Max numbers of assignment, states}, RCTM={ mutual exclusion, Contain, Dependence , Inheritance }. Here is mere description of mutual exclusion relationship of RCTM.

If AS_i and AS_j are mutually exclusive, no roles can have both of the two authorizations at the same time. It is the rule of Separation Rule of Duties. And it is described as follows:

$$Collision(AS_i, AS_j) \rightarrow \neg \exists (r \in RS)(r \in AS_i \wedge r \in AS_j)$$

(2) Task constraints

If task TRS_i and TRS_j are conflicting, no role can execute TRS_i when the role is executing TRS_j. It is described as follows:

$$Collison(TRS_i, TRS_j) \rightarrow \neg \exists (r \in RS)(r \in RS_{TRS_i} \wedge r \in RS_{TRS_j})$$

(3) Role constraints
Similitude with Task constraints, Role constraints contain RCTM and RCTS, the mutual exclusion in RCTS is defined as follows:

$\forall r_i \in RS_{TRS_i}, r_j \in RS_{TRS_j}$, If task TRS_i and TRS_j are conflicting, r_i and r_j are conflicting. It can be marked as $Collison(r_i, r_j)$.

If $\forall r_i, r_j \in CRS \Rightarrow Collision(r_i, r_j)$, CRS is called the set of conflicting roles.

Role constraint rule 1: any user can't belong to two conflicting roles at the same time. The rule is described as follows:

$\forall (u \in US), \forall (r_i, r_j \in CRS) \Rightarrow (u \in r_i \land u \in r_j)$.

Role constraint rule 2: the user number possessed by any role must be not larger than cardinal number of the role. The rule is described as follows:

$\forall r \in RS \Rightarrow (\#us(r) \leq card(r))$

Among this formula, card(r) is cardinal number of the role r, represents the most number of users possessed by the role r; #us(r) represents user number possessed by role r.

4 Role Assignment Algorithm

Security check, authorization assignment and role assignment are the most key algorithms in the role-based access control model[6], this paper only discusses role assignment algorithm[5] as the limit of length.

Definition 3: Let W be a workflow. The constraint base associated with W is named as CB(W), which consists of all the constraints among authorizations, tasks and roles.

In a workflow system, the constraints of role assignment can be categorized into:
(1) Static Constraint: Constraints can be evaluated without executing the workflow.
(2) Dynamic Constraint: Constraints can be evaluated only during executing of workflow.
(3) Hybrid Constraint: Constraints include the former two types of constraints.

Dynamic constraint[3] is the most important authorization constraint, may be described in the following expression C:

$C = ERM(T_i) \mid ERME(T_i) \mid EU(T_i) \mid NULL$

Here, T_i represents a preceding order task of current task T, $ERM(T_i)$ represents that the role executing current task T is higher than the one executing task T_i in role hierarchy, $ERME(T_i)$ represents that the role executing current tasks T is not less than the one executing task T_i in role hierarchy, $EU(T_i)$ represents that the user demanding to execute current task T must be the one executing task T_i, and NULL is empty, represents no dynamic constraints.

In the following, there are some definitions needed in role assignment algorithm.

Definition 4: If $executer_r(R_i, T_j, k)$ is true, then the k-th activation of task T_j is executed by a user belonging to role R_i.

Definition 5: If $must\_executer_r(R_i, T_j)$ is true, then task T_j must be executed by a user belonging to role R_i.

Definition 6: If $cannot\_do_r(R_i, T_j)$ is true, then it indicates that task T_j cannot be assigned to a user belonging to role R_i to execute.

Definition 7:
$Denied\_Role(T_i) = \cup\{R_j \mid cannot\_do_r(R_j, T_i) \in M(CB(W))\}$ represents a RS that cannot execute task T_i according to the constraints of CB (W). M(CB(W)) denotes the meaning of a CB(W) with respect to stable model semantics.

Definition 8:
$Obliged\_Roles(T_i) = \cup\{R_j \mid must\_execute_r(R_j, T_i) \in M(CB(W))\}$
represents a RS that must execute task T_i according to the constraints of CB (W).

Definition 9: (Explicit Rule) An explicit rule is of the form H←, where H is a specification. For example, $executer_r(R_i, T_j, k) \leftarrow$ is a rule that generates depending on whether the k-th activation of task T_j can be executed by a user belonging to role R_i.

Definition 10: Role assignment graph (RAG(W)). A role assignment graph RAG (W) of a workflow is a labeled graph G=(V, E) defined as follows:

(1) Vertex $V = (T_k, R_j), T_i \in TRS_i, R_j \in RS$. The vertex indicates that during execution of a workflow, as long as authorization constraints are not violated, role R_j can be assigned to task T_k.

(2) Edge E. If R_j assigned to T_i and R_k assigned to T_h, and they are in a same workflow, and do not violate authorization constraints, then there is an edge between vertex (T_i, R_j) and vertex (T_h, R_k).

A role-assignment algorithm will be given as follows:

Algorithm name: role-assign(tasknumber, pre_ass)
The tasknumber is current task number in W. The pre_ass is a vector, all role assignment candidates are stored in this vector.
Input: (1) the workflow W = {TRS$_1$, TRS$_2$, ···, TRS$_n$}, n is the number of tasks in W. (2) the constraint base associated with W: CB(W).
Output: Return false if no role assignment satisfying the constraints; Return RAG(W), otherwise.
Procedure role-assign(tasknumber, pre_ass)
{ j:= tasknumber;
  Repeat
    { $Cand\_R := \{R_s \mid R_s \in RS_j \wedge (R_s \text{ is unmarked})\}$;
      If Cand_R≠ϕ
      Then {

```
            For each R_l ∈ Cand _ R Do pre _ ass [ j] := R_l ;
            If j<n Then role_assign(j+1, pre_ass) ;}
         Else {  For i: =1 To n
                 For k: =1 To act_i
                    Insert executer _r (pre _ ass [i], T_i, k) ← To CB (W);
                 If CB (W) is consistent
                 Then { correct:=true;
                       i:=1;
                       While i≤n and correct Do
                          { Calculate Obliged _ Roles (T_i) ;
                            Calculate Denied _ Roles (T_i);
                            If Obliged _ Roles (T_i) ≠ Φ
                               and pre _ ass[i] ∉ Obliged _ Roles (T_i)
                            Then correct:=false;
                            If pre_ ass[i] ∈ Denied _Roles(T_i)
                            Then correct:=false;
                            i:=i+1;          }
                 If correct
                 Then {
                       For i:=1 To n
                          V_i := (T_i, pre _ ass[i]) ;
                       Insert p={V_1, V_2 ,...V_n} Into RAG(W); }
                 }
          }
       }
      Until all roles are in RS_j ;
      IF  RAG(W)= Φ
      Then Output(False)
      Else Output(RAG(W));
    }
```

The function of Output(x) is output "X". The algorithm is initially executed by a call role-assign(1,[,]). In this algorithm, the procedure role-assign is called recursively, which created the role assignment graph of the input workflow W, The candidate role assignments are inserted into vector pre_ass which is incrementally created by recursively calling procedure role-assign.

5 An Application Example Explanation of Role-Assignment Algorithm

In this section, an application example is given to explain the execution of the algorithm role-assign(tasknumber, pre_ass) described in section 4 of this paper.

Supposed that in a workflow W there are four tasks T_1, T_2, T_3 and T_4. T_1 can be executed by market clerk (MC), T_2 by market manager (MM) and general manager (GM), T_3 by MC and GM, and T_4 only be executed by MC.

Constraints in the workflow W are as follows:

Constraint 1(C_1): In this workflow W, at least three roles exist: MC, MM, and GM.

Constraint 2(C_2): The role who executes T_2 must have a higher authorization than those who executes T_1 and T_4.

Constraint 3(C_3): If a user belongs to role MC and has executed T_1, this user cannot execute T_4.

Constraint 4(C_4): If a user has executed task T_2, he (or she) cannot execute task T_3.

Constraint 5(C_5): Each activation of T_2 must be executed by a different user.

Constraint 6(C_6): If a user has executed T_2, then he (or she) cannot execute T_4.

Then, CB(W)={C_1,C_2,C_3,C_4,C_5,C_6}.

According to algorithm role-assign(tasknumber, pre_ass), RAG (W) = {(T_1,MC), (T_2,MM),(T_3,MM),(T_4,MC)},{(T_1,RC),(T_2,MM),(T_3,MM),(T_4,RM)},{(T_1,RC),(T_2,RM),(T_3,RM),(T_4,GM)},{(T_1,RC),(T_2,GM),(T_3, GM), (T_4, RC)}, {(T_1, RC), (T_2, GM), (T_3, GM), (T_4, RM)}, {(T_1, RC), (T2, GM), (T3, GM), (T4, GM)}.

6 The Workflow System Structure of Role-Based Access Control Mechanism

Fig. 2 is a workflow system structure which supports role-based access control mechanism described in above sections of this paper. From Fig.2, we can learn that the workflow system is enriched with Workflow Authorization Mechanism Module, Constraints Analysis and Enforcement Module, Constraint Base, RSH, Role Assignment, User Set, Authorization Management Module an GUI(Graphic User Interface). GUI drives the Authorization Management Module to define RSH, Constraint Base and Role Assignment.

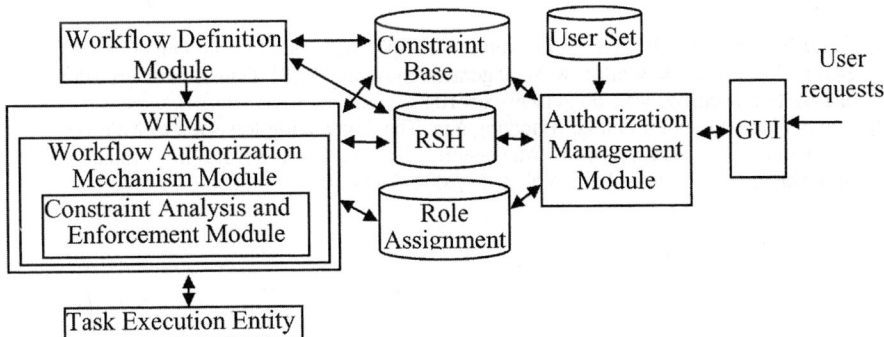

Fig. 2. A workflow system structure

Workflow Authorization Mechanism Module, Constraints Analysis and Enforcement Module are added to the workflow server. Before a workflow task is activated, some constraints and RSH that can execute this task are sent to the Workflow Authorization Mechanism Module. According to Constraints Analysis and history data, the Workflow Authorization Mechanism Module decides the Role Assignment. After this task has assigned to a user, the user and the task will be registered in workflow task list, and then workflow engine sends the task to this user who will perform it. So the workflow system that uses this mechanism not only has a flexible authorization mechanism, but also simplifies the operation complexity of security administrator.

7 Conclusion

The mechanism of role-based access control in WfMS proposed in this paper can be applied in some relevant application domains, such as finance and banking, healthcare, telecommunication, manufacturing, production and office automation system. It has been used in an office automation workflow system, The practice in the system's design and execution has proved that the performance of the access security is improved greatly and it is easy to put into practice. After the workflow system especially distributed workflow system adds the role-based access control mechanism, the execution efficiency of the system must be affected. Considering valid load balance policy can improve the system performance, that load balance policy with high efficiency should be applied to the workflow system is a good idea.

References

1. Ravi Sandhu, Edward Coyne, Hal Feinstein and etc.: Role-Based Access Control Model. IEEE Computer, Vol. 29(2) (1996) 38-47
2. Ying Qiao, De Xu, Guozhong Dai: A New Role-based Access Control Model and It's Implement Mechanism. Journal of Computer Research& Development, Vol. 37(1) (2000) 37-44
3. Weili Han, Gang Chen, Jianwei Yin: Role-Based Constrained Access Control Model and Implement Supported by Constraints among Permissions. Journal of Computer-Aided Design& Computer Graphics, Vol. 14(4) (2002) 333-338
4. Jianxun Liu, Shensheng Zhang, Fenglin Bu: Study on the Application of Role-Based Access Control in Workflow Management System. Mini-Micro Systems, Vol. 24(6) (2003) 1067-1070
5. Elisa Bertino, Elena Ferrari, Vijay Atluri: The Specification and Enforcement of Authorization Constraints in Workflow Management Systems. ACM Transactions on Information and System Security. Vol. 2(1) (1999) 65-104
6. Baoyi Wang, Shaomin Zhang, Xiaodong Xia: The Application Research of Role-Based Access Control Model in Workflow Management System. Lecture Notes in Computer Science of Springer-Verlag Heidelberg, Vol. 3033(2004) 1034-1037

Spatial Query Preprocessing in Distributed GIS

Sheng-sheng Wang and Da-you Liu

College of Computer Science and Technology, Key Laboratory of Symbolic Computing and Knowledge Engineering of Ministry of Education, Jilin University, 130012 Changchun, China
wang_sheng_sheng@163.com, dyliu@mail.jlu.edu.cn

Abstract. This paper focuses on spatial query optimization in distributed GIS. A new qualitative spatial relation model and its consistency problem solution which compose topology, direction, distance and size are proposed. Research integrating the four aspects has not appeared before. A new method to deduce the constraints of spatial query is given, thus saves the query process time in distributed GIS. Finally, The methods and theories are applied to a distributed GIS project, the experiment result is satisfactory.

1 Introduction

Spatial (or geographical) data process was becoming more and more important in the past decade. Nowadays Geographic Information Systems (GIS) and spatial database that use the technology of distributed systems on computer networks are very common [1]. This has represented a current challenge for the several disciplines involved with GIS development. The GIS based on distributed systems or Distributed GIS (DGIS) can be defined as those that have a varied number of autonomous geographic systems, not necessarily homogeneous, interconnected by a computer network and which cooperate in the accomplishment of geographic data processing with the data stored in their local spatial databases.

One of the major features of DGIS is response time. Since many users request spatial queries simultaneously, the spatial data server,which perform every query requires, requires to be more efficient. But query evaluation in geographic databases is often expensive, because the spatial data stored are more complexly structured and data sets are very large. Spatial queries, which are defined by either languages [2] or sketches by hand [3], are usually expressed as a set of spatial objects and a set of spatial constraints among the objects. The spatial constraints may include topology, direction (orientation), distance and size . The goal of a query evaluation is to match the query constraints among the related objects with binary spatial relations between objects that are stored in a spatial database. The set of binary spatial relations (or say the constraints), that need to be checked in spatial database, grows up very quickly with the number of spatial constraints in the query. For m objects in the query and n objects in the database, the query process need to check $m2\binom{n}{!}/(m!\binom{n-m}{!})$ geometry conditions. Such a type of query implies very hard combinatorial computations and presents a challenging problem for spatial information retrieval. This paper focuses on spatial query optimization. To achieve this goal, two steps must be followed:

1. Check the consistency of query constraints. Inconsistency query need not to be performed, directly return NULL set.
2. Reduce the number of constraints in the query by delete the constraint which could be deduced by others.

Another feature of DGIS spatial query is that the requires are variant. Most DGIS systems only support topological or direction, distance, size relation respectively. Most previous GIS and AI researches focused on single spatial aspect. Those were quite inadequate for applications in GIS and DGIS. Some spatial relation models combining topology with direction or distance were proposed[4][5]. Alfonso investigated the spatial reasoning combing topology and size[6]. But combining the four spatial aspects including topology, direction, distance and size has not been studied before. In this paper, a spatial data model which union these four spatial aspect is put forward, so does its SAT problem.

This paper is mainly focused on qualitative spatial relations, but not the metric information. So the major theory and method is derived form qualitative spatial reasoning (QSR). QSR has a wide variety of potential applications in AI (such as spatial information systems, robot navigation, natural language processing, visual languages, qualitative simulation of physical processes and commonsense reasoning) and other fields (such as GIS and CAD)[4][5][7].

2 Representation of Spatial Constraints

Topological Relations

Topological relations are the most important spatial relations which have been studied most. The best known topological theory is Region Connection Calculus (RCC for short) [8]. It is a mereo- topological theory based on spatial region ontology. Many spatial relation models(such as RCC-5, RCC-7, RCC-8, RCC-10, RCC-13) are deduced by RCC theory. Among them, RCC-8 is well-known in state-of-the-art Geographical Information System, spatial database, visual languages and other applied fields . It has one primitive dyadic relation $C(x,y)$ which means "x is connected with y". Eight Jointly Exhaustive and Pairwise Disjoint (JEPD) relations was deduced by $C(x,y)$. {DC,EC,PO,TPP,NTPP,TPPI,NTPPI,EQ} are JEPD basic relations of RCC-8. RCC-8 includes all the possible unions of the basic relations(resulting in 28 relations) .

Composition reasoning is the basic operation for further reasoning works such as CSP. The composition is to determine $R(a,c)$ by $R(a,b)$ and $R(b,c)$. In most cases, $R(a,c)$ can't be unique. The composition of relation A,B is represented by $A \bigcirc B$, and Composition Table is often used. Composition Table is a manual table which couldn't be created automatically in most cases[5][7].

Direction Relations

To Avoid unimportant information which will brings on trivial discussion when composes with other relations, object is abstract to a point (the geometric center of the object) in our qualitative direction model. Space is equivalent divided into four parts by two lines across the reference object. {N,S,W,E,C} are the basic direction relations. Here x {C} y means the geometric centers of object x and y are at the same position, otherwise {N,S,W,E} are the four area that the object related to the reference object may exist. The two lines expect the point 'C' belong to 'N' and 'S' area.

The composition table of this direction model is:

Table 1. Composition table of direction relations

o	N	S	E	W	C
N	N	*	E,N	N,W	N
S	*	S	E,S	S,W	S
E	E,N	E,S	E	*	E
W	W,N	S,W	*	W	W
C	N	S	E	W	C

Distance Relations

Distance is also an important spatial aspect in both qualitative and metric space. Similar to direction our distance model is also depended on the geometric center.

Three basic qualitative distance relations are defined as:

Cnt: x and y are topological connected.

Near: Geometric center distance of x and y is less than or equal to L (a constant value).

Far: Geometric center distance of x and y is more than L.

Although no detail shape information can be inferred by qualitative distance, if we assume that L >> 2*MS (MS is the max diameter of all objects), the composition table of qualitative is got by applying trigonal inequation (Table 2).

Table 2. Composition table of distance relations

o	Cnt	Near	Far
Cnt	Cnt,Near	*	Near,Far
Near	*	*	*
Far	Near,Far	*	*

Size Relations

We will assume that all the spatial regions are measurable sets in R^n. The size of an n-dimensional region corresponds to its n-dimensional measure. For example, the size of a sphere in R^2 corresponds to its area. The size relation of two objects can be qualitative represented by three basic relations $\{<,=,>\}$, and they could be further extended to $\{\emptyset,<,=,>,\leq,\geq,\neq,*\}$. Table 3 is the composition table of size relations.

Table 3. Composition table of size relations

o	<	=	>
<	<	<	<,=,>
=	<	=	>
>	<,=,>	>	>

3 Consistency of Compositive Spatial Constraints

To deal with the consistency problem of compositive constraints, interdependences of spatial aspects defined above must be clarified, then arc consistency algorithm is proposed based on them. All kinds of qualitative spatial relation are some kinds of abstracts of metric geometry interrelations of two objects. So they are inherently interdependent. Topology is the major qualitative spatial relation, other relations such as direction or distance all depend on it. Direction, distance and size relation defined in this paper have no inherently relation. But if we assume that "the geometry center is always inside the object", direction {C} implies the two objects are connected. But the reverse is not sure. Considering all the possibility, interdependence of every two spatial aspects are listed below:

Topology and Direction

Table 4. The dependence from topology to direction

T	DC, EC, PO, TPP, NTPP, TPPI, NTPPI	EQ
TDO(T)	N,S,W,E,C	C

Table 5. The dependence from direction to topology

O	N,S,W,E	C
ODT(O)	DC, EC, PO, TPP, NTPP, TPPI, NTPPI	PO, TPP, NTPP, TPPI, NTPPI, EQ

Topology and Distance

Table 6. The dependence from topology to distance

T	DC	EC, PO, TPP, NTPP, TPPI, NTPPI, EQ
TDD(T)	Near, Far	Cnt

Table 7. The dependence from distance to topology

D	Cnt	Near, Far
DDT(D)	EC, PO, TPP, NTPP, TPPI, NTPPI, EQ	DC

Topology and Size

Table 8. The dependence from topology to size

T	DC	EC	PO	TPP	NTPP	TPPI	NTPPI	EQ
TDS(T)	<,=,>	<,=,>	<,=,>	<	<	>	>	=

Table 9. The dependence from size to topology

S	<	=	>
SDT(S)	DC, EC, PO, TPP, NTPP	DC, EC, PO, EQ	DC, EC, PO, TPPI, NTPPI

Direction and Distance
Similar to the consistency of topology 4, consistency of compositive spatial constraints is defined:

Given a set Θ of spatial constraints of the form x Ry, where x,y are region variables and R is a spatial relation, and R=(T,O,D,S) which represent topology and qualitative direction, distance, size respectively. Deciding the consistency of Θ is a special case of SAT problem.

Every pair of variables is constrained in Θ. If no information is given about the relation holding between two variables x and y, then the universal constraint x{*}y (all the relations are included) is contained in Θ. Whenever a constraint xRy is in Θ, also yR~x (reverse relation) is present.

We propose a new algorithm for dealing with combined topology, direction, distance and size constraints. COMPOSE-CONSISTENCY is a modification of Vilain and Kautz's path-consistency algorithm9. The main novelty of our algorithm is that COMPOSE-CONSISTENCY operates on a graph of groups of constraints. The vertices of the graph are constraint variables, which in our context correspond to spatial regions. Each edge of the graph is labeled by a group(four kinds of) of relations. The function CHECK (i,k,j) has the same role as the function REVISE used.

Table 10. The dependence from direction to distance

O	N,S,W,E	C
ODD(O)	Near, Far, Cnt	Cnt

Algorithm: COMPOSE-CONSISTENCY
Input: A set Θ of spatial constraints, $x_1, x_2, \ldots, x_n$ are region variables in Θ.
Output: true, if Θ is consistent ; fail, otherwise.

1. $Q := \{(i, j) \mid i < j\}; (1 \leq i, j \leq n);$
2. *while* $Q \neq \Phi$ *do*
3.   select and delete an arc (i, j) from Q;
4.   for $k \neq i, k \neq j$ $(k \in \{1, .., n\})$ do
5.   *if* CHECK(i, j, k) *then*
6.   *if* ($T_{ik} = \Phi$ or $O_{ik} = \Phi$ or $D_{ik} = \Phi$ or $S_{ik} = \Phi$) *then* return fail;
7.   *else* add (i,k) to Q;
8.   *if* CHECK(k, i, j) *then*
9.   *if* ($T_{kj} = \Phi$ or $O_{kj} = \Phi$ or $D_{kj} = \Phi$ or $S_{kj} = \Phi$) *then* return fail;
10.   *else* add (k, j) to Q;
11. return true

Function: CHECK(i, k, j)
Input: three region variables x_i, x_k and x_j
Output: true, if R_{ij} is revised; false otherwise.

problem underlies of Grid concept is coordinated resource sharing and problem solving in dynamic, multi-institutional virtual organizations. In Grid computing environment, the computing resources are provided as Grid service. Then, to deal with the complexity in CFD codes application, we can construct a "CFD-Grid" HPC framework, to improve the performance of the CFD codes and gain many new characteristics that is impossible in traditional means for CFD applications. This grid based HPC "CFD-Grid" framework can make the design process be easily controlled and monitored and can be conducted interactively. To construct this novel high performance computational (HPC) framework, three parts work should be done: 1) Every CFD codes have their limitation, the "CFD-Grid" should include every suitable scope provided to users for every CFD codes in this platform; 2) Every CFD codes have various definition for their usage, we should re-define the interface for all CFD codes when transformed into this General "CFD-Grid" platform; 3) Providing general SDK (software develop Kits) for programmers to transplant their CFD codes in "CFD-Grid".

There are three components in this HPC "CFD-Grid" framework: Grid Service module based on IBM Service Domain, general interface standards for various CFD codes and CFD codes for application. All those components in the service domain system are named as "Grid Services" independently. The HPC "CFD-Grid" framework can run on various computers with various platforms, through which various CFD codes and computer resources are provided to the designers in the form of services, no matter where those computers are located. Based on SOA (Service Oriented Architecture), our group present our two-layer "CFD-Grid" framework, which can package existed CFD codes as CFD services with unified interface standards. The upper layer provides unified interfaces for various services and the lower layer provides High Performance Computing environment and SDK (Software Develop Kits) for CFD programmers. By using Service Domain, Registry service, Factory service and Handle-Map service can be constructed.

In the second section, the HPC framework based on Grid technology will be described in details. Then, based on this "CFD-Grid" of Shanghai Jiao Tong University supported by China-Grid project, we can complete aero-crafts aerodynamic simulation and optimization. To demonstrate the ability of this "CFD-Grid" HPC framework, aero-craft simulation and optimization cases was described.

2 "CFD-Grid" HPC Framework Based on Service Domain

2.1 SOA (Service Oriented Architecture) and Service Domain

Service Domain provides tow pieces of layered technologies to enable robust Service Oriented Application and support brokering and marketplace of WSDL services: 1) On Demand Service Grid, the virtualization layer, and 2) Service Domain, the service layer.

The On Demand Service Grid integrates Service Domain and local service hosting technologies to create a virtualization of a Service Broker domain that manages a heterogeneous group of services suppliers to provide services to multiple groups of consumers. The Service Broker establishes itself as a main hub that provides a single logical service image for all the services it represents. The main hub fans out service requests received to multiple secondary hub images, or calls out to other capabilities. Fig 1 illustrated the architecture of ODSG.

The main hub exploits Service Domain for entering contracts with consumers and suppliers separately, for dynamic service mapping, discovery, selection and dispatching, for node topologies, and for rule based operations management. The sub hub administrators provide a supportive role by registering the actual service suppliers locally and promoting selected port types to the main hub, while it is obvious that the sub hubs could also function like a main hub for a local domain. Any service requests distributed from the main hub are managed and charged according to the usage contracts in the main hub.

The Service Domain service layer provides the underneath component technology for brokering and marketplace of WSDL services. In the Service Domain architecture, services sharing and aggregation through a service entry interface, service attachment interface, and service policy interface. Each node, e.g. the main hub, sub hubs, service ports in a hub, etc. in the On Demand Service Grid, is an object with these well-defined interfaces described in WSDL. Service Domain objects can be recursively aggregated using these interfaces to high-level domains thus forming a notion of a larger complex virtualized service.

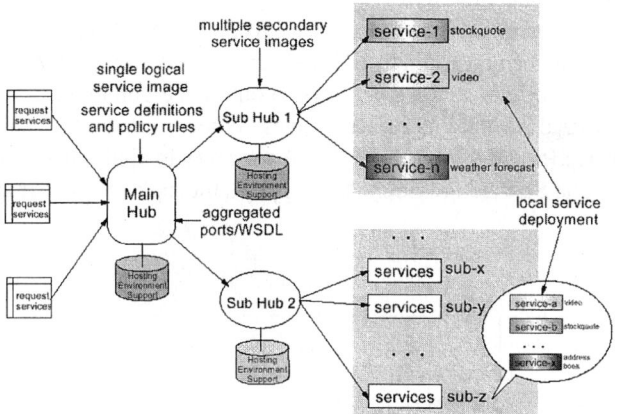

Fig. 1. Flowchart of Service Domain.

2.2 "CFD-Grid" HPC Framework

By using the grid computing technology, a novel high performance framework for "CFD-Grid" that can lessen the complexity or improve performance for CFD simulation and optimization is proposed. How the system has been constructed will be briefly described.

The "CFD-Grid" HPC framework is composed of many Grid services. Each Grid service is built on the web and Service Domain standards. Each service has some standard interfaces to support the registering of service, the querying of service, the communication and the interacting between the client and the service provider. Then, we will discuss the Grid Service features of the HPC framework firstly. Secondly we will discuss all the components of "CFD-Grid" HPC framework.

Grid Service

A Grid service is a web service that conforms to a set of conventions (interfaces and behaviors) that define how a client interacts with a Grid service [1]. Some general features of the Grid service are:

1. Grid service is web service, whose public interfaces and bindings are defined and described using XML. Its definition can be discovered by other software systems. Those systems may interact with the web service in a manner prescribed by its definition, using XML based messages conveyed by internet protocols [2].
2. The Grid service has a set of predefined interfaces, such as Registry, Factory, Handle Map, Primary Key etc [1].
3. Each Grid service has the responsibility to maintain the local policy, which means the Grid service can decide what kind services it will provide and what kind of resources can be shared.

Components of the "CFD-Grid"

The system is composed of two layers as shown in figure 2. The upper layer provides unified interfaces for various services and the lower layer provides High Performance Computing environment and SDK (Software Develop Kits) for CFD programmers. In the upper layer, by using Service Domain, Registry service, Factory service and Handle-Map service can be constructed, and the service requestor are the specific users of this system. The user can submit the tasks, monitor the computing progress and get the middle and final results via the web pages. When received the requests from the users, the Service Provider can query the lower layer to find what kinds of services available now. The Service Providers can supports the user accounts management and

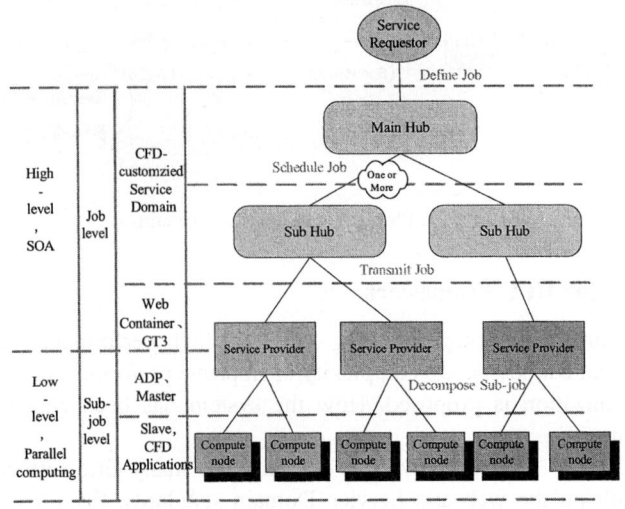

Fig. 2. "CFD-Grid" Framework.

supports the task submitting and parameter adjusting. It can interact with other services to finish the computing work. The Service Provider layer has also the responsibility to act as an information exchange center.

The lower layer of the "CFD-Grid" can complete all computing services, including allocate service and analysis service including various CFD services from every different CFD codes or other analysis code with same defined interface to the main framework of the "CFD-Grid". The Service Providers provide not only a service to control process but also accepting the requests from the users. Then the control process will find the available computing services. Then the connections are built among the upper layer and lower layer.

Different Service Provider provides different CFD codes, named as CFD services. Then, user can select suitable code for their application according to their different computing aims. As demonstrations, in this paper, CFD codes in the "CFD-Grid" can be used for two different applications, aerodynamic simulation for flow field around a wing of an aero-craft and aero-crafts optimization. CFD (computational fluid dynamics) code is a 3-D explicit flow solver in which three-dimensional Navier-Stokes equations are solved by employing a central difference and artificial dissipation finite volume scheme [3]. In order to run all those analysis codes, an automatic grid generation code is required. In the analysis service, we adopt our 3-D overlapping (Chimera) grid generation code [4] and other commercial software to generate computational grids for CFD computations.

3 Applications on Aero-crafts Simulation and Optimization

3.1 Hardware Architecture

To construct the basic hardware environment for our Grid system, we connect computer resources in our university, including a SGI Onyx3800 (64CPUs, 64 Gflops) supercomputer, four set of IBM-1350 Clusters (86 Gflops), Sun E-450 & Series Ultra workstations and many PC connected to our Campus network. The maximum speed of data-flow of our campus network between those computers can reach 1 Gb/s.

3.2 "CFD-Grid" for Aero-crafts Aerodynamic Simulation

The "CFD-Grid" HPC system finally works well by connecting users and the CFD services. For users, if they want to start a new simulation, they only need submit their tasks through the web. Through the task executer web page user can also adjust the computational input parameters. All those web services will help any users complete their whole computing tasks mutually and interactively.

As a demonstrated case, we choice a real aero-craft aerodynamic simulation case to show the performance of the "CFD-Grid". The initial flow condition for this aerodynamic simulation is Mach number is 0.8. The computational grid points for this simulation are 6,000,000. With domain decomposition technology, there are 32 blocks divided for 32 CPUs for the whole flow field computation. The computational grid can be seen from Fig.3 and the flow filed can be shown from the pressure contour Fig.4.

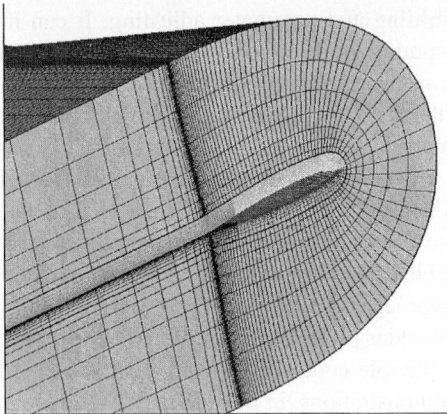

Fig. 3. The computation mesh for aerodynamic simulation around a wing of an aero-craft.

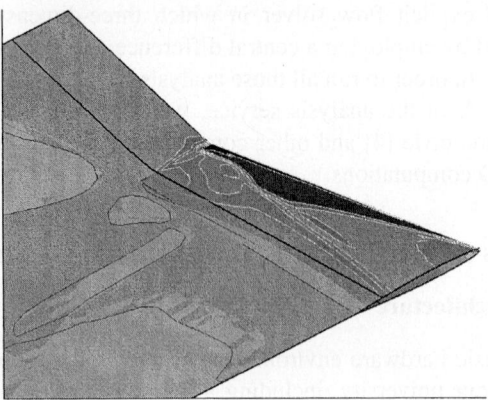

Fig. 4. The pressure contour for the flow field of the wing of the aero-craft.

3.3 "CFD-Grid" for Aero-crafts Optimization Design

The "CFD-Grid" HPC system can also well provide CFD services for users' different application, for example, they can adopt the CFD services in their MDO (Multidisciplinary Design Optimization) task. They also should build an optimization service in service domain and define the connection relationship between the optimization service and CFD services. Then they can also submit their tasks through the web. Through the task executer web page user can also adjust the computational input parameters of optimization design. All those web services will help any users complete their whole computing tasks mutually and interactively, too. To complete a MDO (Multidisciplinary Design Optimization) task, the designer-in-chief also can communicate with other designers from various disciplines through the web. There is a billboard on the web page. The task executor and the co-designers exchange information by it. When the co-designer finds some parameters are not suitable for the task, he leaves messages on the billboard.

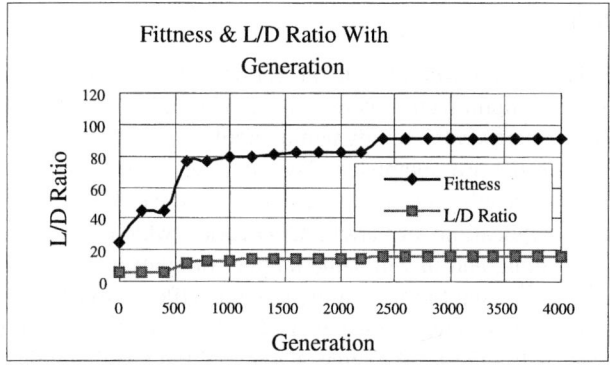

Fig. 5. The Optimization Process: Lift/Drag Ratio with the generation increasing.

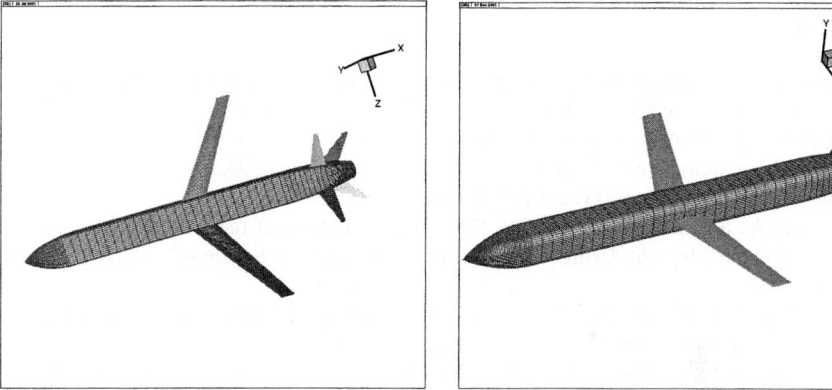

Fig. 6. The aero-craft shape of Generation 500 and Generation 3000, the left one is the shape for 500 generation and the right one is the shape for 3000 generation.

As a demonstration case for the application of "CFD-Grid", a conceptual aero-craft deign is conducted. In the aero-craft conceptual design process, we only consider two basic performances, including aerodynamic high ratio of lift to drag and flight stability. The performances can be computed by CFD services and other CFD-like services respectively. To complete this optimization design, we define an objective function is a function of L/D (ratio of lift to drag) maximization, and set the flight stability as constrain to the optimization problems. For this aero-craft conceptual design, there are 11 design variables to be optimized. The size of population, the mutation rate and crossover rate are the basic control parameters of the optimization service, which can be changed interactively from the web page of the "CFD-Grid". Then we can obtain results, as shown from figure 5 and figure 6. Figure 5 shows all of the optimization process and illustrates that with the generation increasing, Lift/Drag Ratio and the fitness of the evaluation is become larger. Figure 6 is the aero-craft shape of Generation 500 and Generation 3000. As the generation increased, more optimized aero-craft shape can be obtained before the optimal is reached.

4 Conclusion

A "CFD-Grid" HPC framework based on Service Domain for aero-crafts aerodynamic simulation and optimization is described. Through using this "CFD-Grid" an application case for aero-crafts aerodynamic simulation and a computation for conceptual aero-craft design are completed. The advances of this "CFD-Grid" over the traditional parallel computing means can be clearly shown. Firstly, it presents a new framework for the applications of various CFD codes, which can provide users various CFD codes to adopt, and in which the number of analysis service is dynamically changed. Secondly, by using Service Domain in this HPC "CFD-Grid" system we try to define a set of standard interface for every CFD codes. We believe the deeper we investigate on the "CFD-Grid", the more general "CFD-Grid" application standards can be completed.

References

1. Ian Foster, Carl Kesselman, Jeffrey Nick and Steven Tuecke. , "The Physiology of the Grid: An Open Grid Services Architecture for Distributed Systems Integration," 2002.2 http://www.globus.org/research/papers/ogsa.pdf.
2. Web Service Architecture, W3C Working Draft , 14 November 2002., URL: http://www. w3 .org/TR/2002/WD-ws-arch-20021114/
3. Jameson, A., Schmidt, W. and Turkel, "Numerical Solutions of the Euler Equations by the Finite Volume Methods Using Runge-Kutta Time Stepping Schemes," AIAA Paper 81-1259, 1981.
4. Hong Liu, W. Dong, and H.T. Fan, "Conservative and non-conservative Overlapping Grid Generation in Simulating Complex Supersonic Flow," AIAA Paper 99-3305, 1999.
5. Yih-Shin Tan, Brad Topol, Vivekan and Vellanki and Jie Xing, "Business service Grid, Part 1: Introduction, IBM DeveloperWorks," 2003.
   See also ftp://www6.software.ibm.com /software / developer/library/i-servicegrid.pdf.

Power Management Based Grid Routing Protocol for IEEE 802.11 Based MANET

Li Xu[1,2] and Bao-yu Zheng[2]

[1] Dept. of Computer Science, Fujian Normal University, 350007 Fuzhou, China
Xuli@mail.edu.cn
[2] Dept. of Info. Eng., Nanjing University of Post and Telecommunication,
210003 Nanjing, China
Zby@njupt.edu.cn

Abstract. MANET (Mobile Ad Hoc Network) is a collection of wireless mobile nodes forming a temporary communication network without the aid of any established infrastructure or centralized administration. The lifetime of a MANET depends on the battery resources of the mobile nodes. So energy consumption may be one of important design criterions for MANET. With changing the idle model to sleep model, this paper proposes a new energy-aware grid routing protocol. Performance simulation results show that the proposed strategy can dynamic balance the traffic load inside the whole network, extend the lifetime of a MANET without decreasing the throughput ratio.

1 Introduction

MANET (Mobile ad hoc network) is multi-hop wireless network that are composed of mobile nodes communicating with each other through wireless links [1]. MANET is likely to be used in many practical applications, including personal area networks, home area networking, military environments, and search a rescue operations. The wide range of potential applications has led to a recent rise in research and development activities.

Efficient energy conservation plays an important role in the protocol design of each of layer in MANET because mobile nodes in such networks are usually battery-operated [2]. Several studies have addressed energy constraints. In [3], a minimum-power tree is established from source to a destination to support multicast services. In [4], the topology of the whole network is controlled by being adjusted the transmission power of mobile nodes. The goal is to maintain a connected network using minimum power.

Besides reducing transmission power, the energy of a mobile node can be conserved by occasionally turning off its transceiver. A mobile node still consumes much energy even when idle. Turning off the transceiver and entering sleep mode whenever a mobile node is neither transmitting nor receiving is a better way to conserve energy. A longer sleep is preferred to conserve energy. That is, the transceiver should be turned off as soon as possible when idle. However, the problem with turning off the transceiver is that the mobile nodes may fail to receive broadcast or unicast packets. Long sleeping increases probability of losing packets and decreases the throughput ratio. This work considers these two issues together and seeks to maximize energy conservation without decreasing conventional network capacity.

The proposed routing protocol, Power Management based GRID (PMGRID), exploits the concept of a routing protocol call GRID [5] while considering the energy constraints. In GRID, each mobile node has a positioning device such as a Global Positioning System (GPS) receiver to determine its current position. The geographic area of the entire MANET is partitioned into-dimensions logical grid. Routing is performed in a grid-by-grid manner. One mobile node will be elected as the gateway for each grid. This gateway is responsible for:

1. Forwarding route discovery request to neighboring grids
2. Propagating data packets to neighboring grids
3. Maintaining routes for each entry and exit of a node in the grid

No non-gateway nodes are responsible for these jobs unless they are source/destinations of the packets. For maintaining the quality of routes, we also suggest that the gateway node of a grid should be the one more stability and higher remaining energy.

In PMGRID, grid partitioning and grid-by-grid routing are the same as in the GRID routing protocol. The main difference between these two protocols is that PMGRID considers the energy of mobile nodes but GRID does not. Moreover, PMGRID seeks to maximize the lifetime of the networks. For each grid, one mobile node will be elected as the gateway and others can go into sleep mode. The gateway node is responsible for forwarding routing information and propagating data packets as in GRID. Sleeping non-gateway nodes will return to active mode whenever checking data should be sent to them in each beacon interval.

The rest of paper is organized as follows. Section 2 introduces our system environment. Section 3 presents our PMGRID protocol. Section 4 presents simulation result. Section 5 draws conclusion.

2 System Environment

As Figure 1 illustrates, the geographic area of the MANET is partitioned into two-dimensional logical grids. This is exactly the same partition method as described in [5]. Each grid is a square area of size d×d. Grid is numbered (x, y) following the conventional (x, y) coordinate system. Each node still has a unique ID. Each mobile node is made location aware by being equipped with a positioning device, such as a GPS receiver, from which it can read its current location. A predefined mapping should exist from any physical location to its grid coordinate.

As mentioned above, each grid is a square of d×d. Let r be the transmission distance of a radio signal. In this study, the value of d is chose as

$$\frac{\sqrt{2}}{4}r \tag{1}$$

This value setting means that a gateway located at the position can communications with any gateway in its eight neighboring grids.

Each mobile node k periodically calculates its ratio of battery remaining capacity R_k. R_k is defined as

$$R_k = \frac{Battery\ remaining\ capacity\ of\ node\ k}{Battery\ full\ capacity\ of\ node\ k} \tag{2}$$

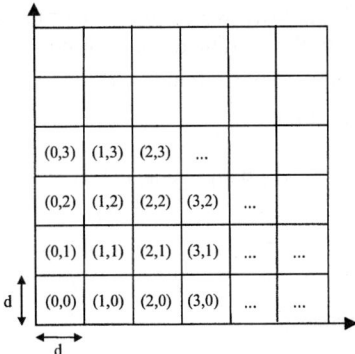

Fig. 1. Logical Grid to Partition a Physical Area

Each node in IEEE 802.11 based MANET is either in active mode or in power-saving mode (PS)[6]. A mobile node in active mode can transmit and receive packets. Only one node (the gateway) is active in each grid. Other non-gateway nodes can turn their transceivers off and enter sleep mode if they do not have packets to transmit or receive. These nodes only wake up periodically to check for possible incoming packet from gateway. Non-gateway always notifies its gateway when changing modes. Periodically, the gateway transmits beacon frames spaced by a fixed beacon interval. A PS node should monitor these frames. In each beacon frame, a traffic indication map (TIM) will be delivered, which contains ID's of those PS nodes with buffered unicast packets in the gateway. A PS node, on hearing its ID, should stay awake for the remaining beacon interval. Under the contention period (i.e., DCF), a wake PS node can issue a PS-POLL to retrieve the buffered packets. While under the contention-free period (i.e., PCF), a PS node will wait for the gateway to poll it.

3 Power Management Based Grid Routing Protocol

In PMGRID protocol, one mobile node in each grid will be elected as the gateway and remains in active mode. Other non-gateway nodes can sleep to conserve battery energy. The gateway node must maintain a node table that stores the node ID and status of all the nodes in the same grid.

A gateway may initiate a new gateway election process when it leaves the grid or seeking to maintain the load balance. To elect a new gateway, all nodes in the same grid must be in active mode. Each grid has a unique broadcast beacon, which is defined as the coordinate of the grid. All nodes must move into active mode when they hear the broadcast beacon. And then a new gateway can be elected according to the gateway election rules. This new gateway will inherit the routing table from the original gateway and remain continuously active.

3.1 Components of Node Priorities

The gateway is responsible for forwarding routing information and transmitting data packets. It plays the most important role in our protocol and consumes much energy. All nodes should take their turn as the gateway to prolong the lifetime of the network. The election of the gateway should take the remaining battery capacity and mobility speed into consideration. A gateway with more remaining energy and lower moving speed is preferred. We denote the weighted value of node k by W_k, the speed of node k by a scalar S_k in term of meters per second, and the remaining energy on node k as R_k. The weighted value function is

$$W_k = 2^{\log_2(R_i*0.9)\log_2(\frac{100}{S_k}+2)} \quad (3)$$

3.2 Gateway Election Process

Gateway election algorithm is executed distributed whenever a new gateway is needed, such as when the network is first initialized or the gateway node initialized a re-election process. The process is as follows:

1. Each node in active mode will periodically broadcast its HELLO message. The HELLO message contains four fields. (1) Id: node ID. (2) Grid: grid coordinate. (3) Gateway flag: set to one when the node is the gateway. (4) Weighted Value: defined as Equation (3).
2. After a HELLO period, which is predefined as the period for nodes to exchange their HELLO messages, all nodes are supposed to receive HELLO message from neighboring nodes in the same grid. Then, Choose that node with the largest W_i as the gateway.
3. The node will declare itself as the gateway by sending a HELLO message with the gateway flag set. The gateway node is responsible for maintaining the node table, which is constructed from the id field of the HELLO message.
4. All other non-gateway nodes receiving the HELLO message from the gateway will move into PS mode if they have no packets to transmit.

The gateway node must stay in active mode. The gateway can move into PS mode only when it releases its gateway responsibility. A non-gateway node may move into PS mode if it is not transmitting /receiving packets. A non-gateway node in PS mode must return to active mode if it receives a request beacon from gateway.

3.3 Grid Maintenance and Update

3.3.1 Controlling the Duty Cycle of Gateway

In order to maintain the stability and avoid oscillation of the routing path, PMGRID strategy tries to prolong the gateway duty cycle. On the other hand, we must re-form the grid before the gateway runs out of energy or leaves its grid region.

So whenever a gateway is former, each node in this grid sets a common wake-up timer that will wake it up in time T_s, which is set to some fraction of the estimated node lifetime and speed prediction of the gateway, which defined as

$$T_s = \min(\frac{R_i}{32}, \frac{d}{4S}) \quad (4)$$

Where *S* and *Ri* respectively represents the gateway current speed and remaining energy. Parameter d is the side of grid.

3.3.2 Normal Node Re-affiliation Policy

Besides a common wake-up timer T_s, each node j also sets a hypo-wake-up timer, which defined as

$$T_h(j) = \begin{cases} T_s & \text{if } S(j) \leq S \\ \dfrac{T_s}{1+\text{int}(\dfrac{S(j)}{S})} & \text{if } 7S \geq S(j) > S \\ \dfrac{1}{6}T_s & \text{if } S(j) > 7S \end{cases} \quad (5)$$

Where $S(j)$ is the current speed of node j. Equation (5) implies that the more fast mobility node the more fast frequency wake-up.

3.3.3 Node Information Update and Maintenance

When node wakes up, the node will see whether it is leaving the current grid. If it is leaving, the non-gateway node will send a unicast message to the gateway node to update the routing and node tables. The node must remain active until if finds another gateway. If it is not leaving, it will recalculate the dwell duration, set the timer, and then enter PS mode again. Two mobility situations should be addressed.

1. Nodes move into a new grid. Nodes will broadcast a HELLO message when they move into a new grid. The gateway node in each grid will also periodically broadcast its HELLO message. After the HELLO message from the gateway is received, the new incoming node will decide whether it should replace the gateway by applying the gateway election rule. In this situation, only a node with a battery level that is higher than that of the original gateway can replace the original gateway. This rule prevents frequent replacement of gateways: such replacement is an overhead of our protocol. If a gateway must be replaced, then the new gateway will declare itself by sending a HELLO message with the gateway flag set. The original gateway, receive this HELLO message, will transmit the routing and node tables to the new gateway. If no gateway is replaced, then the new incoming nodes will enter PS mode to conserve battery energy. If a new node does not receive any HELLO message during a HELLO period, the new node is in an empty grid and will declare itself as the gateway.

2. Nodes move out of a grid. The following considers the case of either a node or a gateway leaving one grid and entering another. A gateway must transfer its routing table to a new gateway before it leaves a grid. The gateway thus first sends a one-hop broadcast packet to wake up all the nodes in the same grid. After waiting for time, T, the gateway will declare its departure by broadcasting a RETIRE message, all non-gateway nodes will store the routing table and apply the gateway election algorithm to elect a new gateway. Then, the new gateway transmits a HELLO message with the gateway flag set to inform all the nodes about its new status. If a non-gateway node leaves a grid, it must notify the gateway about its departure by sending a unicast message to the gateway. The gateway will update its routing and node tables after receiving this unicast message.

4 Simulation and Performance Evaluation

Because it is difficult to capture the detail of PMGRID performance in an analytical model, we implemented PMGRID in the wireless extension NS-2 [7] provided by CMU to evaluate and compare the performance of the AODV, GRID-1 ($d=2r/\sqrt{10}$) and PMGRID. PMGRID is modified from AODV protocol [8]. In PMGRID, the routing table is established in a grid-by-grid manner, instead of in a host-by-host manner. Therefore, the gateway is the only host need to maintain the routing table in a grid, and responsible for the route discovery and route maintenance.

4.1 Simulation Model

We used a network with constant nodes in a rectangular region of size 1000m × 1000m. 15 connections are established at random using CBR traffic. The average duration of each connection is the time that finishes sending 200 packets. RF value is maintained at 281.8mW. Concerning energy consumption, only packets relayed or transmitted consumes a fixed amount of energy from the battery as given as

$$E(packet) = \frac{m \times packet\_size}{Bandwidth} \quad (6)$$

Where m is 1.33W for transmitting packets and 0.97W for receiving packets that correspond to 2400MHz, 2Mbps *WaveLAN* implementation of IEEE 802.11, *packet_size* is 512 bytes and is sent at 4 packets/second.

We set three variables to evaluate PMGRID with Grid and AODV.

Variable 1: Nodes follow random waypoint mobility model. Nodes pause and then move to a randomly chosen location at a fixed speed 5m/s. We consider seven pause times from 0 to 600 seconds.

Variable 2: Change node number ranger from 100 to 300 to change node density.

Variable 3: The nodes initial battery capacity is randomly assigned by a normal Gauss distribution with mean 20.0 joules and variance ranging from 0 to 5.0.

4.2 Network Lifetime

We define the metric lifetime of network as the duration from the beginning of the simulation to the first time a node runs out of energy. We use this metric with different node initial energy variances to measure the load balance and conservation energy performance of our proposed PMGRID.

As shown in Figure 2, PMGRID strategy achieves about 25 percent lifetimes longer than pure GRID and AODV, because the PMGRID strategy can conserve the lower power node. The lower remaining power node can avoid being gateway and turn off his radio. Simulations also show that AODV and GRID protocols are decrease with the network density or initial energy variance increase. By using power management strategy, the PMGRID achieves energy balance in overall network, so the lift time of network almost doesn't vary with the two parameters. The advantage of lifetime of network is more obvious when the network density increase.

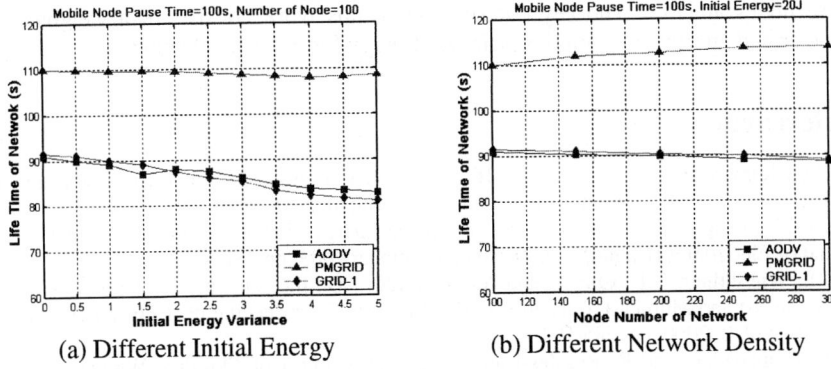

(a) Different Initial Energy   (b) Different Network Density

Fig. 2. Network Lifetime Comparison

4.3 Network Throughput

In the second group experiment, the packet delivery rate is observed at different pause times and different network density. The packet delivery rate is defined as the number of data packets actually received by the destination, divided by the number of packet issued by the corresponding source nodes. Initiate energy of all nodes is 20.0 joules. Figure 3 illustrates that the packet delivery rate exceeds 90% for all three protocols at most of condition. One special case is PMGRID, which has poor delivery rates with sparser nodes, and high rates with denser hosts. The reason is there are too few nodes in each grid when the host density is low.

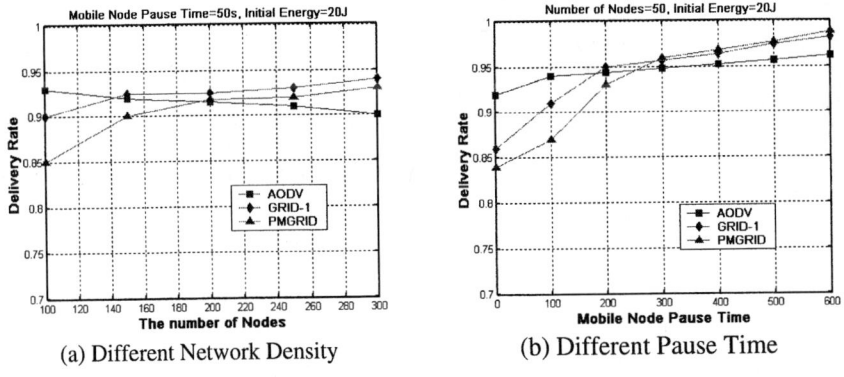

(a) Different Network Density   (b) Different Pause Time

Fig. 3. Packet Delivery Rate

5 Conclusion

The issue of energy conservation is critical in a limited energy MANET. With power management, this study proposes a novel energy-aware routing protocol, PMGRID, for mobile ad hoc networks. PMGRID extends the AODV or GRID protocol to account for energy constraints. One node is elected as a gateway in each grid, according to the gateway election rule. A gateway is responsible for route discovery and packet delivery. Energy conserved by turning the non-gateway node's transceivers off when these nodes are idle. The key contribution of this work consists of a weighted parameter and energy consumption balance strategy for gateway election.

References

1. Ram Ramanathan, Jason Redi, "A Brief Overview of Ad Hoc Networks: Challenges and Directions", IEEE Communications Magazine. Vol.40, No.6. May 2002, pp.48-53.
2. Xu Li, Zheng Baoyu, "Cross layer Coordinated Energy Saving Strategy in MANET", Journal of Electronics (China), Vol.20, No.6, November 2003, pp.451-455.
3. J.E. Wieselthier, G.D.Nguyen, and A.Ephremides, "On the construction of Energy-efficient Broadcast and Multicast Trees in Wireless Networks", Proceeding of IEEE INFOCOM, Tel Aviv, vol. 2, 2000, pp. 585-594.
4. R. Ramanathan and R. Rosales-Hain, "Topology Control of Multihop Wireless Network Using Transmit Power Adjustment", In Proc of the IEEE INFOCOM, Tel Aviv, vol.2, 2000, pp.404-413.
5. W. H. Liao, Y. C. Tseng, and J.P. Shen, "GRID: a fully location-aware routing protocol for mobile ad hoc networks", Telecommunication System, Vol.18, No.1. 2001, pp.37-60.
6. LAN MAN Standards Committee of the IEEE Computer Society, IEEE Std 802.11-1999, Wireless LAN Medium Access Control and Physical Layer specifications, IEEE, 1999.
7. NS-2. http://www.isi.edu/nsnam/ns/doc/index.html
8. C.E.Perkins,S. R. Das, et al.. Ad-hoc on Demand Distance Vector (AODV), Mar 2000; http://www.ietf.org/internet-draft/draft-ietf-manet-aodv-05.txt

On Grid Programming and MATLAB*G

Yong-Meng Teo[1,2], Ying Chen[2], and Xianbing Wang[2]

[1] Department of Computer Science, National University of Singapore, Singapore 117543
[2] Singapore-Massachusetts Institute of Technology Alliance, Singapore 117576
{teoym,cheny,wangxb}@comp.nus.edu.sg

Abstract. This paper discusses the design and implementation of ALiCE object-oriented grid programming template (AOPT). The programming template provides a distributed shared-memory programming abstraction based on JavaSpaces that frees the grid application developer from the intricacies of the underlying grid system. AOPT is designed for developing grid applications and as a programming tool for grid-enabling domain specific software applications such as MATLAB. In this paper, we discuss the design and implementation of MATLAB*G, a grid-enabled MATLAB using AOPT. The performance results indicate that for large matrix sizes MATLAB*G can be a faster alternative to sequential MATLAB.

1 Introduction

The main goal of grid programming is the study of programming models, tools and methods that support the effective development of portable and high-performance algorithms and applications on grid environments [8]. Grid programming will require capabilities and properties beyond that of simple sequential programming or even parallel and distributed programming. A programming model or tool need to present the heterogeneous resources as a common "look-and-feel" to the programmer and hide their differences while allowing the programmer some control over each resource type if necessary.

To reduce the complexity of grid programming, in this paper, we design and implement an object-oriented grid programming template based on the distributed shared-memory model, JavaSpaces [7]. We focus on how to efficiently develop portable and high-performance grid applications. Our propose ALiCE Object-oriented grid Programming Template (AOPT) is implemented on ALiCE (*Adaptive and scaLable Internet-based Computing Engine*), a grid computing *core middleware* designed for secure, reliable and efficient execution of distributed applications on any Java-compatible platform [12]. Our main design goal is to provide grid application developers with a user-friendly programming environment that is transparent of low-level grid infrastructure details, thus enabling them to concentrate solely on the application problems. The middleware encapsulates services for compute and data grids, resource scheduling and allocation, and facilitates application development with a straightforward programming template. Using AOPT, we have demonstrated the ease of programming grid applications [5, 9, 11, 13].

This paper focuses on the use of AOPT as a system programming tool to grid-enabled the domain-specific application package called MATLAB. Performance results indicate that for large matrix sizes MATLAB*G can be a faster alternative to sequential MATLAB. The remainder of this paper is structured as follows. Section 2 presents the ALiCE template-based programming model. Section 3 uses the programming template to implement MATLAB*G. Section 4 presents the performance evaluation of MATLAB*G. Section 5 concludes the paper.

2 Grid Programming

ALiCE is a portable middleware designed for developing and deploying general-purpose grid applications and application programming models, details of ALiCE can be found in [10]. The ALiCE runtime system adopts a three-tiered architecture, and consists of the following three main components: *Consumer, Producer, Resource broker*. In a typical scenario, a user launches an ALiCE application at a consumer, which then submits the application codes to a resource broker in the system. The resource broker schedules the tasks for execution on producers. Results of task execution are then returned to the consumer for visualization.

ALiCE sets out to provide an effective programming model to facilitate the development of grid applications and higher-level specialized programming models. In the ALiCE paradigm, large computations are decomposed into smaller tasks that are then distributed among producers in the network to exploit parallelism as best as possible to achieve a reasonable amount of speedup.

ALiCE adopts the *TaskGenerator-ResultCollector* programming model. This model comprises of four main components: *TaskGenerator, Task, Result* and *ResultCollector*. The consumer first submits the application to the grid system in the form of a .jar file encapsulating the application codes. The *TaskGenerator* running at a task farm manager machine generates a pool of *Task*s belonging to the application. Subsequently, these *Task*s are scheduled for executing by the resource broker and the producers download the tasks from the task pool. The results of the individual executions at the producers are returned to the resource broker as *Result* object. The *ResultCollector*, initiated at the consumer to support visualization and monitoring of data collects all *Result* objects from the resource broker. For batch job, result objects are collected at the resource broker.

Parallel applications development are written using ALiCE programming template. The template allows the programmers to transparently exploit the distributed nature of the ALiCE grid, i.e., without prior knowledge of the underlying technologies for communications, dynamic code linking, etc. The template abstracts methods for generating tasks and retrieving results in ALiCE, leaving the programmers with only the task of filling in the task specifications. Figure 1 shows the ALiCE programming template. Java classes comprising the ALiCE programming template are:

a. *TaskGenerator*. This is run on a task farm manager machine and allows tasks to be generated for scheduling by the resource broker. It provides a method process that generates tasks for the application. The programmer merely needs to specify the circumstances under which tasks are to be generated in the main method.
b. *Task*. This is run on a producer machine, and it specifies the parallel execution routine at the producer. The programmer has to fill in only the execute method with the task execution routine.
c. *Result*. This models a result object that is returned from the execution of a task. It is a generic object, and can contain as many user-specified attributes and methods, thus permitting the representation of results in the form of any data structure that are serializable.

d. *ResultCollector*. This is run on a consumer machine, and handles user data input for an application and the visualization of results thereafter. It provides a method collectResult that retrieves a *Result* object from the resource broker. The programmer has to specify the visualization components and control in the collect method.

```
            TaskGenerator Template

import alice.consumer.*;
import alice.data.*;
public class TASKGEN_CLASSNAME extends TaskGenerator {
  public TASKGEN_CLASSNAME() {}
  public void init() {
    //Place your initialisation code here
  }

  /* Main method - entry point */
  public void main(String args[]) {
    // This is where the tasks are generated, usually in a loop

    // This should be called for each task
    TASK_CLASSNAME t = new TASK_CLASSNAME();
    process(t);

    // To open a data file, read and write from/to it
    DataFile f = Data.openFile("file_name",this);
    READ_BUFF = f.read(POSITION, LENGTH);
    f.write( WRITE_BUFF, POSITION, LENGTH);

    // To send/receive an object
    OBJECT_CLASSNAME obj = new OBJECT_CLASSNAME();
    sendObject(obj, "snd_str_id");
    OBJECT_CLASSNAME rcvObj = (OBJECT_CLASSNAME)
                             requestObject("rcv_str_id");

    // To receive a string message from the result collector:
    String msg = getStringMessage();
  }
}
```

```
                Task Template

import alice.consumer.*;
import java.io.*;
public class TASK_CLASSNAME extends Task {
  // Place variables here
  public TASK_CLASSNAME () {
  }

  public Object execute () {
    // This is where you do your computations. The results can be any kind of
    // objects

    // You can generate and send a new task to be produced
    O_TASK_CLASSNAME t = new O_TASK_CLASSNAME();
    process(t);

    // To open a data file, read and write from/to it
    DataFile f = Data.openFile("file_name",this);
    READ_BUFF = f.read(POSITION, LENGTH);
    f.write( WRITE_BUFF, POSITION, LENGTH);

    // To send/receive an object
    OBJECT_CLASSNAME obj = new OBJECT_CLASSNAME();
    sendObject(obj, "snd_str_id");
    OBJECT_CLASSNAME rcvObj =(OBJECT_CLASSNAME)
                             requestObject("rcv_str_id");
  }
}
```

```
              Result Template

import java.io.*;

public class MyResult implements Serializable {
  public DATA_TYPE var;
  public MyResult() {
    var=NULL;
  }
}
```

```
          ResultCollector Template

import alice.result.*;
public class RESCOL_CLASSNAME extends ResultCollector {
  // Place Variables Here

  public RESCOL_CLASSNAME() {
  }

  public void collect() {
    // Place here the result collection and processing code to obtain
    // number of results ready call
    int resReady = getResultsNoReady();

    // To get a new result call
    RES_CLASSNAME res = (RES_CLASSNAME)collectResult();
  }
}
```

Fig. 1. ALiCE Programming Template

3 MATLAB*G

MATLAB, a widely used mathematical software provides an easy-to-use interface for various science computations. Computation intensive MATLAB applications can benefit from faster execution if parallelism is exploited. With the increasing popularity of distributed computing technology, up to now, at least twenty-seven parallel MATLABs are available [3].

In this paper we present the design, implementation and experimental results of MATLAB*G, a grid-based MATLAB on the ALiCE Grid, which exploits distributed matrix computation using task parallelism and job parallelism. MATLAB*G is an explicitly parallel MATLAB and is similar to MATLAB*P [4] which is designed at

MIT. MATLAB*G handles the communication and synchronization details for the user. However, while users are not required to indicate the matrices to be distributed, they have to explicitly specify the MATLAB computations to be parallelized.

Most parallel MATLABs are built upon *distributed memory architecture* in which each processor has its own memory module, e.g. MATLAB*P, Cornell Multitasking Toolbox for MATLAB [2], etc. The shortcoming is that they can only run on homogenous clusters. MATLAB*G is currently the only parallel MATLAB built on *object-based Distributed Shared Memory* (*DSM*) in which processes on multiple machines share an abstract space filled with shared objects. A shared memory interface is more desirable than a message-passing interface from the application programmer's viewpoint, as it allows the programmer to focus more on algorithmic development rather than on managing communication.

3.1 System Design

MATLAB*G exploits a client-server model using Object-based Distributed-Shared Memory. User submits job interactively from MATLAB environment. The client gets the job, divides it into a number of tasks, sends tasks into the DSM, and polls DSM for the result. A server always tries to get a task from DSM. After receiving a task, the server processes it, and returns the result to DSM. On the client side, after getting all results from servers, it assembles them into a complete result and returns it to the user. The system architecture is shown in Figure 2.

The client side consists of two components: *Extension* and *MGClient*. *Extension* includes a few MATLAB M files. It provides user interfaces for parallelism and links MATLAB with *MGClient*:

a. *ppstart:* This is a MATLAB function introduced by MATLAB*G Extension. When the user calls *ppstart(n)*, *n* servers are initialized and reserved for future computations.

b. *ppstop:* This function releases the reservations by a prior *ppstart*.

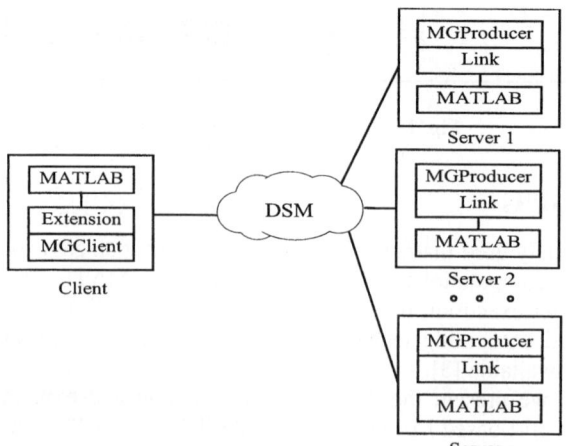

Fig. 2. MATLAB*G Client-Server Architecture

c. *mm:* This function lets the user assign a parallel job. The syntax of *mm* is: A=mm('fname', tasknum, matrices). *fname* is the computation that the user wants to execute, *matrices* are the arguments for this computation and *tasknum* is the task number specified by the user. In the future version, the *tasknum* argument will be removed and the task number will be generated by the system automatically according to certain algorithms. *A* is the result of computation.

MGClient, includes a number of Java classes for communicating with all the servers through DSM and distributing tasks and assembling results.

The Server consists of two main components: *MGProducer* and *Link*. *MGProducer* runs on a server and waits for tasks from DSM. On receiving a *ppstart* from DSM, *MGProducer* starts a MATLAB session at the backend through *Link*. Similarly, on receiving a *ppstop*, *MGProducer* terminates the MATLAB session. Upon receiving a computation task, *MGProducer* performs calculation and sends the result back to DSM. *Link* is used by *MGProducer* to start a MATLAB session, stop a MATLAB session, and execute MATLAB programs. To implement *Link*, we make use of an existing Java interface to the MATLAB engine called JMatLink [7].

A tuplespace is a shared datastore for simple list data structures (tuples) [1]. It provides DSM if every data inside it is an object. In MATLAB*G, communication between processors is handled through a tuplespace where processors post and read objects.

3.2 Mapping MATLAB*G onto ALiCE

A user submits a job through ALiCE Consumer. In response to the submission of a job, the ALiCE Resource Broker will instantiate a MATLAB*G Client, and ALiCE Producers will instantiate MATLAB*G Servers (see figure 3). ALiCE does not provide any interface to allow a user to run a MATLAB*G client directly on the Task Manager or run a MATLAB*G server on a Producer. An application has to be submitted in the form of an ALiCE Program through the Consumer interface.

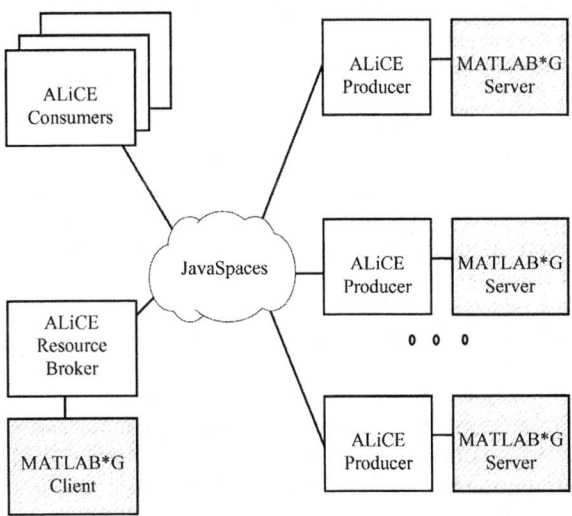

Fig. 3. Mapping MATLAB*G onto ALiCE Grid

To implement MATLAB*G, we added MATLAB*G code into proper templates to generate customized ALiCE program elements.

a. *MGTaskGenerator,* Besides the client side code, the user's MATLAB program is also embedded in the Task Generator template to create *MGTaskGenerator*. *MGTaskGenerator* first starts a MATLAB session, and then initiates n tasks by issuing command *ppstart(n)* to MATLAB. It then asks MATLAB to run the user's MATLAB programs. When finished, it issues a *ppstop* command to terminate tasks. *MGTaskGenerator* sends output it receives from MATLAB to *MGResultCollector* as the result from the computation.

b. *MGTask* is created by adding the server side code into ALiCE Task template. Each *MGTask* instantiates an *MGProducer* and runs it.

c. *MGResultCollector* extends the ALiCE Result Collector template. Running on the ALiCE Resource Broker, it simply waits for the result from JavaSpaces. An ALiCE program is created when we compile these elements together. The structure of such a program is described in Figure 4.

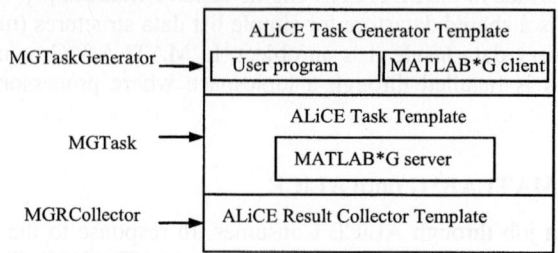

Fig. 4. Structure for a ALiCE Program generated by MATLAB*G

4 Experimental Results

We compare the performance of MATLAB*G with sequential MATLAB on the ALiCE Grid. The experiments are conducted on a ALiCE Grid Cluster with twenty-four nodes connected by 100 Mbps Ethernet. Four nodes are used, each of which is a PIII 866MHz, 256 MB RAM machine running Linux 2.4.

MATLAB*G can exploit two forms of parallelism. The first is task parallelism. To perform computation involving matrices, the computation can be divided into a number of tasks. Tasks are sent to space and each producer gets a task from space and performs computation on its sub-matrices. As an example, we used a simple but compute intensive function that computes for 1000 times the exponential for each element of a matrix A. We compare the execution times by varying the matrix size in Figure 5.

It is observed that for small matrix size (e.g. 100x100), the elapsed time for sequential MATLAB is still less than that of MATLAB*G. This phenomenon is attributed to the communication and partitioning overhead which is much larger than the computation time. However, as matrix size increases, the performance of MATLAB*G improves relative to sequential MATLAB, eventually overtaking it at the cross-point of approximately 500x500.

The second example shows job parallelism. When there are a number of matrix computations (jobs) to be executed one after the other, the jobs can be executed in parallel. $E = pinv(X)$ is the pseudo-inverse function provided by MATLAB. To per-

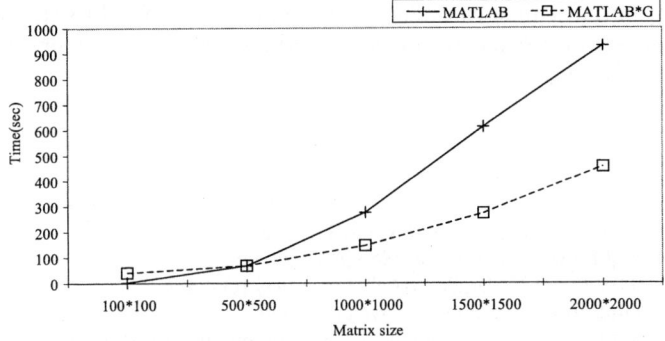

Fig. 5. Task Parallelism - Varying Matrix Size

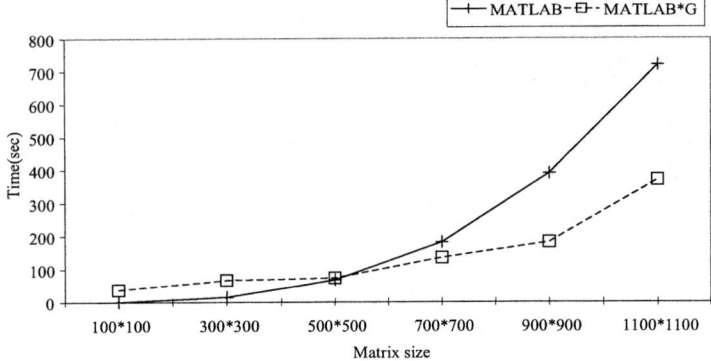

Fig. 6. Job Parallelism - Varying Matrix Size

form *pinv()* on a few matrices such as A1 = A2 = A3 = A4 = randn(1000), the jobs can be executed in parallel as follows:

X(1:1000, :)=A1; X(1001:2000,:)=A2; X(2001:3000, :)=A3; X(3001:4000,:)=A4;
Y=mm('pinv', 4, X);
E1=Y(1:1000,:); E2=Y(1001:2000,:); E3=Y(2001:3000,:); E4=Y(3001:4000,:);

Figure 6 shows the sequential and grid execution times for various matrix sizes. We observed that as matrix size exceeds 500x500, MATLAB*G outperforms sequential MATLAB.

5 Conclusions and Further Works

We discussed the design and implementation of the ALiCE object-oriented grid programming template that supports the distributed-shared memory programming model. We use the grid programming template as a system programming tool to develop a grid parallel MATLAB called MATLAB*G. Currently two types of parallelism for matrix computation are implemented: *task parallelism* and *job parallelism*. Performance results show that for large matrix sizes MATLAB*G can be a faster alternative to sequential MATLAB. Future work includes exploiting MATLAB *for*-loop parallelism, one of the most time-consuming computations in many MATLAB programs. Optimizations are also required to reduce overheads such as communication latency and matrix partitioning.

References

1. N. Carriero, D. Gelernter, "Linda in Context," CACM 32/4, pp. 444-458, 1984.
2. Cornell Multitasking Toolbox for MATLAB. [Online].
   Available: http://www.tc.cornell.edu/Services/Software/CMTM/
3. R. Choy, (2003, Oct. 12). Parallel MATLAB Survey [Online].
   Available: http://theory.lcs.mit.edu/~cly/survey.html.
4. I. Foster, and C. Kesselman, Globus: A Metacomputing Infrastructure Toolkit, *International Journal of Supercomputing Applications*, 11(2), pp 115-128, 1997.
5. D.P. Ho, Y.M. Teo, J.P. Gozali, Solving the N-body Problem on the ALiCE Grid System, 7th Asian Computing Science Conference, Lecture Notes in Computer Science 2250, pp. 87-97, Springer-Verlag, Hanoi, Vietnam, December 2002.
6. S. Hupfer, The Nuts and Bolts of Compiling and Running JavaSpaces Programs, Java Developer Connection, Sun Microsystems, Inc., 2000.
7. JMatLink. [Online]. Available: http://www.held-mueller.de/JMatLink/index.html
8. C. Lee, and D. Talia, "Grid Programming Models: Current Tools, Issues and Directions", in Grid Computing: Making the Global Infrastructure a Reality, F. Berman, G. Fox and T. Hey (eds.), Wiley, chap. 21, pp. 555-578, 2003.
9. Y.M. Teo, S.C. Tay and J.P. Gozalijo, Geo-rectification of Satellite Images using Grid Computing, Proceedings of the International Parallel & Distributed Processing Symposium, IEEE Computer Society Press, Nice, France, April 2003.
10. Y.M. Teo and X. Wang, ALiCE: A Scalable Runtime Infrastructure for High Performance Grid Computing, Proceedings of IFIP International Conference on Network and Parallel Computing, Springer-Verlag Lecture Notes in Computer Science Series, Wuhan, China, October 18-20, 2004.
11. Y.M. Teo, X. Wang, and J.P. Gozali, A Compensation-based Scheduling Scheme for Grid Computing, Proceedings of the 7th International Conference on High Performance Computing, IEEE Computer Society Press, Tokyo, Japan, July 2004.
12. Y.M. Teo, X. Wang and Y.K. Ng, *GLAD: A System for Developing and Deploying Large-scale Bioinformatics Grid,* Technical Report, Department of Computer Science, National University of Singapore, 2004.
13. Y.M. Teo, Y.K. Ng and X. Wang, *Progressive Multiple Biosequence Alignments on the ALiCE Grid,* Proceeding of the 6th International Conference on High Performance Computing for Computational Science, Springer-Verlag Lecture Notes in Computer Science Series, Valencia, Spain, June 28-30, 2004.

A Study on the Model of Open Decision Support System Based on Agent Grid

Xueguang Chen[1], Jiayu Chi[2], Lin Sun[3], and Xiang Ao[1]

[1] Institute of Systems Engineering, Huazhong University of Science and Technology,
430074, P.R. China
xgchen9@mail.hust.edu.cn
[2] Management College, Sun YAT-SAN University, 510275, P.R. China
[3] Linnan College, Sun YAT-SAN University, 510275, P.R. China

Abstract. Decision Support Systems (DSS) are computer-based information systems that help decision-makers solve half-structured or non-structured problems by using data and models. They enable decision-makers to make decisions more effectively. The Web-based DSS have made information sharing on the Internet possible, but they cannot meet the decision-maker's needs in the heterogeneous, autonomic, dynamic and distributed decision support environment, because they only link web pages and lack global mechanism to manage and coordinate decision support resources on the Internet. However, as an advanced technology representing "the third internet revolution", Grid brings about a lot of innovative ideas and technologies for the development of DSS. As an application of Grid in the field of DSS, this paper puts forward an improved model of Agent Grid-based Open DSS (AGBODSS). With the National Economic Mobilization DSS as a practical case, it illustrates how the AGBODSS works in practice.

1 Introduction

DSS have a history of more than 30 years since the beginning of 1970s [1] [2]. In this period, DSS have made a great achievements and its concept has been expanded many times with an integration of computer technology, network technology, database technology, artificial intelligence and decision theory. Nowadays, being more powerful in function, more widely used on the Web, more friendly in man-machine interface, and more intelligent in operation, DSS can provide a great support for the decision-makers.

Currently, there are still some serous problems that DSS developers are facing:

1) The existing DSS model and theory cannot direct DSS development on the Internet environment effectively. They need further expansion to adapt to the great changes in content, form and methods of decision support.
2) There is no effective mechanism to publish, share and reuse decision support resources.
3) Being deficient in openness, it is difficult to integrate new functions and link the other DSS, and it is necessary to establish protocols and standards for DSS.
4) Due to the low level of modular construction, they are difficult to maintain, and the upgrading cost is high.

A lot of researches have been done to find a solution to these problems. Bhargava et al [4] put forward a new framework of electronic market for decision technologies named "Decision Net". The market provides decision technologies as user-based services rather than as products to the consumers. It also assumes a lot of organizing and managing the consumers and providers as well as integrating decision technologies. M.Goul et al [6] proposed a set of protocols for DSS deployment on the Internet called "Open-DSS protocol suite". Consumers and providers are connected via the open DSS protocols. Dong [7] proposed a framework for the architecture of a Web-based DSS Generator utilizing different software agents to enhance the functionalities of the existing DSS. Kwon [8] proposed an Open DSS based on Web in connection with Web technology and ontology metadata web services, in order to enhance the development and integration of the decision services. Sridhar [9] argued about the support of the DSS's three components structure in the Intranet.

All of the studies are valuable to solve the problems stated above, but most of them have stayed in the stage of framework design, and there are still no mature theory and approaches to support the DSS's development on the Internet. As the scale of the Internet keeps growing, the efficient management and utilization of the resources become more challenging.

DSS's development shows that, every step forward of the network technology brings a great influence to the concept, framework and function of DSS. Now, as an advanced technology representing "the third internet revolution", Grid brings about a lot of innovative ideas and technologies for the development of DSS. Grid appears as an effective technology coupling geographically distributed resources for solving large-scale problems in wide area network, which support open standard and dynamic services. In addition, it provides highly intelligent communication between computers and human [11] [12]. These characteristics are very suitable for the constructing need of DSS platform. It will improve DSS greatly, and bring profound revolution to DSS theory and its application.

This article shows the results of the research based on the work in reference [3]. Reference [3] put forward a framework of GBODSS and its operation process that shows some basic characteristics of GBODSS, but the framework is not sufficient for a further study of GBODSS.

MAS technologies, which have been researched for many years and are relatively mature, are very suitable for the modelling and design of the GBODSS. The agent's proactive and responsive characteristics and it ability to cooperate with others makes it appropriate to act as a node in the GBODSS. Moreover, the Agent Grid system introduced below provides a good supporting platform for the implementation of the modeling and design.

In order to give a clear picture of GBODSS for its openness and dynamic characteristic, this article introduces agent technologies into the research of GBODSS, and puts forward an improved model of Agent Grid-based Open DSS (AGBODSS). With the National Economic Mobilization DSS as a real case, it illustrates how this model works.

2 Agent Grid

Agent Grid is a new model of Grid, it was first proposed by the Defense Advanced Research Projects Agency (DARPA) in the Control of Agent Based Systems (CoABS) Program, in which the main aim was to enhance the interoperating, scalability and generalization of military information systems [13].

Manola views the Agent Grid from three different perspectives as below:
1) Agent Grid should be considered as a collection of agent-related mechanisms and protocol.
2) Agent Grid should be considered as a framework for connecting agents and agent systems.
3) Agent Grid should be considered as model entities that might be used to model an org chart, team or other ensemble.

These three views represent different aspects of the Agent Grid, and they can be valid simultaneously.

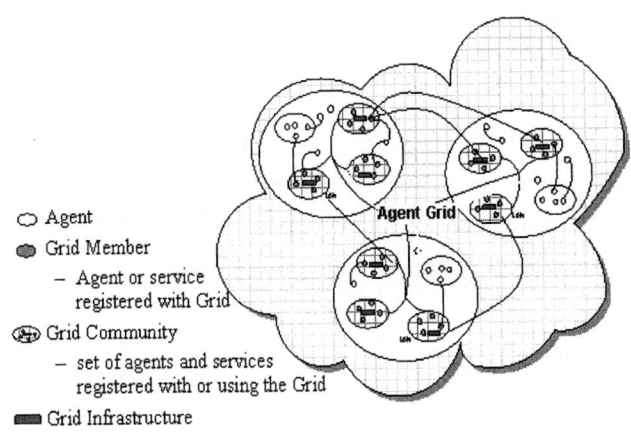

Fig. 1. The Framework of Agent Grid

Figure 1 shows the framework for the CoABS Grid program. Agent Grid is composed of Grid Infrastructure, Grid Community and Grid Member. Agent Grid provides various Grid services to integrate all kinds of distributed and heterogeneous software component on the Internet, such as objects, agents, legacy systems, and so on. CoABS Grid program has offered numerous important grid services, and many other powerful grid services are under construction. All these grid services are elaborated in the reference [5] and [14].

3 A Model of Agent Grid Based Open DSS (AGBODSS)

The concept of Grid system can be divided to different technological layers of computer system for an in-depth study:

1) Computing layer, corresponding to the computing Grid;
2) Data and information layer, corresponding to the information Grid;
3) Software layer, including objects, services and legacy system;
4) Agent layer, namely the Agent Grid. It can be considered as a more intelligent software which can accomplish certain functional tasks;
5) User layer.

Agent Grid layer plays a very important role in whole Grid system. On the one hand, it is this layer that interoperates directly with users. It includes agent and the agent system which perform the main system functions and sub-functions. Moreover, it needs good interfaces for the users, which facilitate the users to utilize the resources in the Grid to solve their problems and make the system integration easier. On the other hand, Agent Grid layer needs to connect the three Grid layers beneath it and manages the corresponding layers and resources. It facilitates the coordination between the different layers to accomplish concrete tasks.

In order to elaborate on the GBODSS, we have improved the model of the GBODSS with the above Grid hierarchy framework based on the Agent Grid. We have established an Agent Grid based open DSS model (AGBODSS) in figure 2. The improved model characterizes the relationship between the components and layers of the GBODSS and illustrates its openness and dynamic characteristics more clearly.

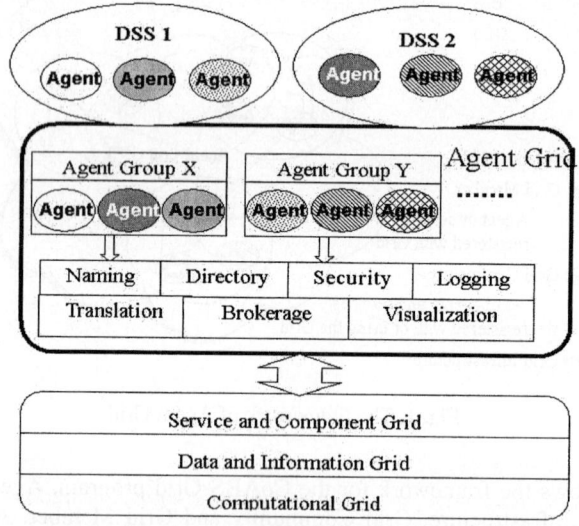

Fig. 2. Open DSS Model Based on the Agent Grid

The Agent Grid layer is the core of the AGBODSS model. This layer has direct connections and interfaces with the other components of the system. The Agent Grid layer is composed of several registered agent groups and grid services, and each agent group is composed of some agents. The Agent Grid provides many Agent Grid services to support the management, interoperation and integration of the agents in this layer. Presently, the Agent Grid services include register service, brokerage service, logging service, security service and visualization service, etc. Agents in the Agent Grid can be divided into several agent groups according to their type to facilitate the management and utilization of these agents.

With practical problems in mind, the designer of the DSS can construct different DSSs easily through a quick integration of the agents in the Agent Grid layer. The designer of the DSS needs firstly to design the structure of the AGBODSS, and then chooses suitable agents in the Agent Grid to play the roles in the AGBODSS.

The three bottom layers of the Grid stack are the basis of the AGBODSS, and can be considered as the Grid infrastructure that can provide all kinds of resource on the Grid. Heterogeneous resources in these layers can be fully utilized in a transparent way with the Agent Grid layer to hide the heterogeneity of them.

The AGBODSS model has many unique advantages:
1) Good intelligence. It handles the problems flexibly through the best utility of the agent's knowledge, autonomy and learning abilities. On the other hand, the agent has a strong communicating ability. Through interoperating and negotiating with the other agents, the multi-agents system can show more intelligent characteristics of the organization.
2) Superior scalability. We can construct a higher-level agent by composing several agents or agent systems, which serves as a basic unit to construct a new DSS.
3) Excellent adaptability to the changing circumstances and tasks. The AGBODSS model is open and dynamic. Each agent can join and leave the DSS dynamically according to the changing circumstances and tasks. And the agent's learning ability can also enhance the adaptive capacity of the DSS.
4) Fine hiding of the heterogeneity of the low-level resources. AGBODSS model can effectively hide the heterogeneity of the low-level resources with the support of the Agent Grid layer.

4 The Design of AGBODSS

The key issue to design an AGBODSS is the design of the Agent Grid layer and building GBODSS with the Agent Grid layer to solve practical problems. We can consider the three bottom layers of the Grid stack (the Computing Grid layer, the Data and Information Grid layer, the Service and Component layer) as Grid infrastructure that can be utilised transparently through the Agent Grid layer. So, here is not to discuss the design issue of these three layers.

4.1 The Design of Agent Grid Layer

As shown in figure 2, the Agent Grid layer is composed of agent groups and grid services. The grid services are developed by the CoABS Grid program and can be used directly. The design issue of the grid services is out of the range of our discussion. This paper focuses on the design issues of the agent and agent groups and building GBODSS with agents in the Agent Grid layer.

The Agent Grid is composed of entity agent, task agent, information agent and resource management agent. The information agent and resource management agent are general-purpose agents for Grid system and can be provided by the Grid system itself. The designer of AGBODSS only needs to care about the design issue of entity agent and task agent.

Each entity agent and task agent can be considered as the model of a practical entity, which has a specific problem solving ability and domain knowledge. The task agent is a specific entity agent that is responsible for the carrying out the entire task and coordinating the process of problem solving. Entity agent and task agent can both be defined as a structure containing the following elements: <Plans, Beliefs, Goals, Intentions, Data system, Reasoning module, Communication module, Sensor module, Act module >.

The agents with similar task type should be aggregated into some agent groups, which will facilitate the management and utilization of these agents. These agent groups can represent the practical teams and organizations in the real world and they can accomplish specific tasks by means of interoperation and negotiation.

4.2 Building DSS with the Agent Grid Layer

An AGBODSS can be built with the agents in the Agent Grid layer. AGBODSS is very similar to the concept of Virtual Organization, and an AGBODSS can be considered as a new agent organization that is composed of agents from several different agent groups to accomplish specific tasks [15]. The following steps should be taken when constructing an AGBODSS:

1) Analyze the problem and task faced by the decision-makers, set up the goal of the AGBODSS; and define decision support requirement.
2) Define the roles needed for the goal, imposing corresponding the responsibilities and rights on them and the relationship between them. The defined roles and their relationships compose the static model of AGBODSS. Generally, there will be a leading role that is responsible for the whole task.
3) Define the dynamic action among the roles in the GBODSS, e.g. defining the agreement of the interaction and coordination among them, to get the dynamic mode of the AGBODSS.
4) Assign the roles to the entity agents in the Agent Grid layer. Especially, the leading role will be assigned to a task agent. This step involves iterative search of eligible agent that satisfy the requirement of the roles in previous steps.
5) With above steps, a primitive AGBODSS has been built. But it still needs to be checked in practical running. There are often some iterative processes to go back to the previous steps for revision and adjustment.

5 A Case Study of AGBODSS

The modeling approach of the AGBODSS is a generic one with wide application. Taking the National Economy Mobilization DSS (NEM-DSS) as a practical case, now we elaborate how to put AGBODSS into practice.

The National Economy Mobilization is a range of government activity to schedule economic and social resources for emergent affairs. This paper uses the automobile mobilization as the case to analyze the constructing process of AGBODSS. The aim of the NEM-DSS is to implement unified and effective management of automobile mobilization activities, including generating plans, simulating plans and executing plans. It helps managers increase the effectiveness of making decisions.

The automobile mobilization involves several kinds of entity units, including components unit, assemble unit, transporting unit, warehouse and mobilization command center. Each entity unit is an autonomic one that needs interoperation and negotiation to form a final plan that should be coordinated in execution.

Based on the above analysis and design steps, we should first create agent model for each entity involved in the automobile mobilization. The design for each entity agent includes its plans, competence, product model, inner spiritual states (e.g. beliefs, goals, intentions), communication module and reasoning module, etc. components unit agent, assemble unit agent, transporting unit agent, warehouse agent and mobilization command center agent are created at this stage.

At this time, the agents can be divided into several groups according to its task type. For example, several components unit agents can be put into a unified agent group and the same is true for the transporting unit agents. These agent groups make up of the Agent Grid layer.

With the development of the Internet and information technology, every practical entity will have a corresponding entity agent on the Grid, so the modeling work of the entity agents and agent groups will not be necessary, and the developing workload to construct an AGBODSS can be reduced greatly.

Then the automobile mobilization DSS can be constructed with the entity agents on the Agent Grid layer:

First, the roles in the mobilization need to be identified according to the requirement of the mobilization goal, including the roles of manufacturing, transporting, storing and commanding. And then the respective goals, responsibilities and rights of these roles and the relationship among them (such as controlling relationship, depending relationship, equal relationship, etc) should be designed. Undoubtedly, the commanding role is the leading role in the mobilization activities.

Secondly, we need to define the dynamic characteristics of the DSS, which include the interaction activities between the roles, the message type, and the rules of coordination.

Thirdly, suitable entity agents should be assigned to the roles in the automobile mobilization DSS. The mobilization command center agent acts the commanding role, and it is the task agent to take charge of the whole mobilization activities. And some suitable agents play other roles according to their competence, reliability, and costs.

By now, a primitive AGBODSS has been built, but it may not be the final one, and needs to be revised and adjusted in practical use.

The AGBODSS has superior openness and excellent scalability. The entity agents can join and leave the AGBODSS dynamically according to the changing circumstances and tasks. The entity agents can sketch out mobilization plans through communication and negotiation, and they can coordinate to solve the problems in the execution of plans. The AGBODSS can efficiently utilize all kinds of decision-making support services and resources on the Grid system.

Notably, the AGBODSS is of great value on the large-scale and distributed simulation of the mobilization plans, which will save a lot of money and time that will be paid in a practical mobilization exercise.

6 Summary and Prospect

This paper proposes an improved Open DSS model based on the Agent Grid and GBODSS model. The improved model can characterize the openness, dynamics and complexity of DSS in Grid circumstance. It effectively hides the heterogeneity of the resources with good intelligence, scalability and adaptability. The paper also shows how to use AGBODSS in a practical case.

It also raises many questions about the AGBODSS, including the performance assessment, the matching of the resource request with the resource providing, the decision resource allocation and the application of AGBODSS in different fields.

References

1. M S Scott Morton, Management decision systems. Computer based support for decision making, Division of Research, Harvard University, Cambridge, Massachusetts, 1971.
2. Sprague.R.H, Carlson.E.D. Building effective decision support systems. Prentice-Hall, Englewood Cliffs, New Jersey, 1982.

3. Chi Jiayu, Chen Xueguang, Sun Ling. A Study on the Framework of Grid Based Decision Support System. In: Lan Hua ed. Proceedings of 2003 International Conference on Management Science & Engineering. Georgia, USA, 2003. Harbin Institute of Technology Press, Harbin, P.R.China. 45-49
4. Bhargava H, Krishnan R, Mueller R. Decision Support on Demand: On Emerging Electronic Markets for Decision Technologies, Decision Support System, 1997, 19(3): 193-214
5. Kettler B. The CoABS Grid: Technical Vision. http://coabs.globalinfotek.com/public /downloads /Grid/documents/ Grid Vision Doc Draft 2-3 2001-09-30.doc, September 2001.
6. Goul M, Philippakis A, Kiang M Y etc. Requirements for the design of a protocol suite to automate DDS deployment on the World Wide Web: A client/server approach. Decision Support System, 1997, 19(3): 151-170
7. DONG C J, LOO G S. Flexible Web-Based Decision Support System Generator (FWDSSG) Utilising Software Agents. DEXA Workshop 2001: 892-897
8. Kwon O B. Meta Web Service: Building Web-based Open Decision Support System Based on Web Services. Expert Systems With Applications, 2003, 24(4): 375-389
9. Sridhar S. Decision Support Using the Intranet, Decision Support Systems, 1998 23(1): 19-28.
10. Foster I, Kesselman C, The Grid: Blueprint for a New Computing Infrastructure, Morgan Kaufmann, San Fransisco, CA, 1999.
11. Hai Jin, Deqing Zou, Hanhua Chen etc. Fault-Tolerant Grid Architecture and Practice. Journal of Computer Science and Technology, Vol.18, No.4, July 2003, pp.423-433
12. Hai Jin, Longbo Ran, Zhiping Wang etc. Architecture Design of Global Distributed Storage System for Data Grid. Exploring the New-generation Computing Technology – Proceedings of the 7th International Conference for Young Computer Scientists (ICYCS'03), International Academic Publishers, August 8-10, 2003, Harbin, China, pp.119-122
13. Manola F, Thompson C. Characterizing the Agent Grid. http://www.objs.com/ agility /techreports, 2001
14. Kahn M, Cicalese C, Brake B, et al. DARPA CoABS Grid Users Manual. http://coabs.globalinfotek.com/public/downloads/Grid/documents/GridUsersManual.v5.0.d oc, June 2003
15. Gao Bo. Study on the Theory and Its Application of Organization Construction Based upon MAS. [D]. 2002, Huazhong University of Science and Technology, Wuhan, China.

Memory Efficient Pair-Wise Genome Alignment Algorithm – A Small-Scale Application with Grid Potential

Nova Ahmed, Yi Pan, and Art Vandenberg

Georgia State University, Atlanta, GA 30303, USA
nahmed2@student.gsu.edu, {yipan,avandenberg}@gsu.edu

Abstract. Grid middleware infrastructure is a distributed environment suitable for many, typically large-scale, applications. An improved genome sequence alignment algorithm is presented that significantly reduces sequence matching computation time. Yet, very long sequences can still present computation challenges. A small-scale application is implemented on a shared memory system, on a cluster, and the grid-enabled cluster. Experimental results show comparable performance of the grid-enabled version, with scalability for large sequences. The grid offers application management, enables dynamic specification of parameters, and enables a choice of distributed computation nodes.

Keywords: Grid computing, distributed computing, computational biology, biological sequence alignment.

1 Introduction

Distributed computing can provide benefits in the combined computation capacity of distributed resources. A grid-enabled computing model assures proper control, reduces overhead of handling shared resources, and provides a service oriented environment. Large or medium-scale applications, typically needing huge amounts of computation, are grid-enabled with good results. To address genome alignment, an important problem in bioinformatics, a small-scale problem, well suited to a shared memory, is grid-enabled and compared with shared memory results.

The basic shared memory algorithm is modified to improve memory usage and implemented for shared memory, cluster, and a grid-enabled cluster. Studying this particular problem is important from two different aspects. First, it involves a lot of communication among the processing elements, challenging the concept of tradeoff expected between computational results versus communications in a grid-enabled environment. Second, it is not a typical, heavily computation intensive problem usually seen in a grid environment, so there may be insights on what kinds of problems may be feasible to implement on a grid. Section 2 describes the shared memory, cluster, and grid-enabled cluster environments. Section 3 treats the genome alignment problem, the algorithm, its memory efficiency improvements, and parallelization. Section 4 presents experimental results and Section 5 gives conclusions and suggests future work.

2 The Distributed Environment

The distributed environments for the algorithm included a shared memory, a cluster and a grid-enabled cluster platform. The shared memory environment is perhaps most desirable from the view point of minimizing the communication but it has limitations on scalability. On the other hand, the cluster environment introduces overhead in terms of communication but has a better scalability. The grid-enabled cluster environment offsets limitations of communication by being very much more scalable. The *Shared Memory* platform was a SGI ORIGIN 2000 machine, with 24 CPUs. The *Cluster environment* was a Beowulf cluster of eight homogenous nodes, each with four 550MHz Pentium III processors with 512 MB RAM. The same Beowulf cluster was then used for the *Grid environment* with implementation of the Globus Toolkit grid software layer. The *Grid environment* is important to enable coordination of resource sharing and problem solving for dynamic, multi-institutional, virtual organizations [1]. It offers a hardware and software infrastructure providing dependable, consistent, pervasive and inexpensive access to the computational resources [2]. The Grid architecture used here follows the Open Grid Systems Architecture (OGSA) standard, providing a user level abstraction, resource management and scheduling as described in OGSA [3].

3 Genome Alignment as a Grid Application

A genome alignment program is considered where, as the genome sequences become very long, the program can become computation and memory intensive. The program is improved to reduce memory and studied in different distributed environments to evaluate the performance of grid.

3.1 The Basic Genome Alignment Algorithm

The sequence alignment of genome structures is different from usual sequence matching as the genomes allow a mismatch, a change of symbol or a gap to be present among two sequences. A match or a mismatch is taken into account by a score assigned to each pair of sequences to be matched [4]. The extremely large size of genomes makes the process of sequence alignment very tedious [5] and the dynamic programming method is a convenient solution for sequence alignment. A two dimensional array is required in this method to store the partial results which has height and width according to the length of the sequences to be matched known as the *similarity matrix*. After the creation of the matrix, the alignments are generated and a *threshold* value is set to determine the best aligned pairs. To generate the aligned sequences a *trace back procedure* is applied, beginning from the threshold value in the matrix to find out the best aligned sequences.

3.2 The Reduced Memory Structure

The basic method requires huge amounts of memory for the similarity matrix, yet there are many zero elements in the matrix that do not contribute to the computation of the genome alignments. The reduced memory structure keeps only the nonzero elements of the matrix and uses the same algorithm for the computation. Memory is dynamically allocated and there are other data structures to maintain the blocks of memory efficiently. The new data structure [6] consists of data blocks to keep the matrix data and location pointers to locate the data blocks. When a data block is filled up, another new data block of fixed size is allocated and its corresponding location pointer is also allocated. There is a static structure defined for each row of the matrix and it contains a single location pointer itself.

3.3 The Parallelization Method

The parallel version maximizes the independent parallel processing taking care of the data dependency involved in the computation. The very long genome sequences are divided among different processors so that each processor computes the part available. A particular processor P_i must wait when it is calculating the elements of its first column for the neighbor element of the last column calculated by P_{i-1}. The parallel program continues in a pipelined fashion. After the similarity matrix is generated, the trace back sequence generation can be initiated by different processors independently. After the threshold value is found out, two events take place. First, processor P_i will get the whole sequence in its own matrix and generate it. Second, a subsequence is found which is to be sent to its neighbor P_{i-1} and in this case the subsequences generated so far are sent to the neighbor so that the neighbor can continue the computation.

4 Experimental Results

The experimental results show very primitive studies of the running time of the algorithm as this particular algorithm is not specially designed for the grid environment. Rather the results indicate that the grid environment is comparable to other ones. For the implementation of the parallel version of the program MPI (the message passing interface) has been used.

The computation times for the shared memory, cluster, and grid-enabled cluster environments are compared in figure 1. The software layer of grid adds overhead for managing the job, so adding some computational time. The fact that the job is done faster in the cluster and grid than the shared memory environment, does not necessarily indicate the superiority of the grid environment, rather it probably signifies that the hardware capabilities are different in two environments. However, it is evident that the computation time in the grid is comparable to other environments even with added communication and overhead introduced for the grid protocols.

In figure 2, the computation time is measured for a much longer genome sequence having genome length of 10,000. (The highest sequence compared was 3,000 in figure 1, due to resource constraints inherent in the shared memory machine compared to the scalable architecture of clusters.)

The speed up performances of the shared memory, cluster, and grid-enabled cluster, are compared in figure 3, where speed up is with respect to running time of the algorithm using multiple processors as compared to a single sequential processor. Figure 3 indicates that the shared memory machine has better speed up for this particular problem, whereas, for the cluster and grid-enable cluster, after a certain extent, the increased the number of processors do not significantly help performance. Indeed, due to the communication overhead of cluster and grid-enabled clusters, it is not perhaps useful to increase the number of processors when the problem is not significantly large. Or, viewed another way, in the cluster and grid-enable cluster environments, allocating processors to a problem is, of itself, an interesting case of finding optimal allocation models. However, the cluster and grid-enabled cluster environment have little difference in the speed up.

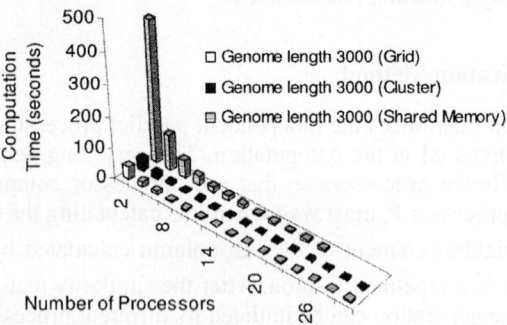

Fig. 1. Comparison of computation time: Shared Memory, Cluster, Grid-enabled Cluster environment

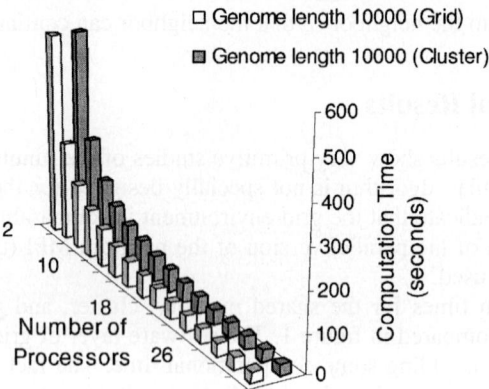

Fig. 2. Comparison of computation time: Cluster, Grid-enabled Cluster environments

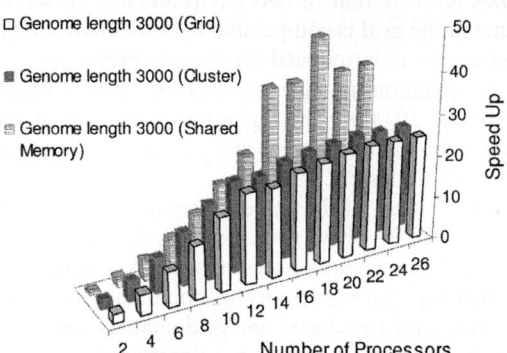

Fig. 3. Comparison of speed up: Shared Memory, Cluster, and Grid-enabled Cluster environments

In figure 4, the speed up is compared for the cluster and grid-enabled cluster environment – the difference is negligible. The grid-enabled environment is comparable to the cluster environment: grid overhead has not affected the performance appreciably and, in fact, introduces additional options for flexibility.

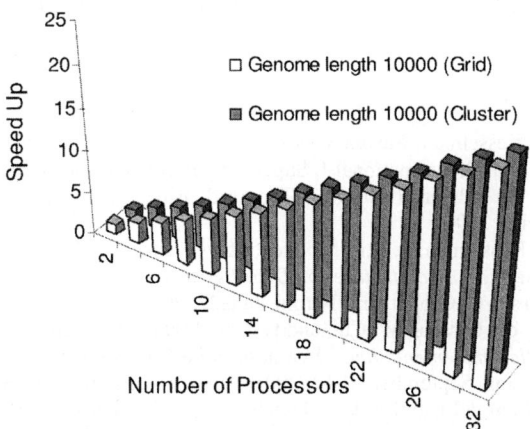

Fig. 4. Comparison of speed up: Cluster, and Grid-enabled Cluster environments

There are other issues related to the shared memory environment compared to the cluster and grid-enabled cluster environments, such as scalability. In the shared memory environment the highest genome length to could be aligned was 3,000, whereas in a grid environment the maximum genome length (given the configuration tested) was 10,000. Certainly, processing elements can be added relatively easily to a grid according to the need, while a shared memory structure is a more fixed structure.

5 Conclusions and Future Work

The Grid environment has opened an efficient way for distributed computing. To utilize this environment best, applications that are well suited for distributed environment can be implemented. The computation required by the pair-wise genome alignment algorithm described here has fewer interdependencies among the processing elements and will be well able to exploit the power of grid.

The future plan is to work on a multiple genome alignment algorithm [7] that uses pair-wise alignments and to adapt it to a distributed environment of grid-enabled clusters. The genome alignment problem uses an algorithm employing fine-grained communication among the processing units. The problem is computation intensive for the fact that the genome sequence can be very long, but it is not very computation intensive compared to problems designed specifically for grid environments. Implementation of such a small-scale problem, with good results, suggests that users can explore the potential of the grid for smaller problems at the same time larger problems are being addressed. A distributed environment is desirable, giving the user greater control over the environment and allowing for scalability of adding resources according to the need.

Acknowledgment

This material is based in part upon work supported by the National Science Foundation Middleware Initiative Cooperative Agreement No. ANI-0123937. Any opinions, findings, conclusions or recommendations expressed herein are those of the author(s) and do not necessarily reflect the views of the National Science Foundation.

References

1. Ian Foster, Carl Kesselman, Steven Tuecke, *"The anatomy of the Grid: Enabling Scalable Virtual Organizations"*, International J. Supercomputer Applications, 15(3), 2001.
2. Ian Foster, Carl Kesselman, (Editors), "The Grid: Blueprint for a New Computing Infrastructure", Elsevier, 2004.
3. Ian Foster, Carl Kesselman, Jeffery M. Nick, Steven Tuecke, *"The physiology of the grid – An Open Grid Services Architecture for Distributed Systems Integration"*, Open Grid Service Infrastructure WG, Global Grid Forum, June22, 2002.
4. W. S. Martins, J. del Cuvillo, W. Cui, and G. Gao, *"Whole Genome Alignment using a Multithreaded Parallel Implementation,"* Symposium on Computer Architecture and High Performance Computing, September 10-12, 2001, Pages 1- 8, Pirenopolis, Brazil.
5. W. S. Martins, Juan del Cuvillo, F. J. Useche, K. B. Theobald, and G. R. Gao, *"A Multithreaded Parallel Implementation of a Dynamic Programming Algorithm for Sequence Comparison,"* Pacific Symposium on Biocomputing. 2001 - January 3 - 7, 2001, Pages 311-322, Big Island of Hawaii.
6. Nova Ahmed, Yi Pan, and Art Vandenberg, *"Memory Efficient Pair-wise Genome Alignment Algorithm -- A Small-Scale Application with Grid Potential,"* Technical Report, Georgia State University, Atlanta, GA, February 2004.
7. W. T. Taylor, *"Multiple sequence alignment by a pair wise alignment"*, 1987, Comp. App. Bio. Sci. 3, 1858-1870.

Research on an MOM-Based Service Flow Management System*

Pingpeng Yuan, Hai Jin, Li Qi, and Shicai Li

Huazhong University of Science and Technology, Wuhan 430074, China
yuanpingpeng@mail.hust.edu.cn, hjin@hust.edu.cn

Abstract. Workflow is an important approach for the specification and management of complex processing tasks. This approach is especially powerful for utilizing distributed service and processing resources in grid systems. Some researches on the combination of workflow and grid exist, leading to many approaches to grid workflow support. We describe our *MOM-based Service Flow Management System* composing complex processing chains. We describe the technologies to enable the execution of these processing chains across wide-area computing systems.

1 Introduction

Grid computing aims to utilize distributed services owned or provided by different organizations as a single, unified service. For example, many computing and data intensive scientific applications require a series of tasks to solve sophisticated problem collaboratively. Thus, how to glue services distributed over Internet and across organizations is the key problem for grid computing. Workflow technique can satisfy these requirements. Workflow is a technique by which a complex process is expressed as an interconnected series of smaller, less complicated tasks. The concept of workflow has successfully been used in many areas.

Due to the great capabilities of workflow to interconnect tasks, grid workflow has emerged. Some goals of grid workflow include: description of the interactions of grid services, composition of grid services and ability to trigger services if necessary. Currently, workflow systems for grid services are evoking a high degree of interest. In this paper we describe our approach to grid workflow.

The remainder of this paper is organized as follows: First, we review some related works. Then, we present our workflow definition language. In section 4, we describe our grid workflow architecture. Finally, we end this paper with conclusion.

2 Related Works

The goal of grid computing is to provide a service-oriented infrastructure that leverages standardized protocols and services to enable pervasive access to, and coordinated sharing of geographically distributed services and resources. New higher-level services and applications can be constructed from the available services through workflow.

---

* This paper is supported by Nature Science Foundation of China under grant 60273076, ChinaGrid project from Ministry of Education, and National 973 Basic Research program under grant 2003CB316903.

There is a limited amount of work related to the issues of the combination of grid and workflow in the grid computing community. Research efforts on grid workflow include Webflow [1], Chimera Virtual Data System (GriPhyN) [2], Symphony [3], DAGMan [4], UNICORE [5], and XCAT [6]. Webflow, one of the earlier workflow systems, supportS application composition in Grid environments. The Chimera Virtual Data System considers grid workflows as static graphs of services. Symphony is a framework for combining existing codes to meta-programs. DAGMan manages the dependencies between jobs and schedule jobs according to their dependency relation. XCAT Application Factories address workflow related issues for grid-based components within the *Common Component Architecture* (CCA) framework [7].

The key issue that differentiates our work from these is that we focus more on service-level support, workflow definition language, distributed workflow management and scheduling, as opposed to some researches at the programming level. The goals are to define a workflow definition language and provide the scheduling solution for services in distributed environment. One advantage of our approach is that services can be selected and scheduled dynamically. This is especially suitable for dynamical grid environments.

3 Workflow Definition Language

In a distributed system, users' requests can be represented as flows of services linked together. An important portion of service flow (workflow) is the definition of the process logic, which is expressed by the usage of a workflow specification language. The *Event-based Service Flow Definition Language* (ESFL) is a workflow specification language defined in our workflow approach.

In ESFL, the definition of a workflow or a process is specified in process element. The process element has a reference attribute to identify the process uniquely. The process elements are important building blocks of ESFL. The process elements include three kinds of sub-elements: *basic element, construction element* and *association element*.

The basic element can be any one of the following: *activity* (A step or task to be performed); *event* (occurrence of a particular situation or condition).

The *activity* types and *event* types can be constructed from construction element. The construction element is as follows: *choice* (exactly one activity (event) out of a set of activities (events)); *and* (all activities (events) out of a set of activities (events)); *or* (any one activity (event) out of two activities (events)).

The association element includes only one element: *OnEvent*. The *OnEvent* element indicates which activity should be executed when an event occurs and which event will occurs after the completion of the activity.

4 The MOM-Based Grid Workflow Management System

Users can describe the interactions between grid services by using of ESFL. Workflow system can orchestrate or control the interactions between services, which are used to perform tasks, such as computing task and storage task.

Our approach to workflow management in grid environments is *MOM-based Service Flow Management Systems* (MSFMS). MSFMS provides high-level middleware to enables transparent access to services and resources distributed over a wide-area network. The fundamental ideas behind the MSFMS are very simple: it consists of a collection of federated servers with hosting MSFMS engine or brokers to applications, or services. The federations of processing resources, which host the MSFMS environment, all make their own placement and scheduling decisions independent of each other. The MSFMS environment provides the necessary framework for the seamless scheduling and execution of the component parts of the users' requests across the distributed system to ensure the request is fulfilled.

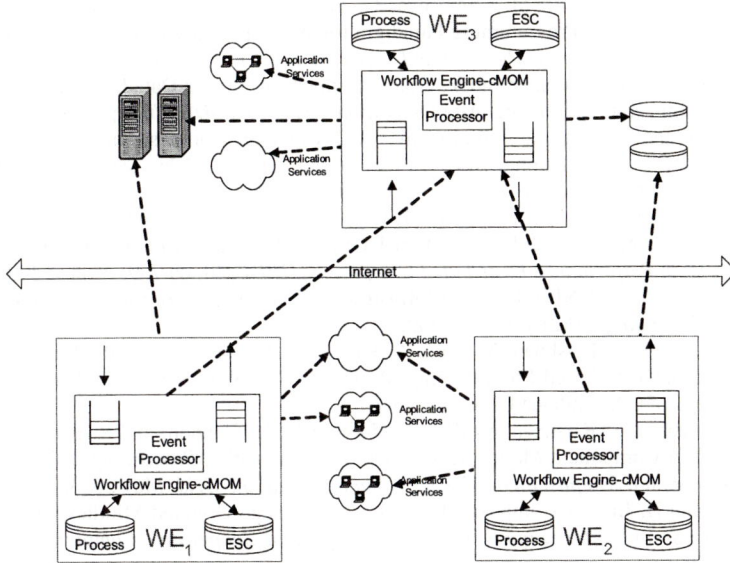

Fig. 1. The Architecture of MSFMS

The architecture of MSFMS adopts a service-oriented perspective, with a high degree of automation that supports flexible collaborations and computations on a large scale, as illustrated in Fig.1. In the architecture, workflow engines are central to the architecture of the MSFMS. Workflow engines are distributed across grid environment. We adopt the cMOM (*Composite-Event-Based Message-Oriented Middleware*) [8] as MSFMS engine. Services or resources can register themselves in one or more workflow engines. MSFMS engine can schedule those services and resources registered in the engine. An ESFL workflow definition can be seen as a template for creating grid services instances, performing a set of operations on the instances and finally destroying them. Therefore, MSFMS engines can also register themselves into other engines. By this means, MSFMS engines can be dynamically assembled into many architectures, such as Peer to Peer architecture, or layer architecture. Due to the dynamic nature of the grid environment, the MSFMS is suitable for grid environment.

5 Conclusion

In this paper, a *MOM-based Service Flow Management System* (MSFMS) is proposed. MSFMS enables the integration of services and resources within and across organizations. In MSFMS, we propose an XML-based workflow definition language,

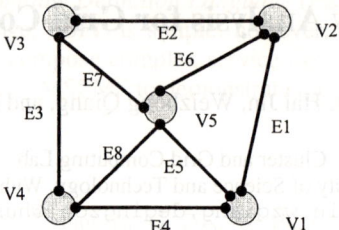

Fig. 1. Graph model for grid environment

For simplicity, Def. 1 does not consider the user space and the access control mechanism of the grid [6], that is, the grid environment can be access by any users.

Definition 2. The **reliability** R(t) of a system at time t is the probability that the system failure has not occurred in the interval [0, t].

Definition 3. A **job element** refers to one activity in the grid environment, such as a file transferring, a program execution on one node. Job element is the basic unit of grid activities.

Definition 4. Period P(i) is the process time for a give job element or a high-level grid activities. P(i) is computed in different ways for different jobs.

Definition 5. A **task** is a collection of series of job elements, that is, $T=(J_1, J_2, \ldots, J_n)$. Any failure occurs during these job elements execution leads to failure of a task.

Definition 6. The **reliability** of the grid, $R_g(t)$, is defined as the probability of successful execution of the tasks running on multiple resources.

We assume that the MTTF of grid resources (software or hardware) follows the exponential distribution, we present our reliability model based on the assumption and definitions above.

We category the task into its serial and parallel job elements, based on the task execution components. For example, if a file transfer job element executes on one storage element, considering fault-tolerance, there are two file severs to serve it, then there are two parallel parts for this task. If the file transfer will be the input file of the next job element, and this creates two serial job elements.

The reliability of grid environment is represented as a *reliability block diagram* (RBD). The reliability is computed as composing the reliability of the job elements. Given a task being partitioned as N job elements, and the failure rate of each job element is donated as $F_i(t)$, ($F_i(t)$ can be computed in different ways for different types of jobs), and F(t), the distribution of the failure time of a task is computed as Eq.1.

$$F(t) = \begin{cases} 1 - \prod_{i=1}^{N}(1 - F_i(t)), & \text{serial parts} \\ \prod_{i=1}^{N} F_i(t), & \text{parallel parts} \end{cases} \quad (1)$$

Assume that the failure rate of the component relative to job element i is λ_i, and exponentially distributed, F(t) can be computed:

$$F(t) = \begin{cases} 1 - \prod_{i=1}^{N} e^{-\lambda_i t} = 1 - \prod_{i=1}^{N} e^{-\lambda_i P(i)}, & \text{serial parts} \\ \prod_{i=1}^{N}(1 - e^{-\lambda_i t}) = \prod_{i=1}^{N}(1 - e^{-\lambda_i P(i)}), & \text{parallel parts} \end{cases} \quad (2)$$

Now we present our analysis model for grid computing. First RBD model is presented specified by Def.5. As depicted in Fig.2, a task is composed as n job elements, some job element have more than one copies for fault-tolerant consideration. In Fig.2 job element 1 has k copies. With Eq.2, the reliability of the task is computed as Eq.3.

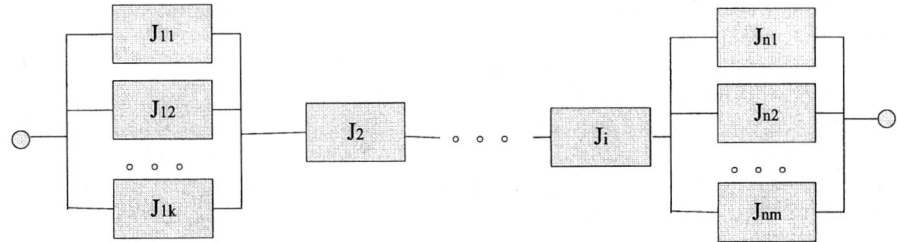

Fig. 2. RBD model for task analysis

$$R(t) = \left[1 - \prod_{l=1}^{k}(1 - e^{-\lambda_{1l}P(1l)})\right] e^{-\lambda_2 P(2)} \ldots e^{-\lambda_i P(i)} \left[1 - \prod_{l=1}^{m}(1 - e^{-\lambda_{nl}P(nl)})\right] \quad (3)$$

where λ_{ij} is failure rate of the j-th copy of i-th job element.

The job element can be separated into two categories: program execution and message exchanging. Analyzing reliability of them have different considerations. For program execution, the reliability binds with the resource where the program is running, so it can be computed as:

$$R_{jp}(t) = e^{-\lambda_p t} \quad (4)$$

For message exchanging, the reliability binds with the resource and the link of resource, shown in Def.1. Reliability of message exchanging between V_1 and V_2 have three RBD blocks, block for V_1, V_2 and E_1, so it can be computed as:

$$R_e(t) = e^{-\lambda_i t} e^{-\lambda_j t} e^{-\lambda_\psi t} = e^{-(\lambda_i + \lambda_j + \lambda_\psi)t} = e^{-(\lambda_i + \lambda_j + \lambda_\psi)\frac{D(i,j)}{S(i,j)}} \quad (5)$$

where λ_i, λ_j denote the failure rate of the two nodes, and λ_ψ donates the failure rate of the link.

MPICH-G [5] is grid-enabled implementation of the *Message Passing Interface* (MPI) that allows the user to run MPI programs across multiple computers at different sites using the same commands that would be used on a parallel computer. MPI programs exchange information among all the nodes during program execution.

With the model presented above, the reliability of MPICH-G program can be computed as:

$$R_{mpi}(t) = e^{-(\sum_{i=1}^{n}\lambda_i + \sum_{k=1}^{\frac{n(n-1)}{2}}\lambda_k)t} \quad (6)$$

where λ_i donates the failure rate of node i, and λ_j donates the failure rate of edge j. From Eq.6, it can be inferred that reliability for MPICH-G program is very low, especially when n is a large number.

3 Uses of Reliability Analysis Model

The analysis model presented above can be used in at least three fields: architecture design, task scheduling, and fault-tolerance policy choosing.

For grid architecture design, availability, scalability, and robustness issues must be considered. Using the model presented above, to deploy more back-ups for the system component where it is used more frequently.

Using the model above, task availability can be increased in two ways. Getting the available recourses which match the task requirements, computing the R(t) for different arrangement, choosing the largest R(t) and scheduling the task to the relative resources. In this way, the system availability for the task can be the highest. Computing R(t) and analyzing the key components which influence the R(t) fiercely, replicating more job elements at these parts to increase system reliability. During task scheduling, the scheduler replicates such job elements automatically, and the scheduler can delete other copies when one of them executed successfully for cost consideration.

Fault-tolerance policy choosing can be separated into two categories: one is for system architecture design; the other is for the users to choose different QoS grid services. For the first category, the system administrators compute $R_i(t)$ for different fault-tolerance policies using the model presented above, and compute the cost for such policies. Based on some economic model, the administrators can choose fault-tolerance polices easily. For the second, the users compute the R(t) based on the task QoS requirements, compute the $R_i(t)$ for different policies, then choose the policy which fits the R(t) well.

4 Conclusions and Future Work

The generic, heterogeneous, dynamic, and high-latency nature of the grid makes it fragile, and how to measure the reliability is a key issue to provide high availability services. This paper presents a general reliability analysis model for grid computing, and uses of the mode are also discussed. In the future, we will use the model to guide our system architecture design and our testbed deployment. More use cases will be considered for the model. In this paper, we do not consider the checkpoint technique and failover technique, such factors will be considered in our future model.

References

1. Foster, C. Kesselman, and S. Tuecke, "The Anatomy of the Grid: Enabling Scalable Virtual Organizations", *International J. Supercomputer Applications*, 2001.
2. V. K. P. Kumar, S. Hariri, and C. S. Raghavendra, "Distributed Program Reliability Analysis", *IEEE Transactions on Software Engineering*, Vol.SE-12, pp.42-50, 1986.
3. S. Hwang and C. Kesselman, "Grid Workflow: A flexible Failure Handling Framework for the Grid", *Proceedings of the 12th IEEE International Symposium on High Performance Distributed Computing*, 2003.
4. J. K. Muppala, R. M. Fricks, and K. S. Trivedi, "Techniques For System Dependability Evaluation", *Computational Probability*, Kluwer Academic Publishers, pp.445-479, 2000.
5. N. Karonis, B. Toonen, and I. Foster, "MPICH-G2: A Grid-Enabled Implementation of the Message Passing Interface", *Journal of Parallel and Distributed Computing*, 2003.
6. L. Pearlman, V. Welch, I. Foster, C. Kesselman, and S. Tuecke, "A Community Authorization Service for Group Collaboration", *Proceedings of the IEEE 3rd International Workshop on Policies for Distributed Systems and Networks*, 2002.

Experiences Deploying Peer-to-Peer Network for a Distributed File System

Chong Wang, Yafei Dai, Hua Han, and Xiaoming Li

CNDS Lab, Dept. of Computer Science, Peking University, Beijing 100871, China
{wangch,dyf,hh,lxm}@net.pku.edu.cn

Abstract. DHT routing algorithms have been studied from various aspects. The advantage of this kind of algorithms is that they can locate objects successfully in polylogarithmic overlay hops. Instead of getting testing results via network simulation, this paper describes experiences deploying Emergint, a DHT based peer-to-peer routing algorithm, in a real circumstance aiming to answer the question whether a DHT algorithm can get its theoretical performance in real-used systems.

1 Introduction

Locating replicated data and services is an important problem in a distributed file system. For the requirement of scalability and performance, peer-to-peer (P2P) networks are constructed as infrastructure of distributed file systems. Thus, object (data and services) lookup mechanism in P2P networks becomes an essential issue. A couple of approaches have addressed this. Among them, DHT routing algorithms using distributed hash tables (DHT) to deterministically locate an object in polylogarithmic overlay network hops, especially Plaxton-like algorithms, such as Plaxton [1], Pastry [2] and Tapestry [3], which take good considerations of underlying physical network characteristics, best meet the requirement of distributed file systems.

YanXing 2.0 [4] is an internet-based distributed file system. The servers of YanXing 2.0 are organized into a P2P network. A client can get the service via any server in the system. Every server has four functional components. They are, from the bottom to the upper application level, P2P routing, erasure coding, caching and upper layer managing user namespaces and providing file services. The latter three all use the object lookup mechanism provided by the P2P routing part. The P2P routing part in YanXing 2.0 employs Emergint, a variant of Tapestry for YanXing 2.0's demands.

2 Emergint Object Lookup Scheme

Emergint shares similarities with Plaxton in object lookup scheme and differs from Plaxton's mainly in the definition of object roots. In Emergint, every node is assigned a globally unique node identifier (nid), every object has a unique object identifier (oid). A hash function maps an object to its root, which is defined as the node with the nid numerically nearest to the 16-bit prefix of the object's oid (we will use the term "destination identifier" to represent the prefix). The location information of an object

---

This work is supported by the NSF (No. 60303002).

is stored in its root. As a result, routing process is for a given object with its identifier oid, delivering the message towards its root and getting its location information there.

Emergint uses local overlay routing tables to perform routing. For each node with a nid of x, the overlay routing table contains i*j*k items. If y is the (i,j,k) item in the overlay routing table of x, y must meet these regulations (RTC): (1) the i-digit prefix of x and that of y are the same; (2) the $(i+1)^{th}$ digit of y is j; (3) among the nodes meeting (1) and (2), y is the kth nearest.

The routing process may take two phases: in the first, matching the destination identifier digit by digit; in the second, if the routing is not finished during the first, delivering the message towards the node with a nid numerically nearest to the destination identifier. Fig. 1 depicts a routing process from node 623A to an object with an oid starting with 3F4A. L1 and L2 are in the first phase. In L3, there being no node with a nid of 3F4*, the process enters the second phase and chooses 3F3*, the numerically nearest one. As 3F3* has set a numerical lower bound, the following step should try to find a node with the next digit as large as possible, that is, 3F3C in L4.

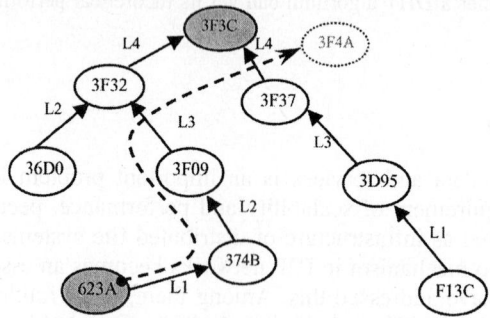

Fig. 1. Object Lookup Scheme in Emergint

3 Node Insertions and Deletions in Emergint

1) Node Insertion: There are three tasks to do to maintain the overlay structure when a node joins: a) generating the overlay routing table of the joining node; b) for the joining node, getting the location information of objects it should be in charge; c) notifying other nodes about its joining. To finish task a), we use the overlay routing table's property: If two nodes' i_0-digit prefixes are the same, they can share the items of (i,j,k) ($0 \leq i \leq i_0$). Thus, a routing from an existing node to the joining node (here we regard the node as the root of some object) is performed to find the nodes sharing the same i_0-digit prefix with the joining node. For task b), according to the definition of object roots, the joining node should retrieve location information of objects whose root is the joining node from two nodes: the one with the nearest but smaller nid (compared with the nid of the joining node) and the one with the nearest but bigger. We call them left neighbor and right neighbor respectively. To do task c), we should inform the new node's joining to a set of nodes (we name the set T), which consists of intermediary nodes during a routing to the joining node. We have already found some (we name these nodes S) in T for task a). The others in T can be found via nodes in S. If the routing message reaches node u, a node that should have delivered the message to node v (v∈ S) in its ith hop before the new node's join, u should have the joining

node as its next hop if the joining node is closer to u than v. So, each node in S informs the nodes that use it as an item in their overlay routing tables. All nodes informed about the new node's joining contact the new node and, if the new node meets the RTC, have it insert into the corresponding item in their routing tables.

2) Node Deletion: There are two kinds of node deletions: voluntary quitting and deletion because of failure. As of failure, the overlay network can discover it by the maintenance mechanism we will discuss later. The process of handling voluntary deletion is contrary to that of node insertions.

4 Emergint Implementation

In our Java-based implementation, we leverage the efficient non-blocking network communications Java has provided in its new package of NIO to reliably deliver messages across the network. For a real system, it is also important to pay attention to the issue of memory resource. We control the number of threads concurrently running and implement a thread pool to manage tasks. Once a message is delivered by NIO, it is decoded into a task and put into the thread pool. When the task is executed, new tasks are generated, encoded into messages and delivered to the NIO.

We have also considered the cases of failure in our implementation. In plaxton's scheme, each item (i,j,k) in which i is the last step in routing has only one node meeting RTC(1) and (2). It is possible that this unique node fails. Instead of changing the Plaxton's structure, we leave the problem to the upper application layers. If a routing fails, the upper application layers change the hash function and route towards another destination identifier. As of node failure in node insertions, we have designed a mechanism to have the left or right neighbor of the failing node act as a surrogate.

In addition, maintenance operations should be done to keep the RTC, especially (3) of it. In so doing, every node sends heartbeat messages at intervals to the nodes in its overlay routing table, measures the network distance from them and adjusts its overlay routing table. Based on a survey of real networks, we have determined the interval to be 3 minutes in our implementation.

5 Emergint Performance: Testing Results in YanXing 2.0

We run Emergint routing in a real system to observe its performance. 8 servers spread at different places in CERNET are deployed to run YanXing 2.0 file system. We have tested for 10 days, during which 1101071 routing queries were received, 175 node insertions and 16 voluntary quits happened. 16428 routing and 18 node insertions failed because of network failure or node failure. The serving time and data explaining the differences in serving time are demonstrated in Table 1.

Table 1. Testing results of Emergint in YanXing 2.0

Host number	1	2	3	4	5	6	7	8
Available memory	5740k	62700k	23580k	5600k	21760k	48230k	9230k	5580k
Bandwidth	10M	10M	10M	100M	100M	100M	100M	100M
CPU	550	800	1G*2	1G*2	1G*2	1G*2	1G*2	1G*2
Number of queries	23689	333057	580727	11631	23665	46867	24004	57431
Avg. serving time (ms)	3.24	7.73	18.3	6.4	8.04	8.48	7.7	6.54
Waiting time in thread pool (ms)	0.38	0.01	0.14	0.26	0.14	0.11	0.08	0.22

The number of intermediary messages every node delivers for each type of service and percentage of message size each service consumes are shown in Fig.2 & Fig 3.

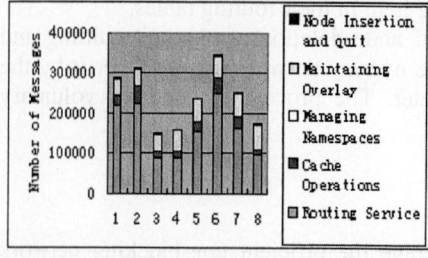

Fig. 2. Number of Intermediary Messages   **Fig. 3.** Percentage of Message Size

6 Conclusion

We have presented the design and implementation of Emergint, a DHT routing algorithm. Observing its performance in YanXing 2.0, we conclude that Emergint performs well in providing routing services for YanXing 2.0. However, from the testing results, we have also found out some problems, such as routing failure and high cost of maintenance. They are also research topics in the realm of P2P networks.

References

1. C.G.Plaxton, R.Rajaraman, and A.W.Richa, Accessing nearby copies of replicated objects in a distributed environment, in proceedings of SPAA, Newport,June 1997
2. B.Y.Zhao, L.Huang, J.Stribling,etc., Tapestry: A Resilient Global-Scale Overlay for Service Deployment, IEEE Journal on Selected Areas in Communications, Vol.22, January 2004
3. A.Rowstron, P.Druschel, Pastry: Scalable, distributed object location and routing for large-scale peer-to-peer systems, in proceedings of Middleware, Heidelberg, Germany, Nov.2001
4. Hua Han, Studies on Internet Oriented Distributed Massive File Storage System, Ph.D. Dissertation, Peking University, 2002

LFC-K Cache Replacement Algorithm for Grid Index Information Service (GIIS)*

Dong Li, Linpeng Huang, and Minglu Li

Department of Computer Science & Engineering,
Shanghai Jiao Tong University
Shanghai 200030, P.R. China
{lidong,huang-lp,li-ml}@cs.sjtu.edu.cn

Abstract. Traditional cache replacement algorithms are not easily applicable to a dynamic and heterogeneous environment. Moreover, the frequently used hit-ratio and byte-hit ratio are not appropriate measures in grid applications, because non-uniformity of the resource object sizes and non-uniformity cost of cache misses in resource information traffic. In this paper, we propose a Least Frequently Cost cache replacement algorithm based on at most K backward references, LFC-K. We define average retrieval cost ratio (ARCR), as the cost saved by using a cache divided by the total retrieval cost if no cache was used. We compare performance of LFC-K with other caching algorithms using ARCR, hit-ratio and byte-hit ratio as performance metrics. Our experimental results indicate that LFC-2 outperforms LRU, LFU and LFU-2.

1 Introduction

A Grid Index Information Service (GIIS) is a Grid middleware component which maintains information about hardware, software, services and people participating in a virtual organization (VO). A key function of the Grid Index Information Service (GIIS) is the management of a large capacity disk cache that it maintains. This paper investigates the problem of cache replacement policies for Grid Index Information Service.

GIIS Cache Management Module generally queues the requests and subsequently makes decisions as to which resource information objects are to be retrieved into its cache. When a decision is made to cache a resource information object it determines which of the resource information objects currently in the cache may have to be evicted to create space for the incoming one. The latter decision is generally referred to as a cache replacement policy and it is the subject of this paper in the context of cache management in the GIIS.

---

* This work is supported by the China Ministry of Science and Technology 863 high Technology program (Grant No 2001AA113160),973 project (No.2002CB312002)of China, and grand project of the Science and Technology Commission of Shanghai Municipality (No. 03dz15027 and No. 03dz15028).

2 Cost Based Caching

We introduce a third measure of goodness which we term the average retrieval cost ratio (ARCR), as the cost by using a cache divided by the total retrieval cost if no cache was used.

$$ARCR = \frac{\sum (H_i * C_i)}{\sum C_i} * 100\% \quad \text{where} \quad H_i = \begin{cases} 1 & \text{if request i is hit} \\ 0 & \text{other} \end{cases} \quad (1)$$

C_i is retrieval cost of resource object.

ARCR gives a better indication of the relative savings in time to retrieve and transfer a resource information object into the cache. It takes into consideration the total delay in caching resource information object of varying sizes, varying source delays and varying network transfer times. Consequently an optimal replacement algorithm based on ARCR, implicitly minimizes the response times of resource requests. This is a more practical objective in designing cache replacement algorithm for on the grid.

3 LRC-K Cache Replacement Algorithm

LRC-K Evaluation Function for Eviction

In LRC-K, a evaluation Function $\theta_i(t)$ value of resource information object i represent its importance in the cache. When the cache is full, the object with the lowest values is replaced.

Given a reference stream, $\omega = r_1, r_2, ..., r_t, ...$, where each reference is for a information object in cache, i.e, $r_t=(o_i), i=1,...,N$, o is information object of cache. We can now consider the reference stream as random variables with a common stationary probability distribution $p_1(t), p_2(t), p_3(t),...,p_n(t),...$ with $Prob(r_i=j)=p_j(t)$ is the probability that r_i references the information object j in cache. For a cache of size S such that the size s_i of ea of each object i is very much less than S, the principle of optimality implies that at the next reference instant t+1 we should always retain in the cache the I objects, considering that the sizes of the cached objects are relatively small compared to the total size S of the cache, the amount of space left after caching the maximum number of objects is negligible. The Cache Space may be restricted to the set I satisfying $\sum_{i \in I} s_i = S$, such that

$$\sum_{i \in I} p_i(t) \quad (2)$$

is maximized subject to

$$\sum_{i \in I} s_i = S \quad (3)$$

The above equation maximizes the hit ratio. Now if we assume the cost of retrieving a information object oi, of size si into the cache is ci(t) and the cost varies from a reference instant to a reference instant then from the reference is maximized subject to instant t to the next instant t+1, we need to retain I objects in cache such that is maximized subject to

$$\sum_{i \in I} p_i(t) * c_i(t) \tag{4}$$

$$\sum_{i \in I} s_i = S \tag{5}$$

We consider the evaluation function is expressed as:

$$\theta_i(t) = \frac{p_i * c_i(t)}{s_i} \tag{6}$$

When a buffer slot of cache is needed for a new information object being read in from network: the information object to be evicted is minimum value of evaluation function $\theta_i(t)$ of all objects in cache.

Evaluation of Re-accessing Probability

The p_i in evaluation function $\theta_i(t)$ indicates the object's future referenced popularity. The computation of p_i is rather complicated and will be discussed in the full paper. To estimate the values of p_i, we utilize the idea applied in developing the *Least Frequently Used Based on the K backward reference(LFU-K)*[2], page replacement policy, The basic idea of LFU-K is to keep track of the times of the last K references to popular database pages, using this information to statistically estimate the arrival times of references on a page by the history basis. Since we do not know the probabilities even more, alternatively, we consider statistical Frequency f_i of LFU-K substitute for p_i because the most frequently accessed object is most likely to be referenced in the future.

$$p_i \approx f_i = \frac{k_i(t)}{t - t_{-k_i}} \tag{7}$$

t_{-k_i} is the time of the k_i backward reference

$k_i(t)$ is number of the most recent references retained in time interval $(t - t_{-k_i})$

We can rewrite equation (6) as

$$\theta_i(t) = \frac{k_i(t)}{t - t_{-k_i}} * \frac{c_i(t)}{s_i} \tag{8}$$

4 Performance Comparison of Some Replacement Policies

We compared the performance metrics of hit ratio, average retrieval cost ratio for a number of cache replacement policies, namely LFU, LRU, LFU-2 and LFC-2. The experiments were carried out on a Grid test bed at the Grid Computing Center of Shanghai Jiao tong University and were based on GT3.

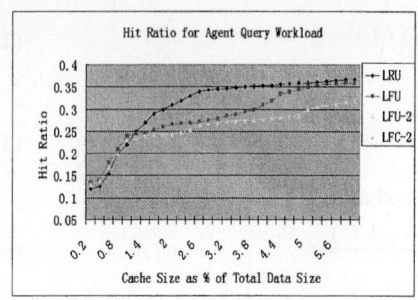 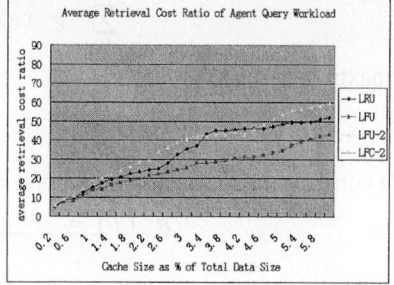

Fig. 1. Hit Ratio of Agent Query Workload   **Fig. 2.** ARCR of Agent Query Workload

Note that there are significantly differences among four caches replace algorithm under the average retrieval cost Ratio performance metric and LFC-2 give the best performance in figure 2.

5 Conclusions and Future Work

In this paper we introduced a new cache replace algorithm named LFC-K. Our experiment results provide evidence that the LFC-K algorithm has significant performance advantages over conventional algorithms in a dynamic and heterogeneous grid environment. We suggest that future research is that develop an analytic model of retrieval cost of resource information object related to the LFC-K algorithm.

References

1. Elizabeth J. O'Neil and Patrick E. O'Neil, UMass/Boston: The LRU–K Page Replacement Algorithm For Database Disk Buffering
2. Leonid B. Sokolinsky: LFU-K: An Effective Buffer Management Replacement Algorithm
3. Zhang, X., Freschl, J., and Schopf, J.M.: 'A performance study of monitoring and information services for distributed systems' Proc. 12th IEEE Int. Symp. on High-performance distributed computing (HPDC- 12), Seattle, WA, 22–24 June 2003, IEEE Press, pp. 270–281

The Design of a Grid Computing System for Drug Discovery and Design*

Shudong Chen[1], Wenju Zhang[1], Fanyuan Ma[1], Jianhua Shen[2], and Minglu Li[1]

[1] The Department of Computer Science and Engineering, Shanghai Jiao Tong University
{chenshudong,zwj03,fyma}@sjtu.edu.cn, li-ml@cs.sjtu.edu.cn
[2] Shanghai Institutes for Biological Sciences, Chinese Academy of Sciences
jhshen@mail.shcnc.ac.cn,

Abstract. This paper presents a novel Grid computing system for drug discovery and design. Utilizing the idle resources donated by the nodes that scatter over the Internet, it can process data-intensive biologic applications. With P2P technologies, the hybrid resource management architecture can avoid some problems, which are inevitable in the traditional master-slave model. Experimental results show the good performance of the proposed system.

1 Introduction

Experience has shown that idle cycles can be utilized to do batch jobs. The goal of this paper is to design a Grid computing system for drug discovery and design (DDG). Utilizing the idle resources donated by the nodes that scatter over the Internet, DDG can process applications of drug discovery and design. Resource management architectures of traditional Grid computing systems, such as SETI@home [1] and BOINC [2], almost adopt a master-slave model. This model may bring some frustrating problems, such as single point of failure and performance bottleneck, etc. So, P2P technologies [3] are introduced into DDG to eliminate these problems.

2 DDG Design Approach

In figure 1, there is a submission node contains a job submission client, which is responsible for submitting requests to the scheduler node. The scheduler node makes scheduling decisions. There is an information repository that maintains the global information about all the idle resources in the Grid.

DDG has a centralized + decentralized P2P network topology. Clusters that would like to contribute their idle cycles will join DDG and become the execution nodes. They will be grouped into several Virtual Organizations (VO) by their structures and interests. VOs communicate with each other through the Resource Management Agents in a P2P mode. If one peer failed, other peers of the network can compensate for the loss of that one. Each local representative Resource Management Agent captures and maintains the resource information of a VO.

---

* This paper is supported by 973 project (No.2002CB312002) of China, and grand project of the Science and Technology Commission of Shanghai Municipality (No. 03dz15027 and No. 03dz15028).

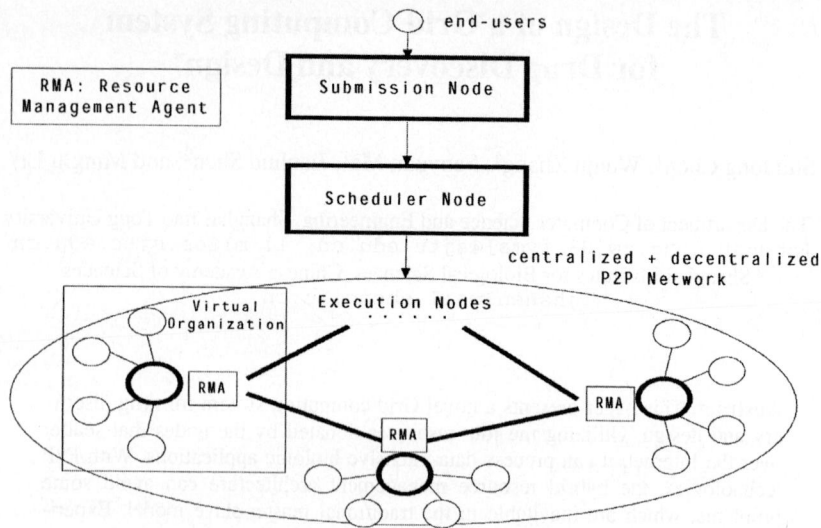

Fig. 1. Hybrid Resource Management Framework of DDG

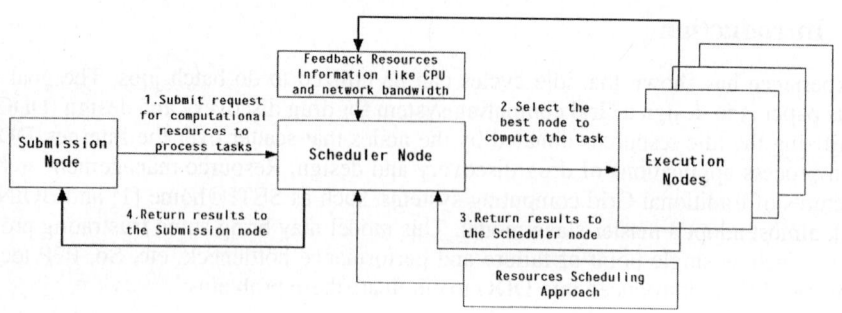

Fig. 2. High-Level Workflow of DDG

Figure 2 gives a high-level workflow of DDG. As an example, the molecular docking experiments will be addressed.

Firstly, the end-users submit protein molecular files to DDG through the web portal. End-users create a job request template for a new application, specifies the desired QoS (quality of service) requirements for the requested applications. Then the job request is submitted to the scheduler node by the submission node.

Secondly, after the scheduler node receives the request, it divides the job into many parallel sub-jobs. DDG provides a distributed repository, where execution nodes can publish their real-time workload information. The scheduler node queries this repository to perform matching of resources in the Grid to satisfy the QoS requirements of the sub-jobs.

Thirdly, these sub-jobs will be allocated to these selected execution nodes. After these sub-jobs are computed, results will be returned to the scheduler node to integrate result fragments into a whole docking result.

Finally, the docking result will be returned to the submission node that will inform the use that job has been completed.

3 Experimental Results

The feasibility and performance of DDG are evaluated by protein molecules docking experiments. Protein molecules docking experiments are to analyze the similarity between protein molecules provided by end-users and those in biological databases. In our experiments, several databases, such as Specs, ACD, CNPP and NCBI are used.

Experiments were carried out on a network with four clusters over the Internet and 20 dedicated PCs connected by LAN in a lab of Shanghai Jiaotong University. Each PC had a 1GHz Pentium IV with 256MB of RAM and 40GB of hard disk space and was connected to a 100Mb/s Ethernet LAN. Four clusters are distributed in Shanghai and Beijing: a SGI Origin 3800 cluster of 64 processors and a SUNWAY cluster of 32 processors were deployed at Shanghai Drug Discovery and Design Center, a SUNWAY cluster of 64 processors was deployed at Shanghai High Performance Computing Center, and one SUNWAY cluster of 256 processors was deployed at Beijing Drug Discovery and Design Center.

3.1 Running Time

Figure 3 gives a comparison between molecules docking time on a SGI Origin 3800 and the DDG. For example, a file with 10,000 protein molecules will be computed approximate 21 days on a SGI Origin 3800 cluster. And it takes 5 days for the Grid to compute the same file. The more number of molecules have, the better performance DDG presents. A conclusion can be drawn that DDG has sped up the process of protein molecules docking greatly.

3.2 Robustness

In this experiment, we evaluated the ability of the Grid to regain consistency after several execution nodes fail simultaneously. There were 20 PCs with each PC receiving 100 protein molecules to dock. Some execution nodes were randomly shut down. After the P2P network stabilized again, we measured the fraction of molecules that could not be processed. Figure 4 shows the effect of the Execution nodes failure on

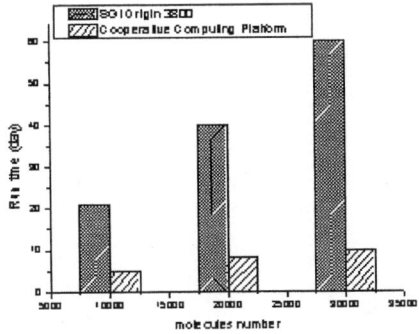
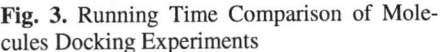
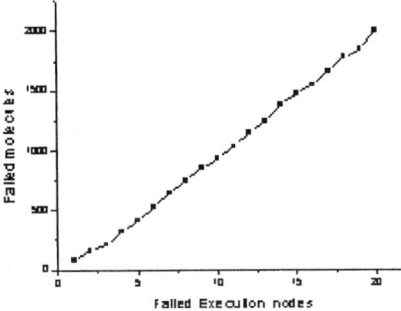

Fig. 3. Running Time Comparison of Molecules Docking Experiments

Fig. 4. The Effect of execution nodes Failure on Molecule Docking Experiments

resource scheduling. The molecules docking failure rate was almost equal to nodes failure rate. This was just the fraction of molecules expected to be failure due to the failure of the responsible execution nodes. That is, there was no significant resource scheduling failure in DDG. Thus it can be concluded that DDG is robust in face of the execution nodes' failure.

4 Conclusions

This paper presentes DDG, a Grid computing system for drug discovery and design. The resource management framework adopts a hybrid architecture, where the scheduler node responsible for the central resource scheduling of DDG and the execution nodes communicate with each other in a P2P manner to eliminate the problems caused by traditional master-slave model. Experimental results show that the robustness of DDG is good and DDG can speed up the process of protein molecules docking greatly.

References

1. D. Anderson, J. Cobb, E. Korpela, M. Lebofsky, and D. Werthimer. Massively distributed computing for SETI. Computing in Science & Engineering, Feb. 2001.
2. http://boinc.berkeley.edu
3. NetSolve Project. Andy Oram(ed). Peer-to-Peer: Harnessing the Power of Disruptive Technologies. O'Reilly & Associates, 2001.

A Storage-Aware Scheduling Scheme for VOD
(Short Version)

Chao Peng and Hong Shen

Graduate School of Information Science
Japan Advanced Institute of Science and Technology
1-1 Tatsunokuchi, Ishikwa, 923-1292, Japan
p-chao@jaist.ac.jp

Abstract. Video-on-demand (VOD) is a service in which a user can view any video program from a server at the time of his choice. In our paper, we present a new effective scheduling scheme, which can intelligently adjust its solution according to available resources and thus achieve an ideal average waiting time. We have realized this protocol in industrial application and have got a good performance.

Keywords: video-on-demand, broadcasting protocols, scheduling.

1 Introduction

Video-on-demand (VOD) refers to video services that allow clients to watch any video content at the time of their choice. Usually a video-on-demand system is implemented by a client-server architecture supported by certain transport networks such as CATV, telecom, or satellite networks. Clients use web browsers or have set-top-boxes (STB) on their television sets. In a pure VOD system, each user is assigned a dedicated video channel so that they can watch the video they chose without delay and many VCR-like functions may be provided.

The bandwidth and storage cost for pure VOD is tremendous, many alternatives have been proposed to reduce it by sacrificing some VCR functions. Broadcasting is one of such techniques and is mostly appropriate for popular or hot videos that are likely to be simultaneously watched by many viewers. The simplest broadcasting scheme is to periodically broadcast the video on several channels, each differentiated by some time. By this method the server need at least K channels in order to keep the waiting time below L/K (here L is the length of the whole video). To enhance the efficiency in channel usage, many schemes have been proposed by imposing a larger enough client receiving bandwidth and an extra buffering space at the client side.

We hereby present a Storage-aware broadcasting scheme, which can intelligently adjust its solution to achieve an ideal waiting time according to available bandwidth and local storage.

2 Modeling Analysis for Video on Demand

The scheduling problem in VoD service can be defined as the following: In a VOD system, if the available bandwidth on the VOD server is B(Mbps); the consumption rate, namely the rate at which the client STB processes data in order to provide video output, is C(Mbps). The minimum memory size in each STB is M(MByte), can we broadcast a video with the length of L(seconds) and satisfy a maximum delay of D(seconds) at the clients end?

Our scheme multiplexes the available bandwidth into B/C virtual channels and divides videos into n segments with equal sizes L/n. A STB will download all the useful segments of a video and store them in its memory, the maximum number of segments can be accommodated is $Max = \lfloor M/(C \times L/n) \rfloor$. Then by playing these segments in order, clients can watch the entire video smoothly. If the available bandwidth on the VOD server satisfies a certain baseline requirement, the server can properly arranging the video segments flow so that the STB can manage to download all the segments needed and store them before the time they should be consumed.

In this scheme, the server will repeatedly broadcast an arranged video program segments flow whose length is $DFSIZE = l \times B/C$ segments. So in the period during which a user watching a single segment, the STB can download at most B/C segments. A user who demand a video can not watch it until the STB find the first segment S_1. So the maximal waiting time is decided by the interval between two consecutive S_1s. Since the flow must satisfy a maximum delay of D seconds at the clients end, so the maximum number of segments between any two consecutive S_1s (we use STEP to denote this number) must satisfy $STEP \leq D/(L/n)$.

To guarantee that any user can download all remained video segments in time after he demands the video and downloads the S_1 segment, we must make sure that the STB can find the required S_i segment in the period between the start point when the S_1 segment is viewed and the point when the S_i segment should be viewed. Thus $DFSIZE$ must satisfy the following requirements:
1) If $STEP \leq B/C$, then:
i) The number of S_1 segments: $num1 = \lceil DFSIZE/STEP \rceil$,
ii) The number of segments S_2 to S_{l-1}:

$$num2 = \lceil \frac{l}{2} \rceil + \lceil \frac{l}{3} \rceil + \lceil \frac{l}{4} \rceil + ... + \lceil \frac{l}{l-1} \rceil = \sum_{i=2}^{l-1} \lceil \frac{l}{i} \rceil,$$

iii) The number of segments S_l to S_n: $num3 = n - L + 1$.
So the total number of segments need to be put in the flow must satisfy:
$\lceil DFSIZE/STEP \rceil + \sum_{i=2}^{l-1} \lceil l/i \rceil + n - L + 1 \leq DFSIZE$.
2) If $STEP > B/C$, then the requirement will be:
$\lceil DFSIZE/STEP \rceil + \sum_{i=2}^{l-1} min(\lceil l/i \rceil, num1) + n - L + 1 \leq DFSIZE$.
If there is vacancy, we can decrease $STEP$ until the table is full.

3 Algorithms for the Broadcasting Scheme

We use the following algorithm to arrange the video program segments flow:

Algorithm 1 The Storage-aware Scheduling Algorithm for VOD
SERVER:
1   Put all S_1s into the flow at the distance of $STEP$;
2   **For** $i = 2$ to $l - 1$
3   **For** $j = 0$ to $DFSIZE - 1$
4   **If** $VideoFlow[j]$ is S_1
5   **If** there is no S_i between $VideoFlow[j]$
                 and $VideoFlow[j+i*RATIO]$,
                 put a S_i into the latest vacant position between them.
6   **If** there is no vacant position, report error, exit;
7   Load S_l to S_{x-1}, report error if no vacancy.

CONSUMER:
1   Start downloading all segments;
2   **If** find segment S_1, start viewing S_1;
3   **For** $i = 2$ to n **do**
4   **If** find segment S_i in the local storage,
5   start viewing S_i;
6   **Else** report error and exit;
7   **For** all segments S_k in the broadcast channel;
8   **If** $k > i$ and S_k is not in the local storage,
9   Download S_k into the local storage;
10   **End for loop**

Claim1: The STB can find the required S_i segment in the period from the beginning to when the S_i segment should be viewed.

Proof: According to Algorithm1, for any S_1, we will put a S_i before $((i-1) \times B/C)$. So for any user who starts at t, the STB will first find a S_1 and then it can find a S_i before $t + (i-1) \times L/n$. Proof Ended.

Claim2: A STB with $(C \times l \times L/n)/8$(MB) local storage can store all the required segments during the whole period of watching the video.

Claim3: The average waiting latency is $(STEP \times (C/u) \times L/n)/2$.

4 Performance Analysis

The the following example: How to broadcast a 120-minute 3Mb/s MPEG-2 video program using a 24 Mbps digital channel and 512MB local storage to achieve a maximum delay of no more than 120 seconds?

We divide the whole video into 67 segments of equal size: $S_1, S_2, S_3 \ldots S_{67}$. So the length of each segment is $7200/67$ seconds. Since the memory size is 512MB, so we can store at most $\lfloor 512*67*8/(7200*3) \rfloor = 12$ segments at the same time. Since the bandwidth is 24Mpbs and the consumption rate is 3Mb/s, we can get 8 sub-channels as the following Table:

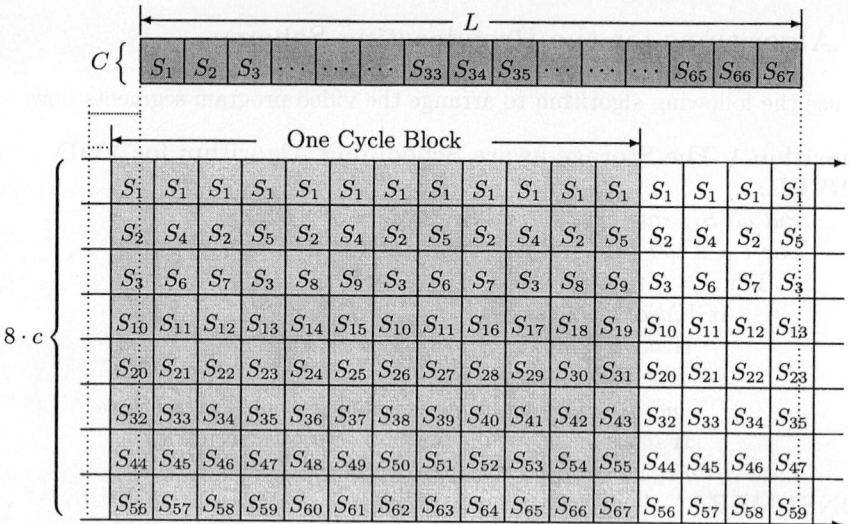

Fig. 1. The Scheduling Table for a video divided into 67 segments.

Now the maximum waiting latency is $7200/67 = 107$ seconds and the average waiting latency is 54 seconds. The used storage is $12*(7200*3/8)/67 = 483.6$MB, which is less than 18% of the whole video. Obviously this solution satisfies the requirements. Compared with some other existing solutions such as the Harmonic broadcasting scheme (HB), our scheme is not efficient in the channel issue. But HB needs 37% of the entire storage size of a video, many other protocols all assume that the user has enough local storage. Our scheme here needs only 18% and we can also smartly adjust the flow schedule to achieve the most efficient bandwidth requirement without breaking the storage restriction.

5 Conclusion

In this paper we present a simple but efficient scheduling scheme, which can be used for VOD service especially when some local STBs have limited storage.

References

1. A. Dan, P. Shabuddin, D. Sitaram and D. Towsley. Channel allocation under batching and VCR control in video-ondemand systems. *Journal of Parallel and Distributed Computing*, 30(2):168-179, Nov. 1994.
2. K. C. Almeroth and M. H. Ammar. The use of multicast delivery to provide a scalable and interactive video-on-demand service. *IEEE Journal on Selected Areas in Communications*, 14(5):1110-1122, Aug 1996.
3. L. Juhn and L. Tseng. Harmonic broadcasting for video-on-demand service. *IEEE Transactions on Broadcasting*, 43(3): 268-271, Sept 1997.

A Novel Approach for Constructing Small World in Structured P2P Systems*

Futai Zou, Yin Li, Liang Zhang, Fanyuan Ma, Minglu Li

Department of Computer Science and Engineering
Shanghai Jiao Tong University, 200030 Shanghai, China
zoufutai@sjtu.edu.cn

Abstract. Many efforts have been made to model small world phenomenon. Kleinberg's recent research models the construction of small world in a grid and shows the possibility to build an efficient search network with a distributed algorithm. However, it needs to use the global information to form the small world structure, which is unpractical for large-scale and dynamic topologies such as the peer-to-peer systems. In this paper, a novel approach is proposed to form the small world structure with only local information. This approach is proved both in theory and simulation. The experimental results show it is more nature and scalable, and can be applied to effectively improve the performance of a class of structured P2P systems.

1 Introduction

A peer-to-peer (P2P) system is a collaborating group of Internet nodes that construct their own special-purpose network on top of the Internet. All nodes attached to the network share any resources (CPU, storage, information etc.) and cooperate with each other. To effectively locate and search these edged Internet resources is the key of P2P systems. In this paper, we present the possibility to realize more efficient key lookups with constant routing tables' size in structured P2P systems.

2 Construction of Small World

Our work is inspired by Kleinberg's construction. Kleinberg [1] recently constructed a two dimensional grid where every point maintains four links to each of its closest neighbor nodes and just one long distance link to a node chosen form a suitable probability function. However, Kleinberg's construction needs to use the global node information to compute probability p for the shortcut. It is very difficult to get these nodes information and is even impossible for a practical network. In this section we purpose a novel approach to **asymptotically form the small world structure in the network with only local information** as more queries are being answered by the system.

Our idea is to use cached long link (we also call it shortcut) instead of fixed long link in Kleinberg's construction. The cache replacing strategy is critical to form the

---

* This paper is supported by 973 project (No.2002CB312002) of China, and grand project of the Science and Technology Commission of Shanghai Municipality (No. 03dz15027 and No. 03dz15028).

small world structure in the system. To be clear, we give an example here. Suppose node T has a shortcut link to node L initially in the cache. When a query request from source node S is forwarded to the target node T, node T replaces L with S with probability $\dfrac{|T-L|^d}{|T-L|^d+|T-S|^d}$[1]. Otherwise it keeps the shortcut link to node L. We have proved that node x would keep node y as its shortcut with the probability p that is the same value defined by Kleinberg's construction with repeating the cache replacing procedure for the shortcut of the node in the system. However, due to the limitation of the space, we omit the detailed proof here.

3 Constructing Small World in Structured P2P Systems

In this paper, we focus on the possibility of constructing small world structure in structured P2P systems, hence we use the well-known structured P2P system CAN to present the possibility based on our construction.

For CAN systems, we assume the key space size is n. Each node in CAN is responsible for a certain portion of the key space. Hence the query request for a key would eventually associate with the answered nodes and we can deal with the cache replacement in the answered node. The steps to form small world structure in CAN are described as follows:

(1) Each node joins the system will contact its adjacent neighbors as its short links. Meanwhile, it also caches the introducing node as its shortcut.

(2) When the system runs and nodes make queries, for example, a query request between two nodes, the target node would cache the source node as its shortcut with the cache replacing strategy described above.

From the construction above, we can learn that the process is very natural and the additional overheads to establish a shortcut link for a node are trivial.

With the construction of small world in P2P systems, there are several merits such as path hops, load balance, fault tolerance and so on. However, as the initial work, we just investigate the path hops as the metric of the performance in this paper. We give the upper bound for path hops in the constructed structured P2P systems as follows:

Theorem 1: *If each node x chooses its random shortcut so that the random shortcut has an endpoint y with probability* $x^d/(x^d+y^d)$ *then the expected number of hops is $O(log^2 n)$. Here n is the number of nodes in the system; d is the dimension of the underlying geometries.*

Proof: The construction of random shortcut follows up Kleinberg's model. Thus the expected number of hops is $O(log^2 n)$ derived from a similar theorem in [1].

CAN have the constant routing table size but has less efficient key lookups. The path hops for CAN is polynomial number of hops as $O(dn^{\frac{1}{d}})$ [2]. According to Theorem 1, CAN would has the poly-logarithmic number of hops as $O(log^2 n)$ with the construction of small world according to the new approach.

---

[1] $|n_1-n_2|$ is the Manhattan distance between node n_1 and node n_2.

4 Simulation

To prove that our new approach can be used in the structured peer-to-peer systems with the improved performance, we have provided the following simulation. For the better understanding our approach applied to structured P2P systems, we use the well-known structured P2P system CAN as the base of our simulation. We use the steps described in Section 3 to build a CAN system. The metrics are average path hops and the average number of queries answered by nodes (cache replacement happened on answering the queries). Our objectives are to test if our approach will drive the shortcut to the desired distribution and to observe the improved performance compared with the original CAN on the metric of path hops.

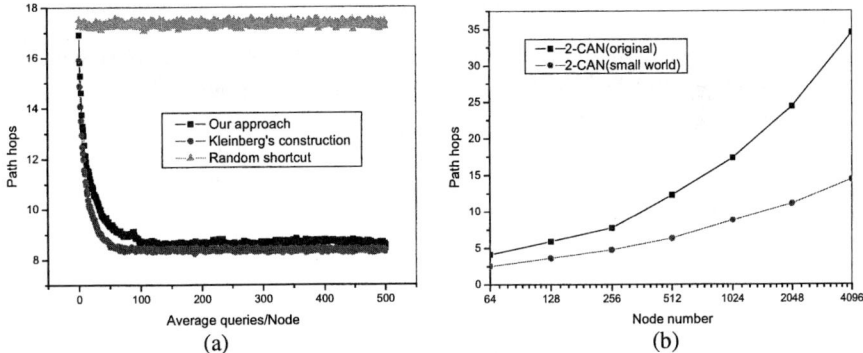

Fig. 1. (a)The effect of the convergence (b) Path hops

To test the effect of the convergence, the simulation runs in the network with 1024 nodes. The result is shown in Fig.1 (a). Our approach shows a similar procedure with Kleinberg's construction. The distribution of shortcut is converged to a stationary state according to our approach and Kleinberg's construction. However, with uniformly random shortcut, the effect is bad. The reason is that random shortcut where shortcuts are chosen uniformly at random shows an emanative process, which have been proved in [1].

Next, we test the path hops. We let the simulation runs in the networks ranging from 64 to 4096 nodes in size. Fig. 1 (b) presents the improved performance by the construction of small world in CAN with our approach. Because of the additional overheads for the construction is trivial (only handled on answering the query request), the improvement is considerable.

5 Conclusion

The routing performance is very important for P2P systems. Structured P2P systems can accurately locate the object in a distributed environment, but they have to burden the high topology maintenance overheads. A low-degree network with efficient search power is the desired topology for structured P2P systems [3] [4]. In this paper, we introduce the basic principle of small world into the routing process and present the improved performance in structured P2P systems. The main merit of our approach is that the routing performance is improved with almost zero additional overheads in the network. Compared with other approach for the improvement of routing performance, it is more nature, scalable. It can be expected to be widely applied to build the low-degree networks with efficient search power for a class of structured P2P systems.

References

1. J. Kleinberg. The small-world phenomenon: an algorithmic perspective. Cornell Computer Science Technical Report 99-1776, 2000.
2. R. Sylvia, F. Paul, H. Mark, K. Richard, S. Scott, 2001. A scalable content-addressable network. In: Proc. ACM SIGCOMM ,San Diego, CA, p. 161–172,2001
3. D. Malkhi, M. Naor, and D. Ratajczak. Viceroy: A scalable and dynamic emulation of the butterfly. In Proc 21st ACM Symposium on Principles of Distributed Computing (PODC 2002), pages 183–192, 2002.
4. G. S. Manku. Routing Networks for Distributed Hash Tables, Proc. 22nd ACM Symposium on Principles of Distributed Computing, PODC 2003, p. 133-142, June 2003.

Design of a Grid Computing Infrastructure Based on Mainframes

Moon J. Kim[1] and Dikran S. Meliksetian[2]

[1] IBM Corporation, 2455 South Road, Poughkeepsie, NY 12601, USA
Moonkim@us.ibm.com
[2] IBM Corporation, 150 Kettletown Road, Southbury, CT 06488, USA
Dikran_Meliksetian@us.ibm.com

Abstract. Businesses are entering the next phase of e-Business, in which they need to be able to sense and respond to fluctuating market conditions in real time. IBM is leading the way toward this on-demand world and a number of initiatives are underway to investigate, design and deploy Grid technologies, which form one of the pillars upon which the on-demand environment can be built. Integral part of these efforts is the integration of existing mainframe resources in various data centers within a Grid environment. In this paper, we present a design that enables the sharing of underutilized capabilities of mainframes in a dynamic way. This approach promises to increase corporate efficiency, by improving return of investment (ROI) on data center investments and opens up the use of data center resources to new applications. An early prototype of the design has already been implemented, and a number of business and engineering applications are being tested on the prototype.

1 Introduction

In an on demand business, the computing environment is integrated to allow systems to be seamlessly linked across the enterprise and across its entire range of customers, partners, and suppliers. It uses open standards such that different systems can work together and link with devices and applications across organizational and geographic boundaries. It is virtualized to hide the physical resources from the application level and make the best use of technology resources, thereby minimizing complexity for users.

A good example of a virtualized model is a grid computing model [1]. In this paper, we describe the design of a grid based on IBM mainframes. The system under consideration is an experimental grid built using the Globus Toolkit version 3 (GT3) [2]. It provides interoperability among zSeries and S/390 [3] processor families. It creates an environment that enables sharing of under-utilized computational resources. Typical mainframe systems are designed and provisioned with a target average utilization that always reserves extra capacity for future growth and to handle unexpected business critical job load demands. The unused capacity on those systems is designated as white space and can be leveraged for on-demand use.

Mainframe systems provide a level of isolation and security that is unparalleled on other platforms. For example, the zOS supports logical partitions and guarantees that there is no leakage of information or execution resources between them. This opens the opportunity to provide better isolation between tasks than available within grids based on other platforms. In current practice, the isolation between grid tasks is based on whatever isolation mechanism an operating system on those platforms can provide.

Usually, grid tasks are run as separate processes within the operating system, thus sharing resources controlled by the OS. This situation might result in intentional or accidental exposure or corruption of the data of one task by another task. This proposed design exploits the capability of running multiple concurrent virtual machines on zSeries and S/390 systems to further isolate task execution in separate virtual machines.

This paper is organized as follows. In the next section, we describe the architecture of the IBM mainframe grid, the components of the system and describe its operation. In Section 3, we identify the benefits of the design and finally in section 4, we present our conclusion and next steps.

2 Architecture and Operations

This infrastructure is built on the Open Grid Services Architecture (OGSA) [4, 5]. It offers services for automated setup, management of users via the Web and synchronization of security files.

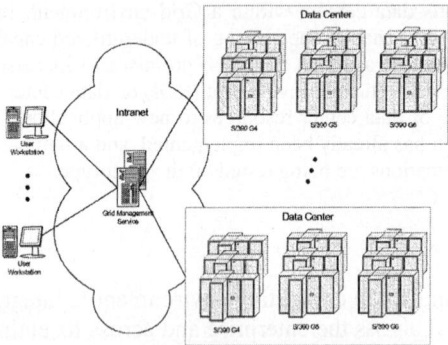

Fig. 1. Mainframe Grid architecture

Fig. 1 is a topological view of various data centers geographically dispersed around the world. Grid users interact with the system through a grid management center. The grid management service includes a job management service as well as a general administration and user management services. Each data center includes one or more zSeries or S/390 nodes. The nodes at a given data center may be homogenous or heterogeneous nodes. In the example of Fig. 1, each data center has a combination of different generation machines.

The mainframe nodes are partitioned via logical partitioning. Each logical partition, called LPAR, functions as a separate system having a host operating system and one or more applications. In general, one or more of the LPARs are dedicated to applications regularly hosted at the data center. However, a special partition can be created for use by the grid infrastructure using the white space. This partition, the Grid LPAR, shares its virtual processors with other partitions at a lower priority than the other LPARs. This ensures that the grid use of the node does not impede the regular operation of the mainframe; it only uses the excess capacity of the mainframe.

Each Grid LPAR runs multiple Linux Virtual Machines. One of those Linux VM instances, the Manager Linux VM, acts as a manager and interfaces with the grid management center. The grid management system allocates jobs to the grid LPARs based on the resources available in that LPAR. The grid management system also migrate jobs from one machine to another when the resources available in a given node are reduced beyond the requirements of the job or jobs running on the system.

A Grid Service hosting environment [6], the *Manager VM Hosting Environment*, is configured to run in this Manager Linux VM. The *Manager VM Hosting Environment* is the equivalent of the *Master Hosting Environment* in GT3. In our experimental system, we used the standalone hosting environment provided in the Globus Toolkit. In addition to the standard grid services, a modified version of the *Globus Master Managed Job Factory Service* [7] is to be deployed. The modification allows the dispatching of the execution of jobs to other Linux instances running in the same LPAR, rather than spawning a *User Hosting Environment* (UHE) in the same instance. The GT3 *HostingEnvironmentStarter* service is replaced by a *Job VM Launcher* service that starts a new virtual Linux virtual machine (Job VM) and communicates with that VM to execute the task using zVM commands.

Multiple Job VM instances are preconfigured. The Globus Toolkit with a modified Local Managed Job Factory Service [7] is deployed on those VMs. The *Job VM Hosting Environment* is the equivalent of the *User Hosting Environment* in GT3, with the modification that it runs in a Linux VM different from the Manager VM Hosting Environment.

The overall sequence of operations is as follows: The Manager VM exposes the capabilities and characteristics of the resources allocated to the LPAR and the current state of the Job VMs running within the LPAR. When a new job request is received, it allocates the necessary resources to a predefined Linux VM and starts the VM. It passes to this new Linux VM the job request and returns to the grid management service a handle to communicate with this VM. The Linux VM executes the tasks. During this period, the grid management service can query for the status of the job and can retrieve the results of the job when ready. Upon completion of the job, the Manager VM shuts down the Job VM, cleans up the used resources and reclaims those resources.

Future extensions of this design will handle more elaborate cases, for example, the user can be replaced by another automated service or program that initiates the task and the initiation of the task might result in the execution of multiple jobs.

3 Benefits

The solution addresses the immediate need for a grid solution for the hosted environment using technologies available today. The solution links mainframe servers from across the company into a single, highly-utilized Grid.

A vast quantity of computing power is wasted due to the under-utilization of resources. In general, planning and sizing for computing requirements are based on peak demand and statistically the actual utilization is of the order of 60%. Harnessing the unutilized compute power will provide immediate economic benefits to any organization that has a large installed base of mainframe servers.

It also provides better isolation between grid jobs, thus enhancing privacy and security in a grid environment. In a generic GT3 environment, the isolation between grid tasks is based on whatever isolation mechanism that an operating system on those platforms can provide. Usually, grid tasks are run as separate processes within the operating system, thus sharing resources controlled by the O/S. This situation might result in intentional or accidental exposure or corruption of the data of one task by the other task. In the proposed design, the isolation mechanisms of GT3 have been replaced by a different model. Instead of running multiple UHE instances in the same Linux instance, each UHE instance is executed in a separate Linux instance. Consequently, jobs belonging to different users never share any resource.

4 Conclusion

The design provides interoperability among the different mainframe processor families. It is based on porting existing grid computing standard architecture components to the zVM virtual Linux environment, adding automated setup/configuration features and exposing the resources available on the IBM mainframe nodes through the grid environment's information indexing infrastructure.

The porting of the Globus Toolkit to the zLinux environment has been completed. The automated configuration of each Manager Linux VM through a registration portal is accomplished. A number of business and engineering applications have been tested on the prototype system. The mechanisms to dynamically start and allocate resources to the Job VM have been designed, implemented and tested. The next step is to integrate these mechanisms with the modified Grid Resource Management services to complete the implementation of the design.

References

1. Foster, C. Kesselman, S. Tuecke, "The Anatomy of the Grid: Enabling Scalable Virtual Organizations," *International J. Supercomputer Applications*, 15(3), 2001.
2. Globus Toolkit Version 3, at http://www-unix.globus.org/toolkit/download.html
3. IBM, *z/Architecure Principles of Operation*, SA22-7832-00, December, 2000
4. I. Foster, C. Kesselman, J. Nick, S. Tuecke, "The Physiology of the Grid: An Open Grid Services Architecture for Distributed Systems Integration," *Open Grid Service Infrastructure WG*, Global Grid Forum, June 22, 2002.
5. Foster, C. Kesselman, J. Nick, S. Tuecke, "Grid Services for Distributed System Integration," *Computer*, 35(6), 2002.
6. Globus Toolkit 3 Core, at http://www-unix.globus.org/core/
7. GT3 GRAM Architecture, at http://www-unix.globus.org/developer/gram-architecture.html

SHDC: A Framework to Schedule Loosely Coupled Applications on Service Networks

Fei Wu and Kam-Wing Ng

Department of Computer Science and Engineering, The Chinese University of Hong Kong,
Shatin, N.T., Hong Kong, SAR, China
{fwu,kwng}@cse.cuhk.edu.hk

Abstract. In this paper, as an attempt to manage Grid resources and schedule Grid applications by a compact and independent system, we propose a framework of Service-based Heterogeneous Distributed Computing (SHDC), which organizes services in the Internet into a P2P network and schedules distributed applications by the cooperation of different peers. In the framework, Grid resources are organized into an addressable service network by low level peers, and Grid applications with a loosely coupled structure can be scheduled intelligently by high-level peers. A Grid Environment Description Language (GEDL) is proposed to compose messages for managing resources and organizing application scheduling in the network. Preliminary results show that the framework can effectively manage large numbers of Grid resources with low cost and schedule loosely coupled applications efficiently and robustly.

1 Introduction

Currently, several research fields such as Grid computing [1] [4] and Web services [2] attempt to bridge the gap between distributed applications and the Internet resources from different points of view. The main goal of Grid computing is to effectively organize various computational resources distributed in the Internet to provide computing facilities to users as a large virtual computer. Some toolkits, such as Globus [6], Legion [5] and ICENI [7], have been produced to manage Grid resources and develop Grid applications. Web services make software application resources available over networks using standard technologies such as XML, WSDL, SOAP, and UDDI. The service-based model is widely accepted as the best to utilize Grid resources in current researches. However, as a heterogeneous and dynamic environment, both the services and the network in a Grid are dynamic. The Grid resources can be available or unavailable at any time without concern of the status of the users' applications. The consideration of dynamic status of Grid resources complicates the developments of Grid applications. In this paper, we propose a framework of Service-based Heterogeneous Distributed Computing (SHDC) to simplify the management of Grid environments and to increase the robustness and schedulability of Grid applications. We organize the available services on the Internet as a Peer-to-Peer network. Services can be published, indexed, registered or requested among the network. Loosely coupled applications can be scheduled onto such a service network with robustness and performance guaranteed by the network.

2 The SHDC Framework

The aim of our SHDC framework is to schedule loosely coupled applications [3][8] on a service network. A loosely coupled application is a distributed application in which different modules of the application communicate by predefined formats of asynchronous messages (loosely coupled messages). We define a loosely coupled module (LCM) as an individual module of a distributed application. It communicates with other parts of the application by loosely coupled messages. A loosely coupled module group (LCMG) is composed of a group of LCMs. Various LCMs in a LCMG communicate with each other by loosely coupled messages. The SHDC framework organizes services in the Internet into a network, and schedules LCMs or LCMGs to complete distributed computations on such a service network.

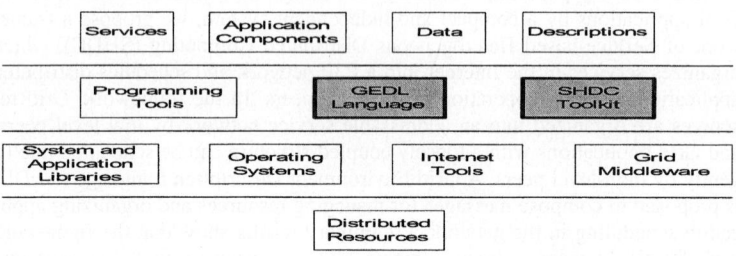

Fig. 1. The architecture of the SHDC framework

Figure 1 shows the architecture of the SHDC framework. On the top layer, services and loosely coupled applications can be developed by various programming tools. For LCMs or LCMGs in the distributed applications, the necessary data and descriptions must be well-defined so that the LCMs or LCMGs can be correctly scheduled onto corresponding resources. Our framework works on the second layer. The SHDC toolkit is responsible for organizing available resources on the Internet into a service network, and scheduling LCMs and LCMGs of a Grid application onto the correct resources. It hides the details of resource mapping and information exchanging between services and applications and makes Grid resources easy to be managed and accessed. The GEDL language is designed to formalize the descriptions of services, LCMs, LCMGs, and loosely coupled messages.

3 Managing a Service Network by SHDC Toolkit

Peer-to-Peer is a central idea to keep the Grid environment open and robust. Various services distributed in the Internet can be organized into a P2P network. There are two kinds of peers to manage the services: local resource managing peers to manage local resources and services, and information peers to deal with service lookup and application scheduling. Also, when a Grid application is to be executed, client-end application manager peers will be created to schedule LCMs and LCMGs of the application onto appropriate services. All these peers are created by the SHDC toolkit.

The P2P network is divided into three layers, as shown in Figure 2. The application manager peers layer includes peers to submit LCM and LCMG jobs to the system as agent of Grid applications. An application manager peer controls the execution of a

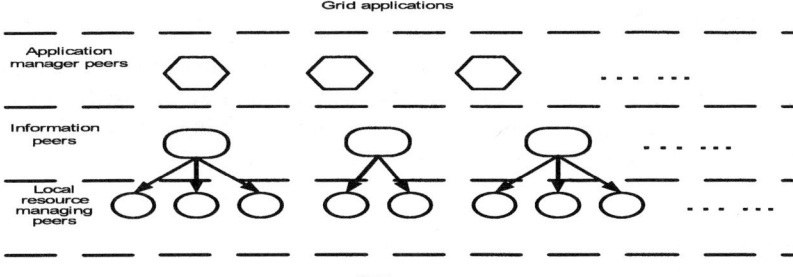

Fig. 2. Structure of the service network

distributed application. It sends out service requests to appointed resource managing peers, coordinate data transfer, and collect results. The other two layers of peers compose the service network. A local service managing peer manages a group of services available on the local machine. It dynamically keeps a service list which contains the features and status of each service. The managing peer watches the status of each service and the local system and then reports such information to high-level information peers. It is also responsible for service registration, service invocation, data caching and data routing. Information peers collect information of services, and provide lookup services to find appropriate services according to the requests.

4 GEDL Descriptions and GEDL Messages in a Service Network

In the SHDC architecture, GEDL acts as a mechanism to formalize the Grid elements including Grid resources, services, applications and messages. GEDL is a daughter-language of the Resource Description Framework (RDF) and it is written in the form of RDF schema. Each description describes a feature or a sub-element of the group. GEDL adopts similar syntax to some prevailing description languages such as WSDL so that it's easy to transform other XML-based description languages into GEDL descriptions or vice versa. But according to the requirement of the framework, some extended properties are introduced. Since the framework is designed to schedule distributed applications, GEDL is concerned more with performance issues such as service capacity and dependency. Moreover, GEDL provides abundant methods to describe complex graph-based structure, which accords with the loosely coupled application model.

GEDL can be used to describe server-end elements such as services and resources. It characterizes heterogeneous resources by resource type, capacity, accessing policy, sharing policy, load and other properties. Services are characterized by service types, platforms, attributes, protocols, production rates, etc. GEDL can also describe client-end LCMs and LCMGs. We represent a LCMG by graph. Every node in the graph represents an individual LCM, and every edge in the graph stands for data dependency or data communication between two adjacent LCMs. And what's more, GEDL is used to format the system messages in the network. Currently, we've designed more than 80 formats of messages to exchange information between various peers in the network.

5 Future Work

Further development on the SHDC framework will be done to improve the functionality of the SHDC toolkit. In the next step, we will try to make the SHDC framework work together with some mature middleware so that some high-level problems such as network security and fault tolerance could be dealt with.

6 Conclusion

In this paper, we have given an introduction to our SHDC framework. We introduced the ideas and architecture of the framework. The working rules of different peers in the service network are discussed. The GEDL language which helps to describe elements in the system and form GEDL messages was described.

References

1. Mark Baker, Rajkumar Buyya and Domenico Laforenza, "Grids and Grid Technologies for Wide-Area Distributed Computing", Software: Practice and Experience, Volume 32, Issue 15, Wiley Press, USA, Nov. 2002.
2. P. Cauldwell, R. Chawla, Vivek Chopra, Gary Damschen, Chris Dix, Tony Hong, Francis Norton, Uche Ogbuji, Glenn Olander, Mark A. Richman, Kristy Saunders, and Zoran Zaev. "Professional XML Web Services". Wrox Press, 2001.
3. Doug Kaye, "Loosely Coupled, The Missing Pieces of Web Services", ISBN 1-881378-24-1, 2003.
4. Foster, I., Kesselman, Jeffrey M. Nick, and Steven Tueche, "The Physiology of the Grid, An Open Grid Services Architecture for Distributed Systems Integration", Open Grid Service Infrastructure WG, Global Grid Forum, 2002.
5. A. Grimshaw, W. Wulf et al., "The Legion Vision of a Worldwide Virtual Computer". Communications of the ACM, vol. 40(1), January 1997.
6. http://www.globus.org
7. Nathalie Furmento, William Lee, Anthony Mayer, Steven Newhouse, John D., 2002, "ICENI: an open grid service architecture implemented with Jini", Supercomputing 2002
8. F. Wu, K.W. Ng, "A Toolkit to Schedule Distributed Applications on Grids", Proc. INC2004, Plymouth, U.K., 2004.

Tree-Based Replica Location Scheme (TRLS) for Data Grids

Dong Su Nam[1,3], Eung Ki Park[1], Sangjin Jeong[2],
Byungsang Kim[3], and Chan-Hyun Youn[3]

[1] Dept. of Information Assurance, National Security Research Institute,
52-1 Hwaam-dong, Yuseong-gu, Daejeon 305-348, Korea
{dsnam,ekpark}@etri.re.kr
[2] Protocol Engineering Center, ETRI
161 Gajeong-dong, Yuseong-gu, Daejeon, 305-350, Korea
sjjeong@etri.re.kr
[3] School of Engineering, Information and Communications University
119 Munji-dong, Yuseong-gu, Daejeon 305-714, Korea
{dsnam,bskim,chyoun}@icu.ac.kr

Abstract. Data grids provide researching groups with data management services such as Grid-Ftp, Reliable File Transfer Service (RFP), and Replica Location Service (RLS). Especially, Replica Location Service could improve data availability by storing multiple data to the distributed location. In this paper, we propose Tree-based Replica Location Scheme (TRLS) that could support users to decide optimally the location of multiple replicas as well as the number of replicas. In addition to, we could decide the number of replicas to satisfy user's requirements by minimizing cost

1 Introduction

The replication scheme is an approach for improving system performance, availability, and quality of service and replicas of particular data can be located at global locations, and keep multiple copies of an important object. Wolfson and Milo considered the communication complexity of maintaining the replicas and proposed a new method in [1]. Heuristics for solving a file location problem with replication are investigated in [2], using a mixed integer programming formulation of the problem. Dowdy and Foster have studied the optimal file assignment problem and they survey a number of Mixed Integer Programming (MIP) models for the file allocation problem [3],[6]. Kalpakis et al proved that the optimal capacitated residence set problem for trees is NP-complete [4]. Two algorithms are presented: one minimizes communications costs, and the other, communications time [5].

In this paper, we propose a Tree-based Replica Location Scheme (TRLS) that minimizes the cost of replica location service. By minimizing both file storage costs and communication costs, we can decide the optimal location of replicas and simultaneously guarantee user's requirements such deadline or budget. In section 2, we propose cost model for tree based grids network environment, and section 3 shows the evaluation results of our proposed model in finding optimal replica locations in the simulation topology. Finally we summarize the paper and discuss future work in section 4.

2   TRLS Model

In this section, we present optimal cost function for locating replicas by using integer-programming model. We consider a tree based graph $G=(V,E)$ where a set $V(G)$ is $V = \{1,2,\cdots,n\}$ and $E(G)$ is $E = \{(v_i,v_j) \mid i,j \in G, i \neq j\}$.

Since we consider tree based graph, the pairs (v_i,v_j) and (v_j,v_i) represent the same edge. For $i,j \in V$, a storage server in node i may transfer a data file to node j; the communication cost of such a transfer is $\overleftrightarrow{\$}_{ij}$. The communication costs $\overleftrightarrow{\$}_{ij}$ form a $n \times n$ triangular matrix Φ. If i,j has intermediate node k, communication cost $\overleftrightarrow{\$}_{ij}$ is represented as $\overleftrightarrow{\$}_{ik} + \overleftrightarrow{\$}_{kj}$. For $i \in V$, a storage server in node i may store a data file to be replicated; the storage cost per unit data is $\$_i$. The weight factor is presented as w_i from node utilization. The storage costs $\$_i$ form a $1 \times n$ vector Γ. Storage capacity constraint for each node i is defined as C_i. Available bandwidth between each pair (v_i,v_j) is denoted by R_{ij} and data file size is defined as L. Moreover, we introduce two additional parameters to represent user's requirements for creating a set of replicas - total budget of user and deadline requirement for creating replicas, namely B and D, respectively.

Our model determines the location of replicas with minimum cost that satisfy user's deadline and budget requirements. A variable x_i has 1, if node i is allocated to a replicated data file. Other case, its value is 0. Since calculating the optimal number of replicas under user's requirements is another research field, presenting an adequate model for determining the number of replicas is not scope of this paper. For simplicity, we assume that we already know the number of replicas, M^f. Also, we only consider read-only storage access and do not consider file update or write access. The proposed scheme is based on zero-one linear programming model. A model that minimizes the sum of file storage costs and communication costs is as follows.

Minimize g

$$g = \sum_{i=1}^{n} \frac{(x_i)(L)(\$_i)}{(w_i)} + \sum_{i=1}^{n-1} \sum_{j=i+1}^{n} (y_{ij})(L)(\overleftrightarrow{\$}_{ij}) \tag{1}$$

Subject to

$$g \leq B \tag{2}$$

$$\sum_{i=1}^{n} x_i = M^f \tag{3}$$

$$\sum_{i=1}^{n} w_i = 1 \tag{4}$$

$$(x_i)(L) \leq C_i \tag{5}$$

$$\frac{(y_{ij})(L)}{R_{ij}} \leq D \tag{6}$$

$$x_i = 0 \text{ or } 1 \tag{7}$$

$$\overset{\leftrightarrow}{\$}_{ij} = \overset{\leftrightarrow}{\$}_{ik} + \overset{\leftrightarrow}{\$}_{kl} + \overset{\leftrightarrow}{\$}_{lm} + \overset{\leftrightarrow}{\$}_{mj} \text{ (if path of } ij \text{ is } i\text{-}k\text{-}l\text{-}m\text{-}j) \tag{8}$$

$$(x_i)(x_j) = y_{ij} \tag{9}$$

$$y_{ij} + (1 - x_j)M \geq x_i \tag{10}$$

$$y_{ij} \leq Mx_j + 0 \quad , \text{ where } M \text{ is large positive integer} \tag{11}$$

In (1), the term $(x_i)(L)(\$_i)/(w_i)$ means the storage cost for each node i when node i stores a data file with size L and weight w. $(x_i)(x_j)(L)(\overset{\leftrightarrow}{\$}_{ij})$ denotes the data transfer cost between node i and j. Therefore, the objective function, i.e., (1), minimizes both storage costs and communication costs. The first constraint, (1), implies the total cost of creating M^f replicas should satisfy user's available budget. The second constraint, (2), limits the number of replicas to M^f and the third constraint, (4), give weight to each node. Equation (5) denotes the capacity constraint for each storage node i. Equation (6) means that the file transfer time from node i to node j is less than user's deadline requirement D. Three contributions of our model are the budget constraint, the deadline constraints and weight constraints.

Since (1) is zero-one nonlinear integer programming model, it is hard to solve this model as the original form. However, we can use a method to reduce zero-one nonlinear integer programming problem to zero-one linear integer programming problem.

3 Evaluation

To evaluate our TRLS, as shown in Fig. 1, we consider the tree topology that consists of 5 edge nodes and 9 intermediate nodes. We assume the respectable value of storage cost vector Γ, communication cost matrix Φ, weighted factor (w_i), deadline D, file size L and budget B. If user's total budget is less than minimum cost for locating replicas, user should increase total available budget or decrease desired availability of a data file. We use CPLEX mathematical package to solve our proposed linear programming model. We simulate total cost for locating replicas when the number of replicas changes from 1 to 14.

Consequently, Fig. 2 shows that the total replication cost have increased exponentially according to the number of replicas. The location of replicas is showed in table 1, where ● indicates that there is a replica at node i. As shown in Fig 3, when we give modest weight to each node, the average number of hop counts is decreased.

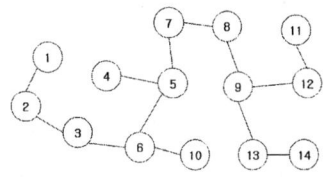

Fig. 1. Network topology for evaluation

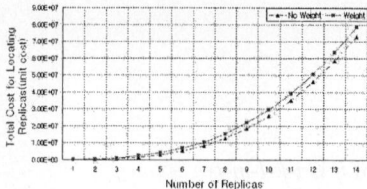
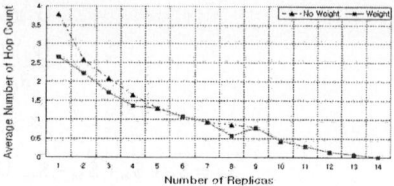

Fig. 2. Minimum cost for 14 nodes topology

Fig. 3. Average number of hop counts for 14 nodes

Table 1. Results of replication

Number of RS	General Replica Location														Replica Location with respectable weights													
	1	2	3	4	5	6	7	8	9	10	11	12	13	14	1	2	3	4	5	6	7	8	9	10	11	12	13	14
(1)					•														•									
(2)			•	•																	•	•						
(3)			•	•		•														•	•	•						
(4)			•	•		•	•												•	•	•	•						
(5)			•	•	•	•	•											•	•	•	•	•						
(6)		•	•	•	•	•	•										•	•	•	•	•	•						
(7)		•	•	•	•	•	•	•								•	•	•	•	•	•	•						
(8)	•	•	•	•	•	•	•	•							•	•	•	•	•	•	•	•						
(9)	•	•	•	•	•	•	•	•	•						•	•	•	•	•	•	•	•	•					
(10)	•	•	•	•	•	•	•	•	•	•					•	•	•	•	•	•	•	•	•	•				
(11)	•	•	•	•	•	•	•	•	•	•	•				•	•	•	•	•	•	•	•	•	•	•			
(12)	•	•	•	•	•	•	•	•	•	•	•	•			•	•	•	•	•	•	•	•	•	•	•	•		
(13)	•	•	•	•	•	•	•	•	•	•	•	•	•		•	•	•	•	•	•	•	•	•	•	•	•	•	
(14)	•	•	•	•	•	•	•	•	•	•	•	•	•	•	•	•	•	•	•	•	•	•	•	•	•	•	•	•

4 Conclusion

In this paper, we evaluated computational analysis using Mixed Integer Programming based on the file assignment problem model. We have reduced the number of hop counts by adapting relative weights to our proposed scheme, and we could be able to reduce the execution time to replicate and to access any important data. Our results also showed that how much of the cost is needed to replicate and this result could be regarded as a basic economic model to decide the location and the number when we replicate any data. Required replication cost augmented exponentially as increasing the number of replicas.

References

1. Wolfson and A. Milo, "The Multicast Policy and Its Relationship to Replicated Data Placement, ACM Transactions on Database Systems, vol. 16, no. 1, pp. 181-205, 1991.
2. Fisher ML, Hochbaum DS, "Database Location in Computer Networks", Journal of ACM, Vol. 27, No. 4, pp. 718-735, 1980.
3. L.W. Dowdy, D.V. Foster, "Comparative Models of the File Assignment Problem", ACM Computing Surveys, Vol. 14, No. 2, pp. 287-313, 1982.
4. K. Kalpakis, et al. "Optimal Placement of Replicas in Trees with Read, Write, and Storage Costs", IEEE Transactions on Parallel and Distributed Systems, 2001.
5. Stephen A. Cook, et al. "The Optimal Location of Replicas in a Network using a READ-ONE-WRITE-ALL Policy", Journal of Distributed Computing, Vol. 15, 2002.
6. Chu W. W., "Optimal File Allocation in a Computer Network", in Computer-Communication Systems, Prentice-Hall, Englewood Cliffs, N. J. , pp. 82-94, 1973.

Evaluating Stream and Disk Join in Continuous Queries

Weiping Wang[1], Jianzhong Li[1,2], Xu Wang[1],
DongDong Zhang[1], and Longjiang Guo[1,2]

[1]School of Computer Science and Technology, Harbin Institute of Technology, China
{wpwang,lijzh,xuwang,zddhit,guolongjiang}@hit.edu.cn
[2] School of Computer Science and Technology, Heilongjiang University, China

Abstract. In many stream applications, a kind of query that join the streams with the data stored in disk is often used, which is termed SDJoin query. To process SDJoin query, two novel evaluating algorithms based on buffer are proposed in this paper, namely BNLJ and BHJ. Since the existed cost models are not suitable for SDJoin evaluating algorithms, a one-run-basis cost model is also presented to analyze the expected performance of proposed algorithms. Theoretical analysis and experimental results show that BHJ are more efficient.

1 Introduction

Recently, a new class of data-intensive applications has become widely recognized: applications in which the data is modeled best not as persistent relations but rather as transient data streams. Examples of such applications include financial applications, network monitoring, web applications, manufacturing, and sensor networks. In many steam applications, it is necessary to process the queries that join the streams with the data stored in disk, which is termed SDJoin query by us.

Processing continuous join query over streams has been widely investigated. The first non-blocking binary join algorithm, symmetric hash join was presented in [1], which was optimized for in memory performance, leading into thrashing on larger inputs. Sliding window joins over two streams were studied by Kang et al in [2], Lukasz et al have discussed sliding window multi-join processing over data streams in [3]. To the best of our knowledge, no research work has discussed the processing of SDJoin. When processing SDJoin in continuous queries, each item from stream must scan the whole data in disk and output the matched join results. There are two possible scenarios for the data in disk when evaluating SDJoin:

(1) Small data set. The whole data in disk can be accommodated in memory.
(2) Large data set. Only part of the whole static data can be stored in memory.

For the first scenario, the SDJoin query is easy to be evaluated. For the second case, we propose two buffer based evaluating algorithms for SDJoin in this paper, which are proved effective, and a new cost model is also given to analyze the proposed algorithms performance.

2 SDJoin Evaluating Algorithms

We model each stream data item as two components <v, t>, where v is a value (or set of values) of the data item, and t is the timestamp that defines the order of the stream sequence. The value of the data item can be a single value or a vector of values. The *timestamp* is considered the sequence number, which is implicitly attached to each item.

The disk data are considered as a set of tuples. We assume that the updates of disk data are very few. Since processing SDJoin needs to access disk frequently, it is more effective to store the disk data in file rather than database. Therefore, in our processing model all the disk data are stored in file.

Two evaluating algorithms, namely BNLJ and BHJ, are presented in this paper. BNLJ algorithm has two buffers that *SBuffer* and *DBuffer*, which stores the items from stream and caches one part of disk data respectively. The structure of *DBuffer* and *SBuffer* both are queues. Each new item from stream is inserted into *SBuffer* as it arrives in, and will be deleted after it has joined the whole disk data. The disk data is sequentially stored in file. The file is divided into k parts, and each part size is equal to the *DBuffer* size. Every part is assigned a part-id ranging from 1 to k.

BNLJ continuously works as follows:

1. Load one part of disk data into *DBuffer*.
2. Process nested loops join over *SBuffer* and DBbuffer.
3. Delete the items in *SBuffer* which have joined the whole disk data.
4. Go to Step 1.

The BNLJ algorithm loads the disk data in a circle. If the last part of disk data is loaded into *DBuffer*, the first part will be read in memory at next time. Since the new item from stream is inserted into *SBuffer* as soon as it arrives in, *SBuffer* size grows dynamically when processing buffers join. To handle this issue, in step 2, the NLJ only consider the tuples in *SBuffer* who have arrived in before the join beginning. The items those arrive in during processing buffers join will be processed in the next run. Every item in *SBuffer* has a bits vector to indicate its status, and the number of bits in vector is equal to the number of disk parts. Initially, all the bits of the vector are set to 0. Suppose that the item has joined the i_{th} part of disk data, the i_{th} bit of its vector is set to 1. The item will be deleted from *SBuffer* if all the bits of its vector have been set to 1.

In BHJ algorithm, both *SBuffer* and *DBuffer* are constructed in hash table. The Disk data are hashed with the same hash function, and the tuples are stored in file with the order of their hash value. That is, the tuples who have the same hash value are stored sequentially. The disk data is divided into k parts, and each part's size is equal to the *DBuffer* size. In general, the hash table entry size varies with its hash value. The large hash table entry may consist of several parts of disk data. Meanwhile, several small entries may belong to the same part.

BHJ algorithm replaces the *DBuffer* in a circle. There is also a bit vector for each item in *SBuffer* to indicate its processing status. The item in *SBuffer* will be deleted after it has joined the disk data whose hash value is same as it. That is, the item can be deleted after it has joined the parts of disk data corresponding to its hash value. We use a table in memory to record the relation between hash table entries and the parts of disk. The join between *SBuffer* and *DBuffer* are processed with hash join algorithm in BHJ.

3 Estimating Cost for Proposed Algorithms

SDJoin query consumes unbounded input stream and outputs query results continuously. Traditional cardinality-based cost model is incapable to estimate the SDJoin evaluating algorithm's cost since it estimates the time needed for a query to be run to completion, however, SDJoin would never stop.

We propose a new cost model to estimate the cost of proposed algorithms. As depicted in section 2, both algorithms repeat to perform three tasks: replace *DBuffer*, process join and delete expired items in *SBuffer*. Processing three tasks once is considered as an execution run. In our cost model, we estimate the cost of an execution run of the algorithms.

A general time cost formula for the i_{th} execution run of algorithms is shown below:

$$T_i = T_{i\text{-load}} + T_{i\text{-join}} + T_{i\text{-invalidate}}$$

The total memory used by algorithm is $M_i = M_{DB} + M_{SB}$, the meanings of M_{DB} and M_{SB} are shown in Table 1. Table 1 lists some symbols that will be used in cost formula.

Table 1. Symbols

Symbol	Meaning
λ_i	Arrival rate of stream during the i_{th} execution run
C_m	Time cost of accessing one item in memory
k	The parts of disk data
M_{DB}	The size of *DBuffer* (bytes)
M_{SB}	The size of *SBuffer* (bytes)
a, β	Tuple size of disk and item size of stream respectively

For the space limitation, the detail of calculating the algorithms cost is ignored in this paper, and can be found in [4]. The time cost of BNLJ in one-run-basis cost model is:

$$T_i = T_0 + C_m * (\left\lfloor \frac{M_{DB}}{\alpha} \right\rfloor + 1) * \sum_{j=i-k+1}^{i}(T_{j-1} * \lambda_j) \quad i > k.$$

The time cost of BHJ is:

$$T_i = T_0 + C_m * (\left\lfloor \frac{M_{DB}}{\alpha} \right\rfloor + 1) * \sum_{j=i-(k+m)/2+1}^{i}(T_{j-1} * \lambda_j) \quad i > k.$$

It is easily to know that BHJ outperform BNLJ algorithm.

4 Experimental Results

To examine the performance of the proposed algorithms, we implement them in C. Experiments are performed in Intel PIV 2Ghz with 256MB of the memory, running windows 2000. We use synthetic stream and disk data in the experiments. The stream item size is 220 Bytes and the size of tuple in disk is 250Bytes. The selectivity of join is 0.04. As is shown in figure 1 and figure 2, BHJ algorithm's performance is always better than BNLJ algorithm.

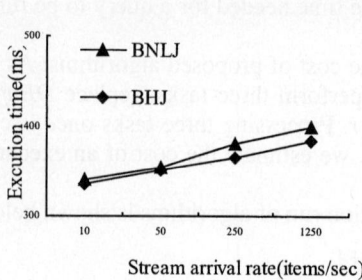 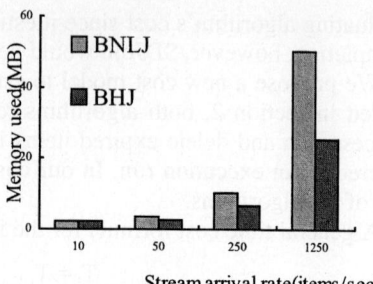

Fig. 1. Execution Time   Fig. 2. Memory used

5 Conclusions

Two novel algorithms BNLJ and BHJ are presented to process SDJoin in this paper. Since the existed cost models are not suitable for evaluating the cost of proposed algorithms, a one-run-basis cost model is also presented. Theoretical analysis and experimental results show that BHJ are more efficient.

References

1. N. Wilschut, P. M. G. Apers: Dataflow Query Execution in a Parallel Main-Memory Environment. PDIS 1991: 68-77
2. Jaewoo Kang. Jeffery F.Naughton. Stratis D. Viglas. Evaluating Window Joins over Unbounded Streams. ICDE Conference 2003.342-352.
3. Lukasz Golab. M.Tamer Ozsu. Processing Sliding Window Multi-Joins in Continuous Queries over Data Streams. VLDB Conference 2003: 500-511
4. Weiping Wang, Jianzhong Li, Xu wang, Dongdong Zhang, Longjiang Guo. Evaluating Stream and Disk Join in Continuous Queries. Technical Report. http://db.cs.hit.edu.cn/~200412.htm.

Keyword Extraction Based Peer Clustering

Bangyong Liang, Jie Tang, Juanzi Li, and Kehong Wang

Knowledge Engineering Group, Department of Computer Science,
Tsinghua University, P.R. China, 100084
liangby97@mails.tsinghua.edu.cn

Abstract. Peer clustering plays an important role in P2P systems like peer discovery, resource sharing and management, etc. Keywords provide rich semantic information about the peers' interests. Keyword extraction from documents is a useful method in topic retrieval and document clustering. Peers exchange resources and some of them are text documents like news and novels. Such documents represent the interests of a peer. This paper proposes a method for clustering peers using the exchange text documents between them. The documents are used for keyword extraction. The keyword extraction is treated as a decision problem and based on Bayesian decision theory. The peers' similarities can be calculated by keyword similarities. And the cluster method is based on the peers' similarities. The experiment gives satisfied results. Finally, the conclusion is discussed.

1 Introduction

Peer to peer network makes resource sharing more adaptive and effective. But due to the growth of the number of peers, the resource discovery and location will be a time consuming task. Peer clustering is a useful method to solve this problem. Peers with same interests can be grouped together. During the peer and resource discovery, other peers can look up peers in specified interest group other than searching in the whole peer society.

There are two methods for keyword extraction. One is based on a predefined controlled vocabulary [1]. This method can't process the unknown phrases. The other doesn't restrict possible keywords to a selected vocabulary [2, 3]. Keywords are picked up from normal words by machine learning algorithm.

In our situation, we can't summarize a keyword vocabulary for the peers before keyword extraction because the modification of the peer society is dynamic. The action of join and leave of a peer may infect the keyword vocabulary of the peer society. So the first kind of method doesn't fit for this situation and the second method has something in common. However, we propose a more adaptive and satisfied method for keyword extraction in dynamic peer to peer environment based on the Bayesian decision theory. Section 2 describes the framework for this clustering method. Section 3 details the clustering algorithm. Section 4 evaluates the experiment result. Finally, the conclusion is discussed.

2 Keyword Based Peer Clustering Framework

We propose the KPCF (Keyword based Peer Clustering Framework) to cluster the peers using the exchange documents between them. The framework is as follows:

The peer information collection robot visits different peers and query them for resources. All resources it collects will hand over to the clustering processor along with their owners. The registration center maintains the record about the peers in the society. The clustering processor uses the resources it can collect to cluster the peers. In the framework, the clustering processor is the core component for clustering the peers. The next section details the clustering algorithm of the processor.

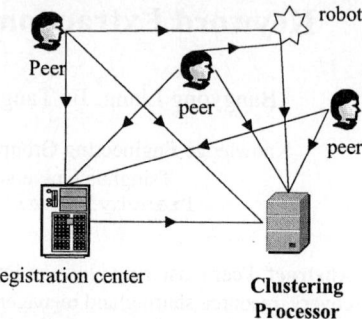

Fig. 1. Keyword based Peer Clustering Framework

3 Clustering Algorithm

The clustering method has four steps. The first step is keyword extraction. The task is to select a set of words as keywords for a peer. Using Bayesian decision theory, the expected loss of the action t is given by

$$R(W \mid RC, R) = \int L(W, \theta, R, RC) p(\theta \mid RC, R) d\theta$$

Where RC is the resource collection; R is the one resource contains the word. $\theta \equiv (\{\theta_i\}_{i=1}^{N})$, θ_i is the model that word is a keyword, the posterior distribution is given by

$$p(\theta \mid RC, R) \propto \prod_{i=1}^{N} p(\theta_i \mid ri, wc, mi)$$

The loss minimization will lead to the following form in the above environment:

$$W^* = \arg_W \min R(W \mid RC, R_j)$$

And the loss function to select W as keyword is as follows:

$$L(W, \theta, R, RC) = \sum_{w \in W} -\delta(w, R)$$

Where $\delta(w, R) = 1$ if the word w is a keyword; or $\delta(w, R) = -1$.

The naïve Bayesian theory gives the method to compute the probability of a word to be a keyword based on the three features. The formula is as follows:

$$p(y \mid ri, wc, mi) = \frac{p(ri \mid y) p(wc \mid y) p(mi \mid y) p(y)}{p(y) + p(n)}$$

The second step is feature definition. Three features are selected to represent the word. The first feature is appearance Count in the Registration Information. Peers need to register to enter the society. The description will be collect and send to the clustering processor. The appearance count of a word is the count divide the whole number of words in the description. This value is in the range of (0,1). The second feature is the TF*IDF weight which is defined as following:

$$w_{ij} = tf_{ij} \times \log_2 \frac{N}{n}$$

Where tf_{ij} is the frequency of word i occurs in resource j. N is the number of resources in the collection. n is the number of resources where i occurs at once. The third feature is the Mutual Information which is defined as follows:

$$w_{ij} = \frac{tf_{ij}/NN}{\sum_i tf_{ij}/NN * \sum_j tf_{ij}/NN} \times factor$$

$$factor = \frac{tf_{ij}}{tf_{ij}+1} \times \frac{\min(\sum_i tf_{ij}, \sum_j tf_{ij})}{\min(\sum_i tf_{ij}, \sum_j tf_{ij})+1}$$

Where tf_{ij} is the word i occurs in the resources j. $NN = \sum_j \sum_i tf_{ij}$ is the frequency counts of all words in all resources. *factor* is used to balance the bias toward infrequent candidates.

The third step is keyword similarity measurement. The similarity of two keywords can be given by WordNet. For w_1 and w_2, the similarity measurement is as follows:

$$Sense(s_1, s_2) = \frac{2 \times \log p(s)}{\log p(s_1) + \log p(s_2)}$$

Where $p(s) = \frac{count(s)}{total}$, $w_1 \in s_1, w_2 \in s_2$. s is a sense node in WordNet. The sense node is the common hypernym for s_1 and s_2. Every peer has three keywords, so the similarity of two peers is to sum up all the keywords' similarities.

$$Sim(peer_1, peer_2) = \frac{\sum_{i=1}^{n} \sum_{j=1}^{m} Sim(w_{1i}, w_{2j})}{n \times m}$$

In our current algorithm, both n and m are three.

The last step is to group peers by similarity. Now we have the similarity about every two peers. Therefore we can form the similarity matrix. The matrix gives the similarity between each two peers in the society. We can obtain the peer groups using equivalence class calculation by similarity matrix with a threshold λ [5]. Currently, the λ is given manually in our framework.

4 Experiments and Evaluations

We simulate a one hundred population of peers in four computers. The peers provide document service. The robot acquires documents from these service providers. All the resources the robot gets will transfer to the clustering processor. The precision in information retrieval is used for evaluating the experiment. Denote *cp* as correctly cluster peers, *wp* as wrong cluster peers, *ap* as all peer numbers. The precision is defined as follows:

$$precision = \frac{cp}{cp + wp} = \frac{cp}{ap}$$

The following figure shows the relation between *precision* and λ.

From the experiment result, we can see that the precision gives preferable result when $\lambda = 0.8$.

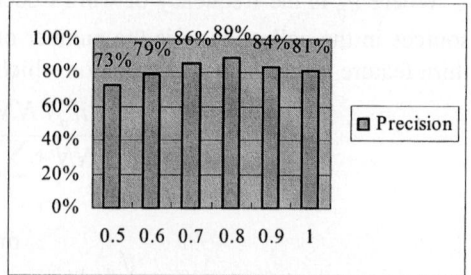

Fig. 2. The relation between precision and λ

5 Conclusions

This paper proposes an approach to deal with the peer clustering in peer to peer network environment. A specific focus is set on the resources the peer exchanges. KPCF treats the keyword extraction as a decision problem. The peers' similarities are calculated by their keywords' similarities. Finally, a similarity matrix is founded and the peer groups are computed by finding the equivalence class from the matrix.

References

1. S.T. Dumais, J. Platt, D. Heckerman, and M. Sahami. Inductive learning algorithms and representations for text categorization. In Proceedings of the 7th International Conference on Information and Knowledge Management, 1998.
2. P.D. Turney. Learning to extract keyphrases from text. Technical Report ERB-1057, National Research Council, Institute for Information Technology, 1999.
3. Eibe Frank and Gordon W. Paynter and Ian H. Witten. Domain-Specific Keyphrase Extraction. Proceedings of the Sixteenth International Joint Conference on Artificial Intelligence (IJCAI-99), Stockholm, Sweden, Morgan Kaufmann. 1999: 668-673.
4. Berger, J. Statistical decision theory and Bayesian analysis. Springer-Verlag. 1985.
5. Zhu Jianying. Non-classical Mathematics for Intelligent Systems. HUST Press, 2001.

A Peer-to-Peer Approach with Semantic Locality to Service Discovery

Feng Yan and Shouyi Zhan

Department of Computer Science and Engineering,
Beijing Institute of Technology, Beijing, 100081, P.R. China
{mryangfeng,Zhan_shouyi}@263.net

Abstract. P2P systems use exact keyword to publish and locate services without semantic support. In this paper, we propose an extension to discover services using the attribute vectors of describing services instead of keyword-hashed identifier, which enables the systems to discover semantically similar services with a semantic locality identifier using locality sensitive hashing. The benefit of such an approach is that it can help identify semantically close services and locate them into the same peers with high probability. Thus this paper represents the initial step towards solving the semantically discover service problem instead of keyword operation.

1 Introduction

In recent years there has been an increasing interest in the efficient service discovery as the number of services grows and becomes more dynamic and larger scale. As a result, there are a number of peer-to-peer (P2P) approaches proposed for building decentralized registries to support service discovery (e.g. [1, 2, and 3]).

One of the most challenges to current P2P approaches is to provide semantic locality discovery mechanism. By "semantic locality" we mean not only the ability to store semantically close service based on content from adverts into same peers (semantic cluster), but the one to find semantically close service and allow requesters to browse similar services.

In this paper we present a P2P discovery approach with semantic search capabilities to support services discovery. The main motivation is that services can be published and located using attribute vectors (AV) composed of a number of values generated from its description or content. Semantically close services are considered to have similar AV. And adverts of similar AV are clustered to the same peers with high probability by using locality sensitive hash functions (LSH) [4]. The approach can be used to index and locate content as a complement for current discovery mechanisms in Grid and Web Services.

Our approach is based on the Chord lookup protocol. Chord and other lookup protocols (CAN, Pastry and Tapestry) only use a unique identifier for lookup. Our approach enhances the lookup protocol with the ability to search similar content.

2 AV of Service

The layered structure is shown in Figure 1. We apply application-specific analysis for given service to generate attribute vector that describe its distinctive features, and to obtain a set of AV from Chord hash function. This ability on attribute vectors can be applied to a variety of contexts.

In application layer, including services publishes and queries, access interfaces are supplied for users to put and get adverts of services. According to different environments such as Grid or Web Services, adverts and queries are mapped into different attribute vectors (AV), such as text fingerprint in the case of P2P file storage system, values of globally defined attributes -memory and CPU frequency- in computing Grid, or port type and operation of WSDL documents in Web Services.

The AV is hashed to the same identifier space using an appropriate locality sensitive hash family to generate semantic identifier (sID). We use Chord to provide the lookup and routing facility.

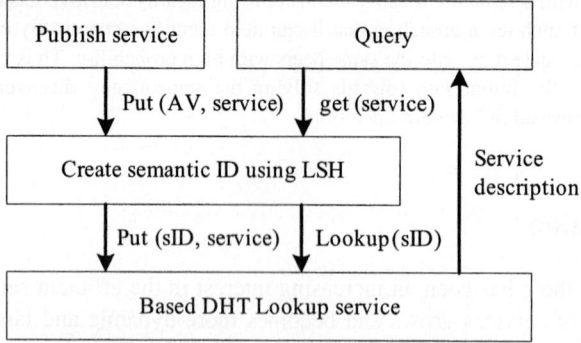

Fig. 1. The system layered structure using LSH

For example, this can be applied to publish such Web Service of automobile price given attributes such as region, business type, and port type. If attributes are canonically ordered as <region, businesstype, porttype>, then AV of this service might be <hash(Wuhan), hash(retail), hash(price)>.

Each member of the vector, an attribute fi, is encoded as a hashed identifier using the same hash function provided by DHT. The AV might be R:=<f1, f2, ...fn>, and n is the dimension of AV. The identifier of peer that publishes this AV is defined srcID. By having each attribute fi as DHT key, an advert {R, srcID} will be stored into a peer that peerID is closest to fi. To publish an advert, we firstly generate an AV for service. For each attribute in the vector, we locate its associated peer whose ID is closest to this attribute, and append the advert to that node's local list.

3 Semantic Locality Discovery

We firstly introduce the locality sensitive hashing (LSH) [5, 6], and then discuss our based on LSH semantic locality discovery in this section.

If A, B are two vectors then a family of hash functions H is said to be locality sensitive if for all $h \in H$ we have:

$$P_{h \in H}[h(A) = h(B)] = sim(A, B) = \frac{|A \cap B|}{|A \cup B|}$$

Where P is the probability, and $sim(A, B) \in [0,1]$ is some similarity function.

Min-wise independent permutations provide an elegant construction of such a locality sensitive hash function family [4]. Given a family of hash functions H, if for any set X and any $x \in X$, we say H is exactly min-wise independent if π is chosen at random in H we have

$$P(\min\{\pi(X)\} = \pi(x)) = \frac{1}{|X|}$$

In other words we require that all the elements of any set X have an equal chance to become the minimum element of the image of X under π, which is said to permutation function. Hence, we can choose say, k independent random permutations $\pi_1, \pi_2, ..., \pi_k$. For each vector, A and B, we store the list

$$A' = (\min\{\pi_1(A)\}, ..., \min\{\pi_k(A)\})$$
$$B' = (\min\{\pi_1(B)\}, ..., \min\{\pi_k(B)\})$$

Then we can readily estimate the similarity of A, B by computing how many corresponding elements in A' and B' are common, that is:

$$P(\min\{\pi(A)\} = \min\{\pi(B)\}) = \frac{|A \cap B|}{|A \cup B|}$$

A family of sensitive hashing functions is composed of permutation functions, such as $h(A) = \min\{\pi(A)\}$. It is very easy to find such π function. Research [6] described a π permutation by using an m-bit key that has exactly m/2 random bits set to 1. The sID is realized with the rough sketch of program code as shown in below.

```
//Algorithm of creating a group of semantic vectors sID.
Hashing each attribute of an advert into AV
for each g[j] do
//g[j] is one of l groups of hashing
    sID[j]=0
    for each h[i] in g[j] do
    //g[j] has k hash functions
        sID[j]^=h(AV)
    endfor
endfor
for each sID[j] do
    querying or publishing the advert by using sID[j] as the
DHT key
endfor
```

For performance of semantic locality, the number of independent random permutations k is a tuning factor. If the value of k increases, the probability of the semantically close advert stored in the same peer may be increased.

4 Conclusions

This paper presented a semantic P2P architecture that can be used for service discovery. Instead of exact keywords used by DHT, our design adapts the attribute vectors of describing services to index and locate services. We have also presented a novel approach to publish and discover of services with a semantic locality identifier mapping the semantically close services into the same peer nodes using locality sensitive hashing.

References

1. M. Schlosser, M. Sintek, and S. Decker. A Scalable and Ontology-Based P2P Infrastructure for Semantic Web servicess. In 2st International Conference on Peer-to-Peer Computing (P2P'02) September 05-07, Linkoping, Sweden 2002.
2. Cristina Schmidt and Manish Parashar. A Peer-to-Peer Approach to Web Service Discovery. In 2st International Peer to Peer System Workshop (IPTPS 2003), Berkeley, CA, USA, February 2003.
3. Feng Yang, Shouyi Zhan, and Fuxiang Shen. PSMI: A JXTA 2.0-Based Infrastructure for P2P Service Management Using Web Service Registries. GCC2003, LNCS 0302, pp. 440-445, Springer-Verlag Berlin Heidelberg 2004.
4. A. Z. Broder, M. Charikar, and A. M. Frieze. Min-wise independent permutations. Journal of Computer and System Sciences, 60(3): 630–659, 2000.
5. A. Gupta, D. Agrawal, and A. El Abbadi. Approximate range selection queries in peer-to-peer systems. In Proceedings of the First Biennial Conference on Innovative Data Systems Research, Asilomar, CA, January 2003.
6. Yingwu Zhu, Honghao Wang and Yiming Hu. Integrating Semantics-Based Access Mechanisms with P2P File System. In 3rd IEEE International Conference on Peer-to-Peer Computing (P2P'03). September, 2003. Linkopings, Sweden.

Implicit Knowledge Sharing with Peer Group Autonomic Behavior*

Jian-Xue Chen[1] and Fei Liu[2]

[1] College of Electronic & Electrical Engineering, Shanghai University of Engineering Science,
Shanghai, 200065, China
cjx@sues.edu.cn

[2] Department of Science & Engineering, Shanghai Jiao Tong University,
Shanghai 200237, China
liufei001@sjtu.edu.cn

Abstract. Autonomic computing is a promising technique in information technology industry. Depending on peer group autonomic computing, one can easily solve those traditional but seem-to-be intractable tasks. This paper introduces the Distributed Hypertext Categorization System (DHCS), in which the Directed Acyclic Graph Support Vector Machines (DAGSVM) for learning multi-class hypertext classifiers is incorporated into peer group autonomic computing. Not like traditional peer-to-peer systems, current behavior among members of DHCS is not confined to resource sharing - data, files, and other concrete types, but also including high-level knowledge sharing. In the prototype DHCS, implicit knowledge sharing among the local learning machines is achieved, which is self-configuring and self-optimizing. The key problems encountered in design and implementations of DHCS are also described with solutions to them.

1 Introduction

Over the years, computer scientists have primarily studied the knowledge discovery process as a single user activity. For example, the research field of automatic text categorization (ATC) has provided us with sophisticated techniques for supporting the information filtering process, but mostly in the context of a single, isolated user's interaction with an information base. Recent research now focuses as well on their cooperative aspects and methods for supporting these. The case study reported in [1] provides insight into the forms of cooperation that can take place during a search process. However, most researchers have studied in depth the kinds of non-autonomic collaboration that can occur in either the physical or digital library [1][2]. Meanwhile, the web mining with autonomic attributes definitely will play a crucial role in the web information acquisition. As a typical application of web information retrieval, automatic hypertext categorization is suffering the large-scale unlabeled web page base. Apparently it is meaningful to extend the state-of-the-art machine learning techniques to cooperative learning even to autonomic computing to solve the problem of distrib-

---

* This paper is supported by the Young Fellowship Fund of Shanghai University of Engineering Science (SUES) under Grant No. 2003Q03.

uted self-managing web information retrieval. Another motivation of this extension is the peer group members are always looking forward to sharing knowledge in one community, but not confined to data, information, or resource sharing.

2 Machine Learning and Hypertext Categorization

Kernel-based learning methods (KMs) are a state-of-the-art class of learning algorithms, whose best-known example is Support Vector Machines (SVMs). SVMs method has been introduced in automated text categorization (ATC) by Joachims [3][4] and subsequently extensively used by many other researchers in information retrieval community. Several methods have been proposed where typically the binary SVMs are combined to construct the multi-class SVMs. Hsu have pointed out that DAGSVM is very suitable for solving practical tasks [5]. Seeing the structure of DAGSVM, we may easily conclude that those binary SVM nodes are very similar to the real nodes in computer networks. First, learning processes are independent with different training samples. Second, the final decision path needs communication among the nodes just like the information transferring in the computer networks. Thus, we may divide the training workload of the whole DAGSVM into several separated groups (each group contains one or more binary SVM nodes).

3 Knowledge-Sharing and Autonomic Computing Attributes

3.1 Implicit Knowledge Sharing in DHCS

We see only those support vectors can affect the decision function, that is, only part of the training samples make contributions to the categorization rules. Researchers have found that the proportion of support vectors in the training set can be very small via proper choice of SVM hyper-parameters. Once the support vector machine has been trained, the decision function is determined so that the margin is enlarged to the max value. And along with the parallel hyper-planes, only those support vectors make contributions to the final categorization rules. It is obvious that the information about the support vectors is enough to represent the knowledge of present training result of SVM. Thus the local knowledge of the categorization can easily be transferred to other computer nodes so that all nodes can obtain the global knowledge of the hypertext categorization.

4 Some Implementation Issues of DHCS

4.1 Self-configuring – Allocation of Computation Load to Computer Nodes

Our system is implemented in a local LAN cooperative environment. However, it is easy to be extended to more complicate circumstances such as the WAN-based peer-to-peer network or even the Internet. In a simple LAN with enough bandwidth, we can ignore the cost of communication between the computer nodes. And all computers in DHCS communicate with each other via simple broadcasting. Every node in the LAN receives all information and takes what it needs as well as skipping the irrelevant information. We have found this simple strategy works very well in DHCS and it can demonstrate some basic autonomic computing functions. Figure 1 shows the simplified DHCS architecture.

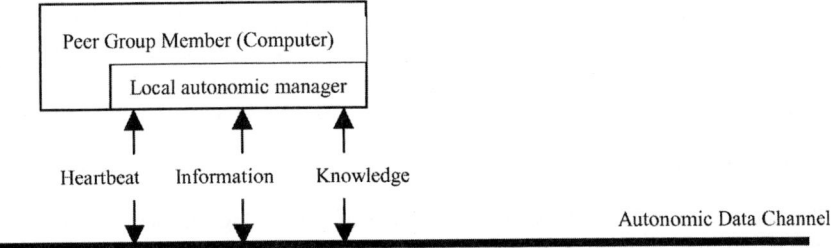

Fig. 1. Simplified architecture of a autonomic peer in DHCS

4.2 Self-updating – Information Exchange in DHCS

To implement dynamic and incremental learning in DHCS, the four computers need to exchange information periodically. Here, we refer to the information but not the knowledge yet, because the computers broadcast the labeled samples periodically. The labeled samples are not the raw hypertext files, but the preprocessed document vectors. In fact, DHCS can update itself in a more intelligent way, e.g. the computer nodes start information exchange respectively when they think they have got enough labeled samples. However, in our experimental DHCS, we just let the computers exchange information simultaneously with a certain frequency. This simple strategy seems efficient enough in the LAN environment.

4.3 Self-optimizing – Iterative Training of Binary SVMs and Adaptive Tuning of Hyper-parameters

In practice, we adopt and implement the modified SMO algorithm in DHCS, which is proved effective and easy to be realized. Empirical choice of SVMs hyper-parameters is always seen as a standard strategy in practical use of DAGSVM. Researchers always manually perform related evaluations based on fixed settings of SVMs, although satisfying experimental results can be obtained then. We have noticed that the optimal hyper-parameters of SVMs can be achieved via minimizing the generalization error of the algorithm. Based on the leave-one-out estimation of the SVMs, one can derive the performance estimators for finding the optimal hyper-parameters. The choosing process is totally automatic without human intervention. Traditional text learning systems are seldom self-optimizing, but it is important in the peer group cooperative environment.

5 Conclusions and Future Work

In this paper, we have introduced a distributed hypertext classification system, which combines the DAGSVM learning architecture and the cooperative learning environment. With little communication cost, knowledge share is achieved in the cooperative learning environment with autonomic attributes. But it should be pointed out here that the knowledge is obtained by machine learning but not by the users directly. The proposed DHCS is based on personal computers in a peer group (typically the LAN). Given the complex and dynamic environment faced by the Web, we believe that an autonomic peer group computing solution should adaptively seek out and local or peer resources, even share the resulting knowledge among the peers.

References

1. Twidale, M. B., D. M. Nichols, G. Smith, J. Trevor: Supporting collaborative learning during information searching. Proceedings of Computer Support for Cooperative Learning(CSCL 95). Bloomington, Indiana: 1995, 367-374
2. Hertaum, M., and Pejtersen, A.M. The information-seeking practices of engineers: Searching for documents as well as for people. Information Processing & Management, 2000, Vol. 36: 761-778
3. Jochims, T. Text categorization with support vector machines: learning with many relevant features. Proceedings of ECML-98, 10th European Conference on Machine Learning. Berlin: 1998, 137-142
4. Jochims, T. Transductive inference for text classification using support vector machines. Proceedings of ICML-99, 16th International Conference on Machine Learning. Bled, Slovenia: 1999, 200-209
5. Hsu, C.W., Lin, C.J. A comparison of methods for multicalsss support vector machines. IEEE Transactions on Neural Networks. March 2002, Vol.13, No.2: 415-425

A Model of Agent-Enabling Autonomic Grid Service System

Beishui Liao, Ji Gao, Jun Hu, and Jiujun Chen

Department of Computer Science and Engineering,
Zhejiang University, Hangzhou 310027, China
{baiseliao,hujun111}@zju.edu.cn, gaoji@mail.hz.zj.cn,
rackycjj@163.com

Abstract. OGSA-compliant Grid Computing has emerged as an important technology for large scale service sharing and integration. However, the management of massive, distributed, heterogeneous, and dynamic Grid services is becoming increasingly complex. This paper proposes a model of agent-enabling autonomic Grid service system to relieve the management burdens such as configuration, deployment, discovery, composition, and monitoring of Grid services. We adopt intelligent agents as autonomic elements to manage Grid services from two aspects: internal management and external transaction and cooperation. On the one hand, within an autonomic element (agent), goal-based planning and scheduling mechanism autonomously (re-) configures, and monitors the enforcement of, managed elements (Grid services or other agents). On the other hand, the automated discovery, negotiation, and contract-based service providing and monitoring among various agents treat with the relationships among autonomic elements.

1 Introduction

In recent years, OGSA-compliant Grid [1, 2] has facilitated the sharing and integration of large scale, heterogeneous resources. However, the management complexity of Grid services is increasing rapidly. The configuration, deployment, discovery, negotiation, composition, and monitoring of Grid services, which are done by human manually now, have posed great challenges. These management issues are tedious and may become unmanageable for human designers and administrators. Fortunately, autonomic computing, pioneered by IBM [3], throws light on treating with this problem. Autonomic computing is defined as computing systems that can mange themselves given high-level objectives from administrator. By self-configuration, self-optimization, self-healing, and self-protection, autonomic computing will relieve human from managing low-level applications and let him concentrate on the definition of high-level policies and goals. On the other hand, with distinguishing characteristics such as autonomy, proactivity, and goal-directed interactivity with their environment, intelligent agents [4] are promising candidates to be used in autonomic computing systems.

This paper proposes a model of agent-enabling autonomic Grid service system. We describe this model from two aspects: autonomic elements and the relationships among them. On the one hand, within an agent that manages the specific Grid services or other agents (managed elements), goal-based planning and scheduling mechanism autonomously (re-) configures, coordinates, invokes, and monitors the enforcement of, these managed elements. On the ther hand, the automated discovery, negotiation, and contract-based service providing and monitoring among various agents treat with the relationships among autonomic elements.

The rest of this paper is structured as follows. In section 2, we introduce the architecture of agent-enabling autonomic Grid service system. The autonomic elements and the relationships among them are represented in section 3 and section 4 respectively. Finally, section 5 is the conclusions.

2 Architecture of Agent-Enabling Autonomic Grid Service System

In this agent-enabling autonomic Grid service system, agents are autonomic managers of autonomic elements. Each agent that governs several Grid services (and/or other agents) is the representative of service provider or service consumer. Agent governs both internal affairs and external interactions (or transactions). Internal affairs include grid services registration, goals-based service planning, enforcing, and monitoring, etc. External interaction involves automated service discovery, negotiation, contract-based service providing and monitoring. The model of agent-enabling autonomic Grid service system (AAGSS) can be defined as a tuple:

$$AAGSS ::= (AE, ER) \tag{1}$$

AE denotes a set of autonomic elements, and ER denotes a set of external relationships between AEs.

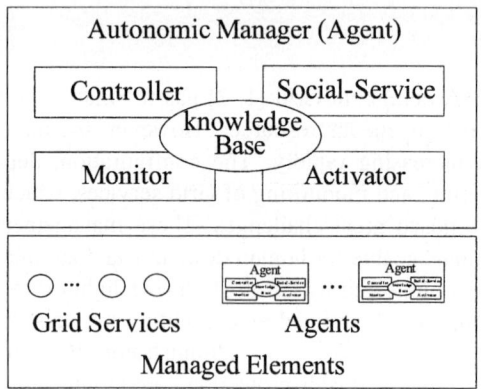

Fig. 1. Structure of autonomic element

The structure of autonomic element (AE) is shown in figure 1. The manager of AE is an agent who governs one or more grid services and/or one or more agents (service providers). AE can be defined as a 3-tuple:

$$AE ::= (AM, ME, IR) \tag{2}$$

AM – autonomic manager that is undertaken by an agent.
ME – managed elements that can be a set of grid services and/or other agents.
IR – a set of internal relationships between AM and ME, including registrations (registering ME to AM), monitoring (monitoring the states of ME), and enforcing (plan-based composition and invocation of MEs).

3 Autonomic Element

Autonomic element is composed of autonomic manager and managed elements. According to the definition (2), AE ::= (AM, ME, IR). The autonomic manager (AM) is an individual agent who governs several managed elements (ME), which can be Grid services or other autonomic elements (agents). The AM is composed of four components: monitor, controller, Social-Service, Activator, and knowledge base (as shown in figure1).

The Monitor is an interface component that monitors both internal and external events related to the autonomic elements, including registration massages from managed elements, negotiation requests from other autonomic elements, state information of managed elements (including run-time states of Grid services or other agents, the situation of agreed-on level of service provisioning), and the invocation messages from higher-level autonomic manager. Social-Service is a special component that provides several services for treating with social issues of autonomic element, including discovery of other autonomic elements (agents), negotiation with other autonomic elements, registrations of managed elements, and contract-based monitoring of services provision. Activator is a component that is in charge of activating the specific managed elements according to the dispatching plan from the Controller. Controller is a component that does goals-based planning and coordinates the behaviors of managed elements.

4 Relationships Among Autonomic Elements

High-level autonomic element is composed of several agents who are themselves autonomic elements. The relationships among them include automated discovery, negotiation, and contract-based service providing and monitoring, of Grid services governed by various agents. The Social-Service component of the autonomic element (as shown in figure 1) deals with the relationships with external agents. There are six steps in realizing the interaction between service-providing agent (SPA) and service-consuming agent (SCA).

Step 1: SPAs register their capability to a middle agent[5].
Step 2: A SCA sends a request to the middle agent.
Step 3: The middle agent matches the capabilities of SPAs with the requirement of SCA.
Step 4: The middle agent returns the capabilities of SPAs (candidates) to the SCA.
Step 5: The SCA negotiates with SPAs in sequence or concurrently, and then a contract between SCA and one of the SPAs forms.

Step 6: The SPA provides its services to the SCA according to the contract, and meanwhile, the service providing process is monitored in terms of contract by both sides.

5 Conclusions

The distributed, heterogeneous, and dynamic grid services pose serious management complexity problem. This paper proposes a model of autonomic grid services system with the supports of the agents that are characteristic of autonomy, proactivity, and sociality. The framework and the principles of this model are described in this paper.

A prototype agent system that supports this model has been developed by our research group in recent years. This agent system is characteristic of automated service discovery based on middle agent[6,7], rational negotiation between agents[8], cooperation and coordination directed by a manager agent[9], and contract-based service provisioning and monitoring among various agents, in the open, distributed, and dynamic environment.

References

1. I.Foster, C. Kesselman, S. Tuecke. The Anatomy of the Grid: Enabling Scalable Virtual Organizations. *International J. Supercomputer Applications*, 15(3), 2001.
2. I. Foster, C. Kesselman, J. Nick, S. Tuecke.The Physiology of the Grid: An Open Grid Services Architecture for Distributed Systems Integration. Open Grid Service Infrastructure WG, Global Grid Forum, June 22, 2002.
3. Jeffrey O. Kephart and David M. Chess.The Vision of Autonomic Computing. IEEE Computer, January 2003.
4. N.R.Jennings.On Agent-Based Software Engineering. Artificial Intelligence, vol.177, no.2, 2000, pp: 277-296.
5. Keith Decker, Katia P. Sycara, Mike Williamson.Middle-Agents for the Internet. IJCAI (1) 1997: 578-583.
6. Zhou Bin. The Systematism of Assistant Service for Agents (pp.31-33) [Thesis of Master degree].Hangzhou: Zhejiang University, 2004.
7. XU Fei. Agent Assistant System [Thesis of Master degree].Hangzhou: Zhejiang University,2003.
8. Ronghua Ye. The Research of Service-Oriented Mechanism of Negotiation Between Agents. [Thesis of Master degree].Hangzhou: Zhejiang University, 2004.
9. HU Jun, GAO Ji, LIAO Bei-shui, CHEN Jiu-jun. An Infrastructure for Managing and Controlling Agent Cooperation. Proceedings of The Eighth International Conference on CSCW in Design, May 26-28, 2004, Xiamen, PR China.

MoDast: A MoD System Based on P2P Networks

Yu Chen and Qionghai Dai

Broadband Network & Multimedia Research Center,
Graduate School in Shenzhen, Tsinghua University, Shenzhen, Guangdong, China, 518055
{cheny,daiqh}@mail.sz.tsinghua.edu.cn

Abstract. This paper presents MoDCast, a MoD servicing system based on P2P network. The hosts in MoDCast cache the media data receiving from streaming media server and forward them to other hosts. In this way, the MoD servicing system can accommodate more users concurrently, which improve the servicing capacity, and reduce the servicing cost.

1 Introduction

Streaming Media on-Demand (MoD) over Internet is one of the main Internet services, and how to improve the servicing capacity of MoD system is a hot topic. IP multicast improves server and networks' resource utilization remarkably, but hasn't been deploy wildly because of the problem of management, dependability, security, and servicing time, etc [1]. Application layer multicast makes hosts to forward data they receive to other host, and needn't dedicated support provided by lower layer. Typical application layer multicast protocols include Narada [2], Gossamer [3] and NICE [4], etc. Based on NICE, Tran, et al develop Zigzag [5] for streaming media broadcast with lower control overhead.

This paper presents a MoD servicing system based on P2P networks, MoDCast. The hosts in MoDCast cache the media data receiving from streaming media server and forward them to other hosts that make the MoD system accommodate more users and reduce the servicing cost. Section 2 introduces the model of MoDCast. Section 3 analyzes its caching strategy. Section 4 analyzes its application layer protocol. In the last section, we give out conclusion.

2 The Model of MoDCast

A basic MoDCast system is composed of a WEB server (ws), streaming media server (ms) that stores a group of clips, and a group of hosts that access those clips. h_1, h_2, h_3 and h_4 are hosts that access clip c_1, and the request of h_1 arriving ws firstly. ws redirects h_1 to ms. Because h_1 can't get service from other hosts, it gets service from ms directly. After this, instead of receiving data from ms, h_2, h_3 and h_4 will receive media data from h_1. In the same way, h_5 get c_2 from ms, and forward data to h_6 and h_8. h_7 receives media data from h_6. Logically, hosts accessing same clip construct a host tree, and ms, is the public root (r) of all host trees. We call the host sending data parent node, and the host receiving data descendant. Host tree is look like application layer multicast tree, but parent node in host tree could send different data to different descendants at same time, and parent node in application layer multicast tree must send same data to different descendants at same time.

The hosts in a host tree could belong to different sub-network. We confine a basic MoDCast system in an autonomous system, such as a campus network, or a network belong to a ISP to control end-to-end delay between hosts, and reduce transmission overhead on backbone.

Web server is the rendezvous of host tree, but not a part of the tree. Besides redirecting hosts to streaming media server, wed server maintains and issues the basic information of the clips, such as clip's introduction, price, etc.

3 Caching Strategy

Caching several seconds or one minute of data makes host be able to tolerate the variation of QoS on network and server, and playback the clip smoothly, but can't satisfy the requirement of MoDCast. Due to the unpredictability of the request' arrival time, the duration time of media data cached on host must long enough to increase the utilization of cached data. Obviously, the longer the data stay on host, the higher the probability of the data used by other hosts. The duration time of cached data is related to the host's cache space. The larger the cache space is, the longer the data's duration time is, but large cache space will occupy host's resource too much. Besides this, MoDCast system hopes the hosts to run as a parent node as long as possible to keep the cached data available to other hosts. But, host has the trend of leaving host tree as soon as possible at the end of playback. So, we design a cache strategy to take advantage of hosts' shared resource efficiently.

Let the duration time of clip is f units, and host can cache l units, $f>l$. Host caches and forwards the data received from parent node. Due to $f>l$, after cache is filled, the oldest data in cache space is given up to reclaim resource for the new arrival data. Let cache space of host h_i is l, and h_i begin to receive data from server at t. Host h_j choose h_i as parent node, and begin to get data from h_i at t'. If $t'-t \leq l$, h_j can get all of the data h_i caches.

Normally, if the interval between the arrival times of two consecutive requests that accessing clip c is always smaller than the duration time that media data cached on a host, one host tree can satisfy all of the users that access clip c. Though it's impossible to assure above statement. Making the host tree as deep as possible can make use of the resource of host sufficiently. But along with the increase of depth of the tree, the number of hosts influenced by the failure of intermediate host in host tree will increase remarkably. So, we define a constant time l for host tree. An host tree created at t when a host become the immediate descendant of MoD server, and other host can only join this tree before $t+l-a$, a is the interval between the time host send request and the time host join host tree. $l-a$ is the duration time that media data cached on a host. MoD server records the deadline of each host tree. For every clip, at any time, at most one host tree available for hosts to join. In this way, we can control the weak consistency of the host tree.

4 Application Layer Protocol for MoDCast

When request is redirected from WEB server to MoD server, MoD server will find whether the user sending this request can join a host tree. If can't, MoD server create a new tree for the user. Hosts in a host tree can be divided into two classes, peer host and selfish host. Peer host forwards the cached data to other host, and selfish host only receive data. New host chooses its parent node from peer hosts in the host tree.

A host tree has several layers. MoD Server is the unique node at the first layer, and second layer only has one host, r. The immediate descendants of r are composed of layer three, H_{L3}, and the immediate descendants of H_{L3} are composed of H_{L4}, etc.

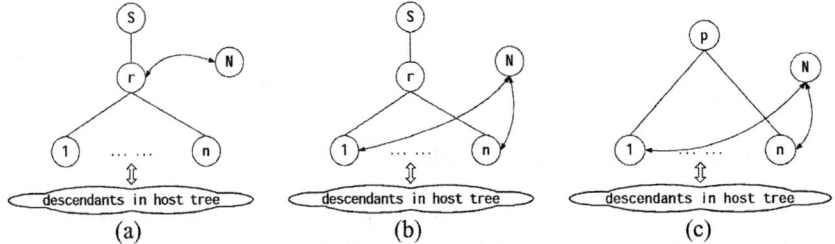

Fig. 1. Join Operation

When host h_n, wants to join in a host tree, server will send the address of r to h_n. h_n contacts with r (in figure 1a). If r has enough shared resource to connect with h_n, h_n become r's immediate descendant. Otherwise, r sends the member address of H_{L3} to h_n. h_n probes the distance between the member of H_{L3} and itself, and chooses the host closest to itself as parent node (in figure 1b). Let p is chosen, but p hasn't enough shared resource, then h_n will choose parent node from p's immediate descendant in H_{L4} (in figure 1c). h_n repeats above operation until find a suitable parent node, or exit because couldn't find parent node. If there are two or more hosts have same distance to h_n, h_n choose the host with lower overhead as parent node.

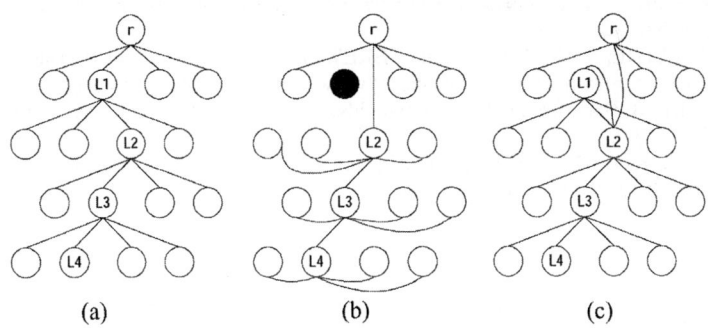

Fig. 2. A example of recovery protocol

We use an example to explain MoDCast's recovery protocol. Figure 2a is a host tree in normal state; $L1$, $L2$, $L3$ and $L4$ are mediate host in the host tree. r is the only host in H_{L2}, and receive data from MoD server. In figure 4b, $L1$ is failed. $L2$ informs its sibling hosts that it can't get media data, and its sibling response with the same message. So, a new parent node should be selected from $L1$'s immediate descendants to replace $L1$. The new parent host should be the first host that chooses $L1$ as parent host, because the data cached in this host can satisfy all other siblings' requirement. In figure 2b, $L2$ is the new parent node, and receives media data from r directly. So, $L2$ is lifted from H_{L4} to H_{L3}. $L2$'s siblings become its descendants, and only one of $L2$'s original descendants continue connect with $L2$. In this example is $L3$, others become $L3$'s descendants.

When $L1$ can't receive data from r, $L1$ exchange information with its siblings to find out whether its parent node is failure or the link to its parent node is failure. If other hosts can still receive data from parent node, the link between $L1$ and r is failed. $L1$ chooses a relay node in its immediate descendants to recover from link failure. As shown in figure 2c, $L2$ is the relay node between $L1$ and r, which receives the date needed by $L1$ from r, and forward them to $L1$. At the same time, $L2$ gets data from $L1$ to playback.

5 Conclusions and Future Works

In this paper, we introduce MoDCast's architecture, and analyze its cache strategy and host tree protocol. We have done some simulation about MoDCast, and the results of simulation demonstrate that, when working in an autonomous system, MoDCast can improve the servicing capability significantly with low control overload. Besides this, we can conclude from simulation that, 1) sharing peer host's resource can reduce the reject ratio, and improve resource utilization, 2) cache more data can improve system's accommodation, but at the cost of increasing hosts' overhead, 3) the number of selfish hosts hasn't significant influence on system performance.

References

1. C. Diot, B. N. Levine and B. Lyles, et al, Deployment Issues for the IP Multicast Service and Architecture. IEEE Network, Jan 2000.
2. Y. Chu, S. Rao, and H. Zhang. A Case For End System Multicast. In Proc. ACM Sigmetrics, June 2000.
3. Y.CHAWATHE, S. MCCANNE and E. BREWER, An architecture for internet content distribution as an infrastructure service. http://www.cs.berkeley.edu/ yatin/papers, 2000.
4. S. Banerjee, B. Banerjee and C. Kommareddy, Scalable Application Layer Multicast, ACM SIGCOMM 2002.
5. D. A. Tran, K. A. Hua and T. T. Do, Zigzag: An Efficient Peer-to-Peer Scheme for Media Streaming, IEEE INFOCOM, 2003.

Maintaining and Self-recovering Global State in a Super-peer Overlay for Service Discovery

Feng Yang[1], Shouyi Zhan[1], and Fouxiang Shen[2]

[1] Department of Computer Science and Engineering Beijing Institute of Technology
Beijing 100081, P.R. China
[2] Communication and Software lab, Institute of Command Automation of Navy
Beijing 100036, P.R. China
{mryangfeng,Zhan_shouyi}@263.net

Abstract. A key problem affecting the routing performance in a Super-Peer Overlay is the fluctuation of network, which makes DHT shifted and generates considerable overhead of maintaining the global state of routing information. In this paper we propose a Power Sorting Multicast algorithm and an adaptive self-recovery mechanism to solve this problem. The experiments show they can efficiently decrease maintenance traffic and restore the system to stable state, holding availability of discovery algorithm.

1 Introduction

The P2P lookup algorithms based on super-peer overlay are seen as an excellent discovery mechanism for large scale distributed systems, such as Grid and Web Services, since superpeers can provide faster, more effective and robust discovery using global state. The powerful nodes are selected as superpeers and take most system tasks. By creating a super-peer overlay network on existing P2P network, the routing tasks can be finished much faster. Using superpeers in P2P systems is a hot research topic recently [1, 2, and 3].

Although P2P approach is a very promising approach, a lot of difficult problems impede their popularity on services discovery.

First issue in this approach is the high overhead of maintaining data structure of superpeer, when nodes joins/leaves the system since superpeer constructs its routing table using all superpeers information, which is called *global state*. In a highly fluctuating and unpredictable environment, the cost of maintaining *global state* of routing is likely to outweigh the advantages of having one. We will likely spend more of our time updating them than performing lookup. It is not only increasing the bandwidth consumption, but also affecting the efficiency of DHT routing algorithms.

Secondly, it lacks effective self-recovery mechanism. A superpeer also stores a large number of indices published on it. If a superpeer is down, the accompanying indices on it disappear. It is not accurate to locate the service and the system has no approach to recover automatically these indices.

2 Related Works

The Project JXTA is a P2P platform based on super-peers network and DHT [4]. JXTA 2 introduces the concept of a rendezvous super-peer network: dynamically and adaptively segregating the P2P network into populations of edge peers plus rendezvous peers. Routing only occurs within the more stable (and smaller population of) rendezvous peers. Rendezvous are superpeers that resolve the queries of routing in JXTA platform, and the JXTA super-peers network is mostly consist of rendezvous.

In this approach, the DHT algorithms require all rendezvous to maintain a consistent view of all rendezvous, namely rendezvous peer view (RPV), which is the list of known rendezvous to the peer, ordered by each rendezvous' unique peer ID by DHT hashed.

Each RDV is able to know the *global state* of rendezvous network, using Gossiping protocol [5] to update RPV. From Gossiping, each rendezvous regularly sends a random subset of its known rendezvous to a random selection of rendezvous in its RPV. Then the rendezvous received the RPV can update itself RPV by using Hash function to create a new order. A super-peer joins /leaves the super-peer overlay, meaning that the RPV is shifted by one position. However, the list of indices has not moved synchronously, causing quires trashing. This problem is called the shift of DHT brought by fluctuant network.

Most super-peer systems [1~4] use Gossiping protocol to maintain the global consistent of RPV in all superpeer. However, the random nature of Gossiping renders the solution costly and with low guarantees. The research [5] has showed that the probability of all rendezvous received the refreshment messages of RPV depends on the number k of random selected nodes in each round and the number of the rounds. If $k = \log n + c$ (n is the total number of nodes in network, c is the constant.), the probability is $e^{(-e^{(-c)})}$. If $c = 2$, only 80 percent of nodes in network may receive this message. With the dilating of network size, the problem "look back" [5] may produce the large number of redundant messages and increase the maintenance overhead.

3 Maintaining the Global State

3.1 Power Sorting Multicast

For the RPV including M superpeers, it is constructed a circular list. For a superpeer s, $s \in list$, define $list(1) = s$. Then, i^{th} neighbor of s represents $list(2^i)$, $i = 1, 2, ..., m$. Let $m = \max\{i \leq \log M\}$. Thus, multicast space of s defines $(list(2^i), list(2^{i+1}))$. The upper bound of multicast space initializing process denotes $upbound = list[M]$. The multicast algorithm is shown in below.

```
//Algorithm of Power Sorting Multicast
s.multicast(update,upbound)
for i=1 to m do
  if list(2^i)∈ (s,upbound) then new = list(2^i);
    if list(2^{i+1})∈ (s,upbound) then new_upbound = list(2^{i+1});
    else new_upbound = upbound;
    end if
    forward(update,new_upbound) to new;
  else exit;
  end if
end for
```

3.2 Discussion of Algorithm

As the DHT constructed a connected graph of superpeers, every peer that receives a multicast message can forward it to all of its neighbors with interval of power length. Therefore, it guarantees eventually every peer in DHT can receive the message.

The other advantage of this algorithm is no redundancy. It ensures that intervals have no overlap for forwarding. Consequently every peer receives the message exactly once, solving the problem of "look back". Compared to Gossiping, the number of selected nodes in each round also equals to $\log n$ and the number of rounds is lower in this algorithm (The number of rounds in Gossiping is $O(\log \sqrt{M})^{1+\varepsilon}$ from [5]).

We denote the number of rounds i, and we get $i = O(\log_{\log M} M) = O\left(\dfrac{\log M}{\log \log M}\right)$.

It is shown that the number of rounds is no more than $O\left(\dfrac{\log M}{\log \log M}\right)$ in our algorithm, lower than $O(\log \sqrt{M})^{1+\varepsilon}$.

3.3 Self-recovery Mechanism of Indices

Even though we can guarantee consistent *global state* of RPV and reduce the maintenance overhead via power sorting multicast, another problem is to recover indices published on disappeared superpeer. One approach is the replication mechanism. Obviously this boost the network consumption, especially in a relatively fluctuant network. The other approach, refresh indices periodically and random walking mechanism, is adopted to ensure certain efficiency when shift occurs. But this approach can't make DHT recover automatically the stable state.

We present an adaptive self-recover mechanism based on random walker algorithm in this section. When the random walker is started to rout the query, the adaptive self-recovery can monitor it and notice the queried edge peer to re-publish its index, and DHT returns automatically to stable state.

The GaV based Method has to be modified whenever a new metadata set (e.g., a heterogeneous resource) is added. Both of these solutions also may suffer from the performance bottleneck and single point of failure because its centralized architecture. Additionally, in GaV method, the modifications on some individual metadata schemas may not be detected by the whole system because they are mapped to a global one and the latter generally is not updated so often.

[1] describes the approach in extending the centralized structure into P2P-based topology. But how to extend the OAI-PMH protocol into the P2P network for DL usage? Currently, the OAI-PMH protocol has been widely adopted in numerous institutes for metadata harvesting purpose because of the low application barrier capability of OAI-PMH protocol. However, few activities work on the extending OAI protocol with P2P concepts except Lagoze's vision in the OAI annual conference [2]. The goal in [2] is to allow users directly request the DPs without dropping into *service providers (SP)* although users still maintain the capabilities to request SPs. Additionally, data are always up-to-date for the users and user queries can reach all available small sized *data providers(DP)*s. Since DPs and SPs are loosely couple, they are allowed to 'join' and 'leave' freely.

Considering that few DPs in the real-world development are willing to add an extra layer to their softwares in terms of the security, reliability issues, we create a wrapper on the user side. Figure 1 illustrates an example for the communication among three roles of user, user wrapper, and DP. In Figure 1, it is assumed that both of the wrapper and DP

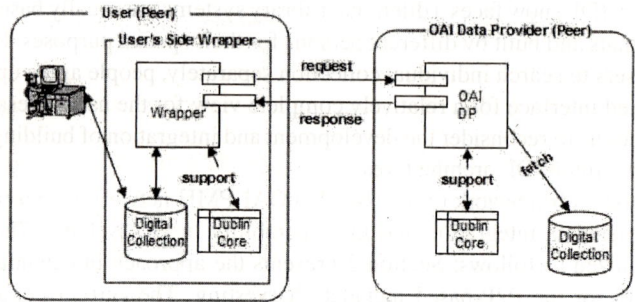

Fig. 1. P2P-based Harvesting Mechanism

support DCMS. The user sends queries as that in the traditional way while the wrapper transform the query to a format understandable by the specific DP. Consequently, the user can always retrieve the 'fresh' data from DPs. Furthermore, the wrapper can also be deployed to harvest data from other DPs into the local repository or cache which acts as the functionalities of SPs.

3 The Design of Metadata Interoperability in P2P Network

Different collections may be built upon various metadata schemas. In the library community, even one specific collection may have several alternative metadata formats to choose from. Actually, having different schemas brings us many conveniences in describing specific collections for specific purposes. For example, if we want to annotate a set of records for scientific publishing purpose, we can have several alternatives, such as, Dublin Core, MARC and EAD, etc. However, problems may arise in combining underlying relevant metadata schemas. It mainly comes from that objectives of some repositories differ from each other heavily.

Figure 2 illustrates a metadata repository which contains three metadata formats indicated by rectangle, square and rhomboid. The dotted lines among 'rectangle' items indicate the relationships of metadata records which share a common schema, while the other set of dotted lines shows the relationships among three items in different schema. We propose to alleviate the aforementioned problems by adopting the concept of *upper-*

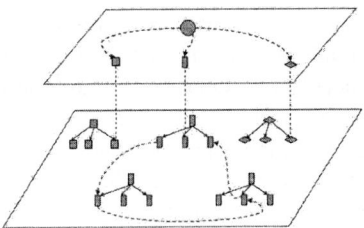

Fig. 2. Relationships among metadata items

level ontology. [3] [4]. Some approaches come recently into play with respect to the upper-level ontology-based application: the ABC (A Boring Core) ontology [5] which is implemented for a domain-specific application on 'hydrogen economy' project [6]; the SUMO (Suggested Upper Merged Ontology) with an ambitious goal of developing a standard upper ontology that will promote data interoperability, information search and retrieval, automated inferencing, and natural language processing [7], and as well as DOLCE (Descriptive Ontology for Linguistic and Cognitive Engineering) [8]. In our approach, we use currently the JENA, an inference engine under the Semantic Web framework, so as to achieve automatic deduction to a certain extent. The general framework is described in Figure 3.

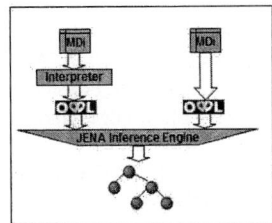

Fig. 3. Metadata Records Integrating by Adopting JENA

The interpreter in Figure 3 is used to transform metadata format into OWL (Web Ontology Language) [9]. Consequently, the explicit relationships among related items can be automatically generated by importing the candidate schemas.

Additionally, the explicit relationships among related terms can be generated according to the predefined rules. And, OWL also defines a special kind of resources called containers for representing collections of things. Combined with the aforementioned method, related records can also be clustered together for further information retrieval.

4 Conclusion and Future Work

This paper has presented the current approaches in the field of integrating/harvesting metadata records in distributed digital libraries. A general framework in harvesting metadata in a super-peer based P2P network was proposed, and the critical issues were indicated and proposed solutions, such as extending OAI protocol for P2P network, mapping heterogeneous metadata schemas by adopting novel inference engine, have also been presented. Since in current applications of OAI-PMH, there is no control of the qualities of the data providers and many trivial but hard to solve problems [10] in the conventional OAI-PMH applications will not avoidable in current system. The expectations, however, from the data providers to re-check and re-annotate their won documents are very unlikely. So, how to automatically annotate the collection with ontology-based semantics will be the future work. Additionally, many efforts will be put into improving the query in order to allow users to choose their favorite data providers.

References

1. Ding, H., Solvberg, I.: A schema integration framework over super-peer based network. In: 2004 IEEE International Conference on Services Computing (SCC 2004), Special Session, to appear, China, China (2004)
2. Carl Lagoze, H.V.d.S.: The oai and oai-pmh: where to go from here? Available at: http://agenda.cern.ch/fullAgenda.php?ida=a035925, last visited:2004/07/02. (2004)
3. S.Staab, R.Studer: Handbook on Ontologies. Springer-Verlag Berlin Heidelberg (2004)
4. Guarino, N.: Formal ontology and information systems. In: In Proceeding of FOIS'98, Trento,Italy (1998) 3–15.
5. Lagoze, C., Hunter, J.: The abc ontology and model (v3.0). Journal of Digital Information **vol.2, no.2** (2001)
6. Hunter, J., Drennan, J., Little, S.: Realizing the hydrogen economy through semantic web technology. IEEE Intelligent System (2004)
7. I.Niles, A.Pease: Towards a standard upper ontology. In: Proceedings of 2nd Int'l Conf. Formal Ontology in Information Systems (FOIS), ACM Press (2001) 2–9.
8. etc., A.: Sweetening ontologies with dolce. In: Proceeding of the 13th International Conference on Knowledge Engneering and Knowledge Management (EKAW 2002),LNCS 2473, Springer-Verlag. (2002) 166–181.
9. Smith, M.K., Welty, C., McGuinness, D.L.: Owl web ontology language. http://www.w3.org/TR/owl-guide/ (2004)
10. Martin Halbert, J.K., Hagedorn, K.: Findings from the melon metadata harvesting initiative. In: European Conference on Digital Library(ECDL), Trondheim, Norway. (2003) 58–69

Service-Oriented Architecture of Specific Domain Data Center*

Ping Ai, Zhi-Jing Wang, and Ying-Chi Mao

School of Computer and Information Engineering,
Hohai University, Nanjing 210098, P.R.China
aip@hhu.edu.cn

Abstract. The application schema based on Web Services is developing quickly, and the component-based software architecture is also researched more and more deeply. The traditional data center architecture based on database service method has been unfit for this technical progress. According to requirements of specific domain data resources development and utilization, a service-oriented architecture of specific domain data center was proposed in this paper. Its essential characteristics are that database and data processing closely combined into data services to be distributed in networks, these services are registered to the service registry center, and data resources are transparently processed by requesting the corresponding data services. The China National Water Resources Data Center is an instance of the service-oriented architecture of specific do-main data center.

1 Service and Domain Data Center

In the specific domain, the main technical difficulty of data integration and sharing based on Internet lies in the heterogeneity of data structure and semantic of databases. Although some techniques can solve some problems in local area, but the service-oriented architecture can offer a new solution based on Internet.

For description of problems, some essential concepts are defined as follows:

- Definition 1 (Specific Domain): Domain is a sphere of activity, concern, or function, e.g. the domain of water resources. Specific domain is a denotation of single domain.
- Definition 2 (Specific Domain Data Center, SDDC): SDDC is the infrastructure used to provide data and services in the specific domain in the centralized way. SDDC can not only integrate the data of various business systems in the domain by a certain business logic relationship, but also implement some functions with some tools to support domain applications.
- Definition 3 (Service): Service is a coarse-grained and discoverable software entity of being a single instance. Service communicates with others by the loosely-coupled message transmission (often is asynchronous communication) [1]. A service can be described with its name, interfaces, functional properties, non-functional properties, specific domain properties and its sub-services.

---

* Supported by the National High-Tech Research and Development Plan of China under Grant No.2001AA113170; the Provincial Natural Science Foundation of Jiangsu Province of China under Grant No.BK2001016; the National Grand Fundamental Research 973 Program of China under Grant No.2002CB312002

- Definition 4 (Service-Oriented Architecture, SOA): SOA is a set of components that can satisfy the business requirements. SOA is made up of several components, ser-vices and procedures [2].

The service is simple logic arrangement of components that satisfy business requirements. Business process is the key part of SOA and components arrangement can meet the requirements of process to make application schema be consistent with the business logic. SOA has many characteristics [3][4], and the Web Service is considered as an implementation of SOA.

2 Service-Oriented Architecture of Specific Domain Data Center

SDDC confronts two important issues: data integration and data sharing. Because of imbalance development of domain applications, the inconsistency between advanced technologies and relatively poor management, data are disordered separated to some extent. As a result, it is an efficient technology strategy to implement scalability integration of separated datasets using the loosely-coupled mechanism of SOA.

2.1 Requirements Analysis of Specific Domain Data Service

There are various application systems in a specific domain because the domain businesses are often implemented in accordance with its own hierarchical structure. In the OLTP era, various specific databases for specific domain applications were established in different level, which results the heterogeneity when data shared within the domain. With the development of OLAP and DS technologies, it is more necessary for data sharing within the domain even among the domains. Furthermore, it has become an impetus to aggregate data into information resources and to make full use of information resources to improve productivity. Therefore, it is key factors for the development of domain applications to integrate all sorts of data resources.

According to the traditional opinions of data integration, it is an approach for solving problems to construct the traditional data center where data can be centralized stored within the domain. Whereas this approach is unpractical in many domains, its main difficulties lie in data maintenance, data service, software reuse and resources optimization. Therefore, SOA with the loosely-coupled characteristic has become the best choice of SDDC. SOA can encapsulate data and processing into service to hide the data heterogeneity, which is quite important in the domain in which data are decentralized stored in the networks and are more frequently updated as well.

2.2 Service-Oriented Architecture of Specific Domain Data Center

In general, the architecture of information systems is mainly multi-level structure based on component and middleware. The databases access function is often separated through adding a data access service level between business logic level and data access level. The business logic not only is separated from the representation level, but also it can be divided into multiple hierarchies. According to the essential thought of SOA and the requirements of domain applications, basic architecture of SDDC based on SOA is illustrated in Figure 1.

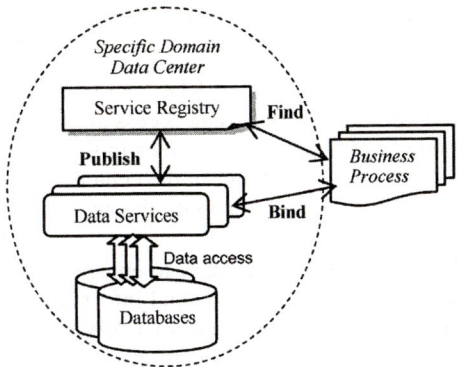

Fig. 1. Basic architecture of SDDC based on SOA

SDDC based on service is made up of three parts: databases, data services and service registry. Databases refer to been established or being established databases in the domain, which can distributed in the networks. Moreover, there is no requirement of the homogeneous database, because the heterogeneity and access security can be solved by data services. Data service is a pure software entity and it can be classified into two class's service based on its own functions. One is technical services to provide various basic data access services. The other is business services to satisfy the common requirements of domain data analysis and processing.

Data services should usually be deployed at the same node that it operates database, in order to lessen the pressure of network transmission and improve the system security. Service registry is used to receive and manage the publication of data services, thus service registry is the core of SDDC and its role is the service locator and the service broker as well. Business process can obtain the service of SDDC by access to service registry, locating the desired data service, binding data service, and obtaining service.

The architecture of SDDC based on SOA takes advantages of the loosely-couple mechanism to encapsulate data access and processing into a service, which makes business applications implement sharing the services of data and data processing by the means of the dynamic service request. In addition, it is unnecessary to reconstruct the existing database, which does not affect the existing business systems, and also unnecessary to change the mode of database maintenance and management.

2.3 Architecture of China National Water Resources Data Center

The China National Water Resources Data Center (CNWRDC) is the center for water resources information collection, storage, management, exchange and service in whole of China and its basic functions are to share the water resources information and to process the common business.

CNWRDC is composed of the center nodes, the basin nodes, and the province nodes, which are geographically distributed. At the same time, CNWRDC is logically composed of three parts: information collection and storage, information service, and support application. Based on the basic technical requirements and constraints of CNWRDC, it can apply the basic architecture of SDDC based on SOA to construct the architecture of CNWRDC.

2 Proposal of an Improved SNMP

We propose an improved SNMP model as follows to solve the network traffic problems mentioned above. Fig.1 shows additional PDUs of the proposed SNMP. Those are added among existing SNMP PDUs. So, 5 PDUs will be used on the proposed model to keep compatibility with the existing SNMP.

GetTiRequest

PDU type	Requested ID	0	0	Start time	End time	Time interval	Variable bindings

GetTiResponse

PDU type	Requested ID	Error status	Error index	Start time	End time	Time interval	Variable bindings

Name 1	Value 1-1	...	Value 1-n	...	Name n	Value n-1	...	Value n-n

Fig. 1. Additional PDUs for measuring TI

A manager can send "GetTiRequest" PDU for measuring TI or "GetRequest" PDU for normal to an agent with management applications. The agent analyzes "PDU type" field of a received PDU, and then processes appropriate operations according to it. In case of the "GetTiRequest" PDU for measuring TI, an agent checks request-id, start time, end time, time interval and variable-binding fields from PDU header, and collects data from the start time to the end time with the time interval. The agent creates a single "GetTiResponse" PDU containing all collected data, and sends it to the manager. Otherwise, an agent receives "GetRequest" PDU and then treats as an original SNMP does.

3 Test Environment and Analysis of Results

The Prototype system environment is as following.
- OS: Manager: UNIX (Solaris2.8), Agent: Linux (RedHat 6.2), Measurement system: Windows XP
- Used SNMP version: SNMP version 2
- Used SNMP program: UCD SNMP v3.4
- Used languages for implementation: C, Shell base CGI-C compiler (GCC v3.1)
- Measuring tool: LANdecoder32, AppDancer FA

An independent test bed is constructed on Ethernet LAN which is unconnected to other network in order to measure network traffics for the proposed SNMP and the existing SNMP. A MIB object is measured for 30 minutes every five seconds by the existing method and the proposed method.

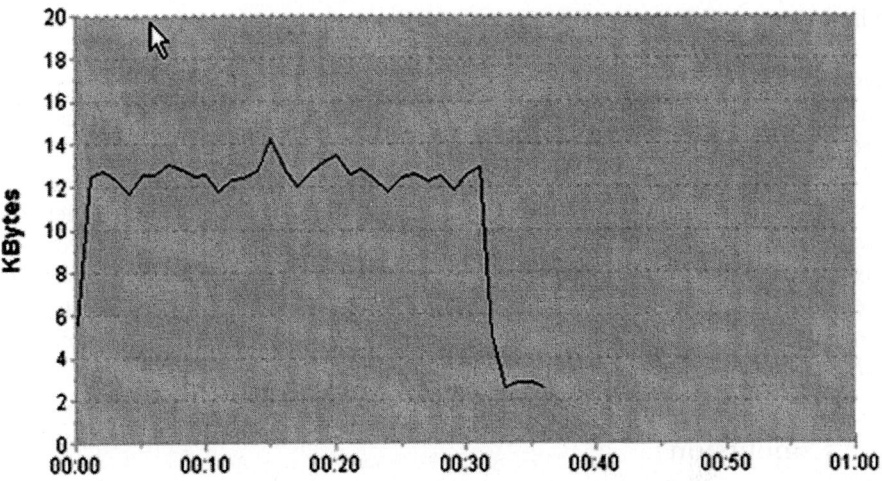

Fig. 2. Network traffic measured by existing SNMP

Fig.2 represents a test result of network traffic measured for 30 minutes according to test condition by the existing SNMP. An average traffic is 12.9Kbytes per minute. However, the traffic is not consistent since there is HTTP for web program to initiate manager from the measurement system, NetBIOS for Windows XP and SNMP in the test bed.

Fig.3 shows that average network traffic is about 0.4Kbytes/ min in the proposed SNMP. The network traffic is relatively higher at starting and ending time. It means the pure traffic of SNMP is mainly produced only twice. The first is when manager transmits GetTIRequest PDU to agent, and the second is

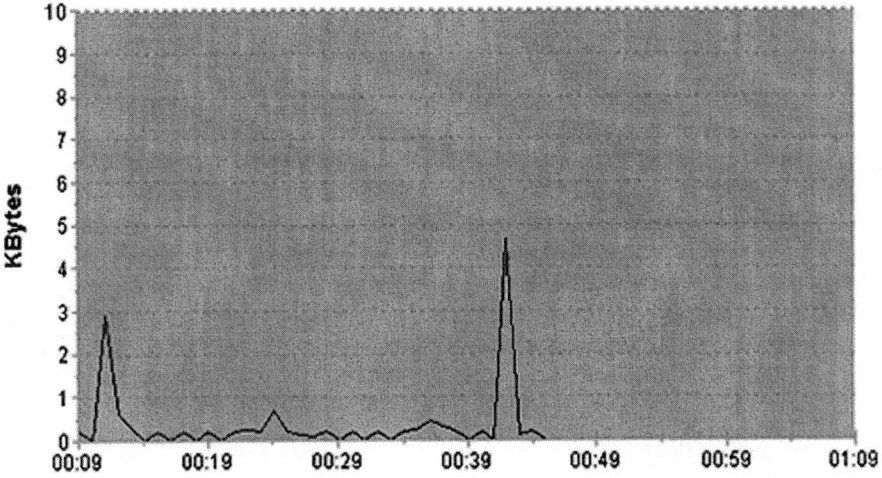

Fig. 3. Network traffic measured by the proposed SNMP

when agent transmits GetTIResponse PDU to manager for response. The other traffics are HTTP for web program to initiate manager from the measurement system and NetBIOS for Windows same as the existing SNMP. In case of single transaction, traffic amount of the existing SNMP is less 83bytes than the proposed SNMP according to the measuring tool. On the other hand, in case of measuring TI, which is interested in this paper, the network traffic of pure SNMP is reduced about 39.3 times in the proposed SNMP. Average network traffic including the other protocols is decreased about 32 times in the proposed SNMP. Conclusively, the network traffic of the proposed SNMP is greatly decreased compare with the existing SNMP as measuring interval gets shorter and measuring term gets longer. And the proposed SNMP keeps compatibility of function with the existing SNMP for measuring TI.

4 Conclusion

This paper is focused on measuring TI for the detail analysis of SNMP network. I proposed an efficient SNMP that can minimize unnecessary network traffic for measuring TI, implement a prototype and test it. The characteristics of the proposed SNMP are as follows. First, it greatly decreases network traffic of measuring TI comparing with the existing SNMP since the number of sending-receiving message between a manager and an agent becomes remarkably smaller. Second, it performs extended functions for measuring TI successfully and keeps compatibility with the existing SNMP perfectly. Therefore, the effect of the proposed SNMP would be great when applied to a NMS that manages wide area network. Of course my proposal has a modification overhead of the original SNMP program, but I think it not so much compare with the contribution of my proposal.

References

1. M. Checkhrouhou and J, Labetoulle, An Efficient polling Layer for SNMP, proceedings of the 2000 IEEE/IFIP Network operations and Management System. pp.447-490, 2000.
2. Min-woo Kim, Seung-kyun, Park, Young-hwan, Oh, A Study on the Polling Mechanism for Optimizing the SNMP Traffics, The jounal of the Korean institute of communication sciences. 06 v.26, n.6A, pp.1051-1059, 2001.
3. Jin-young Cheon, Jin-ha Cheong, Wan-oh Yoon, Sang-bang Choi, Adaptive Network Monitoring Strategy for SNMP-Based Network Management, The Journal of the Korean Institute of Communication Sciences.12 v.27, n.12C, pp.1265-1275 1226-4717, 2002.

A New Grid Security Framework with Dynamic Access Control*

Bing Xie, Xiaolin Gui, Yinan Li, and Depei Qian

Department of Computer Science and Technology, Xi'an Jiaotong University,
710049, Xi'an, China
`xiexiebing@sohu.com, xlgui@mail.xjtu.edu.cn`

Abstract. In this paper, a new Grid security framework based on dynamic access control is introduced to address some security problem in Grid. Based on the dynamic evaluating results about trusts among resources and users, users secure access levels are changed dynamically. The track of users behaviors using resources is criterion for assigning secure levels to different users and for allocating the resources to users in the next application execution. Here, the dynamic access controls are realized by mapping users secure levels to access rights. In our experiment Grid, the evaluation mechanism of users behaviors is applied to support the dynamic access control to Grid resources.

1 Introduction

Grid resources are always distributed in different domains, so security problems in it are more complex and intractable. In network security framework, security techniques are used maturely and widely. Firewall and IDS can prevent most attack from Internet, authentication mechanism are largely used in every security scenario. Although these techniques can assure network runs securely under most conditions, the complex authentication mechanism, plentiful and serious restrictions must result in network resources can't be fully used.

Compared to network security, security problems in Grids are more complex. GSI (Globus Security Framework) [1] is implemented by Globus, which emphasizes traditional security techniques are adequately used in Grid environment. Besides traditional techniques, GSI introduces the concepts of users agent and resources agent specially for the mutual authentication between users and resources. Although this security frame can assure the legal rights of users, it can't guarantee resources are securely used by legal users. Once accredited users attack system or submit the jobs with virus, the Grid system or resource nodes will be wrecked deeply. Another widely used Grid is Legion[2], In Legion, once users are authenticated and assigned with resources, resources nodes can freely change users secure levels and access rights. This framework can secure resources effectively and feasibly, while some problems are inevitable. Such as the Grid dynamic will be satisfied difficultly, at the same time, the Grid scale will be restricted and not easy to extend, and so on.

According to the above analysis, this paper implement a new Grid security framework based on dynamic access control. In this framework, the access rights of users,

---

* This work was sponsored by the National Science Foundation of China (60273085), the state High-tech Research and Development project (Grid Software) and the ChinaGrid project of China (CG2003-GA002).

users history behaviors and resources protection are combined to assure Grid security by using the traditional security techniques such as IDS, resources inspection, etc. By inspecting and saving the users present and history behaviors, the users trust evaluations are calculated as their secure levels and fartherly their access rights are assigned dynamically. And at the same time once the resources inspectors detect the submitted jobs running with virus, the jobs are stopped forcedly for resources security. In the next chapters, this Grid security framework and its application in experiment Grid (Wader [3]) will be given.

2 Grid Security Framework with Dynamic Access Control

In the most popular Grid frameworks, resources access security is implemented by statically assigning users secure levels, and such works are one-off determined. Once trusted users attack the Grid system or computational resources by executing applications or other methods, the results will be fatal. Since any malicious users would emerge some typical characters, the framework can track their behaviors and inspect resource instance to avoid these users illegal options and change their access rights to all protected resources. In this paper, a new Grid security framework based on dynamic access control is introduced to address these problems.

2.1 The Grid Security Framework

The frame of dynamic secure access control based on behaviors evaluation is shown in Fig.1, which includes five components: Grid catalogue, access control list (ACL), behavior inspector, security database and behavior evaluation organization (BEO).

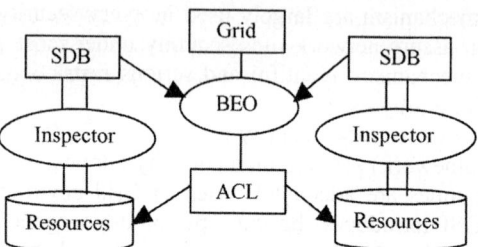

Fig. 1. The framework of the grid security framework with dynamic secure level evaluation

Grid catalogue: mainly responsible for supplying the information to users, and also takes on part of security role, which is to define secure levels as attributes and the relationship between access rights and secure levels.

Inspector: Behavior Inspector, whose main responsibility is to inspect resources used, has two duties:
(1) Inspecting the submitted jobs. Once the jobs are considered as illegal ones, they are stopped immediately. The material option is to inspect network actions, the log files of resources nodes and error report[4],[5],[6] etc. Table1 shows the inspection items by behavior inspector. This function is exactly like IDS, so like IDS, behavior inspector 24-hour monitors possible intrusion or information attack.

(2) Transferring the instance of used resources according to different users to security database for dynamically evaluating the users secure level.

Table 1. Inspection list in Wader

Manager ascertainment	The statistic report of network capability	Log files of resources nodes
ICMP protocol : ping, ports monitor. SNMP protocol: query Statistic report:	The statistic report of users exceptional login and log out	Illegal system transfer, etc.

Security database: responsible for keeping the users present and history behaviors. In another words, it's a tracker of users behaviors. Based on recording users behaviors in it, users themselves are tracked synchronously and the assignment of secure level for users would be more exact and reasonable. In Wader, mysql [7] database server is used as security database. Mysql can be free downloaded, feasibly handled and support several operating system, e.g. Windows, Linux, Solaris, etc, which satisfies the Grid demand for supporting heterogeneous systems.

BEO: behaviors evaluation organization, which is the core component of this frame and responsible for dynamically evaluating the users' secure level by using users present and history behaviors, and even more, the access rights of users will be accordingly and dynamically changed.

ACL: access control list, which is saved in Grid catalogue, is amended owing to the evaluations of BEO. Users are assigned certain right to access corresponding resources. In the following, the description rule of ACL will be given.

The above five components consist of the dynamic evaluation mechanism and secure access control mechanism about the whole Grid. The material rules of access control will be shown in the next part.

2.2 The Access Control Rule

In the above paragraph, the security framework is discussed. Then the factual rules of access control will be shown as the follow.

Table 2. ACL in Wader catalogue

User Domain	Secure Level	Description	Resource Domain	Access Rights
S_0	0	Fresh man	O_0	R
S_1	1	Common user	O_1	R
S_2	2	Regular user	O_2	E
S_3	3	Advanced user	O_3	E
S_4	4	Trustable users	O_4	E and W
S_5	5	Most trustable users	O_5	E and W

In ACL, users are divided into five domains. According to the different secure levels of users, they are assigned with different access rights, where R, E, W desperately express the access rights of *read*, *execute* and *write*. By using ACL, the dynamic secure levels of users are mapped to their access rights, and the resources are protected more effectively.

3 Conclusion

In this paper, a new Grid security framework based on dynamic access control is implemented. In this framework, users behaviors are tracked and then according to these behaviors the secure levels are calculated and the access rights of users are assigned. While in the factual applications, the applied instances of resources are complex and the arithmetic of behavior evaluation will be more exact and material, and it is our next study direction.

References

1. G. Aloisio, M. Cafaro. An introduction to the Globus toolkit. Proceedings of CERN 2000-013, (2000) 117-131
2. J. Bongard. The legion system: A novel approach to evolving heterogeneity for collective problem solving. Lecture Notes in Computer Science, Vol.1802, (2000) 16-28.
3. Gui Xiao-lin, Qian De-pei, He Ge. Design and Implementation of a Campus-Wide Metacomputing System (WADE). Journal of Computer Research and Development, 2002, 39(7): 888~894. (in Chinese)
4. Summaries, http://www.cert.org/summaries
5. Incident_notes, http://www.cert.org/incident_notes
6. Vul_notes, http://www.cert.org/vul_notes

A Dynamic Grid Authorization Mechanism with Result Feedback[*]

Feng Li, Junzhou Luo, Yinying Yang, Ye Zhu, and Teng Ma

Department of Computer Science and Engineering, Southeast University
Nanjing, 210096, P.R. China
{lifengg,jluo}@seu.edu.cn

Abstract. The authorization mechanisms in existing Grids are relatively static, so we present a dynamic authorization mechanism. Through the negotiation between users and resources, the Grid security management entities match the requests of the user and resource according to corresponding stipulations in security policies, form security contracts, and make authorization decisions. In order to reflect whether the security contracts have been well observed, we introduce the trust parameter as the feedback of our authorization mechanism.

Keywords: Grid, Security Contract, Negotiation, Trust, Dynamic Authorization

1 Introduction

As we hope to employ Grid in extensive environments and make it a real strategic infrastructure all over the world, we should consider security problems of the Grid in all aspects in early stage. Further more, because the Grid is dynamic and operates without centralized control, the security demands of Grid cannot be satisfied only by the traditional security measures. In the dynamically changing environment of Grid, one challenge is to authorize user in controllable granularity, and insure that every entity in the Grid complies with the authorization decision.

2 Dynamic Authorization in Grid

User submits his authorization request reflecting the demand of the tasks. On the other hand, resource host puts forward the conditions for the user. And then the management entities consider the Grid's overall situation, user's authorization request and resource's use-condition synthetically, make the authorization decision finally, similar to the service level agreement (SLA) negotiation course. We need to expand SLA's content and design a new negotiation course, fully considering the Grid's security demands. We call the agreements raised by users and resources Security Contract Intents (SCI) and name the authorization decision Security Contract (SC).

---

[*] This work is supported by Key Laboratory for "Network and Information Security" of Jiangsu Province(BM2003201).

2.1 Establish the Functional Entities

Because the users and resources are both dynamic, we design a three-layer trust management model. VO takes the responsibility to manage the user, trust relationship is established between each pair of VO and AD, and AD takes the responsibility to manage the resources in that domain. Two kinds of functional entities are naturally set up: One is responsible for user management considering the VO's overall situation. We call them VO's global security policies (GSP) execution center (GSPEC). Another is mainly responsible for resource management and set up in each AD. We call them AD's local security policies (LSP) execution center (LSPEC).

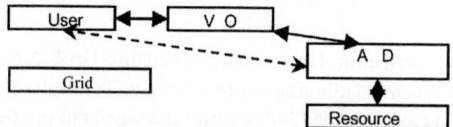

Fig. 1. Trust Relationship in Grid

2.2 Classify the SCIs and SCs

The user and resource negotiate with each other according with cost and their own security demands, and finally reach some kind of agreement.

UTSCIs: Users form UTSCIs according to his demand. A UTSCI is denoted as <I, U_{id}, T_{id}, t_{dead}, j, U_{sig}>. j describes service steps and this task's detailed request.

TSCIs: After the management entities have authenticated the user's identity, they sign the SCI to protect it. A TSCI has the form <I, U_{id}, T_{id}, t_{dead}, j, U_{attr}, M_{id}, M_{sig}>. U_{attr} records the detailed attributes about the corresponding user.

RSCIs: It gives the abstract service ability of the resource and the conditions to use the resource. A complete RSCI has the form<I, R_{id}, t_{dead}, r, U_{req}, R_{sig}>.

BSCs: Represent the final authorization decision. It records the bindings between the task and resources. A complete BSC: <I, T_{id}, [R_{id}], t_{dead}, b, M_{id}, M_{sig}>. b is the description of the binding, it includes the description of user's authority and the resources' commitment about the abstract service ability that they offer to the user.

2.3 The Mutual Course of the Dynamic Authorization

The mutual course is presented in Fig.2. It is the core of the authorization mechanism.

When the Grid system comes into operation, GSPECs and LSPECs should authenticate each other and set up SSL connections. **The first step** is that a user forms UTSCI, and then submits it to the GSPEC. GSPEC authenticate the user according to VO's GSP and then verify the signature in the UTSCI. The GSPEC **then** determines whether to give sanction to the user's authorization request or not considering users' identity, VO's GSP, VO's overall resource acquisition and usage situation and the user's request. If it approves, it will add the user's attribute to the UTSCI, sign this SCI with the GSPEC's private key, and finally form a new TSCI. **In the third step,**

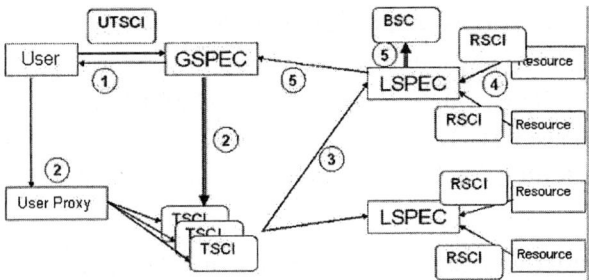

Fig. 2. The mutual course of the dynamic authorization mechanism

GSPEC transmits the TSCI to corresponding LSPEC through SSL connection which has been set up in advance. **The course** that the resource submits its RSCI is actually independent of the 3 steps listed above. Once the resource's situation changes, it will notify the LSPEC by an updated RSCI, LSPEC will update the information it records according to corresponding stipulations in the AD's local security policies. **In the fifth step**, LSPEC should verify the GSPEC's signature. Then it will take four aspects into accounts: the user's attribute in the TSCI, the user's authorization request, the AD's local authorization policies and RSCIs. If the LSPEC determines to refuse this TSCI finally, it will notify the GSPEC directly. If the LSPEC agrees to authorize this user, then it will delete corresponding TSCI and RSCIs, form a BSC to record the binding between the user and the resources. **The sixth step is:** the user and resources audit the implementation of the SC respectively, and submit the auditing result to the GSPEC and LSPEC in the form of trust parameter.

3 Trust Parameter

To realize result feedback, we introduce one security relevant QoS parameter, trust. The same as it is in the realistic society, the concept of trust should include two aspects: direct trust and reputation. The value of direct trust stems from the direct historical mutual record with other entities; however, reputation is the appraisal from other entities in the Grid on the mutual rival. The trust parameter here only reflects whether the user and resource comply with the authorization result strictly.

3.1 The Formulation of Trust Parameter

In Grid, the trust parameter should be directed. User use a post-mortem IDS to audit whether the resource fulfill the task's demand according to regulation in BSC, and submit the auditing result to GSPEC. The GSPEC will gather the results, and compute the direct trust to each AD which offers resource to the VO. The GSPEC also maintains a list which records that VO's intimate degree to others. GSPECs regularly exchange their direct trust about each AD. And a GSPEC calculates the reputation of an AD on the basis of the intimate degree and other VO's direct trust to that AD. The trust parameter's acquisition and updating process can be formalized. GSPEC will gather and compute V_i's direct trust about D_j:

$VDTL(V_i, D_j) = (1-a) \times VDTL(V_i, D_j) + a \times UT(t_{ij})$;

Let $FV(V_i, V_k)$ represent the intimate degree, then the trust parameter from VO_i to D_j is: $VTL(V_i, D_j) = (1-b) \times VDTL(V_i, D_j) + b \times RTL(D_j)$;

D_j's reputation is: $RTL(D_j) = \sum_{k:1\sim n, k\neq i} VDTL(V_k, D_j) \times FV(V_i, V_k)$:

Similarly the LSPEC can compute D_j's trust parameter about user U_l and VO V_i.

3.2 Realize Result Feedback on the Basis of Trust Parameter

Trust parameter can reflect whether the Grid entity obeys the authorization decision in the past. This paper realizes feedback of dynamic authorization through the following two respects: First, to audit the implementation of BSC representing the authorization decision and then update trust parameter based on audit results; Second, using parameter in the following authorization course, both the user and resource can raise requirements of its mutual rival's trust parameter, so the auditing results of the BSC's execution before can have an effect on the following course.

4 Conclusion

In the mechanism described here, different entities are responsible for the execution of VO's GSP and AD's LSP. This ensures the consistent implement of the policies at all levels. Besides, in the negotiation course, GSP and LSP can be formulated and modified totally independently. In traditional network and grid security infrastructure, authorization is a static process. But in fact, users' demands change dynamically. The authorization through negotiation can better meet demands in dynamic environment. In addition, it makes the granularity of authorization finer. Any QoS parameter understood by both sides can be negotiated and controlled. We also take the trust parameter as the feedback result of dynamic authorization. The parameter reflects the historical behavior of users and resources, offers a new basis for following authorization course and will have a kind of deterrent effect on users and resources.

References

1. I.Foster, C.Kesselman, G.Tsudik, S.Tuecke, "A Security Architecture for Computational Grids", Proc.5th ACM Conference on Computer and Communication Security Conferece, 83-92, 1998
2. L.Pearlman,V.Welch,I.Foster,C.Kesselman,S.Tuecke, "A Community Authorization Service for Group Collaboration: Status and Future",
http://www.globus.org/security/CAS/Papers/ CAS_update_CHEP_03-final.pdf , 2003
3. K.Czajkowski, I.Foster, C.Kesselman, "SNAP: A Protocol for Negotiating Service Level Agreements and Coordinating Resource Management in Distributed Systems", LNCS 2537, pp.153-183, Springer, 2002
4. Farag Azzedin and Muthucumaru Maheswaran, "Evolving and Managing Trust in Grid Computing System", Proceeding of the 2002 IEEE Canadian Conference on Electrical & Computer Engeering, 2002

A Dynamic Management Framework for Security Policies in Open Grid Computing Environments

Chiu-Man Yu and Kam-Wing Ng

Department of Computer Science and Engineering, The Chinese University of Hong Kong
Shatin, New Territories, Hong Kong SAR
{cmyu,kwng}@cse.cuhk.edu.hk

Abstract. A computational grid is a kind of open and distributed computing environment enabling heterogeneous resource sharing and dynamic virtual organization (VO) membership. Dynamic security policy management for multiple VOs in grids is challenging due to the heterogeneous nature of grids. Rather than deploying in a centralized VO space to manage the security policies of multiple VOs, we propose a dynamic management framework (DMF) to manage security policies in a decentralized manner. DMF groups VOs under the same security policy framework into a virtual cluster, thus allowing homogeneous conflict analysis to be performed. There is a Policy Processing Unit to coordinate the analysis tasks, but the tasks can be distributed to VOs according to their trust relationships. Heterogeneous conflict analysis for VOs of different policy frameworks takes place at a Principal Policy Processing Unit in the grid environment. Therefore, the homogeneous and heterogeneous policy management tasks are separated.

1 Introduction

Dynamic security policy management is one of the challenges in grid computing. Due to the heterogeneous nature of grids, the security policy frameworks of the Virtual Organizations (VOs) [2] can be different. There may be conflicts between the VOs' security policies. The major consideration of security policies in grid environments is that VOs are dynamic entities which are formed when needed and discontinued when their functions are completed. The setting up of a VO requires each institute to offer certain resources to the shared environment. There are policies to control access to a single machine in a classical organization as well as to a VO. If there is interoperation between multiple VOs, it requires cross-organization policy management [6].

Traditional security policy frameworks [4][6][7] deal with security policy management inside a VO. There is still little research on policy management for multiple VOs. Due to increasing popularity of Grid computing, it is likely that there will be a large number of application systems of Grid computing environments in the future. Each Grid application system forms a VO. With the use of different grid development toolkits, the VOs may deploy different security policy frameworks. To enable secure interaction of heterogeneous VOs, frameworks and mechanisms are needed to manage security policies across the VOs, we propose a Security Policy Dynamic Management Framework to handle this problem.

2 Security Policy Dynamic Management Framework (DMF)

Security Policy Dynamic Management Framework (DMF) is a hierarchical framework which aims to support "dynamic policy management" and "heterogeneous policy conflict analysis" for Grid environments of multiple VOs. It contains a number of "Policy Agents" (PA), "Policy Processing Units" (PPU) and a "Principal Policy Processing Unit" (P-PPU). DMF deploys PAs to divide VOs into virtual clusters according to their security policy frameworks. Conflict analysis can be performed homogeneously in a virtual cluster, or can be performed heterogeneously through the P-PPU. The PA virtual connection architecture inside a virtual cluster is constructed hierarchically according to trust relationship so that policy management tasks can be distributed to PAs.

In the Grid environment model for DMF, there is one Grid Operator and a number of VOs. The Grid Operator coordinates meta-VO interaction or interoperation among the VOs. The Grid Operator also maintains meta-VO level policies. Each VO consists of a number of nodes which can be service providers, or service requesters. Each VO has a policy server. The VOs' policy servers and the Grid Operator are PDPs (Policy Decision Points). The service providers and service requesters on the VOs are PEPs (Policy Enforcement Points). A PEP is the point where the policy decisions are actually enforced; a PDP is the point where the policy decisions are made [5].

2.1 Framework Architecture

Figure 1 illustrates the DMF hierarchical architecture. In DMF, each VO needs to initialize a PA. The Grid Operator initializes a P-PPU. PAs of the same security policy framework would form a virtual cluster. One of the PAs in a virtual cluster would be elected as the PPU. A PPU performs runtime conflict analysis for the same model of security policy framework, whereas the P-PPU is used to perform heterogeneous security policy conflict analysis across different security policy frameworks. In DMF, the service providers are PEPs; whereas the PA, PPU, and P-PPU are PDPs.

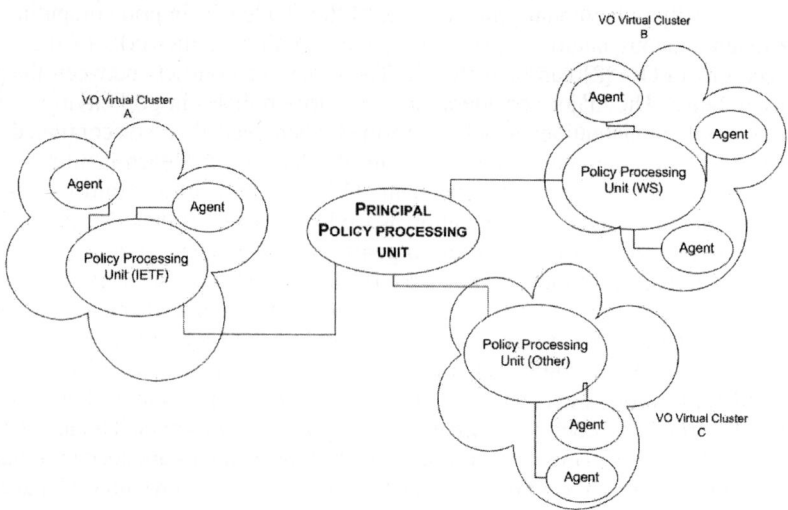

Fig. 1. DMF architecture

All PPUs are connected to one P-PPU. PAs are connected to either PPU or other PA. The virtual connection hierarchy inside a virtual cluster depends on trusts between VOs. PA can access information of local security policies. PPU can access information of local security policies, global security policies, and knowledge of local conflict analysis. P-PPU can access information of global policies, and knowledge of heterogeneous policy conflict analysis. The PA needs to be able to access the VO's security policies. If the VO already has a PDP (or policy server), the PA is likely to be run on the PDP. Otherwise, the PA can be run on a node which has privilege to access the security policy database.

PAs of the same "security policy framework model" would form a virtual cluster. For example, Web Services policy framework (WSPolicy) [1] and IETF security policy framework [3] are different framework models. A newly initialized PA can ask P-PPU for which virtual cluster to join. PAs in a virtual cluster elects one of them to be PPU. PPU is also a kind of PA.

In an ideal trust relationship, if all PAs in the same virtual cluster trust one of the PAs, then the PA can become PPU of the cluster. However, usually the PAs do not fulfill the ideal trust relationship. Some VOs do not trust other VOs, or none of the VOs are trusted by all other VOs in the virtual cluster. Therefore, DMF needs a PPU election mechanism in the non-ideal trust relationship. The election mechanism selects a PA with the most number of supporters to be PPU. The PPU connects to PAs according to the trust relationships of the PAs. The PPU may be untrusted by some PAs such that the PPU is unable to retrieve security policy information from some PAs. Therefore PPU may need to perform conflict analysis with partial information.

2.2 Policy Agent (PA) Hierarchy

In DMF, all PPUs are virtually connected to the P-PPU directly since virtual clusters are deploying different security policy frameworks. In a virtual cluster, not all PAs are virtually connected to the PPU directly. The PA connection hierarchy depends on their trust relationships.

Figure 2 shows possible trust relationships between PAs. The symbol "←" represents direction of trust. The PA in square box means that it has the highest level in that trust relationship. We call it a leader PA. We transform the PA trust relationship diagrams into the following expressions.

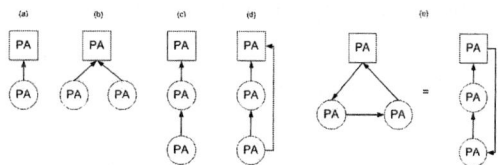

Fig. 2. PA trust relationships

In Figure 2, the PAs in squares are leader PAs. A leader PA can be trusted by other leader PA(s). If PA_1 trusts PA_2, then PA_2 has higher level than PA_1. The PA being trusted has higher level. In the PA hierarchy of DMF, leader PAs with relative highest levels are virtually connected to the PPU directly.

By this PA hierarchy, it is possible for a PPU to delegate to a leader PA to perform conflict analysis. If a conflict analysis only involves PAs in one of the trust relationships as in Figure 2(a), 2(b) and 2(d), the leader PA can get all security policy information from all PAs in the model. Therefore, a PPU can ask the leader PA to perform conflict analysis. This can reduce bottleneck in the PPU.

3 Conclusion and Future Work

DMF (Security Policy Dynamic Management Framework) is a framework to support dynamic policy management and heterogeneous conflict analysis in open grid environments. In DMF, policy agents are used in VOs for exchanging policy information. Policy agents of the same policy framework type would form a virtual cluster, and elect one of the agents to become the policy processing unit. The policy processing unit would handle local policy conflict analysis and policy management. A principal policy processing unit would be responsible to handle heterogeneous conflict analysis. By this framework, only one processing unit needs to handle heterogeneous conflict analysis. Policy agent in individual VO does not need to acquire knowledge of different security policy framework. The virtual cluster formed by agents allows ad hoc update of the cluster topology. It does not require central policy storage. Thus the framework supports dynamic security policy management.

In open environments with heterogeneous VOs, we cannot assume that the VOs can trust another organization. In this case, the PPU may be unable to get all the necessary policy information from PAs to perform conflict analysis and make decisions. Therefore, we need to look into performing conflict analysis with partial information. Besides, at the meta-VO level, DMF needs to perform conflict analysis for the heterogeneous VOs which deploy different policy frameworks. We need to look into heterogeneous policy conflict analysis mechanisms.

References

1. Don Box, Franciso Curbera, Maryann Hondo, Chris Kale, Dave Langworthy, Anthony Nadalin, Nataraj Nagaratnam, Mark Nottingham, Claus von Riegen, John Shewchuk: Specification: Web Services Policy Framework (WSPolicy), available at http://www-106.ibm.com/developerworks/library/ws-polfram/
2. I. Foster and C. Kesselman, J. Nick, and S. Tuecke: The Physiology of the Grid: An Open Grid Services Architecture for Distributed Systems Integration, avaliable at http://www.globus.org, 2002. Version: 6/22/2002.
3. B. Moore, E. Ellesson, J. Strassner, A. Westerinen: Policy Core Information Model – Version 1 Specification, IETF Network Group RFC 3060, February 2001.
4. Gary N. Stone, Bert Lundy, and Geoffery G. Xie, U.S Department of Defense: Network Policy Languages: A Survey and a New Approach, in IEEE Network, Jan/Feb 2001.
5. J. Strassner and E. Ellesson: Terminology for Describing Network Policy and Services, Internet draft draft-strasner-policy-terms-01.txt, 1998.
6. Dinesh Verma, Sambit Sahu, Seraphin Calo, Manid Beigi, and Isabella Chang: A Policy Service for GRID Computing, M. Parashar(Ed.): GRID 2002, LNCS 2536, pp. 243-255.
7. Von Welch, Frank Siebenlist, Ian Foster, John Bresnahan, Karl Czajkowski, Jarek Gawor, Carl Kesselman, Sam Meder, Laura Pearlman, and Steven Tuecke: Security for Grid Services, in Proceedings of the 12th IEEE International Symposium on High Performance Distributed Computing (HPDC'03).

A Classified Method Based on Support Vector Machine for Grid Computing Intrusion Detection

Qinghua Zheng, Hui Li, and Yun Xiao

School of Electronic & Information Engineering, Xi'an Jiaotong University,
710049 Xi'an, P.R. China
qhzheng@mail.xjtu.edu.cn, lihui@sttri.com.cn,
yxiao@sei.xjtu.edu.cn

Abstract. A novel ID method based on Support Vector Machine (SVM) is proposed to solve the classification problem for the large amount of raw intrusion event dataset of the grid computing environment. A new radial basic function (RBF), based on heterogeneous value difference metric (HVDM) of heterogeneous datasets, is developed. Two different types of SVM, Supervised C_SVM and unsupervised One_Class SVM algorithms with kernel function, are applied to detect the anomaly network connection records. The experimental results of our method on the corpus of data collected by Lincoln Labs at MIT for an intrusion detection system evaluation sponsored by the U.S. Defense Advanced Research Projects Agency (DARPA) shows that the proposed method is feasible and effective.

1 Introduction

A Computational Grid is a set of heterogeneous computers and resources spread across multiple administrative domains with the intent of providing users easy access to these resources [1].One goal of software designed as infrastructure supporting Computational Grids is to provide easy and secure access to the Grid's diverse resources. When we enjoy the benefits that Grid computer offers to us, we have to face more security problems. Security is a crucial issue for a grid computing system.

Intrusion Detection System (IDS) is the `burglar alarms' system of the computer security field [2]. IDS can offer a basic safeguard for Grid Computing system. Data acquired in real grid computing network environment is usually highly variable, small sample size and high dimensional. Classification on this data set is always difficult to finish with the conventional machine learning technologies. So the application of the traditional machine learning technologies to IDS is not satisfied, and these technologies show some faults in Grid Computing system.

Support Vector Machine is the core technology of Statistics Learning Theory (SLT) [3]. It aims at small sample size learning and has been successfully used in some small sample size learning problems. These successful stories of SVM prove that SVM is fit for the highly variable, small sample size and high dimensional data. Data acquired in IDS domain (such as network connection records, Unix system logs, etc), takes on these characteristics and is more complex. So we introduce SVM to the research of grid computing intrusion detection (ID) system.

2 SVM and Its Extension on Heterogeneous Dataset for ID Classification

Support vector machines (SVMs) conceptually implement the following idea: input vectors are mapped to a high dimensional feature space through some non-linear mapping chosen a priori. In this feature space a decision surface (a.k.a. hyperplane) is constructed. Special properties of this decision surface, ensures high generalization ability of the learning machine. C-SVM is used for supervised learning which only deals with labeled corpus. Bernhard Scholkopf proposed a new One-Class SVM algorithm which can deal with unlabeled data [4]. It is an unsupervised learning method to be applied in outlier detection and density estimation. The dataset in IDS is often high dimensional, small sample size and heterogeneous. Traditional SVM can only deal with plain dataset and cannot tackle heterogeneous dataset directly, so we do some refinement of kernel function to extend SVM on the heterogeneous dataset.

Suppose two input vector $x, y \in X$, their heterogeneous value difference metric is defined as[5]:

$$H(x,y) = \sqrt{\sum_{a=1}^{m} d_a^2(x_a, y_a)} \qquad (1)$$

Where m is the number of attributes. The function $d_a(x, y)$ returns a distance between the two variables of x and y for attribute a. It is defined as:

$$d_a(x,y) = \begin{cases} 1 & \text{if } x, y \text{ is unknow otherwise} \\ normalized\_vdm_a(x,y) & \text{if } a\text{th attribute is discrete} \\ normalized\_diff_a(x,y) & \text{if } a\text{th attribute is continuous} \end{cases} \qquad (2)$$

$$normalized\_diff_a(x,y) = \frac{|x-y|}{4\sigma_a} \qquad (3)$$

Where σ_a is the standard deviation of the numeric values of the ath attribute of all instances in X.

$$normalized\_vdm_a(x,y) = \sqrt{\sum_{c=1}^{C} \left| \frac{N_{a,x,c}}{N_{a,x}} - \frac{N_{a,y,c}}{N_{a,y}} \right|^2} \qquad (4)$$

Where $N_{a,x}$ is the number of instances in the training set T that have value x for attribute a; $N_{a,x,c}$ is the number of instances in X that have value x for attribute a and output class c; C is the number of output classes in the problem domain.

Based on the HVDM metric of the heterogeneous dataset, we design a new HVDM-based RBF kernel and propose a new HVDM-SVM. The new HVDM-based RBF kernel is as follows (Where H is the HVDM metric):

$$K(x,y) = \exp(-\frac{H(x,y)}{\sigma^2}) \qquad (5)$$

3 Experiments and Results

The experiments dataset for HVDM-SVM ID model is a network connection record set which is restored from the raw data collected by Lincoln Labs at MIT for an intrusion detection system evaluation sponsored by DARPA in 1998 and can be classified into four classes: DOS, Probe, R2L, and U2R. We select two representative datasets "correct" dataset and "10 percent" dataset and construct two groups of datasets: the balanced dataset of DOS and Probe and the unbalanced dataset of U2R and R2L. We employ the C_SVM algorithm to do classification on the balanced dataset of DOS and Probe and one-class SVM on the unbalanced dataset of U2R and R2L. We compute some value related to performance evaluation of intrusion detector such as: 1.Total detection precision rate (P)= the number of sample detected correctly / the number of the total sample; 2.False positive rate (FN) = the number of normal sample that is detected as anomaly / the number of the total normal sample; 3.Detection rate (DR) = the number of anomaly sample which is correctly detected / the number of the total anomaly sample; 4.Average detection time (AT) = overall detection time(ms) / the number of total sample.

We give different combinations of error penalty operator C and regulation operator $G=1/\sigma^2$ of RBF kernel function to test the performance of different model. Due to space constraints the representative results are list in Table1:

Table 1. Results of C_SVM

Class	C	G	10percent	Correct	10per_test_cor	Cor_test_10per
DOS	50	0.5	DR=75.77%	DR=75.39%	DR=73.20%	DR=64.23%
			FN =0.68%	FN=0.49%	FN=0.71%	FN=0.79%
			P=78.55%	P=80.01%	P=75.5%	P=63.57%
			AT=0.27ms	AT=0.25ms	AT=0.27ms	AT=0.13ms
Probe	50	0.5	DR=92.81%	DR=99.84%	DR=96.55%,	DR=87.22%
			FN=0.36%	FN=0.64%,	FN=0.29%	FN=0.47%
			P=90.78%	P=91.86%	P=92.32%,	P=85.57%
			AT=0.27ms	AT=0.5ms	AT=0.4ms	AT=0.3ms
U2R	50	0.5	DR=63.27%	DR=68.19%	DR=60.77%	DR=70.21%
			FN=6.8%	FN=5.1%	FN=8.1%	FN=6.8%
			P=83.16%	P=90.01%	P=78.42%	P=73.55%
			AT=0.27ms	AT=0.13ms	AT=0.13ms	AT=0.27ms
R2L	50	0.5	DR=34.12%	DR=43.39%	DR=37.22%,	DR=34.22%,
			FN=8.4%	FN=3.10%	FN=10.01%	FN=11.79%
			P=65.55%	P=71.01%	P=55.11%	P=63.57%
			AT=0.27ms	AT=0.25ms	AT=0.25ms	AT=0.27ms

We compare the performance of our model with that of Wenke Lee's method. Table 2 gives the results:

Table 2. Performance Comparison Results

	Wenke Lee	C_SVM	One_SVM
DOS	DR=24.3%	DR=76.86% FN=0.57%	
Probe	DR=96.7%	DR=93.24 %FN=0.42%	
U2R	DR=75.0%		DR=66.7 %FN=5.23%
R2L	DR=5.9%		DR=31.58% FN=9.5%

4 Conclusion and Future Work

In this paper, Statistical Learning Theory (SLT) was introduced to the research of intrusion detection and an ID methodology (HVDM-SVM) based upon SVM was proposed. SVM method is generalized on heterogeneous dataset through a refined RBF kernel. Theoretic analysis and experiment on DARPA data corpus prove that the generalization of SVM is reasonable and successful. Experiment results show that our method is feasible and effective than others. The limitation is obvious since our HVDM-SVM method is only tested on the DARPA intrusion dataset, the adaptability and expendability should be verified in further. So the next work is to collect the network data of a real Grid environment from educational resource grid of Xi'an Jiaotong University, and use Kernel principle component analysis to do feature construction. With these selected features, we will design the special HVDM-SVM ID to test and verify. The data of a real Grid environment is similar to the data of common network, so our special HVDM-SVM ID will also show good performance.

References

1. Computer Grids. I. Foster ,C Kesselman: Chapter 2 of "The Grid: Blueprint for a New Computing Infrastructure", Margon-Kaufman,1999
2. S. Axelsson: Intrusion Detection Systems: A Survey and Taxonomy, http://citeseer.nj.nec.com/axelsson00intrusion.html, 2000.
3. Vapnik V N.: The Nature of Statistical Learning Theory,NY:Springer- Verlag, 1995
4. Bernhard Scholkopf, John C. Plattz,Estimating the Support of a High-Dimensional Distribution, Microsoft Research Technical Report MSR-TR-99-87
5. D.Randall Wilson,Tony R.Martinez: Improved Heterogeneous Distance Functions, Journal of Artificial Intelligence Research 6 (1997) 1-34

Real Time Approaches for Time-Series Mining-Based IDS*

Feng Zhao, Qing-Hua Li, and Yan-Bin Zhao

National High Performance Computing Center(WuHan) and
College of Computer Science and Technology,
Hua Zhong University of Science & Technology, 430074, Wuhan, P.R. China
jimmy_zf@sohu.com,liqh@263.net

Abstract. There is often the need to detect currently intrusion and new attacks in existed Intursion Detection System (IDS) due to customers' demands. Since traditional data mining-based IDSs contructed on the basis of historied data, systems are expensive and not real time. In this paper, we present an overview of our research in real time time-series mining-based intrusion detection systems. At first we describe multidimensional spatial model of network events, then present time-series minging-based architecture model and finally discuss real time approaches for systems. We focus on the issues related to deploying an accurate and efficient time-series mining-based IDS in a real time environment.

1 Introduction

Along with the increasingly wide applications of network technologies, the security problems about system and network are more and more remarkable. Thus as an import security mechanism, Intrusion Detection Sytems (IDSs) have become a critical technology to help detect attacks, monitor users' behaviors and protect systems. Traditional IDSs are based on hand-crafted features and fail to detect new attacks or attacks without known signatures.

To avoid shotcomings of taditional IDS, Wenke Lee of Columbia university brings forword a new data mining-based IDSs to generate detection models in a quicker and more automated method [1]. However, a successful time series mining techniques are themselves not enough to create high performance IDSs[2]. Despite the promise of better detection performance and intelligence learning of data minig-based IDSs, there are some inherent weaknesses in the detection process. We can group these difficulties into three general categories: efficiency, accuracy and not real time.

There is a basic premise for our research that we summerize all network stream as time-series data with a set of attributes (such as timestamp, etc.).In this paper, we gave some approaches to improve the accuracy and efficiency of the time-series mining-based network intrusion detection systems.

---

* This paper is supported by National Natural Science Foundation of China (Contract No. 60273075).

2 Multidimensional Model Structure of Network Events/Records

When packet data (in this paper, we just take into account TCP/IP packet) are summarized into network connection events or records, each record/event contains of a set of intrinsic features or attributes. These features are including: timestamp, duration (duration of the connection), service, source host, source port, source bytes, destination host, destination prot, destination bytes, and status flag (such as: SYN, FIN, PUSH, RST) and so on.

Owing to the multidimensional features of network events, we can analysis system security and intrusion detection from network protocol, service type, connection process, pattern features of professional data, network management, etc.. So ,we bring forward multidimensional spatial model to describe the distributed features of network events, each dimension expresses only one attribute of network connection and every network event is a point in this multidimensional space according to it's attribute values' vector.

For more dimensional space, we can use hypercube to describe network events or records. The spatial distance between two points is figuring the similitudity of these two event. When computing the distinct, we can choose different weight according to the significance of each dimension. According to this multidimensional spatial model, we can easily detect intrusion data by time-series mining technology such as frequent itemset mining, classification, sequence analysis, etc..

3 Real Time Approach

A successful time-series mining techniques are themselves not enough to create high performance IDSs. Despite the promise of better detection performance and intelligence learning of time-series minig-based IDSs,there are some inherent weakness in the detection process. We group these difficulties into three general categories: efficiency, accuracy, and not real time. Figure 1 is the real time model for IDS based-on time series mining. Traditional model of time series mining-based IDS architecture always makes detection process unusable in real enviorments because of low efficiency, high misinformation rate, and at the most, it's not real time. The real time model as figured figure 1 can efficiently improve security systems' efficiency, accuracy and timeliness.

Event Process Engineer is used to exact and preprocess raw data from network and system logs. *Data Mining Engineer* (DM Engineer) represents all data mining technologies on time-series, such as sequence analysis, frequent episodes, outlining, association rules mining, classficiation, clustering, etc.. This DM Engineer can product evidence and intuitive rules that can support *Intrusion Detection Engineer* (ID Engineer) [3] to detect intrusions, and can be easily inspected and edited by security experts or network administrators when needed. The learned rules replace the manually detecting intrusion, except from DM Engineer, *Rules Database* (Rules DB) can be from security experts or network users and from *Decision Support System Engineer* (DSS Engineer). The output of ID Engineer is one of terms of DSS. For more interactively and iteratively drive the process of intrusion detection, administrator group in security systems should communicate with DSS when needed.

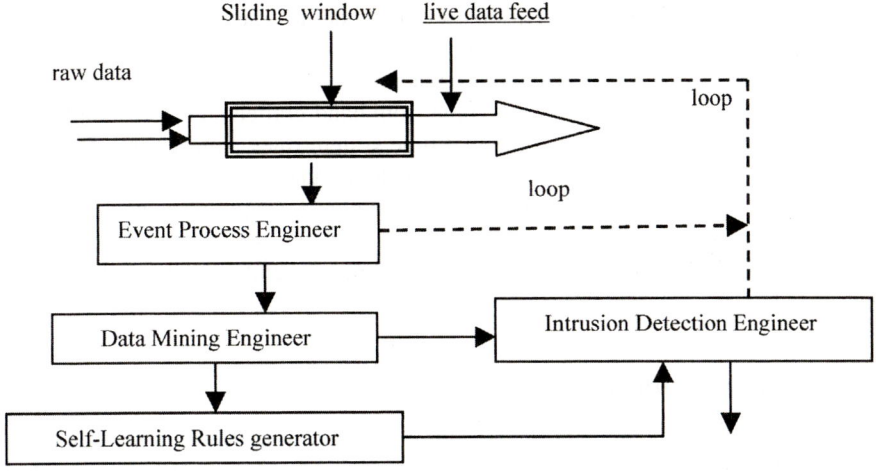

Fig. 1. Real Time Model for IDS Based-On Time-Series Mining

As figured as figure 1, we remodel the time-series mining architecture for IDS by apply some real time strategy to system to improve real time performance of detection. The real time strategy is as following:

The first is the sliding window[3].

A certain user-defined data window size is used as a criterion to snatch network data stream, this size can either be related to a time constraint, or can depend on the number of records wanted in each window, this window is called as a sliding window. A sliding window contains a fixed width span of data elements, the data itmes are implicitly deleted from the sliding window, when it moves out of the window scope. An input item in a net data stream has a value v in the domain D. The domain D can be contiguous or discrete. Each data item associates with a timestamp t_s. There can be more than one data item stamped with the same timestamp or no data item with a particular timestamp.

Let T_c denote the current timestamp. Given the window size w, the domain space for any data item within a sliding window is $D \times \{T_c - w + 1,...,T_c\}$. A net data item d in a sliding window is represented by (v, t_s). When T_c increases by 1 at each time tick, the sliding window moves one step forward. This process can be simulated by a number of insertions of d_i and deletions of d_j, where di is a data item with timestamp T_c and d_j is a data item with timestamp T_c-w.

The sliding window is just like a sticking point which would subsequently choke on the huge amount of data that it has to go through. For example, the window size can be set to contain a fixed-size time frame worth of data. Within a sliding window, we can summarize raw binary data into discrete network connection records with time attributes.

The second one is the loop updating.

After detecting with sliding window, the real time model has additional feedback loops to update the event engineer and intrusion detection engineer based on the re-is to identify a technique for representing complex behavior profiles so that good distance thresholds for anomaly detection can be determined without manual intervention. In this paper, we propose the use of the vector quantization (VQ) technique for representing user profiles.

The rest of the paper is organized as follows. The VQ technique and the intrusion detection model based on which the VQ technique is applied is described in Section 2. Experimental results demonstrating the detection performance of this technique are presented in Section 3. The paper is concluded in Section 4.

2 The Vector Quantization Technique

VQ is an optimization problem in a high dimensional space with a codebook generated as an optimal approximation of the large training data set. Each codeword in the codebook is in fact a representative vector within its partition. In the VQ process, the codebook generation is the primary and the most essential problem to be solved. Many algorithms for optimal codebook design have been proposed. The most popular one is the LBG algorithm which is basically an iterative process and aims to minimize the overall distortion of the resulting final set of codewords from the set of training vectors. For the details of the LBG algorithm please refer to [2].

Once the codebook is generated, the high dimensional space is divided into N partitions, and each of the vectors in the training data set can be approximated by one of the codewords $c_i = (c_{i1}, c_{i2}, \ldots, c_{ip})$, $i = 1, \ldots, N$ in the codebook $C = (c_1, \ldots, c_N)$. In the context of VQ codebook generation, since the squared Euclidean distance $d(x_i, c_j) = \sum_{l=1}^{p}(x_{il} - c_{jl})^2$ between a vector x_i and a codeword c_j is used for measuring their "distance", hence the same measure will be used for the purpose of intrusion detection.

The intrusion detection scheme involves two steps: analysis of correlations and classification of behavior patterns. Correlation analysis transforms the set of user sessions into groups of points in a high dimensional Euclidean space such that sessions within the same group are closely related. Let the measures on p different events from n user sessions, which are extracted from the audit files that are known to be normal, be represented in the form of matrix M. Let x_{ij} denotes the value of measure $j(1 \leq j \leq p)$ in session $i(1 \leq i \leq n)$ of the particular user being monitored. The matrix M also denotes a collection of n vectors (x_i : $i = 1, \ldots, n$) of dimension p. According to the intrusion detection hypothesis sessions belonging to the same user are closely correlated. An efficient method for recognizing groups of closely correlated session vectors is employed to meet the intrusion detection requirements. Correlation analysis is done by transforming the set of p-dimensional user sessions into a p-dimensional unit sphere where points that are closely correlated are clustered together on the surface of the hypersphere. Under this scheme, the original matrix M is transformed to a new matrix $\tilde{M}$ where each element x_{ik} of row i is replaced by a normalized $\tilde{x}_{ik}$. During the profile generation stage, all these n sessions are normal user sessions of the genuine user. Hence, all session vectors form group(s) of closely correlated points in the high dimensional space. To proceed to the classification stage, we need an efficient technique for determining the minimal distance between new test vectors and the profile.

The VQ technique is adopted as an important mechanism in the classification stage that determines the minimal distance between test sessions from the profile. This can be done by treating these n profile vectors as a set of training vectors, and based on them to generate a codebook $C = (c_1, \ldots, c_N)$ using the LGB codebook generation algorithm in VQ. This codebook C is then stored as the user profile of the particular user. For anomaly detection of a new session of the particular user, suppose the event counts of the user session extracted from the audit log is represented by the vector $y = (y_1, \ldots, y_p)$. As mentioned, the codewords generated from the training data is the optimal representative of all the vectors within each partition. According to the intrusion hypothesis, the criterion to determine whether a single session can be labelled as abnormal depends on how "closely" it is related to the group (or groups) of normal sessions in the training data set. In this respect, the "distance" of a new session from codewords is a good measure to determine the abnormality of the session. Within the context of the VQ model, the shortest squared Euclidean distance d of y and C is computed. To determined whether the session is normal or not, d is compared with a pre-determined threshold τ. If $d \leq \tau$, the session is labelled as normal, otherwise it is labelled as intrusive.

3 Experimental Results

In this experiment, the audit data from a Solaris host machine were analyzed. In this study, for easy bench-marking, our experiments used system audit logs from the well-known MIT DARPA data. The experiments used seven weeks of the 1998 audit data set from the MIT Lincoln Laboratory. Since there are about 243 different types of BSM audit event in the audit data, we consider 243 event types in our experiments. We used those data that are labelled as "normal" as the training data set to generate the codebook. During testing, we removed the label of all audit data and use the proposed technique to generate a label. The labels generated in our test are then compared to the given labels to evaluate the performance of our proposed technique.

In order to cater for possible noises in the profile data, it is not unreasonable to set a threshold τ to such a value that allows a small percentage of the profile data to be ignored. In this experiment, we chose 3 different threshold values which are equivalent to 3%, 2% and 1% allowances for noises (i.e. respectively 97%, 98% and 99% of the training data are included in the normal profile). The results on the test data using these 3 threshold values are listed in Table 1.

According to Table 1, the detection performance of the VQ technique is very promising. When we used 4-week data for creating the profile and selected a τ that covers 99% of the training data, the automated detection mechanism achieved 100% detection rate and with false alarm rate at 0.75%, which is close to the expected level of 1%.

Table 1. Detection rate and false alarm rate for different thresholds and codebooks

		2 weeks	3 weeks	4 weeks
97%	τ	0.00249	0.00413	0.00499
	False alarm rate	24.45%	7.23%	2.25%
	Detection rate	100%	100%	100%
98%	τ	0.00303	0.00487	0.00656
	False alarm rate	22.83%	6.59%	1.50%
	Detection rate	100%	100%	100%
99%	τ	0.00447	0.00694	0.00991
	False alarm rate	17.51%	4.74%	0.75%
	Detection rate	100%	100%	100%

The experimental results showed that VQ is an efficient and effective technique for achieving automated intrusion detection and with improved detection performance.

4 Conclusion

An effective intrusion detection scheme in the grid computing environment based on VQ technique was presented. The main contribution of this paper is that it formulates the intrusion detection problem as a VQ problem and investigates the use of the VQ technique as an effective means for representing user profiles with complex groups of correlated activities such that deviations from the profiles can be detected efficiently and reliably. This paper also presented experimental results which demonstrate that our scheme can perform intrusion detection with low false alarm rate and high detection rate. The experimental results indicate that the VQ-based intrusion technique gives a very promising performance in intrusion detection.

Acknowledgement

This research was funded by the National 863 Plan (Projects Numbers: 2003AA148020) and the National Information Security Administration Center, P. R. China.

References

1. K.Y. Lam, L. Hui, S.L. Chung, "Multivariate Data Analysis Software for Enhancing System Security", *Journal of Systems Software*, 1995, 31: 267-275.
2. Y. Linde, A. Buzo, R.M. Gray, "An Algorithm for Vector Quantizer Design", *IEEE Transactions on Communication*, 1980, 28(1): 84-95.

Integrating Trust in Grid Computing Systems

Woodas W.K. Lai, Kam-Wing Ng, and Michael R. Lyu

Department of Computer Science and Engineering
The Chinese University of Hong Kong
Shatin, N.T., Hong Kong
+852-2609-8440
{wklai,kwng,lyu}@cse.cuhk.edu.hk

Abstract. A Grid computing system is a virtual resource framework. Inside the framework, resources are being shared among autonomous domains which can be geographically distributed. One primary goal of such a virtual Grid environment is to encourage domain-to-domain interactions and to increase the confidence of domains to utilize or share resources without losing control and confidentiality. To achieve this goal, a Grid computing system can be viewed more or less as a human community and thus the "trust" notion needs to be addressed. To integrate trust into a Grid, some specific issues need to be considered. In this paper, we view trust in two aspects, identity trust and behavior trust. Further, we briefly present two important issues which help in managing, evolving and interpreting trust. The two issues are grid context and trust tree structure.

1 Introduction

Trust[1] is a complex concept that has been addressed at different levels by many researchers. We classify trust into two categories: identity trust and behavior trust. Identity trust is concerned with verifying the authenticity of an entity and determining the authorizations that the entity is entitled to and is based on cryptographic techniques such as encryption and digital signatures. Behavior trust deals with a wider notion of an entity's "trustworthiness" and focuses more on the behavior of that entity. For example, a digitally signed certificate does not indicate whether the issuer is an industrial spy and a piece of digitally signed code does not show whether the code will perform some malicious actions or not.

In this paper, we will only briefly present and outline the issues that need to be considered when "trust" is being integrated into the Grid Computing Systems. We assume that each Grid service instance has a globally unique id. As stated in [2], for the OGSA architecture, every Grid service instance is assigned a globally unique name, the Grid service handle (GSH).

2 Trust and Reputation

To integrate "trust" into the Grid Computing Systems, first of all, we need to address what "trust" means.

Currently, there is a lack of consensus in the literature on the definition of trust and on what constitutes trust management. In this paper, we propose to modify the definition of trust defined in [3]:

Trust is the firm belief in the competence of an entity to behave as expected such that this firm belief is a dynamic value associated with the entity and it is also subject to the entity's behavior and applies only within a specific context at a given time.

That is, trust is a dynamic value between 0 and 1 inclusively. A value of 0 means very untrustworthy while a value of 1 means very trustworthy. The trust value (TV) is based on the history and is specific to a certain context. For example, entity x might be permitted to use the service s1 of entity y but is not permitted to use the service s2 of entity y at a particular time and context.

Furthermore, to establish a trust relationship, a person will listen to the opinions from others when he wants to make a decision. In Grid computing, when the entities want to make a trust-based decision, the entities may also rely on others for the information and opinion pertaining to a specific entity. For example, if entity x wants to make a decision of whether to call the service s1 of domain y, which is unknown to x, then x can rely on the reputation of the service s1 of domain y. In this paper, we adopt the definition of Reputation as presented in [3] with modification:

The reputation of an entity is an expectation of its behavior based on its identity and other entities' observations or information about the entity's past behavior within a specific context at a given time.

Please note that our trust and reputation definition are both associated with the identity trust and behavior trust while in [3], only behavior trust is concerned.

3 Context in Grid Computing Systems

In the previous section, we mentioned that trust is defined to be context specific. Thus, what is context with respect to a Grid computing system? A service invocation scenario is illustrated in Figure 1.

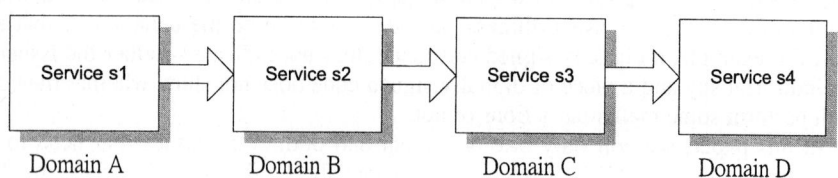

Fig. 1. Service Invocation

In Figure 1, service instance s1 invokes service instance s2, service instance s2 further invokes service instance s3 and then service instance s3 calls service instance s4 finally. In this scenario, we define the Grid context as an ordered four-tuple $(id_{s1}, id_{s2}, id_{s3}, id_{s4})$ where $id_{service}$ is a globally unique service instance id of a particular service.

Therefore, to define context in a Grid computing system, if the service invocation is originated from service $s_1, s_2 \ldots\ldots$ up to s_n, the context will be an ordered n-tuple $(ids_1, ids_2, \ldots\ldots, ids_n)$.

4 Trust Tree

The context of the Grid is service-based. The advantage of the service-based context is that it is highly precise to identity a particular service invocation. However, as there may be many different service instances in a domain, service-based context implies that the number of different contexts can be huge. In a trust system, different contexts should have different trust values. Thus, if we store the trust values in a table and search for the trust value sequentially, it will take quite a long time to do so.

To be efficient, instead of using a simple table to store the direct trust value or reputation from other domains, we propose a trust tree structure. The trust values will be stored in a structure called a trust tree. For each trust value, the associated context tuple will be regarded as a n-dimension record and becomes a node of our trust tree. Other context-based information will become annotations for that node. A trust tree is shown in Figure 2 for illustration.

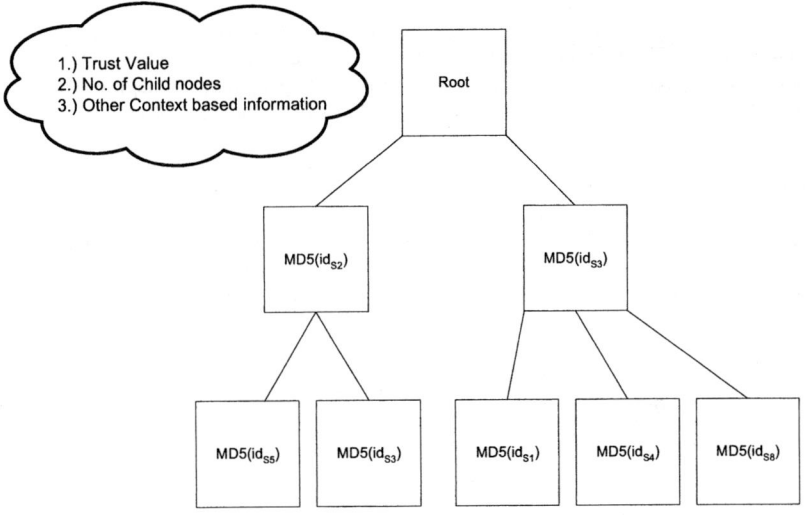

Fig. 2. A sample trust tree

The following statements summarize the structure of a trust tree:
1) Each node of our tree contains the MD5 digest of the corresponding service instance id (128-bit) as the key.
2) Each node may consists of zero to n children nodes.
3) To retrieve a child node efficiently, all the children nodes will be sorted according to the key value and the number of children nodes will be stored in the parent node such that when a key value is given, a binary search could be performed so as to find the matching child node.

Besides, the trust tree provides a *similarity* operation that enables us to search out the trust value of a similar-context node. Consider if we are now making a trust evaluation towards a service request, using the trust tree, we can refer to some similar-context service requests. Undoubtedly, these similar-context service requests do provide a good source for evaluating the trust value of the current service request and leads to a better trust evaluation.

5 Conclusion

In this paper, to integrate "trust" in the Grid Computing Systems, we suggest to address both identity trust and behavior trust. Besides, we give the definition of context in Grid Computing and it makes the meaning of trust to become much more precise and clear. On the other hand, to manage the trust values in an efficient way, we define a trust tree structure. The Trust tree also provides a *similarity* operation that enables us to find out the trust value of other transactions with similar context. It is definitely useful for trust evaluation.

References

1. Alfarez Abdul-Rahman, Stephen Hailes, "Supporting Trust in Virtual Communities," In: Proceedings Hawaii International Conference on System Sciences 33, Maui, Hawaii, 4-7 January 2000.
2. I. Foster et al., "The Physiology of the Grid: An Open Grid Services Architecture for Distributed Systems Integration," Argonne National Laboratory, Argonne, Ill. (2002)
3. Farag Azzedin and Muthucumaru Maheswaran, "Evolving and Managing Trust in Grid Computing Systems," Proceedings of the 2002 IEEE Canadian Conference on Electrical Computer Engineering
4. Geoff Stoker, Brian S. White, Ellen Stackpole, T.J. Highley, and Marty H, Toward Realizable Restricted Delegation in Computational Grids, HPCN Europe 2001, LNCS 2110, pp.32-41, 2001.
5. Ian Foster, Carl Kesselman, and Steve Tuecke, The Anatomy of the Grid, Enabling Scalable Virtual Organizations, International Journal of Supercomputer Applications, 2001.
6. Ian Foster, The Grid: A New Infrastructure for 21st Century Science, Physics Today, February 2002.

Simple Key Agreement and Its Efficiency Analysis for Grid Computing Environments*

Taeshik Shon[1], Jung-Taek Seo[2], and Jongsub Moon[1]

[1] Center for Information Security Technologies, Korea University, Seoul, Korea
{743zh2k,jsmoon}@korea.ac.kr
[2] National Security Research Institute, ETRI, Daejeon, Korea
seojt@etri.re.kr

Abstract. With the increasing use of Grid computing, it is growing to concern a computing power and infrastructure problems of Grid environments. But, it is also important to consider its security problems because of joining many different users. In this paper we first consider satellite environments having a variety of users like Grid environments. We proposed two kinds of key distribution models to achieve a security between satellite terminals. At last, through performance analysis of proposed models, we considered their suitability to satellite environments and Grid computing.

1 Introduction

With the development of Grid technology, users want excellent information communication services provided with high performance, various resources and multimedia. In order to satisfy such requirements, it is pointed out to develop a Grid Infrastructure. In recent times, however, it has begun to change. Many researchers and company get their hands to develop a new Grid technology - Grid Security - to protect their infrastructure already implemented. Among many Grid security we focused on the key agreement to provide their authentication and confidentiality. In this paper, we observe satellite environments for applying to Grid computing, and propose two kinds of key distribution models, and verify them. Key distribution procedure can be classified into the Push and Pull models in accordance with how to get a secret key[1]. We proposed two kinds of key agreement model based on the [1] such as Push typed key distribution model using symmetric or public key encryption algorithm, and Pull typed key distribution model using symmetric or public key encryption algorithm.

2 Simple Key Agreement

Push typed Key distribution model with symmetric key encryption

1. TA sends own ID_A and encrypted message including ID_B , Timestamp and the secret key between A and B to SS

---

* This work was supported (in part) by the Ministry of Information&Communications, Korea, under the Information Technology Research Center (ITRC) Support Program

2. If TA is authenticated, SS sends encrypted message including ID_A, Timestamp and the secret key to TA.
3. TA sends encrypted message including ID_A and the secret key to B.

1. $TA \rightarrow SS : ID_A || E_{AS}[ID_B||K_{AB}||T]$
2. $SS \rightarrow TA : E_{BS}[ID_A||K_{AB}||T]$
3. $TA \rightarrow TB : E_{BS}[ID_A||K_{AB}]$

Push typed key distribution model with public key encryption

1. TA requests communication to SS through satellite. And, TA sends encrypted message with ID of Terminal A, B and the secret key to SS. In this time, the secret key is generated by TA.
2. SS sends encrypted message to TA. This message comprised as follows: Encrypted with public key of TB together with signature to message (m_2) which is comprised with identity of TA, secret key between TA and TB, and timestamp.
3. TA sends this encrypted message to TB.

1. $TA \rightarrow SS : E_{KUss}[m_1||Sign_{KUa}(m_1)]$
2. $SS \rightarrow TA : E_{KUb}[m_2||Sign_{KRss}(m_2)]$
3. $TA \rightarrow TB : E_{KUb}[m_2||Sign_{KRss}(m_2)]$
   $*m_1 = (ID_A||ID_B||K_{AB}), m_2 = (ID_B||K_{AB}||T)$

Pull typed key distribution model using symmetric key encryption

1. TA sends own ID_A and encrypted message including ID_B and timestamp to SS.
2. If TA is authenticated, SS sends encrypted message including ID_B and the secret key to TA. Also, SS sends encrypted message including ID_A , timestamp and the secret key to TB. In this procedure, the secret key for secure channel is generated by SS.

1. $TA \rightarrow SS : E_{AS}[ID_A||ID_B||T]$
2. $SS \rightarrow TA : E_{AS}[ID_B||K_{AB}||T]$
   $SS \rightarrow TB : E_{BS}[ID_A||K_{AB}||T]$

Pull typed key distribution model using public key encryption

1. TA requests communication to SS through satellite. At this time, TA sends encrypted message with ID of Terminal A, B to SS.
2. SS sends encrypted message to TA. This message comprised as follows: Encrypted with public key of A together with signature to message (m_2) which is comprised with identity of A, secret key between TA and TB, and timestamp. Also SS sends encrypted message(m_3) to TB. In this procedure, the secret key for secure channel is generated by SS.

1. $TA \rightarrow SS : E_{KUa}[m_1||Sign_{KUa}(m_1)]$
2. $SS \rightarrow TA : E_{KUa}[m_2||Sign_{KRss}(m_2)]$
   $SS \rightarrow TB : E_{KUb}[m_3||Sign_{KRss}(m_3)]$
   $*m_1 = (ID_A||ID_B), m_2 = (ID_A||K_{AB}||T), m_3 = (ID_B||K_{AB}||T)$

3 Efficiency Analysis

This section analyzes the efficiency of proposed models. To analyze suitability of proposed models, we calculate the sum of delay time related to some parameters such as encryption algorithms, key length. Our model of satellite terminals referred to the security service modeling of normal data communication[2][3][4] and included the characteristic of satellite networks. We assume that packets arrive according to poisson distribution with arrival rate λ and service times have constant values such as μ_1, μ_2 and μ_3. Even if we consider additional information security services to each system such as encryption, decryption, signature and verification, arrival rate is maintained equally, but service rate is added by μ'_1, μ'_2 and μ'_3 respectively. The best efficiency of this systems is $\mu'_b = max(1/\mu'_i)$, i=1,2,3, that is, it is determined by system which has the longest service time. The average delay time of system is same as the sum of the spent time in each system queue. In modeling satellite terminals, because we assume additionally a security service to normal satellite communication, the service time of each system has additional deterministic service time (security service). As according to the addition of security service, the service time of each system also increases deterministically. Thus, among the queuing models, we made modeled satellite communication systems which provide information security service with an M/D/1 queue and derived an equation to calculate the delay time of total systems as follows.

$$\rho_i = \lambda/\mu_i, \ d_i = \mu_i/\mu'_i(\mu'_i = \mu_i/d_i) \tag{1}$$

$$\rho'_i = \lambda/\mu'_i = \lambda * d_i/\mu_i = d_i * \lambda/\mu_i = d_i * \rho_i \tag{2}$$

In equation (3), T is total delay and we can find it from w(M/D/1 queue's average delay time $\mu^{-1} + \rho\mu^{-1}/2(1-\rho)$) plus all systems delay time($\sum_{i=1,i\neq b}^{3} \frac{1}{\mu'_i}$) plus satellite propagation delay(ld*2).

$$T = w + \sum_{i=1,i\neq b}^{3} \frac{1}{\mu'_i} + (ld*2), (\mu'_b = max[1/\mu'_i], i=1,2,3) \tag{3}$$

$$= (1/\mu'_b + \rho'_b\mu'^{-1}_b/2(1-\rho'_b)) + \sum_{i=1,i\neq b}^{3} \frac{1}{\mu'_i} + (ld*2) \tag{4}$$

$$= \rho'_b\mu'^{-1}_b/2(1-\rho'_b) + \sum_{i=1}^{3} \frac{1}{\mu'_i} + (ld*2) \tag{5}$$

$$= \lambda/2\mu'_b(\mu'_b - \lambda) + \sum_{i=1}^{3} \frac{1}{\mu'_i} + (ld*2) \tag{6}$$

In equation (6), we derived the total delay time of systems.

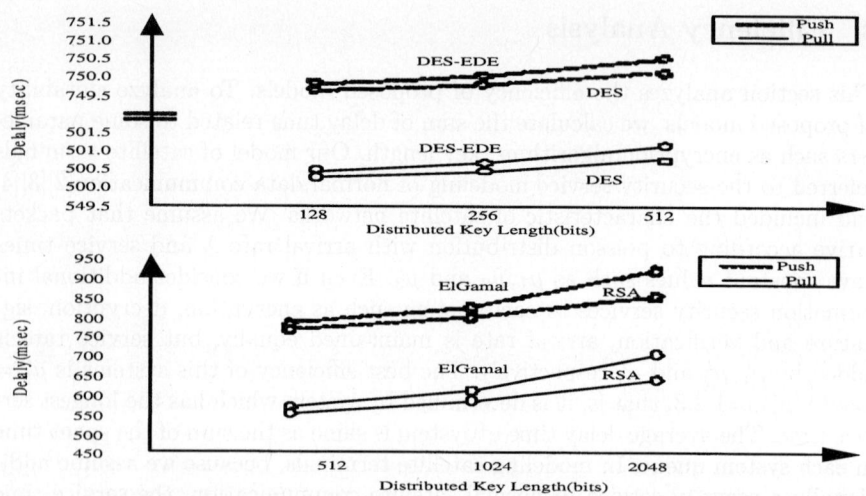

Fig. 1. Delay Comparison according to the key length

4 Conclusion

We first consider satellite environment including many different users because it is similar to Grid computing environment. So, we made a model appropriate to satellite terminals and analyzed its performance. Even though it is not same to Grid environment, we could begin to apply this key agreement ant its result to Grid environment. Through the performance analysis of proposed key distribution models using our simulation equations, we could see that if Pull typed key distribution model using public key algorithm distributes more than the key size of 2048bits, the difference of delay time between the models is to be decreased up to below 87msec, in other words nearly same. Potential future work will adopt additional effectiveness testing with various key distribution models and encryption algorithms to Grid environment.

References

1. ANSI, *X9.17 Financial Institution Key Management Standard*, X9-Secretarait Banker Association, 1985
2. S.W.Kim, *Frequency-Hopped Spread-Spectrum Random Access with Retransmission Cutoff and Code Rate Adjustment*, IEEE, Vol.10, No.2, Feb 1992
3. Kyung Hyune Rhee, *Delay Analysis on Secure Data Communication*, KIISC, Vol.7, No.12, Dec 1997
4. Hyoun K., *Traffic Control and Congestion Control in ATM over Satellite Networks*, IEEK, Vol.4, No.1, Nov 1998
5. Jerry Banks, *Discrete-Event System Simulation*, Prentice-Hall, pp264-265, 1996
6. Bruce Schneier, *Applied Cryptography*, Wiley, pp53 64, 1996
7. Alfred J. Menezes, *Handbook of Cryptography*, pp497 514, CRC Press, 1997

IADIDF: A Framework for Intrusion Detection

Ye Du, Huiqiang Wang, and Yonggang Pang

College of Computer Science and Technology, Harbin Engineering University,
Harbin 150001, China
{duye,hqwang,ygpang}@hrbeu.edu.cn

Abstract. Intrusion detection systems (IDSs) are an important component of defensive measures against system abuse. Firstly, some disadvantages of existing IDSs were analyzed. For solving these problems, an Independent Agents-based Distributed Intrusion Detection Framework – IADIDF was proposed. This paper describes the function of entities, defines the communication and alert mechanism. Each agent operates cooperatively yet independently of the others, providing for efficiency alerts and distribution of resources. The proposed model is an open system, which has good scalability. All the entities of IADIDF were developed in C program under Linux platform. Experiment results indicate that the operating of agents will not impact system performance heavily.

1 Introduction

In recent years, the number of information attacks is increasing and malicious activity becomes more sophisticated. There is clearly a heightened need for information and computer security in all settings. Three main goals of computer security are CIA (Confidentiality, Integrity, Availability). The system designed to detect these types of activities is Intrusion Detection System (IDS).

There are a number of problems with existing IDSs[1]. For solving some existing problems, a distributed IDS framework model based on independent agents – IADIDF was proposed.

2 System Framework

The structure of IADIDF is shown in Figure 1. In IADIDF, every entity of the same host is in the organization of hierarchical structure. Of all these entities, manager is in the supreme level, and detector is in lower level. The cooperating entities among different machines are in the equity position, and there is no control center among these entities. That is, intra-host entities employ hierarchical structure, while inter-host entities are egality.

2.1 Detector

Detector is the basic detection unit in this framework. There may be as many as possible detectors in one host, with their responsibilities for monitoring the operating

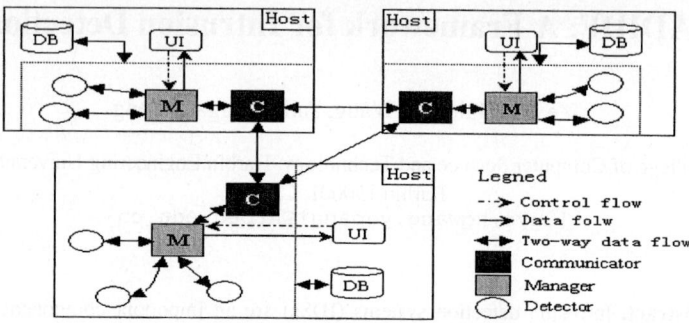

Fig. 1. The architecture of IADIDF

status of host, and reporting abnormal or interesting behaviors to higher-level entity. Detector is a component that operates independently with self-governed data sources and detection modes, and it can be written in any programming language. It is able to fulfill the task of detection alone, or many detectors cooperate one another to take actions. Detector does not produce any alarm. Generally, if the manager receives one or more reports from detectors, it will send an alarm to users.

2.2 Manager

Inside one host, manager is the highest-level entity. It controls the operation of every detector. Furthermore, it plays the important role of transmitting messages between detectors that are in the same host. When a detector getting initialized, it will register itself to its upper entity manager. In this way, manager will hold all IDs of the detectors it controlled. When a detector needs intra-host communication, it merely gives the ID of the aim detector, then manager will redirect the standard input and output of source detector to aim detector. By this mechanism, intra-host communication and cooperation are realized.

2.3 Communicator

If a communicator and a manager are inter-host entities on different hosts, they are in equal position, while not in layer structure on logic. Communicator is responsible for setting up and maintaining communication channels. There is only one communicator on each host, which acts as the bridge of communications between cooperating hosts. Communicator is the intermediary part and offers route service in the course of messages conveying. In the system, two kinds of data package exist: broadcast package and directional package. Communicator must be capable of discerning and receiving these two kinds packages. The inter-host and intra-host communication services of communicators and managers have similar place, but the key difference is that, managers just redirect standard input and output during messages transmitting, while communicators need to choose and update route information, establish and maintain connection channels etc. This is the reason that we select an independent agent for the working of inter-host communication.

2.4 Alert Mechanism

Detectors are responsible for monitoring specific activities and events, while not raising any alarm. This assignment will be done by monitors based on the *State* value reported from detectors. Two main factors are used to describe *State* value: *danger* and *transferability*[3].

The *danger* is defined as the potential damage that can occur if the attack at a particular stage is allowed to continue. *Danger* descriptions (with values in parentheses) are:

(1) The event cannot be deterministically associated with a known attack, but could be an early stage.
(2) The event may lead to a secondary stage of an attack, and the potential damage of the attack is minimal.
(3) The event is an attack with moderate damage.
(4) The event identifies an attack with high damage potential to disrupt service or result in catastrophic losses.

The transferability is defined as the applicability of an attack in other nodes in a network. *Transferability* descriptions (with values in parentheses) are:

(1) Only to some host.
(2) Only to hosts with a common operating environment.
(3) To all hosts across the network.

Then, we get *State* value by the following formula:

$$\text{State} = \text{danger} * \text{transferability} . \tag{1}$$

With the values given above, *State* will range from 1 to 12. Then, the alert level can be evaluated:

√ Normal: *danger*<2;   √ Partial alert: *danger*=2;
√ Alert: *danger*=3;   √ Full alert: *danger*=4;

The higher alert levels should produce alerts. According to the *transferability*, we deploy the scope of alarms (1: local machine; 2: multicasting; 3: broadcasting).

3 Performance Analysis

In this experiment, we use Red Hat Linux 7.2 as the operating system, and hardware devices are CPU: PIII800, Memory: 128M, Harddisk: 20G, and 10M D-link network card. Program language is C, and Libpcap is used to monitor network traffic. In order to receive data continuously, we firstly capture an amount of packets as source data, and then deal with these. Rate of utilization before and after agents running is shown in figure 2. Results indicate that the operating of agents will not impact system performance heavily.

4 Conclusion

We propose architecture for intrusion detection called IADIDF, which is based on independent agents and employs distributed structure. The functionality of entities is

Performance Monitoring-Peer to Peer) and RTPM-CC (Real Time Performance Monitoring-Cluster Computing). RTPM-P2P is for the Peer to Peer Grid(P2P-Grid) testbed. RTPM-CC is for the Cluster Computing Grid(CC-Grid) testbed. It imports most of the performance information from Metacomputing Directory Service (MDS) of Globus in real time mode, processes them internally and graphically displays them.

4 Process Management with Resource Brokering

Users should explicitly specify the computing resources for their parallel processing programs in Globus based computational Grid environment. Thus dynamic load balancing becomes almost impossible. Also the users should check the load information of each computing resources before their job submission on their own risk. This kind of user-dependent static load balancing is un-convenient to users and un-efficient to system. If users can be served by a resource brokering system which automatically checks the load information of all available computing resources in the Globus based computational Grid environment and intelligently selects the best computing resources to the requirements of the users' parallel processing programs, dynamic load balancing of the Grid environment can be done and the users can use the Grid environment with great convenience.

We see many job management systems currently. Some of them are PBS[9], LSF[10] Condor-G[11], Codine, SRB, AppLes, LoadlLeveler, DQS and NQE. However they do not fully support the Globus based testbeds in job management. Thus we developed our own resource brokering system, namely, SGIRBS(Seoul Grid Intelligent Resource Brokering System) and implemented it into the Seoul Grid Portal. SGIRBS supports our Globus based Grid testbed in seamless manner. SGIRBS accommodates any resource management system below it. For example, when we integrated geographically distributed independent cluster systems into the Seoul Grid Testbed using Globus, it allowed each component cluster system to use any of PBS, Condor-G, Nimrod-G and LSF.

SGIRBS collects up-to-date load information of the Grid environment and shows the current available computing resource information to users. Users specify the required computing powers after seeing the complete information of currently available computing resources. According to the user specification, the SGIRBS intelligently selects the best(most idle) computing resources for the user requests, executes the user programs on the selected computing resources and finally returns the execution results to the user.

SGIRBS consists of two main subsystems. They are SGIRBS-CC and SGIRBS-P2P. SGIRBS-CC offers its service to the CC-Grid environment of the Seoul Grid Testbed, which integrated geographically spread independent cluster systems. Each cluster system of our CC-Grid environment can use different job management systems such as PBS, LSF/Clusterware. SGIRBS-CC should care all of these cluster systems. Since it should only consult with Globus for resource information, it uses computing resource information from Metacomputing Directory Service(MDS) of Globus. Users see the general load information of each cluster and specify their required computing powers in our Seoul Grid portal. Then SGIRBS-CC shows the user's selected computing resources. The users finally submit their jobs. After execution, SGIRBS-CC returns the execution results to the users.

SGIRBS-P2P does resource brokering in the P2P-Grid environment of the Seoul Grid Testbed, which aggregated personal computers. MDS of Globus gives very simple internal information about each independent cluster system and SGIRBS-CC restrictively allows users to specify their requirement of computing power. However, SGIRBS-P2P allows users to specify them in more detail as shown in figure 1, since MDS gives more detailed information about peer system.

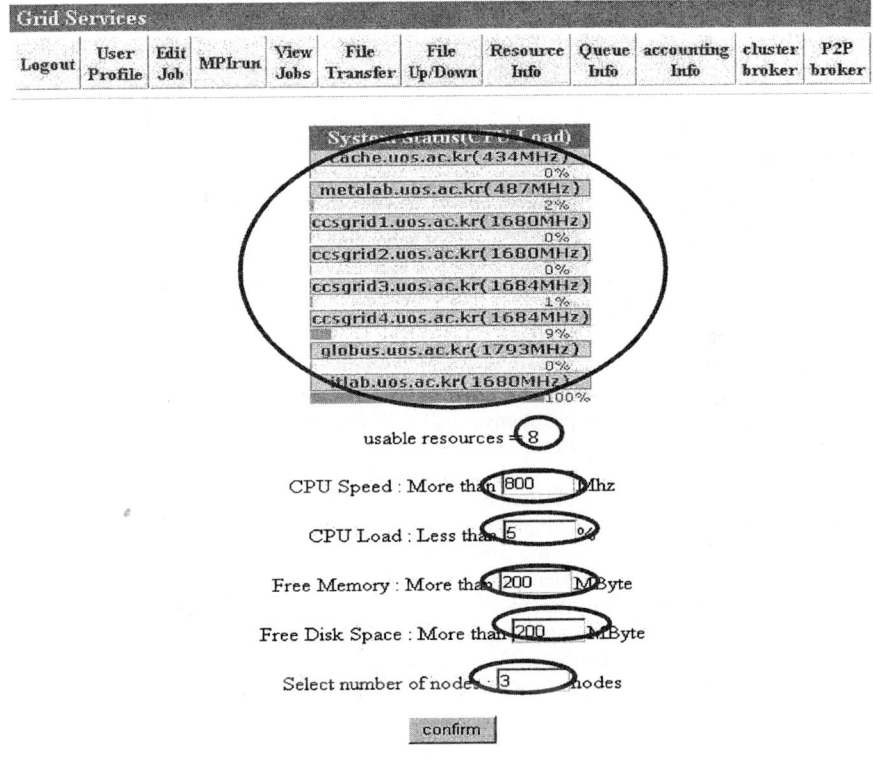

Fig. 1. SGIRBS-P2P allows users to specify their requirement in detail.

5 Account Management

We developed our own user accounting system (SGIUAS: Seoul GRID Institute User Accounting System) using MS-SQL database, matched with the intelligent resource brokering function and implemented it into Seoul Grid portal. Users specify their requests and our user accounting system returns the requested accounting information. Poland VUS(Virtual User Accounting System) [12] and SNUPI [13] by San Diego Supercomputing Center in San Diego, CA, U.S.A. gather user accounting information directly. We collect user accounting information via Globus.

6 Conclusions

In this paper, we present our Grid resource management system which was implemented into the Seoul Grid Portal to support our Seoul Grid Testbed. The Grid resource management system gives three distinctive services to users: the real time performance monitoring service in graphical visualization, the intelligent resource brokering service and the user accounting service. The Seoul Grid portal accommodates file management, user authentication service and on-line manuals of major functions as one stop service as well. The Grid resource management system will evolve to give better services to users as Globus evolves and accommodate more functions such as performance prediction before actual execution in future. This paper has been supported by the 2003 research fund of the University of Seoul. We would like to appreciate its financial support to this paper.

References

1. I. Foster, C. Kesselman and S. Tuecke.: The Anatomy of the Grid: Enabling Scalable Virtual Organizations, International Journal of. Supercomputer Applications, 15(3), (2001)
2. TeraGrid.: http://www.teragrid.org
3. EuroGrid.: http://www.eurogrid.org
4. DAS.: http://www.cs.vu.nl/das
5. NASA IPG.: http://www.ipg.nasa.gov
6. MicroGrid.: http://www-csag.ucsd.edu/projects/grid/microgrid.html
7. I. Foster, C. Kesselman.: Globus: A Metacomputing Infrastructure Toolkit, International Journal of Supercomputer Applications, 11(2), (1997), 115-128
8. DOE Science Grid.: Grid Portal Development Kit, http://doesciencegrid.org/projects/GPDK
9. PBS Homepage.: http://www.pbspro.com
10. Ming Q. Xu.: Effective Metacomputing using LSF MultiCluster, 1st International Symposium on Cluster Computing and the Grid, (2001)
11. J. Frey, T. Tannenbaum, I. Foster, M. Livny, S. Tuecke.: Condor-G: A Computation Management Agent for Multi-Institutional Grids, Cluster Computing, 5(3):237-246, (2002)
12. Miroslaw Kupczyk, et al.: Simplifying Administration and Management Processes in the Polish National Cluster, Distributed Accounting Working Group
13. SNUPI Homepage.: http://snupi.sdsc.edu

PGMS: A P2P-Based Grid Monitoring System

Yuanzhe Yao, Binxing Fang, Hongli Zhang, and Wie Wang

Research Center of Computer Network and Information Security Technology, Harbin Institute
of Technology, Harbin, Heilongjiang, P.R. China, 150001
yyz@pact518.hit.edu.cn

Abstract. A well-designed Grid monitoring system should be able to measure or instrument the performance of Grid resources, the status of Grid applications, and then publish these information for various application aims. P2P-based Grid Monitoring system (PGMS), based on peer-to-peer technology and GMA specification, is designed to monitor large scale and dynamically changed Grid system. In PGMS, the function of directory service is achieved by the P2P-based Grid Distributed Directory Service (PGDDS); in this way, a whole directory service is decentralized into some associated directory services that are peer-to-peer. Universal Resource Description Word (URDW) is a resource description method that adopts the thinking of quantifying and coding, and reduces the storage and communication overhead of resource description information.

1 Introduction

The aim of Grid is to collect usable computation power or resources and then provide ubiquitous and seamless computation power and manifold Grid services to users. It has greater performance than any single computer or site can offer. At same time, Grid system is an extremely complicated distributed computing environment. In Grid, resources are changeful, and the performances of resources are also fluctuating. Therefore, an efficient management of resources and applications running on Grid has become key, difficulty, and also challenge. Moreover, for parallel applications on the Grid, an achievement of high performance is highly dependent upon the effective coordination of their compositive components. Each Grid system has a scheduler system whose aim is to select the most appropriate resources. This choice should be made before the application runs, and based on the actual performance levels at the time the decision is made. In order to implement robust, high available and high performance Grid environment, right and effective monitoring on Grid is a key. A well-designed Grid monitoring system should be able to measure, record, archive, then publish the performance of a Grid resources, status information of Grid applications, these information are used widely, such as fault detection, debugging, performance analysis and tuning, load balancing, task scheduling, security, auditing and accounting, and forecasting service.

2 Grid Monitoring System and General Architecture

A Grid monitoring service should meet several requirements: resources discovery, authorized data access, interoperability, and intrusiveness free [1]. Always, there are many successful monitoring and information systems, such as NWS [2], NetLogger [3] and MDS[4]. However, there are some problems along with these systems, for example, the management of NWS is centralized, so the scale is limited, the scalability is not satisfying. Moreover, any single system can't provide sufficient functions for Grid monitoring. Because of the absence of interoperability, it is a difficulty to make these systems cooperating. It is essential to construct a satisfying Grid monitoring system.

A complete end-to-end monitoring system consists of several levels: instrumentation, presentation, analysis, they separately have one component or several components [1]. Most of these components have been implemented in one form or another in some monitoring systems used for traditional computing environments. However, there seems to be more problems existing in Grid environment. P2P-based monitoring system framework is potential and able to solve many key problems.

3 P2P-Based Grid Monitoring System (PGMS): Components and Functions

Peer-to-peer (P2P) systems can make a convenience of large quantities of decentralized computing and storage resources over the Internet. These resources separately belong to different participants. Some known P2P systems have succeeded in file sharing, distributed computing, and instant messaging. As the result of these successful applications, the researches in P2P also thrive amazingly. Moreover, P2P thinking and technology is introduced into many other fields, as results, many new ideas come into being, which advance the development of the related fields.

P2P-based Grid Monitoring System (PGMS), based on peer-to-peer technology and GMA[5] specification, is designed to monitor large scale and dynamically changed Grid system. In PGMS, each Grid node is named as a *peer*, which runs a copy of PGMS program. In accord with general architecture [1], Components in PGMS are included:

Instrumentation sensors include host sensor, network sensor and software sensor. *Communicating and publication* (CP) module manages local or global information publication directory, communicates with other modules and peers. *Data preprocessing* (DP) module deals with the data acquired from sensors in order to sample useful data and transform them to standard and understandable information forms. *Data cache and archive* (DCA) module saves the historical data and information that are vital to performance forecasting, analysis, tuning, and debugging. *Performance forecasting* (PF) module predicts available resources of a Grid node or peer at next time frame. *Instrumentation management and control* (IMC) module, the core module of PGMS, manages the other modules in same peer.

4 P2P-Based Grid Distributed Directory Service

GMA is an architecture presented and supported by GGF [6]. It aims to constitute standards for grid monitoring and make existing systems interoperate. GMA first adopts a producer-consumer model in Grid monitoring. All monitoring information are events which are basic unit of monitoring data and base on timestamp for storing and transferring. GMA architecture consists of event producer, event consumer and directory service, which is used to publish what event data are available.

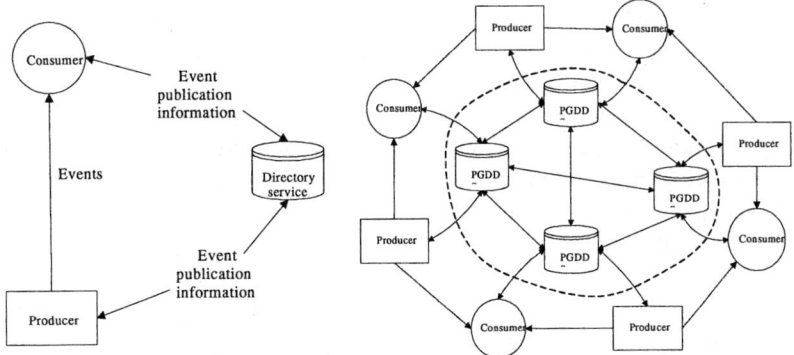

Fig. 1. Implementing GMA directory service using multiple PGDDS cooperating each other. A directory service is decentralized into some associated directory services that are peer-to-peer

In PGMS, the function of directory service is achieved by the P2P-based Grid Distributed Directory Service (PGDDS), in which a whole directory service is decentralized into some associated directory services that are peer-to-peer. The implementation of GMA directory service using multiple PGDDS is showed in Figure 1. Grid Monitoring Architecture components are showed in the left graph. In the right graph, the domain surrounded by broken line may consist of many PGDDS that are not only independent each other but also related nearly. In this way, they cooperate to implement a single directory image. As broken line arrow showed, a producer can provide event data to other consumers as well as original ones.

The copy of PGMS program, cell unit of PGMS, deploys in all peers. Further, some related peers constitute a *peer group* by certain self-organizing or autonomous means. In every peer group, there is a *head peer* and a *backup peer* that is the backup of head peer. Head peer is elected to play a core and key role in PGMS.

A PGDDS consists of two components: a global information publication directory (GIPD), which inhabits in head peer, provides the information of other head peers in other peer groups, and a local information publication directory (LIPD), which provides the information of peer in a same group, is maintained by each peer including head and backup peer. All global directories together compose a virtual, single-image directory. The producers and consumers need not know the details of the directory, the only thing they need do is to interact with local or global directories.

5 Universal Resource Description Word

Universal Resource Description Word (URDW) is a resource description method. It is distinct from general methods that use some resource description language and because of including unprocessed raw data, and the overhead of transform and transfer is too big to satisfy some demand, for example, XML.

In the PGMS, according to the predefined standard levels, the preprocessor of DP module quantify the data acquired by sensors into the performance information. By quantifying, the information of one resource is encoded to some resource description bits (RDB). On all resource information of peer being encoded, they are assembled as a whole word, which is named as Universal Resource Description Word. A URDW includes all key resource information of a peer, so while the other peers obtain a URDW, only knowing the predefined quantifying standard, they can decode it and know the running states of the peer.

Obviously, the use of URDW can sharply reduce the overhead of information communication and the space used for storing and archiving historical data. The disadvantage is that the procedure of encoding and decoding can introduce into some overhead. Because of these, the mechanism more suit for some Grid computing environment on the low bandwidth and high latency network.

6 Future Work

Currently, some application sensors have been implemented and are used to debug parallel program on Grid and the PGMS is implementing. At a specified future date, basing on monitoring, we intend to extend PGMS to PGMMS (P2P-based Grid Monitoring and Management System), which not only can monitor Grid resources but also manage and control their usage so that them can be exploited sufficiently and efficiently.

References

1. Ian Forster, Carl Kesselman, The Grid 2: Blueprint for a New Computing Infrastructure, Morgan Kaufmann; 2nd edition (November 18, 2003): 319-351
2. Rich. Wolski. Dynamically forecasting network performance using the network weather service. Cluster Computing, 1998, (1): 119–132.
3. Brian Tierney, Dan Gunter, NetLogger: A Toolkit for Distributed System Performance Tuning and Debugging, Integrated Network Management 2003: 97-100
4. http://www.globus.org/mds
5. Brian Tierney, Ruth Aydt, Dan Gunter et al. A Grid Monitoring Architecture, GGF Performance Working Group,
http://www-didc.lbl.gov/GGF-PERF/GMA-WG/papers/GWD-GP-16-2.pdf
6. Global Grid Forum: http://www.gridforum.org/

Implementing the Resource Monitoring Services in Grid Environment Using Sensors[*]

Xiaolin Gui, Yinan Li, and Wenqiang Gong

Department of Computer Science and Technology, Xi'an Jiaotong University, Xi'an, China
xlgui@mail.xjtu.edu.cn

Abstract. How to organize and manage the distributed heterogeneous resources in Grid is a challenge. Grid resources may partly come from some dynamic data sources or temporary nodes, so we introduce sensors to monitor the resources changes. Based on sensors, we proposal a novel grid resource monitoring framework (GRMF) for resource share services. In this paper, the architecture of GRMF is presented, and the implementation detail of its groupware also is described.

1 Introduction

Grid [1] need incorporate distributed heterogeneous resources administrated by different organizations into a virtual whole, and provides uniform information services for various clients. How to organize and manage such Grid resources is a challenge. Because these resources may partly come from some dynamic data sources or temporary nodes, to share them is difficult. According to the present Internet-based resource-sharing architecture, however, resource is autonomous and the utility ratio of the resources in common use is at quite a low level. Due to the problems, it is very difficult for users to efficiently share and utilize various resources even in a corporate organization. So in order to improve the utility ratio of resources and facilitate the sharing of resources in a Grid environment, there must be a powerful resource organization and management mechanism. We have developed a sensor-based architecture [2] called Grid Resource Monitoring Framework (GRMF) for resource share services. At present, popular Grid Systems, such as Globus[3] and Legion[4], also support resources share services, but their methods are different from ours.

The Globus project developed the resource management architecture that defines an extensible Resource Specification Language, which is used to communicate requests for resources between components. The lowest level of the resource management architecture is named the LRM. The implementation of this entity is called GRAM [6]. The Legion developed the Resource Management Infrastructure (RMI). It is an object-oriented resource management architecture, which views resource as object. RMI focuses on the scheduling of resources. The resource management model of RMI supports the scheduling philosophy above by allowing user-defined Schedulers to interact with the infrastructure [5].

The GRMF, which was developed using the Service-Oriented Architecture [7], is responsible for organizing and managing the share resources. In GRMF, sensors are

---

[*] This work was sponsored by the National Science Foundation of China (60273085), the state High-tech Research and Development project (2001AA111081) and the ChinaGrid project of China (CG2003-GA002).

used to discover and collect dynamical resources. In this paper, the architecture of GRMF is presented, and the implementation detail of its groupware also is described.

2 The Architecture of GRMF

Fig.1 shows the hierarchical architecture of GRMF. The GRMF is a distributed Grid environment that mainly focuses on providing resources sharing services. The whole system is included the four layers.

(1) RMAI layer: RMAI is the top tier of GRMF that provides the resource management application interfaces such as *Client Tools, API* etc. We have developed a Client Tool called *Wader Grid Shell* that submits all requests from users and returns all responses from Grid environments.

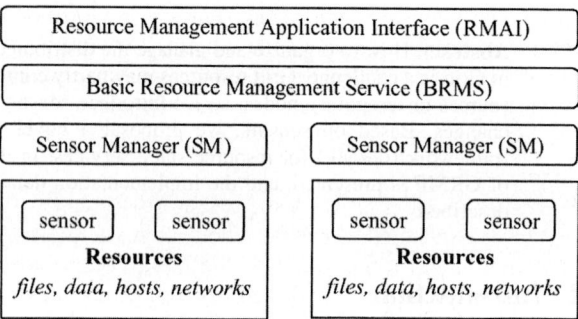

Fig. 1. The architecture of GRMF

(2) BRMS layer: BRMS is a part of middleware. BRMS of the GRMF provides some basic resource management services, these services include resource monitoring, discovering, registering, allocating, searching and naming etc. The resource discovering is implemented using the sensors that distribute in Grid environments. The information about the resource discovered is composed of two major categories: one is the static information, such as the version of OS, type of CPU, number of processors, architecture, memory size, disk space, and the model of network card; the other is the dynamic information, such as file information, load information, CPU utilization ratio, available memory space or free disk space, network bandwidth and latency etc.

(3) SM layer: The sensor managers are also a part of middleware. Each SM manages all sensors in its administration domain. The sensor manager is a daemon process. In general, the sensor manager and the sensors reside in the same host, but we can make the sensor manager and the sensors reside in different hosts by using inter-process communication and the *Uniform Monitoring-Data Message (UMDM)*. The sensor manager is responsible for the management of sensors, such as creating a new sensor process, stopping a running sensor, gathering monitoring-data from all the sensors, and keeping the *Meta Data Catalogue (MDC)* up to date according to the monitoring-data.

(4) Sensors Layer: Grid is a dynamically changed environment, the dynamic changes mainly include the incorporation of new resource into the Grid, the removing of resource from the Grid, and the modification to attributes of resource. In order to efficiently manage all the resources in a Grid, we use sensors to watch over the dynamic changes of resources. Each host is distributed not less than one sensor. Each sensor is stayed in host as a daemon. These daemons collect information such as hosts, files and networks, registers to *MDC*. The layer includes also all physical resources such hosts, files, databases, and networks etc.

3 Implementation of Sensors

The sensor is a daemon process or thread, which resides in the background of each Grid host. The sensors can monitor the frequently dynamic changes of resources such as the incorporation of new resource into the Grid, the removing of resource from the Grid, and the modification to the values of some attributes. Different type of resource has different type of sensor, such as file-system sensor, process sensor, printer sensor, bandwidth sensor, host sensor and so on. Our prototype implementation includes the following sensors.

File-system sensor: File-system sensor is the daemon that is responsible for monitoring the dynamic changes of file system. The changes can be monitored by the sensor include file creation, file deletion, file renaming, directory creation, directory deletion, directory renaming, and attribute modification. We have developed several different file-system sensors for different operating system and architecture, such as Windows and Linux sensors.

Host sensor: Host sensor is the sensor that is responsible for monitoring some important parameters that reflect the architecture and performance of the host and the availability of the host to serve as a Grid node. The parameters mainly include host architecture, host name, network address, operating system type, type and speed of CPU, memory size, total disk space, free disk space, computational performance, and host status.

The most important aspect of implementing sensor is the resource-monitoring algorithm. We have developed a scanning-comparison algorithm. This algorithm periodically (with fixed scanning frequency) scans certain type of resource and records the current scanning result, and then compares the scanning result of last time with the current scanning result to determine whether there are changes of this type of resource occurred. The scanning procedure is to scan all the resources of certain type in order to record changed attributes used to construct *UMDM* message. This scanning-comparison monitoring algorithm is simple and easy implemented. The most important strongpoint of the algorithm is that the scanning frequency can be adjusted in order to be adapted to different types of resources. However, this algorithm is not very efficient mainly because the comparison procedure will cost quite much time especially when the number of resources and the number of attributes to be recorded is very great. In our future work, we must find ways to optimize this algorithm so that it is more efficient.

4 Sensor Manager

The sensor manager assigns an identifier to each newly created sensor to distinct among all the sensors. Each manager is composed of three threads. (1) the management thread is responsible for managing the sensors; (2) the sensor monitoring thread is responsible for monitoring the sensors and gathering monitoring-data from all the sensors; (3) the meta-data updating thread is responsible for creating the resource description and updating all the resource descriptions. The monitoring-data is passed from the sensor monitoring thread to the meta-data updating thread, the meta-data updating thread uses such monitoring-data to construct the resource description and update all the resource descriptions in order to reflect the dynamic changes of resources. The sensor monitoring thread is always waiting around for *UMDM* messages

2 The InstantGrid Framework

InstantGrid works in the client/server mode, where all EE's are stored in and managed by an InstantGrid server from which the compute nodes obtain their EE's on-demand to form the grid platform (Figure 1). The framework allows for replication of the InstantGrid service for better performance and reliability. The framework consists of an *EE management model* and an *EE dissemination service*, as shown in Figure 2.

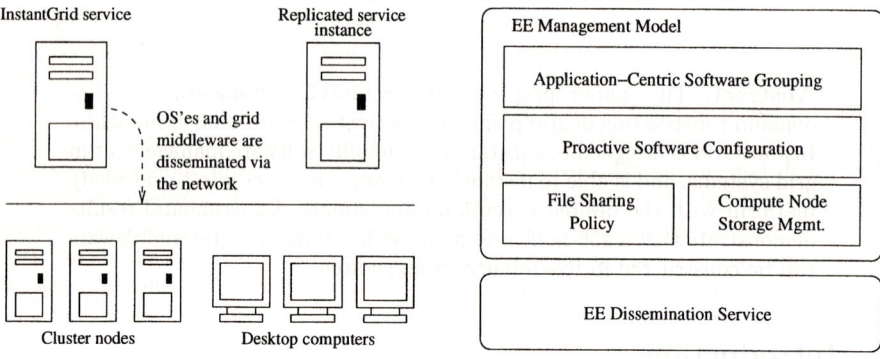

Fig. 1. InstantGrid Servers and Clients   **Fig. 2.** The InstantGrid Framework

Application-centric software grouping. In InstantGrid, an EE is a collection of software components, including an OS, the supporting libraries and applications, grid middleware, cluster middleware, user applications, and the user data. Essentially, we can see what is in an EE by taking a snapshot of all software components in a running system. Each EE is associated with an *EE specification*, which contains a list of software required by a specific application. This application-centric requirement specification facilitates EE management as different users and their applications may require different platforms over time. Once the administrators have defined the list of software in each EE, the system can conveniently switch from one EE to another according to the user requirements, without having to deal with the individual software components.

Proactive software configuration. Traditionally, the OS and system software are installed and configured incrementally. In InstantGrid, by contrast, software belonging to the same EE have to be configured in the central server before being disseminated to the compute nodes. In other words, InstantGrid creates and maintains ready-to-run versions of various EE's. This arrangement saves installation and configuration time during the dissemination process. Nevertheless, some software must perform *local* configuration. For instance, some grid middleware (e.g., Globus Toolkit) require host credentials such as certificates, which have to be handled locally at the respective nodes. In any case, Instant-

Grid takes a "greedy" approach and carries out as many configuration tasks in the central server as possible, leaving minimum work to the compute nodes.

Discriminative file sharing mechanisms. Disseminating the entire EE from the InstantGrid server to the compute nodes is challenging as the size of a typical EE could be in the order of gigabytes, and full replication is therefore impractical. However, to access all files through NFS all the time is also inefficient since many files in an EE are frequently updated—retrieving them through the NFS would result in poor runtime performance as well as heavy loading at the file server. InstantGrid adopts a hybrid approach to the problem: it only replicates files that are likely to be modified (e.g., the files in /etc or /dev), and leaves the others in the file server for sharing via a network file system.

Compute node storage management. InstantGrid allows the files being replicated to a compute node to be stored in the hard disk ("Full-copy to HD") or entirely in the physical memory ("Full-copy to RAM"). The in-memory execution mode enables any computer to participate in a grid without affecting the data stored in its local storage. If the files are stored in the hard disk, InstantGrid supports an additional option of I/O caching ("Copy-if-needed"): before a file is transferred from the InstantGrid server to a compute node, InstantGrid would first check if there is a local version of that file; if so, it would verify whether it is up-to-date, and file transfer would only take place if the file is missing or outdated.

EE dissemination service. This service is offered through a DHCP server, a TFTP server, and an NFS server. When a client machine boots up, it obtains its IP address and the kernel (a Linux kernel in our reference implementation) from the DHCP and TFTP servers, respectively. When the booting process finishes, InstantGrid constructs the pre-defined EE by replicating writable files to local storage and mounting the read-only directories through the NFS.

3 Experiments

We evaluated the performance of InstantGrid using the HKU CS Gideon cluster which consists of 300 Pentium IV machines (we used 256 in the experiments). The EE's are disseminated through a hierarchical Ethernet network in which 13 24-port Fast Ethernet switches are interconnected by a Gigabit Ethernet switch. The InstantGrid server, which connects to the Gigabit Ethernet, is a Pentium IV machine with 512MB RAM and an IDE hard disk. In the experiments, InstantGrid disseminated the Fedora Core 1 OS to the cluster nodes by sharing the /bin, /lib, /sbin, and /usr directories through the NFS, and replicating the remaining directories to the nodes' local storage.

Figure 3(a) presents the time to construct a grid point which is a cluster consisting of a frontend node (Fedora OS, Globus Toolkit 2.4, Ganglia, and PBS) and a number of compute nodes (Fedora OS, Ganglia, PBS, and MPICH-G2). The results show that a 256-node grid point can be constructed from scratch in three ("Copy-if-needed") to five ("Full-copy to HD") minutes. The good performance mainly attributes to the proactive software configuration in InstantGrid,

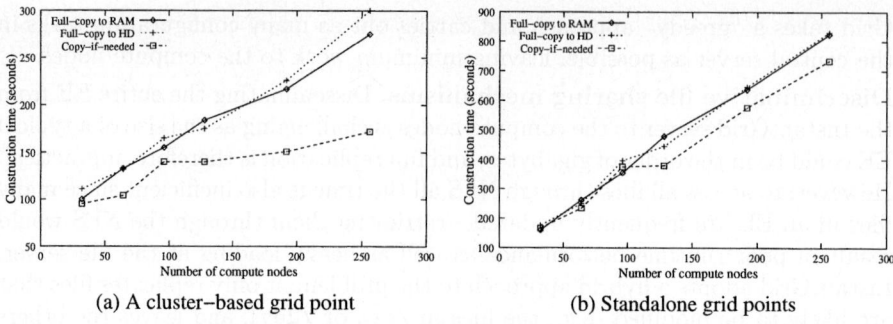

(a) A cluster-based grid point (b) Standalone grid points

Fig. 3. Construction Time of Grid Points

which shortens substantially the dissemination time. Figure 3(b) shows the construction time of standalone service grid points (Fedora OS, Globus Toolkit 3.2, and Ganglia). The reason for the longer construction time compared to that of the previous test is that each compute node is treated as a standalone grid point, i.e., each node requires a unique host certificate; and that the certificates have to be generated *sequentially* in a central certificate authority server. This result suggests that some software configuration processes are indeed time consuming, which should either be avoided or redesigned to allow for more efficient deployment.

4 Conclusion

We have proposed the InstantGrid framework for on-demand construction of grid points. The experimental results show that InstantGrid is able to efficiently construct Linux-based grid points with commodity grid middleware. Future work will be conducted along two dimensions. First, we will devise standard protocols for communicating EE specifications between the InstantGrid servers and compute nodes. Secondly, we will look into possible performance optimizations for InstantGrid in WAN, which could enable remote construction of grid points through broadband networks.

References

1. A. Grimshaw and A. Ferrari and A. Knabe and M. Humphrey. Legion: An Operating System for Wide-Area Computing. *IEEE Computing*, 32(5):29–37, May 1999.
2. I. Foster, C. Kesselman, J.M. Nick, and S. Tueckel. The Physiology of the Grid - An Open Grid Services Architecture for Distributed Systems Integration. In *White Paper, The Globus Project. http://www.globus.org/*.
3. J. Frey, T. Tannenbaum, I. Foster, M. Livny, and S. Tuecke. Condor-G: A Computation Management Agent for Multi-Institutional Grids. *Cluster Computing*, 5:237–246, 2002.
4. S. Zhou. LSF: Load Sharing in Large-scale Heterogeneous Distributed Systems. In *Workshop on Cluster Computing*, 1992.

Scalable Data Management Modeling and Framework for Grid Computing

Jong Sik Lee

School of Computer Science and Engineering
Inha University
#253, YongHyun-Dong, Nam-Ku,
Incheon 402-751, South Korea
jslee@inha.ac.kr

Abstract. There is an increasing demand to execute[1] complex large-scale grid computing systems and to share dispersed data assets and computing resources collaboratively. To perform grid computing within reasonable communication and computation resources, this paper introduces a scalable data management approach and its framework. This paper presents a design and development of the framework which performs the scalable data management approach. For development of this framework, this paper uses a hierarchical and modular object-oriented technology supported by a system modeling theory and specification. The framework provides a superior solution to reduce data traffic loads and local computation demands with a practical application.

1 Introduction

Grid computing system is characterized by numerous interactive data exchanges among multiple distributed entities over a network. Thus, in order to provide a reliable answer in reasonable time with limited communication and computation resources, methodologies for reducing the interactive data exchanges are required in grid computing systems. This paper presents a scalable data management scheme to promote effective reduction of an amount of data communication and develops a framework of Scalable Data Management (SDM). Currently, a major data management concept is interest management [1], [2]. The Data Distribution Management (DDM) [3] service of High Level Architecture (HLA) [4] is an extension of interest management. The scalable data management represents an interest management with a scalability concept and applies the representation to communication data management for data traffic filtering.

The SDM is developed in order to perform a complex and large-scale grid computing system within reasonable communication and computation resources through the scalable data management scheme. This paper presents a design and development of a framework which is able to realize the scalable data management with a practical application, e.g. projectile/missile [5]. In the SDM framework, a system modeling is provided by the DEVS (Discrete Event System Specification) [6] formalism which supports a hierarchical and modular object-oriented technology. Grid computing

---

[1] This work was supported by INHA UNIVERSITY Research Grant. (INHA-31414)

execution of the SDM framework is performed on the HLA middleware. Scalability of the SDM framework is investigated with a projectile/missile application and usefulness of the SDM is demonstrated. The SDM framework provides a superior solution to reduce network data transmission loads and local computations. This paper is organized as follows: Section 2 describes the scalable data management. Section 3 discusses issues of design and development of SDM framework. Section 4 introduces a projectile/missile application and evaluates system performance. The conclusion is in Section 5.

2 Scalable Data Management

The scalable data management is created by representing an interest management concept with a scalability concept and by applying the representation to data management among communication-related components. In conventional interest management scheme, there is only one critical degree of interest to specify the communication relationship between two components and, at any time, this scheme holds or does not hold an interest in all-or-none fashion. This paper classifies a conventional interest management scheme as a semi-scalable data management in a viewpoint of scalability concept. In the scalable data management scheme, there can be more than one critical degree of interest between two components thus allowing communication in a more tunable fashion. A degree of data management is assigned to each interest degree which is created by each critical interest. The degree of data management determines a rule for transferring or discarding messages from sender to receiver. This paper classifies degree of scalable data management: non-scalable, semi-scalable, and scalable. The non-scalable data management does not apply to the scalable data management and assigns a fixed degree of data management for all variation of interest. In the semi-scalable data management, there are, in effect, two degrees of data management: zero and infinity, corresponding to a classification of interest or non-interest. The scalable data management allows multiple degrees of data management, thereby allowing degree of data transmission to be controlled as a smoother function of interest.

3 Design and Development of SDM

The development of scalable data management framework to reduce data traffic among communication-related components has drawn the attention of many researchers as one means of achieving greater scalability. With increasing demand for grid computing systems, methodologies, related to scalable data management with reasonable communication and computation resources, are being noticed. This paper designs and develops a scalable data management framework and performs effective data traffic reduction among communication-related components. This scalable data management framework overcomes the disadvantages of DDM of HLA and performs a reliable grid computing with reasonable communication and computation resources.

The SDM framework is implemented as the upper layer of the DEVS/HLA-Interface layer which supports a portability of models across platforms at a high level of abstraction. Any communication-related model, based on an object-oriented design, can be developed and reused on the SDM layer and it can easily be ported across grid platforms. The SDM layer provides communication specification (e.g. setup of attribute and interaction communications) and component specifications. Data communication among communication-related components distributed in multiple federates are supported by the SDM. The SDM layer supports the DEVS modeling specification for development of communication-related components which contain communication-related attributes. Thus, communication-related components would be specified as DEVS models. The SDM layer takes DEVS coupling information from DEVS models, automatically defines HLA interaction communications using this coupling information, and performs HLA communications. Therefore, the SDM provides a friendly user interface using DEVS modeling and allows that a developer only defines component models with DEVS modeling specification.

4 Experiment and Performance Evaluation

A projectile/missile application, working in a real-world, is used to evaluate performance of the SDM framework. The operation of this application is based on a geocentric-equatorial coordinate system [5]. A projectile is a ballistic flight and accounts for gravitational effects, drag, and motion of rotation of the earth relative to it. A missile is assigned a projectile, and it follows its projectile until it hits its projectile. In modeling the projectile/missile, there are two main models: projectile and missile. The projectile model is the model of a sphere of uniform density following a ballistic trajectory. This projectile model begins at an initial position with an initial velocity, moves, and stops when it meets a missile. When the missile model is close to the assigned projectile model within a certain distance, it stops and this application considers the missile hits its projectile.

Table 1 compares data transmission bits of three systems in varying numbers of projectile/missile pairs; non-SDM, Semi-SDM, and SDM. As the number of projectile/missile pairs increases, data transmission bits of the non-SDM increase significantly and the SDM tremendously reduces data transmission bits. Compared to two SDMs, the SDM shows the more reduction of data transmission bits than that of the Semi-SDM. In addition, Table 1 compares variation of system execution time of the three systems in varying the number of projectile/missile pairs. In the SDM, as the number of projectile/missile pairs increases, system execution time increases slowly and proportionally to data transmission bits. However, system execution time in the non-SDM increases in an exponential manner due to saturation of network transmission. Results of data transmission bits and system execution time in the three systems show that the SDM is very effective in saving inter-federate data and actual system execution time.

5 Conclusion

Data management provides a solution of execution in large-scale grid computing systems and requires understanding of semantic and dynamic characteristics of an

Table 1. Transmission Data Bits (10^6 Bits) and System Execution Time (10^2 Sec) (Non-SDM vs. Semi-SDM vs. SDM)

# of Pairs	Non-SDM		Semi-SDM		SDM	
10	13.926	12.673	9.342	5.003	1.121	1.102
20	14.571	13.932	10.239	6.812	2.453	1.732
40	16.972	16.026	12.635	7.928	3.345	2.327
80	112.037	32.947	14.938	11.394	5.023	3.023

application to tune their parameters for effective filtering with acceptable error. The SDM allows for stratification of degree of interest for a communication-specified attribute, and thereby controls update frequency of the attribute based on the time-varying interest between distributed communicating components.

The SDM framework gives a great promise for system performance improvement and compensates for disadvantages of DDM of HLA mentioned earlier. The SDM framework supports modeling level features inherited from the DEVS, which has a generic dynamic system formalism with a well-defined concept of modularity and coupling of components. The analytical and experimental results from projectile/missile case study show the SDM framework on high level modeling paradigm is very effective in saving inter-federate data transmission and actual system execution time.

References

1. Katherine L. Morse: Interest management in large scale distributed simulations. Tech. Rep. Department of ICS, University of California Irvine (1996) 96-127
2. J. Saville: Interest Management: Dynamic group multicasting using mobile java policies. Proceedings of 1997 Fall Simulation Interoperability Workshop. 97F-SIW-020, (1997)
3. Nico Kuijpers, et al.: Applying Data Distribution Management and Ownership Management Services of the HLA Interface Specification. in SIW, Orlando FL (1999)
4. Defense, D.o.: Draft Standard For Modeling and Simulation (M&S) High Level Architecture (HLA) - Federate Interface Specification, Draft 1. (1998)
5. Erwin Kreyszig: Advanced Engineering Mathematics. 7th Edition. John Wiley& Sons Inc, New York (1993)
6. Zeigler, B.P., T.G. Kim, and H. Praehofer: Theory of Modeling and Simulation. 2nd edn. Academic Press, New York. (2000)

A Resource Selection Scheme for Grid Computing System *META*

KyungWoo Kang[1] and Gyun Woo[2]

[1] Department of Computer and Communication Engineering, Cheonan University,
115, Anseo-dong, Cheonan 330-704, Choongnam, Republic of Korea
kwkang@cheonan.ac.kr
[2] Department of Computer Science and Engineering, Pusan National University,
Busan 609-735, Republic of Korea
woogyun@pusan.ac.kr

Abstract. Our grid system *META* allows CFD users to access computing resources across the network. There are many research issues involved in the grid computing, including fault-tolerance, selection of the computing resource, design of user-interface. In this paper, we propose a resource selection scheme based on the kernel loop of the source code. Also, we developed *META* for executing the parallel SPMD application written in MPI. Also, *META* chooses the resources with the kernel loop model and provides the user with a real-time visualization.

Keywords: CFD, grid computing, resource selection

1 Introduction

The goal of *META* is to help CFD (Computational Fluid Dynamics) users use the supercomputers easily. There may be many research issues involved in the grid system according to the application. The direction of our research is as follows: characterization of the structure of CFD solver, the design of convenient GUI and the automatic resource selection. The selection should be based on user's requirement: shortest response time or least usage cost[1, 2]. The structure of CFD solver is a repetition of same flow work. CFD is a method for finding an approximate solution of the equations like the Navier-Stokes equation using computer. In this paper, we characterize the structure of CFD solvers using Kernel Loop. The Kernel Loop is the loop that contains the repeated work. The elapsed CPU-time of the repeated work is uniform at time. Our selector's decision depends on the expected behavior on current available resources. Therefore, the selector provides realistic estimates of the performance. In this paper, we propose the use of Kernel Loop Model as a general modeling technique for CFD solvers. The Kernel Loop Model is a scheme that entrusts one of the repeated works to all supercomputers and votes on the better machines[2].

This paper is organized as follows: Section 2 provides the properties of CFD analysis program. *META* should choose the supercomputers that might generate the result of CFD solver for shortest CPU time. Section 2 also presents our voting approach to select the faster supercomputers. Section 3 presents the results of this approach for CFD solvers. We conclude in Section 4.

2 The Kernel Loop Model

CFD analysis programs contain a kernel loop whose execution time is the majority of total execution time[2]. In this section, we propose a scheme for selecting some faster supercomputers by using pre-computation of some iterations of kernel loop.

Definition 1 The Kernel loop is the loop that exists in the CFD analysis program, whose CPU-time is the majority of execution time.
Properties of the kernel loop are as following:

1. the number of iteration is between hundreds and thousands.
2. the elapsed-time of one iteration is about some seconds.
3. the elapsed-time of one iteration is uniform regardless of the iteration index.

These properties say that if a supercomputer is faster in the execution of some iteration, the supercomputer is faster in the total execution of the loop.

$META$ transforms a source program, CFD analysis program, into another program that executes one time the iteration of the kernel loop of the program. The transformed program is compiled and executed on every supercomputer. The elapsed-time of each execution is used to select the faster supercomputers.

The target program executes the kernel loop some times and then sends the elapsed-time to $META$ for the vote.

Algorithm 1 Selection of the faster supercomputers by using kernel loop

- Input:
    - CFD analysis program
    - Data files for the execution of the source program
    - Compile command and execution method on each supercomputer
- Output: Execution of the source program on the faster supercomputers
- Steps:
    1. At first, our system checks whether the total number of available nodes can cover the user's job
    2. $META$ searches the source file that contains the kernel loop
    3. the source file is transformed into AST (Abstract Syntax Tree) in order to search begin-statement and end-statement of the kernel loop
    4. the kernel loop is deleted and $META$ inserts a stub code into the source file. The stub measures the elapsed-time that it takes to execute one iteration of the kernel loop.
    5. all source files and data files are transferred to all supercomputers.
    6. every supercomputer compiles the source.
    7. $META$ initiates the executables of all supercomputers.
    8. each execution sends it's elapsed-time to $META$ after executing one iteration of the kernel loop.
    9. $META$ decides the faster supercomputers based on the elapsed-times.
    10. The portions of our job are distributed to the available nodes of the selected supercomputers in order.

3 Implementation and Experiments

Integration of *META* is conducted on top of a massage passing library, PVM (Parallel Virtual Machine) and GMT(Globus Metacomputing Toolkits)[3-5]. Considering the potential advantages of PVM made the selection of PVM. PVM provides the capabilities to link heterogeneous resources and the convenient process control mechanisms such as spawning processes, which are not available in MPI. The tool, PVM-make, is used for automatic file copy, the compilation and the execution in the remote supercomputers. As a grid testbed, we use six parallel systems at the supercomputing centers of Korea, Compaq HPC320, Compaq HPC160, Compaq GS320, Linux-Cluster (KISTI), IBM SP2 (TIT: Tongmyong Univ. of IT) and IBM SP2 (CNU: Chonbuk National Univ.).

The experiment is conducted according to the number of processors required. The elapsed time is measured when the iteration reaches 100. We are not fully able to use the processors of the supercomputers because of the restriction of the management policy. According to the experimental results, the speed-up is proportional to the increase of the number of processors.

4 Conclusion

The goal of *META* is to help CFD users to use the supercomputers easily. Therefore, the direction of our research is as follows: characterization of the

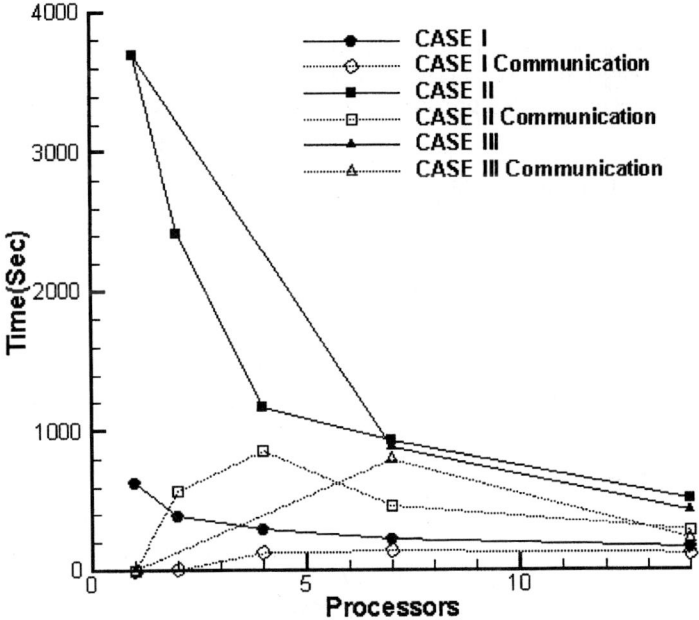

Fig. 1. Performance Analysis

structure of CFD programs, the design of user friendly interface, and the automatic selection of the best supercomputer. The selection should be based on user's request: shortest response time or least cost of usage. We propose the use of Kernel Loop Model as a general modeling technique for CFD programs. The Kernel Loop Model uses the property that the structure of a CFD program is a repetition of same flow work. The Kernel Loop Model is a scheme that entrusts one of the repeated works to all supercomputers and votes on the best one. Our experiment shows two significant results: one is that the supercomputer which shows the fastest response time is selected from the dedicated supercomputers and the other is that the supercomputer which can respond at the least usage cost is selected. These results are based on the properties of kernel loop. $META$ makes many supercomputers' their capabilities appear on the user's PC screen as one system. The user does not have to log into each supercomputer and can make the environment file using a convenient user interface. $META$ automatically selects a supercomputer on which the CFD solver will execute and then makes the solver execute on the supercomputer. $META$ visualizes the output data the solver code generates. The visualization helps the user to confirm the correctness of his or her code.

References

1. V. Sunderam, "Pvm: A framework for parallel distributed computing," *Concurrency: Practice and Experience*, vol. 2, December 1990.
2. K. A. Hoffmann, *Computational Fluid Dynamics for Engineers*. Morgan Kaufmann Publishers, Inc., 1993.
3. C. K. I. Foster, "Globus: A metacomputing infrastructure toolkit," *Intl. J. Supercomputer Applications*, vol. 11, no. 2, 1997.
4. "Creating new information providers," *MDS 2.1 GRIS Specification Document*, May 2002.
5. I. Foster and C. Kesselman, *The Grid: Blueprint for a new Computing Infrastructure*. Morgan Kaufmann Publishers, Inc., 1998.

An Extended OGSA Based Service Data Aggregator by Using Notification Mechanism

YunHee Kang

Dept. of Computer and Communication Engineering, Cheonan University,
115, Anseo-dong,
Cheonan 330-704, Choongnam, Republic of Korea
yhkang@cheonan.ac.kr

Abstract. This paper presents an extended service data aggregator service based on notification mechanism in a Grid environment. To solve scalability problem in its infrastructure, the extended aggregator aperiodically aggregates the Service Data Element(SDE) based on notification scheme about the kinds of data which are gathered. The aggregator parses messages and extracts information about the status of service as well as computing resources. In order to provide the persistent grid information service, we also apply *Xindice* DBMS to maintain SDEs on multiple collections for storing the collection of the resource information as well as its services.

1 Introduction

Grid computing is a kind of distributed system, which is focused on a coordinated use of an assembly of distributed computers that are linked by WAN(Wide Area Network) [1, 2]. But in the other respect the discovery of information about resources including computing resources and services components remain challenging issues.

The Open Grid Services Architecture (OGSA) has been proposed as an enabling infrastructure for systems and applications that require the integration and management of services within distributed, heterogeneous and dynamic "virtual organizations(VO)". OGSA is a service architecture embraces the concepts of Web Services technologies to provide Grid services which is based on the Open Grid Services Infrastructure (OGSI) [3].

We propose an extended service for aggregating SDEs, as part of Grid information system, in Grid environment. To provide highly scalable Grid information system, this system aperiodically aggregates the SDE based on notification scheme about the kinds of data which are gathered. The aggregator also provides persistent Grid information service based on *Xindice* to take over failure problems. We extend the single collection xindice storage scheme into multiple-collection scheme in terms of the context of SDE. Here, we mainly outline our works on aggregating resource information on Grid environment.

This paper is organized as follows: Section 2 describes related works including the characteristics of Grid services as well as OGSA/OGSI. Section 3 shows operational schemes we use in this paper and the result of aggregation. Section 4 concludes this paper.

2 Related Works

The Grid service is a standard basic building block of OGSA structure, built on top of the Web Services standards [4]. Each Grid service is also a Web service and is described by Web Services Description Language(WSDL) [5]. Some extensions to WSDL are proposed to allow Grid-inspired properties to be described, and these may be adopted for wider use in forthcoming standards, called Grid Web Services Description Language(GWSDL) [6]. GWSDL(and WSDL 1.2) had two major improvements: *PortType* inheritance and service data. It allows us to specify the properties of a Grid Service's service data, such as the minimum and maximum amount of SDEs of a certain type that a Grid Service can have.

A Grid service presents some of its properties via SDE. These SDEs may be static or dynamic. Those that are static are invariant for the lifetime of the Grid service they describe, and so may also be available via an encoding in an extension of WSDL. Those that are dynamic present aspects of a Grid service's state. The SDE may be used for introspection and for monitoring to support functions such as performance and progress analysis, fault identification and accounting.

3 The Extended Aggregator Service

Grid environments depend critically on access to information about the underlying networked computing/data storage entity. Information gathering involves searching distributed possibly heterogeneous data sources to answer a query. No one source contains all of the information necessary to answer, so the information must be gathered and integrated in order to respond to a query as if all of these data come from one source.

The OGSI core in GT3 provides a generic interface for mapping Service Data queries and subscriptions for Service Data notification to service implementation mechanisms. The OGSA subscription mechanism is an enhancement over the polling query model of MDS, comparable to an LDAP extension called persistent query.

In OGSA, Grid Service status information is dynamically maintained on SDEs. In order to monitor and control the grid services, service data aggregation is used for gathering information related with the status of services as well as resources in VO. The source of service data, service provider, subscribes to notification by the method *addServiceDataSubscription*, which asynchronously delivers the value of updated SDE by the method *deliverNotification*. To aggregate SDE from Grid services, OGSA services communicate with other using *notificationSources* and *notificationSinks*.

In Figure 1, four services, A, B, C and D constitute the flow of SDEs. Each of services has zero or more objects that represent SDEs. There are five objects O_1,

$O_2 \ldots O_5$. There are notification links connected with two objects. For instance, a single link $O_1 \rightarrow O_2$, which connects the notification source of service A to the notification sink of service B. To deliver the notification, the sink needs to add a subscription to source. Most of these SDEs' values are transferred peer to peer via notifications. The architecture is used to allow communication between services as depicted in Figure 1. O_1 represents SDE_1 in service A. O_2 and O_3 play a role of notification sink of SDE_1, where O_2 and O_3 represent SDE_2 and SDE_3. The notification source of SDE_2 and SDE_3 is O_1. On the other hand O_4 in D is delivered by the value of O_1 via O_2, and O_5 is also delivered by the value of O_1 via O_3.

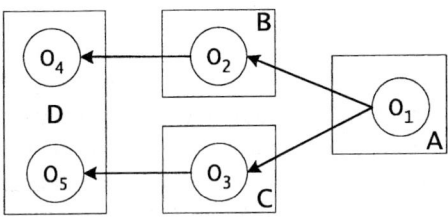

Fig. 1. The architecture for notification based Service Data Aggregation

Notification framework allows clients to register interest in being notified of particular messages and supports asynchronous, one way delivery of such notifications. We extend notification having autonomous and asynchronous characteristics to recover aggregator with service data elements which are kept with in persistent storage. Aggregation scheme should also be extended to the Internet so that infrastructure service can find each other and discover information about up-to-dated and additional services in a VO.

In this aggregator, data is received, stored, and delivered in XML format in any of several Service Data containers, including an XML database named *Xindice* [7]. IndexService in GT3 only provides a single *Xindice* collection with service data container which is used for storing aggregated service data set. But according to the growth of VO in terms of computing resources and Grid Services, single collection has a problem of maintainability and scalability. Basically IndexService based on GT3 does not support multiple-collections.

We extend single *xindiceServiceDataSet* to the multiple-collections based on the context of aggregated service data. To generate dynamically a new collection, the extended service data aggregator parses the root element of the service data when it is delivered from the source of service data, then the service data aggregator determines how to store this service data based on the type of service data element. We show the experimental results including *Host* information which consists of CPU, memory, network bandwidth, etc. The GUI, as part of front-end system, for SDE aggregation displays the result which is delivered from RIPS(see Figure 2).

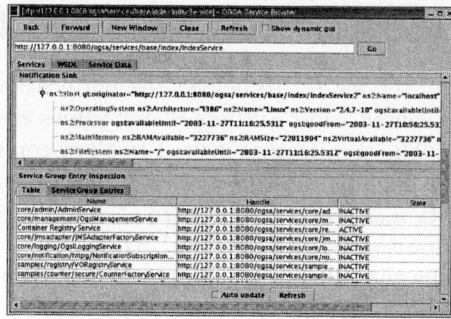

Fig. 2. GUI, a front-end system for the aggregation

4 Conclusion

In this paper, we presented the extended Grid aggregation service based on OGSA. Our system supply the persistent information of the Grid services on the XML database *Xindice*. Main contents of this study consist of the service data aggregator and the manager of XML data. In order to develop the service data aggregator, we extended ServiceDataAggregator class. The aggregator has the notification sink to which the grid service notifies the service data. The received data is classified according to VOs' name and services' kind. And then, the data is stored to the collection of the VO and the service. Our mechanism could support multiple VOs by managing the multiple collections. Our system could help OGSA based resource manager that schedules the resources efficiently and shares their resources in several VOs.

References

1. I. Foster and C. Kesselman. Globus: A metacomputing infrastructure toolkit. *The International Journal of Supercomputer Applications and High Performance Computing*, 11(2):115–128, Summer 1997.
2. I. Foster and C. Kesselman. *The Grid: Blueprint for a new Computing Infrastructure*. Morgan Kaufmann Publishers, Inc., 1998.
3. Ian Foster, Carl Kesselman, Jeffrey M. Nick, and Steven Tuecke. The Physiology of the Grid: An Open Grid Services Architecture for Distributed Systems Integration. Open Grid Service Infrastructure WG, Global Grid Forum, June 2002.
4. S. Tuecke, K. Czajkowski, I. Foster, J. Frey, S. Graham, C. Kesselman, and P. Vanderbilt. Grid service specification. Technical report, Global Grid Forum, October 2002.
5. Erik Christensen, Francisco Curbera, Greg Meredith, and Sanjiva Weerawarana. *Web Services Description Language (WSDL) 1.1*. W3C, 1.1 edition, March 2001. URL: http://www.w3c.org/TR/wsdl.
6. Thomas Sandholm, Rob Seed, and Jarek Gawor. Globus toolkit 3 core - a grid service container framework. Technical report, Globus, January 2003.
7. The Apache Software Foundation. *Xindice*, 2002. URL:http://xml.apache.org/xindice.

Research on Equipment Resource Scheduling in Grids*

Yuexuan Wang, Lianchen Liu, Cheng Wu, and Wancheng Ni

National CIMS Engineering Research Center, Department of Automation,
Tsinghua University, Beijing 100084, China
{wangyx,llc}@cims.tsinghua.edu.cn, wuc@tsinghua.edu.cn,
nwc01@mails.tsinghua.edu.cn

Abstract. In order to resolve equipment resources scheduling problem in grid we develop an equipment resource scheduling system with the equipment grid portal and scheduler including scheduling algorithm, information service and its architecture. The proposed technique provides a method to equipment resources sharing and management on network and will be widely used.

1 Introduction

The distributed equipments sharing via network can improve accessibility and utilization remarkably. Grid technology can establish a sound fundament for building equipments sharing system [1]. Equipment resources including various kinds of machine tools are quite distinct from computing resources or data resources. This particularity increases the complexity of resource management and scheduling in equipment grid (EG), which has been proposed in our previous research [2][3]. So, an equipment grid resource scheduler and portal are proposed in this article to construct the resource management and scheduling system in EG which perform scheduling roles as equipments discovery, selection, and distributing.

2 Equipment Resource Scheduling Portal

To access equipment resources in grid, it is important to utilize *resource management* (RM) interfaces. We have designed and prototyped RM interfaces combined with China Education and Research Network (CERNET). First, we set up a simulated model of the shared equipment as its virtual image to provide an appropriate grid-level active storage resource. Secondly we provide a way that realizes remote access of the equipment resource, management and effective scheduling through the simulated model. Equipment resource scheduler selects and allocates equipment resource. The object of equipment scheduling is to optimize the equipment resource in grids and provide the best quality of service.

In the equipment resource scheduling system, we virtualize or abstract the equipment resource to be used in the task and set up virtual image of the equipment to realize the transparence of the user's interfaces. Under most situations, users do not care where to carry out the task, and how to carry out. What they care about is whether the task can be carried out correctly, how long it will take, how to get the task results, and

---

* This paper is supported by "211 project" "15" construct project: National Universities Equipment and Resource Sharing System and China Postdoctoral Science Foundation (No. 2003034155).

factors like that. And the choice, distribution, scheduling and error processing of the equipment resources that the task uses should be transparent to users. They do not have to handle these with their own efforts. Through the equipment grid portal, users can reserve tasks and control remote equipment easily.

3 Equipment Grid Resource Scheduler

3.1 Equipment Resource Scheduling Algorithm

In this section, a dynamic scheduling algorithm for equipment resource (DSER) is illustrated. First of all, a task sequence called a ring is defined. Then, using a ring of tasks, DSER is described. The tasks in a ring have a total order such that no task has the same order as any other task. A ring has a head which points to a task in the ring. The head in a ring is initialized to point to the task with the lowest order in the ring. The task pointed to by the head is called the current task. The next task in a ring is defined as follows. If the current task is the task with the highest order in the ring, then the next task in the ring is the task with the lowest order in the ring. Otherwise, the next task in a ring is the task with the minimum order of the tasks with higher order than the current task. A head can be moved so that the head points to the next task. Hence, using a head, the tasks in a ring can be scanned in the round-robin fashion. Arbitrary task in a ring can be removed. If the current task is removed, then a head is moved so that the next task is pointed to.

3.2 Equipment Grid Information Service

In this research, we constructed the information service in EG with the help of index service provide by the Globus Toolkit [4], as shown in Fig.1. The functions of its two important components are equipment model and index service.

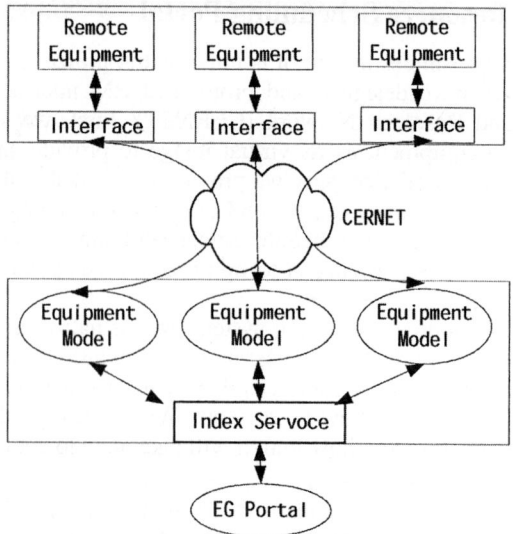

Fig. 1. Equipment Grid Information Service

Equipment resource differs greatly to each other in terms of its nature (physical characteristics) and the demands placed on it (time, quality, cost or service). So, we design equipment simulation models of inhomogeneous resources, which describe the attributes, demands and interfaces of these kinds of resources. Users can access remote equipments by operating equipment simulation model in EG system.

Index service providing collective-level indexing and searching functions. Index service is used to obtain information from multiple resources, and acts as an organization-wide information server for a multi-site collaboration.

3.3 Equipment Resource Scheduling Architecture

The equipment resource scheduling task is analyzing job request, selecting appropriate equipment, managing job, making decision, managing remote equipment and starting up the job based on principles such as cost minimization, profit maximization, distance minimization, response-time minimization, the best of Quality of Service and availability maximization. The architecture of our equipment resource scheduler is shown in Fig. 2. Grid information service functionality is provided by the Monitoring and Discovery Service (MDS) component [5][6] of the Globus Toolkit [4]. MDS provides a uniform framework for discovering and accessing system configuration and status information such as the server configuration and system load. Grid Index Information Service (GIIS) and Grid Resource Information Service (GRIS) [6] components of MDS provide resource availability and configuration information.

The user submits a job to GRAM (Grid Resource Allocation) that is applied to a piece of remote equipment. GRAM gives a notice to GRIS to select required equipment after receiving job request. GRIS gets the required equipment in grid and submits the information of this equipment to GIIS•then GIIS informs the remote equipment. The equipment owner negotiates with GRIS several times then makes sure the equipment will execute this job. GRIS returns the result of equipment scheduling to GRAM. GRAM informs the user and then starts up the system monitor to control the job. As soon as the remote job is finished or failed, scheduling is closed.

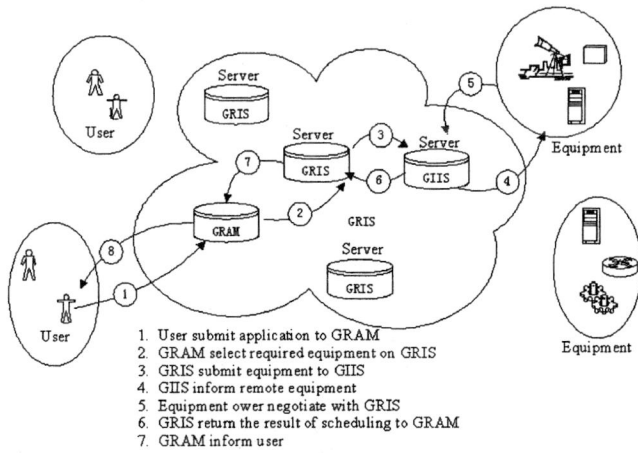

Fig. 2. The Architecture of Equipment Scheduler

4 Conclusions

In order to solve equipment resource management and scheduling problem to handle in equipment grid (EG), we develop an equipment resource management and scheduling system with an algorithm and realization scheme. This scheduling technique in equipment sharing grid is applied in the "211 Project" related National Universities Equipment and Resource Sharing System and indicates that the technique provides a feasible and high-efficiency solution for the realization of equipment resource sharing.

References

1. Foster I, Kesselman C.: The Grid: Blueprint for a future Computing Infrastructure. USA:Morgan Kaufmann (1999)
2. Yuexuan Wang, Lianchen Liu, Xixiang Hu, Cheng Wu: The Study on Simulation Grid Technology for Equipment Resource Sharing System. In Proceedings of the 5th World Congress on Intelligent Control and Automation, (2004) 3235-3239
3. Yuexuan Wang, Cheng Wu, Xixiang Hu, Lianchen Liu: The Study of Equipment Grid Based on Simulation Modeling. Computer Integrated Manufacturing Systems. In press
4. Foster, I. , Kesselman, C: Globus: A Toolkit-Based Grid Architecture. Morgan Kaufmann (1999) 259-278
5. Czajkowski, K., Fitzgerald, S., Foster, I. Kesselman, C.: Grid Information Services for Distributed Resource Sharing. In 10th IEEE International Symposium on High Performance Distributed Computing, IEEE Press (2001) 181-184
6. Fitzgerald, S., Foster, I., Kesselman, C., Laszewski, G.v., Smith, W. and Tuecke, S: A Directory Service for Configuring High-performance Distributed Computations. In Proc. 6th IEEE Symp. on High Performance Distributed Computing (1997) 365-375

A Resource Broker for Computing Nodes Selection in Grid Computing Environments*

Chao-Tung Yang[1], Chuan-Lin Lai[1], Po-Chi Shih[1], and Kuan-Ching Li[2]

[1] High-Performance Computing Laboratory
Department of Computer Science and Information Engineering
Tunghai University
Taichung 407 Taiwan ROC
ctyang@mail.thu.edu.tw

[2] Parallel and Distributed Processing Center
Department of Computer Science and Information Management
Providence University
Shalu, Taichung 433 Taiwan ROC
kuancli@pu.edu.tw

Abstract. As Grid Computing is becoming a reality, there is a need for managing and monitoring the available resources worldwide, as well as the need for conveying these resources to the everyday user. This paper describes a resource broker with its main function as to match the available resources to the user's needs. The use of the resource broker provides a uniform interface to access any of the available and appropriate resources using user's credentials. The resource broker runs on top of the Globus Toolkit. Therefore, it provides security and current information about the available resources and serves as a link to the diverse systems available in the Grid.

Keywords: Resource Broker, Grid Computing, Globus Toolkit, MDS

1 Introduction

Grid computing offers a model for solving massive computational problems using large numbers of computers arranged as clusters embedded in a distributed telecommunications infrastructure. Grid computing has the design goal of solving large problems for any single supercomputer, whilst retaining the flexibility to work on multiple smaller problems. Grid computing involves sharing heterogeneous resources (based on different platforms, hardware/software, computer architecture, computer languages), located in different places belonging to different administrative domains over a network using open standards [1, 2, 3, 4, 5, 6, 7].

As Grid Computing is becoming a reality, there is a need for managing and monitoring the available resources worldwide, as well as the need for conveying these resources to the everyday user. This paper describes a resource broker with its main function being to match the available resources to the user's requests. The use of the resource broker provides a uniform interface to access any of the available and appropriate resources using user's credentials. The resource broker runs on top of the Globus Toolkit. Therefore, it provides security and current information about the

---

* This work is supported by National Center for High-Performance Computing (NCHC), Taiwan under Grant No. NCHC-KING_010200.

available resources and serves as a link to the diverse systems available in the grid. We implement a grid resource broker that considers the network bandwidth and latency for loosely/tightly coupled applications, cluster MDS issues, and application chrematistics. The experiments are conducted on our grid testbed. From the experimental results, we could obtain better performance by applying our resource broker for computing nodes selection.

2 Background Review

The Globus Toolkit is an implementation of these standards, and has become the *de facto* standard for grid middleware [1, 2]. As a middleware component, it provides a standard platform for services to be based upon, it is the only component needed in Grid Computing. Globus has protocols to handle the following three services:

- Resource management: Grid Resource Management Protocol (GRAM).
- Information Services: Monitoring and Discovery Service (MDS).
- Data Movement and management: Global Access to Secondary Storage (GASS) and GridFTP.

The Monitoring and Discovery Service (MDS) is the information service component of the Globus Toolkit. It provides grid information, such as resources that are available and the state of the computational grid. This information may include properties of the machines, computers, and networks in your grid, such as the number of processors available, CPU load, network interfaces, file system information, bandwidth, storage devices, and memory. MDS uses the Lightweight Directory Access Protocol (LDAP), to provide middleware information in a common interface. MDS includes two components: the Grid Resource Information Service (GRIS) and the Grid Index Information Service (GIIS). With the MDS, you can publish information about almost anything in the computational grid [1, 2].

The Grid Resource Information Service (GRIS) provides a uniform mean of querying resources on a computational grid for their current configuration, capabilities, and status. The GRIS is a distributed information service that can answer queries about a particular resource by directing the query to an information provider deployed as part of the Globus services on a grid resource. Examples of information provided by this service include host identity (operating systems and versions) as well as more dynamic information (CPU and memory availability). The Grid Index Information Service (GIIS) combines arbitrary GRIS services to provide exploring and searching capabilities by grid applications. Within a coherent image of the computational grid, GIIS provides the means to identify resources of particular interest, such as the availability of computational resources, storage, data, and networks.

The Java CoG Kit [4] provides access to Grid services through the Java framework. Components providing client and limited server side capabilities are provided. The Java CoG Kit provides a framework on utilizing the many Globus services as part of the Globus metacomputing toolkit. Many of the classes oar provided as pure Java implementation this writing client side applet without installing the Globus toolkit is possible. Nevertheless, some of the components are provided as prototypes in JNI wrappers. If time permits we may work on a pure Java implementation of them.

3 Broker Implementation and Results

Our resource broker is built on top of Globus Toolkit. It makes use of Globus services, such as resource allocation, information, and GridFTP service. Besides our network monitor, measure tool and cluster information provider, our resource broker also includes NWS (Network Weather Service) for forecasting a network bandwidth.

Figures 1 and 2 show the architecture and mechanism of the proposed resource broker respectively. Our resource broker, as shown in Figures 3 and 4, is Java-based implementation by using Java CoG, so it can be executed on different platforms.

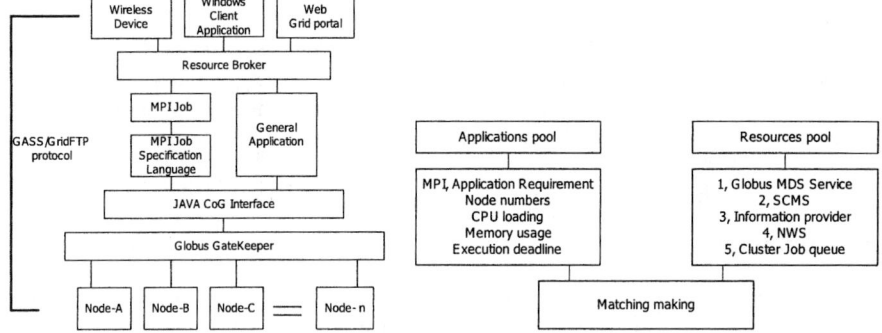

Fig. 1. System architecture of resource broker

Fig. 2. Processing Resource Broker matching

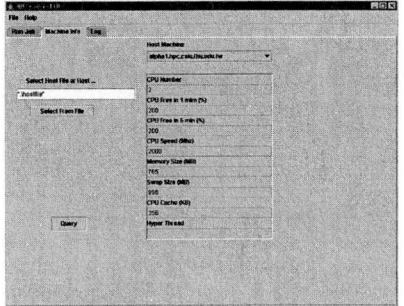

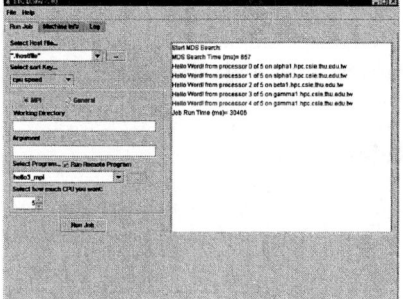

Fig. 3. The information snapshot

Fig. 4. The output snapshot of our resource broker

We performed the experiment using our Grid testbed - THUGrid. The experiments were conducted on 10 (ten) Grid sites, that consists of three PC clusters and seven PCs. Also, Sun Grid Engine is used as local scheduler for the PC-based cluster. We execute three parallel applications - Matrix Multiplication, Prime problem and PI problem.

In Figure 5, Matrix Multiplication is a memory bounded application. Broker strategy changed the memory bound - to select the fastest processors and large memories. The result shows that our broker can achieve better performance under this strategy. To compare with the random selection, if user finds a computing node with small memory size or poor CPU speed. So, it will result in poor computing performance.

Figures 6 and 7 showed the results of Prime and PI problems respectively. These two applications are CPU intensive, and when user submit those types of application, our broker will select computing nodes with the fastest processor speed and maximum CPU available (free idle) to run those jobs. The results also show better performance than random selection.

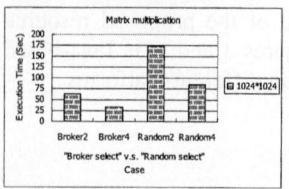 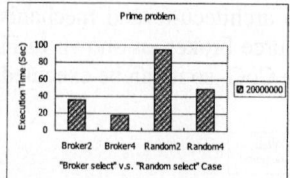 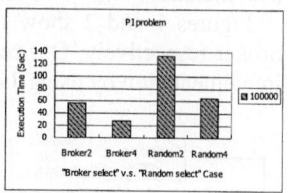

Fig. 5. The performance of matrix multiplication by broker and random select of nodes.

Fig. 6. The performance of Prime problem by broker and random selection of nodes.

Fig. 7. The performance of PI problem by broker and random selection of nodes.

4 Conclusion

As Grid Computing is becoming a reality, there is a need for managing and monitoring the available resources worldwide, as well as the need for conveying these resources to the everyday user. This paper describes a resource broker where its main objective is to match the available resources to a user's needs. The use of the resource broker provides a uniform interface to access any of the available and appropriate computing resources by using user's credentials. The resource broker runs on top of the Globus Toolkit. Thus, it provides security and current information about the available resources and serves as a link to the diverse systems available in the Grid.

References

1. Global Grid Forum, http://www.ggf.org/
2. The Globus Project, http://www.globus.org/
3. B. Allcock, J. Bester, J. Bresnahan, A. L. Chervenak, I. Foster, C. Kesselman, S. Meder, V. Nefedova, D. Quesnal, S. Tuecke, "Data Management and Transfer in High Performance Computational Grid Environments," *Parallel Computing*, Vol. 28 (5), May 2002, pp. 749-771.
4. G. von Laszewski, I. Foster, J. Gawor, P. Lane, "A Java Commodity Grid Toolkit," *Concurrency: Practice and Experience*, 13 (8-9): 645-662, 2001.
5. K. Czajkowski, I. Foster, and C. Kesselman, "Resource Co-Allocation in Computational Grids," *Proceedings of the 8th IEEE International Symposium on High Performance Distributed Computing (HPDC-8)*, pp. 219-228, 1999
6. K. Czajkowski, S. Fitzgerald, I. Foster, and C. Kesselman, "Grid Information Services for Distributed Resource Sharing," *Proceedings of the 10th IEEE International Symposium on High-Performance Distributed Computing (HPDC-10)*, pp. 181-194, 2001
7. Sang-Min Park, Jai-Hoon Kim, "Chameleon: A Resource Scheduler in A Data Grid Environment", *3rd CCGrid International Symposium on Cluster Computing and the Grid*, 2003.

A Model of Problem Solving Environment for Integrated Bioinformatics Solution on Grid by Using Condor

Choong-Hyun Sun[1], Byoung-Jin Kim[1], Gwan-Su Yi[1], and Hyoungwoo Park[2]

[1] School of Engineering, Information and Communications University, 103-6
Munji-Dong, Yusung-Gu, Daejon 305-714, Korea
{chsun,bjkim,gsyi}@icu.ac.kr
http://rosetteer.icu.ac.kr
[2] Grid Technology Research Department, Korea Institute of Science and Technology Information, Yusung, P.O. Box 122, Daejon 305-600, Korea
hwpark@kisti.re.kr

Abstract. To solve the real-world bioinformatics problems on grid, the integration of various analysis tools is necessary in addition to the implementation of basic tools. Workflow based problem solving environment on grid can be the efficient solution for this type of software development. Here we propose a model of simple problem solving environment that enables component based workflow design of integrated bioinformatics applications on Grid environment by using Condor functionalities.

1 Introduction

Bioinformatics field meets inevitable need of high-throughput computing resources as the size of biological data to be managed and analyzed is increasing with the progress of high-throughput biotechnology. Grid computation matches well with this need but it has been applied only to several compute-intensive bioinformatics tools. In addition, there are few examples of integrated bioinformatics solution using grid environment. Many biological analyses need the flexible and diverse integration of basic bioinformatics tools. One of the efficient solutions for this type of software development may be the workflow-based problem solving environment (PSE). There are still heavy overhead, however, to develop and implement this workflow-based model on current grid system or grid middleware. Concerning about this fact, the grid supporting technology in Condor [1] has lots of merit.

In this report, we especially take advantage of simple workflow design functionality of Condor by using a meta-scheduler, DAGMan [1] with many other Condor features for high-throughput computing application on grid. We realized two examples of integrated bioinformatics solutions about sequence searching and orthologous gene finding (OGF) based on our model of workflow based PSE on grid.

2 Model of Integrated Bioinformatics Solution on Grid

We propose a model of workflow-based bioinformatics PSE by using condor functionalities and show practical examples of integrated bioinformatics solutions. Fig. 1 (a) shows the structure of this model. It has basically two layers: components layer and integrated applications layer. The components layer consists of application component and interfacing component. Both of them are grid-enabled executable files. Application component is a typical bioinformatics tool which can perform the separable process of bioinformatics analysis. Interfacing components are all intervening programs necessary to combine application components and related data of the workflow for an integrated solution. The components can be reused or added for new integrated solution. Various workflows matching to specific solutions can be constructed by using simple Condor scripts with proper arrangement of the old and new components. In this way, the system can achieve the maximum flexibility and efficiency for the solution development corresponding to the diverse integrating requests of bioinformatics problems.

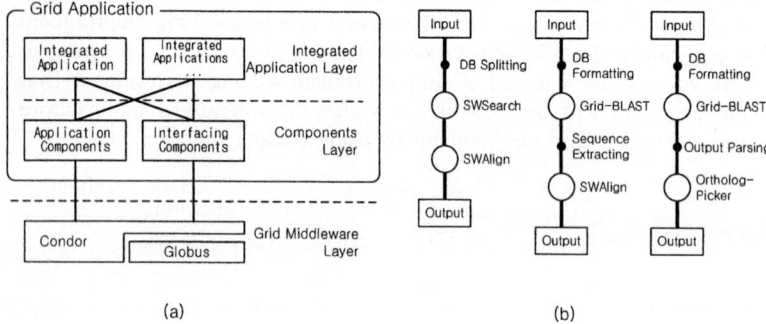

Fig. 1. (a)Structure of bioinformatics problem solving environment on Grid. (b)Examples of workflows for the integrated bioinformatics solutions. Application components and interfacing components are represented by white circles and black dots, respectively

2.1 Components Layer

We selected and implemented sequence search and alignment tools as application components because of their popularity in biological sequence analysis and the need for our integrated solutions. Among them, Grid-BLAST, Grid-FASTA, Grid-SWSearch, Grid-SWAlign, and Ortholog-Picker are briefly introduced here.

Grid-BLAST and Grid-FASTA grid-enabled version of BLAST [2] and FASTA [3]. These tools facilitate searching homologous sequences in multiple sequence databases for various numbers of querying sequences with grid resources. They generate a condor command script file to run the parallel job and submit it into condor pool. The scripts can have the options regarding proper usage of computational resources as well as the original options of BLAST and FASTA.

Grid-SWSearch is a homologous sequence searching program based on Smith-Waterman algorithm [4] that gives more precise results than BLAST or FASTA. In Grid-SWSearch, database and query sequences are divided and assigned to available computing nodes for the parallelization of computing jobs. The computing job is designed not to pass any massage to the other jobs to remove communication overhead and to run each job independently. After bundles of jobs are submitted to condor pool, each job calculates the similarity of all pair of sequences, performs statistical evaluation, and sends back the result. Grid-SWSearch does not show alignment to reduce execution time. We can execute Grid-SWAlign to see sequence alignment.

Grid-SWAlign is a grid enabled sequence alignment program based on Smith-Waterman algorithm. It receives multiple sequences to be aligned and makes a bundle of jobs on grid for all combinations of pair-wise alignment. We could improve the efficiency of both Smith-Waterman algorithm tools with the benefit of grid system.

Ortholog-Picker is an application to pick out orthologous genes from various genome sequence databases. Once Grid-BLAST searches all the gene sequences of query genome for the best hit of all the genes in the other genomes, the corresponding gene IDs are parsed from the BLAST output. In our definition, an ortholog is the group of three or more genes in which all genes are the best hit against the genomes of each other reciprocally.

Interfacing component is the program that process input or output data of application component to facilitate diverse combinations of application components that have their idiopathic input/output format. The examples of the functions of interfacing components are splitting, merging, converting, and formatting files , or extracting and rearranging the content of data.

2.2 Integrated Application Layer

Fig. 1 (b) shows three examples of the workflow for integrated bioinformatics solution. As shown in the left picture of Fig. 1 (b), the searching tool and alignment tool can be linearly arranged as a workflow described in a DAG script file. This work starts by splitting database into fragments. Multiple jobs of SWSearch are submitted and allocated at distributed computers by Condor. Output files are accumulated in the submission computer's directory. Finally, a bundle of jobs for SWAlign are submitted to show aligned sequences. One can easily change the components of this workflow by using DAG script file. For example, as shown in the middle picture of Fig. 1 (b), one may want to choose Grid-BLAST first for whole genome sequence comparison for quick search, then select the sequences of interest and run SWAlign to find more sensitive local alignments that could be missed by Grid-BLAST alone. The third workflow is the other type of example that integrate the sequence comparison tools for the Ortholog Gene Finding (OGF) in multiple genomes. Orthologs [5] are genes retaining the same function in different species that evolved from a common ancestral gene by speciation.

Identification of orthologs is critical for reliable prediction of gene functions in comparative genome analysis. OGF needs high-throughput computing due to the increasing number of sequenced genomes that are more than 60 for microbial genomes and 10 eukaryotic genomes, respectively. In this workflow, genome sequences are converted into query sequence files by FileMerger and converted again into database for BLAST search by BlastDBformatter. Grid-BLAST executes all to all BLAST search and the best hit and gene ID are parsed from BLAST Output files. Ortholog-Picker finds orthologous genes. These series of works are described in DAG script file and are controlled by DAGMan.

3 Conclusions

Most of real world bioinformatics analyses are dealing with heavy computational complexity and subject specific integrated problems. The problem solving environment with simple workflow on grid system can be very efficient model to resolve these problems.

Acknowledgements

This research was partially supported by University IT Research Center Project and by national grid project grant 2004-giban-04 through Korea Institute of Science and Technology Information

References

1. Thain, D., Tannenbaum, T., Livny, M.: Distributed computing in practice: The condor experience. Concurrency and Computation: Practice and Experience (2004)
2. Altschul, S., Gish, W., Miller, W., Myers, E., Lipman, D.: Basic local alignment search tool. J Mol Biol. **215** (1990) 403–410
3. Pearson, W., Lipman, D.: Improved tools for biological sequence comparison. Proceedings of the National Academy of Sciences of the United States of America **85** (1988) 2444–2448
4. Smith, T., Waterman, M.: Identification of common molecular subsequences. J Mol Biol. **147** (1981) 195–197
5. Tatusov, R.L., Koonin, E.V., Lipman, D.J.: A genomic perspective on protein families. SCIENCE **278** (1997) 631–637

MammoGrid: A Service Oriented Architecture Based Medical Grid Application

Salvator Roberto Amendolia[1], Florida Estrella[2], Waseem Hassan[2], Tamas Hauer[1,2], David Manset[1,2], Richard McClatchey[2], Dmitry Rogulin[1,2], and Tony Solomonides[2]

[1] ETT Division, CERN, Geneva, Switzerland
Tel: +41 22 767 8821, Fax: +41 22 7678930
{Salvator.Roberto.Amendolia,Tamas.Hauer,David.Manset,
Dmitry.Rogulin}@cern.ch

[2] CCCS UWE, Frenchay, Bristol BS16 1QY UK
Tel: +44 1179656261, Fax: +44 117 2734
{Florida.Estrella,Waseem.Hassan,Richard.McClatchey,
Tony.Solomonides}@cern.ch

Abstract. The MammoGrid project has recently delivered its first proof-of-concept prototype using a Service-Oriented Architecture (SOA)-based Grid application to enable computing spanning national borders. The underlying AliEn Grid infrastructure has been selected because of its practicality and its emergence as a potential open source standards-based solution for managing and coordinating distributed resources. The resultant prototype is expected to harness the large amounts of medical image data needed to perform epidemiological studies, advanced image processing and ultimately tele-diagnosis over communities of 'virtual organizations'. This paper outlines the MammoGrid approach in managing a federation of Grid-connected mammography databases in the context of the recently delivered prototype and describes the next phase of prototyping.

1 Introduction

The MammoGrid project [1] initiated in 2002 has adopted a lightweight Grid middleware solution (called AliEn [2]) to deliver a set of evolutionary prototypes for Grid-based radiological study through a series of prototypes following a Service Oriented Architecture (SOA) philosophy and with an OGSA interface. MammoGrid uses open source Grid solutions in comparison to the US NDMA [3] project which employs Grid technology on centralized data sets and the UK eDiamond [4] project which uses an IBM-supplied Grid solution to enable applications for image analysis. The current status of MammoGrid is that a single *'virtual organization'* (VO) AliEn solution has been demonstrated using the MammoGrid Information Infrastructure (MII) and images have been accessed and transferred between hospitals in the UK and Italy. The next stage is to provide rich metadata structures and a distributed database in a multi virtual organization environment to enable epidemiological queries to be serviced and the implementation of a service-oriented (OGSA-compliant) architecture for the MII.

The structure of this paper is as follows. In Section 2 we provide an overview of MammoGrid prototypes (P1 and P2). Next, in Section 3 we discuss the advantages of P2 over P1. Finally, in Section 4 we draw conclusions and indicate future work.

2 Description of MammoGrid Prototypes Architecture

2.1 Prototype P1

The MammoGrid prototype P1 enables mammograms to be acquired into files that are distributed across 'Grid-boxes' for simple queries to be executed. The mammogram images are transferred to the Grid-Boxes in DICOM [5] format where AliEn services could be invoked to manage the file catalog and to deal with queries. The medical image (MI) services are directly invoked by authenticated MammoGrid clients and they provide a generic framework for managing image and patient data. The digitized images are imported and stored in DICOM format. The MammoGrid P1 architecture includes a clinician workstation with a DICOM interface to Web Services and the Alien middleware network. There are two sets of Services; one, Java-based comprising the business logic related to MammoGrid services and the other, Perl-based, the AliEn specific services for Grid middleware. As this architecture is Web Services-based, SOAP messages are exchanged between different layers. RPC calls are exchanged from Java specific services to Alien specific services.

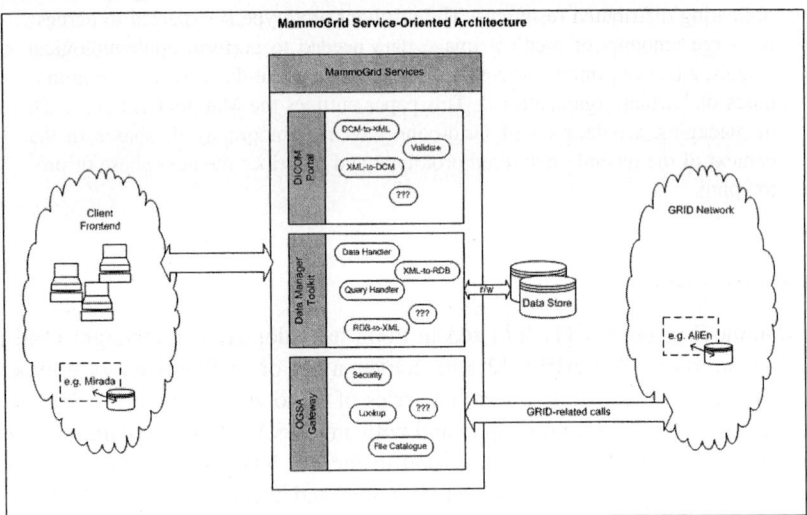

Fig. 1. The MammoGrid Prototype 2 (P2) architecture.

2.2 Prototype P2

Prototype P2 will be based on a Service-Oriented Architecture with clear portals/gateway/interfaces both to external systems such as the MammoGrid image acquisition hardware and to the Grid network. It is diagrammatically shown in Figure 1 and comprises a set of MammoGrid Services including a DICOM portal facilitating the exchange of information between the clinician's workstation and the Grid-Box. This includes validation services and DICOM-to-XML and XML-to-DICOM translators, as well as an OGSA Gateway consisting of Security, Look-up and File Catalogue services. The DICOM front-end will enable any DICOM-compliant external device to exchange information via a set of MammoGrid Services with potentially any OGSA back-end Grid network. Note that the P1 architecture had a tight coupling between different layers of services but P2 architecture will provide loose coupling between different vertical layers of services. P2 architecture will be multi-VO based; its details are given below.

The P2 activities revolve around creating a multi-VO architecture. In essence, the P2 architecture aims to exhibit the following:

1. MammoGrid high level and middleware services that are fully OGSA-compliant. The MammoGrid services will be re-designed and re-deployed as OGSA-compliant Grid services. It is anticipated that AliEn (and the EGEE middleware [6]) and other Grid middleware implementations will move towards compliance in the near future.
2. A set of OGSA-compliant abstract level services to serve as mediator between the MammoGrid services and the middleware services. These abstract services will remove the dependency on Grid-related activities, will allow the use of different Grid implementations as they become available and will create a modular approach in the design and development of the MammoGrid application.
3. Each hospital will belong to a VO which will maintain its own access and authentication rules and its own data. A fourth, central VO, will be responsible for inter-VO communication. A set of inter-VO-related services (currently referred to as VOMS – Virtual Organization Management Services) is anticipated.
4. AliEn will include Virtual Organization Services (VOS). These services will be in charge of authentication/authorization between Virtual Organizations.

2.3 Federation in a Multiple Virtual Organization

The medical sites in a single VO operate within the rules specified by a governing organization. Typically, hospitals have different regulations and governments have different legislations. In order to preserve privacy, patient personal data is – as a first step – partially encrypted in MammoGrid to facilitate anonymization. As a second step, when data is transferred from one service to another through the network, communications are established through a secure protocol HTTPS with encryption at a lower level. Each Grid-box is in fact made part of a VPN (Virtual Private Network) in which allowed participants are identified by a unique host certificate. The general requirements for multi-VO (security) architecture can be found in [7].

The current AliEn design uses a hierarchical LDAP database to describe the static configuration for each VO. This includes People, Roles, Packages, Sites and Grid Partitions as well as the description and configuration of all services on remote sites. The code that is deployed on remote sites or user workstations does not require any specific VO configuration files, everything is retrieved from the LDAP configuration server at run time thus allowing user to select VO dynamically.

3 P1, P2 Architectures: Comparison at a Glance

The MammoGrid requirements analysis process revealed a need for hospitals to be autonomous and to have 'ownership' of all local medical data. The P2 design will be based on a multi-VO architecture where there will be no centralization of data, all metadata will be truly federated and will managed alongside the local medical data records and queries will be resolved at the hospitals where the local data are under governance. Full adoption of the Grid philosophy suggests that the federation should not be confined to the database level but should be realized in all aspects of Grid services (including authentication, authorization, file replication, etc.). A federation of (potentially hierarchical) VOs forms the basis of P2, where the boundaries of the organizations define the access rights and protocols and the service access between 'islands' that provide and consume medical information. Since Prototype P1 was designed to demonstrate client- and middleware-related functionalities, the API was kept as simple as possible using a handful of web-service definitions. The focus of P2 will be the demonstration of full Grid services functionality and its focus will be on interoperability with other OGSA-compliant Grid services.

4 Conclusion and Future Work

This paper outlines the MammoGrid services-based approach in managing a federation of Grid-connected mammography. The current status of MammoGrid is that a single VO AliEn solution has been demonstrated using the MII and images have been accessed and transferred between hospitals in the UK and Italy. The next stage is to provide rich metadata structures and a distributed database in a multi virtual organization environment to enable epidemiological queries to be serviced and the implementation of a service-oriented (OGSA-compliant) architecture for the MII. In doing so, the MammoGrid project is envisioning to become an early adopter of the new EGEE middleware, which will eventually be a seamless transition from the current Grid Middleware to the new EGEE middleware.

References

1. "MammoGrid - A European Federated Mammogram Database Implemented on a Grid Infrastructure". EU Contract IST 2001-37614. See: http://MammoGrid.vitamib.com
2. P. Saiz, L. Aphecetche, P. Buncic, R. Piskac, J.-E. Revsbech, V. Sego., "AliEn - ALICE Environment on the Grid". *Nuclear Instruments & Methods in Physics Resercrh*, A 502 (2003) 339-346 and see: http://alien.cern.ch
3. NDMA: The National Digital Mammography Archive. Contact Mitchell D. Schnall, M.D., Ph.D., University of Pennsylvania. See http://nscp01.physics.upenn.edu/ndma/projovw.htm
4. M Brady, M Gavaghan, A Simpson, M Mulet-Parada & R Highnam. "A Grid-enabled Federated Database of Annotated Mammograms" *in Grid Computing: Making the Global Infrastructure a Reality*, Eds. Bergman F, Fox G and Hey T. Wiley Publishers. 2003.
5. DICOM Digital Imaging and Communications in Medicine. http://medical.nema.org
6. EGEE Project, Available from http://public.eu-egee.org/
7. Ananta Manandhar et al, GRID Authorization Framework for CCLRC Data Portal, Available from www.nesc.ac.uk/events/ahm2003/AHMC

Interactive Visualization Pipeline Architecture Using Work-Flow Management System on Grid for CFD Analysis*

Jin-Sung Park, So-Hyun Ryu,
Yong-Won Kwon, and Chang-Sung Jeong

Department of Electronics Engineering, Korea University, Anamdong 5-ka
Sungbuk-ku, Seoul 136-701, Korea
Fax: +82-2-926-7620, Tel: +82-2-3290-3229
{honong13,messias,luco}@snoopy.korea.ac.kr
csjeong@charlie.korea.ac.kr

Abstract. Since the potential of the impact of complex problem like CFD on design of material that is difficult to test actually has been established, CFD analysis that consist of two parts - simulation part and visualization part is considered as important work. But many researcher think it is difficult to analyze CFD problem because of following two reasons. First reason is that CFD analysis usually needs numerous processing time and much computing power. and Second is that the tasks of simulation and visualization are usually done independently - numerical simulation then visualization with stand-alone visualizing system. we propose solution of two problem. First problem is solved by using parallel/distributed programming on problem solving environment on grid environment - WISE system. and Second problem is solved by interactive visualization pipeline architecture that enables user to conducting simulation and visualization at the same time automatically applying pipeline concept. Therefore we presented interactive visualization pipeline architecture and implemented using WISE system.

1 Introduction

Todays, Computational Fluid Dynamics(CFD) analysis is considered as essential work in engineer and science. But it needs many resources and takes a lot of times to simulate. Therefore most CFD analysis processes in parallel/distributed environment. In the past, CFD analysis was done on expensive supercomputers but advances in high speed network and powerful computers make it possible to construct a large-scale high performance distributed computing environments, called a Grid. In CFD analysis, the tasks of simulation and visualization are usually done independently - numerical simulation then visualization. But if

---

* This work has been supported by a Korea University Grant, KIPA-Information Technology Research Center, University research program by Ministry of Information & Communication, and Brain Korea 21 projects in 2004

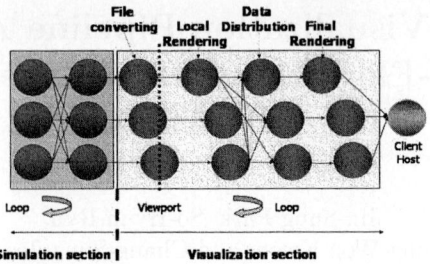

Fig. 1. Overall Architecture - In Standard mode

both steps - simulation and visualization perform at the same time, that is, simulated result visualize immediately, it is more efficient. So, we propose the interactive visualization pipeline architecture on workflow system - WISE for visualizing the CFD result immediately after simulation.

2 Related Work

There are several works that proposed interactive parallel visualization for CFD problem.[1–4] But, in those system, simulation part and visualization part are loosely-coupled : After parallel simulation,all sub-result-data are united into one result data. After that, the result data is sent to visualization system and parallel visualized. But in our approach, simulation part and visualization part are strong-coupled, that is, parallel simulated sub-result-data directly send to visualization parts not combine to one data. Also previous systems have no operations to interact to user or developer. Only things that supports is visualization of simulated CFD data. Therefore it is difficult for user to know how the simulation works. But our approach uses implemented our system on workflow system-WISE, therefore user or developer can monitor how the simulation works and also detect fault that the simulation may occurs.

3 Visualization Pipeline Architecture on WISE

3.1 File-Converting

Most parallel CFD simulation generated unstructured result data. Unstructured data means the topology and geometry of data is unstructured and the cells that compose that data vary in topological dimension. In our system, to use our extended space-leaping algorithm, we present file-converting part that convert from unstructured result data to structured data at each process.

3.2 Local Rendering

After finishing file-converting, user decides viewport which means view-angle that user want to see. And then image screen is divided into number of process and each processes is in charge of each divided image screen. After that, each process executes rendering on active pixel as far as possible with local data

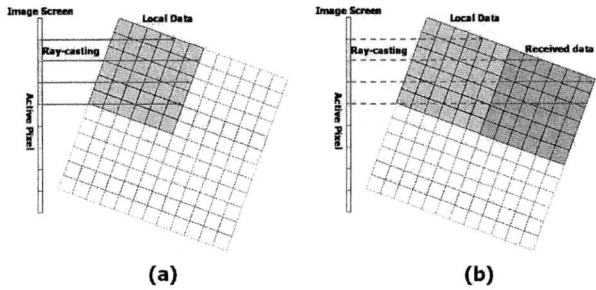

Fig. 2. (a) Local Rendering (b) Final Rendering after Data Distribution

using extended space-leaping algorithm. if ray starts on active pixel and ends with only local data, the process stores its color and opacity. else if ray starts on active pixel but can't continue to the end because the process don't have rest of required data, the process stops ray-casting. And stores calculated color and opacity temporarily, and calculates needed data part. Figure 2-(a) shows local rendering procedure.

3.3 Data Distribution

After finish local rendering, each process has information about data that needs to complete ray-casting of incomplete active pixel. To complete ray-casting, each process send requests to other process to complete ray-casting. And then process that receives the request transfers associated data to sender.

3.4 Final Rendering

After process receives request data, each process finishes rendering with received data as figure 2-(b). Each active pixel that can finish ray-casting continues rendering with color and opacity stored temporarily, calculates final color and opacity finally. Process that finishes rendering of all active pixel sends image data to client host. After all processes finish sending image data to client, client host represents received data to user's image screen. then it's the end of visualization of simulated data for one time step.

4 Experiment

We experiment our visualization pipeline architecture on WISE by analyzing of artificial heart. Simulation is about the change of tension on artificial heart surface as time goes on(figure 3). We experiment it in 3 different mode. First, we simulates and visualization on single machine. Second, we conduct parallel simulation and visualization on homogeneous environment. and Finally we conduct simulation and visualization on heterogeneous environment(table 1). Table 2 shows the execution time and efficiency of system according to number of machines. The efficiency represents the ratio of speed of with respect to expected speed up. As the number of machines increases, our system shows relatively good speed up with efficiency around 70% even on heterogeneous environment.

Table 1. Machine specifications

Machine type	M_1	M_2	M_3
Model	Pentium IV PC	O_2	Octane
CPU	P IV	MIPS	R10000
Clock(MHz)	1740	150	250
Memory(MBytes)	1024	128	512
OS	Linux 2.2	IRIX 6.3	IRIX 6.5

Table 2. Performance results on visualization pipeline architecture

	Mode	$1(M_1)$	$2(M_{1,1,1})$	$4(M_{1,2,3})$
	expected speedup	1.0	3	2.3
	time (min)	308.77	156.69	219.22
performance	speedup	1.0	2.266	1.633
	efficiency (%)	100.0	75.53	71

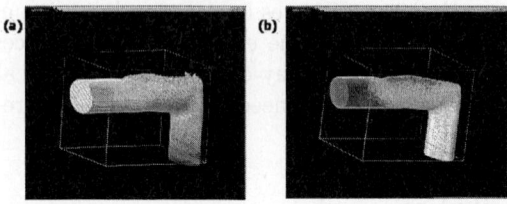

Fig. 3. Artificial Heart after (a) 7steps (b)25steps

5 Conclusion

In this paper we have presented an efficient visualization pipeline architecture on WISE for CFD analysis. we achieve speed up by doing simulation and visualization at same time and conducting visualization using pipeline technique. and because we implemented visualization architecture by using work-flow system-WISE, we can monitor the flow of procedure works and also detect fault of simulation under development.

References

1. S. Schneider, T. May, and M. Schmit. "Parallel Architecture of an interactive scientific visualization system for large datasets" Eurographics Association 2003.
2. A. Gerndt, T. van Reimersdahl, T. Kuhlen, J. Henrichs, C. Bischof, "A parallel approach for VR-based visualization of CFD data with PC clusters", 16th IMACS World Congress, 2000 IMACS.
3. O. Kreylos, A.M. Tesdall, B. Hanmann, J.K. Hunter and K. I. Joy, "Interactive Visualization and Steering of CFD Simulations" The Eurographics Association 2002.
4. T.J.Jankun-Kelly, O. Kreylos, J. M. Shalf, K.-L. Ma, B. Hanmann, K. I. Joy and E. W. Bethel. "Deploying web-based visual exploration tools on the grid." IEEE Computer Graphics and Application, pages 40-50, 2003.

Distributed Analysis and Load Balancing System for Grid Enabled Analysis on Hand-Held Devices Using Multi-agents Systems

Naveed Ahmad[4], Arshad Ali[4], Ashiq Anjum[4], Tahir Azim[4], Julian Bunn[1], Ali Hassan[4], Ahsan Ikram[4], Frank Lingen[1], Richard McClatchey[2], Harvey Newman[1], Conrad Steenberg[1], Michael Thomas[1], and Ian Willers[3]

[1] California Institute of Technology (Caltech), Pasadena, CA 91125, USA
{fvlingen,newman,conrad,thomas}@hep.caltech.edu,
Julian.Bunn@caltech.edu

[2] University of the West of England, Bristol, UK
Richard.mcclatchey@uwe.ac.uk

[3] CERN
Geneva, Switzerland
Ian.Willers@cern.ch

[4] National University of Sciences and Technology, Rawalpindi, Pakistan
{ahsan,ali.hassan,arshad.ali,ashiq.anjum,naveed.ahmad,tahir}
@niit.edu.pk

Abstract. Handheld devices, while growing rapidly, are inherently constrained and lack the capability of executing resource hungry applications. This paper presents the design and implementation of distributed analysis and load-balancing system for hand-held devices using multi-agents system. This system enables low resource mobile handheld devices to act as potential clients for Grid enabled applications and analysis environments. We propose a system, in which mobile agents will transport, schedule, execute and return results for heavy computational jobs submitted by handheld devices. Moreover, in this way, our system provides high throughput computing environment for hand-held devices.

1 Introduction

Handheld computing and wireless networks hold a great deal of promise in the fields of application development and ubiquitous data access. However, inherently constrained characteristics of mobile devices are the main hurdle.

Grids providing computational and storage resource sharing are the best implementation of distributed computing. We present here a brief description of a Grid enabled distributed analysis for handheld devices using mobile agents.

JASOnPDA and WiredOnPDA are the handheld physics analysis clients that we have developed and we will be using these two applications for analysis of our architecture.

2 Selection of Technologies

Grid enabled portal, JClarens [1], is used as a gateway to interact with Grid. For PocketPC we tried various JVMs and technologies, which included IBM Device Developer [5], Personal Profile, Super-Waba [6] and Savaje [7].

In the end we decided to go for PersonalJava and NSICOM's CrEme [8], as they suited our needs more than any other JVM.

For multi agent system we evaluated Aglets [3], DIET [4] and JADE-LEAP [2]. Due to small footprint, Personal Java compatibility and various other features we have chosen Jade-Leap for our research.

3 Architecture Design

Jade-Leap offers three types of containers that include Java based Jade-Leap J2SE containers, Personal Java based Jade-Leap PJava containers and MIDP based Jade-Leap MIDP containers.

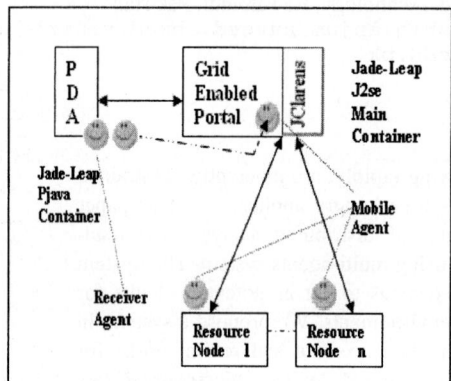
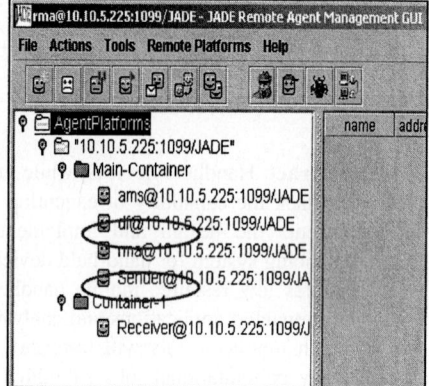

Fig. 1. (a) Architectural Overview, (b) JADE-LEAP Agent Management UI

As shown in Figure-1 (a), our architecture uses a combination of J2SE and PJava containers. Moreover, PJava containers running on handhelds must register with J2SE based Main-Container to communicate with other containers and platform, according to FIPA specifications.

After registering with the Main-Container, the client initiates two agents: the mobile (execution) agent and the receiver agent. While receiver agent resides on handheld device (Figure-1 (b)) waiting for response, mobile agents move to Main-Container (Figure-1 (b)) along with execution ontology and classes. Mobile agents execute job requests on server nodes or available resource nodes. Receiver Agent and Mobile Agent can intercommunicate to check job status, maintain connectivity, receiving results and killing jobs. After completing the job, the mobile agent either brings back the results or directly transfers them to handheld client.

3.1 Load Balancing

Every time an agent moves to a resource node for execution or storage it is provided with the load information of that node. From this information it deduces the load status. Mobile agents checks for the availability of required resources and also caters for the load of that machine. In case the load status notifies that resource node is over utilized the agent withdraws its execution or storage from that machine and looks for a under utilized node available in the farm of that node.

3.2 Fault Tolerance

While agents execute jobs remotely on behalf of handheld devices, they ensure re-connectivity by updating their location and state to their twin agent residing on handheld client.

On the server side architecture has been kept decentralized, there are more than one server nodes and handheld clients can register with all available servers. In case a sever goes down before job submission, handheld client can resubmit job to any other available node.

4 JASOnPDA / WiredOnPDA

JASOnPDA (Figure 2) is a physics analysis tool used for analyzing data obtained from Linear Accelerator in the form of 1D-2D Histograms. Event data to be analyzed is in the form of ROOT files, which stores numerical physics data in a special, quickly accessible hierarchical structure.

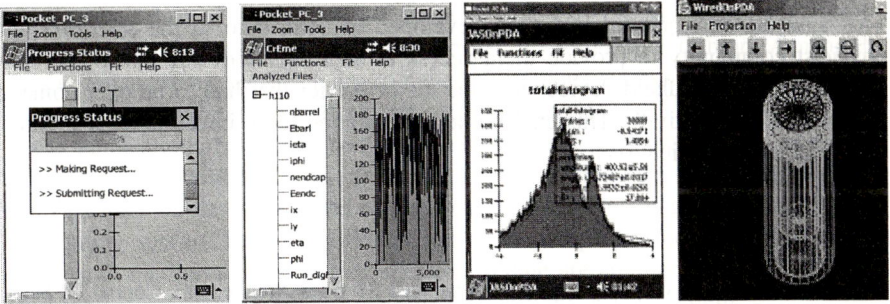

Fig. 2. JASOnPDA and WiredOnPDA Screen Shots

WiredOnPDA (Figure 2) reads data from any source that either provides HepRep XML files or a HepRep-enabled HepEventServer. Files are parsed using a SAX XML parser. "Drawables" are then extracted from the parsed data and are displayed. This procedure is very resource and time consuming and thus makes it difficult to run on handheld devices.

MQAS (Mobile Q&A Agent Server). After creating a clone of the mobile agent, the MQAS dispatches the clone to a mediator. The mediator generates meta-data based on RDFS ontology for the user's question and moves the agent to another MQAS that has answers for the questions. Then the mobile agent returns to the mediator with the most similar answer from the previous RDF meta-data for the questions. The mediator mediates the mobile agent with the answer and meta-data to the original MQAS. Finally, the user can obtain the information of the answer and meta-data through WAP Gateway.

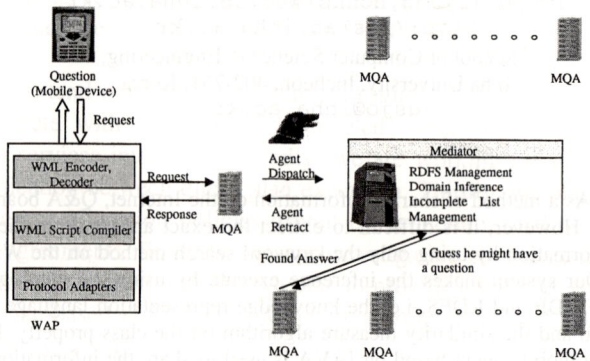

Fig. 1. System Architecture

3   Ontology Based on RDFS and Knowledge Domain Inference

Figure 2 shows a mobile class that is defined as domain knowledge in the mediator. JENA that has the RDF parser makes the user's question an RDF instance document by using the RDFS ontology. The instance document has classes, properties, and information of the super-class it belongs to. Formula 1 expresses the relation between the classes and the properties as a probability function to see if the user's question is in a specific domain.

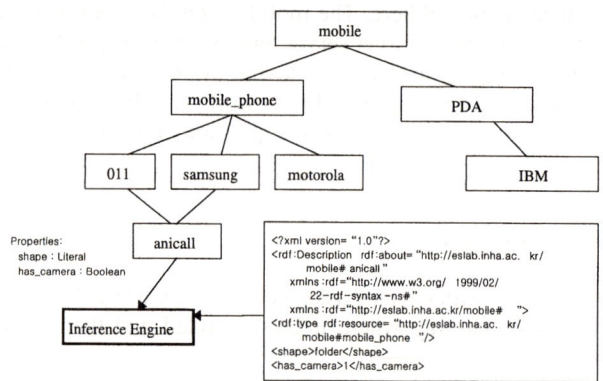

Fig. 2. Domain Inference using RDFS

$$P(C_j \mid B_i) = \sum_{B_k \in children (B_i)} P(C_j \mid B_k) \cdot \frac{P(B_k)}{P(B_i)} \qquad (1)$$

P = Probability that the corresponding class becomes a domain.
C_j = Class that includes the property.
B_i = Property of RDFS that doesn't include leaf node.
B_k = Property that is in the question.

Formula 2 is to obtain the similarity between the documents by using the classes and properties. Here, we consider the condition of existence or non-existence of the class and the arrangement of the property node, not the weight and the frequency of the property.

$$Sim = \sum_{i=1}^{n_c} \frac{1}{n_c} \left(n_p - \sum_{j=1}^{n_p} D_j \right) \qquad (2)$$

Sim = Similarity of the questions
n_c = Number of classes in questions (Only if, no class, $n_c = 1$)
n_p = Total number of the properties defined in RDFS

D_j is a norm to decide the similarity of the properties from the questions as shown in Formula 3.

$$D_j = \left(\frac{D_{qi} - D_{ai}}{n_p + 1} \right)^{|D_{qi} - (D_{ai} + 1)|} \qquad (3)$$

D_{qi} = i'th property node of the class in the current questions
D_{ai} = i'th property node of the class in the past questions

Here, if Sim is the maximum value, we can get the document with the highest similarity for the user's questions.

4 Experimental Evaluation and Conclusion

We obtain the experimentation data from the web site (http://kin.naver.com) and collect the customers' 1040 questions about cellular phone purchases and PDA purchases at random. After extracting each class and property from the collection, we make up RDF documents (Table 1). Figure 3 shows the comparison between the result from our similarity algorithm and the result from the keyword-matching algorithm. Only the results with more than 80% of the similarity are compared here. Then we can see our system has the documents with users' higher satisfaction in spite of relatively smaller number of documents.

Table 1. Number of RDFS classes and properties.

	Cell Phone	PDA
Class 1	12	11
Class 2	22	14
Property	9	10

This paper proposes an information share system for Q&A boards on the Wireless Internet and the semantic web environment by using the mobile agent and the ontol-

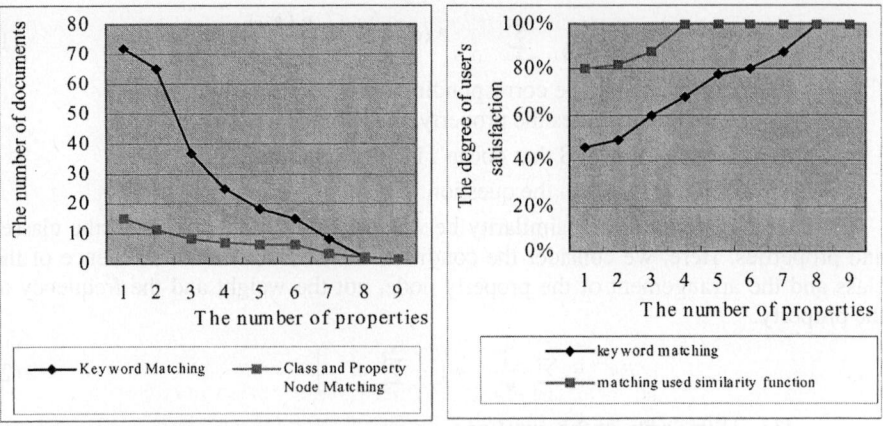

Fig. 3. The data search result with more than 80% of the similarity

ogy. Through the experiment, we represent that the inference by using the properties of classes has higher accuracy and satisfaction than using keyword-matching search method. Hereafter, in terms of evaluations of the weight and the constraint property for the users' questions, studies on the ontology composition with the natural language process should be conducted.

References

1. Jae-Bok Park, Kwang-Yong Lee, Geun-Sik Jo. : Knowledge sharing system using Mobile Agent: A case study of the Q&A board, International Conference on Electronic Commerce, Seoul Korea (2000) 42-46.
2. Kit Hui, Stuart Chalmers, Peter Gray, Alun Preece. : Experience in using RDF in Agent-mediated Knowledge Architectures, AAAI. (2003)
3. Michael N. Huhns, Munindar P.Singh.: Readings in AGENTS, Morgan Kaufmann Publishers Inc. (1998)
4. Robin D. Burke, Kristian J. Hammond, & Vladimir A. Kulyukin. : Question Answering from Frequently-Asked Question Files: Experiences with the FAQ Finder Systems, Technical Report TR-97-05, Chicago Univ. (1997)
5. Roger L. Costello, David B. Jacobs. : Inferring and Discovering Relationships using RDF Schemas, RDFS tutorial, DARPA. (2002)

Semantic Caching Services for Data Grids

Hai Wan[1], Xiao-Wei Hao[2], Tao Zhang[2], and Lei Li[2]

[1]Department of Computer Science and Technology, SUN YAT-SEN University,
Guangzhou 510275, P.R. China
{whwanhai}@163.com

[2]Software Research Institute of SUN YAT-SEN University, Guangzhou, 510275, P.R. China

Abstract. Semantic caching services are very attractive for acting as core services to deliver and share data in Data Grids considering the large sizes of requested data, prolonged data transmission time, and low reliable network. This paper provides semantic caching services for Data Grids and presents the services' formal definitions, modeling, query processing, optimization and coherency control scheme. Finally, the performance of the semantic caching services for Data Grids is examined through a simulation study.

1 Introduction

The Grid approach is an important development in the discipline of computer science and engineering [1]. Compared with other Grid computing, data intensive computing is much more common and comprehensive [2]. However, the disadvantages (such as: narrow bandwidth, frequent disconnection, etc.) of the computing environment, have become the bottleneck of Data Grids. Semantic caching can reduce the network communication cost, and is very attractive and important for use in Data Grids. A storage resource manager, in the context of the Data Grids infrastructure [3], is essentially a middleware component facilitating data sharing and storage. In [4], Qun Ren presented a formal semantic caching definition. We argue that the data processed in Data Grids can be classified as structural data (relation database), semi-structural data, and non-structural data, in which relation database has been used generally and has a logical explanation [5]; hence, semantic caching can be applied in Data Grids.

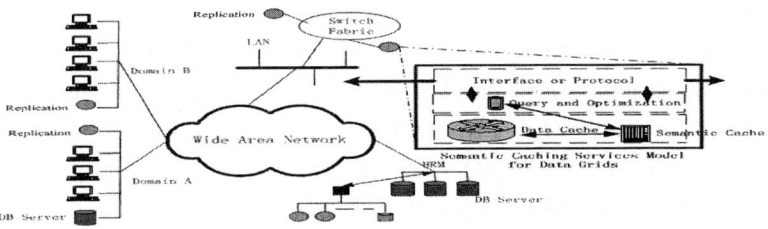

Fig. 1. Semantic caching services model for Data Grids

The rest of this paper is structured as follows: semantic caching services model for Data Grids, query processing, and optimization are formally presented in section 2. Section 4 describes coherency control scheme. The performance is examined in Section 5. Finally, we summarize our work and discuss future research in Section 6.

2 Semantic Caching Services Model and Query Processing

The positional role of semantic caching services model for Data Grids is shown in Fig.1. Semantic caching caches not only data replications but also their semantic descriptions. When there is a query, it would be sent to the semantic caching firstly and the replication reasons from the semantic descriptions to determine whether it can be totally answered. If totally answered from the cache, communication with the server is unnecessary; if partially answered, the original query is trimmed and the trimmed part is sent to the server. The formal definition and concepts are given as follows, with which semantic cache stores and organizes in relation database.

Definition 1. Query Q. Query is <Q_R,Q_P,Q_{rs}>, where Q_R is the relation or view to be queried; Q_P is the semantic description; Qrs= σ $Q_P(Q_R)$, $Q_R \subseteq R$; where R is the replication data.

Definition 2. Query result caching DC. DC is replication data and their semantic descriptions. Each DC_i has the unique semantic description Q_i; DC is <DC_R,DC_P,DC_{rs}>, DC_{rs}= σ $DC_P(DC_R)$, $DC_R \subseteq R$.

Definition 3. Result caching SC. SC is <SC_R,SC_P,SC_{rs}>, where SC_R=∪DC_{iR}, ∩DC_{ip}= φ ; SC_P=∨DC_{ip}, ∧DC_{iR} is unsatisfiable (1≤i≤n); SC_{rs},= σ $SC_p(SC_R)$; and SC is composed of DC_i with the same relation or view and DC_i's cached tuples are all different with each other.

It is Query processing that checks and processes the satisfiability or implication relationship between query semantic Q_P and cached semantic descriptions SC_P. There are three scenarios:1) $Q_P \rightarrow SC_P$, i.e., Qrs⊆SCrs; 2) $Q_P \wedge SC_P$ is satisfiable, i.e., Qrs∩SCrs ≠φ and Qrs∩¬SCrs≠φ; 3) $Q_P \wedge SC_P$ is unsatisfiable, i.e., Qrs∩SCrs = φ.

Both 1) and 3) are easy to process: if 1), Q can be answered without sending to the server; if 3), abandon Q because of no answers in server. The 2) is more complex and Q is divided into two sub-queries: *Probe Query* P_q, retrieving the portion of Q contained in SC_P; *Remainder Query* R_q, retrieving the portion of Q not found in SC_P.

As far as the 3) is concerned, we argue that it is inappropriate to remove query Q_E(unsatisfiable query) simply and we can optimize the evaluation via storing Q_E.

Theorem 1: If Q_E is unsatisfiable in SC_p, for any Q', if Q'→Q_E, Q' is unsatisfiable.

Algorithm 1: Query Processing Optimization Algorithm

Consider query semantic Q_P, caching semantic SC_P and unsatisfiable queries Q_E,

1. If $Q_P \wedge SC_P$ is unsatisfiable then P_q=φ;R_q=Q_P. IsEC=false; otherwise, go 2);
2. If $Q_P \rightarrow SC_P$ then P_q=Q_P,R_q=φ IsEC=true; otherwise, go 3);
3. If $Q_P \wedge SC_P$ is satisfiable, P_q=$Q_P \wedge SC_P$, R_q=$Q_P \wedge \neg SC_P$; IsEC=true. go 4);
4. If IsEC=true then: ①If $P_q \rightarrow Q_E$ then E_q=P_q,R_q=φ; Otherwise:go ②; ②If $P_q \wedge Q_E$≠φ and $P_q \wedge \neg Q_E$≠φ then E_q=$P_q \wedge E_q$,R_q=$R_q \wedge \neg E_q$; otherwise; go ③; ③E_q=φ.

3 Coherency Control Scheme

Semantic caching services for Data Grids may become obsolete and there can be termed our two different conflicts in semantic caching services for Data Grids: *Data error conflict* and *Data non-integrity conflict*.

Definition 4. Semantic Caching Coherency. Given a database server S, S has $<S_R, S_P, S_{rs}>$, $S_{rs} = \sigma S_p(S_R), S_R \subseteq S$; Consider DC is a semantic caching of S, has $<DC_R, DC_P, DC_{rs}>$, and $DC_{rs} = \sigma DC_p(DC_R), DC_R \subseteq S$.

If $\forall$ query q, DC can satisfy:

1. Cached data in semantic caching services can stay on coherency whenever inserting/modifying/updating in S or DC, i.e. $S_{rsq} = \sigma S_q(S_R)$ is satisfiable in S, and $DC_{rsq} = \sigma DC_q(DC_R)$ is satisfiable in DC;
2. Cached data in semantic caching services can match the semantic descriptions whenever *inserting/modifying/updating* in S or DC, i.e., $DC_{rsq} = \sigma DC_q(DC_R)$ is satisfiable;

We say semantic caching coherency is satisfiable.

When the network connection is reliable, each successfully committed operation will be sent to corresponding replications directly. However, when the network is disconnected, all of those operations will store in *update sequence* to wait for the network reconnected again, meanwhile optimize the relevant update operation.

We design Coherency Control Scheme 1 as follows:

Scheme 1: Coherency Control Scheme

1) If the received operation is *delete*, the caching process it directly in SC_R;
2) If the received operation is *insert*, analyze it into atom *insert* operation; i.e. "insert into R tuples $(A_i(i=1,2,...,n))$"; traverse cached semantic descriptions and compare atom *insert* with each SC_P;

   If $\forall DC_{Pj}$, $DC_{Pj} \subseteq SC_P$, $\forall A_i$, and $A_i \subseteq DC_{Pj}$, then process the *insert* operation;
   If not, give up the operation.

3) If the received operation is *update*, then ① Analyze and substitute it with *delete* and *insert*; ② Process *delete* as 1) shows; ③ And then do with *insert* as 2) describes.

4 Performance Study

We design a simulation study to examine the performance of the semantic caching services for Data Grids. In this study, the system is composed of a server, a client, in which the server's OS is Sun Solaris 7 and database is Oracle9i, the client's OS is windows 2000 and we use Ebase designed by Software Research Institute of SUN YAT-SEN University as the embedded database in semantic caching service in client. We use a modified Wisconsin Benchmark [6] to examine the performance, containing one relation R with 5,000 tuples; each tuple is 256 bytes, stored in the server.

1) Effects analysis of semantic caching service (as Fig. 2 shows)

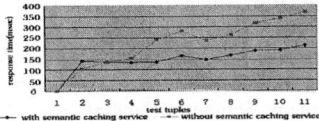

Fig. 2. Effects analysis of semantic caching service

The change in query complexity and the number of query items have impacts on response time, but semantic caching service can improve retrieving process, save response time and make replication more efficient.

2) Effects analysis of query processing and optimization (as Fig. 3 shows)

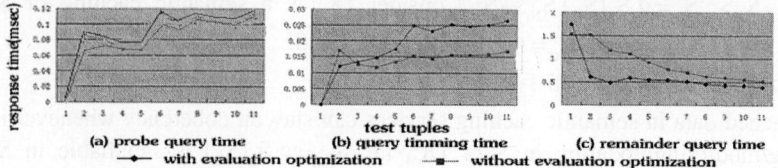

Fig. 3. Effects analysis of semantic caching service with evaluation optimization

Query processing and optimization can save the network bandwidth and send more useful remainder query to server.

3) Effects analysis of coherency control scheme (as Fig. 4 shows)

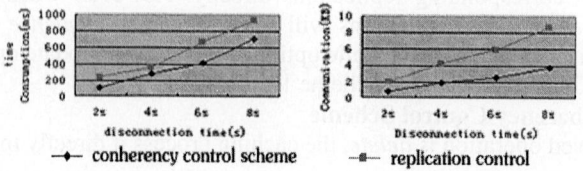

Fig. 4. Effects analysis of coherency scheme and replication control

With the help of coherency control scheme, it is more reliable and utilizable for semantic caching services to be used in *Data Grids*.

5 Conclusions and Future Works

We provide a semantic caching services model for Data Grids with formal definitions, query processing, optimization, and coherency control scheme. Finally, we present and examine its performance through a simulation study.

For the future research, we plan to extend our semantic cache services to include semi-structural data or non-structural data, meanwhile semantic caching services' replacement strategy is another important direction of future research.

References

1. G.von Laszewski. Grid Computing: Enabling a Vision for Collaborative Research. In Conference on Applied Parallel Computing, 3rd CSC Scientific Meeting, Lecture Notes, Espoo, Finland, 15-18 June 2002. Springer.
2. Foster, Kesselman, C., "The Grid: Blueprint for a New Computing Infrastructure," Morgan Kaufmann, San Francisco, 1999.
3. W.Hoschek, J.Jaen-Martinez, A.Samar Data management in an international data grid project. In Proc. 1st IEEE/ACM Int'l. Workshop on Grid Computing, India, 2000.
4. Qun Ren, Margaret H.Dunham, Semantic Caching, and Query Processing, IEEE transactions on knowledge and data engineering,Vol.15,No.1,January/February 2003 192-210
5. S.Ceri, Gottlob and L.Tanca, Logic Programming and Database, Springer-Verlag,1990
6. J. Gray, The Benchmark Handbook. Morgan Kaufmann, 1993.

Managing Access in Extended Enterprise Networks Web Service-Based

Shaomin Zhang[1,2] and Baoyi Wang[1]

[1] School of Computer, North China Electric Power University, Baoding 071003, China
[2] School of Computer, Xidian University, Xi'an 710071, China
zhangshaomin@126.com

Abstract. Most businesses have industrial partnerships with other businesses. We want to develop systems that can automate extended enterprise business processes. BCAs (Bridge Certification Authority), creates a combined multi-enterprise PKI (public key infrastructures) at the cost of increased complexity when evaluating the acceptability of certificates. But today's COTS (Commercial off-the-shelf) products are not entirely prepared to meet the challenges of bridge-connected PKIs. The paper focuses on how to design a secure access control mechanism in extended enterprise networks based on Web service. The characteristics of the two kinds of method are analyzed in detail. The secure access control mechanism has integrated the role-based access control in X.509v4 PMI (Privilege Management Infrastructure) with the XML security. The paper deals with the realization procedure, and some measures to improve the system's running efficiency are suggested also.

1 Introduction

Information security has become a pervasive requirement in nearly all aspects of business, but it becomes difficult when business processes cross-organizational boundaries. Organizations use PKI to support internal business processes, but most businesses have industrial partnerships with other businesses, and these alliances can exploit B2B e-commerce capabilities by connecting corporate PKIs. To develop information systems that can automate extended enterprise business processes, the system must be both secure and flexible. Although there are manifold security issues that affect extended enterprises, we focus on authorization and authentication measures because information must be secured against unauthorized access in order to establish trust, an essential element in B2B transactions.

2 Managing Access in Extended Enterprise Networks with BCA

PKI architectures traditionally fall into three con-figurations: single CA (Certificate Authority), hierarchy of CAs, or mesh of CAs. X.509 public key certificate bind specific identities to specific public keys and usage information via digital signatures.

BCA is a special type of CA that exists solely to connect PKIs, it establishes trust relationships with each participating enterprise PKI. Compared to mesh PKIs, BCA-connected PKIs introduce several important new properties [3].

(1) Certification path discovery might be easier and shorter – because users typically know their path to the BCA and need only construct paths from it to other users' certificates.

(2) Path validation could be more complex. BCAs employ certificate extensions not required in enterprise PKIs.
(3) BCAs are likely to include multiple distribution mechanisms that force applications to implement multiple protocols to retrieve certificates or certificate status information. Even with consistent protocols, each enterprise PKI might organize certificates and CRLs (certificate-revocation lists) differently.

In practice, today's COTS products are not entirely prepared to meet the challenges of bridge-connected PKIs.

3 Managing Access with PKC and AC in Expended Networks

To a trust service system, CA issues PKC (public key certificate). CA is constructed by every e-government application unit or their superior government, and reaches an agreement with national e-government authentication, management centre by business protocol. So CA does not know every person's role, position, experience in social. But in PMI, its primary data structure is an X.509 AC (attribute certificate). An AC strongly binds a set of attribute to its holder (holder, issuer, algorithms, role, the AC validity period, access privilege and various optional extensions)[1], so AA (Attribute Authority) who issues AC must be in user organization. Further, the identity is stable relatively, but the attributes, such as position, department changes frequently, so AC's lifetime is far below that of PKC's. In addition, in a system, everybody can own only one legal PKC, but several applications can use one AC, or AA issues different ACs to different operations in an application, so the usage of AC is more flexible.

When the lifetime of identity and attributes are the same, CA can manage attributes also, that is, PKC can manage authority. To realize authorize by appending attribute fields in extensions. When the performance of servers can not meet the need of an application, one of the methods to realize system's extension suggested by the three greatest suppliers of PKI system Entrust, VeriSign and Baltimore is to assign different CAs to sign the different keys pairs with the extensions in PKC.

4 A Secure Access Control Mechanism Based on Web Service

As defined by the World Wide Web Consortium, XML technology has become a basic architecture in Web services, and XML related security technology is an essential condition for it to apply [5].

X.509v4 PMI supports RBAC (Role-based Access Control). The RBAC model includes the basic RBAC model $RBAC_0$, the hierarchical RBAC model, $RBAC_1$, constrained RBAC model–$RBAC_2$, and consolidated model $RBAC_3$ includes $RBAC_2$ and $RBAC_1$ features [6].

X.509v4 PMI supports $RBAC_1$ by letting AA inserts both roles and privilege as attribute in a role-specification AC, so the role-specification AC inherits the privileges of the encapsulated roles. It supports $RBAC_2$ by letting AA place time constraints on a role-assignment AC's validity period, and it can also restrict the targets at which permission can be used, as well as the policies under which an AC can confer privileges. It support $RBAC_3$ by two kinds of AC:

(1) Role-specification AC holds the permission assignments granted to each role.
(2) Role-assignment AC holds the roles assigned to each user.

The access control structure base on Web service in extended network is shown as figure 1.

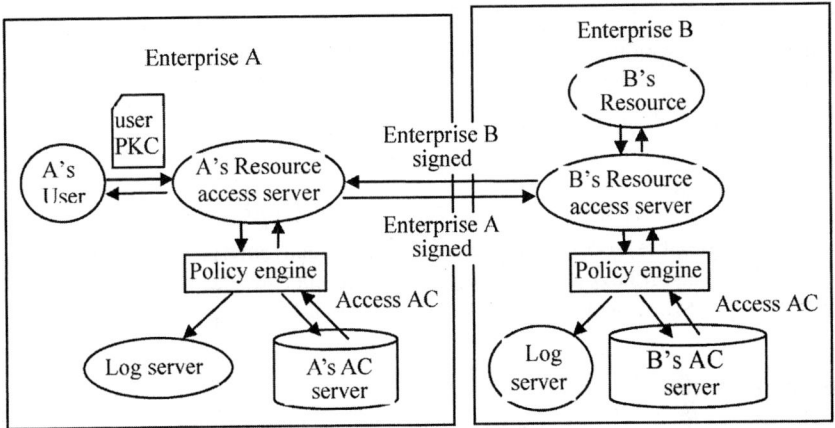

Fig. 1. The access management based on Web service

In figure 1, a user in enterprise A wants to access a target in B, he sends his request and his PKC to resource access server. The latter validates user's PKC, then sends an access request to policy engine, and pushes the access control policy it stored to policy engine also. The policy engine accesses enterprise A's AC server, and get user's AC. At the same time the policy engine records what it has done in log server. The policy engine returns authority result to resource access server. The latter then makes a decision on whether the user's access request should be permitted. If the request is permitted, enterprise A signs and sends the request to B. B's resource access server decrypts the request which signed by A with A's public key, and calculates the hash value with signature algorithm provided in SOAP-Header to validate its integrity. If the message is legal, then B accesses B's AA, obtains the user's AC, validates its authority. Once B's access management component has authorized the request, it accesses the resource and signs the response, then sends it back to enterprise A. A's access management server checks the response's signature and signs it, then transfers it to user. The user obtains corresponding services.

The system is realized by SOAP security token and the related XML security technology [7].

5 The Performance Analysis of the System

In the system, the compliance checker would need to read all role-assignment ACs granted to the user and all role-specification ACs for each granted role, validate the signatures of each AC, and check the various CRLs to ensure that no AC has been revoked since creation. Further, if the policy allows delegation of authority, the compliance checker would need to validate the chain of role-assignment ACs from the user to the trusted source of authority. This is clearly a time-consuming and onerous task.

Several methods can be used to improve system efficiency.

(1) With the PKC extensions, a relative fixed authority model can be realized. It has a relative simple structure, and is easy to realize.
(2) In order to simplify the permission-validation process, all role specifications and permission assignments can be placed into one policy AC[2].
(3) Store all role assignment ACs in LDAP.
(4) Implementing the push model.

6 Conclusion

To realize a flexible access management in extended enterprise networks base on Web service, we still have some problems to overcome, such as how to express policy constraints in XML, how to improve the efficiency of the system etc.

References

1. ITU-T Recommendation X.509, The Directory: Authentication Framework[s], Int'l Telecomm. Union, Geneva, 2000; ISO/IEC 9594-8
2. David W. Chadwick, Alexander Otenko, and Edward Ball. Role-Based access control with X.509 Attribute Certificates [J]. IEEE Internet Computing, 2003,volume 7,No.2, 62-69
3. William T.Polk, Nelson E. Hastings. Public Key Infrastructures that Satisfy Security Goals[J]. IEEE Internet Computing, volume 7(4), 2003, 60-67
4. Karl Furst, Thomas Schmidt, Gerald Wippel. Managing Access in Extended Enterprise Networks[J]. IEEE Internet Computing, volume 6(5), 2003, 67-74
5. Shaomin Zhang, Baoyi Wang, Lihua Zhou. Constructing secure web services based on XML[J]. The Second International Workshop on Grid and Cooperative Computing, Springer, 2003, 12
6. David F. Ferraiolo, Ravi Sandhu, Serban Gavrila, D. Richard Kuhn, Ramaswamy Chandramouli. Proposed NIST Standard for Role-Based Access Control [J]. ACM Transaction on Information and System Security, Vol.4, No.3, August 2001, 224-274
7. Weipeng Shi, Xiaohu Yang: Web Service Security Based on SOAP Protocol. Application Research of Computers, Vol. 20. Chengdu (2003) 100-105

Perceptually Tuned Packet Scheduler for Video Streaming for DAB Based Access GRID*

Seong-Whan Kim[1] and Shan Suthaharan[2]

[1] Department of Computer Science, University of Seoul,
Jeon-Nong-Dong, Seoul, Korea
swkim7@uos.ac.kr
[2] Department of Mathematical Science University of North Carolina,
Greensboro, North Carolina, USA
ssuthaharan@uncg.edu

Abstract. In this paper, we propose a video packet scheduling scheme, which guarantees QoS requirements for video streaming over digital audio broadcast (DAB) IP tunneling networks as a specific access Grid example. Video packet scheduler assumes an architectural framework for multimedia networks based on sub-streams or flows. Each sub-stream has a different QoS requirement, and we can control degree of satisfaction for the required QoS considering network traffic characteristics (e.g. delay, jitter, and packet loss). In this paper, firstly, we designed a classification scheme to partition video data into multiple sub-streams which have their own QoS requirements. Secondly, we designed a management (reservation and scheduling) scheme for sub-streams to support better perceptual video quality such as the bound of end-to-end jitter. We used MPEG-4 advanced visual coding (AVC) for our source coding. We have shown that our video packet scheduling scheme satisfies QoS requirements using real video experiment over DAB networks.

1 Introduction

Access Grid (AG) is exploring the use of large-scale projection based systems as the basis for building room oriented collaboration and semi-immersive visualization systems. A key part of this capability is the efficient transport of audio and video streams over the network. Quality of Service (QoS) guarantee in real time communication for multimedia AG applications is significantly important [1]. The real-time constraint is however not as critical as hard real-time system and is usually classified as soft real time. However, they impose other requirements which are peculiar to multimedia traffic, for example, bounded delay, bounded jitter, bounded packet loss, and bounded synchronization gap. The problem is that how to support QoS in the lossy packet network. In signal processing, source coding and channel coding are used for signal transmission, where source coding removes signal redundancy as well as signal components that are subjectively unimportant, and channel coding adds controlled redundancy so that transmission impairment such as packet loss or bit

---

* This research was supported by AIROIP Networks Inc., and Ministry of Information and Communication (MIC) in South Korea.

errors can be reversed. Video packet scheduler is a way of combining source and channel coding, and it can increase the traffic capacity of a network and also maximize a subjective image quality by exchanging some information between source coding and channel coding [2, 3]. In Sec. 2, we designed video packet scheduling algorithms to maximize perceptual quality in our framework. In Sec. 3, we show simulation results for our proposed scheme, and then compare the results with those of other scheduling schemes. In Sec. 4, we conclude with future research issues.

2 Substream Classification and Scheduling

Packet-based mobile communication networks inevitably introduce three types of impairment: packet loss (failure to arrive), packet corruption (bit errors occurring within the payload), and packet delay. Multimedia services can tolerate some level of loss and corruption without undue subjective impairment, especially if there is an appropriate masking built into the signal decoder. Extending this idea, the substreams model can be developed [3]. In the substreams model, the stream of packets is logically divided into substreams which have their own QoS attributes, and each packet is identified as to its substream, which implicitly specifies the QoS objective for that packet. To support the substream model, the source coder in the service layer should classify source video frames into packets with different QoS objectives, which are based on the HVS knowledge. It should also associate those packets with the appropriate substream [4]. We classified the motion by its motion magnitude relative to the maximum ±12 motion vectors. We classified the motion within ±5 as very low motion, the motion within ±12 as low motion, and the motion outside ±12 as high motion. After extracting the image blocks, each block is packetized with header information. The header gives information on the frame number, the location of the block for this frame, and the timing requirement for each discrete cosine transform (DCT) block. There is also one more parameter which specifies how many coefficients in the DCT block must be displayed.

After substream assignment according to their motion contents, video packets arrive at the destination, and are scheduled to maximize the video quality. We designed the (m, k, w) substream scheduling scheme extending previous real-time scheduling algorithms: Earliest Deadline First [5] and (m, k) scheduling algorithms [6]. EDF scheduler chooses the next customer based on the remained time to deadline [5]. A stream with (m, k)-firm deadlines experiences a dynamic failure if fewer than m customers meet their deadlines in a window of k consecutive customers. A stream gets closer to a failing state when its customer misses its deadline. The closer a stream is to a failing state, the higher the priority assigned to its next customer so as to increase its chances of meeting the deadline and thus moves the stream away from a failing state [6]. Extending these ideas, we assigned (m, k, w) values for each substream. As similar to (m, k) scheduler, (m, k) specifies bounded jitter requirements for each DCT block, and w means the number of DCT coefficients which should be displayed to guarantee the minimum perceptual quality. In the decoder part, the display order of substreams is determined by the *(m, k, w)* scheduler. The *(m, k, w)* scheduler chooses a substream among three (high, low, and very low motion) substreams, and gets a packet to display from the chosen sub-stream's first packet. The *(m, k, w)* scheduler has each substreams' deadline miss history, where a deadline miss can be checked by frame number in the video packets. The substream that has the highest probability of dynamic failure is chosen by the *(m, k, w)* scheduler. In effect, this scheduling will minimize the jitter that causes critical impairment in perceptual quality. The w parameter can be used to manage the congestion that occurs in the client and network.

3 Performance Evaluations

To show real video experiments, we implemented the three algorithms; Earliest Deadline First (EDF), (m, k), and (m, k, w); over the TCP/IP environment.

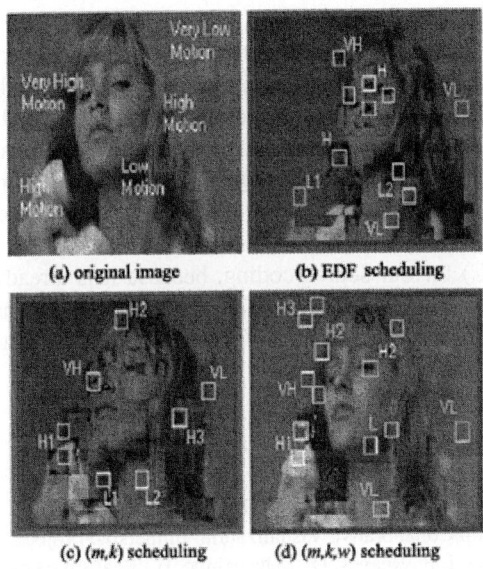

Fig. 1. Video packet scheduler simulation for Susie image sequences

As shown in Fig. 1 (a), a scene can consist of low motion; very low motion, high motion, and very high motion block areas. In EDF, high motion blocks will get the highest priority, and will prevent all the low motion blocks from arriving in the scene until all the high motion blocks arrive. Some high motion blocks will also starve to display in case there are so many high motion blocks. In that case, it will result in the display of past image blocks with respect to the current image blocks as shown in H and VH as shown in Fig. 1 (b), which shows an ugly face because some blocks are new and the other blocks are old. In (m, k), we can decrease the monopoly of high motion blocks which makes the starvation of low motion blocks by allocating allowable timing requirements instead of fixed priority for each high motion and low motion stream. It will result in better image quality as shown in $L1$, $L2$, $H1$, $H2$, and $H3$ of Fig. 1 (c). But some high motion blocks, for example VH, still can be delayed by other high motion blocks in the case there are so many high motion blocks in a scene.

Some low motion blocks can also cause the delay of high motion blocks. In (m, k, w), we can make VH as shown in Fig. 1 (d) visually better if we allocate smaller mandatory display parts for high motion blocks. Almost all the high motion blocks will be displayed and it will not make an unsynchronized face as shown in Fig. 1 (b), because all the delayed high motion blocks will switch to other recent high motion blocks very quickly as shown in VH, $H2$, and $H1$ of Fig. 3. However, it will show that there are some blocks which contain high frequency noise as shown in $H3$ of Fig. 1 (d). It is almost invisible, and it can be post filtered in the receiver side. Image quality in high motion blocks as shown in Fig. 1 (d) can be controlled by allocating more DCT coefficients to the mandatory part of the high motion blocks.

4 Conclusions

Real time multimedia communication over access Grid requires QoS guaranteeing system without bounded delay, bounded jitter, and bounded packet loss. To support these requirements, a loosely coupled video packet scheduler, which is based on the substream concept, is designed. The scheduler classifies each image packet into multiple substreams which have their own QoS requirements. To maximize perceptual video quality such as the bound of end-to-end jitter at the decoder side, we have designed a scheduling scheme (m, k, w) for substreams. We used MPEG-4 advanced visual coding (AVC) for our source coding, because it is already adopted in digital multimedia broadcast (K-DMB) standard in South Korea. We have shown that our video packet scheduler scheme satisfies QoS requirements using real video experiment over DAB networks.

References

1. Childers, L., Disz, T., Olson, R., Papka, M. E., Stevens, R., Udeshi, T.: Access Grid: Immersive Group-to-Group Collaborative Visualization. Immersive Projection Technology (2000)
2. Aras, C. M., Kurose, J. F., Reeves, D., Schulzrinne, H.: Real-time communication in packet-switched networks. Proc. IEEE (1994) 122–138
3. Reason, J., Yun, L. C., Lao, Al. Y., and Messerschmitt, D. G.: Asynchronous Video: Coordinated Video Coding and Transport for Heterogeneous Networks with Wireless Access, Mobile Wireless Information Systems, Kluwer Academic, Dordrecht (1995)
4. Chang, Y. C., Messerschmitt, D. G., Improving network video quality with delay cognizant video coding. IEEE Int. Conf. Image Proc. (1998) 27-31
5. Liu, C. L., Layland, J. W.: Scheduling algorithms for multiprogramming in a hard real-time environment. J. ACM (1973) 46–61
6. Hamdaoui, M., Ramanathan, P.: A dynamic priority assignment technique for streams with (m, k)-firm deadlines. IEEE Trans. Computers, (1995) 1443–1451

Collaborative Detection of Spam in Peer-to-Peer Paradigm Based on Multi-agent Systems

Supratip Ghose[1], Jin-Guk Jung[1], and Geun-Sik Jo[2]

[1] Intelligent E-Commerce Systems Laboratory, School of Computer Science & Engineering,
Inha University, Incheon, 402-751, Korea
{anik,gj4024}@eslab.inha.ac.kr
http://eslab.inha.ac.kr/
[2] School of Computer Science & Engineering,
Inha University, Incheon, 402-751, Korea
gsjo@inha.ac.kr

Abstract. The problem of unsolicited email has continued to increase every month for years. In order to deal with the huge amount of spam received day by day, we combined multiagent systems in a peer-to-peer framework with text categorization to identify spam. The content of the emails is analyzed by the classification algorithm "Support Vector Machines". Information about spam is exchanged between the agents through the networks identification numbers for emails, which where identified as spam, are generated and forwarded to all other agents, connected to the network. These numbers allow agents to identify incoming spam email. Our paper shows that, by this way, powerful email filters with high reliability based on distributed design can be achieved.

1 Introduction

The amount of spam is increasing very rapidly, hitting 64% of email traffic in April 2004 according to a study by Brightmail [1]. With increase in quantity and sophistication of spam, many solutions based on rule based [2] or adaptive statistical filters [3], [4] have been developed and spam filtering for automated message filtering. There have been ongoing research in this stage and there are commercial products with centralized design like Spamnet [9]. The most important benefit of the decentralized design for the spam filtering lies in the fact that effectiveness of the system grows with the number of its users. Additionally, decentralized systems provide higher availability and resilience to failure and attacks than similar centralized solution. A P2P system (e.g. Tapestry) is characterized as a distributed system where the entities at the network end are called peers, and a peer can share and exchange resources with other peers. Each peer in the P2P system is responsible for a range of the hash function's keys. In our proposed method we based our approach on an overlay network that is described in [8]. One of the necessary functions of the antispam agent is the ability to compare two different emails. For reasons of security, a unique identification number for every email is generated locally and send to other agents if the corresponding email is spam. Therefore, it is impossible to conclude from this identification number (hash value) to the content of the corresponding email. One popular

share the fault status amongst them. One of the advantages of our approach is that the change or modification of fault diagnosis algorithm is simple; the replacement of the existing algorithm is feasible through the deployment of mobile agents with a new algorithm.

The ADSD algorithm was developed for detecting and diagnosing faulty nodes in an arbitrary network. Monitoring of a distributed system consisting of a set of nodes is executed in two separate steps as shown in Figure 1; *detection* and *dissemination*, both of which are executed in a distributed way. First, in the detection step, nodes of a system test neighbor nodes one another periodically, forming a ring topology and then, in the dissemination step, pass on the test results to other nodes in the system. The testing results are either *faulty* or *fault-free* and this information is dispersed through a *dissemination tree*, a virtual path which is dynamically formed with the system nodes. The root of this tree is a node that initiates the information dissemination. On receiving the information, nodes determine whether to update their local databases with this information and whether to relay it to next level nodes depending on the recentness of the disseminated information. Once the failure of a node is diagnosed, this failure can be isolated from the system by reconfiguration.

We extend the original ADSD algorithm with *late acknowledgement* and *auditor* both of which are designed to have every node reach consensus on the grid status. We will discuss these mechanisms shortly.

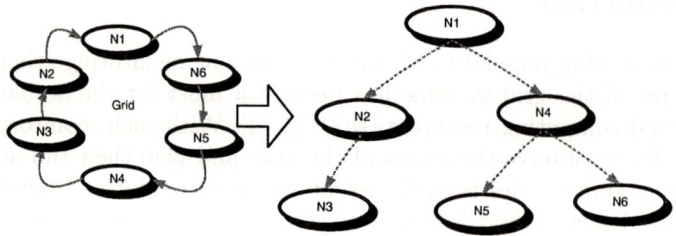

Fig. 1. Fault Detection and Binary Dissemination Tree

This paper is organized as follows. We discuss the extensions to the ADSD for mobile agent based grid monitoring in Section 2. We compare the performance of the mobile agent based approach with that of a traditional centralized scheme in Section 3. Section 4 concludes this paper.

2 Extensions to the ADSD Algorithm for the Fault Monitoring of Grid

The original ADSD algorithm is not able to deal with the failures and joins during the dissemination. In this section, we introduce *late acknowledgement* and *auditor* mechanisms to remedy these drawbacks of the original work.

Late Acknowledgement. It ensures that, once an event dissemination is started, all the fault-free nodes along the binary dissemination tree receive the events even if some of the nodes fail during the dissemination. Moreover, the late acknowledgement is able to detect these failures by timeout.

The acknowledgement is that a mobile agent on a node sends the `info-ack` message to its parent in the binary tree in order to notify that it received an event relayed from the parent. The late acknowledgement is that a mobile agent sends this `info-ack` *late*, i.e. after it confirms that its children nodes in the binary tree received the event. This confirmation is accomplished by the reception of `info-ack` from the children. The correctness of the late acknowledgement can be proved through a space–time diagram. However, the proof is omitted due to the length restriction.

Auditor. It is devised to ensure that newly joining nodes have same knowledge about the grid status as existing nodes have. The auditor mechanism works as follows; whenever a mobile agent on a node receives an event through the dissemination, it ensures that N_{tested} the node that it is testing at the moment receives the same event. To accomplish this, it checks whether N_{tested} is included in a *node list* to which the event is delivered. Note that the node list is conveyed together with the event for this purpose. The correctness of the auditor mechanism that can be proved in a similar way to that of the late acknowledgement is omitted because of the space restriction.

3 Performance Evaluation

We compare the performance of our mobile agent based approach with that of a traditional *central observer* scheme in which a designated object called *central observer* is entirely responsible for the fault detection of the grid in a centralized way. For experiments, we used four groups of grid nodes, each grid consisting of 5, 10, 15, 20 nodes respectively. As the performance metrics for comparison, we used the maximum number of monitoring messages on one node, required time for dissemination, and the maximum elapsed time from a fault detection until the completion of its dissemination.

In the case of mobile agent based approach, the number of monitoring messages is two, remaining as constant regardless of the grid size, while the central observer scheme shows the message numbers increasing linearly proportional to the grid size: $2 * N$ message exchange to monitor all the nodes in a grid, where N is the grid size. The dissemination time matches the results of numerical analysis; with u the time for one message transmission, and N the grid size, the central observer requires the time $2 * N * u$ until the dissemination finishes, while the mobile agent based case needs the time $2 * logN * u$. The maximum elapsed time from a failure detection to its completion also meets the analysis results: $O(N^2)$ for the central observer case, and $O(logN)$ for the mobile agent, where N is the grid size.

4 Conclusions

This paper proposed the mobile agent based approach for the fault monitoring in the grid environment. This approach based upon the ADSD algorithm is adaptive in the sense that participation of grid nodes in the monitoring is dynamic; the faulty nodes are excluded from the monitoring, while repaired nodes are able to join the monitoring again. The monitoring is also distributed; all the operations related with the monitoring are executed by the mobile agents on the grid nodes without the need of a central observer.

We improved the original ADSD algorithm with the late acknowledgement and the auditor; with the former mechanism, failures during the dissemination can be tolerated, the latter one allows the nodes to join a grid without the loss of consistency.

The experiment results showed that the performance of the proposed mobile agent based approach is superior to that of the traditional central observer model because of its distributed and adaptive characteristics. We presented the performance comparisons on the number of monitoring messages, the dissemination time, and the worst case time from the failure detection to the dissemination.

References

1. Fitzgerald, S.: Grid Information Services for Distributed Resource Sharing. In: Proceedings of the 10th IEEE International Symposium on High Performance Distributed Computing (HPDC-10'01). IEEE Computer Society (2001).
2. Brian, T., Crowley, B., Gunter, D., Lee, J., Thompson, M.: A Monitoring Sensor Management System for Grid Environments. Cluster Computing 4 (2001).
3. Lee, J., Gunter, D., Stoufer, M., Tierney, B.: Monitoring Data Archives for Grid Environments. In: Proceedings of the 2002 ACM/IEEE conference on Supercomputing. IEEE Computer Society Press (2002).
4. Lowekamp, B.: Combining Active and Passive Network Measurements to Build Scalable Monitoring Systems on the Grid. SIGMETRICS Perform. Eval. Rev. ACM Press(2003).
5. Tierney, R., et al.: A Grid Monitoring Architecture. Technical Report. Global Grid Forum (2002).
6. Wolski, R., Sprint, N., Hayes, J.: Network Weather Service : a Distributed Resource Performance Forecasting Service for Metacomputing. Future Generation Computer Systems 15 (1999).
7. Byrom, B., et al.: Information and Monitoring Service Architecture : Design, Requirements and Evaluation Criteria. Technical Report. Datagrid (2002).
8. Rangarajan, S., Dahbura, A., Ziegler, E.: A Distributed System-Level Diagnosis Algorithm for Arbitrary Network Topologies. IEEE Transactions on Computers, Special Issue on Fault-Tolerant Computing 44 (1995).

Grid QoS Infrastructure: Advance Resource Reservation in Optical Burst Switching Networks[*]

Yu Hua[1], Chanle Wu[2], and Jianbing Xing[3]

[1] School of Computer, Wuhan University, Wuhan, 430079, China
yhuastar@hotmail.com
[2] National Engineering Research Center of Multimedia Software, Wuhan, 430072, China
[3] The State Key Laboratory of Software Engineering, Wuhan, 430072, China

Abstract. The framework of Grid QoS has been focused widely and researched deeply, so it is necessary and very important to provide a complete and feasible architecture from the Grid application, Grid middleware, to network layer, even the concrete resources. In this paper, a kind of architecture for Grid QoS infrastructure is proposed based on other architectures suggested by Globus project, Global Grid Forum, IETF and so on. In addition, the advance resource reservation is pivotal to realize Grid QoS. At the same time, optical burst switching (OBS) networks with GMPLS support is analyzed and suggested as the Grid infrastructure. Related conclusion is also given in detail.

1 Introduction

Recently, because computational resources cannot satisfy the demands generated by some scientific applications, Grid is considered as a good candidate for providing more computing, storage capacities and high bandwidth shared in distributed circumstances. Nowadays, the grid has been studied widely and deeply from the views of applications and frameworks. In general, Grid applications require large amounts cheap bandwidth provisioned and scheduled on-demand [1].

In order to realize the Grid QoS, resource reservation, advance or on-demand, is necessary. Advance resource reservation may provide better QoS guarantee, but the utilization of corresponding resource is maybe lessened. Optical burst switching (OBS) has combined the advantages of the wavelength routing and optical packet switching in the current WDM networks. The data to be sent are collected at the edge of the OBS networks, and grouped into variable units, called bursts. Then, a control unit is firstly produced and sent towards the destination in order to set up an end-to-end path in advance and make corresponding resources reservation at each node. After the offset time, the data burst is transmitted without any acknowledgment message from the destination node. The essence of OBS is the separation between the control and data burst. Furthermore, the realizations of OBS need the support from the unified control plane, which has complete signaling protocols, routing protocols and link management protocol. Then, GMPLS is an ideal candidate as a unified control plane.

---

[*] Supported by the fund of State Key Lab of Software Engineering and National High Technology Development 863 Program of China under Grant (No.2003AA001032)

Nowadays, with the evolution of MPLS technology, GMPLS can be the unified control plane to provide reliable transportation and efficient resource utilization. In GMPLS networks, the Lambda Label Switch Paths may provide multi-granularity optical flows and a suite of mechanisms that assign a generalized label to bundle consecutive wavelengths, such as a Waveband LSP or a Fiber LSP. The GMPLS-based Grid infrastructure may control the changes by signaling protocols, such as RSVP-TE or CR-LDP.

The main ideas of grid computing include resource sharing and service-oriented methods. The end-to-end QoS realization guaranteed in network-based applications requires dynamic discovery and advance or immediate reservation of resources [2]. But, the resources are heterogeneous in types, independently controlled and administered [1]. Globus Resource Allocation Manager (GRAM) architecture does not address the issue of advance reservations with heterogeneous resource types. Lack of advance reservations, resources cannot be ensured timely when requests arrive. Some Grid network allocation proposals are based on DiffServ configuration and do not take into account the optical layers. Because of the nature of OBS mechanism, the realization and development of advance reservation may be more feasible.

2 Related Works

Currently, several grid frameworks have been provided and discussed widely. GARA (Globus Architecture for Reservation and Allocation) was proposed by Globus project in order to realize the End-to-End QoS guarantee. In particular, the WSRF (Web Service Resource Framework) has been also proposed in order to combine the web service with OGSI (Open Grid Service Infrastructure). In addition, G-QoSM (Grid-QoS Manager) is also a kind of management architecture based on OGSA. The advantages are mainly proper resources reservation and dynamic adaptation. Through OGSA, the Grid user can have a unified network view of its owned resources on top of different autonomous systems. The resources can either be solely owned or shared with other users. At the same time, several methods on resources reservation are also proposed in [3], [4], [5]. Furthermore, Global Grid Forum has begun to research the network plane in order to support the service and demands from application and middleware layers. In addition, many researchers think that web service may play an important role in Grid application. And certain architectures revised reflect the views. So, the research on Grid architecture has been made widely and deeply, especially in application and network layers in order to provide better Grid QoS.

3 Framework of Grid QoS Infrastructure

Now, a kind of complete architecture as Grid Qos infrastructure is proposed in Fig.1.

In OGSA framework for Grid QoS infrastructure, service should be self-contained and modular and can be discovered, registered, monitored, instantiated, created and destroyed with a certain management [2].

In this architecture, the QoS levels are made up of three layers, i.e. the application QoS, middleware QoS, and networks QoS. The application QoS can support all kinds of Gird application and many interfaces. The middleware QoS layer can bridge the service mapping between other two layers. It is made up of performance service and QoS service broker and policy management service. The performance service in-

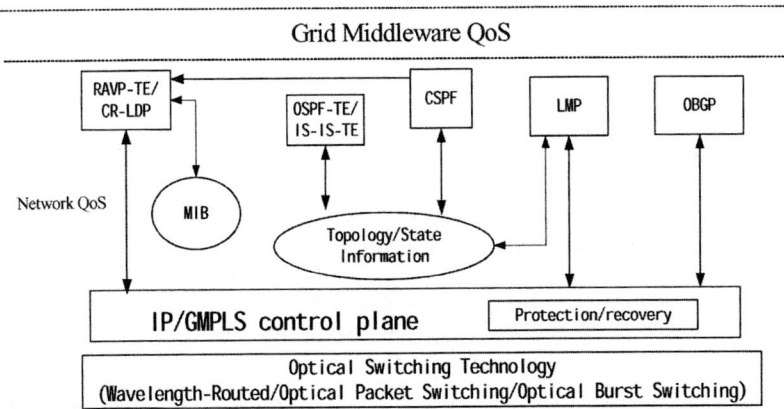

Fig. 1. Architecture for this Grid QoS Infrastructure includes three layers from top to bottom.

cludes performance registry, monitor and adaptation service. And QoS service broker includes service registry, reservation and allocation manager. Then, the policy management service is comprised of policy service, policy repository and admission control. The related services in middleware layer can be realized by corresponding protocols. Thus, the demands from the application layer and resource in the networks layer can be mapped correspondingly. In the layer of network QoS, GMPLS can be the unified control plane based on different types of optical switching. IETF has proposed the signaling protocols (RSVP-TE or CR-LDP), the routing protocols (OSPF-TE or IS-IS-TE), Link Management Protocol (LMP). The routing computation can be based on Constrained-based Shortest Path First (CSPF) with network state information. Thus, based on the architecture, the demands from application QoS layer can be mapped to service constraints in optical transport networks through middleware QoS layer.

With the OBS support, the advance reservation can be realized efficiently. The message of advance reservation may be notified with the control message along the selected path hop by hop. Then, corresponding resources, including computing and storage capacities, will be reserved and scheduled. After the offset time, the tasks of Grid application may arrive with data burst, reserved resources are allocated and tasks can be executed on time. When the tasks are completed, the corresponding resources may be released. The released resources may be notified and broadcasted by OSPF-TE or IS-IS-TE among different domains.

4 Summary

The development and background of current grid technologies have been presented comprehensively. Then, a kind of architecture for Grid QoS infrastructure is proposed based on other architectures. The architecture is complete and consists of three layers from top to bottom. The interior modules and the relationships each other are introduced. In addition, the advance reservation in OBS network layer with GMPLS support is also analyzed and suggested.

Additionally, future works will focus on the requests mapping and message communication in Grid QoS infrastructure. The unified resource manager among different domains will also focused especially.

References

1. Dimitra, S., Reza, N.: Optical Network Infrastructure for Grid. Informational Track, Grid High Performance Networking Research Group in Grid Global Forum (2004)
2. Burchard, L., Heiss, H.: Performance Issues of Bandwidth Reservations for Grid Computing. 15th Symposium on Computer Architecture and High Performance Computing. 11 (2003) 82 – 90
3. Junwei, C., Kerbyson, D.J., Nudd, G.R.: Performance Evaluation of An Agent Based Resource Management Infrastructure for Grid Computing. First IEEE/ACM International Symposium on Cluster Computing and the Grid. 5 (2001) 311 – 318
4. Rui, M., Maheswaran, M.: Scheduling Co-Reservations with Priorities in Grid Computing Systems. 2nd IEEE/ACM International Symposium on Cluster Computing and the Grid. 5 (2002) 250 – 251
5. Ernemann, C., Hamscher, V., Schwiegelshohn, U., Yahyapour, R., Streit, A.: On Advantages of Grid Computing for Parallel Job Scheduling. 2nd IEEE/ACM International Symposium on Cluster Computing and the Grid. 5 (2002) 31 – 38

A Scheduling Algorithm with Co-allocation Scheme for Grid Computing Systems

Jeong Woo Jo and Jin Suk Kim*

School of Computer Science, University of Seoul
90 Cheonnong-dong, Dongdaemun-gu, Seoul, Korea
{jwjo95,jskim}@venus.uos.ac.kr

Abstract. Since the problem of scheduling independent tasks in heterogeneous computational resources is known as NP-complete [2]. Many researchers propose heuristic scheduling algorithms [1, 3, 4, 5]. However, previous algorithms do not support co-allocation to several parallel computer system. In this paper, we propose with co-allocation the scheduling scheme. As a result, we show the performance of scheduling algorithm with co-allocation scheme.

Keyword: scheduling algorithm, co-allocation, Grid

1 Introduction

A Grid computing system is a system which has various machines to execute a set of tasks. In this paper, we propose a scheduling algorithm which assigns tasks to machines in a Grid computing system. The scheduling algorithm determines the execution order of the tasks which will be assigned to machines. Since the problem of allocating independent tasks in heterogeneous computational resources is known as NP-complete [2], an approximation or heuristic algorithm is highly desirable.

We assume that the scheduling algorithms are nonpreemptive, and all the tasks are independent. We can divide scheduling schemes into a static scheduling scheme and a dynamic scheduling scheme [1]. And dynamic scheduling scheme can be divided into an on-line mode scheduling scheme and a batch mode scheduling scheme. In this paper, we consider only the on-line mode scheduling algorithm.

This paper is organized as follows. In Section 2, overviews of the previous algorithms are presented. In Section 3, we examine terminology in scheduling algorithm. In Section 4, we simulate the on-line scheduling algorithm with co-allocation scheme, and finally a summary is given in Section 5.

2 Related Works

The scheduling problem has already been investigated by several researchers [1].

In MET(Minimum Execution Time), the scheduling algorithm assigns a task to the machine which has the least amount of execution time. This algorithm calculates only the minimum one among m machine execution times, and then the algorithm assigns the task to the selected machine.

The MCT(Minimum Completion Time) assigns a task to the machine which has minimum completion time, i.e., the algorithm gets the completion time for each task

---

* Corresponding Author.

```
Algorithm MECT with co-allocation
Input: T_i, {e_i{1}, e_i{2}, ..., e_i{m}}, m=|R|
Output: k (resource index)

I    find b_max = max_{R_j ∈ R} b_j
II   look for one or two cluster combination
III  sort by ready time of each node on the each cluster
IV   find the resources, R̂ that have completion time smaller than b_max
     if (| R̂ |> 0)
V        find k, such that e_i{k} = min_{R_j ∈ R̂} e_i{j}
     else
V'       find k, such that c_i{k} = min_{R_j ∈ R̂} c_i{j}
     endif
VI   return k
```

Fig. 1. MECT scheduling algorithm with co-allocation scheme

by adding begin time and execution time, and calculates the minimum one among m machine completion times. The algorithm assigns the task to the selected machine.

The MECT(Minimum Execution and Completion Time) assigns a task to the machine which has minimum execution time among selected resources [7]. First, compare all the resources of the ready time and calculate the maximum of the ready time. And in that time, it selects the resources with the lowest completion time. Next, a task is allocated to the resource with the lowest amount of execution time among the selected resources. If there isn't any selected resource, a task is allocated to a resource with the least amount of completion time.

Above algorithms are on-line mode scheduling algorithm that handle task assignment in a single cluster. In this paper, we proposed the scheduling algorithm to support the co-allocation in the Grid system. And we analyze the performance of the proposed algorithm.

3   A Scheduling Algorithm

The *execution time* $e_{i\{j\}}$ denotes the amount of time which is taken to execute task T_i on machine M_j [1]. The *completion time* $c_{i\{j,k\}}$ denotes the time at which machine M_j, M_k completes task T_i. Let the *begin time* of T_i on M_j, M_k be $b_{i\{j,k\}}$. We can see that $c_{i\{j,k\}} = b_{i\{j,k\}} + e_{i\{j,k\}}$ from the above definition. $T = \{ T_1, T_2, ... T_n \}$ is defined as the set of tasks. n is the number of tasks and m is the number of machines.

In this paper, we use a *makespan* as a performance metric for scheduling algorithms. The makespan is the maximum completion time when all tasks are scheduled, i.e., the makespan is defined as $\max_{T_i \in T} c_{i\{j\}}$.

Figure 1 shows that the MECT scheduling algorithm with co-allocation scheme. If task is allocated to two machines, scheduling algorithm calculate the new execution time. Because of there are network overhead between two machines. Therefore, the execution time when task (T_k) are performed on two machines (M_i, M_j) can be calculated as follows:

$$e_{k[i,j]} = max\ (e_{k[i]},\ e_{k[j]})\ /\ w_{ij}$$

w_{ij} denotes network weight between two machine and have value between 0 and 1.

By using the numerical formula which we mentioned above, the execution time of the task is longer than the execution time of the slowest machine.

4 Simulation Results

In this section, we experimented scheduling algorithm by simulation program that support co-allocation scheme. In this simulation, we assume that the execution time for each task on each machine is known prior to execution. This assumption is used when studying scheduling algorithm for heterogeneous computing systems [6]. We use two matrixes. One is task-machine matrix which has the execution times, and remainder is matrix that display network weight.

In this simulation, we made task-machine matrixes which have 1000 tasks and 20 machines. Each machine consists of several nodes. And network weight is $0 < w_{ij} < 1$.

We compare the performance of scheduling algorithm with co-allocation scheme with previous algorithms. Previous algorithm is a MET, MCT, KPB and MECT. And we applied the co-allocation scheme in MECT. To analyze the performance of scheduling algorithm in this paper, we have used makespan and response time.

Figures 2 represents the average makespan value of the scheduling algorithms. The makespan of scheduling algorithm with co-allocation scheme is less than other scheduling algorithms. This is because algorithm that have co-allocation scheme uses resources more efficiently.

Figure 3 shows response time for each scheduling algorithm. The response time of scheduling algorithm with co-allocation scheme is less than other scheduling algorithm.

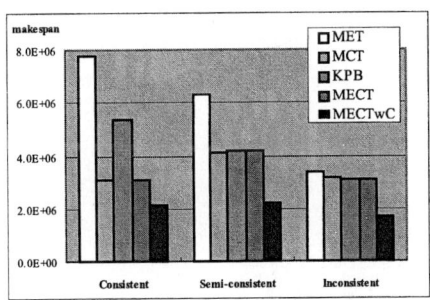

Fig. 2. The makespan value

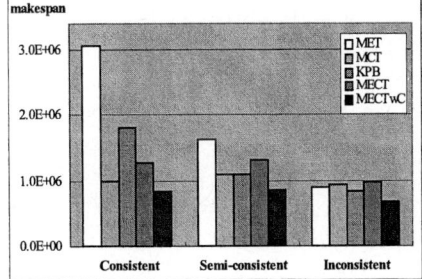

Fig. 3. The response time

5 Conclusion

Previous on-line mode scheduling algorithms are handling task assignment in a single cluster. So it is not support co-allocation scheme. In this paper, we analyzed the performance of on-line scheduling algorithms when tasks are split into multiple machines. We know from the experiment that the scheduling algorithm with co-allocation scheme can use resources more efficiently. So the performance of scheduling algorithm with co-allocation is better than previous scheduling algorithm.

References

1. M. Maheswaran, S. Ali, H. J. Siegel, D. Hensgen, and R. F. Freund, "Dynamic Matching and Scheduling of a Class of Independent Tasks onto Heterogeneous Computing Systems," Proc. of the 8th Heterogeneous Computing Workshop, pp. 30-44, April, 1999.
2. O. H. Ibarra and C. E. Kim, "Heuristic Algorithm for Scheduling Independent Tasks on Nonidentical Processors," Journal of the ACM, vol. 24, no. 2, pp. 280-289, April, 1977.
3. R. Buyya, J. Giddy, and D. Abramson, "An Evaluation of Economy-based Resource Trading and Scheduling on Computational Power Grids for Parameter Sweep Applications," Proc. of the 2nd International Workshop on Active Middleware Services, August, 2000.
4. H. Barada, S. M. Sait, and N. Baig, "Task Matching and Scheduling in Heterogeneous Systems using Simulated Evolution," Proc. of the 15th Parallel and Distributed Processing Symposium, pp. 875-882, 2001.
5. B. Hamidzadeh, Lau Ying Kit, and D.J. Lilja, "Dynamic Task Scheduling using Online Optimization," Journal of Parallel and Distributed Systems, vol. 11, pp. 1151-1163, 2000.
6. T. D. Braun, H. J. Siegel, N. Beck, L. L. Boloni, M. Maheswaran, A. I. Reuther, J. P. Robertson, M. D. Theys, B. Yao, D. Hensgen, and R. F. Freund, "A Comparison Study Mapping Heuristics for a Class of Meta-tasks on Heterogeneous Computing Systems," 8th IEEE Heterogeneous Computing Workshop, pp. 15-29, 1999.
7. Hak Du Kim and Jin S. Kim, "An On-ilne Scheduling Algorithm for Grid Computing Systems," Lecture Note in Computer Science, vol. 3033, pp. 34-39, 2004.

A Meta-model of Temporal Workflow and Its Formalization

Yang Yu, Yong Tang, Na Tang, Xiao-ping Ye, and Lu Liang

Department of Computer Science, Sun Yat-sen University, Guangzhou 510275, China
yuy@zsu.edu.cn

Abstract. Current researches on time constraints in workflow mainly focus on time-relevant process modeling and efficiency analysis based on the temporal attributes of processes and activities. This paper presents the concept of Temporal Workflow. Through extending and modifying the WfMC's Basic Process Definition Meta-model, a modified workflow meta-model is presented, and the temporal attributes of elements and their relations in it are analyzed in detail. Based on this, the primary elements in workflow are formalized, and time is introduced into workflow as a dimension.

1 Introduction

The effect of time element on workflow is extensive and important. Current researches on time in workflow include the following aspects:

1) *Timed Workflow Modeling and Model Verification.* Eder built the Workflow Timed Graph in which time is an attribute of an activity[1]. Sea Ling extended the workflow model based on Petri net(WF-net) to a Time Workflow-net(TWF-net)[2] , mapping the time of activity into a time denotation of transition. Hai Zhuge[3], DU Shuan-zhu[4] made some further researches based on these achievements.
2) *Efficiency of Workflow.* Based on timed workflow process model, many researchers have analyzed the efficiency of workflow, such as J. Leon Zhao[5], LI Jianqiang[6], etc. Further more, Jorge Cardoso presented a Qos model of workflow, considering the time, cost, fidelity, and reliability of workflow synthetically [7].
3) *Temporal Relation in Workflow.* Kafeza reduced the temporal relations of activities to 7 kinds[8]. LIU Xiang-qian made a research on temporal relations between activities in different process instances[9].

In summary, current time-relevant researches on workflow often focus on process modeling and analysis of time constraints and temporal relation constraints based on these models. In fact, not only the activities and processes are in the time dimension, but also the people who participate in these processes, tools (applications) and data are all under the effect of time.

Temporal workflow is the workflow in which time is introduced as a dimension[10]. It researches the temporal attributes of all the elements and their relations in workflow, and their effects on the characteristics and problems of workflow, such as safe, live, sound and exception handling, dynamic evolution, etc.

2 A Modified Workflow Meta-model and Temporal Attributes

A meta-model is a model that defines a language for expressing a model, which is used to define the construct and rules of a semantic model. The workflow meta-model is used to describe the elements, their relations and the attributes of these elements and relations in a workflow system. The first step of Temporal Workflow research is to endue these elements and relations with temporal attributes.

WfMC presented a Basic Process Definition Meta-model [11]. Which elements should be defined in meta-model depends on the requirements of the application domain. We present a meta-model, including a Build-time Meta-model and a Run-time Meta-model. For many applications, the conceptions defined in this meta-model are helpful.

2.1 Build-Time Meta-model

Build-time meta-model consists of 4 sub-models: organization meta-model, information meta-model, application meta-model and process meta-model.

1) *Organization Meta-model*
As shown in Fig.1, organization meta-model describes the resource-relevant conceptions and their relations in workflow. Activities are executed by resources. Resources can be divided into organizational units according to the structure of the organization, or roles according to the functional characteristics.

Considering that resources are often allocated to activities by the intersection of organizational units and roles, a conception "UnitRole" is introduced. An organizational unit consists of several UnitRoles, e.g., market department consists of leaders, supporting engineers, salesmen, etc. At the same time, these leaders belong to a subset of department leader and supporting engineers belong to a subset of engineer, etc. A user can be either a person or an agent [12].

One user can entrust another user with his UnitRoles. "RR" includes "Belong" relation and "Peer" relation[10].

Putting the organization meta-model into time dimension, we can see that an organization has its created time and valid time, for example, a company is founded on a day, and its business license is valid for 10 years; a user is employed on a certain day and the valid time is defined in the labor contract; a person is appointed a role from one day, and the valid time is 3 years; a user will have a holiday for 3 days from the next day, he entrusts his roles to another user, the "entrust" relation has its start time (the next day) and valid time (3 days), etc.

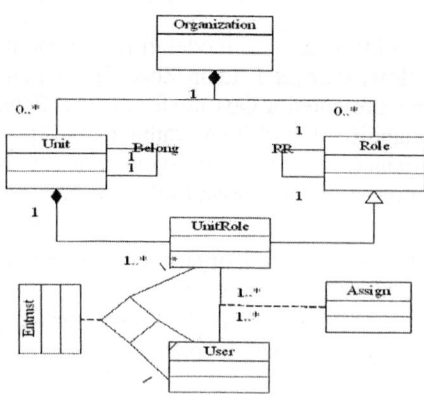

Fig. 1. Organization Meta-model

2) *Information Meta-model*

The data perspective of workflow deals with production data and control data [16]. Production data are information objects (e.g., documents, forms). They are handled by the applications. Control data are data introduced solely for workflow management purposes. They are produced and consumed by workflows, such as the input and output parameters of activities, variables introduced for routing purposes, the index or control parameters for production data (e.g., identifier, priority, deadline), etc. As shown in Fig. 2, information meta-model describes the conceptions and their relations about control data.

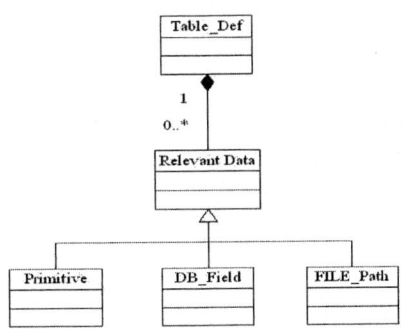

Fig. 2. Information Meta-model

In this meta-model, relevant data can be a primitive type (e.g., parameters, variables), or a DB_Field type which links to a field of database table, or a FILE_Path type which includes the file path. Table_Def is a structured abstract of a group of relevant data, which always expresses a meaning.

The elements in this meta-model also have their own temporal attributes, e.g., in a Claim Handling System, a claim table should be created when receiving a customer's claim, and the table should be handled within 3 days. That means the claim table has a created time and a valid period of 3 days. If the claim table is not handled within its valid period, an exception-handling process should be started.

3) *Application Meta-model*

As shown in Fig. 3, the application meta-model describes the conceptions and their relations about invoked applications. An invoked application can be a common table-handling application, a conventional application, or a URL to a web service.

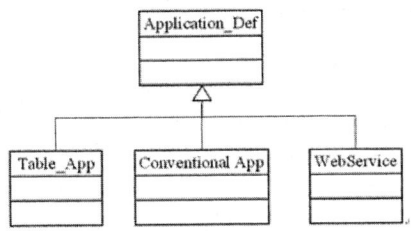

Fig. 3. Application Meta-model

According to the rules specified in ISO 9000, as a tool, an invoked application must be checked with its validity and veracity periodically. Therefore, an application has its own valid period.

4) *Process Meta-model*

As shown in Fig. 4, relevant conceptions are defined in process meta-model to specify which activities need to be executed and in what order (i.e., the routing or control flow). In order to support the structured process definition [13], the conception sub-process is introduced. Profiting from the idea in literature [14], the description of the structures (split, join) and their constraints (AND, OR, XOR) is separated from the description of activity, and a conception connector is introduced to express it. In fact, a connector can be considered as an activity with some special functions. The executing order of activities is determined by "Transition-Rule".

One activity can be described as: one role operates on a table with an invoked application, during which the state of activity changes according to a set of state rules.

The changes of activity states and the transitions among the activities are all inspired by events. Transition rules and state rules consist of events and relevant data.

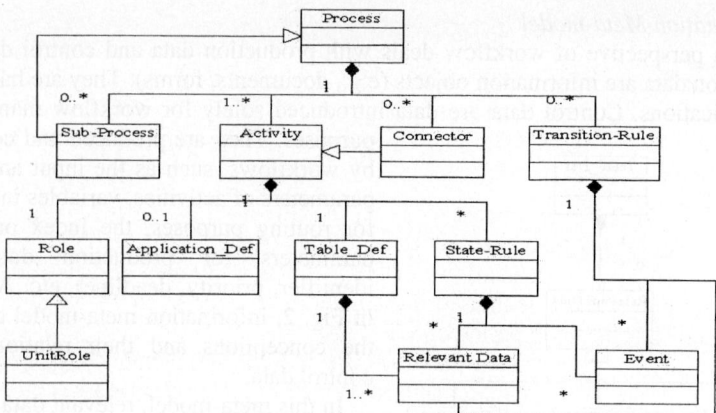

Fig. 4. Process Meta-model

Each version of a process definition has its created time. Once a new version is created, new instances will be generated from it. An activity definition has its created time and valid period too. To meet the needs of workflow management, a process or an activity is always assigned an earliest finish time and a latest finish time [1].

2.2 Run-Time Meta-model

As shown in Fig. 5, run-time meta-model describes the conceptions and their relations in workflow run-time period. In this meta-model, process instances, activity instances, applications and tables are all instantiated from the relevant conceptions in the build-time meta-model. Here, roles have been mapped into concrete users.

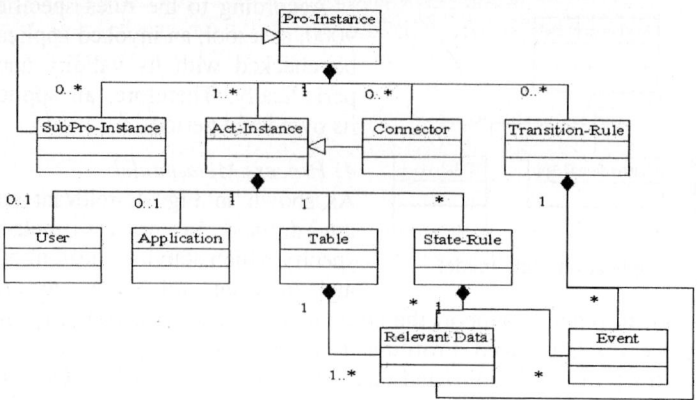

Fig. 5. Run-Time Meta-model

Every process instance or activity instance has its own created time and lifetime. In the build time, the temporal attributes of processes and activities are always defined in relative time. In run time, the temporal attributes of process instances and activity instances are always defined in absolute time.

3 A Formal Model of Temporal Workflow

From the research on workflow meta-model, we get the primary elements and relations, and have analyzed their temporal attributes. Based on these results, we can describe the temporal workflow as a formal model. Here, we only formalize the primary entities.

Definition 1: A workflow process WP = < *Pid, Pn, Ver, TS, AS, F, tp, Tp* >, where: *Pid* is the identifier of the process template; *Pn* is the process name; *Ver* is the version number; *TS* is the set of transition rules. *AS* is the set consisting of activities, connectors and sub-processes; *F* is the set of mapping rules from *AS* to *TS*; *tp* is the created time of WP; *Tp* is the valid period of WP.

From definition 1, we can see that there is a good mapping from the process definition to Petri Net. For example, an activity can be mapped to a TRANSITION, and a set of transition rules can be mapped to a PLACE.

Definition 2: An activity WA = <*Aid, An, Ver, Pid, Rid, Tid, APid, Lim, ta, Ta, State, SRule*>, where: *Aid* is the identifier of WA; *An* is the activity name; *Ver* is the version number; *Pid* is the identifier of the process which WA belongs to; *Rid* is the identifier of the role; *Tid* is the identifier of the table instance; *APid* is the identifier of the application; *Lim* is the limit for activities to access relevant data. *ta* is the created time of WA; *Ta* is the valid period of WA; *State* is a set of activity states; *SRule* is a set of state rules.

Definition 3: A process instance Ip = < *Ipid, Pid, TS, AS, F, PS, tp, Tp* >, where: *Ipid* is the identifier of the process instance; *Pid* is the identifier of the process, which Ip belongs to; *TS* is the set of transition rules. *AS* is the set consisting of instances of activities, connectors and sub-processes; *F* is the set of mapping rules from *AS* to *TS*; *PS* is the current state of Ip; *tp* is the created time of Ip; *Tp* is the valid period of Ip.

Definition 4: An activity instance Ia = < *Iaid, Ipid, Aid, Uid, Tid, APid, ta, Ta, SRule, s* >, where: *Iaid* is the identifier of the activity instance; *Ipid* is is the identifier of process instance which Ia belongs to; *Aid* is the identifier of activity which Ia belongs to; *Uid* is the identifier of the user; *Tid* is the identifier of the table instance; *APid* is the identifier of the application; *ta* is the created time of Ia; *Ta* is the valid period of Ia; *SRule* is a set of state rules; $s \in State$ is the current state of Ia.

Definition 5: A role WR = <*Rid, Rn, C, A, tr, Tr*>, where *Rid* is the identifier of WR; *Rn* is the name of WR; *C* is the set of WR's abilities; *A* is the set of limit rules for WR to access relevant data; *tr* is the created time of WR; *Tr* is the valid period of WR.

Definition 6: A table WT = <*Tid, Tn, Ver, DataList, Frame, td, Td*>, where *Tid* is the identifier of WT; *Tn* is the name of WT; *Ver* is the version number; *DataList* is a list of workflow-relevant data; *Frame* is the definition of table frame; *td* is the created time of WT; *Td* is the valid period of WT.

4 Conclusion and Future Work

This paper presents the concept of Temporal Workflow. Through extending and modifying the workflow meta-model, the temporal attributes of elements and their relations in it are analyzed in detail. Based on this, primary elements in workflow are

described formally, and time is introduced into workflow as a dimension. The future researches on temporal workflow will include the following aspects:
1) Temporal workflow process modeling based on meta-model and Petri Net.
2) Analyzing the efficiency and validity of temporal workflow.
3) Exception handling and flexibility of temporal workflow.
4) The implementation of temporal workflow.

References

1. Eder, J., Panagos, E., Pozewaunig, H., et al. Time management in workflow systems. In: Proceedings of the 3rd International Conference on Business Information Systems. Heidelberg, London, Berlin: Springer-Verlag, 1999. 265~280.
2. Ling, S., Schmidt, H. Time Petri nets for workflow modeling and analysis. In: Proceedings of the IEEE International Conference on System, Man and Cybernetics. 2000.4:3039-3044.
3. Hai Zhuge,To-yat Cheung,Hung-keng Pung. A timed workflow process model. The Journal of Systems and Software, 2001.55: 231-243
4. Du Shuan-zhu, Tan Jian-rong, Lu Guo-dong. An Extended Time Workflow Model Based on TWF-net and Its Application, Journal of Computer Research and Development, 2003. 40(4):524-530 (Chinese).
5. J. Leon Zhao,Edward A. Stohr. Temporal workflow management in a claim handling system, Software Engineering Notes, March 1999.24(2):187-195.
6. Li Jian-qiang, Fan Yu-shun. A Method of Workflow Model Performance Analysis, Chinese Journal of Computers, 2003.26(5):513-523(Chinese).
7. Jorge Cardoso, Amit Sheth, Krys Kochut. Implementing QoS Management for Workflow Systems. Technical Report, LSDIS Lab, Computer Science, University of Georgia, July 2002.
8. Kafeza, E., Kamalakar, K. Temporally constrained workflows. In: Internet Applications. Lecture Notes in Computer Sciences 1749, Springer-Verlag, 1999. 246~265.
9. Liu Xiangqian, Wang Xiaolin, Zeng Guangzhou. Multi-workflow Combining Method Based on Coordination Mechanisms, Computer Engineering, 2003.29(2):118-119,194 (Chinese).
10. Yang Yu, Yong Tang, Lu Liang, Zhi-sheng Feng. Temporal extension of workflow meta-model and its application. In: Proceedings of The 8th International Conference on CSCW in Design (CSCWD'2004) Volume II, Xiamen, P. R. China, 2004, 293-297.
11. Hollingsworth D. The workflow reference model. Workflow Management Coalition. 1995.
12. Dickson K.W. Chiu, Qing Li, Kamalakar Karlapalem. A Meta Modeling Approach To Workflow Management Systems Supporting Exception Handling, Information Systems, 1999.24(2):159-184.
13. Johann Eder, Wolfgang Gruber. A Meta Model for Structured Workflows Supporting Workflow Transformations, LNCS 2435, Springer-Verlag Berlin Heidelberg,2002:326-339.
14. Zhao Wen, Hu Wen-Hui, Zhang Shi-Kun, Wang Li-Fu. Study and Application of a Workflow Meta-Model, Journal of Software, 2003.14(6)1052-1059(Chinese).

Commitment in Cooperative Problem Solving

Hong-jun Zhang[1,2], Qing-Hua Li[1], and Wei Zhang[1]

[1] Department of Computer Science and Technology,
Huazhong University of Science and Technology, Wuhan China 430074
[2] Information Department of the 161 Hospital, The General Logistics Department,
Wuhan 430010
zhanghjun@hotmail.com

Abstract. In this paper we aim to describe commitment attitudes in teams of agents involved in Cooperative Problem Solving (CPS). Particular attention is given to the collective commitment. First, a logical framework is sketched in which a number of relevant social and collective attitudes is formalized, leading to the plan-based definition of collective commitments. Then, a social structure component is added to this framework in order to capture how the structure-relation concerning the group can influence the collective commitment and CPS.

1 Introduction

In *multi-agent systems* (MAS), a paradigmatic example of joint activity is *cooperative problem solving* (CPS) in which a group of autonomous agents choose to work together, both in advancement of their own goals as well as for the good of the system as a whole. As in human societies, cooperation is an intricate and subtle activity in multi-agent systems. Many researchers have attempted to formalize it[1,2,3]. There is an agreement about "shared mind" view in these studies. The collective commitment is accounted for the crucial notion for stable group formation and activity. It is concerns itself with individual mental concepts, such as the mutual beliefs and intentions of the participating agents. However, as open agent societies (such as GRID, The server-oriented computing) are becoming ubiquitous, recent developments recognize that the modeling of cooperation in MAS cannot simply rely on the agent's own architectures and capabilities. Cooperation can be affected and constrained by some social structures [4]. In this paper, we discuss and formalize collective commitment by using several concepts as social structure and plan, we gave a first attempt to formalize how the social structure can affect collective commitment and then CPS process.

2 The Logical Framework

The logical framework L is an extension of the BDI paradigm in [2], which based on a branching-time logic. This means that formulae are interpreted over tree-type branching structures that represent all conceivable ways the system can evolve. Due to space limitations we will only describe the basic ideas and the extensions made to the original framework briefly.

Definition 1. The language is based on three sets: atomic propositions symbols set Φ, a finite set A of agents. a finite set Ac of atomic actions. A motivational attitude can appear in two forms: with respect to propositions, or with respect to actions.

Definition 2. The class Sp of social plan expressions is defined inductively, $stit(G,\varphi)$, $confirm(\varphi)$, $<\alpha;\beta>$ (sequential composition), $<\alpha \| \beta>$ (parallelism), $<\alpha | \beta>$ (choice) are social plan expressions.

Definition 3. (semantics model) Based on Kripke's possible world model, the model is a structure $M =< W, D_A, D_{Ac}, B, G, I, Tr, Cap, Will, Val >$

We adopt the KD45 axiom for belief, KD axiom for goal and intention.

3 Commitment and Social Structure

Commitment is a mental states which intuitively reflects the relations and directly leads to agent actions. An agent can make commitment to an action, to a goal, or to another agent about an action or a goal. Commitment results in rights and obligations so it can represent the social attitudes of an agent. In CPS, commitment attitudes are considered on the following three levels: *individual*, *social* and *collective*.

3.1 Collective Commitment

The collective commitment is the strongest motivational attitude to be considered in teamwork [2]. A collective intention may be viewed as an inspiration for team activity, whereas the collective commitment reflects the concrete manner of achieving the intended goal by the group. Planning provides this concrete manner. The social plan should result from the main goal by task division, means-end analysis, and action allocation, as reflected in constitute(φ,p). The constitute(φ,p) means P is a plan to achieve goal ϕ. In addition, for every one of the subgoal φ that occur in plan P, there should be one agent in the team who is socially committed to at least one agent in the team to fulfill the subgoal. It is defined as below:

$$CCOMM\ (G,\varphi,P) \leftrightarrow CINT\ (G,\varphi) \wedge constitute\ (\varphi,P) \wedge \bigwedge_{\varphi \in P} \bigvee_{i,j \in G} SCOMM\ (i,j,\varphi) \wedge CBEL\ (G, \bigwedge_{\varphi \in P} \bigvee_{i,j \in G} SCOMM\ (i,j,\varphi))$$

3.2 Social Structure and Commitment

Commitment is important for cooperation activity and society. However, Recent developments recognize that the agents are not so free to commit themselves as they like: they are conditioned by some social structure. A social structure is a set of relations that hold between agents in a society, and determine the rights and responsibilities of each agent in the society with respect to its peers. There are several structures in any MA system: The interdependence and power structure; the acquaintance structure; the communication structure and the commitment structure etc. Those structure has been seen as a natural way of improving coordination and cooperation in mutiagent systems because they can restrict, and make more predictable, the behavior of agents. For example, using dependence network can determines and predicts partnerships and coalitions formation [5].

We will represent the social structure of the MAS by Socstr(G):

$$Socstr\ (G) := \{relation\ (i_l, i_l', \varphi, ty) \mid i_l, i_l' \in G\}$$

Meaning that Socstr (G) is the set of all existing relations between two agents and in the group G. ty is relation types. The relation relation(j,k, ,power) means informally that agent j has the power to get k to achieve φ. Power relations are reflexivity and transitive. The relation relation(j,k, ,auth) means that agent j is authorized by k to achieve φ. Authorization relations can be established by mutual agreement [6]. We have follow axiom:

$$\models relation\ (i, j, \varphi, power) \to relation\ (i, j, \varphi, auth)$$
$$\models request\ (i, j, \varphi) \land relation\ (i, j, \varphi, power) \to SCOMM\ (j, i, \varphi)$$
$$\models request(i, j, \varphi) \land relation(i, j, request(i, j, \varphi), auth) \to SCOMM(i, j, \varphi)$$

Under social structure, the collective commitment is defined as above can be modified as:

$$CCOMM\ (G, \varphi, P \mid Socstr\ (G)) \leftrightarrow CINT\ (G, \varphi) \land constitute\ (<i_1 : \varphi_1, ..., i_n : \varphi_n>, \varphi, P, G)$$
$$\land \bigwedge_{\varphi_l \in P} \bigvee_{i_l, j \in G} \{(request\ (j, i_l, \varphi_l) \land relation\ (j, i_l, \varphi_l, power) \to SCOMM\ (i_l, j, \varphi_l)) \lor$$
$$(request\ (i_l, j, \varphi_l) \land relation\ (i_l, j, request\ (i_l, j, \varphi_l), auth) \to SCOMM\ (i_l, j, \varphi_l))\}$$
$$\land CBEL\ (G, \bigwedge_{\varphi_l \in P} \bigvee_{i_l, j \in G} SCOMM\ (i, j, \varphi))$$

4 The Commitment in CPS

In [1], Wooldridge and Jennings presented a model of CPS that divided the process into four stages. The third stage is plan formation. Here the team divides the overall goal into subtasks, associates these with actions and allocates these actions to team members. In terms of motivational attitudes the end results of this stage is a collective commitment to perform the social plan that realizes the overall goal. Formally:

$$[confirm\ (succ\ (division\ (\varphi, \sigma); means - end\ (\sigma, \tau); allocation\ (\tau, P)]$$
$$CCOMM\ (G, \varphi, P \mid Socstr\ (G))$$

5 Conclusion

This paper based a logical framework, formalized a number of relevant social and collective attitudes in CPS, leading to the plan-based definition of collective commitments. Then, a social structure component is added to this framework in order to capture how the structure-relation concerning the group can influence the collective commitment and CPS.

The research work is still coarse, and many detailed are to be investigated. Future work includes formalize social structure model and upon which account for how cooperation can arise, and how it can proceed.

References

1. M. Wooldridge and N. R. Jennings. Cooperative problem solving. In *Journal of Logic and Computation*, 9(4), 1999.
2. B. Dunin-Keplicz and R. Verbrugge. Collective motivational attitudes in cooperative problem solving. In V.Gorodetsky et al. (eds.), Proceedings of the First International workshop of Central and Eastern Europe on Multi-agent Systems(CEEMAS'99), St. Petersburg,1999,pp.22-41
3. F. Dignum, B. Dunin-Keplicz and R. Verbrugge. Dialogue in Team Formation, In F. Dignum and M. Greaves (eds.) *Issues in Agent Communication (LNCS-1916)*, Springer-Verlag, 2000, pages 264-280.
4. C.Castelfranchi, Modelling social action for AI agents, Artificial Intelligence, Volume: 103,Issue: 1-2, August, 1998, pp. 157-182
5. J.S. Sichman, R.Conte, Y.Demazeau, and C.Castelfranchi, A social reasoning mechanism based on dependences networks, Proceeding of the 11th European Conference on Artificial Intelligence, T.Conte editor, Amsterdam, the Netherlands, 182-192, 1994.
6. V.Dignum , J.-J.Meyer, F.Dignum , and H.Weigand, Formal Specification of Interaction in Agent Societies. In: Hinchey, M., Rash, J.,Truszkowski, W., Rouff, C., Gordon-Spears, D., (Eds.): *Formal Approaches to Agent-Based Systems* (FAABS'02). LNAI 2699, Springer 2003.

iOmS: An Agent-Based P2P Framework for Ubiquitous Workflow

Min Yang[1], Zunping Cheng[2], Ning Shang[2], Mi Zhang[2], Min Li[1], Jing Dai[2], Wei Wang[2], Dilin Mao[1], and Chuanshan Gao[1]

[1] Department of Computer Science and Engineering, Fudan University,
Shanghai 200433, China
{m_yang,leemin,dlmao,cgao}@fudan.edu.cn
[2] Department of Computer Information and Technology, Fudan University,
Shanghai 200433, China
{032021175,032021171,012021158,daijing,weiwang1}@fudan.edu.cn

Abstract. Workflow is referred to automatic business process, and current involving methods mainly depend on traditional centralized C/S infrastructure and email systems. As far as we know, the existing methods have following weaknesses in nature: 1) The abilities of workflow processing is difficult to scale up; 2) Workflow execution is not flexible; 3) Workflow transfer process is not reliable; 4) Handheld devices in mobile environments are seldom considered. However, peer-to-peer technology is characterized by decentralized control, large scale, and extreme dynamism, while agent technology can construct overlay network more easily and process workflow automatically. In this paper, we present a novel framework based on office-secretary model, which utilizes concept of peer-to-peer, agent and transaction mechanism. In particular, we believe that iOmS framework could easily support scalable, dynamic, reliable, and ubiquitous workflow applications.

1 Introduction

The workflow concept emanates from the notion of commercial processes and social office activities. Some researchers have concentrated on the routine aspects of work activities to improve the processes efficiency[9].

Workflow has two types [12]: Internal organizational; External organizational. Workflow is ubiquitous. In order to pursue this need, lots of products have been developed [10]. Nearly all of the existing business products are characterized by Client-Server mode, email systems or message queues. There are several intrinsic limitations [8]: It is difficult to scale up the abilities of workflow processing; Workflow execution is not flexible; Workflow transfer process is not reliable; handheld devices in mobile environments are seldom considered.

As such, we present a novel architecture and framework – iOmS based on office-secretary model for end-to-end applications. We construct our infrastructure with the consideration of decentralized control, large scale, self-organization, resilience and usability.

This paper is organized as the following. Section 2 presents related work, section 3 describes iOmS model, and section 4 offers conclusion and future work.

2 Related Work

Workflow is a concept closely related to reengineering and automating business and information processes in an organization. [3] [4] [5] [6] [9] [10] provide several typical workflow management systems.

Peer-to-peer is a technology concept applied at different levels of the systems architecture. [1] [2] give an overview of the different areas peer-to-peer technology is used and introduces the main characteristics of peer-to-peer systems. [11] introduces a framework for the development of agent-based peer-to-peer systems. [13] proposes to evaluate agents as one enabling technology for active wireless / mobile networking. [7] proposes to develop middleware services that additionally provide services for information sharing in mobile ad-hoc networks, because the possibility to share information is mission critical for many ad-hoc network applications.

3 The iOmS Model

Overview of iOmS

An iOmS system is structurally composed of a self-organizing P2P overlay network of interconnected intelligent Offices (iOffice) and mobile Secretaries (mSecretary), as illustrated in Figure 1. iOmS systems can be more scalable and robust since working load is respectively assigned to each iOffice. Joint devices, no matter whether connected to Internet, or joined up with each other through an ad-hoc protocol, or organized in any other form of interconnection, as long as running iOmS-based application, can be regarded as iOffices. The network is characterized by the absence of any fixed structure, as iOffices come, go and discover each other on top of a communication substrate. Thus, iOmS systems can put up with variform connectivity, which is the base of the Internal and External organizational workflow.

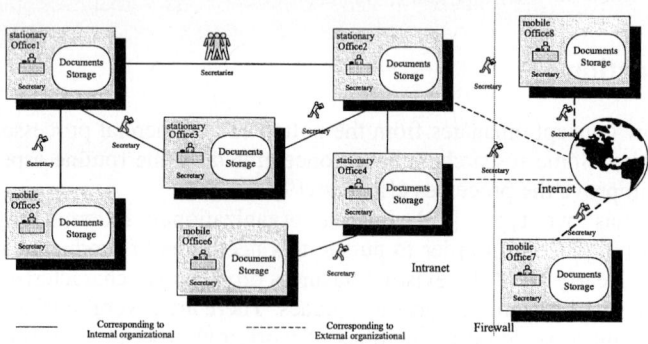

Fig. 1. Overview of our iOmS network

Intelligent Office (iOffice)

An iOffice corresponds to a peer node, which is the abstraction of a user's desktop. The principle characteristic of iOffice is to accomplish workflows automation management and to complete tasks such as search, statistic, pushing and pulling information, etc. by means of manipulating mSecretaries that adhere to it. Also, iOffice must keep track of information about the overlay network for mSecretaries to locate and route. For an end user, an iOffice may directly offer services to mSecretaries, or serve other upper-layer applications that directly interact with the user.

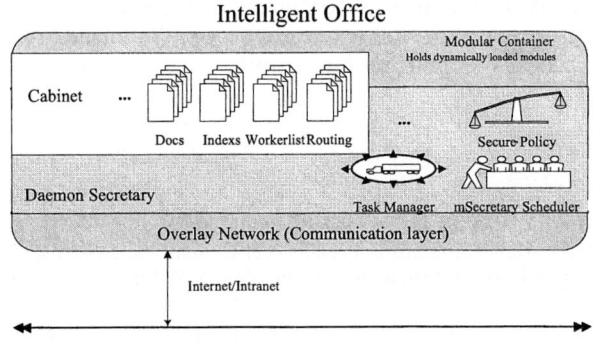

Fig. 2. Structure of an iOffice

Figure 2 illustrates the structure of an iOffice composed of several logical modules. The dSecretary is one of the most important roles in the iOffice, which provided with a "sandbox" so that we can limit the documents available to them and prohibit them from performing potentially dangerous actions. The task of a dSecretary is to serve upper-layer applications, to interact with external mSecretaries, to interact with local mSecretaries, to operate on all kinds of cabinets, and to carry out secure polices.

Mobile Secretary (mSecretary)
Mobile Secretary is generated by iOffice in response to user's request. Those mSecretaries that cannot accomplish their mission within a time-to-live (TTL) parameter will be terminated and automatically removed. A mSecretary is generally composed of three modules which are route map, tasks, briefcase(including all documents).

Documents
It is the abstractions of business archives and configures information, which includes two components, a header and a body. Header is made up of meta data, such as moving routes, business rules, timestamps, signatures, and so on. Body is a container to comprize the real business content.

4   Conclusions and Future Work

Current workflow applications are restricted to Client/Server (Browser/Server) infrastructure, email system and message queue in common networks. On the one hand, C/S (B/S) systems limit scalability and robustness, while email system and message queue lead to unreliable and unpredictable; on the other hand, ubiquitous workflow is required in e-business applications. Considering characteristics of the peer-to-peer and agent, we propose a novel framework to overcome the weaknesses of traditional workflow applications. In this paper, we present an academic model and implement primary iOmS prototype on heterogeneous hardware and platform. We are currently performing emulations based on a more realistic network environment, which will prove that the novel architecture is scalable, flexible, reliable and mobile-fit in terms of experimental results. Also we believe that programmers can develop upper-layer applications for battlefield information system, emergency rescue communication system, etc. Because of the complexity of workflow and business process, there are still a lot of things to do, such as concurrent mechanism, instantaneous message, workflow management engine, failover recovery, etc. We intend to implement a more complex and practical system in our future work.

References

1. D. Milojicic, V. Kalogeraki, R. Lukose, K. Nagaraja, J. Pruyne, B. Richard, S. Rollins, Z. Xu: Peerto-Peer Computing, hp Technical Report(2002).
2. Andreas Mauthe, David Hutchison: Peer-to-Peer Computing: Systems, Concepts and Characteristics, Praxis in der Informationsverarbeitung & Kommunikation (PIK), K. G. Sauer Verlag, Special Issue on Peer-to-Peer, 26(03/03)(June 2003).
3. Gregory, Alan, Bolcer: Magi: An Architecture for Mobile and Disconnected Workflow, IEEE Internet Computing, (May·June 2000).
4. Robert Tolksdorf: Workspaces: A Web-Based Workflow Management System, IEEE Internet Computing, (September 2002).
5. Sea Ling, Seng Wai Loke: Advanced Petri Nets for Modelling Mobile Agent Enabled Interorganizational Workflows, the Ninth Annual IEEE International Conference and Workshop on the Engineering of Computer-Based Systems (ECBS.02)(April 2002).
6. Junwei Cao, Stephen A. Jarvis, Subhash Saini, Graham R. Nudd: GridFlow: Workflow Management for Grid Computing, 3st International Symposium on Cluster Computing and the Grid(May 2003).
7. Thomas Plagemann, Vera Goebel, Carsten Griwodz, Pål Halvorsen: Towards Middleware Services for Mobile Ad-Hoc Network Applications, The Ninth IEEE Workshop on Future Trends of Distributed Computing Systems (FTDCS'03)(May 2003).
8. Kwang-Hoon Kim: Workflow Dependency Analysis and Its Implications on Distributed Workflow Systems, 17th International Conference on Advanced Information Networking and Applications (AINA'03)(March 2003).
9. Diimitrios Georgakopoulos, Mark Hornick, Amit Sheth: An Overview of Workflow Management: From Process Modeling to Workflow Automation Infrastructure, Distributed and Parallel Databases, vol. 3, no. 2, 119-153(1995).
10. Martin Ader: Workflow Comparative Study 2004, http://www.waria.com, Volume III, Workflow Comparative Analysis(2004).
11. Özalp Babao, Hein Meling, Alberto Montresor: Anthill: A Framework for the Development of Agent-Based Peer-to-Peer Systems, 22 nd International Conference on Distributed Computing Systems (ICDCS'02)(July 2002).
12. Domino Application Security & Workflow (Lotus Domino Designer Release 5), Lotus Authorized Education(2003).
13. Henning Sanneck, Michael Berger: Application of agent technology to next generation wireless/mobile networks, http://www.citeseer.com(2002).

MobiMessenger: An Instant Messenger System Based on Mobile Ad-Hoc Network

Ling Chen, Weihua Ju, and Gencai Chen

College of Computer Science
Zhejiang University, Hangzhou 310027, P.R. China
lingchen@cs.zju.edu.cn

Abstract. Mobile ad-hoc network offers many new ways for CSCW resolution. So we develop an ad-hoc instant messenger, MobiMessenger, which will be used as a platform for our future research work on ad-hoc collaboration. In this paper, we present the design and implement of MobiMessenger, which is based on AODV algorithm, including the system architecture and implementation. Also present our future work on ad-hoc collaboration.

1 Introduction

An ad-hoc network is a collection of wireless mobile hosts forming a temporary network without the aid of any established infrastructure or centralized administration.

It offers new ways for CSCW resolution. Various applications on ad-hoc collaboration have been developed, such as: Hocman [1] and RoamWare [2].

Our research goal is to develop an application to support mobile interaction, collaboration, and transparent data exchange used in ad-hoc network. Up to date, we have designed and implemented MobiMessenger which is a mobile ad-hoc instant messenger. This is a work in progress, and it will be a platform for our future research work on CSCW in an ad-hoc environment.

The remainder of this paper is organized as follows. In section 2, the system architecture is given. Section 3 describes the system implementation in detail. Section 4 gives some conclusions and the further work.

2 System Architecture

The architecture of the MobiMessenge is derived from AODV [3] algorithm, so it is very simple, involving four modules: Chat Routing, Hello and Transceiver as shown in Fig.1.

The Chat module handles the user interface with Windows Form Controls. Users use GUI to input the message to send and destination IP, and also use GUI to display the received messages.

The Routing module handles all tasks of path discovery including reverse path set up and forward setup, next-hop route table management and path maintenance. All the routing information is stored in a next-hop table.

The Hello module handles the local connectivity management through the maintenance of Potential table and Neighbor table. The Potential table contains all nodes from which it has received packet directly. The Neighbor-table contains all neighbors which it has directly received packet from and send packet to successfully.

The Transceiver module handles all tasks of sending and receiving packets using UDP socket API. It also packets the message and route information before sending, and extracts message to be displayed from a packet after receiving.

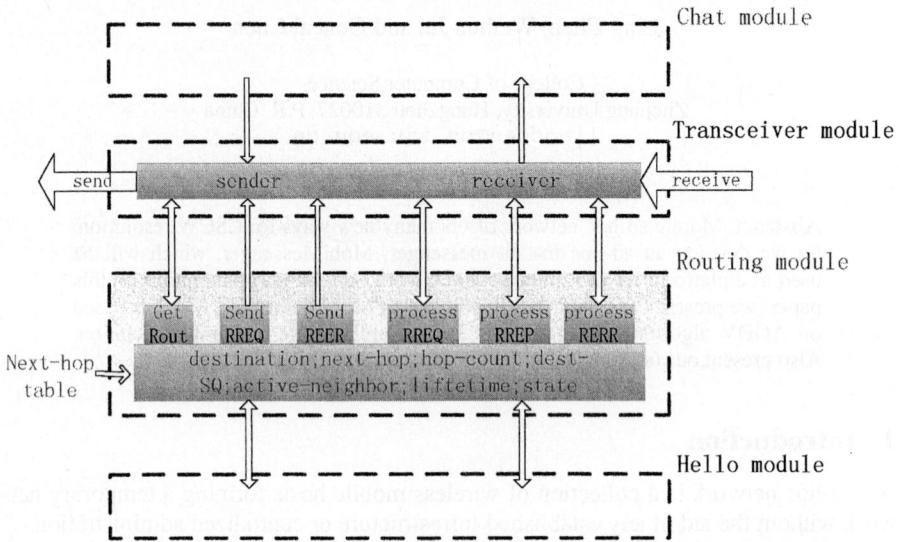

Fig. 1. The architecture of MobiMessenger

Data flow among modules is also shown in Fig 1. In one hand, when a user inputs the message to be sent and the destination IP address, and clicks the "send" button, the Chat module sends the data (the message) to the Transceiver module. The Transceiver module tries to get a rout for the destination from the Routing module using the GetRout method. If succeed, the Transceiver module packets the message and route information, and sends the packet. Else, the Transceiver module waits the Routing class for the route information. And the Routing module sends RREQ. When it receives the RREP packet, the processRREP method first updates the next-hop table using the information in the RREP packet, then send the route information to the Transceiver module.

In the other hand, when a Routing class receives RREQ packet from its neighbor, it first updates its next-hop table about its neighbor and the source originator. Then if it is the destination of the RREQ or can answer the RREQ using its own next-hop table, answers the RREQ. Else, forwards the RREQ packet.

3 Implementation

The main development platform is a workstation installed with Windows 2000 operating system. The development software components include Microsoft Visual Studio .Net(C#), Microsoft Pocket PC 2003 SDK and Microsoft .Net Compact Framework, especially using UDP socket API from the .Net Compact Framework.

There was no established physical infrastructure when we evaluated the MobiMessenger. All Pocket PCs communicated with each other using 802.11b wireless LAN cards. In order to distinguish different Pocket PCs, different IP addresses were used to assign to Pocket PCs.

We first ran the MobiMessenger on the Pocket PC 2003 Emulator in .Net Compact Framework environment, and scouted the data flow between modules. The interface of MobiMessenger is shown in Fig.2.

Fig. 2. Interface of MobiMessenger

We evaluated MobiMessenger in an outdoor environment within a fixed-size 100m×50m area. The evaluation involved four participants, and each of them was equipped with a HP iPAQ 5550 Pocket PC (400MHz PXA255 XScale processor with a 200MHz system bus, 128 MB of RAM, 802.11b Wireless LAN card). Participants could communicate with each other freely whenever they were standing or moving.

4 Conclusions and Further Work

In summary, we have presented a mobile ad-hoc network instant messenger: MobiMessenger. It is a practical instant messenger. Users that hold a Pocket PC installed IEEE 802.11b wireless LAN card can communicate with each other freely without any established infrastructure or centralized administration. Also, it is extensible for our future research work on CSCW in an ad-hoc environment.

We will improve the MobiMessenger by adding it a buddy list, Then users can select some friends from the buddy list to have a temporary conference. Also other friends can ask for joining the conference, if agreed, they do. To make such service come true, we will improve the MobiMessenger to support section control and multicast.

We will make some future researching work on ad-hoc collaboration on the base of MobiMessenger, including consistency maintenance, section control and authority maintenance in an ad-hoc network environment. We will develop ad-hoc network conferencing and some other ad-hoc collaboration applications on the base of MobiMessenger.

References

1. M. Esbjornsson, M. Ostergren. Hocman: Supporting Mobile Group Collaboration. In Proc of CHI Extended Abstracts on Human Factors in Computing Systems, 2002:838-839.
2. M. Wiberg. RoamWare: An Integrated Architecture for Seamless Interaction In between Mobile Meetings. In Proc of International ACM SIGGROUP Conference on Supporting Group Work, 2001:288-297.
3. C.E. Perkins, P. Bhagwat. Highly Dynamic Destination Sequenced Distance-Vector Routing (DSDR) for Mobile Computers. In Proc of ACM Special Interest Group on Data Communication, 1994:234-244.

Temporal Dependency for Dynamic Verification of Temporal Constraints in Workflow Systems

Jinjun Chen and Yun Yang

CICEC – Centre for Internet Computing and E-Commerce
School of Information Technology
Swinburne University of Technology
PO Box 218, Hawthorn, Melbourne, Australia 3122
{jchen,yyang}@it.swin.edu.au

Abstract. The current workflow management systems do not take into consideration the mutual dependency between temporal constraints, which is affecting the effectiveness and efficiency of the temporal verification. In this paper, we investigate this dependency and its effects on the verification of temporal constraints. Furthermore, based on these analyses on the temporal dependency, we develop some new methods for more effective and efficient temporal verification. These analyses and new methods further reinforce the time management of the current workflow management systems.

1 Introduction

To control the temporal correctness of the workflow execution, explicit temporal constraints are set at the build-time stage and verified at the build-time, run-time instantiation and run-time execution stages. The mutual dependency between these constraints is affecting the effectiveness and efficiency of the temporal verification. Although there has some time management work which has been done, for example, [3] uses the modified Critical Path Method (CPM) to calculate temporal constraints. [5] presents a method for dynamic verification of absolute deadline constraints and relative deadline constraints. However, all these work together with other work such as [1, 4] do not take into consideration the temporal dependency. In this paper, we investigate this temporal dependency and its relationships with the temporal verification. Furthermore, we develop some new temporal verification methods by which we can achieve better verification effectiveness and efficiency.

2 Timed Workflow Representation

Considering an arbitrary execution path in a timed acyclic workflow graph, we denote minimum duration, maximum duration, run-time start time, run-time end time and run-time real completion duration of ith activity a_i as $d(a_i)$, $D(a_i)$, $S(a_i)$, $E(a_i)$, and $Rcd(a_i)$ respectively [3, 5]. And we denote maximum, minimum and run-time real completion durations between a_i and a_j ($j \geq i$) as $D(a_i, a_j)$, $d(a_i, a_j)$ and $Rcd(a_i, a_j)$ respectively [3, 5]. Meanwhile, upper bound constraint $upb(a_i, a_j)$ means the duration between a_i and a_j should not exceed it. Lower bound constraint $lob(a_i, a_j)$ means the duration between a_i and a_j should not be lower than it. Deadline constraint at a_i means a_i must be completed by a certain time, denoted as $rdl(a_i)$ at build-time and $adl(a_i)$ at run-time [3]. And we have: $rdl(a_i)=adl(a_i)-S(a_1)$, $Rcd(a_i, a_j)=E(a_j)-S(a_i)$. In addition, we can see each parallel or selective structure as a composite activity. Then, $D(a_i, a_j)$ is equal to the sum of all D values between a_i and a_j, and $d(a_i, a_j)$ is equal to the sum of all d values between a_i and a_j. At the same time, checkpoints are selected at the run-time execution stage for more efficient temporal verification. And we give:

Definition 1. An upper bound constraint between a_i and a_j ($j \geq i$) is consistent at the build-time stage if and only if $D(a_i, a_j) \leq upb(a_i, a_j)$, and the corresponding consistency condition for a lower bound constraint is if and only if $d(a_i, a_j) \geq lob(a_i, a_j)$.

Definition 2. An upper bound constraint between a_i and a_j is consistent at checkpoint a_p between a_i and a_j ($j \geq p$, $p \geq i$) at the execution stage if and only if $Rcd(a_i, a_p) + D(a_{p+1}, a_j) \leq upb(a_i, a_j)$, and the corresponding consistency condition for a lower bound constraint is if and only if $Rcd(a_i, a_p) + d(a_{p+1}, a_j) \geq lob(a_i, a_j)$.

Definition 3. The deadline constraint at a_i is consistent at the build-time stage if and only if $D(a_1, a_i) \leq rdl(a_i)$.

Definition 4. The deadline constraint at a_i is consistent at the instantiation stage if and only if $D(a_1, a_i) \leq adl(a_i)-S(a_1)$, and is consistent at checkpoint a_p by a_i ($p \leq i$) at the execution stage if and only if $Rcd(a_1, a_p)+D(a_{p+1}, a_i) \leq adl(a_i)-S(a_1)$.

3 Temporal Dependency Between Temporal Constraints

Upper bound constraints without mutual nesting relationships are relatively independent. For those with nesting relationships, based on Allen's interval logic [2], we can conclude three kinds of basic nesting relationships between upper bound constraints A, B, C, and a basic nesting extension, as depicted in Fig. 1.

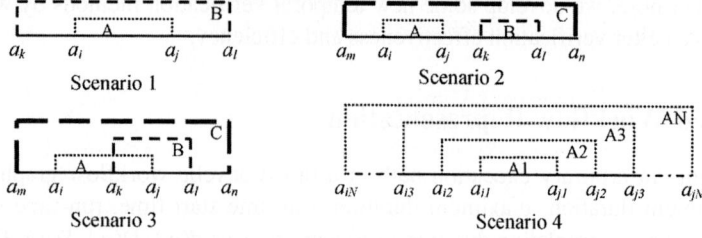

Fig. 1. Basic nesting relationships between upper bound constraints

In Fig. 1, through analysing the nesting relationships, we can deduce that we must consider the temporal dependency between upper bound constraints, and we have:

Definition 5. The temporal dependency between A, B and C is consistent in scenario 1 if and only if $D(a_k, a_{i-1})+upb(A)+D(a_{j+1}, a_l) \leq upb(B)$; in scenario 2 if and only if $D(a_m, a_{i-1})+upb(A)+D(a_{j+1}, a_{k-1})+upb(B)+D(a_{l+1}, a_n) \leq upb(C)$; and in scenario 3 if and only if $D(a_m, a_{i-1})+upb(A)+upb(B)-D(a_k, a_j)+D(a_{l+1}, a_n) \leq upb(C)$.

Furthermore, for scenario 4, an extension of scenario 1, we can prove:

Theorem 1. In scenario 4, if the dependency between any two adjacent upper bound constraints is consistent, the dependency between any two non-adjacent upper bound constraints must be consistent.

The proof process is straightforward, so, we simply omit it. If we apply scenario 4 to scenarios 2 and 3 in Fig. 1, the corresponding dependency discussion is similar.

The dependency between lower bound constraints is similar to that between upper bound constraints. For deadline constraints, because of their mutual nesting relationships, similarly, we must consider their mutual dependency and we have:

Definition 6. The dependency between two adjacent deadline constraints respectively at a_i and a_j ($j>i$) is consistent if and only if $D(a_{i+1}, a_j) \leq rdl(a_j)-rdl(a_i)$.

Furthermore, for two non-adjacent deadline constraints, we have:

Theorem 2. If the dependency between any two adjacent deadline constraints is consistent, the dependency between any two non-adjacent deadline constraints must be consistent.

Again, we simply omit the proof process like theorem 1.

4 Build-Time Temporal Verification

At build-time, according to section 3, for the effectiveness of the temporal verification, we have to verify the dependency between temporal constraints.

For upper bound constraints, on one hand, we conduct verification computations according to definition 1. On the other hand, for scenarios 1, 2 and 3 in Fig. 1, we verify the temporal dependency according to definition 5. For scenario 4, according to definition 5 and theorem 1, we only need to verify the dependency between any two adjacent upper bound constraints. For lower bound constraints, the corresponding verification is similar. For deadline constraints, on one hand, we verify them based on definition 3. On the other hand, based on definition 6 and theorem 2, we verify the dependency between any two adjacent deadline constraints.

5 Run-Time Temporal Verification

At the instantiation stage, because we still do not have specific execution times, the temporal constraint and dependency verification is the same as that of the build-time.

At the execution stage, for those upper bound constraints without mutual nesting relationships, we conduct the temporal verification according to definition 2. For those nested one another, for scenario 1 in Fig. 1, we can have:

Theorem 3. In scenario 1 of Fig. 1, at checkpoint a_p between a_i and a_j, if $Rcd(a_k, a_{i-1}) \leq D(a_k, a_{i-1})$, then, if A is consistent, B must be consistent.

Similar to theorem 1, we simply omit the corresponding proof process. According to theorem 3, for scenario 1 in Fig. 1, if A is consistent and $Rcd(a_k, a_{i-1}) \leq D(a_k, a_{i-1})$, B is consistent. Under all other conditions, we verify A and B directly based on definition 2. Clearly, the temporal verification based on the temporal dependency described in theorem 3 is more efficient than that only based on definition 2 because under the above condition, if based on definition 2, we still need to conduct significant extra computations described in definition 2. Theorem 3 can be extended for scenarios 2 and 3 in Fig. 1, and similar conclusions can be drawn.

For lower bound constraints, the corresponding verification is similar. For deadline constraints, we can prove the following theorem. Again, the proof process is omitted.

Theorem 4. At the execution stage, at checkpoint a_p, if a deadline constraint D after a_p is consistent, any deadline constraint after D must be consistent.

According to theorem 4, at a checkpoint, we need not verify any deadline constraints after a consistent one. Obviously, this will improve the verification efficiency.

6 Conclusions

In this paper, the dependency between temporal constraints and its effects on the temporal verification are investigated. Furthermore, based on the temporal dependency and relevant concepts and principles provided by this paper, some new verification methods are presented, which enable us to conduct more effective and efficient temporal verification. All these discussions, relevant concepts, principles and new verification methods strengthen the current workflow time management.

Acknowledgements

The work reported in this paper is partly supported by Swinburne Vice Chancellor's Strategic Research Initiative Grant 2002-2004.

References

1. Bussler, C.: Workflow Instance Scheduling with Project Management Tools. In Proc. of the 9th Workshop on Database and Expert Systems Applications (DEXA'98). Vienna, Austria (1998) 753-758
2. Chinn, S., Madey, G.: Temporal Representation and Reasoning for Workflow in Engineering Design Change Review. IEEE Transactions on Engineering Management 47(4) (2000) 485-492
3. Eder, J., Panagos, E., Rabinovich, M.: Time Constraints in Workflow Systems. In Proc. of the 11th International Conference on Advanced Information Systems Engineering (CAiSE'99). Lecture Notes in Computer Science, Vol. 1626. Springer-Verlag, Germany (1999) 286-300
4. Li, H., Yang, Y., Chen, T.Y.: Resource Constraints Analysis of Workflow Specifications. The Journal of Systems and Software, Elsevier, in press
5. Marjanovic, O.: Dynamic Verification of Temporal Constraints in Production Workflows. In Proc. of the Australian Database Conference. Canberra, Australia (2000) 74-81

Integrating Grid Services for Enterprise Applications

Feilong Tang and Minglu Li

Department of Computer Science and Engineering,
Shanghai Jiao Tong University, Shanghai 200030, China
{tang-fl,li-ml}@cs.sjtu.edu.cn

Abstract. Integrating heterogeneous systems in the "plug and play" way is a goal of distributed computing researchers. In this paper[1], we propose a general platform to integrate Grid services for enterprise applications, and present implementation mechanisms of its core components. The platform can handle transaction, workflow and security to satisfy enterprise application requirements. Moreover, the platform is fully built on open standards so that it has some good properties such as scalability, adaptability and portability.

1 Introduction

Goals of Grid computing are sharing large-scale resources geographically distributed in different organizations and controlled in different security policies, and facilitating cooperative tasks. This may be achieved when applications are integrated by the integration platform, which has to hide the heterogeneous nature of the resources and provides (1)uniform interfaces for creation, publication and discovery of services, (2)uniform mechanisms to handle transaction and workflow, and (3)security control, including authentication, authorization, delegation and encryption.

This paper presents such an integration platform, by which various resources are encapsulated into Grid services [1] and integrated together in the "Plug and Play" way.

2 Architecture

Our integration platform is based on the WS-Resource Framework [2] and structured in Fig.1. It manages the resources within an organization and acts as a gateway among different enterprises. The key components include (1) a discovery service, together with the registry and handleResolver, for finding appropriate Grid resources, (2) a workflow service to composite multiple Grid services into a logic service, (3) a transaction service for reliable business activities, and (4) a security service to guarantee secure access to autonomous Grid resources.

---
[1] This paper is supported by 973 project (No.2002CB312002) of China and grand project of the Science and Technology Commission of Shanghai Municipality (No.03dz15027).

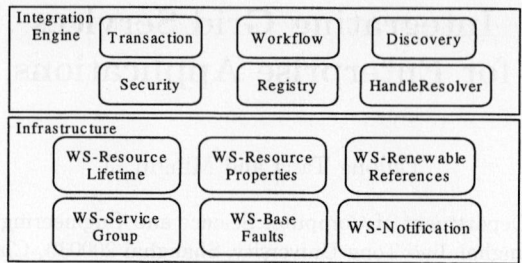

Fig. 1. The architecture of the integration platform.

3 Implementation

3.1 Discovery

Discovery service contacts the global UDDI, the registry and handleResolver in a local integration platform to find the service reference, as described in Fig.2.

- A client discovers qualified service and URL of its local registry in the UDDI.
- The client queries service handle in the registry, then resolves its service reference in the handleResolver. For a transient service, its service factory registered in registry creates a new service instance.

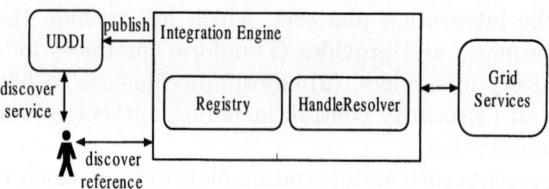

Fig. 2. Discovery of Grid services.

Service discovery operations include (1) lookupUDDI() to find desired services and the URL of its local registry, (2) lookupHandle() to discover the handle of the service instance in registry or create a new instance for a transient service by using its factory, and (3) lookupReference() to resolve the reference of the service instance by the handleResolver.

3.2 Workflow

The Workflow service serves as the base for specifying and executing collaborative tasks within Grids. A workflow involves the following procedures [3,4]. (1) A client defines the specification to specify the workflow rules. (2) The client call workflow factory to create an instance and submit the workflow description document to the instance. (3) The workflow instance controls workflow execution by resolving the specifications, creating service instances involved in the workflow, statically or dynamically submitting sub-tasks to those instances according to dependencies among them until either the work is completed or is terminated, and monitoring the status of submitted sub-tasks. (4) Invoked services execute specified sub-tasks and return results to the workflow instance, which destroys the invoked service instances and itself after receiving acknowledgements.

```
ActionOfParent{
    sends CC to all participants;
    wait Response from participants;
    send Prepare to all P_i;
    if (receive N Prepared messages){
        record commit in log;
        send Commit to all P_i;
        if (not receive N Committed)
            send Rollback to all P_i;
    } else   send Abort to all P_i; }
```
(a) Coordinator algorithm of AT

```
ActionOfChild{
    send Response to TS;
    wait Prepare from TS;
    if (reserves resources successfully){
        send Prepared to TS;
        if (receive a Commit){
            allocate reserved resources;
            commit transaction;
            send Committed to TS;
    } else cancel reservation; } }
```
(b) Participant algorithm of AT

```
ActionOfSuperior{
    while (transaction doesn't complete){
        sends CC to all candidates;
        wait Response from candidates;
        send Enroll to all candidates;
        while (t ≤T) {
            if (receive Committed)
                if (user selects some candidates)
                    send Confirm messages to them;
                else
                    send Cancel messages to them;
    } } }
```
(c) Coordinator algorithm of CT

```
ActionOfInferior{
    send Response to TS;
    wait Enroll from TS;
    reserve and allocate resources;
    commit transaction;
    if (commit successfully){
        send Committed to TS;
        while (t ≤T){
            if (receive a Cancel)
                invoke compensation transaction;
            else  if (receive a Confirm)
                send Confirmed; } } }
```
(d) Candidate algorithm of CT

Fig. 3. Coordination algorithms

3.3 Transaction

The transaction service (TS), which acts as a coordinator, interacts with the participant/candidate to manage a transaction. It may handle two types of transactions: atomic transaction (AT) to coordinate short-lived operations and cohesion transaction (CT) to coordinate long-lived business activities. An atomic transaction requires all participants to commit synchronously while a cohesion transaction allows candidates independently to commit and can undo the committed sub-transactions by performing compensation transactions. Their coordination algorithms are shown in Fig. 3.

Transaction service provides the following operations: (1) startTransaction() to send CoordinationContext(CC) message to participants/candidates, (2) prepareTransaction() to reserve resources in AT, (3) commitTransaction() to commit transaction in AT, (4) rollback() to notify participants in AT to undo operations taken previously, (5) enroll() to notify candidates commit transaction in CT, (6) confirm() to confirm committed sub-transactions in CT, and (7) cancel() to start a compensation transaction in CT.

3.4 Security

The security service works like this:

- Authentication. A client creates a proxy credential signed by user's private key, using a user proxy. Then the security service checks the user's identity using its authentication algorithm.
- Authorization. The security service maps the proxy credential into a local user name, then checks local policy to determine whether the user is allowed to access to local resource. If the user is authorized, it allocates a credential C_p.
- Delegation. Security service promulgates credential C_p to invoke other remote services. By tracing back along the certificate chain to check the original user certificate, services started on separate sites by the same user can authenticate one another, enabling user to sign once, run anywhere [5].
- Encryption. SOAP messages may be forwarded through multiple intermediate nodes. If these intermediate nodes are trusted, secure communication can be ensured by means of the SSL. Otherwise, the security service uses SOAP encryption and SOAP digital signature to carry out the confidentiality.

4 Conclusion

This paper presents a Grid service integration platform. In order to implement information integration for enterprise applications, the platform provides the support for workflow, transaction and security. In particular, the platform is open, distributed, and modular. Thus, it can be easily added into new modules to adapt to more complex requirements.

References

1. I. Foster, C. Kesselman, J. M. Nick and S. Tuecke. The Physiology of the Grid-An Open Grid Services Architecture for Distributed Systems Integration. June, 2002.
2. K. Czajkowski, D. F. Ferguson, I. Foster et.al.. The WS-Resource Framework. March, 2004. http://www.globus.org/wsrf/.
3. D. Cybok. A Grid Workflow Infrastructure. http://www.extreme.indiana.edu/groc/ggf10-ww/grid_workflow_infrastructure_dieter_cybok/GWI.pdf.
4. S. Hwang, C. Kesselman. Grid Workflow: A Flexible Failure Handling Framework for the Grid. Proceedings of the HPDC'03. June, 2003, pp. 126-137.
5. R. Butler, V.Welch, D. Engert, I. Foster, S.Tuecke, J. Volmer and C. Kesselman. A national-scale authentication infrastructure. Computer, December, 2000.

A Middleware Based Mobile Scientific Computing System – MobileLab[*]

Xi Wang, Xu Liu, Xiaoge Wang, and Yu Chen

Department of Computer Science & Technology,
Tsinghua University, Beijing, P.R. China
LX2000@mails.tsinghua.edu.cn

Abstract. In this paper we present a scientific computing system named MobileLab which is designed for pervasive computing on the Internet. A middleware named TOM (stands for Tsinghua Operation Middleware) is designed and employed in MobileLab as a key to access the grid. Based on TOM, MobileLab can access the computational resource transparently; distribute the components to optimize the computation procedure and computation can be moved from one node to another. To use a middleware model in the architecture of scientific computing system is our new approach corresponding to the grid computing and mobile computing.

1 Introduction

Grid computing and the Internet are revolutionizing scientific and technological computing which has gained extremely huge capability from the computational grid. However, there has not been a computing environment for non-expert users to benefit the computational power from the gird, especially on the mobile equipments. There has been large amount of work dealing with networking, data management and power management issues on mobile computing but few focuses on the software architecture of mobile scientific computing. Related work named "Science Pad" [1] was published by Purdue University during the rush of "Pen-Like Computers" in mid 1990s. Flexibility and usability is highly required which grid computing requires the computational resources on the internet to be re-coordinated. Thus a middleware model which bridges the mobile device and super computers is designed as a novel feature of our system. The middleware model acts as the "software glue" [2] which enables supercomputers on the Internet work cooperatively to gain a much higher capability than the traditional parallel computer. Although recent technology of embedded system has advanced the computational ability of mobile equipments most of the scientific computing problems can not be solved on these "weak" equipments. These problems often require super computers (cluster) and a long solving period. Unfortunately, scientists who need to run these computational tasks know little about programming on high-performance computers while a lot of computational abilities on super computers were left idle. Mobile Lab is designed for this purpose.

MobileLab is a robust computing environment for high performance computing based on middleware technique. A middleware model named TOM [7] is designed and implemented for the Mobile Lab environment. Distribution of resources, includ-

---

[*] The work is supported by the 863 national high technology plan (under the contract 2003AA1Z2090) and by Tsinghua University research fund (Jc2003021).

ing computation time, is transparent in MobileLab. MobileLab can handle failure gracefully, including network failures, hardware failures and software bugs. Mobile Lab would achieve its best performance when distributing tasks to peers.

2 Architecture of MobileLab

The following picture shows the system architecture of MobileLab:

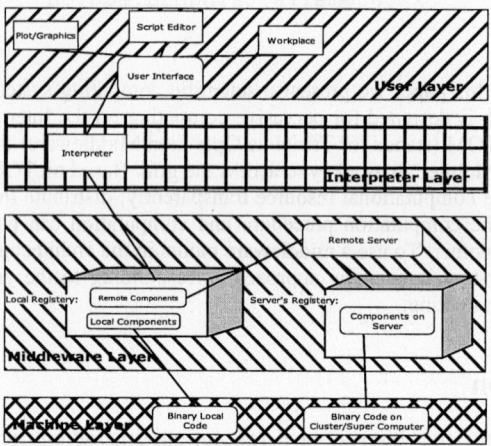

Fig. 1. System Architecture of MobileLab

The MobileLab uses a server-client model: The server runs on PC/cluster. When receiveing a requirement to create a component it starts a new thread to maintain it.

As the MobileLab system is designed as an integration of computational resources on grid [5], the server and client must not be closely coupled. Thus when a server is temporarily unavailable the computation can be moved to another node.

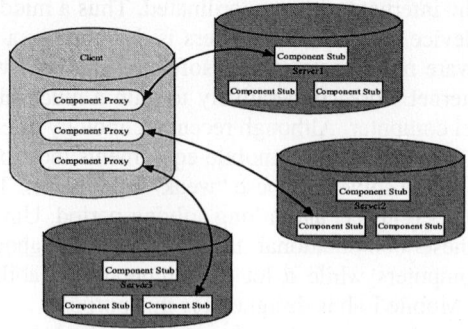

Fig. 2. Client-Server relationship of MobileLab

MobileLab has a lightweight client which runs on PC/Laptop/PDA. The client maintains a local registry recording the location of components, both local and remote, for the the interpreter to load the corresponding components when necessary.

There are 3 ways to load a component:

(1) the component is local : query its location by UUID and load it into process
(2) the component is remote: query its URL by UUID then we have 2 choices: download and load it into process or call it remotely.

A client may use several components on several different servers. In whichever way the component is loaded it is transparent to the interpreter.

3 Middle-Ware Layer

Most research on distributed and parallel computing is based on an existing and fixed network infrastructure and therefore leads to fairly static solutions such as PVM, MPI, CORBA, RMI, etc. With the emerging ubiquitous computing paradigm, certain significant limits relevant to current middleware are addressed such as: middleware usually tied up with a central naming mechanism.

Middleware are usually built for use of workstations or servers, meaning most current middlewares are heavy and therefore unsuitable for use on lightweight devices[6]. Unlike CORBA or DCOM, our middleware model TOM is designed specifically for MobileLab, moveable and lightweight, and is also compatible with Microsoft COM.

As the component is location transparent, TOM employs a proxy-stub structure. If the component is instantiated inside local process the proxy exposes the actual interface directly; parameters and results are transferred via stack. If the component is remote the proxy marshals the parameters to a consecutive data block and transfers it to the stubs which will un-marshal the parameters and start a new thread. Inside the thread the un-marshaled parameters are passed to the actual components. A call back address is also passed to the thread, when the computation is finished, the thread uses this address to call back and pass back the result. The stub then marshals the result which will be sent back to the proxy and un-marshaled by the proxy [8].

4 Example and Performance Analysis

In the following example a components which generates the Mandelbrot–set is employed. We also use this example to compete the efficiency between local (serial) and remote (parallel) components.

```
a=mandelbrot([[-1.4] [-1] [.01] [200] [200] [10000] [100000]]);
plot(a);
```

The script contains only 2 lines but it covers all. The first line requires a component named "mandelbrot" to calculate the `mandelbrot set`When the interpreter reads the first line, it searches the local registry for a component named "mandelbrot" and tries to instantiate it. Then it inquires its "`I_Complexity`" interface to estimate approximately how much time it will cost. If the complexity is acceptable then do the computation locally, otherwise the interpreter searches another remote component with the same name "mandelbrot" on the parallel computer and try to compute remotely. When the computation is finished the user interface runs a "plot" function to show the result. All these procedures are done automatically.

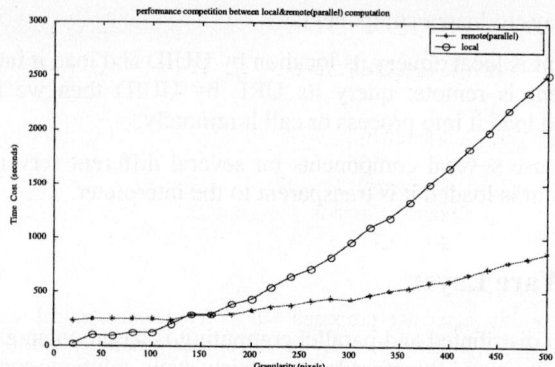

Fig. 3. Performance Comparison between Local and Remote Computation

Client: AMD Duron 800, 256 Mb memories, Windows XP
Server: 4-nodes cluster with Intel P4 1.4 on every node, 1 GB memories, Red-hat Linux 9.0

From this comparison result we can see that when the pixel number is small, the computation load is light, the local computation is faster that remote computation, because the network overhead costs more time than the computation itself [10]. While the pixel number is growing, the advantage of remote computing is distinctive since it cost much less time. It is especially superior when the client is on a PDA-like system.

References

1. Joshi, S. Weerawarana, T.T. Drashansky and E.N. Houstis, SciencePad: An inteligent electronic notepad for ubiquitous scientific computing, Proc. Int. Conf. on Inteligent Information Management Systems (June 1995) pp. 107-110.
2. http://www.gnu.org/software/glue/glue.html
3. A. Athan and D. Duchamp, Agent-mediated message passing for constrained environments, Proc. Mobile and Location-Independent Computing, USENIX, Cambridge, MA (August1993) pp. 103-107.
4. Joshi, S.Weerawarana, R.A.Weerasinghe, T.T. Drashansky, N. Ramakrishnan and E.N. oustis, A survey of mobile computing technologies and applications, Tech. Report TR-95-050, Dept. of Computer Sciences, Purdue University (1995).
5. Ian Foster, Carl Kesselman Jeffrey M. Nick Steven Tuecke, the Physiology of the Grid an Open Grid Services Architecture for Distributed Systems Integration, Feb. 2002. http://www.globus.org
6. Luc Hogie, Christian Hutter, Middleware Support for Ubiquitous Computing, Contents ERCIM News No. 54 Special Theme: Applications and Service Platforms for the Mobile User, July 2003
7. http://hpclab.cs.tsinghua.edu.cn/~liuxu/mobilelab/doc/tom/
8. Guy/Henry Eddon, Inside DCOM, Microsoft Press, 1997
9. J. Xu, B. Randell and A. F. Zorzo, "Implementing Software Fault Tolerance in C++ and Open C++: An Object-Oriented and Reflective Approach", in Proc. Int. Workshop on Computer-Aided Design, Test, and Evaluation for Dependability (CADTED'96), (Beijing, China), pp.224-9, International Academic Publishing, 1996.
10. Ian Foster, Designing and Building Parallel Programs, Chapter 8.7,19951999 DARPA Intrusion Detection Evaluation Dataset. http://www.ll.mit.edu/IST/ideval/index.html

Towards Extended Machine Translation Model for Next Generation World Wide Web

Gulisong Nasierding, Yang Xiang, and Honghua Dai

School of Information Technology, Deakin University
Melbourne Campus, Burwood 3125, Australia
{gn,yxi,hdai}@deakin.edu.au

Abstract. In this paper, we proposed a Data Translation model which potentially is a major promising web service of the next generation world wide web. This technique is somehow analogy to the technique of traditional machine translation but it is far beyond what we understand about machine translation in the past and nowadays in terms of the scope and the contents. To illustrate the new concept of web services based data translation, a multilingual machine translation electronic dictionary system and its web services based model including generic services, multilingual translation services are presented. This proposed data translation model aims at achieving better web services in easiness, convenience, efficiency, and higher accuracy, scalability, self-learning, self-adapting.

1 Introduction

With the strong increasing of data complexity both computationally and descriptively, a creative technique which can greatly simplify the process, storage, transmission and communication of data becomes essential and in great demanding. The World Wide Web is increasingly used for application to application communication. The programmatic interfaces made available are referred to as Web services. An increasing number of software systems extend their capability to encompass web services technology. Web services provide a platform to develop automatic application systems based on cross-organizational and heterogeneous software components. This characteristic of web-services is intrinsically similar with Machine Translation (MT) and Data Translation technology.

Machine translation refers to the use of computer systems to translate texts among various languages and help human translators on their translation work. Machine Translation has been researched for decades. However, today's MT system is still facing challenges to provide reliable, high-accuracy and generic portal translations. Laurie [1] stated in her article that majority people in worldwide area couldn't satisfy simply translating web site. Since customers need to communicate with each other by using different language channels or exchange ideas through preferred type of data. As a result, web services based data translation has become important and necessary as we proposed in this paper.

2 Machine Translation Systems

2.1 Review of MT Systems

According to Andy Way [2], machine translation methods can be classified to two major branches: (1) Rule-based MT and (2) Data-Driven MT [3]. In Arul's research work, machine-learning technique is used to attain translation knowledge from bilingual datasets, which improved the quality of translation. Example-Based Machine Translation via the Web system is proposed by Nano [4]. Contemporary web-based MT systems use HTML pages transmission to translate large amount of data. For example, in UNIX system, 'wget' function is used to pass the source web pages to translation system. After complete translation, the web server returns the translated web pages to the request. In the IBM, LMT machine translation system has been implemented based on rules of word formation rather than rely on entries in the bilingual core lexicons. Our system belongs to rule-based MT systems which implicitly contains rule bases in the translation main modal. An instance model of MT system can be seen from the previously presented work [6].

2.2 A Model of Web Services Based Multilingual Translation System

The system model is shown in the following Figure 1. It includes five layers, Generic Multilingual Translation Web Services layer, Specific Translation Web Services layer, Multilingual Translation Engine layer, Knowledge Base layer, and Database and Rule Base layer. The main function of multilingual web services model is to provide distance services to a client's request of multilingual translation focusing on words, terms or phrase translation with phonetic notation, then approaching to translate sentences and paragraphs in the future work. The generic multilingual translation web services layer is an open portal to bridge different translation. According to the requirements, specific request is directed to different translation web services layer. Then the request is passed to translation engine.

The key role of the multilingual translation engine is to translate between every two different languages, and add notation to the translated ones based on production rules. Machine learning techniques are used for searching, extracting and retrieving translated corresponding words and phrases. The different language translation rules are different upon to the characters or scripts of the languages and formation of the words. All relevant language translation rules are remained in the rule bases and all different lexical information are stored in different data bases. In order to retrieve the translated words efficiently, the lexical data can be indexed and associated retrieval techniques are adopted.

Our multilingual translation web services model can be applied in different scenarios. For example, staff in different company branches have to use the same management information system belongs to the same group. They need to manage the same source information, but face to different language interface. By using our system, they don't need to purchase different language versions of the system so that it could bring financial benefits to the user if they register to our web services.

3 Data Translation (DT)

The concept of data translation is an 'ISO 9001-certified supplier to the data acquisition, imaging and machine vision markets with expertise in the design of high-accuracy, high-quality and reliable analogue-to-digital products' [7]. We extend the concept of data translation to translating or transforming data within different types, i.e. it refers to different pattern of visualized information, such as text, images/pictures (still or moving) or audio data (the spectrum of it can be visualized).

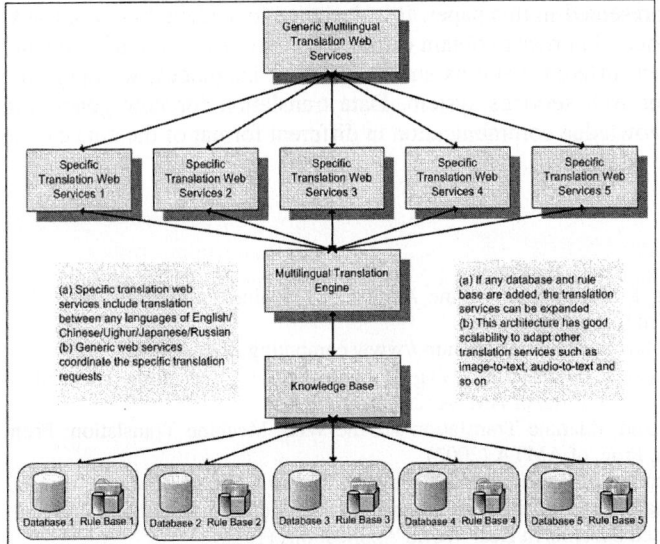 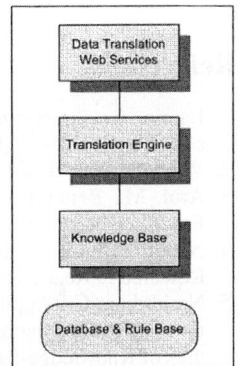

Fig. 1. A model for multilingual translation web services

Fig. 2. Data translation Web services model

The data translation web services model is an extension of the multilingual translation web services model. The main methods and techniques we designed to apply here is similar to previous one. But it expands to the databases not only containing the lexical database but also comprised by image databases. Knowledge base builds up with databases and rule bases. Descriptions and annotations of the image shapes are to be stored in the data bases. The rules of searching and retrieving specific images from image database and matching rules between images and its text-annotations or descriptions are stored in the rule bases. The function of data translation engine is to be a reasoning machine based on production rules. Decision tree learning algorithms need to be applied in order to realize accurate and efficient translation within different types of visual data. We also could consider object recognition as machine translation when translating from images or moving pictures to text or vice versa in our further work.

Currently the web-based translation services are mainly for website translation, and can only handle uncomplicated context. Our approach in fact is to build an open architecture for data translation web services. It is not only faced to the translate websites, but also to handle all the software systems, file systems and even operating systems. This is a great advantage for people with minority languages because it costs too much to translate system into every minority language; but with this web services, it is easy for the people to access the information out of their language knowledge. All kinds of information in the world such as text, image, audio and video can be shared with people all around the world by the data translation web services.

4 Conclusions and Future Work

In this paper, we proposed a data translation web service model which is extended from machine translation. As an instance of data translation, a multilingual translation web services model is presented in this paper. It is designed to achieve high accuracy, scalability and intelligence. In order to obtain distance communication, retrieving and extracting information in different patterns and in different languages, we apply the data translation into our web services system. Data translation for new generation world wide web with knowledge communication in different format of the data can be a direction of our future work.

References

1. Laurie, G.: Supporting a Multilingual Online Audience, Machine Translation: From Research to Real Users, Proc. of AMTA (2002)
2. Way, A.: Web-Based Machine Translation, http://www.computing.dcu.ie/~away, 2003
3. Arul, M.: Better Contextual Translation Using Machine Learning, Machine Translation: From Research to Real Users, Proc. of AMTA (2002)
4. Nano, G.: Example-Based Machine Translation via the Web, Machine Translation: From Research to Real Users, Proc. of AMTA (2002)
5. Nasierding, G.: Application of Artificial Intelligence in Uighur-Chinese Bilingual Translation Electronic Dictionary, Proc. of the First International Conference on Success and Failures of Knowledge-Based Systems in Real-World Applications (1996)
6. Nasierding, G., Tu, X.: Research on Uighur, Chinese, English, Japanese and Russian multilingual two-way translation Electronic Dictionary, Proc. of The Second Chinese World Congress on Intelligent Control and Intelligent Automation (1997)
7. DataTranslation.com, http://www.datatranslation.com (2003)
8. W3C, Extensible Markup Language (XML), http://www.w3.org/XML (2003)
9. Sergei, N.: Machine Translation: Theoretical and methodological issues (1987)

Author Index

Ahmad, Naveed 947
Ahmed, Nova 777
Ai, Ping 855
Ali, Arshad 947
Amendolia, Salvator Roberto 939
Anjum, Ashiq 947
Ao, Xiang 769
Astfalk, Greg 5
Azim, Tahir 947

Bao, Feng 317
Bogdanski, Maciej 175
Bunn, Julian 947

Cao, Jian 50
Cao, Jiannong 8, 277, 535, 543, 665, 687
Cao, Lei 448
Cao, Liqiang 129
Cao, Qiang 364
Carter, Adam C. 696
Catlett, Charlie 4
Chen, Daoxu 535, 665
Chen, Gencai 1001
Chen, Guihai 225
Chen, Huajun 372
Chen, Jian 671
Chen, Jian-Xue 835
Chen, Jinjun 1005
Chen, Jiujun 839
Chen, Lin 301
Chen, Ling 1001
Chen, Ming 551
Chen, Minghua 277
Chen, Shudong 799
Chen, Xiaolin 225
Chen, Xiaowu 720
Chen, Xinuo 217
Chen, Xueguang 769
Chen, Ying 761
Chen, Yu 167, 634, 843, 1013
Chen, Zhigang 472
Cheng, Zunping 997
Chi, Jiayu 769
Choi, Ju-Ho 81
Chung, Ilyong 859
Chung, Siu-Leung 883

Chung, Young-Jee 121

Dai, Guanzhong 464
Dai, Honghua 1017
Dai, Jing 997
Dai, Qionghai 634, 843
Dai, Yafei 791
Deng, Qianni 269
Deng, Robert H. 317
Deng, Zhiqun 464
Ding, Hao 851
Ding, Wenkui 404, 601
Dong, Haitao 511, 519
Dong, Xiaomei 348
Dongarra, Jack 2
Du, Ye 895
Du, Zhi-Hui 167, 249
Duan, Hai-Xin 325

Estrella, Florida 939

Fang, Binxing 113, 903
Fang, Lina 65
Feng, Boqin 341, 389, 583
Foster, Ian 1
Fu, Cheng 73
Fu, Wei 105
Fu, Xianghua 389

Gao, Chuanshan 997
Gao, Ji 839
Ge, Sheng 161
Ghose, Supratip 971
Gong, Wenqiang 907
Gu, Ming 883
Guan, Jianbo 26
Gui, Xiaolin 285, 863, 907
Guo, Longjiang 823
Guo, Minyi 679

Han, Dezhi 364
Han, Hua 791
Han, Jincang 341
Han, Weihong 26
Han, Yanbo 89
Hao, Xiao-Wei 959
Hassan, Ali 947

Hassan, Waseem 939
Hauer, Tamas 939
He, Chuan 167
He, Ligang 217
He, Tao 495
Hey, Tony 3
Ho, Roy S.C. 911
Hong, Feng 626
Hong, Ye 193
Hong, Yoon Sik 712
Hu, Jinfeng 519
Hu, Jun 839
Hu, Mingzeng 201
Hua, Yu 979
Huang, Hai 431
Huang, Jianhua 618
Huang, Joshua 448
Huang, Linpeng 657, 795
Huang, Min 277
Huang, Yingchun 720
Hume, Alastair C. 696
Hung, Daniel H.F. 911

Ikram, Ahsan 947

Jarvis, Stephen A. 217
Jeong, Chang-Sung 81, 943
Jeong, Sangjin 819
Ji, Yuefeng 183
Jia, Weijia 502
Jia, Yan 26
Jiang, Yichuan 356
Jin, Hai 293, 479, 671, 783, 787
Jo, Geun-Sik 575, 955, 971
Jo, Jeong Woo 983
Joo, Su-Chong 121
Ju, Weihua 1001
Jun, Kyungkoo 975
Jung, Hyung Soo 567
Jung, Jason J. 955
Jung, Jin-Guk 971

Kang, KyungWoo 919
Kang, Seokhoon 975
Kang, YunHee 923
Kankanhalli, Mohan S. 317
Kim, Byoung-Jin 935
Kim, Byungsang 819
Kim, Heung-Nam 575
Kim, Hyun-Jun 575
Kim, Jin Suk 983

Kim, Moon J. 811
Kim, Seong-Whan 967
Kosiedowski, Michal 175
Kwon, Yong-Won 81, 943

Lai, Chuan-Lin 931
Lai, Woodas W.K. 887
Lam, Kwok-Yan 883
Lam, Wai-kin 167
Lau, Francis C.M. 167, 249, 301, 911
Law, Kincho 50
Le, Jia-Jin 951
Lee, David C.M. 911
Lee, Do-Hoon 333
Lee, Jong Hyuk 712
Lee, Jong Sik 915
Lee, Yong Il 712
Lee, Yong Woo 899
Li, Bo 161
Li, Chunjiang 233
Li, Dong 795
Li, Feng 867
Li, Gang 89
Li, Hui 183, 875
Li, Jianzhong 823
Li, Juanzi 827
Li, Kuan-Ching 931
Li, Lei 959
Li, Lian 261
Li, Min 591, 997
Li, Ming 519
Li, Minglu 50, 73, 137, 269, 448, 626, 657, 745, 795, 799, 807, 1009
Li, Ping 591
Li, Qing-Hua 879, 993
Li, Sanli 167, 249
Li, Shanping 650
Li, Shicai 783
Li, Sikun 642
Li, TieYan 317
Li, Wei 129
Li, Xiang 495
Li, Xiaoming 791
Li, Xing 325
Li, Yin 137, 807
Li, Yinan 863, 907
Li, Ying 626, 671
Li, Yun-Jie 456
Li, Zupeng 618
Liang, Bangyong 827

Liang, Lu 987
Liao, Beishui 839
Liao, Xiaofei 479
Lin, Chuang 193
Lin, Hongchang 720
Lin, Xin 650
Lin, Xin-hua 745
Lingen, Frank 947
Liu, Aibo 183
Liu, Anfeng 472
Liu, Da-you 737
Liu, Fei 657, 835
Liu, Hao 89
Liu, Hong 745
Liu, Jian-Wei 951
Liu, Li 261
Liu, Lianchen 927
Liu, Lingxia 58
Liu, Wu 325
Liu, Xinpeng 42, 97, 145
Liu, Xu 1013
Liu, Xuezheng 551
Liu, Yun-Jie 397
Liu, Zhicong 464
Liu, Zhong 34, 423
Lloyd, Ashley D. 696
Long, Guoping 472
Lu, Jian 535
Lu, Sanglu 8, 225, 687
Lu, Xicheng 105, 487
Lu, Xinda 269, 626, 745
Lu, Yueming 183
Luo, Hong 464
Luo, Junzhou 867
Luo, Xiangfeng 381
Luo, Xixi 720
Luo, Yingwei 42, 97, 145, 153
Lv, ZhiHui 440
Lyu, Michael R. 887

Ma, Dian-fu 161
Ma, Fanyuan 137, 559, 657, 799, 807
Ma, Teng 867
Ma, Tianchi 301, 650
Ma, Zhaofeng 389
Manset, David 939
Mao, Dilin 997
Mao, Ying-Chi 855
Mao, Yuxing 372
Mazurek, Cezary 175

McClatchey, Richard 939, 947
Meliksetian, Dikran S. 811
Moon, Jongsub 891
Mu, Dejun 464

Nam, Dong Su 819
Nasierding, Gulisong 1017
Newman, Harvey 947
Ng, Kam-Wing 815, 871, 887
Ni, Jun 495
Ni, Lionel M. 209
Ni, Wancheng 927
Niu, Guangcheng 527
Noh, Kyoung-Sin 955
Nudd, Graham R. 217

Ou, Haifeng 720

Pan, Yi 679, 777
Pan, Zhangsheng 720
Pang, Yonggang 895
Park, Eung Ki 819
Park, Hyoungwoo 935
Park, Jin-Sung 943
Pellicer, Stephen 679
Peng, Chao 803
Peng, Gang 650

Qi, Li 783
Qi, Yang 745
Qi, Zhengwei 73
Qian, Depei 863
Qiang, Weizhong 293, 787
Qiu, Jie 129
Qu, Xiangli 348, 704

Rankich, Greg 6
Rao, Weixiong 559
Rao, Yuan 341
Ren, Ping 325
Ren, Yi 26
Rogulin, Dmitry 939
Rong, Henry 448
Ryou, Jae-Cheol 333
Ryu, So-Hyun 81, 943

Seo, Jung-Taek 333, 891
Sere, Kaisa 527
Serra, Moisés Ferrer 527
Sha, Edwin H.-M. 543
Shang, Ning 997
Shen, Fouxiang 847

Author Index

Shen, Hong 803
Shen, Jianhua 799
Shi, Dongyu 73
Shi, Wei 650
Shi, Xuanhua 293, 787
Shi, Yao 249
Shih, Po-Chi 931
Shin, Chang-Sun 121
Shon, Taeshik 891
Sloan, Terence M. 696
Solchenbach, Karl 7
Solomonides, Tony 939
Song, Hui 559
Spooner, Daniel P. 217
Steenberg, Conrad 947
Stroinski, Maciej 175
Sun, Choong-Hyun 935
Sun, Hong-Wei 883
Sun, Jia-Guang 883
Sun, Lin 769
Sun, Yuzhong 129
Sung, Mee Young 712
Suthaharan, Shan 967

Tan, Zhangxi 193
Tang, Feilong 1009
Tang, Hongji 464
Tang, Jie 827
Tang, Na 987
Tang, Rui-Chun 249
Tang, Shengqun 65
Tang, Yong 987
Tang, Yu 704
Tao, Muliu 241
Teo, Yong-Meng 17, 761
Thomas, Michael 947
Tu, Wanqing 502

Vandenberg, Art 777

Wan, Hai 959
Wang, Baoyi 729, 963
Wang, Bin 404
Wang, Cho-Li 167, 301, 911
Wang, Chong 791
Wang, Dongsheng 511, 519
Wang, Ge 495
Wang, Hui 591
Wang, Huiqiang 895
Wang, Jie 50
Wang, Kehong 827

Wang, Lina 348
Wang, Qiang 277
Wang, Qingjiang 285
Wang, Shaofeng 413, 431
Wang, Shaowen 495
Wang, Sheng-sheng 737
Wang, Wei 113, 903, 997
Wang, Weiping 823
Wang, Wenjun 42, 97, 145, 153
Wang, Xi 1013
Wang, Xianbing 761
Wang, Xiaoge 167, 1013
Wang, Xiaolin 42, 97, 145, 153
Wang, Xingwei 277
Wang, Xu 823
Wang, Yi 626
Wang, Yijie 642
Wang, Yongwen 487
Wang, Yuelong 153
Wang, Yuexuan 927
Wang, Zhi-Jing 855
Weng, Chuliang 269
Willers, Ian 947
Wo, Tian-yu 161
Woo, Gyun 919
Wu, Chanle 241, 979
Wu, Cheng 927
Wu, Fei 815
Wu, Jian-Ping 325
Wu, Libing 241
Wu, Quanyuan 26, 58
Wu, Wei 413, 431
Wu, Xiaobo 591
Wu, Yongwei 551
Wu, Zhaohui 372

Xia, Zhengyou 356
Xiang, Guang 348
Xiang, Yang 309, 1017
Xiao, Bin 543
Xiao, Lijuan 209
Xiao, Nong 105, 233, 704
Xiao, Yun 875
Xie, Bing 285, 863
Xie, Changsheng 364
Xie, Li 8, 687
Xie, Yong 17
Xing, Jianbing 241, 979
Xu, Cheng 404
Xu, Li 753

Xu, Zhenning 34
Xu, Zhi 720
Xu, Zhiwei 209
Xu, Zhuoqun 42, 97, 145, 153, 404, 601
Xue, Guangtao 269
Xue, Tao 583

Yan, Feng 831
Yan, Lu 527
Yan, Xiaolang 591
Yang, Chang 225
Yang, Chao-Tung 931
Yang, Dongsheng 34
Yang, Feng 847
Yang, Guangwen 551
Yang, Min 997
Yang, Xuejun 233, 704
Yang, Yan 65
Yang, Yinying 867
Yang, Yun 1005
Yao, Yuanzhe 113, 903
Ye, Baoliu 8, 665
Ye, Xiao-ping 987
Yeom, Heon Y. 567
Yi, Gwan-Su 935
Yin, Hao 193
Yin, K.K. 911
Yin, Yanbin 404
You, Jinyuan 73
Youn, Chan-Hyun 819
Youn, Chunkyun 859
Yu, Chiu-Man 871
Yu, Ge 348
Yu, Haiyan 129
Yu, Huashan 404, 601
Yu, Jiadi 626
Yu, Yang 987
Yu, Young-Hoon 955
Yuan, Pingpeng 783

Zha, Li 129
Zhan, Jian 261
Zhan, Shouyi 831, 847
Zhang, Cheng 89
Zhang, DongDong 823
Zhang, Fan 397, 456
Zhang, Hong-jun 993
Zhang, Hongli 113, 903
Zhang, Huyin 241
Zhang, Liang 807

Zhang, Mi 997
Zhang, Qin 671
Zhang, Shaomin 729, 963
Zhang, Shensheng 50
Zhang, ShiYong 440
Zhang, Tao 959
Zhang, Wei 993
Zhang, Weiming 34
Zhang, Weizhe 201
Zhang, Wenju 799
Zhang, Wentao 65
Zhang, Xinjia 464
Zhang, Yan 431
Zhang, Yang 8, 665, 687
Zhang, Yun-Yong 397, 456
Zhang, Zhi-Jiang 397, 456
Zhang, Zhigang 583
Zhao, Feng 879
Zhao, Menglian 591
Zhao, Qinping 413
Zhao, Xiubin 618
Zhao, Yan-Bin 879
Zhao, Yue 610
Zheng, Bao-yu 753
Zheng, Qianbing 487
Zheng, Qinghua 875
Zheng, Ran 671
Zheng, Shouqi 285
Zheng, Weimin 511, 519
Zhong, YiPing 440
Zhou, Aoying 610
Zhou, Bin 58
Zhou, Haifang 704
Zhou, Jingyang 535, 665, 687
Zhou, Shuigeng 610
Zhou, Wanlei 309
Zhou, Xing-Ming 423
Zhou, Xinrong 527
Zhou, Zhong 413
Zhu, Bo 317
Zhu, Cheng 34
Zhu, Guangxi 193
Zhu, HuaFei 317
Zhu, Weiwei 535
Zhu, Yanmin 209
Zhu, Ye 867
Zhuge, Hai 381
Zou, Deqing 293, 787
Zou, Futai 137, 807

Lecture Notes in Computer Science

For information about Vols. 1–3186

please contact your bookseller or Springer

Vol. 3305: P.M.A. Sloot, B. Chopard, A.G. Hoekstra (Eds.), Cellular Automata. XV, 883 pages. 2004.

Vol. 3293: C.-H. Chi, M. van Steen, C. Wills (Eds.), Web Content Caching and Distribution. IX, 283 pages. 2004.

Vol. 3287: A. Sanfeliu, J.F.M. Trinidad, J.A. Carrasco Ochoa (Eds.), Progress in Pattern Recognition, Image Analysis and Applications. XVII, 703 pages. 2004.

Vol. 3286: G. Karsai, E. Visser (Eds.), Generative Programming and Component Engineering. XIII, 491 pages. 2004.

Vol. 3284: A. Karmouch, L. Korba, E.R.M. Madeira (Eds.), Mobility Aware Technologies and Applications. XII, 382 pages. 2004.

Vol. 3280: C. Aykanat, T. Dayar, İ. Körpeoğlu (Eds.), Computer and Information Sciences - ISCIS 2004. XVIII, 1009 pages. 2004.

Vol. 3274: R. Guerraoui (Ed.), Distributed Computing. XIII, 465 pages. 2004.

Vol. 3273: T. Baar, A. Strohmeier, A. Moreira, S.J. Mellor (Eds.), <<UML>> 2004 - The Unified Modelling Language. XIII, 454 pages. 2004.

Vol. 3271: J. Vicente, D. Hutchison (Eds.), Management of Multimedia Networks and Services. XIII, 335 pages. 2004.

Vol. 3270: M. Jeckle, R. Kowalczyk, P. Braun (Eds.), Grid Services Engineering and Management. X, 165 pages. 2004.

Vol. 3269: J. López, S. Qing, E. Okamoto (Eds.), Information and Communications Security. XI, 564 pages. 2004.

Vol. 3266: J. Solé-Pareta, M. Smirnov, P.V. Mieghem, J. Domingo-Pascual, E. Monteiro, P. Reichl, B. Stiller, R.J. Gibbens (Eds.), Quality of Service in the Emerging Networking Panorama. XVI, 390 pages. 2004.

Vol. 3265: R.E. Frederking, K.B. Taylor (Eds.), Machine Translation: From Real Users to Research. XI, 392 pages. 2004. (Subseries LNAI).

Vol. 3264: G. Paliouras, Y. Sakakibara (Eds.), Grammatical Inference: Algorithms and Applications. XI, 291 pages. 2004. (Subseries LNAI).

Vol. 3263: M. Weske, P. Liggesmeyer (Eds.), Object-Oriented and Internet-Based Technologies. XII, 239 pages. 2004.

Vol. 3262: M.M. Freire, P. Chemouil, P. Lorenz, A. Gravey (Eds.), Universal Multiservice Networks. XIII, 556 pages. 2004.

Vol. 3261: T. Yakhno (Ed.), Advances in Information Systems. XIV, 617 pages. 2004.

Vol. 3260: I.G.M.M. Niemegeers, S.H. de Groot (Eds.), Personal Wireless Communications. XIV, 478 pages. 2004.

Vol. 3258: M. Wallace (Ed.), Principles and Practice of Constraint Programming – CP 2004. XVII, 822 pages. 2004.

Vol. 3257: E. Motta, N.R. Shadbolt, A. Stutt, N. Gibbins (Eds.), Engineering Knowledge in the Age of the Semantic Web. XVII, 517 pages. 2004. (Subseries LNAI).

Vol. 3256: H. Ehrig, G. Engels, F. Parisi-Presicce, G. Rozenberg (Eds.), Graph Transformations. XII, 451 pages. 2004.

Vol. 3255: A. Benczúr, J. Demetrovics, G. Gottlob (Eds.), Advances in Databases and Information Systems. XI, 423 pages. 2004.

Vol. 3254: E. Macii, V. Paliouras, O. Koufopavlou (Eds.), Integrated Circuit and System Design. XVI, 910 pages. 2004.

Vol. 3253: Y. Lakhnech, S. Yovine (Eds.), Formal Techniques, Modelling and Analysis of Timed and Fault-Tolerant Systems. X, 397 pages. 2004.

Vol. 3251: H. Jin, Y. Pan, N. Xiao, J. Sun (Eds.), Grid and Cooperative Computing - GCC 2004. XXII, 1025 pages. 2004.

Vol. 3250: L.-J. (LJ) Zhang, M. Jeckle (Eds.), Web Services. X, 301 pages. 2004.

Vol. 3249: B. Buchberger, J.A. Campbell (Eds.), Artificial Intelligence and Symbolic Computation. X, 285 pages. 2004. (Subseries LNAI).

Vol. 3246: A. Apostolico, M. Melucci (Eds.), String Processing and Information Retrieval. XIV, 332 pages. 2004.

Vol. 3245: E. Suzuki, S. Arikawa (Eds.), Discovery Science. XIV, 430 pages. 2004. (Subseries LNAI).

Vol. 3244: S. Ben-David, J. Case, A. Maruoka (Eds.), Algorithmic Learning Theory. XIV, 505 pages. 2004. (Subseries LNAI).

Vol. 3243: S. Leonardi (Ed.), Algorithms and Models for the Web-Graph. VIII, 189 pages. 2004.

Vol. 3242: X. Yao, E. Burke, J.A. Lozano, J. Smith, J.J. Merelo-Guervós, J.A. Bullinaria, J. Rowe, P. Tiňo, A. Kabán, H.-P. Schwefel (Eds.), Parallel Problem Solving from Nature - PPSN VIII. XX, 1185 pages. 2004.

Vol. 3241: D. Kranzlmüller, P. Kacsuk, J.J. Dongarra (Eds.), Recent Advances in Parallel Virtual Machine and Message Passing Interface. XIII, 452 pages. 2004.

Vol. 3240: I. Jonassen, J. Kim (Eds.), Algorithms in Bioinformatics. IX, 476 pages. 2004. (Subseries LNBI).

Vol. 3239: G. Nicosia, V. Cutello, P.J. Bentley, J. Timmis (Eds.), Artificial Immune Systems. XII, 444 pages. 2004.

Vol. 3238: S. Biundo, T. Frühwirth, G. Palm (Eds.), KI 2004: Advances in Artificial Intelligence. XI, 467 pages. 2004. (Subseries LNAI).

Vol. 3236: M. Núñez, Z. Maamar, F.L. Pelayo, K. Pousttchi, F. Rubio (Eds.), Applying Formal Methods: Testing, Performance, and M/E-Commerce. XI, 381 pages. 2004.

Vol. 3235: D. de Frutos-Escrig, M. Nunez (Eds.), Formal Techniques for Networked and Distributed Systems – FORTE 2004. X, 377 pages. 2004.

Vol. 3232: R. Heery, L. Lyon (Eds.), Research and Advanced Technology for Digital Libraries. XV, 528 pages. 2004.

Vol. 3231: H.-A. Jacobsen (Ed.), Middleware 2004. XV, 514 pages. 2004.

Vol. 3230: J.L. Vicedo, P. Martínez-Barco, R. Muñoz, M. Saiz Noeda (Eds.), Advances in Natural Language Processing. XII, 488 pages. 2004. (Subseries LNAI).

Vol. 3229: J.J. Alferes, J. Leite (Eds.), Logics in Artificial Intelligence. XIV, 744 pages. 2004. (Subseries LNAI).

Vol. 3226: M. Bouzeghoub, C. Goble, V. Kashyap, S. Spaccapietra (Eds.), Semantics for Grid Databases. XIII, 326 pages. 2004.

Vol. 3225: K. Zhang, Y. Zheng (Eds.), Information Security. XII, 442 pages. 2004.

Vol. 3224: E. Jonsson, A. Valdes, M. Almgren (Eds.), Recent Advances in Intrusion Detection. XII, 315 pages. 2004.

Vol. 3223: K. Slind, A. Bunker, G. Gopalakrishnan (Eds.), Theorem Proving in Higher Order Logics. VIII, 337 pages. 2004.

Vol. 3222: H. Jin, G.R. Gao, Z. Xu, H. Chen (Eds.), Network and Parallel Computing. XX, 694 pages. 2004.

Vol. 3221: S. Albers, T. Radzik (Eds.), Algorithms – ESA 2004. XVIII, 836 pages. 2004.

Vol. 3220: J.C. Lester, R.M. Vicari, F. Paraguaçu (Eds.), Intelligent Tutoring Systems. XXI, 920 pages. 2004.

Vol. 3219: M. Heisel, P. Liggesmeyer, S. Wittmann (Eds.), Computer Safety, Reliability, and Security. XI, 339 pages. 2004.

Vol. 3217: C. Barillot, D.R. Haynor, P. Hellier (Eds.), Medical Image Computing and Computer-Assisted Intervention – MICCAI 2004. XXXVIII, 1114 pages. 2004.

Vol. 3216: C. Barillot, D.R. Haynor, P. Hellier (Eds.), Medical Image Computing and Computer-Assisted Intervention – MICCAI 2004. XXXVIII, 930 pages. 2004.

Vol. 3215: M.G.. Negoita, R.J. Howlett, L.C. Jain (Eds.), Knowledge-Based Intelligent Information and Engineering Systems. LVII, 906 pages. 2004. (Subseries LNAI).

Vol. 3214: M.G.. Negoita, R.J. Howlett, L.C. Jain (Eds.), Knowledge-Based Intelligent Information and Engineering Systems. LVIII, 1302 pages. 2004. (Subseries LNAI).

Vol. 3213: M.G.. Negoita, R.J. Howlett, L.C. Jain (Eds.), Knowledge-Based Intelligent Information and Engineering Systems. LVIII, 1280 pages. 2004. (Subseries LNAI).

Vol. 3212: A. Campilho, M. Kamel (Eds.), Image Analysis and Recognition. XXIX, 862 pages. 2004.

Vol. 3211: A. Campilho, M. Kamel (Eds.), Image Analysis and Recognition. XXIX, 880 pages. 2004.

Vol. 3210: J. Marcinkowski, A. Tarlecki (Eds.), Computer Science Logic. XI, 520 pages. 2004.

Vol. 3209: B. Berendt, A. Hotho, D. Mladenic, M. van Someren, M. Spiliopoulou, G. Stumme (Eds.), Web Mining: From Web to Semantic Web. IX, 201 pages. 2004. (Subseries LNAI).

Vol. 3208: H.J. Ohlbach, S. Schaffert (Eds.), Principles and Practice of Semantic Web Reasoning. VII, 165 pages. 2004.

Vol. 3207: L.T. Yang, M. Guo, G.R. Gao, N.K. Jha (Eds.), Embedded and Ubiquitous Computing. XX, 1116 pages. 2004.

Vol. 3206: P. Sojka, I. Kopecek, K. Pala (Eds.), Text, Speech and Dialogue. XIII, 667 pages. 2004. (Subseries LNAI).

Vol. 3205: N. Davies, E. Mynatt, I. Siio (Eds.), UbiComp 2004: Ubiquitous Computing. XVI, 452 pages. 2004.

Vol. 3204: C.A. Peña Reyes, Coevolutionary Fuzzy Modeling. XIII, 129 pages. 2004.

Vol. 3203: J. Becker, M. Platzner, S. Vernalde (Eds.), Field Programmable Logic and Application. XXX, 1198 pages. 2004.

Vol. 3202: J.-F. Boulicaut, F. Esposito, F. Giannotti, D. Pedreschi (Eds.), Knowledge Discovery in Databases: PKDD 2004. XIX, 560 pages. 2004. (Subseries LNAI).

Vol. 3201: J.-F. Boulicaut, F. Esposito, F. Giannotti, D. Pedreschi (Eds.), Machine Learning: ECML 2004. XVIII, 580 pages. 2004. (Subseries LNAI).

Vol. 3199: H. Schepers (Ed.), Software and Compilers for Embedded Systems. X, 259 pages. 2004.

Vol. 3198: G.-J. de Vreede, L.A. Guerrero, G. Marín Raventós (Eds.), Groupware: Design, Implementation and Use. XI, 378 pages. 2004.

Vol. 3196: C. Stary, C. Stephanidis (Eds.), User-Centered Interaction Paradigms for Universal Access in the Information Society. XII, 488 pages. 2004.

Vol. 3195: C.G. Puntonet, A. Prieto (Eds.), Independent Component Analysis and Blind Signal Separation. XXIII, 1266 pages. 2004.

Vol. 3194: R. Camacho, R. King, A. Srinivasan (Eds.), Inductive Logic Programming. XI, 361 pages. 2004. (Subseries LNAI).

Vol. 3193: P. Samarati, P. Ryan, D. Gollmann, R. Molva (Eds.), Computer Security – ESORICS 2004. X, 457 pages. 2004.

Vol. 3192: C. Bussler, D. Fensel (Eds.), Artificial Intelligence: Methodology, Systems, and Applications. XIII, 522 pages. 2004. (Subseries LNAI).

Vol. 3191: M. Klusch, S. Ossowski, V. Kashyap, R. Unland (Eds.), Cooperative Information Agents VIII. XI, 303 pages. 2004. (Subseries LNAI).

Vol. 3190: Y. Luo (Ed.), Cooperative Design, Visualization, and Engineering. IX, 248 pages. 2004.

Vol. 3189: P.-C. Yew, J. Xue (Eds.), Advances in Computer Systems Architecture. XVII, 598 pages. 2004.

Vol. 3188: F.S. de Boer, M.M. Bonsangue, S. Graf, W.-P. de Roever (Eds.), Formal Methods for Components and Objects. VIII, 373 pages. 2004.

Vol. 3187: G. Lindemann, J. Denzinger, I.J. Timm, R. Unland (Eds.), Multiagent System Technologies. XIII, 341 pages. 2004. (Subseries LNAI).